An Introduction to

Classical Electrodynamics

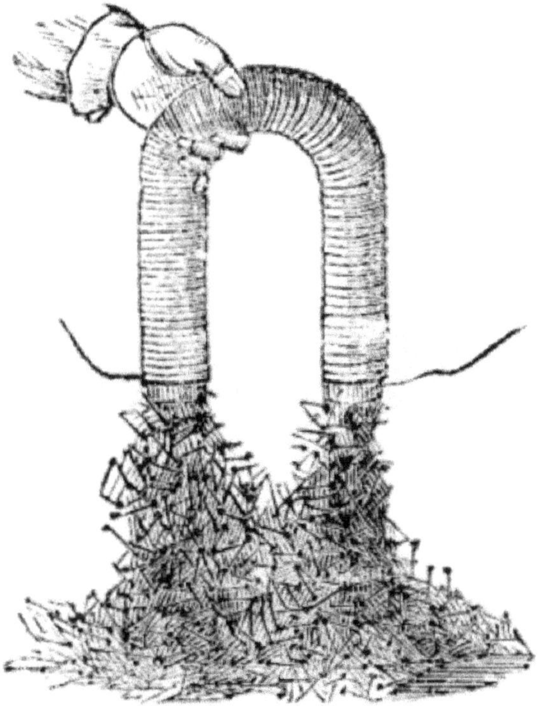

Jonathan W. Keohane
and
Joseph P. Foy

Maricourt Academic Press

An Introduction to Classical Electrodynamics

Jonathan W. Keohane and Joseph P. Foy

Illustrated by Jonathan W. Keohane

Maricourt Academic Press
Farmville, VA USA
https://maricourt.press

Edition 1.1

The authors can be contacted by E-mail at jkeohane@hsc.edu and jpfoy@asu.edu.

ISBN: 978-1-949942-00-2

Library of Congress Control Number: 2018913008

Library of Congress Catalog Number: QC631.K46

Dewey Decimal Classification Number: 537.6

Library of Congress Cataloging-in-Publication Data:
Keohane, Jonathan W., 1966-, Foy, Joseph P., 1962-
An Introduction to Classical Electrodynamics / Jonathan W. Keohane, Joseph P. Foy
755 pages, 650 illustrations
1. Electrodynamics 2. Physics:Electricity and Magnetism 3. Physics:Electromagnetism 4. Physics:Special Relativity
5. Electrodynamics:History 6. Mathematical Physics

Short Table of Contents

Table of Fundamental Laws

Table of Plates

Long Table of Contents

Preface

A good popular science book tells a story. A good instructional manual includes many examples. And, finally, a good treatise develops a persuasive argument backed by evidence. Our goal is to combine the best practices of all three genres: the story telling of a popular science book, the worked examples of an instructional manual, and the persuasiveness of a treatise. We will let you decide if we have succeeded in this endeavor.

Over the past decade we have finished this textbook many times, after which it was sent it out for peer review. Each time half of the reviewers loved our novel approach, while the other half were somewhat skeptical. After each iteration we reorganized the text to please the skeptics, while still keeping the spark that makes the text interesting. While our drafts have improved with each revision, there are inevitably errors that we have missed. Every time we edit, we find typographical errors, such as misspellings, missing words, and copy and paste errors (especially in dates and equations). We also have purposely not numbered our equations and figures. Rather all figures are placed in context by hand, but they sometimes move. We apologize in advance for such things, and we would very much appreciate your letting us know via e-mail when you find errors, so we can correct them for future readers. This book is printed on demand, so corrections can be implemented fairly quickly.

Since we have reorganized the presentation of material multiple times, our assumed prior knowledge has changed. At times we may assume that something new has already been introduced, or we may introduce something more than once. We hope that our comprehensive table of contents, biographical index, and many cross-references helps the reader navigate to the right place. Please let us know about these oversights, or any others, so we can correct them for the second edition.

We have learned much over the past decade, and in the process, we have discovered many holes in our prior knowledge, and even found that we each held some fundamental physical misconceptions. If you believe that anything slipped through the cracks, please let us know. Just keep in mind that we have attempted to present Maxwell's theory, and the story of its discovery, in a way that should be accessible to undergraduate students of either physics or electrical engineering. Moreover, different fields of physics and engineering do things differently, and use different terminology. We are both astrophysicists, so we came to electrodynamics with that particular perspective. Over time, we have learned to appreciate other perspectives as well.

This is a long book, but do not be intimidated. The length is designed to add clarity and perspective, not to be encyclopedic. Many texts skip algebra steps, or leave out important context, simply for brevity. We have used the space to explain how concepts actually developed, so that you can make the connections required of a highly educated professional.

We believe that nobody, least of all a student, should be priced out of reading our book. We also would like for practicing physicists and electrical engineers to keep a copy of our book for reference. For this reason, we have priced the book close to the marginal cost of production, without the usual markup that a major publisher would apply. Our goal, after all, is to convey a topic of physics that we love.

We also want to thank everyone who helped in the creation of this book. We owe many thanks to Alistair Kwan for his detailed and insightful comments on our draft chapters. Isaac Keohane put in many hours helping with programming on the numerical exercises, and Jacob Keohane was a superb copy editor. We also owe thanks to Beckie Smith for typing up our first draft and

Robert Keohane for making helpful suggestions. We would like to thank Dennis Strauss for notifying us about parts of the text which were unclear or contain errata. We also thank to Michael Brabanski who told us about Michelson's time in Berlin, and the U.S. Naval Observatory library for providing high resolution scans of some of Benjamin Franklin's figures. We also owe thanks to Mitchell Willis and Bruce Drain for reading over the problems that refer to their research.

We would also like to thank the National Radio Astronomy Observatory for kindly hosting a year of sabbatical, and our two institutions which generously provided support through Hampden-Sydney College Summer Research Funds and the Barrett Honors College Drescher Award.

We would like to thank Joseph Calamia for all the time he spent finding suitable experts to review our work, and of course the anonymous reviewers themselves, whose frank and insightful comments were invaluable to our ability to produce this book.

Jonathan W. Keohane — Hampden-Sydney College

Joseph P. Foy — Arizona State University

To Laura and Emalee

Plate 1: Albert Michelson's Interferometer[1]

[1] The interferometer in the photo is a replica on display in Michelsonhaus (Michelson house), on Telegrafenberg (Telegraph mountain), in Potsdam Germany, where Michelson carried out his experiment due to city vibrations in his Berlin optics lab. (Photo: Michael Brabanski).

Overview: From Newton to Einstein

Sir Isaac Newton produced two great works that revolutionized natural philosophy. One was theoretical, the other experimental. One was written in Latin, the other English. One is now considered his greatest work; the other is remembered for advancing a failed theory. One seemed to answer all the questions of the universe, the other asked more questions than it answered. These works were Philosophiæ Naturalis Principia Mathematica (Mathematical Principles of Natural Philosophy) and Opticks or A Treatise of the Reflections, Refractions, Inflections and Colours of Light.

The Principia presented Newton's three laws of motion, and his Universal Law of Gravitation. Written in Latin, the universal language of scholarship, it was embraced by scientists across Europe and sparked a mathematical revolution. Unifying physics and astronomy, and providing the link between many disparate disciplines, many in Europe regarded the *Principia* as presenting the ultimate solution to the laws of nature.

Despite its success, Newton was still worried about several lingering philosophical questions, especially the issue of *action at a distance*. Newton's primary theme concerns cause and effect. When one object applies a force on another, according to his second law, the other object accelerates by a proportionate amount. This principle was easy to appreciate in the case of contact forces, such as collisions, where one body directly interacts with another. However, gravity works differently. The moon orbits the earth, and water in the earth's oceans moves in response. How does this happen? What is it that links the seas to the moon, despite their vast distance? Presumably some *mechanism* conveys the gravitational force, but Newton did not know what it was, and so he did not speculate about it formally.

Like gravity, light travels through vast distances of free space, but unlike gravity we can see light. Moreover, the intensity of each diminishes in proportion to the inverse square of the distance. So, while Newton may have liked to investigate the invisible gravitational mechanism, he could study light. Not only could he study it, but he could investigate it experimentally, and write about it in a way that any educated person could understand.

While most of Opticks describes the geometrical optics of lenses, mirrors, and prisms, now called *Newtonian optics*, Newton's most startling experiments investigated color. Prior to Newton, scientists reasoned that colored light contains some sort of pigment that gives it color, much like dye-soaked cloth or stained glass. Under this hypothesis, overlapping projected colors should combine in the same way as paints mix. Newton, however, shocked the scientific community when he used two prisms to separate light into a rainbow and then "compound Whiteness by mixing their Colours."

To explain his experimental results, Newton proposed that light is composed of particles, or *corpuscles*, which travel in straight lines over great distances through free space. He believed the color of light varies with the mass of a light particle. By passing white light through prisms, Newton found that red light bends less than blue light, which he attributed to red light corpuscles being more massive than blue ones. Newton's particle model sharply contrasted with another theory, proposed by his Dutch contemporary Christiaan Huygens, which modeled light as a wave though a medium, much like sound traveling through air. Newton reasoned, however, that if such a medium exists, it could not be so dense as to degrade planetary orbits, but must be elastic enough to convey forces over long distances, which was highly unlikely. However, Newton also reasoned, that if future experiments disproved his particle model, then this medium, called the *aether*, must exist. The alternative, after all, would be *action at a distance* across vast distances of truly empty space, something he could not abide.

Along with gravity, two other forces also appeared magically to act from a distance: electricity and magnetism. To study the mechanism that conveys these forces, Newton's intellectual descendants had to separate cause from effect. They did this by defining a new concept called a *force field.*

Consider, for example, the force with which the earth pulls down on you, commonly known as your *weight.* If you traveled, say, to the moon your weight would be different, but you would still be the same. There must be some property intrinsic to the earth that causes some *gravitational field*, which, in turn, interacts with some property of you to pull you downward. We call this intrinsic property *mass*, and we define the *gravitational field* at your location to be the ratio of your weight to your mass, or in introductory physics notation:

$$\vec{g} \equiv \frac{\vec{F}_g}{m}.$$

By defining the gravitational field in this way, natural philosophers could circumvent the pernicious problem of *action at a distance.* Someone, some day, would figure out what this elusive gravitational field really is, scientists reasoned. For the time being, however, the gravitational field is defined *operationally* based on how it is measured.

The force field appropriate for electric charges can be defined in a similar way to gravity. If you rub two identical balloons with, say, your hair, they will repel. One balloon causes something, call it the *electric field*, which in turn interacts with the other balloon to push it away. By defining the electric field as the ratio of the force on an object to some governing quantity, call it *charge*, we can say the electric field is pushing the balloon. Or, in introductory physics notation:

$$\vec{E} \equiv \frac{\vec{F}_E}{q}.$$

And, finally, if you hold a compass, the compass needle twists until it points approximately in the north-south direction. The torque on the compass needle must be caused by the interaction between some *magnetic field* and some intrinsic property of the magnet, called the *magnetic moment*, or $\vec{\tau} = \vec{m} \times \vec{B}$, where $\vec{\tau}$ is the torque on the compass needle, and $\vec{m}$ and $\vec{B}$ are the magnetic moment of the compass needle and the local magnetic field respectively.

Not everyone was spooked by *action at a distance.* After all, Newton's law of gravity was so successful that it made sense to search for similar laws relating the forces between charged objects and magnets. Charles Agustin Coulomb, for example, found inverse square force laws relating charged objects and the poles of compass needles. While his famous law involving charged objects is still considered a great achievement, his analogous law involving magnetic poles was later shown to be an overgeneralization of reasoning by analogy, as magnetic poles fundamentally do not exist.

Electricity and magnetism merged when, in 1820, Hans Christian Ørsted discovered that a current carrying wire deflects a compass needle. This led to a flurry of experimental research, where scientists raced each other to understand this new phenomenon. Within only a short time, scientists had carefully measured the forces between magnets and wires to develop more mathematical laws in the same style as Newton's law of gravity. By the middle of the nineteenth century, Continental physicists had worked out a comprehensive *action at a distance* theory of *electromagnetism.*

In England, Michael Faraday was not so sanguine about *action at a distance,* so he took a different tack. Using iron filings, Faraday mapped out the direction of the magnetic field around magnets and through wire loops. Like the flow of water, and electric current, these magnetic field lines circulated, always making closed loops. Even more startling was Faraday's discovery that moving a magnet through loops of wire caused electricity to flow. Faraday's fluid approach made a great deal of sense conceptually, but Faraday was no mathematician.

After successfully modeling Saturn's rings, James Clerk Maxwell moved to London and began expressing the known laws of electricity and magnetism in terms of the electric and magnetic fields, using the mathematics of fluid mechanics. In free space, Maxwell discovered, the electric and magnetic fields oscillate between each other, and propagate at a speed numerically equivalent to the speed of light. Thus, light was now an electromagnetic phenomenon. Moreover, the problem of *action at a distance* was largely put to rest. A changing current causes an oscillating electric field. This, in turn, causes an oscillating magnetic field, which causes an electric field, which, in turn, causes a magnetic field yet again. Thus, all electromagnetic information travels through space at the characteristic speed of the medium.

Maxwell had now successfully combined electricity, magnetism, and optics into a single branch of physics called *electrodynamics.* The last frontier for physics would be to determine the nature of the medium that links cause to effect: from battery to motor, from magnet to steel, and from star to telescope, as Maxwell's final sentence of his Treatise on Electricity and Magnetism so eloquently states:

> Hence all these theories lead to the conception of a medium in which the propagation takes place, and if we admit this medium as an hypothesis, I think it ought to occupy a prominent place in our investigations, and that we ought to endeavor to construct a mental representation of all the details of its action, and this has been my constant aim in this treatise.

Maxwell's theory quickly passed the rigorous experimental tests of Heinrich Hertz, leaving little doubt about its correctness and the existence of the aether. A natural next step was to measure the velocity of the earth with respect to the aether.

Consider taking a round-trip airplane trip on a windy day, with a headwind going toward your destination, and a tailwind on the way back. As it turns out, the additional speed of the return flight cannot make up for the slower flight out. Thus, round-trip flights take longer on windy days. The wind velocity, therefore, could, in principle, be calculated from the differences in round-trip flight times of planes traveling in different directions. This is the approach to finding the speed of the *aether wind* that a young American physicist took while spending a semester in Berlin.

With his interferometer (photo on p. 0 above), Albert Michelson attempted to measure differences in the round-trip travel time of light with direction. Michelson found no measureable time differences, no matter the time of year. Therefore, he published the following bold conclusion:

> The result of the hypothesis of stationary ether is thus shown to be incorrect, and the necessary conclusion follows that the hypothesis is erroneous.[2]

---

[2] Albert A. Michelson, "The Relative Motion of the Earth and the Luminiferous Ether", American Journal of Science, **22** (1881), 120-129.

At the time, however, this result was seen as support for the idea, proposed by George Gabriel Stokes about thirty years before, that the earth drags the aether along as it moves about the sun. While the significance of his null result remained unappreciated, it did launch Michelson's career. He moved to a larger university, where he and his colleague Edward Morley built the most accurate optical interferometer to that date. Alas, he failed again to measure any time differences, and at this he concluded:

> the relative velocity of the earth and the ether is probably less than one sixth the earth's orbital velocity, and certainly less than one-fourth.[3]

By the turn of the century even true believers in the aether's existence were having doubts. For example, the elderly Lord Kelvin wrote:

> The beauty and clearness of the dynamical theory, which asserts heat and light to be modes of motion, is at present obscured by two clouds. I. The first came into existence with the undulatory theory of light, and was dealt with by Fresnel and Dr. Thomas Young; it involved the question, How could the earth move through an elastic solid, such as essentially is the luminiferous ether? II. The second is the Maxwell-Boltzmann doctrine regarding the partition of energy.[4]

What could possibly explain the propagation of electromagnetic waves without a medium? It just made no sense for a cause one place to produce an effect somewhere else, without either a particle traveling or a medium carrying the signal.

Albert Einstein cleared Kelvin's first cloud in 1905 by crafting a new aetherless theory that unified kinematics and electrodynamics. Einstein's theory of relativity did not modify the equations of classical electrodynamics, but rather our fundamental concepts of time and space.[5]

Einstein's theory of relativity, while bold, was not considered reckless to his colleagues because it so beautifully unified all of classical physics under a single paradigm. Despite its classical beauty, and immediate acceptance, special relativity was not the end of the story for the nature of light.

By dismissing the aether, Einstein's theory of relativity opened the door for *action at a distance* to, once again, cast a pall over the scientific world. This was not lost on Einstein, who, also in 1905, reincarnated the long-dead particle model of light to address whether light from a hot source could be consistent with "the Maxwell-Boltzmann doctrine regarding the partition of energy."

Even after the acceptance of relativity theory, few people, least of all Robert Millikan, were persuaded that light was a particle; the evidence for the wave model was too overwhelming. This all changed in 1916, when Millikan finished a research program designed to disprove Einstein's particle model of light once and for all. He wrote:

> It was in 1905 that Einstein made the first coupling of photo effects with any form of quantum theory by bringing forward the bold, not to say the reckless, hypothesis of an electro-magnetic light corpuscle of energy, $h\nu$, which energy was transferred upon absorption to an electron.[6]

---

[3] Albert A. Michelson & Edward W. Morley, "On the Relative Motion of the Earth and the Luminiferous Ether", American Journal of Science **34** (1887), 333–345.

[4] W. Thomson, "Nineteenth century clouds over the dynamical theory of heat and light", Philosophical Magazine Series 6, **2**:7, (1901), 1-40.

[5] A. Einstein, "On the Electrodynamics of Moving Bodies," Annalen der Physik **17** (1905): 891-921, Translated by Anna Beck, in *The Collected Papers of Albert Einstein, Volume 2, The Swiss Years, 1900-1909*. John Stachel, ed. (Princeton: Princeton University Press). English Translation, ©1989 by the Hebrew University of Jerusalem.

[6] Millikan, R.A., "A Direct Photoelectric Determination of Planck's 'h'," Physical Review, **7** (1916), 355-388.

However, rather than sending Einstein's particle model of light back to the land of the dead, Millikan's experimental results directly verified Einstein's ideas, which opened the door to the quantum electrodynamic theories of the twentieth century.

What, after all, are these electric and magnetic fields that will dominate this 755-page book? Are they really just measures of the density of these electromagnetic particles, called *photons*? And, if they are, how do they communicate to act collectively? Electric and magnetic fields are simply different manifestations of an overall electromagnetic field, but what about the electromagnetic field? Is it really some sort of quantum field, and if so, what is that?

As far as *classical* electrodynamics is concerned, the electric field is simply a force per charge; similarly, the magnetic field is simply a torque per magnetic moment—no more and no less. Through 17 chapters of this book, this is exactly what they are. In the last chapter, however, we will come back to Einstein's reckless hypothesis that light might behave as a collection of particles, or even that all fields are really collections of particles. Or, perhaps the opposite is true, that all particles are actually manifestations of fields. And if they are, what is a field anyway other than something we construct to explain observed phenomena?

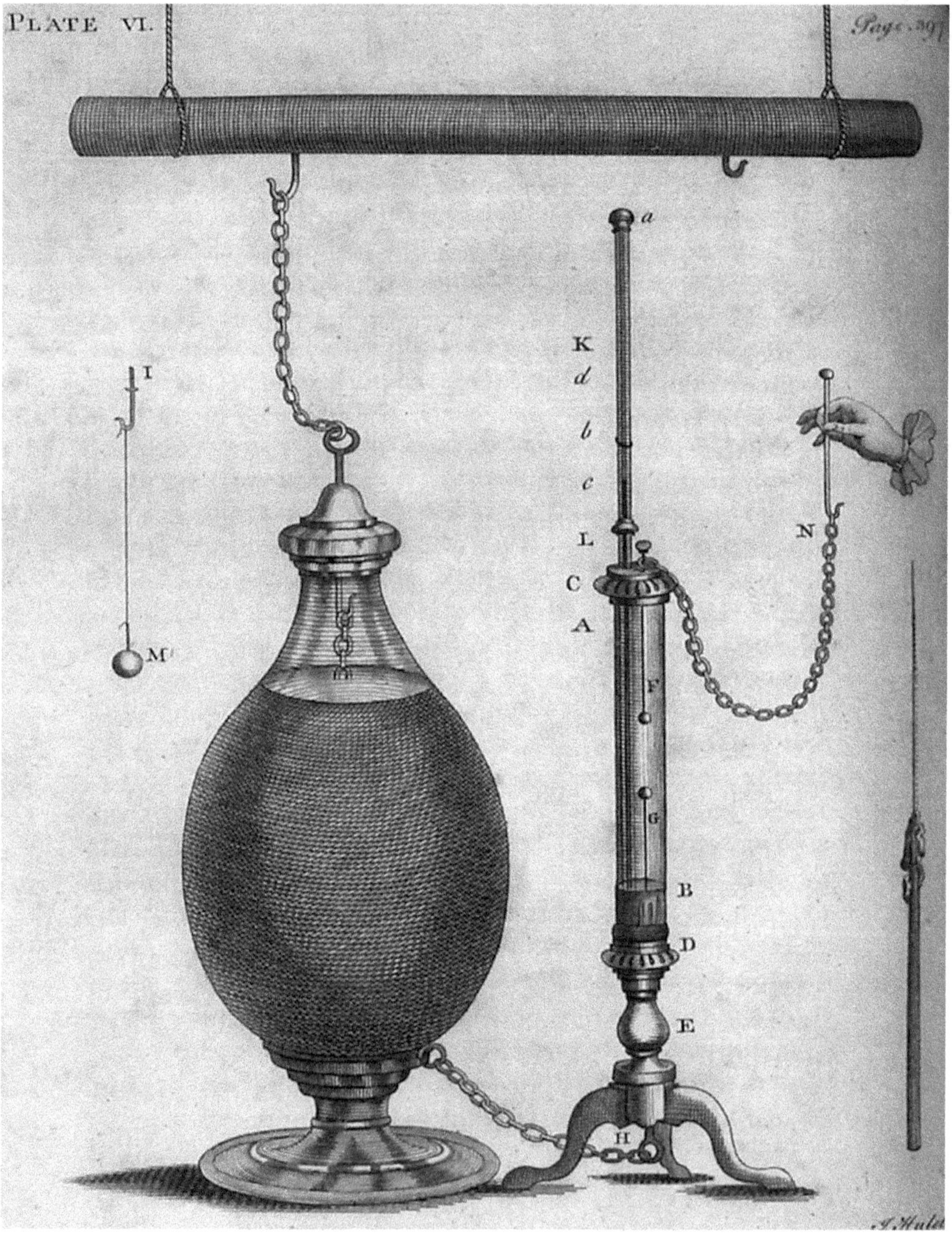

Plate 2: Benjamin Franklin's Leyden Jar[7]

[7] B. Franklin, <u>Experiments and Observations on Electricity</u>, illustrated by I. Hulett, 5th Ed., (London: F. Newberry at the Corner of St. Paul's Church-Yard, 1774), p. 397, plate 6.

Part I: Electricity

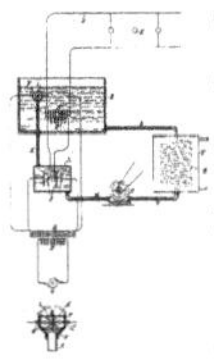

For two and a half millennia, the fields of electricity and magnetism were separate and distinct. This all changed in 1820 when Hans Christian Ørsted discovered electromagnetism. However, in order to observe an electric current deflect a compass needle, Ørsted must have had access to a powerful battery of electrochemical cells. For someone to build a powerful battery, someone else had to invent the electrochemical cell, which required access to electrical wires and knowledge of conductivity. Thus, knowledge begets knowledge, and that is how science progresses over time.

This *positivist* view of the history of science predicts that the rate of knowledge increases in proportion to the existing knowledge. The more we know the faster we learn, so we would expect an exponential increase in knowledge over time. For example, let K be the amount of knowledge. If $\frac{d}{dt}K \propto K$, then by separation of variables $K = K_0 e^{(t-t_0)/\tau} = K_0 2^{(t-t_0)/\tau_2}$, where K_0 represents the amount of knowledge at time t_0, and τ_2 represents the doubling time. For example, the American chemist, electrical engineer, and entrepreneur, Gordon Moore, predicted that the number of transistors per microchip, at fixed area and cost, would double every two years as shown in his graph.[8] If we quantify *knowledge of electricity* as Moore did, it has increased exponentially for over half a century. Moore's law is somewhat unique, as it became the driving goal of the microchip industry, so one could argue that it became its own cause—a *self-fulfilling prophecy*[9]. Regardless of the details, the conceptual idea that knowledge begets knowledge is firmly etched in our scientific mythos.

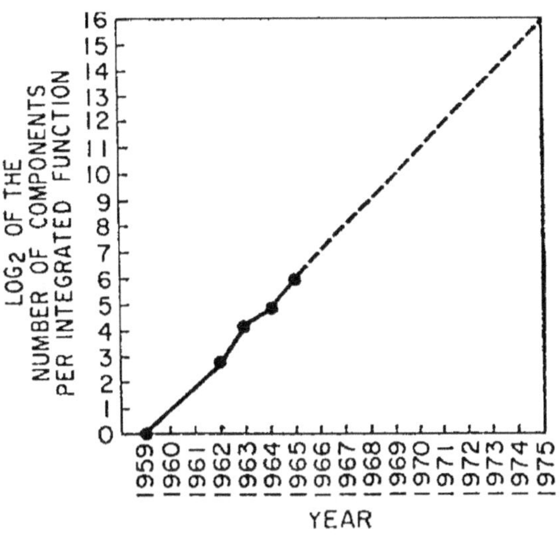

If any field of science actually followed this cartoon history, it was electricity. For the past 2,500 years, the Western understanding of electricity progressed in a monotonically increasing manner from Thales's discovery of static electricity to Moore's law today. This is likely because, as opposed to many other fields[10], physicists mostly got electrical theory right the first time around.

We begin our book with the work of an Englishman, although he later revolted against his king. This experimentalist not only developed the most precisely verified law in all of physics, *the conservation of charge*, but developed a framework for understanding electricity as a fluid of particles that we still use today. Or, in his words:

---

8 G.E. Moore, *Cramming More Components onto Integrated Circuits*, Electronics, April 19, 1965, pp. 114-117.

9 Had Moore not come up with his law, the speed of computers probably would have still increased pretty much exponentially, but not spot on as it has for the past 50 years. Afterall it is great for marketing if they can keep it up, as we can see from the Intel website, which says about Moore's law: "A Forecast and a Challenge" and "Expoential Growth that Continues Today."

10 This contrasted sharply with the history of both magnetism and optics, as we will discuss in the introductions to Parts II and III (pp. 255 and 521 respectively).

> The electrical matter consists of particles extremely subtle, since it can permeate common matter, even the densest metals, with such ease and freedom as not to receive any perceptible resistance.[11]

Thus, with the work of Benjamin Franklin, the fluid model of electricity was born.

Over time, Franklin's attention turned from physics to politics, which eventually led him to Paris to persuade the French to declare war on Britain. While there, Franklin met a scientist named Charles-Augustin de Coulomb. Soon afterwards, Coulomb measured the force law between charged objects, which is the topic of our second chapter.

Meanwhile in Italy, Alessandro Volta showed that a pile of electrodes, made of different metals soaked in salt water, causes charge to flow. This invention, then, led to whole batteries of these voltaic cells, which, when placed in series or parallel, could produce steady electrical current that could we used for further study. This, in turn, allowed Ørsted to pass a steady current near a magnetized compass needle.

In the hands of the top European mathematicians, Franklin's ideas and Coulomb's law, turned into a complete theory of electrical charge, electrical forces, and electrical potential energy. This is the topic of our first 5 chapters.

---

[11] Benjamin Franklin, "Opinions and Conjectures, Concerning the Properties and Effects of the Electrical Matter, etc.," letter to Peter Collinson, July 29, 1750, reprinted in The Papers of Benjamin Franklin, vol. 4 (Yale University Press, 1961), p. 9. See also, B. Franklin, Experiments and Observations on Electricity, illustrated by I. Hulett, 5th Ed., (London: F. Newberry at the Corner of St. Paul's Church-Yard, 1774), p. 54.

Plate 3: The Colossus Code Breaking Computer at Bletchley Park[12]

12 This is a 1943 picture of the first programmable electronic computer, which was used by the British to decode German military messages during the second world war. This picture is part of the National Archives of the United Kingdom, record number FO 850/234.

Plate 4: Entertaining Electrical Demonstrations[13]

---

[13] See the enlightening paper by Paola Bertucci, "Sparks in the dark: the attraction of electricity in the eighteenth century," Endeavour, **31:3** (2007), 88-93. The figures were made for the French translation of William Watson's Experiments and Observations tending to illustrate the Nature and Properties of Electricity (London: The Royal Society, 1746). The French translation is titled Expériences et observations pour server à l'explication de la nature et des propriétés de l'électricité (Paris, 1748).

Chapter 1 Charge

A boy appears to hover in midair, but closer inspection reveals fine silk threads suspending him above the ground. A turning wheel touches the soles of the boy's bare feet and his hand reaches out to a little girl standing on a stool. Her dress puffs out and her hands begin to attract bits of finely chopped straw. A pretty young lady stands on the stool, and invites a suitor to kiss her in the dark. Much to his chagrin, the gentleman is thwarted by a bright spark of fire to his lips. The woman retains her virtue, and the man must reclaim his masculinity by igniting a vat of spirits with a metal sword.

In the eighteenth century, such crowd-pleasing demonstrations not only showed that human beings could hold quite a bit of what's now known as electric charge, but also that electrical effects could travel—from wheel, to boy, to girl, to woman, to man, to sword, to spirits. Electric charge is the subject of this chapter.

The "flying boy" demonstration is only one example of Stephen Gray's extensive experimentation with threads, wires, moist fibers and electrical fire. Gray's work established that electrical effects could travel great distances—800 feet in once case—and cast doubt on the earlier notion that electricity was firmly attached to specific materials. During the eighteenth century, investigators showed that, like heat or liquid, electricity could flow from one object to another, sometimes without direct contact. Electricity, it seemed, was like a fluid.

Charles Francois de Cisternay du Fay was among those who continued where Gray's work left off. He noted that a glass tube could charge and then repel a light gold leaf, while resins such as amber could attract the same leaf. Such experiments led him to propose a model of electricity that consisted of two fluids: *vitreous* (relating to the Latin word for glass) and *resinous* (for resin). Du Fay further noted the difference between conductors, which he called (somewhat counterintuitively) *non-electrics* and insulators, which he called *electrics*. Du Fay reported that bodies *electrified* (we would now say *charged*) with vitreous electricity attract bodies electrified with resinous electricity but repel other bodies electrified with vitreous electricity. Du Fay hadn't helped at all to answer the question of how electricity could act through the air. Instead, he raised another question: where did the two apparent types of electricity come from?

Benjamin Franklin built on this work and showed that while objects contain charges, they are usually neutral. His use of the less-colorful terms *positive* and *negative* for the two types of electric charge replaced du Fay's *vitreous* and *resinous* electric fluids and conveyed the idea that a neutral body contains an equal amount of both. In Franklin's view, a body becomes electrically charged because of a transfer of electricity. A negatively charged body, for example, can be thought to have an excess of *negative* charge, or a deficiency of *positive* charge.

Coulomb subsequently proposed a perfectly good rival theory to Franklin's one-fluid theory, which was the *two-fluid theory of electricity*. In Franklin's model there is one common electrical fluid, and the two types of charge merely reflect a deficit or surplus of this common fluid. On the other hand, Coulomb contended that there are separate positive and negative electric fluids, and that each fluid cancels the other when present in equal amounts. Both models have some merit, along with deficiencies, but their best features are now contained within our understanding of the atomicity of charge.[14]

---

[14] Roller and Roller, The Development of the Concept of Electric Charge: Electricity from the Greeks to Coulomb (Cambridge, Mass.: Harvard University Press, 1954), 80-1.

In this chapter, we will:

- Develop the ideas of charge conservation, electric current, and current density,
- Discuss the continuity equation, and some of the simpler solutions to the partial differential equations that govern the flow of electric charge.

1.1 *Charge Conservation*

This section introduces the principle of the conservation of charge and the experiments of eighteenth century scientists such as Benjamin Franklin that led to its discovery. Franklin's earliest work involved charging various objects by rubbing them with buckskin, and studying how charge moves between different bodies. Franklin interpreted his results by conceiving of electric charge as a conserved quantity, and neutral matter as composed of equal amounts of positive and negative charge. A charged body could then be explained as having an imbalance of electric charge. An object which is positively charged, for example, has a net surplus of positive charge, or what amounts to the same thing, a net deficit of negative charge (Discussion 1.1, p. 12).

Franklin was then in a position to explain the action of the Leyden Jar, which is a device that stores electric charge. From a series of Leyden Jar experiments (Thought Experiment 1.2, p. 15) Franklin concluded that charge is neither created nor destroyed, but rather that it can only be transferred from one body to another. This fundamental law of nature has been confirmed in every experiment performed since, including modern nuclear and particle experiments, including a test that conserves charge to better than one part in 100 septillion (Discussion 1.3, p. 19).

Conservation of electric charge is one of the most important laws of nature, and it is hard to imagine how radically electrodynamics, and the rest of physics, would be altered from its present form if it were not true. For example, charge conservation places a fundamental restriction on the building blocks of normal matter and the electromagnetic forces that operate between them.

Discussion 1.1: Franklin's Sign Convention for Charge

Franklin's earliest work involved the electrification of various objects and the study of how charge transfers between different bodies. Franklin interpreted the results of these experiments by conceiving of electric charge as a conserved quantity and neutral matter as composed of equal amounts of positive and negative charge. He could then explain a charged body as an imbalance of electric charge. An object which is positively charged, for example, has a net surplus of positive charge, or what amounts to the same thing, a net deficit of negative charge.

The question naturally arises that if we are to call one kind of charge positive and the other kind negative, how should we define positive and negative? This choice is arbitrary, but it is important that scientists agree on which is positive and which is negative for efficient communication. In twenty-first century hindsight, we say that common matter is composed of positively charged protons and negatively charged electrons, but Franklin did not know of such things. He did know how to charge various rods by rubbing, so he defined positive and negative charge by rubbing buckskin and green glass together as he described in a letter:

> We suppose, as aforesaid, that electrical fire is a common element, of which every one of the three persons above-mentioned has his equal share, before any operation is begun with tube. *A*, who stands on wax and rubs the tube, collects the electrical fire from himself into the glass; and his communication with the common stock being cut off by the wax, his body is not again immediately supply'd. *B*, (who stands on wax likewise) passing his knuckle along near the tube, receives the fire which was collected by the glass from *A*; and his communication with the common stock being likewise cut off, he retains the additional quantity received. - - - To *C*,

> standing on the floor, both appear to be electrised: for he having only the middle quantity of electric fire, receives a spark upon approaching *B*, who has an over quantity; but gives one to *A*, who has an under quantity. If *A* and *B* approach to touch each other, the spark is stronger, because the difference between them is greater: After such touch there is no spark between either of them and *C*, because the electrical fire in all is reduced to the original equality. If they touch while electrising, the equality is never destroy'd, the fire only circulating. Hence have arisen some new terms among us: we say *B*, (and bodies like circumstanced) is electrised *positively*; *A*, *negatively*. Or rather, *B* is electrised *plus*; *A*, *minus*. And we daily in our experiments electrise bodies *plus* or *minus*, as we think proper.[15]

Franklin did not include a figure in the letter, so we will ask you to make the pictures yourself in Problem 1.1.

Franklin developed the idea that charge ("electrical fire") is conserved ("a common element") and is contained in matter in equal quantities of positive and negative charge. Moreover, he clearly explains how to charge ("electrise") a body as either positive or negative. We retain this sign convention for electric charge today.

Franklin's sign convention was a choice that he made arbitrarily. There was no experiment that Franklin, or anyone else at the time, could perform to know whether electricity was actually a fluid, and in which direction it flows. There were no one-way valves, such as the diodes we now take for granted, that could distinguish between the flow of positive charge in one direction, negative charge in the other, nor the case where both kinds of charges flow in opposite directions. For most electrical appliances it does not even mater which way the electricity flows, much less whether the substance that flows is actually positive or negative. In fact, even in some branches of physics, such as solid state, it is often easier to count positive *holes* rather than negative *electrons*.

In 1879, Edwin Hall discovered that negative charge flows in most wires,[16] which means, unfortunately, that Franklin made the wrong choice for his sign convention. Of course, by 1879 Franklin's sign convention had been established for over a century, and it was impractical to change it. Today's physics students learn that negative charge actually flows through wires backwards from the direction of the current.

Problem 1.1: Franklin's Letter

On page 12 we included an extensive quote from Benjamin Franklin that defined what we consider to be positive and negative charge. For this problem, you will need to reread the quote carefully and clearly diagram persons A, B, and C and the experiments they conduct.

(a) Clearly explain, in drawings and text, the flow of positive charge in each of Franklin's scenarios.

(b) Explain how "we can daily in our experiments electrise bodies plus or minus as we think proper." In other words, how could one charge an object using only a green glass rod and a piece of buckskin. If you had access to such things, could you reproduce the experiments?

---

[15] Letter to Peter Collinson of May 25, 1747, <u>The Papers of Benjamin Franklin</u>, vol. 3 (New Haven: Yale University Press, 1961), 126-34.

[16] In some wires positive charge appeared to flow, which we now know is due to a subtle quantum mechanical effect.

(c) The first conclusive experiment showing that negative charge flows rather than positive charge was conducted over a century after Franklin's work. Draw diagrams, as in part (a), but showing Franklin's thought experiment as a flow of negative charge rather than positive charge.

(d) Were there any experiments that could possibly have been conducted in the eighteenth century to distinguish between the three following models of charge flow: (1) only positive charge flows, (2) only negative charge flows, or (3) both charges flow equally?

Thought Experiment 1.1: Charging an Electroscope

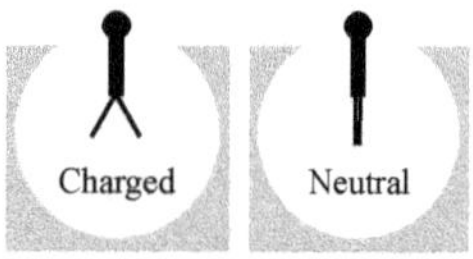

An electroscope is a device that indicates whether it is charged and approximately how much charge is on it. The simplest electroscope, seen in most physics demonstrations, consists of a metal knob connected to two gold foil leaves that spread apart when they are charged (either positively or negatively) and relax when they are neutral. The foil leaves are inside a conducting container that is insulated from the knob and leaves.

In this thought experiment, we show two methods of charging an electroscope with a charged wand. After showing the experiments, we will discuss how to interpret the results. It is important to realize that many ideas that the modern student takes for granted, like the existence of moveable electrons, were not discovered until well over a century after Franklin and his colleagues conducted their experiments.

The diagram shows the charging of an electroscope by *contact* with a positively charged rod. The steps shown are: (a) The electroscope is initially neutral. (b) A charged tube approaches; the leaves separate. (c) Upon contact, the leaves have maximum separation. (d) When the rod is removed, the electroscope remains charged. (e) Now the rod and electroscope are both charged in same sense (positive).

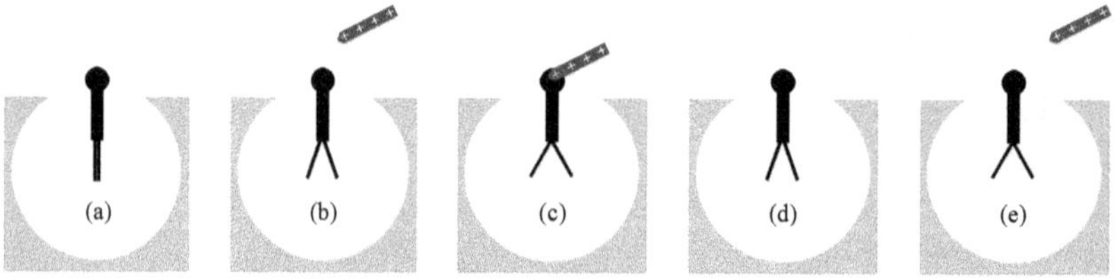

This next diagram shows the charging of an electroscope by *induction*, again with a positively charged rod. The steps shown are: (1) The electroscope is initially neutral. (2) When a charged rod approaches the knob, the leaves separate. (3) Ground the electroscope. (4) Remove the ground wire before the rod. (5) When the rod is removed, the leaves separate. (6) Now the electroscope is charged in the opposite sense (negative) as the rod.

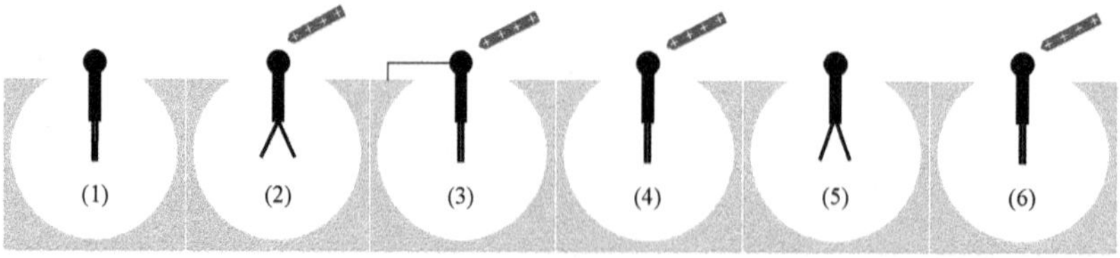

We will now interpret these experiments using a fluid model of charge flow. First notice that holding a charged rod near the knob separates the leaves. We can explain this as the leaves having the same kind of charge, so they repel each other. This could be caused by positive charge moving away from the positive rod and into the leaves, or negative charge moving toward the rod and out of the leaves. Regardless of what really moves, the quantity *charge* flows away from the rod and toward the leaves.

Until contact is made, no charge can enter or leave the electroscope as a whole: charge can only move between the knob and the leaves. Thus if no contact at is made, the leaves relax when we remove the wand.

When contact is made, as in *charging by contact*, charge flows from the rod to the electroscope. This would suggest that charge is a substance that somehow distributes itself throughout the rod and the electroscope. After removing the wand, the electroscope retains a net *like* charge.

On the other hand, in the case of *charging by induction*, holding the rod near the grounded electroscope sends like charge out of the electroscope through the grounding wire. When the wire is disconnected, before the rod is removed, a net *opposite* charge remains.

With these two charging techniques, and the sign convention discussed on page 12, eighteenth-century scientists could produce objects of either positive or negative charge at will, allowing for even more sophisticated experiments like the ones by Benjamin Franklin we discuss below.

Thought Experiment 1.2: Franklin's Leyden Jar Experiments

Franklin conducted many experiments using a "wonderful bottle" that the Dutch scientist Pieter van Musschenbroek invented in the city of Leiden. This *Leyden jar* consists of an insulating bottle filled with salt water and surrounded on the outside by a thin conductor. A conducting knob sticks out and connects to the inside of the bottle but is insulated from the outside. We encourage you to create your own Leyden jar out of everyday materials, such as a soda bottle, a metal skewer, and aluminum foil.

By choosing whether to set the bottle on blocks of wax (drawn as blocks) or lead (drawn as cylinders), Franklin was able to isolate the bottle from its surroundings. This allowed him to complete the following experiments, described in a 1747 letter.[17]

In his Figure I, Franklin shows that the outside of the bottle, which is isolated from ground, must always keep an equal and opposite charge as the inside, so the bottle remains net neutral.

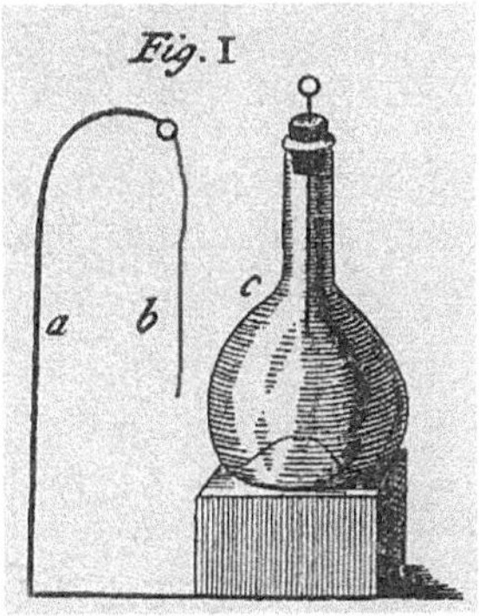

> (Fig. I) Form a bent wire (a) sticking in the table. Let a small [conducting] linen thread (b) hang down within half an inch from the electrised phial (c). Touch the wire [on top] of the phial repeatedly with your finger, and at every touch you will see the thread instantly attracted by the bottle. (This is best done by a vinegar cruet, or some belly'd bottle). As soon as you draw any fire out from the upper part, by touching the wire, the lower part of the bottle draws an equal quantity in by the thread.

Next, Franklin shows how a bobber is repeatedly attracted and repelled, as it oscillates between having a positive and negative charge:

> (Fig. II) Fix a wire in the lead, with which the bottom of the bottle is armed (d) so as that bending upwards, its ring-end may be level with the top of the ring-end of the wire in the [insulating] cork

---

[17] Letter to Peter Collinson of July 28, 1747, The Papers of Benjamin Franklin, vol. 3 (New Haven: Yale University Press), 156. The figures are excerpted from B. Franklin, Experiments and Observations on Electricity, illustrated by I. Hulett, 5th Ed., (London: F. Newberry at the Corner of St. Paul's Church-Yard, 1774).

(e) and at three or four inches distance. Then electricise the bottle, and place it on wax.

If a cork is suspended by [insulating] silk thread (f) hang between these two wires, it will play incessantly from one to the other, 'til the bottle is no longer electrised; that is, it fetches and carries fire from the top to the bottom of the bottle, 'till the equilibrium is restored.

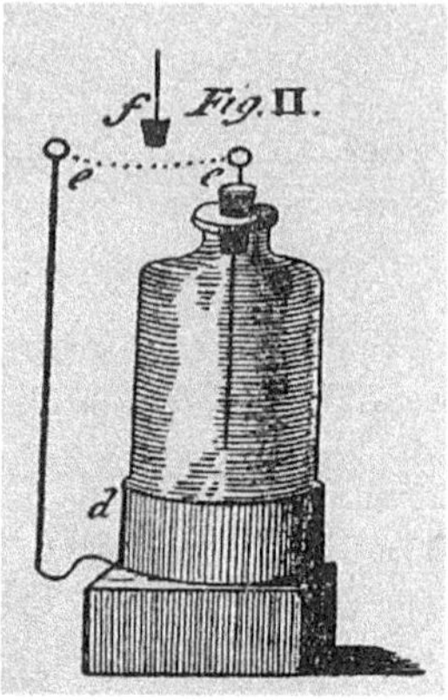

Franklin now insulates the whole Leyden jar from ground, and shows that it can be discharged without any connection to the Earth. This is probably his simplest and most compelling demonstration that charge is a conserved quantity, rather than simply a natural property of the Earth.

(Fig. III) Place an electrised phial on wax [to isolate it from the earth]; take a wire (g) in form of a C, the ends at such a distance when bent, as that the upper may touch the wire of the bottle, when the lower touches the bottom: stick the outer part on a stick of sealing-wax (h), which will serve as a handle; then apply the lower end to the bottom of the bottle, and gradually bring the upper end near the wire in the cork. The consequence is, spark follows spark till the equilibrium is restored. Touch the top first, and on approaching the bottom with the other end, you have a constant stream of fire from the wire entering the bottle. Touch the top and bottom together, and the equilibrium will instantly be restored: the crooked wire forming the communication.

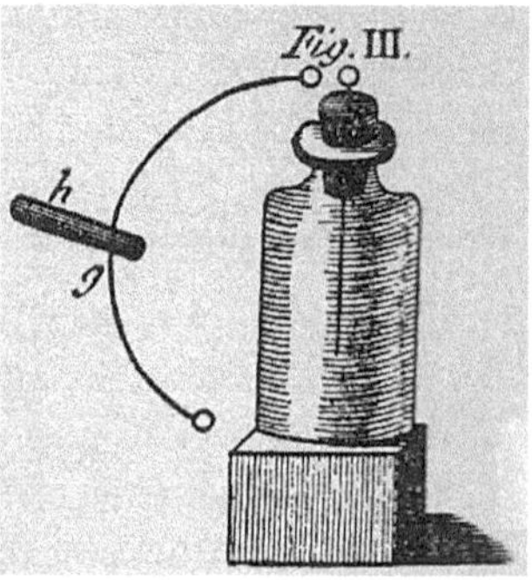

Extending the preceding experiment, Franklin describes discharging the jar.

(Fig. IV) Let a ring of thin lead, or paper, surround a bottle (i) even at some distance from or above the bottom. From that ring, let a wire proceed up, till it touch the wire of the cork (k). A bottle so fixt cannot be any means be electrised: the equilibrium is never destroyed: for while the communication between the upper and lower parts of the bottle is continued by the outside wire, the fire only circulates: what is driven out at bottom, is constantly supplied from the top. Hence a bottle cannot be electrised that is foul or moist on the outside, if such moisture continue up to the cork or wire.

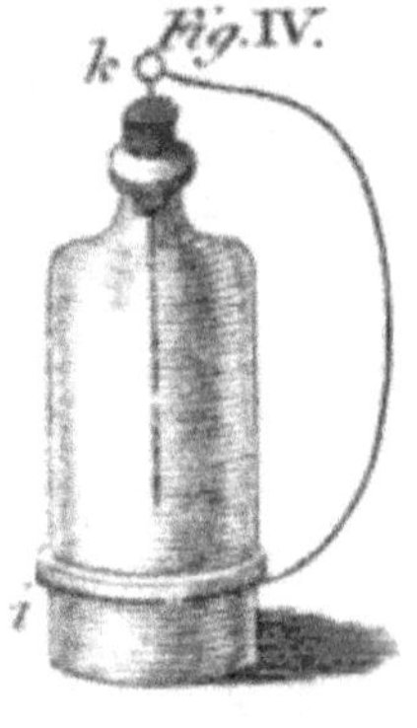

And finally Franklin describes a dramatic demonstration of this using a book decorated with gold foil in a darkened room. Pay particular attention to his footnote; notice how he reinforces the result of the experiment in his Figure III above.

(Fig. V) The passing of the electrical fire from the upper to the lower part of the bottle, to restore the equilibrium, is rendered strongly visible by the following pretty experiment. Take a book whose covering is filletted with gold; bend a wire of eight or ten inches long in the form of (m). Slip it on the end of the cover of the book, over the gold line, so as that the shoulder of it may press upon one end of the gold line, the ring up, but leaning towards the other end of the book. Lay the book on a glass or wax*, and on the other end of the gold lines set the bottle electrised: then bend the springing wire, by pressing it with a stick of wax till its ring approaches the ring of the bottle wire, instantly there is a strong spark and stroke, and the whole line of gold, which completes the communication, between the top and bottom of the bottle, will appear a vivid flame, like the sharpest lightning. The closer the contact between the shoulder of the wire and the gold at one end of the line, and between the

bottom of the bottle and the gold at the other end, the better the experiment succeeds. The room should be darkened. If you would have the whole filletting round the cover appear in fire at once, let the bottle and wire touch the gold in the diagonally opposite corners.

* Placing the book on glass or wax is not necessary to produce the appearance; it is only to show that the visible Electricity is not brought up from the common stock of the earth.

These experiments demonstrated to Franklin that the total charge on the bottle never changes. Notice how he stresses that it is unnecessary to ground the bottle in order to bring it back to equilibrium. This shows that charge is not unique to the Earth, but rather the Earth is so large that it has the ability to neutralize either positive or negative objects without significantly changing its own neutrality.

This leads us to our first fundamental law, the conservation of charge.

Fundamental Law 1: The Conservation of Charge

In any closed system, charge cannot be created or destroyed.

Definition 1.1: The Coulomb

The *Système international* (SI) unit of charge is called a *coulomb* with the symbol C. The 26th General Convention on Weights and Measures (CGPM) recently redefined the coulomb quantum-mechanically such that "the elementary charge e is 1.602 176 634 × 10^{-19} C." [18]

In other words, one coulomb is now defined to be 6,241,509,074,460,762,608 elementary charges, or about six pentillion elementary charges.

Problem 1.2: Franklin's Bells

In the picture shown,[19] the top bar is connected to the outside of a charged Leyden Jar. When the bottom chain is connected to the inside of the same Leyden Jar, the bells ring. Notice in the picture that the bar and chains are conducting and the threads are insulating.

a) Discuss qualitatively with diagrams why the bells ring. What is the mechanism for discharging the Leyden Jar?

b) Imagine that each ball can hold a charge q and that the period of oscillation for each clapper is P. What is the total *current* (see Definition 1.2, p. 21) I flowing out of the Leyden jar?

Problem 1.3: Franklin's Tower

In a letter dated June 29, 1755, Benjamin Franklin wrote about a church that was struck by lightning. In this church, the bell in its tower was connected via a metal wire to a clock on a wall located two floors below the bell. "The wire was not bigger than a common knitting needle," Franklin wrote. When the lightning struck, the spire above the bell was "split to all pieces by the lightning," and below the clock "the building was exceedingly rent and damaged, and some stones in the foundation-wall torn out, and thrown to a distance of twenty or thirty feet." Yet in the region between the bell and the clock there was no damage to the church. While the clock

---

[18] From Resolution 1, of the resolutions adopted at the 26th CGPM, Versailles, France, 13-15 of Novermber, 2018.

[19] Figure from George Adams' Lectures on Natural and Experimental Philosophy (first American printing 1806).

pendulum was still intact, "no part of the afore-mentioned long small wire, between the clock and the hammer, could be found, except about two inches that hung to the tail of the hammer, and about as much that was fastened to the clock; the rest being exploded, and its particles dissipated in smoke and air, as gunpowder is by common fire."[20]

a) Based on Franklin's observations, estimate the ratio of the current density flowing through the wire to the current density flowing through the clock's pendulum, which was "about the thickness of a goose quill." (Hint: estimate the size of these objects.)

b) Presumably, the current density flowing through the spire was less than the current density flowing through the pendulum. Discuss, qualitatively, why the spire sustained so much more damage than the clock.

c) Look up the total current of a typical lightning strike, and cite your source. Estimate the current density flowing through the long small wire and the clock pendulum. Express your answer in SI units.

Problem 1.4: A Pointed Lightning Rod

Benjamin Franklin conducted a number of experiments that demonstrated that the more pointed an object is the better it draws the "electrical fire." To illustrate this, consider a conical lightning rod of height h and half opening angle α. Next assume that the charge on the lightning rod is evenly distributed on its lateral surface with a constant charge per area σ. For lightning rods of equal heights, find the average charge per volume of the cone as a function of α.

Problem 1.5: The Dissipation Law of Coulomb

According to the so-called dissipation law of Coulomb, "the loss of charge in unit time of a conductor placed in still air or moving air is proportional at any moment to the charge still remaining."[21]

(a) Write down this law in terms of the charge remaining on the conductor Q, its time derivatives, and a constant (τ) with units of time.

(b) Assuming that a conductor has an initial charge Q_0, find the charge on the conductor as a function of time.

(c) Discuss qualitatively some physical properties on which you would expect the characteristic time to depend.

Discussion 1.2: Fundamental Laws and Charge Conservation

Physicists have often discovered laws of nature only to find later that these "laws" are restricted in scope and apply only to particular situations. In physics, we strive to find the fundamental laws of nature. These laws must, by definition, apply in every situation, everywhere, and without exception. This book is about *classical* electrodynamics, so when we introduce a fundamental law we mean that it is a fundamental law so far as physical understanding stood until about 1900. (Some of these fundamental laws are still considered valid today.)

---

[20] Letter to Thomas-François Dalibard of June 29, 1755, The Collected Papers of Benjamin Franklin, vol. 6 (New Haven: Yale University Press, 1961), 97-100.

[21] See Hess V.F. 1928, The Electrical Conductivity of the Atmosphere and its Causes, 16, where he sites C.A. Coulomb, *Mém. de l' Acad. Paris*, 1785, 616.

To be more specific, in 1905 Albert Einstein wrote four revolutionary papers while holding a day job as a patent examiner. Two of these papers offered new interpretations of electrodynamics, another presented a mathematically sophisticated argument that atoms exist based on the kinetic theory of heat, and the last of the four papers was a brief two page argument justifying one of the most famous formulas in twentieth century physics. (The predictions of the kinetic theory paper had already been observed as the so-called Brownian motion of pollen grains suspended in solution.)

The third of these papers, *On the Electrodynamics of Moving Bodies*, suggested that the laws of classical electrodynamics were more fundamental than Newton's laws of motion. Thus, in areas where the two theories were in conflict, Newton's laws must yield to electrodynamics. Einstein's *Theory of Special Relativity* (as it was eventually named) soon became widely accepted, and it was also assumed to apply everywhere without exception, while Newton's laws of motion apply only to objects moving at relative velocities much less than the speed of light.

Einstein's first 1905 electrodynamics paper proposed that instead of being an electromagnetic wave, light is actually composed of particles or *photons*. This new photon idea was not readily accepted, mostly because it wasn't at all clear how a light "quantum" or "particle" could also display wave properties like diffraction, interference, and polarization. Einstein's light particles contradicted Maxwell's wave theory of light as well as Newton's classical laws of mechanics.

When we introduce a fundamental law in this book, we mean that it is fundamental for everything, everywhere, in all contexts that do not relate to quantum mechanics and particle physics. Thus, *classical* includes Einstein's theory of relativity. The opposite of *classical* is *quantum.*

That said, charge conservation holds good—everywhere and at all times. It is completely compatible with quantum electrodynamics, and has been tested to unprecedented accuracy as we discuss next. If there is anything we know about the universe, it is that charge is conserved.

Discussion 1.3: Modern Tests of Charge Conservation

In modern experiments, an electron may interact with its anti-particle, the positron, which results in annihilation and other particles called gamma ray photons. Thus two particles with opposite charge interact and result in particles with no charge. Where did the charge go?

In this case, the negative electric charge of the electron exactly balances the positive electric charge of the positron. The net charge of the system (electron plus positron) is zero before the interaction, and it remains zero after annihilation. This thought experiment discusses additional examples of charge conservation in modern experiments.

The status of conservation of charge as a fundamental law is rooted in experiment: we have yet to observe any violations of this law. Since Franklin's day, scientists have conducted innumerable experiments to test charge conservation. Here we first describe a series of common particle decay reactions that conserve charge followed by a recent experimental limit on charge conservation.

We begin with the radioactive decay of uranium into thorium, which is also used to find the age of the oldest rocks in the solar system:

$$ {}^{238}_{92}\mathrm{U} \rightarrow {}^{234}_{90}\mathrm{Th} + {}^{4}_{2}\mathrm{He}. $$

In this equation the letters indicate the elements, while the superscript number (the mass number) indicates the total number of neutrons and protons in the nucleus and the subscript number (the atomic number) is the number of protons.

The radioactive nucleus of the uranium atom, which contains 92 positively charged protons, spontaneously decays by emitting an alpha particle (^{4}He). A thorium nucleus, which contains 90 protons, remains. Since the alpha particle comprises two protons, the net electric charge present before the radioactive decay is identical to the net charge after the decay.

Another example is positron emission (a form of beta decay), where a proton decays into a neutron, releasing a positron and a neutrino:

$$p^+ \rightarrow n^0 + e^+ + \nu_e .$$

This process, which is crucial in powering stars, conserves electric charge, because the proton and the positron have the same charge, and the neutrino and neutron are electrically neutral. Ordinary beta decay, in which an electrically neutral neutron is converted into a proton, an electron, and an anti-neutrino, also conserves charge:

$$n^0 \rightarrow p^+ + e^- + \overline{\nu}_e .$$

Stronger evidence for charge conservation comes from looking for particle decays that would occur if electric charge were not conserved. One example of such a charge non-conserving decay is given by the hypothetical decay of a neutron into a proton, a neutrino, and antineutrino:

$$n^0 \rightarrow p^+ + \nu_e + \overline{\nu}_e .$$

Such a decay has never been observed, but not for want of searching. In 1996, Eric Norman, John Bachall, and Maurice Goldhaber tested this in what would be an otherwise favorable decay of Gallium-71 to Germanium-71 to better than one part in 100 septillion (10^{26}).[22]

The law of the conservation of charge has been exhaustively tested since its inception over two and a half centuries ago. Even when using the most sophisticated experimental techniques, researchers have yet to observe any process where charge is not conserved.

1.2 *Current*

Franklin presented charge as a substance that can be neither created nor destroyed, and this is borne out by modern experiment to unprecedented accuracy. Of course, charge can move from one place to another, and so we define *current*, I, as the rate of charge flow. We use the letter I for the historical reason that André-Marie Ampère originally called it *intensité de courant* (current intensity). Current is not analogous to a canoeist's notion of river current (velocity), but is similar to the total water flow past a position on the bank—huge for the Mississippi river but small for a swift mountain stream. Current flows in a particular direction, making it seem to be a vector quantity in the direction of positive charge flow. However, the vector nature of current becomes more subtle when the flow of charge is not in a constant direction. For this reason, we will define the quantity *current density*, $\vec{J}$, to be the current per area. Unlike current, the current density is a vector with a well-defined direction, and is analogous to velocity from hydrodynamics.

---

[22] Norman, Bachall, Goldhaber, "Improved Limit on Charge Conservation derived from 71Ga Solar Neutrino Experiments," Phys. Rev. D, **53** (1996), 556.

We begin this section by defining the current density as the current per perpendicular area, and then considering the current associated with a moving charge and charge flow in a neutral medium (Thought Experiment 1.4, p. 23). We elaborate on the idea of current density by considering the flow of charge in simple situations, such as in a neutral wire of constant cross-sectional area, in a wire with a tapering cross-section (Example 1.3, p. 32), on the plates of a capacitor (Thought Experiment 1.5, p. 29), and surrounding a charged sphere (Thought Experiment 1.6, p. 30).

Definition 1.2: Current

Current is the quantity of charge that passes through a surface per unit time.

The SI unit of current is the *ampere*, which was recently redefined as:

> The ampere, symbol A, is the SI unit of electric current. It is defined by taking the fixed numerical value of the elementary charge e to be $1.602\ 176\ 634 \times 10^{-19}$ when expressed in the unit C, which is equal to A s, where the second is defined in terms of $\Delta \nu_{Cs}$.[23]

Discussion 1.4: Units of Charge

Units of charge have been the subject of particular controversy over the years, and physics students are sometimes caught in the middle of these disagreements. The unit conventions of various fields of physics, and of many texts, are so different that they even express the equations of electrodynamics differently. The primary debate has been whether to define a unit of charge electrostatically (based of the forces between charges) or electromagnetically (based on the forces between wires). While this may seem like an arcane detail, in the mid twentieth century it became a proxy war as to the nature of physics itself. Is physics a quest for the simplest, and most elegant, universal laws governing the universe? Or, is physics a set of applicable mathematical tools for understanding how things work? Is elegance more important than practicality, or the other way around?

Before the discovery of electromagnetism, the unit of charge had to be defined electrostatically, as the charge needed to produce a unit of force (say one dyne) between two identical objects separated by one unit of length (say on centimeter). With the discovery of electromagnetism, it became possible to measure current flow using magnets, but only if the instruments were calibrated based on the forces between wires. This led to an electromagnetic definition of the ampere, as the current needed to produce two dynes of force between parallel wires one meter long and one centimeter apart. When the two definitions were eventually reconciled, and put in consistent units, it turned out that one unit of electromagnetic charge was greater than one unit of electrostatic charge by a factor of the speed of light.

In Part I of this text we follow the school of thought that favored the electrostatic system of units, while in Part II we develop the approach that favored the electromagnetic system of units. In Part III we reconcile the two. If it were particularly important for students to learn both unit systems equally well, we would use one in Part I and the other in Part II. Thankfully, science has mostly standardized around one system, the SI, so that is what we will use. Nevertheless, you will sometimes see equations that look like the ones in this book, but with factors of the speed of light or 4π that you do not recognize. This is because they are using the electrostatic, or Gaussian, system of units.

---

[23] Appendix 3 of the resolutions adopted at the 26th CGPM, Versailles, France, 13-15 of Novermber, 2018.

By the late 1800s, the meter-kilogram-second (MKS) system was in development. In 1901, the Italian electrical engineer, Giovanni Giorgi, proposed a coherent set of units based on the MKS system, and the electrical engineering unit of resistance—the *ohm*. This system, with four base units, was called the MKSΩ system. This was replaced by a similar system that used the electromagnetic definition of the ampere as the base unit, called the MKSA system, and was adopted by the General Convention on Weights and Measures (CGPM) in 1948. However, since the MKSA system requires an additional base unit, and an additional constant in Maxwell's equations, it is considered less elegant so many physicists still use electrostatic units.

We have adopted the SI, which is based on the MKSA system, as our unit system of choice, primarily because it is now the international standard for all branches of science and engineering. Thus, while many scientists and engineers prefer to work in specialized units, all can speak SI. Moreover, the SI is the most rigorously defined unit system in the world.

Since you will encounter electrostatic units in your physics career, we have included a description of them, and tips on translating between unit systems, in Appendix D (p. 719).

Thought Experiment 1.3: A Charged Arrow Shaft

Consider an arrow that, after being rubbed with a cloth, contains a charge per length of λ (S.I. units: $\mathrm{C/m}$). Imagine that it is shot off at a velocity $\vec{v}$. What is the current associated with the moving arrow?

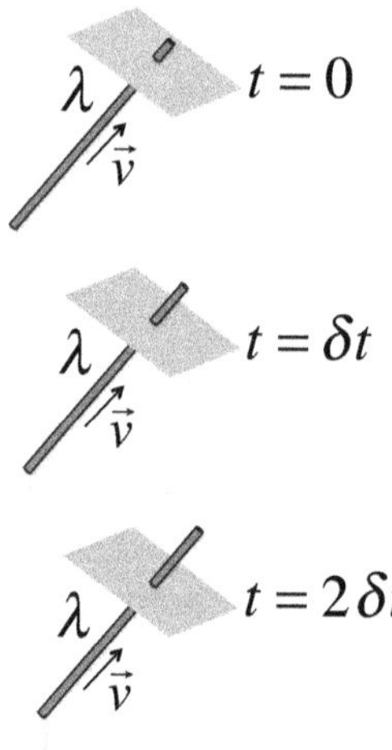

First, we draw an imaginary surface that the arrow will pass through. Now, in a given time interval δt, the arrow will have traveled a distance $\delta \ell = v\delta t$, so the corresponding small charge that will have passed through the imaginary plane is given by:

$$\delta Q = \lambda \delta \ell = \lambda v \delta t .$$

Thus, the total charge per time (current) that crosses the perpendicular surface is given by:

$$I = \frac{\delta Q}{\delta t} = \lambda v .$$

This leads us to a common question, is current a vector or a scalar? Here, the velocity is a vector, and charge is moving in the same direction of the arrow if $\lambda > 0$, or in the opposite direction if $\lambda < 0$. So, we would think that we can just put a vector sign on the current above.

While this makes sense in this case, what if the current diverged, so that at different places it flowed in different ways, such as at the junction of a Y-connector? What is the direction of the current there? Do we really care what it is right there, or do we only care about how much current flows in and how much flows out? This is the important question, so we make a distinction between the scalar current, I, and the vector *current density*, $\vec{J}$ (Definition 1.4, p. 25).

If we care which direction the charge actually flows, at any infinitesimal point in space, then we must consider the current density. On the other hand, if we are interested in the rate that charge flows—in, out, or through—something of finite size, then we are interested in the current. Thus, the current, has one of two directions, in or out, while the current density has an absolute direction.

Thought Experiment 1.4: Current Flow in a Neutral Medium

Consider the arrow in Thought Experiment 1.3 (p. 22) traveling at velocity $\vec{v}$ with linear charge density λ. Now imagine this arrow is moving through a nonconducting stationary shaft of equal and opposite linear charge density. What is the rate of current flow?

The shaft is stationary, so no current is associated with the shaft, only the moving arrow, so the current is:

$$I = I_{Arrow} + I_{Shaft} = \lambda v + 0 = \lambda v .$$

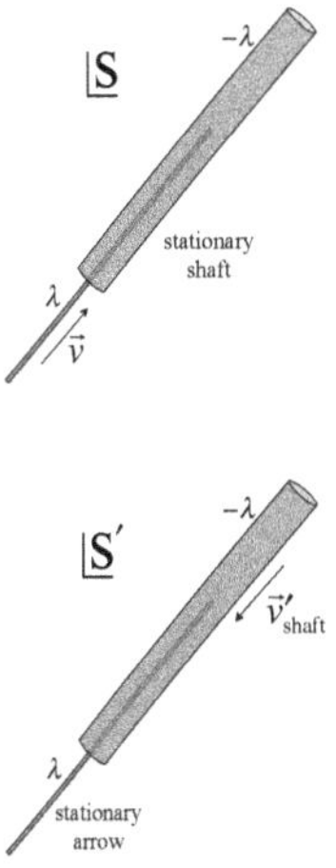

Now imagine an observer is also moving at velocity $\vec{v}$, so that the arrow appears to be stationary. What is the current in the observer's reference frame?

$$I' = I'_{Arrow} + I'_{Shaft} = 0 + (-\lambda)(-v) = \lambda v = I .$$

This illustrates that when current flows in a neutral medium ($\lambda_{\text{arrow}} + \lambda_{\text{shaft}} = 0$), the current is independent of the velocity of the observer.

This is clearly not the case for a moving charge, such as the arrow alone, because the two observers would measure different currents.

Problem 1.6: A Wheatstone Bridge I

A *Wheatstone bridge* can measure very small differences in current using a sensitive current meter called a *galvanometer.* It is considered to be "balanced" if no current flows through the galvanometer. Consider four electrical components configured as in the diagram, with a known total current (I) flowing into the circuit.

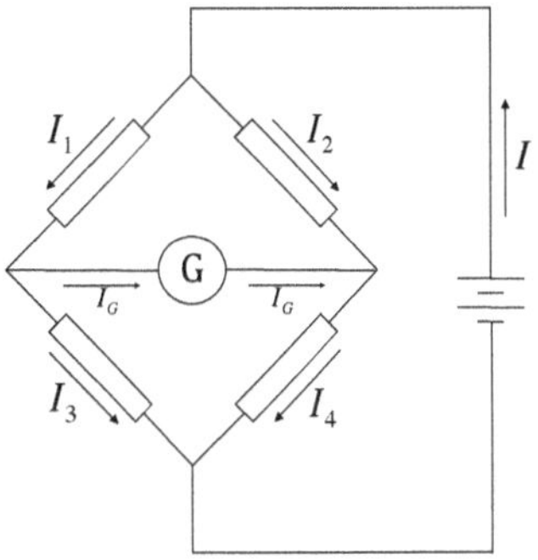

a) Consider a Wheatstone bridge circuit that is built in such a way that a well known fraction of the current, α, flows through each top component, so $I_1 = \alpha I$ and $I_2 = (1-\alpha)I$. When it is balanced, how much current flows through I_3 and I_4?

b) Next imagine that there are two knobs, one to control α and the other to control I_3 directly. How much current flows through the galvanometer in terms of α, I_3 and I?

c) Now imagine that a piece of lab equipment needs exactly I_0 of current for it to work properly, and any fluctuations will increase the experimental error. No current source is perfect, so your job is to design a monitoring program. The maximum fluctuation expected is $\delta I_{\max}$. You find an old galvanometer strip-chart recorder in the lab whose pen deflects the maximum amount when a current $\pm I_G^{\max}$ flows through it. Find the optimum knob settings of α and I_3 for your application.

d) Consider an application requires a steady flow of exactly one amp, but your current supply is only guaranteed to produce within $1\,\text{mA}$ of the set output. Your strip-chart recorder can record a maximum current of $\pm 100\,\mu\text{A}$, and has a 100:1 sensitivity ratio. How would you set I_3 and α for this application?

Problem 1.7: The Discovery of Cosmic Radiation

Measuring the sources of ionizing radiation was a topic of interest in the early 20th century. Since each ion contains a known charge, the radiation rate could be measured by charging an electroscope and measuring its discharge rate. In the years 1911-1912, Victor Hess made measurements of ionizing radiation from a hot air balloon so he could measure the decrease in naturally occurring radiation with distance from the earth. Hess did measure the expected decrease with altitude close to the earth, but he also measured an unexpected increase in radiation as the balloon rose higher. We credit Hess with the discovery of cosmic rays, which we now believe mostly come from the sun and interstellar shock waves.

Hess describes his electroscope as follows:[24]

> Fig. 9 shows a vertical section of Wulf's radiation apparatus. The distance of the two fibres, which is proportional to the [charge on] the system, is measured by means of the ocular micrometer in the microscope *F*. The fibres are illuminated from behind by mirror *a*. In order to allow the insulation loss to be determined separately, a metal cylinder *J*, fitting fairly closely round the fibre system, is built into the instrument in such a way that it may be let down over the fibres during in insulation test so as to separate them from the rest of the ionisation chamber. The apparatus, when closed, is completely air and watertight, so that is may be employed for absorption experiments under water. The fibre system is charged by means of a lever, not visible from the diagram, which can be operated from the outside.

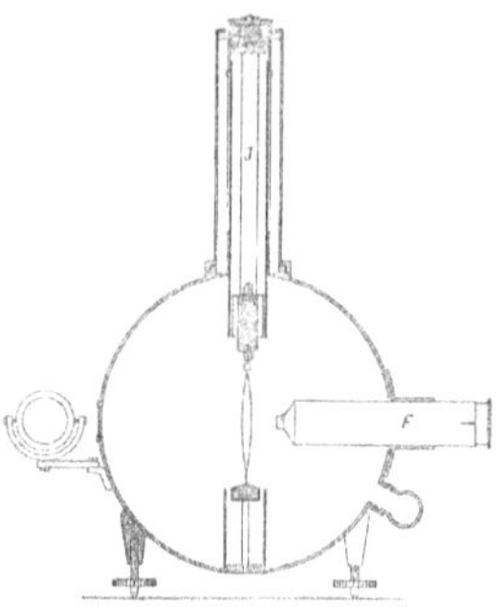

FIG. 9.—Wulf's radiation apparatus.

By measuring the charge as a function of time $Q(t)$, Hess was able to work out the number of ionizing rays that entered his chamber as a function of time:

> An ionising power q = 10 I (i.e. ten pairs of ions produced per cubic centimeter per second) corresponds, in an ionisation vessel of 1 litre volume, to a current of only 1·6 . 10^{-15} amp.

(a) Derive a relationship between Hess' ionizing power q, and the charge on the electroscope Q.

(b) Confirm Hess's relationship between the ionizing power and the current.

Definition 1.3: The Area Vector

The area $\vec{A}$ of a flat surface is defined as a vector whose direction is *the normal* to that surface. Notice that there are two equal, and opposite, perpendicular directions to a flat planar surface. One of these is considered positive, and the other negative, based on the following conventions:

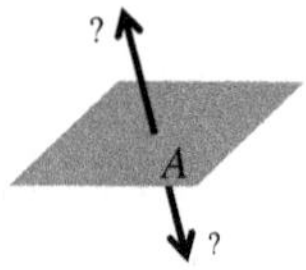

I. On the outside of a closed three-dimensional surface, the area vector points in the outward normal direction.

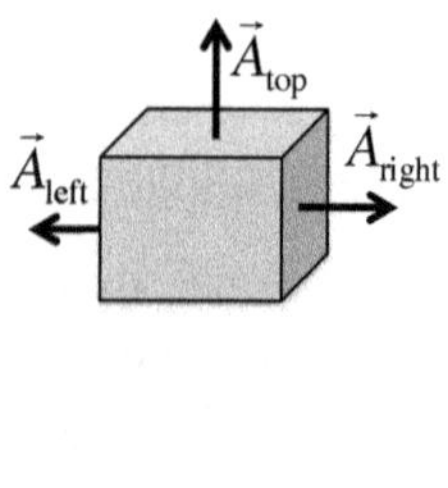

[24] Figure from Hess V.F., The Electrical Conductivity of the Atmosphere and its Causes, 119, Translated by L.W. Codd, published in 1928 by D. Van Nostrad Company, New York.

II.   On a flat surface that is surrounded by a loop, the area vector points in the direction defined using the right-hand-rule, based on the positive direction around the loop.

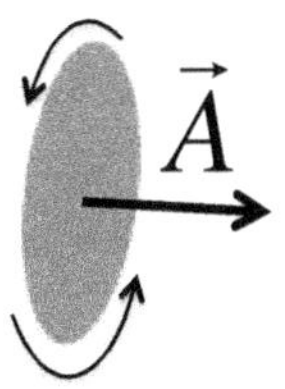

III.   If the plane is the cross-section of a tube that forms a loop, the positive direction forms a closed loop inside the tube.

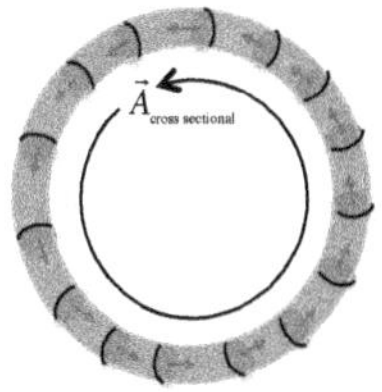

Sometimes, more than one of these rules apply. When this is the case, if you can make two apply then do so.

For example, imagine two coils of wire wrapped around an iron ring, as is a common design for an inductor (picture p. 284). In that case, pick a positive direction for each wire, such that both of the last two criteria hold.

Sometimes, however, these rules explicitly conflict. When they do, simply use common sense when picking a positive direction, and make a clear drawing. If you apply an equation that assumes the other sign convention than you picked, pay special attention to make sure you put the correct negative signs in by hand.

Definition 1.4: Current Density

We define the current density, $\left(\vec{J}\right)$ as the vector whose magnitude is the current per perpendicular area, and whose direction is in the direction that the charge flows. The S.I. unit is: A/m^2.

$$I = \vec{J} \cdot \vec{A} = J\,A\cos\theta$$

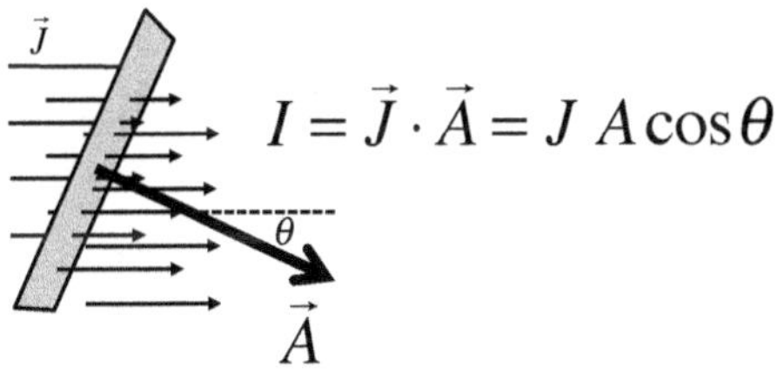

Now that we have defined the current density, we can write the current through any surface as:

$$I = \int\limits_{surface} \vec{J} \cdot d\vec{A}.$$

Notice that the current, I, depends on both the current density and the surface through which the charge flows.

Example 1.1: Unidirectional Current Density

First let us consider a unidirectional current density $\left(\vec{J} = J_{(x)}\hat{x}\right)$ inside a neutral wire. Given the current density at one point $\left(J_0 = J_{(x = 0)}\right)$, our goal is to find the current density at any other position $\vec{r} = x\hat{x} + y\hat{y} + z\hat{z}$.

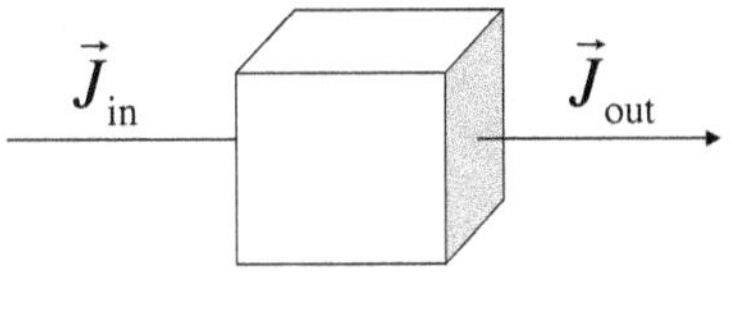

To accomplish this, we draw an imaginary rectangular box of dimensions $b \times a \times a$, with its left hand face at location $x = 0$, and the right hand side located at position $\vec{r} = b\hat{x}$. So, the area vectors of the six faces of the box are: $-a^2\hat{x}, -ab\hat{y}, -ab\hat{z}, a^2\hat{x}, ab\hat{y}, ab\hat{z}$.

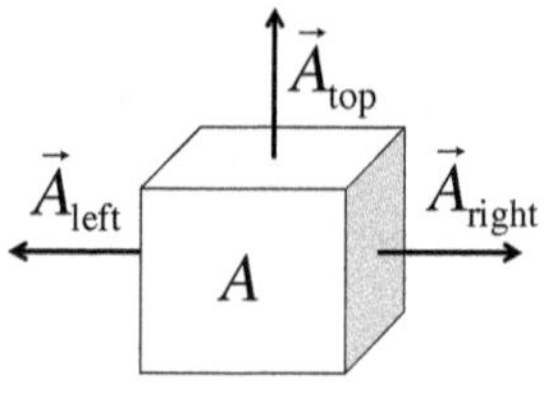

We also note that charge is conserved, and that the wire must remain neutral, so an equal amount of current must flow in as flows out.

First we look at the current density flowing through the left hand face:

$$I_{\text{left}} = \int_{A_{\text{left}}} \vec{J}_0 \cdot \mathrm{d}\vec{A} = \int_{A_{\text{left}}} (J_0\hat{x}) \cdot (-\mathrm{d}A\,\hat{x}) = -(\hat{x}\cdot\hat{x}) \int_{A_{\text{left}}} J_0\,\mathrm{d}A = -(\hat{x}\cdot\hat{x}) J_0 \int_0^a\!\!\int_0^a \mathrm{d}x\,\mathrm{d}y = -J_0 \int_0^a\!\!\int_0^a \mathrm{d}x\,\mathrm{d}y = -a^2 J_0.$$

Notice that the current is negative because the current density is opposite the direction of positive area, so this current is flowing *into* the box.

Next we look at the current density flowing through the top face:

$$I_{\text{top}} = \int_{A_{\text{top}}} \vec{J} \cdot \mathrm{d}\vec{A} = \int_{A_{\text{top}}} (J_0\hat{x}) \cdot (\mathrm{d}A\,\hat{y}) = J_0(\hat{x}\cdot\hat{y}) \int_0^a\!\!\int_0^b \mathrm{d}x\,\mathrm{d}y = 0.$$

Notice that the current is zero because the current density is parallel to the plane, so no current flows across the plane. The dot product between the area vector and the current density vector nicely accounts for this physical fact. Clearly, the current flowing through the other lateral sides is also zero for the same reasons.

Often integrals can be trivially determined by considering their physical meaning. Notice how we wrote the above equation as an integral, but we never needed to evaluate it explicitly. In fact, had we tried to solve the integral we would have obtained:

$$\int_{A_{\text{top}}} J\,\mathrm{d}A = \int_0^a\!\!\int_0^a J(x)\,\mathrm{d}x\,\mathrm{d}y = a\int_0^a J(x)\,\mathrm{d}x\,.$$

Because we do not know, *a priori*, the current density as a function of position, we cannot explicitly evaluate this integral. We do know, however, that no current flows out of the top of the box, and this allows us to solve the problem despite our not knowing $J(x)$. A finite quantity multiplied by zero equals zero, after all, even if we do not know the value of that quantity.

Finally, we look at the current flowing out through the right hand face:

$$I_{\text{right}} = \int_{A_{\text{right}}} \vec{J}(x) \cdot \mathrm{d}\vec{A} = \int_{A_{\text{right}}} (J(a)\hat{x}) \cdot (\mathrm{d}A\,\hat{x}) = (\hat{x}\cdot\hat{x}) \int_0^a\!\!\int_0^a J(b)\,\mathrm{d}x\,\mathrm{d}y = J(b) \int_0^a\!\!\int_0^a \mathrm{d}x\,\mathrm{d}y = a^2 J(b).$$

We then come back to conservation of charge, which says that the charge contained in the box must remain constant, and find the total current *leaving* the box as:

$$I_{\text{box}} = I_{\text{left}} + I_{\text{right}} + I_{\text{top}} + I_{\text{bottom}} + I_{\text{front}} + I_{\text{back}} = -a^2 J_0 + a^2 J(b) + 0 + 0 + 0 + 0 = a^2\left(-J_0 + J(b)\right).$$

Furthermore, conservation of charge requires that whatever current flows in, must also flow out:

$I_{\text{box}} = 0$.

Now recall the sign convention for area on page 24, where the positive direction was defined as facing outward from a closed surface. While current flows in (negative) and out (positive), these must add up to zero in order for the box to remain net neutral.

We can therefore write: $J(b) = J_0$.

Now comes another conceptual leap. The length, b, of our box was arbitrarily chosen. We could have made b anything, because the box we have drawn is imaginary. So, for example, we may now replace b with our spatial variable x:

$$J(x) = J_0 .$$

Notice that this is actually an interesting statement about the physics of the problem. It says that, if the current density is unidirectional in a neutral wire, then the magnitude of the current density must be constant and uniform.

The imaginary box we employed in this problem is an example of a *Gaussian surface*, which is simply an imaginary closed three-dimensional surface that is used for integration.

We can now write the conservation of charge in integral form as:

$$\oint_{\text{surface}} \vec{J} \cdot \mathrm{d}\vec{A} = -\frac{\mathrm{d}Q}{\mathrm{d}t},$$

where Q is the total charge contained inside the same Gaussian surface.

Example 1.2: Current Through a Constriction

A current is flowing through a wire of cross sectional area A_{in} with current density $\vec{J}_{\text{in}} = J_{\text{in}}\hat{x}$. The wire narrows to a cross sectional area A_{out}. What is the current density at the position in the wire after the cross section narrows?

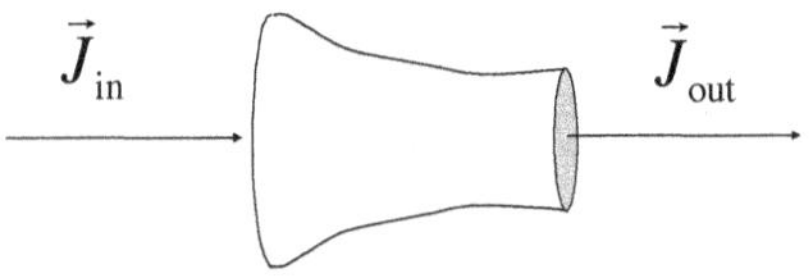

To answer this question, we proceed as we did in the previous thought experiment by accounting for the charge entering the surface on the left, and the charge exiting the surface on the right (as shown in the figure). The total current out of the closed surface must be zero:

$$I_{\text{surface}} = \int_{A_{\text{in}}} \vec{J}_{\text{in}} \cdot \mathrm{d}\vec{A} + \int_{A_{\text{out}}} \vec{J}_{\text{out}} \cdot \mathrm{d}\vec{A} = \left(\hat{x}\cdot(-\hat{x})\right) \int_{A_{\text{in}}} \vec{J}_{\text{in}} \cdot \mathrm{d}\vec{A} + \left(\hat{x}\cdot\hat{x}\right) \int_{A_{\text{out}}} \vec{J}_{\text{out}} \cdot \mathrm{d}\vec{A} = 0.$$

So the relationship between the average current density in, and out, is simply:

$$J_{\text{out}} = J_{\text{in}}\left(\frac{A_{\text{in}}}{A_{\text{out}}}\right).$$

This is similar to a nozzle on a hose, which can greatly increase the velocity of the water leaving the hose. The key difference, however, between this and a nozzle is that charge-carriers have negligible momentum, so the flow is not always the same as one would have from the flow of a fluid where the conserved quantity (mass) also carries the inertia.

Problem 1.8: Drift Velocity

We now know that the electrical fluid is made up of particles, called electrons, which each have a charge $q_e = -e$, where e represents one elementary charge. Let n_e be the number of free electrons per unit volume in the wire and let $\vec{J}$ be the current density.

a) Derive a formula for the velocity of the electrons in the wire. This is called the *drift velocity.*

b) Sixteen-gauge copper wire has a diameter of 1.29 mm. Assuming there are two free electrons per copper atom, what is the drift velocity of the electrons in the wire when it is carrying a current of $1\,\text{A}$? (The density of copper is $8.92\ {}^{\text{g}}\!/_{\text{cm}^3}$ and the atomic mass is $63.55\,{}^{\text{g}}\!/_{\text{mol}}$.

c) Look up, in an external source, typical thermal velocities of electrons in copper. Cite your source. Are these velocities similar to, much greater than, or much less than, the drift velocity? Discuss the relationship, if any, between the drift velocity and the thermal velocity.

Problem 1.9: Electrolysis of Water

In 1800, William Nicholson, Anthony Carlisle, and William Cruickshank separated hydrogen and oxygen from water using the current from the newly invented voltaic pile (i.e. a battery).[25] Consider a current, I, flowing through the water.

a) Derive an expression for the rate of diatomic Oxygen (O_2) and Hydrogen (H_2) produced as a function of the current.

b) If one ampere of current is flowing, what is this rate in units of moles per hour?

c) What is the mass of water per hour that is being converted into these gasses? Express your answer in grams per hour.

Problem 1.10: A Fuel Cell Car

In a fuel cell, hydrogen and oxygen combine to produce water and an electric current that can be used to do work. A 2015 Honda FCX car has an efficiency rating of 60 miles travelled per kg of hydrogen fuel, so it can go about 240 miles on one tank. How many coulombs of charge must flow total through the fuel cells per mile driven?

Problem 1.11: The Large Hadron Collider

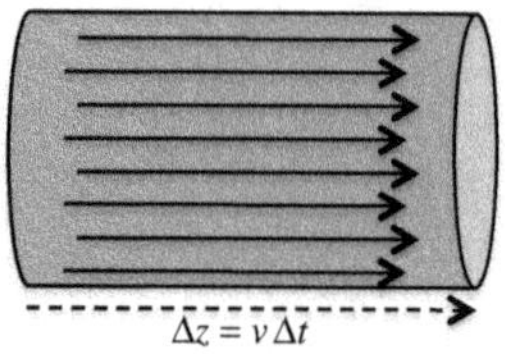

The Large Hadron Collider accelerates a beam of protons to nearly the speed of light. The total current in the beam is $I = 0.53\,\text{A}$. This beam can be squeezed down to a diameter of only 15 microns.

(a) Find the flux of the proton beam in units of $\dfrac{\text{protons}}{\text{cm}^2\,\text{s}}$.

(b) Estimate the number density of the protons in the beam.

---

[25] Nicholson, Carlisle and Cruickshank, "Experiments in galvanic electricity," Philosophical Magazine, Series 1, Volume 7, Issue 28, (1800), 337-347.

Thought Experiment 1.5: Charge of a Parallel Plate Capacitor

Two initially neutral parallel conducting plates are placed next to each other, but insulated from one another, to form a *capacitor*. Like the Leyden jar (Thought Experiment 1.2), as charge builds up on one plate, an equal and opposite charge builds up on the other plate, but no charge can directly flow between the plates. What is the relationship between the current flow and the charge on each plate?

Taken together, both plates always remain net neutral despite each having a charge. Since charge is conserved we conclude that: $I_{\text{in}} = I_{\text{out}}$.

We could have also applied the concept of a Gaussian surface (p. 27), which we will draw around both plates. Since the total charge on both plates remains constant (zero), we can write the conservation of charge in integral form as:

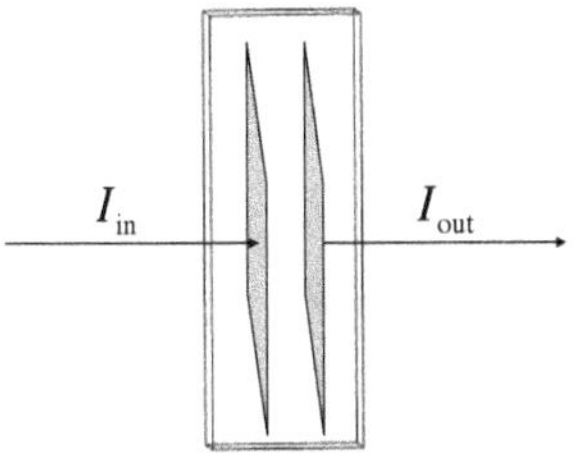

$$\oint_{\text{surface}} \vec{J}\cdot \mathrm{d}\vec{A} = -\frac{\mathrm{d}Q}{\mathrm{d}t} = 0\,.$$

Using the area vector sign convention (p. 24), and calling the right-hand direction $\hat{x}$, this becomes:

$$0 = \oint_{\text{surface}} \vec{J}\cdot \mathrm{d}\vec{A} = \int_{\text{left}} (J_{\text{in}}\,\hat{x})\cdot(-\hat{x}\,\mathrm{d}A) + \int_{\text{right}} (J_{\text{out}}\,\hat{x})\cdot(\hat{x}\,\mathrm{d}A) = -\int_{\text{left}} J_{\text{in}}\,\mathrm{d}A + \int_{\text{right}} J_{\text{out}}\,\mathrm{d}A = I_{\text{out}} - I_{\text{in}}\,.$$

Next we consider just the left plate, where the same Gaussian surface method gives us:

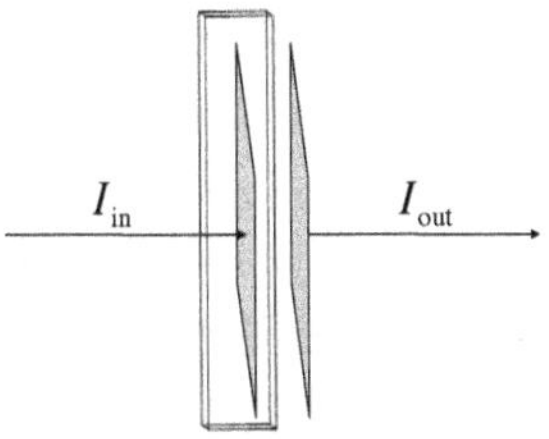

$$-\frac{\mathrm{d}Q}{\mathrm{d}t} = \oint_{\text{surface}} \vec{J}\cdot \mathrm{d}\vec{A} = \int_{\text{left}} (J_{\text{in}}\,\hat{x})\cdot(-\hat{x}\,\mathrm{d}A) = -I_{\text{in}}\,,$$

so that

$$Q_{\text{left}} = \int_0^t I_{\text{in}}\,\mathrm{d}t\,.$$

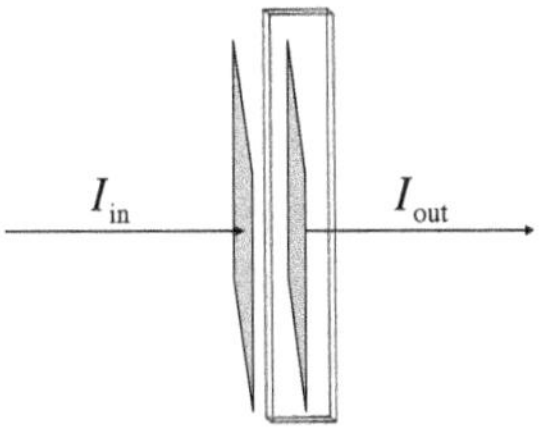

Conversely, considering only the right plate, we see that current is only exiting so its charge must be decreasing with time. Thus:

$$-\frac{\mathrm{d}Q}{\mathrm{d}t} = \oint_{\text{surface}} \vec{J}\cdot \mathrm{d}\vec{A} = \int_{\text{right}} (J_{\text{out}}\,\hat{x})\cdot(\hat{x}\,\mathrm{d}A) = I_{\text{out}} \quad\rightarrow\quad Q_{\text{right}} = -\int_0^t I_{\text{out}}\,\mathrm{d}t\,.$$

Again, if both plates taken together are to always remain net neutral then $I_{\text{in}} = I_{\text{out}}$.

We have not yet addressed the question of why the two plates together remain net neutral. For now we will assume that there is some force that tends to neutralize objects. Whether this force is between two different fluids, as du Fay suggested, or a single property of one kind of fluid as Franklin suggested, does not matter. We can still assume that the fluid, or fluids, will rearrange itself so as to be as neutral as possible, within the constraints of the system. This is much like the way water always flows downhill whenever possible.

Problem 1.12: A Leaky Parallel Plate Capacitor

Two infinitely large, and oppositely charged, thin parallel plates are located a distance d apart, and each contains a surface charge density (charge per area) of $\sigma(t)$ and $-\sigma(t)$ respectively. In between the plates is a poorly conducting medium which discharges at a rate proportional to the charge buildup $\vec{J} = K\sigma$. Find the surface charge density as a function of time, assuming an initial surface charge density of σ_0.

Thought Experiment 1.6: Charge Leakage from a Balloon.

A rubber balloon of radius R is rubbed with rabbit fur, giving it a total negative charge Q_0. Over time the charge on the balloon is lost due to ions in the surrounding air. On average, negative ions will be repelled and positive ions attracted, thereby neutralizing the balloon slowly over time. We will assume that we know this rate of total charge accretion as: $\frac{d}{dt}Q$. What, then, is the current density in the air surrounding the balloon?

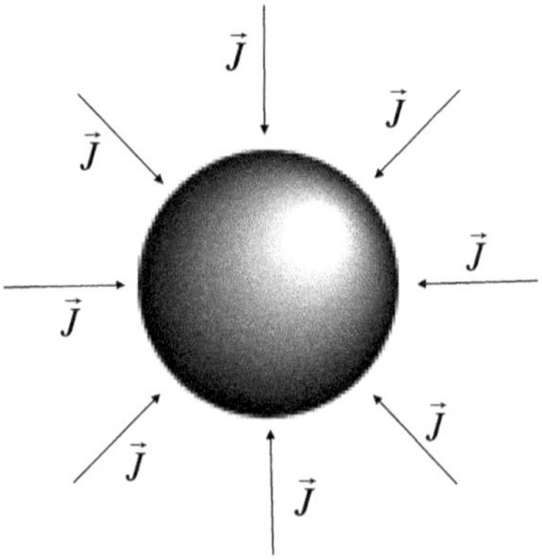

The rate of charge loss per area is a current density, which can be written in spherical coordinates as:

$$\vec{J} = J(r, t)\,\hat{r}.$$

We see from the definition of $\vec{J}$ that:

$$I_{\text{surface}} = \oint_{\text{surface}} \vec{J}\cdot d\vec{A} = -\frac{dQ}{dt}.$$

Now, if we draw an imaginary sphere of radius r co-centered with the balloon, we can account for the negative charge flowing out of this imaginary sphere.

If the sphere is smaller than the balloon $(r < R)$, then it remains net neutral so:

$$I_{\text{surface}} = \oint_{\text{surface}} \vec{J}\cdot d\vec{A} = 0, \text{ or } (J\hat{r})\cdot(4\pi r^2\hat{r}) = 0, \text{ or } J = 0.$$

So no current flows within the sphere. On the other hand, if we consider space outside the balloon $(r > R)$ we must include the charge of the balloon. Thus,

$$I_{\text{surface}} = \oint_{\text{surface}} \vec{J}\cdot d\vec{A} = -\frac{dQ}{dt}, \text{ or } (J\hat{r})\cdot(4\pi r^2\hat{r}) = -\frac{dQ}{dt}.$$

Solving for J, we solve for the surrounding current density:

$$\vec{J} = -\frac{1}{4\pi r^2}\frac{dQ}{dt}\hat{r}.$$

Notice how spherical symmetry implies that the current density must fall off as $1/r^2$.

1.3 *The Continuity Equation*

Conserved quantities, such as charge, mass, and energy, are conserved not only for an entire system, but also at every point in space. This allows us to apply the law of the conservation of charge continuously at every point in the system, rather than on a particular scale of interest.

In the eighteenth century, Leonhard Euler devised the *continuity equation*, which expresses the continuous conservation of a scalar quantity in a fluid. In hydrodynamics, for example, the rate of constant-density fluid flow (e.g. gallons per minute) must be continuous, because whatever mass flows into a particular place must also flow out of that same place.

The integral form of current continuity, which we have already developed, simply expresses charge conservation as a global fact. Whatever the messy details of the current flow at the microscopic level, the rate at which charge accumulates in a region of space must be accounted for by the flow of current into, and out of, that region. It is indeed a powerful and useful tool to express conservation of charge as global fact, but it is important and often advantageous to note that charge must also be conserved at every point in space. This is a *local* requirement that we express as a relation between the time rate of change of the charge density at a particular point in space, and the *divergence* of the current density at the same point. In Thought Experiment 1.8 we geometrically introduce the concept of the divergence of a vector field and the divergence theorem,[26] which connects the local and global formulations of charge conservation. The divergence theorem is of fundamental importance in nearly all branches of physics, so we present several examples to demonstrate the usefulness of this local formulation of charge conservation.

Since charge is the primary conserved quantity in electrodynamics, depending on the geometry involved you will encounter a variety of different charge densities. When moving, these are often paired with a corresponding current density so as to conserve charge. Equations that ensure that charge is conserved at all scales are called *continuity equations.*

Thought Experiment 1.7: Linear Charge Continuity

Recall Thought Experiment 1.3 (p. 22), with the charged arrow. In this example the charge per length, λ, called the *linear charge density*, is the governing charge density. In one-dimensional cases, with a uniform linear charge density, the current is given by:

$$I = \frac{\delta Q}{\delta t} = \lambda v .$$

If the medium is neutral, current can still flow, so long as either opposite charge flows in opposite directions, or only one kind of charge flows as in Thought Experiment 1.4 (p. 23).

However, if the current is not uniform, parts of the wire would become more, or less, charged as we discussed in Thought Experiment 1.5 (p. 29) not all of the charge flowing into something would necessarily also flow out. So, for a given length element $\delta\ell$ the criterion of charge conservation gives us:

$$\frac{\delta Q}{\delta\ell} = \frac{I_{\text{in}}\,\delta t - I_{\text{out}}\,\delta t}{\delta\ell} = -\frac{\left(I_{\text{out}} - I_{\text{in}}\right)}{\delta\ell}\delta t = \frac{\left(I\left(\ell + \tfrac{1}{2}\delta\ell\right) - I\left(\ell - \tfrac{1}{2}\delta\ell\right)\right)}{\delta\ell} .$$

---

[26] The divergence theorem is also called Gauss's Theorem, however we avoid that terminology in this text because of confusion with Gauss's Law.

Taking the limit as $\delta\ell$ approaches zero, while holding the time constant, gives us:

$$\lim_{\delta\ell\to 0}\frac{\delta Q(\ell,t)}{\delta\ell}=(-\delta t)\lim_{\delta\ell\to 0}\left(\frac{\left(I\left(\left(\ell+\frac{1}{2}\delta\ell\right),t\right)-I\left(\left(\ell-\frac{1}{2}\delta\ell\right),t\right)\right)}{\delta\ell}\right)=(-\delta t)\left(\frac{\partial I(\ell,t)}{\partial\ell}\right).$$

The partial symbol, ∂, means that we hold the other independent variables fixed (i.e. treat them as constants). Notice also that the term on the left is not the charge density λ, but rather the change in charge density $\delta\lambda$, over a time interval δt. Dividing through by this time and taking its limit, while holding the position constant, gives us, for the left hand side:

$$\lim_{\delta t\to 0}\frac{\delta\lambda(\ell,t)}{\delta t}=\lim_{\delta t\to 0}\frac{\lambda\left(\ell,\left(t+\frac{1}{2}\delta t\right)\right)-\lambda\left(\ell,\left(t-\frac{1}{2}\delta t\right)\right)}{\delta t}=\frac{\partial\lambda(\ell,t)}{\partial t}.$$

Thus, we can finally equate these, to write one partial differential equation relating the current gradient to the rate of change of the linear charge density.

$$\frac{\partial I}{\partial\ell}=-\frac{\partial\lambda}{\partial t}.$$

Notice that the units work out, since each side has units of charge per length per time.

Example 1.3: A Current Pulse

Consider a current pulse flowing through a wire with the following functional form:

$$I(x,t)=I_0e^{-(x-vt)^2},$$

where x and t represent the position and time variables. I_0 and v represent constants with units of current and velocity respectively.

What is the linear charge density in the wire as a function of position and time?

Since charge is always conserved everywhere and at all times, the continuity equation must hold. Since we are given the current, we will start by taking the spatial derivative of the current, while holding the time constant:

$$\tfrac{\partial}{\partial x}I=\tfrac{\partial}{\partial x}\left(I_0e^{-(x-vt)^2}\right)=I_0\left(e^{-(x-vt)^2}\right)\tfrac{\partial}{\partial x}\left(-(x-vt)^2\right)=-2I_0(x-vt)e^{-(x-vt)^2}.$$

Now, applying conservation of charge everywhere, to the linear charge density give us:

$$\tfrac{\partial}{\partial t}\lambda=-\tfrac{\partial}{\partial x}I=2I_0(x-vt)e^{-(x-vt)^2},$$

so we will need to integrate this, with respect to time, to find the charge density. To do this we will make the variable substitution $x'=x-vt$, so:

$$t=\frac{x-x'}{v}\text{ and }\mathrm{d}t=-\frac{\mathrm{d}x'}{v},\text{ thus:}$$

$$\lambda=\int 2I_0(x-vt)e^{-(x-vt)^2}\,\mathrm{d}t=\int 2I_0x'e^{-(x')^2}\left(-\frac{\mathrm{d}x'}{v}\right)=\frac{1}{v}\int -2I_0x'e^{-(x')^2}\,\mathrm{d}x'.$$

Notice this is simply the inverse of one of the derivatives we just took, so:

$$\lambda = \frac{1}{v}\int -2I_0 x' e^{-(x')^2}\,\mathrm{d}x' = \frac{1}{v}\left(I_0 e^{-(x')^2}\right) + C(x) = \frac{I_0}{v}\left(e^{-(x-vt)^2}\right) + C(x).$$

Assuming that the wire was initially neutral, the charge density is therefore:

$$\lambda = \frac{I_0}{v} e^{-(x-vt)^2}.$$

We could have guessed that this would be the solution, had we noticed beforehand the following relationship with our same variable substitution, making it only a function of a single variable:

$$I = I_0 e^{-(x-vt)^2} = I_0 e^{-(x')^2}, \text{ so: } \quad \frac{\partial\lambda}{\partial t} = \frac{\mathrm{d}\lambda}{\mathrm{d}x'}\frac{\partial x'}{\partial t} = -v\frac{\mathrm{d}\lambda}{\mathrm{d}x'} = -\frac{\mathrm{d}(v\lambda)}{\mathrm{d}x'} = -\frac{\partial I}{\partial x} = -\frac{\mathrm{d}I}{\mathrm{d}x'}\frac{\partial x'}{\partial x} = -\frac{\mathrm{d}I}{\mathrm{d}x'}.$$

Therefore, the simplest solution is for $v\lambda = I$.

Discussion 1.5: Intensive and Extensive Properties

Unlike charge, the charge density of a uniform substance does not change with the volume selected. Thus, charge is an *intensive* quantity. The current density, $\vec{J}$, is an intensive vector quantity for the same reason.

Charge is different. It is an *extensive* quantity. In a uniform substance, charge is proportional to the volume. The ratio of any two extensive quantities is an intensive quantity, because the volume divides out. For example, charge density is the ratio of charge and volume, both of which are extensive quantities.

Physicists like to work in intensive quantities, because any classical theory should work on any scale. This means that we will always express our fundamental laws of electrodynamics, such as the conservation of charge, as relationships among intensive quantities, such as charge density and current density.

We express this use of intensive quantities in a number of ways. For example, we often say that we write something in its *differential form*, or that we are expressing it *locally*. On the other hand, we can write an equation in its *integral form*, which is to say that we are using a *global* representation.

The main reason we prefer expressing equations in their differential form is simply this: representing an intensive quantity as a function of position gives information that is left out by the corresponding extensive quantity for the system as a whole. For example, the charge density of a net neutral water molecule is more positive on one side, and negative on the other. The global representation of charge would be the same as with any other neutral molecule, but the dipole nature of water is one of the things that make life on Earth possible.

Many quantities are neither intensive, nor extensive. For example, consider a current I flowing through a wire. It is related to the cross-section of the wire $\vec{A}$, and the current density $\vec{J}$, through the relationship $I = \vec{J} \cdot \vec{A}$. Consider two scenarios: (1) we double the length of the wire and hold the cross section constant, or (2) we double the cross section while holding the length constant. In both cases, we double the volume of the wire. However, in case (1) the current remains the same, but in case (2) the current doubles. So, the current not only depends on the volume, but the shape of its container, making it neither intensive nor extensive.

Definition 1.5: Charge Density

Charge density, ρ, is defined as the charge per unit volume. The SI unit of the charge density is the coulomb per meter cubed.

Thought Experiment 1.8: Charge Continuity

Recall Thought Experiment 1.6 (p. 30) with the charge leaking from a spherical balloon. Now, instead of a balloon, we have some arbitrary charge density distribution $\rho(\vec{r},t)$, and we draw an imaginary Gaussian surface centered at any location $\vec{r}$ with a volume V and a surface area A. Then, again, the charge contained in this surface must change based on the total current that passes out of the surface, so we again have:

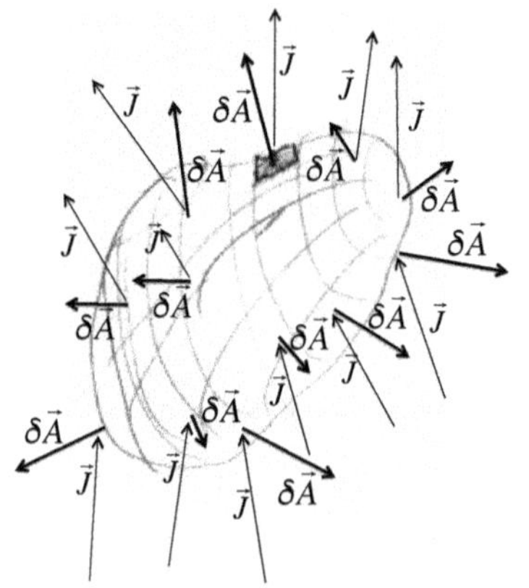

$$I_{\text{surface}} = \oint_{\text{surface}} \vec{J}\cdot \mathrm{d}\vec{A} = -\frac{\mathrm{d}Q}{\mathrm{d}t} = -\frac{\mathrm{d}}{\mathrm{d}t}\left(\oint_{\text{volume}} \rho\cdot \mathrm{d}V\right).$$

Now, in the limit that the volume of the Gaussian surface becomes very small, the charge density will be uniform, so we could factor it out. Therefore:

$$-\frac{\mathrm{d}\rho}{\mathrm{d}t}\left(\lim_{V\to 0}\oint_{\text{volume}} \mathrm{d}V\right) = \lim_{V\to 0}\left(\oint_{\text{surface}} \vec{J}\cdot \mathrm{d}\vec{A}\right) = \lim_{V\to 0}\left(\frac{\oint_{\text{surface}} \vec{J}\cdot \mathrm{d}\vec{A}}{\oint_{\text{volume}} \mathrm{d}V}\right)\left(\lim_{V\to 0}\oint_{\text{volume}} \mathrm{d}V\right).$$

Therefore:

$$-\frac{\mathrm{d}\rho}{\mathrm{d}t} = \lim_{V\to 0}\left(\frac{\oint_{\text{surface}} \vec{J}\cdot \mathrm{d}\vec{A}}{\oint_{\text{volume}} \mathrm{d}V}\right) = \lim_{V\to 0}\left(\frac{1}{V}\oint_{\text{surface}} \vec{J}\cdot \mathrm{d}\vec{A}\right).$$

This is actually quite profound! We just took a global idea, that in any closed system charge is neither created nor destroyed, and turned it into a local condition that can be solved as function of position.

Since the right hand side is a limit as the volume goes to zero, it is clearly some sort of derivative, but of a vector function. This type of derivative is called the *divergence*, which is defined as follows for any spatial vector function:

$$\vec{\nabla}\cdot\vec{J} \equiv \lim_{V\to 0}\left(\frac{1}{V}\oint_{\text{surface}} \vec{J}\cdot \mathrm{d}\vec{A}\right).$$

Thus conservation of charge can be written, in general, with the following equation:

$$\vec{\nabla}\cdot\vec{J} = -\frac{\partial\rho}{\partial t}.$$

$\vec{\nabla}$ has two names, *del* or *nabla*, and it is a vector operator that represents a three dimensional spatial derivative.

Physicists write the divergence as the dot product $(\vec{\nabla}\cdot)$ because in Cartesian coordinates we can represent it as:

$$\vec{\nabla}\cdot\vec{J}=\frac{\partial J_x}{\partial x}+\frac{\partial J_y}{\partial y}+\frac{\partial J_z}{\partial z}=\left(\hat{x}\frac{\partial}{\partial x}+\hat{y}\frac{\partial}{\partial y}+\hat{z}\frac{\partial}{\partial z}\right)\cdot\left(J_x\hat{x}+J_y\hat{y}+J_z\hat{z}\right),$$

which we will derive in Thought Experiment 1.9 (p. 36). For this reason, the del vector operator is often written as:

$$\vec{\nabla}=\left(\hat{x}\tfrac{\partial}{\partial x}+\hat{y}\tfrac{\partial}{\partial y}+\hat{z}\tfrac{\partial}{\partial z}\right).$$

However you must be very careful not to overgeneralize! In cylindrical and polar coordinates it is not this simple!

An example of this is the solution to our balloon example (Thought Experiment 1.6, p.30), where we found the current density outside of the balloon to be:

$$\vec{J}=-\frac{1}{4\pi r^2}\frac{\mathrm{d}Q}{\mathrm{d}t}\hat{r}.$$

We also know that the charge density outside of the balloon is zero, so:

$$\vec{\nabla}\cdot\vec{J}=0=\vec{\nabla}\cdot\left(-\frac{1}{4\pi r^2}\frac{\mathrm{d}Q}{\mathrm{d}t}\hat{r}\right)=-\frac{1}{4\pi}\frac{\mathrm{d}Q}{\mathrm{d}t}\vec{\nabla}\cdot\left(\frac{\hat{r}}{r^2}\right).$$

While this is true, it is not obvious from the nomenclature.

Finally we want to emphasize the difference between the differential expression of charge conservation and the integral expression. In the differential expression, charge is conserved on all scales no matter how small, while the integral expression simply states that charge is conserved within a particular volume.

For example, if the charge density increases at any position (x,y,z), then at the same time net current must be flowing to that point from the surrounding points, and the same thing must be true for each of those points in turn *continuously*. Thus, $\vec{\nabla}\cdot\vec{J}=-\frac{\partial}{\partial t}\rho$ is called the *equation of charge continuity*, or simply *the continuity equation.*

1.4 *The Divergence in Curvilinear Coordinates*

In this section, we will derive the divergence in our three standard coordinate systems. While this may seem somewhat abstract, you will find it worth your while to make sure you understand how we got the vector calculus relationships in the covers of the book.

Definition 1.6: The Divergence

Consider any Gaussian surface enclosing a volume V. At every point on the surface there is a small area vector $\delta\vec{A}(\vec{r})$ with a direction normal to the surface and pointing outward. Next consider any vector field, $\vec{J}(\vec{r})$, that passes through the surface. We first take the limit $\delta\vec{A}\rightarrow\mathrm{d}\vec{A}$ so that we can integrate $\vec{J}$ over the whole Gaussian surface, then we take the limit as the volume of the surface become infinitesimal.

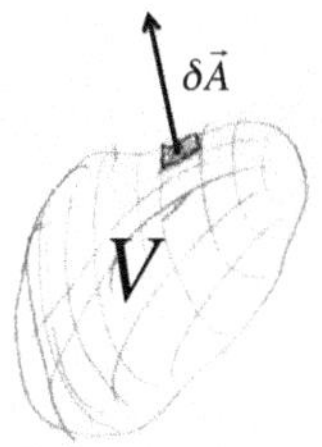

Now we can define the *divergence* of a vector field $\vec{J}(\vec{r})$ as:

$$\vec{\nabla}\cdot\vec{J} \equiv \lim_{V\to 0}\frac{1}{V}\oint_{\text{surface}} \vec{J}\cdot d\vec{A}.$$

We can think of the units of the del operator as inverse length:

$$\left[\vec{\nabla}\cdot\vec{J}\right] = \left[\lim_{V\to 0}\left(\frac{\oint_{\text{surface}}\vec{J}\cdot d\vec{A}}{V}\right)\right] = \frac{[\vec{J}][d\vec{A}]}{[V]} = \frac{[\vec{J}]\text{m}^2}{\text{m}^3} = \frac{1}{\text{m}}[\vec{J}] = [\vec{\nabla}][\vec{J}].$$

Thought Experiment 1.9: The Divergence in Cartesian Coordinates

In this thought experiment, we assume that we know the current density vector as a function of position in Cartesian coordinates ($\vec{J} = \vec{J}(\vec{r}) = \vec{J}(x,y,z)$) and that it is rendered also in the same Cartesian coordinate directions. Therefore:

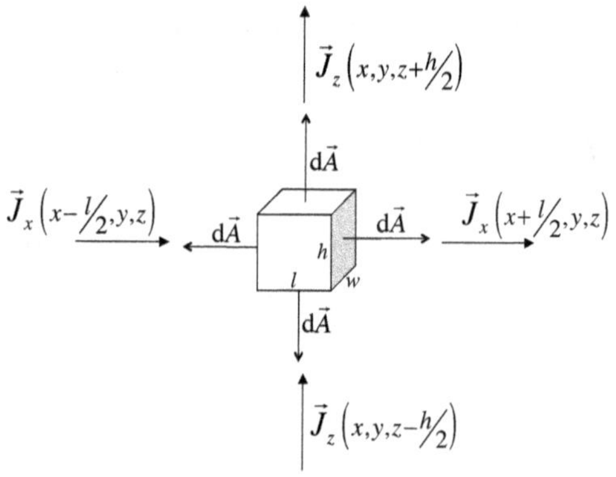

$$\vec{J}(\vec{r}) = J_x(x,y,z)\,\hat{x} + J_y(x,y,z)\,\hat{y} + J_z(x,y,z)\,\hat{z}.$$

Now we will draw a rectangular box, centered at coordinates (x,y,z) with dimensions $l\times w\times h$. For this box, we will find the net current flowing out of this box by calculating the surface integral $\oint_{\text{box}}\vec{J}\cdot d\vec{A}$, by summing over all six sides.

$$\oint_{\text{box}}\vec{J}\cdot d\vec{A} = \int_{\text{left}}\vec{J}\cdot d\vec{A} + \int_{\text{right}}\vec{J}\cdot d\vec{A} + \int_{\text{front}}\vec{J}\cdot d\vec{A} + \int_{\text{back}}\vec{J}\cdot d\vec{A} + \int_{\text{bottom}}\vec{J}\cdot d\vec{A} + \int_{\text{top}}\vec{J}\cdot d\vec{A}$$

$$\oint_{\text{box}}\vec{J}\cdot d\vec{A} = \int_{A=wh} J_x\left(x-\tfrac{1}{2}l,y,z\right)\hat{x}\cdot\left(-\hat{x}\,dA\right) + \int_{A=wh} J_x\left(x+\tfrac{1}{2}l,y,z\right)\hat{x}\cdot\left(\hat{x}\,dA\right)$$

$$+ \int_{A=hl} J_y\left(x,y-\tfrac{1}{2}w,z\right)\hat{y}\cdot\left(-\hat{y}\,dA\right) + \int_{A=hl} J_y\left(x,y+\tfrac{1}{2}w,z\right)\hat{y}\cdot\left(\hat{y}\,dA\right)$$

$$+ \int_{A=lw} J_z\left(x,y,z-\tfrac{1}{2}h\right)\hat{z}\cdot\left(-\hat{z}\,dA\right) + \int_{A=lw} J_z\left(x,y,z+\tfrac{1}{2}h\right)\hat{z}\cdot\left(\hat{z}\,dA\right)$$

$$\oint_{\text{box}}\vec{J}\cdot d\vec{A} = \int_{A=wh}\left(J_x\left(x+\tfrac{1}{2}l,y,z\right) - J_x\left(x-\tfrac{1}{2}l,y,z\right)\right)dA$$

$$+ \int_{A=hl}\left(J_y\left(x,y+\tfrac{1}{2}w,z\right) - J_y\left(x,y-\tfrac{1}{2}w,z\right)\right)dA$$

$$+ \int_{A=lw}\left(J_z\left(x,y,z+\tfrac{1}{2}h\right) - J_z\left(x,y,z-\tfrac{1}{2}h\right)\right)dA$$

Now we use the definition of the partial derivative to simplify each integrand, in the limit that the box is small:

$$\oint_{\text{box}}\vec{J}\cdot d\vec{A} \approx \int_{A=wh}\left(\tfrac{\partial}{\partial x}J_x\right)l\,dA + \int_{A=hl}\left(\tfrac{\partial}{\partial y}J_y\right)w\,dA + \int_{A=lw}\left(\tfrac{\partial}{\partial z}J_z\right)h\,dA$$

$$\approx \left(\tfrac{\partial}{\partial x}J_x\right)l(wh) + \left(\tfrac{\partial}{\partial y}J_y\right)w(hl) + \left(\tfrac{\partial}{\partial z}J_z\right)h(lw) \approx lwh\left(\tfrac{\partial}{\partial x}J_x + \tfrac{\partial}{\partial y}J_y + \tfrac{\partial}{\partial z}J_z\right).$$

Notice that the volume of the box is simply $V = lwh$, and all the approximations are exact in the limit of an infinitesimally small box. Therefore:

$$\vec{\nabla}\cdot\vec{J} \equiv \lim_{V\to 0}\left(\frac{1}{V}\oint_{\text{surface}} \vec{J}\cdot \mathrm{d}\vec{A}\right) = \lim_{V\to 0}\left(\frac{\cancel{lwh}\left(\frac{\partial}{\partial x}J_x + \frac{\partial}{\partial y}J_y + \frac{\partial}{\partial z}J_z\right)}{\cancel{V}}\right) = \frac{\partial J_x}{\partial x} + \frac{\partial J_y}{\partial y} + \frac{\partial J_z}{\partial z}.$$

Now we see why the vector operator ($\vec{\nabla}$) is often written as a dot product:

$$\vec{\nabla} = \frac{\partial}{\partial x}\hat{x} + \frac{\partial}{\partial y}\hat{y} + \frac{\partial}{\partial z}\hat{z}.$$

While this makes the notation very handy in Cartesian coordinates, it can be confusing in curvilinear coordinates, as the divergence does not follow the same mnemonic.

Thought Experiment 1.10: The Divergence in Cylindrical Coordinates

Consider now a system represented by cylindrical coordinates s, φ, and z, and corresponding orthogonal unit vectors $\hat{s}$, $\hat{\varphi}$ and $\hat{z}$. Notice that each unit vector always points in the direction that its corresponding spatial coordinate increases. Thus the position vector is given by $\vec{r} = s\hat{s} + z\hat{z}$ and a vector field can be written:

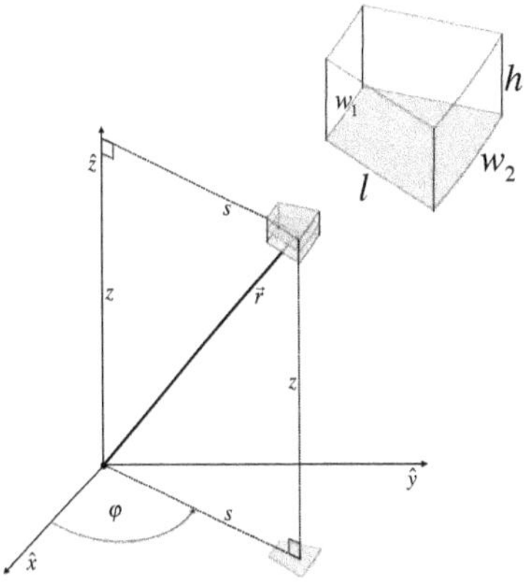

$$\vec{J}(\vec{r}) = J_s(s,\varphi,z)\,\hat{s} + J_\varphi(s,\varphi,z)\,\hat{\varphi} + J_z(s,\varphi,z)\,\hat{z}.$$

Now we will draw a pseudo-rectangular box, centered at coordinates (s,φ,z) with dimensions $l \times w \times h$. However, because of the angular coordinate φ, the inner width w_1 and outer width w_2 differ, and we will need to account for this. Consider a small change in azimuthal angle, $\delta\varphi$, the corresponding width at a given radius from the $\hat{z}$ axis is given by $w = s\delta\varphi$. We can therefore write the inner and outer widths as $w_1 = (s - \frac{1}{2}l)\delta\varphi$ and $w_2 = (s + \frac{1}{2}l)\delta\varphi$ respectively.

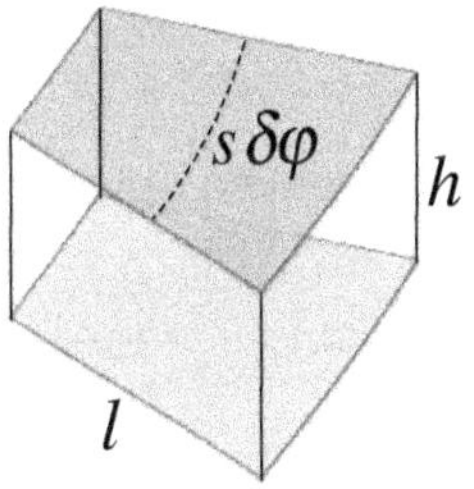

For this pseudo-box, we will calculate the total surface integral $\oint_{\text{box}} \vec{J}\cdot \mathrm{d}\vec{A}$, by summing over all six sides.

$$\oint_{\text{box}} \vec{J}\cdot \mathrm{d}\vec{A} = \int_{\text{inner}} \vec{J}\cdot \mathrm{d}\vec{A} + \int_{\text{outer}} \vec{J}\cdot \mathrm{d}\vec{A} + \int_{\text{front}} \vec{J}\cdot \mathrm{d}\vec{A} + \int_{\text{back}} \vec{J}\cdot \mathrm{d}\vec{A} + \int_{\text{bottom}} \vec{J}\cdot \mathrm{d}\vec{A} + \int_{\text{top}} \vec{J}\cdot \mathrm{d}\vec{A}$$

The areas of the inner and outer faces are given by $(s - \frac{1}{2}l)\delta\varphi h$ and $(s + \frac{1}{2}l)\delta\varphi h$ respectively. The area of the side faces are each lh, and the top and bottom faces each have an area given by:

$$A_{\text{top}} = A_{\text{bottom}} = \left(\frac{\delta\varphi}{2\cancel{\pi}}\right)\left(\cancel{\pi}\left(s + \tfrac{1}{2}l\right)^2 - \cancel{\pi}\left(s - \tfrac{1}{2}\right)^2\right) = \left(\frac{\delta\varphi}{\cancel{2}}\right)\left(\cancel{4}s\left(\frac{1}{\cancel{2}}\right)\right) = sl\delta\varphi$$

Thus, the total integral around the box is given by:

$$\oint_{\text{box}} \vec{J}\cdot \mathrm{d}\vec{A} = \int_{A=(s-\frac{1}{2})\delta\varphi h} J_s\left(s-\tfrac{1}{2}l,\varphi,z\right)\hat{s}\cdot\left(-\hat{s}\,\mathrm{d}A\right) + \int_{A=(s+\frac{1}{2})\delta\varphi h} J_s\left(s+\tfrac{1}{2}l,\varphi,z\right)\hat{s}\cdot\left(\hat{s}\,\mathrm{d}A\right)$$
$$+\int_{A=hl} J_\varphi\left(s,\varphi-\tfrac{1}{2}\delta\varphi,z\right)\hat{\varphi}\cdot\left(-\hat{\varphi}\,\mathrm{d}A\right) + \int_{A=hl} J_\varphi\left(s,\varphi+\tfrac{1}{2}\delta\varphi,z\right)\hat{\varphi}\cdot\left(\hat{\varphi}\,\mathrm{d}A\right)$$
$$+\int_{A=sl\delta\varphi} J_z\left(s,\varphi,z-\tfrac{1}{2}h\right)\hat{z}\cdot\left(-\hat{z}\,\mathrm{d}A\right) + \int_{A=sl\delta\varphi} J_z\left(s,\varphi,z+\tfrac{1}{2}h\right)\hat{z}\cdot\left(\hat{z}\,\mathrm{d}A\right)$$

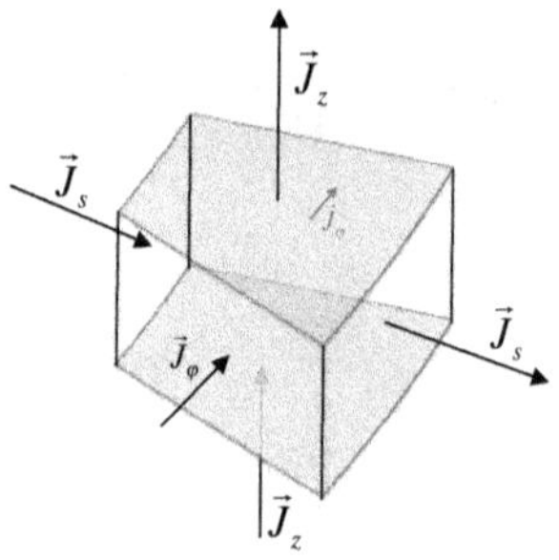

$$\oint_{\text{box}} \vec{J}\cdot \mathrm{d}\vec{A} = \int_{A=(s+\frac{1}{2})\delta\varphi h} J_s\left(s+\tfrac{1}{2}l,\varphi,z\right)\mathrm{d}A - \int_{A=(s-\frac{1}{2})\delta\varphi h} J_s\left(s-\tfrac{1}{2}l,\varphi,z\right)\mathrm{d}A$$
$$+\int_{A=hl}\left(J_\varphi\left(s,\varphi+\tfrac{1}{2}\delta\varphi,z\right) - J_\varphi\left(s,\varphi-\tfrac{1}{2}\delta\varphi,z\right)\right)\mathrm{d}A$$
$$+\int_{A=sl\delta\varphi}\left(J_z\left(s,\varphi,z+\tfrac{1}{2}h\right) - J_z\left(s,\varphi,z-\tfrac{1}{2}h\right)\right)\mathrm{d}A$$

Now we will assume the box is small, so we can take a first order approximation for each integral:

$$\oint_{\text{box}} \vec{J}\cdot \mathrm{d}\vec{A} \approx \delta\varphi h\left(\left(s+\tfrac{1}{2}l\right)J_s\left(s+\tfrac{1}{2}l,\varphi,z\right) - \left(s-\tfrac{1}{2}\right)J_s\left(s-\tfrac{1}{2}l,\varphi,z\right)\right)$$
$$+hl\left(J_\varphi\left(s,\varphi+\tfrac{1}{2}\delta\varphi,z\right) - J_\varphi\left(s,\varphi-\tfrac{1}{2}\delta\varphi,z\right)\right)$$
$$+sl\delta\varphi\left(J_z\left(s,\varphi,z+\tfrac{1}{2}h\right) - J_z\left(s,\varphi,z-\tfrac{1}{2}h\right)\right)$$

$$\oint_{\text{box}} \vec{J}\cdot \mathrm{d}\vec{A} \approx \delta\varphi h\frac{\partial\left(sJ_s(s,\varphi,z)\right)}{\partial s}\left(\left(s+\tfrac{1}{2}l\right)-\left(s-\tfrac{1}{2}l\right)\right)$$
$$+hl\frac{\partial J_\varphi(s,\varphi,z)}{\partial\varphi}\left(\left(\varphi+\tfrac{1}{2}\delta\varphi\right)-\left(\varphi-\tfrac{1}{2}\delta\varphi\right)\right)$$
$$+sl\delta\varphi\frac{\partial J_z(s,\varphi,z)}{\partial z}\left(\left(z+\tfrac{1}{2}h\right)-\left(z-\tfrac{1}{2}h\right)\right)$$

$$\oint_{\text{box}} \vec{J}\cdot \mathrm{d}\vec{A} \approx l\delta\varphi h\frac{\partial\left(sJ_s(s,\varphi,z)\right)}{\partial s} + l\delta\varphi h\frac{\partial J_\varphi(s,\varphi,z)}{\partial\varphi} + l\delta\varphi h s\frac{\partial J_z(s,\varphi,z)}{\partial z}$$

The volume of the pseudo-box is $V = ls\delta\varphi h$, so:

$$\oint_{\text{box}} \vec{J}\cdot \mathrm{d}\vec{A} \approx \frac{V}{s}\frac{\partial\left(sJ_s(s,\varphi,z)\right)}{\partial s} + \frac{V}{s}\frac{\partial J_\varphi(s,\varphi,z)}{\partial\varphi} + V\frac{\partial J_z(s,\varphi,z)}{\partial z}$$

and the approximations become exact in the limit of a very small volume. Therefore:

$$\vec{\nabla}\cdot\vec{J} \equiv \lim_{V\to 0}\left(\frac{\oint_S \vec{J}\cdot\mathrm{d}\vec{A}}{V}\right) = \lim_{V\to 0}\left(\frac{\frac{\cancel{V}}{s}\frac{\partial\left(sJ_s(s,\varphi,z)\right)}{\partial s} + \frac{\cancel{V}}{s}\frac{\partial J_\varphi(s,\varphi,z)}{\partial\varphi} + \cancel{V}\frac{\partial J_z(s,\varphi,z)}{\partial z}}{\cancel{V}}\right).$$

Finally, the divergence in cylindrical coordinates is:

$$\vec{\nabla}\cdot\vec{J} = \frac{1}{s}\frac{\partial\left(sJ_s(s,\varphi,z)\right)}{\partial s} + \frac{1}{s}\frac{\partial J_\varphi(s,\varphi,z)}{\partial\varphi} + \frac{\partial J_z(s,\varphi,z)}{\partial z}.$$

Thought Experiment 1.11: The Divergence in Spherical Coordinates

Consider now a system represented by spherical coordinates r, θ, and φ and their corresponding orthogonal unit vectors $\hat{r}$, $\hat{\theta}$, and $\hat{\varphi}$. These are referred to as the *radial*, *polar*, and *azimuthal* coordinates.

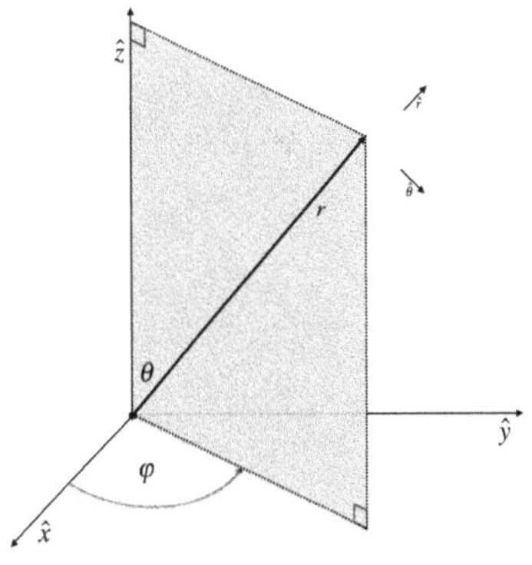

Notice that each unit vector always points in the direction that its corresponding spatial coordinate increases. Thus the position vector is now given by $\vec{r} = r\hat{r}$, and a vector field can be written:

$$\vec{J}(\vec{r}) = J_r(r,\theta,\varphi)\,\hat{r} + J_\theta(r,\theta,\varphi)\,\hat{\theta} + J_\varphi(r,\theta,\varphi)\,\hat{\varphi}.$$

Now we will draw a pseudo-rectangular box, centered at coordinates (r,θ,φ). The figures below show this pseudo-box from two projections, followed by a detailed view of the box.

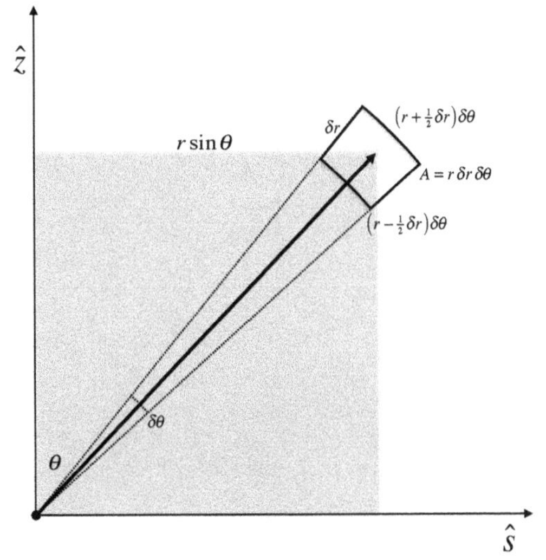

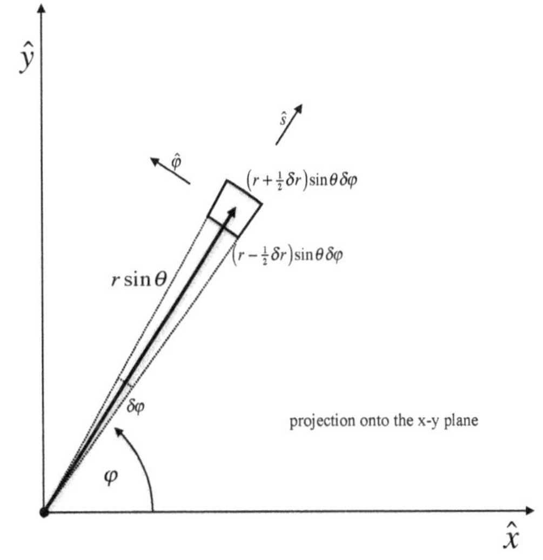

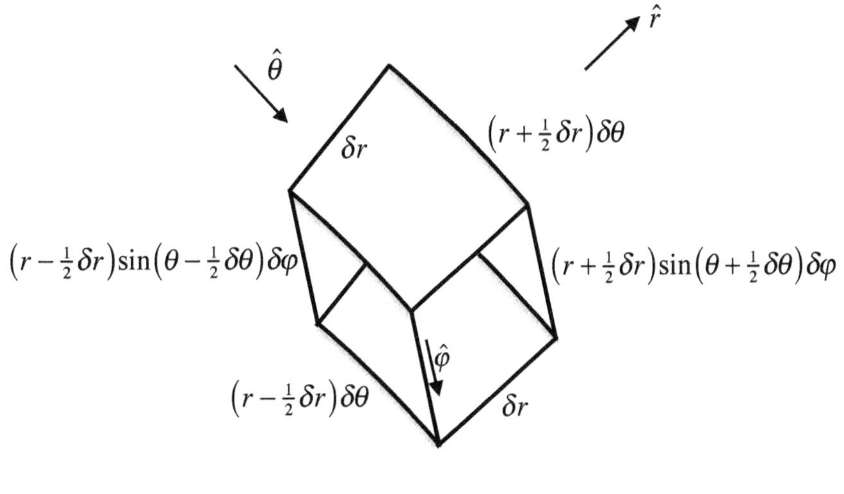

The total surface integral around the pseudo-box is given by:

$$\begin{aligned}\oint_{\text{box}} \vec{J}\cdot d\vec{A} = &\int_{A=(r-\frac{1}{2}\delta r)^2\sin\theta\,\delta\theta\,\delta\varphi} J_r\left(r-\tfrac{\delta r}{2},\theta,\varphi\right)\hat{r}\cdot(-\hat{r}\,dA) + \int_{A=(r+\frac{1}{2}\delta r)^2\sin\theta\,\delta\theta\,\delta\varphi} J_r\left(r+\tfrac{\delta r}{2},\theta,\varphi\right)\hat{r}\cdot(\hat{r}\,dA)\\ &+\int_{A=r\sin(\theta-\frac{1}{2}\delta\theta)\,\delta r\,\delta\varphi} J_\theta\left(r,\theta-\tfrac{\delta\theta}{2},\varphi\right)\hat{\theta}\cdot(-\hat{\theta}\,dA) + \int_{A=r\sin(\theta+\frac{1}{2}\delta\theta)\,\delta r\,\delta\varphi} J_\theta\left(r,\theta+\tfrac{\delta\theta}{2},\varphi\right)\hat{\theta}\cdot(\hat{\theta}\,dA)\\ &+\int_{A=r\,\delta\theta\,\delta\varphi} J_\varphi\left(r,\theta,\varphi-\tfrac{\delta\theta}{2}\right)\hat{\varphi}\cdot(-\hat{\varphi}\,dA) + \int_{A=r\,\delta\theta\,\delta\varphi} J_\varphi\left(r,\theta,\varphi+\tfrac{\delta\theta}{2}\right)\hat{\varphi}\cdot(\hat{\varphi}\,dA)\end{aligned}$$

$$\begin{aligned}\oint_{\text{box}} \vec{J}\cdot d\vec{A} = &\int_{A=(r+\frac{1}{2}\delta r)^2\sin\theta\,\delta\theta\,\delta\varphi} J_r\left(r+\tfrac{\delta r}{2},\theta,\varphi\right)dA - \int_{A=(r-\frac{1}{2}\delta r)^2\sin\theta\,\delta\theta\,\delta\varphi} J_r\left(r-\tfrac{\delta r}{2},\theta,\varphi\right)dA\\ &+\int_{A=r\sin(\theta+\frac{1}{2}\delta\theta)\,\delta r\,\delta\varphi} J_\theta\left(r,\theta+\tfrac{\delta\theta}{2},\varphi\right)dA - \int_{A=r\sin(\theta-\frac{1}{2}\delta\theta)\,\delta r\,\delta\varphi} J_\theta\left(r,\theta-\tfrac{\delta\theta}{2},\varphi\right)dA +\\ &+\int_{A=r\,\delta\theta\,\delta r} J_\varphi\left(r,\theta,\varphi+\tfrac{\delta\theta}{2}\right)dA - \int_{A=r\,\delta\theta\,\delta r} J_\varphi\left(r,\theta,\varphi-\tfrac{\delta\theta}{2}\right)dA\end{aligned}$$

Again we will make a first order approximation of each integral for a small box:

$$\begin{aligned}\oint_{\text{box}} \vec{J}\cdot d\vec{A} \approx\; &J_r\left(r+\tfrac{\delta r}{2},\theta,\varphi\right)\left(r+\tfrac{1}{2}\delta r\right)^2\sin\theta\,\delta\theta\,\delta\varphi - J_r\left(r-\tfrac{\delta r}{2},\theta,\varphi\right)\left(r-\tfrac{1}{2}\delta r\right)^2\sin\theta\,\delta\theta\,\delta\varphi\\ &+J_\theta\left(r,\theta+\tfrac{\delta\theta}{2},\varphi\right)r\sin\left(\theta+\tfrac{1}{2}\delta\theta\right)\delta r\,\delta\varphi - J_\theta\left(r,\theta-\tfrac{\delta\theta}{2},\varphi\right)r\sin\left(\theta-\tfrac{1}{2}\delta\theta\right)\delta r\,\delta\varphi\\ &+J_\varphi\left(r,\theta,\varphi+\tfrac{\delta\theta}{2}\right)r\,\delta\theta\,\delta r - J_\varphi\left(r,\theta,\varphi-\tfrac{\delta\theta}{2}\right)r\,\delta\theta\,\delta r\end{aligned}$$

$$\begin{aligned}\oint_{\text{box}} \vec{J}\cdot d\vec{A} \approx\; &\sin\theta\,\delta\theta\,\delta\varphi\left(\left(r+\tfrac{1}{2}\delta r\right)^2 J_r\left(r+\tfrac{\delta r}{2},\theta,\varphi\right) - \left(r-\tfrac{1}{2}\delta r\right)^2 J_r\left(r-\tfrac{\delta r}{2},\theta,\varphi\right)\right)\\ &+r\,\delta r\,\delta\varphi\left(\sin\left(\theta+\tfrac{1}{2}\delta\theta\right)J_\theta\left(r,\theta+\tfrac{\delta\theta}{2},\varphi\right) - \sin\left(\theta-\tfrac{1}{2}\delta\theta\right)J_\theta\left(r,\theta-\tfrac{\delta\theta}{2},\varphi\right)\right)\\ &+r\,\delta\theta\,\delta r\left(J_\varphi\left(r,\theta,\varphi+\tfrac{\delta\theta}{2}\right) - J_\varphi\left(r,\theta,\varphi-\tfrac{\delta\theta}{2}\right)\right)\end{aligned}$$

And now we apply the definition of the partial derivative, and find that for our small pseudo-box:

$$\oint_{\text{box}} \vec{J}\cdot d\vec{A} \approx \sin\theta\,\delta r\,\delta\theta\,\delta\varphi\tfrac{\partial}{\partial r}\left(r^2 J_r(r,\theta,\varphi)\right) + r\,\delta r\,\delta\theta\,\delta\varphi\tfrac{\partial}{\partial r}\left(\sin\theta\, J_\theta(r,\theta,\varphi)\right) + r\,\delta r\,\delta\theta\,\delta\varphi\tfrac{\partial}{\partial\varphi}\left(J_\varphi(r,\theta,\varphi)\right).$$

The volume of the pseudo-box is $V = r^2\sin\theta\,\delta r\,\delta\theta\,\delta\varphi$, so:

$$\oint_{\text{box}} \vec{J}\cdot d\vec{A} \approx \frac{V}{r^2}\frac{\partial}{\partial r}\left(r^2 J_r(r,\theta,\varphi)\right) + \frac{V}{r\sin\theta}\frac{\partial}{\partial\theta}\left(\sin\theta\, J_\theta(r,\theta,\varphi)\right) + \frac{V}{r\sin\theta}\frac{\partial}{\partial\varphi}J_\varphi(r,\theta,\varphi).$$

Taking the limit as the volume goes to zero, we have the divergence:

$$\vec{\nabla}\cdot\vec{J} \equiv \lim_{V\to 0}\left(\frac{1}{V}\oint_{\text{surface}}\vec{J}\cdot d\vec{A}\right) = \frac{1}{r^2}\frac{\partial}{\partial r}\left(r^2 J_r(r,\theta,\varphi)\right) + \frac{1}{r\sin\theta}\frac{\partial}{\partial\theta}\left(\sin\theta\, J_\theta(r,\theta,\varphi)\right) + \frac{1}{r\sin\theta}\frac{\partial}{\partial\varphi}\left(J_\varphi(r,\theta,\varphi)\right).$$

1.5 *Solving The Continuity Equation*

The continuity equation guarantees that charge is always conserved, at all times and all places. However, we have not discussed the forces on a charged fluid, except to say that they must exist.

In this next section, we will make reasonable, common sense assumptions to guess *particular solutions* for the current density as a function of position and time, so long as they satisfy the continuity equation. However, there often exists more than one solution to a given problem that satisfies the continuity equation. Without additional constraints, such as knowledge of the forces involved, the techniques we describe in this section provide possible solutions that conserve charge, but do not necessarily produce unique solutions.

Example 1.4: The Spherical Continuity Equation

In Thought Experiment 1.6 (p. 30), we introduced a negatively charged balloon that gains charge at a known rate $I = \frac{d}{dt}Q$. Here we will again solve for the current density, $\vec{J}$, outside the sphere, but we will do so more formally, beginning with the continuity equation in differential form:

$$\vec{\nabla}\cdot\vec{J} = -\frac{\partial\rho}{\partial t}.$$

As we did in Thought Experiment 1.6 (p. 30), we draw a concentric sphere, with a finite radius, and then take the volume integral of both sides of the continuity equation:

$$\oint_{\text{volume}} \left(\vec{\nabla}\cdot\vec{J}\right) dV = \oint_{\text{volume}} \left(-\tfrac{\partial}{\partial t}\rho\right) dV .$$

Now, looking at the left side of the equation we find:

$$\oint_{\text{volume}} \left(\vec{\nabla}\cdot\vec{J}\right) dV = \oint_{\text{volume}} \left(\lim_{\delta V\to 0}\left(\frac{\oint_{\text{surface}} \vec{J}\cdot d\vec{A}}{\delta V}\right)\right) dV = \oint_{\text{surface}} \vec{J}\cdot d\vec{A} = I,$$

which is simply an application of a fundamental theorem of calculus, called the divergence theorem. This theorem was discovered by the mathematician, physicist, and astronomer, Carl Friedrich Gauss. As such, the surface of integration one chooses is called a Gaussian surface, which is a term that will be used throughout this book. We will discuss Gauss and his other remarkable contributions to the study of electricity and magnetism in Chapter 5.

Now looking at the right hand side of the equation, we see that:

$$\oint_{\text{volume}} \left(-\frac{\partial\rho}{\partial t}\right) dV = -\frac{d}{dt}\left(\oint_{\text{volume}} \rho\, dV\right) = -\frac{dQ}{dt} = I.$$

Finally, we can write:

$$I = \oint_{\text{surface}} \vec{J}\cdot d\vec{A} = -\frac{dQ}{dt}.$$

We have "recovered" the integral form of the continuity equation from its local, or differential, form. The divergence theorem, which we have obtained along the way, connects the differential and integral forms of the continuity equation and, hence, the local and global versions of the law of charge conservation. We will use the divergence theorem throughout this book, even when we are covering topics other than charge conservation.

Because we know that the current density is pointing radially inward, and is therefore normal to the spherical Gaussian surface, it follows that $\vec{J}\cdot d\vec{A} = -J\,dA$. So, as in Thought Experiment 1.6:

$$I = \oint_{\text{surface}} \vec{J} \cdot \mathrm{d}\vec{A} = -\frac{\mathrm{d}Q}{\mathrm{d}t}, \quad \text{and} \quad -J \cdot \left(4\pi r^2\right) = -\frac{\mathrm{d}Q}{\mathrm{d}t}.$$

Now, recall that the current density was inward, and the charge was increasing (becoming less negative), so we can write the current density vector as:

$$\vec{J} = \frac{I}{4\pi r^2}\hat{r} = -\frac{\mathrm{d}Q}{\mathrm{d}t}\frac{1}{4\pi r^2}\hat{r}.$$

Let us check our work using the differential form of the continuity equation:

$$\vec{\nabla} \cdot \vec{J} = -\frac{\partial \rho}{\partial t}.$$

Outside of the balloon, $\rho = 0$, so $\vec{\nabla} \cdot \vec{J} = 0$. In spherical coordinates we can calculate $\vec{\nabla} \cdot \vec{J}$ as:

$$\vec{\nabla} \cdot \vec{J} = \frac{1}{r^2}\frac{\partial}{\partial r}\left(r^2 J_r\right) + \frac{1}{r\sin\theta}\frac{\partial}{\partial \theta}\left(\sin\theta\, J_\theta\right) + \frac{1}{r\sin\theta}\frac{\partial J_\phi}{\partial \phi}.$$

Now in our case, $J_r = {}^{I}\!/_{4\pi r^2}$ and $J_\theta = J_\phi = 0$, because $\vec{J} = J_r\,\hat{r}$.

Thus:

$$\vec{\nabla} \cdot \vec{J} = \frac{1}{r^2}\frac{\partial}{\partial r}\left[r^2\left(\frac{I}{4\pi r^2}\right)\right] = \frac{1}{r^2}\frac{\partial}{\partial r}\left(\frac{I}{4\pi}\right) = 0,$$

so our solution checks out.

Problem 1.13: The Current Around a Star

Around a star, the current density is sometimes modeled with the following function:

$$\vec{J} = A\frac{\sin(2\theta)}{r^3}\hat{r} + B\frac{\sin^2(\theta)}{r^3}\hat{\theta}.$$

a) Solve for the ratio of the constants $\frac{A}{B}$, assuming that the charge density remains constant (and presumably neutral) with respect to time.

b) Research current flow around stars. Report on the evidence for current flows around stars, their causes, and any interesting effects. Clearly cite your references.

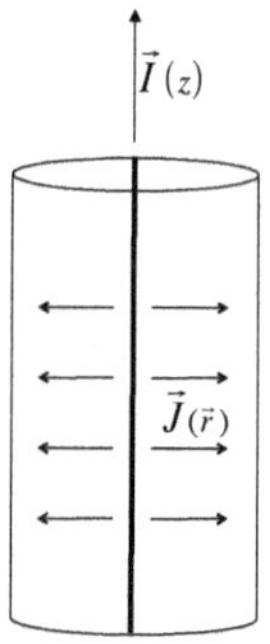

Problem 1.14: The Leaky Cable

Current flows through a long piece of poorly insulated coaxial cable with an end at the origin, so that a small amount of current leaks out. The current in the center of the cable is given by $\vec{I} = I_0 e^{-kz}\,\hat{z}$, where $\hat{z}$ is the direction along the wire. Solve for $\vec{J}(\vec{r}) = J(s,z)\,\hat{s}$, which is the current density that is leaking out.

Problem 1.15: An AC Transmission Line

A wave of AC current is propagating down a linear transmission line at a phase speed v with a current given by the following relationship:

$$I(z,t) = I_0 \sin\left(\omega\left(t - \frac{z}{v}\right)\right).$$

What is the linear charge density λ as a function of position and time?

Problem 1.16: A Cylindrical AC Current Wave

A cylindrically symmetrical wave of AC current emanates sideways from a long rod located on the $\hat{z}$ axis. This wave propagates with a phase velocity v, so the current density is given by the following relationship:

$$\vec{J}(s) = \frac{C}{s}\sin\left(\omega\left(t - \frac{s}{v}\right)\right)\hat{s}.$$

What is the charge density as a function of space and time?

Problem 1.17: A Point AC Current Source

A small spherical ball is attached to a slowly oscillating current source with $I = I_0 \sin(\omega t)$, which propagates radially outward.

a) If the ball is located in a conductive and neutral medium, what is the current density outside the ball as a function of position and time?

b) Now consider that rather than presuming a continuously neutral medium, we assume a particular propagation speed of v in the medium. Find the surrounding charge and current densities as a function of position and time.

Thought Experiment 1.12: A Two-Pole Circuit

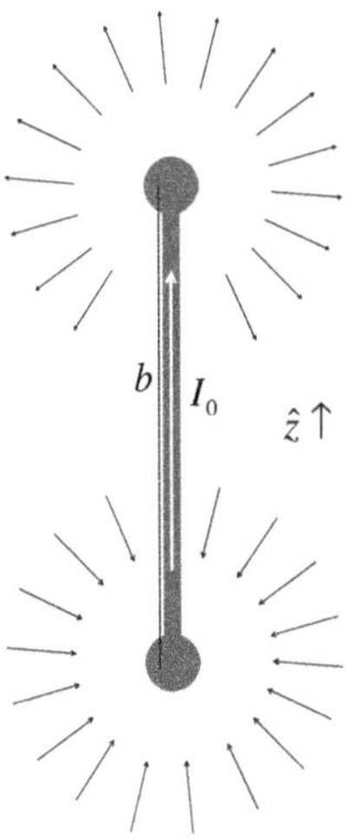

Consider two small conducting spheres, separated by a distance b, and connected at opposite ends of a conducting rod of cross-sectional radius $a \ll b$, which contains a battery that produces a current I_0.

One sphere becomes a current source, the *anode*, and the other a current sink, the *cathode*. The whole system is located inside a neutral ion bath that conducts electricity. What will be the current density as a function of position?

First we require charge to be conserved everywhere, so we can write the continuity equation as:

$$\vec{\nabla} \cdot \vec{J} = 0.$$

In the last thought experiment, we solved this for a single sphere, but here we have two spheres. However, the current continuity equation is a *linear partial differential equation*, so it must follow the superposition principle.

In other words, we know that by translating the prior solution up by half of the separation, that near the top sphere the current density is:

$$\vec{J}_{\text{top}} = \frac{I_0}{4\pi\left(\left|\vec{r} - \tfrac{1}{2}b\hat{z}\right|\right)^2}\left(\frac{\left(\vec{r} - \tfrac{1}{2}b\hat{z}\right)}{\left|\vec{r} - \tfrac{1}{2}b\hat{z}\right|}\right).$$

and translating down, we know that near the bottom sphere it is:

$$\vec{J}_{\text{bottom}} = -\frac{I_0}{4\pi\left(\left|\vec{r}+\tfrac{1}{2}b\hat{z}\right|\right)^2}\left(\frac{\left(\vec{r}+\tfrac{1}{2}b\hat{z}\right)}{\left|\vec{r}+\tfrac{1}{2}b\hat{z}\right|}\right).$$

Both of these solutions satisfies $\vec{\nabla}\cdot\vec{J}=0$, and therefore their vector sum must as well, because:

$$\vec{\nabla}\cdot\vec{J} = \vec{\nabla}\cdot\left(\vec{J}_{\text{top}}+\vec{J}_{\text{bottom}}\right) = \vec{\nabla}\cdot\vec{J}_{\text{top}} + \vec{\nabla}\cdot\vec{J}_{\text{bottom}} = 0+0=0\,.$$

Therefore the best guess for the solution is going to be simply the sum of the top and bottom solutions. This is called a *dipole* vector field, as there are two poles of equal and opposite strengths.

Notice that this predicts that the current density would be in the negative $\hat{z}$ direction on the axis between the two poles, when in reality it is:

$$\vec{J}_{\text{rod}} = \frac{I_0}{\pi a^2}\hat{z}\,,$$

due to the battery between the two poles.

It is important to always remember that a divergence-free vector field cannot have any sources, nor sinks, unless there is some sort of hidden pipe from the sink and back to the source.

Example 1.5: Electroplating a Cylindrical Object

Conducting objects can be coated with a thin metal surface, through a process called *electroplating*. As charge flows through a ion bath, oxidation and reduction reactions take place and metal is transferred from the anode to the cathode.

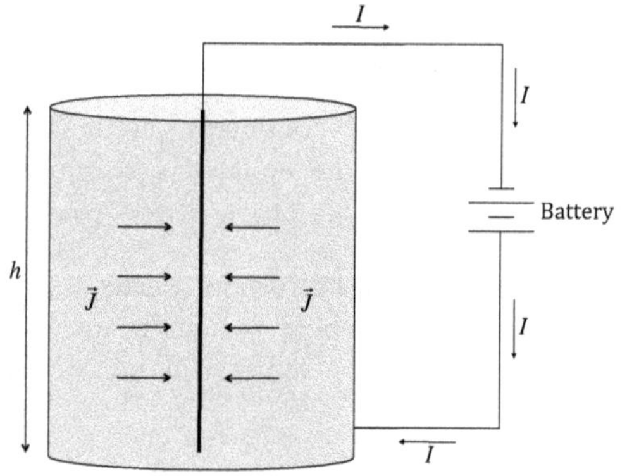

In this example, we consider a rod being electroplated in a cylindrical metal jar of height h and radius R that is filled with an ion bath. A steady current I flows through the wire and power source. What is the current density $\vec{J}$ inside the jar?

To answer this, first we render the system in cylindrical coordinates, since cylindrical symmetry allows us to write $\vec{J} = J_s\,\hat{s}$ and easily apply the conservation of charge:

$$\vec{\nabla}\cdot\vec{J} = \frac{-\partial\rho}{\partial t}\,.$$

We solve this in integral form by drawing a cylindrical Gaussian surface of radius s < R, and a height equal to the height (h) of the jar. We integrate both sides of the above equation with respect to volume:

$$\oint_{\text{volume}}\left(\vec{\nabla}\cdot\vec{J}\right)dV = \oint_{\text{volume}}\left(\frac{-\partial\rho}{\partial t}\right)dV\,.$$

By applying the definition of the divergence we find:

$$\oint_{\text{volume}} \left(\vec{\nabla}\cdot\vec{J}\right)\mathrm{d}V = \oint_{\text{volume}} \left(\lim_{\delta V\to 0}\left(\frac{1}{\delta V}\oint_S \vec{J}\cdot \mathrm{d}\vec{A}\right)\right)\mathrm{d}V = \oint_{\text{surface}} \vec{J}\cdot\mathrm{d}\vec{A},$$

and, from the continuity equation, we find:

$$\oint_{\text{surface}} \vec{J}\cdot\mathrm{d}\vec{\mathrm{A}} = \oint_{\text{volume}} \left(\frac{-\partial\rho}{\partial t}\right)\mathrm{d}V = -\frac{\mathrm{d}Q}{\mathrm{d}t}.$$

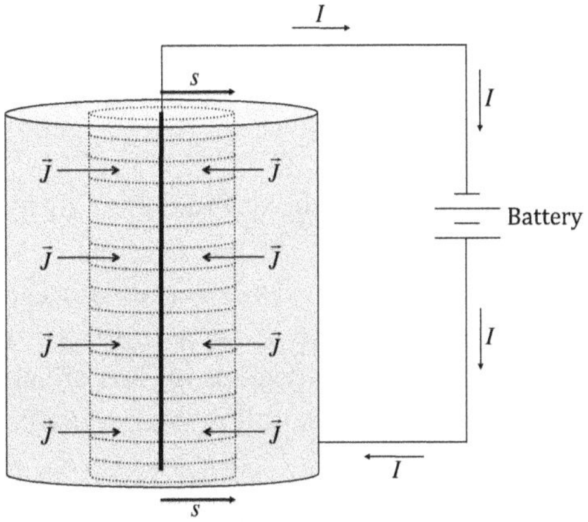

Everything is neutral, so:

$$\frac{\mathrm{d}Q}{\mathrm{d}t} = 0.$$

We can now evaluate the surface integral for our cylindrical Gaussian surface:

$$\oint_{\text{surface}} \vec{J}\cdot\mathrm{d}\vec{\mathrm{A}} = \int_{\text{top}} \vec{J}\cdot\mathrm{d}\vec{\mathrm{A}} + \int_{\text{bottom}} \vec{J}\cdot\mathrm{d}\vec{\mathrm{A}} + \int_{\text{sides}} \vec{J}\cdot\mathrm{d}\vec{\mathrm{A}}.$$

First we recall that the current density is perpendicular to the axis of the rod ($\vec{J} = J_s\hat{s}$), except inside the wire leaving the top of the Gaussian surface. Next we recall that the differential area vector points normal to the surface. Thus, for the top, bottom, and sides, $\mathrm{d}\vec{A} = \hat{z}\,\mathrm{d}A$, $\mathrm{d}\vec{A} = -\hat{z}\,\mathrm{d}A$, $\mathrm{d}\vec{A} = \hat{s}\,\mathrm{d}A$ respectively. So,

$$\oint_{\text{surface}} \vec{J}\cdot\mathrm{d}\vec{\mathrm{A}} = \int_{\text{top}} \left(J_{\text{top}}\hat{z}\right)\cdot\left(\hat{z}\,\mathrm{d}A\right) + \int_{\text{bottom}} (0)\cdot\left(-\hat{z}\,\mathrm{d}A\right) + \int_{\text{sides}} \left(J_s\,\hat{s}\right)\cdot\left(\hat{s}\,\mathrm{d}A\right),$$

$$0 = I + 0 + \int_0^h \left(J_s\right)\left(2\pi s\,\mathrm{d}z\right) = I + 2\pi s h J_s.$$

Solving for $\vec{J}(\vec{r}) = J_s\hat{s}$, our final answer becomes:

$$\vec{J}(\vec{r}) = \left(\frac{-I}{2\pi\,h}\right)\frac{\hat{s}}{s}.$$

Now we will check our work by calculating the divergence of $\vec{J}$ in cylindrical coordinates:

$$\vec{\nabla}\cdot\vec{J} = \frac{1}{s}\frac{\partial\left(sJ_s\right)}{\partial s} + \frac{1}{s}\frac{\partial J_\phi}{\partial\phi} + \frac{\partial J_z}{\partial z} = \frac{1}{s}\frac{\partial}{\partial s}\left(s\left(\frac{-I}{2\pi\,s}\right)\right) = \frac{1}{s}\frac{\partial}{\partial s}\left(\frac{-I}{2\pi}\right) = 0,$$

which is, again, the continuity equation when there is no build up of charge.

Throughout your study of electrodynamics relatively simply solutions arise for problems with lots of symmetry, so the direction of the vector field can be assumed. This is true of current flow in a fluid, but it will also be the case with other vector fields. Unfortunately the reverse is also true; systems without obvious symmetry are usually impossible to solve analytically. Since the universe is filled with asymmetrical systems, we must also learn how to solve these equations numerically. And, finally, we cannot stress our first chapter caveat too much. Applying only the conservation of charge, even with numerical algorithms, only produce unique solutions in symmetrical systems.

Example 1.6: Current Density Inside a Box

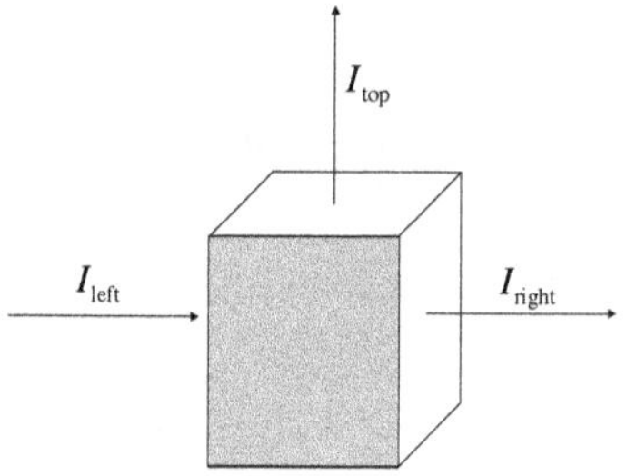

A box of known dimensions $(l \times w \times h)$ contains a uniform substance and is electrically neutral throughout. Three of the six sides are insulators (front, back, bottom) and three are conductors (left, right, top) but insulated from each other. The box is connected to a battery and ammeters (ammeters are devices which measure current), so that a known current flows into the left side (I_{left}) and out of the top and right sides (I_{top} and I_{right}) of the box. Find $\vec{J}(\vec{r})$ inside the box.

We will assume that $\vec{J}$ is constant on each conducting side and perpendicular to that particular face. This may seem to be a simplistic assumption, but it will suffice for illustrative purposes here. Such simplifying assumptions are always necessary when we model real physical processes. We therefore have the following boundary conditions:

$$\vec{J}_{left} = \frac{I_{left}}{wh}\hat{x}, \quad \vec{J}_{right} = \frac{I_{right}}{wh}\hat{x} \text{ and } \vec{J}_{top} = \frac{I_{top}}{hl}\hat{z}.$$

If we consider the whole box as our Gaussian surface, we have:

$$\oint_{surface} \vec{J} \cdot d\vec{A} = -\frac{dQ}{dt} = 0,$$

because it is neutral throughout. Let us now explicitly evaluate the surface integral on the left side of this equation:

$$\oint_{box} \vec{J} \cdot d\vec{A} = \int_{left} \vec{J} \cdot d\vec{A} + \int_{right} \vec{J} \cdot d\vec{A} + \int_{top} \vec{J} \cdot d\vec{A} = \int_{left} \left(\frac{I_{left}\hat{x}}{wh}\right) \cdot (-\hat{x}\, dA) + \int_{right} \left(\frac{I_{right}\hat{x}}{wh}\right) \cdot (\hat{x}\, dA) + \int_{top} \left(\frac{I_{top}\hat{z}}{lw}\right) \cdot (\hat{z}\, dA).$$

Now, we know that:

$$\oint_{box} \vec{J} \cdot d\vec{A} = -I_{left} + I_{right} + I_{top} = 0,$$

and this confirms our global constraint that: $I_{left} = I_{right} + I_{top}$.

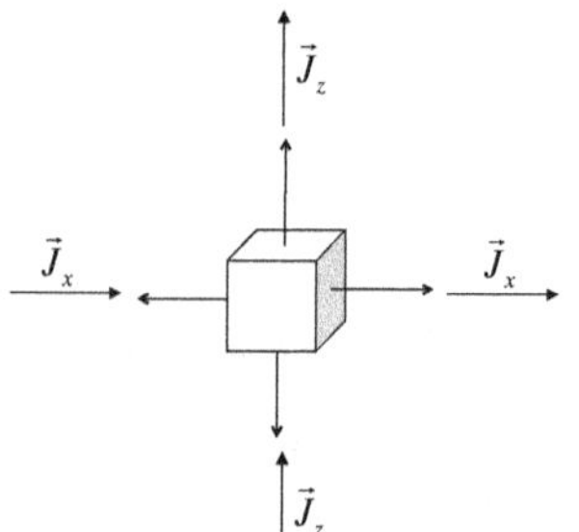

Next we will look at an infinitesimal cell inside at position $\vec{r} = x\hat{x} + y\hat{y} + z\hat{z}$ inside the box. Because no current flows into or out of the box in the $\hat{y}$ direction, and nothing depends on the y coordinate, we can write: $\vec{J} = J_x(x,z)\hat{x} + J_z(x,z)\hat{z}$. Then we can write the continuity equation in differential form as:

$$\vec{\nabla} \cdot \vec{J} = \frac{\partial J_x}{\partial x} + \frac{\partial J_z}{\partial z} = \frac{-\partial \rho}{\partial t} = 0, \quad \text{or simply} \quad \frac{\partial J_x}{\partial x} = -\frac{\partial J_z}{\partial z}.$$

The simplest non-trivial solution to this equation is a constant:

$$K = \frac{\partial J_x}{\partial x} = -\frac{\partial J_z}{\partial z}.$$

We will take this solution and then integrate and apply our boundary conditions to find K in terms of the known quantities I_{left}, I_{right}, and I_{top}.

$$J_x = \int_0^x \left(\frac{\partial J_x}{\partial x}\right) \mathrm{d}x + J_x(x=0) = \int_0^x (K)\,\mathrm{d}x + J_{\text{left}} = J_{\text{left}} + K\,x,$$

$$J_z = \int_0^z \left(\frac{\partial J_z}{\partial z}\right) \mathrm{d}z + J_z(z=0) = \int_0^z (-K)\,\mathrm{d}z + J_{\text{bottom}} = -K\,z.$$

Now, applying the top and right boundary conditions, we can write:

$$J_{\text{right}} = J_x(x=l) = J_{\text{left}} + K\,l \quad \text{and} \quad J_{\text{top}} = J_z(z=h) = -K\,h, \text{ so}$$

$$K = \frac{J_{\text{right}} - J_{\text{left}}}{l} = \frac{I_{\text{right}} - I_{\text{left}}}{lwh} \quad \text{and} \quad K = \frac{-J_{top}}{h} = \frac{-I_{top}}{lwh}.$$

This checks out, because we already established that overall conservation of charge implies:

$$I_{\text{left}} = I_{\text{right}} + I_{\text{top}}.$$

So, our final solution becomes:

$$\vec{J} = J_x\hat{x} + J_z\hat{z} = \frac{1}{lwh}\left(\left(I_{\text{left}}(l-x) + I_{\text{right}}\,x\right)\hat{x} + \left(I_{\text{top}}\,z\right)\hat{z}\right).$$

This is plotted to the right, assuming that half the current flows out the side and half out the top.

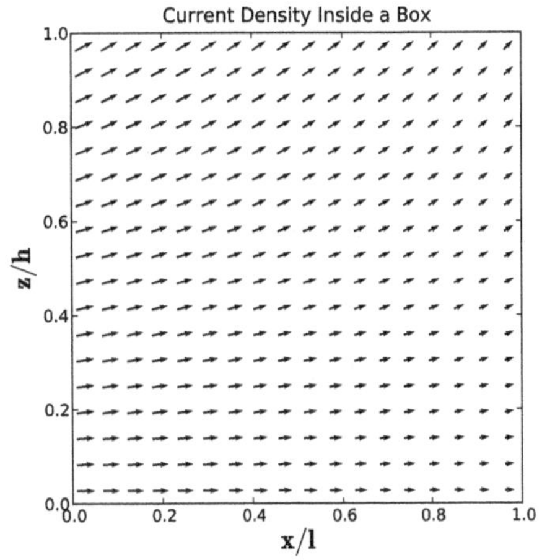

Now let us check our work, knowing that $\vec{\nabla}\cdot\vec{J} = 0$:

$$\vec{\nabla}\cdot\vec{J} = \frac{\partial J_x}{\partial x} + \frac{\partial J_y}{\partial y} + \frac{\partial J_z}{\partial z} = \left(\frac{1}{lwh}\right)\left(-I_{\text{left}} + I_{\text{right}} + I_{\text{top}}\right) = 0.$$

We emphasize again that current continuity is a necessary criterion, but that, depending on the details of the boundary conditions and other assumptions, there may be multiple solutions that satisfy charge conservation. However, all solutions must satisfy the continuity equation.

Problem 1.18: A Discharging Long Rod

A long thin rod has a charge per length of $\lambda(t) = \lambda_0\left(1 - \frac{t}{\tau}\right)$, within the time range $0 \le t \le \tau$. Find the current density $\vec{J}(\vec{r},t)$ surrounding the rod during the same time range.

Problem 1.19: An AC Spherical Shell

Consider a spherical shell of radius R, located inside a conductive medium and connected via a wire to a slowly varying AC current source, so it has a uniform, but time dependent, surface charge density of $\sigma(t) = \sigma_0 \cos(\omega t)$.

a) Assuming everywhere else remains continuously neutral, find the current density as a function of radial position and time outside the sphere.

b) Show that the current density is zero inside the sphere. Fully justify your results.

Problem 1.20: A Discharging Cloud

A cloud has a charge density given by $\rho(r,t) = \frac{C}{r^2}\left(1 - \frac{t}{\tau}\right)$ within the time range $0 \le t \le \tau$. Find the current density as a function of position and time.

Problem 1.21: A Plasma Globe

In a plasma globe a spherical clear ball contains low pressure inert gases, such as neon and argon. In the center is a metal ball connected to a high frequency AC current source. When enough current flows through the gas, it glows. Plasma globes are fun, because they do not glow uniformly, but rather have moving strands of light extending approximately radially outward.

a) My plasma globe has a total current of 3 milliamperes, and under normal conditions has about 50 glowing tubes extending radially outward from a central ball that is 3.5 cm in diameter. Each of these tubes has a diameter of about 2 millimeters. Estimate the current density through each of the glowing filaments.

b) Touching the plasma globe attracts these filaments and, when I touch it on top with my gold ring, only a single brighter filament extends to my ring finger. This filament appears to be the same diameter as each of the others. Estimate the current density in this brighter strand.

c) The frequency of the AC current in my plasma globe is 35 kHz. Find the surface charge density on the central metal ball as a function of time. Clearly discuss any assumptions you make.

Problem 1.22: The Heliospheric Current Sheet

In the solar wind there is an inward current density flowing in a spiral pattern. This is offset by a current density flowing out of the poles. Here we will very crudely model each term as:

$$\vec{J}_{\text{spiral}} = J_1 \sin^2\theta \, e^{-(\varphi - kr - \omega t)^2} \left(\Phi(r)\hat{\varphi} - R(r)\hat{r}\right) \quad \text{and} \quad \vec{J}_{\text{poles}} = J_2\left(\frac{\cos^4\theta}{(kr)^2}\right)\hat{r},$$

where J_1, J_2, and k are constant and $\Phi(r)$ and $R(r)$ are unitless functions of r. The net current density is therefore modeled as: $\vec{J} = \vec{J}_{\text{spiral}} + \vec{J}_{\text{poles}}$.

The total current flowing out the poles is a few billion amperes $\left(I_0 \approx 3\times10^9\,\text{A}\right)$.

The spiral structure is caused by the rotation of the sun. The sun's rotation period P is about 25 days, and the speed of the solar wind v_w is usually about 300 km/s.

a) Find the sun's angular velocity ω in terms of P and its value in units of radians per day.

b) Find the constant k in terms of v_w and P and estimate its value in units of radians per AU $\left(1\,\text{AU} = 1.5\times10^8\,\text{km}\right)$.

c) Find the constant J_2 in terms of I_0 and k, and find its value in SI units.

d) Assuming that the current is in a direction parallel to the spiral, find the ratio of $\Phi(r)$ to $R(r)$. Notice that for every circle of circumference $2\pi r$, the spiral shifts radially by a distance $\frac{2\pi}{k}$, so the ratio of the azimuthal to radial components of the current density must have the same ratio as the corresponding spiral distances.

e) Find solutions for $\Phi(r)$ and $R(r)$, which are unitless functions of r that satisfy the continuity equation. These may also be a function of the constant k.

f) Find the constant J_1 in terms of I_0, and k, and find its value in SI units.

g) Write down the total current density $\vec{J}$ in terms of the known constants I_0 and k.

h) Estimate the magnitude of the current density near the earth as a function of time, and plot it. Make sure you remember that the Earth is moving in an approximate circle around the sun's equator, with a radius of one astronomical unit and a period of one year.

Problem 1.23: Electrotyping a Sculpture

Often works of sculpture are carved in a soft material such as clay, painted with a conducting paint, and electrotyped using copper. This, among other methods, is often referred to as being "cast in bronze." This is done by dipping the sculpture and a copper bar into an aqueous solution of copper sulfate ($CuSO_4$) and sulfuric acid (H_2SO_4), and running electricity through the solution. How much total charge must flow per kilogram of copper transferred?

Problem 1.24: Electroplating a Globe

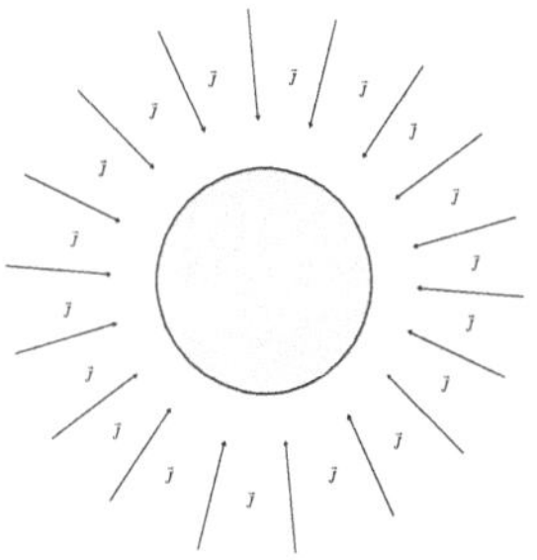

A spherical mold is being electrotyped on the outside to make a copper globe of radius a, with a total current I flowing through the device.

(a) Find the current density as a function of distance from the globe center, $\vec{J}(\vec{r})$.

(b) If the concentration of Cu^{2+} is n ions per unit volume, find the systemic velocity of the copper ions in the solution as a function of the total current and the distance from the globe center.

Problem 1.25: Electroplating a Cone

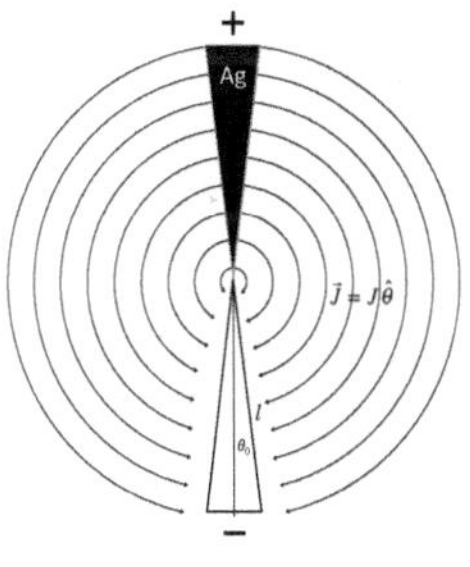

A small decorative Christmas tree is being electroplated evenly on its outside surface using a solution of silver cyanide. The Christmas tree is a cone of side length l and half opening angle θ_0, and the electroplating machine carries a total DC current I_0.

Assuming the ionic current density is only in the $\hat{\theta}$ direction, find $\vec{J}(\vec{r}) = J(r,\theta)\hat{\theta}$ in the ion bath.

Thought Experiment 1.13: Algorithmically Solving the Continuity Equation

Consider a similar problem as the last thought experiment, but without such simple constraints. We can use the same idea to iteratively ensure that charge is always conserved in a steady state, two dimensional flow of current.

Consider a small conducting square, of size $2a \times 2a \times h$. Knowing the current density at the eight surrounding points, what is the current density in the middle?

We start this problem by setting up a Cartesian coordinate grid, as in the diagrams, and noticing that we are looking for the following current density:

$\vec{J}(a,a) = J_x(a,a)\,\hat{x} + J_y(a,a)\,\hat{y}$.

Next we draw a Gaussian surface around the left half the cell, as shown. Conservation of charge implies that the current into the square must equal the flowing out, so:

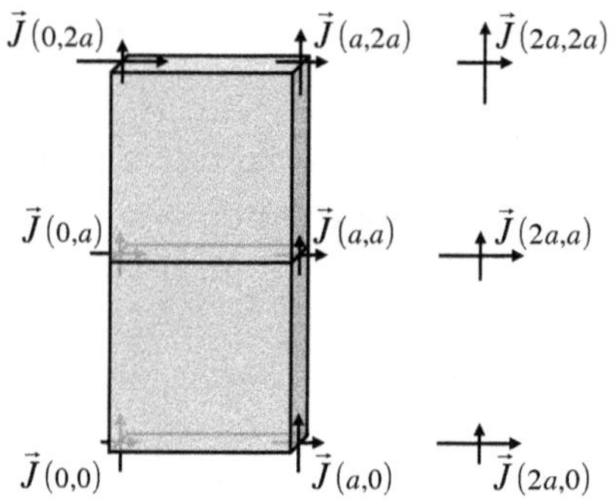

$$I_{\text{net}} = \left(I_{\text{right}} - I_{\text{left}}\right) + \left(I_{\text{top}} - I_{\text{bottom}}\right) = 0 .$$

And in integral form this is:

$$0 = h\int_0^{2a} J_x\big|_{x=a}\,\mathrm{d}y - h\int_0^{2a} J_x\big|_{x=0}\,\mathrm{d}y + h\int_0^{a} J_y\big|_{y=2a}\,\mathrm{d}x - h\int_0^{a} J_y\big|_{y=0}\,\mathrm{d}x .$$

Each of these integrals is the average current density multiplied by the cross-sectional area. So the current flowing out of the top would be:

$$I_{\text{top}} \approx ha\left(\frac{J_y(0,2a) + J_y(a,2a)}{2}\right).$$

Similarly the current flowing into the bottom would be:

$$I_{\text{bottom}} \approx ha\left(\frac{J_y(0,0) + J_y(a,0)}{2}\right).$$

The sides, on the other hand, have three sample points, with the one in the middle being most important. Assuming it is twice as important as the other two combined, we can write the current flowing into the left side as:

$$I_{\text{left}} \approx 2ha\left(\frac{J_x(0,0) + 4J_x(0,a) + J_x(0,2a)}{6}\right),$$

and out of the right as:

$$I_{\text{right}} \approx 2ha\left(\frac{J_x(a,0) + 4J_x(a,a) + J_x(a,2a)}{6}\right).$$

Now the net current becomes:

$$I_{\text{net}} \approx \frac{ha}{2}\left(J_y(0,2a) + J_y(a,2a) - J_y(0,0) - J_y(a,0)\right) + \frac{ha}{3}\left(J_x(a,0) + 4J_x(a,a) + J_x(a,2a) - J_x(0,0) - 4J_x(0,a) - J_x(0,2a)\right)$$

which must be zero for steady state flow. Setting this to zero, and solving for $J_x(a,a)$ we obtain:

$$J_x(a,a) \approx J_x(0,a) + \tfrac{1}{4}\left(J_x(0,0) + J_x(0,2a) - J_x(a,0) - J_x(a,2a)\right) + \tfrac{3}{8}\left(J_y(0,0) - J_y(0,2a) + J_y(a,0) - J_y(a,2a)\right).$$

First notice that if the current density is uniform, then $J_x(a,a)$ would simply return this same value. Next notice that what flows in, must also flow out. All the positive terms represent inward flow and the negative terms represent outward flow.

This expression is promising, but we have not used all of the data that we have to find the horizontal component at the center. To do this, we must perform a similar analysis using a rectangular box with the center point entering, rather than exiting. You can now show for yourself that:

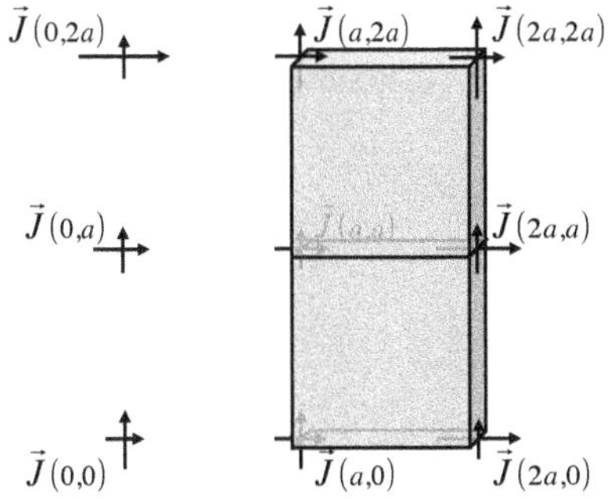

$$\begin{aligned} J_x(a,a) \approx\ & J_x(2a,a) \\ & +\tfrac{1}{4}\left(J_x(2a,0)+J_x(2a,2a)-J_x(a,0)-J_x(a,2a)\right) \\ & +\tfrac{3}{8}\left(J_y(2a,2a)+J_y(a,2a)-J_y(a,0)-J_y(2a,0)\right). \end{aligned}$$

We average the two results to get a more robust result:

$$\begin{aligned} J_x(a,a) \approx\ & \tfrac{1}{2}\left(J_x(0,a)+J_x(2a,a)\right) \\ & +\tfrac{1}{8}\left(J_x(0,0)+J_x(0,2a)-J_x(a,0)-J_x(a,2a)+J_x(2a,0)+J_x(2a,2a)-J_x(a,0)-J_x(a,2a)\right) \\ & +\tfrac{3}{16}\left(J_y(0,0)-J_y(0,2a)+\cancel{J_y(a,0)}-J_y(a,2a)+J_y(2a,2a)+J_y(a,2a)-\cancel{J_y(a,0)}-J_y(2a,0)\right). \end{aligned}$$

This simplifies to:

$$\begin{aligned} J_x(a,a) \approx\ & \tfrac{1}{2}\left(J_x(0,a)+J_x(2a,a)\right)-\tfrac{1}{4}\left(J_x(a,0)+J_x(a,2a)\right) \\ & +\tfrac{1}{8}\left(J_x(0,0)+J_x(0,2a)+J_x(2a,0)+J_x(2a,2a)\right) \\ & +\tfrac{3}{16}\left(J_y(0,0)-J_y(0,2a)-J_y(2a,0)+J_y(2a,2a)\right). \end{aligned}$$

And rotating the system yields the vertical component:

$$\begin{aligned} J_y(a,a) \approx\ & -\tfrac{1}{4}\left(J_y(0,a)+J_y(2a,a)\right)+\tfrac{1}{2}\left(J_y(a,0)+J_y(a,2a)\right) \\ & +\tfrac{3}{16}\left(J_x(0,0)-J_x(0,2a)-J_x(2a,0)+J_x(2a,2a)\right) \\ & +\tfrac{1}{8}\left(J_y(0,0)+J_y(0,2a)+J_y(2a,0)+J_y(2a,2a)\right). \end{aligned}$$

Using this relationship we can interpolate the current density, and we can iteratively refine values in a grid as a function of each surrounding point. We simply replace the value at every point in the grid with this function of the surrounding points, and then subject it to any boundary conditions and other constraints we place on a system. We can often do this by solving a very simple model first, and then use the result in making a grid of points. We then refine the model by putting in more realistic boundary conditions. Algorithms that use this type of technique are called *relaxation methods*, and this approach is common in computer solutions to steady-state partial differential equations with known boundary conditions. The only catch is that current continuity does not *guarantee* a unique solution, so the final result can depend on the initial guess. The more additional physical constraints you can place on a system, the better relaxation methods work.

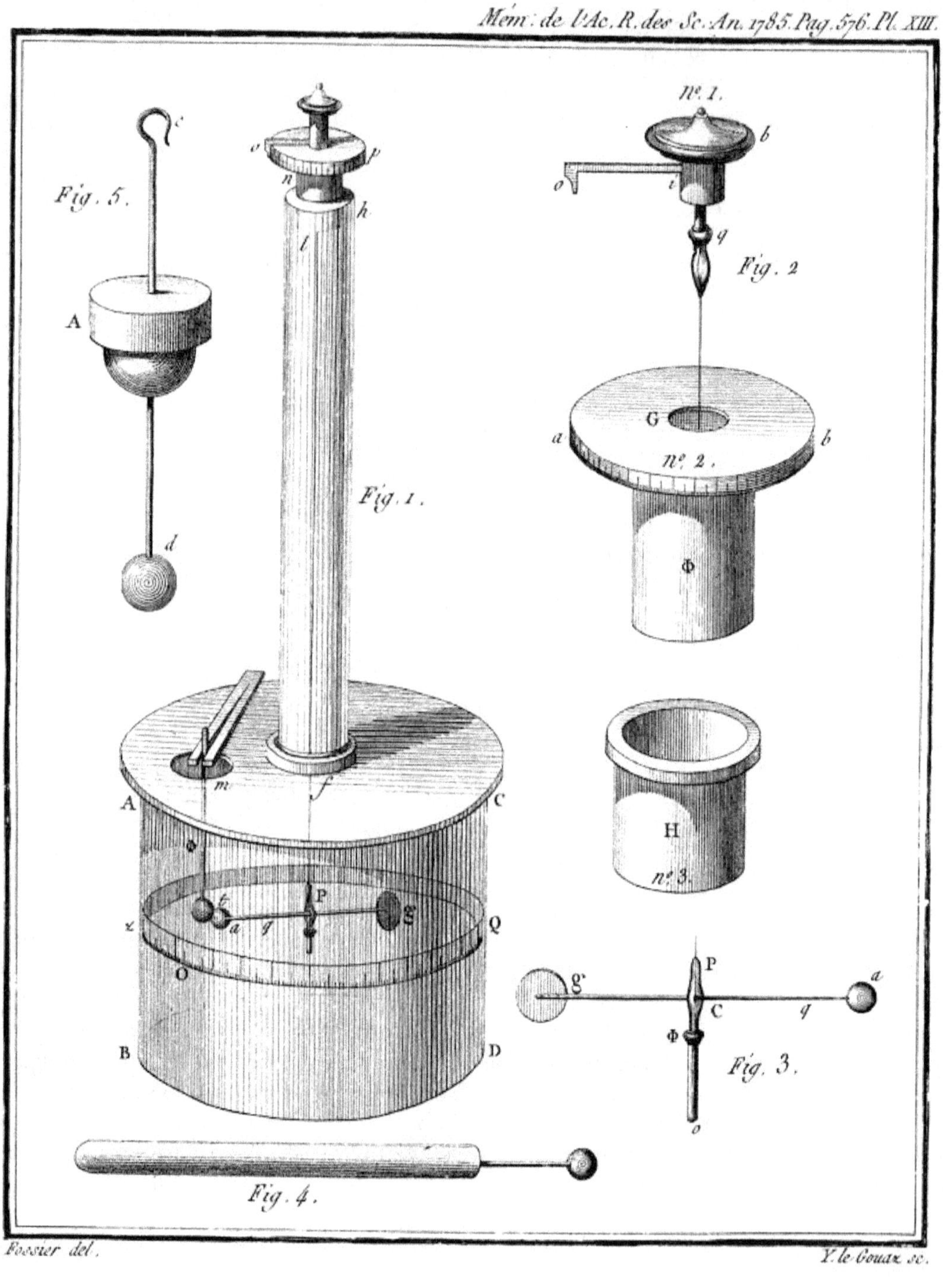

Plate 5: Coulomb's Torsion Balance[27]

---

[27] Coulomb, Mémoires des Académie des Sciences, (Paris, 1785), p. 576, plate 8.

Chapter 2 The Electrostatic Force

In 1687, Sir Isaac Newton published his three laws of motion and his universal law of gravitation, arguably the most successful theory in the whole history of science. Philosophiæ Naturalis Principia Mathematica was written in Latin, was embraced by scientists across all of Europe, and sparked a mathematical revolution. Unifying physics and astronomy, and providing the link between disparate disciplines, many in Europe regarded the *Principia* as presenting the ultimate solution to the laws of nature.

Despite its success, Newton was worried about several lingering philosophical questions, especially the issue of action at a distance. The primary theme of Newton's laws of motion, for instance, concerns cause and effect. When one object applies a force on another, according to Newton's second law, the other object accelerates by an amount proportional to the force applied and inversely proportional to its own mass. This was easy enough to appreciate in the case of contact forces, such as collisions, where one body directly interacts with another. However, gravity works differently. The moon orbits the earth, and the earth's oceans move in response to the moon. How does this happen? What is it that links the seas to the moon, despite the vast distance between them? Presumably some mechanism conveys this gravitational force, but Newton did not know what it was and would not speculate about it in his formal works, writing:

> I have not as yet been able to discover the reason for these properties of gravity from phenomena, and I do not feign hypotheses. For whatever is not deduced from the phenomena must be called a hypothesis; and hypotheses, whether metaphysical or physical, or based on occult qualities, or mechanical, have no place in experimental philosophy. In this philosophy particular propositions are inferred from the phenomena, and afterwards rendered general by induction.[28]

While Newton's disdain for action at a distance permeated Cambridge, Continental physicists appreciated the mathematical beauty and were not disturbed by action at a distance. After all, God surely made an elegant universe, but why should he be constrained by issues of communication over great distances? More importantly, eighteenth century French mathematicians had the skills, and government backing, to develop intricate theories of celestial mechanics, which were eventually laid out by Pierre-Simon Laplace in his 1798 Traité de Mechanique Céleste (Treatise on Celestial Mechanics).

Meanwhile, in Colonial America, Benjamin Franklin had joined a radical political faction, which, by 1776, was about to loose a revolution against the British Crown. Something had to be done. In a last ditch effort, Benjamin Franklin snuck over to Paris to ask King Louis XVI for help. Let us not forget that, all through the eighteenth century, France and Britain were enmeshed in a great struggle for world dominance. Much like today's great powers, France's government used scientific funding to project power and improve their military technology.

Every day, the local magnetic field changes slightly, but by how much and why? Can it be predicted and accounted for by naval navigators? This was the question of the annual scientific *Gran Prix* of the *Académie des Sciences* when Franklin arrived in the French capital. This *Gran Prix* question also launched the second career of a first lieutenant in the French corps of engineers, who, after the Seven Years War, built a new fort on the island of Martinique.

---

[28] Isaac Newton, The Principia: Mathematical Principles of Natural Philosophy (Philosophiae Naturalis Principia Mathematica) Third ed. (1726), trans. I. Bernard Cohen and Anne Whitman (California: University of California Press, 1999), p. 943.

When Charles-Augustin de Coulomb returned to France he continued his military service domestically, where he had time also to apply his craft to his scientific interests, such as how to measure finely the terrestrial magnetic field. Coulomb suspected that the frictional torque of a suspended compass needle could be made negligible if he used fine silk threads. This made the new compass like a pendulum, and Coulomb realized that it would twist back and forth in a similar manner as a pendulum oscillates. Just as the period of a pendulum remains constant with small amplitude, the suspended needle also swings back and forth isochronously. Furthermore, the torque should be directly proportional to the angle of twist, provided that the angle of twist does not exceed the elastic limit of the thread.

This new torsion magnetometer allowed the little known engineer to share the 1777 *Gran Prix* with an established compass designer. Meanwhile, Franklin showed himself to be a brilliant diplomat, as France went to war the next year and he remained in Paris as the American ambassador. Franklin may have attended the 1781 scientific Gran Prix, which Coulomb won outright with a detailed study of friction in simple machines.

Franklin's job in France was to be a diplomat, but he was also an eminent scientist. The king put him to work in this capacity, perhaps in the hope that it would help the war effort. In 1784, Franklin chaired a scientific commission that evaluated the technical designs of lightning rods to protect French military outposts.[29] Also authoring the report was Charles-Augustin de Coulomb.

Also in 1784, Coulomb published a report titled Theoretical Research and Experimentation on Torsion and the Elasticity of Metal Wire, which paved the way for perhaps the most fundamental electrical discovery since Franklin's work as a young man.

As in his diagram (p. 52), Coulomb replaced the silk thread in his compass with a piano wire, as it had a more linear rotational spring constant. He then suspended a lead cylinder in a pot of water and observed the rate at which the amplitude of the torsional motions decreased. In this manner, Coulomb determined the force of resistance of the water to the cylinder, equivalent to a torque of $3\times10^{-9}\,\mathrm{N\cdot m}$, which was beyond the sensitivities of previous instruments.[30]

Pursuing an analogy with Newton's Law of Gravity, Coulomb applied his technique to measure the force between electric charges, by using a waxed silk thread to suspend a bar holding small charged spheres at each end. In spite of significant experimental error, Coulomb announced his conclusion that the force between the charges obeyed an inverse square law similar to Newton's universal law of gravity.[31] In this way, Coulomb demonstrated experimentally that the force between two stationary point charges acts along the line joining them and is inversely proportional to the square of their mutual distance, or in symbols:

$$\vec{F} \propto \frac{q_1 q_2}{r^2}\hat{r}.$$

---

[29] MM. Franklin, Le Roy, Coulomb, Delaplace and l'Abbé Rochon, *Académie Royale des Sciences*, minutes, tome CIII, fol. 90 v°-95 r° (session for Saturday April 24, 1784). From The Papers of Benjamin Franklin, Yale University Press, posted on franklinpapers.org.

[30] J.L. Heilbron, Electricity in the17th and 18th centuries: a study of early Modern physics. (University of California Press, 1979): 469-70.

[31] Coulomb, Mémoires des Académie des Sciences, (1785): 229-638. These are a series of papers. It is interesting to note that the English scientist Henry Cavendish (1731-1810) obtained the same result, but using a different apparatus. Unlike Coulomb, Cavendish did not publish his result. J.C. Maxwell published this, along with other pioneering work done by Cavendish, nearly a hundred years later in 1879.

Before the recent ascent of computational integration techniques, applying Coulomb's law was quite difficult in practice. However, since Coulomb's law follows the same inverse square law as Newton's Law of Gravity, Laplace and others could apply the physics of celestial mechanics to the field of electrostatics. This led to what was known as *the astronomical model of electricity*, and is now the basis for the field of electrostatics.

In the end, Franklin gained the help of Louis XVI and the French government. Thanks to France's involvement, the war against Britain was a great success for the American revolutionaries. France got little out of it, however, except a worsening financial situation and perhaps a strong whiff of Franklin's subversive political ideas. In 1785 Franklin returned to America and by 1789 political unrest took over France. In 1793, Louis XVI lost his head and the young Napoleon Bonaparte sieged Toulon.

Coulomb retired from the military and laid low during the French Revolution. After the French Revolution, Coulomb came out of retirement to help with the development of the metric system and public education. He remained active in research until his death at age 70.

Meanwhile, across the English channel, Michael Faraday picked up where Franklin's Leyden jar experiments had left off. In particular, Faraday conducted experiments using spherical Leyden jars with interchangeable insulators made of various materials. To explain his experiments, Faraday suggested that the positive and negative charge in the insulator remain bound together, but become somewhat separated, so he called these materials *dielectrics*. Written in clear English without mathematical complexity, his well illustrated three volume treatise[32] documents each experiment, along with how they fit together. Eventually he forms a coherent theory involving long chains of microscopic dipoles linking cause to effect.

Regardless of the microscopic details, Faraday could map out these *lines of electric force* from positive to negative charge. The electric force imparted on another charge would simply be related to the number of lines of force per area, which he eventually called *the electric field*. His idea also nicely explained the inverse-square nature of Coulomb's law, as the number of radial field lines per area would decrease as the square of the distance from a spherical charge.

Using his spherical Leyden jar, Faraday also showed that the constant in Coulomb's law varied as a function of each dielectric material. Thus, every substance would have a corresponding property, called the *permittivity*, that measured how much the internal charges are displaced when exposed to an electric field, which in turn would convey the electric force.

While it did not have anything to do with gravity directly, the idea of an analogous gravitational field, which was somehow conveyed through matter, gave hope to the British physics community that it was only a matter of time until mechanisms would be found for every force that appears to act at a distance.

2.1 *Coulomb's Law*

As we discussed in the introduction, Coulomb demonstrated that the force between two spherical charges is inversely proportional to the square of their mutual distance, or:

$$\vec{F}_{1,2} \propto \frac{Q_1 Q_2}{r_{1,2}^2} \hat{r}_{1,2},$$

---

[32] Michael Faraday, Experimental Researches in Electricity, 3 vols. (London: Taylor, 1839, 1844; Taylor and Francis, 1855)

where $\vec{F}_{1,2}$ is the force between the two spherical charges, Q_1 and Q_2 are the magnitudes of the charges, $r_{1,2}$ is their mutual distance from one another, and $\hat{r}_{1,2}$ is the unit vector along the line that connects the charges.

Electrical charges obey a mathematical relationship that is analogous to Newton's universal law of gravity in two key ways. First, two point charges which are initially at rest with respect to one another are attracted or repelled along the line that joins them, which makes the force a central force just like the gravitational force between point masses in Newton's expression. Second, Coulomb's law predicts that the mutual acceleration of the two charges is inversely proportional to the square of the distance between them.

That Coulomb's electrostatic force law resembles Newtonian gravity is remarkable, especially if we note that the strong and weak nuclear forces, the other two fundamental forces found in nature, are not central forces and do not conform to such a simple mathematical relationship as an inverse square law. The most obvious difference, however, between the two force laws is that the gravitational force is exclusively attractive.

To most scientists, the fact that Coulomb's expression for the electrical force between point charges and Newton's universal law of gravity both obey inverse square laws must be more than mere coincidence. The history of physics is replete with examples of apparent coincidences, such as the equality of "inertial" and "gravitational" mass, that later turn out not to be coincidences at all and which reveal unsuspected physical truths.

Einstein, among others, puzzled over the mathematical similarity between the fundamental force laws of Coulomb and Newton for many decades. He hoped to find the answer in a field theory that would provide a unified description of gravity and electricity. Although Einstein did not succeed in his quest, scientists still seek an explanation for the remarkable resemblance between Coulomb's and Newton's force laws.

In this section, we examine electric forces between charges using Coulomb's law.

Discussion 2.1: Gaussian Units

As we have mentioned, electrodynamics developed in parallel east and west of the English Channel. One significant legacy of this, as we mentioned in Discussion 1.4 (p. 21), are two competing representations of electrodynamics each with their own metric system of units, where charge was defined either electrostatically or electromagnetically. Physics has lived with these competing representations for well over a century, and depending on the field you choose you may need to translate between them.

The metric base units of the gram, meter, and second did not seem to work very well for engineering, as most objects that one would measure in meters are much more massive than a gram, while those measured in grams tend to be quite small. After all, the density of water is one gram per cubic centimeter. Therefore two popular systems of measure arose. One used the meter, kilogram and second (MKS) as it base units, while the other used the centimeter gram and second (cgs). The SI is an updated MKS system, but the cgs system is still used in some fields such as astrophysics.

For any mechanical quantity, the conversion factor between MKS and cgs units is simply an easily calculable power of ten, and the equations of physics will be the same regardless of unit system. For example, the cgs unit of energy is called an *erg*, which is related to the joule as:

$$1\ \text{J} = 1\frac{\text{kg}\cdot\text{m}^2}{\text{s}^2} \quad \text{so} \quad 1\ \text{erg} = 1\frac{\text{g}\cdot\text{cm}^2}{\text{s}^2}$$

$$1\ \text{J} = \frac{(1{,}000\ \text{g})\cdot(100\ \text{cm})^2}{\text{s}^2} = 10^7\frac{\text{g}\cdot\text{cm}^2}{\text{s}^2} = 10^7\,\text{ergs}.$$

Similarly the cgs unit for force is the *dyne*, which is related to the newton as:

$$1\ \text{N} = 1\frac{\text{kg}\cdot\text{m}}{\text{s}^2} \quad \text{so} \quad 1\ \text{dyne} = 1\frac{\text{g}\cdot\text{cm}}{\text{s}^2}$$

$$1\ \text{N} = \frac{(1{,}000\ \text{g})\cdot(100\ \text{cm})}{\text{s}^2} = 10^5\frac{\text{g}\cdot\text{cm}}{\text{s}^2} = 10^5\,\text{dynes}.$$

However, things are not so obvious when it comes to electrodynamics, as the units of charge are defined completely differently. The cgs unit of charge is called the *electrostatic unit*, *esu*, *franklin* or *statcoulomb*, and is defined such that the repulsive force between two stationary unit point charges, separated by one centimeter, is exactly one dyne. Therefore, Coulomb's law becomes simply:

$$\underset{\text{Gaussian}}{\vec{F}_{1,2}} = \frac{Q_1 Q_2}{r_{1,2}^2}\hat{r}_{1,2},$$

and we can write the esu in terms of the other base units as:

$$\left[\underset{\text{Gaussian}}{\vec{F}_{1,2}}\right] \propto \frac{[Q_1][Q_2]}{[r_{1,2}]^2}\left[\not{\hat{r}_{1,2}}\right] = \frac{(\text{esu})^2}{(\text{cm})^2}$$

$$1\ \text{esu} = 1\,\text{cm}\cdot\sqrt{\text{dyne}} = 1\,\text{cm}\cdot\sqrt{\frac{\text{g}\cdot\text{cm}}{\text{s}^2}} = 1\ \text{cm}^{3/2}\,\text{g}^{1/2}\,\text{s}^{-1}.$$

Therefore, in SI units Coulomb's law requires a constant, K, analogous to Newton's constant G, however in cgs units $K \equiv 1\frac{\text{dyne cm}^2}{\text{esu}^2}$. Appendix D (p. 719) explains these units in more detail and discusses how to translate between them.

Fundamental Law 2: Coulomb's Electrostatic Force Law

The electric force acts along the line between any two small charged objects, and is repulsive if the two charges have like charges, but attractive for unlike charges. The magnitude of this force is proportional to each charge, Q_1 and Q_2, and inversely proportional to the square of the distance $r_{1,2}$ between them. Or, in mathematical symbols:

$$\vec{F}_{1,2} = K\frac{Q_1 Q_2}{r_{1,2}^2}\hat{r}_{1,2}.$$

Only if the space between the two points is a vacuum will the constant of proportionality, known as *Coulomb's constant*, be a universal constant. In SI units, Coulomb's constant is:

$$K = \left(299{,}792{,}458\tfrac{\text{m}}{\text{s}}\right)^2 \times 10^{-7}\tfrac{\text{N}}{\text{A}^2} \approx 8.99\times 10^9\tfrac{\text{N m}^2}{\text{C}^2}.$$

Problem 2.1: The Franklin and the Coulomb

a) Based on Discussion 2.1, show that one nanocoulomb is approximately equal to 3.00 electrostatic units, abbreviated esu or Fr, as it is also called the franklin.

b) Show that one nanocoulomb is exactly 2.99792458 Fr.

c) Look up the 2018 revision of the SI. Will one nanocoulomb always be exactly 2.99792458 Fr, or is it now subject to experimental error? Explain and cite your sources.

Discussion 2.2: Spooky Action at a Distance

Like Newton's law of gravity, Coulomb's law raises a disturbing philosophical question: how can distant charges affect each other?

Newton's law of universal gravitation is the canonical example of *action at a distance*, since we have the sun pulling on the earth from a distance of a hundred and fifty million kilometers! How can this be since there is no rope attaching them? If the sun suddenly disappeared, would the earth immediately be affected? And, if not, how would it take for the Earth know about it? Clearly there must be some *mechanism* relating cause and effect, but we just do not know what it is yet.

Theories that do not involve *local* interactions are referred to as *action at a distance.* According to an action at a distance interpretation two bodies can act on one another instantaneously no matter how far apart they are. For this reason, such theories are often derided as *spooky action at a distance,* since they appear to invoke ideas similar to those of supernatural occultists. Isaac Newton, in particular, expressed serious doubts about such action at a distance theories, despite his law of universal gravity being the canonical example of such a theory.

The working solution to this quandary it to posit a *force field.* Not force fields like the invisible barriers in science fiction movies, but rather something whose cause is one object (the sun), which then affects another object (the earth). So, for example, the force on the earth can be thought of as an interaction between it and some gravitational field, $\vec{g}$. This gravitational field, in turn, must somehow be ultimately caused by the sun, even if we do not yet know how. Thus, by supposing that this *field* exists and mediates the interaction between the two distinct bodies, it no longer seems quite like the hocus-pocus of action at a distance.[33]

Assuming the existence of some sort of force field was not the only approach to the question of action at a distance. Another approach, which we will discuss much later in this text (Chapters 16 & 17), rather postulates that there is a finite speed v at which electromagnetic information can travel through the universe. Using this approach, the electric force between any two charges will also depend on this communication time delay. In this case, Coulomb's law would need to be modified as follows:

$$\underset{\text{of 1 on 2}}{\vec{F}_{1,2}(t)} = K \frac{Q_1\left(t-\frac{r}{v}\right)Q_2(t)}{\left|\vec{r}_2(t)-\vec{r}_1\left(t-\frac{r}{v}\right)\right|^2}\hat{r}_{1,2} \quad \text{and} \quad \underset{\text{of 2 on 1}}{\vec{F}_{2,1}(t)} = K \frac{Q_1(t)Q_2\left(t-\frac{r}{v}\right)}{\left|\vec{r}_1(t)-\vec{r}_2\left(t-\frac{r}{v}\right)\right|^2}\hat{r}_{1,2}.$$

This approach, however, does not address the question of how the information travels, making action at a distance almost as spooky.

---

33 See A. Hobson, "There are no particles, there are only fields," American Journal of Physics, **81** (2013), 211-223 for a fascinating discussion of these ideas.

2.2 *The Electric Field*

How can one charged object affect another one far away? There must be some mechanism mediating cause and effect, but what is it? Is it matter with separable positive and negative charge, as Faraday thought, or is it something else? While Faraday knew there must be some mechanism, and speculated on what it might be, he was careful not to rush to conclusions. Rather, as he knew for certain that nothing in one place can directly affect something somewhere else, he defined an, as yet unknown, intermediary that we call a *field*. Moreover, this field must be measureable at any particular location. Thus, the idea of the *electric field* stems directly from the philosophical quandary of action at a distance (Discussion 2.2).

This is similar to the surface gravity on earth, which is defined *operationally* as the ratio of an object's weight to its mass:

$$\vec{g} \equiv \tfrac{1}{m}\vec{F}_g .$$

The electric field is also defined operationally, as the ratio of the electric force on an object to the charge of that same object. That way, by definition, it can be measured *in situ.*

Definition 2.1: The Electric Field

The electric field is a vector quantity, defined as the ratio of the electric force per charge:

$$\vec{E} \equiv \frac{\vec{F}_E}{q} ,$$

so long as the size and charge of the object are small.

We can make this definition more robust by considering Franklin's fluid model of electric charge. Consider a small quantity of charge δq, which is experiencing a force $\delta\vec{F}_E$, then the electric field at that location is defined as:

$$\vec{E} \equiv \lim_{\delta q \to 0} \frac{\delta\vec{F}_E}{\delta q} = \frac{\mathrm{d}\vec{F}_E}{\mathrm{d}q} .$$

Thus, whatever their causes, electric fields are defined by the effects they have on a small charge. The SI unit of $\vec{E}$ is simply a newton per coulomb and it has no special name.

Problem 2.2: The Electric Body Force

Consider a fluid with a charge density ρ in the presence of an electric field $\vec{E}$. Show that the electric force per volume $\vec{f}_E$, or *body force*, is given by:

$$\vec{f}_E = \rho\vec{E} .$$

Example 2.1: A Charge on a Spring

Two large thin metal plates are attached to a battery that charges the top plate positive and the bottom plate negative. A small block with positive charge q and mass m is attached to a nonconducting idealized spring of spring constant k inside the region between the plates. When the plates are charged, there is an equilibrium stretch length y_0, such that:

$k y_0 + mg = qE$.

The effect of the charged plates on the charged object can therefore be determined simply by measuring the stretch of the spring. Thus the electric field can be measured as:

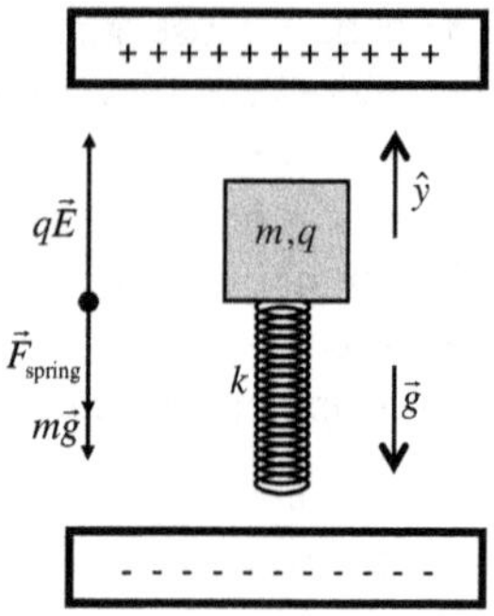

$$\vec{E} \equiv \frac{\vec{F}_{\text{electric}}}{q} = \frac{(mg + k y_0)\hat{y}}{q}.$$

If we slightly displace the object from equilibrium, what will be the frequency of oscillation ω_0 of the object about its equilibrium position?

The equation of motion is:

$$\vec{F}_{\text{net}} = -mg\hat{y} - ky\hat{y} + qE\hat{y} = m\vec{a} \quad \rightarrow \quad \frac{d^2y}{dt^2} = -\frac{k}{m}y - g + \frac{q}{m}E.$$

The solution to this differential equation is in the form:

$$y = y_0 + A\sin(\omega_0 t + \phi),$$

where A and ϕ depend on the initial conditions, and:

$$y_0 = \frac{qE - mg}{k} \quad \text{and} \quad \omega_0 = \sqrt{\frac{k}{m}}.$$

Now imagine hooking the plates to an AC current source, rather than a battery, so the electric field is given by $\vec{E} = E_0 \sin(\omega t)\hat{y}$. The new equation of motion is then:

$$\frac{d^2y}{dt^2} = \omega_0^{\,2}\, y + \frac{q}{m}E_0 \sin(\omega t) - g.$$

This is the equation for a driven oscillator, which is an interesting problem in classical mechanics.

Problem 2.3: The Net Force on an Electric Dipole

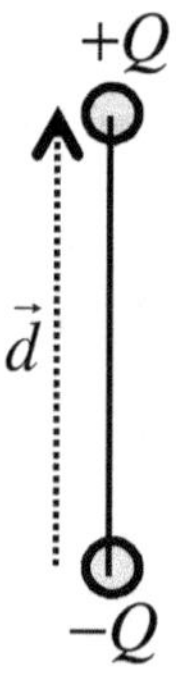

Consider an electric field that increases linearly in the $\hat{z}$ direction $(\vec{E} = Cz\hat{z})$ acting on an electric dipole that consists of two equal, but opposite, point charges $\pm Q$, separated by a fixed displacement $\vec{d}$ pointing from the negative to the positive charge.

(a) What is the net force on the dipole as a function of the given parameters C, Q, and d, as well as the polar angle θ.

(b) From your result in part a, show that, in this particular case, the force on the dipole is: $\vec{F} = C(\vec{p}\cdot\hat{z})\hat{z}$., where $\vec{p} = Q\vec{d}$.

(c) Consider a small electric dipole at angle θ. Show that for any electric field in the form of $\vec{E} = E(z)\hat{z}$, the force on the dipole is given by: $\vec{F} = (\vec{p}\cdot\hat{z})\frac{d\vec{E}}{dz}$.

(d) A common school chemistry demonstration involves rubbing a rubber rod with rabbit fur and holding it by a stream of water; the stream bends toward the rod. This is touted as evidence that water molecules are polar in nature. Explain clearly why the rod attracts the water, assuming that each molecule has a *dipole moment* $\vec{p}$.

(e) There is an alternative explanation for the same experimental result. If ions were dissolved in the water, then those with the same type of charge as the rod would move away from it, and those of opposite charge would move closer. Thus, there would also be a net attractive force on the rod. What simple experiment could one do to distinguish between these two explanations for why a stream of tap water bends toward a charged rod.

(f) Either conduct the experiment yourself, or research and cite a piece of scientific work investigating this question.

Problem 2.4: The Torque on an Electric Dipole

An electric dipole consists of two equal but opposite point charges $\pm Q$ separated by a fixed displacement $\vec{d}$.

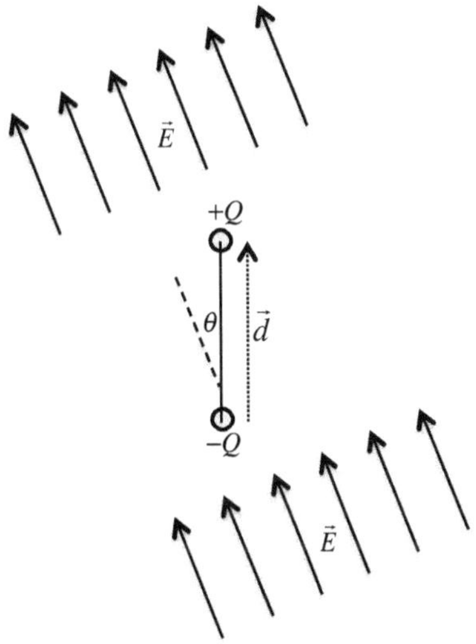

a) Find a vector expression for the torque $\vec{\tau}$ on a dipole located in an electric field $\vec{E}$ in terms of the quantities E, Q, d and the angle θ.

b) The electric dipole moment $\vec{p}$ is defined as $\vec{p} \equiv Q\vec{d}$. Show that the torque on a dipole is given by $\vec{\tau} = \vec{p} \times \vec{E}$.

c) How much work is required to twist the dipole from its equilibrium position to an angle θ?

d) Show that the potential energy of the dipole, due to twisting, can be written as $\mathbb{U} = -\vec{p} \cdot \vec{E}$, so long as zero potential energy is defined to be when $\theta = 90°$.

e) If the mass of each charge is M, what is the frequency of oscillation about the equilbrium position assuming small angles? Express this in terms of E, Q, d and M.

2.3 *The Electric Field Surrounding Point Charges*

In the last section we defined the electric field in terms of how it affects charges. We also discussed Coulomb's law, and the forces between charges in the section before. In this section we put these two ideas together to find the electric field surrounding a point charge. We then use this relationship to find the electric field surrounding multiple small point charges and continuous charge distributions.

Electric and gravitational forces, but not all forces, obey the principle of superposition: the net force exerted on a small *test charge* is simply the vector sum of all the individual electric forces that act on it. The net electric force on single test charge, due to a finite number of other charges, can therefore be directly computed from Coulomb's law merely by taking the vector sum of all the forces between the test charge and each one of the other charges. This procedure can be easily extended to continuous distributions of charge.

Consider a spherical charge Q with a very small radius—a point charge. What is the electric field at a distance r from the charge?

First we consider a charge element, δq, placed at the location of interest, with a force on it of:

$$\delta \vec{F}_E = K\frac{Q\delta q}{r^2}\hat{r}.$$

The electric field is therefore:

$$\vec{E} \equiv \lim_{\delta q\to 0}\frac{\delta \vec{F}_E}{\delta q} = \lim_{\delta q\to 0}\left(\frac{1}{\delta q}K\frac{Q\delta q}{r^2}\hat{r}\right) = K\frac{Q}{r^2}.$$

To find the electric field surrounding collections of charges, or even a continuous distribution, we take advantage of the principle of superposition, and add them up vectorally.

Thought Experiment 2.1: Coulomb's Law for a Collection of Charges

Assume there exists a collection of N point charges located at known positions, so the total electric field is simply the vector sum of the electric field from each charge.

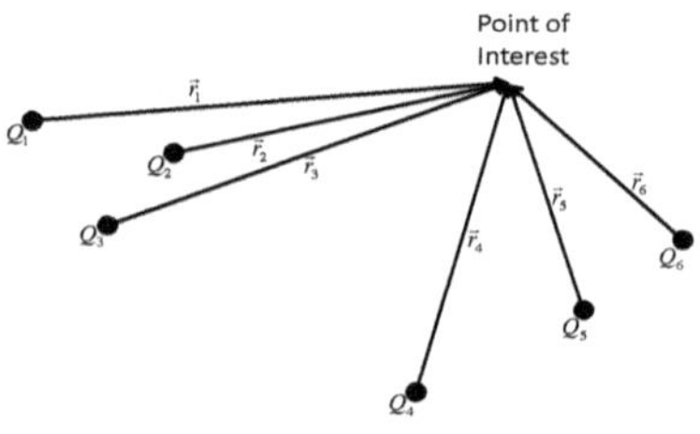

$$\vec{E} = \sum_{i=1}^{N}\vec{E}_i = K\sum_{i=1}^{N}\frac{Q_i\hat{r}_i}{r_i^2} = K\sum_{i=1}^{N}\frac{Q_i\vec{r}_i}{r_i^3}.$$

However, there is a catch; each vector $\vec{r}_i$ originates at each charge Q_i, but terminates at the point of interest.

Once we define an origin, the position of each charge (Q_i) can be written as $\vec{R}_i$. The position of the point of interest, with respect to the origin, is therefore $\vec{r} = \vec{R}_i + \vec{r}_i$. Thus the net electric field at the point of interest becomes:

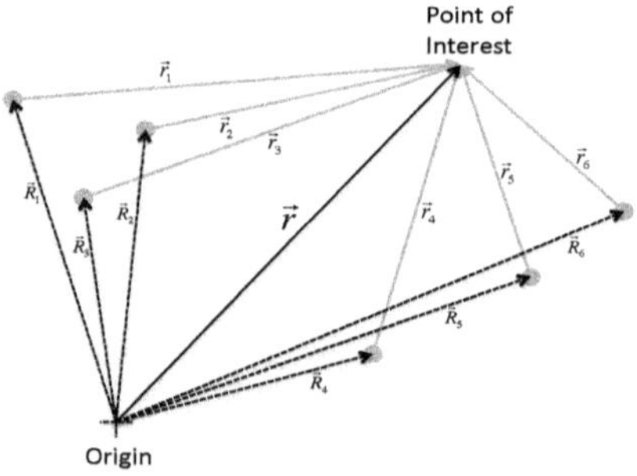

$$\vec{E} = K\sum_{i=1}^{N}\frac{Q_i\left(\vec{r}-\vec{R}_i\right)}{\left|\vec{r}-\vec{R}_i\right|^3}.$$

If we now consider the electric field due to a continuous charge distribution, $\rho(\vec{R})$, the summation is replaced by an integral over the distribution:

$$\vec{E} = K\int_{\vec{R}}\frac{\left(\vec{r}-\vec{R}\right)}{\left|\vec{r}-\vec{R}\right|^3}dQ = K\int_{Volume}\frac{\left(\vec{r}-\vec{R}\right)\rho(\vec{R})}{\left|\vec{r}-\vec{R}\right|^3}dV.$$

Notice that, for any point of interest, $\vec{r}$ is a constant while $\vec{R}$ varies over the charge distribution.

Example 2.2: A Single Hoop of Wire

Consider a hoop of radius R and constant linear charge density λ. What is the electric field at a height h above the center of the hoop?

We first render the problem in cylindrical coordinates, and apply Coulomb's law:

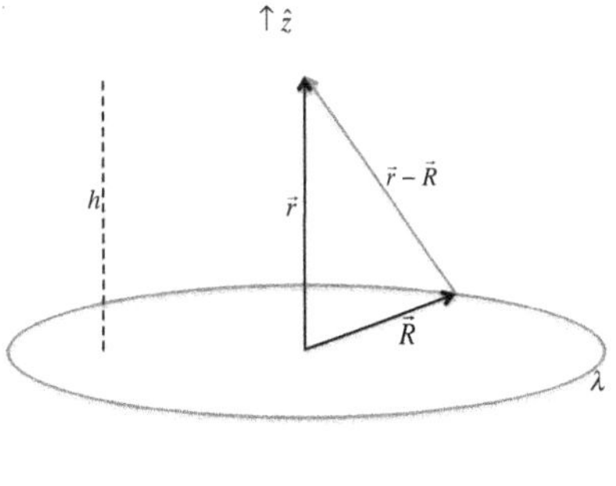

$$\vec{E} = K\int_{\vec{R}} \frac{\vec{r}-\vec{R}}{\left|\vec{r}-\vec{R}\right|^3} dQ,$$

and simplifying:

$$\vec{E} = K\int_0^{2\pi} \frac{\vec{r}-\vec{R}}{\left|\vec{r}-\vec{R}\right|^3}(\lambda R d\varphi) = \frac{K\lambda R}{\left(R^2+h^2\right)^{3/2}}\left(h\int_0^{2\pi}\hat{z}\,d\varphi - R\int_0^{2\pi}\hat{s}\,d\varphi\right).$$

For every $\hat{s}$ in one direction there is a corresponding one in the opposite direction, so the second integral is zero. Therefore:

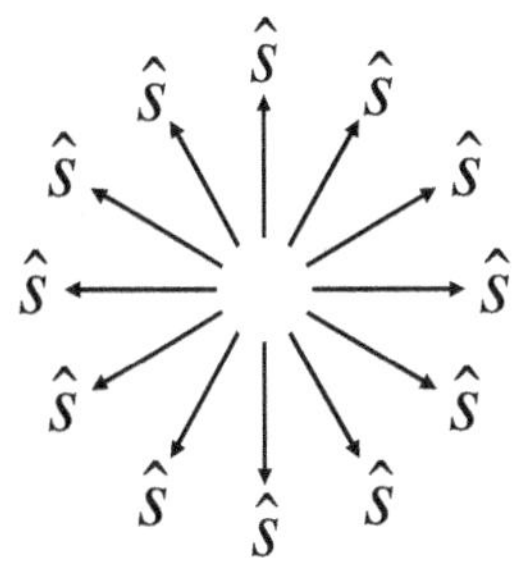

$$\vec{E} = \frac{K\lambda R}{\left(R^2+h^2\right)^{3/2}}\left(h\int_0^{2\pi}\hat{z}\,d\varphi - R\int_0^{2\pi}\hat{s}\,d\varphi\right) = \frac{2\pi K\lambda R h}{\left(R^2+h^2\right)^{3/2}}\hat{z}.$$

Notice that for large observation distances, for which $h \gg R$, the field is approximately that due a point charge:

$$\vec{E} = \frac{2\pi K\lambda R h}{\left(R^2+h^2\right)^{3/2}}\hat{z} = \frac{K(\lambda 2\pi R)h}{h^3\left(1+\left(\frac{R}{h}\right)^2\right)^{3/2}}\hat{z} \approx \frac{KQ}{h^2}\hat{z},$$

which is precisely what we expect if we evaluate the field at a distance h from a point charge. Now does our result for the field of a single charged hoop also make sense when we go to the opposite extreme and set $h=0$? To answer this we should rewrite our result in powers of h/R, and evaluate the result for $h=0$:

$$\vec{E} = \frac{2\pi K\lambda R h}{R^3\left(1+\frac{h^2}{R^2}\right)^{3/2}}\hat{z} = 2\pi K\frac{\lambda h}{R^2}\frac{1}{\left(1+\left(\frac{h}{R}\right)^2\right)^{3/2}}\hat{z} \approx \left(2\pi K\frac{\lambda}{R^2}\hat{z}\right)h = 0, \text{ when } h=0.$$

Now, does this make good physical sense? You should have little trouble using the symmetry of the situation, the principle of superposition, and the vector character of Coulomb's law to see that the net force on a test charge placed at the center of the charged ring is zero. (Hint: sketch a free body diagram of the forces acting on such a test charge.)

Problem 2.5: A Charged Bead on the Axis of a Charged Hoop

Consider a small bead with mass m, and charge q constrained to the axis of a hoop with a large radius R and a uniform linear charge density. The whole system is net neutral.

a) What is the electric force on the bead? Express you answer in terms of q, R, and z, where z is the axial distance from the center of the hoop.

b) Show that if $z \ll R$, the bead will undergo simple harmonic motion. What would be the angular frequency of this bead's oscillations?

c) Consider a hoop of one meter in radius and a charge of -1 μC, along with a one-gram bead with a charge of 1 μC. What would be its period of oscillation?

d) Using your answer in part (a), express the acceleration of the bead, in general, as a function the axial position z. Use a spreadsheet, or a programming language, to numerically calculate the position of the bead in part (c) as a function of time, if it starts from rest at an initial position h. The easiest way to do this is to: (1) calculate the acceleration from the position using Newton's second law, (2) find the velocity after a short time step using $v = v_0 + a\,\Delta t$, (3) find the position after the same short time step using $z = z_0 + \frac{1}{2}(v_0 + v)\Delta t$, and repeat.[34] Use a time step of 1% of your answer in part (c), and let the simulation run for a length 10 times your answer in part (c). Plot your results for 10 different initial conditions and show that it agrees with simple harmonic motion for small amplitudes. Does the period increase, or decrease, with amplitude?

Example 2.3: The Electric Field Near a Flat Disk

Consider a flat disk of radius a, and uniform surface charge density σ. What is the electric field at height h above the middle of the disk?

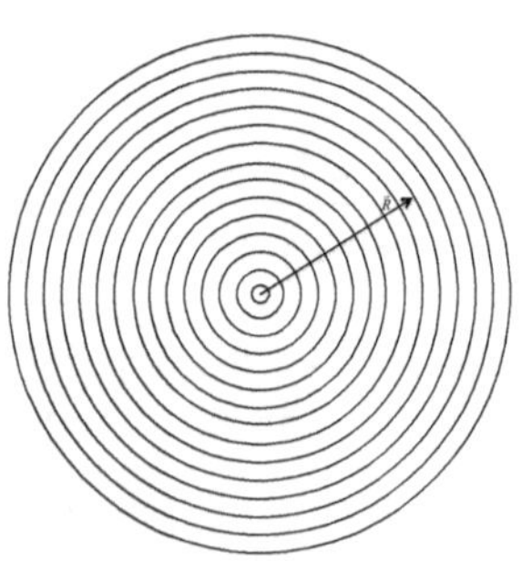

To solve this problem we model our disk as a collection of concentric hoops, each of radius R and width δR. Thus the charge per length of each infinitesimal hoop is given by:

$$\delta\lambda = \frac{\delta Q}{2\pi R} = \frac{\sigma\,\delta A}{2\pi R} = \frac{\sigma\left(\cancel{2\pi R}\,\delta R\right)}{\cancel{2\pi R}} = \sigma\,\delta R.$$

We now use our result from Example 2.2 (p. 62) to find the contribution from each hoop:

$$\delta\vec{E} = \frac{2\pi K\,\delta\lambda\,R h}{\left(R^2+h^2\right)^{3/2}}\hat{z} = \frac{2\pi K\sigma h R\delta R}{\left(R^2+h^2\right)^{3/2}}\hat{z}.$$

Integrating over the disk yields the net electric field,

$$\vec{E}(h) = \int_0^a \frac{2\pi K\sigma h R\,\mathrm{d}R}{\left(R^2+h^2\right)^{3/2}}\hat{z} = \pi K\sigma h\hat{z}\int_0^{a^2}\left(R^2+h^2\right)^{-3/2}\mathrm{d}\left(R^2\right),$$

which simplifies to:

$$\vec{E}(h) = 2\pi K\sigma h\left(\frac{1}{h} - \frac{1}{\sqrt{a^2+h^2}}\right)\hat{z} = 2\pi K\sigma\left(1 - \frac{h}{\sqrt{a^2+h^2}}\right)\hat{z}.$$

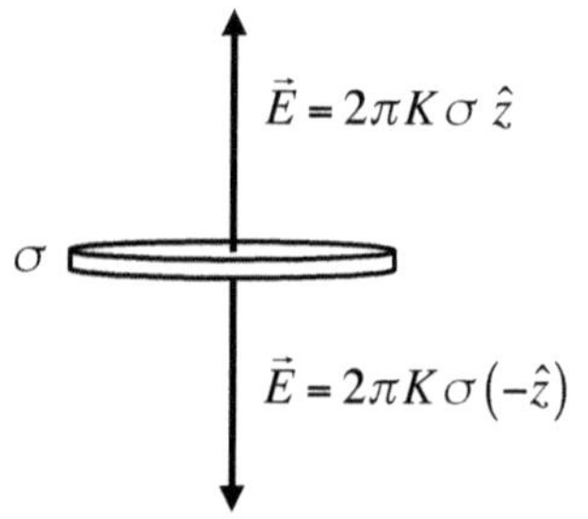

In the limit that $h \gg a$, we get:

$$\lim_{a\to\infty}\vec{E} = 2\pi K\sigma\,\hat{z}.$$

Notice that the situation is symmetrical: since the charge is evenly distributed on both sides of the disk, the field is also on both sides of the disk as indicated in the figure.[35]

---

[34] This is called Euler's method, but you can improve on it by estimating the average acceleration in the middle, rather than the beginning, of the next time step as: $\frac{1}{2}\left(a_{n+1}+a_n\right) \approx \frac{3}{2}a_n - \frac{1}{2}a_{n-1}$, as in the flow chart on p. 316.

[35] William Thomson (Lord Kelvin) reproduced Coulomb's result that the electric force on a test charge near a charged plane is perpendicular to that surface and proportional to the surface charge density in his famous paper (which he wrote under the pseudonym P.Q.R), "On the uniform motion of heat and its connection with the mathematical theory of electricity", Cambridge Mathematical Journal **3**, (1842), 71–84.

Problem 2.6: A Coil of Charge

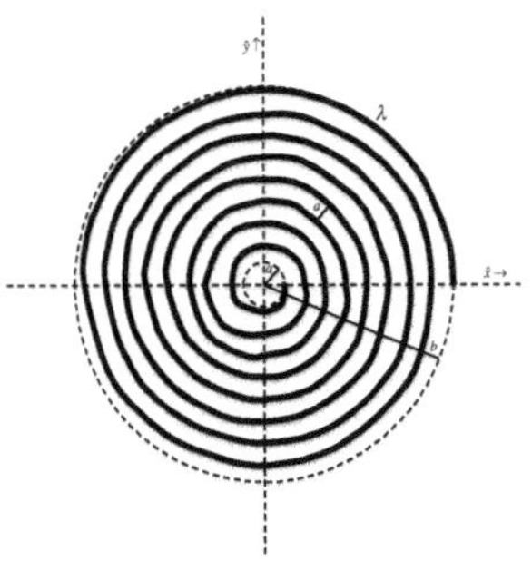

Consider a coil of wire with a constant linear charge density λ. The inner coil starts at a radius a, and continues out increasing in radius by a distance a again with each turn until finally terminating at a radius b after an integral number of turns.

a) Calculate the electric field on the axis of the coil, a height z from the coil's center.

b) Show that this reproduces the result for a flat disk of radius b in the limit that $a \to 0$.

Example 2.4: An Idealized Parallel Plate Capacitor

Let us next consider two large disks with equal and opposite uniform surface charge densities $+\sigma, -\sigma$. As the plates are placed close to each other, the electric field from each plate will add vectorally.

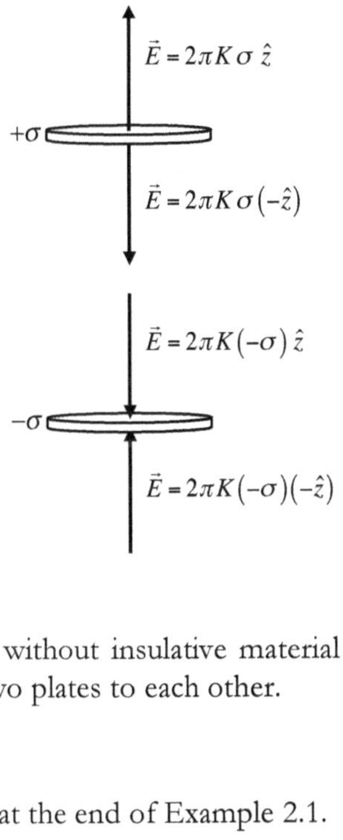

Above both plates the electric field will become:

$$\vec{E} = 2\pi K \sigma \, \hat{z} + 2\pi K (-\sigma) \, \hat{z} = 0 ,$$

while between the plates the net field is given by:

$$\vec{E} = 2\pi K \sigma (-\hat{z}) + 2\pi K (-\sigma) \, \hat{z} = -4\pi K \sigma \, \hat{z} ,$$

and below both plates it is:

$$\vec{E} = 2\pi K \sigma \, (-\hat{z}) + 2\pi K (-\sigma)(-\hat{z}) = 0 .$$

This shows that, so long as the gap is small compared to the distance to the edge of the plate, the electric field is uniform in the gap between a parallel plate capacitor and zero outside of the capacitor. If the gap is not small compared to the dimensions of the plates, then the field will not be uniform at the edges of the plate, a condition that scientists commonly refer to as *edge effects*.

In practice, however, it would be impossible to make such a device without insulative material sandwiched between the plates because of the attractive force of the two plates to each other.

Problem 2.7: A Driven Oscillator

Consider a mass on a spring in an oscillating electric field as discussed at the end of Example 2.1.

a) Using a spreadsheet application, or your favorite programming language, build a simulator which will plot the position as a function of time for the mass on the spring, given a set of appropriate input values such as the initial conditions, mass, charge, spring constant, E_0 and ω.

b) Investigate the solutions when:

$$\omega << \sqrt{k/m} , \quad \omega \lesssim \sqrt{k/m} , \quad \omega = \sqrt{k/m} , \quad \omega \gtrsim \sqrt{k/m} \text{ and } \omega >> \sqrt{k/m} .$$

Example 2.5: A Spherical Shell

Consider a uniformly charged spherical shell of radius a and constant surface charge density σ. Find the electric field at position $\vec{r}$ from the center.

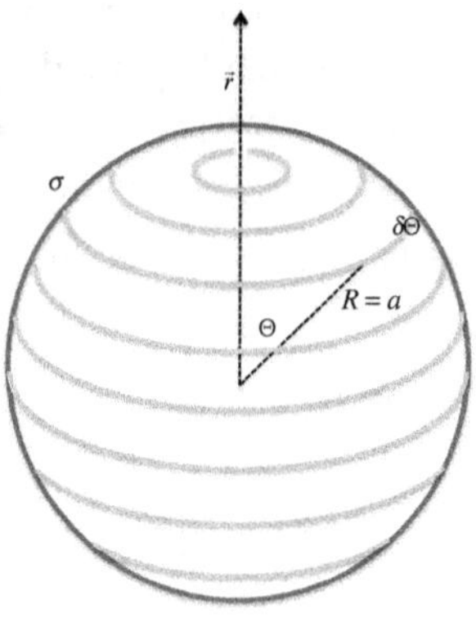

To solve this problem, we will again use our solution to Example 2.2 (p. 62) and model the shell as a series of stacked hoops. We can take advantage of symmetry and place our point of interest a distance r along the $\hat{z}$ axis, so that $\vec{r} = r\hat{z}$.

We will now represent each point on the charged sphere in polar coordinates, with each hoop having constant R and Θ coordinates. Moreover, each hoop has $R = a$, and the domain of Θ is between 0 and π.

Our point of interest is now located at a height $h_{\text{hoop}} = r - a\cos\Theta$ above each hoop. Each hoop has a radius $R_{\text{hoop}} = a\sin\Theta$, and a linear charge density of $\delta\lambda_{\text{hoop}} = \sigma\, a\,\delta\Theta$.

Now applying our result from Example 2.2, we see that the contribution to the electric field of each hoop is given by:

$$\delta\vec{E} = \frac{2\pi K\,\delta\lambda_{\text{hoop}}\,R_{\text{hoop}}\,h_{\text{hoop}}}{\left(R^2_{\text{hoop}} + h^2_{\text{hoop}}\right)^{3/2}}\hat{z} = \frac{2\pi K\,\sigma\,a^2\sin\Theta(r - a\cos\Theta)\delta\Theta}{\left((a\sin\Theta)^2 + (r - a\cos\Theta)^2\right)^{3/2}}\hat{z} = \frac{2\pi K\,\sigma\,a^2\sin\Theta(r - a\cos\Theta)\delta\Theta}{\left(a^2 - 2ar\cos\Theta + r^2\right)^{3/2}}\hat{z}.$$

We now integrate this from $\Theta = 0$ to $\Theta = \pi$, to find the total contribution for the whole sphere.

$$\vec{E} = \int_0^{\pi}\frac{2\pi K\,\sigma\,a^2\sin\Theta(r - a\cos\Theta)}{\left(a^2 + r^2 - 2ar\cos\Theta\right)^{3/2}}\hat{z}\,d\Theta = \left(\frac{2\pi K\sigma\,a^2}{r^2}\hat{z}\right)\left(\frac{a - r\cos\Theta}{\sqrt{a^2 - 2ar\cos\Theta + r^2}}\right)\Bigg|_0^{\pi}$$

$$= \left(\frac{2\pi K\,\sigma\,a^2}{r^2}\right)\left(\frac{a+r}{\sqrt{(a+r)^2}} - \frac{a-r}{\sqrt{(a-r)^2}}\right)\hat{z} = \left(\frac{2\pi K\,\sigma\,a^2}{r^2}\right)\left(1 - \frac{a-r}{|a-r|}\right)\hat{z}.$$

Notice that the problem is spherically symmetrical, so we can replace $\hat{z}$ with $\hat{r}$. Next we solve the problem separately for points inside, and outside, of the charged sphere.

If our point of interest is inside the sphere then $r < a$, and $a - r = |a - r|$ so the electric field is:

$$\vec{E} = \left(\frac{2\pi K\,\sigma\,a^2}{r^2}\right)\left(1 - \frac{a-r}{|a-r|}\right)\hat{r} = \left(\frac{2\pi K\,a^2}{r^2}\right)(1-1)\hat{r} = 0,$$

while outside the sphere $r > a$ so $a - r = -|a - r|$, so:

$$\vec{E} = \left(\frac{2\pi K\,\sigma\,a^2}{r^2}\right)\left(\frac{a+r}{|a+r|} - \frac{a-r}{|a-r|}\right)\hat{r} = \left(\frac{2\pi K\,\sigma\,a^2}{r^2}\right)(1-(-1))\hat{r} = \frac{4\pi K\,\sigma\,a^2}{r^2}\hat{r}.$$

Finally substituting for the surface charge density $\left(\sigma = \frac{Q}{4\pi a^2}\right)$ the electric field is given by:

$$\vec{E} = \begin{cases} 0 & : r < a \\ \dfrac{KQ}{r^2}\hat{r} & : r > a. \end{cases}$$

This result is actually quite profound. Inside the sphere there is no electric field at all, while the field outside the sphere is exactly the same as if all the charge were simply located at a point in the center of the sphere.

Problem 2.8: The Torque on an Electric Dipole

An electric dipole consists of two equal but opposite point charges $\pm Q$ separated by a fixed displacement $\vec{d}$.

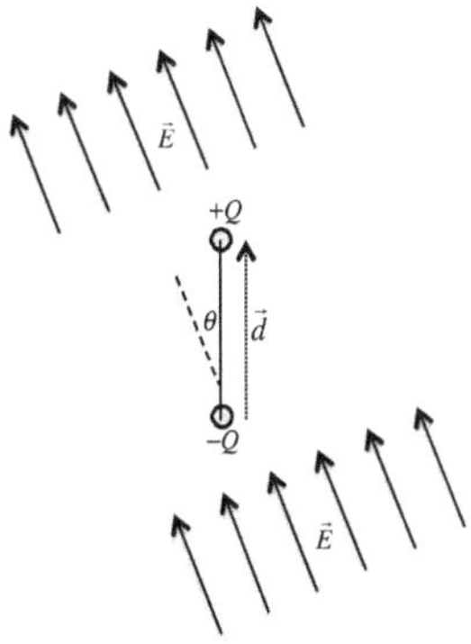

a) Find a vector expression for the torque $\vec{\tau}$ on a dipole located in an electric field $\vec{E}$ in terms of the quantities E, Q, d and the angle θ.

b) The electric dipole moment $\vec{p}$ is defined as $\vec{p} \equiv Q\vec{d}$. Show that the torque on a dipole is given by $\vec{\tau} = \vec{p} \times \vec{E}$.

c) How much work is required to twist the dipole from its equilibrium position to an angle θ?

d) If the mass of each charge is M, what is the frequency of oscillation about the equilbrium position assuming small angles? Express this in terms of E, Q, d and M.

Problem 2.9: The Net Force on an Electric Dipole

Consider an electric field that increases linearly in the $\hat{z}$ direction $\vec{E} = Kz\hat{z}$ acting on an electric dipole that consists of two equal but opposite point charges $\pm Q$, separated by a fixed displacement $\vec{d}$.

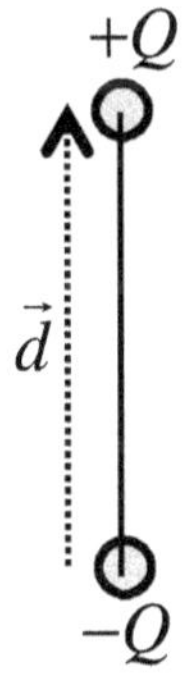

(a) What is the net force on the dipole as a function of the angle θ between $\vec{d}$ and $\hat{z}$? Show your work.

(b) Consider a small electric dipole with $\vec{d} = d\hat{z}$. Show that for any electric field in the form of $\vec{E} = E(z)\hat{z}$ the force of the dipole is given by: $\vec{F} = Qd\dfrac{d\vec{E}}{dz} = p\dfrac{d\vec{E}}{dz}$, where $\vec{p} = Q\vec{d}$.

(c) This force is can also be written in vector notation using the *directional derivative* as: $\vec{F} = \left(\vec{p}\cdot\vec{\nabla}\right)\vec{E}$. Show that this is consistent with both parts (a) and (b) above.

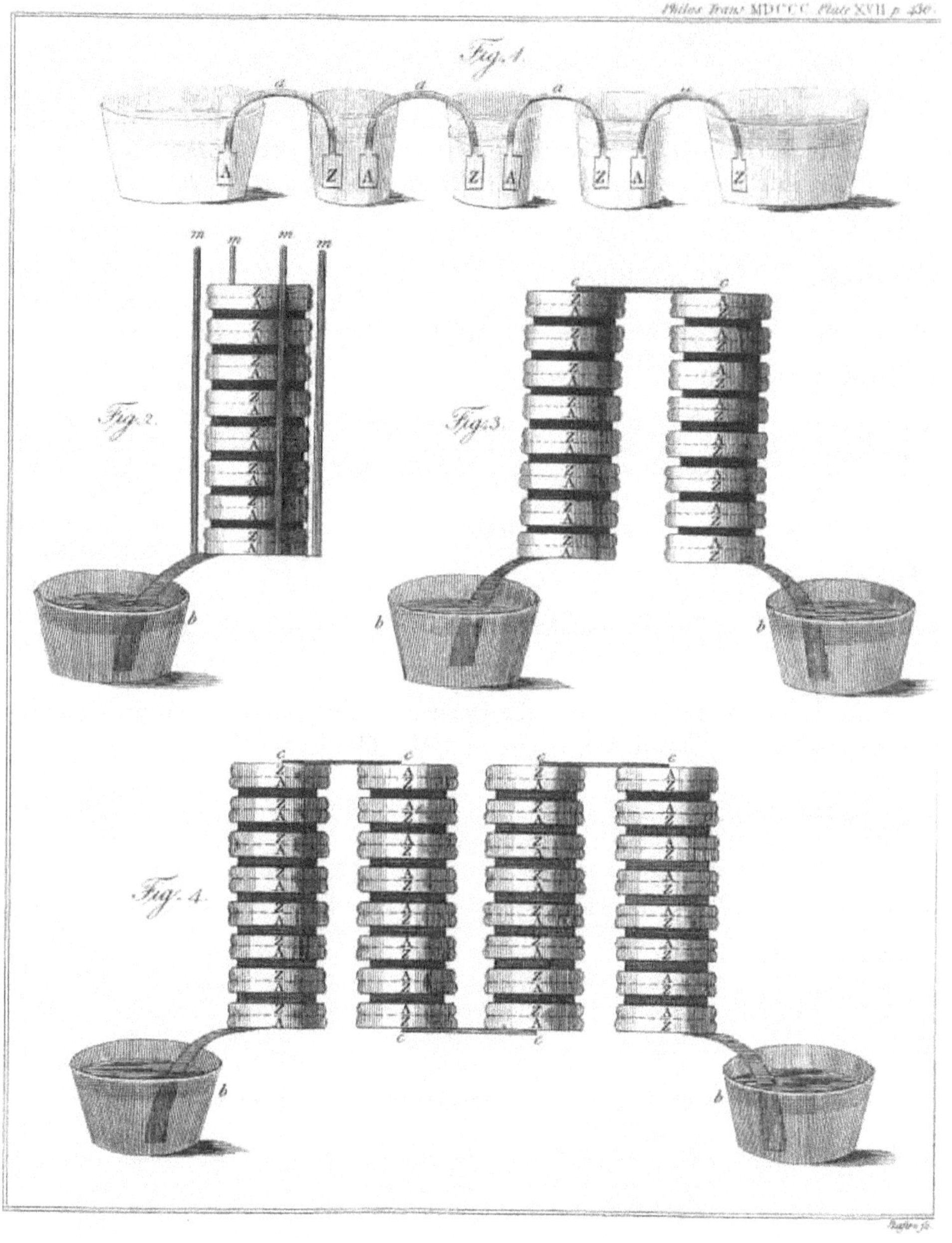

Plate 6: Volta's Pile of Electrochemical Cells[36]

[36] Alessandro Volta, "On the Electricity Excited by the Mere Contact of Conducting Substances of Different Kinds," Phil. Trans. R. Soc. Lond. **90** (1800), p. 430, plate 17.

Chapter 3 Electrical Potential Energy

Connected to two different metals, a dead frog's legs twitch. Did energy flow to or from the frog? What is unique about recently dead tissue that make it appear to come back to life?

This experiment was first conducted by the Italian scientist Luigi Galvani who in 1791 took this discovery as evidence that the frog's leg muscle, like a Leyden jar, retained some electric charge even after death.[37] By touching the frog, he asserted, this *animal electric fluid* becomes discharged and the leg twitches.

Mary Wollstonecraft Shelley spent the summer of 1816 vacationing in Switzerland, where she read Galvani's report of latent animal electric fluid making the frog's legs twitch. At the same time, she composed an idea for a novel about a scientist bringing dead tissue back to life. As Shelley's focus was on her protagonist, Victor Frankenstein, she did not specifically mention electricity as the means by which the dead corpse is revived.[38] In our modern interpretations of both Galvani's experiment and Shelley's story, electricity is the cause of both the dead frog's twitching leg and the creature's resuscitation.

Alessandro Volta coined the term *galvanism* in honor of his colleague Galvani, but had a different opinion as to the source of electricity. He maintained that the source of the electricity was due to the contact of the dissimilar metals. To prove his point, Volta built a device to store electrical energy, with no animal tissue.

Volta stacked bimetallic disks of copper and zinc plates, layered with salt-water soaked paper. This *voltaic pile* became the first practical source of steady electrical current.[39] Like the frog's leg in Galvani's experiment, the salt-water provides an electrolytic solution for ions to travel from one plate to the other. Once the voltaic pile was invented, it became relatively straightforward to stack them to arbitrary height, which made it possible to supply fairly steady electrical current to wires forming an electric circuit.

Volta's achievement spurred remarkable progress in the subjects of electricity and magnetism. In 1820, the Danish physicist Hans Christian Ørsted used a voltaic pile to pass a steady current near a magnetized compass needle causing the needle to deflect, and therefore demonstrated that electricity and magnetism are related. Using Ørsted's result, French mathematician and physicist Andre Marie Ampere and German physicist and chemist Johann Schweigger independently developed the earliest *galvanometer* to measure current in terms of the torque exerted on a compass needle suspended near the current. In the same year that Ørsted announced his discovery of electromagnetism, Michael Faraday employed the steady currents produced by Volta's cells, or *batteries,* to construct the first electric motor.

---

[37] See, for example, J.L. Heilbron, Electricity in the17th and 18th centuries: a study of early Modern physics. (University of California Press, 1979), 491.

[38] Mary Shelley, Frankenstein (1818 text), ed. Marilyn Butler (Oxford: Oxford University Press, 1993), p. 195. See also the following article that situates Shelley's novel within the debate between Volta and Galvani over the nature of so-called animal electricity: R. C. Sha, "Volta's Battery, Animal Electricity, and *Frankenstein*," European Romantic Review, **23**(1) (2012), 21-41.

[39] A. Volta, "On the Electricity Excited by the Mere Contact of Conducting Substances of Different Kinds," Phil. Trans. R. Soc. Lond. **90** (1800), 403-431. This article is in French, but an English translation of it may be found under the same title in Philosophical Magazine, **7:2** (1800), 289-311. The figure on this page is a plate from near the end of Volta's original paper.

Beginning in 1825, Georg Ohm used voltaic cells and a galvanometer to begin his investigation of how different metals conduct electric current. In his first two sets of experiments, he inserted metal wires of different lengths and materials between the terminals of a voltaic cell made of copper and zinc, and varied the number of plates to control the strength of the current. Among other things, he determined that the current flowing through a wire is related to the length of the wire and the type of metal from which it is made. But the voltaic cells of the early 1820's were far from perfect, and Ohm found that he could not obtain a sufficiently steady current to make his measurements as precise as he felt they needed to be. For the third experiment, Ohm replaced the voltaic cell with a thermocouple, for which a temperature difference between two dissimilar metals causes current to flow, so that he could produce the steadier and more reliably measurable currents required to establish a relationship between current flowing in a conductor and the length of the conductor. He found that the current was inversely proportional to the length of the wire plus a constant and directly proportional to the temperature difference in the thermocouple source.[40]

Ohm leaned heavily on the work of Joseph Fourier, whose The Analytical Theory of Heat proved so influential for nineteenth century physics, and proposed that the flow of electrical charge in a conductor was analogous to the flow of heat. He also suggested that the cause of current flow was analogous to the temperature difference or gradient that in Fourier's theory is responsible for heat flow. Ohm asserted that the length of the wire, plus the constant term that he conceived of as the "effective length" of the voltaic cell or the thermocouple source, is proportional to the electrical *resistance* of the wire plus the "internal resistance" of the current source, much like the thermal resistance to heat flow in a conductor.[41]

Although Ohm was guided by an analogy with Fourier's theory of heat flow, the connection between his discovery and other electrical quantities only become clear as the concept of energy became more developed. For example, Gustav Kirchhoff pointed out in 1850 that it is the difference in *voltage* or *electrostatic potential*, which is defined as the work per unit charge to transport a unit charge from one location to another in an electric field, between the terminals of the wire that is ultimately responsible for the flow of current. This is true whether the current source is a voltaic cell or a thermocouple. Therefore, Kirchhoff asserted, the electric current density is proportional to another vector derivative, the negative *gradient*, of the electrostatic potential, which in turn is directly proportional to the electric field. With this, Kirchhoff helped link electrostatics with electrodynamics.

Of course, Kirchhoff also used conservation of charge and conservation of energy to develop his famous rules for currents flowing in a circuit. His *junction rule* states that the sum of the currents entering the point where two or more wires meet must equal the sum of the currents leaving that point or junction. His second rule, often referred to as the *loop rule*, requires that the total energy around a closed loop of current carrying wire must be conserved. It is impressive that he

---

[40] Details of Ohm's experimental set up and analysis can be found in Joseph F. Keithley, The Story of Electrical Measurements from 500 BC to the 1940s, (New York: IEEE Press, 1999), 92-103.

[41] While the analogy between current and heat flow was useful to Ohm in formulating the law that now bears his name, it also misled him into making erroneous predictions. For example, based on the formal analogy with Fourier's theory of heat, Ohm predicted that a conductor with charge would be in equilibrium if the charge were uniformly distributed throughout the volume of the conductor. It is easy to show (using Gauss's law, for example, see Chapter 4) that this is not the case. In static equilibrium, the excess or net charge resides on the surface of the conductor.

formally derived these two rules, which are now a standard part of introductory physics and electrical engineering courses, while he was still an undergraduate student![42]

In this chapter we make a distinction between *conservative* and *nonconservative* forces. If you do work against a conservative force, the force will be able to do the same work on something else in the future. However, the work done against a nonconservative force goes into the random kinetic energy we call heat. In that case, we say *the energy is dissipated.*

We explore energy concepts by first reviewing work and energy in mechanical systems. Next we define *voltage*, and then we cover circuits with batteries and resistors. This, in turn, leads to Kirchhoff's loop and junction rules for an electric circuit, which are a consequence of energy and charge conservation respectively.

3.1 *Work and Energy in Mechanical Systems*

> The conservation laws are in a sense not laws at all, but postulates which we insist must hold in any physical theory. If, for example, for moving charged particles, we find that the total energy, defined as $(\mathbb{T}+\mathbb{U})$, is not constant, we do not abandon the law, but change its meaning by redefining energy to include electromagnetic energy in such a way as to preserve the law. We prefer always to look for quantities which are conserved, and agree to apply the names 'total energy,' 'total momentum,' 'total angular momentum' only to such quantities. The conservation of these quantities is then not a physical fact, but a consequence of our determination to define them in this way. It is, of course, a statement of physical fact, which may or may not be true, that such definitions...can always be found. This assertion, has so far been true...
>
> -K. Symon[43]

Unlike charge conservation, the idea of conservation of energy arose concurrently with electrodynamics and thermodynamics. In fact, many of the same physicists, such as Hermann von Helmholtz and William Thomson (later Lord Kelvin), played a significant part in both the development of electrodynamics and thermodynamics. From a modern perspective the conservation of energy is one of the fundamental tenets of physics. We are usually introduced to energy conservation in conjunction with Newtonian mechanics or thermodynamics, where we learn about forces doing work, conservative forces, and the mechanical equivalent of heat that was demonstrated by James Prescott Joule in the 1840s. The idea of energy conservation and fields is usually deferred until gravity is discussed in terms of field physics, or until we begin our study or electricity and magnetism for the first time.

If you lift a ball from the floor and place it on the surface of a table, you have done work against the gravitational field. If you then knock the ball off of the table so that it falls back to the ground, the gravitational field does work on the ball. The work done by a force in displacing an object is easily defined and, since the electric field is just the electric force per charge, we can speak of the work done by the electric field in displacing a test charge. This can be generalized to the notion that the electromagnetic field is capable of doing work, as we will see when we consider Faraday's law of induction.

Now, we performed zero *total* work in lifting the ball from the ground and then letting gravity return it to the ground. It does not matter by which path the ball traveled in going from its initial

---

[42] Kalil T. Swain Odham, The doctrine of description: Gustav Kirchhoff, classical physics, and "the purpose of all science" in 19th-century Germany, (PhD., University of California, Berkeley), 137-38.

[43] K. Symon, Mechanics, 3rd ed. (Reading, Mass.: Addison-Wesley, 1971), 172.

to its final location. It matters only that the ball ended where it returned without dissipating energy in the process. This is the defining feature of conservative forces. Like gravity, the electric force is a conservative one, at least for those instances where the electric field is not generated by a time-varying magnetic field, and the work done by such a static electric field in displacing a charge depends only on the initial and final position of the charge with respect to the field. As far as the work done by the field on the charge is concerned, it matters not one bit whether the path taken by the charge carried it from the surface of the earth to the moon and back or simply moved it in the shortest trajectory from its initial to its final point.

Thought Experiment 3.1: The Work-Energy Theorem

We accept Newton's laws as the basis of a rational (albeit classical) theory of mechanics and work out the consequences, the result of which we can check with experiment (at least in principle). If we consider the cumulative effect a net force has on an object, we obtain two interesting and useful results. The first comes from considering what happens to something that is subject to a force for a definite time. The *impulse* is defined as the time integral of the net force acting on an object, and we can use Newton's second law to derive the *impulse-momentum* theorem:

$$\int_{t_i}^{t_f} \vec{F}\,dt = \int_{t_i}^{t_f} \left(\tfrac{\mathrm{d}}{\mathrm{d}t}\vec{p}\right)\mathrm{d}t = \int_{\vec{p}_i}^{\vec{p}_f} \mathrm{d}\vec{p} = \vec{p}_f - \vec{p}_i\,,$$

where $\vec{F}$ is the net force acting on the object, and $\vec{p}$ is the object's momentum. The subscripts i and f indicate initial and final values respectively. Classically, the relationship between an object's momentum and its velocity is $\vec{p} = m\vec{v}$, where m is the object's mass. If we assume that the mass of the object is constant, then the impulse-momentum theorem can be written as:

$$\int_{t_i}^{t_f} \vec{F}\,\mathrm{d}t = m\vec{v}_f - m\vec{v}_i\,.$$

The impulse-momentum theorem has many applications, the most important is its use in deriving conservation of momentum for a system of freely moving objects that collide with one another. Now we turn our attention to the question of what happens to an object that is subjected to an applied force over a definite distance.

The *work* done by a force $\vec{F}$ in displacing a body from an initial position given by $\vec{r}_i$ to the final position $\vec{r}_f$ is defined by the following path integral of the force:

$$\mathrm{W}_{if} = \int_{\vec{r}_i}^{\vec{r}_f} \vec{F}\cdot\mathrm{d}\vec{r}\,.$$

Now consider the case where $\vec{F}$ is the net force acting on an object of constant mass. We may invoke Newton's second law to write:

$$\mathrm{W}_{if} = \int_{\vec{r}_i}^{\vec{r}_f} \vec{F}\cdot\mathrm{d}\vec{r} = m\int_{\vec{r}_i}^{\vec{r}_f} \left(\tfrac{\mathrm{d}}{\mathrm{d}t}\vec{v}\right)\cdot\mathrm{d}\vec{r}\,.$$

Now, $\mathrm{d}\vec{r} = \frac{\mathrm{d}}{\mathrm{d}t}\vec{r}dt = \vec{v}\,dt$, so we may rewrite the expression for work in the following form:

$$\mathrm{W}_{if} = \int_{\vec{r}_i}^{\vec{r}_f} \vec{F}\cdot\mathrm{d}\vec{r} = m\int_{\vec{r}_i}^{\vec{r}_f} \left(\tfrac{\mathrm{d}}{\mathrm{d}t}\vec{v}\right)\cdot\vec{v}\,\mathrm{d}t = m\int_{\vec{v}_i}^{\vec{v}_f} \vec{v}\cdot\mathrm{d}\vec{v} = m\int_{v_i}^{v_f} v\,\mathrm{d}v = \tfrac{1}{2}mv_f^2 - \tfrac{1}{2}mv_i^2\,.$$

We say that a body of mass m, moving at velocity $\vec{v}$, has a *kinetic energy* $\mathbb{T}$ where $\mathbb{T} = \frac{1}{2}mv^2$. The expression we just derived is called the *work-energy theorem* and is sometimes written simply:

$$\mathbb{T}_f - \mathbb{T}_i = \mathbb{W}_{if} ,$$

where $\mathbb{T}_i$ is the initial kinetic energy, $\mathbb{T}_f$ is the final kinetic energy, and $\mathbb{W}_{if}$ is the work done along a specified path from the initial point *i* to the final point *f*.

The work done in raising a boulder ten meters is the same whether it takes an hour or a day, but taking an entire day to raise a boulder isn't a very effective use of time. We often need to know just how quickly work is performed. For that reason, power $\mathbb{P}$ is defined as the instantaneous rate of doing work:

$$\mathbb{P} = \tfrac{\mathrm{d}}{\mathrm{d}t}\mathbb{W} .$$

The SI unit of power is the watt, which is defined as one joule per second: $1\mathrm{W} = 1\,{}^{\mathrm{J}}\!/_{\mathrm{s}}$. By the work-energy theorem, the power tells us time rate of change of the body's kinetic energy:

$$\mathbb{P} = \tfrac{\mathrm{d}}{\mathrm{d}t}\mathbb{T} = \tfrac{\mathrm{d}}{\mathrm{d}t}\left(\tfrac{1}{2}m\vec{v}\cdot\vec{v}\right) = \tfrac{1}{2}m\left(\vec{v}\cdot\left(\tfrac{\mathrm{d}}{\mathrm{d}t}\vec{v}\right) + \left(\tfrac{\mathrm{d}}{\mathrm{d}t}\vec{v}\right)\cdot\vec{v}\right) = m\vec{a}\cdot\vec{v} = \vec{F}\cdot\vec{v} .$$

Thus, according to Newton's second law, the rate at which an object's kinetic energy changes (i.e. the power) equals the dot (or scalar) product of the *net external force* on it and its velocity.

If we the integrate the power over time, we find that the change in the object's kinetic energy:

$$\int_{t_i}^{t_f} \mathbb{P}\,\mathrm{d}t = \int_{t_i}^{t_f} \vec{F}\cdot\vec{v}\,\mathrm{d}t = \int_{\vec{r}_i}^{\vec{r}_f} \vec{F}\cdot\mathrm{d}\vec{r} = \mathbb{T}_f - \mathbb{T}_i .$$

The work-energy theorem is not a new discovery that is somehow independent of Newtonian physics. All we have done is define two quantities, kinetic energy and work, and used Newton's second law to relate the change in an object's kinetic energy to the work done on the object. The importance of the work-energy theorem lies in its utility: for a variety of problems it is relatively easy to determine the work done, without having to worry about too many details..

Example 3.1: Lifting and Dropping a Mass

Consider a mass, initially at rest, lifted vertically by a rope of tension $\vec{F}_T = F_T\,\hat{z}$ from the ground to a height h, where it is placed at rest. How much work is done on the mass?

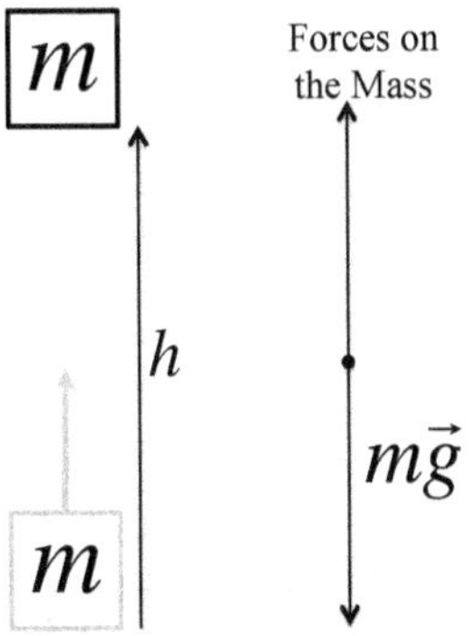

Notice first that the initial and final kinetic energy of the mass is zero, so according to the work energy theorem the total work done on the mass is zero. Thus:

$$\mathbb{T}_f - \mathbb{T}_i = 0 = \int_{\vec{r}_i}^{\vec{r}_f} \vec{F}\cdot\mathrm{d}\vec{r} .$$

Next, draw a free body diagram of the forces acting on the mass as it transits upward. Two forces act on it: the rope pulling it up, and the earth pulling it down. Given the gravitational field of the earth, $\vec{g} = -g\,\hat{z}$, the total work done on the mass is:

$$\int_{\vec{r}_i}^{\vec{r}_f} \vec{F}\cdot d\vec{r} = \int_{\vec{r}_i}^{\vec{r}_f} \left(\vec{F}_T + m\vec{g}\right)\cdot d\vec{r} = \int_{\vec{r}_i}^{\vec{r}_f} \left(F_T\hat{z}\right)\cdot\left(\hat{z}dz\right) + \int_{\vec{r}_i}^{\vec{r}_f} \left(-mg\hat{z}\right)\cdot\left(\hat{z}dz\right) = \int_0^h F_T\, dz - \int_0^h mg\, dz = \int_0^h F_T\, dz - mgh .$$

The work done by the gravitational field of the earth is simply a function of the mass and its initial and final positions. Because the total work is zero, we can solve for the work done by the rope even if we do not know the force of tension on the mass as a function of position. Thus:

$$\int_0^h F_T\, dz = mgh .$$

Now imagine that the mass is subsequently dropped, and it falls toward the ground. The work done on the mass during the fall is given by:

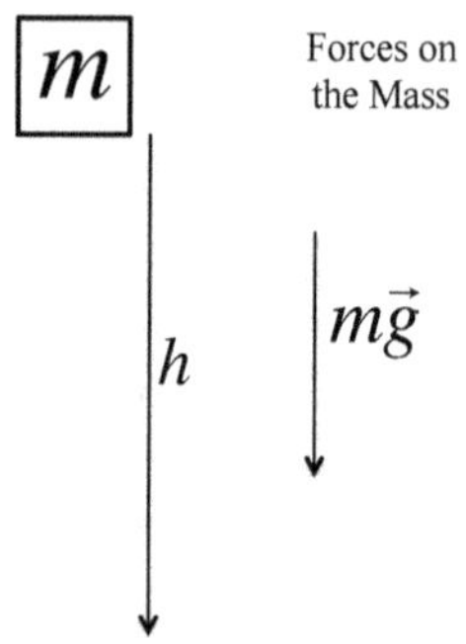

$$\int_{\vec{r}_i}^{\vec{r}_f} \vec{F}\cdot d\vec{r} = \int_{\vec{r}_i}^{\vec{r}_f} \left(m\vec{g}\right)\cdot d\vec{r} = \int_h^0 \left(-mg\hat{z}\right)\cdot\left(\hat{z}dz\right) = 0 - \left(-mgh\right) = mgh .$$

The kinetic energy, just before hitting the ground, is therefore:

$$\mathbb{T}_f = mgh .$$

Notice something profound: the same work done against gravity to lift the mass was also done by gravity when the mass fell. Forces that have this property are called *conservative forces.* Accounting for the work such forces do is greatly simplified by defining a function of position called the *potential energy.*

Thought Experiment 3.2: The Potential Energy

Consider the same scenario as in the last problem, but now let's account for the work done by gravity using the concept of potential energy.

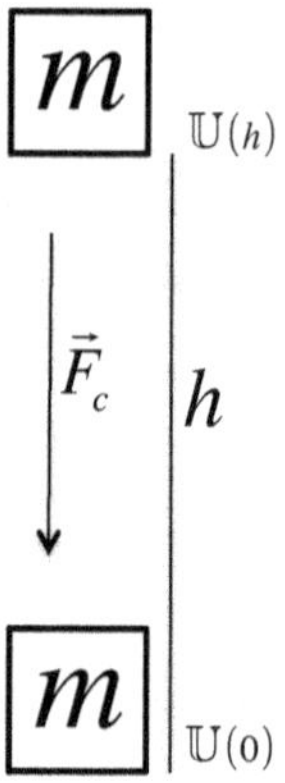

We wish to define a function of position, $\mathbb{U}(z)$, whose difference represents the "potential" for the conservative force to do work. We begin by writing a modified version of work-energy theorem that substitutes this potential energy function for the line integral of the corresponding conservative force. For the case of lifting the mass, we now can write:

$$\left(\mathbb{U}(h) + \mathbb{T}\right) - \left(\mathbb{U}(0) + \mathbb{T}_0\right) = \int_{\vec{r}_i}^{\vec{r}_f} \vec{F}_T \cdot d\vec{r} = \int_0^h F_T \cdot dz ,$$

and in our case, we know from the last example that the work done by the rope, against gravity, was simply mgh. Recall that the last problem assumed the initial and final kinetic energies were zero, so we have:

$$\mathbb{U}(h) - \mathbb{U}(0) = \int_0^h F_T \cdot dz = mgh .$$

Now consider the second part of the problem, when the mass falls. In this case, the only work is done by gravity, so we will ignore this work because we are accounting for it via the gravitational potential energy. So we can simply write:

$$\left(\mathbb{U}(h) + \cancel{\mathbb{T}(h)}\right) - \left(\mathbb{U}(0) + \mathbb{T}(0)\right) = 0 .$$

Solving for the kinetic energy immediately before the mass strikes the ground yields:

$$\mathbb{T}(0) = \mathbb{U}(h) - \mathbb{U}(0) = mgh\,.$$

Now we can move toward a more general definition of the potential energy. We showed already that for a given conservative force, the work done against it would be somehow stored as potential energy. Thus, for any conservative force $\vec{F}_c$, the force required to counteract this force is simply $-\vec{F}_c$. So, the work done against the force from position $\vec{r}_0$ to $\vec{r}$ is the corresponding potential energy difference:

$$\mathbb{U}(\vec{r}) - \mathbb{U}(\vec{r}_0) = \int_{\vec{r}_0}^{\vec{r}} \left(-\vec{F}_c\right)\cdot \mathrm{d}\vec{r} = -\int_{\vec{r}_0}^{\vec{r}} \vec{F}_c \cdot \mathrm{d}\vec{r}\,.$$

Problem 3.1: A Point Charge and a Flat Sheet

Consider a point charge, q, a height h from a large flat sheet of surface charge density σ.

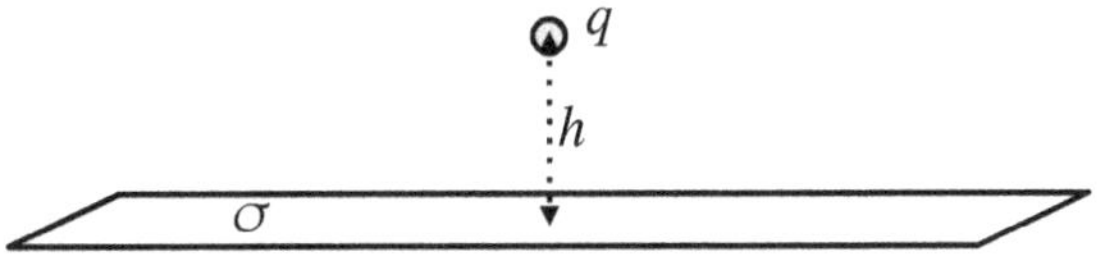

Use our result from Example 2.3 (p.64) for the electric field from the sheet in order to find the potential energy of the point charge. Clearly show where you have defined zero potential energy, which is referred to as *ground*.

Problem 3.2: Energy of a Bead and a Hoop

Recall Example 2.2 (p. 62) and Problem 2.5 (p. 63) with a small bead of charge q on a rod constrained to the axis of a large hoop with linear charge density λ and radius R.

a) Find the electrical potential energy of the charge as a function of axial position z, defining the middle of hoop as ground.

b) What would be the potential energy function if, rather, $z = \pm\infty$ is defined as ground.

c) In the special case of the net-neutral system, as in Problem 2.5, what is the difference in potential energy between the bead being in the middle, and the bead being very far away.

Discussion 3.1: Energy Concepts in Electrodynamics

The nineteenth century witnessed the development of thermodynamics and electrodynamics, both of which were important for the technological revolution that was taking place in the western world. Engineers who were primarily interested in power transmission and telecommunications pioneered much of electrodynamics and thermodynamics.

In the past century and a half, energy has become an ever more fundamental concept in physics. This is not merely a matter of convenience, but rather because even when Newton's laws of motion have failed, experiment sides with conservation of energy. An important example of this is quantum physics, where force is a "derived" concept and energy methods are taken as fundamental.

From the modern perspective, also, the electromagnetic field is the fundamental quantity from which one derives forces. This inverts the order in which we learn about such things. For example, Coulomb's law, which expresses the electric force of attraction or repulsion between two point charges, is typically taken as an experimental fact, and we then abstract from that to construct the notion of an electrostatic force field. The electric and magnetic fields add vectorally, just as with forces, and the associated energy is conceived as somehow residing in the fields. So, in short, much of this book involves understanding the effects and causes of the electric and magnetic fields. On the other hand, the fields can contain and carry energy from one place to another in the form of light.

In Part I of this book, the energy related to the electromagnetic field can be considered potential energy, that is, we conceive of electromagnetic energy as stored in the fields. But a light wave carries energy as it travels, and according to Maxwell's theory this energy flux is given by the *Poynting vector*, which we introduce in Section 3.3. As we discussed in the preface, in modern physics light is conceived of as energy quanta, or photons, and the energy associated with freely moving light particles is treated as kinetic energy.

In principle, once *all* forms of kinetic and potential energy are taken into account, energy then becomes a universally conserved quantity. This is easy to state but becomes quite subtle in practice since there are usually energy losses from non-conservative forces. These losses can be accounted for, however, by an increase of the thermal (random) energy in a system, which is measured as an increase in temperature.

Fundamental Law 3: The Conservation of Energy

In any closed system, energy can neither be created nor destroyed, but may change from one form to another.

Thought Experiment 3.3: Force Laws from Potential Energy

We've discussed how to find the potential energy by knowing a conservative force law, where the potential energy difference is given by the following path integral:

$$\mathbb{U}(\vec{r}) - \mathbb{U}(\vec{r}_0) = \int_{\vec{r}_0}^{\vec{r}} -\vec{F}_c \cdot d\vec{r}\,.$$

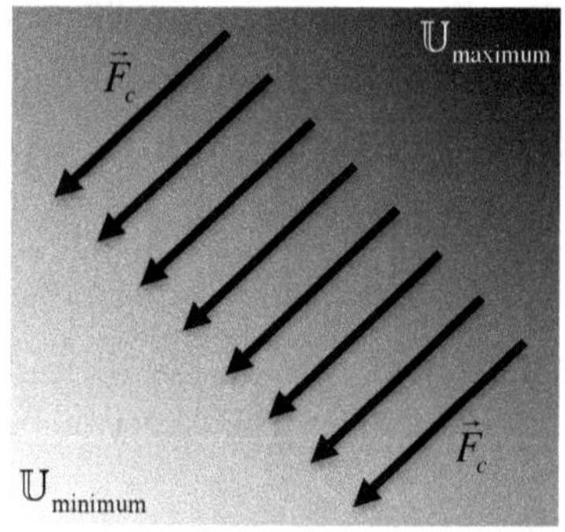

Naturally, this only works if the integral is the same regardless of which path is used for integration.

Consider a differential change in position, so we can the potential energy as:

$$\mathbb{U}\left(\vec{r}+\tfrac{1}{2}\delta\vec{r}\right) - \mathbb{U}\left(\vec{r}-\tfrac{1}{2}\delta\vec{r}\right) = \int_{\vec{r}-\frac{1}{2}\delta\vec{r}}^{\vec{r}+\frac{1}{2}\delta\vec{r}} -\vec{F}_c \cdot d\vec{r} \approx -\vec{F}_c \cdot \int_{\vec{r}-\frac{1}{2}\delta\vec{r}}^{\vec{r}+\frac{1}{2}\delta\vec{r}} d\vec{r} \approx -\vec{F}_c \cdot \delta\vec{r}\,.$$

If we pick a path over which to integrate such that $\delta\vec{r}$ is parallel to the force, then we can write the force in the limit as δr becomes infinitesimal:

$$\vec{F}_c = -\lim_{\delta r \to 0} \frac{\mathbb{U}\left(\vec{r}+\frac{1}{2}\delta\vec{r}\right) - \mathbb{U}\left(\vec{r}-\frac{1}{2}\delta\vec{r}\right)}{\delta r}\delta\hat{r} = -\vec{\nabla}\mathbb{U}(\vec{r})\,.$$

This is our second vector derivative, the first being the divergence (p. 35). In the case of divergence, the derivative was a scalar. In this case, the *gradient* of a scalar function returns a vector field. Now, if we calculate the potential energy directly, we can find the corresponding force by taking the negative gradient.

3.2 *The Gradient in Curvilinear Coordinates*

A scalar function of the position, such as the potential energy, has a magnitude and direction of steepest ascent. The most common example is the gravitational potential energy on the surface of Earth, which is proportional to one's elevation. The direction "straight up the mountain" is the direction of the gradient of your elevation, while the slope of this steepest ascent is the magnitude of the gradient. Similarly, water flows in the opposite direction as the gradient of the elevation, since that is the direction of the sum of the gravitational and normal forces $\left(\vec{F}_g + \vec{F}_N\right)$.

In this section, we derive the gradient of a scalar function in Cartesian, cylindrical, and spherical coordinate systems.

Definition 3.1: The Gradient

The gradient of a scalar function is a vector field, which is defined as:

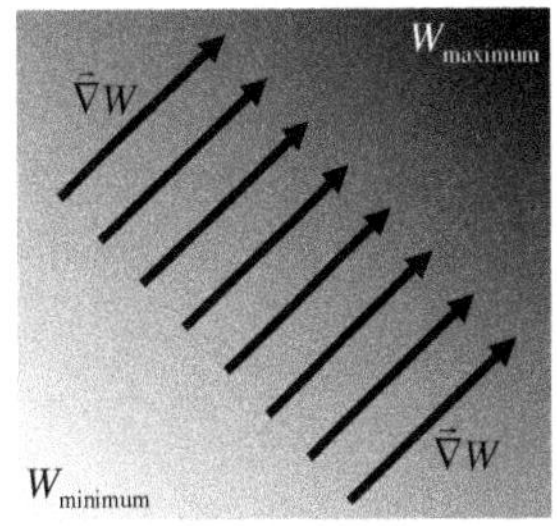

$$\vec{\nabla} W \equiv \lim_{\delta r \to 0} \frac{W\left(\vec{r} + \frac{1}{2}\delta\vec{r}\right) - W\left(\vec{r} - \frac{1}{2}\delta\vec{r}\right)}{\delta\vec{r}} = \lim_{\delta r \to 0} \frac{\delta W}{\delta \vec{r}} .$$

Notice that the gradient can be interpreted as simply the derivative with respect to the position vector, with the vector pointing in the direction where W is increasing the fastest.

This definition can also be written as:

$$\vec{\nabla} W = \lim_{\delta r \to 0} \frac{\delta W}{\delta \vec{r}} = \lim_{\delta r \to 0} \frac{\delta W}{\delta \vec{r}} \left(\frac{\delta \vec{r} \cdot \delta \vec{r}}{\delta \vec{r} \cdot \delta \vec{r}} \right) = \lim_{\delta r \to 0} \delta W \frac{\delta \vec{r}}{\left(\delta r\right)^2} = \lim_{\delta r \to 0} \frac{\delta W}{\delta r} \frac{\delta \vec{r}}{\delta r} = \lim_{\delta r \to 0} \frac{\delta W}{\delta r} \delta \hat{r} .$$

And we can use the fundamental theorem of calculus to write this in integral form as:

$$W(\vec{r}) = \int_0^{W(\vec{r})} \mathrm{d}W = W(\vec{r}_0) + \int_{\vec{r}_0,\ \text{path}}^{\vec{r}} \left(\vec{\nabla} W\right) \cdot \mathrm{d}\vec{r} .$$

Derivation 3.1: The Gradient in Cartesian Coordinates

Consider now a position vector, $\vec{r}$, and another position $\vec{r} + \delta\vec{r}$. We will first find the vector difference in Cartesian coordinates as in the picture:

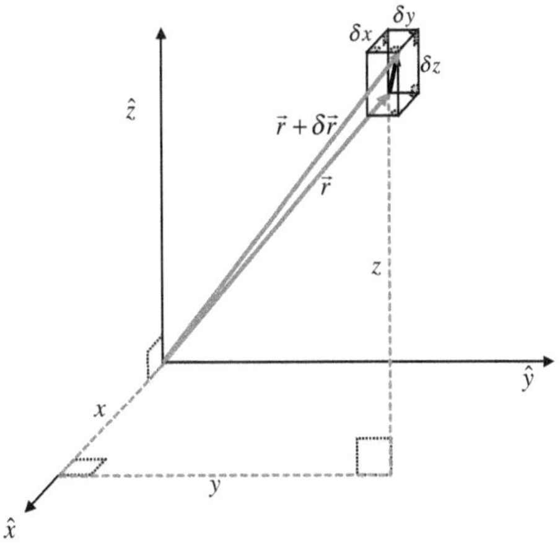

$$\left(\vec{r} + \delta\vec{r}\right) - \left(\vec{r}\right) = \delta\vec{r} = \delta x\, \hat{x} + \delta y\, \hat{y} + \delta z\, \hat{z}$$

Now consider a scalar function of position, $W(\vec{r})$, and the definition of the gradient $\vec{\nabla} W$:

$$\vec{\nabla} W \equiv \lim_{\delta r \to 0} \frac{\delta W}{\delta \vec{r}} .$$

Now we assume the gradient is constant over small differences, so that:

$$\delta W \approx \left(\vec{\nabla} W\right) \cdot \delta\vec{r} = \left(\vec{\nabla} W\right)_x \delta x + \left(\vec{\nabla} W\right)_y \delta y + \left(\vec{\nabla} W\right)_z \delta z$$

Now, we can expand W in a Taylor series using Cartesian coordinates and obtain:

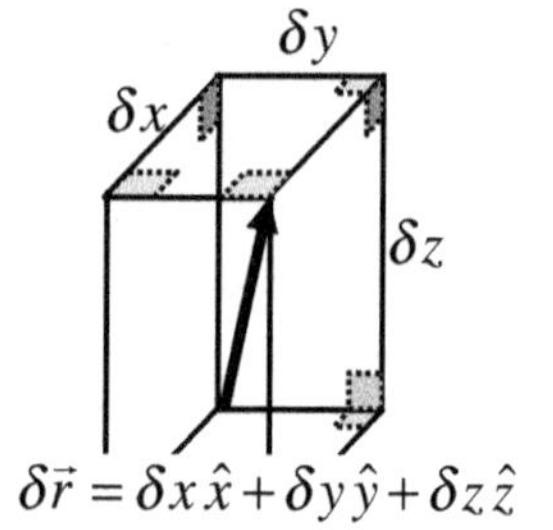

$$\delta W \approx \left(\frac{\partial W}{\partial x}\right)\delta x + \left(\frac{\partial W}{\partial y}\right)\delta y + \left(\frac{\partial W}{\partial z}\right)\delta z$$

Equating these two expressions we find:

$$\delta W \approx \left(\vec{\nabla} W\right)_x \delta x + \left(\vec{\nabla} W\right)_y \delta y + \left(\vec{\nabla} W\right)_z \delta z \approx \left(\frac{\partial W}{\partial x}\right)\delta x + \left(\frac{\partial W}{\partial y}\right)\delta y + \left(\frac{\partial W}{\partial z}\right)\delta z$$

Taking the limit as $\delta r \to 0$, we can take advantage of the orthogonality of our coordinate system to equate each component of the gradient:

$$\vec{\nabla} W = \left(\frac{\partial W}{\partial x}\right)\hat{x} + \left(\frac{\partial W}{\partial y}\right)\hat{y} + \left(\frac{\partial W}{\partial z}\right)\hat{z} = \left(\frac{\partial}{\partial x}\hat{x} + \frac{\partial}{\partial y}\hat{y} + \frac{\partial}{\partial z}\hat{z}\right)W.$$

As with the divergence in Chapter 1, it is important to realize that while defining the vector operator $\vec{\nabla}$ algebraically as:

$$\vec{\nabla} \equiv \left(\frac{\partial}{\partial x}\hat{x} + \frac{\partial}{\partial y}\hat{y} + \frac{\partial}{\partial z}\hat{z}\right)$$

is a wonderful memory aid, and it works well in Cartesian coordinates, however it has significant flaws in curvilinear coordinates where the unit vectors themselves point in different directions at different positions. We will show this in the next two derivations.

Derivation 3.2: The Gradient in Cylindrical Coordinates

We now turn our attention to the derivation of the gradient in cylindrical coordinates.

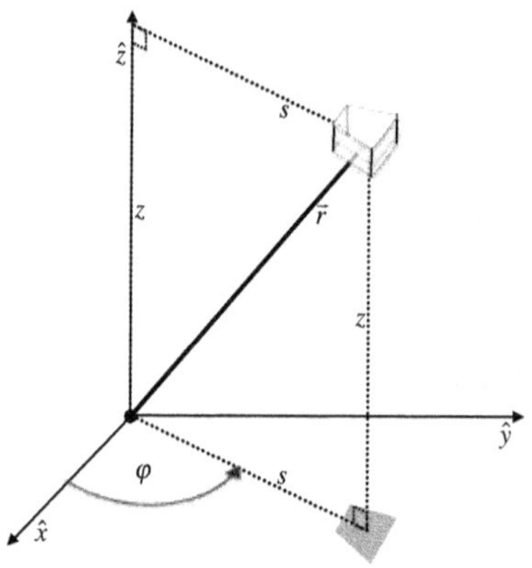

We begin with the difference of two positions as shown in the picture as:

$$\left(\vec{r} + \delta\vec{r}\right) - \left(\vec{r}\right) = \delta\vec{r} \approx \delta s\,\hat{s} + s\,\delta\varphi\,\hat{\varphi} + \delta z\,\hat{z}$$

Again consider a scalar function of position, $W(\vec{r})$, and the definition of the gradient $\vec{\nabla} W$:

$$\delta W = W(\vec{r} + \delta\vec{r}) - W(\vec{r}) = \int_{\vec{r}}^{\vec{r}+\delta\vec{r}} \left(\vec{\nabla} W\right) \cdot d\vec{r}$$

Again we assume the gradient is constant over small differences, so that:

$$\delta W \approx \left(\vec{\nabla} W\right) \cdot \delta\vec{r} \approx \left(\vec{\nabla} W\right)_s \delta s + \left(\vec{\nabla} W\right)_\varphi s\,\delta\varphi + \left(\vec{\nabla} W\right)_z \delta z,$$

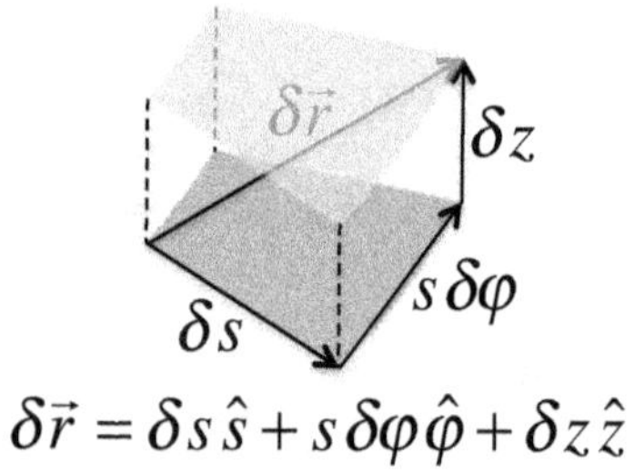

and we can expand W in a Taylor series using cylindrical coordinates to obtain:

$$\delta W \approx \left(\frac{\partial W}{\partial s}\right)\delta s + \left(\frac{\partial W}{\partial \varphi}\right)\delta\varphi + \left(\frac{\partial W}{\partial z}\right)\delta z.$$

Equating these two expressions we find:

$$\delta W \approx \left(\vec{\nabla} W\right)_s \delta s + \left(\vec{\nabla} W\right)_\varphi s\,\delta\varphi + \left(\vec{\nabla} W\right)_z \delta z \approx \left(\frac{\partial W}{\partial s}\right)\delta s + \left(\frac{\partial W}{\partial \varphi}\right)\delta\varphi + \left(\frac{\partial W}{\partial z}\right)\delta z$$

Finally taking the limit as $\delta r \to 0$, we can take advantage of the orthogonally of our coordinate system to equate each component of the gradient:

$$\left(\vec{\nabla} W\right)_s \cancel{\delta s} = \left(\frac{\partial W}{\partial s}\right)\cancel{\delta s},\quad \left(\vec{\nabla} W\right)_\varphi s\,\cancel{\delta\varphi} = \left(\frac{\partial W}{\partial \varphi}\right)\cancel{\delta\varphi},\quad \text{and}\quad \left(\vec{\nabla} W\right)_z \cancel{\delta z} = \left(\frac{\partial W}{\partial z}\right)\cancel{\delta z}.$$

Thus the gradient in cylindrical coordinates is:

$$\vec{\nabla} W = \left(\frac{\partial W}{\partial s}\right)\hat{s} + \frac{1}{s}\left(\frac{\partial W}{\partial \varphi}\right)\hat{\varphi} + \left(\frac{\partial W}{\partial z}\right)\hat{z}.$$

Derivation 3.3: The Gradient in Spherical Coordinates

First we consider the small difference between two position vectors as in the figure, so:

$$\delta\vec{r} \approx \delta r\,\hat{r} + r\,\delta\theta\,\hat{\theta} + r\sin\theta\,\delta\varphi\,\hat{\varphi}.$$

Again consider a scalar function of position, $W(\vec{r})$, and the definition of the gradient $\vec{\nabla} W$:

$$\delta W = W(\vec{r}+\delta\vec{r}) - W(\vec{r}) = \int_{\vec{r}}^{\vec{r}+\delta\vec{r}} \left(\vec{\nabla} W\right)\cdot \mathrm{d}\vec{r} \approx \left(\vec{\nabla} W\right)\cdot\delta\vec{r}.$$

Again we assume the gradient is approximately constant over small displacements, so that:

$$\delta W \approx \left(\vec{\nabla} W\right)_r \delta s + \left(\vec{\nabla} W\right)_\theta s\,\delta\theta + \left(\vec{\nabla} W\right)_\varphi r\sin\theta\,\delta\varphi.$$

Expanding W in spherical coordinates, we find:

$$\delta W \approx \left(\frac{\partial W}{\partial r}\right)\delta r + \left(\frac{\partial W}{\partial \theta}\right)\delta\theta + \left(\frac{\partial W}{\partial \varphi}\right)\delta\varphi.$$

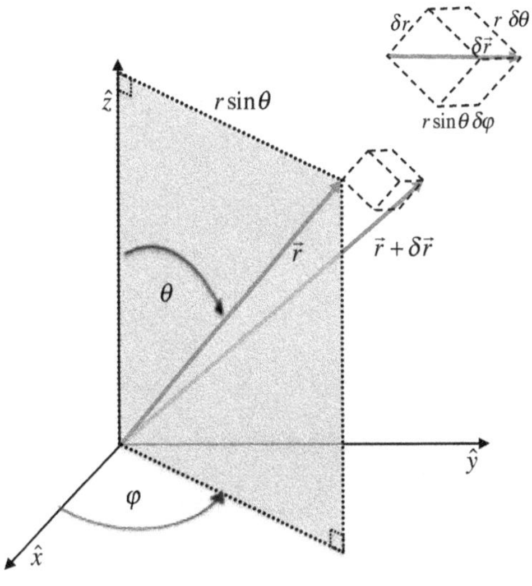

$\delta\vec{r} = \delta r\,\hat{r} + r\,\delta\theta\,\hat{\theta} + r\sin\theta\,\delta\varphi\,\hat{\varphi}$

Equating these two expressions, we find:

$$\delta W \approx \left(\vec{\nabla} W\right)_r \delta s + \left(\vec{\nabla} W\right)_\theta s\,\delta\theta + \left(\vec{\nabla} W\right)_\varphi r\sin\theta\,\delta\varphi \approx \left(\frac{\partial W}{\partial r}\right)\delta r + \left(\frac{\partial W}{\partial \theta}\right)\delta\theta + \left(\frac{\partial W}{\partial \varphi}\right)\delta\varphi.$$

Finally taking the limit as $\delta r \to 0$, we can equate each component of the gradient:

$$\left(\vec{\nabla} W\right)_r \cancel{\delta r} = \left(\frac{\partial W}{\partial r}\right)\cancel{\delta r},\quad \left(\vec{\nabla} W\right)_\theta r\,\cancel{\delta\theta} = \left(\frac{\partial W}{\partial \theta}\right)\cancel{\delta\theta},\quad \text{and}\quad \left(\vec{\nabla} W\right)_\varphi r\sin\theta\,\cancel{\delta\varphi} = \left(\frac{\partial W}{\partial \varphi}\right)\cancel{\delta\varphi}.$$

Thus the gradient in cylindrical coordinates is:

$$\vec{\nabla} W = \left(\frac{\partial W}{\partial r}\right)\hat{r} + \frac{1}{r}\left(\frac{\partial W}{\partial \theta}\right)\hat{\theta} + \frac{1}{r\sin\theta}\left(\frac{\partial W}{\partial \varphi}\right)\hat{\varphi}$$

Discussion 3.2: Directional Derivatives

The directional derivative generalizes the notions of partial derivative and gradient by giving the derivative of a function in an arbitrary direction. The directional derivative of a function in a specific direction tells us the initial rate of change of the function in that direction. It has many applications in physics, especially in fluid mechanics.

Definition 3.2: The Directional Derivative of a Scalar

The *directional derivative*, $\hat{n}\cdot\vec{\nabla}$, of a scalar valued function is:

$$\nabla_{\hat{n}}\psi = \left(\hat{n}\cdot\vec{\nabla}\right)\psi(\vec{r}) \equiv \lim_{h\to 0}\tfrac{1}{h}\left(\psi\left(\vec{r}+\tfrac{h}{2}\hat{n}\right)-\psi\left(\vec{r}-\tfrac{h}{2}\hat{n}\right)\right) = \hat{n}\cdot\vec{\nabla}\psi(\vec{r}).$$

Problem 3.3: The Directional Derivative and Unit Vectors

(a) Show that for Cartesian unit vectors, $\hat{x},\hat{y},$ and $\hat{z}$, $\nabla_{\hat{x}}\psi = \frac{\partial}{\partial x}\psi$, $\nabla_{\hat{y}}\psi = \frac{\partial}{\partial y}\psi$, and $\nabla_{\hat{z}}\psi = \frac{\partial}{\partial z}\psi$.

(b) Find the directional derivative for the other six standard curvilinear unit vectors.

Problem 3.4: Directional Derivative Identities

(a) Show, using diagrams, like the one shown, the following holds for any divergence free unit vector field:

$$\nabla_{\hat{n}}\psi = \lim_{\delta V\to 0}\tfrac{1}{\delta V}\left(\oint_{\text{surface}}\psi(\vec{r})\left(\hat{n}\cdot d\vec{A}\right)\right) = \vec{\nabla}\cdot\left(\psi(\vec{r})\,\hat{n}\right).$$

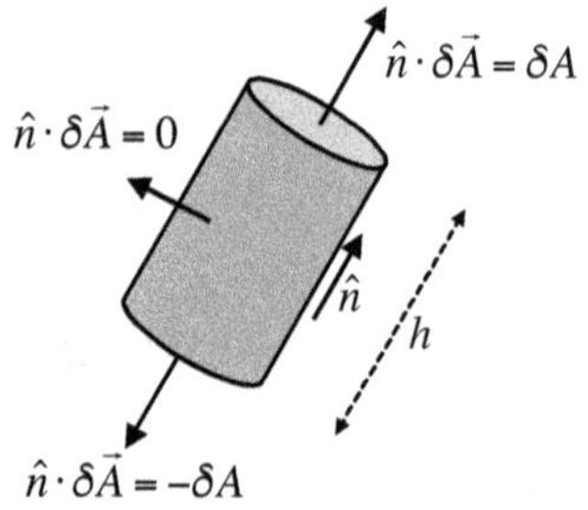

(b) Using the appropriate vector identities, show that:

$$\vec{\nabla}\cdot\left(\psi\hat{n}\right) = \psi\left(\vec{\nabla}\cdot\hat{n}\right) + \nabla_{\hat{n}}\psi.$$

(c) Verify your last derivation using $\hat{s}$ as the unit vector.

Problem 3.5: The Directional Derivative in Curvilinear Coordinates

(a) Derive, in spherical coordinates, the directional derivative in the $\hat{r}$ direction.

(b) Derive, in cylindrical coordinates, the same directional derivative in the $\hat{r}$ direction.

(c) Derive, in Cartesian coordinates, the same directional derivative in the $\hat{r}$ direction.

(d) Show that they are all equivalent.

3.3 *Energy Continuity*

Thought Experiment 3.4: The Energy Continuity Equation

For any closed system, energy cannot be created or destroyed, however it can transform from one form to another. Our goal here is to express this as a continuity equation, in much the same way as we did for charge conservation in Chapter 1.

In Chapter 1, we considered the conservation of charge, Q, and wrote this in terms of the *flux* of charge, which we called the current density $\vec{J}$. Notice that we are using the word flux in this context to mean the *rate of flow per area.* This definition of a *flux* presupposes that we are discussing a conserved quantity, because otherwise the idea of a flow would have little meaning. As with charge, we make an analogy to the flow of water and the conservation of mass in a fluid.

To solve conservation of energy problems, we must first define *the system*, inside of which we account for the total energy. Let's imagine that our system is surrounded by a Gaussian surface, which marks an imaginary boundary between the system and the rest of the universe. We then define the total energy flux vector, $\vec{\mathbb{S}}_{tot}$, as the flow of energy per time per area. We can relate the total energy, $\mathbb{E}_{tot}$, contained inside this Gaussian surface to the total power, $\mathbb{P}$, that passes through the surface via the surface integral of the energy flux:

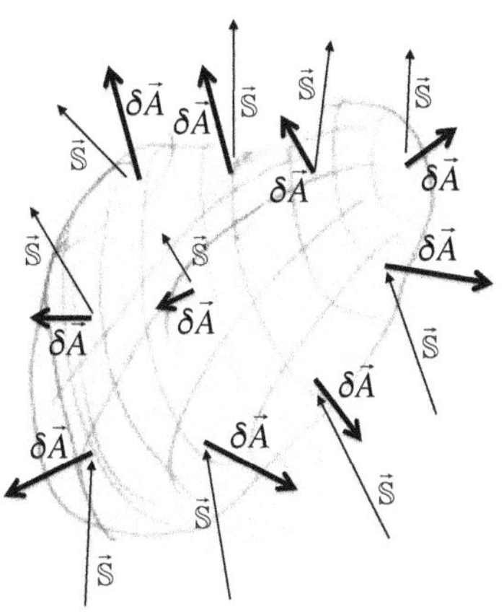

$$\mathbb{P} = \oint_{surface} \vec{\mathbb{S}}_{tot} \cdot d\vec{A} = -\frac{d\mathbb{E}_{tot}}{dt}.$$

The more energy that leaves the surface, the less energy is left inside.

We can extend this concept from a global (extensive) form, to a local (intensive) form, by applying the divergence theorem:

$$\lim_{\delta V \to 0} \frac{\oint_{surface} \vec{\mathbb{S}}_{tot} \cdot d\vec{A}}{\delta V} = -\lim_{\delta V \to 0} \frac{\frac{d}{dt}\mathbb{E}_{tot}}{\delta V} = -\frac{d}{dt} \lim_{\delta V \to 0} \frac{\mathbb{E}_{tot}}{\delta V}, \quad \text{or more compactly:} \quad \vec{\nabla} \cdot \vec{\mathbb{S}}_{tot} = -\frac{d\mathbb{u}_{tot}}{dt}.$$

where $\vec{\mathbb{S}}_{tot}$ is the energy flux and $\mathbb{u}_{tot}$ is the energy density.

Example 3.2: The Luminosity of the Sun

Consider a spherical source of light, such as the sun, that gives off total power $\mathbb{P}$, which astronomers call the *bolometric luminosity*.[44] What is its brightness (energy flux) $\vec{\mathbb{S}}$ as a function of distance? Or, conversely, if we measure the brightness of the sun, at what rate is its internal energy decreasing?

We first draw a spherical Gaussian surface, centered on the sun, with a radius r. We imagine that the observer is located at the surface. So, r, would be the distance from the earth to the sun (one *astronomical unit*).

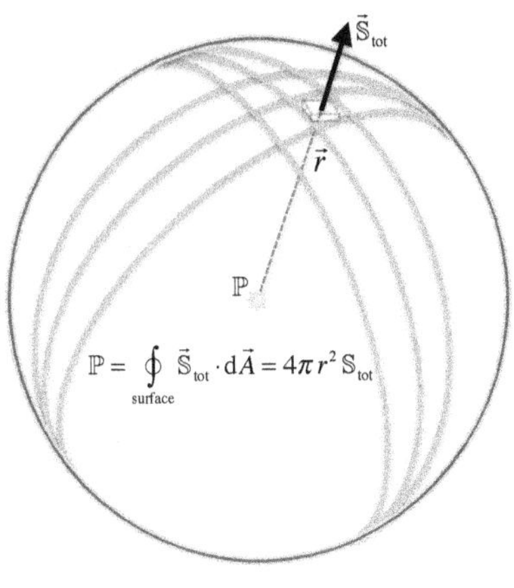

According to conservation of energy:

$$\mathbb{P} = \oint_{surface} \vec{\mathbb{S}}_{tot} \cdot d\vec{A} = -\frac{d\mathbb{U}_{tot}}{dt}.$$

But we know that the sun emits light equally in all directions, so this becomes:

$$\mathbb{P} = \mathbb{S}_{tot}\left(4\pi r^2\right) = -\frac{d\mathbb{U}_{tot}}{dt}.$$

---

[44] The term *bolometric* emphasizes that this is frequency-integrated luminosity.

This has a number of important implications for us on Earth. First, we can find the total power emitted by the sun, by measuring the solar constant, $\mathbb{S}_{\text{tot}} = 1{,}361\frac{\text{W}}{\text{m}^2}$, and the astronomical unit. $r = 1\,\text{AU} = 1.50\times10^{11}\,\text{m}$, so the luminosity of the sun is:

$$\mathbb{P} = \mathbb{S}_{\text{tot}}\left(4\pi r^2\right) = \left(1{,}361\tfrac{\text{W}}{\text{m}^2}\right)4\pi\left(1.50\times10^{11}\,\text{m}\right)^2 = 3.84\times10^{26}\,\text{W}.$$

This also means that the sun is losing energy at this same rate, which is presumably coming from some sort of potential energy source inside the sun. The sun has a finite lifetime, therefore, which depends on its energy source.

The oldest objects in the solar system are just under five billion years old, as measured through radioactive decay measurements. This would imply that the sun has used up about the following potential energy so far:

$$\Delta\mathbb{U} \approx \mathbb{P}t \approx \left(4\times10^{26}\,\text{W}\right)\left(1.5\times10^{17}\,\text{s}\right) \approx 6\times10^{43}\,\text{J}.$$

Problem 3.6: Visual Brightness of a Camera Flash

The SI of visual brightness (*illuminance*) is the lux, which is defined as the visible brightness thrown by one standard candle a distance of one meter. A *standard candle* has the same visual luminosity as does a 540 THz monochromatic isotropic light source emitting exactly $\frac{4\pi}{683}$ watt of radiant power.

The brightness of a typical cloudy day is about 1,000 lux, which is about the right amount of ambient light for a $\frac{1}{250}$s exposure with a $f/_{5.6}$ focal ratio and ISO 200 film speed.

(a) How many standard candles would be required to light a photography subject to at 1,000 lux at a distance of 3 meters?

(b) The SI unit of *luminous power* is the *lumen*. The luminous power of one standard candle is 4π lumens. How many lumens are required to light the photography subject in part (a)?

(c) The *luminous efficacy* is the ratio of the luminous power output, to the power input, of a light source as measured in lumens per watt. What is the luminous efficacy of a perfectly efficient 540 THz light source?

(d) Cree Inc., in a 2014 press release, boasted:

> Cree reports thet the LED efficacy was measured at 303 lumens per watt, at a correlated color temperature of 5150 K and 350 mA. Standard room temperature was used to achieve the results.

What is the efficiency, in percent, of this LED?

(e) How much power would be needed to light the photography subject in part (a) with the LEDs in part (d).

(f) The Canon EOS T6i can take about 550 still pictures without using the flash, or 440 shots using the flash half of the time, on one fully charged 7.5 watt-hour battery pack. Estimate the luminous efficacy of its camera flash.

Example 3.3: A Hydropower Dam

Much of electrodynamics is taken by analogy from hydrodynamics, and so the purpose of this example is to make the connection to a simple hydrodynamics problem.

Consider water flowing at a velocity $\vec{v} = -v\hat{z}$, down through a turbine from a lake a height h above the river below.

Assuming the water has a constant mass density, ρ_{m}, the total mass per time per area that flows past a particular place (such as the turbine) is:

$$\vec{J}_{\text{m}} = \rho_{\text{m}}\vec{v} = -\rho_{\text{m}} v\hat{z} .$$

So, the total mass of water that flows down through a pipe of cross section $\vec{A} = -A\hat{z}$ is given by:

$$I_{\text{m}} = \int_A \vec{J}_{\text{m}} \cdot \mathrm{d}\vec{A} = \int_A \left(-J_{\text{m}}\hat{z}\right)\cdot\left(-\mathrm{d}A\hat{z}\right) = \int_A J_{\text{m}} \cdot \mathrm{d}A = AJ_{\text{m}} = A\rho_{\text{m}} v .$$

Now we can ask the important question for hydropower, how much work per time is the falling water doing on the turbine? To answer this, we calculate the total potential energy of the river, both above and below the dam, as:

$$\mathbb{U} = \oint_{M_{\text{total}}} gz\mathrm{d}m = \oint_{\text{above dam}} gz\mathrm{d}m + \oint_{\text{below dam}} gz\mathrm{d}m .$$

Now we assume a steady state flow of water above and below the dam, so in a given time δt, a mass of $\delta m = I_m \delta t$ is transferred from above to below the dam, but all other water is assumed to be the same. Thus the potential energy of the dam changes by an amount:

$$\Delta\mathbb{U} = \oint_{\delta m \text{ below dam}} gz\mathrm{d}m - \oint_{\delta m \text{ above dam}} gz\mathrm{d}m = -\delta m \int_0^h \vec{g}\cdot\mathrm{d}\vec{r} .$$

Notice something interesting. Regardless of the details of the dam, the potential energy per mass can be simply given by:

$$\mathbb{V}_m = \frac{\delta\mathbb{U}}{\delta m} = -\int_0^h \vec{g}\cdot\mathrm{d}\vec{r} = -g\int_0^h \left(-\hat{z}\right)\cdot\mathrm{d}\vec{r} = g\int_0^h \mathrm{d}z = gh .$$

We refer to the potential energy per unit quantity as simply the *potential*, so the quantity $\mathbb{V}_m$ would be called the *gravitational potential.*

Finally the work per time (power) that the water does against the turbine can be written as:

$$\mathbb{P} = -\frac{\delta\mathbb{U}}{\delta t} = -\left(\frac{\delta m}{\delta t}\right)\left(\frac{\delta\mathbb{U}}{\delta m}\right) = -\left(-I_m\right)\left(\mathbb{V}_m\right) = I_m\mathbb{V}_m . .$$

There is a direct analogy between basic fluid flow and charge flow in circuits. In both cases, there is a conserved quantity (mass or charge), which means that the same amount of this quantity which enters a pipe (or wire) also exits the same pipe (or wire). Moreover, it can do work if it is pushed by something (the gravitational or electric field). In the case of water, the rate of energy flow (power) is simply the product of the total mass flow rate $\left(I_m\right)$ and the potential energy difference per mass $\left(\mathbb{V}_m = hg\right)$, therefore the total power of water flowing through the dam is simply given by $\mathbb{P} = I_m\mathbb{V}_m$.

This relationship should look familiar from basic electronics, where the power through a circuit is simply the current multiplied by the drop in voltage. We developed the idea of current in Chapter 1 and electric fields in Chapter 2. We will develop voltage in the next section.

3.4 *Energy and Work in DC Circuits*

As we saw in Chapter 1, Franklin and his contemporaries used the Leyden charge to provide the "electrical fire" in their experiments. Benjamin Franklin considered the idea of using the Leyden jar as an energy source. In his revolutionary vigor, he imagined hooking together a whole *battery* of Leyden jars, likening it to a battery of cannon, which is probably the reason why modern electrochemical cells are called *batteries.*

Despite this history, the Leyden jar was not a primitive battery, but rather a primitive capacitor. To see the difference, consider the experiments you may have performed in school using lemons to power small lights. In those demonstrations, the student pokes two different metals into the lemon and completes the circuit through a light bulb, which lights up.

This contrasts with another common science demonstration, that of the soda bottle Leyden jar. A plastic bottle, filled with salt water, is coated with aluminum foil and fitted with a one-hole rubber stopper, with a protruding metal rod. The rod is charged by an external source, such as a *van de Graaff generator*, and shocking fun is had by all.

The key distinction is that of a *chemical*, as opposed to *physical*, reaction. In the case of the lemon battery, differences in chemical potential energy powers the source, while in the Leyden jar charged particles physically rearrange themselves, but the electron structure of the particular atoms remain intact.

Alessandro Volta invented the first electrochemical cell in 1800, and it worked in much the same way as the lemon battery. As shown in the figure from his work (p. 68), this was a pile of copper (A) and zinc (Z) plates, separated by brine soaked paper disks.

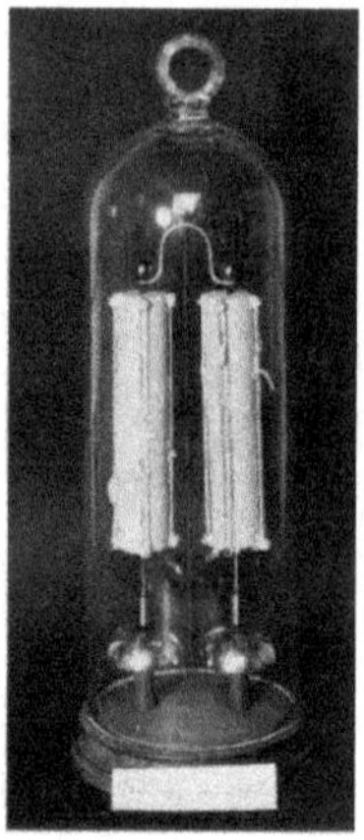

In this *voltaic pile*, zinc becomes ionized as it dissolves in the solution, while copper becomes neutralized and collects on the other end. Notice that current flows through the battery, but the charged particles in both the wire (electrons) and the brine (ions) move slowly away from one end toward the other.

Once the voltaic pile was invented, it became relatively straightforward to stack them to arbitrary height, thus allowing for some very high voltage circuits such as the Oxford electric bell (pictured).[45]

In this section, we will investigate DC circuits, which contain simple components such as batteries, resistors, and capacitors. In all of these cases, we will discuss the relationship between the voltage and the current.

Definition 3.3: Voltage

The *voltage*, or *potential difference*, $(\Delta\mathbb{V})$ is the work per charge between two points:

$$\Delta\mathbb{V} \equiv \tfrac{1}{q}\Delta\mathbb{U}.$$

The unit of voltage, named after Alessandro Volta, is the volt, symbol V:

$$1\,\mathrm{V} = 1\,\mathrm{J}/\mathrm{C} = 1\,\mathrm{W}/\mathrm{A}.$$

When ground is defined, the voltage with respect to ground is called the *electrostatic potential.*

---

[45] A.J. Croft, "The Oxford Electric Bell," European Journal of Physics (1984), 193-94.

In circuit diagrams, the symbol for a constant voltage source is a stack of long and short lines, which is indicative of the pile of electrochemical cells. The positive terminal has a long line, while the negative terminal has a short line. Usually the voltage is written next to the diagram.

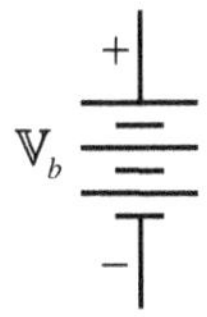

Discussion 3.3: Electrochemical Cell Chemistry

Consider Alessandro Volta's electrochemical cell (p 68). We now know this consists of two separate chemical half-reactions take place in the ion bath—near the positive, and negative, terminals respectively:

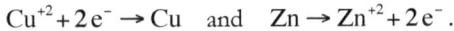

$$Cu^{+2} + 2e^- \rightarrow Cu \quad \text{and} \quad Zn \rightarrow Zn^{+2} + 2e^- .$$

When the circuit is closed, a net negative charge moves through a wire from the negative to the positive terminal, while a net positive charge moves through the brine. This does not mean that every electron actually makes it from one place to the other, since each one moves very slowly (Problem 1.8, p. 28). This is similar in the brine, as often other positive ions are present, such as sodium in salt water or hydrogen in an acid.

When studying batteries, the terms *anode* and *cathode* can sometimes be a source of confusion. This is because the anode is defined as a current source, while the cathode is a current sink. Thus, when considering a battery in a circuit, the positive end is the anode, while the negative end is the cathode. However, inside the ion bath, these roles are reversed. For a chemist, the positive terminal (zinc) is the cathode, and the negative terminal (copper) is the anode, which is exactly the opposite sense an electrical engineer would use.

Problem 3.7: The Oxford Electric Bell

The Oxford Electric Bell is an experimental electric bell that has continued ringing since it was set up in 1840. The experiment consists of two brass bells each connected to a *voltaic pile.* A metal sphere of about 4 mm in diameter is suspended by an insulating thread between the bells. As the clapper touches one bell it is charged, and then electrostatically repelled and attracted to the other bell. On hitting the other bell, the process repeats. The ball has a diameter of 4 mm and its oscillation frequency is 2 hertz.

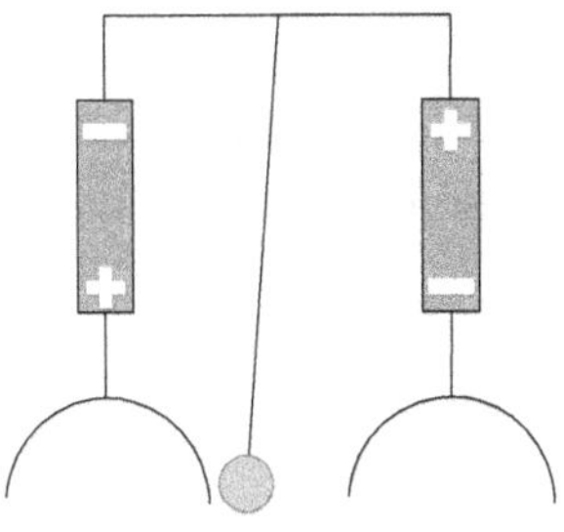

a) Estimating that the clapper holds about 100 nC of charge, what would be the average current flowing through the piles?

b) The voltage between the bells is about 2 kV, and each cell has an estimated voltage of 0.8 V. How many cells are in the pile?

c) How much energy is used each time the clapper rings? From this, give a reasonable estimation of the speed of the clapper hitting the bell.

d) Research this bell, and discuss the experiment in further detail.

Thought Experiment 3.5: Power in a Circuit

Consider a circuit containing a battery attached to some device, like a portable music player, as in circuit diagram. The battery has a voltage difference $\mathbb{V}_b$, and a current I flowing through the device. How much energy per time is being transferred from the battery to the device?

Now we will draw a Gaussian surface around the battery in a circuit, and we must conserve two quantities, charge and energy. The battery remains net neutral, so whatever charge leaves the anode, must return at the cathode (i.e. for every electron that leaves the cathode, another electron must enter the anode). In addition the energy must be conserved, so for each unit of charge that passes out, and in, the battery the chemical potential energy of the battery decreases by an amount $\Delta \mathbb{U} = q\,\mathbb{V}_b$. Thus the rate that energy leaves the battery must be:

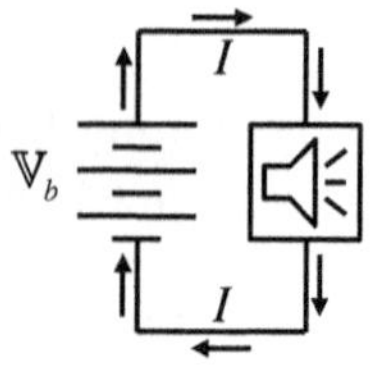

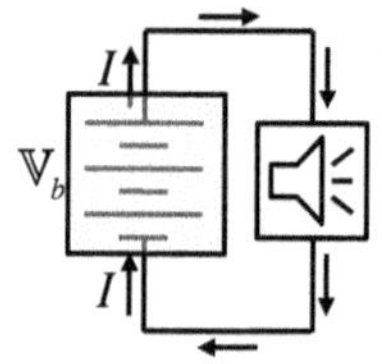

$$\mathbb{P} = I\,\mathbb{V}_b .$$

In other words, the total work per time (power) is the current *through* the battery times the voltage *across* the battery.

Such a definition implies that the voltage from one place to another is the work done by the battery per charge, which we can write in terms of the force on each charge:

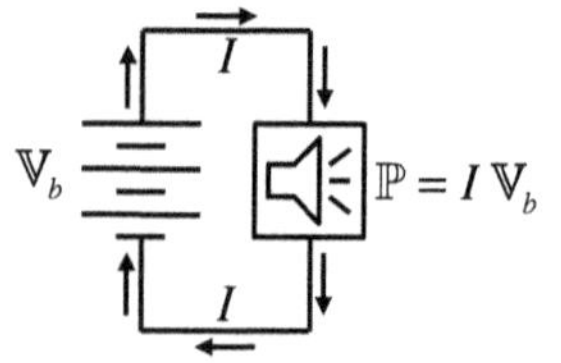

$$\mathbb{V}_B - \mathbb{V}_A = -\tfrac{1}{q}\int_A^B \vec{F}\cdot d\vec{r} = -\int_A^B \left(\vec{E} + \vec{v}\times\vec{B}\right)\cdot d\vec{r}.$$

and in the case of stationary circuits, this simply becomes:

$$\mathbb{V}_B - \mathbb{V}_A = -\int_A^B \vec{E}\cdot d\vec{r}.$$

So, whatever circuits are in the media player, the total work done, per time, by the battery on them is the amount of potential energy that the battery lost.

Notice that a battery does not lose charge, but rather it loses potential energy. The battery starts out in a neutral state, and remains in a neutral state at all times. Technically speaking, then, you do not "charge" your phone, but rather you "energize" it.

Finally, we will check the units to verify that everything works out as it should:

$$[\mathbb{P}] = [I][\mathbb{V}] = \mathrm{A}\,\mathrm{V} = \left(\frac{\cancel{\mathrm{C}}}{\mathrm{s}}\right)\left(\frac{\mathrm{J}}{\cancel{\mathrm{C}}}\right) = \frac{\mathrm{J}}{\mathrm{s}} = \mathrm{W}.$$

Problem 3.8: The O'Shaughnessy Dam

The O'Shaughnessy Dam is the oldest tall dam in the United States, and was built under great controversy between 1914 and 1923 closing up the Hetch Hetchey Valley, which was considered a twin to California's famous Yosemite valley.

Water flows from the Hetch Hetchey through tunnels to its first powerhouse after dropping an elevation of 440 meters from the surface of the lake. The powerhouse contains three turbines, producing a maximum of 234 megawatts of power to northern California.

For this problem, you can make the naive assumption that everything is 100% efficient.

a) What is the potential energy difference, per unit mass, between the surface of the reservoir and the first powerhouse?

b) What is the rate of water flowing through the turbines in units of cubic meters per second?

c) The power flows out of the powerhouse at a voltage of 2,300 volts, what is the total current flowing from the powerhouse?

d) Soon the power is stepped-up to a voltage of 230 kV. What is the current flowing from this transformer?

e) Eventually the power enters homes and businesses throughout Northern California at 220 volts. How much total current flows to all of the customers?

Problem 3.9: A Tandem Van de Graaff Accelerator

A tandem van de Graaff (Plate 7, p. 113) particle accelerator is used for nuclear physics experiments. In these accelerators negative ions are injected into one end where they are attracted toward the middle, which is held at a high voltage V_0. When they reach the middle, they fly through a screen that strips the negative ions of their electrons, making them now positive ions. They continue to be accelerated as they travel to the other side.

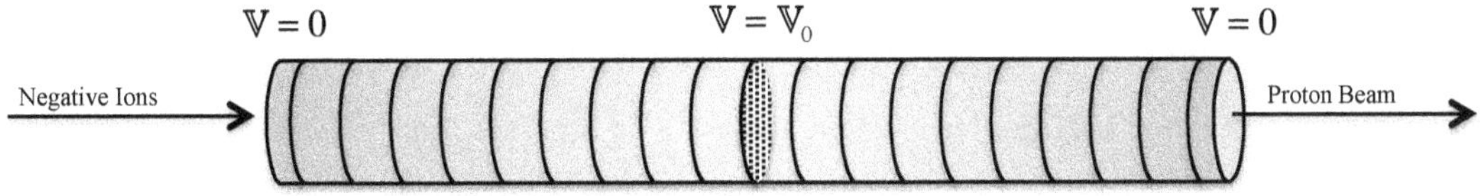

a) Find an expression for the energy per proton as they leave the accelerator. Assume that they are injected as hydride ions (H^-).

b) What is the proton flux (protons per area per time) in terms of the total current I of the beam?

c) Find the total kinetic energy of the proton beam that passes by a point per time interval. Express this in terms of the central voltage V_0 and the total current I in the beam.

d) Ignoring relativistic effects, what is the speed of each proton in the beam?

e) The largest van de Graff accelerators have central voltages of about 20 MV, with total beam currents of 20 μA, and can be as long as 30 meters. What is the kinetic energy per proton that these machines produce (express your answer in both eV and joules)? What is the total power of the beam? What is the electric field inside the tube?

f) Research, and write a short report on, one particular tandem van de Graff accelerator of your choice. What kind of research is it used for? What ions does it produce? At what voltage does it typically run?

Discussion 3.4: Electromotive Force

In a steady state situation, the work done per unit charge between any two points cannot depend on the path taken by the charges, so the potential energy is well defined in this case. For situations in which there is no time dependence, in other words, the electric force behaves as a conservative one. From this we can derive Kirchhoff's loop rule for a circuit: the sum of the voltage drops around a closed electric circuit must add to zero.

Imagine a closed loop around which a group of identical charge carriers is transported. Clearly a force is required to transport these charges and it takes work to bring them from one point to another, let alone to complete a circuit. The line integral, taken around a complete loop or circuit, of this force divided per unit charge is the *work done per unit charge* and is commonly referred to as the *electromotive force* or *EMF*. The name, *EMF*, is misleading since it is *not* a *force*, but rather the *line integral of a force* divided by the magnitude of a charge. You should avoid using the term EMF, but you need to be aware what others mean when they refer to *sources of EMF*.

Historically the term goes back to early days of electrical experimentation, when Ohm, and others, thought of the *EMF* as analogous to pressure in a fluid system, so it was seen as the thing that pushes the current through the circuit.

Sources of EMF are devices, such as batteries, fuel cells, or electrical generators, which add electrical energy to a system from outside. For example, a battery acts as a source of EMF that converts chemical energy into electrical energy. Mechanical energy is transformed into electrical energy via the source of EMF known as an electrical generator. Although it is usually thought of as causing a drop in voltage or dissipation, a resistor in a circuit is technically a "sink of EMF" because it converts electrical energy to thermal energy (also known as heat) that leaves the system. When people speak of EMF sources, they mean a source of electrical energy at constant voltage, in the case of DC circuits, or constant peak amplitude voltage in the case of alternating currents (AC). A power supply that works like a battery, as opposed to a current supply, which supplies energy at constant current (rather than voltage), is a good example of this.

The *voltage* is the work required per unit charge to transport a charge from one location to another. In other words, it is the integral of the force per unit charge taken between two points A and B:

$$\Delta\mathbb{V} = -\frac{1}{q}\int_A^B \vec{F}\cdot d\vec{r}\,.$$

When we define a particular zero point for voltage, which is called *ground*, we refer to the voltage difference with respect to ground as the *electric potential*. This is very convenient as we often encase circuits inside of metal boxes, which are in turn connected to a grounding wire, which is eventually connected physically to the actual ground of the earth.

In the case of a steady state situation, the electric field is conservative and the voltage is therefore identical to the difference in electrical potential between points A and B:

$$\Delta\mathbb{V} = \mathbb{V}_B - \mathbb{V}_A = -\int_A^B \vec{E}\cdot d\vec{r}\,.$$

Consider a loop of wire connected to a battery. If we place the two probes of a voltmeter to the same point we measure zero potential difference, as expected. Now if we place the same probes at the terminal ends of a battery, for example, we will measure a voltage.

In a two-terminal device, such as an electrochemical cell or generator, the EMF can be measured as the voltage across the two open-circuited terminals.

Electromotive force (EMF) in other words, is really the voltage developed by any source of electrical energy such as a battery, fuel cell, or generator. A source of EMF is a source of electrical energy; a potential difference, or voltage drop, represents a sink in electrical energy where electrical energy is converted to other forms of energy. As a charge traverses a closed loop, the total energy it delivers to all the elements (e.g. a resistor) in the loop (i.e. the sum of the potential differences) is equal to the total electrical energy that was supplied to it (the net EMF).

In general, we prefer the terms *voltage rise* and *voltage drop* rather than *source* or *sink* of *EMF*. The terms sources and sinks of energy, or power, are perhaps even better, because they reinforce the connection with the continuity equation for energy.

Thought Experiment 3.6: Ohm's Law

In a series of experiments conducted between 1825 and 1826, Georg Simon Ohm investigated the relationship between current and voltage through a series of metals of different sizes, shapes, and cross sections.[46] He concluded that the flow of electrical current is analogous to Joseph Fourier's law of heat conduction.

Heat always flows from a higher temperature to a lower temperature, and similarly charge always flows from a high voltage to a low voltage. Ohm could also have made a similar analogy with fluid flow, which now makes more sense as temperature is not a measure of potential energy. Nevertheless, Fourier's work was timely, and seemed to provide a fruitful analogy between the flow of heat and the flow of electricity. Gustav Kirchhoff first identified the important quantity furnished by the source in Ohm's experiment, which Ohm had termed "electroscopic force" and "tension," as the change in electrostatic potential.[47]

In modern terms, *resistance* is defined as the ratio of the electric potential drop $(-\Delta\mathbb{V})$ to the current flowing through an electrical device:

$$R \equiv \frac{-\Delta\mathbb{V}}{I},$$

which Ohm claimed was constant for a given device held at a constant temperature.

The SI unit of resistance is the ohm, and it is defined as:

$$\Omega = [R] = \frac{[\mathbb{V}]}{[I]} = \frac{\mathrm{V}}{\mathrm{A}}.$$

An inverse ohm used to be called the *mho*, but is now called a siemens (S).

Next he showed that the resistance primarily depended on three things: (1) the conducting material involved, (2) the length of the material that the current transverses, and (3) the cross-sectional area of the resistor. For a given material, the resistance is directly proportional to the length, and inversely proportional to the area. Thus he defined the electrical conductivity, κ, in an analogous way as the relationship between thermal conductivity, κ_T, and the thermal resistance (the R factor) in heat flow such that:

$$\kappa \equiv \frac{l}{A\,R}.$$

Material	Conductivity at 20°C (S/m)
Silver	6.3E+07
Copper	5.8E+07
Gold	4.1E+07
Aluminum	3.5E+07
Calcium	3.0E+07
Tungsten	1.8E+07
Nickel	1.4E+07
Iron	1.0E+07
Tin	9.2E+06
Platinum	9.1E+06
Lead	4.5E+06
Mercury	1.0E+06
Carbon	2.9E+04
Germanium	2.2E+00
Silicon	1.6E-03

[46] G.S. Ohm, The Galvanic Circuit Investigated Mathematically (Berlin: 1827); Translated by William Frances (New York: D. Van Nostrad Company, 1891)

[47] G. Kirchhoff, "On a deduction of Ohm's Laws, in connexion with the theory of electrostatics," Philosophical Magazine, **37** (1850), 463-68.

The unit of electrical conductivity is, thus, a siemens per meter, because:

$$[\kappa] = \frac{[l]}{[R][A]} = \frac{\text{m}}{\Omega\text{m}^2} = \frac{1}{\Omega\text{m}} = \frac{1/\Omega}{\text{m}} = \frac{\text{S}}{\text{m}}.$$

Next we apply the definition of voltage difference, and write the resistance in terms of the current and voltage as:

$$R \equiv \frac{-\Delta V}{I} = -\frac{V_B - V_A}{I_{AB}} = -\frac{1}{I_{AB}}\left(-\int_A^B \vec{E}\cdot d\vec{r}\right) = \frac{1}{I_{AB}}\left(\int_A^B \vec{E}\cdot d\vec{r}\right).$$

Next we recall the definition of current density (p. 25), and write the current in terms of the area vector. Thus the resistance becomes:

$$R = \frac{1}{\vec{J}\cdot\vec{A}}\left(\int_A^B \vec{E}\cdot d\vec{r}\right) = \int_A^B \frac{\vec{E}}{\left(\vec{J}\cdot\vec{A}\right)}\cdot d\vec{r} \ .$$

We will now assume that the electric field and the current density are parallel, as we would expect for steady state flow. So, for a homogeneous material, we find the resistance as:

$$R = \int_A^B \frac{\vec{E}}{\left(\vec{J}\cdot\vec{A}\right)}\cdot d\vec{r} = \int_A^B \frac{E}{(JA)}dr = \frac{E\,l}{J\,A} \ .$$

Thus, for steady state flow, the current density becomes:

$$\vec{J} = \left(\frac{l}{RA}\right)\vec{E} = \kappa\vec{E} \ ,$$

which is the local (i.e. continuous) version of Ohm's law.

Thought Experiment 3.7: The Heaviside Interpretation of Ohm's Law

In 1847, Ohm worked out an empirical relationship between the negative voltage gradient (the equivalent of $\vec{E}$ under the pressure gradient interpretation) and the current through a circuit.[48] Ohm's law, in its modern intensive form, says that the current density is proportional to the electric field ($\vec{J} \propto \vec{E}$), and applies to conducting metals where frictional forces resist the flow of current. The point of this thought experiment is to show how a simple model for the average drag force density leads to Ohm's law.

In 1883, Oliver Heaviside described a laminar flow model to explain Ohm's law.[49] He wrote:

> Now let the frictional resistance be exactly proportional to the velocity, as it is said to be approximately for low velocities in thin pipes. Then, if C' be the current, R'C', where R' is some constant multiplier, will be the fractional resistance, and C' will increase until R'C'=E', which gives the steady current corresponding to the motive force E', after which there will be no further acceleration of velocity.

---

[48] G.S. Ohm, Die galvanische Kette mathematisch bearbeitet, (Berlin: T.H. Riemann 1827).

[49] Oliver Heaviside, "The Energy of the Electric Current," published in The Electrician June 30, (1883) 149, republished in Electrical Papers, 2nd ed. (Bronx, NY: Chelsea Publishing, 1970) Vol. 1, 285.

In Heaviside's model, the charged fluid flowing through a copper wire is like the water flowing through a copper pipe. To conserve charge, the velocity of the negatively charged fluid must be:

$$\vec{v}_- = -\frac{\vec{J}}{(-\rho_-)} = \frac{\vec{J}}{\rho_-},$$

where $\vec{J}$ is the current density, and ρ_- is the charge density of the moving negative charge. Notice that $\rho_- < 0$, so the velocity is in the opposite direction as the current density.

The drag force $\vec{F}_D$ on the moving fluid of velocity $\vec{v}_-$ of viscosity η, due to a stationary spherical object of radius a, is approximately[50]:

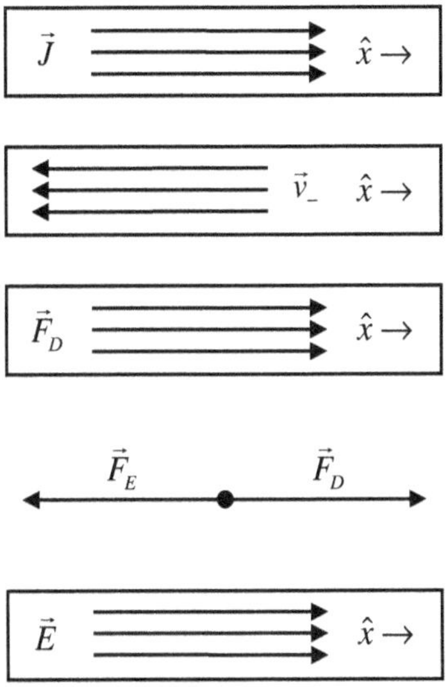

$$\vec{F}_D \approx -6\pi\eta a\vec{v}_- \approx -6\pi\eta a\left(\frac{\vec{J}}{\rho_-}\right) \approx \left(\frac{6\pi\eta a}{(-\rho_-)}\right)\vec{J}.$$

Assuming that there are n_+ ions per volume, and each has a radius a, the average drag force per volume becomes:

$$\vec{f}_D \sim n_+\vec{F}_D \sim n_+\left(\frac{6\pi\eta a}{-\rho_-}\right)\vec{J}.$$

The electric field is also acting on the fluid, with a body force:

$$\vec{f}_E = \rho_-\vec{E}.$$

Because the fluid is moving at a constant velocity, the net body force must be zero, so:

$$0 = \vec{f}_D + \vec{f}_E \approx -\frac{6\pi\eta a n_+\vec{J}}{\rho_-} + \rho_-\vec{E}.$$

Solving for the current density, and grouping the constants:

$$\vec{J} \approx \left(\frac{\rho_-^2}{6\pi\eta a n_+}\right)\vec{E} = \kappa\,\vec{E},$$

where the constant κ is the conductivity.

Discussion 3.5: Classical Models of Conductivity

We have employed reasoning by analogy between metals and fluids from Franklin's fluid model in Chapter 1 to Heaviside's derivation of Ohm's law here. This analogy breaks down, however, when we consider the temperature dependence of conductivity. The conductivity of a metal increases as it cools, implying that cold wires are less viscous than warm ones. Such behavior is completely opposite that of common liquids like honey or molasses, which become less viscous when heated, but it is similar to that of gases, whose viscosity increases with temperature.

Paul Drude sought to explain the thermal and electrical properties of metals by modeling them as a gas of free electrons. This classical free electron, or so-called Drude-Lorentz, model had

[50] This drag force on a spherical object is called Stoke's law, and we will derive it in Chapter 10 (p. 354).

notable if limited success. The classical free electron model accounts for Ohm's law, as did Heaviside's simple laminar flow model. Unlike the laminar flow model, the Drude-Lorentz model also predicts that metallic conductivity (κ) decreases with temperature (T) as the inverse square root $\left(\kappa \propto T^{-\frac{1}{2}}\right)$, just like the viscosity of an ideal gas. Although the Drude-Lorentz model correctly predicted that a metal's conductivity will increase as the temperature decreases, it forecasted the wrong functional dependence. According to experiment, the conductivity of a metal actually decreases *linearly* with temperature $\left(\kappa \propto T^{-1}\right)$. This fact is something for which Drude's classical model simply cannot account.

In fact, both conductivity and viscosity involve processes at the atomic scale and therefore require quantum mechanics for their correct explanation. While a simple classical model accounts fairly well for the temperature dependence of viscosity in gases, it fails for liquids. Arnold Sommerfeld corrected the deficiencies of Drude's model by applying quantum mechanical Fermi-Dirac statistics, in place of classical Maxwell-Boltzmann statistics, to the valence electrons.

Problem 3.10: The Earth's Electric Field

The downward component of the electric field of the Earth is approximately 100 to 300 N/C and there is an associated current density of about $2\times10^{-16}\,\frac{\text{A}}{\text{cm}^2}$, also in the downward direction.[51]

(a) Estimate the average conductivity of the Earth's atmosphere.

(b) Estimate the total downward current across the whole Earth.

(c) Discuss how charge is conserved between the Earth and the atmosphere; cite your sources.

Example 3.4: A Cylindrical Resistor

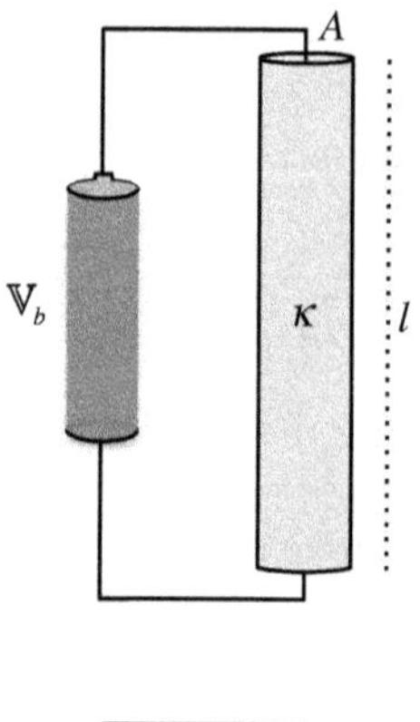

Consider the example of a cylinder made of some material with an electrical conductivity, κ, cross-sectional area A, and length l, which is connected to a battery of voltage $\mathbb{V}_b$.

What is the total current flowing through the battery?

First we notice that charge is conserved, so the current must be constant throughout the circuit. Next we conserve energy, so that the power from the battery is equal to the rate of work done on the resistor:

$$\mathbb{P}_b = I\,\mathbb{V}_b = \int_{+}^{-} I\,\vec{E}\cdot d\vec{\ell} = \int_{+}^{-} \vec{E}\cdot d\vec{\ell} = I\int_{\text{top}}^{\text{bottom}} \vec{E}\cdot d\vec{\ell}.$$

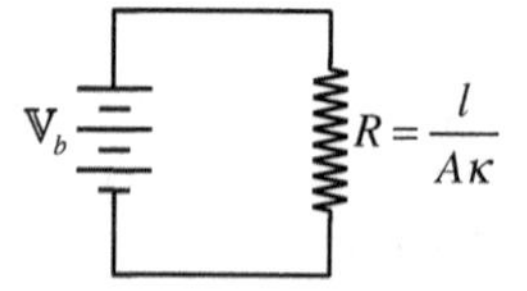

Now, canceling out the current, and applying Ohm's law to the cylinder, yields the voltage between the terminals of the battery:

$$\mathbb{V}_b = \int_{\text{top}}^{\text{bottom}} \vec{E}\cdot d\vec{\ell} = \int_{\text{top}}^{\text{bottom}} \left(\frac{\vec{J}}{\kappa}\right)\cdot d\vec{\ell} = \frac{I\,l}{A\kappa} = I\,R$$

---

[51] V. F. Hess, The Electrical Conductivity of the Atmosphere and its Causes, trans. L.W. Codd, (New York: D. Van Nostrand Company, 1928), 3.

Finally we can solve for the current in terms of the voltage as:

$$I = \frac{\mathbb{V}_b A\kappa}{l} = \frac{\mathbb{V}_b}{R},$$

which is simply the global version of Ohm's law.

Problem 3.11: A Conical Resistor

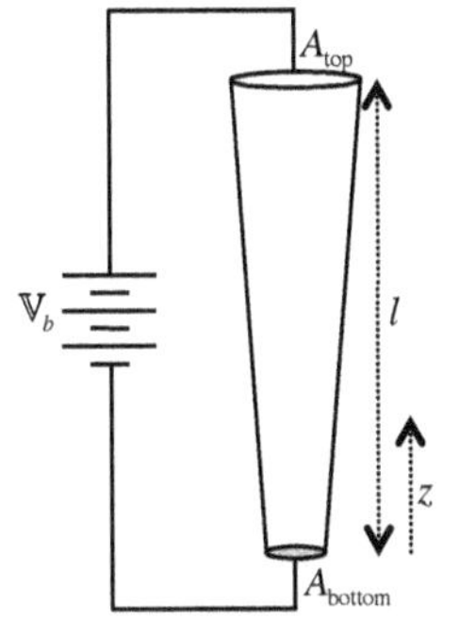

A fez-shaped resistor with conductivity κ, length l, and end areas A_{bottom} and A_{top} is connected to a battery of voltage $\mathbb{V}_b$. Find:

a) The electric field in the resistor $\vec{E}(s,z)$.

b) The current density in the resistor $\vec{J}(s,z)$.

c) The overall resistance R.

d) The total power dissipated by the resistor.

Thought Experiment 3.8: Voltmeters and Ammeters

A galvanometer is a very sensitive measure of current, and we will discuss how they work later in the book (p. 249). In practice, how do we measure larger currents through circuits? And, how can we measure the voltage between two points? To do this, we combine resistors with a galvanometer, as is done in a typical analog multimeter.

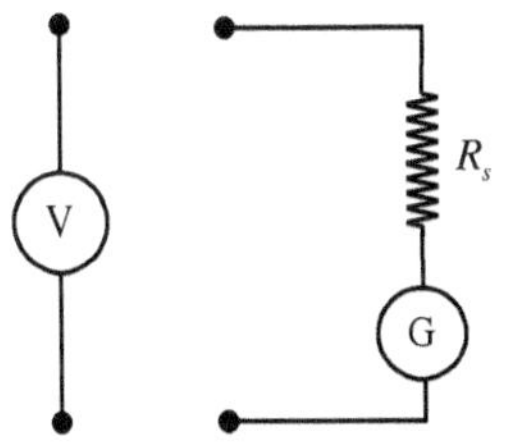

Consider first how one would hook up a voltmeter in such a way as not to disturb an existing circuit. We would like to place two probes on our circuit, record the voltage between the two points, but not have the circuit affected in the process. Thus, we want the internal resistance of a voltmeter to be as large as possible. In order to do this, we connect the galvanometer in series with a large resistor, R_s. For a given voltage difference across the voltmeter, $\Delta\mathbb{V}$, the current that flows through the voltmeter will be given by Ohm's law to be:

$$I = \frac{-\Delta\mathbb{V}}{R_s}.$$

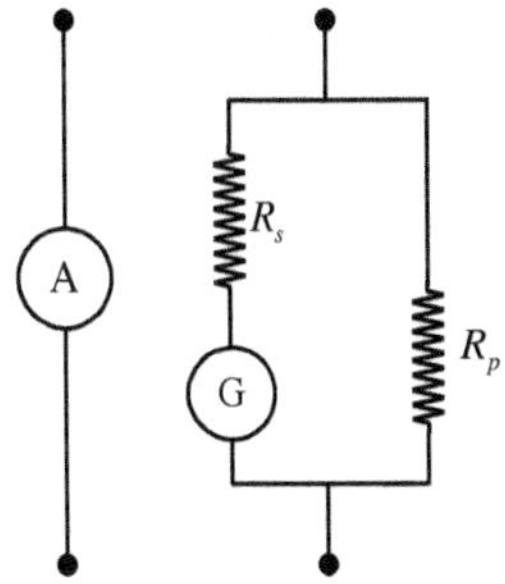

We will choose a series resistor whose resistance corresponds to our desired voltage sensitivity, given the sensitivity of the galvanometer. An idealized voltmeter, of course, has an infinite internal resistance.

On the other hand, charge is conserved. Whatever current enters our ammeter must also exit it, and, in the process, we do not want to disturb the circuit. Therefore, ammeters are hooked up in series to the wire we wish to measure, and we would like the internal resistance to be as small as possible. In order to build an ammeter, we essentially put the voltmeter above in parallel with a very small resistor, R_p. Thus the total current can be found in terms of the current through the galvanometer as:

$$(I - I_G)R_p = I_G R_s \quad \text{and} \quad I R_p = I_G(R_s + R_p) \quad \text{so} \quad I = \left(1 + \frac{R_s}{R_p}\right) I_G.$$

We, naturally, choose our two resistors to corresponds to the sensitivity of the galvanometer. Thus, for a perfectly sensitive galvanometer, we can ignore the resistance of the ammeter.

It is standard laboratory practice to always leave multimeters in voltmeter mode, because mistaking a voltmeter for an ammeter creates a short circuit, while mistaking an ammeter for a voltmeter simply acts as an open circuit.

Example 3.5: Kirchhoff's Loop and Junction Rules

In many practical circuit problems, we are interested in the current as a function of the voltage. The first thing to do, then, is to apply the conservation of charge, and the conservation of energy, to the circuit. For example, consider a circuit in which two resistors (R_1, R_2) are placed in a parallel circuit connected to a battery of voltage V_b. Find the current through the battery, and through each resistor.

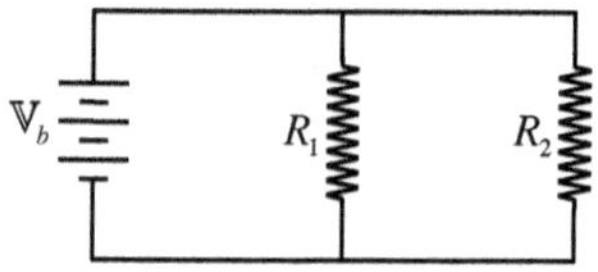

First let I_b, I_1, and I_2 represent the current through the battery, and each resistor.

Now, because the electrostatic force is conservative, the work is path independent. Therefore, around any loop, the work per charge must add to zero:

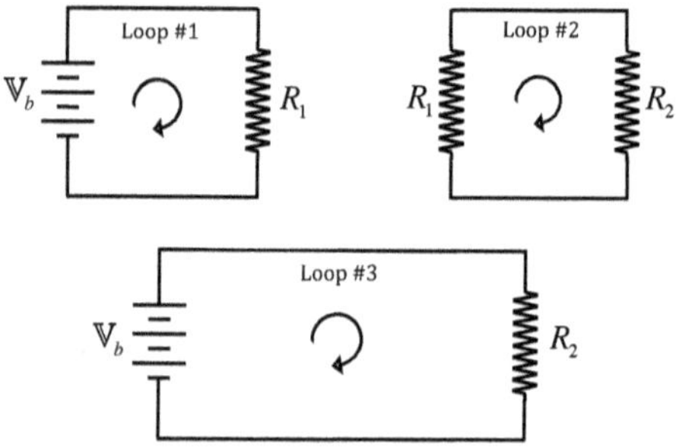

Loop #1: $V_b - I_1R_1 = 0$

Loop #2: $I_1R_1 - I_2R_2 = 0$

Loop #3: $V_b - I_2R_2 = 0$

And solving for the current through each resistor:

Loop #1 $V_b = I_1R_1$ so: $I_1 = V_b/R_1$

Loop #3 $V_b = I_2R_2$ so: $I_2 = V_b/R_2$

Now we apply the conservation of charge to either junction, so:

$$I_b = I_1 + I_2 = V_b\left(\frac{1}{R_1} + \frac{1}{R_2}\right).$$

Since energy must be conserved, we can check to make sure that this is so:

$$P_{in} = P_{out}$$
$$I_bV_b = I_1(I_1R_1) + I_2(I_2R_2)$$
$$I_bV_b = \left(\frac{V_b}{R_1}\right)(I_1R_1) + \left(\frac{V_b}{R_2}\right)(I_2R_2)$$
$$I_bV_b = V_b(I_1 + I_2)$$
$$I_bV_b = I_bV_b$$

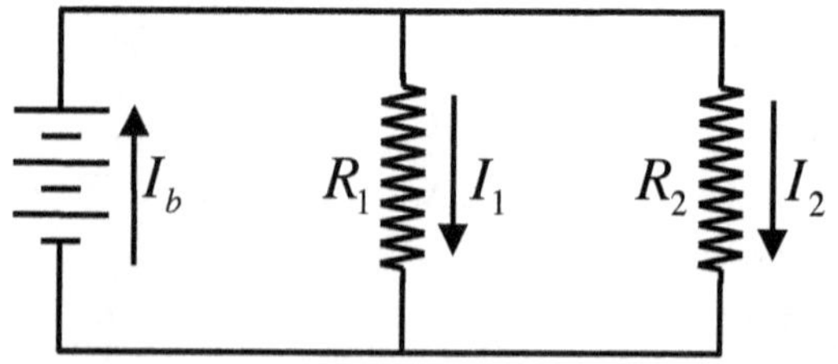

So, our solution is consistent with conservation of energy for the system.

Problem 3.12: A Wheatstone Bridge II

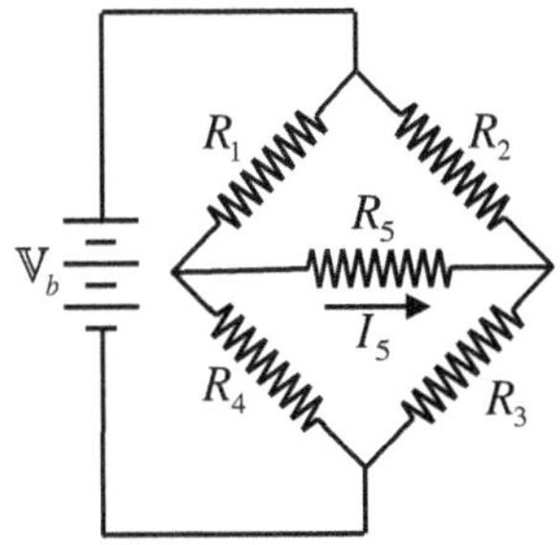

A Wheatstone bridge circuit consists of 5 resistors configured as in the diagram.

a) Find the current flowing through the middle resistor (I_5) as a function of the resistances $(R_1, R_2, R_3, R_4, R_5)$ and the voltage of the battery (V_b).

b) What is the relationship between the resistors when $I_5 = 0$?

Problem 3.13: An Infinite Resistor Ladder I

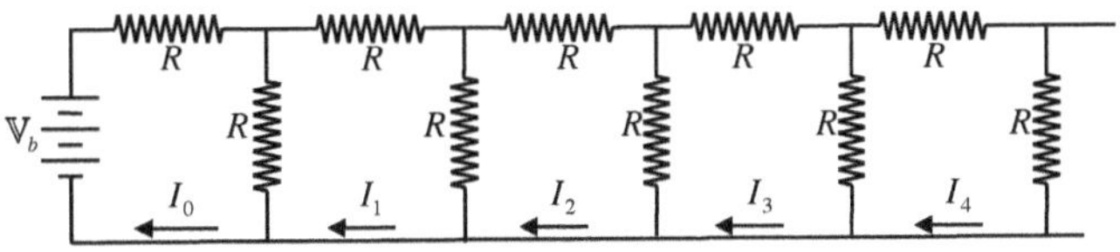

Consider the long chain of identical resistors in the diagram. From one end, you can measure the voltage V_b and current I_0.

a) Find the current I_1 in terms of V_b, each resistance R, and I_0.

b) Find the current I_2 also in terms of V_b, R, and I_0.

c) Find the current I_n in terms of I_{n-1} and I_{n-2}.

d) Find the current I_n in terms of V_b, R, I_0, and n.

e) Argue that the effective resistance $\left(\frac{V_b}{I_0}\right)$ must be between R and $2R$.

f) Calculate, and plot as a function of N, $\frac{V_b}{I_0 R}$ for a chain of $2N$ total resistors.

g) Find the effective resistance of an infinite resistor chain in terms of R.

Problem 3.14: An Infinite Resistor Ladder II

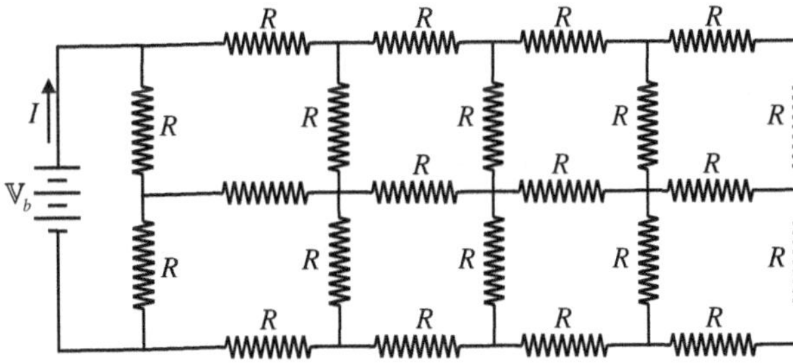

Consider the infinite chain of identical resistors in the diagram.

a) Given a constant voltage on the battery V_b, what is the current I through the circuit?

b) What is the overall effective resistance of the chain?

Discussion 3.6: Non-linear Electronic Components

The resistor is clearly an idealized linear component, and in practice carbon resistors have a constant resistance over a large range of applied voltages. On the other hand, if the voltage is too great across the resistor, the dissipated power will burn out the resistor. At this point, it is no longer linear. A common example of a non-linear system is an incandescent light bulb, because the conductivity of a metal depends on its temperature.

Here we will consider another example, the diode. An idealized diode is simply a one-way door, with the resistance given by:

$$R = \begin{cases} 0 & : \Delta V > 0 \\ \infty & : \Delta V < 0 \end{cases} .$$

This would mean that either no current flows through a diode, or there is no voltage drop, so either way no energy can be dissipated. Some diodes, however, emit light, which dissipates energy, making them far from ideal.

In the field of practical electronics it is also important to remember the distinction between fundamental physics and what is merely a convenient design specification of each electronic component. Ohm's law for a resistor, the capacitor relationship, and the diode rule, are design specifications, and require some knowledge of the materials involved. On the other hand, conservation of charge and energy are among the most fundamental laws of nature. As such, they must always apply—regardless of materials, temperatures, or other circumstances. Thus, in steady-state circuits, Kirchhoff's loop and junction rules must hold for all components, regardless of whether or not the manufacturer kept high standards.

Problem 3.15 An Incandescent Light Bulb

The voltage as a function of the current through a circuit need not be linear, as in an idealized resistor. A common example is the tungsten filament of an incandescent light bulb.

a) A typical North American 60 watt incandescent light bulb has a Tungsten double-coiled filament that is approximately 580 mm long and 220 μm in radius. Knowing that standard American wall voltage is 110 Volts, what is the conductivity, κ, of the tungsten filament?

b) Look up the conductivity of tungsten, how close are your results? Explain why the looked up conductivity is so different from the conductivity, κ, you found in part a.

c) All objects at a given temperature radiate light, called *blackbody radiation.* The total power radiated is related to the temperature via Stefan's law:

$$\mathbb{P} = \oint_{\text{surface}} \sigma T^4 \, dA \ ,$$

where $\sigma = 5.67 \times 10^{-8} \frac{\text{W}}{\text{m}^2\,\text{K}^4}$ is called the Stefan-Boltzmann constant. What is the temperature of the filament of the light bub from part (a)?

3.5 *Electrostatic Potential Energy*

Since the mathematical form of Coulomb's law is identical to Newton's law of universal gravitation, French and German natural philosophers could apply mathematics previously developed for understanding gravity to understanding electrostatics. Therefore, in this section, we begin with a review of gravitational potential energy, before we tackle the Coulomb potential.

There are two differences, however, between Newton's gravity and Coulomb's electrostatics: (1) Gravity is always attractive, so the maximum potential energy between two masses is always when they are furthest apart; and (2) Gravity is governed by the mass, which is the same quantity that measures inertia, so the motion of small objects become mass independent.

Thought Experiment 3.9: Universal Gravitational Potential Energy

We will now consider the universal law of gravitational potential energy, so we can better understand the implications of Coulomb's law. Rather than the one object being much less massive than the other, we will consider a system of two arbitrary objects, such as stars for example, interacting under the influence of their mutual gravitational attraction.

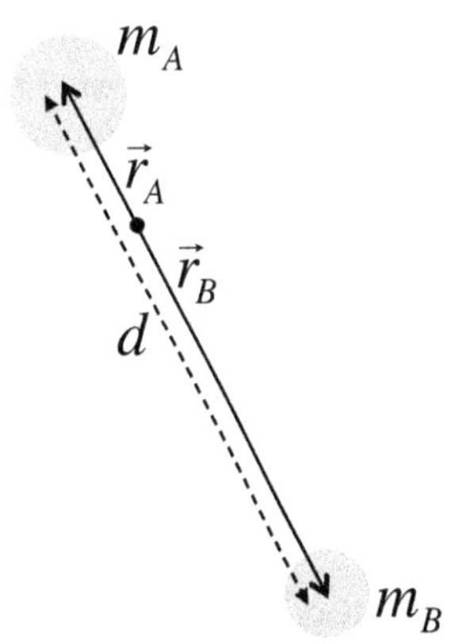

Consider two stars with masses, m_A and m_B, initially at rest and separated by a distance d_0 such that at time zero $d_0 = |\vec{r}_{A0} - \vec{r}_{B0}| = r_{A0} + r_{B0}$, where $\vec{r}$ represents the position measured from the center of mass. The center of mass is always between the two stars, so at any time we have that:

$$\frac{r_A}{d} = \frac{m_B}{m_A + m_B} \quad \text{and} \quad \frac{r_B}{d} = \frac{m_A}{m_A + m_B} \, .,$$

where d is the (variable) distance between the two stars. The gravitational force between them is:

$$\vec{F}_{\text{of B on A}} = -\frac{G m_A m_B}{d^2} \hat{r}_A \quad \text{and} \quad \vec{F}_{\text{of A on B}} = -\frac{G m_B m_A}{d^2} \hat{r}_B \, .$$

Notice that Newton's law of gravity conforms to Newton's third law. Thus:

$$\vec{F}_{\text{of B on A}} = -\frac{G m_A m_B}{d^2} \hat{r}_A = \frac{G m_B m_A}{d^2} \hat{r}_B = -\vec{F}_{\text{of A on B}} \, .$$

Let's calculate the work done on mass A by mass B as the separation between the objects decreases from d_0 to d:

$$\mathrm{W}_{\text{of B on A}} \equiv \int_{\vec{r}_{A0}}^{\vec{r}_A} \vec{F}_{\text{of B on A}} \cdot d\vec{r} = \int_{\vec{r}_{A0}}^{\vec{r}_A} \left(-\frac{G m_A m_B}{(r_A + r_B)^2} \hat{r}_A \right) \cdot d\vec{r}_A = -G m_A m_B \int_{r_{A0}}^{r_A} \left(\frac{dr_A}{(r_A + r_B)^2} \right) = G \frac{m_A m_B^2}{m_A + m_B} \left(\frac{1}{d} - \frac{1}{d_0} \right) .$$

And, the system is symmetrical under the swapping of A and B, so:

$$\mathrm{W}_{\text{of A on B}} = G \frac{m_A^2 m_B}{m_A + m_B} \left(\frac{1}{d} - \frac{1}{d_0} \right) .$$

Thus the total work done, on the whole system (both A and B) is therefore:

$$\mathrm{W} = G \frac{m_A m_B^2}{m_A + m_B} \left(\frac{1}{d} - \frac{1}{d_0} \right) + G \frac{m_A^2 m_B}{m_A + m_B} \left(\frac{1}{d} - \frac{1}{d_0} \right) = \frac{G m_A m_B}{d} - \frac{G m_A m_B}{d_0} .$$

Notice that work was done on the system, despite the absence of external forces. Notice something else: the total work done depends only on the initial and final relative distance between the two stars, not on the direction or the details of the integration.

We understood that work was the transfer of energy from one system to another, but that is clearly not the case here. We have a single isolated system, yet the total work done on this system is much greater than zero. How can this be?

The answer to this question comes only if we reframe the problem. Rather than consider the work done by this conservative force, we account for it via energy concepts. We now define the difference in gravitational potential energy of the system (both A and B) to be the work that an external force would need to do in order make this particular change in state. In other words:

$$\Delta\mathbb{U} = -\mathbb{W} = -\left(\frac{G m_A m_B}{d} - \frac{G m_A m_B}{d_0}\right) = \left(-\frac{G m_A m_B}{d}\right) - \left(-\frac{G m_A m_B}{d_0}\right).$$

Notice, also, that we have freedom to set an arbitrary zero point, even as $d \to \infty$, so we then define the gravitation potential energy as:

$$\mathbb{U}(d) = -\frac{G m_A m_B}{d} \ .$$

This is the familiar relationship from classical mechanics with which we can account for the total energy of the system. As such, we do not have to worry about the details of the interactions in order to find the kinetic energy of the system as a function of distance. Instead, we have only to invoke conservation of energy.

Problem 3.16: Universal Electrostatic Potential Energy

Show from Coulomb's law that the electrostatic potential energy between two point-like charges, q_1 and q_2 is:

$$\mathbb{U}(d) = \frac{K q_1 q_2}{d},$$

where d is the separation distance and K is Coulomb's constant.

Example 3.6: The Impact of Two Astronomical Objects

The current explanation for the existence of our moon, the tilt of the earth's axis, and the original location of the continents, is that a Mars-sized object impacted the earth about four and a half billion years ago. Let us find the total energy released, if we make the assumption that the object eventually became part of the earth and moon system.

Assume that there were two objects initially, one of which was the impacting body, of mass m_I (I for impactor), and the other being the earth, whose mass we will designate as $M_\oplus$. We also assume that the impacting object came from a long distance away from Earth, so the initial energy of the system was:

$$\mathbb{E}_0 = \mathbb{T}_0 + \mathbb{U}_0 = 0 - \frac{G m_I M_{0\oplus}}{\infty} = 0 \ .$$

At the end, we also have two objects, the moon and the earth, so we can write the final energy of the system at any time after the collision as:

$$\mathbb{E}_f = \mathbb{T}_f + \mathbb{U}_f = \tfrac{1}{2} m_{\text{moon}} v_{\text{moon}}^2 + \tfrac{1}{2} m_\oplus v_\oplus^2 - \frac{G m_{\text{moon}} M_\oplus}{d} \ .$$

We know about the earth and moon now, so we will calculate the final energy. Using modern values for these quantities, gives us the current energy of the Earth moon system of:

$$\mathbb{E}_f = \tfrac{1}{2}\left(7.34\times10^{22}\,\text{kg}\right)\left(1{,}076\tfrac{\text{m}}{\text{s}}\right)^2 + \tfrac{1}{2}\left(5.97\times10^{24}\,\text{kg}\right)\left(13.2\tfrac{\text{m}}{\text{s}}\right)^2 - \frac{\left(6.67\times10^{-11}\,\text{J}\cdot\text{m}/\text{kg}^2\right)\left(7.34\times10^{22}\,\text{kg}\right)\left(5.97\times10^{24}\,\text{kg}\right)}{\left(3.63\times10^{8}\,\text{m}\right)}.$$

So, $\mathbb{E}_f = 4.25\times10^{28}\,\text{J} + 5.21\times10^{26}\,\text{J} - 8.05\times10^{28}\,\text{J} = -3.75\times10^{28}\,\text{J} \approx -4\times10^{28}\,\text{J}$.

Therefore, the total non-conservative work done on the system must have been:

$$\mathbb{W} = \mathbb{E}_f - \mathbb{E}_0 = -4\times10^{28}\,\text{J}.$$

The total energy that went into deforming, and heating, the earth must have been about 4×10^{28} J. This is approximately the amount of energy the sun emits in a couple of minutes, or the energy given off by about ten trillion modern nuclear warheads.

Problem 3.17: Skew Line Charges

Consider two long cylindrical line charges, each with charge per length λ and radius a. They are aligned in a *skew* configuration, so that they are perpendicular but do not cross, with a closest approach b.

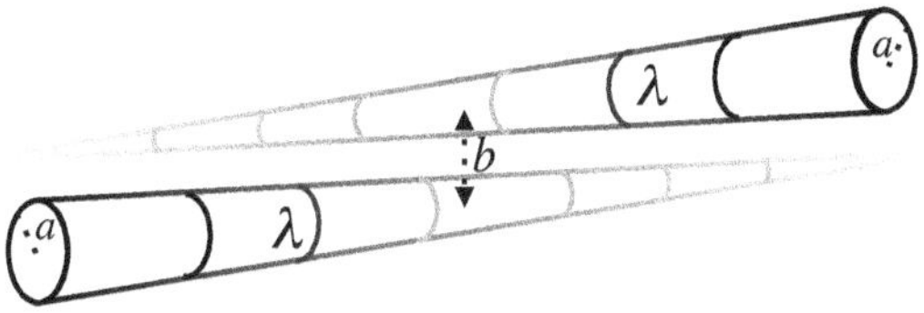

a) Find the mutual potential energy of the two line charges. You may make the assumption that $b \gg a$.

b) What is the mutual force of repulsion between the two rods?

Example 3.7: The Self-Gravity of the Sun

In this example we calculate the gravitational potential energy of a constant density sphere with a total mass M and a radius R.

We begin by imagining the sun was formed as a spherical mass that was built up layer by layer over time. Whether it actually happened this way is immaterial, because potential energy is path independent, so whatever any historical scenario we assume, so long as it is conservative, will result in the same current energy state.

Consider a spherical mass m, and imagine wrapping it with a small layer of atmosphere of mass δm brought from very far away. The potential energy of the system decreased, as gravity is an attractive force, so the change in potential energy due to the new layer of mass is:

$$\delta\mathbb{U} = -Gm\delta m\left(\frac{1}{r} - \frac{1}{\infty}\right) = -G\frac{m\delta m}{r}.$$

Now, we are assuming a constant density, so the mass and volume must be directly proportional to one another. That means that the mass ratio of the forming to final sun is given by:

$$\frac{m}{M} = \frac{\cancel{\tfrac{4\pi}{3}}r^3}{\cancel{\tfrac{4\pi}{3}}R^3} = \left(\frac{r}{R}\right)^3.$$

The change in potential energy due to the addition of mass δm can therefore be written as:

$$\delta\mathbb{U} = -G\frac{m\delta m}{r} = -Gm\delta m\sqrt[3]{\frac{M}{m}}\frac{1}{R} = -\frac{GM^{\frac{1}{3}}}{R}m^{\frac{2}{3}}\delta m\,.$$

We can integrate this expression to find the total potential energy that was released during our hypothetical formation of the sun. So:

$$\mathbb{U} = -\frac{GM^{\frac{1}{3}}}{R}\int_0^M m^{\frac{2}{3}}\,\mathrm{d}m = -\frac{GM^{\frac{1}{3}}}{R}\left(\tfrac{3}{5}M^{\frac{5}{3}} - 0\right) = -\frac{3GM^2}{5R}\,.$$

This is the total energy of formation of a constant density object of mass M and radius R. Most astronomical objects, however, are denser toward the center than toward the outside.

We expect to get similar results by assuming a more realistic density profile. To see this, imagine replace the constant density in the previous analysis with one given by the profile $\rho \propto e^{-\left(\frac{r}{R}\right)^3}$. Integrating this from zero to r gives us:

$$m \propto \int_0^r 4\pi r^2 e^{-\left(\frac{r}{R}\right)^3}\mathrm{d}r \propto \int_0^r r^2 e^{-\left(\frac{r}{R}\right)^3}\mathrm{d}r \propto r^3\left(1 - e^{-\left(\frac{r}{R}\right)^3}\right),$$

and normalizing this gives us:

$$\frac{m}{M} = \frac{r^3\left(1 - e^{-\left(\frac{r}{R}\right)^3}\right)}{R^3\left(1 - e^{-\left(\frac{R}{R}\right)^3}\right)} = \frac{e}{e-1}\left(\frac{r}{R}\right)^3\left(1 - e^{-\left(\frac{r}{R}\right)^3}\right) \approx 1.6\left(\frac{r}{R}\right)^3\left(1 - e^{-\left(\frac{r}{R}\right)^3}\right)$$

This function is more difficult to invert, so we will integrate with respect to r instead of m:

$$\delta\mathbb{U} = -G\frac{m\delta m}{r} = -\frac{G}{r}m\frac{\mathrm{d}m}{\mathrm{d}r}\delta r\,.$$

And so we can substitute for the mass and the mass gradient:

$$m = M\left(\tfrac{e}{e-1}\right)\left(\frac{r}{R}\right)^3\left(1 - e^{-\left(\frac{r}{R}\right)^3}\right)$$

$$\frac{\mathrm{d}m}{\mathrm{d}r} = M\left(\tfrac{e}{e-1}\right)\left(3\left(\frac{r^2}{R^3}\right)\left(1 - e^{-\left(\frac{r}{R}\right)^3}\right) + \left(\frac{r}{R}\right)^3\left(3\frac{r^2}{R^3}e^{-\left(\frac{r}{R}\right)^3}\right)\right) = M\left(\tfrac{e}{e-1}\right)3\left(\frac{r^2}{R^3}\right)$$

$$\delta\mathbb{U} = -\frac{G}{r}m\frac{\mathrm{d}m}{\mathrm{d}r}\delta r = -G\frac{M^2}{R}\left(3\left(\tfrac{e}{e-1}\right)^2\right)\left(\frac{r}{R}\right)^4\left(1 - e^{-\left(\frac{r}{R}\right)^3}\right)\frac{\delta r}{R}$$

Next we make a unitless variable substitution, where $\eta \equiv \frac{r}{R}$, so:

$$\mathbb{U}_\odot = -\frac{GM^2}{R}\int_0^1\left(3\left(\tfrac{e}{e-1}\right)^2\right)\eta^4\left(1 - e^{-\eta^3}\right)\mathrm{d}\eta \approx -0.67\frac{GM^2}{R}\,.$$

The 0.67 is slightly higher than the $\frac{3}{5} = 0.6$ from the constant density model, so as the density profile becomes more centralized the potential energy decreases (i.e. becomes more negative).

Assuming that this second scenario represents the sun, the total gravitational potential energy released from its formation would have been:

$$\mathbb{U}_{\odot} = 0.67\frac{GM^2}{R} \approx 2.5\times10^{41}\,\text{J}\,.$$

In Chapter 3 we found that the current bolometric luminosity of the sun is $\mathbb{P}_{\odot} = \left(1361\tfrac{\text{W}}{\text{m}^2}\right)4\pi\left(1.50\times10^{11}\text{m}\right)^2 = 3.84\times10^{26}\,\text{W}$. William Thomson (later Lord Kelvin) considered every energy source imaginable, and he found that the sun's self-gravity provided the most energy for the sun's heating. Thus, he argued in 1862, that the age of the sun must be:

$$t = \int_0^{\mathbb{U}}\left(\frac{dt}{d\mathbb{U}}\right)d\mathbb{U} = \int_0^{\mathbb{U}}\frac{d\mathbb{U}}{(-\mathbb{P})} \approx -\frac{\mathbb{U}}{\mathbb{P}} \approx -\frac{2.5\times10^{41}\,\text{J}}{4\times10^{26}\,\text{W}} \approx 7\times10^{14}\,\text{s} \approx 2\times10^{7}\,\text{years},$$

or as he put it:

> It seems, therefore, on the whole most probable that the sun has not illuminated the earth for 100,000,000 years, and almost certain that he has not done so for 500,000,000 years. As for the future, we may say, with equal certainty, that inhabitants of the earth cannot continue to enjoy the light and heat essential to their life, for many million years longer, unless sources now unknown to us are prepared in the great storehouse of creation.[52]

William Thomson's conclusion contradicted the age of the earth, and life on it, which was concurrently being determined by biologists and geologists of the day. At the time, the validity of Thomson's argument above, and that of a similar result he obtained by estimating the time it takes for the Earth to cool, seemed beyond reproach. In both cases, the solution was that differences in the binding energy of the nuclei of atoms was the source in "the great storehouse of creation," a fact that was totally unknown (and unknowable) to Thomson.

Problem 3.18: Building up a Sphere of Charge

Consider making a sphere of constant charge density, from far away charged material. This will take work in order to squeeze the like charges together. By calculating the work one would do to make a sphere of charge Q and radius a, you can calculate the electrostatic potential energy in the sphere $U(Q,a)$.

a) Clearly show that $U(Q,a) = \frac{3}{5}\frac{KQ^2}{a}$..

b) Once you have made this sphere, what is the electric field $\vec{E}(r)$ inside the sphere?

c) In the creation of Neutron star, 1.4 solar masses ($3\times10^{30}\,\text{kg}$) of material collapses from a radius of about 5000 km down to about 12 km. How much gravitational potential energy is released in such an implosion?

Example 3.8: The Electrostatic Self-Energy of a Disk

Now we will consider making a disk of uniform charge density with total charge Q, radius a, and thickness h such that $h \ll a$. This disk would be very unstable unless there is some very strong glue holding it together. So we ask, how much work did it take to make the disk in the first place? Or, similarly, if whatever forces were holding it together suddenly let go, how much energy would be released in the explosion?

---

52 W. Thomson, "On the Age of the Sun's Heat," Macmillan's Magazine, **5**, (1862), 388-393.

To simplify this problem, we will ignore edge effects on the disk and assume that the electric field is the same as in the middle of a circular disk (p. 64). We will imagine building up this disk layer by layer, where each layer adds an additional charge δQ. Thus, the amount of work done to add an additional layer of charge must be:

$$\delta W = -\int_{r=\infty}^{z=h} \delta Q\,\vec{E}(\vec{r})\cdot d\vec{r} = -\delta Q \int_{\infty}^{h} \left(2\pi K\sigma\left(1-\frac{z}{\sqrt{a^2+z^2}}\right)(-\hat{z})\right)\cdot(\hat{z}\,dz)$$

$$= \delta Q(2\pi K\sigma)\int_{\infty}^{h}\left(1-\frac{z}{\sqrt{a^2+z^2}}\right)dz = \delta Q(2\pi K\sigma)\left(z-\sqrt{a^2-z^2}\right)\Big|_{z=\infty}^{z=h}.$$

Since the disk has a constant volume charge density, each additional layer of charge with thickness δh will have charge $\delta Q = \rho\pi a^2\delta h$. Thus the work required to add the layer will be:

$$\delta W = \left(\rho\pi a^2\delta h\right)(2\pi K\sigma)\left(h-\sqrt{a^2-h^2}\right).$$

The surface charge of the existing disk, σ, is given by:

$$\sigma = \frac{Q}{\pi a^2} = \frac{Qh}{\pi a^2 h} = \rho h,$$

so in terms of the thickness h, additional thickness δh, and values independent of h, the additional work required for each layer is:

$$\delta W = \left(\rho\pi a^2\delta h\right)(2\pi K\rho h)\left(h-\sqrt{a^2-h^2}\right) = 2\pi^2 K\rho^2 a^2\left(h^2-h\sqrt{a^2-h^2}\right)\delta h.$$

Finally the total work required to make the disk must be:

$$W = 2\pi^2 K\rho^2 a^2\int_0^h\left(h^2-h\sqrt{a^2-h^2}\right)dh = \tfrac{2}{3}\pi^2 K\rho^2 a^2\left(\left(\sqrt{a^2-h^2}\right)^3+h^3\right).$$

Recall that we justified our ignoring of edge effects by assuming a thin disk, in the limit that $a \gg h$ the work required to assemble a disk of charge Q, radius a, and thickness h becomes:

$$W \approx \tfrac{2}{3}\pi^2 K\rho^2 a^5 = \left(\frac{Q}{\pi a^2 h}\right)^2\frac{2\pi^2 K a^5}{3} = \frac{Q^2 2\cancel{\pi^2} K a^5}{3\cancel{\pi^2} a^4 h^2} = \frac{2KQ^2 a}{3h^2}.$$

We can now reinterpret this result in light of Example 3.7 above. In that example we used the potential energy of a small object to build up the build up a sphere the size of the sun layer by layer. In this example, we did the same thing for a disk, but we calculated the work explicitly from the electrostatic force.

3.6 *The Coulomb Potential*

Here we combine the concepts of voltage, potential energy, and Coulomb's law. This is a very powerful technique, not only for finding the electric potential due to point charges, or a collection thereof, but also for continuously charged media.

Definition 3.4: The Coulomb Potential

The Coulomb potential is the potential energy, per charge, due to a single point charge. As is the case of gravity, it is defined to be zero at a distance of infinity. Thus, the Coulomb potential due to a positive charge is always positive, and that due to a negative charge is negative. Applying your result from Problem 3.16 (p. 98), the Coulomb potential is:

$$\mathbb{V}(\vec{r}) \equiv \lim_{\delta q \to 0} \frac{\mathbb{U}_E(\vec{r})}{\delta q} = \lim_{\delta q \to 0} \frac{K Q \cancel{\delta q}}{\cancel{\delta q}\, r} = \frac{K Q}{r}.$$

As with Coulomb's law, the potentials due to different charges add. Unlike Coulomb's law, the potential is a scalar quantity, so there is no need for vector addition.

Problem 3.19: The Lattice Energy of Table Salt

When salt dissolves in water, it gives up some potential energy, which is measured by chemists to be $\Delta\mathbb{U} \approx -786\,\frac{\text{kJ}}{\text{mol}} = -8.15\,\text{eV}$. In 1918 Max Born and Alfred Landé published a model to calculate lattice energy, and the point of this exercise is for you to both derive and use what is now known as the Born-Landé equation.

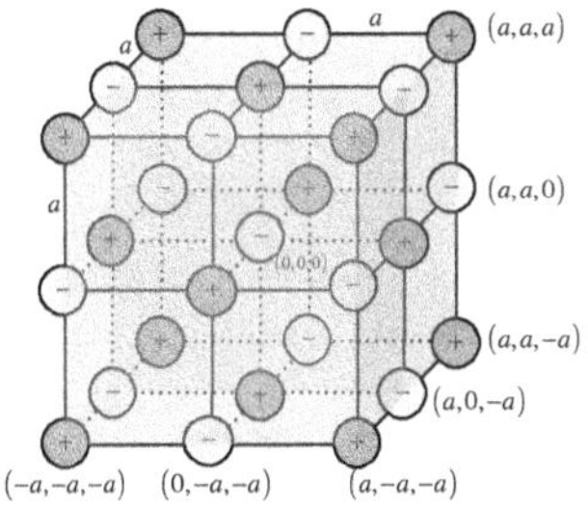

a) Consider the molecule NaCl, which forms a cubic lattice structure, as shown. Assigning the origin to one of the negative ions, show that the electric potential of the ion at the origin, due to all the other ions, can be given by the following relationship:

$$\mathbb{V}_1 = \frac{K e}{a} \sum_i \sum_j \sum_k \frac{(-1)^{i+j+k}}{\sqrt{i^2 + j^2 + k^2}} = \frac{K e}{a} M,$$

where i, j and k, span all integers except where $i = j = k = 0$. The unitless sum, M, is called the Madelung constant after the German physicist Erwin Madelung.

b) Born and Landé suggested that there was another potential energy having to do with quantum mechanical effects. This, they suggested, should be given by an inverse power law, which for NaCl is:

$$\mathbb{U}_2 \propto \frac{1}{a^8}.$$

Show that when the total potential energy is minimized, it is given by the Borne-Landé equation, which for NaCl is:

$$\mathbb{U}_{\min} = -\frac{7}{8} \frac{K M e^2}{a_{\min}}.$$

c) Find the Madelung constant, M, using the measured value above for the lattice energy, and the independently measured lattice spacing of: $a = 564\,\text{pm} = 5.64\,\text{Å}$.

d) Notice that each nearby ion is closer to the origin, so each one exerts more influence over the potential. However, there are more ions that are far away, so it is not obvious which will have more influence.

Thus, if a cube whose nearest face is a distance na contributes m_n towards the Madelung constant, then the Madelung constant would be given by:

$$M = \sum_{n=1}^{\infty} m_n \,.$$

Find the first contribution, m_1, which would be the Madelung constant due to the just the ions shown in the figure.

e) Find m_n as a function of the n.

f) Use a computer to calculate the Madelung constant for NaCl. Plot the current Madelung constant, after each iteration, as a function of n the grid size. Also plot a horizontal line representing your result in part c? Does your program converge? Does it appear to be converging to the measured value, or at least oscillating around it? Run it for as many iterations as you believe to be practical.

Thought Experiment 3.10: The Monopole Electric Potential

Here we derive the Coulomb potential, again, using the *gradient*. In the case of a point source at the origin, Coulomb's law gives the electric field:

$$\vec{E} = \frac{KQ}{r^2}\hat{r} = -\vec{\nabla}\mathrm{V} = -\frac{\partial \mathrm{V}}{\partial r}\hat{r} \,.$$

By collecting terms, we find that:

$$\frac{\mathrm{dV}(r)}{\mathrm{d}r} = -\frac{KQ}{r^2} \,.$$

Notice that the charge density is infinite at the location of the point charge, but we are interested only in knowing the field at some finite distance r. As in the gravitational analogies, we set the potential to zero at infinity:

$$\mathrm{V}(r) = \int_0^{\mathrm{V}(r)} \mathrm{dV} = \int_{\infty}^{r} \left(\frac{-KQ}{r^2}\right) \mathrm{d}r = KQ\frac{1}{r}\bigg|_{\infty}^{r} = KQ\left(\frac{1}{r} - \frac{1}{\infty}\right) = \frac{KQ}{r} \,.$$

This is a powerful result, because the potential of a collection of point charges is simply given by the scalar sum of the potential of each charge.

To check our work, we reproduce Coulomb's law as:

$$\vec{E} = -\vec{\nabla}\mathrm{V}(r) = -\frac{\partial}{\partial r}\left(\frac{KQ}{r}\right)\hat{r} = \frac{KQ}{r^2}\hat{r} \,.$$

Thought Experiment 3.11: The Electric Dipole

Now we will consider the potential at a point of interest that is at distances $r_\pm \equiv |\vec{r}_\pm|$ from equal and opposite point charges $\pm Q$:

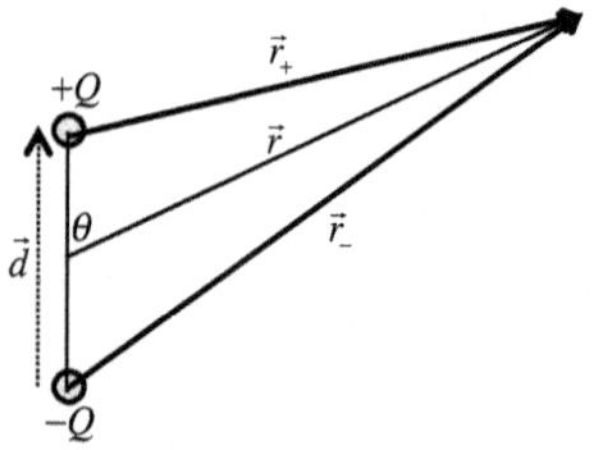

$$\mathrm{V}(\vec{r}) = \frac{KQ}{r_+} + \frac{-KQ}{r_-} = KQ\left(\frac{1}{r_+} - \frac{1}{r_-}\right).$$

Now we write the square of the distances $r_\pm$ in terms of the position $\vec{r}$ and the dipole separation vector $\vec{d}$:

$$r_\pm^2 = \vec{r}_+ \cdot \vec{r}_+ = \left(\vec{r} \mp \tfrac{1}{2}\vec{d}\right)\cdot\left(\vec{r} \mp \tfrac{1}{2}\vec{d}\right) = r^2 \mp \vec{r}\cdot\vec{d} + \left(\tfrac{1}{2}d\right)^2 = r^2 + \tfrac{1}{4}d^2 \mp r\,d\cos(\theta)$$

But, what we are interested in is $r_+^{-1} - r_-^{-1}$, so we write this in terms of the unitless ratios:

$$\frac{r}{r_\pm} = \left(1 + \left(\frac{d}{2r}\right)^2 \mp \frac{d\cos(\theta)}{r}\right)^{-\frac{1}{2}} = \left(1 + x_\pm\right)^{-\frac{1}{2}}, \text{ where } x_\pm \equiv \left(\frac{d}{2r}\right)^2 \mp \frac{d\cos(\theta)}{r}.$$

Since this is small if $r \gg d$, we can Taylor expand it as:

$$\left(1 + x_\pm\right)^{-\frac{1}{2}} \approx 1 - \tfrac{1}{2}x_\pm + \tfrac{3}{8}x_\pm^2 + \ldots,$$

so the difference $r_+^{-1} - r_-^{-1}$ can be written to first order as:

$$\frac{1}{r_+} - \frac{1}{r_-} = r\left(\left(1 + x_+\right)^{-\frac{1}{2}} - \left(1 + x_-\right)^{-\frac{1}{2}}\right) \approx r\left(\left(1 - \tfrac{1}{2}x_+\right) - \left(1 - \tfrac{1}{2}x_-\right)\right) \approx -\tfrac{1}{2}r\left(x_+ - x_-\right)$$

$$\approx -\tfrac{1}{2}r\left(\left(\left(\frac{d}{2r}\right)^2 - \frac{d\cos(\theta)}{r}\right) - \left(\left(\frac{d}{2r}\right)^2 + \frac{d\cos(\theta)}{r}\right)\right) \approx 2d\cos(\theta).$$

Now that we have done this mathematical prep-work, we can return to finding the electric potential in the far limit as:

$$\mathbb{V}(\vec{r}) \approx KQ\frac{d\cos(\theta)}{r^2}.$$

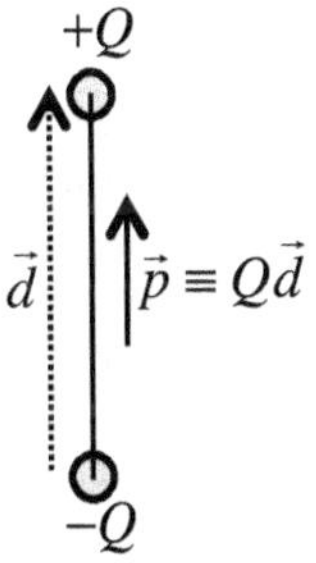

Defining the *electric dipole moment* as $\vec{p} \equiv Q\vec{d}$, we have:

$$\mathbb{V}(\vec{r}) \approx K\frac{p\cos(\theta)}{r^2} = K\frac{\vec{p}\cdot\hat{r}}{r^2}.$$

Note that the dipole potential is proportional to the inverse square of the distance, but unlike the potential for a point charge which varies simply as the inverse distance.

Now that we have an expression for the potential of a dipole, we can take its gradient to find an expression for the electric field in any coordinate system of choice. In this instance, it is most straightforward to employ spherical coordinates:

$$\vec{E} = -\vec{\nabla}\mathbb{V}(\vec{r}) = -\frac{\partial \mathbb{V}}{r}\hat{r} - \frac{1}{r}\frac{\partial \mathbb{V}}{\partial\theta}\hat{\theta} \approx \frac{Kp}{r^3}\left(2\cos(\theta)\hat{r} + \sin(\theta)\hat{\theta}\right).$$

The gradient could also have been found in coordinate-free vector form using the identity for the gradient of the dot product of two vectors:

$$\vec{E} = -\vec{\nabla}\frac{K\vec{p}\cdot\hat{r}}{r^2} = -K\left(\vec{p}\times\left(\cancel{\vec{\nabla}\times\frac{\hat{r}}{r^2}}\right) + \left(\vec{p}\cdot\vec{\nabla}\right)\frac{\hat{r}}{r^2}\right) = -K\left(\vec{p}\cdot\vec{\nabla}\right)\frac{\vec{r}}{r^3}.$$

To see how this works, using the product and chain rules:

$$\vec{E} \approx -K\left(\vec{p}\cdot\vec{\nabla}\right)\left(r^{-3}\vec{r}\right) \approx -K\left(\vec{r}\left(\vec{p}\cdot\vec{\nabla}\right)r^{-3} + r^{-3}\left(\vec{p}\cdot\vec{\nabla}\right)\vec{r}\right)$$

$$\approx -K\left(-3r^{-4}\vec{r}\left(\vec{p}\cdot\vec{\nabla}\right)r + r^{-3}\left(\vec{p}\cdot\vec{\nabla}\right)\vec{r}\right) \approx \frac{K}{r^3}\left(3\frac{\vec{r}}{r}\left(\vec{p}\cdot\vec{\nabla}\right)r - \left(\vec{p}\cdot\vec{\nabla}\right)\vec{r}\right).$$

If $\vec{p} = p\hat{z}$ then the operator:

$$\left(\vec{p}\cdot\vec{\nabla}\right) = p\left(\hat{z}\cdot\vec{\nabla}\right) = p\tfrac{\partial}{\partial z},$$

thus we will investigate the two derivatives:

$$\frac{\partial r}{\partial z} = \frac{\partial}{\partial z}\left(\sqrt{s^2+z^2}\right) = \frac{z}{\sqrt{s^2+z^2}} = \frac{z}{r} \quad , \quad \frac{\partial \vec{r}}{\partial z} = \frac{\partial}{\partial z}\left(s\hat{s} + z\hat{z}\right) = \hat{z},$$

so the electric field in cylindrical coordinates becomes:

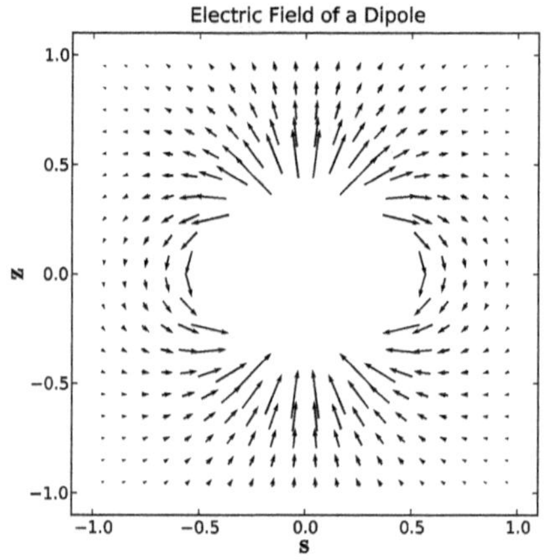

$$\vec{E} \approx K\frac{p}{r^3}\left(3\frac{\vec{r}}{r}\frac{z}{r} - \hat{z}\right) \approx K\frac{p}{r^3}\left(\frac{3z\vec{r} - r^2\hat{z}}{r^2}\right).$$

Evaluating the terms in the numerator of the final term on the right yields:

$$\vec{E} \approx K\frac{p}{r^3}\left(\frac{3sz}{r^2}\hat{s} + \frac{2z^2 - s^2}{r^2}\hat{z}\right).$$

Note that field of a dipole falls off more severely with increasing distance than does the field of a single point charge. The potential and field of a collection of charges can be approximated by the sum of a monopole term and a dipole term. We will generalize this idea later, by showing that the electric potential of an arbitrary charge distribution can be expressed as an infinite sum of multipoles.

Thought Experiment 3.12: The Electric Quadrupole

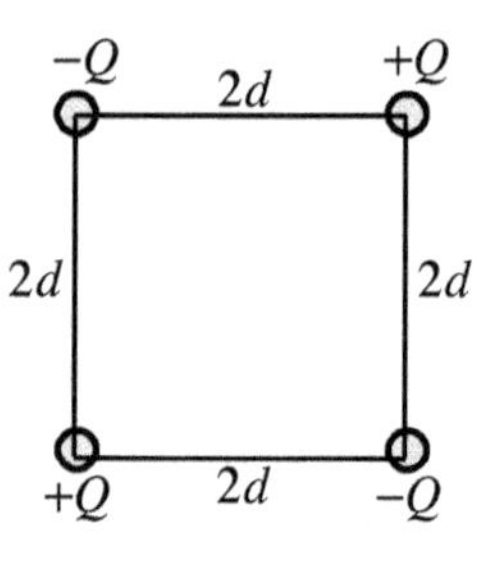

An electric quadruple is a neutral collection of charges with no net dipole moment. Since the monopole moment is the overall charge, and the dipole moment is a vector, we simply align two dipoles in opposite directions as in the figure. With our experience from the electric dipole, we will make the spacing between charges $2d$, as we expect the algebra simpler that way.

Let us label each charge with a number, and use Cartesian coordinates to write out the position vectors of the charges, r_1, r_2, r_3, and r_4 as well as the position vector $\vec{r}$ to our point of interest. We choose the origin to be at the center of the square, and obtain the following:

$$\vec{r} = x\hat{x} + y\hat{y} + z\hat{z}$$
$$\vec{r}_1 = (x-d)\hat{x} + (y-d)\hat{y} + z\hat{z}$$
$$\vec{r}_2 = (x-d)\hat{x} + (y+d)\hat{y} + z\hat{z}$$
$$\vec{r}_3 = (x+d)\hat{x} + (y+d)\hat{y} + z\hat{z}$$
$$\vec{r}_4 = (x+d)\hat{x} + (y-d)\hat{y} + z\hat{z}$$

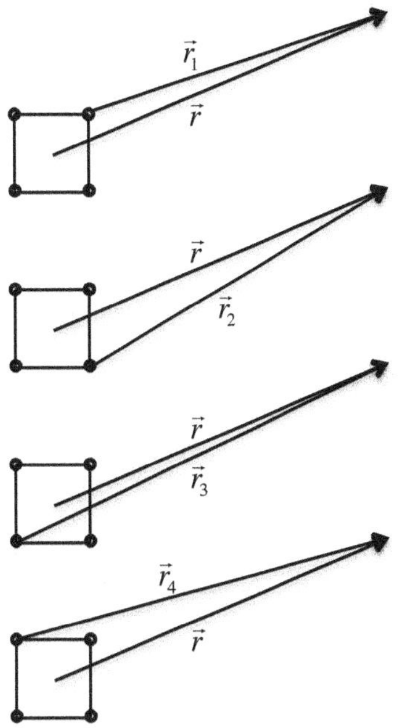

The electric potential due to a quadrupole is therefore:

$$\mathbb{V}(\vec{r}) = \frac{KQ}{r_1} - \frac{KQ}{r_2} + \frac{KQ}{r_3} - \frac{KQ}{r_4} = \frac{KQ}{r}\left(\frac{r}{r_1} - \frac{r}{r_2} + \frac{r}{r_3} - \frac{r}{r_4}\right).$$

Now let us look at one term of the potential, and rewrite it in terms of Cartesian coordinates:

$$\frac{r}{r_1} = \frac{r}{\sqrt{(x-d)^2 + (y-d)^2 + z^2}} = \frac{r}{\sqrt{x^2 + y^2 + z^2 - 2xd - 2yd + 2d^2}}$$
$$= \frac{1}{\sqrt{1 - \frac{2xd}{r^2} - \frac{2yd}{r^2} + \frac{2d^2}{r^2}}} \approx \frac{1}{\sqrt{1 + \frac{2d}{r^2}(-x-y)}}.$$

Applying a Taylor's expansion as we did with the dipole:

$$\frac{r}{r_1} \approx 1 - \frac{d}{r^2}(-x-y) + \frac{3d^2}{2r^4}(-x-y)^2.$$

By repeating the same analysis for the other three charges, we find, the approximate distance ratio for each charge:

$$\frac{r}{r_1} \approx 1 + \frac{d}{r^2}(x+y) + \frac{3d^2}{2r^4}(x^2 + y^2 + 2xy) \quad \frac{r}{r_2} \approx 1 + \frac{d}{r^2}(x-y) + \frac{3d^2}{2r^4}(x^2 + y^2 - 2xy)$$
$$\frac{r}{r_3} \approx 1 - \frac{d}{r^2}(x+y) + \frac{3d^2}{2r^4}(x^2 + y^2 + 2xy) \quad \frac{r}{r_4} \approx 1 - \frac{d}{r^2}(x-y) + \frac{3d^2}{2r^4}(x^2 + y^2 - 2xy)$$

Using these approximate results allows us to express the electric potential in a useful form:

$$\mathbb{V}(\vec{r}) = \frac{KQ}{r}\left(\frac{r}{r_1} - \frac{r}{r_2} + \frac{r}{r_3} - \frac{r}{r_4}\right) \approx KQ\frac{12d^2xy}{r^5}.$$

We then calculate the electric field by taking the gradient:

$$\vec{E} = \vec{\nabla}\mathbb{V}(\vec{r}) \approx \vec{\nabla}\left(KQ\frac{12d^2xy}{r^5}\right) = 12KQd^2\vec{\nabla}\left(\frac{xy}{r^5}\right)$$
$$= 12KQd^2\left[xy\vec{\nabla}\left(r^{-5}\right) + \frac{\vec{\nabla}(xy)}{r^5}\right],$$

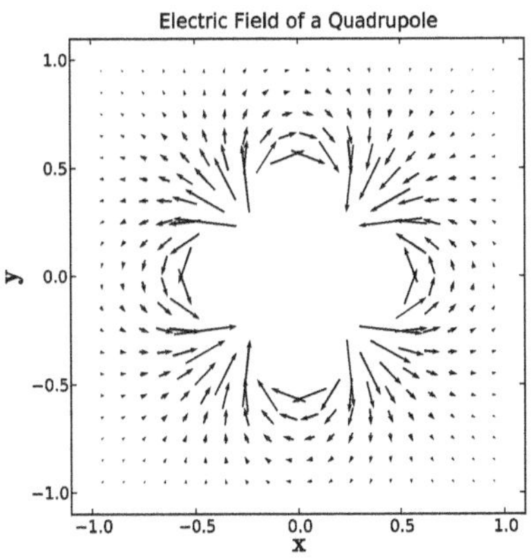

which yields,

$$\vec{E} \approx 12KQ\frac{d^2}{r^4}\left(\left(\frac{y}{r} - \frac{5x^2y}{r^3}\right)\hat{x} + \left(\frac{x}{r} - \frac{5xy^2}{r^3}\right)\hat{y} - \frac{5xyz}{r^3}\hat{z}\right).$$

Definition 3.5: The Quadrupole Moment Tensor

In the case of a single point charge, the magnitude of the charge is the only parameter needed to describe the far electric field, so a scalar quantity is sufficient to describe a monopole. A dipole has a particular direction, however, so we need a vector to describe it. A quadrupole moment needs yet more parameters. It is described by a *second order tensor* which can be represented by a 3x3 matrix. As we have already taken into account both the monopole and dipole terms, this tensor is traceless, as you can check by summing the diagonal elements below.

The quadrupole moment is a tensor, $\ddot{\mathbf{Q}}$, which yields a potential such that:

$$\mathrm{V}(\vec{r}) = \frac{K}{r^5}\left(\tfrac{1}{2}\vec{r}\cdot\ddot{\mathbf{Q}}\cdot\vec{r}\right) = \frac{K}{r^3}\left(\tfrac{1}{2}\hat{r}\cdot\ddot{\mathbf{Q}}\cdot\hat{r}\right).$$

From the result of Thought Experiment 3.12 (p. 106), we see that:

$$\ddot{\mathbf{Q}} = 12Qd^2\begin{pmatrix} 0 & 1 & 0 \\ 1 & 0 & 0 \\ 0 & 0 & 0 \end{pmatrix}.$$

We can represent the quadrupole moment tensor for a collection of n discrete charges located at positions $\vec{R}_i = X_i\hat{x} + Y_i\hat{y} + Z_i\hat{z}$ with the following matrix:

$$\ddot{\mathbf{Q}} = \begin{pmatrix} \sum_{1=1}^{n} Q_i\left(2X^2_i - Y_i^2 - Z_i^2\right) & 3\sum_{1=1}^{n} Q_iX_iY_i & 3\sum_{1=1}^{n} Q_iZ_iX_i \\ 3\sum_{1=1}^{n} Q_iX_iY_i & \sum_{1=1}^{n} Q_i\left(2Y^2_i - Z_i^2 - X_i^2\right) & 3\sum_{1=1}^{n} Q_iY_iZ_i \\ 3\sum_{1=1}^{n} Q_iZ_iX_i & 3\sum_{1=1}^{n} Q_iY_iZ_i & \sum_{1=1}^{n} Q_i\left(2Z^2_i - X_i^2 - Y_i^2\right) \end{pmatrix}$$

The electric field is the gradient of the potential and may therefore be expressed as:

$$\vec{E} = \vec{\nabla}\mathrm{V}(\vec{r}) = K\vec{\nabla}\left(\frac{\frac{1}{2}\vec{r}\cdot\ddot{\mathbf{Q}}\cdot\vec{r}}{r^5}\right) = K\left[\tfrac{1}{2}\vec{r}\cdot\ddot{\mathbf{Q}}\cdot\vec{r}\;\vec{\nabla}\left(r^{-5}\right) + \frac{\vec{\nabla}\left(\frac{1}{2}\vec{r}\cdot\ddot{\mathbf{Q}}\cdot\vec{r}\right)}{r^5}\right] = \frac{K}{r^4}\left[r\left(\vec{\nabla}\left(\tfrac{1}{2}\hat{r}\cdot\ddot{\mathbf{Q}}\cdot\hat{r}\right)\right) - \tfrac{5}{2}\left(\hat{r}\cdot\ddot{\mathbf{Q}}\cdot\hat{r}\right)\hat{r}\right].$$

Thought Experiment 3.13: Multipole Expansion I

Consider an arbitrary finite distribution of charge, $\rho(\vec{r})$. To first approximation, the distant electric potential should simply follow Coulomb's law (Thought Experiment 3.10). The total charge, or monopole moment, and the approximate potential are given by:

$$Q = \int_{\text{volume}} \rho(\vec{r})\,\mathrm{d}V \;\rightarrow\; \mathrm{V}_1 \approx KQ\,\frac{1}{r}.$$

The monopole approximation becomes increasingly accurate the closer the center of the charge distribution is to the origin, i.e. where $\int_{\text{volume}} |\rho(\vec{r})|\vec{r}\,\mathrm{d}V = 0$. For a point charge at the origin, of course, this ceases being an approximation and is the exact expression for the potential.

We can improve this first-order approximation by finding the dipole moment (Thought Experiment 3.11) and adding the electric potentials:

$$\vec{p} = \int_{\text{volume}} \vec{r}\, dQ = \int_{\text{volume}} \rho(\vec{r})\vec{r}\, dV \quad \rightarrow \quad \mathbb{V}_2 = K\frac{p\cos(\theta)}{r^2} = K\,\vec{p}\cdot\hat{r}\frac{1}{r^2}.$$

In a similar manner, the quadrupole moment (Thought Experiment 3.12) can be calculated, thus improving the approximation even more:

$$\ddot{\mathbf{Q}} = \begin{pmatrix} \int_{\text{volume}} \rho(\vec{r})\left(2x^2 - y^2 - z^2\right)dV & \int_{\text{volume}} 3\rho(\vec{r})\,xy\,dV & \int_{\text{volume}} 3\rho(\vec{r})\,zx\,dV \\ \int_{\text{volume}} 3\rho(\vec{r})\,xy\,dV & \int_{\text{volume}} \rho(\vec{r})\left(2y^2 - z^2 - x^2\right)dV & \int_{\text{volume}} 3\rho(\vec{r})\,yz\,dV \\ \int_{\text{volume}} 3\rho(\vec{r})\,zx\,dV & \int_{\text{volume}} 3\rho(\vec{r})\,yz\,dV & \int_{\text{volume}} \rho(\vec{r})\left(2z^2 - x^2 - y^2\right)dV \end{pmatrix}$$

$$\mathbb{V}_3 = K\left(\tfrac{1}{2}\hat{r}\cdot\ddot{\mathbf{Q}}\cdot\hat{r}\right)\frac{1}{r^3}.$$

Thus, to a very good approximation, we can model the electric potential away from any charge distribution as the sum of these three terms.

$$\mathbb{V}(\vec{r}) \approx \mathbb{V}_1 + \mathbb{V}_2 + \mathbb{V}_3$$

$$\mathbb{V}(\vec{r}) \approx K\left(Q\frac{1}{r} + \vec{p}\cdot\hat{r}\frac{1}{r^2} + \tfrac{1}{2}\hat{r}\cdot\ddot{\mathbf{Q}}\cdot\hat{r}\frac{1}{r^3}\right).$$

Problem 3.20: Quadrupole Electric Potential

Show that a quadrupole moment of $\ddot{\mathbf{Q}} = 12Qd^2\begin{pmatrix} 0 & 1 & 0 \\ 1 & 0 & 0 \\ 0 & 0 & 0 \end{pmatrix}$, will yield the electric potential

$$\mathbb{V}(\vec{r}) = \frac{12KQd^2xy}{r^5}.$$

Problem 3.21: Linear Quadrupole

The linear quadrupole is comprised of 3 charges configured as shown.

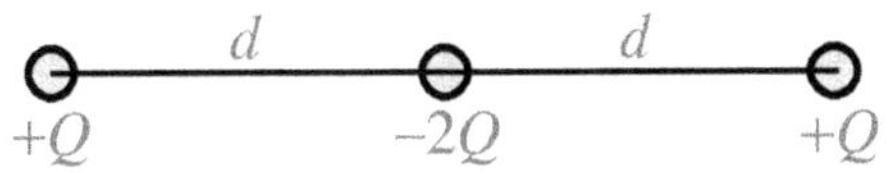

In spherical coordinates:

a) Find the electric potential $\mathbb{V}(r,\theta)$

b) Find the electric field $\vec{E}(r,\theta)$

c) Show that the electric field lines follow the relation $r^2 \propto \sin^2\theta\cos\theta$.

Example 3.9: The Water Molecule

The water molecule is often modeled as a triangular configuration of charges.

Here we will make the *a priori* assumption that it is a configuration of 3 point charges as shown.

The empirical dipole and quadrupole moments are:

$\vec{p} = 0.3862\, e\, \text{Å}\, \hat{y}$, and

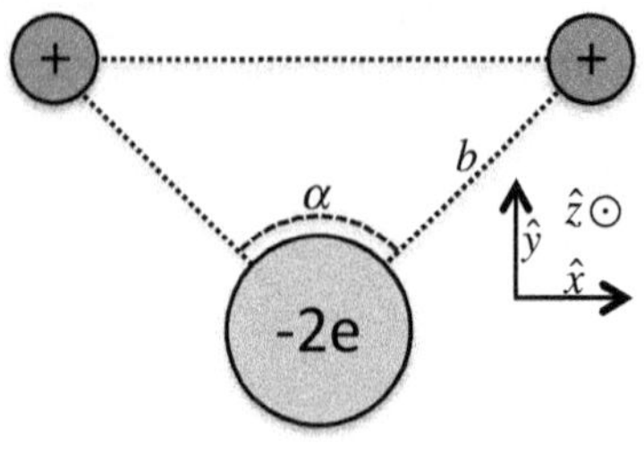

$$\ddot{\mathbf{Q}} = \begin{pmatrix} 0.548 & 0 & 0 \\ 0 & -0.027 & 0 \\ 0 & 0 & -0.520 \end{pmatrix} e\text{Å}^2 .$$

Find the opening angle α and spacing b of the charges.

We begin by using our toy model to find the moments in terms of α and b.

We will first define our origin at the charge center, so that the single positive charges are located at positions $(\pm x_+, y_+)$, and the negative charge is located at $(0, -y_-)$. Thus, in terms of our parameters we have:

$$b = \sqrt{x_+{}^2 + (y_+ + y_-)^2} \quad \text{and} \quad \alpha = 2\arctan\left(\frac{x_+}{(y_+ + y_-)}\right).$$

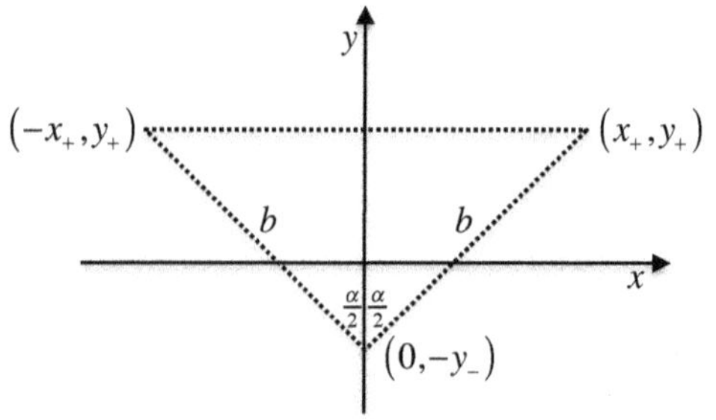

Now, the dipole moment components are given by:

$$p_x = e(-x_+) + e(x_+) = 0$$

$$p_y = ey_+ + ey_+ + (-2e)(-y_-) = 2e(y_+ + y_-)$$

$$p_z = 0$$

And the quadrupole moment is given by:

$$\mathbf{Q}_{xx} = \sum_{1=1}^{n} Q_i\left(2X^2{}_i - Y_i^2 - Z_i^2\right) = e\left(2(-x_+)^2 - y_+{}^2\right) + e\left(2(x_+)^2 - y_+{}^2\right) + (-2e)\left(0 - (-y_-)^2\right)$$

$$= 2e\left(2x_+{}^2 - y_+{}^2 + y_-{}^2\right)$$

$$\mathbf{Q}_{yy} = \sum_{1=1}^{n} Q_i\left(2Y^2{}_i - Z_i^2 - X_i^2\right) = e\left(2y_+ - (-x_+)^2\right) + e\left(2y_+ - x_+{}^2\right) + (-2e)\left(2(-y_-)^2\right)$$

$$= 2e\left(2y_+{}^2 - x_+{}^2 - 2y_-{}^2\right)$$

$$\mathbf{Q}_{zz} = \sum_{1=1}^{n} Q_i\left(2Z^2{}_i - X_i^2 - Y_i^2\right) = e\left(0 - (-x_+)^2 - y_+{}^2\right) + e\left(0 - x_+{}^2 - y_+\right) + (-2e)\left(0 - 0 - (-y_-)^2\right)$$

$$= 2e\left(y_-{}^2 - x_+{}^2 - y_+{}^2\right).$$

Since the tensor is traceless, we can check our work:

$$\mathbf{Q}_{xx} + \mathbf{Q}_{yy} + \mathbf{Q}_{zz} = 2e\left(2x_+{}^2 - y_+{}^2 + y_-{}^2 + 2y_+{}^2 - x_+{}^2 - 2y_-{}^2 + y_-{}^2 - x_+{}^2 - y_+{}^2\right) = 0.$$

We can then solve our system of equations for the coordinate positions in terms of the moments:

$$\mathbf{Q}_{yy} + 2\mathbf{Q}_{zz} = 2e\left(\cancel{2y_+^2} - x_+{}^2 - \cancel{2y_-^2}\right) + 2e\left(\cancel{2y_-^2} - 2x_+{}^2 - \cancel{2y_+{}^2}\right) = -6ex_+{}^2$$

so, we now can solve for x_+ :

$$x_+ = \sqrt{-\frac{\mathbf{Q}_{yy} + 2\mathbf{Q}_{zz}}{6e}}.$$

And, from the dipole moment we have:

$$(y_+ + y_-) = \frac{p}{2e}.$$

So, the parameters we are interested in are:

$$b = \sqrt{x_+^2 + (y_+ + y_-)^2} = \sqrt{\left(-\frac{\mathbf{Q}_{yy} + 2\mathbf{Q}_{zz}}{6e}\right) + \left(\frac{p}{2e}\right)^2}$$

$$\alpha = 2\arctan\left(\frac{x_+}{(y_+ + y_-)}\right) = 2\arctan\left(\left(\frac{2e}{p}\right)\sqrt{\left(-\frac{\mathbf{Q}_{yy} + 2\mathbf{Q}_{zz}}{6e}\right)}\right).$$

Now in terms of numbers, for the H_2O molecule:

$$b = \sqrt{\left(\cancel{-}\frac{(\cancel{-}0.027 + 2(\cancel{-}0.520))\cancel{e}\,\text{Å}^2}{6\cancel{e}}\right) + \left(\frac{0.386\cancel{e}\,\text{Å}}{2\cancel{e}}\right)^2} = 0.464\,\text{Å}$$

$$\alpha = 2\arctan\left(\left(\frac{2\cancel{e}}{0.386\cancel{e}\,\text{Å}}\right)\sqrt{\left(-\frac{(-0.027\cancel{e}\,\text{Å}^2) + 2(-0.520\cancel{e}\,\text{Å}^2)}{6\cancel{e}}\right)}\right) = 131°.$$

To put this in perspective, the corresponding nuclear distance and angle are $0.958\,\text{Å}$ and $104.5°$ respectively, so these results make sense given the covalent bonding nature of water.

Problem 3.22: The Geometry of Hydrogen Cyanide

The Hydrogen Cyanide molecule has a linear geometry as shown in the diagram. The dipole and quadrupole moments are given in Cartesian coordinates by:

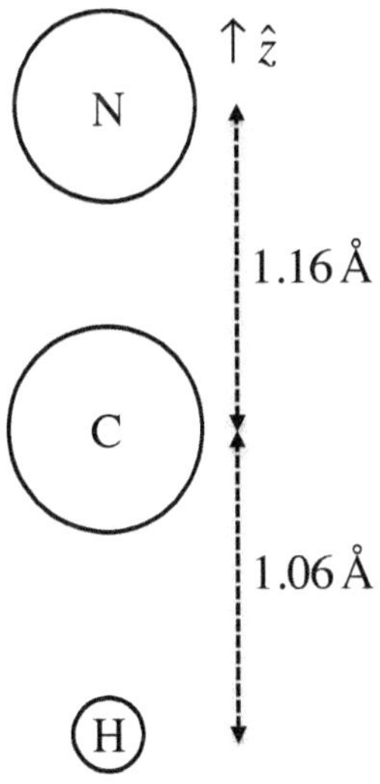

$$\vec{p} = -2.98\,\text{D}\ \hat{z} \qquad \ddot{\mathbf{Q}} = 0.34\,\text{D}\cdot\text{Å}\begin{pmatrix} -1 & 0 & 0 \\ 0 & -1 & 0 \\ 0 & 0 & 2 \end{pmatrix}.$$

A common unit of dipole moment is the Debye, where $1\,\text{D} \approx 0.208\,e\text{Å}$.

Model this molecule as three point charges, q_H, q_C, q_N with charge separations of 1.06Å and 1.16Å for H-C and C-N respectively. Express your answer for each of these effective charges in elementary charge units.

Problem 3.23: The Octopole

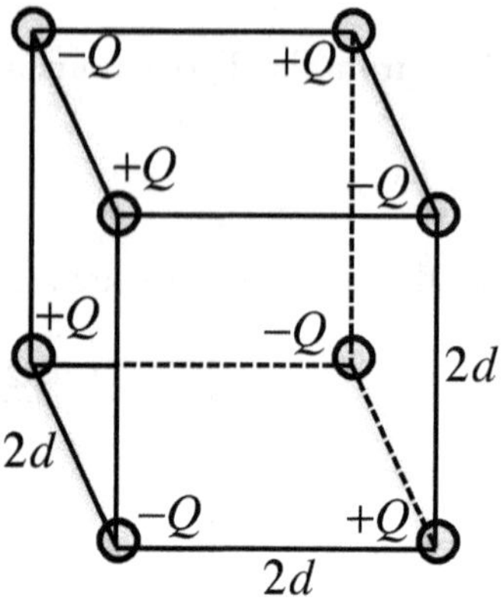

The classic octopole consists of a cube of two quadrupoles stacked.

a) Find the electric potential $\mathbb{V}(\vec{r})$, over all space.

b) Find the electric potential in the limit that $r \gg d$.

c) Find the electric field $\vec{E}(\vec{r})$ in the far limit.

d) The monopole moment was simply the charge, which could also be considered a "rank 0 tensor." A vector is a first order tensor, or we say its rank is 1. The second order tensor, or rank 2, consisted of a 3×3 matrix. Following this pattern, how should we mathematically express a general octopole moment? Discuss how this works.

Problem 3.24: Four Charges

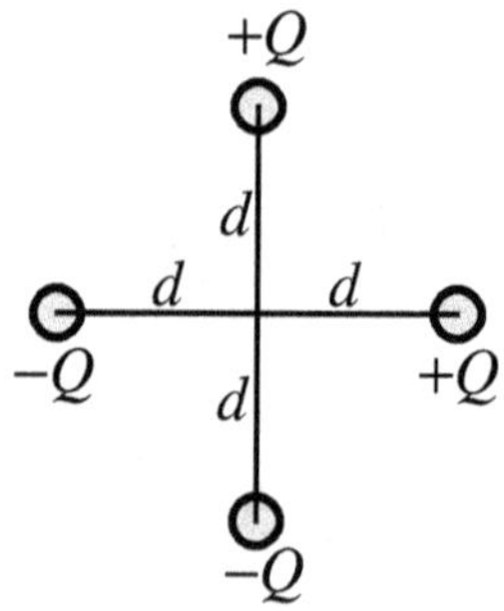

Four charges are configured in a square as shown.

a) Find the monopole and dipole moments of the configuration, in terms of d and Q.

b) Find the electric potential in the far limit.

c) Find the far electric field.

d) Using a computer, plot the electric potential you found in (b), the exact electric potential from Coulomb's law, and the difference between the two. These plots may be two dimensional images or contour maps limited to the in the plane that holds the charges. Discuss the results.

Problem 3.25: Six Charges

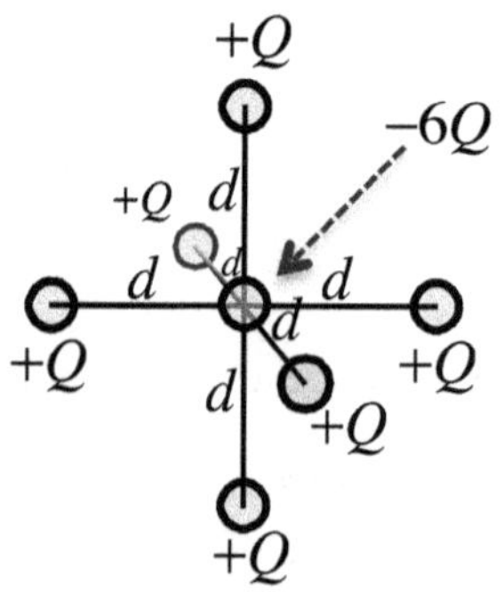

Six equal positive charges are configured on the faces of a cube, with a negative charge in the center, as shown as shown. Note the configuration in net neutral.

a) Find the electric potential at a large distance from the configuration.

b) At large distances, what is the electric field?

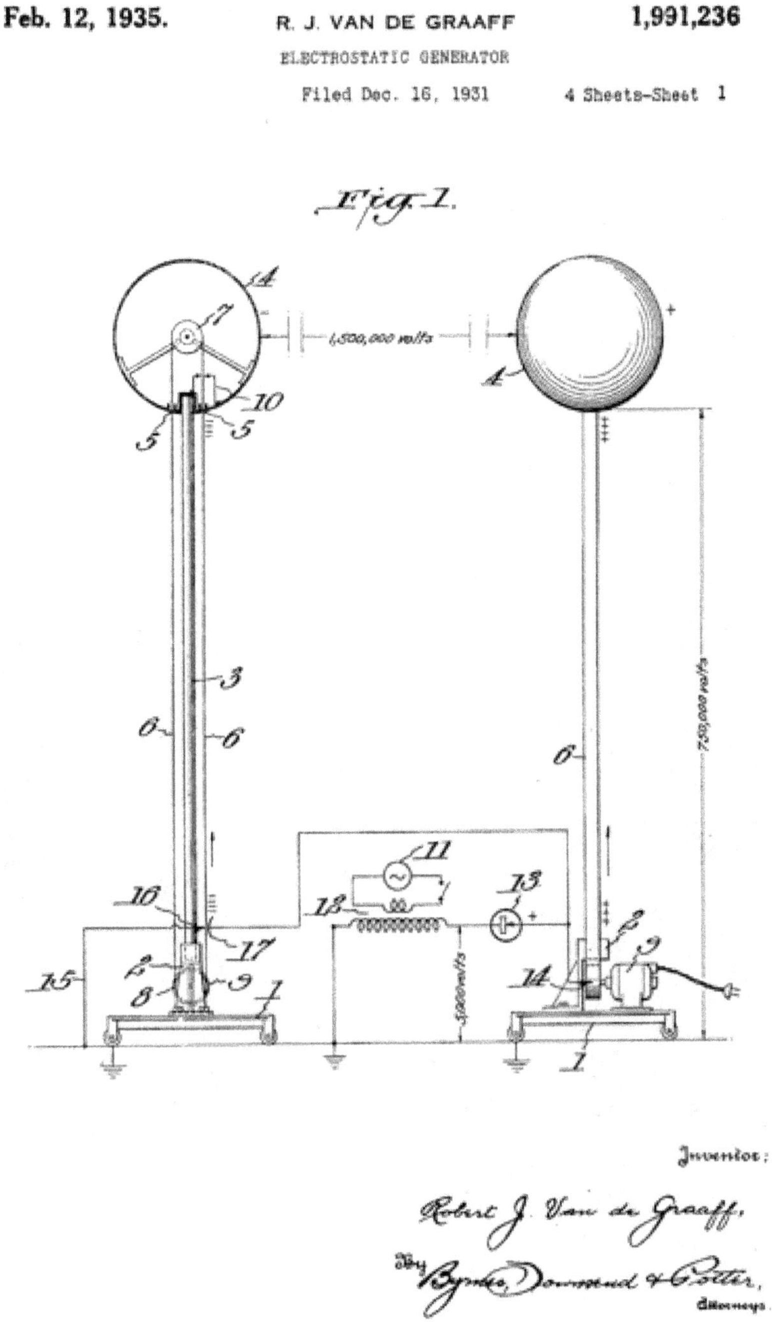

Plate 7: Robert van de Graaff's Invention[53]

[53] Figure from patent US1991236A filed by R.J. van de Graaff on Dec. 16, 1931, and awarded to the Massachusets Institute of Technology. Photo of one of the towers is on the next page.

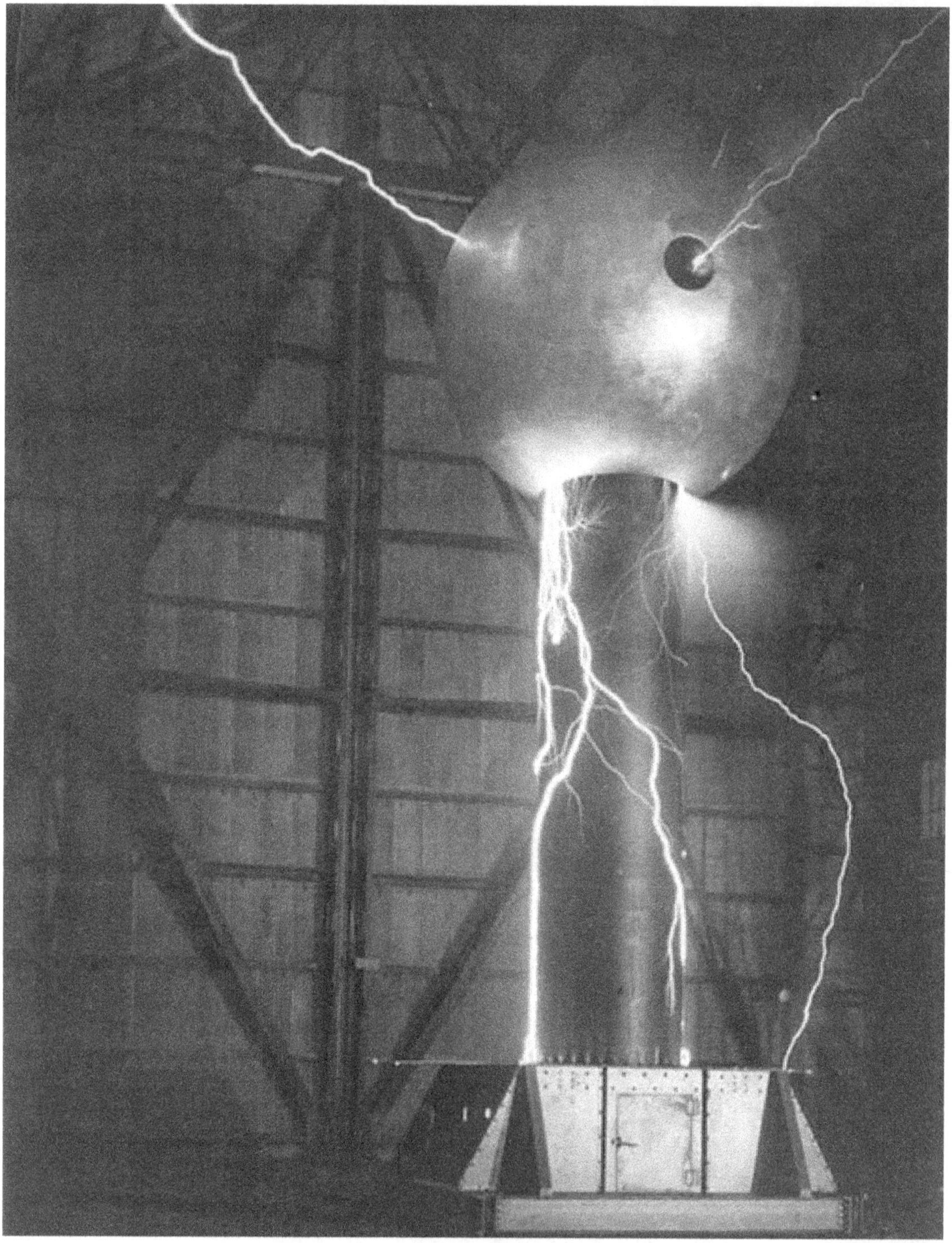

Plate 8: The MIT van de Graaff Electrostatic Generator[54]

[54] This is a 1933 time exposure of half of the MIT van de Graaff high voltage generator. It still works and is on permanent display at the Museum of Science in Boston. Image from the MIT Institute Archives & Special Collections.

Chapter 4 Gauss's Law

In 1813, the great German physicist, mathematician, and astronomer, Carl Friedrich Gauss, reformulated Newton's law of gravity into an equivalent and elegant statement, which relates the total mass enclosed by an arbitrary closed surface to the surface integral of the gravitational field through the same surface,[55] and in 1839 he applied this to inverse square laws in general.[56] In 1861, Maxwell applied Gauss's law to electricity,[57] relating the total charge enclosed within an arbitrary surface to the electric field through the same surface. The divergence theorem of vector calculus, which Gauss also discovered, states that the surface integral of a vector field (such as the electric field) through a closed surface equals the volume integral of the divergence of that vector field taken over the volume bounded by the closed surface. Gauss's law can thus be stated locally as well as globally: the divergence of the electric field at a point is proportional to the charge density at that point. Thus, despite being physically equivalent to Coulomb's law, Gauss's law is mathematically similar to the continuity equation.

To see Gauss's law in action, consider Robert van de Graaff's electrostatic generator (pp. 113-114). Since the electric charge distributes itself on the outside of the metal sphere, a Gaussian surface matching the inside of the sphere contains no charge, so it is easy to show, with Gauss's law, that the electric field inside of the sphere is zero. Thus, the inside of the sphere can accumulate more and more charge, so long as it remains insulated from the ground. Moreover, even though the electric potential of the sphere is 1.5 MV above ground, it is still safe for both people and electronics inside it even when it is charged.

Not only is Gauss's law usually more useful in the practical matter of directly computing the electric field, it is almost always more useful in formulating general statements of principle in electrodynamics. We will use Gauss' law to determine the electric field between the surfaces of a charged capacitor and a coaxial cable, and introduce the notion of capacitance in an example on charging a parallel plate capacitor through a resistor.

Recall that in Chapter 1 we made a distinction between the global principle of the conservation of charge, and the local condition of charge being conserved on all scales. Similarly, we use the integral formulation of Gauss's law to find the electric field surrounding a known charge distribution, but its differential form applies continuously to every point in space. Just as marrying Franklin's experiments to Euler's mathematics produced the continuity equation, the marriage of Coulomb's experiment to Gauss's mathematics leads to Gauss's law.

4.1 *Capacitors*

In our discussion to this point, we have ignored electric fields in insulative matter, except for our discussion of action at a distance in the introduction to Chapter 2 and Discussion 6.2. We now

---

[55] C.F. Gauss, "Theoria attractionis corporum sphaeroidicorum ellipticorum homogeneorum methodo nova tractata," Werke, (Gottingen, 1867) **5**: 1-22.

[56] C.F. Gauss, "General Propositions relating to Attractive and Repulsive Forces acting in inverse ratio of the square of the distance," Scientific Memoirs, Selected from the Transactions of Foreign Academies and Learned Societies and from Foreign Journals, ed. R. Taylor (London: R. and E. Taylor, 1841), Vol. III, Part X, 151-196.

[57] Gauss's law in differential form is equation (115) in Maxwell's 1861 paper "Lines of Physical Force." Maxwell, subsequently, states the integral version of Gauss's law in Chapter II, articles 75 and 76 of his Treatise on Electricity and Magnetism, Vol. I (Oxford: Clarendon, 1873), 76-79, and uses the differential version of the law in the course of deriving Poisson's equation (section 77, page 79 of Volume I of the Treatise).

treat *dielectrics*, as insulators are called, whose behavior differs considerably from conductors because charges within insulating materials are only slightly mobile. Electrons and protons are bound together within the molecules that comprise dielectrics, and these molecules can only stretch or rotate through small distances in response to an external electric field. After dealing with conductors, it may come as a surprise that electric fields have much effect on insulators. But Faraday found that the capacitance of a capacitor increases considerably when dielectric material is inserted between the parallel plates of the capacitor.

When the region between the plates is entirely filled with an insulator, in fact, the capacitance increases by a factor which depends only on the nature of the dielectric material itself. This factor is called the *dielectric constant* or dielectric strength of the material and can be measured directly in the lab using a capacitor and a voltmeter. The net electric field between the plates decreases by the same dielectric constant, which means that it is easier to store charge on a capacitor with a dielectric than without one. To understand why fields behave this way inside dielectric material, it is necessary to consider a toy model of matter and think about how an atom might respond to an applied electric field.

Example 4.1: Faraday's Spherical Leyden Jar

Consider a thin metal sphere of radius b, with a small central metal ball of radius a in the center. If we place a known charge Q on the central ball, what voltage, between the inner sphere and outer shell, what does Coulomb's law predict the voltage difference to be?

Recall from Thought Experiment 3.10 (p. 104) that the electric potential at distance r from a spherical charge is given by:

$$\mathbb{V} = \frac{KQ}{r},$$

so the potential difference from just outside the small sphere, to just inside the large sphere, should be:

$$\Delta\mathbb{V} = \frac{KQ}{a} - \frac{KQ}{b} = KQ\left(\frac{1}{a} - \frac{1}{b}\right).$$

However, when Michael Faraday conducted similar experiments, but with various spherical insulators between the plates, the voltage difference decreased by an amount that depended on the inserted material!

Thus, instead of finding Coulomb's relationship, he found:

$$\Delta\mathbb{V} \leq KQ\left(\frac{1}{a} - \frac{1}{b}\right).$$

Now, recall, the philosophical problem of spooky action at a distance (pp. 1, 58). Coulomb's law, and Newton's law of gravity for that matter, did not sit well with Faraday because there was no mechanism to relate cause to effect. Therefore, Faraday concluded that it is matter that conveys the electric field. And, much like elasticity conveys shock forces, there must be a similar intensive property of matter, called the *permittivity*, that conveys electric forces. The higher the permittivity, the lower the resultant voltage difference.

Thought Experiment 4.1: Capacitance

In Chapter 1 we discussed many thought experiments involving the Leyden jar, and how it was used to store charge. Technically speaking, the jar really stores potential energy, not charge. As charge moves into the jar, an equivalent amount of opposite charge moves as close as possible to the charged area. For example, if the inside of the jar is net positive, the surrounding outside becomes net negative, making the jar as a whole (both inside and out) net neutral. Moreover, the jar's capacity to hold charge increases when it is filled with salt water, but the salt water is also conductive, so there has to be another insulation layer.

The main point here is this: as charge flows into the Leyden jar, the jar remains net neutral, but charge separates and this separation increases the potential energy. If $\mathbb{V}$ is the voltage across the terminals of the Leyden Jar, then the work it takes to push a charge Q into the jar is given by:

$$\mathbb{U}_E = \int_0^Q \mathbb{V}\,dQ .$$

Clearly, the voltage across the terminals increases with charge build-up, and experiment has shown that the voltage tends to be proportional to the charge. Modern equivalents of the Leyden jar are now called *capacitors* (but were once called *condensers*), and manufacturers make them as linear as possible. The measure of a capacitor's ability to store charge is called its *capacitance*, C, and is defined as the ratio of the charge separation $\pm Q$ to the voltage across the terminals, or:

$$C \equiv \frac{Q}{\mathbb{V}} .$$

where C is the capacitance. The SI unit of capacitance is the farad, named after Michael Faraday, as:

$$1\,\text{F} = [C] = \frac{[Q]}{[\mathbb{V}]} = \frac{\text{C}}{\text{V}} = \frac{\text{C}^2}{\text{J}} .$$

In general, a one-farad capacitor would be very large. Typically, large capacitors are rated in units ranging from picofarads to microfarads. The photo-flash capacitor shown in the picture is rated for a capacitance of 100 microfarads, up to a maximum voltage of 330 volts. This capacitor is shaped like a can, not because it is a cylindrical Leyden jar, rather because inside is a layered roll containing the anode, cathode and insulators.

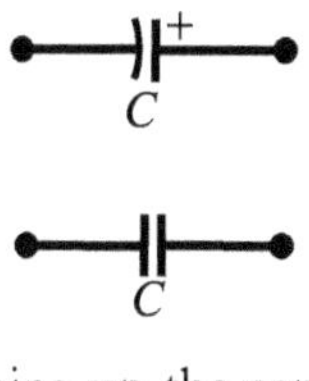

Modern capacitors are often made with two different materials, one for the anode and the other for the cathode. If the leads are reversed, then current can flow from one plate to the other, usually burning up the capacitor. In circuit diagrams, these *polarized capacitors* are shown with a curved cathode, while symmetrical capacitors are drawn with parallel lines, as either side can be the anode or the cathode depending on the voltage difference. Clearly, one can use a non-polarized capacitor when a polarized one is called for, but not the other way around.

The total potential energy stored in a capacitor is:

$$\mathbb{U}_E = \int_0^Q \mathbb{V}\,dQ = \int_0^Q \frac{Q}{C}\,dQ ,$$

so for an ideal linear capacitor:

$$\mathbb{U}_E = \frac{1}{C}\int_0^Q Q\,\mathrm{d}Q = \frac{Q^2}{2C} = \tfrac{1}{2}Q\mathbb{V} = \tfrac{1}{2}C\mathbb{V}^2.$$

Thus, for example, the photoflash capacitor shown could hold potential energy up to:

$$\mathbb{U}_E = \tfrac{1}{2}C\mathbb{V}^2 = \tfrac{1}{2}\left(100\times10^{-6}\,\mathrm{F}\right)\left(330\,\mathrm{V}\right)^2 = 5.4\,\mathrm{J}.$$

Example 4.2: An RC Circuit

Now we consider a circuit consisting of a battery, a resistor and a capacitor. Imagine that the circuit is closed at time zero. First we will consider a capacitor, with capacitance C, which as been charged using a battery of voltage $\mathbb{V}_0$. At time zero the switch is closed, and the capacitor discharges through a resistor with resistance R. What is the current I as a function of time?

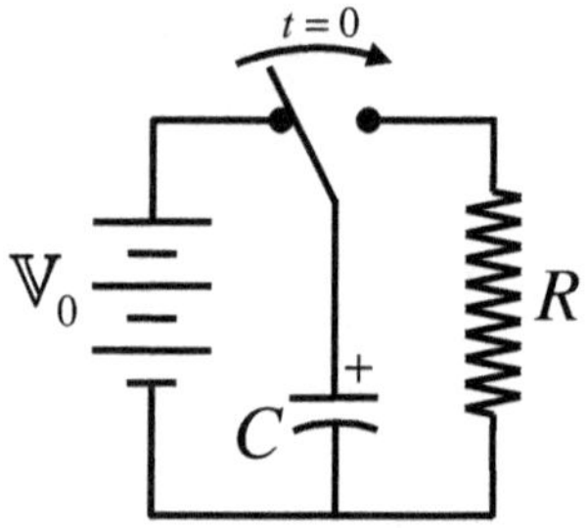

First we consider the charge continuity equation. Drawing a Gaussian surface around the positive plate of the capacitor, and applying charge continuity, allows us to determine the current flowing from the positive plate:

$$\vec{\nabla}\cdot\vec{J} = -\frac{\partial\rho}{\partial t} \;\rightarrow\; \oint_{\text{surface}} \vec{J}\cdot\mathrm{d}\vec{A} = \oint_{\text{volume}} \left(-\frac{\partial\rho}{\partial t}\right)\mathrm{d}V \;\rightarrow\; I = -\frac{\mathrm{d}Q}{\mathrm{d}t}.$$

The same current must be flowing to the negative plate since the capacitor is net neutral, though not continuously neutral. This charge is related to the voltage difference across the capacitor via:

$$\mathbb{V}_{\text{capacitor}} = \frac{Q}{C}.$$

The voltage across the capacitor must equal the voltage across the resistor $\left(\mathbb{V}_{\text{resistor}} = \mathbb{V}_{\text{capacitor}}\right)$.

Putting these together with Ohm's law, we can now write:

$$\mathbb{V}_{\text{resistor}} = \mathbb{V}_{\text{capacitor}} \;\rightarrow\; IR = \frac{Q}{C} \;\rightarrow\; \left(-\frac{\mathrm{d}Q}{\mathrm{d}t}\right)R = \frac{Q}{C},$$

so we can separate the differentials and integrate:

$$\int_{Q_0}^{Q}\frac{\mathrm{d}Q}{Q} = -\int_0^t \frac{\mathrm{d}t}{RC}.$$

Solving this for the charge, we get:

$$\int_{Q_0}^{Q}\frac{\mathrm{d}Q}{Q} = -\int_0^t \frac{\mathrm{d}t}{RC} \;\rightarrow\; \ln\left(\frac{Q}{Q_0}\right) = -\frac{t}{RC} \;\rightarrow\; Q = Q_0\,e^{\frac{-t}{RC}}.$$

We are interested in the finding current through the resistor. Noting that the capacitor's voltage difference is $\mathbb{V}_0$, initially, we have that:

$$I = -\frac{dQ}{dt} = -Q_0\left(-\frac{1}{RC}\right)e^{-\frac{t}{RC}} = \cancel{C}\mathbb{V}_0\left(\frac{1}{R\cancel{C}}\right)e^{-\frac{t}{RC}} = \frac{\mathbb{V}_0}{R}e^{-\frac{t}{RC}}.$$

Notice that the current also falls off exponentially with a time constant of RC. The power dissipated by the resistor would then be given by:

$$\mathbb{P} = I\mathbb{V} = I^2R = \left(\frac{\mathbb{V}_0}{R}e^{-\frac{t}{RC}}\right)^2 R = \frac{\mathbb{V}_0^2}{R}e^{-\frac{2t}{RC}} = \frac{\mathbb{V}_0^2}{R}e^{-\frac{t}{\left(\frac{1}{2}RC\right)}},$$

so we see that the power falls off twice as fast as the voltage and the current individually, so the power time constant is $\tau_{\mathbb{P}} = \frac{1}{2}RC$.

We often model a light bulb as a resistor, so this could be a photo-flash circuit. For example, if we were interested in a flash bulb with a power time constant of one millisecond that runs off the capacitor shown on page 117 so that $C = 100\,\mu\text{F}$ and $\mathbb{V}_0 \le 330\,\text{V}$, then we would pick a light bulb with an effective resistance given by:

$$R = \frac{2\tau_{\mathbb{P}}}{C} = \frac{2\left(10^{-3}\,\text{s}\right)}{10^{-4}\,\text{F}} = 20\,\Omega.$$

How much total energy would it put out during a typical 10 ms camera shutter time?

The power output would then be given by:

$$\mathbb{P} = \frac{\mathbb{V}_0^2}{R}e^{-\frac{2t}{RC}} = \frac{(330\,\text{V})^2}{(20\,\Omega)}e^{-\frac{t}{1\text{ms}}} = (5400\,\text{W})e^{-\frac{t}{1\text{ms}}},$$

and the total energy emitted would then be:

$$\mathbb{E} = \int_0^{10\text{ ms}} \mathbb{P}\,dt = \int_0^{10\text{ ms}} (5.4\,\text{kW})e^{-\frac{t}{1\text{ms}}}\,dt = (5.4\,\text{kW})(1\,\text{ms})\int_0^{10} e^{-\left(\frac{t}{1\text{ms}}\right)}\,d\left(\tfrac{t}{1\text{ms}}\right)$$

$$= (5.4\,\text{J})\int_0^{10} e^{-\left(\frac{t}{1\text{ms}}\right)}\,d\left(\tfrac{t}{1\text{ms}}\right) = (5.4\,\text{J})\left(-e^{-10} + e^0\right) = 5.4\,\text{J}.$$

This, of course, is the same as the energy stored in the capacitor on page 117.

Problem 4.1: A Photoflash

Recalling Problem 3.6 (p. 82) about flash photography, estimate the peak visual luminosity of an LED powered by the photoflash capacitor above. Express your answer in lumens.

Thought Experiment 4.2: The Permittivity of Dielectrics

Imagine charging a parallel plate capacitor with a charge on the top and bottom plates of $+Q$ and $-Q$ respectively. The amount of charge, $\pm Q$, that each plate can hold, given a particular surface electric field, should be proportional to the surface area of the plates A, as the charge will distribute itself evenly on the plate surface. Franklin demonstrated this, and so he advocated the use of sharply pointed lightning rods. The capacitance should therefore increase linearly with plate area A, and decrease with the plate separation h:

$$C \equiv \frac{Q}{\mathbb{V}} \propto \frac{A}{h}.$$

This is certainly the correct relationship, but it isn't the whole story. For example, a soda bottle Leyden jar that is filled with water holds much more charge than one that is empty, and one filled with salt water holds even more charge. Thus the insulating material between charged conductors also has a significant effect on the capacitance.

In the late 1830s, Faraday clearly demonstrated this with his electrical induction experiments. He surmised, mostly correctly, that the molecules in the medium were somehow charged positive on one side, and negative on the other, so he called them *dielectrics*.

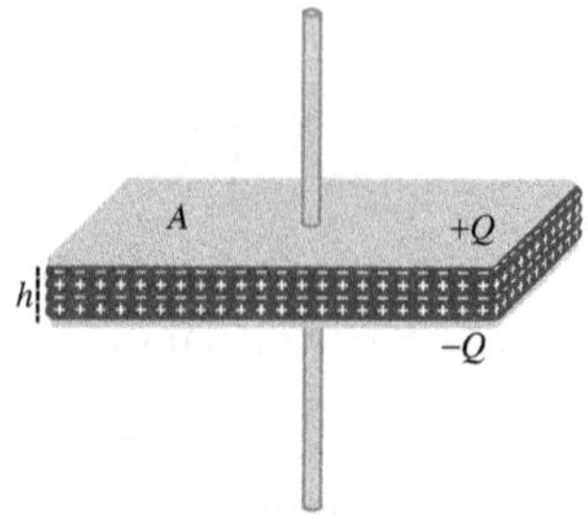

The constant of proportionality in the relation above is an intensive property of the material. For a parallel plate capacitor:

$$C \equiv \frac{Q}{\mathbb{V}} = \varepsilon \frac{A}{h}.$$

ε is called the *permittivity*, which can be measured by simply putting a sample between, or even near, a pair of capacitor plates, and measuring the charge and voltage. The permittivity of typical insulating materials in capacitors is about $\varepsilon \gtrsim 10\frac{\text{pF}}{\text{m}}$. The permittivity of air is $\varepsilon_{\text{air}} \approx 8.86\frac{\text{pF}}{\text{m}}$, and water, being a polar molecule, has a fairly large permittivity of about $\varepsilon_{\text{H}_2\text{O}} \approx 700\frac{\text{pF}}{\text{m}}$.

The measured permittivity in a vacuum is called the *permittivity of free space*, and has the symbol ε_0.

Problem 4.2: The Electric Field in a Capacitor

Show, from the fundamental concepts of work and energy, and the discussion of capacitance, that the electric field between the plates of a parallel plate capacitor is given by:

$$\vec{E} = \frac{Q}{\varepsilon A}\hat{z},$$

where the $\hat{z}$ direction is toward the positive plate.

Thought Experiment 4.3: The Permittivity of Free Space

In Example 2.4 (p. 65) we found, using Coulomb's law, the electric field between the plates of an idealized parallel plate capacitor is given by:

$$\vec{E} = -4\pi K \sigma\, \hat{z} = -\frac{4\pi K Q}{A}\hat{z},$$

where the $\hat{z}$ direction is toward the positive plate.

The work required, per charge, to push a positive test charge from the positive to the negative plate is, by definition, the voltage across the two plates. Thus:

$$\mathbb{V}_{\text{capacitor}} = \int_{z=h}^{z=0} \vec{E}\cdot d\vec{\ell} = -\int_0^h E\, dz = -\int_0^h \left(-\frac{4\pi K Q}{A}\right) dz = \frac{4\pi K Q}{A} h.$$

The capacitance is, therefore:

$$C \equiv \frac{Q}{\mathbb{V}} = \frac{Q}{\left(\frac{4\pi K h Q}{A}\right)} = \frac{A}{4\pi K h} = \left(\frac{1}{4\pi K}\right)\left(\frac{A}{h}\right) = \varepsilon_0 \left(\frac{A}{h}\right).$$

Thus the permittivity of an idealized capacitor is a simple function of the Coulomb constant. In practice, most physicists use ε_0, rather than Coulomb's constant, to express the laws of electrodynamics. It's value is:

$$\varepsilon_0 = \frac{1}{4\pi K} \approx 8.854 \tfrac{\text{pF}}{\text{m}} .$$

Between 1983 and 2019, ε_0 has been a derived (defined) constant in the SI, based on the accepted definition of the speed of light. However, since the 2018 General Conference on Weights and Measures (CGPM), the standard value of ε_0 is now subject to measurement error. The currently accepted value of the permittivity of free space is:

$$\varepsilon_0 = 8.8541878 \times 10^{-12} \tfrac{\text{C}^2}{\text{N}\cdot\text{m}^2} .$$

Therefore, Coulomb's constant can be written $K = \frac{1}{4\pi\varepsilon_0}$, so Coulomb's law becomes:

$$\vec{F}_{1,2} = \frac{Q_1 Q_2}{4\pi\varepsilon_0\, r_{1,2}^2}\hat{r}_{1,2} .$$

As we mentioned in Discussion 2.1 (p. 56), many physicists, including many astronomers and particle physicists, work in Gaussian units. In these units, Coulomb's constant is defined to be unity. In fact, many theorists work in units where a number of other fundamental constants are also set to unity, such as the speed of light, Newton's constant, and Planck's constant.

In particle physics units, where both the speed of light and Coulomb's constant are unity, the permittivity of free space simply becomes $\varepsilon_0 = \frac{1}{4\pi}$. However, if the speed of light is not also set to unity, converting from one system to another requires a little more attention to detail when translating equations involving magnetism as we discuss in Appendix D.

Discussion 4.1: Permittivity, the Aether, and Toy Models of Matter

All quantities with real physical meaning must be measurable *in situ*, otherwise we are left with spooky action at a distance (p. 58). We defined the electric field, therefore, based on its measurability. Simply defining a field to relate cause to effect, however, simply kicks the can down the road. After all, the field itself may simply be a figment of our imagination.

This leads to the question: if the electric field is to be a real quantity, what physical medium must convey it? In every other context vector fields represent something tangible, such as the velocity of water in a stream. Why should the electric field be any different? Free space must, therefore, be filled with a substance, called *the aether* after Aristotle's fifth element,[58] with a measurable permittivity. As you already know (p. 1), the aether turned out not to exist after all, and modern scientists have dropped the word *aether* from their lexicon.

The historian of science E.T. Whittaker argues that physicists are still investigating the same question of how forces are conveyed across vast distances of seemingly empty space. Thus, the aether of the nineteenth century simply became the virtual particles, dark matter, and dark energy, which we take seriously today. Had physicists followed Whittaker's suggestion, much of cosmology might still be called aether theory today—perhaps it should.[59]

---

58 Aristotle, De Caelo, Translated by J.L. Stocks (London: Oxford, 1922), Book III.

59 See the Whittaker quote on the top of page 425.

Consider the counter-example of chemistry. Nature was thought to be comprised of four elements: earth, air, fire, and water. When this theory was debunked, chemists simply replaced them with other elements. Today no chemist believes in the four traditional elements, but they do believe that matter is comprised of some substances that are more basic than others.

Keep this analogy in mind, because one of the most important goals of nineteenth century physics was to model matter on the smallest scales. What is it that makes conductors different from dielectrics? What makes glass transparent and silver shiny? Are all good conductors reflective, and if so why? After such great success in explaining conductors as fluid conduits, perhaps the same ideas can be turned to insulators as conduits of the electric field?

Classical models of electricity, by and large, both explain the observed phenomena and provide a reasonable analog to the underlying quantum physics. This is true for conductors, dielectrics, and even superconductors. However, the opposite is true for magnetic materials, such as permanent magnets, where every classical toy model has been completely debunked!

No classical model, regardless of how successfully it may reproduce certain observations, truly applies to matter at atomic scales. Classical models regarding electricity have fared better than those regarding magnetic properties of matter, however, and are sometimes useful as rough and ready approximations to reality. Despite their poor reputation, nineteenth century aether theories were successful enough to pave the way for twentieth century particle physics.

Definition 4.1: The Polarization Vector

In Problem 2.8 (p. 67) you showed that the torque on an electric dipole is given by:

$$\vec{\tau} = qd \times \vec{E} = \vec{p} \times \vec{E},$$

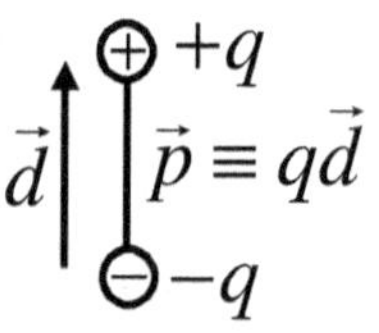

and you noticed that the torque only depended on the product of the charge and the separation vector $\vec{d}$, so we defined a new vector quantity $\vec{p} \equiv q\vec{d}$, called the *electric dipole moment*, or simply *the dipole moment*. We imagine that neutral atoms could consist of a bunch of dipoles, which may, or may not, be lined up. To represent this property of matter, we define a quantity called the polarization, $\vec{P}$, as the dipole moment per volume. Thus, according to our toy model:

$$\vec{P} = \frac{1}{\delta V} \sum_{\text{over volume } \delta V} \vec{p}_{\text{each molecule}} \,.$$

Or, to put it in macroscopic terms, the dipole moment of a volume of insulative material is:

$$\vec{p} = \oint_{\text{volume}} \vec{P} \mathrm{d}V \,.$$

Thought Experiment 4.4: The Dumbbell Model of Dielectrics

Consider a trapezoidal dielectric with a dipole moment per volume $\vec{P}$, as shown in the figure. What is the surface molecular charge on the outside of the dielectric?

This seems like a silly question, because the net charge is zero, as even a polarized medium is neutral on all scales. Of course, this is correct because we defined *all scales* as larger than a molecule, but smaller than everything else.

Consider, for example, a very simple toy model of matter, where we imagine that it is composed of small dumbbells each of length d with equal and opposite charges attached to each end. For each negative charge of a single dipole there will be an adjacent positive charge from the next dipole over. The negative and positive charges, therefore, cancel each other so that there is no net charge in the immediate vicinity of this pair. This will be the case everywhere except for the very top and bottom layers of dipoles, as there is no adjacent dumbbell to cancel out the charge.

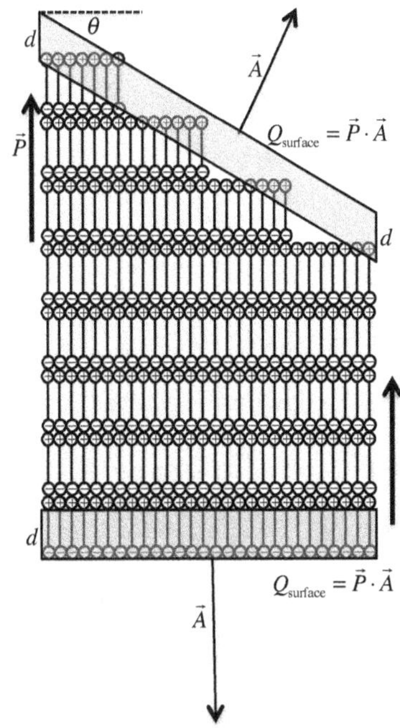

In this simplistic model, we will assume n perfectly aligned dipoles per unit volume, so the polarization is given by:

$$\vec{P} = n\vec{p} = nq\vec{d} = nqd\hat{z}\,.$$

First consider the bottom surface, which is perpendicular to the polarization. When we draw a Gaussian surface containing only the negative charges of the dipoles, we obtain a total charge given by:

$$Q_{\text{bottom}} = -nqV = -nqAd = -PA = \vec{P}\cdot\vec{A}\,.$$

Notice how the surface charge is negative when the polarization and area vectors are antiparallel.

Now consider the top surface. The total surface charge is also equal to the charge density times the volume, but now the volume of a parallelogram prism of height $\vec{d}$ and area $\vec{A}$ is:

$$V = \vec{A}\cdot\vec{d} = Ad\cos\theta\,.$$

Therefore the surface charge on the top is:

$$Q_{\text{top}} = nqV = nq\vec{d}\cdot\vec{A} = \vec{P}\cdot\vec{A}\,,$$

which is equal and opposite the charge on the bottom, because charge is always conserved.

Problem 4.3: Polarization Vector in a Capacitor

Consider a parallel plate capacitor with charge $\pm Q$ on the plates, with a polarized dielectric of uniform polarization $\vec{P} = -P\hat{z}$ sandwiched between the plates, like the one in Thought Experiment 4.2 (p. 119).

(a) As we did in Thought Experiment 1.5 (p. 29), draw pictures, with Gaussian surfaces, to show that the total charge near the positive plate is given by $Q_{\text{total}} = Q - PA$, where A is the area of each plate.

(b) Use this result, and the result from Example 2.4 (p. 65), to find the electric field between the plates in terms of Q, P, and A.

(c) Refer to your result from Problem 4.2 (p. 120), to show that the polarization of the dielectric can be written in terms of the permittivity as $\vec{P} = -\sigma\left(1 - \frac{\varepsilon_0}{\varepsilon}\right)\hat{z}$, where σ is the surface charge density on the positive plate.

(d) Show that $\vec{P} = (\varepsilon - \varepsilon_0)\vec{E}$.

(e) Suppose that the capacitor's potential energy were stored uniformly within the dielectric. Show that the energy density would be: $u_E = \frac{1}{2}\varepsilon E^2 = \frac{1}{2}\vec{P}\cdot\vec{E} + \frac{1}{2}\varepsilon_0 E^2$.

It seems reasonable that the electric field would be the cause of the polarization, so one would expect this last two results to be true in any linear medium.

Thought Experiment 4.5: The Electric Field in a Polarized Medium

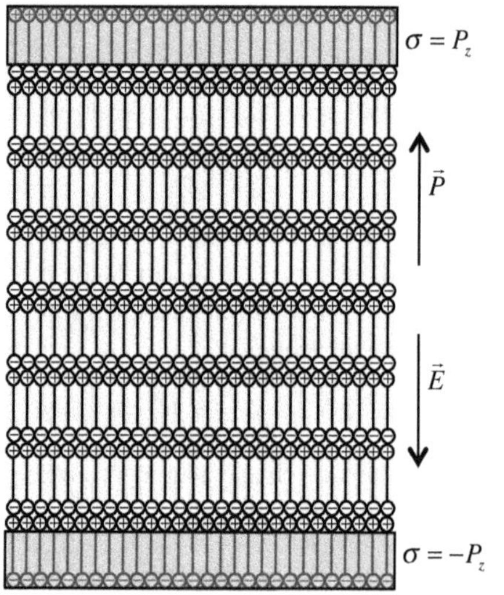

The purpose of this thought experiment is to find the electric field inside a uniformly polarized substance due solely to the dipole moment per volume $\vec{P}$. To do this, we must imagine that the material is inherently polarized and that there is no external electric field.

To do this we will return to the simplistic dumbbell toy model we introduced in Chapter 4. We assume there are n, perfectly aligned, dipoles per unit volume, so the polarization is given by:

$$\vec{P} = nq\vec{d} = nqd\hat{z}$$

For each negative charge of a single dipole, there will be an adjacent positive charge from the next dipole over. The negative and positive charges therefore cancel each other so that there is no net charge in the immediate vicinity of this pair. This will be the case everywhere except for the very top and bottom layers of dipoles, where the charge densities will be $\rho = +nq$ and $\rho = -nq$ respectively.

Now consider either Gaussian surface shown in the figure, and assume that the electric field $\vec{E}_P$ is contained in the dielectric. Then the electric field inside the capacitor must be given by:

$$\oint_{\text{surface}} \vec{E}_P \cdot d\vec{A} = \frac{1}{\varepsilon_0}\oint_{\text{volume}} \rho\, dV \rightarrow E_P \cancel{A} = \frac{nq\cancel{A}d}{\varepsilon_0}.$$

The electric field due to the polarization is in the opposite direction as the dipole moments, and is therefore opposite the polarization vector. Now $\vec{P} = nqd\hat{z}$, so we can write the electric field due to the polarization of the dielectric, in terms of the polarization vector $\vec{P}$:

$$\vec{E}_P = \frac{1}{\varepsilon_0} nq\, d(-\hat{z}) = -\frac{1}{\varepsilon_0}\vec{P}..$$

Thought Experiment 4.6: The Continuity of Molecular Charge

Now consider a region inside a neutral dielectric with a dipole moment per volume $\vec{P}(\vec{r})$. There may be a corresponding atomic scale charge density depending on the orientation of the small-scale polarization vector.

To see this, we draw an arbitrary Gaussian surface inside the dielectric, and notice that the leftover charge, just inside the surface, will be given by:

$$Q_{\text{inside}} = -\oint_{\text{surface}} \vec{P} \cdot \mathrm{d}\vec{A}\,.$$

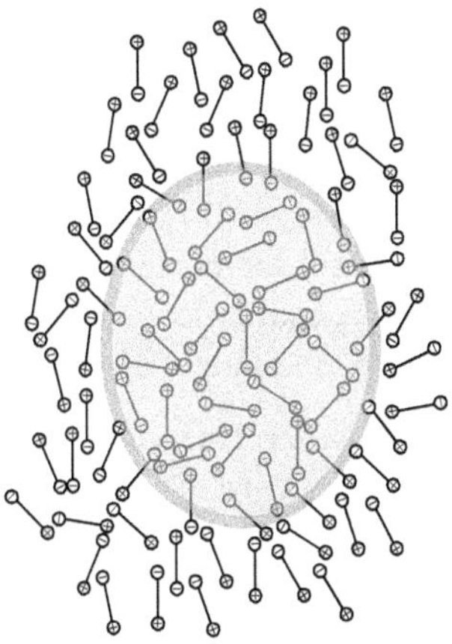

$$Q_{\text{surface}} = \oint_{\text{surface}} \vec{P} \cdot \mathrm{d}\vec{A}$$

This is simply the negative of the leftover charge, just outside of the surface, which we had above. Now taking the limit as the volume of the Gaussian surface approaches zero yields:

$$\rho_{\text{bound}} = \lim_{\delta V \to 0} \frac{Q_{\text{inside}}}{\delta V} = -\lim_{\delta V \to 0} \tfrac{1}{\delta V} \oint_{\text{surface}} \vec{P} \cdot \mathrm{d}\vec{A} = -\vec{\nabla} \cdot \vec{P}\,,$$

so we see that the molecular charge density is simply the negative of the divergence of the polarization vector. This is often called the *bound charge density*, because the charge is bound to the atoms, and therefore cannot flow even if the medium is conductive.

Next consider our first, and most fundamental, law that charge must be conserved. We expressed this before in the form of the continuity equation for charge that could flow, which is still correct on any scale where classical physics applies. However, it is also correct to write the continuity equation for *all* of the charge, whether bound or not, as:

$$\vec{\nabla} \cdot \vec{J}_{\text{total}} = -\tfrac{\partial}{\partial t} \rho_{\text{total}} \;\rightarrow\; \vec{\nabla} \cdot \left(\vec{J} + \vec{J}_{\text{bound}} \right) = -\tfrac{\partial}{\partial t} \left(\rho + \rho_{\text{bound}} \right) \;\rightarrow\; \left(\vec{\nabla} \cdot \vec{J} + \tfrac{\partial}{\partial t} \rho \right) + \left(\vec{\nabla} \cdot \vec{J}_{\text{bound}} + \tfrac{\partial}{\partial t} \rho_{\text{bound}} \right) = 0.$$

Since the continuity equation applies for free charge, therefore, it must hold for bound charge:

$$\vec{\nabla} \cdot \vec{J}_{\text{bound}} = -\tfrac{\partial}{\partial t} \rho_{\text{bound}}\,.$$

However, you should remember that the concept of bound charge is really a construct of our toy model. The atomic world may not be so simple. Classical physics, however, does work on scales larger than atoms, so we write this in terms of the polarization:

$$\vec{\nabla} \cdot \vec{J}_{\text{bound}} = -\tfrac{\partial}{\partial t} \left(-\vec{\nabla} \cdot \vec{P} \right) \;\rightarrow\; \vec{\nabla} \cdot \vec{J}_{\text{bound}} = \vec{\nabla} \cdot \tfrac{\partial}{\partial t} \vec{P} \;\rightarrow\; \vec{J}_{\text{bound}} = \tfrac{\partial}{\partial t} \vec{P}\,.$$

And now we can write down the total current density as:

$$\vec{J}_{\text{total}} = \vec{J} + \vec{J}_{\text{bound}} = \vec{J} + \tfrac{\partial}{\partial t} \vec{P}\,,$$

so $\frac{\partial \vec{P}}{\partial t}$ is called the *polarization current density*.

The total charge continuity equation therefore reverts to:

$$\vec{\nabla} \cdot \left(\vec{J} + \tfrac{\partial}{\partial t} \vec{P} \right) = -\tfrac{\partial}{\partial t} \left(\rho - \vec{\nabla} \cdot \vec{P} \right) \;\rightarrow\; \vec{\nabla} \cdot \vec{J} + \cancel{\vec{\nabla} \cdot \tfrac{\partial}{\partial t} \vec{P}} = -\tfrac{\partial}{\partial t} \rho + \cancel{\tfrac{\partial}{\partial t} \vec{\nabla} \cdot \vec{P}} \;\rightarrow\; \vec{\nabla} \cdot \vec{J} = -\tfrac{\partial}{\partial t} \rho\,.$$

Keep in mind that we only expect classical models of matter to work on scales much larger than atoms, where the bound charge density is always zero. However, continuously neutral matter can still have an overall dipole moment!

To stress that classical physics does not necessarily apply on atomic scales, our notation slightly differs from some other authors. In this book ρ and $\vec{J}$ always represent the free charge density and the free current density. In many other texts, ρ and $\vec{J}$ represent the sum of the free and bound charge and current densities.

4.2 *Gauss's Law*

In Chapter 1 we applied the concept of the divergence to conserved scalar quantities, including charge and energy. We drew Gaussian surfaces around charging objects, and related this to the total current flowing into the same Gaussian surface. Moreover, when the charging object was spherical the current density followed an inverse square law with distance (p. 30). In Chapter 3 we applied the concept of continuity to energy (p. 80), where we found that sunlight also follows an inverse-square law with distance (p. 81).

In Chapter 2 we discussed the inverse-square force laws of Newton and Coulomb. Might these also relate to some conserved quantity? If so, what? Maybe there is something, be in a disturbance in the aether or some governing particle, that flows from one place to another to make these laws work? Whether or not any of these ideas pan out, the force laws of Newton and Coulomb invoke action at a distance, so there must be a better way of representing them.

Joseph-Louis Lagrange and Carl Friedrich Gauss represented Newton's law of gravity by analogy with fluid dynamics, showing mathematically that an inverse-square force law can be represented locally in terms of the mass density and the gravitational field. This representation turns out to be very handy in astronomy, where the fluids are gravitationally important. James Clerk Maxwell applied this formulation to Coulomb's law, called it Gauss's law, and made it one of his four equations of electrodynamics.

Example 4.3: Gauss's Law of Gravity

Imagine that the earth were a uniform sphere with mass M and radius R then according to Newton's Law of Gravity, the gravitational field $\left(\vec{g} \equiv \frac{\vec{F}}{m}\right)$ above the Earth's surface, a distance r from its center is:

$$\vec{g} = -\frac{GM}{r^2}\hat{r}\,.$$

Notice that this is similar to the leaking balloon on page 30, where the current density also followed a similar inverse square law. Recall in that case, the divergence of the current density outside the balloon was zero. Now we will take the divergence of the gravitational field:

$$\vec{\nabla}\cdot\vec{g} = \vec{\nabla}\cdot\left(-\frac{GM}{r^2}\hat{r}\right) = -\frac{GM}{r^2}\frac{\partial}{\partial r}\left(r^2\frac{1}{r^2}\right) = 0\,.$$

Now let us investigate the gravitational field inside our uniform Earth. In this case:

$$\vec{g} = -\frac{GM_{\text{inside}}}{r^2}\hat{r} = -\frac{G\left(\rho_m\frac{4\pi}{3}r^3\right)}{r^2}\hat{r} = -\tfrac{4\pi}{3}G\rho_m r\,\hat{r}\,,\,.$$

where ρ_m represents the uniform mass density: $\rho_m = \dfrac{M}{\frac{4\pi}{3}R^3}$.

The divergence now is:

$$\vec{\nabla}\cdot\vec{g} = \vec{\nabla}\cdot\left(-\tfrac{4\pi}{3}G\rho_m r\,\hat{r}\right) = \frac{1}{r^2}\tfrac{\partial}{\partial r}\left(-\tfrac{4\pi}{3}G\rho_m r^3\right) = -4\pi G\rho_m\frac{1}{r^2}\tfrac{\partial}{\partial r}\left(\tfrac{1}{3}r^3\right) = -4\pi G\rho_m\,.$$

So, by example, we can make a conjecture that Newton's law of gravity can also be written:

$$\vec{\nabla}\cdot\vec{g} = -4\pi G\rho_m .$$

Lagrange in 1773 mathematically proved this relationship, followed by Gauss in 1813.[60]

Thought Experiment 4.7: Gauss's Law

Since Coulomb's law and Newton's law of gravity follow the same mathematical form, the following law must hold in free space:

$$\vec{\nabla}\cdot\vec{E} = 4\pi K\rho = \frac{\cancel{4\pi}\rho}{\cancel{4\pi}\varepsilon_0} = \frac{\rho}{\varepsilon_0}.$$

However, as Faraday demonstrated (p. 116), Coulomb's law also depended on the permittivity of the medium involved. From our experience thus far, we would guess that in a linear medium we would simply replace ε_0 with ε. While this may be so, we can do better by returning to the toy model of dielectrics and the concept of the bound charge.

Recall from Thought Experiment 4.6 (p. 124) that the total charge density, including the molecularly bound charge, is given by:

$$\rho_{\text{total}} = \rho - \vec{\nabla}\cdot\vec{P}.$$

Since the forces between charges can depend on small variations in total charge, so must the resulting electric field. Therefore, taking matter into account, Gauss's law becomes:

$$\vec{\nabla}\cdot\vec{E} = \frac{\rho_{\text{total}}}{\varepsilon_0} = \frac{\rho - \vec{\nabla}\cdot\vec{P}}{\varepsilon_0} = \frac{\rho}{\varepsilon_0} - \frac{1}{\varepsilon_0}\vec{\nabla}\cdot\vec{P}.$$

Fundamental Law 4: Gauss's Law

$$\vec{\nabla}\cdot\vec{E} = \frac{\rho}{\varepsilon_0} - \frac{1}{\varepsilon_0}\vec{\nabla}\cdot\vec{P}.$$

Problem 4.4: Gauss's Law in Integral Form

Use the definition of the divergence to show that Gauss's law can be written in integral form as:

$$\oint_{\text{surface}} \vec{E}\cdot \mathrm{d}\vec{A} = \tfrac{1}{\varepsilon_0}Q - \tfrac{1}{\varepsilon_0}\oint_{\text{surface}} \vec{P}\cdot \mathrm{d}\vec{A}.$$

Example 4.4: The Parallel Plate Capacitor and Gauss's Law

Consider the a parallel plate capacitor, with two plates each of surface area A, opposite surface charge densities $-\sigma,+\sigma$, and a linear medium of permittivity ε sandwiched in the narrow gap.

To solve this problem, start with Gauss's law, take the volume integral of both sides of it, then apply the divergence theorem. The result is Gauss's law stated in integral form:

$$\vec{\nabla}\cdot\vec{E} = \tfrac{1}{\varepsilon_0}\rho - \tfrac{1}{\varepsilon_0}\vec{\nabla}\cdot\vec{P} = \frac{\rho}{\varepsilon} \quad\rightarrow\quad \oint_{\text{surface}} \vec{E}\cdot d\vec{A} = \oint_{\text{volume}} \left(\frac{\rho}{\varepsilon}\right) dV = \frac{1}{\varepsilon}Q_{\text{inside}}.$$

---

[60] J.L. Lagrange, "Sur l'attraction des spheroides elliptiques" ("On the Attraction of Elliptical Spheroids"), Mémoires de l'Académie Royale des Sciences et Belles-Lettres de Berlin (1773) **4**: 121-48.

Now draw a Gaussian surface around the entire capacitor, and apply Gauss's law to calculate the total electric flux through the whole surface:

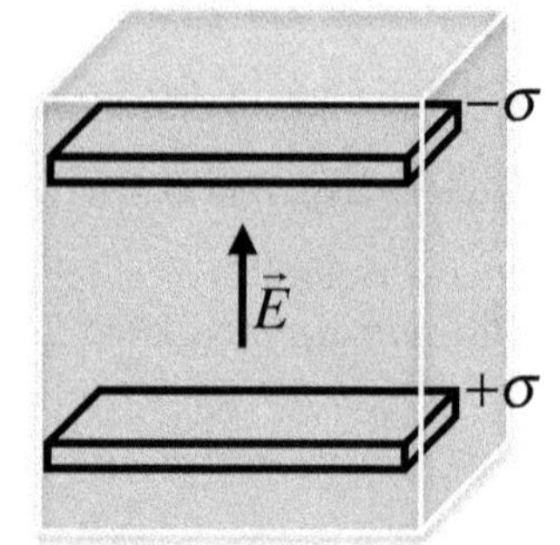

$$\oint_{\text{surface}} \vec{E}\cdot d\vec{A} = \oint_{\text{volume}} \left(\frac{\rho}{\varepsilon}\right) dV = -\frac{\sigma}{\varepsilon}A + \frac{\sigma}{\varepsilon}A = 0$$

Expanding this out for each face gives us:

$$0 = \oint_{\text{surface}} \vec{E}\cdot d\vec{A} = \int_{\text{top}} \vec{E}\cdot d\vec{A} + \int_{\text{bottom}} \vec{E}\cdot d\vec{A} + \cancel{\int_{\text{sides}} \vec{E}\cdot d\vec{A}}\ .$$

The last term is zero for the idealized capacitor because the electric field is perpendicular to the sides of our Gaussian surface. In fact, that is why we drew the Gaussian surface this way.

Investigating the top and bottom faces of the Gaussian surface, we see the following:

$$0 = \vec{E}_{\text{above}} \cdot A_{\text{top}} (+\hat{z}) + \vec{E}_{\text{below}} \cdot A_{\text{bottom}} (-\hat{z})$$

$$\vec{E}_{\text{above}} = \vec{E}_{\text{below}}$$

Thus, Gauss's law states that if there is an electric field above or below the plates, those fields must be equal and in the same direction. From symmetry, we will assume that $E_{\text{above}} = E_{\text{below}} = 0$.

Now consider a Gaussian surface surrounding only the top plate. Applying Gauss's law, we obtain the following result:

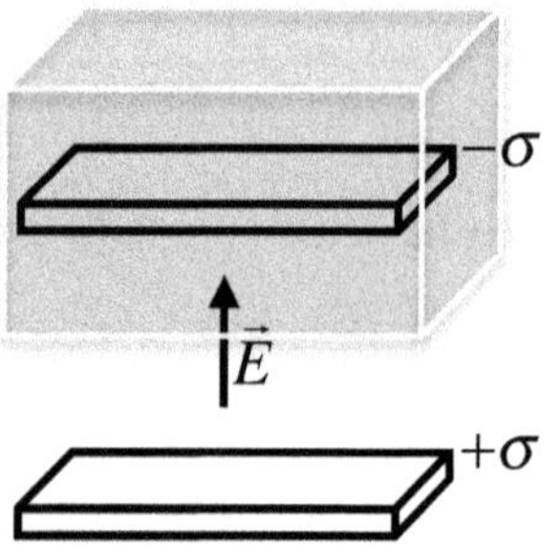

$$\oint_{\text{surface}} \vec{E}\cdot d\vec{A} = \oint_{\text{volume}} \left(\frac{\rho}{\varepsilon}\right) dV$$

$$\int_{\text{top}} \cancel{\vec{E}}\cdot d\vec{A} + \int_{\text{bottom}} \vec{E}\cdot d\vec{A} + \cancel{\int_{\text{sides}} \vec{E}\cdot d\vec{A}} = \frac{-\sigma A}{\varepsilon}.$$

Here we have used our prior knowledge from symmetry that the electric field above the plate is zero and the electric field has no sideways component anywhere. This leaves only the face located between the plates (the "bottom" surface in this instance) to contribute to the total electric flux throughout the surface. We can now find the electric field between the plates:

$$\vec{E}\cdot\left(\cancel{-}\cancel{A}\,\hat{z}\right) = \frac{\cancel{-}\sigma \cancel{A}}{\varepsilon} \quad \rightarrow \quad \vec{E}\cdot\hat{z} = \frac{\sigma}{\varepsilon}.$$

Knowing that there are no sideways components to the electric field, because the gap is narrow, we can write:

$$\vec{E} = \frac{\sigma}{\varepsilon}\hat{z}.$$

This agrees with our previous result, but we have followed a very different route to obtain it.

What about the bottom plate? If it did not exist, the resulting electric field would not agree with the expression that we have just derived. Yet we did not take the bottom plate into account when obtaining our result. How is this possible? The answer to this question comes in our assumption that there is no electric field above the plate. Had there not been a bottom plate, we would have included a different set of a priori assumptions based on a different symmetry.

We now apply a Gaussian surface to the bottom plate, for completeness, and we find the same thing as did above:

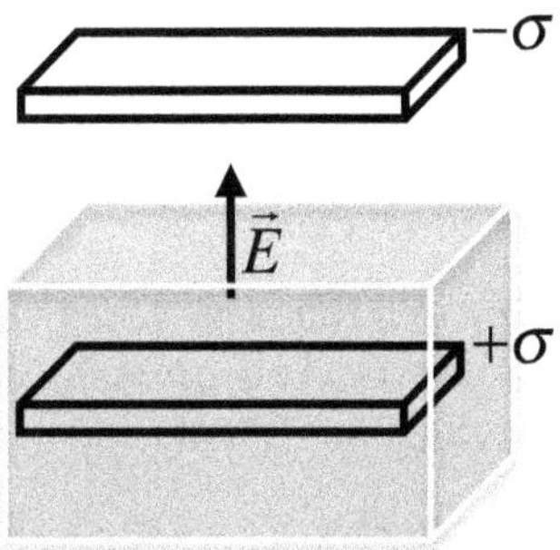

$$\oint_{\text{surface}} \vec{E}\cdot d\vec{A} = \oint_{\text{volume}} \left(\frac{\rho}{\varepsilon}\right) dV$$

$$\int_{\text{top}} \vec{E}\cdot \mathrm{d}\vec{A} + \int_{\text{bottom}} \cancel{\vec{E}}\cdot \mathrm{d}\vec{A} + \cancel{\int_{\text{sides}} \vec{E}\cdot \mathrm{d}\vec{A}} = \frac{\sigma A}{\varepsilon}$$

Now, solving for the electric field between the plates we have:

$$\vec{E}\cdot\left(\cancel{A}\,\hat{z}\right) = \frac{\sigma \cancel{A}}{\varepsilon} \quad\rightarrow\quad \vec{E}\cdot\hat{z} = \frac{\sigma}{\varepsilon}$$

Since there are no sideways components to the electric field:

$$\vec{E} = \frac{\sigma}{\varepsilon}\hat{z}.$$

This agrees with our prior results for the electric field between the plates.

Finally we will consider a surface between the plates that contains no charge. How can this be consitent with the presence of an electric field in between the plates?

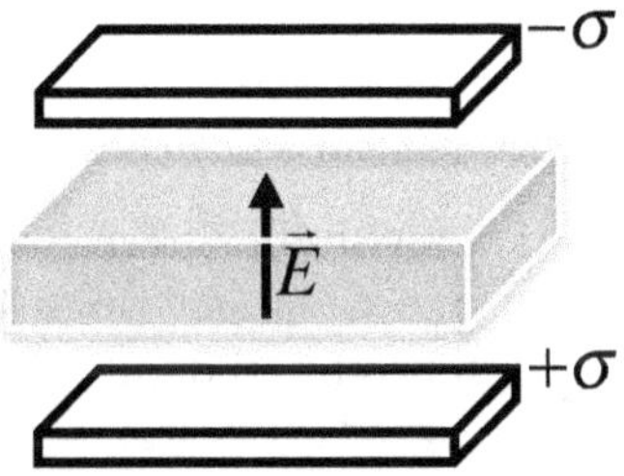

$$\oint_{\text{surface}} \vec{E}\cdot d\vec{A} = \oint_{\text{volume}} \left(\frac{\rho}{\varepsilon}\right) dV$$

$$\int_{\text{top}} \vec{E}\cdot \mathrm{d}\vec{A} + \int_{\text{bottom}} \vec{E}\cdot \mathrm{d}\vec{A} + \cancel{\int_{\text{sides}} \vec{E}\cdot \mathrm{d}\vec{A}} = 0$$

$$\vec{E}_{\text{top face}}\cdot\left(\mathrm{d}A\,\hat{z}\right) + \vec{E}_{\text{bottom face}}\cdot\left(-\mathrm{d}A\,\hat{z}\right) = 0$$

$$\vec{E}_{\text{top face}} = \vec{E}_{\text{bottom face}}$$

Thus, the electric field between the plates must be uniform.

Thought Experiment 4.8: A Hollow Metal Sphere

Now consider a hollow metal sphere located inside of an external electric field. What is the electric field inside of its cavity?

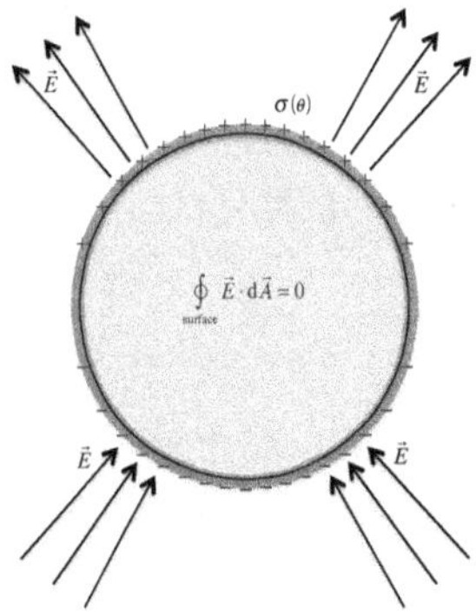

First consider what would happen if there were an electric field inside the conductor itself—current would flow. However, without a loop the current cannot flow forever. The builds up until there is no electric field. The only exception is normal to the surface, where the charge is prevented from moving. Thus, inside of all static conductors, $\vec{E} = 0$, and on the surface $\vec{E}\times\vec{A} = 0$.

We now draw a Gaussian surface such that the whole surface is contained inside the conductor. Now, because there is no net field within the conductor, it must also follow that:

$$\oint_{\text{surface}} \vec{E}\cdot \mathrm{d}\vec{A} = \frac{Q}{\varepsilon_0} = 0 .$$

We therefore determine that all of the charge separation takes place on the outer surface of the conductor. In other words, the inner surface of the conductor is net neutral.

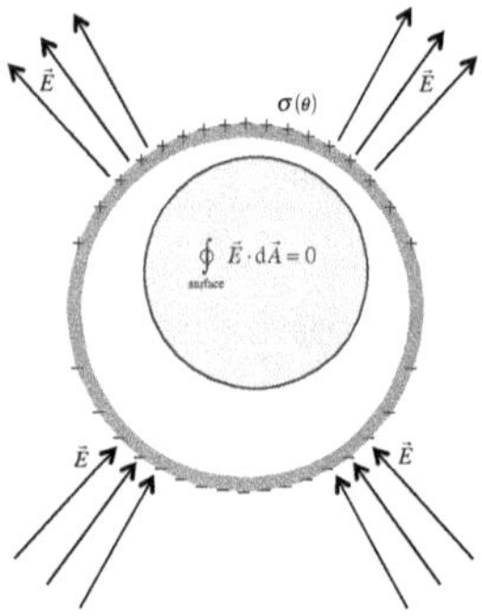

We can now draw any arbitrary surface inside the shell, and the enclosed charge will be zero. Thus, because the electric field was zero on the inner surface of the conductor, it is also zero everywhere inside the conductor.

This is referred to as a Faraday cage, after Michael Faraday, and it is the reason you are safe during a thunderstorm inside your car, a building, or even inside of a wire mesh cage.

This is also why van de Graaff's machine works. No matter how much charge builds up on the outside of the sphere, the electric field inside the conducting sphere will be zero, and therefore the electric field inside the whole sphere will be zero.

Problem 4.5: A Faraday Cage

a) Show that there will be no electric field inside of any hollow conductor.

b) The MIT van de Graaff generator (pp. 113-114) is now the main attraction at the Theater of Electricity at the Museum of Science in Boston. The presenter sits in what looks like a large birdcage. At some point during the show, the presenter rises and lightning strikes the cage. Clearly explain why neither the presenter, nor the electronic controls, are hurt.

c) If you visit the Green Bank Observatory, which has some of the largest radio antennas in the world, the computer lab is completely sealed with copper screening. To enter you must go through two metal doors, which, like an airlock, may not be open at the same time. Explain why they designed it this way.

Thought Experiment 4.9: Electric Flux

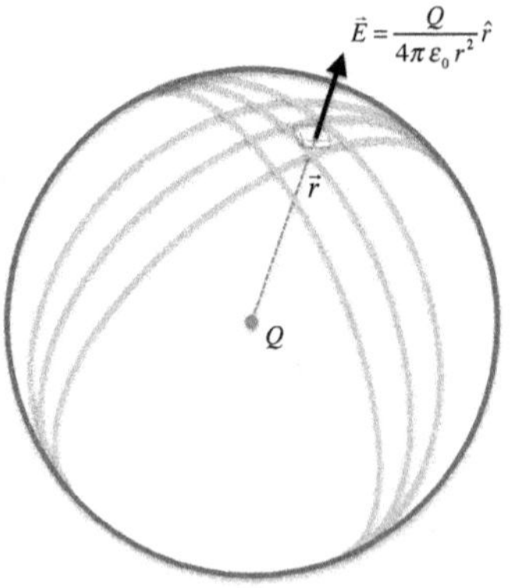

Consider an imaginary spherical shell of radius r surrounding a point charge located at the shell's center, and imagine the electric field vector poking through this spherical shell. Now consider the surface integral of the electric field through the surface of the shell, which is given by:

$$\Phi_E \equiv \oint_{\text{surface}} \vec{E}\cdot d\vec{A} .$$

where Φ_E is called the *electric flux*, and has SI units of $\frac{\text{N}\cdot\text{m}^2}{\text{C}}$.

The relationship between electric flux and the electric field is, therefore, mathematically similar to the relationship between current and current density. For this reason positive, and negative, charges are considered *sources*, and *sinks*, of electric flux—in much the same way as anodes, and cathodes are considered sources, and sinks, of current.

We can also write Gauss's law as:

$$\Phi_E \equiv \oint_{\text{surface}} \vec{E}\cdot d\vec{A} = \tfrac{1}{\varepsilon_0}Q - \tfrac{1}{\varepsilon_0}\oint_{\text{surface}} \vec{P}\cdot \mathrm{d}\vec{A}.$$

Notice that, in free space $(\vec{P}=0)$, the total electric flux through the spherical shell is only a function of the enclosed charge Q.

Discussion 4.2: The Term *Flux* in Physics

One of the most confusing terms in physics is *flux*. In this book we have used the term *flux* for two very different concepts:

- The vector rate of flow of a conserved quantity per area. For example, $\vec{J}$ is the charge flux and $\vec{\mathbb{S}}$ is the energy flux.
- The multiplication of a field vector with an area vector. For example, $\Phi_E = \int \vec{E}\cdot \mathrm{d}\vec{A}$.

This is very confusing, because these are really mathematically opposite concepts. In the first case, flux is a local vector quantity that represents the flow of a conserved scalar quantity. However in the second case, flux is a global scalar quantity that comes from integrating a local vector. Had we the choice of terminology, we would have picked a different word to mean one of these two ideas.

Thought Experiment 4.10: The Electric Flux Through a Cube

Consider the simple free space Coulomb problem, but, instead of drawing an imaginary spherical shell around the point charge, we now surround the charge with an imaginary cube such that the center of each face is a distance a from the charge. What is the total electric flux through the cube?

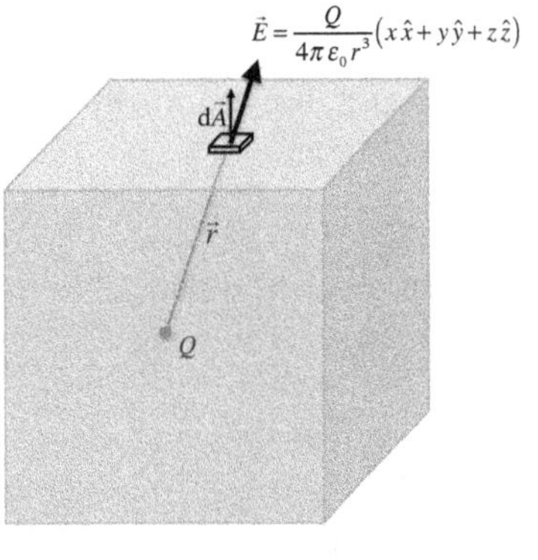

Due to symmetry, the electric flux through each face of the cube should be equal, so we should be able to calculate one and multiply by six. Solving for the face located at $x=a$ we see:

$$\Phi_E \equiv \oint_{\text{surface}} \vec{E}\cdot d\vec{A} = \oint_{\text{surface}} \left(\frac{Q}{4\pi\varepsilon_0 r^2}\hat{r}\right)\cdot d\vec{A} = \oint_{\text{surface}} \left(\frac{Q}{4\pi\varepsilon_0 r^3}\vec{r}\right)\cdot d\vec{A}$$

$$= 6\int_{-a}^{a}\int_{-a}^{a} \frac{Q}{4\pi\varepsilon_0 r^3}\vec{r}\cdot(\hat{x}\,\mathrm{d}y\mathrm{d}z) = 6\int_{-a}^{a}\int_{-a}^{a} \frac{Qa}{4\pi\varepsilon_0\left(a^2+y^2+z^2\right)^{3/2}}\mathrm{d}y\mathrm{d}z.$$

Notice that this integral is even in both x and y, so:

$$\Phi_E = 24\int_0^a\int_0^a \frac{Qa}{4\pi\varepsilon_0\left(a^2+y^2+z^2\right)^{3/2}}\mathrm{d}y\mathrm{d}z = \frac{6Q}{\pi\varepsilon_0}\int_0^a\int_0^a \frac{a}{\left(a^2+y^2+z^2\right)^{3/2}}\mathrm{d}y\mathrm{d}z.$$

We will now make a unitless variable substitution of $\gamma = y/a$ and $\xi = z/a$, so:

$$\Phi_E = \frac{6Q}{\pi\varepsilon_0}\int_0^1\int_0^1 \frac{1}{\left(1+\gamma^2+\xi^2\right)^{3/2}}\mathrm{d}\gamma\,\mathrm{d}\xi.$$

Evaluating the double integral, we get:

$$\int_0^1\int_0^1 \frac{1}{\left(1+\gamma^2+\xi^2\right)^{3/2}}\,\mathrm{d}\gamma\,\mathrm{d}\xi = \int_0^1 \frac{\mathrm{d}\xi}{\left(\xi^2+1\right)\sqrt{\xi^2+2}} = \frac{\pi}{6}.$$

So we obtain:

$$\Phi_E \equiv \oint_{\text{surface}} \vec{E}\cdot d\vec{A} = \frac{Q}{\varepsilon_0}.$$

Note the significance of this result. It suggests that the total electric flux through a closed surface does not depend upon the shape of the surface. The flux will be the same whether the closed surface, surrounding the charge, is a cube, a sphere, or an ellipsoid. With this example as our guide, we tentatively assert the following: the total electric flux through any closed surface depends only upon the total amount of charge that is enclosed by that surface.

Before we prove this, we will consider one more special case.

Thought Experiment 4.11: Electric Flux Through an Empty Surface

Consider a point charge Q located at a point in free space that we will take as the origin of some system of coordinates—just as we did in the previous two examples. Only now we will draw our imaginary closed surface so that it does not enclose the source Q of the electric field. What is the total flux through our imaginary surface?

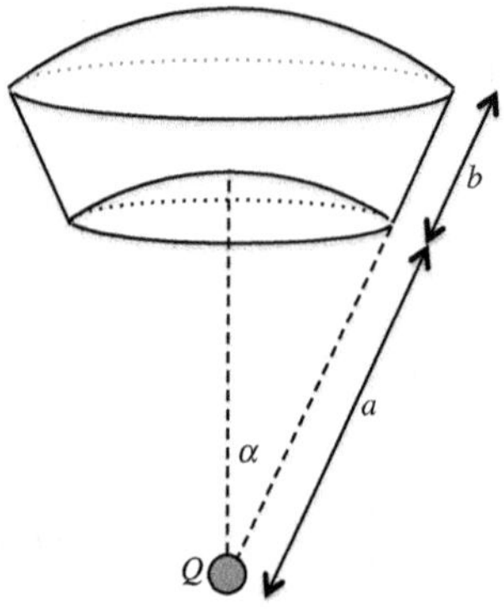

For simplicity, let us make a wedge surface with a range in spherical polar coordinates of $a<r<a+b$ and $\theta<\alpha$, as shown in the diagram.

There is no electric flux through the side of the surface, because the electric field is parallel to the surface (i.e. $\vec{E}\perp \mathrm{d}\vec{A}$), so the total electric flux is:

$$\Phi_E \equiv \oint_{\text{surface}} \vec{E}\cdot\mathrm{d}\vec{A} = \int_{\text{Bottom}} \vec{E}\cdot\mathrm{d}\vec{A} + \int_{\text{Top}} \vec{E}\cdot\mathrm{d}\vec{A} + \cancel{\int_{\text{Side}} \vec{E}\cdot\mathrm{d}\vec{A}}$$

Using Coulomb's law to calculate the field and inserting the result in the remaining integrals on the right hand side of this expression yields the total flux through the surface:

$$\Phi_E = \int_{\text{Bottom}} \vec{E}\cdot\mathrm{d}\vec{A} + \int_{\text{Top}} \vec{E}\cdot\mathrm{d}\vec{A} = \int_{\text{Bottom}} \left(\frac{Q\,\hat{r}}{4\pi\varepsilon_0 a^2}\right)\cdot\mathrm{d}\vec{A} + \int_{\text{Top}} \left(\frac{Q\,\hat{r}}{4\pi\varepsilon_0 (a+b)^2}\right)\cdot\mathrm{d}\vec{A}.$$

This integral simplifies to zero:

$$\Phi_E = \int_0^\alpha\int_0^{2\pi} \left(\frac{Q\hat{r}}{4\pi\varepsilon_0}\right)\cdot\left(\sin\theta\,\mathrm{d}\varphi\,\mathrm{d}\theta(-\hat{r})\right) + \int_0^\alpha\int_0^{2\pi} \left(\frac{Q\hat{r}}{4\pi\varepsilon_0}\right)\cdot\left(\sin\theta\,\mathrm{d}\varphi\,\mathrm{d}\theta\,\hat{r}\right)$$

$$= \frac{Q}{4\pi\varepsilon_0}\int_0^\alpha\int_0^{2\pi}\left(-\sin\theta\,\mathrm{d}\varphi\,\mathrm{d}\theta + \sin\theta\,\mathrm{d}\varphi\,\mathrm{d}\theta\right) = 0.$$

This is another example where t electric flux through a closed surface will be zero, if there is not enclosed charge—even if a charge is located just outside the surface.

Derivation 4.1: Gauss's Law in Free Space

Each one of the surfaces we constructed in the previous three thought experiments are examples of a Gaussian surface. Like an inflated balloon that has been tied off, a Gaussian surface is always closed. Moreover, a Gaussian surface need not coincide with any physical surface. In most applications, in fact, it is an *imaginary surface* that is used to evaluate the flux of the electric field at points in space which happen to coincide with the location of the surface.

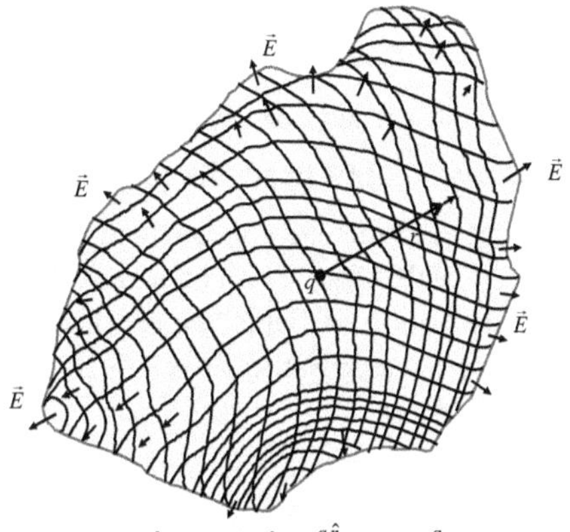

$$\Phi_E = \oint_{\text{surface}} \vec{E}\cdot d\vec{A} = \oint_{\text{surface}} \frac{q\hat{r}}{4\pi\varepsilon_0 r^2}\cdot d\vec{A} = \frac{q}{\varepsilon_0}$$

The results of these three previous thought experiments suggest that an equivalent way of expressing Coulomb's law is the following: *the electric flux through any Gaussian surface is equal to the net charge enclosed by that surface divided by* ε_0. This is another statement of Gauss's law in free space, but we only showed this for particular Gaussian surfaces rather than for *any* Gaussian surface.

First consider a point charge q located at the origin completely surrounded by any Gaussian surface. Then let's calculate the total flux through the Gaussian surface Φ_E due to the point charge q using Coulomb's law:

$$\Phi_E = \oint_{\text{surface}} \vec{E}\cdot d\vec{A} = \oint_{\text{surface}} \frac{q}{4\pi\varepsilon_0 r^2}\hat{r}\cdot d\vec{A} = \frac{q}{4\pi\varepsilon_0}\oint_{\text{surface}} \frac{\hat{r}\cdot d\vec{A}}{r^2},$$

where $\vec{r}$ is the position of the Gaussian surface with respect to the point charge.

Now consider a patch of the Gaussian surface $\delta\vec{A}$, the dot product with $\hat{r}$ is area of the shadow that $\delta\vec{A}$ would cast in the $\hat{r}$. We can now express the projected area $\hat{r}\cdot d\vec{A}$ in spherical coordinates, as shown in the diagram. We see that an element of outward surface area, projected onto the plane normal to, can be found as:

$$\hat{r}\cdot\delta\vec{A} = (r\,\delta\theta)(r\sin\theta\,\delta\varphi) = r^2\sin\theta\,\delta\theta\,\delta\varphi.$$

We can now write the total electric flux through an arbitrary closed surface surrounding the point charge as:

$$\Phi_E = \frac{q}{4\pi\varepsilon_0}\oint_{\text{surface}} \frac{\cancel{r^2}\sin\theta\,d\theta\,d\varphi}{\cancel{r^2}} = \frac{q}{4\pi\varepsilon_0}\int_0^{2\pi}\int_0^{\pi}\sin\theta\,d\theta\,d\varphi.$$

Notice how essential it is that the Coulomb field is a central force (i.e. that the field lies in the direction of $\hat{r}$) and that it exactly obeys an inverse square law. These two properties allow for the radial distance r to completely cancel out, so that we can evaluate the integral over every direction without regard to the distance between the point source and the Gaussian surface, so the flux becomes:

$$\Phi_E = \frac{q}{\cancel{4\pi}\varepsilon_0}\cancel{4\pi} = \frac{q}{\varepsilon_0}.$$

In the case of a single point charge, the total electric flux evaluated over any closed surface is proportional to the net charge enclosed by that surface.

Notice that we can move our Gaussian surface and as long as the point charge is inside of it and get the same result. Similarly we could move the charge around inside the surface and the total flux would still not change.

What if instead of merely placing the charge off-center inside the surface, we move the charge outside the surface altogether? Notice that the part of the surface in the foreground has negative flux, and the background has positive flux. On one hand there is more projected area on the far side of the Gaussian surface than the near side, on the other hand the Coulomb field is greater on the near side of Gaussian surface than the far side. These will completely cancel each other out, because the magnitude of the flux does not depend on the distance from the charge.

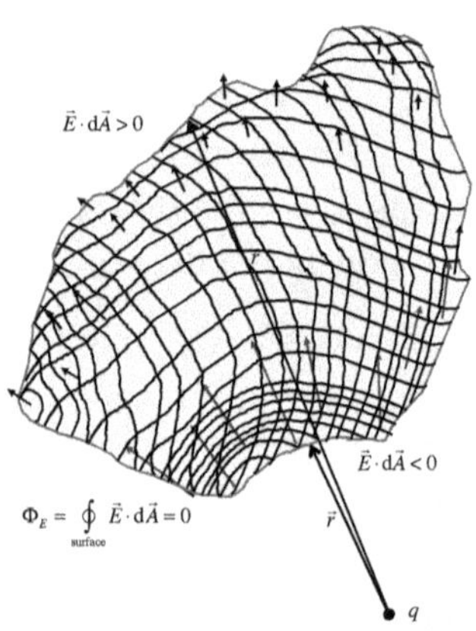

We have argued that for any point charge q inside the surface it will contribute a fixed electric flux $\frac{1}{\varepsilon_0}q$ through the surface, while a charge outside of the surface will not contribute at all to the electric flux. Now consider that there are many charges. By linear superposition of fields, we can now state the total flux through the surface as simply the flux contribution from each of them added together, or:

$$\Phi_E = \sum_{\text{inside}} \frac{q_i}{\varepsilon_0}.$$

Next let us consider an arbitrary charge density distribution $\rho(\vec{r})$, each charge element is simply a point charge so via the same superposition principle we can simply integrate over all the charge inside the Gaussian surface, and ignore any charge that lies outside of it, so:

$$\Phi_E = \sum_i \frac{q_i}{\varepsilon_0} = \frac{1}{\varepsilon_0} \oint_{\text{volume}} \rho \, dV = \frac{Q_{\text{enclosed}}}{\varepsilon_0}.$$

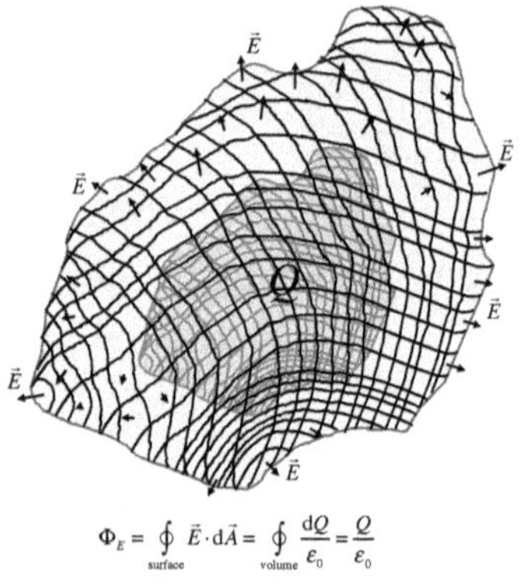

If we now take the limit as our volume of integration becomes small, we can use the definition of the divergence to recover the differential form of Gauss's law in free space:

$$\vec{\nabla}\cdot\vec{E} = \lim_{\delta V \to 0} \frac{\oint_{\text{surface}} \vec{E}\cdot d\vec{A}}{\delta V} = \lim_{\delta V \to 0} \frac{\frac{1}{\varepsilon_0}\oint_{\text{volume}} \rho \, dV}{\delta V} = \frac{\rho}{\varepsilon_0}.$$

We should be careful not to misinterpret Gauss's law. It asserts that $\vec{\nabla}\cdot\vec{E} = \frac{1}{\varepsilon_0}\rho$ only at those points in space where electric charge is actually present. The divergence of the electric field is zero everywhere else. The divergence of the electric field produced by a point charge, for example, is undefined at the location of the charge (this shouldn't surprise us since the charge density of a point charge is infinite) and zero everywhere else. If you sketch the electric field vectors produced by a positive charge, the vectors spread out with increasing distance r from the location of the charge. In other words, the vectors "diverge" from the charge in the radial direction. But the field also diminishes in magnitude as $1/r^2$, and the little vectors in the sketch will be smaller farther out from the charge. This latter effect just cancels the increased spreading to give a zero divergence everywhere outside the charge.

Discussion 4.3: Gauss's Law vs. Coulomb's Law

Which is more fundamental, Gauss's law or Coulomb's law? They are in fact equivalent statements, so far as electrostatics is concerned, and thus make identical predictions about the same physical quantities: charges and fields. But we have seen that Gauss's law is often easier to use. Does that make it the more fundamental, at least in some sense, than Coulomb's law?

Gauss's law, rather than the explicit form of Coulomb's law, is listed among the complete set of Maxwell's equations, and Gauss's law dovetails nicely with Maxwell's fluid representation of electrodynamics by allowing us to think about sources and sinks of the electric field. Ultimately, Maxwell's equations describing the fundamental laws of electrodynamics are mathematical statements of experimentally determined facts.

When Hertz succeeded in confirming Maxwell's theory of light, physicists the world over accepted Maxwell's ideas. These included all of Maxwell's ideas—even those that we now know to be wrong. When Einstein showed the fallacy of the aether, he reinterpreted electrodynamics, but did not change the physically measurable predictions of Maxwell's theory. As Hertz put it:

> To the question, "What is Maxwell's theory?" I know of no shorter or more definite answer than the following: "Maxwell's theory is Maxwell's system of equations."[61]

Most physicists treat Gauss's law as more fundamental than Coulomb's law for any number of theoretical and practical reasons. For one, Gauss's law holds true even when the source charge is in motion. This is not the case for Coulomb's law, which must be modified so as to account explicitly for the retarded time in the case of moving charges. Gauss's law is manifestly consistent with special relativity, but so is the modified version of Coulomb's law. Gauss's law, especially in its integral form, is particularly useful because it is by far the easiest way to handle electrostatic problems in situations involving a high degree of symmetry. And it is more useful, both practically and theoretically, especially when dielectric media are involved.

Problem 4.6: A Finite Sheet of Charge

Consider a flat rectangular sheet of dimensions $a \times b$ with a uniform surface charge density σ.

a) Use Coulomb's law to directly find an expression for the electric field a height z above the middle of the sheet.

b) Find an expression for the potential as a function height above the center of the sheet.

c) Take the negative gradient of the potential function from part (b) and show that it gives the electric field from part (a).

d) In the limit that the sheet is infinitely large, use Gauss's law to find the electric field above the middle of the sheet. Is this consistent with your finite answers above?

e) In the limit that $b \to \infty$ and $a \to 0$, show that you reproduce the line charge solution with $(\sigma a) \to \lambda$.

---

[61] H. Hertz, Electric Waves, trans. D. Jones (London: MacMillan and Co., 1893), 21.

Problem 4.7: The Disk of the Milky Way

The sun orbits the Milky Way at a distance called the *solar circle* of radius:

$r_\odot \approx 27{,}000\,\text{Light Years} \approx 8300\,\text{pc} \approx 8.3\,\text{kpc} \approx 2.5\times 10^{20}\,\text{m.},$

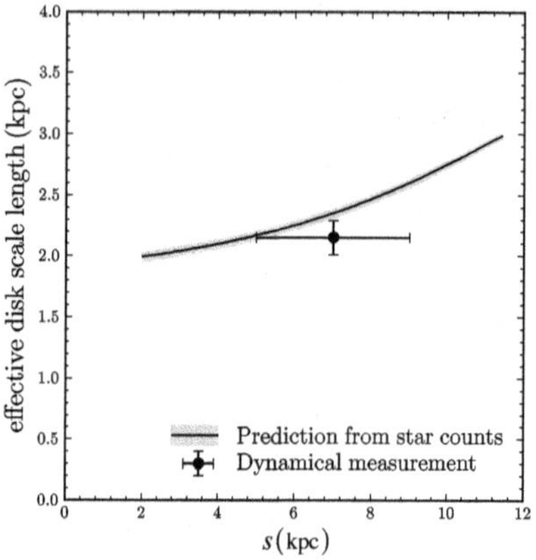

with an orbital speed $v_\varphi \approx 220\,{}^{\text{km}}\!/\!{}_{\text{s}}$. A recent paper[62] fit the mass density profile of the Galactic Disk within the range $4\,\text{kpc} \lesssim s \lesssim 9\,\text{kpc}$ to the following functional form:

$$\rho_{\text{disk}}(s,z) \approx \rho_\odot \exp\left(-\left(\tfrac{(s-r_\odot)}{R}+\tfrac{|z|}{h}\right)\right),$$

where they assumed a scale height of $h \approx 300\,\text{pc}$, and measured a radial scale length of $R \approx 2.15 \pm 0.14$ kpc and the surface density at the location of the solar circle to be:

$$\sigma_\odot \equiv \int_{-1.1\text{kpc}}^{1.1\text{kpc}} \left(\rho(r_\odot,z)\right)dz \approx \int_{-1.1\text{kpc}}^{1.1\text{kpc}} \rho_\odot e^{-\left(\frac{(s-r_\odot)}{R}+\frac{|z|}{h}\right)} dz \approx 68 \pm 4\,{}^{M_\odot}\!/\!{}_{\text{pc}^2}.$$

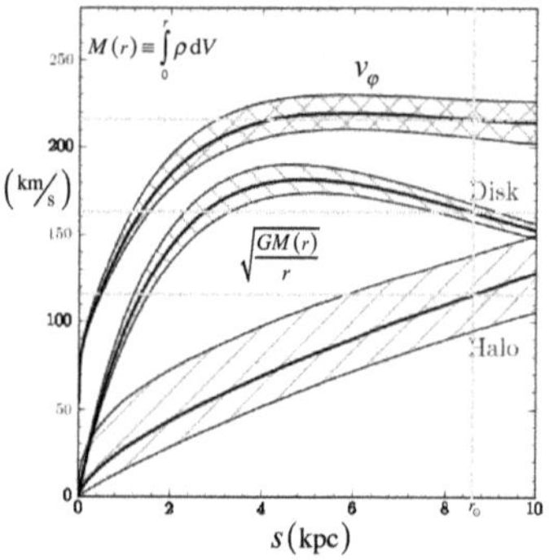

a) What is the density of the galaxy, $\rho_\odot$, near us? Express your answer in solar masses per cubic parsec.

b) Using Gauss's formulation of Newton's Law of Gravity, $\vec{\nabla}\cdot\vec{g} = 4\pi G \rho_m$, find the vertical component of the gravitational field, g_z, in the galactic plane. Express your result in units of ${}^{\text{N}}\!/\!{}_{\text{kg}}$. How does this compare to the gravitational field on Earth's surface ($g_\oplus = 9.8\,{}^{\text{N}}\!/\!{}_{\text{kg}}$)? (note $1M_\odot = 2.0\times 10^{30}\,\text{kg}$).

c) The Sun moves up and down in an approximately sinusoidal pattern. If the amplitude of the up and down motion is small compared to h, what is the period of this vertical motion? How often does the sun cross the plane of the Galaxy? How does this compare to the periodicity of mass extinctions on Earth, which have happened about every 26 million years on average?[63]

d) Estimate the $\hat{s}$ component of the gravitational field due to the Galactic Disk.

e) Assuming that their surface density profile holds all the way from the center of the Galaxy to its far reaches, what is the total mass of the Galactic disk? Express you answer in solar masses.

f) The Milky Way's disk is not the only contribution to its mass. It also has a spherically symmetrical component called the *halo*. The lower plot shows the orbital velocity of stars in the galaxy as a function of azimuthal radius s, and the relative contributions from the disk and the halo. From the plot, estimate the total mass, mass of the disk, and mass of the halo, inside of the solar circle.

---

[62] J., Bovy and H.W. Rix, "A direct dynamical measurement of the Milky Way's disk surface density profile, disk scale length, and dark matter profile at $4\,\text{kpc} \lesssim R \lesssim 9\,\text{kpc}$," The Astrophysical Journal, (2013) **779**, 115-145.

[63] D.M. Raup and J.J. Sepkoski, "Periodicity of extinctions in the geologic past," Proceedings of the National Academy of Sciences (1984), **81**, 801–805.

g) Their best-fit density profile for the halo is:

$$\rho_{\text{halo}} \approx \left(0.008 \pm 0.0025 \tfrac{M_\odot}{pc^3}\right)\left(\tfrac{r}{r_\odot}\right)^{-\alpha} : 0 < \alpha < 1.53 \cdot$$

where the best fit was at $\alpha \sim 0.5$. Considering just the halo, what would the galactic gravitational field be as a function of r in units of kiloparsecs?

h) What is the combined gravitational field taking into account both the halo and the disk? Render this in cylindrical coordinates.

i) Stars in the disks of other Galaxies orbit their galactic center at approximately the same orbital speed. Assuming this is also the case with the Milky Way, estimate the centripetal acceleration of these stars.

Example 4.5: A Conductor Inside a Capacitor

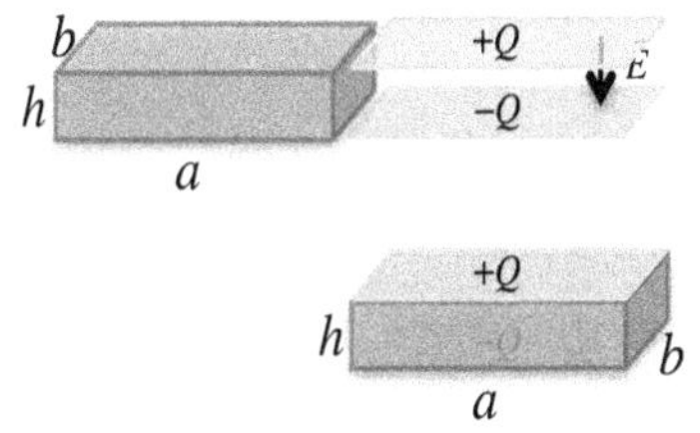

Imagine a parallel plate capacitor, whose plates have dimensions $a \times b$, a gap separation h and a charge $\pm Q$. What is the total potential energy due to the charge separation? To answer this, we simply imagine draining the capacitor of charge and ask how much work would be done. We first determine the work needed to transport each element of charge, dQ, from the upper to the lower plate, and then sum over all charge to find the work required to drain the capacitor of charge:

$$\mathbb{U} = \int_0^Q \left(\int_0^h \vec{E} \cdot d\vec{r} \right) dQ = \int_0^Q Eh dQ = \int_0^Q \frac{\sigma}{\varepsilon_0} h dQ = \frac{h}{ab\varepsilon_0} \int_0^Q Q dQ = \frac{hQ^2}{2ab\varepsilon_0}.$$

This is the potential energy due to the charge separation between the plates. Now let us insert a conducting slab, with a thin insulating coating and also of dimensions $a \times b \times h$, into the region between the charged plates. Now, the potential energy contained in the capacitor is zero, because the electric field is zero inside the conductor.

$$\mathbb{U} = \int_0^Q \left(\int_0^h \vec{E} \cdot d\vec{r} \right) dQ = 0$$

So where did the energy go when the conductor was inserted? Before the conductor was inserted, the charges rearranged themselves so there was a net force of attraction between the conductor and the capacitor. Clearly conservation of energy must imply that:

$$\frac{hQ^2}{2ab\varepsilon_0} = \int \vec{F}_{Net} \cdot d\vec{r}.$$

So, in this case, the average force on the conductor could be estimated as:

$$\left\langle \vec{F}_{Net} \right\rangle \approx \frac{\int_{-\infty}^{0} \vec{F}_{Net} \cdot d\vec{r}}{a} = \frac{hQ^2}{2a^2 b\varepsilon_0}.$$

In other words, the field between the plates does work by attracting the conducting slab and, while the charge residing on the plates remains the same in this example, the energy stored in the field is reduced after the conducting slab is introduced between the plates.

Example 4.6: A Spherically Symmetrical Charge Distribution

Consider a spherically symmetrical charge distribution in free space, with total charge Q and charge density given by:

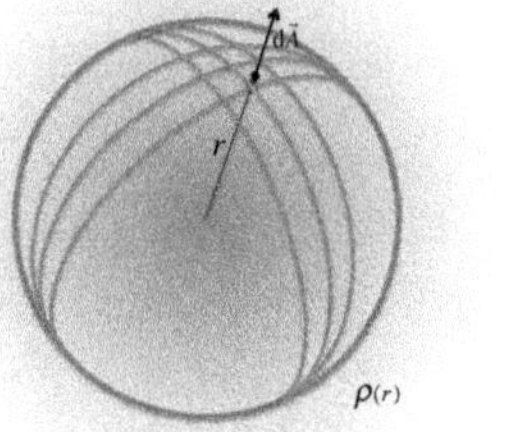

$$\rho = k\frac{e^{-(r/a)^2}}{r},$$

where k and a are constants. What is the electric field $\vec{E}(\vec{r})$?

We will first normalize the charge density distribution in order to express the arbitrary constant k in terms of the total charge Q:

$$Q=\int_0^\infty \rho 4\pi r^2\,\mathrm{d}r = \int_0^\infty \frac{k e^{-(r/a)^2}}{r}4\pi r^2\,\mathrm{d}r = k4\pi a^2\int_0^\infty e^{-(r/a)^2}\left(\tfrac{r}{a}\right)\mathrm{d}\left(\tfrac{r}{a}\right) = 2\pi k a^2$$

So, the charge density is given by:

$$\rho = \frac{Q}{2\pi a^2}\frac{e^{-(r/a)^2}}{r}$$

We now apply Gauss's Law to get the field:

$$\vec{\nabla}\cdot\vec{E} = \frac{\rho}{\varepsilon_0} \rightarrow \oint_{\text{surface}} \vec{E}\cdot\mathrm{d}\vec{A} = \oint_{\textit{volume}} \frac{Q}{2\pi\varepsilon_0 a^2}\frac{e^{-(r/a)^2}}{r}\mathrm{d}V \rightarrow E4\pi \mathrm{r}^2 = \frac{Q}{2\pi\varepsilon_0 a^2}\int_0^{\mathrm{r}}\frac{e^{-(r/a)^2}}{r}4\pi r^2\mathrm{d}r.$$

Now solving for the electric field we find:

$$E = \frac{Q}{2\pi\varepsilon_0 a^2 \mathrm{r}^2}\int_0^{\mathrm{r}} e^{-(r/R)^2} r\,\mathrm{d}r = \frac{-Q}{4\pi\varepsilon_0 \mathrm{r}^2}\int_0^{r/R} e^{-(r/a)^2}(-2)\left(\frac{r}{a}\right)\mathrm{d}\left(\frac{r}{a}\right) = \frac{-Q}{4\pi\varepsilon_0 \mathrm{r}^2}\left[e^{-(r/a)^2}-1\right].$$

As the field must be in the radial direction, we finally have:

$$\vec{E} = \frac{Q}{4\pi\varepsilon_0 \mathrm{r}^2}\left(1-e^{-(r/a)^2}\right)\hat{r}.$$

Notice that this reverts to Coulomb's law when $\mathrm{r} \gg a$.

Example 4.7: The Motion of a Dielectric in a Capacitor

Consider a parallel plate capacitor with dimensions $l\times w\times h$ held at a constant charge $\pm Q$, and containing a frictionless dielectric slab of permittivity ε and mass m. If we slide the dielectric slab a small distance x_0, what is the force on the dielectric slab? If it will oscillate, what is the period of this oscillation?

To find the force on the dielectric as a function of the position x, let us write down the total energy in the electric field as a function of position. Ignoring edge effects, this is:

$$\mathbb{U}_E = \oint_{\text{volume}} \mathbb{u}_E\, dV = \oint_{\text{volume inside dielectric}} \mathbb{u}_E\, dV + \oint_{\text{volume outside dielectric}} \mathbb{u}_E\, dV .$$

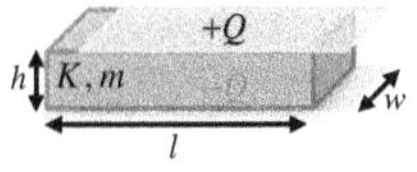

We start by recalling your result from Problem 4.3 (p. 123) for the energy density of a capacitor:

$$\mathbb{u}_E = \tfrac{1}{2}\vec{P}\cdot\vec{E} + \tfrac{1}{2}\varepsilon_0 E^2 = \tfrac{1}{2}\varepsilon E^2 .$$

And since electric field between the plates is given by $E = \frac{\sigma}{\varepsilon}$:

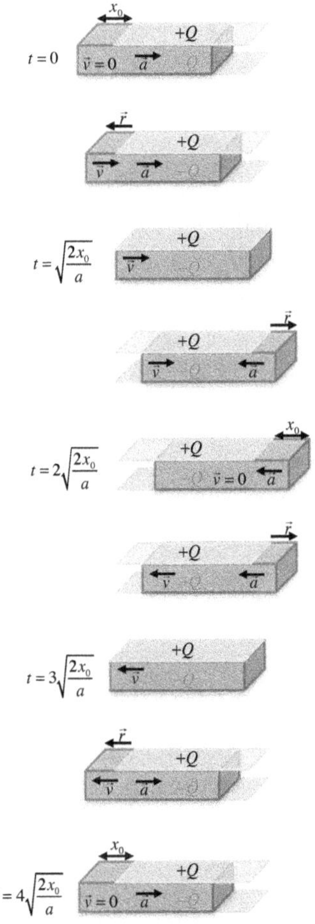

$$\mathbb{u}_E = \tfrac{1}{2}\varepsilon E^2 = \tfrac{1}{2}\varepsilon\left(\tfrac{\sigma}{\varepsilon}\right)^2 = \tfrac{1}{2}\tfrac{\sigma^2}{\varepsilon},$$

where σ is the surface charge density on the positive plate, which we will assume is uniform. Therefore, we will keep the energy density in terms of σ until the end of the problem.

We will now write the total energy, considering the permittivities involved:

$$\mathbb{U}_E = \oint_{\text{volume inside dielectric}} \left(\tfrac{1}{2}\tfrac{\sigma^2}{\varepsilon}\right) dV + \oint_{\text{volume outside dielectric}} \left(\tfrac{1}{2}\tfrac{\sigma^2}{\varepsilon_0}\right) dV .$$

From the diagram, the volumes inside and outside of the dielectric yield:

$$V_{\text{total}} = l \times w \times h, \quad V_{\text{inside}} = \left(l - |x|\right)\times w \times h, \text{ and } V_{\text{outside}} = |x| \times w \times h.$$

So the potential energy is given by:

$$\begin{aligned}\mathbb{U}_E &= \left(\tfrac{1}{2}\tfrac{\sigma^2}{\varepsilon}\right)\left(\left(l-|x|\right)wh\right) + \left(\tfrac{1}{2}\tfrac{\sigma^2}{\varepsilon_0}\right)\left(|x|wh\right)\\ &= \left(\tfrac{1}{2}\sigma^2 wh\right)\left(\tfrac{1}{\varepsilon}\left(l-|x|\right) + \tfrac{1}{\varepsilon_0}\left(|x|\right)\right)\\ &= \left(\tfrac{1}{2}\sigma^2 wh\right)\left(\tfrac{1}{\varepsilon}l + \left(\tfrac{1}{\varepsilon_0} - \tfrac{1}{\varepsilon}\right)|x|\right).\end{aligned}$$

Notice that when the slab is fully inside the capacitor, the total potential energy is:

$$\mathbb{U}_E\big|_{x=0} = \left(\tfrac{1}{2\varepsilon}\sigma^2 whl\right),$$

and totally outside it is similarly:

$$\mathbb{U}_E\big|_{|x|=l} = \left(\tfrac{1}{2\varepsilon_0}\sigma^2 whl\right).$$

Thus, our limits check out. We can now find the force on the dielectric as:

$$\vec{F} = -\vec{\nabla}\mathbb{U} = -\frac{\partial}{\partial x}\left(\left(\tfrac{1}{2}\sigma^2 wh\right)\left(\tfrac{1}{\varepsilon}l + \left(\tfrac{1}{\varepsilon_0} - \tfrac{1}{\varepsilon}\right)|x|\right)\right)\hat{x} = -\left(\tfrac{1}{2}\sigma^2 wh\right)\left(\tfrac{1}{\varepsilon_0} - \tfrac{1}{\varepsilon}\right)\frac{\partial|x|}{\partial x}\hat{x} = \left(\tfrac{1}{2}\sigma^2 wh\right)\left(\tfrac{1}{\varepsilon_0} - \tfrac{1}{\varepsilon}\right)\left(-\tfrac{|x|}{x}\hat{x}\right).$$

Thus the magnitude of the force is constant, but the force reverses direction in the center. So the absolute value of its motion will be more like a super ball bouncing than a mass on a spring. Using Newton's second law, we can find the magnitude of the acceleration to be:

$$a = \frac{F}{m} = \left(\frac{\sigma^2 wh}{2m}\right)\left(\tfrac{1}{\varepsilon_0} - \tfrac{1}{\varepsilon}\right) = \left(\frac{Q^2 wh}{2m(wl)^2}\right)\left(\tfrac{1}{\varepsilon_0} - \tfrac{1}{\varepsilon}\right) = \left(\frac{Q^2 h}{2mwl^2}\right)\left(\tfrac{1}{\varepsilon_0} - \tfrac{1}{\varepsilon}\right).$$

Since the acceleration is constant, until the dielectric passes through the center when it switches direction. The period of oscillation must be simply four times the free-fall motion time (see picture) as:

$$T = 4\sqrt{\frac{2x_0}{a}} = 4\sqrt{\frac{2x_0}{\left(\frac{Q^2 h}{2mwl^2}\right)\left(\frac{1}{\varepsilon_0} - \frac{1}{\varepsilon}\right)}} = 4\sqrt{\frac{2x_0}{\left(\frac{1}{\varepsilon_0} - \frac{1}{\varepsilon}\right)}\left(\frac{2mwl^2}{Q^2 h}\right)} = 8\left(\frac{l}{Q}\right)\sqrt{\frac{mwx_0}{h\left(\frac{1}{\varepsilon_0} - \frac{1}{\varepsilon}\right)}}.$$

And, finally, we will check our units using the SI system:

$$[T] = \left(\frac{\mathrm{m}}{\mathrm{C}}\right)\sqrt{\frac{\mathrm{kg\,m}^2}{\mathrm{m}}\left(\frac{\mathrm{F}}{\mathrm{m}}\right)} = \sqrt{\left(\frac{\mathrm{m}}{\mathrm{C}}\right)^2 \mathrm{kg}\left(\frac{\mathrm{C}}{\mathrm{V}}\right)} = \sqrt{\left(\frac{\mathrm{kg\,m}^2}{\mathrm{J}}\right)} = \sqrt{\left(\mathrm{s}^2\right)} = \mathrm{s}.$$

Problem 4.8: An Atmospheric Electric Field

Consider a relatively flat surface of the Earth with a decreasing electric field with height given by:

$$\vec{E} = E_0 e^{-z/h}\,\hat{z},$$

where $\hat{z}$ is in the upward direction and h is a constant.

What is the charge density distribution $\rho(z)$?

Problem 4.9: A Dielectric Sphere in an Electric Field

A sphere of radius a and permittivity ε is placed in an external electric field that had been constant and uniform. Find the following:

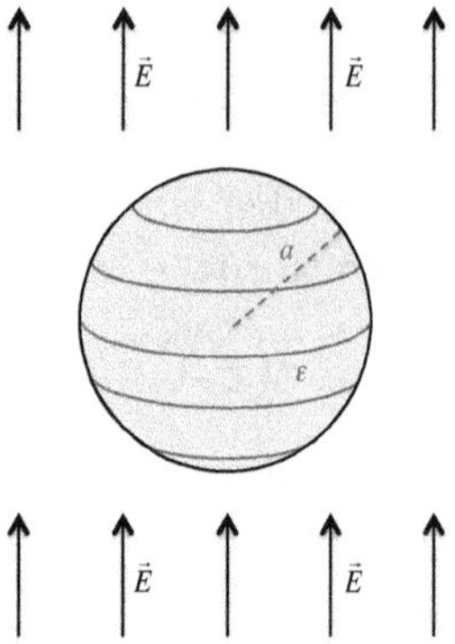

a) The polarization $\vec{P}(r,\theta)$ of the dielectric material.

b) The effective (bound) surface charge $\sigma_b(\theta)$ on the outside of the cylinder.

c) The electric field $\vec{E}(r,\theta)$ far away from the cylinder.

d) Show that, in the center of a uniformly polarized sphere, $\vec{E} = -\frac{1}{3\varepsilon_0}\vec{P}$.

Example 4.8: The Capacitance of a Coaxial Cable

Consider a long coaxial cable with a central wire of cross-sectional radius a, and a charge per length λ on the central wire and $-\lambda$ on the outer mesh, which is at a radius b. Between the outer mesh and the central wire is an insulator with permittivity ε. What is the electric field surrounding the wire?

First let us consider the space between the coaxial cable and the wire mesh. By drawing a Gaussian cylinder of radius s, and noting that $\vec{E} = E\,\hat{s}$, we write:

$\vec{\nabla}\cdot\vec{E} = \frac{\rho}{\varepsilon_0} - \frac{1}{\varepsilon_0}\vec{\nabla}\cdot\vec{P} = \frac{\rho}{\varepsilon}$, so in integral form

$$\oint_{\text{surface}} \vec{E}\cdot d\vec{A} = \oint_{\text{volume}} \left(\frac{\rho}{\varepsilon}\right) dV = \frac{Q_{\text{inside}}}{\varepsilon}.$$

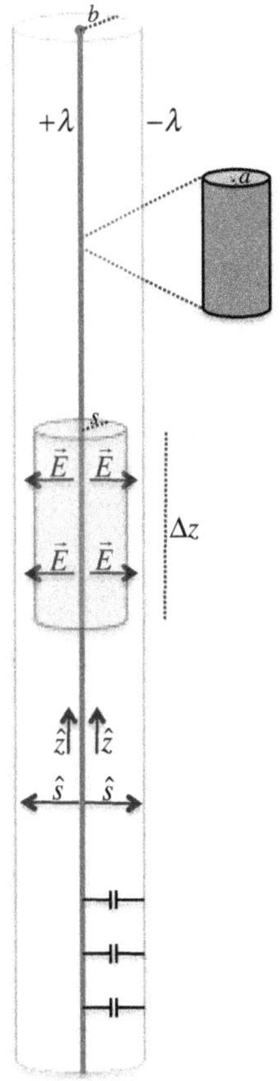

Therefore, over a length Δz this becomes:

$$E\left(2\pi s \cancel{\Delta z}\right) = \frac{\lambda \cancel{\Delta z}}{\varepsilon} \quad \rightarrow \quad \vec{E} = \frac{\lambda}{2\pi\varepsilon s}\hat{s}.$$

However, outside of the cable the electric field is zero, because the total charge contained inside the Gaussian cylinder would be zero. This is an example of shielding, where the electric field is contained between the central wire and the outside mesh. Notice that everywhere, except on the actual wires, $\vec{\nabla}\cdot\vec{E} = 0$.

We are now tempted to ask, what is the capacitance of the coaxial cable?

$$C \equiv \frac{Q_{\text{capacitor}}}{\mathbb{V}_{\text{capacitor}}} = \frac{(\lambda\Delta z)}{\int_a^b E\cdot ds} = \frac{(\cancel{\lambda}\Delta z)}{\int_a^b \left(\frac{\cancel{\lambda}}{2\pi\varepsilon s}\right) ds} = \frac{2\pi\varepsilon}{\ln(b/a)}\Delta z.$$

Since the capacitance is proportional to the length of the wire, it makes sense instead to simply find the capacitance per unit length c such that:

$$\mathsf{c} = \frac{dC}{dl} = \frac{2\pi\varepsilon}{\ln(b/a)} \quad \rightarrow \quad \mathbb{V}(z) = \frac{\lambda(z)}{\mathsf{c}}.$$

Problem 4.10: The Capacitance of Two Parallel Wires

(a) Consider two parallel line charges each of a small radius a with linear charge densities $+\lambda$ and $-\lambda$, which are separated by a distance b and held together by a rigid insulator with permittivity ε. Find the electric field in the plane between the two wires as a function of the distance from the midpoint of the wires.

(b) Find the potential difference $\mathbb{V}$ between the two line charges.

(c) Find the capacitance per length c of the two long parallel wires.

(d) We now embed the wires inside a dielectric with permittivity ε. What is the capacitance per length of these wires now?

4.3 *Classical Models of Dielectrics*

Ordinary matter is composed of atoms. A convenient picture of a single atom, which is physically valid for many purposes, consists of a small positively charged nucleus surrounded by an electron cloud with a spherically symmetrical probability distribution of location. The probability distribution is governed by the rules of quantum physics, so the electron cloud tells us the likelihood of where the electron is within the atom.

With its positive charge concentrated at the center, and an equal but opposite charge in the surrounding volume, the atom is electrically neutral and thus has no net charge Q. Nor does it possess a dipole moment $\vec{p}$, since the centers of negative and positive charge coincide within the atom.

Now suppose we apply an electric field to the atom; what happens then? The field separates the negative from the positive charge by pulling the nucleus in the direction of the field and the negatively charged electron cloud in the opposite direction.

The electrostatic attraction between the positive nucleus and the negatively charged cloud attempts to counter this separation, and an equilibrium is reached when the competing forces exactly balance. The centers of negative and positive charge no longer coincide in this new state of equilibrium, however, and the atom has therefore acquired a dipole moment.

We can model this situation and obtain a rough estimate of the size of the shift and the magnitude of the dipole moment to be expected when an electric field is applied to, say, a hydrogen atom. Neither the size of the displacement nor the magnitude of the dipole moment is very large. The shift is hardly larger than the diameter of the nucleus.

We also find that the induced dipole moment $\vec{p}$ is proportional to the field $\vec{E}$: $\vec{p} = \alpha\vec{E}$, where the constant of proportionality α is called the *atomic polarizability*. This is a measure of how easily any atom stretches or deforms under the action of an electric field, and it varies considerably across the periodic table.

The atomic polarizability is fairly high for hydrogen, as well as the alkali metals. Within a column on the periodic table, it increases with atomic number. For example, for hydrogen, $\alpha_H = 7.42\times10^{-41}\dfrac{C^2s^2}{kg}$, but for lithium, $\alpha = 18\alpha_H$, and $\alpha = 41\alpha_H$ for sodium.

On the other hand, noble gases such as helium and neon have polarizability of only $\alpha = 0.32\alpha_H$ and $\alpha = 0.61\alpha_H$ respectively. The more loosely bound is the outer, or *valence*, electron, the more easily an electric field can induce a dipole moment.

An electric field can induce an atomic dipole moment, but this induced moment typically lasts for only as long as the field is applied. However, molecules often a have permanent dipole moment. These are called *polar molecules*. Of course there are molecules, such as oxygen or carbon dioxide, which have mirror symmetry and therefore do not possess a permanent dipole moment.

On the other hand, diatomic molecules formed by dissimilar atoms are unsymmetrical even in their ground state. Hydrogen and chlorine atoms, which are spherically symmetric, for example, become connected by a single covalent bond when they combine to form hydrogen chloride. Since chlorine has more of tendency to become negative than hydrogen, the electron of the hydrogen atom shifts partially to the chlorine atom and leaves behind a partial positive charge due to the hydrogen nucleus. The result is a permanent dipole moment that points toward the hydrogen atom.

Water molecules have an extremely large dipole moment owing to their bent shape. The individual dipole moments of the O-H bonds make an angle with each other and therefore add to give a net dipole moment. Here again, there is a partial shift of negative charge toward the oxygen side of the molecule.

This results in water having a permanent dipole moment that points away from the oxygen atom. The measured value for the magnitude of the dipole moment of water is $p = 6.17 \times 10^{-30}\ \mathrm{C \cdot m}$. If we treat the H_2O molecule as a simple dipole system of negative and positive charge, the effective separation distance between the centers of charge is:

$$d_{eff} = \frac{p}{10e} \approx 3.9 \times 10^{-12}\ \mathrm{m} \approx 4\,\mathrm{pm}\,.$$

For comparison, the atomic polarizability of the hydrogen is $\alpha = 7.42 \times 10^{-41} \frac{\mathrm{C^2 s^2}}{\mathrm{kg}}$, which means that even for an extremely large field such as $E = 3 \times 10^{6}\ \frac{\mathrm{N}}{\mathrm{C}}$, the induced dipole moment is only $p_H = \alpha E = 2 \times 10^{-34}\ \mathrm{C \cdot m}$. The effective shift is then less than the diameter of a gold atom's nucleus:

$$d_{eff} \approx 1.4 \times 10^{-15}\ \mathrm{m} \approx 0.0014\,\mathrm{pm}!$$

Permanent dipole moments are much larger than dipole moments induced in nonpolar molecules by laboratory electric fields (which are typically $\sim 10^{3}\ \frac{\mathrm{N}}{\mathrm{C}}$). To induce a dipole moment in a neutral hydrogen atom equal to that of water, for example, would require an electric field of $E \sim 8 \times 10^{10}\ \frac{\mathrm{N}}{\mathrm{C}}$. This is not only beyond our present technology, but pushes things to the breaking point of the atom itself.

Internal electric fields in atoms and molecules are typically on the order of $e^2/(4\pi\varepsilon_0 r^2) = e^2/((4\pi\varepsilon_0)\cdot(1\times10^{-10}\mathrm{m})^2) \approx 10^{11}\ \frac{\mathrm{N}}{\mathrm{C}}$, so applying such field strengths to matter would threaten the disintegration of the very atoms which compose matter!

The dielectric strengths of polar materials are generally much greater than dielectric strengths for nonpolar media. For example, the permittivities of water and ethyl alcohol are $\varepsilon_{\mathrm{H_2O}} = 80\,\varepsilon_0$ and $\varepsilon_{\mathrm{C_2H_5OH}} = 28\,\varepsilon_0$, respectively, while a typical value for nonpolar substances is $\varepsilon \approx 2\,\varepsilon_0$.

As we have shown, it takes a substantial electric field to induce an electric dipole moment in each molecule of a nonpolar substance. The polar molecules that make up polar dielectric substances have tremendous individual dipole moments, but they are usually randomly oriented. When an electric field is applied to a polar material, however, the field causes these polar molecules to at least partially align and the effect of this can be quite significant. We close this section by finding an expression for the average effective dipole moment induced in a polar material subject to an external electric field.

Thought Experiment 4.12: A Single Atom in an Electric Field

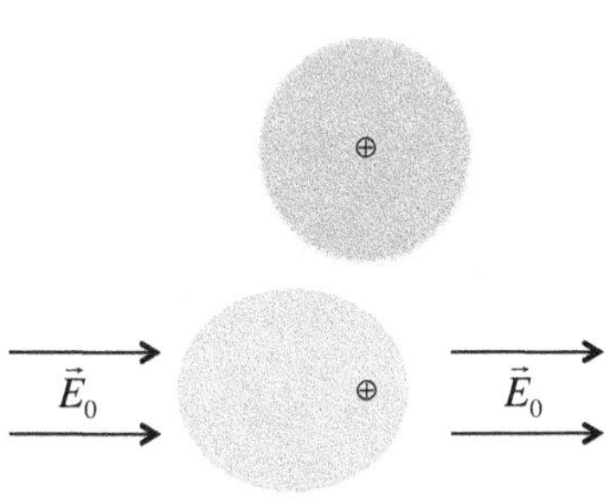

The observation of dielectric phenomena leads us to the question of how to account for their behavior. We begin to answer this by considering a simple semi-classical model of an atom in an external electric field.

We start with a single atom and assume that at its center is a positively charged point nucleus, which is surrounded by an electron cloud with a spherically symmetrical probability distribution. Therefore, the atom is both net neutral (no monopole moment), and negative charge is symmetrical (no dipole moment).

Now consider the same atom inside an electric field. The electron cloud will shift, creating a dipole moment in the direction of the external field. The actual shape of the electron probability distribution will be governed by quantum mechanics, but we can apply a simple model to find the displacement of the nucleus from the charge center of the electron cloud.

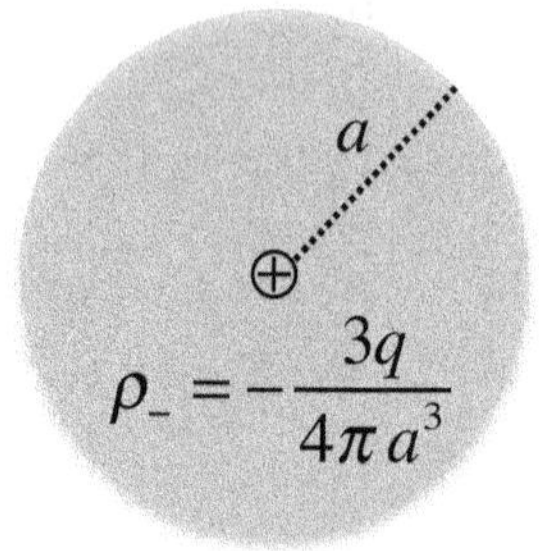

First we consider the body force density on the electron cloud by the nucleus, as well as other electron-electron forces. When there is no external electric field, the net body force density must be zero, so:

$$0 = \vec{f}(\vec{r})\Big|_{E_0=0} = \left(\frac{q\rho_-}{4\pi\varepsilon_0|\vec{r}|^3}\right)(\vec{r}) + f_{\text{electrons}} \quad \text{so} \quad f_{\text{electrons}} = \left(\frac{-q\rho_-}{4\pi\varepsilon_0|\vec{r}|^3}\right)(\vec{r}),$$

where q is the nuclear charge and ρ_- is the electron cloud charge density.

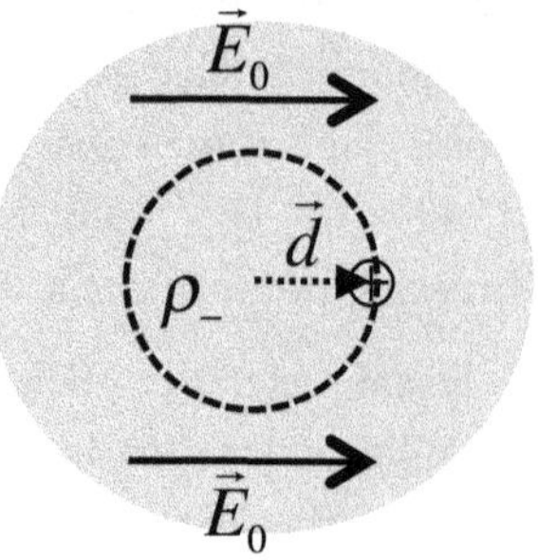

Now when an external electric field $\vec{E}_0$ is applied, the net body force on the electron cloud becomes:

$$\vec{f}(\vec{r}) = \left(\frac{q\rho_-}{4\pi\varepsilon_0|\vec{r}-\vec{d}|^3}\right)(\vec{r}-\vec{d}) + \left(\frac{-q\rho_-}{4\pi\varepsilon_0|\vec{r}|^3}\right)(\vec{r}) + \rho_-\vec{E}_0 .$$

This should be true everywhere in the cloud including the edge $r = a$, and assuming that $a \gg d$ this becomes:

$$0 \approx \left(\frac{q\cancel{\rho_-}}{4\pi\varepsilon_0 a^3}\right)(a\,\hat{r} - \vec{d}) + \left(\frac{-q\cancel{\rho_-}}{4\pi\varepsilon_0 a^3}\right)(a\,\hat{r}) + \cancel{\rho_-}\vec{E}_0 \quad \rightarrow \quad 0 \approx \cancel{qa\hat{r}} - q\vec{d} - \cancel{qa\hat{r}} + \vec{E}_0 4\pi\varepsilon_0 a^3 .$$

The dipole moment in terms of the applied field is therefore given by:

$\vec{p} = q\vec{d} = 4\pi\varepsilon_0 a^3 \vec{E}_0$.

This linear relationship between the induced dipole moment and the electric field can be measured in the lab, for which purpose it is convenient to introduce the *atomic polarizability*, α, defined as $\alpha \equiv \frac{p}{E_0}$, and express the induced dipole-field relationship as:

$$\vec{p} = 4\pi\varepsilon_0 a^3 \vec{E}_0 = \alpha\,\vec{E}_0 .$$

For Hydrogen, $\alpha = 7.42 \times 10^{-41}\,\frac{\text{C}^2\text{s}^2}{\text{kg}} = 4.5\left(4\pi\varepsilon_0 a_0^{\,3}\right)$, where a_0 is the Bohr radius.

Now we find the amount of work $\mathbb{W}$ of the field on the atom:

$$\mathbb{W} = \int_0^d \vec{F}\cdot d\vec{r} = -\frac{q^2}{4\pi\varepsilon_0 a^3}\int_0^d z\,dz = -\frac{q^2 d^2}{8\pi\varepsilon_0 a^3} = -\frac{p^2}{2\alpha} .$$

We now invoke the relationship between the applied field and the induced dipole moment:

$$\mathbb{W} = -\frac{p^2}{2\alpha} = -\frac{p}{2}\left(\frac{p}{\alpha}\right) = -\tfrac{1}{2}\vec{p}\cdot\vec{E}\,.$$

Thought Experiment 4.13: The Clausius-Mossotti Model

In Thought Experiment 4.12 we defined the atomic polarizability, α, as the ratio of the atomic dipole moment to the electric field that that atom experiences. Our goal now is to predict the permittivity in terms of the number density of atoms n and each atom's polarizability α.

The tricky part of this thought experiment is figuring out the electric field that each atom actually experiences. This is not a trivial problem, because we cannot take into account the field produced by the same atom that we are talking about.

Imagine a block of dielectric material placed in an external field $\vec{E}_0$. The total field inside the block $\vec{E}$ is the vector sum of the external field and the field caused by the induced polarization $\vec{E}_P$, so:

$$\vec{E} = \vec{E}_0 + \vec{E}_P\,.$$

By applying the dumbbell model of a dipole with two charges $\pm q$ separated by a distance d, we can use Coulomb's Law to find the electric field at midpoint of the dipole itself:

$$\vec{E}_{\text{self}} = \frac{q}{4\pi\varepsilon_0\left(\frac{d}{2}\right)^2}\left(-\hat{d}\right) + \frac{-q}{4\pi\varepsilon_0\left(\frac{d}{2}\right)^2}\left(\hat{d}\right) = -\frac{2q}{\pi\varepsilon_0 d^2}\hat{d} = -\frac{2q}{\pi\varepsilon_0 d^3}\vec{d}\,.$$

Recall from Thought Experiment 4.5 (p. 124) that the electric field due to the polarization, in terms of the number density of the material, is:

$$\vec{E}_P = -\frac{nq\,d}{\varepsilon_0}\hat{z}\,.$$

Therefore, the field at the midpoint of a single dumbbell dipole in terms of the total field:

$$\vec{E}_{\text{self}} = \left(\frac{2}{\pi n d^3}\right)\vec{E}_P\,.$$

Our goal now is to calculate the measurable permittivity in terms of the number density and the atomic polarizability, so we must now calculate the dipole moment per atom.

$$\vec{p} = \alpha\left(\vec{E} - \vec{E}_{\text{self}}\right) = \alpha\left(\vec{E} - \left(\tfrac{2}{\pi n d^3}\vec{E}_P\right)\right) = \alpha\left(\vec{E} - \tfrac{2}{\pi n d^3}\left(-\tfrac{\vec{P}}{\varepsilon_0}\right)\right) = \alpha\left(\vec{E} + \tfrac{2}{\pi n d^3 \varepsilon_0}\vec{P}\right).$$

The polarization vector is simply the dipole moment per volume, so:

$$\vec{P} = n\vec{p} = n\alpha\left(\vec{E} + \frac{2\vec{P}}{\pi n d^3\varepsilon_0}\right) = n\alpha\vec{E} + \frac{2\cancel{n}\alpha\vec{P}}{\pi\cancel{n}d^3\varepsilon_0} = \frac{n\alpha}{\left(1 - \frac{2\alpha}{\pi d^3\varepsilon_0}\right)}\vec{E}\,.$$

And remembering that $\vec{P} = (\varepsilon - \varepsilon_0)\vec{E}$, we can find the permittivity in this model by:

$$\vec{P}=\frac{n\alpha}{1-\frac{2\alpha}{\pi d^3\varepsilon_0}}\vec{E}=(\varepsilon-\varepsilon_0)\vec{E} \quad \text{so} \quad \varepsilon=\frac{n\alpha}{\left(1-\frac{2\alpha}{\pi d^3\varepsilon_0}\right)}+\varepsilon_0 .$$

In 1850 Ottaviano-Fabrizio Mossotti came up with a similar relationship that was independently found in 1879 by Rudolf Clausius. In their case, however, they assumed a spherical cavity surrounding N atoms. The radius of the sphere is $R=\frac{1}{2}d\sqrt[3]{N}$ so the number density of dipoles is:

$$n=\frac{N}{V}=\frac{N}{\frac{4\pi}{3}\left(\frac{d}{2}\sqrt[3]{N}\right)^3}=\frac{6}{\pi d^3}=3\left(\frac{2}{\pi d^3}\right).$$

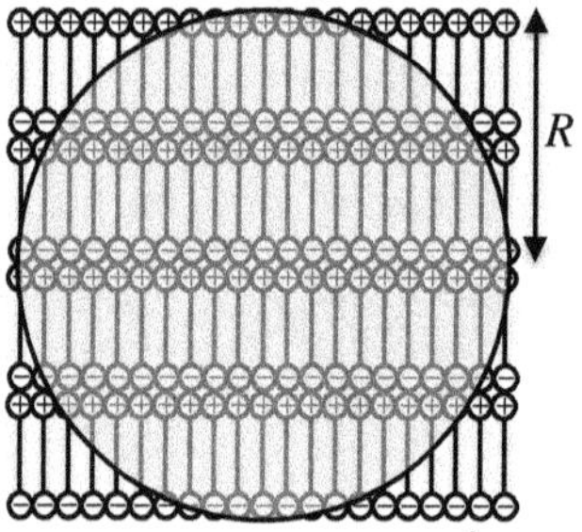

$V_{\text{cube}}=(2R)^3$ $\quad V_{\text{sphere}}=\frac{4\pi}{3}R^3$

With this assumption the *relative permittivity*, also called the *dielectric constant*, is given by the Clausius-Mossotti formula:

$$\frac{\varepsilon}{\varepsilon_0}\approx\frac{n\alpha}{\varepsilon_0-\frac{1}{3}n\alpha}+1 .$$

Thought Experiment 4.14: The Langevin Formula for polar dielectrics

In the absence of an electric field, a sample of polar dielectric is not polarized since the individual (permanent) dipoles are randomly oriented. In such a case, the polarization $\vec{P}$, which is just the sum of all dipole moments $\vec{p}$ per volume (δV), will be zero:

$$\vec{P}=\frac{1}{\delta V}\sum\vec{p}=0 .$$

If an electric field is applied to the dielectric, the individual dipoles will tend to align with the field because each dipole experiences a torque exerted by the applied field. In the limit of a very strong field, the permanent dipoles could possibly become completely aligned with the external field:

$$\vec{P}=N\vec{p},$$

in which case we have saturation. This is unlikely, however, since the thermal agitation of the dipoles destroys their alignment with the field to some extent. For water at room temperature ($T=293\text{K}$) we have that $p_{H_2O}\approx 6\times10^{-30}\,\text{C}\cdot\text{m}$. So for a very strong electric field, say $E\approx10^6\,\frac{\text{V}}{\text{m}}$, we have an energy of only $U=p_{H_2O}E\approx6\times10^{-24}\,\text{J}$. Now at room temperature, $k_BT\approx\,3\times10^{-21}\,\text{J}$, where $k_B=1.38\times10^{-23}\,\frac{\text{J}}{\text{K}}$ is the Boltzmann constant. So U is on the order of $\frac{1}{500}k_BT$ in this extreme example. At ordinary field strengths, therefore, the polarization is far less than its saturation value, and it becomes less still if the temperature of the dielectric increases. We wish to determine the polarization of a sample of polar dielectric (i.e. a sample composed entirely of permanent dipoles). In order to do this, we need an expression for the average *effective* dipole moment per molecule. By effective, we mean that component of the dipole moment that lies in the direction of the electric field: $\vec{p}_0\cdot\vec{E}=p_0E\cos\theta$, where $\vec{p}_0$ is the permanent dipole moment of the molecule and θ is the angle between $\vec{p}_0$ and the electric field $\vec{E}$.

At room temperature T, the probability of a molecule having a particular energy

$$U = -\vec{p}_0 \cdot \vec{E} = -p_0 E \cos\theta$$

is given by the Boltzmann factor $e^{-U/k_B T}$.

We ignore the molecular kinetic energies in this instance because these do not depend upon the applied electric field. Now the average energy is given by

$$\langle U \rangle = \frac{\int_{U_{min}}^{U_{max}} U e^{-U/k_B T}\, dU}{\int_{U_{min}}^{U_{max}} e^{-U/k_B T}\, dU},$$

and the average effective dipole moment can be obtained from its relationship to $\langle U \rangle$ and the electric field:

$$\langle p_0 \cos\theta \rangle = -\frac{1}{E}\langle U \rangle.$$

The minimum energy occurs when the dipole moment is completely aligned with the field, so we have that:

$$U_{min} = -p_0 E \cos(0^\circ) = -p_0 E, \quad U_{max} = -p_0 E \cos(180^\circ) = +p_0 E,$$

and therefore,

$$\langle U \rangle = \frac{\int_{-p_0 E}^{p_0 E} U e^{-U/k_B T}\, dU}{\int_{-p_0 E}^{p_0 E} e^{-U/k_B T}\, dU} = \frac{(k_B T)^2 e^{-U/k_B T}\left[-(U/k_B T) - 1\right]\Big|_{-p_0 E}^{p_0 E}}{-(k_B T) e^{-U/k_B T}\Big|_{-p_0 E}^{p_0 E}}$$

$$\langle U \rangle = k_B T - p_0 E\left[\frac{e^{p_0 E/k_B T} + e^{-p_0 E/k_B T}}{e^{p_0 E/k_B T} - e^{-p_0 E/k_B T}}\right] = k_B T - p_0 E \coth\left(\frac{p_0 E}{k_B T}\right),$$

where $\coth(x) = \dfrac{e^x + e^{-x}}{e^x - e^{-x}} = \dfrac{\cosh x}{\sinh x}$ is the hyperbolic cotangent. The average effective dipole moment is therefore given by

$$\langle p_0 \cos\theta \rangle = -\frac{k_B T}{E} + p_0 \coth\left(\frac{p_0 E}{k_B T}\right).$$

It is convenient to define

$$x = \frac{p_0 E}{k_B T},$$

so that we may write

$$\langle p_0 \cos\theta \rangle = p_0 \left[\coth x - \frac{1}{x} \right] .$$

The expression on the right hand side of this equation is referred to as the *Langevin function* after the French physicist Paul Langevin who studied the response of molecules to fields in matter during the early part of the twentieth century. Having performed the necessary mathematical gymnastics, we can now ask if our result makes sense physically. For example, in the limit that the electric field vanishes, $E \to 0$, we have that $x \to 0$, and the Langevin function vanishes which makes sense since we would expect that $\langle p_0 \cos\theta \rangle = 0$ in the absence of an external field. We would obtain that same result in the limit that the temperature $T \to \infty$, as we would expect. On the other hand, as $T \to 0$ (or, in the extremely strong field limit as $E \to \infty$ for a finite temperature) the Langevin function approaches unity, i.e. $\langle p_0 \cos\theta \rangle \to p_0$, as we obtain the ideal case of perfect alignment or saturation.

Note the linear behavior of the Langevin function at ordinary values of field and temperature, i.e. values for which $x \ll 1$ or $p_0 E \ll k_B T$ (see our earlier discussion about dipole energies at room temperature). We expand the Langevin function for $x \ll 1$ by first noting that

$$\coth x \approx \frac{1}{x} + \frac{x}{3}$$

for small values of x. Therefore:

$$\langle p_0 \cos\theta \rangle \approx \frac{1}{3} p_0 x = \frac{p_0^{\,2} E}{3k_B T} ,$$

a result referred to as the *Langevin-Debye formula.* Following our earlier discussion of dipole moments temporarily induced in spherically symmetric atoms and nonpolar molecules, we can define the orientational polarizability, α, which is the molecular dipole moment per unit polarizing field, as:

$$\alpha \equiv \frac{\langle p_0 \cos\theta \rangle}{E} = \frac{p_0^{\,2}}{3k_B T} .$$

This takes into account only the orientation of the permanent dipoles with respect to the applied electric field. But even molecules with permanent dipole moments will experience the deformation we discussed earlier, and we should therefore define a total molecular polarizability by combining the deformation polarizability and orientational polarizability derived here:

$$\alpha_T \equiv 4\pi\varepsilon_0 a^3 + \frac{p_0^{\,2}}{3k_B T} = \alpha_0 + \frac{p_0^{\,2}}{3k_B T} .$$

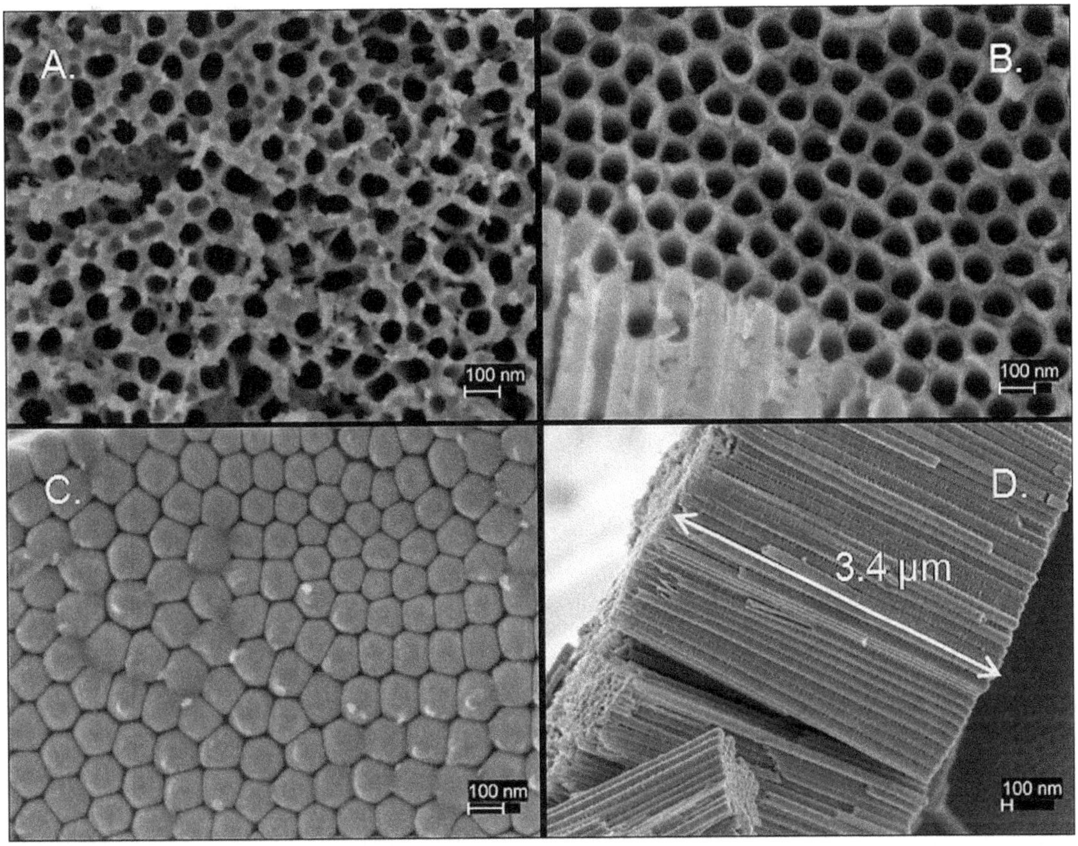

Plate 9: Superdielectric Nanotubes[64]

---

[64] Scanning electron microscope images of TiO nanotubes., which have dielectric constants greater than ten million $\left(\varepsilon > 10^7 \varepsilon_0\right)$. F.J.Q. Cortes and J. Phillips, "Tube-Super Dielectric Materials: Electrostatic Capacitors with Energy Density Greater than 200 J cm^{-3}", Materials (2015) **8**, 6208-6227, doi: 10.3390/ma8095301.

Plate 10: Adrien-Marie Legendre and Joseph Fourier[65]

[65] A watercolor portrait of French mathematicians Adrien-Marie Legendre and Joseph Fourier: Boilly, Julien-Leopold. (1820). Album de 73 Portraits-Charge Aquarelle's des Membres de I'Institute (watercolor portrait #29). Biliotheque de l'Institut de France. In the original, the faces are colored in.

Chapter 5 The Equations of Laplace and Poisson

Mathematicians love general tools, which can be applied to a variety of different problems. This has its great advantages, since much of the difficult work can be recycled from one field to the next. It also has its disadvantages, however, as it biases reasoning by analogy over other ways of solving physical problems. As a physicist, it is important for you to understand different ways of approaching similar problems. Some methods will be more mathematically efficient, others may better elucidate the physics. Often there are compelling reasons to pick one way over another. This is a textbook, so in the preceding chapters we chose methods of solving problems primarily to elucidate the physics, and somewhat to stress the scientific history of the subject. This chapter covers common mathematical techniques to solve similar problems more efficiently.

There were two formulations of electrodynamics in the nineteenth century, each based on its own analogy, which were: the fluid model of electrodynamics and the astronomical model of electrodynamics. These were not incompatible, but each had its own focus.

The fluid model not only applied the continuity equation to charge conservation, but also imagined the electric field vectors much like velocity flow in a liquid. This allowed for the recycling of the mathematics derived in the context of fluid flow to electrodynamics. Maxwell's equations are the ultimate achievement of this analogy.

The astronomical model, on the other hand, emphasized the parallels between Newton's law of gravity and Coulomb's law of electrostatics. This allowed for the recycling of the prior century's worth of work on celestial mechanics by such French powerhouses of mathematical physics as: Joseph-Louis Lagrange), Pierre-Simon Laplace, Adrien-Marie Legendre, and Siméon Denis Poisson. Perhaps the discovery of gravitational waves is a recent success of this analogy, since they are often thought of as an analogy to electromagnetic radiation. Of course, Einstein developed his idea for gravitational waves in analogy to the way Maxwell's equations predict electromagnetic waves, so the recent detection of gravitational waves can also be seen as a triumph for the fluid point of view!

We found in Chapter 3 that it is usually easier to solve for the electric potential, rather than the electric field directly. As long as we find the potential as a function of position, we can then determine the electric field by taking the gradient of the electric potential. In Chapter 4 we discussed how Gauss's law combines these two analogies, by applying Lagrange's divergence theorem to Coulomb's law of electrostatics. When written in potential form, it becomes:

$$\vec{\nabla}\cdot\vec{E}=\tfrac{1}{\varepsilon_0}\rho-\tfrac{1}{\varepsilon_0}\vec{\nabla}\cdot\vec{P}\approx\tfrac{1}{\varepsilon}\rho \quad\rightarrow\quad -\vec{\nabla}\cdot\left(\vec{\nabla}\mathbb{V}\right)=\tfrac{1}{\varepsilon}\rho .$$

This is called Poisson's equation, and it can be numerically solved given the appropriate boundary conditions and knowing the charge density as a function of position. In a neutral medium, this simplifies to $\vec{\nabla}\cdot\vec{\nabla}\mathbb{V}=0$, which is called Laplace's equation. While Laplace's equation is not easy to solve analytically, certain standard techniques to do so are known.

Solving differential equations comes down to making educated guesses as to the solution, and then checking to see if one's guess satisfies the equation and the initial or boundary conditions appropriate to the problem. But Joseph Fourier inaugurated a new phase of mathematical approaches to solving boundary value problems, in his famous 1822 work The Analytical Theory of Heat. In particular, he introduced the technique of representing a function by an infinite trigonometric series to solve a partial differential equation describing heat flow. Fourier's heat equation, as this partial differential equation is known, reduces to the mathematical form of Laplace's equation in the case of a stationary temperature without a heat source, or to the form of Poisson's equation when heat sources are included. Fourier's techniques are therefore capable of being extended to solve Laplace's equation for the electrostatic potential.

Yet Fourier did more than provide new mathematical tools to solve partial differential equations with prescribed boundary values; his work on heat theory suggested a new approach to the subject of electricity. William Thomson (later Lord Kelvin) took Fourier's theory of heat flow as an analogy to electricity, finding that the electrostatic potential behaves like the temperature in Fourier's theory for the time-independent case.[66]

By assuming point sources of heat continuously distributed on the surface of a solid, he derived an expression for the temperature mathematically identical to the expression for the electrostatic potential due to a surface charge density. By considering the case of a solid body with an isothermal surface, Thomson reasoned that if a solid body has a uniform surface potential, then the potential is also uniform everywhere inside the solid. Since a conductor forms a surface of constant potential, it follows that the electric field is zero everywhere inside.

Thomson then reproduced Coulomb's result that the electric force on a test charge, near a charged closed surface, is perpendicular to that surface and proportional to the surface charge density. He then applied this to Fourier's theory of heat to deduce that the density of heat sources which sustain a constant temperature on the surface of a body is proportional to the heat flux across the surface. The temperature outside of any isothermal surface depends only on the temperature on that surface, and on the heat flux at every point of the surface, so long as all heat sources are within or on the surface. Applied to electrostatics, his result reads:

> the electric force due to any distribution of electric charge is the same as the force due to a fictitious distribution of charge on a surface of constant potential containing all real charges, the surface density being proportional to the electric force created on the surface by the real charges.[67]

Thomson used this to show that the electric field due to a point charge placed above an infinite conducting plane is equivalent to the field of the point charge alone plus the field due to an imaginary equal but opposite mirror image charge placed at a specific distance below the plane.[68] This is now called the *method of images*, which we discuss on page 173.

Gauss, whose contributions to the study of electricity and magnetism were substantial, independently discovered many of Thomson's results as part of his program to make electrodynamics more rigorous. Having worked with German astronomer and mathematician Friedrich Wilhelm Bessel to improve precision and accuracy in astronomical measurements, Gauss was not impressed when he visited the magnetic observatory built by the scientist-philosopher Alexander von Humboldt in Berlin in the 1830s. Gauss, who was equally adept with theory and practical experimentation, demanded more rigorous standards of measurement. Collaborating with the physicist Wilhelm Weber, he soon did just that.[69]

---

[66] William Thomson, later Lord Kelvin, (under the pseudonym P.Q.R), "On the uniform motion of heat and its connection with the mathematical theory of electricity", Cambridge Mathematical Journal **3**, (1842), 71–84. Maxwell would later replace Thomson's analogy of heat flow with "an imaginary incompressible fluid" because it would make for a less abstract analogy.

[67] Olivier Darrigol, Electrodynamics from Ampère to Einstein, (London: Oxford University Press, 2000), 114-15.

[68] Lord Kelvin, Reprint of Papers on Electrostatics and Magnetism, (London: Macmillan, 1872), 52-85.

[69] Olivier Darrigol, Electrodynamics from Ampère to Einstein, (London: Oxford University Press, 2000), 49.

To implement his research agenda, which would culminate in the first serious efforts at mapping the Earth's magnetic field, Gauss first developed potential theory from the scalar function $\mathbb{V}$ of Lagrange, Laplace, and Poisson, naming it *potential* independently of George Green, the English physicist who had coined the term in an obscure paper published in 1828.[70] In his earlier research into the gravitational force near a massive ellipsoid, Gauss had obtained the result that the potential on a closed surface surrounding all masses will determine the gravitational potential everywhere outside of the surface. He applied this result to electrostatics in his 1839 paper on potential theory.[71]

Certain arrangements of discrete charges are so common in physics and chemistry that they are standard, and we will derive the potential and field for some of these configurations such as the electric dipole and quadrupole. We then find an expression for the potential due to an arbitrary, but localized, distribution of charge in terms of equivalent expressions for the potential due to these standard arrangements: a dipole, quadrupole, octopole, etc. While the resulting multipole expansion is exact, its utility lies in allowing us to calculate a suitable approximation to the electrostatic potential of an arbitrary charge distribution at large distances.

Although many of the fundamental ideas of electrostatics were formulated during the latter half of the eighteenth, and the very beginning of the nineteenth centuries, astronomers, mathematicians, and physicists spent the first half of the nineteenth century developing highly specialized mathematical techniques to solve specific problems involving stationary charges and the effects that they produce. In the early development of such sophisticated techniques, investigators looked to the analogy between electrostatics and Newtonian gravity.

The mathematician and astronomer Joseph Louis Lagrange, for example, introduced the idea of a scalar potential in the context of studying the gravitational attraction of the moon. He demonstrated that if the potential of a body at a point external to that body were known, the gravitational force it exerted on other masses could be obtained directly from differentiation.[72] Legendre polynomials, which will be used in an example later in this chapter, were first introduced as coefficients of the Newtonian gravitational potential by the French mathematician Adrien-Marie Legendre.[73]

---

[70] Ibid., 49-54, and G. Green, "An Essay on the Application of Mathematical Analysis to the Theories of Electricity and Magnetism," reprited in Mathematical Papers of the Late George Green, ed. N.M. Ferrers, (London: McMillian & co., 1871).

[71] "General Theory of Terrestrial Magnetism," and "General Theory of attraction and repulsion forces acting inversely proportional to the square of the distance," (1839) published in Scientific Memoirs, Selected from the Transactions of Foreign Academies and Learned Societies and from Foreign Journals, ed. R. Taylor (London, 1841), Vol. II, Part VI, 184–251, and Vol. III, Part X, 151-196, respectively. Elizabeth Sabine, an accomplished linguist who happened to be the wife of physicist Edward Sabine, translated both papers from the original German. A new translation of the first paper has been recently given by Glassmeier and Tsurutani, Hist. Geo Space Sci., **5**, (2014): 11–62. Gauss emphasized that potential theory could be applied to gravitational, electrostatic, and magnetic forces despite the apparent differences between each phenomenon.

[72] J.L. Lagrange, "Sur l'Equation Séculaire de la Lune", L'Académie Royale des Sciences de Paris, VII (1773); Prix pour l'année 1774; Oeuvres, VI, 335; "Theorie de la Libration de la Lune", Mémoires de l'Académie royale des Sciences et Belles-Lettres de Berlin, (1780) in Oeuvres, V, 5; and A. S. Hathaway, "Early History of the Potential", Am. Bull. New York Math. Soc. 1, no. 3, (1891): 66-74. The name "potential" was apparently first used in an essay by George Green in 1828.

[73] Adrien-Marie Le Gendre, "Recherches sur l'attraction des sphéroïdes homogènes," Mémoires de Mathématiques et de Physique, présentés à l'Académie royale des sciences (Paris) par sçavants étrangers, vol. 10, pages 411-435.

Shortly after Legendre introduced his polynomials, the famous mathematician Pierre-Simon Laplace generalized Legendre's results to three dimensions (the co-called spherical harmonics), discussed the gravitational potential function of Lagrange, and presented the equation, which now bears his name, for the gravitational potential between masses.[74]

Since the mathematical form of the gravitational potential due to a point mass is identical to the electrostatic potential due to a point charge, it is natural that Lagrange's potential, Legendre's polynomials, and Laplace's partial differential equation were applied to electrostatic problems. In 1812 Siméon Denis Poisson, who had been a student of Lagrange and a disciple of Laplace, took the gravitational potential of Laplace and Lagrange, and applied to it to electrostatics. Poisson extended Laplace's equation to include the charge density, and solved it for several simple cases.[75]

The introduction of Bessel functions, which can be used to solve Laplace's equation when expressed in cylindrical coordinates, provides yet another instance in which the study of gravity supplied tools for handling electrostatic problems. These functions were developed and studied by Friedrich Bessel in 1817 during an investigation of the *three-body problem*: to determine the motion of three masses acting only under the influence of their mutual gravitational attraction. Bessel later extended his work on these functions in a study of planetary perturbations, and the optics of telescopes. This last application allowed Bessel, in 1838, to measure the parallax of the star 61 Cygni to be 0.3 seconds of arc, which corresponds to a distance of 10 light years.[76]

In practice, specialized functions such as Bessel functions and Legendre polynomials were defined by infinite series and numerically tabulated. In the absence of modern computers, which we now take for granted, scientists like Bessel and Legendre developed numerical methods to solve differential equations for problems where the solution could not be expressed in the simple closed form of a simple polynomial, exponential or sine function. These special functions are still tabulated, or calculated on the fly, and included in modern data analysis packages.

Applied mathematicians and engineers will look to you, the physicist, to know which physical laws to apply to which problem. This ability to keep the fundamental principles in mind, while also understanding the gist of the details, will be your relative strength in the modern diverse workplace. Keep in mind Mark Twain's remark: "To a man with a hammer, everything looks like a nail." Your job will be to select the correct tool to use for the problem under investigation.

5.1 *Equations for the Electrostatic Potential*

As you will see in Chapter 14, all of electrodynamics can be boiled down to only one equation, called the *wave equation*, making it look deceptively simple. In this Chapter, we solve this equation

---

[74] Simon-Pierre Laplace, "Théorie des attractions des sphéroïdes et de la figure des planets", *Paris Mémoires* but is also reprinted in the third volume of Laplace's Méchanique céleste (1799).

[75] J.L. Heilbron, Electricity in the17th and 18th centuries: a study of early Modern physics (Berkley: University of California Press, 1979), 499.

[76] Credit for the first successful measurement of stellar parallax should perhaps belong to F. von Struve, who published his measured value for the parallax of Vega a few years before Bessel's announcement. Struve's result for Vega was preliminary, and he later refined it in light of Bessel's work on 61 Cygni, and ended up with a reported value that was twice his initial result. Interestingly, Struve's original value of 0.125 seconds of arc for the parallax of Vega is nearly identical with the modern, accepted value of 0.129 seconds of arc, which corresponds to a distance of about 25 light-years. See Suzanne Débarbat, "The First Successful Attempts to Determine Stellar Parallaxes in the Light of the Bessel/Struve Correspondence" in Mapping the Sky: Past Heritage and Future Directions, Volume 133 of the IAU Symposium (Dorddrecht: Kluwer, 1988).

in the electrostatic case, and introduce you to a number of standard techniques for solving second order partial differential equations. Here we begin by defining the equations of Laplace and Poisson, and discussing the solutions that we already found.

Thought Experiment 5.1: Gauss's Law in Potential Form

As discussed in 0, the electric potential $\mathbb{V}$, in the special case of a static situation, the electric field and the vector potential by the relation:

$$\vec{E} = -\vec{\nabla}\mathbb{V}.$$

Applying Gauss's law yields:

$$\vec{\nabla}\cdot\vec{E} = \tfrac{1}{\varepsilon_0}\rho - \tfrac{1}{\varepsilon_0}\vec{\nabla}\cdot\vec{P} \quad\rightarrow\quad -\vec{\nabla}\cdot\left(\vec{\nabla}\mathbb{V}\right) = \frac{\rho}{\varepsilon_0} - \tfrac{1}{\varepsilon_0}\vec{\nabla}\cdot\vec{P}.$$

This is called *Poisson's equation*, and can be difficult to solve depending on the charge distribution. In net-neutral regions between charges, Poisson's equation simplifies to $\vec{\nabla}\cdot\vec{\nabla}\mathbb{V} = 0$, (evaluate operators from right to left) in which case it is called *Laplace's equation.*

For this reason, the mathematical operator $\vec{\nabla}\cdot\vec{\nabla}$ is referred to as the *Laplacian operator*, and is written as ∇^2. We, therefore, refer to $\nabla^2\mathbb{V}$ as "the Laplacian of the electrostatic potential."

The equations of Poisson and Laplace, in a linear medium, become:

$$\nabla^2\mathbb{V} = -\tfrac{1}{\varepsilon}\rho - \tfrac{1}{\varepsilon_0}\vec{\nabla}\cdot\vec{P} \quad \text{and} \quad \nabla^2\mathbb{V} = 0.$$

As you may have guessed, Laplace's equation is usually easier to solve than Poisson's equation.

These equations are general tools which can be applied to diverse problems. You can use them to find the electric field around a conductor, but also the current density inside of a conductor and the magnetic field surrounding permanent magnets and superconductors, as we will do in Chapters 7 & 12.

Thought Experiment 5.2: Linearity and Superposition

Recall using Coulomb's law to find the electrostatic potential surrounding a small charge. Everywhere, except at the location of the charge, the charge density is zero, so Laplace's equation must hold. Therefore, Coulomb's law must be a solution to Laplace's equation.

Now we take this one step further. The electrostatic potential surrounding any collection of charges is simply the sum of the Coulomb potential. Therefore, regardless of how we choose to solve the problem, the principle of superposition must hold.

Consider any two functions of position, ψ_1 and ψ_2, both of which are particular solutions of Laplace's equation, the principle of superposition states:

$$\nabla^2(\psi_1 + \psi_2) = \nabla^2\psi_1 + \nabla^2\psi_2.$$

The most important property of Laplace's equation is its linearity, which lets you add multiple particular solutions together. Nonlinear partial differential equations, as in fluid dynamics, are usually impossible to solve analytically, and very expensive to solve numerically for reasons we will discuss in Chapter 12 .

Problem 5.1: Non-Linearity and Poisson's Equation

Consider a purely mathematical form of Poisson's equation, $\nabla^2\psi = f(\vec{r})$.

a) Show that this equation in non-linear when $f(\vec{r}) \neq 0$.

b) Discuss what this says about analytical solutions to Poisson's equation.

5.2 *Vector Second Derivatives*

So far we have discussed two ways of taking vector derivatives, the divergence (p. 35) and the gradient (p. 77). One of these receives a scalar function and returns a vector function, while of other does the opposite. What happens, then, if we perform both in succession?

Definition 5.1: The Laplacian

The Laplacian of a scalar function is defined as the divergence of the gradient, so:

$$\nabla^2 W \equiv \vec{\nabla}\cdot\left(\vec{\nabla}W\right).$$

This is essentially the second derivative with respect to the position vector.

The Laplacian can be applied to a vector function. It is defined, such that, if we render the vector field $\vec{F}$ into Cartesian coordinates with unit vectors $\hat{x},\hat{y},\hat{z}$, then the Laplacian becomes:

$$\nabla^2\vec{F} = \nabla^2\left(F_x\,\hat{x} + F_y\,\hat{y} + F_z\,\hat{z}\right) = \left(\nabla^2 F_x\right)\hat{x} + \left(\nabla^2 F_y\right)\hat{y} + \left(\nabla^2 F_z\right)\hat{z}.$$

Be careful, however, as this is not always true in curvilinear coordinates!

Thought Experiment 5.3: The Laplacian in Cartesian Coordinates

Consider a scalar function of position in Cartesian coordinates, $W(x,y,z)$. What is its Laplacian?

In order to derive this, we will use the definition of the Laplacian as $\vec{\nabla}\cdot\left(\vec{\nabla}W\right)$, so:

$$\nabla^2 W = \vec{\nabla}\cdot\left(\vec{\nabla}W\right) = \vec{\nabla}\cdot\left(\frac{\partial W}{\partial x}\hat{x} + \frac{\partial W}{\partial y}\hat{y} + \frac{\partial W}{\partial z}\hat{z}\right) = \frac{\partial}{\partial x}\left(\frac{\partial W}{\partial x}\right) + \frac{\partial}{\partial x}\left(\frac{\partial W}{\partial y}\right) + \frac{\partial}{\partial x}\left(\frac{\partial W}{\partial z}\right).$$

Thus, in Cartesian coordinates:

$$\nabla^2 W = \frac{\partial^2 W}{\partial x^2} + \frac{\partial^2 W}{\partial y^2} + \frac{\partial^2 W}{\partial z^2}$$

Thought Experiment 5.4: The Laplacian in Cylindrical Coordinates

Consider a scalar function of position in cylindrical coordinates, $W(s,\varphi,z)$, what is its Laplacian?

In order to derive this, we will use the definition of the Laplacian as $\vec{\nabla}\cdot\left(\vec{\nabla}W\right)$, so:

$$\nabla^2 W = \vec{\nabla}\cdot\left(\vec{\nabla}W\right) = \vec{\nabla}\cdot\left(\frac{\partial W}{\partial s}\hat{s} + \frac{1}{s}\frac{\partial W}{\partial \varphi}\hat{\varphi} + \frac{\partial W}{\partial z}\hat{z}\right)..$$

Now applying the divergence in cylindrical coordinates, we get:

$$\vec{\nabla}\cdot\left(\frac{\partial W}{\partial s}\hat{s}+\frac{1}{s}\frac{\partial W}{\partial \varphi}\hat{\varphi}+\frac{\partial W}{\partial z}\hat{z}\right)=\frac{1}{s}\frac{\partial}{\partial s}\left(s\left(\frac{\partial W}{\partial s}\right)\right)+\frac{1}{s}\frac{\partial}{\partial \varphi}\left(\frac{1}{s}\frac{\partial W}{\partial \varphi}\right)+\frac{\partial}{\partial z}\left(\frac{\partial W}{\partial z}\right).$$

The Laplacian in cylindrical coordinates is therefore:

$$\nabla^2 W=\frac{1}{s}\frac{\partial}{\partial s}\left(s\left(\frac{\partial W}{\partial s}\right)\right)+\frac{1}{s^2}\frac{\partial^2 W}{\partial \varphi^2}+\frac{\partial^2 W}{\partial z^2}.$$

Sometimes it is more convenient to expand this expression using the product rule:

$$\nabla^2 W=\frac{\partial^2 W}{\partial s^2}+\frac{1}{s}\frac{\partial W}{\partial s}+\frac{1}{s^2}\frac{\partial^2 W}{\partial \varphi^2}+\frac{\partial^2 W}{\partial z^2}.$$

Thought Experiment 5.5: The Laplacian in Spherical Coordinates

Consider a scalar function of position in spherical coordinates, $W(r,\theta,\varphi)$, what is its Laplacian?

In order to derive this, we will use the definition of the Laplacian as $\vec{\nabla}\cdot\left(\vec{\nabla}W\right)$, so:

$$\nabla^2 W=\vec{\nabla}\cdot\left(\vec{\nabla}W\right)=\vec{\nabla}\cdot\left(\frac{\partial W}{\partial r}\hat{r}+\frac{1}{r}\frac{\partial W}{\partial \theta}\hat{\theta}+\frac{1}{r\sin\theta}\frac{\partial W}{\partial \varphi}\hat{\varphi}\right).$$

Now applying the divergence in spherical coordinates, we get:

$$\vec{\nabla}\cdot\left(\frac{\partial W}{\partial r}\hat{r}+\frac{1}{r}\frac{\partial W}{\partial \theta}\hat{\theta}+\frac{1}{r\sin\theta}\frac{\partial W}{\partial \varphi}\hat{\varphi}\right)=\frac{1}{r^2}\frac{\partial}{\partial r}\left(r^2\frac{\partial W}{\partial r}\right)+\frac{1}{r\sin\theta}\frac{\partial}{\partial \theta}\left(\frac{\sin\theta}{r}\frac{\partial W}{\partial \theta}\right)+\frac{1}{r\sin\theta}\frac{\partial}{\partial \varphi}\left(\frac{1}{r\sin\theta}\frac{\partial W}{\partial \varphi}\right).$$

The Laplacian in spherical coordinates is therefore:

$$\nabla^2 W=\frac{1}{r^2}\frac{\partial}{\partial r}\left(r^2\left(\frac{\partial W}{\partial r}\right)\right)+\frac{1}{r^2\sin\theta}\frac{\partial}{\partial \theta}\left(\sin\theta\left(\frac{\partial W}{\partial \theta}\right)\right)+\frac{1}{r^2\sin^2\theta}\frac{\partial^2 W}{\partial \varphi^2}.$$

It is occasionally more convenient to expand this result and write it out term by term:

$$\nabla^2 W=\frac{\partial^2 W}{\partial r^2}+\frac{2}{r}\frac{\partial W}{\partial r}+\frac{1}{r^2}\frac{\partial^2 W}{\partial \theta^2}+\frac{1}{r^2\tan\theta}\frac{\partial W}{\partial \theta}+\frac{1}{r^2\sin^2\theta}\frac{\partial^2 W}{\partial \varphi^2}.$$

5.3 *Multipole Solutions to Laplace's Equation*

We discussed in Thought Experiment 5.1 (p. 155) that since Laplace's equation holds in between charges, and that is where Coulomb's law applies, then the Coulomb potential must be a solution to Laplace's equation. Furthermore, we devoted much of Section 3.6 (p. 102) to the idea of approximating the electric potential due to an arbitrary collection of charge as the sum of a monopole, dipole, and quadrupole. Here we generalize this *multipole expansion* to an arbitrary number of terms.

Problem 5.2: The Monopole Electric Potential

a) Show that the potential due to a point source, located at the origin, satisfies Laplace's equation.

b) Argue that, for any finite collection of charge centered at the origin:

$$\lim_{r\to\infty} \mathbb{V}(r) = \frac{Q_{\text{tot}}}{4\pi\varepsilon_0 r}, \quad \text{where} \quad Q_{\text{tot}} = \oint_{\text{volume}} \rho \, dV.$$

Problem 5.3: The Electric Dipole

Show that, for any vector dipole moment $\vec{p}$, the electric dipole solution:

$$\mathbb{V}(\vec{r}) = \frac{\vec{p}\cdot\hat{r}}{4\pi\varepsilon_0 r^2},$$

is a solution to Laplace's equation.

Example 5.1: A Conducting Ball in an Electric Field

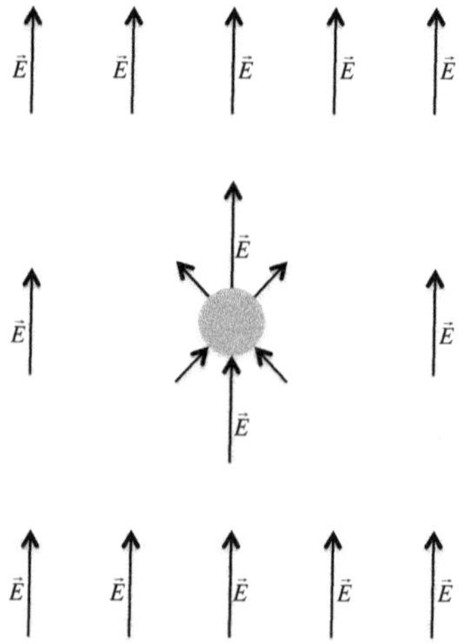

Consider a neutral conducting sphere of radius a located inside a uniform electric field $\vec{E}_0$. What is the electric potential everywhere?

Notice that as long as there is an electric field inside the conductor, charge will move until the field goes to zero. Therefore, the potential must be constant throughout the ball. Since, we can set ground arbitrarily, we will define the potential inside the sphere to be zero: $\mathbb{V}(r\leq a) = 0$.

Next, notice that far away from the conducting sphere the potential must be consistent with a constant and uniform field. In other words:

$$-\vec{\nabla}\mathbb{V}_{r\gg a} = \vec{E}_0 = E_0\hat{z}.$$

Integrating this, we can find the far away boundary condition as:

$$\mathbb{V}(r\gg a) = \int -E_0 \, dz = -E_0 z + C = -E_0 r\cos\theta + C.$$

We set $C = 0$, because then it is zero at the location of the sphere.

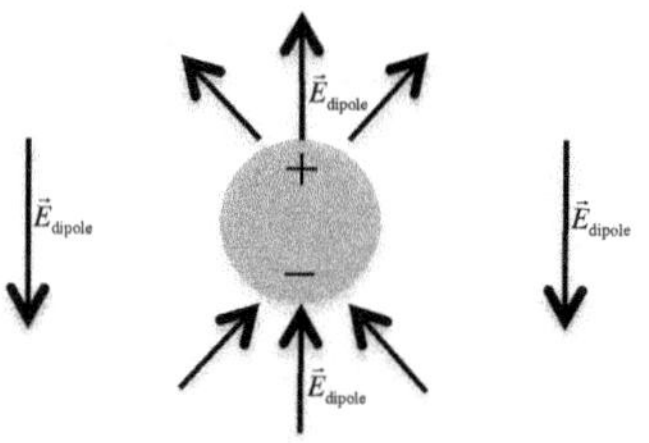

Now that we understand the boundary conditions, we can do a multipole expansion of the sphere. Since it is net neutral, it will have no monopole moment. Therefore we will start with the dipole moment, $\vec{p} = p\hat{z}$. Summing both potentials, gives:

$$\mathbb{V}(r\gg a) = \mathbb{V}_{\text{dipole}} - E_0 r\cos\theta \approx \frac{\vec{p}\cdot\hat{r}}{4\pi\varepsilon_0}\frac{1}{r^2} - E_0 r\cos\theta$$

$$\approx \frac{p\cos\theta}{4\pi\varepsilon_0 r^2} - E_0 r\cos\theta \approx \left(\frac{p}{4\pi\varepsilon_0 r^2} - E_0 r\right)\cos\theta.$$

Setting the potential on the sphere to zero gives us the dipole moment:

$$\mathbb{V}(r=a) = 0 \approx \left(\frac{p}{4\pi\varepsilon_0 ra^2} - E_0 a\right)\cos\theta \quad \rightarrow \quad \vec{p} = 4\pi\varepsilon_0 E_0 a^3\,\hat{z}.$$

Notice that the dipole solution happens to match our boundary condition perfectly, as we could factor out the cosine term. If it did not match, we would need to include a quadrupole moment term. Now, substituting back in the dipole moment, we obtain the solution:

$$\mathbb{V}(\vec{r}) = \left(\frac{p}{4\pi\varepsilon_0 r^2} - E_0 r\right)\cos\theta = \left(\frac{\cancel{4\pi\varepsilon_0} E_0 a^3}{\cancel{4\pi\varepsilon_0} r^2} - E_0 r\right)\cos\theta = \left(\frac{a^2}{r^2} - \frac{r}{a}\right)E_0 a\cos\theta.$$

To check our work, we verify that this is a solution to Laplace's equation, which, in spherical coordinates, is given by:

$$\nabla^2\mathbb{V} = \frac{1}{r^2}\frac{\partial}{\partial r}\left(r^2\frac{\partial\mathbb{V}}{\partial r}\right) + \frac{1}{r^2\sin\theta}\frac{\partial}{\partial\theta}\left(\sin\theta\frac{\partial\mathbb{V}}{\partial\theta}\right) + \cancel{\frac{1}{r^2\sin^2\theta}\frac{\partial^2\mathbb{V}}{\partial\varphi^2}} = 0 \ ,$$

where we disregard the last term because of azimuthal symmetry.

$$0 \stackrel{?}{=} \frac{\frac{\partial}{\partial r}\left(r^2\frac{\partial}{\partial r}\left(\cancel{(E_0 a)}\left(\frac{a^2}{r^2} - \frac{r}{a}\right)(\cos\theta)\right)\right)}{\cancel{r^2}} + \frac{\frac{\partial}{\partial\theta}\left(\sin\theta\frac{\partial}{\partial\theta}\left(\cancel{(E_0 a)}\left(\frac{a^2}{r^2} - \frac{r}{a}\right)(\cos\theta)\right)\right)}{\cancel{r^2}\sin\theta}.$$

and we simplify:

$$0 \stackrel{?}{=} (\cos\theta)\tfrac{\partial}{\partial r}\left(r^2\left(\tfrac{-2a^2}{r^3} - \tfrac{1}{a}\right)\right) + \left(\tfrac{a^2}{r^2} - \tfrac{r}{a}\right)\frac{1}{\sin\theta}\tfrac{\partial}{\partial\theta}\left(-\sin^2\theta\right) = -(\cos\theta)\tfrac{\partial}{\partial r}\left(\tfrac{-2a^2}{r} - \tfrac{r^2}{a}\right) - \left(\tfrac{a^2}{r^2} - \tfrac{r}{a}\right)\frac{1}{\sin\theta}\tfrac{\partial}{\partial\theta}\left(\sin^2\theta\right)$$

$$0 \stackrel{?}{=} -\left(\cancel{\cos\theta}\right)\left(\frac{-\cancel{2}a^2}{r^2} + \frac{\cancel{2}r}{a}\right) - \left(\frac{a^2}{r^2} - \frac{r}{a}\right)\frac{1}{\cancel{\sin\theta}}\left(\cancel{2}\cancel{\sin\theta}\cancel{\cos\theta}\right) = \cancel{\left(\frac{a^2}{r^2} - \frac{r}{a}\right)} - \cancel{\left(\frac{a^2}{r^2} - \frac{r}{a}\right)} = 0.$$

Now that we have verified that our solution conforms to the boundary condition and solves Laplace's equation, we will take the negative gradient to find the electric field:

$$\begin{aligned}\vec{E}(r,\theta) &= -\vec{\nabla}(E_0 a)\left(\tfrac{a^2}{r^2} - \tfrac{r}{a}\right)(\cos\theta) = -E_0\vec{\nabla}\left(\tfrac{a^3}{r^2} - r\right)(\cos\theta) = -E_0(\cos\theta)\vec{\nabla}\left(\tfrac{a^3}{r^2} - r\right) - E_0\left(\tfrac{a^3}{r^2} - r\right)\vec{\nabla}(\cos\theta) \\ &= -E_0(\cos\theta)\tfrac{\partial}{\partial r}\left(\tfrac{a^3}{r^2} - r\right)\hat{r} - E_0\left(\tfrac{a^3}{r^2} - r\right)\tfrac{1}{r}\tfrac{\partial}{\partial\theta}(\cos\theta)\hat{\theta} = -E_0(\cos\theta)\left(\tfrac{-2a^3}{r^3} - 1\right)\hat{r} - E_0\left(\tfrac{a^3}{r^3} - 1\right)(-\sin\theta)\hat{\theta} \\ &= E_0\left(1 + \tfrac{2a^3}{r^3}\right)(\cos\theta)\hat{r} + E_0\left(1 - \tfrac{a^3}{r^3}\right)(\sin\theta)\hat{\theta}.\end{aligned}$$

Now checking our conclusion for large r:

$$\lim_{r\to\infty}\vec{E}(r,\theta) = E_0\left(1 + \cancel{\tfrac{2a^3}{r^3}}\right)(\cos\theta)\hat{r} + E_0\left(1 - \cancel{\tfrac{a^3}{r^3}}\right)(\sin\theta)\hat{\theta} = E_0\left((\cos\theta)\hat{r} - (\sin\theta)\hat{\theta}\right) = E_0\hat{z}.,$$

which agrees with the far boundary condition.

Now we consider the near boundary condition, when $r = a$:

$$\vec{E}(a,\theta) = E_0(\cos\theta)(1+2)\hat{r} - E_0(1-1)(\sin\theta)\hat{\theta} = 3E_0\cos\theta\,\hat{r}.$$

Thus the electric field is normal to the surface of the conductor!

Finally, let us ask ourselves how the charge is distributed on the surface of the sphere. In other words, what is the surface charge density $\sigma(\theta)$?

To answer this, we draw a Gaussian surface, on the surface of the conductor $(r = a)$, so it encloses a small charge δQ, such that:

$$\delta Q = \sigma(\theta)\delta A = \sigma(\theta)(a\delta\theta)(a\sin\theta\,\delta\varphi) = \sigma(\theta)a^2\sin\theta\,\delta\theta\,\delta\varphi\,.$$

Now, Gauss's law states:

$$\vec{\nabla}\cdot\vec{E} = \frac{\rho}{\varepsilon_0} \rightarrow \oint_{\text{surface}} \vec{E}\cdot \mathrm{d}\vec{A} = \frac{\delta Q}{\varepsilon_0}\,.$$

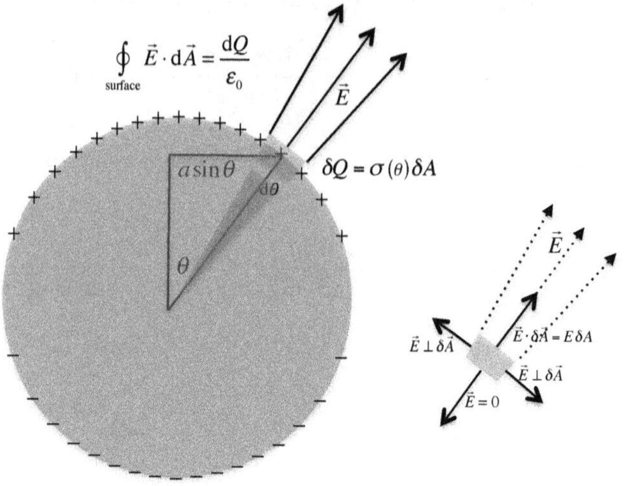

Inside the conductor yields zero, because there is no electric field, and the electric field at the sides is parallel to the Gaussian surface, so the only term that remains is the one outside of the conducting surface:

$$E\cancel{\delta A} = \frac{\sigma(\theta)\cancel{\delta A}}{\varepsilon_0}$$

$$\sigma(\theta) = \varepsilon_0 E(a,\theta) = 3\varepsilon_0 E_0 \cos\theta\,.$$

Problem 5.4: The Electric Quadrupole

Show that, for any tensor quadrupole moment $\ddot{\mathbf{Q}}$, the corresponding electric potential:

$$\mathbb{V}(\vec{r}) = \frac{1}{8\pi\varepsilon_0 r^3}\left(\hat{r}\cdot\ddot{\mathbf{Q}}\cdot\hat{r}\right),$$

is a solution to Laplace's equation.

Thought Experiment 5.6: Multipole Expansion with Legendre Polynomials

Now that we can approximate the electric potential surrounding a charge distribution as a sum of monopole, dipole and quadrupole terms, we would like to generalize the technique to an arbitrarily order of multipoles. In principle, if the number of poles were infinite, this would yield an exact solution for the electric potential for any static charge distribution.

Consider a distribution of charge $\rho(\vec{R})$. Integrating Coulomb's Law over $\vec{R}$ gives us the exact value for the electric potential:

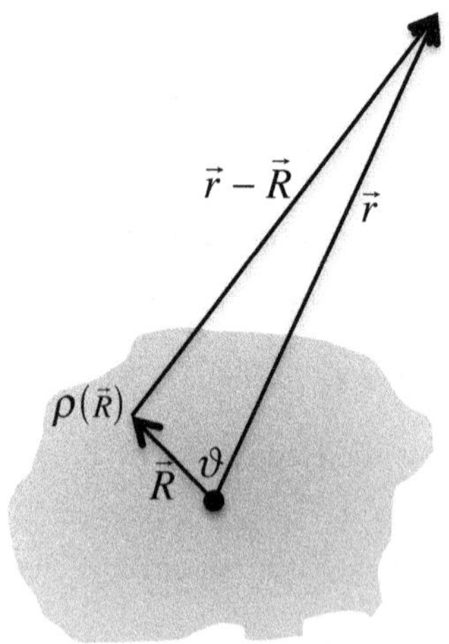

$$\mathbb{V}(\vec{r}) = \frac{1}{4\pi\varepsilon_0}\int_{\text{volume}} \frac{\rho(\vec{R})}{\left|\vec{r}-\vec{R}\right|}\mathrm{d}V\,.$$

We apply the law of cosines, so that:

$$\left|\vec{r}-\vec{R}\right|^2 = r^2 + R^2 - 2rR\cos\vartheta\,,$$

where $\cos\vartheta = \frac{\vec{r}\cdot\vec{R}}{|\vec{r}||\vec{R}|}$, then we can write:

$$\frac{1}{\left|\vec{r}-\vec{R}\right|} = \frac{1}{r\sqrt{1+\left(\frac{R}{r}\right)^2 - 2\frac{R}{r}\cos\vartheta}}\,.$$

Recall the series:

$$\frac{1}{\sqrt{1+x}}=1-\tfrac{1}{2}x+\tfrac{3}{8}x^2-\tfrac{5}{16}x^3+\ldots,$$

setting $x=\left(\frac{R}{r}\right)^2-2\frac{R}{r}\cos\vartheta$, we can now write:

$$\frac{1}{\left|\vec{r}-\vec{R}\right|}=\frac{1}{r}\left(1-\frac{1}{2}\left(\frac{R}{r}\right)^2+\frac{R}{r}\cos\vartheta+\frac{3}{8}\left(\frac{R}{r}\right)^2\left(\frac{R}{r}-2\cos\vartheta\right)^2+\ldots\right)=\frac{1}{r}\left(1+\frac{R}{r}\cos\vartheta+\left(\frac{R}{r}\right)^2\left(\frac{3\cos^2\vartheta-1}{2}\right)+\ldots\right)$$

The electric potential is therefore given by:

$$\mathbb{V}(\vec{r})=\frac{1}{4\pi\varepsilon_0}\int_{\text{volume}}\frac{\rho(\vec{R})}{r}\left(1+\frac{R}{r}\cos\vartheta+\left(\frac{R}{r}\right)^2\left(\frac{3\cos^2\vartheta-1}{2}\right)+\ldots\right)\mathrm{d}V.$$

Note that the first 3 terms of this series reproduces the monopole, dipole, and quadrupole:

$$\mathbb{V}(\vec{r})\approx\frac{1}{4\pi\varepsilon_0}\left(Q\frac{1}{r}+\vec{p}\cdot\hat{r}\frac{1}{r^2}+\tfrac{1}{2}\hat{r}\cdot\ddot{\mathbf{Q}}\cdot\hat{r}\frac{1}{r^3}\right).$$

Legendre polynomials were first found in a similar context involving gravity, and provide us with a convenient way to express the multipole expansion as a function of the spherical coordinates r and ϑ. Though named in honor of Legendre, who first introduced them, these polynomials are often defined using the formula that bears the name of another French mathematician, Benjamin Olinde Rodrigues (1795-1851):

$$P_n(x)=\frac{1}{2^n n!}\left(\frac{\mathrm{d}}{\mathrm{d}x}\right)^n\left(x^2-1\right)^n.$$

Thus, the electric potential becomes:

$$\mathbb{V}(\vec{r})=\frac{1}{4\pi\varepsilon_0}\sum_{n=0}^{\infty}\frac{1}{r^{n+1}}\int_{\text{volume}}\rho(\vec{R})R^nP_n(\cos\vartheta)\mathrm{d}V,$$

where the integral represents each moment of the expansion.

Evaluating Rodrigues' formula to obtain the Legendre polynomials quickly becomes tedious. In practice, it is easier to calculate each Legendre polynomial using Bonnet's recursion relation:

$$P_{n+1}(x)=\frac{(2n+1)xP_n(x)-nP_{n-1}(x)}{n+1}.$$

So, the first four Legendre polynomials are given by:

$$P_0(x)=1,\quad P_1(x)=x,\quad P_2(x)=\frac{3x^2-1}{2},\quad P_3(x)=\frac{5x^3-x}{2}.$$

Problem 5.5: Legendre Polynomials

a) Show that the first three Legendre polynomials $\left(P_0(\cos\theta),P_1(\cos\theta),P_2(\cos\theta)\right)$ are a solution to polar dependence of Laplace's equation:

$$\left(\frac{1}{\Theta}\frac{1}{\sin\theta}\frac{\mathrm{d}}{\mathrm{d}\theta}\left(\sin\theta\frac{\mathrm{d}\Theta}{\mathrm{d}\theta}\right)=-n(n+1)\right).$$

b) Show that any Legendre polynomial is a solution to this equation using the Rodrigues formula.

Problem 5.6: A Dielectric Sphere in an Electric Field

An insulative sphere of radius a and permittivity ε is placed in an uniform external electric field. Find the following:

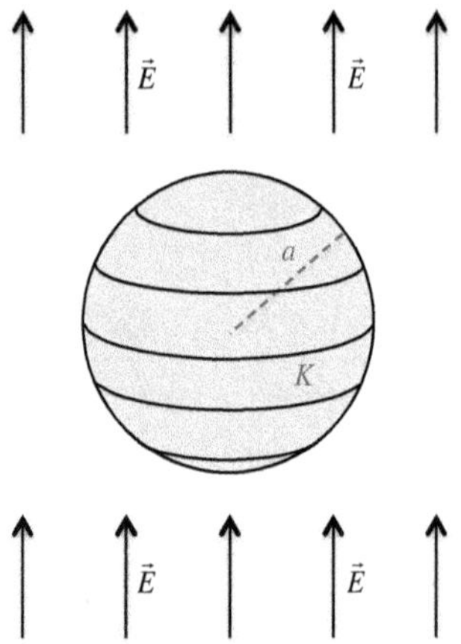

a) The initial potential function $\mathbb{V}_0(r,\theta)$ that corresponds to the uniform initial field $\vec{E}_0 = E_0\,\hat{z}$.

b) The polarization $\vec{P}(r,\theta)$ of the dielectric material.

c) The effective surface charge $\sigma(\theta)$ on the outside of the cylinder.

d) The electric potential $\mathbb{V}(r,\theta)$ throughout all space outside the dielectric.

e) The electric field $\vec{E}(r,\theta)$ outside of the sphere.

5.4 *Separation of Variables*

Separation of vaiables in the contex of partial differential equations means something different than it does in calculus and in simple first order differential equations. If an equation is seperable, then the solution can be written as a product of functions of a single variable. For example, in Cartesian coordinates, we may write the potential as:

$$\mathbb{V}(x,y,z) = X(x)Y(y)Z(z).$$

Then, by applying the chain rule, you can solve for each function using Laplace's equation and the boundary conditions.

Example 5.2: The Potential in a Long Coaxial Cable

First we will solve a problem without having to actually separate any variables. Consider a long coaxial cable held at a constant voltage $\mathbb{V}_0$. We will set the electric potential of the center wire to $\mathbb{V}_0$, and the outside to zero as is common with coaxial cables. Between the central wire and the outside sheath the potential must obey Poisson's equation, which must only depend on the radial coordinate s, so:

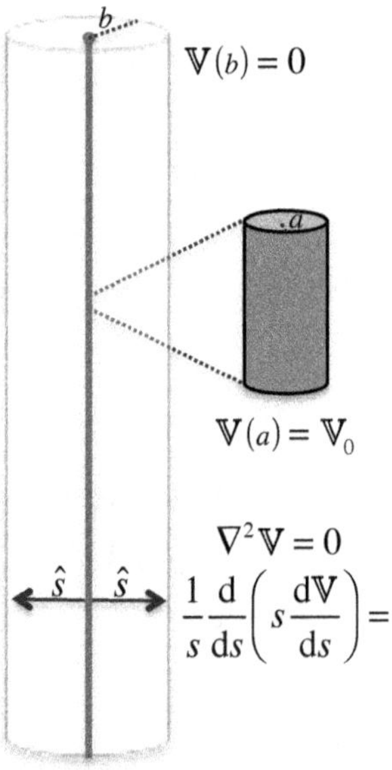

$$\frac{1}{s}\frac{\mathrm{d}}{\mathrm{d}s}\left(s\frac{\mathrm{d}\mathbb{V}}{\mathrm{d}s}\right) = 0.$$

Thus $s\frac{\mathrm{d}}{\mathrm{d}s}\mathbb{V}$ must be a constant C, so:

$$\mathrm{d}\mathbb{V} = C\frac{\mathrm{d}s}{s} \quad \rightarrow \quad \mathbb{V}(s) - \mathbb{V}(b) = C\int_b^s \frac{1}{s}\mathrm{d}s = C\ln\left(\frac{s}{b}\right).$$

Moreover, the potential difference is $\mathbb{V}_0$, so we have that:

$$\mathbb{V}_0 = \mathbb{V}(a) - \mathbb{V}(b) = C\int_b^a \frac{1}{s}\,\mathrm{d}s = C\ln\left(\frac{a}{b}\right).$$

Thus, solving for C, our solution is:

$$\mathbb{V}(s) = \mathbb{V}_0 \frac{\ln\left(\frac{s}{b}\right)}{\ln\left(\frac{a}{b}\right)}.$$

The inner cable is a conductor, so it is an equipotential: $\mathbb{V}(s<a) = \mathbb{V}_0$.

Problem 5.7: A Spherical Capacitor

Consider, again, a metal sphere of radius a surrounded by a larger concentric spherical shell of radius b. Each sphere is connected to opposite sides of a battery with constant voltage difference $\mathbb{V}$, and the whole system is net neutral, as in Faraday's experiments (p. 119).

a) Find the charge $+Q$ and $-Q$ of the two spheres.

b) What is the capacitance of the system?

c) Now we fill the space between the spheres with a dielectric of permittivity ε. What is the capacitance now of the system?

d) Find the potential as a function of r between the two shells using Laplace's equation.

Example 5.3: The Electric Potential Inside a 2D Conducting Box

A rectangular box of known dimensions $l \times w \times h$ consists of five metal plates that are grounded, and a metal top that is insulated with respect to the box and held at a voltage $\mathbb{V}_0$ with respect to ground. Find the electric potential inside the box.

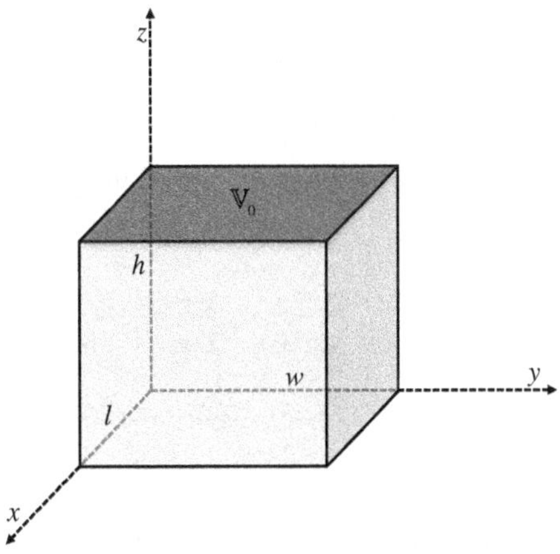

First we notice that inside the box Laplace's equation must hold, and given the rectangular symmetry it is clearly easiest to solve this problem in Cartesian coordinates:

$$\nabla^2\mathbb{V} = \frac{\partial^2\mathbb{V}}{\partial x^2} + \frac{\partial^2\mathbb{V}}{\partial y^2} + \frac{\partial^2\mathbb{V}}{\partial z^2} = 0.$$

Since we know the potential on the surfaces of the box, our boundary conditions are clearly specified:

$$\mathbb{V}(0,y,z) = \mathbb{V}(l,y,z) = \mathbb{V}(x,0,z) = \mathbb{V}(x,w,z) = \mathbb{V}(x,y,0) = 0$$
$$\mathbb{V}(x,y,h) = \mathbb{V}_0$$

This *partial differential equation* can be solved by the method of *separation of variables*, by employing the natural symmetry of the system to guess that the three dimensions are independent. Thus we will guess that the potential can be written as a product of three functions, each of one spatial variable:

$$\mathbb{V}(x,y,z) = \mathbb{V}_0\, X(x)Y(y)Z(z).$$

Now we use the product rule:

$$\vec{\nabla}\mathbb{V}(x,y,z)=\mathbb{V}_0\left(\left(\vec{\nabla}X(x)\right)Y(y)Z(z)+X(x)\left(\vec{\nabla}Y(y)\right)Z(z)+X(x)Y(y)\left(\vec{\nabla}Z(z)\right)\right)$$

$$\frac{\vec{\nabla}\cdot\vec{\nabla}\mathbb{V}(x,y,z)}{\mathbb{V}(x,y,z)}=\left(\frac{\nabla^2X(x)}{X(x)}+\frac{\nabla^2Y(y)}{Y(y)}+\frac{\nabla^2Z(z)}{Z(z)}\right)+\frac{\left(\vec{\nabla}X(x)\right)\cdot\vec{\nabla}\left(Y(y)Z(z)\right)}{X(x)Y(y)Z(z)}$$

$$+\frac{\left(\vec{\nabla}Y(y)\right)\cdot\vec{\nabla}\left(X(x)Z(z)\right)}{X(x)Y(y)Z(z)}+\frac{\left(\vec{\nabla}Z(z)\right)\cdot\vec{\nabla}\left(X(x)Y(y)\right)}{X(x)Y(y)Z(z)}$$

Clearly the gradient of $X(x)$ is in the $\hat{x}$ direction, and the gradient of $Y(y)Z(z)$ has only $\hat{y}$ and $\hat{z}$ components, thus their dot products are zero, as is true for the other directions as well. Therefore, we can write:

$$\frac{\nabla^2\mathbb{V}(x,y,z)}{\mathbb{V}(x,y,z)}=\frac{\nabla^2X(x)}{X(x)}+\frac{\nabla^2Y(y)}{Y(y)}+\frac{\nabla^2Z(z)}{Z(z)}=\frac{1}{X(x)}\frac{\mathrm{d}^2X(x)}{\mathrm{d}x^2}+\frac{1}{Y(y)}\frac{\mathrm{d}^2Y(y)}{\mathrm{d}y^2}+\frac{1}{Z(z)}\frac{\mathrm{d}^2Z(z)}{\mathrm{d}z^2}=0\,.$$

This is where we notice that each term is simply a constant, which must sum to zero. Therefore we can write our partial differential equation as three ordinary differential equations:

$$\frac{1}{X(x)}\frac{\mathrm{d}^2X(x)}{\mathrm{d}x^2}=C_x,\quad \frac{1}{Y(y)}\frac{\mathrm{d}^2Y(y)}{\mathrm{d}y^2}=C_y,\quad \frac{1}{Z(z)}\frac{\mathrm{d}^2Z(z)}{\mathrm{d}z^2}=C_z$$

where $C_x+C_y+C_z=0$.

Now we can solve each equation in turn, by applying the known boundary conditions. The solution to the equation $\frac{\mathrm{d}^2X(x)}{\mathrm{d}x^2}=C_xX(x)$ depends on whether C_x is positive or negative. If C_x is positive the solutions are exponential, but if C_x is negative the solutions will be sinusoidal. Because $X(0)=X(l)=0$, the solutions must be sinusoidal, with a solution of the form:

$$X(x)=A_x\sin\left(k_x x\right).$$

Now, $X(0)=A_x\sin(0)=0$ and $X(l)=A_x\sin\left(k_xl\right)=0$, so $n_x\pi=k_xl$, or $k_x=\frac{n_x\pi}{l}$ where n_x is an integer. The box is symmetrical between X and Y, so the solutions to the first two equations must be:

$$X(x)=A_x\sin\left(\frac{n_x\pi}{l}x\right),\quad Y(y)=A_y\sin\left(\frac{n_y\pi}{w}y\right).$$

Before going any further, let's check our work, by differentiating:

$$\frac{\mathrm{d}^2X(x)}{\mathrm{d}x^2}=-\left(\frac{n_x\pi}{l}\right)^2A_x\sin\left(\frac{n_x\pi}{l}x\right),\quad \frac{\mathrm{d}^2Y(y)}{\mathrm{d}y^2}=-\left(\frac{n_y\pi}{w}\right)^2A_y\sin\left(\frac{n_y\pi}{w}y\right).$$

So we find that:

$$\frac{1}{X(x)}\frac{\mathrm{d}^2X(x)}{\mathrm{d}x^2}=C_x=-\left(\frac{n_x\pi}{l}\right)^2,\quad \frac{1}{Y(y)}\frac{\mathrm{d}^2Y(y)}{\mathrm{d}y^2}=C_y=-\left(\frac{n_y\pi}{w}\right)^2.$$

We finally have the tools to solve the last differential equation. We begin by recalling the original differential equation:

$$0=\frac{\frac{d^2}{dx^2}(X(x))}{X(x)}+\frac{\frac{d^2}{dy^2}(Y(y))}{Y(y)}+\frac{\frac{d^2}{dz^2}(Z(z))}{Z(z)}=-\left(\frac{n_x\pi}{l}\right)^2-\left(\frac{n_y\pi}{w}\right)^2+\frac{\frac{d^2}{dz^2}(Z(z))}{Z(z)}.$$

So we can write:

$$\frac{\frac{d^2}{dz^2}(Z(z))}{Z(z)}=\left(\frac{n_x\pi}{l}\right)^2+\left(\frac{n_y\pi}{w}\right)^2 \quad\rightarrow\quad \frac{d^2Z(z)}{dz^2}=\left(\left(\frac{n_x\pi}{l}\right)^2+\left(\frac{n_y\pi}{w}\right)^2\right)Z(z).$$

Now, recalling the lower boundary condition $(Z(0)=0)$, the solution to this equation is the difference of two exponentials called the hyperbolic sine function:

$$\frac{d^2Z(z)}{dz^2}=\left(\left(\frac{n_x\pi}{l}\right)^2+\left(\frac{n_y\pi}{w}\right)^2\right)Z(z) \rightarrow Z(z)=A_z\sinh\left(\left(\sqrt{\left(\frac{n_x\pi}{l}\right)^2+\left(\frac{n_y\pi}{w}\right)^2}\right)z\right).$$

Finally we take the product, so we have particular solutions:

$$\frac{\mathbb{V}_{n_x,n_y}(x,y,z)}{\mathbb{V}_0}=C_{n_x,n_y}\sin\left(\frac{n_x\pi}{l}x\right)\sin\left(\frac{n_y\pi}{w}y\right)\sinh\left(\left(\sqrt{\left(\frac{n_x\pi}{l}\right)^2+\left(\frac{n_y\pi}{w}\right)^2}\right)z\right).$$

To find the general solution, we sum over all particular solutions:

$$\mathbb{V}(x,y,z)=\mathbb{V}_0\sum_{n_y=1}^{\infty}\sum_{n_x=1}^{\infty}\mathbb{V}_{n_x,n_y}(x,y,z).$$

The next step is in applying the top boundary condition $\mathbb{V}(x,y,h)=\mathbb{V}_0$, so that we can constrain the coefficients C_{n_x,n_y}:

$$\sum_{n_y=1}^{\infty}\sum_{n_x=1}^{\infty}C_{n_x,n_y}\sin\left(\frac{n_x\pi}{l}x\right)\sin\left(\frac{n_y\pi}{w}y\right)\sinh\left(\left(\sqrt{\left(\frac{n_x}{l}\right)^2+\left(\frac{n_y}{w}\right)^2}\right)\pi h\right)=1.$$

In order to solve for C_{n_x,n_y} we will multiply both sides by $\sin\left(\frac{n'_x\pi}{l}x\right)\sin\left(\frac{n'_y\pi}{w}y\right)$, where n'_x and n'_y are arbitrary integers, and integrate over the whole box.

This right hand side integrates to:

$$\int_0^l\int_0^w\sin\left(\tfrac{1}{l}n'_x\pi x\right)\sin\left(\tfrac{1}{w}n'_y\pi y\right)dx\,dy=\int_0^l\sin\left(\tfrac{1}{l}n'_x\pi x\right)dx\int_0^w\sin\left(\tfrac{1}{w}n'_y\pi y\right)dy$$

$$=\frac{l}{n'_x\pi}\left(1-\cos\left(n'_x\pi\right)\right)\frac{w}{n'_y\pi}\left(1-\cos\left(n'_y\pi\right)\right).$$

and note that $\cos(n'\pi)=\pm1$, so $1-\cos(n'\pi)=0$ for even integers and $1-\cos(n'\pi)=2$ for odd integers.

The left side becomes:

$$\sum_{n_y=1}^{\infty}\sum_{n_x=1}^{\infty} C_{n_x n_y}\left(\int_0^l \sin\left(\tfrac{1}{l}n_x'\pi x\right)\sin\left(\tfrac{1}{l}n_x\pi x\right)\mathrm{d}x\right)\left(\int_0^w \sin\left(\tfrac{1}{w}n_y'\pi y\right)\sin\left(\tfrac{1}{w}n_y\pi y\right)\mathrm{d}y\right)\sinh\left(\pi h\sqrt{\left(\tfrac{1}{l}n_x\right)^2+\left(\tfrac{1}{w}n_y\right)^2}\right).$$

Now, if $n_x' \neq n_x$ or $n_y' \neq n_y$ then the integrals go to zero, on the other hand if $n_x' = n_x$ or $n_y' = n_y$ then the corresponding integral becomes simply $\frac{1}{2}l$ or $\frac{1}{2}w$ respectively. So we can finally simplify our relation as follows:

$$\sum_{n_y=1}^{\infty}\sum_{n_x=1}^{\infty} C_{n_x n_y}\left(\tfrac{1}{2}l\right)\left(\tfrac{1}{2}w\right)\sinh\left(\pi h\sqrt{\left(\tfrac{1}{l}n_x\right)^2+\left(\tfrac{1}{w}n_y\right)^2}\right)=\underset{n_y \text{ and } n_x \text{ are both odd}}{\sum_{n_y=1}^{\infty}\sum_{n_x=1}^{\infty}}\left(\frac{2l}{n_x\pi}\right)\left(\frac{2w}{n_y\pi}\right).$$

Thus for each odd term, we can solve for the coefficient:

$$\underset{\text{both odd}}{C_{n_x,\,n_y}} = \frac{16}{\pi^2 n_x n_y \sinh\left(\left(\sqrt{\left(\frac{n_x}{l}\right)^2+\left(\frac{n_y}{w}\right)^2}\right)\pi h\right)}.$$

And finally we can write the electric potential as:

$$\mathbb{V}(x,y,z) = \mathbb{V}_0 \underset{\text{odd } n_x\, n_y}{\sum_{n_y=1}^{\infty}\sum_{n_x=1}^{\infty}}\left(\frac{4\sin\left(\frac{n_x\pi}{l}x\right)}{\pi n_x}\right)\left(\frac{4\sin\left(\frac{n_y\pi}{w}y\right)}{\pi n_y}\right)\left(\frac{\sinh\left(\pi z\sqrt{\left(\tfrac{1}{l}n_x\right)^2+\left(\tfrac{1}{w}n_y\right)^2}\right)}{\sinh\left(\pi h\sqrt{\left(\tfrac{1}{l}n_x\right)^2+\left(\tfrac{1}{w}n_y\right)^2}\right)}\right).$$

Notice that the boundary conditions are shaped like a square well potential, but the solutions follow sine functions. One can also express any repeating function as an infinite sum of sine functions, which is referred to as a *Fourier series* after Jean-Baptiste Joseph Fourier who first found the series in order to understand heat flow. Below is a plot of the first six terms of the series (first parentheses of the solution above), and the cumulative result, holding the second and third terms to unity. Notice that with only a few terms it approximates the boundary condition more-or-less correctly toward the center of the box, but still yields a poor result toward the edges.

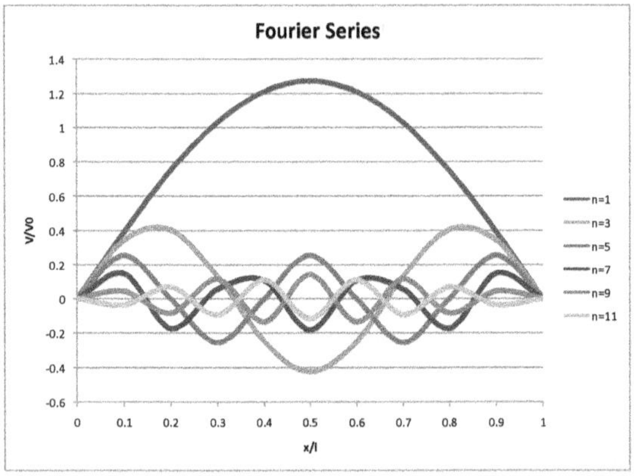

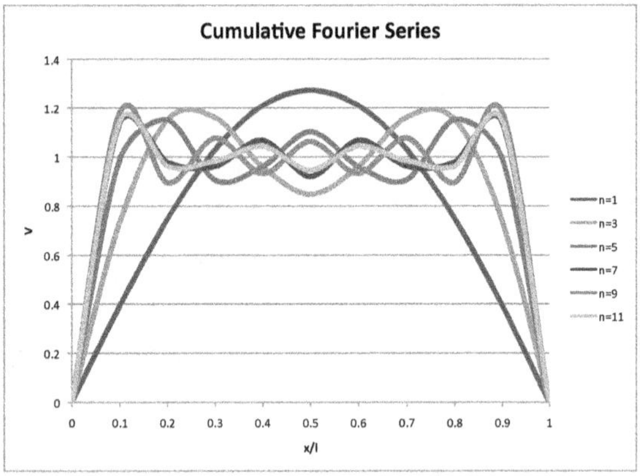

Example 5.4: A Conducting Ball in an Electric Field Revisited

Here we revisit the example of a neutral conducting sphere of radius a located inside a constant electric field, $\vec{E}_0$ (Example 5.1, p. 158). However, we now solve the same problem more formally using the technique of separation of variables. This technique should work, in principle, for most azimuthally symmetrical solutions in polar coordinates.

We assume a potential $\mathbb{V} = R(r)\Theta(\theta)$, so Laplace's equation becomes:

$$\nabla^2\mathbb{V} = \frac{\partial}{\partial r}\left(r^2\left(\frac{\partial \mathbb{V}}{\partial r}\right)\right) + \frac{1}{\sin\theta}\frac{\partial}{\partial\theta}\left(\sin\theta\frac{\partial\mathbb{V}}{\partial\theta}\right) = 0$$

$$0 = \frac{\partial}{\partial r}\left(r^2\left(\frac{\partial(R(r)\Theta(\theta))}{\partial r}\right)\right) + \frac{1}{\sin\theta}\frac{\partial}{\partial\theta}\left(\sin\theta\frac{\partial(R(r)\Theta(\theta))}{\partial\theta}\right)$$

$$0 = \frac{1}{R(r)}\frac{\partial}{\partial r}\left(r^2\left(\frac{\partial(R(r))}{\partial r}\right)\right) + \frac{1}{\Theta(\theta)}\frac{1}{\sin\theta}\frac{\partial}{\partial\theta}\left(\sin\theta\frac{\partial(\Theta(\theta))}{\partial\theta}\right).$$

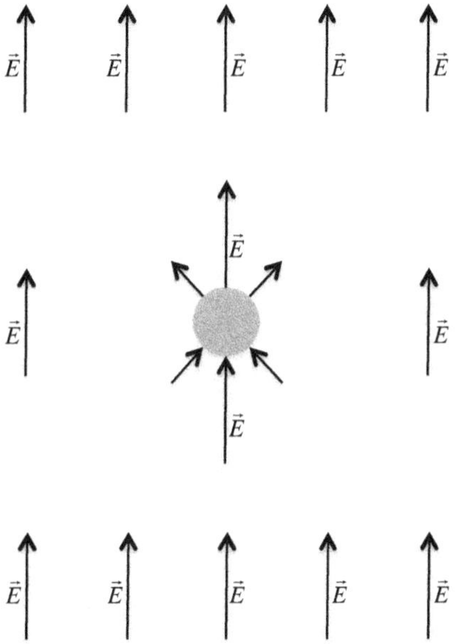

Now, we see that each term must simply equal a constant, and they must add to zero. Thus:

$$\frac{1}{R}\frac{\mathrm{d}}{\mathrm{d}r}\left(r^2\left(\frac{\mathrm{d}R}{\mathrm{d}r}\right)\right) = C \quad \text{and} \quad \frac{1}{\Theta}\frac{1}{\sin\theta}\frac{\mathrm{d}}{\mathrm{d}\theta}\left(\sin\theta\frac{\mathrm{d}\Theta}{\mathrm{d}\theta}\right) = -C.$$

We can now solve each differential equation in turn, beginning with the radial equation:

$$\frac{\mathrm{d}}{\mathrm{d}r}\left(r^2\left(\frac{\mathrm{d}R}{\mathrm{d}r}\right)\right) = CR.$$

We will guess a solution, and then realize that the general solution is a superposition of all particular solutions. That said, how do we guess the solution?

In principle we could perhaps solve this in terms of multipole expansion, because it is a solution to Laplace's equation in spherical coordinates. The potential for multipole expansion should be in the form of:

$$\mathbb{V}(\vec{r}) = \left(\frac{1}{4\pi\varepsilon_0}\right)\sum_{n=0}^{\infty}\left(\frac{1}{r^{n+1}}\right)\left(\int_{\text{volume}} \rho(\vec{R})R^n P_n(\cos\vartheta)\mathrm{d}V\right).$$

So, we will guess a power law solution $\left(R = \frac{A}{r^{n+1}} = Ar^{-(n+1)}\right)$ and check it:

$$\frac{\mathrm{d}}{\mathrm{d}r}\left(r^2\left(\frac{\mathrm{d}R}{\mathrm{d}r}\right)\right) = \frac{\mathrm{d}}{\mathrm{d}r}\left(r^2\left(\frac{\mathrm{d}\left(Ar^{-(n+1)}\right)}{\mathrm{d}r}\right)\right) = \frac{\mathrm{d}}{\mathrm{d}r}\left(r^2A(-(n+1))r^{-(n+2)}\right)$$

$$= -A(n+1)\frac{\mathrm{d}}{\mathrm{d}r}\left(r^{-n}\right) = n(n+1)\left(Ar^{-(n+1)}\right) = CR.$$

Now we see that our guess was right, so long as:

$$C = n(n+1), \quad n^2 + n - C = 0.$$

Since this equation is quadratic it has two possible solutions:

$$n_1 = \frac{-1+\sqrt{1+4C}}{2} = -\frac{1}{2}+\sqrt{\tfrac{1}{4}+C}, \quad n_2 = \frac{-1-\sqrt{1+4C}}{2} = -\frac{1}{2}-\sqrt{\tfrac{1}{4}+C},$$

with the property that:

$$n_1 + n_2 = -1 \quad \rightarrow \quad n_2 = -(n_1+1).$$

One solution is what we expected from multipole expansion, and the other increases with distance from the origin. These solutions are unphysical for most situations, but not in this case, as we will see.

For a given n we can write:

$$R_n = \frac{A_1}{r^{n+1}} + \frac{A_2}{r^{-(n+1)+1}} = \frac{A_1}{r^{n+1}} + \frac{A_2}{r^{-n}} = \frac{A_1}{r^{n+1}} + A_2 r^n,$$

and $C_n = n(n+1)$.

Let us now turn our attention to the polar component, where:

$$\frac{1}{\Theta}\frac{1}{\sin\theta}\frac{\mathrm{d}}{\mathrm{d}\theta}\left(\sin\theta\frac{\mathrm{d}\Theta}{\mathrm{d}\theta}\right) = -C = -n(n+1).$$

Again, we will use our multipole expansion solution to guess a solution to this differential equation. Thus, by looking at our solution in Thought Experiment 5.6 (p. 160), we can take the θ dependent factor as:

$$\Theta = \int_{\text{volume}} \rho(\vec{R}) R^n P_n(\cos\theta)\mathrm{d}V = B_n\, P_n(\cos\theta).$$

Where P_n are the Legendre polynomials, as the solutions to the θ dependence of multipole expansion. In fact, it can be shown with some mathematical effort that these are the solutions to our polar differential equation here (see Problem 5.5, p. 161).

We can now write the general solution in spherical coordinates, with azimuthal symmetry, as:

$$\mathbb{V}(\vec{r}) = \mathbb{V}_0 \sum_{n=0}^{\infty}\left(\frac{A_{1,n}}{r^{n+1}} + A_{2,n} r^n\right)\left(B_n\, P_n(\cos\theta)\right).$$

Notice that introducing an overall normalization allows for all of the coefficients to be unitless.

Notice that we have the following boundary conditions:

$$\mathbb{V}(r{=}a) = 0 \quad \text{and} \quad \mathbb{V}(r \gg a) = -E_0 z = -E_0 r\cos\theta.$$

In our problem, the quantity with units of volt that we can make from known constants will specify our overall normalization parameter. So, let:

$$\mathbb{V}_0 = E_0 a.$$

Next we will take the limit that $r \gg a$, so:

$$\lim_{r\gg a}\mathbb{V}(\vec{r}) = \lim_{r\gg a}\mathbb{V}_0 \sum_{n=0}^{\infty}\left(\frac{A_{1,n}}{r^{n+1}} + A_{2,n} r^n\right)\left(B_n\, P_n(\cos\theta)\right) = -E_0 r\cos\theta.$$

Thus, $A_{2,n}=0$ if $n>1$, $P_0(\cos\theta)=1$ and $P_1(\cos\theta)=\cos\theta$. So we can write just the first two terms at large distance as:

$$\cancel{E_0}\, a\left(\left(A_{2,0}\right)\left(B_0\right)+\left(A_{2,1} r\, B_1 \cos\theta\right)\right)=-\cancel{E_0}\, r\cos\theta \,.$$

We now see that $A_{2,0}B_0=0$, so our far away boundary condition becomes:

$$a\left(A_{2,1}\cancel{r}\, B_1 \cancel{\cos\theta}\right)=-\cancel{r}\,\cancel{\cos\theta} \quad\rightarrow\quad A_{2,1}B_1=-\frac{1}{a}\,.$$

Now let us apply the boundary condition at the surface of the sphere:

$$0=\cancel{\mathbb{V}_0}\sum_{n=0}^{\infty}\left(\frac{A_{1,n}}{a^{n+1}}+A_{2,n}a^n\right)\left(B_n\, P_n(\cos\theta)\right).$$

We can simplify this by evaluating the first two terms explicitly:

$$0=\left(\frac{A_{1,0}}{a^{0+1}}+A_{2,0}a^0\right)\left(P_0(\cos\theta)\right)+\left(\frac{A_{1,1}}{a^{1+1}}+A_{2,1}a^1\right)\left(B_1\, P_1(\cos\theta)\right)+\sum_{n=2}^{\infty}\left(\frac{A_{1,n}}{a^{n+1}}+A_{2,n}a^n\right)\left(P_n(\cos\theta)\right).$$

At the surface of the sphere, this simplifies to:

$$0=\left(\frac{B_1A_{1,1}}{a^2}+B_1A_{2,1}a\right)\cos\theta=\left(\frac{B_1A_{1,1}}{a^2}+\left(-\frac{1}{\cancel{a}}\right)\cancel{a}\right)\cos\theta$$

$$\frac{B_1A_{1,1}}{a^2}=1 \quad\rightarrow\quad B_1A_{1,1}=a^2\,.$$

Thus, we can finally write down the expression for the potential as:

$$\mathbb{V}(\vec{r})=\mathbb{V}_0\sum_{n=0}^{\infty}\left(\frac{A_{1,n}}{r^{n+1}}+A_{2,n}r^n\right)\left(B_n\, P_n(\cos\theta)\right)=E_0a\left(\frac{a^2}{r^2}-\frac{r}{a}\right)\cos\theta\,.$$

This result agrees with our result from Example 5.1. (p. 158).

Example 5.5: A Capped Coaxial Cable

We will now tackle a more mathematically complicated example, that of a capped coaxial cable. Imagine that we place a hollow conductive cap of length h on the end of the coaxial cable from Example 5.2. What is the electric potential inside the cavity created by the cap?

First we notice the azimuthal symmetry, so:

$$\mathbb{V}=\mathbb{V}(s,z)\,.$$

Also note that at the conductive cap surface $\mathbb{V}=0$, so:

$$\mathbb{V}(z=0)=\mathbb{V}(s=b)=0$$

And we now assume that the potential at the end of the cable must match Example 5.2, so:

$$\mathbb{V}(s,z=h)=\begin{cases} s\le a & \mathbb{V}_0 \\ s\ge a & \mathbb{V}_0\dfrac{\ln\left(\frac{b}{s}\right)}{\ln\left(\frac{b}{a}\right)}\end{cases}$$

We can now tackle Laplace's equation inside the cap:

$$\nabla^2\mathbb{V}=\frac{1}{s}\frac{\partial}{\partial s}\left(s\frac{\partial\mathbb{V}}{\partial s}\right)+\cancel{\frac{1}{s^2}\frac{\partial^2\mathbb{V}}{\partial\varphi^2}}+\frac{\partial^2\mathbb{V}}{\partial z^2}=0$$

$$\frac{1}{s}\frac{\partial}{\partial s}\left(s\frac{\partial\mathbb{V}(s,z)}{\partial s}\right)+\frac{\partial^2\mathbb{V}(s,z)}{\partial z^2}=0.$$

We assume the solution is separable, so it can be written in the form:

$$\mathbb{V}(s,z)=S(s)Z(z),$$

which allows Laplace's equation to be written as:

$$0=\frac{1}{s}\frac{\partial}{\partial s}\left(s\frac{\partial\left(S(s)Z(z)\right)}{\partial s}\right)+\frac{\partial^2\left(S(s)Z(z)\right)}{\partial z^2}$$

$$=\frac{1}{S(s)}\frac{1}{s}\frac{\partial}{\partial s}\left(s\frac{\partial S(s)}{\partial s}\right)+\frac{1}{Z(z)}\frac{\partial^2 Z(z)}{\partial z^2}.$$

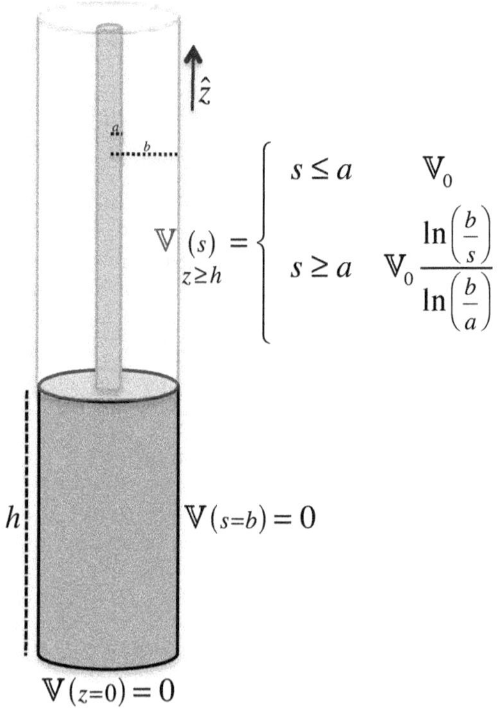

Thus we have two ordinary differential equations to solve:

$$\frac{1}{Z(z)}\frac{\mathrm{d}^2Z(z)}{\mathrm{d}z^2}=C \quad\text{and}\quad \frac{1}{S(s)}\frac{1}{s}\frac{\mathrm{d}}{\mathrm{d}s}\left(s\frac{\mathrm{d}S(s)}{\mathrm{d}s}\right)=-C.$$

Notice that, as in the case of the box (p. 163) one solution will be oscillatory and the other exponential in form. Clearly, by analogy with the box, and because each side of the cap is held at ground, $S(s)$ must be oscillatory. Thus $C\ge 0$, and noticing that the units of C are inverse length squared, we can replace C with a new constant k of units inverse length, such that $k=\sqrt{C}$.

Not every oscillatory function is a standard trigonometric function, and the reader can verify that a function in the form $S(s)=A\sin(ks)+B\cos(ks)$ does not satisfy the above differential equation. However, this differential equation can be solved with another oscillatory function called the Bessel function, after the German astronomer and mathematician Friedrich Wilhelm Bessel (p. 154). To solve this, we will put the equation in *Bessel's form*, so that we can apply his solution:

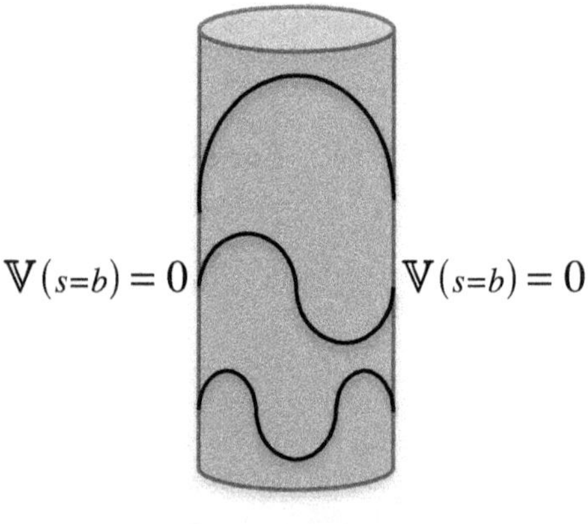

$$u^2\frac{\mathrm{d}^2S}{\mathrm{d}u^2}+u\frac{\mathrm{d}S}{\mathrm{d}u}+\left(u^2-m^2\right)S=0\,.$$

Our equation becomes:

$$\frac{1}{S(s)}\frac{1}{s}\frac{\mathrm{d}}{\mathrm{d}s}\left(s\frac{\mathrm{d}S(s)}{\mathrm{d}s}\right)=-k^2=\frac{1}{S(s)}\frac{1}{s}\left(s\frac{\mathrm{d}^2S(s)}{\mathrm{d}s^2}+\frac{\mathrm{d}S(s)}{\mathrm{d}s}\right)$$

$$s^2\frac{\mathrm{d}^2S(s)}{\mathrm{d}s^2}+s\frac{\mathrm{d}S(s)}{\mathrm{d}s}=-(ks)^2S(s)$$

$$(ks)^2\frac{\mathrm{d}^2S(ks)}{\mathrm{d}(ks)^2}+(ks)\frac{\mathrm{d}S(ks)}{\mathrm{d}(ks)}+(ks)^2S(ks)=0\,,$$

and with the substitution $u=ks$, we can write:

$$u^2\frac{\mathrm{d}^2S(u)}{\mathrm{d}u^2}+u\frac{\mathrm{d}S(u)}{\mathrm{d}u}+\left(u^2-0^2\right)S(s)=0\,.$$

The independent and orthogonal solutions to this linear second order differential equation are Bessel and Neumann functions, $J_m(u)$ and $Y_m(u)$ respectively. In our case, $m=0$. Moreover, our boundary conditions require a real solution on-axis, but $\lim_{u\to 0}Y_m(u)=-\infty$, so we can disregard the Neumann functions:

$$S(s)=A\,J_0(ks).$$

Applying the boundary condition that the outer wall is grounded:

$$S(b)=A\,J_0(kb)=0\,.$$

Thus, kb are the roots of J_0, or we can give each root a subscript:

$$k_n=\frac{J^{-1}{}_0(0)}{b}\quad\rightarrow\quad S(s)=\sum_{n=1}^{\infty}A_n\,J_0(k_ns)\,.$$

Consider, now, the original separable equation:

$$\mathbb{V}(s,z)=S(s)Z(z),$$

which we solve for the function $Z(z)$:

$$\frac{1}{Z(z)}\frac{\mathrm{d}^2Z(z)}{\mathrm{d}z^2}=k^2\,.$$

We solved this differential equation in Example 5.3 (p. 163) as the hyperbolic sine and cosine functions. Given that $Z(0)=0$ the solution is in the form of:

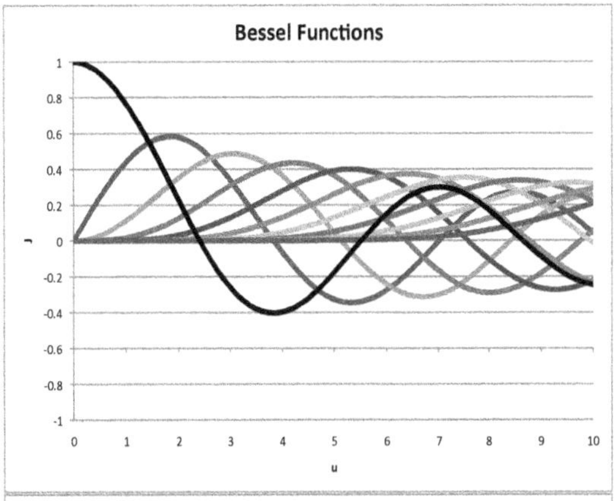

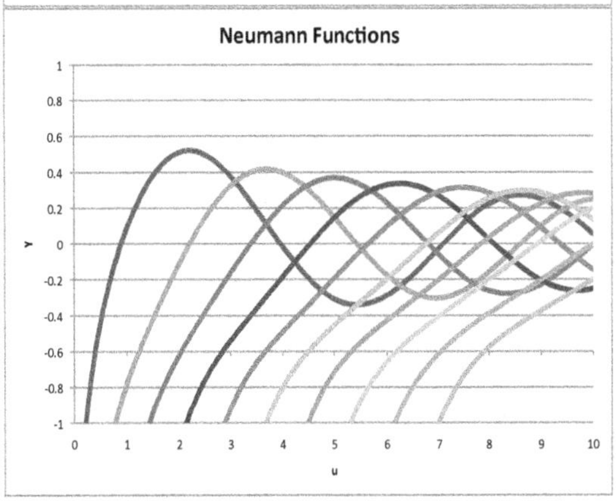

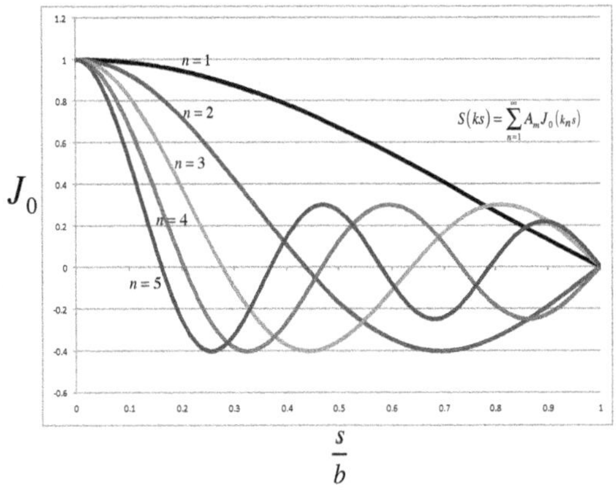

$Z(z) \propto \sinh(kz)$

Now, write the potential as a linear combination of the two solutions:

$$\mathbb{V}(s,z) = S(s)Z(z) = \mathbb{V}_0 \sum_{n=1}^{\infty} C_n J_0(k_n s)\sinh(k_n z).$$

Now we can finally apply the top boundary condition:

$$\mathbb{V}(s,h) = \mathbb{V}_0 \sum_{n=1}^{\infty} C_n J_0(k_n s)\sinh(k_n h).$$

We can solve for C_n by noting that, like sine functions, Bessel functions are orthogonal:

$$\int_0^b J_0(k_n s)\mathbb{V}(s,h)\, s\, \mathrm{d}s = \int_0^b J_0(k_{n'} s)\mathbb{V}_0 \sum_{n=1}^{\infty} C_n J_0(k_n s)\sinh(k_n h)\, s\, \mathrm{d}s = \mathbb{V}_0 C_n \sinh(k_n h)\int_0^b J_0(k_{n'} s) J_0(k_n s)\, s\, \mathrm{d}s .$$

Bessel functions have an orthogonality property, such that:

$$\int_0^b J_0(k_{n'} s) J_0(k_n s)\, s\, \mathrm{d}s = \frac{b^2}{2}\left(J_1(k_n b)\right)^2, \text{ so:}$$

$$\int_0^b J_0(k_n s)\mathbb{V}(s,h)\, s\, \mathrm{d}s = \mathbb{V}_0 C_n \sinh(k_n h)\frac{b^2}{2}\left(J_1(k_n b)\right)^2 .$$

Therefore, we can write the coefficients as:

$$\begin{aligned} C_n &= \frac{2}{b^2 \sinh(k_n h)\left(J_1(k_n b)\right)^2}\frac{1}{\mathbb{V}_0}\int_0^b J_0(k_n s)\mathbb{V}(s,h)\, s\, \mathrm{d}s \\ &= \frac{2}{b^2 \sinh(k_n h)\left(J_1(k_n b)\right)^2}\left(\int_0^a J_0(k_n s)\, s\, \mathrm{d}s + \int_a^b J_0(k_n s)\frac{\ln\left(\frac{b}{s}\right)}{\ln\left(\frac{b}{a}\right)}\, s\, \mathrm{d}s\right). \end{aligned}$$

And finally, we can write the potential as:

$$\mathbb{V}(s,z) = \mathbb{V}_0 \sum_{n=1}^{\infty} \frac{\sinh(k_n z)}{\sinh(k_n h)}\frac{J_0(k_n s)}{\left(J_1(k_n b)\right)^2}\frac{2}{b^2}\left(\int_0^a J_0(k_n s)\, s\, \mathrm{d}s + \int_a^b J_0(k_n s)\frac{\ln\left(\frac{b}{s}\right)}{\ln\left(\frac{b}{a}\right)}\, s\, \mathrm{d}s\right).$$

5.5 *The Method of Images*

We have now shown that superpositions of Coulomb's law are solutions to Laplace's equation. We have also shown that the electric field must be normal to any conductor's surface. Recall from introductory physics how light bounces off of mirrors, in a way that is symmetrical about the normal. Putting these ideas together, we can visualize Thomson's method of images.

If a charge is near a conducting plane placing an equal, but opposite, charge at the location of its virtual image (behind the mirror) gives the same electric potential as having the conductor.

Example 5.6: A Point Charge Near a Conducting Plane

Consider a point charge Q a distance h above a large grounded conducting plane. What is the electric potential surrounding the charge?

Notice the boundary conditions on the surface of the conducting plane are such that the electric field must be normal to the surface, and near the charge is simply Coulomb's Law:

$$\vec{E}(z=0) = -E(z=0)\hat{z} \quad \text{and} \quad \mathbb{V}_{\text{near charge}} \approx \frac{Q}{4\pi\varepsilon_0\sqrt{s^2+(z-h)^2}}.$$

Notice that we have rendered our system in cylindrical coordinates with the origin on the surface of the plane directly below the point charge.

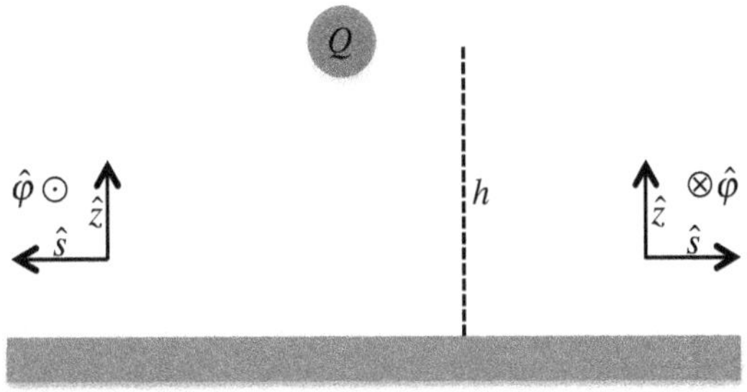

Now, by placing an imaginary, equal but opposite, mirror image charge below the plane, the boundary conditions are identical to those posed in the problem. Since Coulomb's law must satisfy Laplace's equation, we can therefore replace the original problem with an equivalent dipole whose potential is given by:

$$\mathbb{V} = \frac{Q}{4\pi\varepsilon_0}\left(\frac{1}{\sqrt{s^2+(z-h)^2}} - \frac{1}{\sqrt{s^2+(z+h)^2}}\right).$$

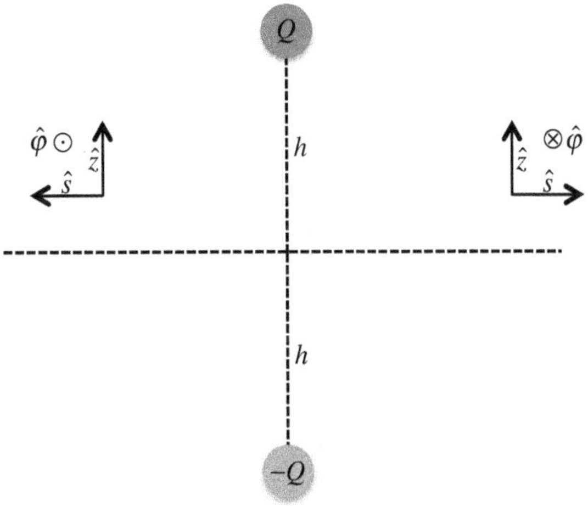

This conforms to the boundary conditions of the original problem, and satisfies Laplace's equation. It is only valid above the conducting plane, as in the conductor $\mathbb{V}=0$.

The reasoning we have employed here is an example of the *method of images*, which is useful for solving problems involving point charges near conductors.

Problem 5.8: A Line Charge Near a Conducting Plane

Consider a infinitely long line charge of charge per length λ a distance h from a infinite conducting plane.

a) Find a fictitious set of charges which will hopefully yield the same electric potential function as the real charge and the conducting plane.

b) Show that your fictitious charge distribution reproduces the correct boundary conditions at the location of the conducting plane.

c) Find the electric potential associated with the scenario given in the problem.

d) Show explicitly that this satisfies Laplace's equation everywhere except at the charge.

Problem 5.9: A Point Charge and Two Conducting Planes

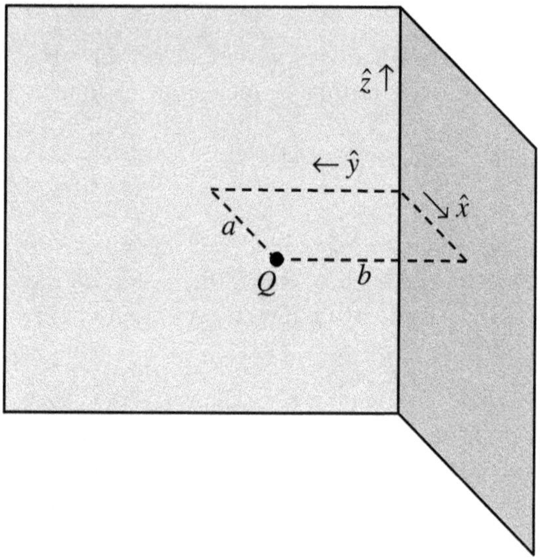

Consider a point charge near two large perpendicular grounded conducting planes. One plane is located at position $x=0$, and the other at position $y=0$. The point charge is located at position $\vec{r}=a\hat{x}+b\hat{y}$.

a) Find a fictitious set of charges which will hopefully yield the same electric potential function as the charge and conducting planes shown.

b) Show that your fictitious charges reproduce the correct boundary conditions at the location of the conducting planes.

c) Find the electric potential associated with the scenario given in the problem for positive x and y positions.

d) Show explicitly that this satisfies Laplace's equation everywhere except at the charge.

Problem 5.10: A Point Charge and a Conducting Paraboloid

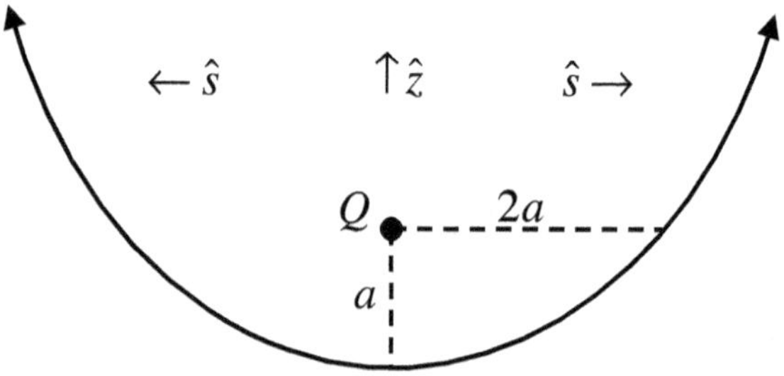

Consider a large metal paraboloid whose surface is given in cylindrical coordinates by the equation $z=\dfrac{s^2}{4a}-a$, and a point charge Q located at the origin.

a) Find a fictitious set of charges which will hopefully yield the same electric potential function as the charge and paraboloid.

b) Show that your fictitious charges reproduce the correct boundary conditions at the location of the conducting planes.

c) Find the electric potential function that is valid for the combined real and fictitious charge distribution.

d) Show explicitly that this satisfies Laplace's equation at positions such that $z\geq\dfrac{s^2}{4a}-a$.

5.6 *The Method of Relaxation*

With the advent of modern computers, the approach to solving Poisson's or Laplace's equations has shifted from specialized classical standard techniques to modern algorithms used to obtain numerical solutions to boundary value problems. While we briefly discussed relaxation methods in Chapter 1 (p. 49), here we develop the method of relaxation for solving both Poisson's equation and Laplace's equation.

Thought Experiment 5.7: Poisson's Equation with Relaxation

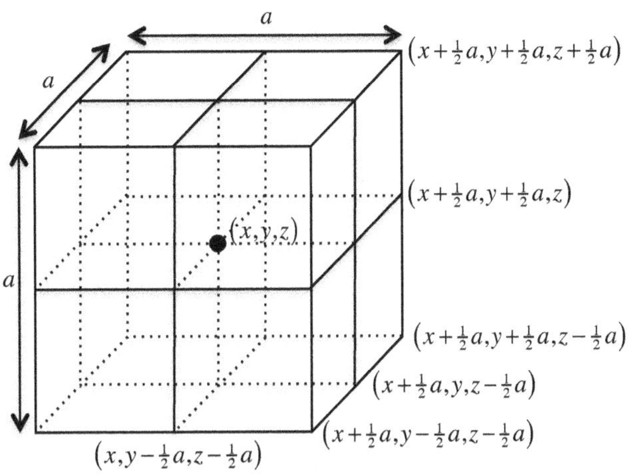

We now turn to a class of problem with both a continuous charge distribution, $\rho(\vec{r})$, and a boundary at a fixed voltage. These involve solving Poisson's equation, which can be quite difficult analytically. However, if we apply Gauss's law to each cell in a computer grid, we can use the method of relaxation to solve arbitrarily complex problems in electrostatics numerically.

Consider now that there is some grid resolution, a, that is the smallest scale of interest, so we draw a cubic Gaussian surface of dimensions $a \times a \times a$ centered at a point of interest (x,y,z), and we will assume that a is sufficiently small so that the charge contained in the box is given simply by:

$$Q(x,y,z) = a^3 \rho(x,y,z).$$

Applying Gauss's Law to this box the electric field at each face becomes:

$$\oint_{\text{surface}} \vec{E}\cdot d\vec{A} = \tfrac{1}{\varepsilon_0}Q = \tfrac{1}{\varepsilon_0}a^3\rho(x,y,z) = \begin{array}{l} a^2E_x(x+\tfrac{1}{2}a,y,z) - a^2E_x(x-\tfrac{1}{2}a,y,z) + a^2E_y(x,y+\tfrac{1}{2}a,z) \\ -a^2E_y(x,y-\tfrac{1}{2}a,z) + a^2E_z(x,y,z+\tfrac{1}{2}a) - a^2E_z(x,y,z-\tfrac{1}{2}a). \end{array}$$

Now we apply the definition of the derivative, so:

$$\vec{E} = -\vec{\nabla}\mathbb{V} \quad \text{so} \quad E_x(x',y',z') \approx -\tfrac{1}{a}\left(\mathbb{V}(x'+\tfrac{1}{2}a,y',z') - \mathbb{V}(x'-\tfrac{1}{2}a,y',z')\right).$$

And a similar relationship holds for the other two directions, so substituting $x' = x + \frac{1}{2}a$:

$$E_x(x+\tfrac{1}{2}a,y,z) \approx -\tfrac{1}{a}\left(\mathbb{V}(x+\tfrac{1}{2}a+\tfrac{1}{2}a,y,z) - \mathbb{V}(x+\tfrac{1}{2}a-\tfrac{1}{2}a,y,z)\right) \approx -\tfrac{1}{a}\mathbb{V}(x+a,y,z) + \tfrac{1}{a}\mathbb{V}(x,y,z)$$

and applying the substitution $x' = x - \frac{1}{2}a$ the following relationship holds:

$$E_x(x-\tfrac{1}{2}a,y,z) \approx -\tfrac{1}{a}\left(\mathbb{V}(x-\tfrac{1}{2}a+\tfrac{1}{2}a,y,z) - \mathbb{V}(x-\tfrac{1}{2}a-\tfrac{1}{2}a,y,z)\right) \approx -\tfrac{1}{a}\mathbb{V}(x,y,z) + \tfrac{1}{a}\mathbb{V}(x-a,y,z).$$

and with the corresponding substitutions in the two other directions, we can write:

$$\begin{array}{l} a^2\left(-\tfrac{1}{a}\mathbb{V}(x+a,y,z) + \tfrac{1}{a}\mathbb{V}(x,y,z)\right) - a^2\left(-\tfrac{1}{a}\mathbb{V}(x,y,z) + \tfrac{1}{a}\mathbb{V}(x-a,y,z)\right) + a^2\left(-\tfrac{1}{a}\mathbb{V}(x,y+a,z) + \tfrac{1}{a}\mathbb{V}(x,y,z)\right) \\ -a^2\left(-\tfrac{1}{a}\mathbb{V}(x,y,z) + \tfrac{1}{a}\mathbb{V}(x,y-a,z)\right) + a^2\left(-\tfrac{1}{a}\mathbb{V}(x,y,z+a) + \tfrac{1}{a}\mathbb{V}(x,y,z)\right) - a^2\left(-\tfrac{1}{a}\mathbb{V}(x,y,z) + \tfrac{1}{a}\mathbb{V}(x,y,z-a)\right) \\ = \tfrac{1}{\varepsilon_0}a^3\rho(x,y,z). \end{array}$$

Solving for $\mathbb{V}(x,y,z)$ in terms of the 6 surrounding points, we obtain:

$$\mathbb{V}(x,y,z) = \frac{\frac{1}{\varepsilon_0}a^3\rho(x,y,z)}{6a} + \frac{\cancel{a}}{6\cancel{a}}\begin{pmatrix} \mathbb{V}(x+a,y,z) + \mathbb{V}(x-a,y,z) + \mathbb{V}(x,y+a,z) \\ +\mathbb{V}(x,y-a,z) + \mathbb{V}(x,y,z+a) + \mathbb{V}(x,y,z-a) \end{pmatrix}$$

Or in terms of the average of the surrounding points, $\langle \mathbb{V}(x\pm a, y\pm a, z\pm a)\rangle$:

$$\mathbb{V}(x,y,z) = \langle \mathbb{V}(x\pm a, y\pm a, z\pm a)\rangle + \frac{a^2\rho(x,y,z)}{6\varepsilon_0}.$$

If the charge density is zero, this becomes particularly simple. For a small box, the potential in the middle is simply the average of the surrounding points.

It is more robust to use all the points on a 3×3 grid, rather than just the four closest points. Like a Rubik's cube, a 3×3×3 grid has 6 face points, 12 edge points, and 8 corner points. Each edge point is $\sqrt{2}$ times further from the center than each face point, and each corner point is $\sqrt{3}$ times further from each face point. By analogy with the two dimensional case, it would be reasonable to use an algorithm that weighted each face:edge:corner point in a $1:\frac{1}{2}:\frac{1}{3}$ ratio, or in terms of averages a $\frac{6}{1}:\frac{12}{2}:\frac{8}{3}$ ratio would make sense, thus:

$$\mathbb{V}(x,y,z) \approx \tfrac{9}{22}\langle \mathbb{V}(\text{faces})\rangle + \tfrac{9}{22}\langle \mathbb{V}(\text{edges})\rangle + \tfrac{2}{11}\langle \mathbb{V}(\text{corners})\rangle + \frac{a^2\rho(x,y,z)}{6\varepsilon_0}.$$

Example 5.7: A Flat Stack of Capacitors

Consider now a thin stack of capacitors of height h, sandwiched between two grounded plates. Every-other plate is has opposite charge. We will model this as if we had a stack of alternating charge densities, as in the top figure. Our goal is to find the electric field throughout the stack,

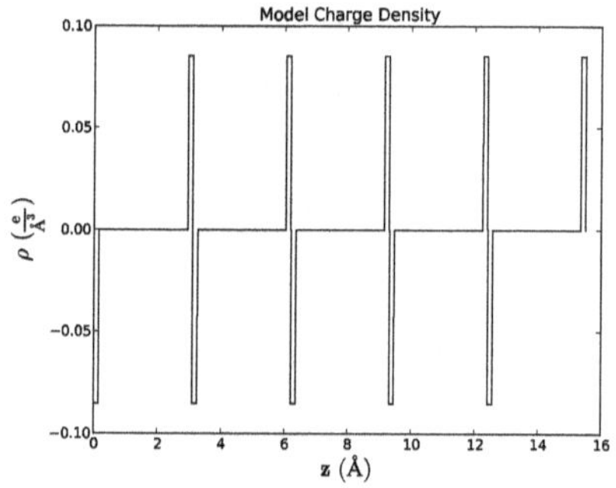

Given our assumption of horizontal symmetry, the method of relaxation can be written simply as:

$$\mathbb{V}(z) \approx \tfrac{1}{2\varepsilon_0}a^2\rho(z) + \tfrac{1}{2}\left(\mathbb{V}(z+a) + \mathbb{V}(z-a)\right).$$

We begin with an initial guess of zero potential everywhere, and insist on neutrality at the top and bottom. The potential after convergence is shown.

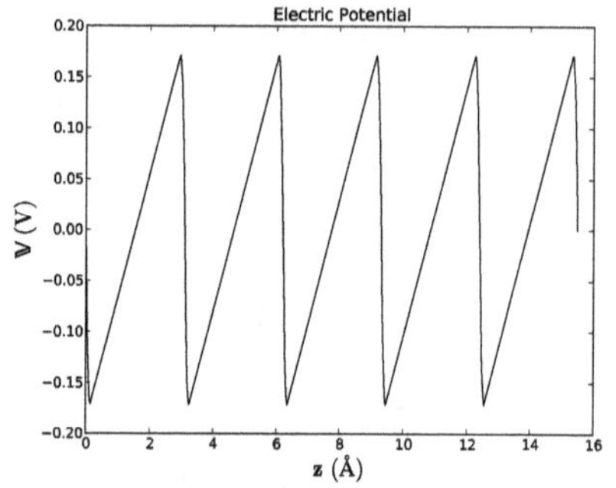

Once we have calculated the potential, we differentiate it using the central difference to find the electric field using the relationship:

$$E(z) \approx -\frac{\mathbb{V}(z+a) - \mathbb{V}(z-a)}{2a}.$$

Notice from the plot how between each capacitor the electric field is constant and in the negative direction, which was opposite the direction of the dipole moments.

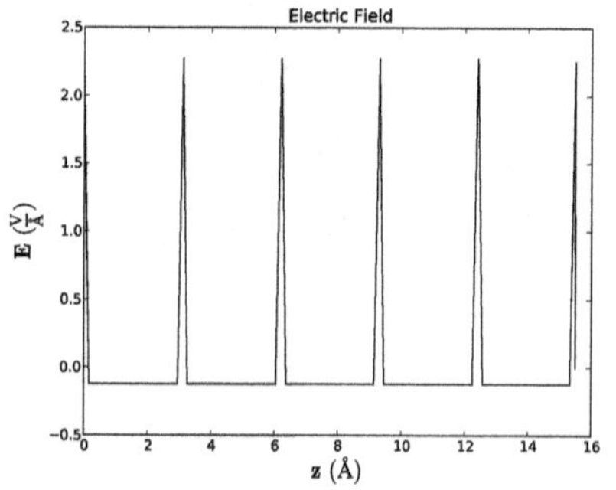

In the plots of the charge density, electric potential, and electric field we used values appropriate if each capacitor had one elementary charge, and each gap width were the size of a typical molecule. In other words, we applied this to a classical toy model of polarized matter.

Example 5.8: A 2D Dipole in a Grounded Box

Consider taking a grounded square metal box of dimension $l \times l \times h$, with one long side $(h \gg l)$, so that $\mathbb{V} = \mathbb{V}(x,y)$. Inside the box are two cylindrical metal posts with voltages $\mathbb{V}_0$ and $-\mathbb{V}_0$ respectively. Our goal is to find the electric potential inside the box.

Inside the box, and not at the posts, Laplace's equation $\nabla^2 \mathbb{V} = 0$ must hold, so we will apply the method of relaxation in two dimensions to the inside of our box, where we replace each pixel with the weighted average of the surrounding points, so for the nth iteration with cell spacing a:

$$\mathbb{V}_n(x,y) = \tfrac{2}{3}\langle \mathbb{V}_n(\text{sides})\rangle + \tfrac{1}{3}\langle \mathbb{V}_n(\text{corners})\rangle, \quad \text{where:}$$

$$\langle \mathbb{V}_n(\text{sides})\rangle = \tfrac{1}{4}\left(\mathbb{V}_n(\text{x+a, y}) + \mathbb{V}_n(\text{x-a, y}) + \mathbb{V}_n(\text{x, y+a}) + \mathbb{V}_n(\text{x, y-a})\right)$$

$$\langle \mathbb{V}_n(\text{corners})\rangle = \tfrac{1}{4}\left(\mathbb{V}_n(\text{x+a, y+a}) + \mathbb{V}_n(\text{x-a, y+a}) + \mathbb{V}_n(\text{x+a, y-a}) + \mathbb{V}_n(\text{x-a, y-a})\right).$$

The three plots show: the initial guess, the potential after convergence, and the corresponding electric field lines. Notice how the electric field is normal to the surface of all conductors.

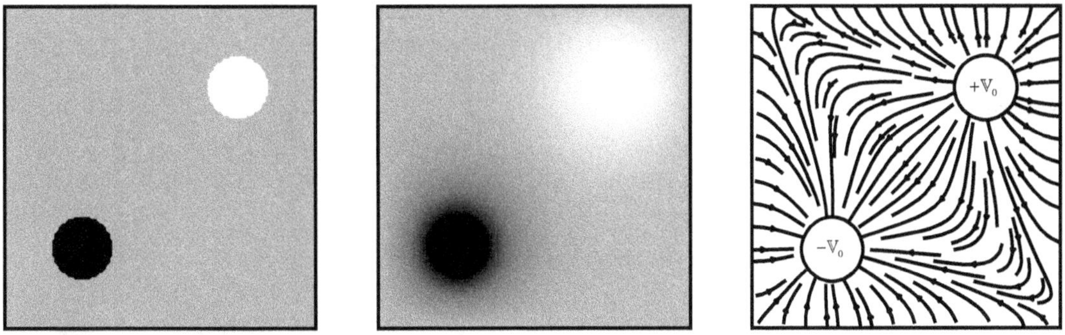

Problem 5.11: A 2D Quadrupole in a Grounded Box

a) Reproduce the 2D dipole simulation from Example 5.8, producing similar data plots.

b) Simulate a 2D quadrupole, with four cylindrical metal posts arranged like the dots on a die, with alternating voltages of $\mathbb{V}_0$ and $-\mathbb{V}_0$, arranged such that the net dipole moment is zero.

c) Simulate a similar scenario with 3 charged posts, again arranged like the dots on a die, but with each outside post being held at a voltage of $-\mathbb{V}_0$, and the middle post at a voltage of $\mathbb{V}_0$.

d) Again, simulate a similar scenario but with 5 charged posts, again arranged like the dots on a die with each outside post being held at $-\mathbb{V}_0$, and the middle post at a voltage of $\mathbb{V}_0$.

Example 5.9: A 3D Octopole in a Grounded Box

Consider a grounded conducting box of inside dimensions $L \times L \times L$ with a fixed charge density distribution inside the box given by:

$\rho = \rho_0 \sin(\frac{2\pi}{L}x)\sin(\frac{2\pi}{L}y)\sin(\frac{2\pi}{L}z)$.

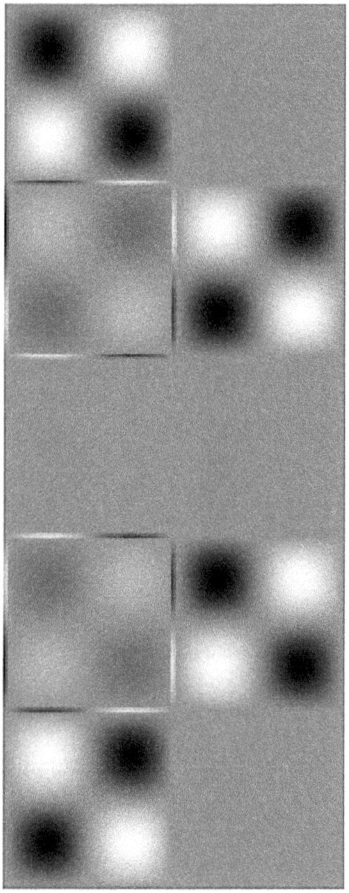

In order to solve this problem we first need to consider the question of grid spacing. On one hand, the finer the grid spacing (a) the better the solution, but on the other hand the finer the grid the more calculations the computer must perform. Thus, if $N = \frac{1}{a}$ then the total number of calculations per iteration is given by N^3, but we will need to run it at least N times, so the total number of calculations is proportional to N^4. Thus a small difference in grid spacing can make a huge difference in the time for the computer to run.

We begin solving this problem by setting up an array of electric potential values that contains the very simple initial guess of a uniform electric potential. Next we define an array for the source term for the potential as:

$$\mathbb{V}_0(x,y,z) \equiv \frac{a^2\rho(x,y,z)}{6\varepsilon_0} = \frac{a^2\rho_0}{6\varepsilon_0}\sin(\tfrac{2\pi}{L}x)\sin(\tfrac{2\pi}{L}y)\sin(\tfrac{2\pi}{L}z).$$

At each iteration, we will calculate the more relaxed value of:

$$\mathbb{V}_2(x,y,z) = \mathbb{V}_0(x,y,z) + \langle \mathbb{V}_1 \rangle,$$

where $\mathbb{V}_2$ represents the more refined potential and $\langle \mathbb{V}_1 \rangle$ represents the average of the adjacent cells of the last iteration. Except for the boundaries, which we will keep at zero potential because the box is grounded and conducting.

The figure shows x,y slices of the charge density (left column) and the potential (right column) at five evenly spaced z positions from top to bottom of 0, $\frac{1}{4}L$, $\frac{1}{2}L$, $\frac{3}{4}L$, and L respectively.

Thought Experiment 5.8: Relaxation in 2D Cylindrical Coordinates

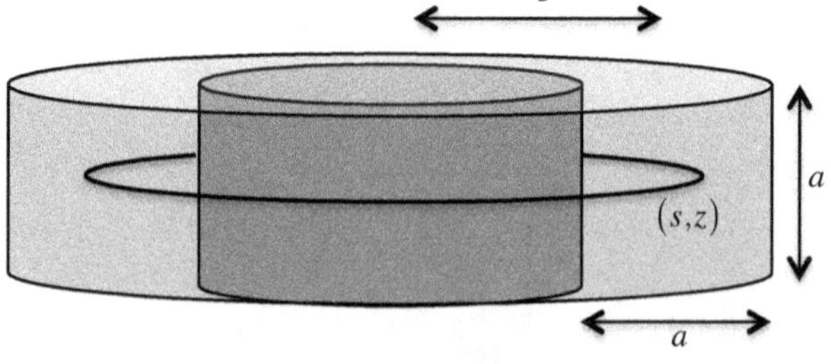

In three dimensions there is really no advantage in using cylindrical coordinates, rather than Cartesian coordinates. However, in azimuthally symmetrical systems there is a great advantage in using cylindrical coordinates.

Consider a circle in a cylindrical grid located at position s,z, inside a circular Gaussian surface of radial width a, and a corresponding height also a. The total charge enclosed by this surface will be defined as $Q(s,z) \approx 2\pi s a^2 \rho(s,z)$, making the assumption that the charge density is uniform inside the surface. We now apply Gauss's law to this surface, so:

$$\oint_{\text{surface}} \vec{E} \cdot d\vec{A} = \frac{Q}{\varepsilon_0}.$$

Now we substitute our the electric field at the edge of each cell:

$$\begin{aligned} &E_s\left(s+\tfrac{1}{2}a,z\right)\left(2\pi\left(s+\tfrac{1}{2}a\right)a\right) - E_s\left(s-\tfrac{1}{2}a,z\right)\left(2\pi\left(s-\tfrac{1}{2}a\right)a\right) + E_z\left(s,z+\tfrac{1}{2}a\right)\left(\pi\left(s+\tfrac{1}{2}a\right)^2 - \pi\left(s-\tfrac{1}{2}a\right)^2\right) \\ &-E_z\left(s,z-\tfrac{1}{2}a\right)\left(\pi\left(s+\tfrac{1}{2}a\right)^2 - \pi\left(s-\tfrac{1}{2}a\right)^2\right) = \tfrac{1}{\varepsilon_0}\rho(s,\varphi,z)\left(2\pi s a^2\right). \end{aligned}$$

Simplifying, this becomes:

$$E_s\left(s+\tfrac{1}{2}a,z\right)\left(\tfrac{s+\frac{1}{2}a}{s}\right) - E_s\left(s-\tfrac{1}{2}a,z\right)\left(\tfrac{s-\frac{1}{2}a}{s}\right) + E_z\left(s,z+\tfrac{1}{2}a\right) - E_z\left(s,z-\tfrac{1}{2}a\right) = \tfrac{a}{\varepsilon_0}\rho(s,\varphi,z)$$

Next we will investigate the gradient of the electric potential, where:

$$\begin{aligned} &\vec{E} = -\vec{\nabla}\mathbb{V} = -\tfrac{\partial}{\partial s}\mathbb{V}\,\hat{s} - \cancel{\tfrac{1}{s}\tfrac{\partial}{\partial\varphi}\mathbb{V}\,\hat{\varphi}} - \tfrac{\partial}{\partial z}\mathbb{V}\,\hat{z} \\ &E_s(s',\varphi',z') \approx -\tfrac{1}{a}\left(\mathbb{V}\left(s'+\tfrac{1}{2}a,z'\right) - \mathbb{V}\left(s'-\tfrac{1}{2}a,z'\right)\right) \\ &E_z(s',\varphi',z') \approx -\tfrac{1}{a}\left(\mathbb{V}\left(s',z'+\tfrac{1}{2}a\right) - \mathbb{V}\left(s',z'-\tfrac{1}{2}a\right)\right). \end{aligned}$$

Now substituting for the primed coordinates on each face of the Gaussian surface, in much the same way as we did in Thought Experiment 5.7 (p. 175), we can write:

$$\begin{aligned} \tfrac{a}{\varepsilon_0}\rho(s,z) = &\left(-\tfrac{1}{a}\left(\mathbb{V}(s+a,z) - \mathbb{V}(s,z)\right)\right)\left(\tfrac{s+\frac{1}{2}a}{s}\right) - \left(-\tfrac{1}{a}\left(\mathbb{V}(s,z) - \mathbb{V}(s-a,z)\right)\right)\left(\tfrac{s-\frac{1}{2}a}{s}\right) \\ &+\left(-\tfrac{1}{a}\left(\mathbb{V}(s,z+a) - \mathbb{V}(s,z)\right)\right) - \left(-\tfrac{1}{a}\left(\mathbb{V}(s,z) - \mathbb{V}(s,z-a)\right)\right), \end{aligned}$$

which simplifies to:

$$\mathbb{V}(s,z) = \left(\tfrac{s+\frac{1}{2}a}{4s}\right)\mathbb{V}(s+a,z) + \left(\tfrac{s-\frac{1}{2}a}{4s}\right)\mathbb{V}(s-a,z) + \tfrac{1}{4}\mathbb{V}(s,z+a) + \tfrac{1}{4}\mathbb{V}(s,z-a) + \tfrac{a^2}{4\varepsilon_0}\rho(s,z).$$

This is simply the weighted average of the surrounding voltages plus a source term based on the charge density. The outer voltage is more important than the inner voltage because the circumference is greater.

So, as long as an azimuthally symmetrical problem is divided into square cells, this method can be used to solve Poisson's equation.

Example 5.10: A Coaxial Cable with Charge

Consider now a cylindrical section of coaxial cable of radius R and height h, and a circular top and bottom located at $z = 0$ and $z = h$. The center wire, and the outside are also grounded.

We now imagine that it has a fixed charge density given by the following equation:

$$\rho = \rho_0\left(\frac{R}{s}\right)\sin\left(\tfrac{\pi}{R}s\right)\sin\left(\tfrac{2\pi}{h}z\right).$$

Our goal is to calculate the electric potential as a function of position inside the coaxial cable, as well as the electric field $\vec{E}$.

In order to apply the method of relaxation to this model, we must decide on a grid size a. Clearly, the finer the grid size the more accurate the results. Remember that for the method to work, the grid must be square, so if we use N_s slices in the radial direction, then $a = R/N_s$, and therefore the number of slices in the axial direction must be:

$$N_z = \frac{h}{a} = \frac{h}{R} N_s .$$

We begin with a zero potential everywhere, and apply the method of relaxation in cylindrical coordinates in order to find the potential as a function of s and z.

Now to find the electric field, we need to take the cylindrical gradient:

$$\vec{E} = -\vec{\nabla}\mathbb{V} = -\tfrac{\partial}{\partial s}\mathbb{V}\,\hat{s} - \tfrac{\partial}{\partial z}\mathbb{V}\,\hat{z} .$$

To take a derivative numerically, we take a central difference, so the electric field is:

$$\begin{pmatrix} E_s \\ E_z \end{pmatrix} \approx \begin{pmatrix} -\tfrac{1}{2a}\left(\mathbb{V}(s+a,z) - \mathbb{V}(s-a,z)\right) \\ -\tfrac{1}{2a}\left(\mathbb{V}(s,z+a) - \mathbb{V}(s,z-a)\right) \end{pmatrix} .$$

The figure shows the charge distribution, the potential, and the electric field, with a grid size of 250×1000 cells and parameters: $R = 1\,\text{cm}$, $h = 4\,\text{cm}$, and $\rho_0 = 1\,{}^{\text{C}}\!/\!_{\text{m}^3}$. The simulation required 542,000 iterations until the maximum voltage difference between iterations was only 1 nV.

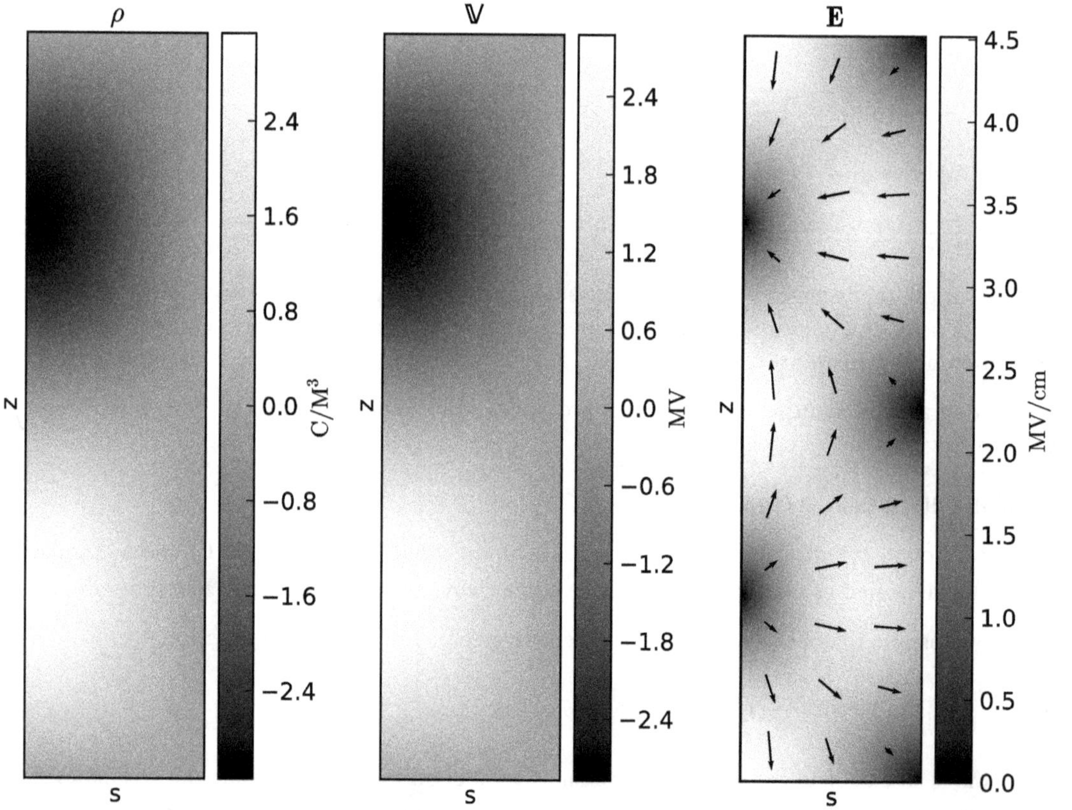

And finally we apply Gauss's law to the boundaries, so we can calculate the surface charge density on the top and bottom plate and the linear charge density along the central cable and outer sheath that is needed to keep each boundary at zero voltage. These are shown in the plots below.

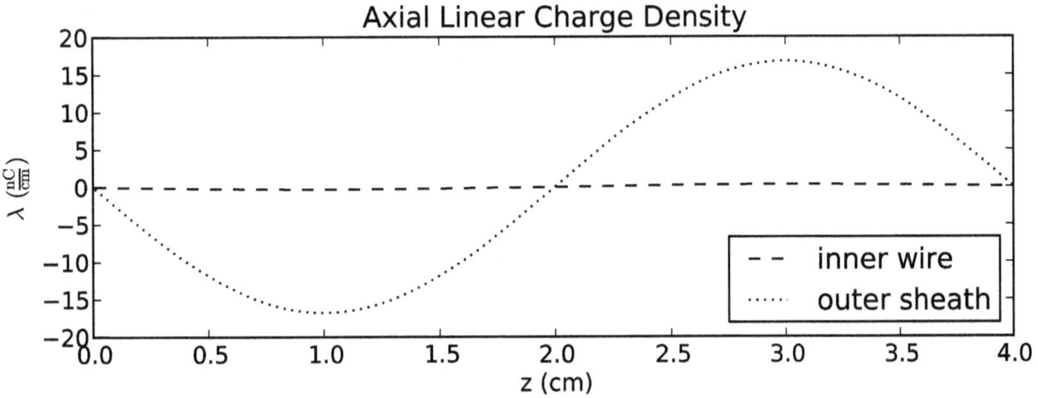

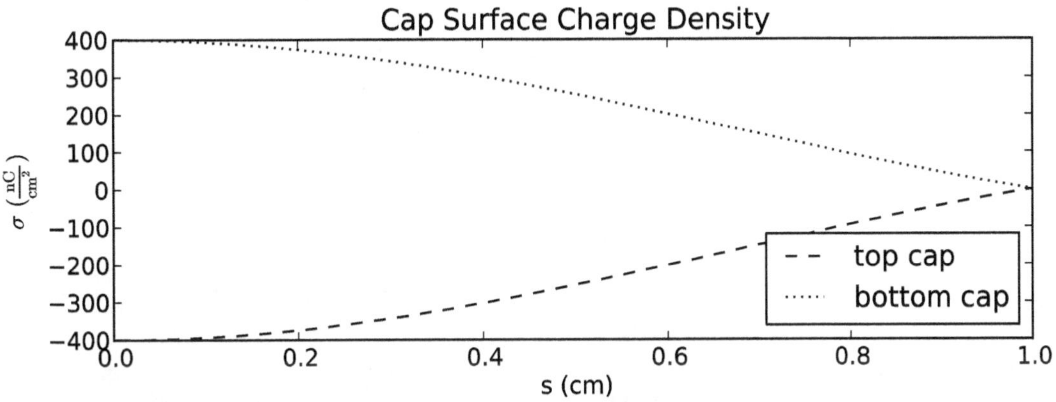

Plate 11: South Pointing Chinese Compass[77]

[77] This is a replica of a first century Chinese compass, with a magnetic spoon that balances on a nonmagnetic plate.

Part II: Magnetism

Growing fields of study specialize, ideally leading to ever deeper understanding. Since, over time, each new field develops its own culture, this fragmentation can lead to isolated groups with incommensurate views.

In principle, physics is immune from balkanization because we follow a strict scientific method of experimental hypothesis testing. Theorists compete to explain each new experimental result, and experimentalists design new experiments to distinguish among these competing explanations. Eventually one theory passes every experimental test and everyone moves on. What used to be cutting edge is now accepted knowledge.

The great philosopher of science Thomas Kuhn called the most entrenched theories *paradigms*.[78] Franklin's law of the conservation of charge, for example, became a paradigm during the eighteenth century. It has been continuously applied ever since, so there is no reason for a physicist to actually test it.[79]

Sometimes, however, scientists come to consensus too quickly, and the wrong theory becomes a paradigm. The more entrenched the paradigm, the more likely that a *physical misconception* will be propagated to the next generation of scientists. Over time, experimentalists ignore clearer and clearer evidence, and theorists come up with odder and odder theories, all to preserve the existing paradigm. Finally someone presents an alternative theory that elegantly explains the existing observations. At this point, the field is forced to go through a *paradigm shift*, after which a great burst of productivity usually follows.

The old paradigm can, however, linger—often for good reasons. For example, Galileo's observations of the phases of Venus debunked the geocentric model of the solar system,[80] throwing the field of astronomy into turmoil. By this time, however, systems of time keeping and celestial coordinates were already well established, so astronomers still use the geocentric representation for those applications—even though nobody today actually believes that the sun revolves around the earth. Like the geocentric model of the universe, the pole model of magnetism lasted for centuries before being debunked.

Coulomb not only developed his law of electrostatics, but also a parallel law for magnetism. Coulomb's pole model posited an inverse square law between magnetic poles, exactly analogous to electrostatics and gravity. Magnetization, therefore, was due to the separation of northness from southness, in exactly the same way that separation of positive from negative charge causes electrical phenomena.

Luckily, André-Marie Ampère developed a competing model of magnetization, involving microscopic current loops rather than the separation of northness from southness. Since current

---

[78] Kuhn, Thomas S., The Structure of Scientific Revolutions, (Chicago: The University of Chicago Press 1962).

[79] The astute reader will protest that charge conservation has been tested more accurately than any other law of nature (p. 49). Yes, but the purpose of that experiment was to observe neutrinos from space. Confirmation of charge conservation was simply a fun byproduct of an existing experiment. Why spend resources to confirm something we already know?

[80] Aristarchus of Samos proposed the heliocentric model of the solar system in the third century BC. However, critics pointed out that it predicted very fast prevailing winds and annual variations in the brightness and location of stars, none of which were observed.

loops are the classical analog to quantized angular momentum, we credit Ampère with finding the correct paradigm.

However, Ampère's classical picture completely failed to predict why magnets stick to steel. Without quantum mechanics, it actually predicted a repulsive force, just as we now observe in superconductors. Thus, for very good scientific reasons, Ampère failed to convince most scientists, especially those working with permanent magnets.

James Clerk Maxwell avoided the controversy by making his field equations work with both representations. He did this by introducing two electric fields, $\vec{D}$ and $\vec{E}$, and two magnetic fields, $\vec{H}$ and $\vec{B}$. Physicists pretty much agreed that $\vec{E}$ was the fundamental electric field, which could be measured *in situ*. Which magnetic field one considered more fundamental, however, depended on one's favored model of magnetic matter.

Under the pole model interpretation, $\vec{H}$ was called *the magnetic field* and considered more fundamental, while $\vec{B}$ was *the magnetic induction.* Under Ampère's current loop paradigm, however, $\vec{B}$ is considered the true magnetic field, while $\vec{H}$ has little intrinsic meaning. Since both interpretations predict the same macroscopic phenomena, it made little difference which model one chose to apply. Thus, scientists and engineers would use whichever paradigm simplified their particular analysis or worked better at explaining their own experiments.

The beauty of the pole model is that the same intricate mathematics of Newtonian gravity and electrostatics can be recycled into solving magnetostatics problems. The heroes of electrostatics, including Gauss and Poisson, defined a magnetic pole density ρ_{M}, analogous to the electric charge density. Next they applied a pseudo-Gauss's law, $\vec{\nabla}\cdot\vec{H}=\rho_{\mathrm{M}}$, and introduced a corresponding magnetostatic scalar potential ϕ_M that satisfies $\vec{\nabla}\phi_M=\vec{H}$. Using these laws of *magnetostatics*, they successfully characterized the measured magnetic field surrounding any permanent magnet.

The only catch is that magnetic monopoles do not exist. Moveable magnetic "poles" are simply convenient fictions, which have no real physical meaning. However, the definitive experiments debunking the pole model did not take place until 1915 (Discussion 11.4, p. 369).

Like timekeeping after Galileo, some fields of magnetics that matured in the nineteenth century continue their existing practices to this day. This is primarily true of geomagnetism, because the magnetic scalar potential does an excellent job of empirically modeling the magnetic field surrounding the earth. Just as no modern timekeeper believes that the sun orbits the earth, no modern geophysicist actually believes that poles move around inside of magnets. However, there is a significant cost to changing existing practice, with no additional benefit in geological predictive power.

There are additional costs, however, of continuing to use unphysical legacy models, which are borne by you, the student. You must not only learn how the natural world works to the best of our twenty-first century knowledge, but you must also learn the history of your own field of study. Only through historical context can you distinguish practices based on current physics from those based on past physics.

Plate 12: Measuring the Geomagnetic Field in 1919[81]

[81] This is a 1919 picture of Captain J.P. Ault and his assistant measuring the earth's magnetic field. The measurements were so successful that the same scientific department at the Carnegie Instutition of Washington is still called the Department of Terrestrial Magnetism, even though they no longer measure the Earth's magnetic field. Rather they concentrate on astrophysics, geophysics, and planetary science. Photo kindly provided by the Department of Terrestrial Magnetism of the Carnegie Instutition of Washington.

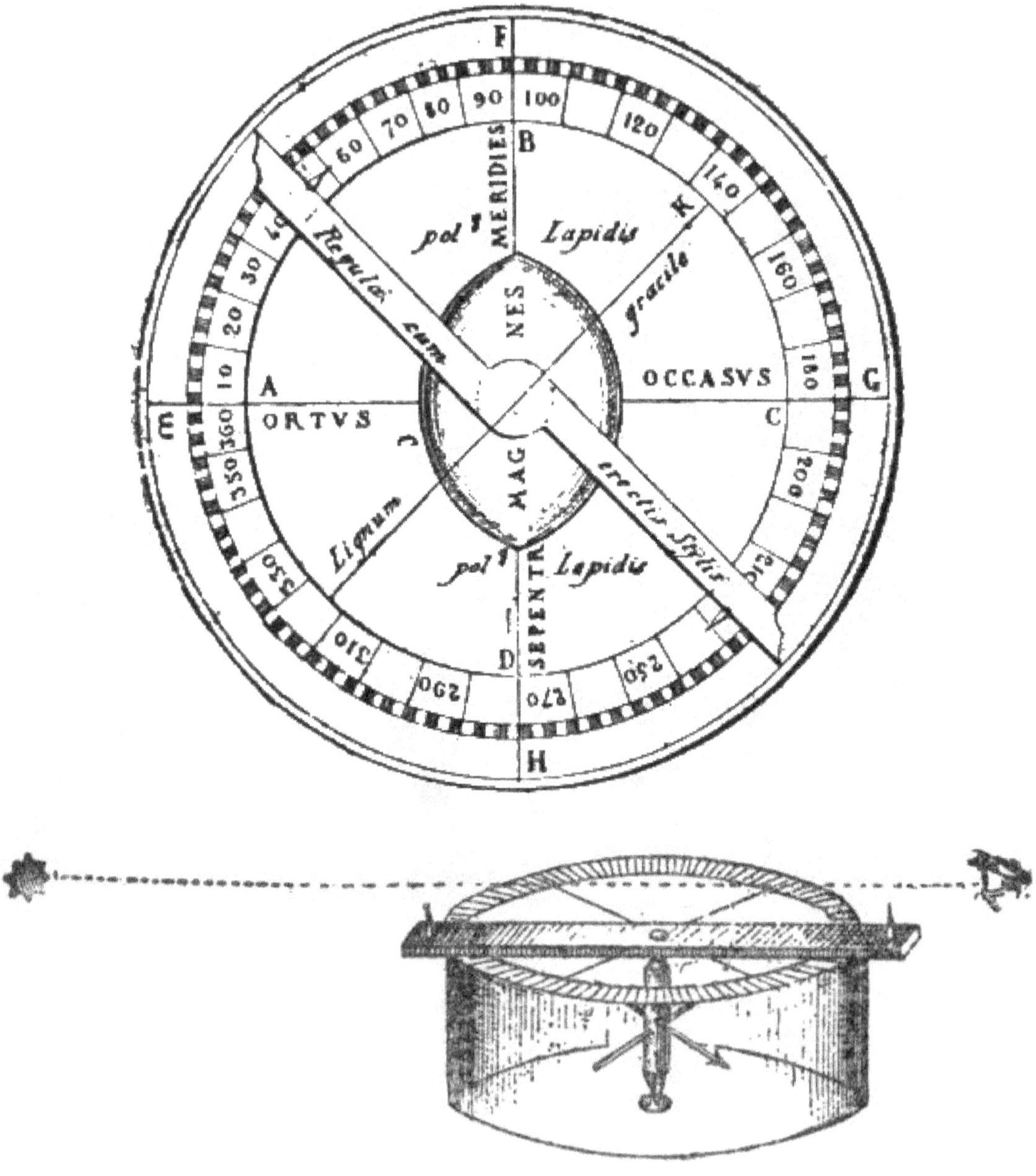

Plate 13: Petrus Peregrinus's Compass

Chapter 6 Permanent Magnets

In 1264, the French Pope Urban IV gave southwestern Italy to the French prince Charles of Anjou. There was a catch, however; Charles would have to take it by force. Despite the valiant efforts of a band of Muslim archers from the town of Lucera, he won the Battle of Benevento in 1266. Two years later, the German mayor of Lucera led a revolt against the rule of, now, King Charles I of Sicily, and a year-long siege began. Petrus Peregrinus (a.k.a. Peter of Mericourt and Pierre de Mericourt) was a knight serving under King Charles during the siege of Lucera.

A siege is really a waiting game, so Peregrinus had plenty of time on his hands for scientific experiment and theory, and authoring the earliest known extensive treatise on magnetism, Epistola de magnete, which he sent to his best friend via a letter in 1269.[82] In this letter, he described how to make better compasses (see facing page), discussed the attraction of iron by magnets, the magnetization of iron by permanent magnets, and the ability to reverse the polarity in such induced magnets.

Peregrinus also formed a spherical magnet out of the mineral magnetite (also called lodestone) and used a small compass needle to map out the region in the vicinity of the sphere. He discovered that the lines mapped along the surface of the magnetite ball resembled the meridional lines of a globe and converged at two opposite points of the sphere. He called these antipodal points *poles*, in analogy with the geographic poles of the Earth, implying that the Earth itself was a large magnet.

By convention, the north pole of a magnetic needle points northward. Thus, Peregrinus noticed that if one models the Earth as a large magnet it is actually the *south* magnetic pole of the spherical magnet that must correspond to the earth's *north* pole. The magnetic axis of the earth is more or less antiparallel to the geographic axis, and the poles are the points where the axis intersects with the surface of the earth.

Peregrinus also came up with a thought experiment that crystalized the principle behind one of the fundamental field equations of electrodynamics. Consider a bar magnet with a north and a south pole. If we break it into two pieces, we get two smaller bar magnets each of which possesses its own north and south magnetic pole. Therefore, Peregrinus reasoned, this could be done no matter how small the magnetic fragment.

Peregrinus's principle, that neither northness, nor southness, independently exist, still holds true today. Despite the most modern particle accelerators searching in vain for many decades, there have been no reproducible detections of any magnetic monopoles. For the next 500 years, no other original work on magnets matched the scientific insight of the brilliant knight from Mericourt.

We do not know what happened to Sir Peter of Mericourt after his letter of 1269. Presumably he would have completed more scientific works had he returned to Picardy, but one never knows. He may have fallen in battle, quietly joined a monastery, or settled down somewhere in Italy.

What we do know, however, is that in 1270 the king granted a number of the knights funds to build a church and hospital in Naples (St. Eligio) to tend to the wounded. We also know that the

---

[82] Pierre de Maricourt, a.k.a. Petrus Peregrinus, Epistola de magnete (1269) published as The Letter of Petrus Peregrinus On The Magnet, A.D. 1269, trans. Brother Arnold, M.Sc. Principal of La Salle Institute, Troy (New York: McGraw Publishing Company, 1904).

king brought as much French culture as possible to southwestern Italy, and personally oversaw the building of two Cistercian monasteries in the French Gothic style.[83]

After the siege of Lucera, King Charles taxed the Islamic city heavily as punishment for their revolt, but otherwise let them live in peace for the rest of his reign.[84] Alas, his son (King Charles II) sold the Muslims of Lucera into slavery in a deliberate attempt to ethnically cleanse his kingdom.

After Peregrinus's friend published Epistola de magnete, the work greatly influenced later scholars, both directly and indirectly—especially William Gilbert. Gilbert reproduced (and expanded upon) the experiments, and in turn published his own treatise.[85] Gilbert's work influenced other scientists, whose work influenced others, and so on, until some of Peregrinus's ideas eventually influenced Michael Faraday .

Among his many accomplishments, Faraday reformulated Peregrinus's thought experiment in terms of his concepts of magnetic field, magnetic flux, and lines of magnetic force—showing "that the lines of force are continuous through the body of the magnet, and with that continuity gives the necessary reason why no absolute charge of northness and southness is found." (see p. 197)

6.1 *The Magnetic Field*

Ever since the invention of the compass, the horizontal direction of the terrestrial magnetic field has been measured and mapped. With Robert Norman's invention of the dip needle, by around 1600 the three dimensional magnetic field direction could be mapped out over the known earth (at least in principle). This meant that the fields surrounding permanent magnets could also be measured once a standard (such as the Earth's magnetic field at a particular observatory) was defined. Since iron filings act like little compasses, the suggestive patterns they make near a permanent magnet display the contours of its field.

Faraday did not actually use the term "magnetic field" until 1845, but he defines the concept in a footnote to a paper published in 1831:[86]

> By magnetic curves, I mean the lines of magnetic forces, however modified by the juxtaposition of the poles, which could be depicted by iron filings; or those to which a very small magnetic needle would form a tangent.[87]

In the following thought experiments, we introduce the magnetic field through just this type of *operational definition*, and demonstrate how to model the magnetic field of permanent magnets in much the same way as we modeled the current density in Chapter 1.

---

[83] Caroline A. Bruzelius, "ad modum francia': Charles of Anjou and Gothic Architecture in the Kingdom of Sicily," Journal of the Society of Architectural Historians Vol. 50, No. 4 (1991), 402-420.

[84] Taylor, Julie, Muslims in Medieval Italy: The Colony at Lucera, (Lexington Books, 2005)

[85] William Gilbert of Colchester, On the Magnet, Magnetick Bodies also, and on the great magnet the earth; a new Physiology, demonstrated by many arguments & experiments. (Latin version: London: 1600, 1628, 1633; English translation: London: Chadwick, 1900)

[86] "If a man could be in the Magnetic field, like Mahomet's coffin, he would turn until across the magnetic line" quoted from Faraday's diary on page 98 of Olivier Darrigol, Electrodynamics from Ampère to Einstein, (Oxford: Oxford University Press, 2000).

[87] Faraday, Experimental Researches in Electricity, 3 vols. (London: Taylor, 1839, 1844; Taylor and Francis, 1855) **I**: First Series, para.114, note [A], 32.

Thought Experiment 6.1: Magnetic Field Lines

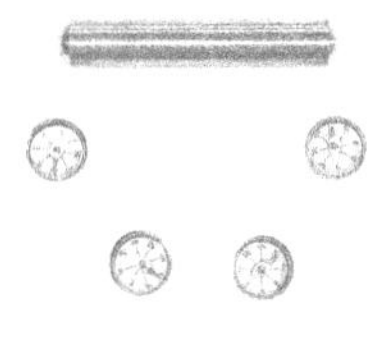

Imagine that you are out at sea and sailing toward magnetic north by continually taking a bearing with a magnetic compass. If you map your trip based on the readings of the compass, you will be following the horizontal component of the Earth's magnetic field. Faraday defined these curves of magnetic force, now called magnetic field lines, as follows:

> The term *line of magnetic force* is intended to express simply the direction of the force in any given place, and not any physical direction or notion of the manner in which the force may be exerted; as by actions at a distance, or pulsations, or waves, or a current, or what not. A line of magnetic force may be defined to be that line which is described by a very small magnetic needle, when it is so moved in either direction correspondent to its length, that the needle is constantly a tangent to the line of motion; or, it is that line along which, if a transverse wire be moved in either direction, there is no tendency to the formation of an electric current in the wire, whilst if moved in any other direction there is such a tendency. The direction of these lines is easily represented in a general manner by the well-known use of iron filings.[88]

It also seems evident that: (1) magnetic field lines never cross each other, and, (2) for every magnetic field line entering a magnet there must be a corresponding one exiting the same magnet, as described by Faraday in the same paper:

> The lines of force already described will, if observed by iron filings or a magnetic needle or otherwise, be found to start off from one end of a bar-magnet, and after describing curves of different magnitudes through the surrounding space, to return to and set on at the other end of the magnet; and these forces being regular, it is evident that if a ring, a little larger than the magnet, be carried from a distance toward the magnet and over one end until it has arrived at the equatorial part, it will have intersected *once* all the external lines of force of that magnet.

Naturally, it was impossible to map out the field lines inside the magnet this way, but outside the magnet it became clear to Faraday that the best way to describe magnetism was by these lines of magnetic force.

Thought Experiment 6.2: The Torque on a Permanent Magnet

In Thought Experiment 6.1 we discussed how Faraday defined his *lines of magnetic force*. These magnetic field lines clearly indicate the direction of the magnetic field, although the question remains of how to measure the magnetic field strength. The goal of this thought experiment is to provide an operational definition of the magnetic field.

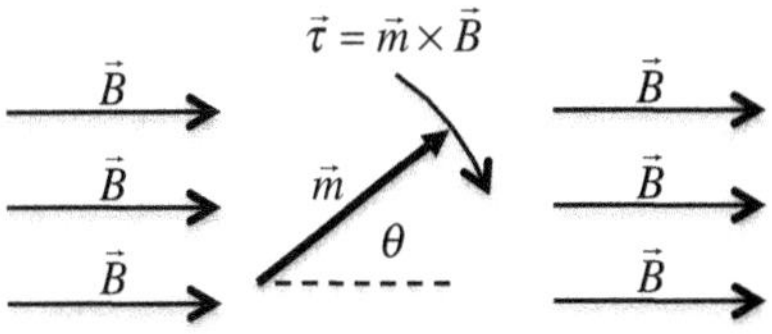

A simple magnet is characterized by its *magnetic moment* $(\vec{m})$, which is a vector pointing from the south end of a magnet to the north and has a magnitude proportional to the overall strength of the magnet. In the presence of an externally applied magnetic field, the small magnet experiences a torque that aligns it along the field. Through simple experiment, the torque on the magnet can be shown to be proportional to $\sin\theta$, where θ is the angle between the magnetic

---

[88] M. Faraday, "On the Lines of Magnetic Force," Royal Institution Proceedings, 23rd January, 1852, reprinted in M. Faraday, Experimental Researches in Electricity: Series 19-29, 3 vols. (London: Taylor and Francis, 1839, 1844, 1855) **III**, 403-404. Faraday's remark about the formation of a current in a wire that moves through a magnetic field refers to the phenomenon of electromagnetic induction, which is the topic of Chapter 11. A current is induced in a wire moving through a magnetic field, which is evidence that a potential difference has been created in the wire.

moment and the direction of the magnetic field. The facts are economically expressed by the following vector cross product relationship:

$$\vec{\tau} = \vec{m} \times \vec{B},$$

where $\vec{\tau}$ represents the torque on the needle, $\vec{m}$ the magnetic moment of the magnet, and $\vec{B}$ the external magnetic field. We now have a definition of the magnetic field vector: the orientation of the compass needle defines its direction, and its strength is proportional to the torque it exerts on a magnet of known magnetic moment.

Despite knowing the units of torque $(\mathrm{N \cdot m})$, we cannot derive both units of magnetic moment and magnetic field from this thought experiment alone. We can, at least, name our unit of magnetic field, even if we have yet to completely specify it. The SI unit of magnetic field is called the tesla (T), after the Serbian-American electrical engineer Nikola Tesla (, and the unit of magnetic moment is therefore a newton-meter per tesla. The magnetic field at the Earth's surface ranges from about 30 to 60 microtesla, while a typical compass needle has a magnetic moment on the order of one newton-meter per tesla.

Definition 6.1: The Magnetic Field

The local magnetic field, $\vec{B}$, is defined such that a small permanent magnet, with a magnetic moment, $\vec{m}$, will experience a torque given by: $\vec{\tau} = \vec{m} \times \vec{B}$. Thus, a permanent magnet will align itself with the direction of the magnetic field, and the magnetic field strength is the torque per unit magnetic moment when the field and moment are perpendicular.

Notice that this definition requires a standard unit of magnetic moment, or a standard unit of magnetic field, depending on which quantity we wish to treat as fundamental. This is a common difficulty that arises in working definitions of quantities. For example, consider the challenge in defining units of force through Newton's second law, which allows us to determine the unit of force only after the unit of mass has been specified.[89] Later, when we discuss the connection between electricity and magnetism, we will define the tesla in terms of SI base units.

Thought Experiment 6.3: The Potential Energy of a Magnet

Consider a compass needle, with a magnetic moment $\vec{m}$, which experiences a torque in an external magnetic field, so that $\vec{\tau} = \vec{m} \times \vec{B}$. It must take some work $\mathbb{W}$ to twist it against this torque. Since the applied force is not dissipative, this work should correspond to a potential energy difference $\Delta\mathbb{U} = \mathbb{W}$. Thus, by calculating the work done against the field, we should be able to find the potential energy of a magnetic dipole in a magnetic field.

Consider a dipole moment that is aligned with the field $(\theta = 0)$ so that there is no torque on the magnetic dipole and it is in stable equilibrium. Now apply the minimum external torque $\vec{\tau}_{ext}$

---

[89] Mass is a fundamental quantity and we therefore cannot describe it in terms of simpler physical quantities. Instead, we must define mass by a set of operations that can be used to measure masses in terms of an arbitrary standard mass. This is what we mean by providing an *operational definition* of a physical quantity such as mass. Ernst Mach proposed such a definition of mass in terms of the mutual accelerations of two interacting bodies, and this is often used in modern physics textbooks. For a discussion of some of the difficulties with Mach's approach, see Eugene Hecht, "There Is No Really Good Definition of Mass," The Physics Teacher, **44** (2006), 40-45.

needed to twist the dipole until it is at rest at an angle θ with respect to the magnetic field. The work done by this external torque is given by:

$$\mathbb{W} = \int_0^\theta \tau_{ext}\, d\theta .$$

Now we propose a function $\mathbb{U}(\theta)$ to represent the potential energy we stored (presumably in the system defined by the combination of the given magnetic field and the field due to the dipole magnet) by applying an external torque:

$$\mathbb{U}(\theta) - \mathbb{U}(0) = \mathbb{W} = \int_0^\theta \tau_{ext}\, d\theta .$$

Recall that the applied external torque was the minimum necessary to twist the magnetic needle from its alignment with the field. The applied torque is thus equal and opposite to the magnetic torque exerted on the needle, because we are supposing no acceleration of the needle. The difference in potential energy is therefore given by:

$$\mathbb{U}(\theta) - \mathbb{U}(0) = \int_0^\theta \tau_{ext}\, d\theta = \int_0^\theta |\vec{m} \times \vec{B}|\, d\theta = \int_0^\theta mB\sin\theta\, d\theta = -mB(\cos\theta - 1).$$

Now we must choose the zero point of potential energy. When the magnetic dipole is oriented perpendicular to the prescribed magnetic field, we have that $(\theta = \pi/2)$ and $\cos\frac{\pi}{2} = 0$, so we define $\mathbb{U}(\frac{\pi}{2}) = 0$. We can therefore write the potential energy as follows:

$$\begin{aligned}\mathbb{U}(\theta) &= \mathbb{U}(\theta) - \mathbb{U}(\tfrac{\pi}{2}) = (\mathbb{U}(\theta) - \mathbb{U}(0)) - (\mathbb{U}(\tfrac{\pi}{2}) - \mathbb{U}(0)) \\ &= (-mB(\cos\theta - 1)) - (-mB(\cos\tfrac{\pi}{2} - 1)) = (-mB\cos\theta + mB) - (mB) \\ &= -mB\cos\theta = -\vec{m}\cdot\vec{B}.\end{aligned}$$

Thus, the potential energy of a magnetic dipole in the presence of a magnetic field is given by:

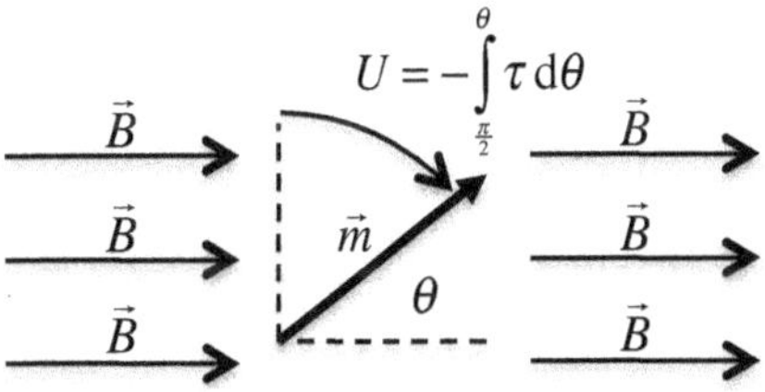

$$\mathbb{U}(\theta) = -\vec{m}\cdot\vec{B}.$$

Had we known from the beginning that $\theta = \frac{\pi}{2}$ would be the most convenient zero point, we could have found this directly from the definition of potential energy:

$$\mathbb{U}(\theta) = \int_{\pi/2}^\theta \tau_{ext}\, d\theta = -\int_{\pi/2}^\theta \tau\, d\theta = \int_{\pi/2}^\theta mB\sin\theta\, d\theta = -mB\cos\theta = -\vec{m}\cdot\vec{B}.$$

Problem 6.1: Zeeman Splitting

A gas discharge tube, or a neon sign, emits (nearly) monochromatic light at specific wavelengths that appears as a series of unique and discrete colored lines when viewed through a prism or a diffraction grating. This light is produced by the internal transitions electrons make between the quantized energy levels of the gas atom to which they belong. If the discharge tube is placed in a strong magnetic field, some of these energy levels split into two closely spaced energy levels.

Let's apply a *semi-classical model* to explain this apparent orbital splitting. In this context that means we will accept the experimentally demonstrated fact that the electrons of an atom behave like tiny magnets and possess a magnetic moment, and that according to experiment (and the modern quantum mechanical theory of matter) an electron in a magnetic field can only have its magnetic moment aligned, or anti-aligned, with the magnetic field. Denote the magnetic field as $\vec{B} = B\hat{z}$, and let $\vec{m} = \pm\mu_e\hat{z}$, where $\mu_e = 9.285\times10^{-24}\ \mathrm{J}/\mathrm{T}$ is the magnitude of the electron magnetic moment.

a) Pieter Zeeman discovered this effect by observing splitting of the blue Cadmium spectral line when he applied a very large magnetic field of 3.2 tesla.[90] What is the energy difference you would expect classically between the two energy states with this magnetic field?

b) The Zeeman effect is of particular importance in solar physics, because it is used to measure magnetic fields. Report on some recent solar physics observations that utilize this effect. To what resolution can they measure the magnetic fields? Can they measure them in all three dimensions, or do they only make two dimensional maps?

Thought Experiment 6.4: The Force on a Magnet in a Non-Uniform Field

A magnet experiences no net force in a uniform magnetic field, although there is a torque that tends to align it with the uniform field. Now, we will consider a magnet in a non-uniform magnetic field where there is indeed a net force.

First, write down the potential energy of the magnet in an external magnetic field as:

$$\mathbb{U}(\vec{r}) = -\vec{m}\cdot\vec{B}(\vec{r}).$$

Note that the potential energy of this system depends on position through the spatial dependence of the magnetic field. Since the potential energy depends on the position $\vec{r}$, and since there is no energy dissipation term here, the force is conservative and given by the negative gradient of this potential energy:

$$\vec{F} = -\vec{\nabla}U(\vec{r}).$$

We can therefore write down the force on the magnet as:

$$\vec{F} = -\vec{\nabla}U(\vec{r}) = -\vec{\nabla}\left(-\vec{m}\cdot\vec{B}(\vec{r})\right) = \vec{\nabla}\left(\vec{m}\cdot\vec{B}(\vec{r})\right).$$

Notice that it is not the magnetic field itself that applies a force on the permanent magnet, but rather the *gradient of the magnetic field strength*. This can be seen after the magnet has aligned itself with the magnetic field:

$$\vec{F} = \vec{\nabla}\left(\vec{m}\cdot\vec{B}(\vec{r})\right) = \vec{\nabla}\left(mB(\vec{r})\right) = m\vec{\nabla}B(\vec{r}).$$

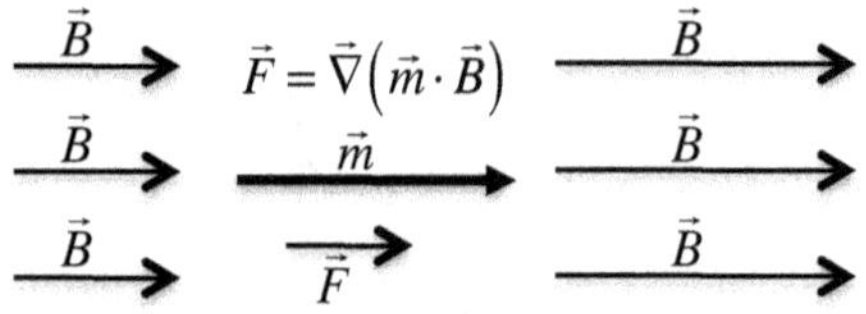

If the magnetic moment is held in place, so it does not align itself with the magnetic field, then the force need not be in the direction of increasing field strength. For example, if the dipole moment is aligned opposite that of the external field, then the force would act in the direction of the decreasing magnetic field strength.

---

[90] P. Zeeman, "Doubles and triplets in the spectrum produced by external magnetic forces," Phil. Mag. **44** (1897), 55.

Problem 6.2: A Magnet on a Torsion Spring

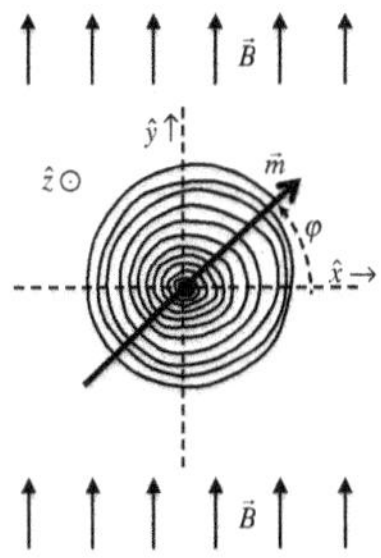

Consider a permanent magnet with magnetic moment $\vec{m}$, attached to a torsion spring with a constant k, so that $\tau = -k\varphi$ if φ is the angle from the spring's rest position. The whole thing is placed in a constant and uniform magnetic field $\vec{B} = B\hat{y}$.

a) This device can be used to measure the magnetic field $\vec{B} = B\hat{y}$. Find the magnetic field as a function of the observed equilibrium angle φ_0?

b) Now the magnet will have a mass and length, so it must have a moment of inertia $\mathbf{I}$. Write down the equations of motion (i.e. find the equation relating φ and its time derivatives) in terms of the given quantities $(k, B, \mathbf{I})$.

c) Find the frequency of oscillation about the equilibrium angle. You can assume small angles ($\delta\varphi = \varphi - \varphi_0$, $|\delta\varphi| \ll 1$).

Discussion 6.1: The Stern-Gerlach Experiment

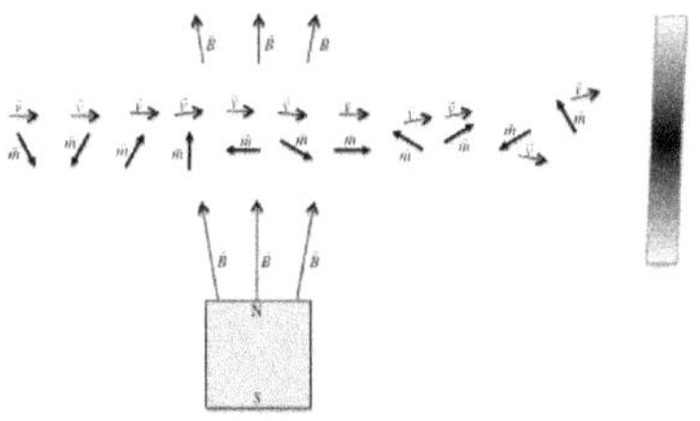

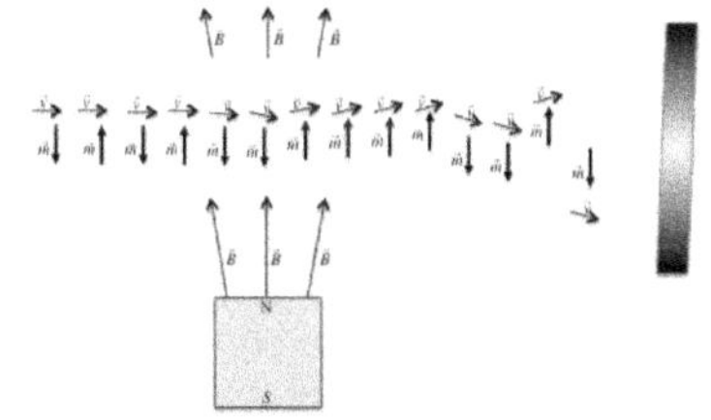

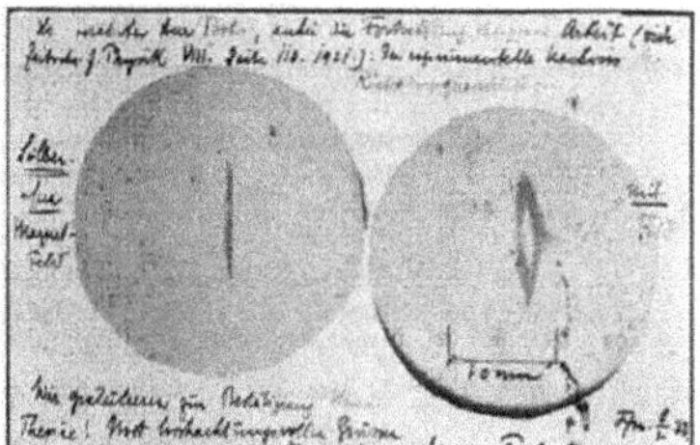

In a famous 1922 experiment conducted by Otto Stern and Walter Gerlach, vaporized silver was injected into a non-uniform magnetic field so as to measure the distribution of the atomic magnetic moments. Classically, since one would expect that atoms would have randomly oriented magnetic moment vectors, a deflection of neutral atoms by a non-uniform magnetic field should be continuously distributed (top figure). However, this was not observed. Instead, the magnetic moments appeared to be always aligned with the detector, regardless of the detector direction, with 50% pointing along in one direction called "up" and 50% of the magnetic moments in the "down" direction, but none in between (bottom figure). The postcard was sent by Gerlach to Niels Bohr with the message: "Attached the experimental proof of directional quantization."

When characterizing magnetic moments on atomic scales, therefore, it is important to note that the effects are inherently governed by quantum mechanics. Classical physics was developed primarily during the eighteenth and nineteenth centuries, and it should not come as a surprise that many phenomena that were once explained with classical models ultimately require quantum mechanical ideas for their correct description.

Problem 6.3: A Spherical Magnet in an External Field

Consider a small spherical magnet with a mass M and a magnetic moment $\vec{m}$, inside a magnetic field given in cylindrical coordinates as:

$\vec{B} = K(s\hat{s} - 2z\hat{z})$,

where K is a constant.

a) For this problem, you can assume that the magnetic moment quickly aligns itself parallel to the field. Discuss in what conditions this would be a reasonable assumption, and when it would not.

b) What is the force on the magnet as a function of s and z?

c) If the sphere is placed at a location $\vec{r}_0 = z_0\,\hat{z}$, what is the position of the magnetic ball as a function of time?

d) If the sphere is placed at $\vec{r}_0 = s_0\,\hat{s}$, what would its position be?

e) In the two special cases you worked out, the motion of the magnet was in the $\hat{r}$ direction. Is this always the case, regardless of the initial position? If not, how will it deviate from the radial direction? Will it deviate in the polar or equatorial direction? Find a relationship between the polar position angle, θ, and the direction of acceleration $\theta_{\vec{F}}$. Draw qualitative pictures of trajectories from various initial positions.

f) Are your answers in parts (d) and (e) stable to perturbations in the initial position or angle?

g) Write a computer simulation program that will plot the trajectory, as a function of initial position. Does it agree with your qualitative assessment.

Problem 6.4: Random Spherical Magnets

Imagine that you load a BB gun with spherical magnets of mass M and dipole moment $\vec{m} = m\cos\varphi\,\hat{x} + m\sin\varphi\,\hat{y}$, where φ is the angle $\vec{m}$ makes with the x-axis. You shoot these with an initial velocity $\vec{v}_0 = v_0\,\hat{x}$ into a magnetic field $\vec{B} = ky\hat{y}$, where k is a constant, and put the target a distance x_0 away from the gun. For simplicity ignore gravity an assume that the magnetic moment is confined to the $\hat{x},\hat{y}$ plane (i.e. $\vec{m}\cdot\hat{z} = 0$). Also assume that the direction φ of each magnetic moment does not change while in transit.

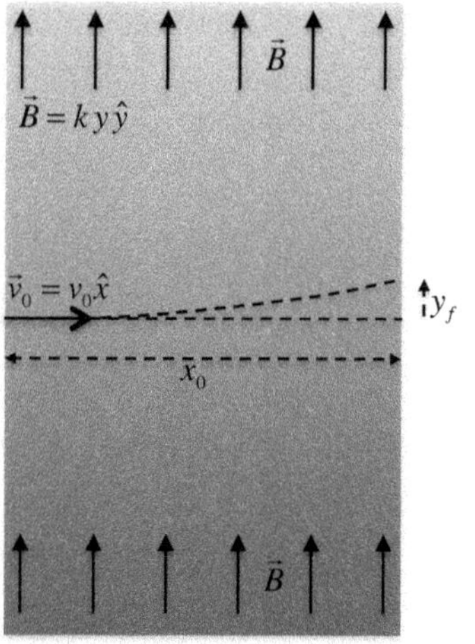

a) Find an expression for the force on each BB while it is interacting with the magnetic field.

b) Find the deflection y_f.

c) We will now assume that when a bunch of BBs are shot, each BB has the same velocity and magnetic moment strength. However, the direction of the magnetic moment will now be randomly orientated. Plot the hole pattern distribution (i.e. y_f) for 100 BBs with randomly oriented magnetic moments from $\varphi = 0$ to $\varphi = 2\pi$.

d) Now imagine that instead of being randomly oriented, 50 BB's have a magnetic moment $\vec{m} = m\hat{y}$ and 50 have $\vec{m} = -m\hat{y}$, but no dipoles are pointed at other angles. In other words, they are either pointed up or down. Plot the hole pattern you would observe in this case, and compare it to part c.

e) Reread Discussion 6.1 (p. 193) and discuss the image on the postcard sent to Niels Bohr.

Discussion 6.2: Operational Definitions and Fields

We have just defined the magnetic field by how it affects a small permanent magnet such as an iron filing, compass, or dip needle. This definition allows us to measure the magnetic field using permanent magnets if we know the value of the magnetic moment. First, we place a compass and dip needle in a magnetic field and with these find the direction of the magnetic field. Next, we place a permanent magnet perpendicular to the magnetic field and measure the torque on it. The definition of the magnetic field is a direct result of how it affects observables.

In general, we seek to define physical quantities based on how we measure them. If something is not measureable, at least in principle, then we do not know if it is actually "physical" or "real." In the case of the magnetic field, we presume there is some cause for the field's existence. We say that this cause creates a magnetic field, which in turn applies a torque on our compass. So the magnetic field is measurable, by definition, since we can observe its effect on the compass.

But why do we need a magnetic field at all? If the only way a magnetic field can be measured is by how it affects objects, and something caused the field (say another magnet), why not just write a relationship that describes the forces and torques between the magnets? For example, the Earth is a magnet and therefore causes the torque on a compass needle. Why do we need some seemingly imaginary intermediary, such as a field, to describe this phenomenon?

The reason is that, without such an intermediary, the description of forces between objects violates our ideas of cause and effect. We understand forces between magnets (or charges, or masses) as acting or propagating in some way through the space that separates these bodies. How can something from far away directly affect something here? Theories that do not involve *local* interactions are referred to as *action at a distance*. According to an action at a distance interpretation, for example, two bodies can act on one another instantaneously no matter how far apart they are. For this reason, among others, such theories are often derided as *spooky action at a distance*, as they appear to invoke ideas similar to those of supernatural occultists (see overview on p. 0 and Discussion 2.2 on page 58).

6.2 *Peregrinus's Principle*

In the historical introduction, we told the story of Petrus Peregrinus of Maricourt, the thirteenth century French knight who wrote a brilliant treatise while laying siege to the Italian town of Lucera.

When a small magnet is balanced, one end, or pole, will point north, and the other south, so Peregrinus named them accordingly. He also observed that if the Earth were modeled as a large magnet, its north magnetic pole would be located in the geographic south, and south magnetic pole located in the geographic north. Consider a magnetic compass needle that is allowed to pivot freely on its axis. One end will swing and point geographically northward, and the other end will point toward the geographical south. We call the north seeking end of the magnet the *north pole*, and the southward pointing end the *south pole*. So it is that a compass needle's north magnetic pole is attracted to the south magnetic pole of the Earth, which ironically lies geographically northward.

The Earth's geographic north and south poles are located where its rotation axis crosses the surface, like the holes for the pole in the middle of a globe. Similarly, you should imagine the poles of a magnet as the two points where the magnetic axis of symmetry emerges from the magnet's surface.

Peregrinus's brilliant theoretical insight was that, no matter how you slice it, for every north pole there must also be a south pole. In other words, magnets do not act like gravitating masses or electrical charges. Thus, there is no intrinsic property of northness nor southness, they are just the names for each end of a magnet's axis of symmetry.

Maxwell incorporated this principle into the four canonical field equations that collectively bear his name: Maxwell's equations. Each equation has a separate name, such as Gauss's law, but the statement that there are no magnetic monopoles has traditionally had no name. We shall call it *Peregrinus's principle* since he was the originator of the idea. And, of course, "Peregrinus's principle" is alliterative, so it sounds catchy in English. We hope the name sticks.

Thought Experiment 6.5: There Are No Magnetic Monopoles

Peregrinus wrote the following text to discuss the nature of magnets.[91] We have added the illustrations to help visualize Peregrinus's discussion.

> Take a lodestone which you may call AD, in which A is the north pole and D the south; cut this stone into two parts, so that you may have two distinct stones; place the stone having the pole A so that it may float on water and you will observe that A turns towards the north as before; the breaking did not destroy the properties of the parts of the stone, since it is homogeneous; hence it follows that the part of the stone at the point of fracture, which may be marked B, must be a south pole; this broken part of which we are now speaking may be called AB. The other, which contains D, should then be placed so as to float on water, when you will see D point towards the south because it is a south pole; but the other end at the point of fracture, lettered C, will be a north pole; this stone may now be named CD.

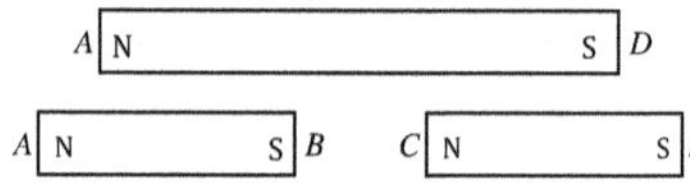

> If we consider the first stone as the active agent, then the second, or CD, will be the passive subject. You will also notice that the ends of the two stones, which before their separation were together, after breaking will become one a north pole and the other a south pole. If now these same broken portions are brought near each other, one will attract the other, so that they will again be joined at the points B and C, where the fracture occurred. Thus, by natural instinct, one single stone will be formed as before. This may be demonstrated fully by cementing the parts together, when the same effects will be produced as before the stone was broken. As you will perceive from this experiment, the active agent desires to become one with the passive subject because of the similarity that exists between them. Hence C, being a north pole, must be brought close to B, so that the agent and its subject may form one and the same straight line in the order AB, CD and B and C being at the same point. In this union the identity of the extreme parts is retained and preserved just as they were at first; for A is the north pole in the entire line as it was in the divided one; so also D is the south pole as it was in the divided passive subject, but B and C have been made effectually into one.

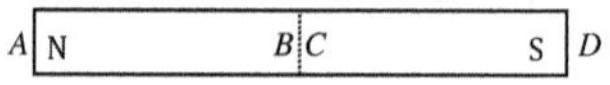

> In the same way it happens that if A be joined to D so as to make the two lines one, in virtue of this union due to attraction in the order CD AB, then A and D will constitute but one point, the identity of the extreme parts will remain unchanged just as they were before being brought together, for C is a north pole and B a south, as during their separation. If you proceed in a different fashion, this identity or similarity of parts will not be preserved; for you will perceive that if C, a north pole, be joined to A, a north pole,

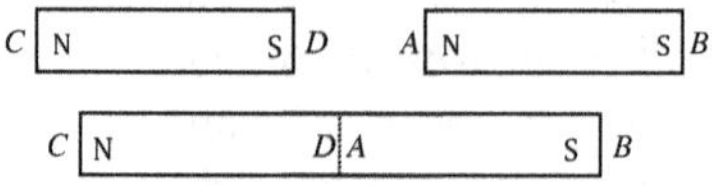

[91] Petrus Peregrinus, The Letter of Petrus Peregrinus On The Magnet, A.D. 1269, trans. Brother Arnold, M.Sc. Principal of La Salle Institute (New York: McGraw Publishing Company, 1904), Chapter IX, 14-18.

> contrary to the demonstrated truth, and from these two lines a single one, BACD, is formed, as D was a south pole before the parts were united, it is then necessary that the other extremity should be a north pole, and as B is a south pole, the identity of the parts of the former similarity is destroyed.

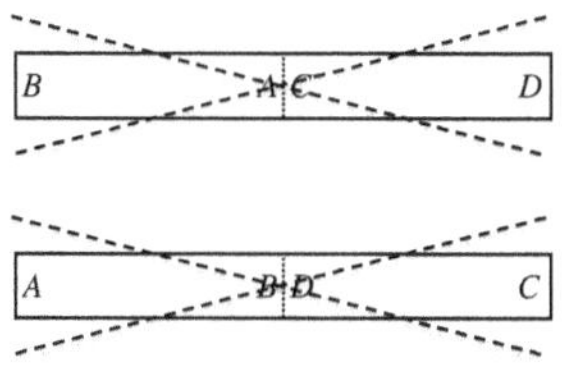

> If you make B the south pole as it was before they united, then D must become north, though it was south in the original stone; in this way neither the identity nor similarity of parts is preserved. It is becoming that when the two are united into one, they should bear the same likeness as the agent, otherwise nature would be called upon to do what is impossible. The same incongruity would occur if you were to join B with D so as to make the line ABDC, as is plain to any person who reflects a moment. Nature, therefore, aims at being and also at acting in the best manner possible; it selects the former motion and order rather than the second because the identity is better preserved. From all this it is evident why the north pole attracts the south and conversely, and also why the south pole does not attract the south pole and the north pole does not attract the north.

Peregrinus understands that, after their merger, points B and C (or D and A) are one and the same. No matter how you slice the magnet, there can never be only a north pole or only a south pole. There is always one of each "type" of pole. As of this writing, no magnetic monopole has ever been reproducibly observed in nature. Therefore, *Peregrinus's principle,* that there are no magnetic monopoles, still holds universally, everywhere, and in all contexts, at least to the best of our knowledge, 750 years after his work.

Thought Experiment 6.6: The Continuity of Magnetic Flux

In 1852, Michael Faraday expressed Peregrinus's principle in the following way:

> In the magnet such a division does develop new external lines of force; which being equal in amount to those dependent on the original poles, shows that the lines of force are continuous through the body of the magnet, and with that continuity gives the necessary reason why no absolute charge of northness and southness is found in the two halves.[92]

As Faraday articulates so well, there is no experimental evidence for the existence of magnetic "charge." The lines of magnetic field are thus mathematically similar to streamlines of current, in that the field lines are continuous and circulatory. Therefore, it makes sense to define something, called the *magnetic flux*, which is denoted by Φ_B, as the surface integral of the magnetic field in much the same way as the current is the surface integral of the current density:

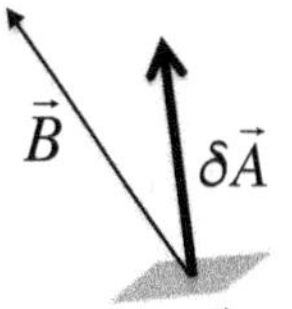

$\delta\Phi_B = \vec{B}\cdot\delta\vec{A}$

$$\Phi_B \equiv \int \vec{B}\cdot \mathrm{d}\vec{A}\,.$$

The SI unit of magnetic flux is the weber, after the German physicist Wilhelm Weber, and is defined as a tesla times a square meter.

Faraday considered the question of the *sources* or *sinks* of magnetic flux, in much the same way as there are sources and sinks of current. Just as the *anode* is defined as a source of current, the *north*

[92] Faraday, M. (1852), Experimental Researches in Electricity 3 vols. (London: Taylor, 1839, 1844, Taylor and Francis, 1855) **III**, Series 19-29: , para. 3264, 418.

pole of a magnet is defined as a source of magnetic flux. The *cathode* and *south pole* are sinks of current and magnetic flux respectively.[93]

Combining the definition of magnetic flux with Peregrinus's principle, we can consider *any* closed surface. Since neither northness nor southness exist, all entering magnetic flux must be balanced by the exiting magnetic flux.

Like Peregrinus did in his slicing thought experiment of 750 years ago, we can take the limit as our Gaussian surface becomes arbitrarily small. Thus:

$$\lim_{V\to 0}\left(\frac{\Phi_B}{V}\right)=\lim_{V\to 0}\frac{\oint_{\text{surface}}\vec{B}\cdot d\vec{A}}{V}=0.$$

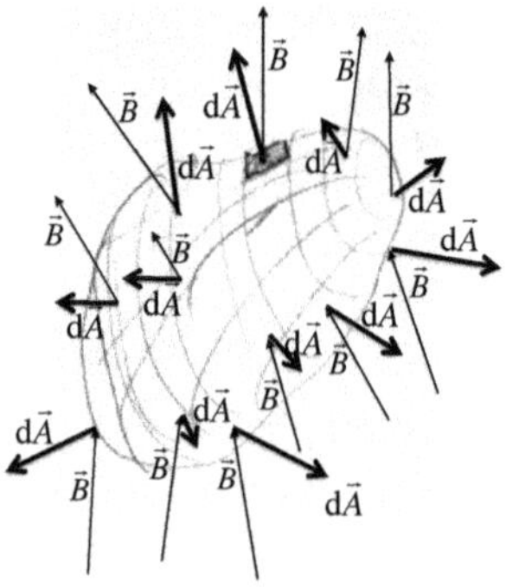

And now, we obtain the following by applying the definition of the divergence to the magnetic field:

$$\vec{\nabla}\cdot\vec{B}\equiv\lim_{V\to 0}\frac{\oint_{\text{surface}}\vec{B}\cdot d\vec{A}}{V}=0.$$

This result expresses the continuity of magnetic flux in its differential form.

Fundamental Law 5: Peregrinus's Principle

$$\vec{\nabla}\cdot\vec{B}=0.$$

This is the mathematical statement of Petrus Peregrinus of Maricourt's thought experiment (p. 196). This law has traditionally had no name, and so various authors have called it different things, including: *Gauss' law of magnetism,*[94] *the continuity of magnetic flux*, *Faraday's other law*, or simply *the "there are no magnetic monopoles" law*. In practice, people sometimes just say, "del dot B equals zero," and leave it at that. We will call it *Peregrinus's principle* since he so clearly articulated the concept half a millennium prior to Faraday's remarks.

Thought Experiment 6.7: Faraday's Magnetic Field Lines

Michael Faraday conducted a number of experiments using iron filings and weak permanent magnets. In the presence of a magnet, the iron filings become magnetized and act as tiny compasses that indicate the direction $(\hat{B})$, though not the strength (B), of the magnet's field at each filing's location. Faraday used this fact to map out the fields of various magnets. The top figure is a reproduction of one of Faraday's figures[95], with the magnetic field emanating from the left side of the magnet (north) and returning into the right side (south). Faraday had no direct measurement of the magnetic field inside this magnet.

---

93 Chemists define the anode and cathode in the opposite manner, as they are concerned with the internal workings of a battery rather than what the battery does in a circuit.

94 Gauss used $\vec{\nabla}\cdot\vec{H}=4\pi\rho_M$ to model the earth's magnetic field, not $\vec{\nabla}\cdot\vec{B}=0$.

95 M. Faraday, Experimental Researches in Electricity: Series 19-29, 3 vols. (London: Taylor and Francis, 1839, 1844, 1855) **III**, plate III, Fig. 1, p. 592 of the 1855 edition. (The full plate is shown on p. 240.)

Here we apply Peregrinus's principle to Faraday's magnetic field lines in order to find the magnetic field vector, including both its magnitude and direction using the method of relaxation that we discussed in Chapter 5.

To digitize Faraday's data, we first created an ad-hoc model of the magnetic field direction based on what it looks like to the eye. The middle figure shows our ad hoc model unit vectors, overlaid on top of Faraday's data to scale. As you can see, they line up pretty well, but not perfectly.

Recall from Chapter 1 that algorithmically solving the continuity equation only produces unique results with significant additional information. In this case, we have Faraday's data, so we constrain the *direction*, but not the strength, of the magnetic field to match the ad hoc unit vector model exterior to the magnet itself.

We also make two additional assumptions: (1) the *strength*, though not the direction, of the magnetic field is constant inside the magnet, and (2) the strength is much smaller (zero) at the outside boundaries of the box. This latter assumption is necessary because we assume the field strength falls off at far distances.

The bottom figure shows our final model of the magnetic field. The greyscale image shows the magnetic field strength, and the unit vectors show the magnetic field direction both inside and outside of the magnet.

Notice how much stronger the magnetic field is inside, rather than outside, the magnet.

Modeling Faraday's Magnetic Field Lines

	data	assumptions	model
inside the magnet	n/a	$B = B_0$	$\hat{B}$
surrounding the magnet	$\hat{B}$	n/a	B
on the outside border	$\hat{B}$	$B = 0$	n/a

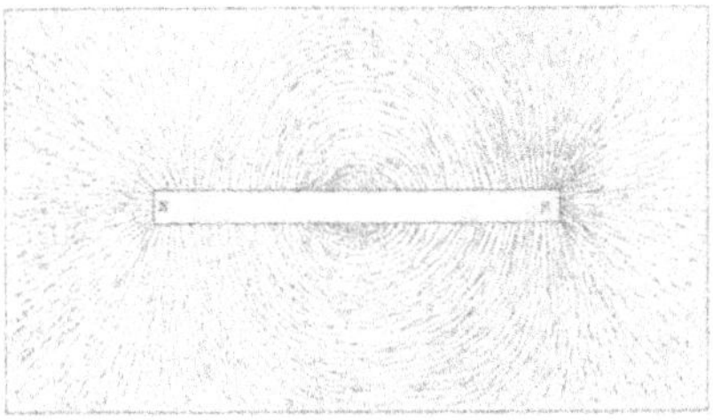

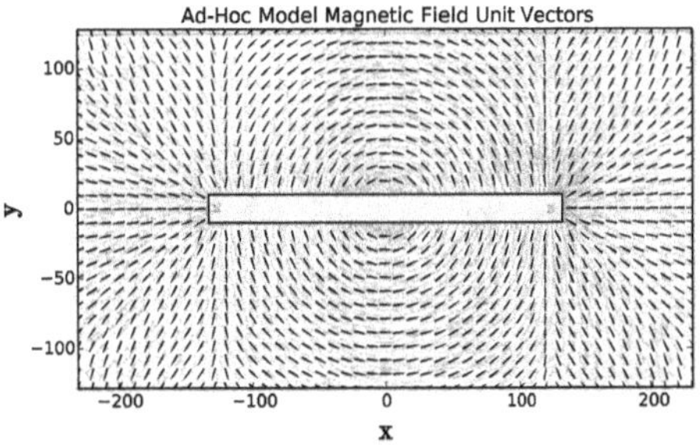

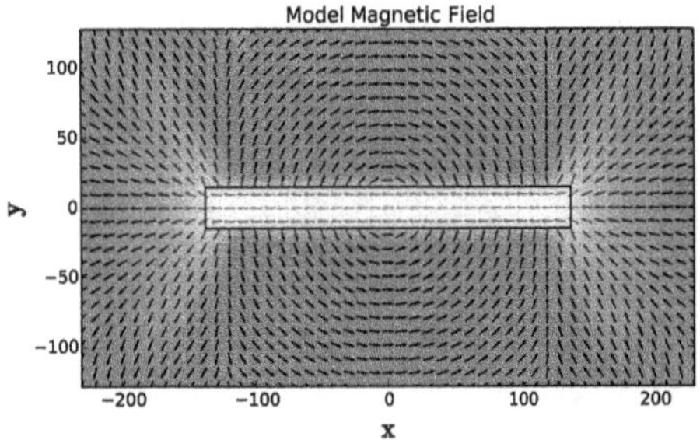

Problem 6.5: An Inverse Square Magnetic Field

Consider a magnetic field given in spherical coordinates by:

$$\vec{B} = K\frac{\sin\theta}{r^2}\hat{\varphi},$$

where K is a suitably chosen constant.

Show that $\vec{\nabla}\cdot\vec{B} = 0$ for this magnetic field.

Problem 6.6: A Thin Magnetic Needle

The magnetic field outside a thin magnetic needle of length l can be modeled by the following relationship in cylindrical coordinates (p. 231):

$$\vec{B} \propto \frac{1}{s}\left(\frac{1}{\sqrt{s^2+\left(z-\frac{1}{2}\right)^2}}-\frac{1}{\sqrt{s^2+\left(z+\frac{1}{2}\right)^2}} + \frac{\left(z+\frac{1}{2}\right)^2}{\left(s^2+\left(z+\frac{1}{2}\right)^2\right)^{3/2}}-\frac{\left(z-\frac{1}{2}\right)^2}{\left(s^2+\left(z-\frac{1}{2}\right)^2\right)^{3/2}}\right)\hat{s}$$

$$-\left(\frac{z+\frac{1}{2}}{\left(s^2+\left(z+\frac{1}{2}\right)^2\right)^{3/2}}-\frac{z-\frac{1}{2}}{\left(s^2+\left(z-\frac{1}{2}\right)^2\right)^{3/2}}\right)\hat{z}.$$

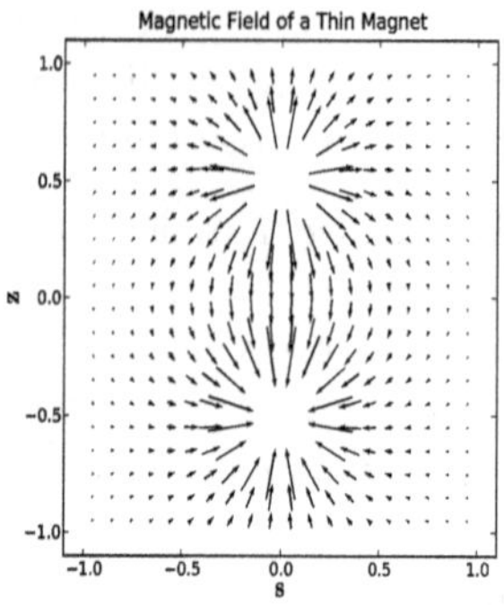

a) Show that Peregrinus's principle holds everywhere outside the magnet.

b) Show that the normalization factor is $\frac{1}{4\pi}\Phi_B$, where Φ_B is the magnetic flux through the needle.

6.3 *The Field Surrounding Permanent Magnets*

Ever since the compass was invented, people have mapped the direction of the magnetic field surrounding permanent magnets—especially the field surrounding the Earth. Without a paradigm, mapping these field lines could reveal little about the magnetic field inside the magnet.

Petrus Peregrinus's thought experiment provided one model for the internal workings of a permanent magnet. In fact, in Thought Experiment 6.7 (p. 198) we algorithmically apply his principle to Faraday's drawings and show that the internal magnetic field must be very strong and mostly parallel to the magnetic moment of the magnet.

Eighteenth century physicists who pioneered the field of electrostatics took a different tack. They assumed that matter is made up of little poles, north and south, which behave much like electric charges. Given the success of Coulomb's inverse square law of electrostatic forces, it seemed only natural to investigators such as Coulomb, Jean-Baptiste Biot, Siméon Denis Poisson, and Carl Friedrich Gauss to assume that magnetic interactions obeyed a similar force law. This allowed them to use the tools developed for gravity and electricity to study permanent magnets. Yet, although magnetic poles appeared to follow an inverse square law like Coulomb's, it proved impossible to isolate magnetic poles as had been done for electric charges. Scientists were unable to produce an excess of magnetic "charge" by, say, rubbing, nor could they find any way of magnetizing a body with an excess of one kind of magnetic pole. Furthermore, there was no magnetic analogue to the electrical phenomenon of conduction. The pole model for magnetism was misleading.

Wrong models, however, can sometimes produce similar, or even identical, observational predictions as correct ones. Moreover, when an incorrect model adds mathematical simplicity, it can linger for quite a long time. This was the case with the pole model. As one author has written, "it is a remarkable feature of magnetism that the fiction (as it is now understood) of magnetic end-poles obeying an inverse-square law can predict correct results for the forces between magnets."[96]

---

[96] John J. Roche, *The Mathematics of Measurement: A Critical History* (London: The Athlone Press, 1998), 141. For a discussion of Coulomb's attempt to deduce a force law for magnets, especially in the context of how a model may appear to be mathematically successful yet turn out to be wrong, see pages 140-42 of Roche's book.

The competing classical model of magnetic matter, championed by A André-Marie Ampère, was based on the observation, first made by Danish physicist Hans Christen Ørsted in 1819, that electric currents give rise to magnetic effects. From this, and later experiments by Ampère himself, arose the modern view that *magnetic fields are due to the relative motion of electric charges*. In fact, all magnetic effects are now believed to be the result of interactions between charges in relative motion. It is also an experimental fact that a planar loop of steady electric current placed in a magnetic field experiences a torque just like a tiny magnet! Furthermore, a loop of current produces a magnetic field whose magnetic moment equals the product of the current and the area enclosed by the loop. These facts led Ampère to suggest that magnetism in matter is caused by microscopic circular electric currents.

Ampère's hypothesis that molecular currents are the ultimate cause of magnetism in matter had obvious advantages over the pole model of magnetism. The fact that electric currents and permanent magnets produce magnetic effects suggested the possibility of two very different causes for the same phenomenon, but Ampère's hypothesis united these two causes into a single explanation. According to his idea, moving electric charges generate magnetic fields inside matter just as they generate magnetic fields in a lab. The proper alignment of tiny molecular magnets, created by atomic or molecular current loops inside a piece of magnetic material, plausibly explains why magnets have inseparable north and south poles. Material in which the molecular magnetic moments are aligned randomly would not be magnetic because the individual fields cancel one another. Ampère's hypothesis easily accommodates the basic facts of Peregrinus's observations that magnetic poles are inseparable, while the pole model cannot even suggest why moving electric charges should have anything to do with magnetism.

Early in the twentieth century, such scientists as Albert Einstein, Wander deHaas, Owen Richardson, and Samuel Barnett confirmed some key predictions of the Ampèrian current loop hypothesis by studying ferromagnets, which are materials, such as iron, that can be easily magnetized and retain their magnetic fields in the absence of an external, magnetizing field. Einstein and deHaas demonstrated in the lab that magnetizing a magnet can induce a mechanical rotation, which Einstein and Richardson independently predicted would be a consequence of the Ampère model. After reading Richardson's theoretical prediction, Barnett successfully detected the inverse effect that spinning a ferromagnet could change its magnetization.

Nevertheless, Ampère's model also predicts that magnets should repel rather than attract matter. The simple classical picture of an atomic current loop caused, say, by an electron or a bunch of electrons orbiting the nucleus of an atom does not reproduce one of the most obvious features of magnets. We now know that it is impossible to explain ferromagnetism microscopically by naively applying Ampère's picture of current loops or spinning charged spheres. The correct description of ferromagnets requires quantum rather than classical physics (p. 264).

Over a century after Ampère's death, Paul Dirac resurrected the pole model when he predicted the existence of a quantum of magnetic monopole.[97] He argued that if charge is quantized, so must be pole strength. Dirac also showed that the force holding two quanta of monopoles together must be about 4700 times stronger than that holding a proton and electron of the same distance, so it would be difficult to separate these monopoles if they did actually exist.

---

[97] Dirac, P.A.M., "Quantised Singularities in the Electromagnetic Field," Proc. Roy. Soc. A., **133** (1931), 60-72.

Despite much experimental effort, magnetic monopoles have yet to be reproducibly found in nature.[98] Thus, the principle of Petrus Peregrinus still holds 750 years after the siege of Lucera.

Problem 6.7: Magnetic Monopoles

Despite Peregrinus's arguments, and 750 years of experimental verification, the question of whether magnetic monopoles exist remains of current interest. In particular, there is a phase of matter called a *spin ice*, in which some researchers claim to have found separated magnetic poles.[99]

For this problem, research this topic using primary sources, and write a short paper arguing either that magnetic monopoles do, or do not, exist. Remember to base your argument both on the current experimental evidence, and on the interpretation of said evidence. What hypothetical future experimental result, or theoretical explanation, would cause you to change your mind?

Thought Experiment 6.8: The Magnetic Field Surrounding Point-like Magnets

Peregrinus's thought experiment tells us that slices of permanent magnets, no matter how small, appear to have both a north and a south pole. Our last clue is this: the debunked pole model could successfully predict the fields surrounding permanent magnets.

We, therefore, posit the following classical toy model of magnetic matter:

(1) Each atom has an intrinsic magnetic moment. In aggregate, these magnetic moments add vectorally to produce the overall magnetic moment of the magnet involved.

(2) The far magnetic field surrounding an atom follows the same mathematical form as does the far electric field of an electric dipole (p. 104).

Therefore, we expect that the magnetic field surrounding a tiny magnet with a magnetic moment $\vec{m} = m\hat{z}$ would obey the standard form for a dipole field:

$$\vec{B} \propto \frac{m}{r^3}\left(\frac{3sz}{r^2}\hat{s} + \frac{2z^2 - s^2}{r^2}\hat{z}\right).$$

The constant of proportionality, in SI units, is by definition[100] $1/10{,}000{,}000\,\frac{\text{T}^2\text{m}^3}{\text{J}}$. For reasons having to do with the Maxwellian interpretation of the aether, this constant is labeled $\mu_0/4\pi$, where μ_0 is called the *permeability of free space.*

The field of a magnetic dipole can also be written in vector notation as follows:

$$\vec{B} = \left(10^{-7}\,\tfrac{\text{T}^2\text{m}^3}{\text{J}}\right)\left(\frac{3(\vec{m}\cdot\hat{r})\hat{r} - \vec{m}}{r^3}\right) = \frac{\mu_0}{4\pi}\left(\frac{3(\vec{m}\cdot\vec{r})\vec{r}}{r^5} - \frac{\vec{m}}{r^3}\right).$$

The magnetic moment of an electron, μ_e, has a value, in SI units, of $9.28\times10^{-27}\,\frac{\text{J}}{\text{T}}$.

---

98 In 1982 one monopole was claimed to be found, but no others were ever detected, (B. Cabrera, "First Results from a Superconductive Detector for Moving Magnetic Monopoles," Phys. Rev. Lett., **48**, (1982) 1378–1381).

99 B. Castelnovo, R. Moessner, and S.L. Sondhi, "Magnetic monopoles in spin ice," Nature, **451**, (2008) 42-45.

100 This is because of the pre-2018 definition of the Ampere, which in turn defines the Tesla (see p. 280).

Problem 6.8: The Divergence of a Dipole Field

a) Show, in cylindrical coordinates, that the divergence of a dipole field is zero at far distances.

b) Is the divergence also zero in the limit when r becomes very small? Is this relevant?

Thought Experiment 6.9: A Collection of Small Magnets

Consider n small magnets each at known locations $\vec{R}_i$ with magnetic moment vectors $\vec{m}_i$ with respect to some common origin.

In order to find the magnetic field at a given point of interest $\vec{r}$, we simply add the magnetic field at that point from each of these magnets.

As in the diagram, you can see that we must evaluate the magnetic field function at the relative position between each magnet and the point of interest. Therefore the resultant magnetic field is:

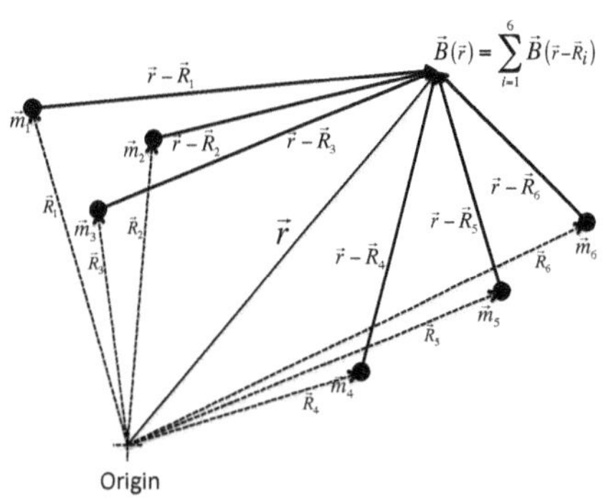

$$\vec{B}(\vec{r})=\sum_{i=1}^{n}\vec{B}(\vec{r}-\vec{R}_i)=\left(\frac{\mu_0}{4\pi}\right)\sum_{i=1}^{n}\left(\frac{3\left(\vec{m}_i\cdot\left(\vec{r}-\vec{R}_i\right)\right)\left(\vec{r}-\vec{R}_i\right)}{\left|\vec{r}-\vec{R}_i\right|^5}-\frac{\vec{m}_i}{\left|\vec{r}-\vec{R}_i\right|^3}\right).$$

Problem 6.9: Two Small Magnets

Consider two identical small magnets, located at positions $\vec{R}_\pm=\pm a\,\hat{y}$, and aligned anti-parallel, so that $\vec{m}_\pm=\pm m\,\hat{z}$.

(a) Show that the total magnetic moment adds to zero.

(b) Show that $(r_\pm)^n=\left|\vec{r}-\vec{R}_\pm\right|^n\approx r^n\left(1\pm n\tfrac{a}{r}\hat{y}\cdot\hat{r}\right)$, when $a\ll r$.

(c) Find the far magnetic field, in unit vector notation, surrounding both magnets.

(d) Find the far magnetic field in spherical coordinates.

(e) Compare and contrast this to the electric quadruple in Thought Experiment 3.12 (p. 106).

Thought Experiment 6.10: The Attractive Force Between Magnets

One of our first experiences with magnets is that like poles repel and opposite poles attract. We can easily chain permanent magnets together, and it takes work to separate them. How do we explain this force in terms of the magnetic field theory we've presented so far?

Empirically, the on-axis field surrounding a small magnet is proportional to the inverse cube of the distance, which is consistent with Thought Experiment 6.8 above.

$$\vec{B}_1(s=0,z\gg a)\propto\frac{\hat{z}}{z^3}\approx\frac{\mu_0}{4\pi z^3}\left(3(m\hat{z}\cdot\hat{z})\hat{z}-m\hat{z}\right)\approx\frac{\mu_0 m\,\hat{z}}{2\pi z^3}.$$

Consider another identical small magnet, with a magnetic moment given by $\vec{m} = m\hat{z}$ that is located at a position z from the first magnet. We expect the two magnets to attract as the north pole of one is close to the south pole of the other.

What is the force exerted on the second magnet by the first magnet?

As in Thought Experiment 6.4 (p. 192), we begin by finding the potential energy of the second dipole magnet in the field caused by the first one:

$$U = -\vec{m}_2 \cdot \vec{B}_1 = -(m\hat{z}) \cdot \left(\frac{\mu_0 m \hat{z}}{2\pi z^3}\right) = -\frac{\mu_0 m^2}{2\pi z^3}.$$

The work-energy relation implies that the force between the magnets is given by the negative gradient of the potential energy function:

$$\vec{F}_{\text{of 1 on 2}} = -\vec{\nabla} U = -\vec{\nabla}\left(-\frac{\mu_0 m^2}{2\pi z^3}\right) = \frac{\mu_0 m^2}{2\pi} \frac{\mathrm{d}}{\mathrm{d}z} z^{-3}\hat{z} = -\frac{3\mu_0 m^2}{2\pi z^4}\hat{z}.$$

The forces between magnets become very strong when they are close together, but can be quite weak when pulled apart.

Problem 6.10: The Force Between Two Repulsive Magnets

Consider two identical dipole magnets arranged north to north on a line, separated by a distance z, as shown in the diagram.

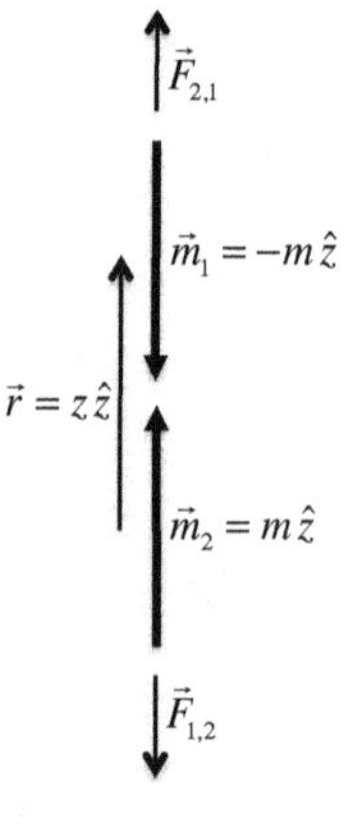

a) Show that the interaction energy is:

Therefore, the interaction energy is given by:

$$U(\vec{r}) = \frac{\mu_0 m^2}{2\pi z^3}.$$

b) And the force on magnet 1, by magnet 2, is:

$$\vec{F} = \frac{3\mu_0 m^2}{2\pi z^4}\hat{z}.$$

Problem 6.11: The Force Between Magnetic Rods

Consider two thin uniform magnetic rods, each of length l, and total magnetic moment $\vec{m}$. If they are stuck together, end to end, how much force would be required to pull them apart?

Thought Experiment 6.11: The Interaction Energy of Two Magnets

Consider the case of two magnets, each possessing a magnetic moment $\vec{m}$, which interact with one another. In order to understand the forces and torques on each dipole, we need to first write down the interaction energy between them.

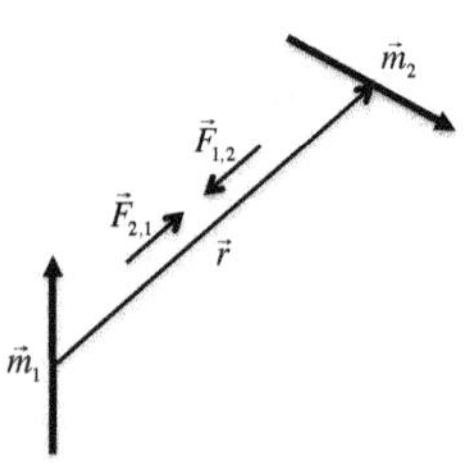

We begin by assigning one, $\vec{m}_1$, as the cause of the magnetic field, and the other, $\vec{m}_2$, as the object being affected by the field. Moreover, we will set $\vec{m}_1$ at the origin and have it aligned with the $\hat{z}$ axis, so that $\vec{m}_2$ is located at a position $\vec{r}$. The magnetic field due to $\vec{m}_1$ is given by:

$$\vec{B}_1(\vec{r}) = \left(\frac{\mu_0}{4\pi}\right)\frac{3(\vec{m}_1\cdot\hat{r})\hat{r}-\vec{m}_1}{r^3}.$$

Thus, the interaction energy is:

$$U(\vec{r}) = -\vec{m}_2\cdot\vec{B}_1(\vec{r}) = -\vec{m}_2\cdot\left(\frac{\mu_0}{4\pi}\right)\frac{3(\vec{m}_1\cdot\hat{r})\hat{r}-\vec{m}_1}{r^3}$$

$$= -\left(\frac{\mu_0}{4\pi}\right)\frac{3(\vec{m}_1\cdot\hat{r})(\vec{m}_2\cdot\hat{r})-\vec{m}_1\cdot\vec{m}_2}{r^3}.$$

Notice that the expression is symmetrical with regard to the reversal of $\vec{m}_1$ and $\vec{m}_2$, that is:

$$U(\vec{r}) = -\vec{m}_1\cdot\vec{B}_2(-\vec{r}) = -\vec{m}_1\cdot\left(\frac{\mu_0}{4\pi}\right)\frac{3(\vec{m}_2\cdot(-\hat{r}))(-\hat{r})-\vec{m}_2}{r^3}$$

$$= -\left(\frac{\mu_0}{4\pi}\right)\frac{3(\vec{m}_1\cdot\hat{r})(\vec{m}_2\cdot\hat{r})-\vec{m}_1\cdot\vec{m}_2}{r^3}.$$

In Problem 6.12 below you will show that the interaction force, in general, is:

$$\vec{F} = -\vec{\nabla}U = \frac{3\mu_0}{4\pi r^4}\left((\vec{m}_1\cdot\hat{r})\vec{m}_2+(\vec{m}_2\cdot\hat{r})\vec{m}_1+(\vec{m}_1\cdot\vec{m}_2)\hat{r}-\frac{5(\vec{m}_1\cdot\hat{r})(\vec{m}_2\cdot\hat{r})\hat{r}}{r}\right).$$

Note that the symmetry of the situation guarantees that we get the same result if we reverse the roles of each dipole and assign $\vec{m}_2$ as the field source and $\vec{m}_1$ as the test magnet.

Problem 6.12: The Interaction Force Between Two Magnets

Show that the general force law between magnets mathematically follows from the potential energy in Thought Experiment 6.11 above.

Problem 6.13: The Force Between Flat Bar Magnets

Consider two thin uniform bar magnets, each of width a, length b, and total magnetic moment $\vec{m}$, pointing lengthwise.

a) If they are stuck together, end to end, with parallel magnetic moments, how much force would be required to pull them apart?

b) If they are stuck together sideways, with antiparallel magnetic moments, how much force would be required to separate them?

Problem 6.14: Hyperfine Structure

Consider two magnets with magnetic moments $\vec{m}_1$ and $\vec{m}_2$ placed parallel to each other and separated by a distance a.

a) Find the energy difference between having the two magnets aligned compared to anti-aligned. For example, if $\vec{m}_1 = m_1\hat{z}$ and is located at the origin, then $\vec{m}_2$ would be located at a position in cylindrical coordinates $s = a$. Find the difference in energy between when $\vec{m}_2 = m_2\hat{z}$ and $\vec{m}_2 = -m_2\hat{z}$.

b) The most important energy transition in radio astronomy is the 21 cm line, which is due to a spin flip between the proton and electron in the ground state of hydrogen. The electron magnetic moment is $\mu_e = 9.285\times10^{-24}\,\frac{\text{J}}{\text{T}}$, the proton magnetic moment is $\mu_p = 1.411\times10^{-26}\,\frac{\text{J}}{\text{T}}$, and the distance between the electron and the proton in the Bohr model is $5.29177\times10^{-11}\,\text{m}$. Calculate the energy difference of this semi-classical hyperfine transition, and compare it to the measured value of $5.551\times10^{-6}\,\text{eV}$.

c) The 21 cm line of Hydrogen was first discovered in 1951. Report on its discovery, and how it is used in radio astronomy today?

Problem 6.15: The Torque Between Magnets

Consider two magnets with magnetic moments pointed in different, known, directions so that $\vec{m}_1$ does not necessarily equal $\vec{m}_2$.

(1) Find the torque about an axis parallel to $\vec{m}_1$, but through magnet 2, using the definition of the magnetic moment from Thought Experiment 6.2 (p. 189).

(2) Now, find the torque on the center of mass of magnet 2, about the axis defined by magnet 1, using the basic definition of torque and the two magnet force law.

(3) What is the net torque on magnet 2 about the center of magnet 1?

(4) Discuss how this relates to the parallel axis theorem from introductory physics.

Thought Experiment 6.12: Magnetization

Consider a chunk of magnetic material. Clearly the larger the magnet, the larger its overall magnetic moment. So, how do we characterize the strength of the material itself, rather than the overall magnet?

Most magnet manufactures simply specify the strength of the residual magnetic field inside of the magnet, which is typically about one tesla for strong magnets.

Most physicists, however, consider the magnetic moment per volume. This magnetic moment density is called the *magnetization*. Since dipole moments add vectorally, the magnetic moment of a permanent magnet is simply the vector sum of all the atomic dipole magnets that it is made of:

$$\vec{m}_{\text{tot}} = \sum \vec{m}_i \,.$$

Whether these atomic magnets reinforce or cancel each other will depend on how they are aligned and the magnitude of the atomic magnetic moments. We expect a material sample in which the moments are aligned purely at random to exhibit no net magnetic moment because the moments will cancel each other when they are summed. Adding the magnetic moments up over scales much larger than atoms, but much smaller than those of macroscopic interest, we can write the magnetization as:

$$\vec{M}(\vec{r}) \equiv \frac{1}{\delta V}\sum_{\delta V} \vec{m}_i \,,$$

where δV is a volume element.

The SI unit of magnetization is given by:

$$[M]=\frac{[\text{magnetic moment}]}{[\text{volume}]}=\frac{\left(\frac{\text{J}}{\text{T}}\right)}{\text{m}^3}=\frac{\left(\text{J}/\text{m}^3\right)}{\text{T}}=\frac{\text{T}}{\left(\frac{\text{T}^2\cdot\text{m}^3}{\text{J}}\right)}.$$

Finally, the total magnetic moment of a permanent magnet is given by:

$$\vec{m}_{\text{tot}}=\oint_{\text{volume}}\vec{M}\,\mathrm{d}V.$$

It may be tempting to think that finding the magnetic moment of a permanent magnet is sufficient to determine the surrounding magnetic field. Although this is true in the far limit, it is far from true in general. The correct way to find the overall magnetic field is to calculate the magnetic field contribution from each small volume element, and then integrate over the whole magnet.

Thought Experiment 6.13: The Magnetic Moment of a Spherical Magnet

The strongest permanent magnets presently used in industry are called *rare earth magnets* because they contain alloys of the rare earth element neodymium, though it is neither rare nor unique to Earth. These magnets have a high permanence, so after applying a very strong magnetic field in one direction, the magnetization remains strong and the same throughout the magnet.

Now we consider a spherical magnet of radius a and constant magnetization $\vec{M}$, and ask, "What is its total magnetic moment $\vec{m}$?" As we discussed earlier, translated magnetic moments simply add vectorally, so the solution to this problem is:

$$\vec{m}=\oint_{\text{volume}}\vec{M}\,\mathrm{d}V=\vec{M}\oint_{\text{volume}}\mathrm{d}V=\frac{4\pi}{3}a^3\vec{M}.$$

At distances far from the material, we expect the magnetic field to be that of a magnetic dipole:

$$\vec{B}(\vec{r})\approx\left(\frac{\mu_0}{4\pi}\right)\frac{3(\vec{m}\cdot\hat{r})\hat{r}-\vec{m}}{r^3}.$$

In Problem 6.18 (p. 208), you will show that the field outside of a spherical magnet turns out to be exactly that of a magnetic dipole.

Problem 6.16: The Force Between Cubic Magnets

Consider two uniform cubic magnets, each with side length a, and total magnetic moment $\vec{m}$.

a) What is the magnetization of each cube?

b) If they are stuck together, north end to south end, how much force would be required separate them?

c) If they are stuck together sideways, with antiparallel magnetic moments, how much force would be required to pull them apart?

d) What is the ratio of these two forces?

Thought Experiment 6.14: The Field Outside a Magnet

Consider now a permanent magnet with a magnetization vector $\vec{M}(\vec{R})$, how do we find the magnetic field at a position $\vec{r}$?

In Thought Experiment 6.9 (p. 203) we found the magnetic field due to six small magnets. Here we will find a magnetic field element $\delta\vec{B}$ as a function of a volume element δV:

$$\delta\vec{B} = \left(\frac{\mu_0}{4\pi}\right)\left(\frac{3\left(\delta\vec{m}\cdot(\vec{r}-\vec{R})\right)(\vec{r}-\vec{R})}{\left|\vec{r}-\vec{R}\right|^5} - \frac{\delta\vec{m}}{\left|\vec{r}-\vec{R}\right|^3}\right) = \left(\frac{\mu_0}{4\pi}\right)\left(\frac{3\left(\vec{M}\cdot(\vec{r}-\vec{R})\right)(\vec{r}-\vec{R})}{\left|\vec{r}-\vec{R}\right|^5} - \frac{\vec{M}}{\left|\vec{r}-\vec{R}\right|^3}\right)\delta V.$$

Taking the limit, and writing this as an integral over the magnetic object, the magnetic field becomes:

$$\vec{B}(\vec{r}) = \left(\frac{\mu_0}{4\pi}\right)\oint_{volume}\left(\frac{3\left(\vec{M}(\vec{R})\cdot(\vec{r}-\vec{R})\right)(\vec{r}-\vec{R})}{\left|\vec{r}-\vec{R}\right|^5} - \frac{\vec{M}(\vec{R})}{\left|\vec{r}-\vec{R}\right|^3}\right)\mathrm{d}V.$$

The integral above is very difficult to solve analytically, except in a few special situations. In most applications, however, the scientist or engineer will write a computer application to calculate the magnetic field from the magnetic material of interest. Given the linearity of these equations, this process is relatively straightforward.

Problem 6.17: A Cylindrical Magnet

Consider a long cylindrical magnet of length h, radius a, and uniform magnetization $\vec{M} = M\,\hat{z}$.

(a) Show that the magnetic field at the center of the magnet is given by $\vec{B} = \mu_0\vec{M}\left(\frac{h}{\sqrt{(2a)^2+h^2}}\right)$.

(b) In the limit that the magnet is long and thin, show that the on-axis magnetic filed is simply given by $\vec{B} = \mu_0\vec{M}$.

(c) The magnetization of permanent magnets are often measured in magnetic field units, with an typical strong magnet having a magnetization of $\mu_0\vec{M} = 1$ T. What would the magnetic moment be, in SI units, of this 1 tesla magnet?

(d) Imagine you want to make a flat disc shaped refrigerator magnet with a diameter of about one inch $(2a = 1\,\text{in} = 2.5\,\text{cm})$, out of some material with a magnetization of about 100 mT. How thick would it need to be in order to have a magnetic field of about 5 mT?

(e) Typical magnetic recording tape has a thin layer of the same magnetic material on top of plastic tape. If your goal were to have a field strength of about 50 μT in the tape, how thick would you need to make the layer of magnetic material?

Problem 6.18: A Magnetized Sphere

Consider the magnetized sphere with uniform magnetization $\vec{M}$ and a radius a, as discussed in Thought Experiment 6.13 above.

a) Show that the magnetic field in the middle of the sphere is given by: $\vec{B}\Big|_{r=0} = \dfrac{2\mu_0 \vec{M}}{3}$.

b) Show that outside the sphere the magnetic field is simply that of an equivalent point magnet with the same a total magnetic moment: $\vec{m} = \left(\tfrac{4}{3}\pi a^3\right)\vec{M}$.

Problem 6.19: The Neocube™

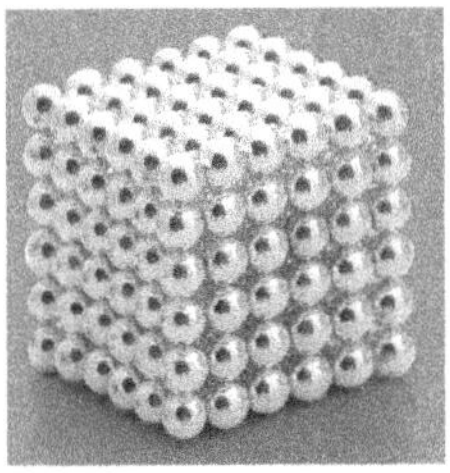

A Neocube™ is a desk toy with 216 neodymium spherical magnets, stuck together arranged as a $6\times 6\times 6$ cube. Once they are taken apart, it takes some practice and thought to put them back together again. Assume that each magnet has a radius a and magnetic moment magnitude m.

a) How much work is required to separate two spherical magnets aligned north to south (end to end)?

b) How much work is required to separate two anti-aligned spherical magnets that are lying sideways to each other?

c) Clearly draw the direction of each magnetic moment when they are arranged as a cube.

d) How much work is required to separate all the magnets in the cube. You can ignore those magnets which are not touching.

e) Imagine making a long chain with all 216 magnets. Is this more, or less, energetically stable than the cube? Clearly explain.

Problem 6.20: A Toroidal Magnet

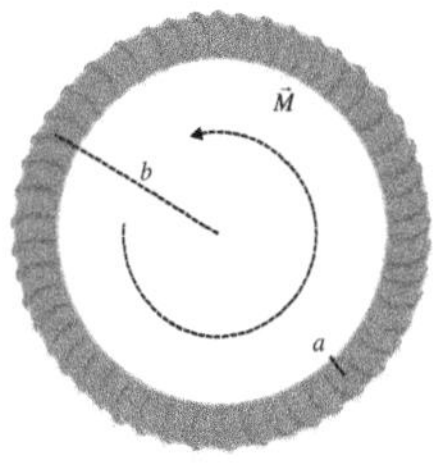

Consider a toroidal magnet with constant counter-clockwise magnetization $\vec{M} = M\,\hat{\varphi}$, cross-sectional radius a, and ring radius $b \gg a$, as shown.

a) Write down an expression for the differential magnetic field $\delta\vec{B}(\phi,s,z)$, where ϕ is the location of the differential magnetic moment $\delta\vec{m} = M\,b\,\delta\phi\,\hat{\varphi}$, and $\vec{r} = s\,\hat{s} + z\,\hat{z}$ is the location of interest.

b) Express, in integral form, the magnetic field $\vec{B}(s,z)$, over all of space.

c) Show that the magnetic field is zero on the axis of symmetry.

d) Express in integral form the magnetic field, $\vec{B}(s)\Big|_{z=0}$, in the plane of symmetry.

e) Taylor expand the differential magnetic field $\delta\vec{B}(\phi,s)\Big|_{z=0}$, in the equatorial plane and in the far limit so that $s \gg b$.

f) Find the magnetic field, $\vec{B}(s)\Big|_{z=0,s\gg b}$, in the far limit. Discuss how quickly the magnetic field falls off with distance. Based on Section 5.3 (p. 157), what would this be called in the language of multipole expansion (e.g. dipole, quadrupole, etc. ..)?

g) Argue that, in the limit that the magnet becomes infinitesimally small, in such a way that $b \gg a$, the magnetic field goes to: $\vec{B} \rightarrow \mu_0 \vec{M}$ inside the magnet.

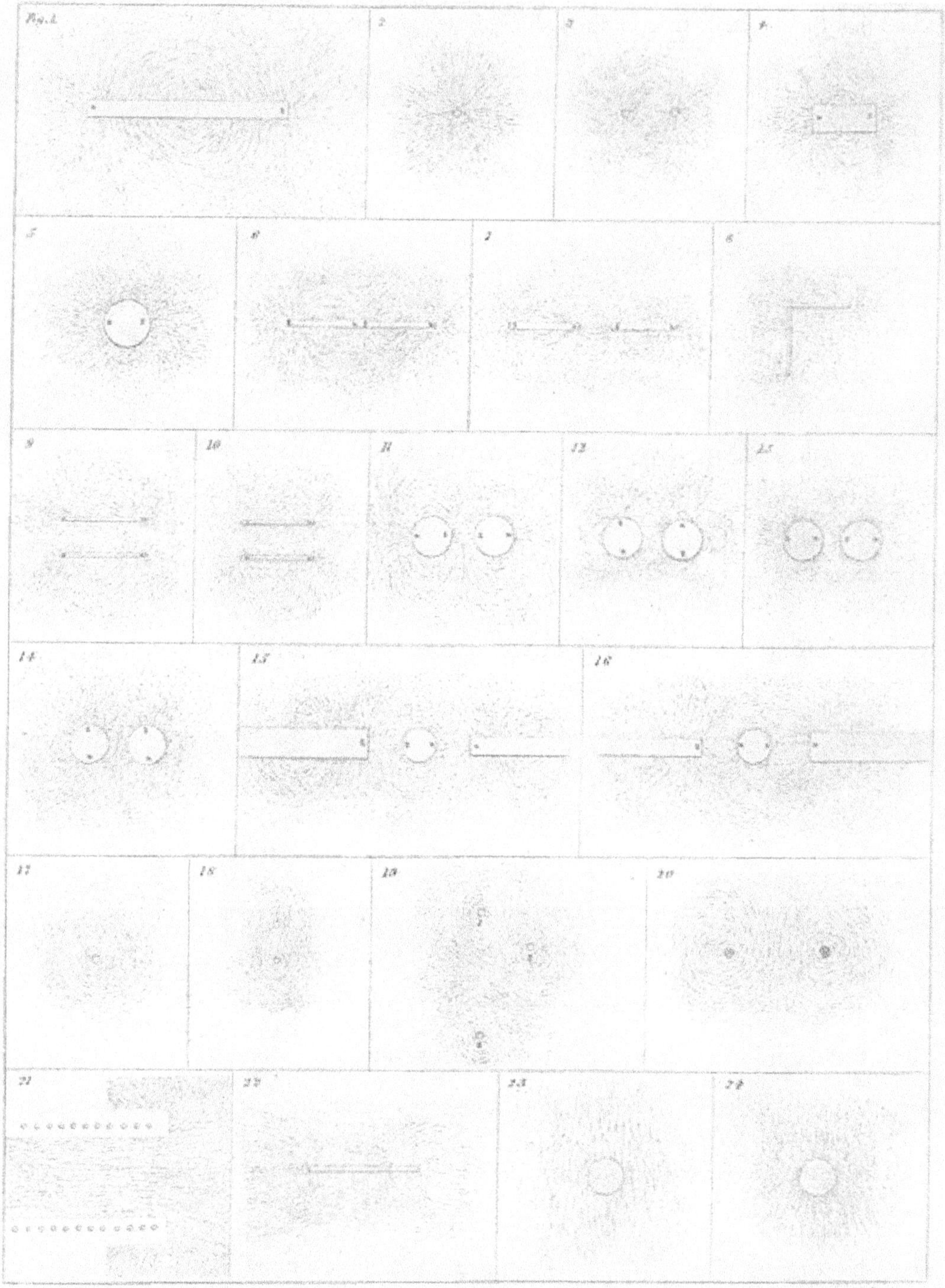

Plate 14: Michael Faraday's Drawings of Magnets and Iron Filings[101]

[101] M. Faraday, Experimental Researches in Electricity: Series 19-29, 3 vols. (London: Taylor and Francis, 1839, 1844, 1855) **III**, plate III, p. 592 of the 1855 edition.

Chapter 7 The Vector Potential and the Curl

How do we measure openings in surfaces? Say, for example, you close up a draw-string bag. Do you think of the decreasing area or circumference of the hole? This distinction is essentially the relationship between magnetic field and the vector potential.

Consider an open drawstring bag; the magnetic flux through the opening is:

$$\Phi_B = \int_{\text{surface}} \vec{B}\cdot \mathrm{d}\vec{A}.$$

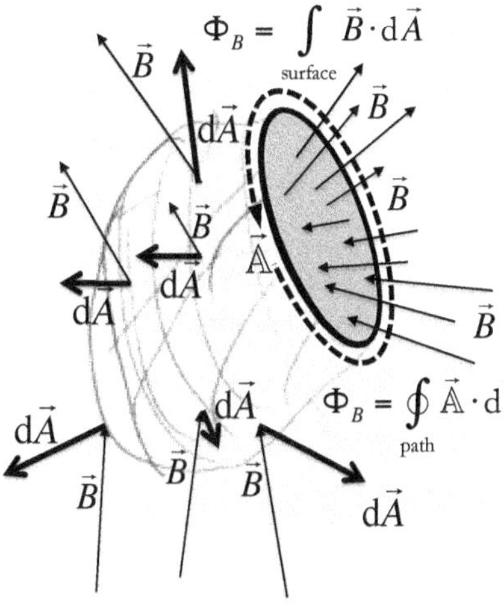

Since the hole has both a perimeter and an area, we could have represented the magnetic flux as a line integral around the opening, rather than as a surface integral through the hole. The vector potential is just this, the quantity which when integrated around an open surface, represents the magnetic flux through the surface. Thus:

$$\Phi_B = \int_{\text{path}} \vec{\mathbb{A}}\cdot \mathrm{d}\vec{\ell}.$$

To continue with the bag analogy, the open surface need not be flat. We can wrap the bag around anything and calculate the flux through the hole.

Imagine, hypothetically, that we instead wanted to calculate the electric, rather than magnetic, flux. We could choose to wrap the bag around some charge, or not, and we would get different answers either way, some sort of electric vector potential would be meaningless. Luckily we do not have that problem, since no absolute charge of northness nor southness exists. The magnetic flux through our bag only depends on what passes through the opening, and does not depend on the shape of the surface of choice. This is what makes the vector potential such a powerful tool.

Franz Ernst Neumann introduced an expression for the vector potential in 1845. Wilhelm Weber soon suggested a different expression in 1846. In 1847, William Thomson suggested an expression for the vector potential in a number of cases, including the vector potential of a small magnet in terms of the magnetic moment, which we use extensively in this chapter. [102]

The vector potential was controversial throughout the 1800s because of its apparent fatal flaw of non-uniqueness. Recall that both Neumann and Weber published different expressions, which were both physically equivalent, as later shown by Hermann von Helmholtz . By the end of the nineteenth century, most physicists agreed with Oliver Heaviside and Heinrich Hertz that it added no additional physical significance.

Only a few decades later, however, Albert Einstein and others found that it fit beautifully within the theory of relativity because $\vec{\mathbb{A}}$ and $\mathbb{V}$ together form a relativistically covariant four dimensional space-time vector from which the electric and magnetic fields can be derived.

[102] See the review article: A.C.T. Wu and C.N. Yang, *Evolution of the Concept of the Vector Potential in the Description of Fundamental Interactions* International Journal of Modern Physics A **21** (2006) 3235-3277, World Scientific Publishing Company

This rapid change in fate of $\vec{\mathbb{A}}$ coincided with a fall in the prestige of the vector $\vec{H}$, and the downfall of the pole model. If magnetic monopoles exist, than $\vec{\mathbb{A}}$ simply becomes a function of convenience. If, however, Petrus Peregrinus was correct and magnetic monopoles do not exist, than it is $\vec{H}$, not $\vec{\mathbb{A}}$, that becomes the function of convenience.

Yakir Aharonov and David Bohm addressed this question in 1959, when Aharonov was Bohm's doctoral student. They developed thought experiments which would yield results that depended on the vector potential, even when there were no fields present. As they put it:

> In classical mechanics, we recall that potentials cannot have such significance because the equation of motion involves only the field quantities themselves. For this reason, the potentials have been regarded as purely mathematical auxiliaries, while only the field quantities were thought to have a direct physical meaning.
>
> In quantum mechanics, the essential difference is that the equations of motion of a particle are replaced by the Schrodinger equation for a wave. This Schrodinger equation is obtained from a canonical formalism, which cannot be expressed in terms of the fields alone, but which also requires the potentials.[103]

In this Chapter we introduce the *magnetic vector potential*, $\vec{\mathbb{A}}$, and use it to simplify the calculation of magnetic fields surrounding permanent magnets. In order to do this, however, we must define a new vector derivative, called *the curl*.

The curl's corresponding fundamental theorem of calculus is called *Stokes Theorem*. It was developed by William Thomson and sent to George Gabriel Stokes in a letter. Stokes then put it as a question on the 1854 Smith's Prize Exam for graduate students at the University of Cambridge, where James Clerk Maxwell sat for the exam. Maxwell won the prize having done particularly well on that question (see Problem 7.4, p. 223).

7.1 *Magnetic Flux and the Vector Potential*

In Thought Experiment 6.6 (p. 197) we discussed Faraday's analogy between magnetic flux and current in a neutral medium. In both cases, what appears to flow in must also flow out, because of Peregrinus's principle and charge conservation respectively. In fact, since Peregrinus's principle is always true, with absolutely no caveats, the circulatory nature of the magnetic flux is also equally fundamental.

What does this mean for a magnetic needle, or a string of paper clips? In Problem 6.17 (p. 208), you found that the central magnetic field inside of a long thin cylindrical magnet is given by:

$$\vec{B} = \lim_{a \ll h}\left(\mu_0\,\vec{M}\,\frac{h}{\sqrt{(2a)^2+h^2}}\right) = \mu_0\,\vec{M}\,.$$

The magnetic flux inside the magnet is, therefore, expressed as:

$$\Phi_B = \int_{\text{surface}} \vec{B}\cdot \mathrm{d}\vec{A} \approx \left(\mu_0\vec{M}\right)\cdot\vec{A}\,.$$

---

[103] Y. Aharonov and D. Bohm, "Significance of Electromagnetic Potentials in the Quantum Theory," Physical Review, **115** (1959), 485-491.

Each of the imaginary planes, $\vec{A}$, that the flux passes through is surrounded by an imaginary loop, whose positive direction is related to the plane's area vector by the right hand rule (Definition 1.3, p. 24). These loops are called *Amperian loops*, because they are very common when applying Ampère's law (p. 253).

Since the magnetic field is circulatory $\left(\vec{\nabla}\cdot\vec{B}=0\right)$, we can define a new vector $\vec{\mathbb{A}}$, called the *vector potential*, such that for any given Amperian loop, the magnetic flux through the corresponding enclosed plane is given by:

$$\Phi_B = \int_{\text{surface}} \vec{B}\cdot d\vec{A} = \oint_{\text{path}} \vec{\mathbb{A}}\cdot d\vec{\ell}.$$

The SI unit of the vector potential has no particular name, so it is simply called a meter-tesla or a weber per meter.

Note that here $\Phi_B \neq 0$ since the flux is calculated over an *open surface*, which is bounded by the closed loop over which $\vec{\mathbb{A}}\cdot d\vec{\ell}$ is evaluated.

Classically, the vector potential cannot be measured *in situ*, so all that really matters is that it predicts the correct magnetic field vector $\vec{B}$. For this reason, depending on how we solve a problem, we have some freedom of choice in terms of zero point, and something else called *gauge*, which we will discuss later (p. 300).

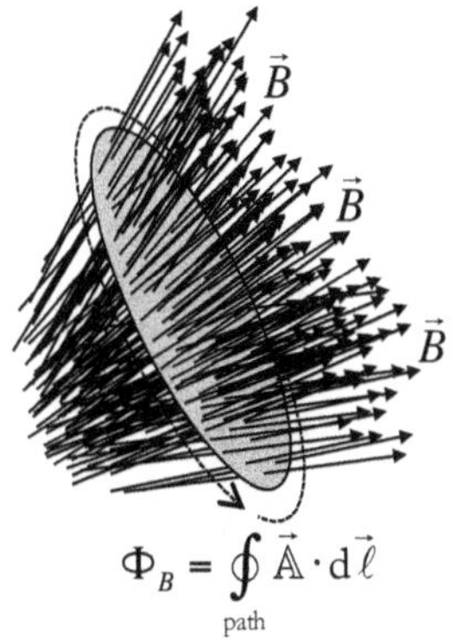

Problem 7.1: The Vector Potential of a Very Long Magnet

Consider a very long cylindrical magnet with radius a and uniform internal magnetic field $\vec{B}_0$.

(a) Clearly draw a picture of the magnet and the direction of the magnetic field. Define your coordinate system so that inside the magnet $\vec{B}=B_0\hat{z}$. Also draw the direction of the magnetization and the vector potential.

(b) Draw an Amperian loop of radius a so it just encloses the magnet. Show that the magnetic flux through the magnet is $\Phi_m = \pi a^2 B_0$.

(c) Show that, in cylindrical coordinates, the vector potential at the edge of the magnet is: $\vec{\mathbb{A}} = \frac{1}{2} a B_0 \hat{\varphi}$.

(d) Show that, in cylindrical coordinates, the vector potential outside the magnet is: $\vec{\mathbb{A}} = \dfrac{a^2}{2s} B_0 \hat{\varphi}$.

(e) Show that, inside the magnet, the vector potential is: $\vec{\mathbb{A}} = \frac{1}{2} s B_0 \hat{\varphi}$.

(f) Find the divergence of the vector potential both inside and outside of the wire. If this is zero, then you are using the *Coulomb gauge*. Is this the case here?

Thought Experiment 7.1: The Vector Potential of a Tiny Magnet

Consider a simple tiny magnet located at the origin with a magnetic moment $\vec{m}=m\hat{z}$. What is the surrounding vector potential?

First we recall Thought Experiment 6.8 (p. 202) where wrote the external magnetic field, due to a small magnet, to be:

$$\vec{B} = \left(10^{-7}\,\tfrac{\mathrm{T^2 m^3}}{\mathrm{J}}\right)\left(\frac{3(\vec{m}\cdot\hat{r})\hat{r}-\vec{m}}{r^3}\right) = \frac{\mu_0}{4\pi}\left(\frac{3(\vec{m}\cdot\vec{r})\vec{r}}{r^5}-\frac{\vec{m}}{r^3}\right).$$

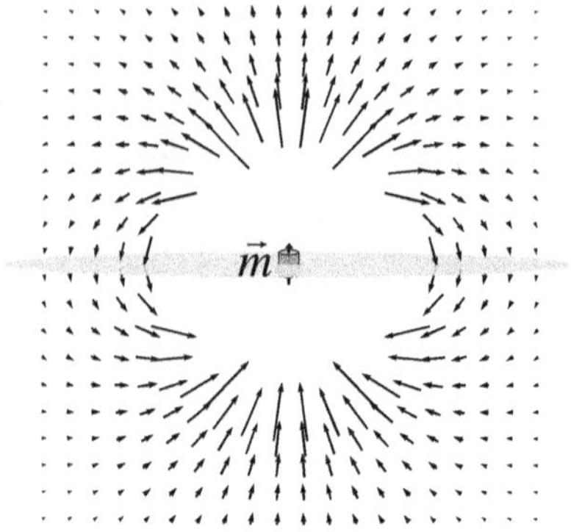

According to Peregrinus's principle the total flux through the equator must be zero. This is because we could, in principle, draw a Gaussian surface through the equation, but of infinite size, so only the equatorial flux would matter as the magnetic field goes to zero at far distances. Thus, the total flux inside the magnet must be equal, and opposite to, the total flux through the equator outside of the magnet.

The total flux through the equator outside of a magnet of radius a, as a function of position is:

$$\underset{\text{outside}}{\Phi_B} = \int_{\text{surface}} \vec{B}\cdot \mathrm{d}\vec{A} = \int_a^\infty \tfrac{\mu_0 m}{4\pi}\left(\tfrac{3sz}{r^5}\hat{s}+\tfrac{2z^2-s^2}{r^5}\hat{z}\right)\cdot\left(2\pi s\,\mathrm{d}s\,\hat{z}\right) = \left(\frac{\mu_0 m}{2}\right)\int_a^\infty \frac{2z^2 s - s^3}{r^5}\,\mathrm{d}s.$$

On the equator we can set $z = 0$ and $r = s$, so:

$$\underset{\text{outside}}{\Phi_B} = \left(\frac{\mu_0 m}{2}\right)\int_a^\infty\left(\frac{-s^3\,\mathrm{d}s}{r^5}\right) = \left(\frac{\mu_0 m}{2}\right)\frac{1}{s}\bigg|_a^\infty = \left(\frac{\mu_0 m}{2}\right)\left(\frac{1}{\infty}-\frac{1}{a}\right) = -\frac{\mu_0 m}{2a}.$$

Next we find the net magnetic flux inside any circular equatorial loop:

$$\Phi_B = \underset{\text{inside}}{\Phi_B} + \underset{\text{outside}}{\Phi_B} = \frac{\mu_0 m}{2a} + \left(\frac{\mu_0 m}{2}\right)\left(\frac{1}{s}-\frac{1}{a}\right) = \frac{\mu_0 m}{2s}.$$

Thus the equatorial vector potential is:

$$\oint_{\text{path}} \mathbb{A}\cdot \mathrm{d}\vec{\ell} = 2\pi s\,\mathbb{A} = \frac{\mu_0 m}{2s} \rightarrow \vec{\mathbb{A}} = \frac{\mu_0\, m}{4\pi\, s^2}\hat{\varphi}.$$

Now that we see what we are doing in a specific example, we can do this in general.

Let us again draw an Amperian loop about the axis, because of the axial symmetry, so we expect that the vector potential will again be in the azimuthal direction.

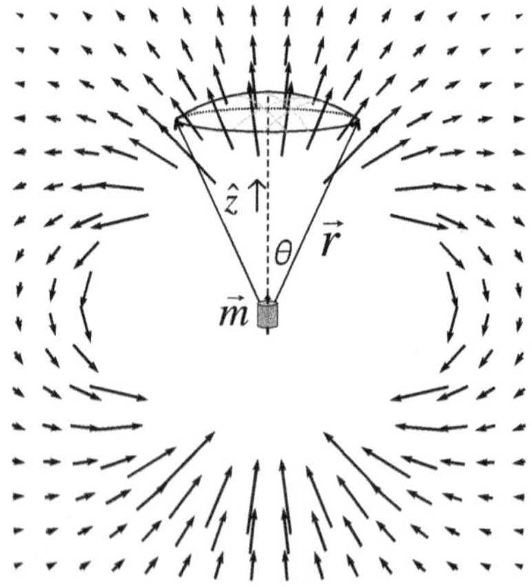

The surface does not need to be flat. In fact, because through any closed surface the total magnetic flux is zero, it can be any shape, so long as the loop is the opening. So we can make the surface dome shaped with constant radius, and render the magnetic field into spherical coordinates, so $\hat{z}\cdot\hat{r} = \cos(\theta)$ and $\hat{z} = \cos(\theta)\,\hat{r} - \sin(\theta)\,\hat{\theta}$, therefore:

$$\vec{B}(\vec{r}) = \left(\frac{\mu_0 m}{4\pi}\right)\frac{3(\hat{z}\cdot\hat{r})\hat{r}-\hat{z}}{r^3} = \left(\frac{\mu_0 m}{4\pi r^3}\right)\left(2\cos(\theta)\hat{r}+\sin(\theta)\hat{\theta}\right).$$

The flux through the cap of this surface is:

$$\Phi_B = \int_{\text{surface}} \vec{B}\cdot \mathrm{d}\vec{A} = \int_0^{\theta}\left(\left(\frac{\mu_0 m}{4\pi r^3}\right)\left(2\cos(\theta)\hat{r}+\sin(\theta)\hat{\theta}\right)\right)\cdot\left(2\pi s r \mathrm{d}\theta(\hat{r})\right) = \left(\frac{\mu_0 m}{2r^3}\right)\int_0^{\theta}\left(2r^2\cos(\theta)\sin(\theta)\right)\mathrm{d}\theta$$

$$= \left(-\frac{\mu_0 m}{4r}\right)\int_0^{\theta} 2\left(-\sin(2\theta)\right)\mathrm{d}\theta = \left(-\frac{\mu_0 m}{4r}\right)\cos(2\theta)\Big|_0^{\theta} = \left(-\frac{\mu_0 m}{4r}\right)\left(\cos(2\theta)-1\right) = \left(\frac{\mu_0 m}{4r}\right)\left(1-\cos(2\theta)\right).$$

In the case of a full sphere this is zero $\left(1-\cos\left(2\times 180^{\circ}\right)=0\right)$, thus complying with Peregrinus's principle. Now, applying the definition of the vector potential yields:

$$\oint_{\text{path}} \mathbb{A}\cdot \mathrm{d}\vec{\ell} = \left(\mathbb{A}\cdot\hat{\varphi}\right)\left(2\pi r\sin(\theta)\right).$$

Given the axial symmetry in the problem, we are assuming that the vector potential will be in the azimuthal direction. Putting these together:

$$\Phi_B = \left(\frac{\mu_0 m}{4r}\right)\left(1-\cos(2\theta)\right) = \left(\frac{\mu_0 m}{4r}\right)\left(2\sin^2(\theta)\right), \text{ so:}$$

$$\vec{\mathbb{A}} = \left(\frac{\mu_0 m\sin(\theta)}{4\pi r^2}\right)\hat{\varphi} = \vec{\mathbb{A}} = \left(\frac{\mu_0 m}{4\pi r^2}\right)\left(\hat{z}\times\hat{r}\right) = \frac{\mu_0 \vec{m}\times\hat{r}}{4\pi r^2}.$$

This is particularly important because by integrating this result, we can find the vector potential, and in turn the magnetic field, surrounding permanent magnets.

Problem 7.2: Odd Shaped Surfaces

For any loop, there are an infinite number of open surfaces with that as a perimeter. Consider, for example, blowing a soap bubble through a circular wand. Until it detaches, all these surfaces have the same perimeter, but the surface area keeps growing as you blow.

a) Show that the magnetic flux through all surfaces, whether curved or flat, bounded by the same perimeter, are equal.

b) Is this also, necessarily, true of electric flux? Explain why or why not.

Thought Experiment 7.2: The Vector Potential Everywhere

We defined the vector potential using the magnetic flux, where:

$$\Phi_m \equiv \int_{\text{surface}} \vec{B}\cdot \mathrm{d}\vec{A} = \oint_{\text{path}} \mathbb{A}\cdot \mathrm{d}\vec{\ell}.$$

Consider a circular Amperian loop, with surface area $\vec{A} = \pi a^2 \hat{B}$ parallel to the magnetic field, so the magnetic flux simplifies to:

$$\Phi_m \equiv \int_{\text{surface}} \vec{B}\cdot \mathrm{d}\vec{A} = \int_{\text{surface}} \vec{B}\cdot\hat{n}\,\mathrm{d}A = \oint_{\text{path}} \mathbb{A}\cdot \mathrm{d}\vec{\ell}.$$

If we now take the limit as the area goes to zero, the magnetic field strength becomes:

$$\vec{B}\cdot\hat{n} = \lim_{\delta A\to 0}\left(\frac{1}{\delta A}\oint_{\text{path}} \mathbb{A}\cdot \mathrm{d}\vec{\ell}\right).$$

However, there is a catch! What if we don't know the magnetic field's direction in advance?

The solution is to evaluate this expression along loops around three orthogonal axes (rather than only one), with each axis aligned with each coordinate. In the case of a Cartesian coordinate system, we would choose $\hat{n}_1 = \hat{x}$, $\hat{n}_2 = \hat{y}$, and $\hat{n}_3 = \hat{z}$ in succession. This way each component of the magnetic field can be calculated.

This vector derivative is called the *curl*, and the relationship above can be written as:

$$\vec{B} = \vec{\nabla} \times \vec{\mathbb{A}}.$$

Discussion 7.1: Peregrinus's Principle and the Vector Potential

Vector second derivatives are very important in electrodynamics, as there are many different ways to take more than one derivative. There are two second derivatives that are always mathematically equal to zero, regardless of the physical meaning of the functions in question. These are:

$$\vec{\nabla} \cdot \left(\vec{\nabla} \times \vec{\mathbb{A}}\right) = 0 \quad \text{and} \quad \vec{\nabla} \times \left(\vec{\nabla} \mathbb{V}\right) = 0.$$

Thus, according to Peregrinus's principle, and the definition of the vector potential, we can write:

$$\vec{\nabla} \cdot \vec{B} = \vec{\nabla} \cdot \left(\vec{\nabla} \times \vec{\mathbb{A}}\right) = 0,$$

by mathematical definition. Therefore the simple act of using the vector potential guarantees that the magnetic field is divergence free.

The same thing can be done for any divergence free vector field.

For example, consider substituting Gauss's law into the continuity equation:

$$\vec{\nabla} \cdot \vec{J} = -\tfrac{\partial}{\partial t} \rho = -\tfrac{\partial}{\partial t}\left(\varepsilon_0 \vec{\nabla} \cdot \vec{E} + \vec{\nabla} \cdot \vec{P}\right) = -\vec{\nabla} \cdot \left(\tfrac{\partial}{\partial t}\left(\varepsilon_0 \vec{E} + \vec{P}\right)\right),$$

so we can write the following statement:

$$\vec{\nabla} \cdot \vec{J} + \vec{\nabla} \cdot \left(\tfrac{\partial}{\partial t}\left(\varepsilon_0 \vec{E} + \vec{P}\right)\right) = \vec{\nabla} \cdot \left(\vec{J} + \tfrac{\partial}{\partial t}\left(\varepsilon_0 \vec{E} + \vec{P}\right)\right) = 0.$$

Similarly, in this case too we can define another vector field, call it $\vec{H}$, such that:

$$\vec{\nabla} \times \vec{H} = \vec{J} + \tfrac{\partial}{\partial t}\left(\varepsilon_0 \vec{E} + \vec{P}\right).$$

Notice the contrast between an operational definition (p. 195) and a mathematical definition. When something is defined operationally, it means that you know how to measure it. You may, or may not, know what causes it or why it may be important, but you can measure it *in situ*.

On the other hand, when something is defined mathematically, it is known only as well as the operationally defined quantities on which it depends. So, in our example here, $\vec{\mathbb{A}}$ is only defined if $\vec{B}$ is. Similarly $\vec{H}$ would only has physical meaning if $\vec{J}$, $\vec{E}$, and $\vec{P}$ do. Mathematically defined quantities sometimes turn out to have physical content in addition to mathematical convenience, but whether they do, or not, depends on the underlying physical theory.

7.2 *The Curl in Curvilinear Coordinates*

By now you have become accustomed to using the divergence to take spatial derivatives of vector fields. In the first chapter we defined the divergence geometrically in order to ensure that a conserved quantity, such as charge, is continuously conserved. Thus, they are conserved locally at every point in space.

Later we applied the same mathematics to the gravitational force per mass, $\vec{g}$, surrounding a spherical object. This allowed us to write Newton's law of gravity in continuous form as $\vec{\nabla}\cdot\vec{g} = -4\pi G\rho_m$. Since Coulomb's law follows the same mathematical form as Newton's, we then found the electric field, in free space, to be $\vec{\nabla}\cdot\vec{E} = \frac{1}{\varepsilon_0}\rho$. And, finally, since there are no magnetic monopoles, we wrote Peregrinus's principle as $\vec{\nabla}\cdot\vec{B} = 0$.

Now, the divergence represents the spatial derivative in the same direction as the vector field itself. Thus, if a vector field changes in the perpendicular direction, it will not affect the divergence. But what if we are interested in the perpendicular gradients to a field? For example, how much faster does water flow in the middle, as opposed to the bank, of a river? Do winds blow faster at high or low altitude? What makes it unsafe for an airplane to takeoff? To address these questions, and many like them, we need to be able to take a perpendicular derivative of a vector field. This is called *the curl.*

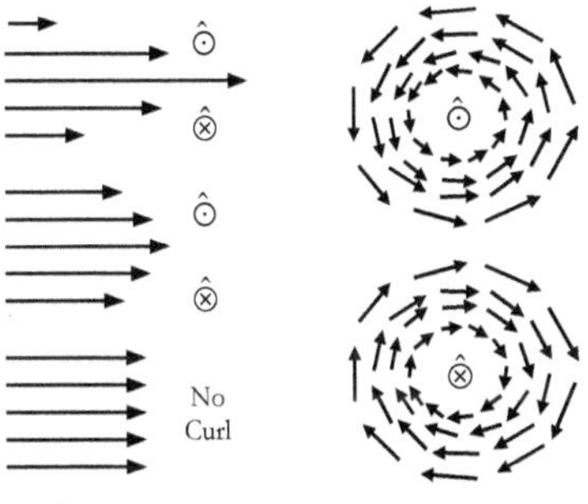

In the figure, the top-left vector field has the most curl, while the bottom-left has none. Like the area vector, the direction of the curl is always normal to the relevant plane. If, as in the figure, the $\hat{x}$ direction is in the direction of the vector, and the magnitude of the vectors are decreasing the $\hat{y}$ direction, then the curl will be in the $\hat{z}$ direction. Similarly, if the vector increases the $\hat{y}$ then the curl will be in the $-\hat{z}$ direction.

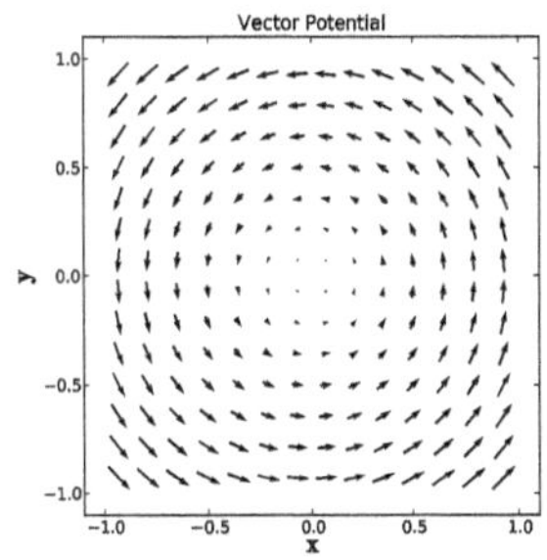

This sign convention seems odd when looking at a bunch of vectors all in the same direction. On the other hand if, say, the quantity were circular, as in the right figures, the sign convention follows the right hand rule as we would expect. For example, in Problem 7.1 (p. 213), you found the vector potential, $\vec{\mathbb{A}}$, of a uniform magnetic field, $\vec{B} = B_0\,\hat{z}$, to be $\vec{\mathbb{A}} = \frac{1}{2}s\,B_0\,\hat{\varphi}$, which we have plotted here. Since $\vec{B} = \vec{\nabla}\times\vec{\mathbb{A}}$, we know that $\vec{\nabla}\times\left(\frac{1}{2}s\,B_0\,\hat{\varphi}\right) = B_0\,\hat{z}$. Notice that simply taking the derivative with respect to s would not give the correct result. Why? Because $\hat{\varphi}$, itself, is not curl free.

In this section, we will derive, geometrically, the curl in curvilinear coordinates. As with the divergence, it is particularly important to keep in mind the geometrical definition of the curl so you always know when it is applicable and you can check your work.

Problem 7.3: Azimuthal Vectors and the Curl

a) Use one of your results from Problem 7.1 (p. 213) to show that $\vec{\nabla}\times\left(s\,\hat{\varphi}\right) = 2\,\hat{z}$.

b) Use another one your results to show that $\vec{\nabla}\times\left(\frac{1}{s}\,\hat{\varphi}\right) = 0$.

c) In the second case the vector field is clearly rotating counter-clockwise around the $\hat{z}$ axis, yet is curl free. Discuss why that can be?

d) In what direction is $\vec{\nabla}\times\left(\frac{1}{s^2}\hat{\varphi}\right)$? Why?

e) Some students are tying to remember the functional form of $\vec{\nabla}\times\hat{\varphi}$, one person guesses $-\frac{\hat{z}}{s}$, while someone else suggests $s\hat{z}$. How do you know that they are both wrong? Without calculating it, what would you guess its functional form to be? Why?

Thought Experiment 7.3: Conservative Forces and the Curl of the Force

Consider the work done, for example, during a roller coaster ride. The work done to lift the train is counterbalanced by the work done to stop the train, and the total work done by gravity is zero because ride ends where it begins. We can therefore write the following integral relationship:

$$\begin{aligned} W &= \oint_{\text{path}} \vec{F}\cdot d\vec{\ell} = \oint_{\text{path}} \left(\vec{F}_{\text{push on train}} + \vec{F}_{\text{gravity on train}} + \vec{F}_{\text{friction on train}}\right)\cdot d\vec{\ell} \\ &= \oint_{\text{path}} \left(\vec{F}_{\text{push on train}}\right)\cdot d\vec{\ell} + \oint_{\text{path}} \left(\vec{F}_{\text{gravity on train}}\right)\cdot d\vec{\ell} + \oint_{\text{path}} \left(\vec{F}_{\text{friction on train}}\right)\cdot d\vec{\ell} \\ &= W_{\text{push on train}} + W_{\text{gravity on train}} + W_{\text{friction on train}} \,. \end{aligned}$$

Since the train starts and stops with the same energy, the total work around the closed loop must be zero. Similarly, since the net work done by gravity only depends on the difference in height, it must also be zero because the train starts and stops at the same place. Thus:

$$W_{\text{push on train}} > 0\,,\quad W_{\text{gravity on train}} = 0\,,\quad W_{\text{friction on train}} < 0\,.$$

In the case of gravity, or any other *conservative force*, the path integral around any closed loop must be zero—including loops that are infinitely small. The curl calculates the path integral, per area, around not only one tiny loop per point, but rather three orthogonal tiny loops per point—one per dimension. For this reason, the curl returns a vector which is mutually perpendicular to the vector field and its perpendicular rate of change. For two dimensional systems, the curl always has a single component that points in the third direction.

The work done on an object, as we have discussed, is the path integral of the force taken between the initial and final position. We have emphasized that to define the potential energy of an object it must not matter which path we choose to integrate over. We now use this criterion to derive a test for whether a force law is conservative or not.

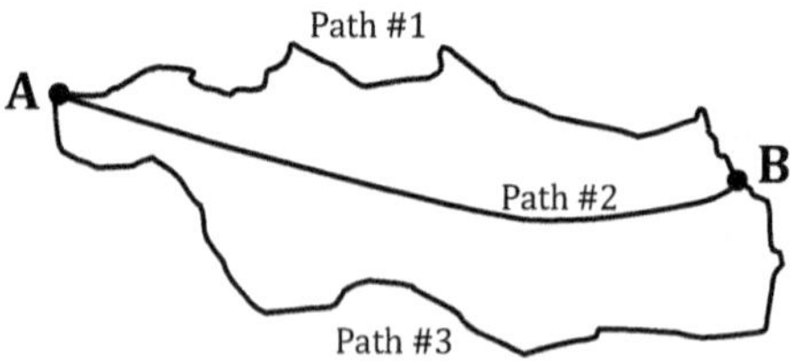

Consider a force that is a function of position, $\vec{F}(\vec{r})$. It is a conservative force if the work it does on an object that travels from point A to point B is the same regardless of which path the object takes to get from A to B. If this is so, the work is said to be *path independent.* Alternatively:

$$W_{AB\,\text{Path\#1}} = \int_{A\,\text{Path \#1}}^{B} \vec{F}\cdot d\vec{r} = W_{AB\,\text{Path\#2}} = \int_{A\,\text{Path \#2}}^{B} \vec{F}\cdot d\vec{r} = W_{AB\,\text{Path \#3}} = \int_{A\,\text{Path \#3}}^{B} \vec{F}\cdot d\vec{r}\,.$$

In other words, the work done around any closed path is zero for a conservative force:

$$0 = \underset{\text{Path\#1}}{W_{AB}} - \underset{\text{Path\#2}}{W_{AB}} = \underset{\text{Path\#1}}{W_{AB}} + \underset{\text{Path\#2}}{W_{BA}} = \underset{\text{Path\#1}}{\int_A^B \vec{F}\cdot \mathrm{d}\vec{r}} + \underset{\text{Path\#2}}{\int_B^A \vec{F}\cdot \mathrm{d}\vec{r}} = \oint_{\text{path}} \vec{F}\cdot \mathrm{d}\vec{r}\,,$$

Thus, if $\oint_{\text{any path}} \vec{F}\cdot \mathrm{d}\vec{r} = 0$ around any closed loop, then $\vec{F}$ is a conservative force.

Checking every path seems like a tall order, but there is a relatively simple test, similar to taking a derivative, which we can apply to determine whether or not a particular vector function is a suitable candidate to be a conservative force. In general, the curl is a differential vector operation that measures the change in a vector field *perpendicular* to the direction of the vector. The curl is defined in the following way: the component of the curl of the vector $\vec{F}(\vec{r})$, in the direction defined by the unit vector $\hat{n}$, is defined as the limiting value of the (closed) path integral of $\vec{F}(\vec{r})$ in the plane perpendicular to $\hat{n}$, divided by the area the path encloses:

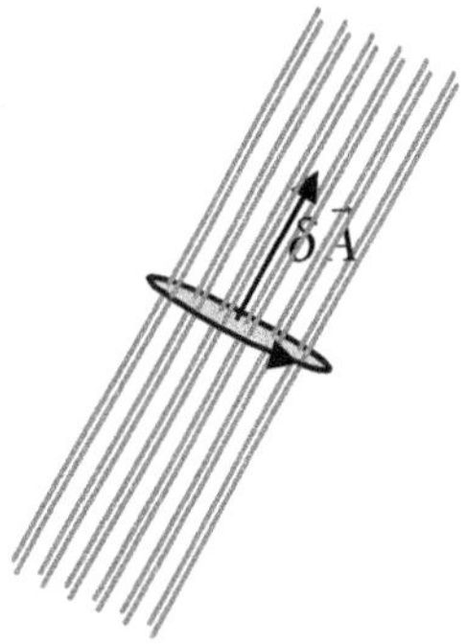

$$\left(\vec{\nabla}\times\vec{F}(\vec{r})\right)_n \equiv \lim_{\delta A_n\to 0}\left(\frac{\oint_{\text{path}} \vec{F}(\vec{r})\cdot(\mathrm{d}\vec{r})}{\delta A_n}\right)$$

A vector field is path independent if, and only if, the curl of the vector field is zero everywhere.

Definition 7.1: The Curl

The curl of a vector field is also a vector field, and it is defined such that for any set of three orthogonal unit vectors, each component of the curl of $\vec{F}(\vec{r})$, is given by:

$$\left(\vec{\nabla}\times\vec{F}(\vec{r})\right)_n \equiv \lim_{\delta A_n\to 0}\left(\frac{1}{\delta A_n}\oint_{\text{path}} \vec{F}(\vec{r})\cdot \mathrm{d}\vec{r}\right).$$

The path of integration is the right-handed perimeter of a two dimensional surface, normal to the relevant unit vector, in the limit that the area becomes infinitesimal.

Thought Experiment 7.4: The Curl on a 3x3 Numerical Grid

The curl for two-dimensional flow measures the rotation of a vector about a particular axis on the smallest scale. For a velocity field in a fluid, this is called the *vorticity*, and it is a measure of turbulence. We define the curl normal to a plane as:

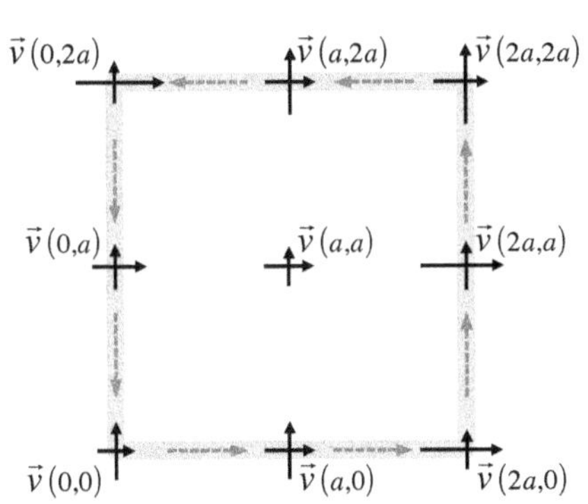

$$\left(\vec{\nabla}\times\vec{v}\right)_\perp \equiv \lim_{\delta A\to 0}\left(\frac{\oint_{\text{path}} \vec{v}\cdot \mathrm{d}\vec{r}}{\delta A}\right),$$

where the positive direction of integration is counter-clockwise. In a right-handed Cartesian coordinate system, this can be approximated as:

$$\left(\vec{\nabla}\times\vec{v}\right)_z(a,a)\approx\frac{1}{4a^2}\left(\int_0^{2a}v_x\Big|_{y=0}\mathrm{d}x+\int_0^{2a}v_y\Big|_{x=2a}\mathrm{d}y+\int_{2a}^{0}v_x\Big|_{y=2a}\mathrm{d}x+\int_{2a}^{0}v_y\Big|_{x=0}\mathrm{d}y\right),$$

because the area enclosed by the loop is $4a^2$. By reversing the limits of the last two integrals, this can also be written:

$$\left(\vec{\nabla}\times\vec{v}\right)_z(a,a)\approx\frac{1}{4a^2}\left(\int_0^{2a}v_x\Big|_{y=0}\mathrm{d}x-\int_0^{2a}v_x\Big|_{y=2a}\mathrm{d}x-\int_0^{2a}v_y\Big|_{x=0}\mathrm{d}y+\int_0^{2a}v_y\Big|_{x=2a}\mathrm{d}y\right).$$

We use Simpson's rule, which is a parabolic interpolation, to solve for the integrals. For example:

$$\int_0^{2a}v(x,y)\,\mathrm{d}x\approx\frac{a}{3}\left(v_0+4v_1+v_2\right),$$

where $v_0=v(0,y)$, $v_1=v(a,y)$, and $v_0=v(2a,y)$.

We therefore have the following results:

$$\begin{aligned}\left(\vec{\nabla}\times\vec{v}\right)_z(a,a)\approx&\frac{1}{12a}\left(v_x(0,0)+4v_x(a,0)+v_x(2a,0)\right)-\frac{1}{12a}\left(v_x(0,2a)+4v_x(a,2a)+v_x(2a,2a)\right)\\&-\frac{1}{12a}\left(v_y(0,0)+4v_y(0,a)+v_y(0,2a)\right)+\frac{1}{12a}\left(v_y(2a,0)+4v_y(2a,a)+v_y(2a,2a)\right),\end{aligned}$$

and rearranging we obtain:

$$\begin{aligned}\left(\vec{\nabla}\times\vec{v}\right)_z(a,a)\approx&\frac{1}{3a}\left(v_x(a,0)-v_x(a,2a)-v_y(0,a)+v_y(2a,a)\right)\\&+\frac{1}{12a}\left(v_x(0,0)-v_x(0,2a)+v_x(2a,0)-v_x(2a,2a)\right)\\&+\frac{1}{12a}\left(-v_y(0,0)-v_y(0,2a)+v_y(2a,0)+v_y(2a,2a)\right).\end{aligned}$$

This measures the rotation of the flow.

Derivation 7.1: The Curl in Cartesian Coordinates

We begin this thought experiment by first considering a point of interest, given by $\vec{r}(x,y,z)=x\hat{x}+y\hat{y}+z\hat{z}$, and a vector field given by $\vec{F}(\vec{r})=\vec{F}(x,y,z)=F_x\hat{x}+F_y\hat{y}+F_z\hat{z}$. Our goal is to derive an convenient differential expression in Cartesian coordinates for the curl of $\vec{F}(\vec{r})$ using the geometric definition of curl:

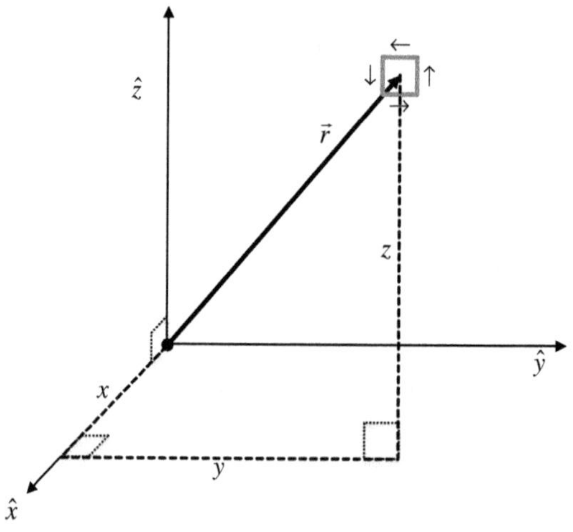

$$\left(\vec{\nabla}\times\vec{F}(\vec{r})\right)_n\equiv\lim_{\delta A_n\to 0}\left(\frac{1}{\delta A_n}\oint_{\text{path}}\vec{F}(\vec{r})\cdot\mathrm{d}\vec{r}\right).$$

To do this, we will consider each direction independently and find an expression for each component of the curl. First consider a loop about the $\hat{x}$ axis, as in the figure, so that $\delta\vec{A} = \delta A_x\,\hat{x}$.

We then find the integral of the vector $\vec{F}$ around the loop, starting at the lower left of the figure, so that:

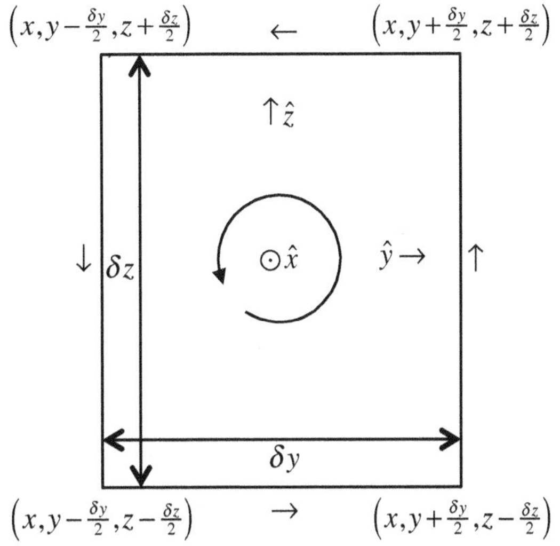

$$\left(\oint_{\text{path}} \vec{F}\cdot d\vec{l}\right)_x = \int_{y-\delta y/2}^{y+\delta y/2} F_y\left(x, y, z-\tfrac{\delta z}{2}\right)dy + \int_{z-\delta z/2}^{z+\delta z/2} F_z\left(x, y+\tfrac{\delta y}{2}, z\right)dz$$
$$+ \int_{y+\delta y/2}^{y-\delta y/2} F_y\left(x, y, z+\tfrac{\delta z}{2}\right)dy + \int_{z+\delta z/2}^{z-\delta z/2} F_z\left(x, y-\tfrac{\delta y}{2}, z\right)dz.$$

Reversing the limits of the lower integrals and combining terms yields:

$$\left(\oint_{\text{path}} \vec{F}\cdot d\vec{l}\right)_x = \int_{y-\delta y/2}^{y+\delta y/2} \left(F_y\left(x, y, z-\tfrac{\delta z}{2}\right) - F_y\left(x, y, z+\tfrac{\delta z}{2}\right)\right)dy$$
$$+ \int_{z-\delta z/2}^{z+\delta z/2} \left(F_z\left(s, y+\tfrac{\delta y}{2}, z\right) - F_z\left(s, y-\tfrac{\delta y}{2}, z\right)\right)dz$$

Now we apply the definition of the derivative to the small loop:

$$\left(\oint_{\text{path}} \vec{F}\cdot d\vec{l}\right)_x \approx \int_{y-\delta y/2}^{y+\delta y/2} \left(-\frac{\partial F_y}{\partial z}\delta z\right)dy + \int_{z-\delta z/2}^{z+\delta z/2} \left(\frac{\partial F_z}{\partial y}\delta y\right)dz \approx -\frac{\partial F_y}{\partial z}\delta z\,\delta y + \frac{\partial F_z}{\partial y}\delta y\,\delta z.$$

Now we write the area of the loop as:

$\delta A_x = \delta y\,\delta z.$

Now, by the definition of the curl, we can write down its $\hat{x}$ component as:

$$\left(\vec{\nabla}\times\vec{F}(\vec{r})\right)_x \equiv \lim_{\delta A_x\to 0}\left(\frac{\oint_{\text{path}} \vec{F}(\vec{r})\cdot d\vec{r}}{\delta A_x}\right) = \lim_{\delta A_x\to 0} \frac{-\frac{\partial F_y}{\partial z}\cancel{\delta z}\,\cancel{\delta y} + \frac{\partial F_z}{\partial y}\cancel{\delta y}\,\cancel{\delta z}}{\cancel{\delta y}\,\cancel{\delta z}} = -\frac{\partial F_y}{\partial z} + \frac{\partial F_z}{\partial y}.$$

While we could do a simple cyclic permutation to find the curl in the other two orthogonal directions, we cannot stress too much the importance of the geometrical definition of the curl. We will therefore proceed with finding the curl about the other two axes using the geometric definition.

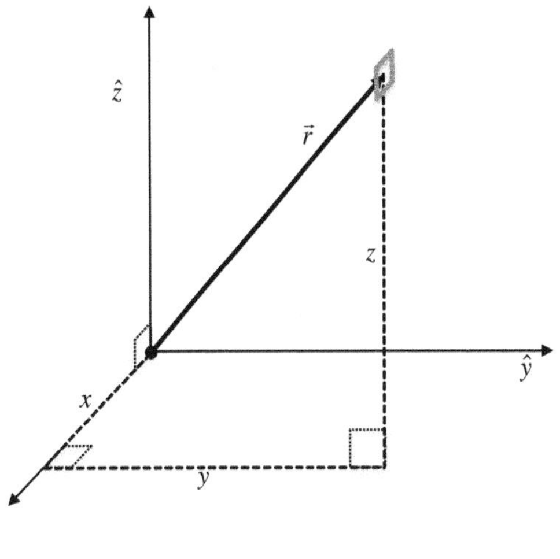

To find the curl about the $\hat{y}$ direction, we must consider a small loop in the $\hat{z}\times\hat{x}$ plane. Now we can follow the same method we used for the $\hat{x}$ component of the curl, but making a cyclic substitution in axis labels:

$$\left(\oint_{\text{path}} \vec{F}\cdot \mathrm{d}\vec{l}\right)_y = \int_{z-\delta z/2}^{z+\delta z/2} F_z\left(x-\tfrac{\delta y}{2}, y, z\right)\mathrm{d}z + \int_{x-\delta x/2}^{x+\delta x/2} F_x\left(x, y, z+\tfrac{\delta z}{2}\right)\mathrm{d}x$$

$$+\int_{z+\delta z/2}^{z-\delta z/2} F_z\left(x-\tfrac{\delta y}{2}, y, z\right)\mathrm{d}z + \int_{x+\delta x/2}^{x-\delta x/2} F_x\left(x, y, z-\tfrac{\delta z}{2}\right)\mathrm{d}x$$

$$\left(\oint_{\text{path}} \vec{F}\cdot \mathrm{d}\vec{l}\right)_y \approx -\frac{\partial F_z}{\partial x}\delta x\,\delta z + \frac{\partial F_x}{\partial z}\delta z\,\delta x.$$

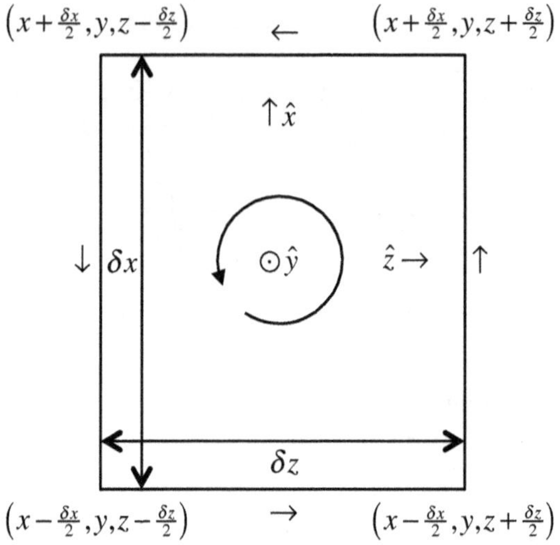

We now write the area of the loop as:

$$\delta A_y = \delta z\, \delta x$$

So the $\hat{y}$ component of the curl becomes:

$$\left(\vec{\nabla}\times\vec{F}(\vec{r})\right)_y \equiv \lim_{\delta A_y \to 0}\left(\frac{\oint_{\text{path}} \vec{F}(\vec{r})\cdot \mathrm{d}\vec{r}}{\delta A_y}\right) = -\frac{\partial F_z}{\partial x} + \frac{\partial F_x}{\partial z}.$$

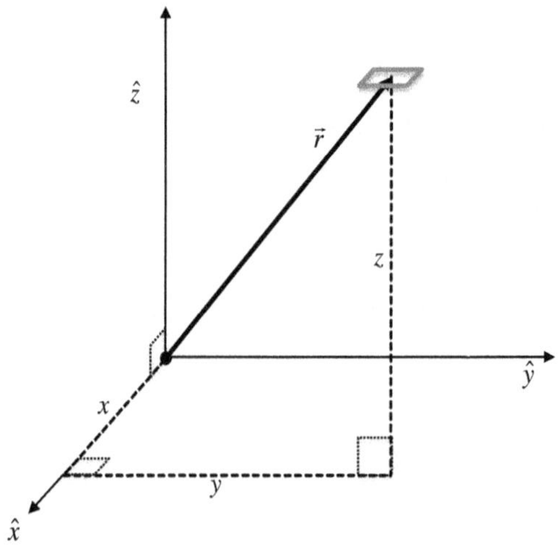

Finally, we will find the $\hat{z}$ component of the curl in a similar manner as:

$$\left(\oint_{\text{path}} \vec{F}\cdot \mathrm{d}\vec{l}\right)_z = \int_{x-\delta x/2}^{x+\delta x/2} F_x\left(x, y-\tfrac{\delta y}{2}, z\right)\mathrm{d}x + \int_{y-\delta x/2}^{y+\delta y/2} F_y\left(x+\tfrac{\delta x}{2}, y, z\right)\mathrm{d}y$$

$$+\int_{z+\delta z/2}^{z-\delta z/2} F_x\left(x, y+\tfrac{\delta y}{2}, z\right)\mathrm{d}x + \int_{x+\delta x/2}^{x-\delta y/2} F_y\left(x-\tfrac{\delta x}{2}, y, z\right)\mathrm{d}y,$$

Again, this implies:

$$\left(\oint_{\text{path}} \vec{F}\cdot \mathrm{d}\vec{l}\right)_z \approx -\frac{\partial F_x}{\partial y}\delta y\,\delta x + \frac{\partial F_y}{\partial x}\delta y\,\delta x.$$

Here, we write the area of the loop as:

$$\delta A_z = \delta x\, \delta y$$

and the $\hat{z}$ component of the curl is therefore:

$$\left(\vec{\nabla}\times\vec{F}(\vec{r})\right)_z \equiv \lim_{\delta A_z \to 0}\left(\frac{\oint_{\text{path}} \vec{F}(\vec{r})\cdot \mathrm{d}\vec{r}}{\delta A_z}\right) = -\frac{\partial F_x}{\partial y} + \frac{\partial F_y}{\partial x}.$$

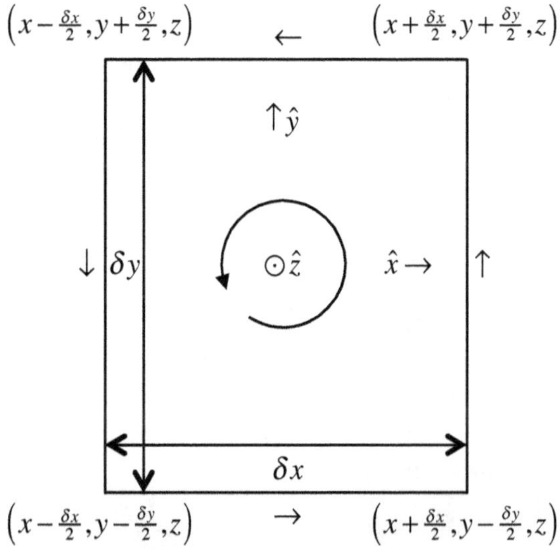

We can now put all three components together and write the curl as:

$$\vec{\nabla}\times\vec{F}(x,y,z)=\left(\frac{\partial F_z}{\partial y}-\frac{\partial F_y}{\partial z}\right)\hat{x}+\left(\frac{\partial F_x}{\partial z}-\frac{\partial F_z}{\partial x}\right)\hat{y}+\left(\frac{\partial F_y}{\partial x}-\frac{\partial F_x}{\partial y}\right)\hat{z}\,.$$

Notice that in Cartesian coordinates you can simply take the cross-product of the vector derivative operator and obtain the correct result for the curl.

$$\vec{\nabla}\times\vec{F}=\left(\hat{x}\frac{\partial}{\partial x}+\hat{y}\frac{\partial}{\partial y}+\hat{z}\frac{\partial}{\partial z}\right)\times\left(F_x\hat{x}+F_y\hat{y}+F_z\hat{z}\right)=\vec{\nabla}\times\vec{F}\,.$$

As with the divergence and the gradient, the curl in cylindrical and spherical coordinates are both more complicated than it is in Cartesian coordinates. Thus it is especially important to always keep in mind the geometrical definition of the curl, rather than this algebraic expression.

Problem 7.4: The 1854 Smiths Exam Problem 8

Solve the exam question mentioned on page 212, which stated:

> 8. If X,Y,Z be functions of the rectangular co-ordinates x,y,z, $\mathrm{d}S$ an element of any limited surface, l,m,n the cosines of the inclinations of the normal at $\mathrm{d}S$ to the axes, $\mathrm{d}s$ an element of the bounding line, shew that
>
> $$\iint\left\{l\left(\frac{\mathrm{d}Z}{\mathrm{d}y}-\frac{\mathrm{d}Y}{\mathrm{d}x}\right)+m\left(\frac{\mathrm{d}X}{\mathrm{d}z}-\frac{\mathrm{d}Z}{\mathrm{d}x}\right)+n\left(\frac{\mathrm{d}Y}{\mathrm{d}x}-\frac{\mathrm{d}X}{\mathrm{d}y}\right)\right\}\mathrm{d}S=\int\left(X\frac{\mathrm{d}x}{\mathrm{d}s}+Y\frac{\mathrm{d}y}{\mathrm{d}s}+Z\frac{\mathrm{d}z}{\mathrm{d}s}\right)\mathrm{d}s,$$
>
> the differential coefficients of X,Y,Z being partial, and the single integral being taken all round the perimeter of the surface.

Derivation 7.2: The Curl in Cylindrical Coordinates

We begin this derivation by considering the position vector rendered in cylindrical coordinates, so that $\vec{r}=\vec{r}(s,\varphi,z)=s\,\hat{s}+z\,\hat{z}$. As in the previous thought experiment, we will find the curl by separately applying the geometric definition of the curl to each orthogonal component, $\hat{s},\hat{\varphi},\hat{z}$.

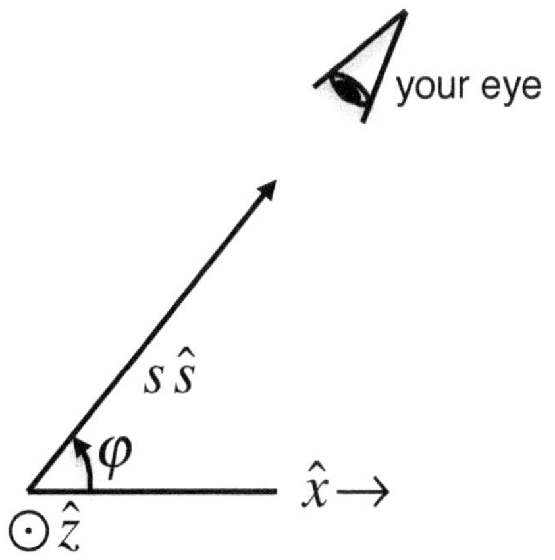

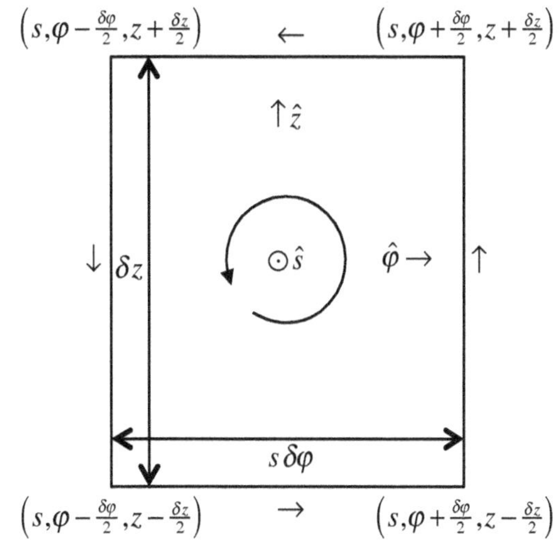

If we first observe this along the $\hat{s}$ direction, as shown in the figure, we need to investigate a loop surrounding the position $\vec{r}$ in the $\hat{\varphi}\times\hat{z}$ plane.

The path integral of $\vec{F}(\vec{r})$, starting at the lower left of the figure, is:

$$\left(\oint_{\text{path}}\vec{F}\cdot\mathrm{d}\vec{l}\right)_s = \int_{\varphi-\delta\varphi/2}^{\varphi+\delta\varphi/2} F_\varphi\left(s,\varphi,z+\delta z/2\right)\left(s\,\mathrm{d}\varphi\right)+\int_{z-\delta z/2}^{z+\delta z/2} F_z\left(s,\varphi+\delta\varphi/2,z\right)\mathrm{d}z+\int_{\varphi+\delta\varphi/2}^{\varphi-\delta\varphi/2} F_\varphi\left(s,\varphi,z-\delta z/2\right)\left(s\,\mathrm{d}\varphi\right)+\int_{z+\delta z/2}^{z-\delta z/2} F_z\left(s,\varphi-\delta\varphi/2,z\right)\mathrm{d}z.$$

Reversing the limits of the lower integrals yields:

$$\left(\oint_{\text{path}}\vec{F}\cdot\mathrm{d}\vec{l}\right)_s = s\int_{\varphi-\delta\varphi/2}^{\varphi+\delta\varphi/2}\left(F_\varphi\left(s,\varphi,z+\delta z/2\right)-F_\varphi\left(s,\varphi,z-\delta z/2\right)\right)\mathrm{d}\varphi+\int_{z-\delta z/2}^{z+\delta z/2}\left(F_z\left(s,\varphi+\delta\varphi/2,z\right)-F_z\left(s,\varphi-\delta\varphi/2,z\right)\right)\mathrm{d}z.$$

Now we apply the definition of the partial derivative to the loop:

$$\left(\oint_{\text{path}}\vec{F}\cdot\mathrm{d}\vec{l}\right)_s \approx s\int_{\varphi-\delta\varphi/2}^{\varphi+\delta\varphi/2}\left(-\frac{\partial F_\varphi}{\partial z}\delta z\right)\mathrm{d}\varphi+\int_{z-\delta z/2}^{z+\delta z/2}\left(\frac{\partial F_z}{\partial\varphi}\delta\varphi\right)\mathrm{d}z\approx -s\frac{\partial F_\varphi}{\partial z}\delta z\,\delta\varphi+\frac{\partial F_z}{\partial\varphi}\delta\varphi\,\delta z.$$

Notice that the area of the loop is given by: $\delta A_s = s\,\delta\varphi\,\delta z$, so we now apply the definition of the curl around the $\hat{s}$ direction as:

$$\left(\vec{\nabla}\times\vec{F}(\vec{r})\right)_s = \lim_{\delta A_s\to 0}\frac{-s\frac{\partial E_\varphi}{\partial z}\cancel{\delta z}\,\cancel{\delta\varphi}+\frac{\partial E_z}{\partial\varphi}\cancel{\delta\varphi}\,\cancel{\delta z}}{s\,\cancel{\delta\varphi}\,\cancel{\delta z}} = -\frac{\partial E_\varphi}{\partial z}+\frac{1}{s}\frac{\partial E_z}{\partial\varphi}.$$

To find the $\hat{\varphi}$ component of the curl, we observe the position of interest from the $\hat{\varphi}$ direction, and investigate the path integral about $\hat{\varphi}$:

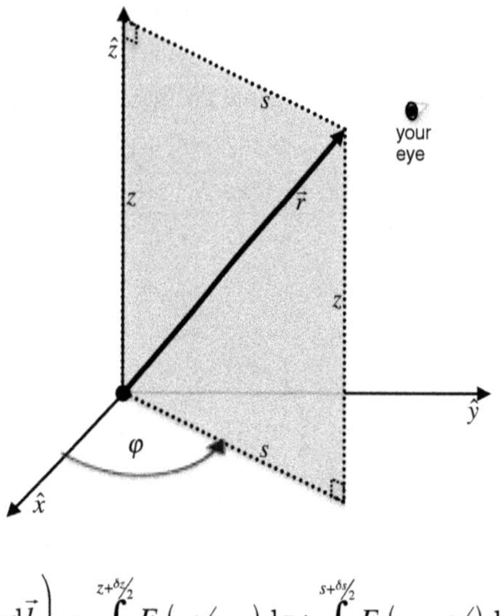

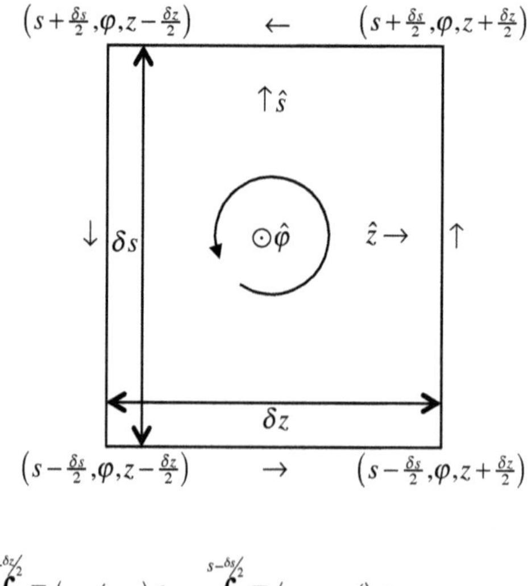

$$\left(\oint_{\text{path}}\vec{F}\cdot\mathrm{d}\vec{l}\right)_\varphi = \int_{z-\delta z/2}^{z+\delta z/2} F_z\left(s-\delta s/2,\varphi,z\right)\mathrm{d}z+\int_{s-\delta s/2}^{s+\delta s/2} F_s\left(s,\varphi,z+\delta z/2\right)\mathrm{d}s+\int_{z+\delta z/2}^{z-\delta z/2} F_z\left(s+\delta s/2,\varphi,z\right)\mathrm{d}z+\int_{s+\delta s/2}^{s-\delta s/2} F_s\left(s,\varphi,z-\delta z/2\right)\mathrm{d}s.$$

Again, reversing the limits of the lower integrals and combining terms yields:

$$\left(\oint_{\text{path}} \vec{F}\cdot\mathrm{d}\vec{l}\right)_{\varphi} = \int_{z-\delta z/2}^{z+\delta z/2}\left(F_z\left(s-\tfrac{\delta s}{2},\varphi,z\right)-F_z\left(s+\tfrac{\delta s}{2},\varphi,z\right)\right)\mathrm{d}z + \int_{s-\delta s/2}^{s+\delta s/2}\left(F_s\left(s,\varphi,z+\tfrac{\delta z}{2}\right)-F_s\left(s,\varphi,z-\tfrac{\delta z}{2}\right)\right)\mathrm{d}s\,.$$

Now we, again, apply the definition of the partial derivative:

$$\left(\oint_{\text{path}} \vec{F}\cdot\mathrm{d}\vec{l}\right)_{\varphi} \approx \int_{z-\delta z/2}^{z+\delta z/2}\left(-\frac{\partial F_z}{\partial s}\delta s\right)\mathrm{d}z + \int_{s-\delta s/2}^{s+\delta s/2}\left(\frac{\partial F_s}{\partial z}\delta z\right)\mathrm{d}s \approx -\frac{\partial F_z}{\partial s}\delta s\,\delta z + \frac{\partial F_s}{\partial z}\delta z\,\delta s\,.$$

Now the area of this loop is given by:

$$\delta A_{\varphi} = \delta z\,\delta s\,.$$

The curl about the $\hat{\varphi}$ direction becomes:

$$\left(\vec{\nabla}\times\vec{F}(\vec{r})\right)_{\varphi} = \lim_{\delta A_{\varphi}\to 0}\frac{-\frac{\partial E_z}{\partial s}\cancel{\delta s}\,\cancel{\delta z}+\frac{\partial E_s}{\partial z}\cancel{\delta z}\,\cancel{\delta s}}{\cancel{\delta z}\,\cancel{\delta s}} = -\frac{\partial E_z}{\partial s}+\frac{\partial E_s}{\partial z}$$

Now we observe the system along the $\hat{z}$ direction. Then for some point $\vec{r}$ we imagine a wedge-shaped loop of radial length δs and arc length $s\,\delta\varphi$ surrounding that point, shown in the figure. Notice how the angular dimension has curved sides, so its length will depend on the radial component.

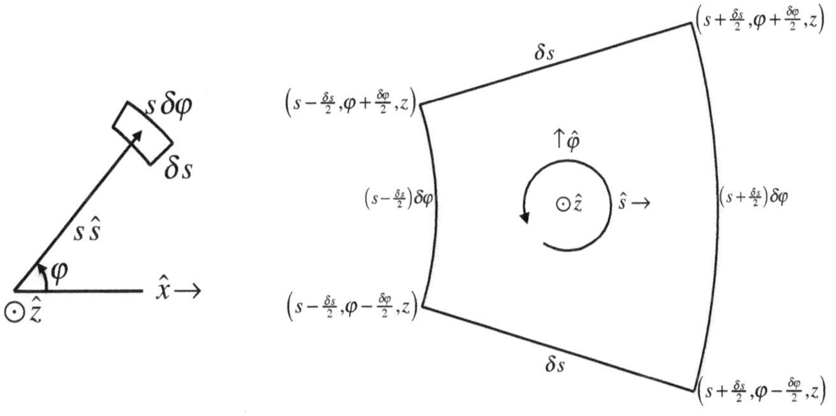

The path integral counter-clockwise around the loop is:

$$\left(\oint_{path}\vec{F}\cdot\mathrm{d}\vec{\ell}\right)_{z} = \int_{s-\delta s/2}^{s+\delta s/2} F_s\left(s,\varphi-\frac{\delta\varphi}{2}\delta\varphi,z\right)\mathrm{d}s + \int_{\varphi-\frac{1}{2}\delta\varphi}^{\varphi+\frac{1}{2}\delta\varphi} F_{\varphi}\left(s+\delta s/2,\varphi,z\right)\left(s+\delta s/2\right)\mathrm{d}\varphi$$
$$+\int_{s+1/2}^{s-1/2} F_s\left(s,\varphi+\frac{\delta\varphi}{2},z\right)\mathrm{d}s + \int_{\varphi+\frac{1}{2}\delta\varphi}^{\varphi-\frac{1}{2}\delta\varphi} F_{\varphi}\left(s-\delta s/2,\varphi,z\right)\left(s-\delta s/2\right)\mathrm{d}\varphi\,.$$

This simplifies to:

$$\left(\oint_{path}\vec{F}\cdot\mathrm{d}\vec{\ell}\right)_{z} \approx \int_{s-\delta s/2}^{s+\delta s/2}\left(-\frac{\partial F_s}{\partial\varphi}\delta\varphi\right)\mathrm{d}s + \int_{\varphi-\frac{1}{2}\delta\varphi}^{\varphi+\frac{1}{2}\delta\varphi}\left(\frac{\partial\left(sF_{\varphi}\right)}{\partial s}\delta s\right)\mathrm{d}\varphi \approx -\frac{\partial F_s}{\partial\varphi}\delta\varphi\,\delta s + \frac{\partial\left(sF_{\varphi}\right)}{\partial s}\delta s\,\delta\varphi$$

Now, let us consider carefully the area associated with the loop. To obtain an expression for the area of the loop, we can subtract the areas of two circles and reduce the result by the ratio of the angle to that of a full circle:

$$A_z = \left(\pi\left(s+{}^{\delta s}\!/_{2}\right)^2 - \pi\left(s-{}^{\delta s}\!/_{2}\right)^2\right)\left(\frac{\delta\varphi}{2\pi}\right) = \left(\cancel{\pi}\,\cancel{2}\,s\,\delta s\right)\left(\frac{\delta\varphi}{\cancel{2}\,\cancel{\pi}}\right) = s\,\delta s\,\delta\varphi .$$

Finally, we again apply the definition of a component of the curl to obtain:

$$\left(\vec{\nabla}\times\vec{F}(\vec{r})\right)_z = \lim_{\delta A_z\to 0}\frac{-\dfrac{\partial E_s}{\partial\varphi}\cancel{\delta\varphi}\,\cancel{\delta s}+\dfrac{\partial\left(sE_\varphi\right)}{\partial s}\cancel{\delta s}\,\cancel{\delta\varphi}}{s\,\cancel{\delta\varphi}\,\cancel{\delta s}} = \frac{-\dfrac{\partial E_s}{\partial\varphi}+\dfrac{\partial\left(sE_\varphi\right)}{\partial s}}{s}$$

We can now put each of these together to find the curl of the vector field $\vec{F}$ expressed in cylindrical coordinates:

$$\vec{\nabla}\times\vec{F}(s,\varphi,z) = \left(\frac{1}{s}\frac{\partial F_z}{\partial\varphi}-\frac{\partial F_\varphi}{\partial z}\right)\hat{s}+\left(\frac{\partial F_s}{\partial z}-\frac{\partial F_z}{\partial s}\right)\hat{\varphi}+\frac{1}{s}\left(\frac{\partial\left(sF_\varphi\right)}{\partial s}-\frac{\partial F_s}{\partial\varphi}\right)\hat{z} .$$

Problem 7.5: Algebraic Derivation of the Cylindrical Curl

Derive the expression for the curl in cylindrical coordinates from the expression of the curl in Cartesian coordinates, and the appropriate coordinate transformations.

Derivation 7.3: The Curl in Spherical Coordinates

To find the curl in spherical coordinates of a vector $\vec{F}(\vec{r})=\vec{F}(r,\theta,\varphi)$, we must consider three perpendicular loops about a position $\vec{r}$.

We first consider the projection along the $\hat{r}$ direction, and a loop surrounding the $\hat{r}$ direction located at the position $\vec{r}$ as in the figure.

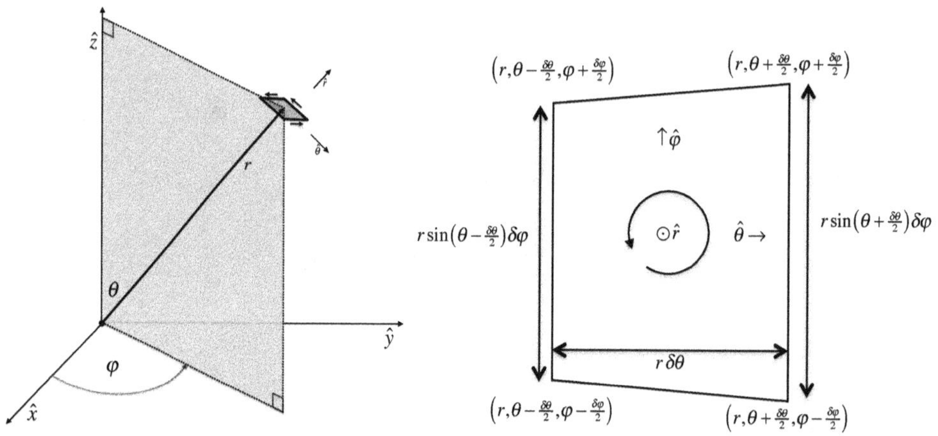

Again, we find the path integral around the loop. Starting at the bottom left-hand corner, and proceeding counter-clockwise, this is:

$$\left(\oint_{\text{path}} \vec{F}\cdot \mathrm{d}\vec{l}\right)_r = \int_{\theta-\delta\theta/2}^{\theta+\delta\theta/2} F_\theta\left(r,\varphi-\tfrac{\delta\varphi}{2},\theta\right)\left(r\,\mathrm{d}\theta\right) + \int_{\varphi-\delta\varphi/2}^{\varphi+\delta\varphi/2} F_\varphi\left(r,\varphi,\theta+\tfrac{\delta\theta}{2}\right)\left(r\sin\left(\theta+\tfrac{\delta\theta}{2}\right)\mathrm{d}\varphi\right)$$
$$+\int_{\theta+\delta\theta/2}^{\theta-\delta\theta/2} F_\theta\left(r,\varphi+\tfrac{\delta\varphi}{2},\theta\right)\left(r\,\mathrm{d}\theta\right) + \int_{\varphi+\delta\varphi/2}^{\varphi-\delta\varphi/2} F_\varphi\left(r,\varphi,\theta-\tfrac{\delta\theta}{2}\right)\left(r\sin\left(\theta-\tfrac{\delta\theta}{2}\right)\mathrm{d}\varphi\right).$$

Now we reverse the integration limits, and combine like integrals, so the path integral around the loop is given by:

$$\left(\oint_{\text{path}} \vec{F}\cdot \mathrm{d}\vec{l}\right)_r = r\int_{\theta-\delta\theta/2}^{\theta+\delta\theta/2}\left(F_\theta\left(r,\theta,\varphi-\tfrac{\delta\varphi}{2}\right)-F_\theta\left(r,\theta,\varphi+\tfrac{\delta\varphi}{2}\right)\right)\mathrm{d}\theta + r\int_{\varphi-\delta\varphi/2}^{\varphi+\delta\varphi/2}\left(F_\varphi\left(r,\theta+\tfrac{\delta\theta}{2},\varphi\right)\sin\left(\theta+\tfrac{\delta\theta}{2}\right)-F_\varphi\left(r,\theta-\tfrac{\delta\theta}{2},\varphi\right)\sin\left(\theta-\tfrac{\delta\theta}{2}\right)\right)\mathrm{d}\varphi .$$

Finally, we use the definition of the partial derivative to simplify the prior result to:

$$\left(\oint_{\text{path}} \vec{F}\cdot \mathrm{d}\vec{l}\right)_r \approx r\int_{\theta-\delta\theta/2}^{\theta+\delta\theta/2}\left(-\frac{\partial F_\theta}{\partial\varphi}\delta\varphi\right)\mathrm{d}\theta + r\int_{\varphi-\delta\varphi/2}^{\varphi+\delta\varphi/2}\left(\frac{\partial\left(F_\varphi\sin(\theta)\right)}{\partial\theta}\delta\theta\right)\mathrm{d}\varphi \approx -r\frac{\partial F_\theta}{\partial\varphi}\delta\varphi\,\delta\theta + r\frac{\partial\left(F_\varphi\sin(\theta)\right)}{\partial\theta}\delta\theta\,\delta\varphi .$$

Now, taking another look at the figure, the loop is a trapezoid whose area is given by:

$$\delta A_r = \left(r\delta\theta\right)\left(\tfrac{1}{2}\left(r\sin\left(\theta+\tfrac{\delta\theta}{2}\right)\delta\varphi + r\sin\left(\theta-\tfrac{\delta\theta}{2}\right)\delta\varphi\right)\right) \approx r^2\sin(\theta)\delta\theta\,\delta\varphi .$$

Applying the definition of the curl:

$$\left(\vec{\nabla}\times\vec{F}(\vec{r})\right)_r \equiv \lim_{\delta A_r\to 0}\frac{1}{\delta A_r}\oint_{\text{path}}\vec{F}(\vec{r})\cdot\mathrm{d}\vec{r} = \frac{-r\frac{\partial}{\partial\varphi}F_\theta\,\cancel{\delta\varphi\,\delta\theta} + r\frac{\partial}{\partial\theta}\left(F_\varphi\sin(\theta)\right)\cancel{\delta\theta\,\delta\varphi}}{r^2\sin(\theta)\,\cancel{\delta\theta\,\delta\varphi}}$$
$$= \frac{1}{r\sin(\theta)}\left(-\frac{\partial F_\theta}{\partial\varphi}+\frac{\partial\left(F_\varphi\sin(\theta)\right)}{\partial\theta}\right).$$

Of course, this is only one component of the curl. We still need to complete a similar exercise for the $\hat{\theta}$ and $\hat{\varphi}$ directions.

Next, we consider the view from the $\hat{\theta}$ direction, and consider the path integral around a wedge-shaped loop surrounding the position $\vec{r}$.

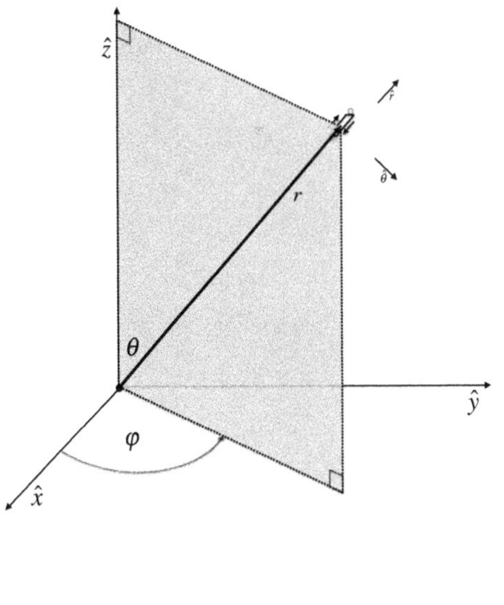

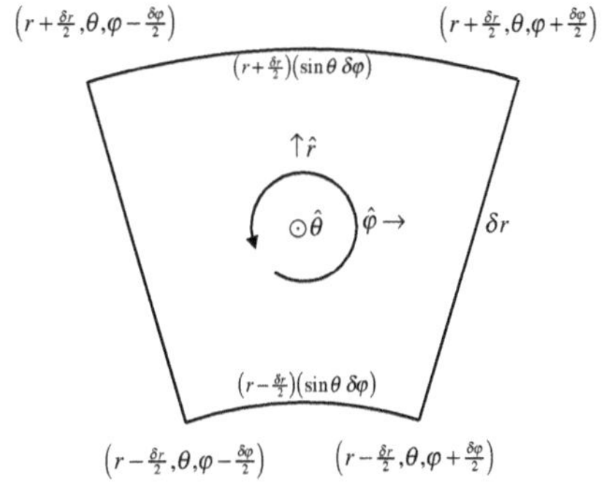

$$\left(\oint_{\text{path}} \vec{F}\cdot \mathrm{d}\vec{l}\right)_\theta = \int_{\varphi-\delta\varphi/2}^{\varphi+\delta\varphi/2} F_\varphi\left(r-\tfrac{\delta r}{2},\theta,\varphi\right)\left(r-\tfrac{\delta r}{2}\right)\sin\theta\,\mathrm{d}\varphi + \int_{r-\delta r/2}^{r+\delta r/2} F_r\left(r,\theta,\varphi+\tfrac{\delta\varphi}{2}\right)\mathrm{d}r$$
$$+\int_{\varphi+\delta\varphi/2}^{\varphi-\delta\varphi/2} F_\varphi\left(r+\tfrac{\delta r}{2},\theta,\varphi\right)\left(r-\tfrac{\delta r}{2}\right)\sin\theta\,\mathrm{d}\varphi + \int_{r+\delta r/2}^{r-\delta r/2} F_r\left(r,\theta,\varphi-\tfrac{\delta\varphi}{2}\right)\mathrm{d}r\,.$$

Again, we combine terms and reverse the limits of integration, so:

$$\left(\oint_{\text{path}} \vec{F}\cdot \mathrm{d}\vec{l}\right)_\theta = -\sin\theta \int_{\varphi-\delta\varphi/2}^{\varphi+\delta\varphi/2} \left(\left(r-\tfrac{\delta r}{2}\right)F_\varphi\left(r-\tfrac{\delta r}{2},\theta,\varphi\right) - \left(r+\tfrac{\delta r}{2}\right)F_\varphi\left(r+\tfrac{\delta r}{2},\theta,\varphi\right)\right)\mathrm{d}\varphi + \int_{r-\delta r/2}^{r+\delta r/2}\left(F_r\left(r,\theta,\varphi+\tfrac{\delta\varphi}{2}\right) - F_r\left(r,\theta,\varphi-\tfrac{\delta\varphi}{2}\right)\right)\mathrm{d}r$$
$$= -\sin\theta \int_{\varphi-\delta\varphi/2}^{\varphi+\delta\varphi/2}\left(\frac{\partial(rF_\varphi)}{\partial r}\delta r\right)\mathrm{d}\varphi + \int_{r-\delta r/2}^{r+\delta r/2}\left(\frac{\partial F_r}{\partial\varphi}\delta\varphi\right)\mathrm{d}r = -\sin\theta\frac{\partial(rF_\varphi)}{\partial r}\delta r\,\delta\varphi + \frac{\partial F_r}{\partial\varphi}\delta\varphi\,\delta r\,.$$

Now the area of the loop is once again simply the area of a trapezoid:

$$\delta A_\theta = (r\sin\theta\,\delta\varphi)(\delta r)\,.$$

If we once again apply the definition of the curl, we obtain the following:

$$\left(\vec{\nabla}\times\vec{F}(\vec{r})\right)_\theta \equiv \lim_{\delta A_\theta\to 0}\frac{1}{\delta A_\theta}\oint_{\text{path}}\vec{F}(\vec{r})\cdot\mathrm{d}\vec{r} = \frac{-\sin\theta\frac{\partial}{\partial r}(rF_\varphi)\cancel{\delta r\,\delta\varphi} + \frac{\partial}{\partial\varphi}F_r\cancel{\delta\varphi\,\delta r}}{r\sin\theta\cancel{\delta\varphi\,\delta r}}\,.$$

The theta component is:

$$\left(\vec{\nabla}\times\vec{F}(\vec{r})\right)_\theta = -\frac{1}{r}\frac{\partial(rF_\varphi)}{\partial r} + \frac{1}{r\sin\theta}\frac{\partial F_r}{\partial\varphi}$$

Finally, we need to consider a loop at position $\vec{r}$ around the $\hat{\varphi}$ axis.

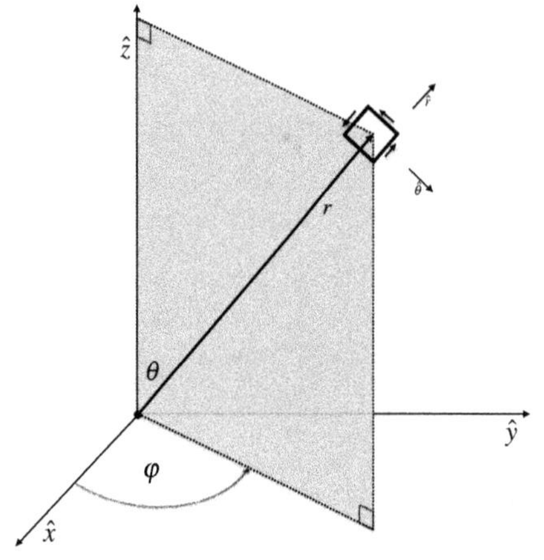

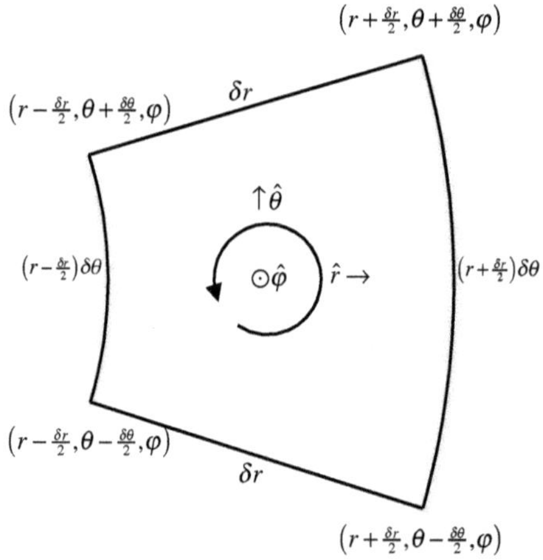

By taking the path integral around this loop, we find that:

$$\left(\oint_{\text{path}} \vec{F}\cdot \mathrm{d}\vec{l}\right)_\varphi = \int_{r-\delta r/2}^{r+\delta r/2} F_r\left(r,\theta-\tfrac{\delta\theta}{2},\varphi\right)\mathrm{d}\theta + \int_{\theta-\delta\theta/2}^{\theta+\delta\theta/2} F_\theta\left(r-\delta r/2,\theta,\varphi\right)\left(r-\delta r/2\right)\mathrm{d}\theta + \int_{r+\delta r/2}^{r-\delta r/2} F_r\left(r,\theta+\tfrac{\delta\theta}{2},\varphi\right)\mathrm{d}\theta + \int_{\theta+\delta\theta/2}^{\theta-\delta\theta/2} F_\theta\left(r+\delta r/2,\theta,\varphi\right)\left(r+\delta r/2\right)\mathrm{d}\theta\,.$$

Again, we reverse the limits of integration and apply the definition of the partial derivative:

$$\left(\oint_{\text{path}} \vec{F}\cdot\mathrm{d}\vec{l}\right)_{\varphi} = \int_{r-\delta r/2}^{r+\delta r/2}\left(F_r\left(r,\theta-\tfrac{\delta\theta}{2},\varphi\right)-F_r\left(r,\theta+\tfrac{\delta\theta}{2},\varphi\right)\right)\mathrm{d}r+\int_{\theta-\delta\theta/2}^{\theta+\delta\theta/2}\left(\left(r-\delta r/2\right)F_\theta\left(r-\delta r/2,\theta,\varphi\right)-\left(r+\delta r/2\right)F_\theta\left(r+\delta r/2,\theta,\varphi\right)\right)\mathrm{d}\theta$$

$$= \int_{r-\delta r/2}^{r+\delta r/2}\left(-\frac{\partial F_r}{\partial\theta}\delta\theta\right)\mathrm{d}r+\int_{\theta-\delta\theta/2}^{\theta+\delta\theta/2}\left(\frac{\partial\left(rF_\theta\right)}{\partial r}\delta r\right)\mathrm{d}\theta = -\frac{\partial F_r}{\partial\theta}\delta\theta\,\delta r+\frac{\partial\left(rF_\theta\right)}{\partial r}\delta r\,\delta\theta\,.$$

And, again, we will find the area of the loop by subtracting circles:

$$\delta A_\varphi = \frac{\delta\theta}{2\pi}\left(\pi\left(r+\tfrac{1}{2}\delta r\right)^2-\pi\left(r-\tfrac{1}{2}\delta r\right)^2\right) = \frac{\delta\theta}{2\pi}\left(2\pi r\,\delta r\right) = r\,\delta r\,\delta\theta\,.$$

The φ component of the curl is therefore:

$$\left(\vec{\nabla}\times\vec{F}(\vec{r})\right)_\varphi \equiv \lim_{\delta A_\varphi\to 0}\frac{1}{\delta A_\varphi}\oint_{\text{path}}\vec{F}(\vec{r})\cdot\mathrm{d}\vec{r} = \frac{-\frac{\partial}{\partial\theta}F_r\,\cancel{\delta\theta\,\delta r}+\frac{\partial}{\partial r}\left(rF_\theta\right)\cancel{\delta r\,\delta\theta}}{r\,\cancel{\delta r\,\delta\theta}} = \frac{1}{r}\left(-\frac{\partial F_r}{\partial\theta}+\frac{\partial\left(rF_\theta\right)}{\partial r}\right).$$

Finally, the curl in spherical coordinates becomes:

$$\vec{\nabla}\times\vec{F} = \frac{1}{r\sin(\theta)}\left(\frac{\partial\left(F_\varphi\sin(\theta)\right)}{\partial\theta}-\frac{\partial F_\theta}{\partial\varphi}\right)\hat{r}+\frac{1}{r}\left(\frac{1}{\sin\theta}\frac{\partial F_r}{\partial\varphi}-\frac{\partial\left(rF_\varphi\right)}{\partial r}\right)\hat{\theta}+\frac{1}{r}\left(\frac{\partial\left(rF_\theta\right)}{\partial r}-\frac{\partial F_r}{\partial\theta}\right)\hat{\varphi}\,.$$

Problem 7.6: Algebraic Derivation of the Spherical Curl

Derive the expression for the curl in spherical coordinates from the expression of the curl in cylindrical coordinates, and the appropriate coordinate transformations.

Thought Experiment 7.5: Newton's Law of Gravity is Path Independent

Newton's law of gravity expresses the force on object A, by object B, as:

$$\vec{F}_\mathrm{A} = -\frac{G m_\mathrm{A} m_\mathrm{B}}{\left(r_A+r_B\right)^2}\hat{r}_\mathrm{A}\,.$$

Substituting, we find:

$$\vec{F}_\mathrm{A} = -\frac{G m_\mathrm{A} m_\mathrm{B}}{\left(r_A+r_B\right)^2}\hat{r}_\mathrm{A} = -\frac{G m_\mathrm{A} m_\mathrm{B}}{\left(r_\mathrm{A}+\dfrac{m_\mathrm{A}}{m_\mathrm{B}}r_\mathrm{A}\right)^2}\hat{r}_\mathrm{A} = -G\frac{m_\mathrm{A} m_\mathrm{B}^3}{\left(m_\mathrm{B}+m_\mathrm{A}\right)^2}\frac{\hat{r}_\mathrm{A}}{r_\mathrm{A}^2}\,.$$

Now we can take the curl of the force on body A, to test whether it is path independent:

$$\vec{\nabla}\times\vec{F}_\mathrm{A} = \vec{\nabla}\times\left(-G\frac{m_\mathrm{A} m_\mathrm{B}^3}{\left(m_\mathrm{B}+m_\mathrm{A}\right)^2}\frac{\hat{r}_\mathrm{A}}{r_\mathrm{A}^2}\right) = -G\frac{m_\mathrm{A} m_\mathrm{B}^3}{\left(m_\mathrm{B}+m_\mathrm{A}\right)^2}\vec{\nabla}\times\left(\frac{\hat{r}_\mathrm{A}}{r_\mathrm{A}^2}\right).$$

This is already rendered in spherical polar coordinates, so:

$$\vec{\nabla}\times\left(\frac{\hat{r}_\mathrm{A}}{r_\mathrm{A}^2}\right) = \frac{1}{r_A\sin\theta}\frac{\partial}{\partial\varphi}\left(\frac{1}{r_A^2}\right)\hat{\theta}-\frac{1}{r_A\,\mathrm{s}}\frac{\partial}{\partial\theta}\left(\frac{1}{r_A^2}\right)\hat{\varphi} = 0\,.$$

Therefore $\vec{\nabla}\times\vec{F}_A = 0$, so the force is path independent. As we discussed in Chapter 3. path independence is a necessary condition for the application of potential energy concepts.

Problem 7.7: The Electrostatic Force is Path independent

Show that Coulomb's electrostatic force is path independent.

Problem 7.8: Conservative Force Fields

Consider the following forces $\vec{F}(\vec{r})$. By taking the curl, determine whether each is a conservative or non-conservative force. If you recognize an application of any of these forces, discuss it.

a) $\vec{F}\propto\frac{\hat{z}}{z^4}$, b) $\vec{F}\propto\frac{\hat{r}}{r}$, c) $\vec{F}\propto yz\hat{x}+xz\hat{y}+xy\hat{z}$, d) $\vec{F}\propto s(\hat{s}+\hat{\varphi})$, e) $\vec{F}\propto\frac{x\hat{x}+y\hat{y}+z\hat{z}}{\left(x^2+y^2+z^2\right)^{3/2}}$,

f) $\vec{F}\propto\frac{\hat{\varphi}}{s}$, g) $\vec{F}\propto z\sin\varphi\,\hat{s}+z\cos\varphi\,\hat{\varphi}+s\sin\varphi\,\hat{z}$, h) $\vec{F}\propto Ks\hat{s}-\frac{s\hat{s}+z\hat{z}}{\left(\sqrt{s^2+z^2}\right)^3}$, i) $\vec{F}\propto 2r\theta\hat{r}+r\hat{\theta}$,

j) $\vec{F}\propto\frac{\hat{r}}{r^2\sin\theta}+\frac{\hat{\theta}}{r^2\tan\theta\sin\theta}$, k) $\vec{F}\propto y\hat{x}+x\hat{y}$, l) $\vec{F}\propto z\hat{x}+x\hat{y}+y\hat{z}$, m) $\vec{F}\propto\hat{r}$. n) $\vec{F}\propto\hat{\varphi}$.

7.3 *The Vector Potential and Permanent Magnets*

In this section, we investigate the magnetic field surrounding permanent magnets by finding the vector potential via direct integration, and then taking its curl.

Thought Experiment 7.6: The Vector Potential of a Permanent Magnet

The magnetic moments in a permanent magnet add vectorally, but the total magnetic field is not exactly that of a dipole. Simply finding the total magnetic moment will determine the field at points far from a magnet, but it will not necessarily determine the field nearby. We can work around this if we are prepared to integrate directly all the pure dipole fields that each moment in the magnet contributes to the net magnetic field.

Consider a small magnetic moment $\delta\vec{m}$ at a position $\vec{R}$ from the origin, with the associated vector potential $\delta\vec{\mathbb{A}}$ evaluated at a position $\vec{r}$ from the origin. Using the formula we derived in Thought Experiment 7.1, we see that:

$$\delta\vec{\mathbb{A}}=\frac{\mu_0}{4\pi}\frac{\delta\vec{m}\times\left(\vec{r}-\vec{R}\right)}{\left|\vec{r}-\vec{R}\right|^3}.$$

The vector potential due to a permanent magnet is then just the integral of this expression over the volume of the magnet.

The magnetic field, which is the curl of the vector potential, is also a linear vector: it too obeys superposition. One thus has a choice of integrating the vector potential and taking its curl, or directly integrating the corresponding magnetic field (see pp. 208-209).

Example 7.1: A Long Thin Magnetic Needle

Now we will investigate the magnetic field surrounding a long thin needle, whose magnetization is aligned with the needle. We will assume a length l and cross-sectional radius $a \ll l$, and a constant magnetization $\vec{M}$.

We will solve for the magnetic field as a function of position from the middle of the magnet.

First we write down the vector potential for a magnetic moment element:

$$\delta\vec{\mathbb{A}} = \frac{\mu_0}{4\pi}\frac{\delta\vec{m}\times\left(\vec{r}-\vec{R}\right)}{\left|\vec{r}-\vec{R}\right|^3} = \frac{\mu_0}{4\pi}\frac{\hat{z}\times\vec{r}-\hat{z}\times\vec{R}}{\left|\vec{r}-\vec{R}\right|^3}\delta m\,.$$

And notice that:

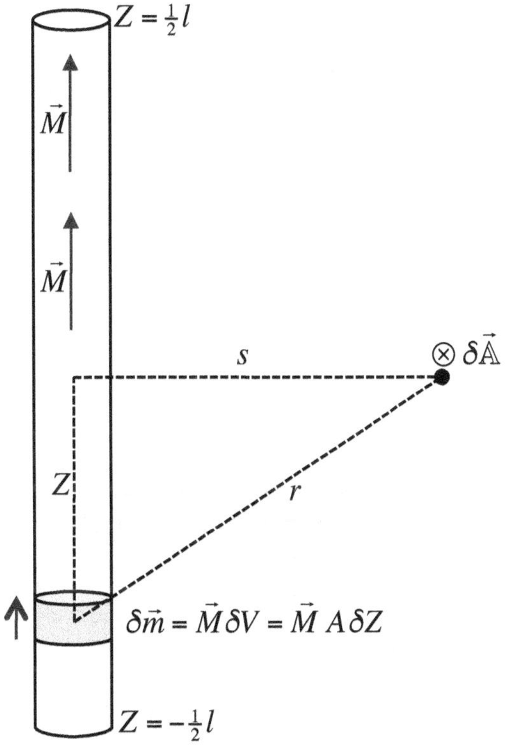

$$\frac{\hat{z}\times\vec{r}-\hat{z}\times\vec{R}}{\left|\vec{r}-\vec{R}\right|^3}\delta m = \frac{\hat{z}\times\vec{r}-Z\,\cancel{\hat{z}\times\hat{z}}}{\left|\vec{r}-\vec{R}\right|^3}\delta m$$

$$= \frac{\hat{z}\times\vec{r}}{\left|\vec{r}-\vec{R}\right|^3}\delta m$$

$$= \frac{s\,\delta m}{\left|\vec{r}-\vec{R}\right|^3}\hat{\varphi}\,.$$

Moreover, $\left|\vec{r}-\vec{R}\right| = \sqrt{s^2+\left(z-Z\right)^2}$

and $\delta m = M\,\delta V = M\,\pi a^2\,\delta Z$, so:

$$\delta\vec{\mathbb{A}} = \frac{\mu_0}{4\pi}\frac{s\,\delta m}{\left|\vec{r}-\vec{R}\right|^3}\hat{\varphi} = \frac{\mu_0}{4\pi}\frac{s\,\delta m}{\left(s^2+\left(z-Z\right)^2\right)^{3/2}}\hat{\varphi}\,.$$

The vector potential is therefore given by:

$$\vec{\mathbb{A}} = \frac{\mu_0 m}{4\pi l}\int_{-1/2}^{1/2}\frac{s\,\mathrm{d}Z}{\left(s^2+\left(z-Z\right)^2\right)^{3/2}}\hat{\varphi}\,.$$

By evaluating the integral, we obtain:

$$\vec{\mathbb{A}} = \frac{\mu_0 m}{4\pi l}\left(\frac{z+1/2}{s\sqrt{s^2+\left(z+1/2\right)^2}} - \frac{z-1/2}{s\sqrt{s^2+\left(z-1/2\right)^2}}\right)\hat{\varphi}\,.$$

Now, we take the curl of the vector potential to find the magnetic field. Because there is only a $\hat{\varphi}$ term, the curl is simply:

$$\vec{B} = \vec{\nabla}\times\vec{\mathbb{A}} = -\tfrac{\partial}{\partial z}\mathbb{A}_\varphi\hat{s} + \tfrac{1}{s}\left(\tfrac{\partial}{\partial s}\left(s\,\mathbb{A}_\varphi\right)\right)\hat{z}\,.$$

Differentiating each term separately yields:

$$-\frac{\partial \mathbb{A}_\varphi}{\partial z} = -\frac{\partial}{\partial z}\frac{\mu_0 m}{4\pi l}\left(\frac{z+\frac{1}{2}}{s\sqrt{s^2+\left(z+\frac{1}{2}\right)^2}} - \frac{z-\frac{1}{2}}{s\sqrt{s^2+\left(z-\frac{1}{2}\right)^2}}\right)$$

$$= \frac{\mu_0 m}{4\pi l s}\left(\frac{1}{\sqrt{s^2+\left(z-\frac{1}{2}\right)^2}} - \frac{1}{\sqrt{s^2+\left(z+\frac{1}{2}\right)^2}} + \frac{\left(z+\frac{1}{2}\right)^2}{\left(s^2+\left(z+\frac{1}{2}\right)^2\right)^{3/2}} - \frac{\left(z-\frac{1}{2}\right)^2}{\left(s^2+\left(z-\frac{1}{2}\right)^2\right)^{3/2}}\right).$$

And the next term:

$$\frac{1}{s}\left(\frac{\partial\left(s\,\mathbb{A}_\varphi\right)}{\partial s}\right) = \frac{1}{s}\frac{\partial}{\partial s}\left(\cancel{s}\frac{\mu_0 m}{4\pi l}\left(\frac{z+\frac{1}{2}}{\cancel{s}\sqrt{s^2+\left(z+\frac{1}{2}\right)^2}} - \frac{z-\frac{1}{2}}{\cancel{s}\sqrt{s^2+\left(z-\frac{1}{2}\right)^2}}\right)\right)$$

$$= \frac{\mu_0 m}{4\pi l s}\frac{\partial}{\partial s}\left(\frac{z+\frac{1}{2}}{\sqrt{s^2+\left(z+\frac{1}{2}\right)^2}} - \frac{z-\frac{1}{2}}{\sqrt{s^2+\left(z-\frac{1}{2}\right)^2}}\right)$$

$$= \frac{\mu_0 m}{4\pi l \cancel{s}}\left(-\frac{1}{\cancel{2}}\frac{z+\frac{1}{2}}{\left(s^2+\left(z+\frac{1}{2}\right)^2\right)^{3/2}}\cancel{2}\cancel{s} + \frac{1}{\cancel{2}}\frac{z-\frac{1}{2}}{\left(s^2+\left(z-\frac{1}{2}\right)^2\right)^{3/2}}\cancel{2}\cancel{s}\right)$$

$$= -\frac{\mu_0 m}{4\pi l}\left(\frac{z+\frac{1}{2}}{\left(s^2+\left(z+\frac{1}{2}\right)^2\right)^{3/2}} - \frac{z-\frac{1}{2}}{\left(s^2+\left(z-\frac{1}{2}\right)^2\right)^{3/2}}\right).$$

Adding these components gives the magnetic field:

$$\vec{B} = \frac{\mu_0 m}{4\pi l s}\left(\frac{1}{\sqrt{s^2+\left(z-\frac{1}{2}\right)^2}} - \frac{1}{\sqrt{s^2+\left(z+\frac{1}{2}\right)^2}} + \frac{\left(z+\frac{1}{2}\right)^2}{\left(s^2+\left(z+\frac{1}{2}\right)^2\right)^{3/2}} - \frac{\left(z-\frac{1}{2}\right)^2}{\left(s^2+\left(z-\frac{1}{2}\right)^2\right)^{3/2}}\right)\hat{s}$$

$$-\frac{\mu_0 m}{4\pi l}\left(\frac{z+\frac{1}{2}}{\left(s^2+\left(z+\frac{1}{2}\right)^2\right)^{3/2}} - \frac{z-\frac{1}{2}}{\left(s^2+\left(z-\frac{1}{2}\right)^2\right)^{3/2}}\right)\hat{z}$$

We plotted this, for a unit length, back on page 199. The magnetic field inside the magnet was not plotted, but it is strong and must be in the positive $\hat{z}$ direction so that the divergence of the magnetic field is zero.

Problem 7.9: The Long Thin Needle and the Dipole Field

Consider Example 7.1 above, where we found that the magnetic field due to a thin needle. Show, in the following cases, the answer agrees with that of a perfect magnetic dipole with the same magnetic moment:

a) When in the equatorial plane $(z=0)$.

b) In the limit that $l \to 0$.

c) Find the total magnetic flux through the magnet in terms of m and l.

Example 7.2: A Flat Magnetic Disk

Consider a flat disk of radius a and height $h \ll a$, which is uniformly magnetized parallel to its axis with a magnetization $\vec{M} = M\,\hat{z}$. What is the vector potential?

In this example we will solve for $\vec{\mathbb{A}}$ in Cartesian coordinates for a point in the $x-z$ plane, and when all is finished we will take advantage of the azimuthal symmetry to generalize our result. Thus we have:

$$\vec{r} = x\hat{x} + \cancel{y\hat{y}} + z\hat{z} \quad , \quad \vec{R} = X\hat{x} + Y\hat{y} + \cancel{Z\hat{z}} \quad \rightarrow \quad \vec{r} - \vec{R} = (x - X)\hat{x} - Y\hat{y} + z\hat{z}\,.$$

Applying our result from Thought Experiment 7.1 (p. 189), the element of vector potential due to $\delta\vec{m}$ is given by:

$$\delta\vec{\mathbb{A}} = \frac{\mu_0}{4\pi}\frac{\delta\vec{m}\times(\vec{r}-\vec{R})}{\left|\vec{r}-\vec{R}\right|^3} = \frac{\mu_0}{4\pi}\frac{\delta m\left(\hat{z}\times\vec{r}-\hat{z}\times\vec{R}\right)}{\left|\vec{r}-\vec{R}\right|^3}\,.$$

Now we can write each factor as:

$$\left(\hat{z}\times\vec{r}-\hat{z}\times\vec{R}\right) = x\hat{y} - \cancel{y\hat{x}} - X\hat{y} + Y\hat{x}\,, \quad \left|\vec{r}-\vec{R}\right|^3 = \left((x-X)^2 + Y^2 + z^2\right)^{3/2}$$

and the magnetic moment element is:

$$\delta m = M\,\delta V = M\,h\,\delta X\,\delta Y\,,$$

so:

$$\delta\vec{\mathbb{A}} = \frac{\mu_0 M\,h}{4\pi}\frac{x\hat{y} - X\hat{y} + Y\hat{x}}{\left((x-X)^2 + Y^2 + z^2\right)^{3/2}}\delta X\,\delta Y\,.$$

Notice that $a^2 \geq X^2 + Y^2$, because we are considering contributions δm inside the magnet, which implies that the vector potential is given by the following integral:

$$\vec{\mathbb{A}} = \frac{\mu_0 M\,h}{4\pi}\int_{-a}^{a}\int_{-\sqrt{a^2-X^2}}^{\sqrt{a^2-X^2}}\frac{(x-X)\hat{y} + Y\hat{x}}{\left((x-X)^2 + Y^2 + z^2\right)^{3/2}}\,.$$

Now the $\hat{x}$ and $\hat{y}$ terms of the inner integral are even and odd functions of Y respectively, so:

$$\int_{-\sqrt{a^2-X^2}}^{\sqrt{a^2-X^2}}\frac{(x-X)\hat{y} + Y\hat{x}}{\left((x-X)^2 + Y^2 + z^2\right)^{3/2}}\mathrm{d}Y = 2\int_{0}^{\sqrt{a^2-X^2}}\frac{(x-X)\hat{y}}{\left((x-X)^2 + Y^2 + z^2\right)^{3/2}}\mathrm{d}Y + 0\cdot\hat{x}\,.$$

This last integral can be evaluated:

$$2\int_{0}^{\sqrt{a^2-X^2}}\frac{(x-X)\hat{y}}{\left((x-X)^2 + Y^2 + z^2\right)^{3/2}}\mathrm{d}Y = \frac{2(x-X)\sqrt{a^2-X^2}\;\hat{y}}{\left(x^2 - 2xX + X^2 + z^2\right)\sqrt{x^2 - 2xX + z^2 + a^2}}\,.$$

The vector potential is therefore given by:

$$\vec{\mathbb{A}} = \frac{\mu_0 M h}{4\pi} \int_{-a}^{a} \frac{2(x-X)\sqrt{a^2-X^2}}{\left(x^2-2xX+X^2+z^2\right)\sqrt{x^2-2xX+z^2+a^2}} \mathrm{d}X\,\hat{y}.$$

Given the azimuthal symmetry of the problem, we can replace x with s:

$$\vec{\mathbb{A}} = \frac{\mu_0 M h}{4\pi} \int_{-a}^{a} \frac{2(s-X)\sqrt{a^2-X^2}}{\left(s^2+z^2-2sX+X^2\right)\sqrt{s^2+z^2-2sX+a^2}} \mathrm{d}X\,\hat{\varphi}.$$

Outside the magnet, where $r \gg a$, this can be approximated by:

$$\vec{\mathbb{A}} \approx \frac{\mu_0 M h}{4\pi} \int_{-a}^{a} \frac{2(s-\cancel{X})\sqrt{a^2-X^2}}{\left(s^2+z^2-\cancel{2sX+X^2}\right)\sqrt{s^2+z^2\cancel{-2sX+a^2}}} \mathrm{d}X\,\hat{\varphi}$$

$$\approx \frac{\mu_0 M h s}{4\pi\left(s^2+z^2\right)^{3/2}} \int_{-a}^{a} 2\sqrt{a^2-X^2}\,\mathrm{d}X\,\hat{\varphi} \approx \frac{\mu_0 M h s}{4\pi\left(s^2+z^2\right)^{3/2}} (\pi a^2)\,\hat{\varphi},$$

so, for large distances $(r \gg a)$:

$$\vec{\mathbb{A}} \approx \frac{\mu_0 m s}{4\pi\left(s^2+z^2\right)^{3/2}}\,\hat{\varphi}.$$

It is now a straightforward, if somewhat tedious, task to compute the magnetic field by taking the curl of the vector potential, using:

$$\vec{B} = \vec{\nabla}\times\vec{\mathbb{A}} = -\tfrac{\partial}{\partial z}\mathbb{A}_\varphi \hat{s} + \tfrac{1}{s}\tfrac{\partial}{\partial s}\left(s\mathbb{A}_\varphi\right)\hat{z}.$$

First, let us evaluate the radial component of the curl:

$$B_s = -\frac{\partial}{\partial z}\left(\frac{\mu_0 M h}{4\pi} \int_{-a}^{a} \frac{2(s-X)\sqrt{a^2-X^2}}{\left(s^2+z^2-2sX+X^2\right)\sqrt{s^2+z^2-2sX+a^2}} \mathrm{d}X \right).$$

We first reverse the order of integration and differentiation:

$$B_s = -\frac{\mu_0 M h}{4\pi} \int_{-a}^{a} \frac{\partial}{\partial z}\left(\frac{2(s-X)\sqrt{a^2-X^2}}{\left(s^2+z^2-2sX+X^2\right)\sqrt{s^2+z^2-2sX+a^2}} \right) \mathrm{d}X,$$

to obtain the following expression:

$$B_s = \frac{\mu_0 M h}{4\pi} \int_{-a}^{a} \left(\begin{array}{c} \dfrac{4z(s-X)\sqrt{a^2-X^2}}{\left(s^2+z^2-2sX+X^2\right)^2\sqrt{s^2+z^2-2sX+a^2}} \\ + \dfrac{2z(s-X)\sqrt{a^2-X^2}}{\left(s^2+z^2-2sX+X^2\right)\left(s^2+z^2-2sX+a^2\right)^{1.5}} \end{array} \right) \mathrm{d}X$$

We determine the axial component of the field in the same way:

$$B_z = \frac{1}{s}\frac{\partial}{\partial s}\left(s\frac{\mu_0 M\, h}{4\pi}\int_{-a}^{a} \frac{2(s-X)\sqrt{a^2-X^2}}{\left(s^2+z^2-2sX+X^2\right)\sqrt{s^2+z^2-2sX+a^2}}\mathrm{d}X \right)$$

where we again reverse the order of integration:

$$B_z = \frac{\mu_0 M\, h}{4\pi s}\left(\int_{-a}^{a} \frac{\partial}{\partial s}\left(\frac{2s(s-X)\sqrt{a^2-X^2}}{\left(s^2+z^2-2sX+X^2\right)\sqrt{s^2+z^2-2sX+a^2}} \right)\mathrm{d}X \right)$$

and obtain:

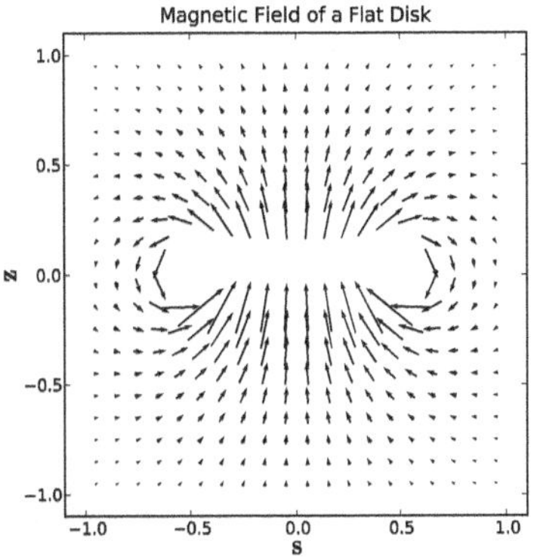

$$B_z = \frac{\mu_0 M\, h}{4\pi s}\int_{-a}^{a} \left(\begin{array}{c} \dfrac{2(2s-X)\sqrt{a^2-X^2}}{\left(s^2+z^2-2sX+X^2\right)\sqrt{s^2+z^2-2sX+a^2}} \\ -\dfrac{4s(s-X)^2\sqrt{a^2-X^2}}{\left(s^2+z^2-2sX+X^2\right)^2\sqrt{s^2+z^2-2sX+a^2}} \\ -\dfrac{2s(s-X)^2\sqrt{a^2-X^2}}{\left(s^2+z^2-2sX+X^2\right)\left(s^2+z^2-2sX+a^2\right)^{1.5}} \end{array} \right)\mathrm{d}X$$

These integrals appear daunting, but all is far from lost because both can be solved numerically using the trapezoidal method.

The magnetic field due to a thin disk is plotted here, as a function of the coordinates s and z, for a thin disk of radius 0.5 in these units. The magnetic field is much stronger inside than outside the disk, so we have excluded the interior field from our plot.

Example 7.3: The Magnetic Field of a Very Thick Magnet

Let us now consider the magnetic field of a magnet which is infinitely thick and uniform in the $\hat{z}$ direction, but of arbitrary size and shape in the other two dimensions. Moreover, we will assume that the magnetization vector has no $\hat{z}$ component, so:

$$\vec{M} = M_x(x,y)\hat{x} + M_y(x,y)\hat{y}\,,\quad \vec{\mathbb{A}} = \vec{\mathbb{A}}(x,y),\ \text{ and }\ \vec{B} = \vec{B}(x,y).$$

Now consider a volume element, $\delta V = \delta \mathrm{X}\,\delta \mathrm{Y}\,\delta \mathrm{Z}$, so a magnetic moment volume element is:

$$\delta\vec{\mathrm{m}} = \vec{M}\,\delta V = \left(M_x(x,y)\hat{x} + M_y(x,y)\hat{y}\right)\delta \mathrm{X}\,\delta \mathrm{Y}\,\delta \mathrm{Z}.$$

Again we apply the result from Thought Experiment 7.1 (p. 189):

$$\delta\vec{\mathbb{A}} = \frac{\mu_0}{4\pi}\frac{\delta\vec{m}\times\left(\vec{r}-\vec{R}\right)}{\left|\vec{r}-\vec{R}\right|^3} = \frac{\mu_0}{4\pi}\frac{\left(M_x(x,y)\hat{x}+M_y(x,y)\hat{y}\right)\times\left(\vec{r}-\vec{R}\right)}{\left|\vec{r}-\vec{R}\right|^3}\delta V$$

and we expand out the position vectors in Cartesian coordinates as:

$$\delta\vec{\mathbb{A}} = \frac{\mu_0}{4\pi}\left|\vec{r}-\vec{R}\right|^{-3}\left(M_x(x,y)\hat{x}+M_y(x,y)\hat{y}\right)\times\left((x-X)\hat{x}+(y-Y)\hat{y}+(z-Z)\hat{z}\right)\delta V\,.$$

Evaluating the unit vectors, we have:

$$\delta\vec{\mathbb{A}} = \frac{\mu_0}{4\pi}\frac{(z-Z)(M_y\hat{x}-M_x\hat{y})+M_x(y-Y)\hat{z}-M_y(x-X)\hat{z}}{\left|\vec{r}-\vec{R}\right|^3}\delta V .$$

Assume a thickness h, and integrating over Z we find:

$$\delta\vec{\mathbb{A}} = \frac{\mu_0}{4\pi}\int_{-h/2}^{h/2}\frac{(z-Z)(M_y\hat{x}-M_x(z-Z)\hat{y})+M_x(y-Y)\hat{z}-M_y(x-X)\hat{z}}{\left((x-X)^2+(y-Y)^2+(z-Z)^2\right)^{3/2}}\delta Z\delta X\delta Y$$

Noticing that $\mathrm{d}(z-Z)=-\mathrm{d}Z$, we make the substitution $\zeta = z-Z$:

$$\delta\vec{\mathbb{A}} = -\tfrac{\mu_0}{4\pi}\delta X\,\delta Y\int_{z+h/2}^{z-h/2}\frac{(M_y\hat{x}-M_x\hat{y})\zeta+M_x(y-Y)\hat{z}-M_y(x-X)\hat{z}}{\left((x-X)^2+(y-Y)^2+\zeta^2\right)^{3/2}}\mathrm{d}\zeta$$

$$\delta\vec{\mathbb{A}} = -\tfrac{\mu_0}{4\pi}\delta X\,\delta Y\,(M_y\hat{x}-M_x\hat{y})\left.\frac{1}{\sqrt{(x-X)^2+(y-Y)^2+\zeta^2}}\right|_{z+h/2}^{z-h/2}$$

$$-\tfrac{\mu_0}{4\pi}\delta X\,\delta Y\left(M_x(y-Y)-M_y(x-X)\right)\hat{z}\left.\frac{(\zeta)}{\left((x-X)^2+(y-Y)^2\right)\sqrt{(x-X)^2+(y-Y)^2+\zeta^2}}\right|_{z+h/2}^{z-h/2}$$

Now, we assumed a thick magnet, so $x,y,z \ll h$, so:

$$\vec{\mathbb{A}}(x,y) = \frac{\mu_0}{2\pi}\int_{\text{magnet}}\frac{M_x(x,y)(y-Y)-M_y(x,y)(x-X)}{(x-X)^2+(y-Y)^2}\mathrm{d}X\,\mathrm{d}Y\;\hat{z}$$

In principle, one could integrate this over the volume for any magnet. In practice, it is usually much easier to integrate this numerically rather than analytically, which we illustrate in the following examples.

Example 7.4: A Square 2D Magnet

Consider a magnetic bar with a square cross-section of dimension $a \times a \times h$, where $h \gg a$. The magnetization is given by $\vec{M} = M\,\hat{y}$ in the magnet. What is the magnetic field, $\vec{B}(x,y)$ inside and outside the magnet?

In order to solve this problem, we use the 2D vector potential relationship:

$$\vec{\mathbb{A}}(x,y) = \frac{\mu_0}{2\pi}\int_{\text{magnet}}\frac{\cancel{M_x(x,y)}(y-Y)-M_y(x,y)(x-X)}{(x-X)^2+(y-Y)^2}\mathrm{d}X\,\mathrm{d}Y\;\hat{z} .$$

Putting in the limits, we find:

$$\vec{\mathbb{A}}(x,y) = \frac{\mu_0 M}{2\pi}\int_{-a/2}^{a/2}\int_{-a/2}^{a/2}\frac{-(x-X)}{(x-X)^2+(y-Y)^2}\mathrm{d}X\,\mathrm{d}Y\;\hat{z} .$$

While we can obtain an exact solution (see Problem 7.10, below), the functional form of the solution is not particularly enlightening unless we plot it. It is possible, but again not terribly enlightening, to take the curl of this exact expression to find the magnetic field. We therefore solve this problem numerically rather than analytically. The magnitude of the vector potential is plotted in greyscale, after computing the double integral using the trapezoidal rule. The magnetic field is shown as a vector plot (left).

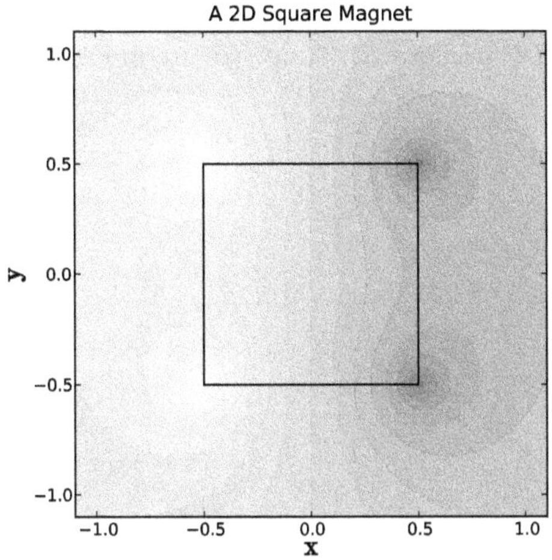

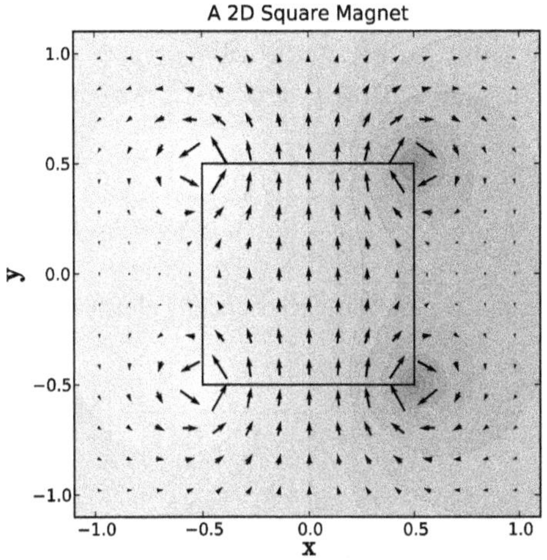

Notice that the vector equipotentials map out the magnetic field lines. This is expected because the curl of a unidirectional vector field will be perpendicular to the gradient. We can see this as follows:

$$\vec{B}(x,y) = \frac{\partial \mathbb{A}_z}{\partial y}\hat{x} - \frac{\partial \mathbb{A}_z}{\partial x}\hat{y} = \left(\frac{\partial \mathbb{A}_z}{\partial x}\hat{x} + \frac{\partial \mathbb{A}_z}{\partial y}\hat{y}\right) \times \hat{z}.$$

For small values of δx and δy, the magnetic field is given by:

$$B_x = \frac{\partial \mathbb{A}_z}{\partial y} \approx \frac{\mathbb{A}_z(y+\delta y) - \mathbb{A}_z(y-\delta y)}{2\delta y}, \quad B_y = -\frac{\partial \mathbb{A}_z}{\partial x} \approx -\frac{\mathbb{A}_z(x+\delta x) - \mathbb{A}_z(x-\delta x)}{2\delta x}.$$

As we discussed in Section 5.6 (p. 174), taking a numerical derivative can magnify small numerical errors, so it is best to utilize multiple adjacent points whenever possible, as we did in Thought Experiment 7.4 (p. 219).

Problem 7.10: A 2D Square Magnet

Consider the long rectangular magnet with a square face from Example 7.4 above with a constant magnetization in the $\hat{y}$ direction. We solved this numerically, but it can also be solved analytically.

a) Find the vector potential $\vec{\mathbb{A}}(x,y)$.

b) Plot your results, and compare them to our plot of the numerical solution for the same problem. Do they agree?

c) Inside of the magnet much of our discussion would suggest that the magnetic field is simply equal to $\mu_0\vec{M}$. Is that the case in your solution? Discuss why or why not. Which model is more simplistic?

d) How does the Maxwell-Ampère law apply to this situation, and under what conditions would it suggest that the magnetic field inside a permanent magnet is simply equal to $\mu_0\vec{M}$?

Example 7.5: The Magnetic Field Inside a Cylindrical Magnet

Let us consider a cylindrical permanent magnet of radius a and height h, with magnetization $\vec{M} = M\,\hat{z}$. What is the magnetic field at the center of the magnet?

To solve this problem, we will calculate the magnetic field directly, since we are interested in the magnetic field only at a particular point:

$$\delta\vec{B} = \left(\frac{\mu_0}{4\pi}\right)\left(\frac{3\left(\delta\vec{m}\cdot(\vec{r}-\vec{R})\right)(\vec{r}-\vec{R})}{\left|\vec{r}-\vec{R}\right|^5} - \frac{\delta\vec{m}}{\left|\vec{r}-\vec{R}\right|^3}\right).$$

Setting $\vec{r} = 0$, each magnetic field element becomes:

$$\delta\vec{B} = \left(\frac{\mu_0 M}{4\pi}\right)\left(\frac{3\left(\hat{z}\cdot(-\vec{R})\right)(-\vec{R})}{\left|-\vec{R}\right|^5} - \frac{\hat{z}}{\left|-\vec{R}\right|^3}\right)\delta V$$

$$= \left(\frac{\mu_0 M}{4\cancel{\pi}}\right)\left(\frac{3Z(S\,\hat{s}+Z\,\hat{z})}{R^5} - \frac{\hat{z}}{R^3}\right)2\cancel{\pi}S\,\delta Z\,\delta S$$

$$= \mu_0 M\left(\frac{3ZS\,\hat{s}}{2\left(S^2+Z^2\right)^{5/2}} + \frac{3Z^2\,\hat{z}}{2\left(S^2+Z^2\right)^{5/2}} - \frac{\hat{z}}{2\left(S^2+Z^2\right)^{3/2}}\right)S\,\delta Z\,\delta S\,.$$

And integrating over the whole cylinder we have:

$$\vec{B} = \mu_0 M\int_0^a\int_{-h/2}^{h/2}\left(\frac{3ZS\,\hat{s}}{2\left(S^2+Z^2\right)^{5/2}} + \frac{3Z^2\,\hat{z}}{2\left(S^2+Z^2\right)^{5/2}} - \frac{\hat{z}}{2\left(S^2+Z^2\right)^{3/2}}\right)\mathrm{d}Z\,S\,\mathrm{d}S\,.$$

Notice that the first term is an odd function with respect to Z; its integral must therefore equal zero. The other two are even functions, and remembering that $\vec{M} = M\,\hat{z}$, we find:

$$\vec{B} = \mu_0\vec{M}\int_0^a\int_0^{h/2}\left(\frac{3Z^2}{\left(S^2+Z^2\right)^{5/2}} - \frac{1}{\left(S^2+Z^2\right)^{3/2}}\right)\mathrm{d}Z\,S\,\mathrm{d}S\,.$$

Notice that that this definite integral is unitless.

First we reverse the order of integration, and evaluate one integral, and simplify:

$$\vec{B} = \mu_0\vec{M}\int_0^{h/2}\int_0^a\left(\frac{3Z^2 S}{\left(S^2+Z^2\right)^{5/2}} - \frac{S}{\left(S^2+Z^2\right)^{3/2}}\right)\mathrm{d}S\,\mathrm{d}Z = \mu_0\vec{M}\int_0^{h/2}\left(\left.\frac{-Z^2}{\left(S^2+Z^2\right)^{3/2}}\right|_0^a + \left.\frac{1}{\left(S^2+Z^2\right)^{1/2}}\right|_0^a\right)\mathrm{d}Z$$

$$= \mu_0\vec{M}\int_0^{h/2}\left(\frac{-Z^2}{\left(a^2+Z^2\right)^{3/2}} + \cancel{\frac{Z^2}{\left(Z^2\right)^{3/2}}} + \frac{1}{\left(a^2+Z^2\right)^{1/2}} - \cancel{\frac{1}{\left(Z^2\right)^{1/2}}}\right)\mathrm{d}Z = \mu_0\vec{M}\int_0^{h/2}\left(\frac{-Z^2}{\left(a^2+Z^2\right)^{3/2}} + \frac{1}{\left(a^2+Z^2\right)^{1/2}}\right)\mathrm{d}Z\,.$$

Now, integrating over Z gives:

$$\vec{B} = \mu_0 \vec{M}\left(\left.\frac{Z}{\sqrt{a^2+Z^2}}\right|_0^{h/2} - \cancel{\left.\ln\left(2Z+2\sqrt{a^2+Z^2}\right)\right|_0^{h/2}} + \cancel{\left.\ln\left(2Z+2\sqrt{a^2+Z^2}\right)\right|_0^{h/2}} \right)$$

$$= \mu_0 \vec{M}\left(\frac{h/2}{\sqrt{a^2+\left(h/2\right)^2}} - \frac{0}{\sqrt{a^2+0^2}} \right) = \mu_0 \vec{M}\left(\frac{h}{\sqrt{4a^2+h^2}} \right).$$

In the limit that we have a flat disk as in Example 7.2 (p. 233)the magnetic field at the center is:

$$\vec{B} \approx \mu_0 \vec{M}\left(\frac{h}{2a} \right).$$

And in the limit of a long thin cylinder as in Example 7.1 (p. 231), then $h \gg a$, so:

$$\vec{B} \approx \mu_0 \vec{M} .$$

This latter limit is of particular importance, because it is often assumed inside magnetized matter.

Problem 7.11: A Magnetized Sphere

Consider the magnetized sphere with uniform magnetization $\vec{M}$ and a radius a, as discussed in Thought Experiment 6.13.

Show that, outside the sphere, the magnetic field is simply that of an equivalent point magnet located at the origin, with the same total magnetic moment: $\vec{m} = \left(\frac{4\pi}{3}a^3\right)\vec{M}$.

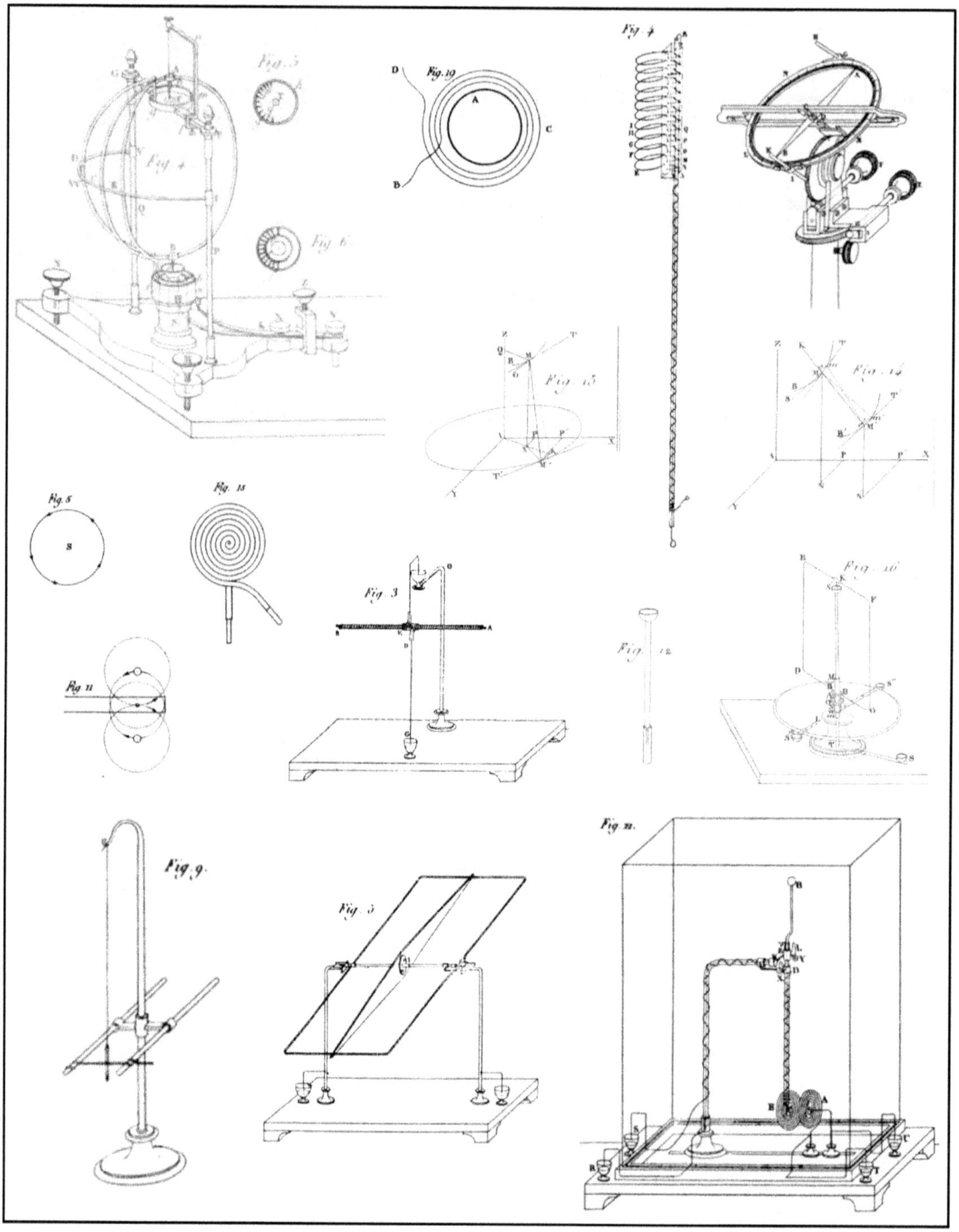

Plate 15: Ampère's Experiments[104]

104 Selected figures from: André-Marie Ampère, Recueil d'observations électro-dynamiques, (Paris: Crochard, 1822)

Chapter 8 Electromagnetism

The beheading of the great French chemist Antoine-Laurent de Lavoisier was swift, but its effects were long lasting. Lavoisier's head was not the only one to roll during the Reign of Terror of 1792-1794, led by Maximilien de Robespierre and his fellow Jacobins. The city of Lyon was particularly contentious, as it was a stronghold of the competing Girondists who controlled the city even after Robespierre had secured power in Paris and executed the Parisian Girondists *en masse*. Caught in the middle of this power struggle was a well-meaning local magistrate named Jean-Jacques Ampère.

Ten years earlier, Ampère moved his family to his country estate so he could concentrate on homeschooling his children. He filled his house with books and encouraged his children to grow up with a love of ideas. As an example of his great teaching, his son, André-Marie, wrote a treatise of mathematical proofs and submitted them to the local academy at age 13. While this work was not new to mathematics, it did teach André-Marie how to produce original ideas and articulate them using mathematics and the written word. In this fashion, André-Marie developed a keen mind and strong academic skills.[105]

The Reign of Terror brought this to a complete halt, when a kangaroo court of Jacobin nationalists sentenced and executed his father. Thus, at the age of 17, André-Marie Ampère suffered his first bout of depression, which would punctuate the rest of his mostly unhappy personal life.

Ampère started teaching high school mathematics in 1795, fell in love in 1796, and fathered a son in 1800. In 1802 he got a college teaching job in another town, and spent time there alone concentrating on his research. His hard work paid off, as he secured a college teaching position in Lyon in 1803, where he could be with his family and close friends. Unfortunately, his wife had been ill for some time, and died. Ampère was soon offered a teaching position at the prestigious *École Polytechnique* in Paris.

Meanwhile, during the Reign of Terror, the Royal Academy of Sciences was abolished, and most of its members quietly stayed out of the spotlight doing their work. After Robespierre went to the guillotine himself, things greatly improved. The new leader of the Academy, Lazare Carnot, was an engineer and great military tactician, who had studied at Coulomb's alma mater and published a paper on kinetic energy. In 1795 the new National Institute of Sciences and Arts was formed, and divided into three classes for the following subjects: (1) the natural sciences, (2) the social sciences and (3) the arts. Thus, the *First Class of the Institute* succeeded the old Royal Academy retaining the same members, except for Lavoisier of course. As the change was largely in name, we will refer to it simply as *the Academy*.

Pierre-Simon Laplace took the primary leadership role in the mathematical sciences of the First Class, where he dictated the research agenda over the next two decades. Laplace's clear and forceful leadership, with an emphasis on careful experimental techniques, led to great advances in those areas of physics where Laplace's ideas were correct. Thus, Laplace ran the first truly modern research laboratory with exacting standards for mathematical rigor, experimental precision, and the empirical testing of theory.

---

[105] James R. Hofmann, André-Marie Ampère, (Oxford: Blackwell, 1995), which is one of the Blackwell science biographies.

Laplace also promoted *the astronomical view of nature*, which supposed that molecules act on each other from a distance, in much the same way that gravity acts from a distance between astronomical objects. The Laplacians made a distinction between ordinary *ponderable* matter, and the *imponderable fluids* of heat, light, electricity and magnetism. These fluids were thought to comprise mutually repulsive particles, so they would spread out, and then interact with ordinary ponderable matter. The primary goal of early nineteenth century physics was to carefully characterize these forces and fluids experimentally, or as Laplace put it:

> All terrestrial phenomena depend on forces of this kind, just as celestial phenomena depend on universal gravitation. It seems to me that the study of these forces should now be the chief goal of mathematical philosophy. I even believe that it would be useful to introduce such a study in proofs in mechanics, laying aside abstract considerations of flexible or inflexible lines without mass and of perfectly hard bodies. A number of trials have shown me that by coming closer to nature in this way one could make these proofs no less simple and far more lucid than by the methods used hitherto.[106]

One problem with Laplacian physics was that its followers did not allow for dissent. So long as a young scientist agreed with Laplace in general, his work was considered worthy and he was given more opportunities to advance his career. As it turns out, the Laplacian paradigm was correct for electricity, but incorrect for heat, light, and magnetism.

In 1799 Napoleon Bonaparte took power in a *coup d'état*, and soon appointed Laplace as his minister of the interior. While Laplace was forceful and decisive for a scientist, he was not for a Napoleonic executive—so Bonaparte promoted him. As a member of the senate, Laplace received a handsome salary, a great deal of influence over scientific policy, and time for research. It was the perfect position for an ambitious mid-career scientist such as Laplace. The chemist Claude Louis Berthollet and the elderly Joseph-Louis Lagrange also received senatorial posts.

In 1806, Laplace bought an estate next door to Berthollet. Each summer, Laplace, Berthollet, and their German friend Alexander von Humboldt, would invite a small number of the most promising young scientists to spend their summers in an idyllic environment of intellectual interchange. The group called themselves the Society of Arcueil, as they were located in the small town of Arcueil about five kilometers south of central Paris. Society of Arcueil participants who contributed to electrodynamics included: Dominique François Jean Arago, Jean-Baptiste Biot, Joseph Louis Gay-Lussac, Étienne-Louis Malus, and Siméon Denis Poisson.

On July 21 of 1820, Hans Christian Ørsted published a short Latin paper summarizing his discovery that a current carrying wire deflects a compass needle in a circular pattern around the wire. But it was not until late summer that, while visiting Geneva, Arago learned of the discovery. As the news was received with disbelief when Arago reported it on the first Monday in September, he experimentally demonstrated it the following Monday. This sparked a race for an explanation, primarily between Biot and Ampère.

Ampère got off to a running start with a demonstration that a compass points in the tangential direction surrounding the wire. And at the last Monday in September, Ampère demonstrated that coils of current carrying wires behave like permanent magnets. Following on the heels of Ampère's success, Arago wrapped an iron needle with wire, and discovered that the poles of this new electromagnet reverse when the current reverses.

---

[106] P.S. Laplace, Traité de mécanique céleste, 5 vols, (Paris 1799-1825), 5, 99. Translation and discussion of Laplacian physics from the book chapter by Robert Fox, "The Rise and Fall of Laplacian Physics," in Historical Studies in the Physical Sciences **4**, edited by R. McCormmach, (Princeton: Princeton University Press, 1974).

By the meeting of October 2, Ampère presented a draft paper describing the electrodynamic forces between current-carrying wires, and in the next meeting he demonstrated these forces using an apparatus. By the very next meeting, he had demonstrated the force on a current-carrying wire due to the Earth's magnetic field.

After all of these demonstrations, and discussions, Biot presented his analysis of the same problem on the last Monday in October. Biot, and his protégé Félix Savart, had made very careful measurements of the torque on a magnetized needle as a function of the distance from the wire, which they reported as:

> Biot and Savart were led to the following result which rigorously expresses the action experienced by a molecule of astral or boreal magnetism located at an arbitrary distance from an extended, very fine, cylindrical wire magnetized by a voltaic current. Draw a perpendicular to the axis of the wire from the point where the is molecule resides: the force which influences the molecule is perpendicular to the line and to the axis of the wire. Its intensity is inversely proportional to the simple distance.[107]

In November, Ampère demonstrated that the magnetic field adds with the number of current carrying wires in the same direction, and had outlined the main points of his eventual force law.

On the last Monday before Christmas, Biot and Savart presented measurements with bent wires, with the goal of finding the force between infinitesimal magnetic fluid elements. Now they found that the force is proportional to the distance squared between their supposed poles of magnetic fluid, just like Coulomb's law for electricity.

Thus, by the end of the calendar year of 1820, Biot and Savart had characterized empirically the torque on a magnetic needle, but interpreted their result using the existing Laplacian paradigm which supposed that magnetic forces were caused by forces between the southern (astral) or northern (boreal) molecules which make up the magnetic fluid. Despite a revolutionary experiment that completely contradicted the Laplacian paradigm, Biot over-interpreted the data to conform with Laplacian two-fluid theory of magnetism.

It was the provincial André-Marie Ampère who first correctly explained the magnetic forces between current-carrying wires. While Ampère's explanation was clearly correct, the tenacious Biot would not concede defeat. So, Ampère and Biot published rival papers, and were given joint credit for the discovery of the force law between wires.

In general, the Society of Arcueil members supported Biot, with the notable exception of François Arago. The careers of Arago and Biot were launched together by Laplace, who gave them the task of measuring the meridian arc through Paris, as the meter was defined as one ten millionth (10^{-7}) of the distance from the north pole to the equator through Paris. As it turns out, after taking some measurements near the Spanish border, Biot returned to Paris. Arago stayed to finish the work, was captured, and became a prisoner of war. Eventually he escaped and through a series of great adventures returned to France, while somehow keeping all of the data safe and intact.

When the heroic Arago sided with the awkward Ampère over the popular Biot, it finally made other members of the Academy more comfortable voicing their true thoughts on scientific ideas. Laplacian hegemony was beginning to end, as now the two interpretations of magnetism could be

---

[107] James R. Hofmann, André-Marie Ampère, Blackwell science biographies, (Oxford: Blackwell, 1995), 232.

freely debated.[108] This opened the door for Fourier and Augustin-Jean Fresnel, whose ideas regarding heat and light were now taken seriously, but this new openness also meant that the rivalry between Biot and Ampère grew more intense over the next few years.

Around the same time, Poisson developed a theory of magnetism in the Laplacian mold, which explained the new electromagnetic phenomena in terms of the imponderable (i.e. massless) fluid of magnetic poles. This theory was consistent with the measurements of both Biot and Ampère, and at the same time preserved Laplacian magnetism. While Poisson's theory was mathematically elegant, it was physically wrong.

Ampère, for his part, obtained the services of the brilliant young scientist Félix Savary (not to be confused with Biot's young collaborator Félix Savart) whose work propelled the Ampèrian theory of magnetic forces further, until Ampère eventually wrote his complete treatise fully documenting the forces between wires. Ampère and his collaborators also promoted the idea that permanent magnets contain small circular currents.

In the hands of James Clerk Maxwell, Ampère's force law became the basis for defining the magnetic field. In this chapter we develop Ampère's idea of equilibrium forces between wires, but do so within the context of Maxwell's field theory and the mathematics of vector calculus.

We begin with a presentation of a fundamental law that is mistakenly attributed to Ampère: the circulation of a magnetic field around a closed loop is directly proportional to the total current that flows through any surface bounded by that loop. Maxwell first stated this theorem in a letter to William Thomson in 1854.[109] The local, or differential, version of this law is that the curl of a magnetic field is proportional to the current density. Maxwell presented this local version in differential form in his 1855 paper, "On Faraday's lines of force."[110]

8.1 *Hans Christian Ørsted's Discovery*

In 1819, Hans Christian Ørsted discovered that a compass needle moves when current flows through a nearby wire, and mapped out the direction of the magnetic field caused by the current in the wire. Ørsted's experiments involved placing a horizontal wire along the magnetic north-south direction. He summarized his findings thus:

> In order that this may be more easily remembered, we may use this formula: The pole above which the negative electricity enters turns to the west, below to the east.[111]

---

[108] See the book chapter by Robert Fox, "The Rise and Fall of Laplacian Physics," in Historical Studies in the Physical Sciences **4**, edited by R. McCormmach, (Princeton: Princeton University Press, 1974), 89-136.

[109] The Scientific Letters and Papers of James Clerk Maxwell, ed. P. M. Harman. 2 vols. (Cambridge: Cambridge University Press, 1990-95), **1**: 256-7. This is discussed in some detail in O. Darrigol, Electrodynamics from Ampère to Einstein (Oxford: Oxford University Press, 2000), 139-42. Maxwell's letter of November 13, 1854 is also reprinted in full in "The origins of Clerk Maxwell's Electric ideas, as described in familiar letters to William Thomson," **32** Proc. Cambridge Philos. (1936), 701- 705 (see especially p. 702). For a clear explanation of the evolution of Maxwell's ideas, beginning with his letter to Thomson, see M. Norton-Wise, "The Mutual Embrace of Electricity and Magnetism," Science **203** (30 March 1979), 1310-18. Norton-Wise nicely translates Maxwell's terminology into modern language.

[110] J.C. Maxwell, "On Faraday's Lines of Force," Transactions of the Cambridge Philosophical Society, **10**, no. 1, (1856), 27-83. See the equations on p. 56.

[111] H.C. Ørsted, "Experiments on the Effect of the Electric Conflict on the Magnetic Needle" Vindenskabernes Selskabos Overskrifter **3** (1820-21): 12-21, Translated by K. Jelved, A.D. Jackson, and Ole Knudsen, in *Selected Scientific Works of Hans Christian Ørsted* (Princeton: Princeton University Press). ©1998 by Princeton University Press

We illustrate his "formula" in the diagram. He also noted that when the wire is in the east-west direction, the compass needle does not turn, but rather pitches with the north end pointing down, or up, depending on the direction of the current.

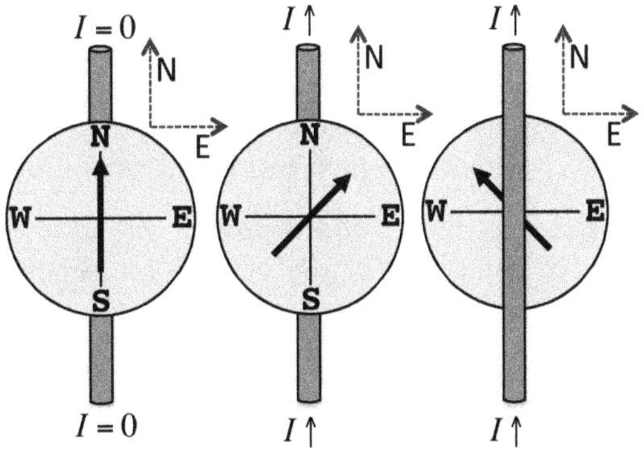

Ørsted also experimented with trying to shield the effect by placing different materials between the wire and the compass, and found that all non-magnetic materials had no effect whatsoever. He concluded that the magnetic field ("electric conflict"), due to the wire, is circular and drops off quickly with distance. As he explained it:

> From all this, we may make a few observations towards an explanation of these phenomena.
>
> The electric conflict can act only on the magnetic particles of matter. All non-magnetic bodies seem to be penetrable by the electric conflict, whereas magnetic bodies, or rather their magnetic particles, seem to resist the passage of this conflict; hence, they can be moved by the impetus of the contending powers.
>
> It is sufficiently evident from the preceding observations that the electric conflict is not confined to the conductor but, as mentioned above, is dispersed quite widely into the circumjacent space.
>
> From what we have observed we may likewise conclude that this conflict moves in circles; for without this condition it seem impossible that the same part of the connecting wire moves the magnetic needle towards the east when placed below the magnetic pole, but towards the west when placed above it; for it is the nature of circles that motion in opposite parts must have opposite directions.

We now remember the equivalent "formula" using the "right hand rule," where we point our thumb in the direction of the current and the magnetic field follows our fingers.

Thought Experiment 8.1: The Magnetic Field Surrounding a Vertical Wire

Ørsted did consider vertical wires, but he mostly investigated the horizontal ones. However, a vertical wire makes for a very nice thought experiment.

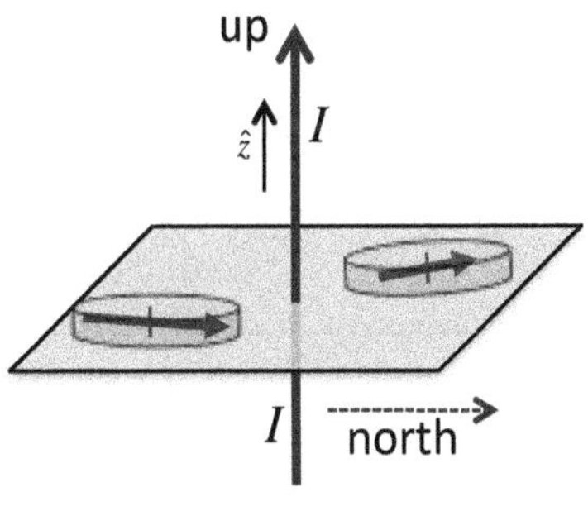

Consider stringing a wire vertically through a hole in a non-magnetic table, and measuring the horizontal direction of the magnetic field surrounding the wire, as in the figure.

With the current turned off the compass needle points north. However, when current flows, the needle deflects. The following observations are made:

- If the compass is placed north of the wire, as the current increases the needle deflects towards the west. The greater the current, the higher the deflection angle, φ, with the following proportionality: $\tan\varphi \propto I$.
- If the compass is placed due east of the wire, the current does not affect the direction of the compass needle.

- A compass due south of the wire deflects to the east, as the current increases the deflection angle again increases with the proportionality: $\tan\varphi \propto I$.
- A compass west of the wire points north, until the current is greater than some threshold current, at which point it starts pointing south.

Looking at these observations, we conclude that the magnetic field, due to the wire, points azimuthally around the long wire.

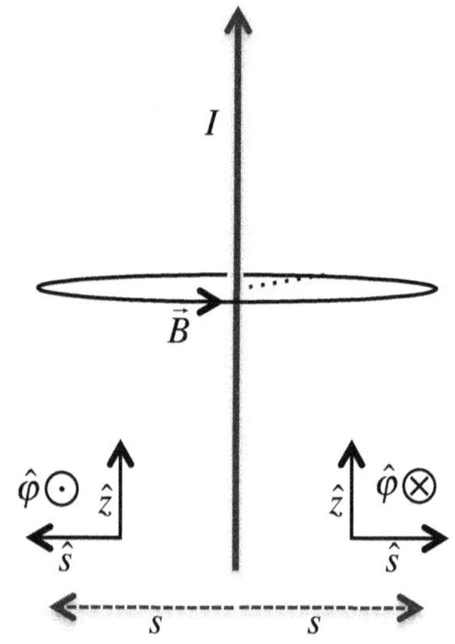

By placing a compass north, or south, of the wire we can measure the magnetic field, in ratio to that of the Earth's horizontal component, as:

$$B_{\text{wire}} = B_{\oplus}\tan(\varphi).$$

Once we measure the magnetic field, we can go one step further and find a relationship between the distance and the current. By doing these experiments, you can verify that the magnetic field is inversely proportional to the distance from the wire.

Putting these observations together, we can write the following:

$$\vec{B}_{\text{wire}} \propto \frac{I}{s}\hat{\varphi}.$$

Recall that in Chapter 6 we named the SI unit of the magnetic field, but we never gave the official definition. Actually, the tesla is defined in term of the ampere, which is defined based on the force between two long wires—at least up through 2018 as we discussed back on page 21.

Thought Experiment 8.2: Parallel Wires

As Ampère first observed: two parallel wires with current have an attractive force between them if the currents are in the same direction, and a repulsive force if in opposite directions.

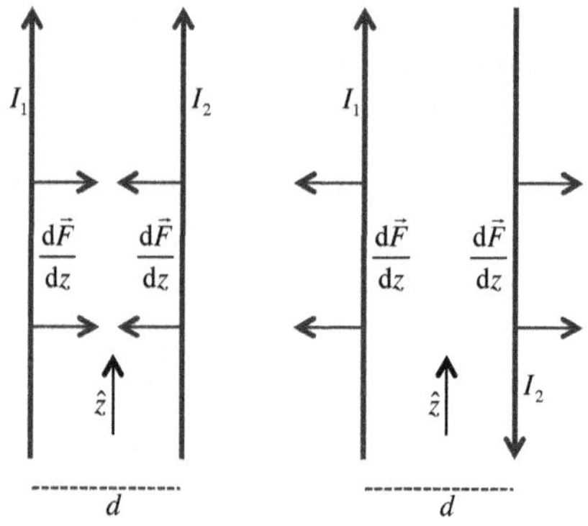

When two current carrying wires were observed interacting, it was noted that the force between them increased proportionally whenever the current in either wire was increased. Moreover, the force was observed to decrease inversely with distance when the separation between the wires was increased. Thus:

$$\frac{\mathrm{d}F}{\mathrm{d}z} \propto \frac{I_1 I_2}{d}.$$

Definition 8.1: The Ampere

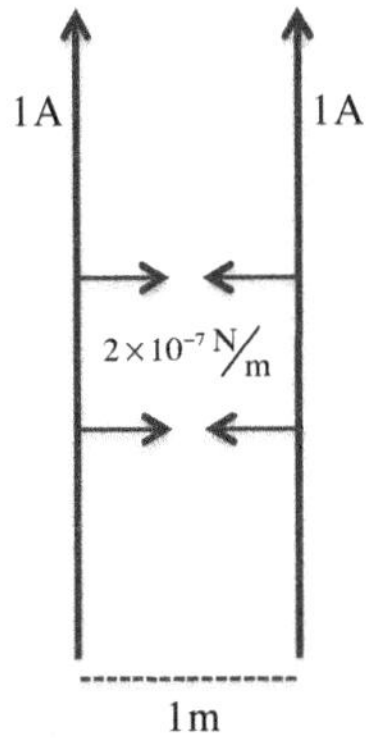

As we discuss in Discussion 1.4 (p. 21), Discussion 2.1 (p. 56), and Appendix D (p. 719), since the 1948 General Conference on Weights and Measures (CGPM), the ampere has been the accepted electrical base unit, with a definition of:

> The ampere is that constant current which, if maintained in two straight parallel conductors of infinite length, of negligible circular cross section, and placed 1 meter apart in vacuum, would produce between these conductors a force equal to 2×10^{-7} newton per meter of length.[112]

Based on this definition, *the permeability of free space*, μ_0, is exactly $4\pi\times10^{-7}\,{}^{\mathrm{N}}\!/_{\mathrm{A}^2}$, which is equivalent to the magnetostatic definition we introduced as experimental fact in Thought Experiment 6.8 (p. 202).

8.2 *The Law of Laplace*

Within a year after Ørsted's discovery of electromagnetism, the great experimentalist Michael Faraday used a similar concept to set a current carrying wire into a constant rotation in the presence of a strong permanent magnet—and the first electric motor was born.[113] Also in 1820, back in France, Jean-Baptiste Biot and Félix Savart measured the force on an electrical wire near a permanent magnet. A combination of their work and Faraday's field concepts, is what we now call the *law of Laplace*,[114] and led, in principle, to a more rigorous definition of the magnetic field. By applying the fluid model of current flow, we can also derive an expression for the magnetic force per volume as a function of current density in an external magnetic field.

Thought Experiment 8.3: The Magnetic Force on a Wire

Consider a wire carrying a current I, in an external magnetic field $\vec{B}$, what is the force on a small length of wire $\delta\vec{\ell}$?

The main results of the experiments we discussed are:

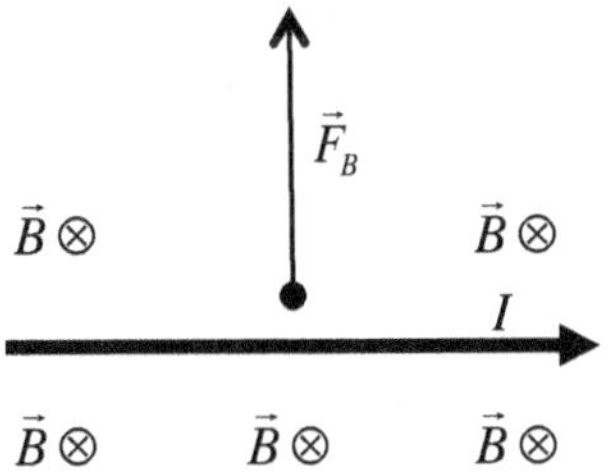

- If the magnetic field, whose direction can be defined using a compass, is perpendicular to the current, then the force on the wire is greatest. The force vanishes when the field and the current are aligned parallel, or anti-parallel, to one another.
- The strength of the force on the wire is proportional to the current flowing through the wire, the length of the wire, and the strength of the external magnetic field.

---

[112] Language from the 1948 General Conference on Weights and Measures.

[113] M. Faraday, "On Some New Electro-Magnetical Motions, and on the Theory of Magnetism," Quarterly Journal of Science and the Arts (Royal Institution of Great Britain), XII (1821), 74-96.

[114] This should not be confused with a different "Laplace's law" that arises in fluid mechanics.

- The force on the wire is mutually perpendicular to the direction of current flow through the wire and the external magnetic field.

We can summarize these results as the *law of Laplace*. In vector notation, this physical law can be expressed as:

$$\vec{F}_B \propto \int_{\text{wire}} \left(I\,\mathrm{d}\vec{\ell} \times \vec{B}\right) \propto I \int_{\text{wire}} \left(\mathrm{d}\vec{\ell} \times \vec{B}\right) \propto -I \int_{\text{wire}} \vec{B} \times \mathrm{d}\vec{\ell},$$

where I is the current flowing through the wire, $\vec{B}$ is the external magnetic field, and $\mathrm{d}\vec{\ell}$ is a differential length along the wire in the direction of positive current. The integration is along the wire.

Now, with the law of Laplace, we can define the tesla in terms of base units, such that the proportionality above becomes an equation:

$$\vec{F}_B = -I \int_{\text{wire}} \vec{B} \times \mathrm{d}\vec{\ell}\ .$$

Definition 8.2: The Tesla

A magnetic field of one tesla exerts a force of one newton on one meter of wire, carrying one ampere of current aligned perpendicular to the magnetic field. Therefore, a tesla can be represented in terms of other SI units as:

$$1\,\mathrm{T} = 1\frac{\mathrm{N}}{\mathrm{A\,m}} = 1\frac{\mathrm{N \cdot s}}{\mathrm{C\,m}} = 1\frac{\mathrm{kg}}{\mathrm{C\,s}}.$$

The tesla is an excellent unit for strong magnetic fields, such as those inside permanent magnets, modern loudspeakers, or used in magnetic resonance imaging. The magnetic field on the surface of the earth ranges from 22 to 67 microtesla, with lower fields in equatorial regions (Brazil) and higher fields in more northerly regions (Canada).

Problem 8.1: Raising the Bar

A cylindrical conducting bar of mass m, radius a, and length b is connected to two frictionless conducting vertical guides, so it is free to move vertically while remaining in electrical contact. The whole apparatus is located in a perpendicular horizontal magnetic field $\vec{B} = B\hat{y}$, and a gravitational field $\vec{g} = -g\hat{z}$, such as one might find on Earth. Each of the supports is connected to a source of constant current I.

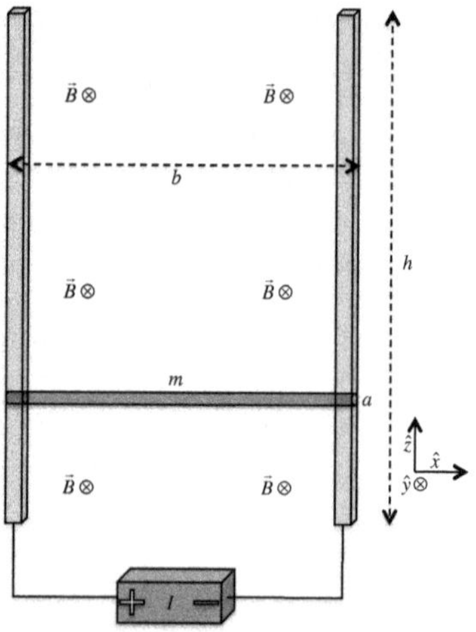

a) What current must flow through the loop for it to be held at a constant height z?

b) At time zero, the bar is at rest on the ground. Then a constant current source turns on with a current I, which is larger than your answer in part a. Solve for the velocity, $\vec{v} = v\hat{z}$, and height, z, as a function of time.

Example 8.1: A Loudspeaker

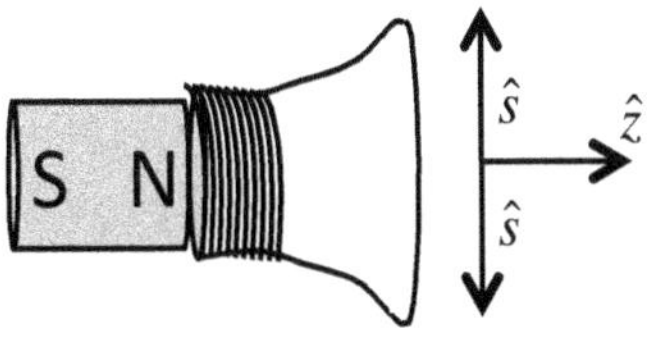

A typical loudspeaker is made with a permanent magnet and a coil of wire attached to a baffle of area A. The coil has N turns with a radius s, and the magnet has a magnetic field emanating from the north pole of the magnet and entering the south pole of $\vec{B} = B_s\hat{s} + B_z\hat{z}$. The amplifier sends a current through the speaker, which we will represent by $I(t)$. What is the pressure the speaker baffle exerts on the air?

We are given a cylindrically symmetrical system so a small length of wire is given by:

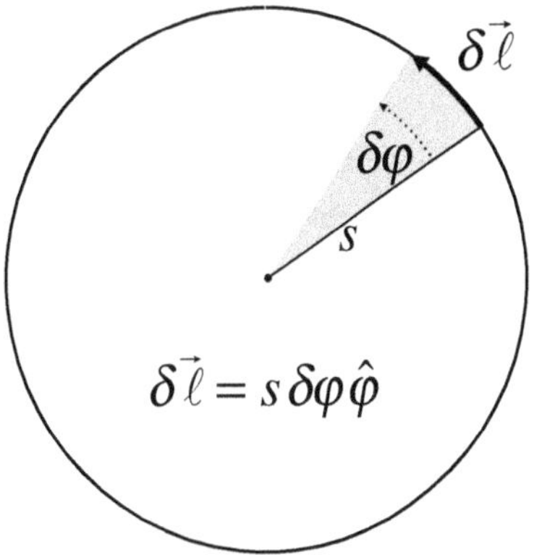

$$\delta\vec{\ell} = s\,\delta\varphi\,\hat{\varphi}.$$

So the force per length of wire is given by:

$$\begin{aligned}\delta\vec{F} &= I\,\delta\vec{\ell}\times\vec{B} = I(t)\left(s\,\delta\varphi\right)\left(\hat{\varphi}\times\vec{B}\right)\\ &= I(t)\left(s\,\delta\varphi\right)\hat{\varphi}\times\left(B_s\hat{s}+B_z\hat{z}\right)\\ &= I(t)\left(s\,\delta\varphi\right)\left(-B_s\hat{z}+B_z\hat{s}\right).\end{aligned}$$

The total force is the sum of loops:

$$\vec{F} = \int \mathrm{d}\vec{F} = \int_0^{2\pi N}\left(I(t)\left(s\,\mathrm{d}\varphi\right)\left(-B_s\hat{z}+B_z\hat{s}\right)\right).$$

Now let us assume that the speaker is designed to resist sideways expansion or contraction ($\hat{s}$ direction), but not to resist the axial force ($\hat{z}$ direction). Assuming that the magnetic field and coil radius are constant, we find that:

$$\vec{F}_z(t) = -2\pi N s B_s I(t).$$

Now the bellow has an area A, so the pressure exerted by the bellow is given by:

$$P(t) = \left(F_z(t)\big/A\right) = \left(\frac{-2\pi N s B_s}{A}\right)I(t).$$

Example 8.2: A Galvanometer

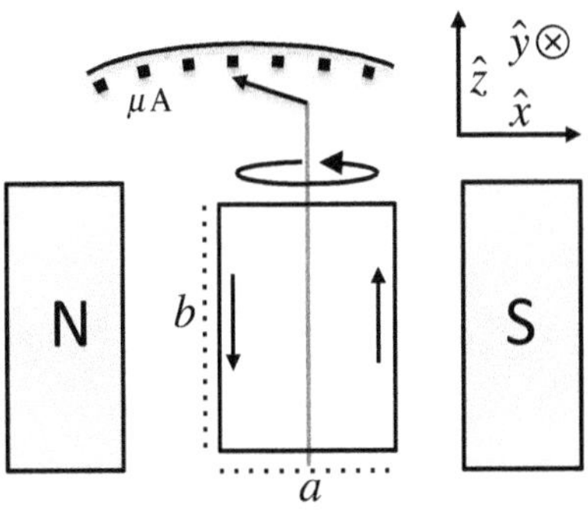

A *galvanometer* is used to measure small currents flowing through a circuit. Invented circa 1820, it is named for Luigi Galvani, the eighteenth-century Italian physician who discovered that an electric current can make a dead frog legs kick (p. 69). A galvanometer consists of a coil of wire placed in a magnetic field and attached to a torsion spring and an arrow indicator. The indicator arrow is set to zero, usually in middle of the dial, when no current flows.

Consider rectangular coil of N wires, with dimensions $a\times b$, located in a constant magnetic field $\vec{B} = B\,\hat{x}$ as shown. This is allowed to pivot about the $\hat{z}$ axis, where it

$$\vec{\tau} = \oint_{\text{path}} \vec{r}\times\mathrm{d}\vec{F} = I\oint_{\text{path}} \vec{r}\times\left(\mathrm{d}\vec{\ell}\times\vec{B}\right)$$

pushes, or pulls, against a torsion spring with spring constant k_s.

Since the galvanometer is in static equilibrium, the net torque must be zero, so the torque on the wire coil, by the magnetic field, must be given by:

$$\vec{\tau} = -k_s\,\varphi\,\hat{z},$$

where φ is the angle of rotation from equilibrium.

Next consider the force on a small piece of wire, $\delta\vec{\ell}$, which is given by:

$$\delta\vec{F} = I\,\delta\vec{\ell}\times\vec{B},$$

and so the torque on the same current element is:

$$\delta\vec{\tau} = \vec{r}\times\delta\vec{F} = \vec{r}\times\left(I\,\delta\vec{\ell}\times\vec{B}\right) = I\,\vec{r}\times\left(\delta\vec{\ell}\times\vec{B}\right).$$

Before we integrate around the loop, let us simplify the torque in two special cases: (1) The wire is perpendicular to the axis, or $\delta\vec{\ell} = \pm\delta\ell\,\hat{s}$; and (2) the wire is parallel to the axis, or $\delta\vec{\ell} = \pm\delta\ell\,\hat{z}$.

Case #1 ($\delta\vec{\ell} = \pm\delta\ell\,\hat{s}$):

$$\begin{aligned}\delta\vec{\tau} &= I\,\vec{r}\times\left(\delta\vec{\ell}\times\vec{B}\right) = I\,B\,\delta s\,\vec{r}\times\left(\left(\pm\hat{s}\right)\times\hat{x}\right) = I\,B\,\delta s\,\vec{r}\times\left(\mp\sin\varphi\,\hat{z}\right) = I\,B\,\delta s\left(s\,\hat{s}+z\hat{z}\right)\times\left(\mp\sin\varphi\,\hat{z}\right)\\ &= I\,B\,\delta s\;s\sin\varphi\left(\hat{s}\times\mp\hat{z}\right) = B\,\delta s\;s\sin\varphi\left(\hat{s}\times\mp\hat{z}\right) = \pm B\,\delta s\,s\sin\varphi\;\hat{\varphi} = \pm\delta\tau\;\hat{\varphi}.\end{aligned}$$

Case #2 ($\delta\vec{\ell} = \pm\delta\ell\,\hat{z}$):

$$\begin{aligned}\delta\vec{\tau} &= I\,\vec{r}\times\left(\delta\vec{\ell}\times\vec{B}\right) = I\,B\,\delta z\,\vec{r}\times\left(\left(\pm\hat{z}\right)\times\hat{x}\right) = I\,B\,\delta z\,\vec{r}\times\left(\mp\hat{y}\right) = I\,B\,\delta z\left(s\,\hat{s}+z\hat{z}\right)\times\left(\mp\hat{y}\right) = I\,B\,\delta z\left(s\,\hat{s}\right)\times\left(\mp\hat{y}\right)\\ &= \mp I\,B\,\delta z\;s\left(\hat{s}\times\hat{y}\right) = \mp I\,B\,\delta z\;s\,\cos\varphi\;\hat{z} = \mp\delta\tau\;\hat{z}.\end{aligned}$$

Now we integrate around the loop N times to find the total torque:

$$\begin{aligned}\vec{\tau} &= N\oint_{\text{path}}\mathrm{d}\vec{\tau} = N\left(\int_{s=0}^{s=a/2}\mathrm{d}\vec{\tau}\Big|_{z=-\frac{1}{2}b} + \int_{-b/2}^{b/2}\mathrm{d}\vec{\tau}\Big|_{s=\frac{1}{2}a} + \int_{s=a/2}^{s=0}\mathrm{d}\vec{\tau}\Big|_{z=\frac{1}{2}b} + \int_{s=0}^{s=a/2}\mathrm{d}\vec{\tau}\Big|_{z=\frac{1}{2}b} + \int_{b/2}^{-b/2}\mathrm{d}\vec{\tau}\Big|_{s=\frac{1}{2}a} + \int_{s=a/2}^{s=0}\mathrm{d}\vec{\tau}\Big|_{z=-\frac{1}{2}b}\right)\\ &= N\left(+\cancel{\int_{s=0}^{s=a/2}\mathrm{d}\tau\Big|_{z=-\frac{1}{2}b}} - \cancel{\int_{s=a/2}^{s=0}\mathrm{d}\tau\Big|_{z=\frac{1}{2}b}} + \cancel{\int_{s=0}^{s=a/2}\mathrm{d}\tau\Big|_{z=\frac{1}{2}b}} - \cancel{\int_{s=a/2}^{s=0}\mathrm{d}\tau\Big|_{z=-\frac{1}{2}b}}\right)\hat{\varphi} + N\left(-\int_{z=-b/2}^{z=b/2}\mathrm{d}\tau\Big|_{s=\frac{1}{2}a} - \int_{z=-b/2}^{z=b/2}\mathrm{d}\tau\Big|_{s=\frac{1}{2}a}\right)\hat{z}\\ &= -2N\left(\int_{z=-b/2}^{z=b/2}\mathrm{d}\tau\Big|_{s=\frac{1}{2}a}\right)\hat{z} = -2NI\,B\left(\int_{z=-b/2}^{z=b/2}\left(\tfrac{1}{2}a\right)\cos\varphi\;\mathrm{d}z\Big|_{s=\frac{1}{2}a}\right)\hat{z} = -NI\,B\left(ab\right)\cos\varphi\;\hat{z}.\end{aligned}$$

Now, returning to the galvanometer, the relationship therefore between the dial angle and the current flow is:

$$\tau = -k_s\,\varphi\,\hat{z} = -NI\,B\left(ab\right)\cos\varphi\;\hat{z}\quad\text{and}\quad\varphi = \frac{NIBab}{k_s}\cos\varphi \approx \frac{NIBab}{k_s}\left(1-\varphi^2\right) \approx \left(\frac{NBab}{k_s}\right)I.$$

Next we will write the torque in terms of the area vector of the loop, $\vec{A} = -ab\,\hat{\varphi}$:

$$\vec{\tau} = \left(NIA\right)\left(-\cos\varphi\;\hat{z}\right)B = \left(NIA\right)\left(\hat{\varphi}\times\hat{x}\right)B = \left(NIA\,\hat{\varphi}\right)\times\left(B\,\hat{x}\right) = \left(NI\vec{A}\right)\times\vec{B}.$$

Notice how the coil of wire responds to a magnetic field in much the same way as a permanent magnet. On page 189 we defined the magnetic moment $\vec{m}$ such that $\vec{\tau} = \vec{m} \times \vec{B}$. We can therefore write the magnetic moment of a coil of wire as simply $\vec{m} = N I \vec{A}$. You will show this in general in Problem 8.3 below.

Problem 8.2: The Magnetic Moment of a Current Loop

Use the definition of the magnetic moment $\left(\vec{\tau} = \vec{m} \times \vec{B}\right)$ to show that the magnetic moment of any current loop is given by $\vec{m} = I \vec{A}$.

Problem 8.3: A Pivoting Wire Loop

Consider an $a \times b$ wire loop with linear mass density λ_m attached to a frictionless hinge on the bottom edge. A current I flows through the loop from a constant current source. It is placed in a horizontal magnetic field $\vec{B} = B\hat{x}$, perpendicular to the axis of rotation.

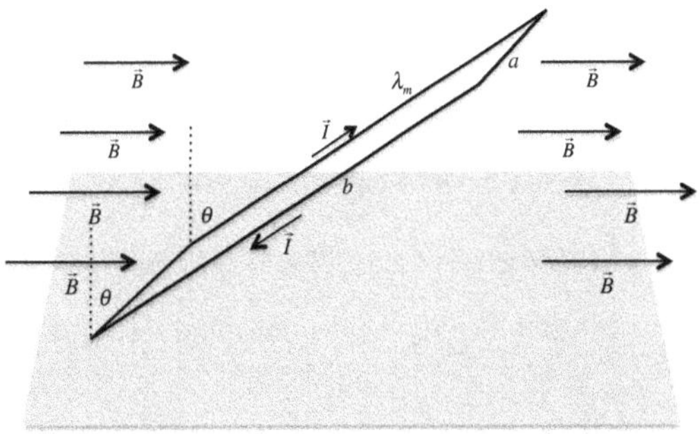

a) Find the angular acceleration, $\ddot{\theta} \equiv \frac{d^2\theta}{dt^2}$, as a function of θ.

b) What minimum current must flow through the loop for it to be in stable equilibrium when vertical? Now assume that I is greater than this.

c) The loop is now displaced a small angle θ_0 and let go to oscillate about the top. What is the characteristic frequency ω_0 of this oscillation? Solve the equations of motion for small displacements about the vertical.

Problem 8.4: A Leaning Wire Loop

Consider a wire loop of dimensions $a \times b$ with a linear mass density λ_m located in a constant magnetic field in the vertical direction. The wire loop is placed at an angle θ from the vertical. The surfaces between the wire and the floor, and the wire and the wall, each have different coefficients of static friction μ_F, and μ_W, respectively. Next you lower the current I until the wire loop just begins to slip.

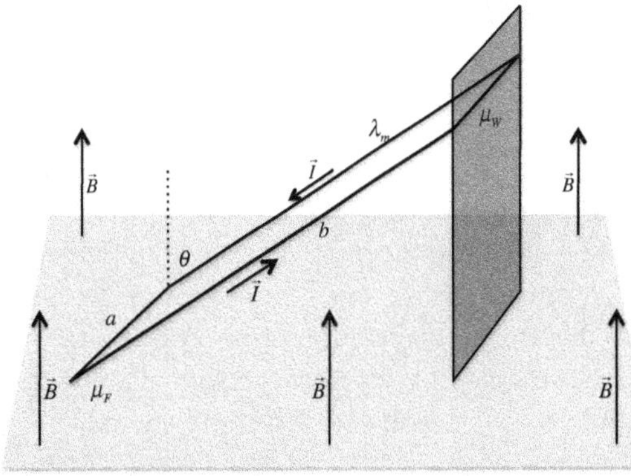

a) Assuming the wall is slippery $\left(\mu_W \approx 0\right)$, find an inequality expression for the current in the wire, so long as the wire does not slip on the floor.

b) Now assume that the floor is slippery $\left(\mu_F \approx 0\right)$, but the wall is not. Find an inequality expression for the current in the wire so long as the wire does not slip on the wall.

c) Now assume neither the floor nor the wall is slippery and again find an inequality expression for the current in the wire so long as the wire does not slip.

d) Does a higher current mean that is more or less likely to slip? What if the current is reversed? Explain.

Hints: (i) Recall from introductory physics that the force of static friction can only hold an object without slipping when $F_s \le \mu_s F_N$. (ii) Also recall that, for an object to be in static equilibrium, all the forces, and the torques, must be in balance.

Problem 8.5: A Spinning Charged Sphere

Consider a solid sphere of constant charge density. The ball has a radius a and charge Q. It is spinning with a uniform angular velocity $\vec{\omega} = \omega\hat{z}$ and is in a constant magnetic field $\vec{B} = B\hat{x}$ perpendicular to its axis of rotation.

a) Find the net torque on the sphere in terms of the variables given.

b) What is the magnetic moment $\vec{m}$ of the sphere?

Thought Experiment 8.4: The Magnetic Field Surrounding a Current Loop

In Problem 8.2 (p. 251) you showed that a current loop exhibits a magnetic moment given by:

$$\vec{m} = I\,\vec{A},$$

where I and $\vec{A}$ are the current and area vector respectively.

Since magnetic fields affect current loops in much the same way as they do permanent magnets, it seems reasonable to hypothesize that the corollary is also true: the magnetic field caused by current loops should mimic the field surrounding permanent magnets.

Under this logic, the magnetic field surrounding a loop of current, in the far limit, should be given in by the dipole formula:

$$\vec{B}_{r\to\infty} = \left(\frac{\mu_0}{4\pi}\right)\frac{3(\vec{m}\cdot\hat{r})\hat{r} - \vec{m}}{r^3} = \left(\frac{\mu_0 I}{4\pi}\right)\frac{3(\vec{A}\cdot\hat{r})\hat{r} - \vec{A}}{r^3} = \left(\frac{\mu_0 I A}{4\pi}\right)\frac{1}{r^3}\left(\frac{3sz}{r^2}\hat{s} + \frac{2z^2 - s^2}{r^2}\hat{z}\right).$$

This is what is experimentally observed for distances much greater than the loop radius. It turns out, however, that the most general expression for a magnetic field produced by a current loop is not exactly that of dipole, nor is it simple. For example, the formula for such a field measured at some point not along the current loop's axis of symmetry, or very far away from the loop, is a complicated function of so-called elliptic integrals (which must be evaluated numerically).[115] Since physics is the study of the universal laws of nature, we concentrate on the simplest system that illustrates the phenomenon in question. In this case, this phenomenon is the production of magnetic fields by current carrying wires. Ampère began with the force of interaction between parallel wires, which is much more tractable than the current loop and to which we now turn.

---

[115] J. Simpsons, J. Lane, C. Immer, R. Youngquist, "Simple analytic expressions for the magnetic field of a circular current loop," NASA technical documents, 2001. Simpson et al. obtain closed-form expressions for the field due to a "flimantary" current loop that are "exact" everywhere outside the current-carrying wire. Despite the descriptor of "exact," the formulas are in term of elliptic integrals, which must be evaluated numerically.

8.3 *Ampère's Law*

André-Marie Ampère was obsessed with finding an analog to Coulomb's law relating the forces between differential elements of current, and he succeeded in finding this relationship. The equivalent relationship in modern field notation is not called Ampère's law, but rather the law of Biot and Savart, which we will cover in Chapter 11 (p. 354).

Given that Ampère contributed the most to electrodynamics of all of the mathematical physicists in the French Academy, it is only fitting that James Clerk Maxwell would name his equation for the source of magnetic fields after him.

Thought Experiment 8.5: The Magnetic Field Surrounding a Wire

Consider now Wire 2 from Thought Experiment 8.2 (p. 246). There must be something at its location, say a magnetic field, which we can say causes the force on Wire 2. Since the cause of this magnetic field was Wire 1, we will call this $\vec{B}_1$. Here, we will use this cause and effect argument to find the relationship between the magnetic field $\vec{B}_1$ and the current I_1 that caused it in the first place. Recall that $\vec{B}_1$ was caused by Wire 1, and is the cause of the force on Wire 2.

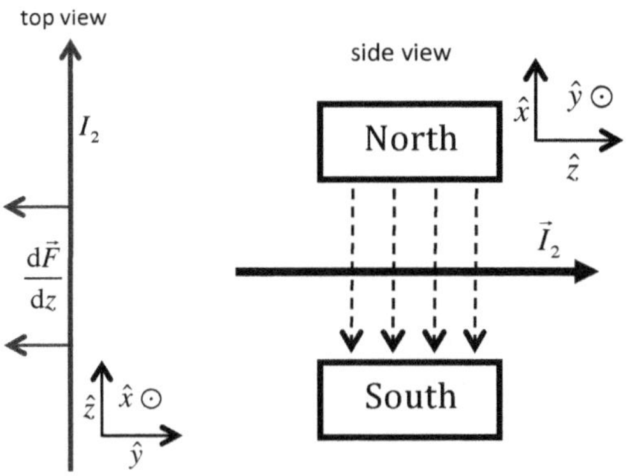

We now equate the Laplace force per length on Wire 2:

$$\left(\frac{d\vec{F}}{dz}\right)_{\text{on wire 2}} = \vec{I}_2 \times \vec{B}_1 = \frac{\mu_0}{2\pi}\frac{I_1 I_2}{d}(-\hat{y}) = \cancel{I_2}\,\hat{z} \times \vec{B}_1 = \frac{\mu_0}{2\pi}\frac{I_1 \cancel{I_2}}{d}(-\hat{y}),$$

which becomes:

$$\vec{B}_1 = -\frac{\mu_0 I_1}{2\pi d}\hat{x}.$$

Now looking at the other wire, we can see:

$$\left(\frac{d\vec{F}}{dy}\right)_{\text{on wire 1}} = \vec{I}_1 \times \vec{B}_2 = \frac{\mu_0}{2\pi}\frac{I_1 I_2}{d}\hat{y}$$

$$\cancel{I_1}\,\hat{z} \times \vec{B}_2 = \frac{\mu_0}{2\pi}\frac{\cancel{I_1} I_2}{d}\hat{y}$$

$$\hat{z} \times \vec{B}_2 \times \hat{z} = \frac{\mu_0}{2\pi}\frac{I_2}{d}\hat{y} \times \hat{z} \quad \rightarrow \quad \vec{B}_2 = \frac{\mu_0 I_2}{2\pi d}\hat{x}.$$

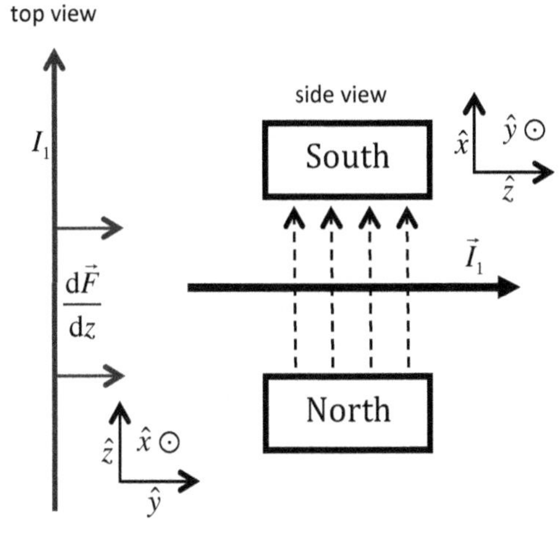

We now use cylindrical coordinates to express the magnetic field caused by a current along the $\hat{z}$ axis:

$$\vec{B} = \frac{\mu_0 I}{2\pi s}\hat{\varphi}.$$

As we mentioned before, Biot and Savart made careful measurements of the torque surrounding a wire, so as to find the equivalent relationship in their unit system.

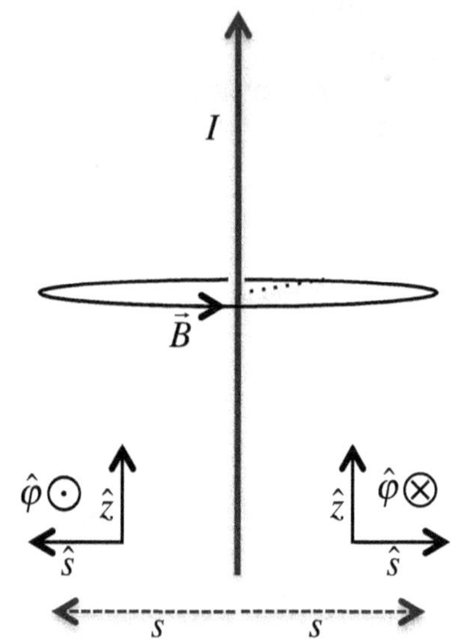

Recall from Peregrinus that all magnetic field lines are closed loops, and, since these loops are circular, each has a length of $2\pi s$, so the denominator above is simply the length of this circular loop, so this equation could be written as:

$$\oint_{\text{path}} \vec{B}\cdot \mathrm{d}\vec{\ell} = \mu_0 I \,.$$

By considering the forces that current carrying wires exert on each other, we have recovered the results of Ørsted and Ampère: electric currents act as sources of magnetic field.

There is some irony in the fact that Ampère never wrote this relationship down! Maxwell published derivations of it in the context of a fluid analogy to magnetism in two seminal papers: "On Faraday's Lines of Force" and "On the Physical Lines of Force."[116]

In Chapter 1 we discussed Franklin's fluid model of charge, which led to our discussing the body force exerted by an electric field on a charged fluid in Chapter 2. Here we will discuss the force exerted by a magnetic field on a moving charged fluid.

Problem 8.6: Ørsted's Current Source

> If the distance of the connecting wire from the magnetic needle does not exceed ¾ inch, the deviation will amount to an angle of approximately 45°. If the distance is augmented, the angles will decrease as the distances increase. Moreover, the deviation varies with the power of the apparatus. – Hans Christian Ørsted (Copenhagen, 1820)[117]

In Demark in 1820, the horizontal component of the Earth's magnetic field was 16 μT.[118] How much current was flowing through Ørsted's wire?

Thought Experiment 8.6: The Magnetic Field Around a Capacitor

Let us now imagine that we attach a parallel plate capacitor to a long thin wire, as shown. At some point in time a constant current I flows through the wire. What is the magnetic field surrounding the capacitor?

No charge flows between the plates, so the current equals zero there. But we might expect, for reasons of continuity alone, that the magnetic field should not suddenly vanish between the capacitor plates. Perhaps the field is the same whether we draw our loop just above the upper plate, or between the plates, despite the fact that there is no current flowing through this loop.

---

[116] J.C. Maxwell, "On Physical Lines of Force," Philosophical Magazine, **21**:139, (1861), 161-75. Specifically, note Maxwell's equation (9) on p. 171, and his Fig. (6) on p. 172.

[117] H.C. Ørsted, "Experiments on the Effect of the Electric Conflict on the Magnetic Needle" Vindenskabernes Selskabos Overskrifter **3** (1820-21): 12-21, Translated by K. Jelved, A.D. Jackson, and Ole Knudsen, in *Selected Scientific Works of Hans Christian Ørsted* (Princeton: Princeton University Press). ©1998 by Princeton University Press

[118] According to NOAA's National Centers for Environmental Information, at www.ngdc.noaa.org/geomag-web.

If this is indeed the case, we might ask ourselves, "If not the current, then what seems to be flowing through the loop?" Our answer is the electric flux due to the charge that builds up on the capacitor plates! So, we can use Gauss's law to rewrite the magnetic field in terms of the electric flux:

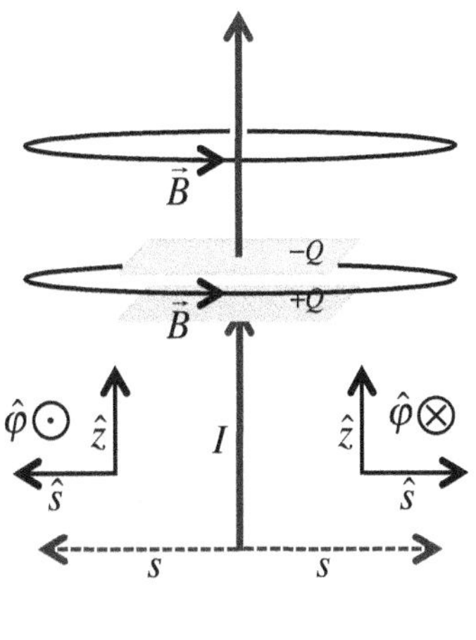

$$\vec{B}=\frac{\mu_0 I}{2\pi s}\hat{\varphi}=\frac{\mu_0\left(\frac{\partial Q}{\partial t}\right)}{2\pi s}\hat{\varphi}=\frac{\mu_0\varepsilon_0}{2\pi s}\frac{\mathrm{d}}{\mathrm{d}t}\left(\int\vec{E}\cdot\mathrm{d}\vec{A}\right)\hat{\varphi}=\frac{\mu_0\varepsilon_0}{2\pi s}\frac{\mathrm{d}\Phi_E}{\mathrm{d}t}\hat{\varphi},$$

where Φ_E is the electric flux through the Amperian loop. Again noticing that $2\pi s$ is the path around the Amperian loop, we can write this as:

$$\oint_{\text{path}}\vec{B}\cdot\mathrm{d}\vec{\ell}=\mu_0\varepsilon_0\frac{\mathrm{d}\Phi_E}{\mathrm{d}t}.$$

This is clearly not a steady-state situation. If the electric field were static the charge density would not vary with time, so the time rate of change of the electric flux would equal zero.

Our calculation here is an educated guess, and similar to one that James Clerk Maxwell made about a century and a half ago. Maxwell, who first introduced the term, referred to $\varepsilon_0\frac{\mathrm{d}}{\mathrm{d}t}\Phi_E$ as the *displacement current* in analogy with the conventional conduction current I.

Thought Experiment 8.7: The Maxwell-Ampère Law in Free Space

Consider a long wire, of finite cross section, that carries a constant current I_{in}. Let the wire now have a small gap where it is interrupted by a pair of capacitor plates and a very thin wire that carries a small current I between the capacitor plates, so that some current charges the capacitor and some passes through to the other side of the gap. The magnetic field would therefore simply be the sum of that due to the small current I and that due to the changing electric flux from the capacitor:

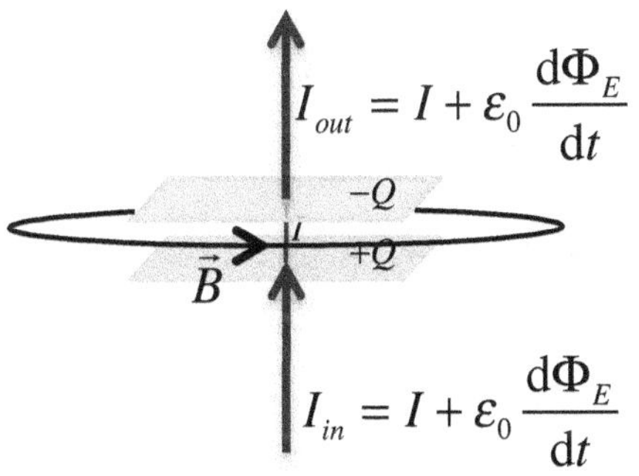

$$\vec{B}=\frac{\mu_0 I}{2\pi s}\hat{\varphi}+\frac{\mu_0\varepsilon_0}{2\pi s}\frac{\mathrm{d}\Phi_E}{\mathrm{d}t}\hat{\varphi}=\frac{\mu_0}{2\pi s}\left(I+\varepsilon_0\frac{\mathrm{d}\Phi_E}{\mathrm{d}t}\right)\hat{\varphi}.$$

Extrapolating this relationship to non-constant magnetic fields, or non-circular loops, we can rewrite this as:

$$\oint_{\text{path}}\vec{B}\cdot\mathrm{d}\vec{l}=\mu_0\left(I+\varepsilon_0\frac{\mathrm{d}\Phi_E}{\mathrm{d}t}\right)=\mu_0\oint_{\text{surface}}\left(\vec{J}+\varepsilon_0\frac{\mathrm{d}\vec{E}}{\mathrm{d}t}\right)\cdot\mathrm{d}\vec{A}$$

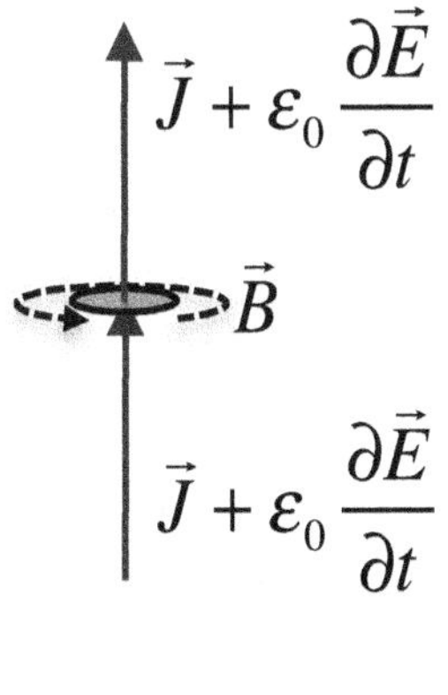

Next we take the limit as the area of this loop becomes tiny:

$$\lim_{\delta A\to 0}\frac{\oint_{\text{path}}\vec{B}\cdot\mathrm{d}\vec{l}}{\delta A}=\lim_{\delta A\to 0}\frac{\oint_{\text{surface}}\mu_0\left(\vec{J}+\varepsilon_0\frac{\mathrm{d}\vec{E}}{\mathrm{d}t}\right)\cdot\mathrm{d}\vec{A}}{\delta A}.$$

This is the definition of the curl in the direction of the current density, so:

$$\vec{\nabla} \times \vec{B} = \mu_0 \left(\vec{J} + \varepsilon_0 \tfrac{\partial}{\partial t} \vec{E} \right).$$

This is the Maxwell-Ampère Law in free space.

Thought Experiment 8.8: The Magnetic Field Inside an Empty Solenoid

Consider, a long solenoid with n turns per unit length, and a steady current I flowing through the wire, wrapped around a non-magnetic tube. What is the magnetic field inside of the solenoid?

From symmetry we can assume that the magnetic field is along the axis of the solenoid, so $\vec{B} = B\hat{z}$, and that all time derivatives are zero.

Next we draw an Ampèrian loop of length l so from the Maxwell-Ampère law:

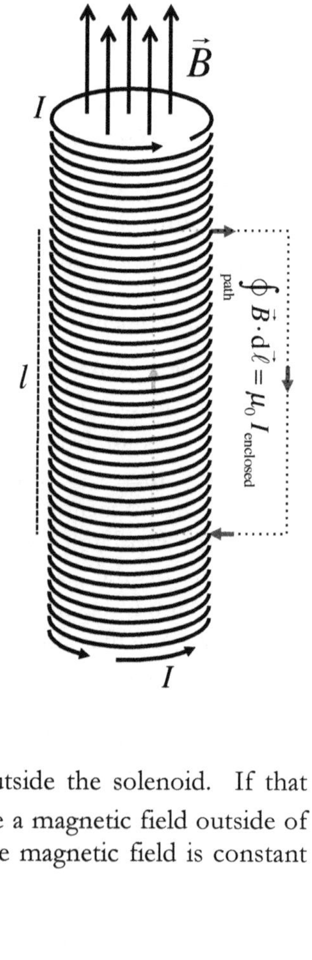

$$\vec{\nabla} \times \vec{B} = \mu_0 \left(\vec{J} + \varepsilon_0 \frac{\partial \vec{E}}{\partial t} \right) = \mu_0 \vec{J}$$

$$\oint_{\text{surface}} \left(\vec{\nabla} \times \vec{B} \right) \cdot d\vec{A} = \oint_{\text{path}} \vec{B} \cdot d\vec{\ell} = \mu_0 \oint_{\text{surface}} \vec{J} \cdot d\vec{A}.$$

Now integrating around the loop shown in the diagram, we find:

$$\oint_{\text{path}} \vec{B} \cdot d\vec{\ell} = \int_{0}^{l} \underset{s=s_1}{B\, dz} - \int_{l}^{0} \underset{s=s_2}{B\, dz}$$

where s_1 and s_2 are at the inner and outer integration paths.

Noticing that the field must fall off quickly as $s_2 \to \infty$, we see:

$$\oint_{path} \vec{B} \cdot d\vec{\ell} = \int_{0}^{l} \underset{s=s_1}{(B)\, dz} = (B)_{\text{Inside}}\, l.$$

Now consider the right hand side of the original equation:

$$\oint_{\text{surface}} \vec{J} \cdot d\vec{A} = nlI.$$

Notice that as long as $\vec{B} = B\hat{z}$, the magnetic field must be zero outside the solenoid. If that criterion is no longer valid, say for a short solenoid, then there can be a magnetic field outside of the solenoid. Notice also that our result is independent of s, so the magnetic field is constant and uniform inside the solenoid.

Problem 8.7: The Magnetic Field in a Long Solenoid Magnet

Consider, a long solenoid with n turns per unit length, and a steady current I flowing through the wire. However, rather it is now wrapped around a magnetic cylinder with uniform magnetization $\vec{M} = M\hat{z}$.

Argue, using the Maxwell-Ampere law, that the magnetic field on the axis of the cylinder is: $\vec{B} = \mu_0 nI\,\hat{z} + \mu_0 M\,\hat{z}$. (See Problem 6.17, p. 208).

Problem 8.8: An Large Current Sheet

Consider an infinitely large thin sheet of uniform current density $\vec{J} = J\,\hat{x}$ with small thickness h.

a) What is the magnetic field a distance z normal to the current sheet?

b) Report on the Heliospheric current sheet, its shape, and the associated magnetic field. Verify that the current, and the associated magnetic field, are related via the relationship you found in part a.

Thought Experiment 8.9: A Hollow Magnetic Ring

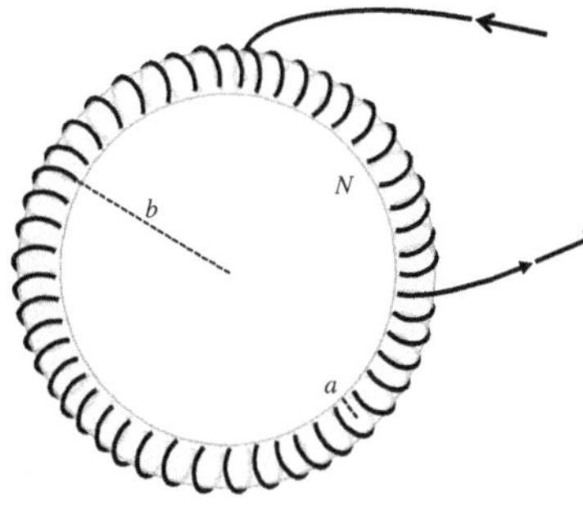

Consider a hollow non-magnetic toroidal tube with N coils of wire wrapped around it as shown, with the cross-sectional radius a and ring radius $b \gg a$, as shown in the picture.

Since this is simply the long solenoid wrapped upon itself, we can assume that the magnetic field is counter-clockwise and given by the relation:

$$\vec{B} = \mu_0 n I\,\hat{\varphi} = \frac{\mu_0 N I}{2\pi b}\,\hat{\varphi}.$$

Dotting both sides by the circumference, gives us:

$$\vec{B}\cdot(2\pi b\hat{\varphi}) = \left(\frac{\mu_0 N I}{\cancel{2\pi b}}\,\hat{\varphi}\right)\cdot\left(\cancel{2\pi b}\,\hat{\varphi}\right) = \mu_0 N I.$$

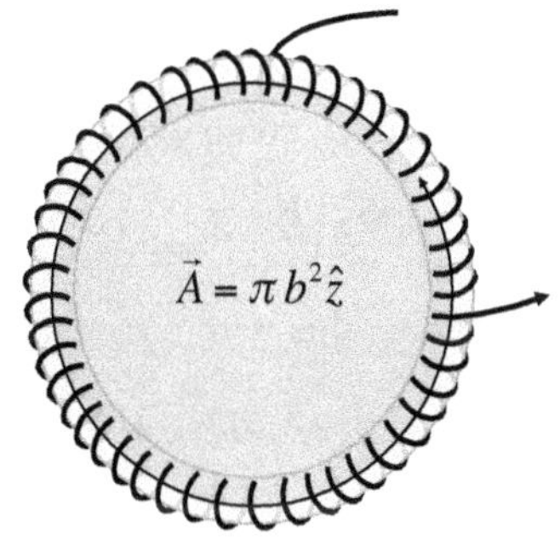

Now we draw an Amperian loop around the ring as shown. Notice that the inside part of each coil of wire passes through the surface, while none of the outer parts do. Therefore, the total amount of current passing through the surface is given by:

$$I_{\text{tot}} = \int_{\text{surface}} \vec{J}\cdot d\vec{A} = N I.$$

Now, writing this in integral form, we have the following relationship:

$$\oint \vec{B}\cdot d\vec{l} = \int_{\text{surface}} \vec{J}\cdot d\vec{A}.$$

Taking the limit as $A \to 0$, yields the curl of the magnetic field as:

$$\vec{\nabla}\times\vec{B} = \lim_{A\to 0}\oint_{\text{path}} \vec{B}\cdot d\vec{\ell} = \mu_0 \lim_{A\to 0}\int_{\text{surface}} \vec{J}\cdot d\vec{A} = \mu_0\vec{J}.$$

Next, imagine that we added a capacitor to some of the wires in such a way that our Amperian surface is between the plates. As in Thought Experiment 8.7 (p. 255), we would expect the magnetic field to stay the same, so our relationship would then become:

$$\vec{\nabla}\times\vec{B} = \mu_0\vec{J} + \mu_0\varepsilon_0 \tfrac{\partial}{\partial t}\vec{E},$$

which is, again, the Maxwell-Ampère Law in free space.

Thought Experiment 8.10: A Magnetic Ring

Consider the toroidal permanent magnet from Problem 6.20 (p. 209), with uniform counter-clockwise magnetization $\vec{M} = M\,\hat{\varphi}$. You found that inside the ring the magnetic field is simply $\vec{B}_{\text{inside}} = \mu_0 \vec{M}$, and outside the ring the magnetic field falls off rather quickly with distance.

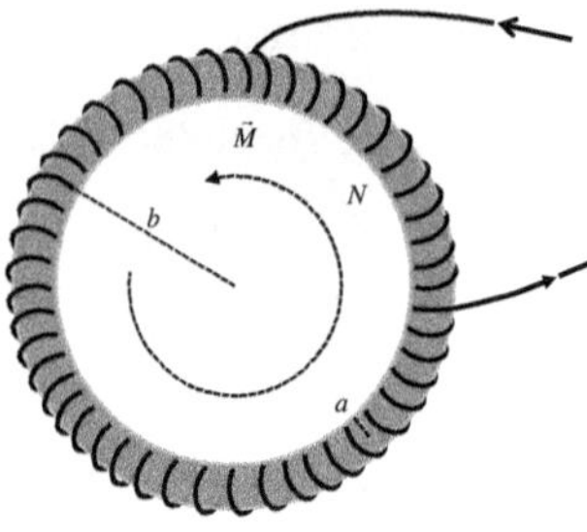

Let us, now, wrap a wire around the torus, as in the diagram, like in Thought Experiment 8.9 above with a free space donut. Putting these two concepts together, we can superimpose the magnetic field:

$$\vec{B} = \vec{B}_{\text{magnet}} + \vec{B}_{\text{coil}}, \quad \text{or} \quad \vec{B}_{\text{coil}} = \vec{B} - \vec{B}_{\text{magnet}}.$$

Now we take the curl of both sides, so:

$$\vec{\nabla} \times \vec{B}_{\text{coil}} = \vec{\nabla} \times \left(\vec{B} - \vec{B}_{\text{magnet}}\right) = \lim_{A\to 0} \tfrac{1}{A} \oint_{\text{path}} \left(\vec{B} - \vec{B}_{\text{magnet}}\right) \cdot \mathrm{d}\vec{\ell} = \lim_{A\to 0} \tfrac{1}{A} \oint_{\text{path}} \vec{B} \cdot \mathrm{d}\vec{\ell} - \lim_{A\to 0} \tfrac{1}{A} \oint_{\text{path}} \vec{B}_{\text{magnet}} \cdot \mathrm{d}\vec{\ell}.$$

The first term is simply the curl of the magnetic field. What about the second term? You argued in Problem 6.20 that, while not true in the large limit, in the small limit the magnetic field of a toroidal magnet is simply:

$$\vec{B}_{\text{magnet}} = \mu_0 \vec{M} \quad \text{everywhere.}$$

Since we have taken exactly this same limit to find the curl, the following must be true:

$$\vec{\nabla} \times \vec{B}_{\text{coil}} = \vec{\nabla} \times \left(\vec{B} - \vec{B}_{\text{magnet}}\right) = \vec{\nabla} \times \vec{B} - \vec{\nabla} \times \vec{B}_{\text{magnet}} = \vec{\nabla} \times \vec{B} - \mu_0 \vec{\nabla} \times \vec{M}.$$

Combining this with our result from Thought Experiment 8.9 above:

$$\vec{\nabla} \times \vec{B}_{\text{coil}} = \mu_0 \vec{J} + \mu_0 \varepsilon_0 \tfrac{\partial}{\partial t} \vec{E},$$

gives us the Maxwell-Ampère Law with magnetic materials:

$$\vec{\nabla} \times \vec{B} = \mu_0 \vec{J} + \mu_0 \varepsilon_0 \tfrac{\partial}{\partial t} \vec{E} + \mu_0 \vec{\nabla} \times \vec{M}.$$

Thought Experiment 8.11: The Maxwell-Ampère Law in General

As of now we have found the Maxwell-Ampère Law in free space, and in a magnetic material. What about a dielectric with polarization $\vec{P}$?

Given our experience thus far, let us guess that working in a dielectric will simply add another term that acts like a current to the Maxwell-Ampère Law .

To find this last term, we define a vector function, $\vec{f}(\vec{P})$, such that:

$$\vec{\nabla} \times \vec{B} = \mu_0 \left(\vec{J} + \varepsilon_0 \tfrac{\partial}{\partial t} \vec{E} + \vec{\nabla} \times \vec{M} + \vec{f}(\vec{P})\right).$$

Our goal, now, is to apply the other fundamental laws one at a time. Since many of them involve the divergence, we will take the divergence of both sides and distribute:

$$\vec{\nabla}\cdot\left(\vec{\nabla}\times\vec{B}\right)=\mu_0\left(\vec{\nabla}\cdot\vec{J}+\varepsilon_0\vec{\nabla}\cdot\tfrac{\partial}{\partial t}\vec{E}+\vec{\nabla}\cdot\left(\vec{\nabla}\times\vec{M}\right)+\vec{\nabla}\cdot\vec{f}(\vec{P})\right).$$

Notice that the divergence of the curl is zero, and we can swap the order of integration, so:

$$0=\vec{\nabla}\cdot\vec{J}+\varepsilon_0\tfrac{\partial}{\partial t}\left(\vec{\nabla}\cdot\vec{E}\right)+\vec{\nabla}\cdot\vec{f}(\vec{P}).$$

Now we can substitute in the continuity equation and Gauss's law, so:

$$0=\left(-\frac{\partial\rho}{\partial t}\right)+\cancel{\varepsilon_0}\frac{\partial}{\partial t}\left(\frac{\rho}{\cancel{\varepsilon_0}}-\frac{\vec{\nabla}\cdot\vec{P}}{\cancel{\varepsilon_0}}\right)+\vec{\nabla}\cdot\vec{f}(\vec{P})=-\vec{\nabla}\cdot\left(\tfrac{\partial}{\partial t}\vec{P}\right)+\vec{\nabla}\cdot\vec{f}(\vec{P}).$$

Therefore, we see that the simplest solution is: $\vec{f}(\vec{P})=\frac{\partial}{\partial t}\vec{P}$, so with all microscopic intensive quantities, the Maxwell-Ampère law becomes:

$$\vec{\nabla}\times\vec{B}=\mu_0\vec{J}+\mu_0\varepsilon_0\tfrac{\partial}{\partial t}\vec{E}+\mu_0\tfrac{\partial}{\partial t}\vec{P}+\mu_0\vec{\nabla}\times\vec{M}.$$

Fundamental Law 6: The Maxwell-Ampère Law

$$\vec{\nabla}\times\vec{B}=\mu_0\left(\vec{J}+\varepsilon_0\tfrac{\partial}{\partial t}\vec{E}+\tfrac{\partial}{\partial t}\vec{P}+\vec{\nabla}\times\vec{M}\right).$$

8.4 *Electromagnets*

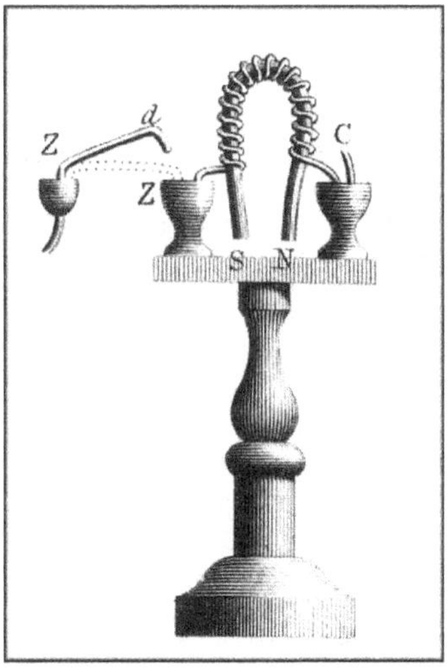

The 1824 electromagnet in the figure[119] was made by the English self-taught electrical engineer William Sturgeon, who also made some of the first practical electric motors. Notice that his wire was wrapped loosely because it was not insulated. However, as we discussed in the introduction to this chapter, Arago and others had already made electromagnets as part of their scientific research.

The key feature of an electromagnet is that the coil is wrapped around an iron bar. If the bar is made of a material that otherwise would not attract a permanent a magnet, rather, the electromagnet does not work. Why is this so? What is so unique about iron?

As we mentioned before (p. 183), attempts to explain the microphysics of ferromagnetism by the French physicists in the nineteenth century failed miserably. Ampère's model of microscopic current loops predicted that magnets would repel metal, while the pole model predicted the existence of magnetic monopoles (p. 196), neither of which agrees with experiment or experience! However, this did not prevent us, with the benefit of hindsight, from making our own semi-classical toy model. In this modern toy model, the electron has its own *intrinsic magnetic moment* which can be aligned, or anti-aligned, with an external magnetic field called *spin down*, or *spin up*, respectively. This was shown in the famous Stern-Gerlach experiment (p. 193), but that was not until 1922—a full century after the events just discussed.

---

[119] W. Sturgeon, W. "Improved Electro Magnetic Apparatus", Trans. Royal Society of Arts, Manufactures, and Commerce, (1824) **43**. Figure is Plate 3, fig.13 and downloaded from Wikimedia commons.

Thought Experiment 8.12: The Magnetizing Field

Consider the simple elementary school electromagnet experiment with a wire wrapped around a nail. This does not work if we wrap our wire around a pencil instead, so what makes the nail special? Since we now know that classical physics does not work on atomic scales, we will not attempt to answer this question, but we will characterize the observed phenomena.

We first suppose that the *current in the coil* creates some *magnetizing field* (see the introduction to section 6.3 starting on p. 200), which we will call $\vec{H}$, inside of the nail. This field, in turn, somehow causes the nail to become magnetized. Ideally this magnetizing field will be directly proportional to the magnetization $\left(\vec{M}\right)$ in common magnetic materials, like iron and its alloys. Moreover, we want materials that are more *susceptible* to becoming magnetized to have a greater ratio of $\vec{M}$ to $\vec{H}$.

Now consider Thought Experiment 8.10 (p. 258), where we found the Maxwell-Ampère law in a magnetic material (ignoring and dielectric effects) as:

$$\vec{\nabla} \times \vec{B} = \mu_0 \vec{J} + \mu_0 \varepsilon_0 \tfrac{\partial}{\partial t} \vec{E} + \mu_0 \vec{\nabla} \times \vec{M}.$$

Now, doing some rearranging, gives us:

$$\vec{\nabla} \times \vec{B} - \mu_0 \vec{\nabla} \times \vec{M} = \vec{\nabla} \times \left(\vec{B} - \mu_0 \vec{M}\right) = \mu_0 \vec{J} + \mu_0 \varepsilon_0 \tfrac{\partial}{\partial t} \vec{E}.$$

Officially the *magnetizing field* is defined via the *magnetic constitutive relation*, which is:

$$\vec{H} \equiv \tfrac{1}{\mu_0} \vec{B} - \vec{M}.$$

This is simply a mathematical relationship, which only has relevance in terms of its curl. To find $\vec{H}$ both $\vec{B}$ and $\vec{M}$ would need to be measured. Why do physicists and engineers use $\vec{H}$? Moreover, what is the best way to think of it?

The answer to the first question is simply because current is easy to measure. Consider a typical electromagnet, like the wrapped nail. The current through the wire can be measured using an ammeter, and based on the geometry the magnetizing field can be found using:

$$\vec{\nabla} \times \vec{H} = \vec{J} + \varepsilon_0 \tfrac{\partial}{\partial t} \vec{E}.$$

The more difficult question is how to imagine $\vec{H}$. Simply, $\mu_0 \vec{H}$ is what the magnetic field world be, had the experiment taken place in free space. Another answer is that $\vec{H}$ is the effective magnetization due to currents. Recall from Thought Experiment 6.12 (p. 206) that the magnetization of a permanent magnet is simply its magnetic moment per volume. Moreover, the magnetic moment of a loop of wire is simply equal to $I\vec{A}$, as you showed in Problem 8.2 (p. 251). Thus, a stack of N loops, aligned in the same direction, would have a magnetic moment:

$$\vec{m} = N I \vec{A}.$$

Now, imagine drilling a hole of area A and depth l inside an insulator. You then thread the hole using a tap with n turns per length, and paint the threads with a good conductor. By running a current I through the threads, the magnetic moment of this hole is therefore:

$$\vec{m} = N I \vec{A} = (nl) I \vec{A} = nI(lA)\hat{z} = nIV\hat{z},$$

where V is the volume of the hole, so the effective magnetization inside the hole is given by:

$$\frac{\vec{m}}{V} = nI\,\hat{z} = \vec{H}.$$

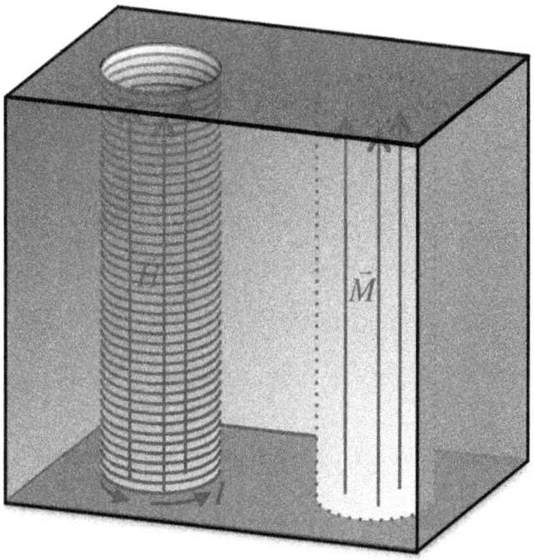

For many materials, the ratio of the magnetization to the magnetizing field is approximately constant, and is called the *magnetic susceptibility* (χ_M):

$$\chi_M \equiv \frac{M}{H}.$$

$\vec{M}$ can often be a complicated function, not only of $\vec{H}$, but of what $\vec{M}$ was before. Even in these cases, this relationship approximately holds for small values of $\vec{H}$.

Problem 8.9: Magnetic Potential Energy

Consider a small permanent magnet of volume V, and magnetization $\vec{M} = M_0\hat{y}$, aligned perpendicular with an external uniform magnetic field $\vec{B}_{\text{external}} = \mu_0 H_0 \hat{z}$.

(a) Draw a diagram of the initial scenario, and call its potential energy $\mathbb{U}_a$.

(b) Imagine that you let the magnet twist 90°, but you have it do work in the process. How much work did the magnet do while twisting? What is the potential energy difference $\mathbb{U}_b - \mathbb{U}_a$?

(c) Now imagine that, after completing part (b), you take out the permanent magnet, and return it with a chuck of iron with susceptibility $\chi_M = \frac{H_0}{M_0}$. What is the potential energy difference $\mathbb{U}_c - \mathbb{U}_b$?

(d) Now imagine that you twist this piece of iron by 90°. How much work was required?

(e) Clearly explain the difference between twisting magnets or iron, in a magnetic field.

Problem 8.10: The Vector Potential of a Solenoid

a) Find the vector potential inside, and outside, of a long, thin, empty solenoid of radius a with n turns of wire per unit length.

b) Calculate the same thing for a long solenoid with an iron core of susceptibility χ_M.

c) What is the ratio of these two?

Thought Experiment 8.13: The Magnetizing Field in a Permanent Magnet

Recall Problem 6.18 (p. 208) where you showed that the magnetic field in the middle of a uniform magnetized sphere is: $\vec{B} = \frac{2}{3}\mu_0 \vec{M}$. What is the magnetizing field in the center of the spherical magnet in terms of $\vec{M}$?

The solution is mathematically quite simple. We simply apply the *magnetic constitutive relation*, equate $\vec{B}$ and solve for $\vec{H}$. Thus:

$$\vec{H} = \left(\tfrac{2}{3} - 1\right)\vec{M} = -\tfrac{1}{3}\vec{M}.$$

Conceptually this makes no physical sense because there is neither an external field nor any current, so we would expect that there were no magnetizing field. And, moreover, why would it be pointing backwards compared to the magnetization?

However, this makes excellent historical sense. The debunked pole model of matter suggested that magnetization is caused by the separation of some sort of magnetic analogy to charge. Thus, the internal field due to this charge separation must be backwards, in much the same way that the electric field is opposite the electrical polarization in a dielectric. In fact, you showed in Problem 4.9 (p. 140) that the electric field in the middle of a uniformly polarized sphere is given by a similar relationship $\left(\vec{E} = -\frac{1}{3\varepsilon_0}\vec{P}\right)$. Thus you should avoid using $\vec{H}$, except as a known externally applied field, as it leads to physical misconceptions.

Discussion 8.1: The Confusing Magnetizing Field

In the introduction to Section 6.3 (p. 200) we discussed the failed pole model of magnetism, where north and south poles were thought to be analogous to electrical charges and separable from one another. The magnetizing field, $\vec{H}$, despite being quite useful at times, is a legacy of this debunked theory, so it can be very confusing to the contemporary student, who rightfully wants to apply a modern semi-classical toy model of matter. For this reason, most physics instructors recommend using $\vec{H}$ as little as possible. Unfortunately $\vec{H}$ is sill widely used, especially in engineering, so we recommend that you understand its history, so you will be able to put it in context when it becomes confusing.

William Thomson introduced $\vec{H}$ in the context of Poisson's polar theory of magnetism, as a magnetic analogue of sorts to the electric field $\vec{E}$ in a dielectric. The fact that Thomson himself referred to $\vec{H}$ as the magnetic field "according to the polar definition" and to $\vec{B}$ as the field "according to the electromagnetic definition" might have given him and other scientists sufficient reason to be careful in applying this new quantity.[120] But the mathematical convenience that $\vec{H}$ allows when describing magnetic fields in matter led to its widespread use and also to a great deal of confusion about its status as a fundamental quantity in electricity and magnetism.

As we discussed in Thought Experiment 8.11 (p. 258), it is also the equivalent magnetic moment per volume *due to* the motion of charge, rather than *due to* the material involved. Therefore, inside any particular medium, the local effective magnetization, from both atomic magnetic moments and the current, is simply the sum of $\vec{M}$ and $\vec{H}$. It is this sum, $\vec{M} + \vec{H}$, that is used to predict the local magnetic field $\vec{B}$ inside the medium.

There is a significant problem, however, to this way of thinking. Unlike $\vec{M}$, $\vec{H}$ is not an intrinsic property of matter. Since we have defined $\vec{H}$ in terms of causes, rather than the effects, it is not directly measureable *in situ* (i.e. at the particular location of a detector). In other words, we measure the current in the coil, and the geometry of the system, and then we calculate $\vec{H}$. However, we can not measure $\vec{H}$ directly from an unknown source. How would you know what causes it in the first place? Was it due to currents or atoms? Is there any way to know?

---

120 E.T. Whittaker, A History of the Theories of Aether and Electricity, Vols. 1&2 (1951; rpt. New York: Dover Publications, 2017), **1**: 219. See also, W. Thomson, Phil. Trans. Roy. Soc. Lond. **141** (1851), 243-268 and Phil. Trans. Roy. Soc. Lond. **141** (1851), 269-285.

To complicate things further, physicists used to call $\vec{H}$ the *magnetic field*, and $\vec{B}$ was called the *magnetic induction*. The idea being that $\vec{H}$ causes $\vec{B}$ in a material—much like we have just discussed $\vec{M}$. In many electrical circuits, $\vec{H}$ is a function of the current, and the voltage is a function of $\vec{B}$, so it is often useful to correlate $\vec{H}$ with $\vec{B}$. In many cases, but not all, these are directly proportional. The constant of proportionality is called the *permeability*, and is defined as:

$$\mu \equiv \frac{B}{H}.$$

In a linear magnetic medium, this is:

$$\mu = \frac{B}{H} = \frac{\mu_0 (H + M)}{H} = \mu_0 \left(1 + \frac{M}{H}\right) = \mu_0 (1 + \chi_M),$$

where χ_M is the *magnetic susceptibility*. Notice that in free space, $\mu = \mu_0$.

Relationships relating how materials respond to an applied generalized force are called *constitutive relations*. These include equations of state and stress and strain functions. This expression is called the *magnetic constitutive relation,* because nineteenth century physicists interpreted these relationships mechanically, in much same way as stress and strain.

Thought Experiment 8.14: Magnetic Hysteresis

Materials that can retain a magnetic moment are called *ferromagnetic*. The most common ferromagnetic material is iron, and the many different alloys that are made from iron, such as steel. Each of these materials not only becomes magnetized when exposed to an external magnetic field, but can retain a magnetization even after the external field is removed.

By placing the material within a magnetic coil, and measuring the magnetic moment as a function of the current in the coil, we can measure the properties of these magnetic materials. Thus the magnetization can be found as a function of the magnetizing field $\vec{H}$. These plots are called *magnetic hysteresis curves.*

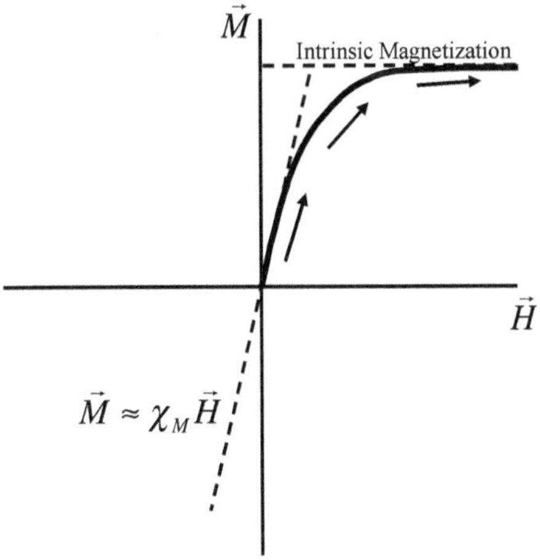

As one slowly increases the current in the coil, the magnetization of most ferromagnetic substances increases approximately linearly, with a unitless slope, χ_M, called the *magnetic susceptibility*. Ferromagnetic substances have large magnetic susceptibilities, with typical values in the hundreds or thousands.

Eventually, the magnetization no longer responds to increases in the applied field and remains constant. The value for which this happens is called the *intrinsic magnetization*. Ignoring frictional forces, and for weak paramagnetic substances, the magnetization as a function of a constant magnetizing field is well-approximated by:

$$M \approx M_I \tanh\left(\frac{\chi_M H}{M_I}\right),$$

where M_I and χ_M are the intrinsic magnetization and magnetic susceptibility respectively. This is commonly derived theoretically in courses on statistical mechanics, but it was measured well before atoms were understood.

By heating and cooling the material inside of a strong applied magnetic field, permanent magnets can be produced. The magnetization that remains, when we lower the applied magnetic field to zero, is called the *remanence*, which is the strength of the resulting permanent magnet.

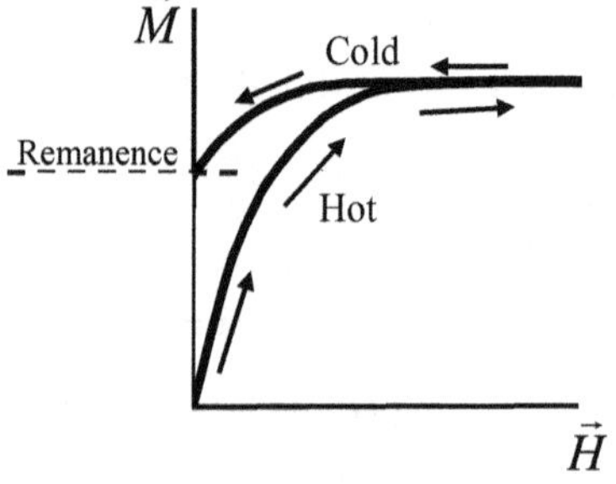

In applications, such as transformers and inductors, these are driven with a sinusoidal applied magnetic field. If the driving field is strong, at some point the magnetization, as a function of the applied field, becomes irreversible; the magnet is now *saturated*. This phenomenon of irreversibility is referred to as *hysteresis*, which comes from the Greek for "lagging behind." After saturation, the magnetization increases only slightly with increasing values of the applied magnetic field.

When this happens, we must reverse the applied field to a level called the *coercivity*, to make the magnetization go to zero. By increasing the reversed field more, it will saturate again in the opposite direction, and the cycle will continue.

These hysteresis curves can sometimes be modeled as two shifted hyperbolic tangent function as:

$$M \approx M_I \tanh\left(\frac{\chi_M (H \mp H_C)}{M_I}\right),$$

where M_I, χ_M, and H_C are the intrinsic magnetization, magnetic susceptibility, and the coercivity respectively. In practice, they are published by vendors who make magnetic devices.

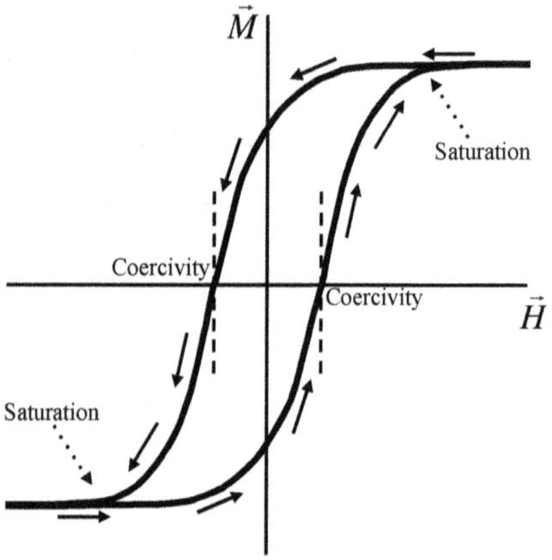

Discussion 8.2: Models of Magnetic Matter

Although classical, or, more precisely, semi-classical, models can serve as useful heuristics, magnetism in matter is fundamentally a quantum mechanical phenomenon. Atoms contain electrons in orbitals about the nucleus and these are often thought of as current loops arising from the motion of electric charge about the nucleus. The dominant contribution to the magnetic moment of a one-electron atom, however, is not due to this orbital angular momentum as much as it is because each electron contains an intrinsic magnetic moment related to its quantum spin state. So long as there are unpaired electrons inside of an atom, electron spin will dominate the magnetic moment and contributions to magnetic effects.

Magnetization of matter is divided into three broad categories: diamagnetism, paramagnetism, and ferromagnetism; however there are many features of quantum magnetism that do not fall into any of these three categories. Diamagnetism and paramagnetism involve the inducement of a magnetic moment in an otherwise nonmagnetic material. If the nonmagnetic material is repelled by a permanent magnet it is considered *diamagnetic*, but if it is attracted it is considered *paramagnetic*. Neither diamagnetism, nor paramagnetism, are observed in everyday life, but they are both routinely measured by chemists in the laboratory. If, however, a material can retain a magnetic moment in isolation, it is considered *ferromagnetic*. All useful magnets, including electromagnets and the cores of power transformers, are ferromagnetic.

Diamagnetism can be qualitatively explained as resulting from the change of the electrons' orbital motion due to an applied magnetic field, and the Lorentz force (p. 310), or, from the point of view of Faraday's law (p. 271) and Lenz's law (p, 347), due to an induced electric field that causes a current that is equivalent to a magnetic moment opposite to the applied field. This induced magnetic moment, which is ultimately responsible for a diamagnetic specimen's repulsion from a non-uniform magnetic field, disappears the moment the applied field is removed. This classical model is simple and therefore appealing, but only a fully quantum mechanical calculation yields the correct values for the diamagnetic susceptibility.

The quantum mechanical origin of paramagnetism is more obvious than with diamagnetism, since paramagnetic phenomena are due to the electrons' spins and the coupling of these spins to the orbital angular momentum of these electrons. Nonetheless, three French Physicists, Pierre Curie, Paul Langevin, and Pierre Weiss, made significant advances using classical, and semi-classical, molecular models.

Curie found that the magnetization in a paramagnetic specimen is directly proportional to the externally applied magnetic field but inversely proportional to the temperature, so long as the material is warm enough that magnetic moments do not interact.[121] This relationship is now called *Curie's law*, and the temperature where The temperature at which certain materials undergo an abrupt change in their magnetic properties, presumably due to magnetic moment interactions, is called *the Curie point.*

Langevin developed a semi-classical model that reproduced Curie's law and the Curie point. He simply assumed that each atom or molecule possessed a magnetic moment, and investigated how a system of such magnetic moments, at a finite temperature, would respond to an applied magnetic field. As we might expect, he found that thermal agitation would inhibit spins from aligning with the applied field, thereby keeping the magnetic susceptibility small (though positive).

Paramagnetic substances retain their magnetization only while the external field is being applied, just like the case with diamagnetism. However, some materials retain their magnetization even when the field is removed. Among these are ferromagnets, in which the spins align in the same direction. Ferromagnetic materials also display spontaneous magnetization: at temperatures below the Curie point and in the absence of an applied magnetic field, ordered spin states cause the specimen to be magnetized. At temperatures above the Curie point, ferromagnetic materials exhibit paramagnetic behavior.

Weiss extended Langevin's classical model of paramagnetism to explain ferromagnetism by positing the existence of a *mean internal magnetic field* that would act to keep atomic and electronic spins aligned in the absence of an applied external magnetic field. In the Weiss theory, or *mean field approximation*, we assume that each atom is subject to a magnetic field that is proportional to the magnetization. Once again, we are presented with an appealing classical model, but one that is even more unphysical than those used as heuristics for diamagnetism and paramagnetism: this *mean magnetic field* must exceed that of any expected magnetic interaction by a factor on the order of 500 or more! This mean field is, therefore, entirely fictitious, yet it remains a useful empirical parameter to engineers, chemists, and applied physicists.

---

[121] Pierre Curie Propriétés magnétiques des corps a diverses temperatures, Gauthier-Villars et fils (1895)

The currently accepted explanation for ferromagnetism came about in 1928 with the advent of quantum theory and the work of one of its founders, Werner Heisenberg, who showed that only the electron spin magnetic moments contribute to the effect. More subtlety, he showed that the alignment of spins is energetically favored because of a combination of electrostatic forces and the Pauli exclusion principle.

Another important feature of matter is the magnetization structure on microscopic scales that are nevertheless larger than an atom (Plate 16, p. 269). These *magnetic domains*, for example, are small regions within a magnetic material that have uniform magnetization. That is, individual magnetic moments of the atoms are aligned with one another and point in the same direction. When cooled below the Curie temperature, the magnetization of a piece of ferromagnetic material, such as iron, spontaneously divides into magnetic domains. The magnetization within each domain points in a uniform direction, but the magnetization of different domains generally point in different directions. That is, while the magnetization within each domain is saturated, that of the entire specimen is not because the domains do not all align in one direction. In 1935, the Russian physicists Lev Landau and Evgeny Liftshitz showed that magnetic domain structure is ultimately due to the fact that a specimen with a domain configuration minimizes magnetic energy compared to one with a saturated configuration.[122]

Magnetic domain structure is responsible for the magnetic behavior of ferromagnetic materials like iron, nickel, cobalt, and their alloys. In regions separating magnetic domains, the magnetization vectors rotate coherently from the direction in one domain to that in the next domain, and are called domain or Bloch walls after the Swiss physicist Felix Bloch.

Problem 8.11: Classically Modeling Ferromagnetism

As we discussed above, microscopy of ferromagnetic materials shows that ferromagnetic materials contain *magnetic domains*, each with their own magnetization. These regions are on the scale of microns, so they are much bigger than atoms, yet much smaller than macroscopic magnets, so they are ripe for analysis using classical electrodynamics. On the other hand, we need to be very careful, because applying classical physics within these domains led nineteenth century physicists astray, leading to theories of magnetization which have since been debunked. Atoms, after all, are quantum systems.

a) Do some research on the imaging of magnetic domains. As of this writing multiple research groups have posted images and videos showing ferromagnets as the external magnetic field increases.

b) For simplicity, assume that each domain has the same magnetization strength and that there are N domains total, so the magnetic moment strength of each domain is: $m = |\vec{m}_i| = M_I \delta V$. Explain, clearly using diagrams, what it means for a ferromagnetic material to be demagnetized.

c) Now imagine the other extreme, when the material is completely saturated. Explain, clearly using diagrams, what it means for a ferromagnetic material to be completely magnetized.

d) Express the fraction, f, of domains that are aligned with the magnetic field in terms of the externally applied field, the intrinsic magnetization, magnetic susceptibility, and the coercivity.

---

[122]Landau, L. and Lifshitz, E., "Theory of the dispersion of magnetic permeability in ferromagnetic bodies," Phys. Z. Sowietunion, 8, 153-169, which has been translated and published in: Collected papers of L. Landau. ed. D. Ter Haar, 2nd ed. (New York: Gordon and Breach, Science Publishers; London: Pergamon Press Ltd, 1967), 101-114.

e) Show that this fraction is always between negative one and one. What does a negative fraction mean?

Thought Experiment 8.15: The Magnetic Moment of an Electromagnet

An electromagnet has both a magnetic core and a coil of wire, therefore we will need to sum up the magnetic moments of the two.

$$\vec{m} = N I \vec{A} + \oint_{\text{volume}} \vec{M}\, \mathrm{d}V .$$

If the electromagnet has a length l and area A, then the total magnetic moment is now:

$$\vec{m} = N I \vec{A} + \oint_{\text{volume}} \vec{M}\, \mathrm{d}V = N I A\, \hat{z} + M(lA)\hat{z} = (nI + M)(lA)\,\hat{z} .$$

Recalling the magnetizing field, we see that the magnetic moment of an electromagnet can be written as:

$$\vec{m} = N I \vec{A} + \oint_{\text{volume}} \vec{M}\, \mathrm{d}V = \oint_{\text{volume}} \vec{H}\, \mathrm{d}V + \oint_{\text{volume}} \vec{M}\, \mathrm{d}V = \oint_{\text{volume}} \left(\vec{H} + \vec{M}\right) \mathrm{d}V .$$

And, finally, in terms of the magnetic field inside of the electromagnet we can simply write:

$$\vec{m} = \oint_{\text{volume}} \left(\vec{H} + \vec{M}\right) \mathrm{d}V = \tfrac{1}{\mu_0} \oint_{\text{volume}} \vec{B}\, \mathrm{d}V .$$

Problem 8.12: Work Done by an Electromagnet

Consider a cylindrical electromagnet, of height h and cross sectional radius a, attached to a constant current source with I flowing through it, with N turns of wire surrounding an iron core with magnetic susceptibility χ_M.

a) What is the magnetic moment of the electromagnet?

b) What is the on axis magnetic field, as a function of a distance z, below the center of the electromagnet?

c) What would be the magnetic moment of a small chunk same material (i.e. χ_M), with volume δV, at the on-axis distance z below the magnet center.

d) What would be the force $\delta\vec{F}$ of the electromagnet on a small chunk of the same material as its core with volume δV at the on-axis distance z from the center of the electromagnet.

e) What would be the total force on a cylinder of radius a and height h, with its near end a distance z from the electromagnet?

f) How much work would it take to pull the metal rod completely away from the electromagnet?

g) If the rod were pulled away at a constant velocity, v, what would be the voltage across the electromagnet as a function of time? Why?

Thought Experiment 8.16: A Chain of Paperclips and Magnetic Flux

Consider a permanent magnet holding a chain of paperclips, as in the drawing. Why does the bottom paperclip stick, even if the magnet is too weak to lift it from that distance?

To understand forces we must understand the energetics of the magnetic field and its interaction with the steel paperclips. First consider the top paperclip next to the magnet. The external magnetic field of the magnet acts like a magnetizing field to the paperclip, so it induces a magnetization into the paperclip:

$$\vec{M} = \chi_M \vec{H}.$$

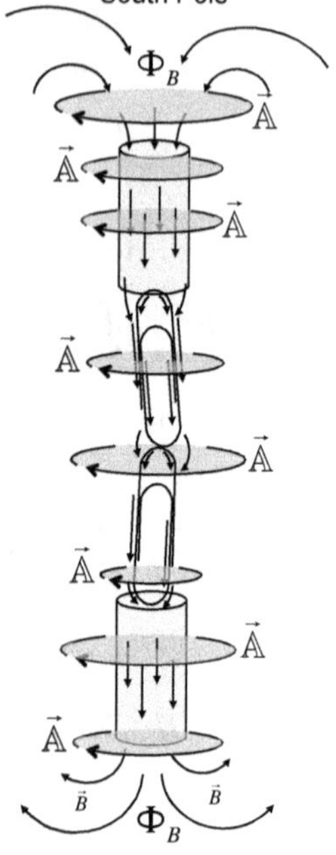

Now the magnetic moment of the paper clip is in the same direction as the magnetic moment of the magnet, so they are attractive. The same effect will, in turn, apply to the next paper clip down the line, inducing a magnetic moment per volume in this paper clip as well.

Now consider that $\vec{\nabla} \cdot \vec{B} = 0$, implying that there is total magnetic flux leaving the north end of the magnet, and returning on the south end. Once we induce a magnetization, it in-turn intensifies the magnetic field, increasing $\vec{B}$ inside the paperclip. Since there is a finite magnetic flux leaving the magnet, as the magnetic field inside the paper clip increases, the magnetic field outside must decrease accordingly. Thus, the magnetic flux appears to flow through the paperclips and spill out the other end.

This led to the concept of a *magnetic circuit*, in analogy to electrical circuits, with analogies to resistance (reluctance), electromotive force (magnetomotive force) and current (magnetic flux). Despite being useful, this analogy had no additional physical meaning whatsoever. There is no substance that flows through the paperclips, but it is a common analogy due to the mathematically circulatory magnetic field.

Problem 8.13: A Magnetic Circuit

Consider a toroidal piece of electrical steel, wound with N turns of wire carrying current I, with magnetic susceptibility χ_M, cross sectional area $A = \pi a^2$, and circumference $\ell = 2\pi b$.

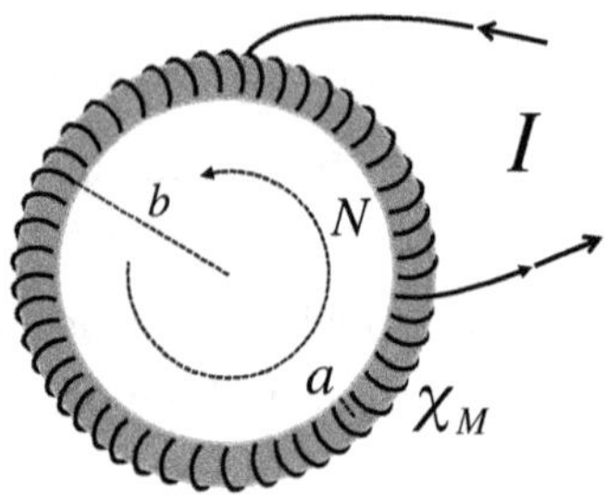

a) Find the magnetizing field, $\vec{H}$, magnetization, $\vec{M}$, magnetic field, $\vec{B}$, and the magnetic flux, Φ_B, inside the torus.

b) Show that your solutions follow a similar mathematical form to the electrical conductivity κ, with the following substitutions: $\Phi_B \Leftrightarrow I$, $NI \Leftrightarrow V$, and $\chi_M + 1 \Leftrightarrow \kappa$.

c) Reluctance, R_m, is the analogous quantity to resistance. Express the reluctance in terms of the geometry of the torus and χ_M.

d) Show that the magnetizing field and the electric field are analogous: $\vec{H} \Leftrightarrow \vec{E}$.

e) What electrical quantity is analogous to the magnetic field, $\vec{B}$, in the analogy, and what fundamental equation of electricity, therefore, would be analogous to $\vec{\nabla} \cdot \vec{B} = 0$?

f) Show that the magnetizing field is the electrical analogy to the vector potential $\vec{A} \Leftrightarrow \vec{H}$.

g) Discuss how the concept of a magnetic circuit reinforces the debunked pole model.

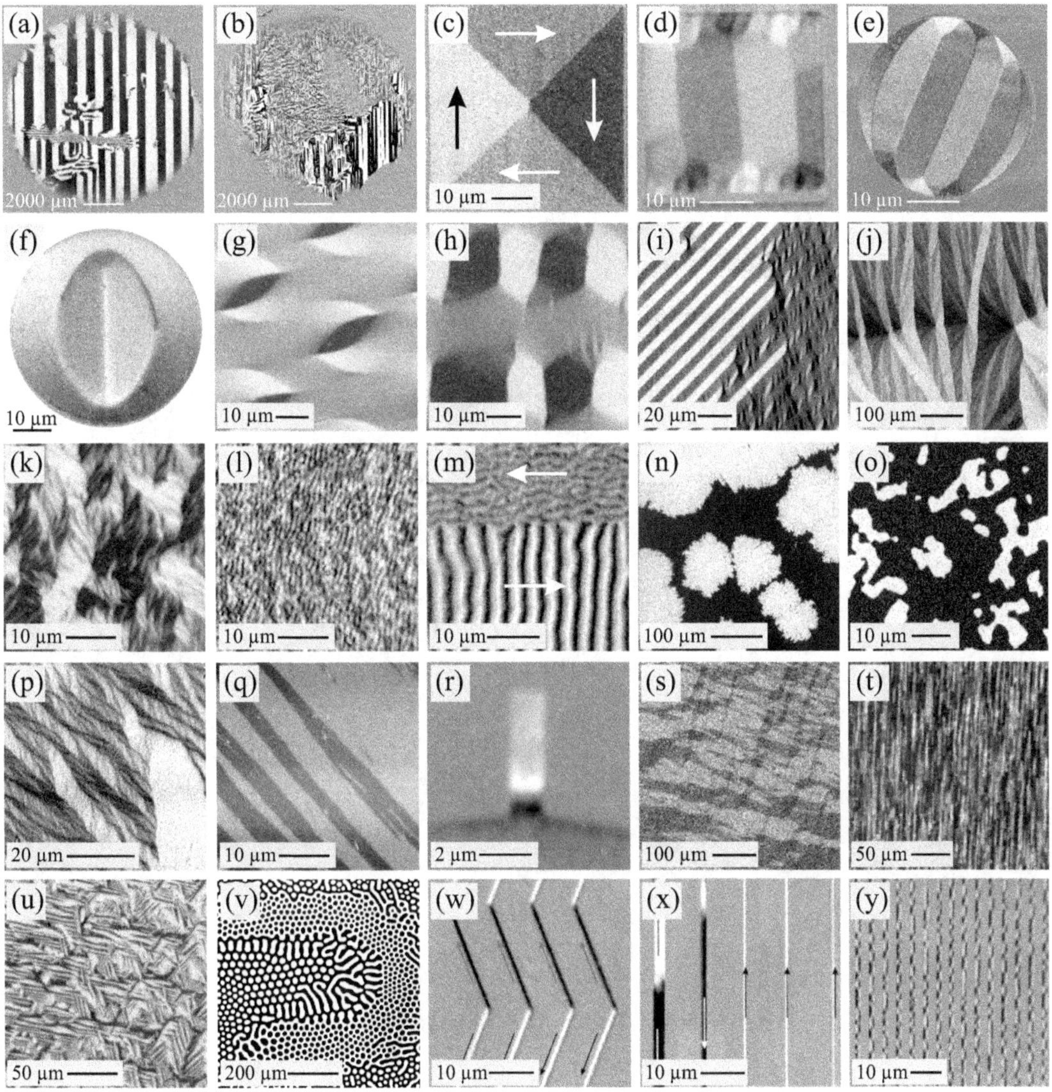

Plate 16: Magnetic Domains in Electrical Steel[123]

---

[123] Jeffrey McCord, "Progress in magnetic domain observation by advanced magneto-optical microscopy," Journal of Physics D Applied Physics, **48** (IOP Publishing, 2015), 3330001, Figure 29.

Plate 17: Nikola Tesla in his Colorado Laboratory[124]

[124] This picture is a double exposure taken by the photographer Dickenson V. Alley around 1889. Tesla was not actually sitting calmly at the same time as the equipment was turned on. Rather, two pictures were taken and later superimposed in the darkroom. This copy is inscribed: "To my illustrious friend Sir William Crookes of whom I always think and whose kind letters I never answer! Nikola Tesla June 17, 1901" (Wellcome Collection. CC BY 4.0)

Chapter 9 Faraday's Law of Induction

After Ørsted discovered that electric currents give rise to magnetic effects, it was only natural to ask if the converse is also true: can magnetism somehow create electricity? After nearly a decade of work, Faraday obtained an answer in 1831 when he demonstrated that a time changing magnetic field gives rise to an electric current in a wire loop placed in a changing magnetic field.

As an historical aside, it appears that the American physicist Joseph Henry actually discovered electromagnetic induction in 1830—a full year before Faraday! Henry's results were not published until after Faraday's announcement, however, so the discovery of induction is properly credited to Faraday.

Henry is now recognized for his discovery that a current in a loop induces a field in the circuit itself, an effect that is called self-inductance. Henry's further work in electromagnetism, especially his work on the electromagnetic relay, by which a weak current can operate a powerful local electromagnet over very long distances, contributed to the development of the electrical telegraph. Among other honors, Henry served as the first secretary of the Smithsonian Institute between 1846 and 1878.[125]

We develop Faraday's idea of magnetic induction in this chapter. For example, by rotating a magnet in the vicinity of a wire coil, or a wire coil near a magnet, electric current is induced to flow. Now mechanical energy can be converted into electrical energy! This is essentially how Hoover Dam works to deliver 4.2 trillion watt-hours of energy every year to hundreds of thousands of homes in the U.S. Southwest, and the alternator in your car delivers power to recharge the car's battery and power its electrical system.

9.1 ***Faraday's Law***

Michael Faraday conducted an experiment where he wound a long wire around a spool and inserted a magnet.

Fig. 4.

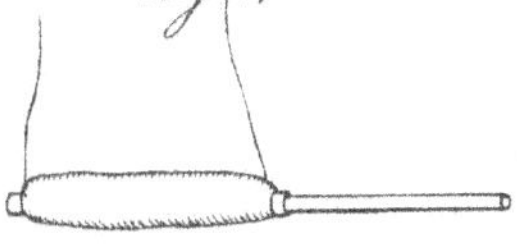

> All the similar ends of the compound hollow helix were bound together by copper wire, forming two general terminations, and these were connected with the galvanometer. The soft iron cylinder was removed, and a cylindrical magnet, three quarters of an inch in diameter and eight inches and a half in length, used instead. One end of the magnet was introduced into the axis of the helix (fig. 4), and then, the galvanometer-needle being stationary, the magnet was suddenly thrust in; immediately the needle was deflected ... Being left in, the needle resumed its first position, and then, the magnet being withdrawn, the needle was deflected in the opposite direction.[126]

---

125 The figure shows a painting titled "Professor Henry Posts Daily Weather Map in Smithsonian Institution Building, 1858," which was commissioned for the 1933 Chicago Century of Progress Exposition (Hoover, Louise Rochon, 1933, Smithsonian Archives - History Div, 84-2074 and 31052-E and 94-12563).

126 Michael Faraday, Experimental Researches in Electricity, 3 vols. (London: Taylor and Francis, 1839, 1844, 1855) **I**: Paragraph 39.

In modern classical mechanics courses you have learned to routinely change reference frames solve problems. To us it would, therefore, seem most natural to analyze this problem from the point of view of the coil, and separately from the point of view of the rod, and compare our results. This works wonderfully in electrodynamics, and we will do this in Chapter 11 (p. 331). However, neither Faraday nor Maxwell followed this relational approach. What they did, however, works equally well and, because of Peregrinus's principle, is mathematically equivalent.

Like current, magnetic flux is continuous and circular. Why not, then, interpret the magnet as some sort of conduit for magnetic flux, like a wire is a conduit for current. This analogy worked well in helping us to understand inductors and conductors, and it illustrated one of Maxwell's fundamental equations. As we discussed in Thought Experiment 8.16 (p. 267), this analogy is still used, but unfortunately also fosters physical misconceptions.

Here we follow Faraday's style of analysis, but with modern mathematical notation, in order to understand his law of induction.

Thought Experiment 9.1: The Magnetic Flux through a Magnetic Rod

First we consider a long magnetic rod with length l, radius a, and constant magnetization $\vec{M} = M\,\hat{z}$, like the one we analyzed in Example 7.1 (p. 231). Applying the solution we found, the vector potential at radius a is:

$$\vec{\mathrm{A}} = \tfrac{1}{4}\mu_0 a M \left(\frac{z + l/2}{\sqrt{a^2 + (z + l/2)^2}} - \frac{z - l/2}{\sqrt{a^2 + (z - l/2)^2}} \right) \hat{\varphi} \, .$$

Notice that the vector potential points azimuthally around the magnet. From this vector potential we can find the magnetic flux through each plane along the rod, which will be a function of the lateral surface area, as the flux can only escape sideways. The flux is therefore:

$$\Phi_B = \int_{\text{surface}} \vec{B} \cdot \mathrm{d}\vec{A} = \oint_{\text{path}} \vec{\mathrm{A}} \cdot \mathrm{d}\vec{\ell} = \tfrac{\pi}{2}\mu_0 a^2 M \left(\frac{z + l/2}{\sqrt{a^2 + (z + l/2)^2}} - \frac{z - l/2}{\sqrt{a^2 + (z - l/2)^2}} \right).$$

The figure shows the magnet with the magnetic field overlaid. Using his iron filings Faraday understood this magnetic field structure, even if he did not calculate it mathematically.

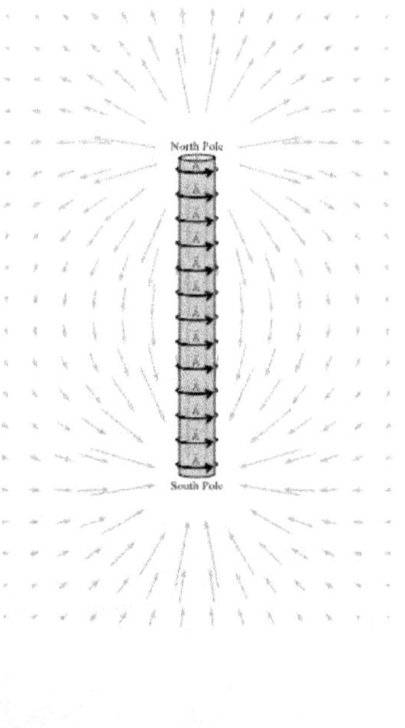

Consider the axial flux entering the magnet at one end, in the center, and exiting at the other end, in the limit that $a \ll l$:

$$\Phi_B \approx \begin{cases} \text{south } \tfrac{1}{2}\pi a^2 \mu_0 M \\ \text{middle } \pi a^2 \mu_0 M \\ \text{north } \tfrac{1}{2}\pi a^2 \mu_0 M \, . \end{cases}$$

This also means that half of the flux enters through the south end, while the other half enters through the sides.

Faraday noticed, again non-mathematically, that the flux gradient along the magnet is greatest near the ends. Differentiating, you will show that:

$$\tfrac{d}{dz}\Phi_B \approx \begin{cases} \text{south } \tfrac{\pi}{4} a\mu_0 M \\ \text{middle } 0 \\ \text{north } \tfrac{-\pi}{4} a\mu_0 M . \end{cases}$$

Problem 9.1: The Flux Gradient through a Long Cylindrical Magnet

Consider Thought Experiment 9.1 above.

a) Verify the expression for the flux through the magnet starting with the result for the vector potential from Example 7.1 (p. 231).

b) Show that in the limit $a \ll l$ the expressions for the magnetic flux at the ends, and in the middle, are correct.

c) What is the average magnetic field inside the magnet in the middle and at the ends.

d) Differentiate the expression for the magnetic flux to find an expression for the gradient of the magnetic flux. Show that this is consistent with what we reported in Thought Experiment 9.1.

Thought Experiment 9.2: Faraday's Wire Spool Experiment

When conducting his wire spool experiment (page 271), Faraday observed that the galvanometer needle moved one way when one end of the magnet was thrust in, and the other way when pulled out. From this he hypothesized that a changing magnetic flux induces a voltage in the surrounding coil. Experimentally, he tested his hypothesis, and refined hypotheses, until he came up with a law consistent with all of his experiments, which states that the voltage increase around a loop is proportional to the rate of change of the flux through the loop. Here we will use this hypothesis to predict the voltage across his wire coil as a function of the velocity and position of the magnetic rod.

If the center of the rod is at position z_0, then the magnetic flux through a loop located at position z will be:

$$\Phi_B = \tfrac{\pi}{2}\mu_0 a^2 M \left(\frac{z - z_0 + \frac{1}{2}l}{\sqrt{a^2 + \left(z - z_0 + \frac{1}{2}l\right)^2}} - \frac{z - z_0 - \frac{1}{2}l}{\sqrt{a^2 + \left(z - z_0 - \frac{1}{2}l\right)^2}} \right).$$

When the rod moves, the middle position changes, but the position of interest does not, so the rate of change of flux is given by:

$$\tfrac{\partial}{\partial t}\Phi_B = \left(\tfrac{\partial}{\partial z_0}\Phi_B\right)\left(\tfrac{\partial}{\partial t} z_0\right) = v\left(\tfrac{\partial}{\partial z_0}\Phi_B\right).$$

Since we replaced our old z with our new $z - z_0$, we can use the flux gradient that we calculated to find:

$$\tfrac{\partial}{\partial t}\Phi_B = \left(\tfrac{\partial}{\partial z_0}\Phi_B\right)\left(\tfrac{\partial}{\partial t} z_0\right) = v\left(\tfrac{\partial}{\partial z_0}\Phi_B\right) = -v\left(\tfrac{\partial}{\partial z}\Phi_B\right).$$

Near the locations we discussed, with the north pole going in first:

$$\frac{\partial}{\partial t}\Phi_B \approx \begin{cases} \text{north pole} & \frac{\pi}{4} a v \mu_0 M & t=0 \\ \text{middle} & 0 & t=\frac{L}{2v} \\ \text{south pole} & -\frac{\pi}{4} a v \mu_0 M & t=\frac{L}{v} \end{cases}.$$

so according to Faraday's hypothesis, if we sent a very long rod through a short coil of wire at a constant velocity, the galvanometer would deflect one way when the north pole entered, and then the opposite way when the south pole exits.

Interestingly, Faraday noticed that the galvanometer needle moved in the same direction if the south pole were thrust in first. Seeing this, he concluded that the voltage difference was in one direction if the flux increased, and the opposite direction if it decreased.

By observation, the voltage induced is clockwise for increased flux. Given our sign convention for loops and areas, the voltage increase around the loop, often called EMF, equals $-\frac{\partial}{\partial t}\Phi_B$.

Thought Experiment 9.3: Faraday's Law in Various Forms

Consider Faraday's observation that the observed voltage, per loop of wire, due to the changing magnetic flux, equals $-\frac{\partial}{\partial t}\Phi_B$. What would this imply about the electric field?

Due to the work/energy relationship, the electric field around the loop must be the voltage rise divided by the length of the wire. Therefore the induced electric field is:

$$\vec{E} = \frac{\left(-\frac{\partial}{\partial t}\Phi_B\right)}{2\pi a}\hat{\varphi} = -\frac{\partial}{\partial t}\frac{\Phi_B}{2\pi a}\hat{\varphi} = -\frac{\partial}{\partial t}\vec{\mathbb{A}}.$$

The electric field has an operational definition, so its value must not depend on whether it was created through a difference in electrostatic potential or a changing magnetic field. Therefore, the electric field should be the sum of the results from Part I and here, or:

$$\vec{E} = -\vec{\nabla}\mathbb{V} - \frac{\partial}{\partial t}\vec{\mathbb{A}}.$$

This is the electric field in terms of both potentials. Notice that it is the integral of the electric field that we measure as voltage differences, so when working with time dependent systems you will need to take both potentials into account.

Now, taking the curl of both side, we can write this as a relationship between the electric and magnetic fields:

$$\vec{\nabla}\times\vec{E} = \vec{\nabla}\times\left(-\vec{\nabla}\mathbb{V} - \frac{\partial}{\partial t}\vec{\mathbb{A}}\right) = -\frac{\partial}{\partial t}\left(\vec{\nabla}\times\vec{\nabla}\mathbb{V} + \vec{\nabla}\times\vec{\mathbb{A}}\right) = -\frac{\partial}{\partial t}\left(\vec{\nabla}\times\vec{\mathbb{A}}\right) = -\frac{\partial}{\partial t}\vec{B}.$$

Problem 9.2: A Time Dependent Magnet

Consider an electromagnet with a time dependent magnetic moment, $\vec{m} = m_0 \sin(\omega t)\,\hat{z}$, with a wire coil of radius a in the equatorial plane, with N loops. As in Thought Experiment 7.1 (p. 213) the outside vector potential is given by:

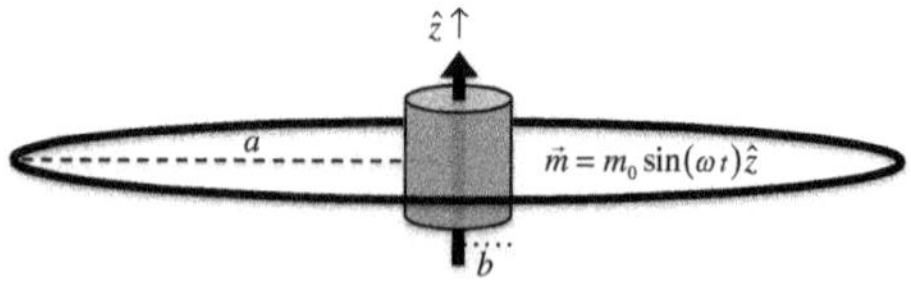

$$\vec{\mathbb{A}} = \frac{\mu_0 \vec{m}\times\hat{r}}{4\pi r^2}.$$

a) Find the vector potential inside the wire.

b) Find the electric field in the wire.

c) Express the voltage across the coil, as a function of time.

d) Find the magnetic flux through the loops.

e) If the coil of wire were attached to a load resistor R, what would be the total power dissipated? Where did this energy come from?

Problem 9.3: Heating a Ring

A gold wedding ring is placed in a spatially constant, but time-dependent, magnetic field $\left(\vec{B}(t)\right)$, which is at an angle θ from the axis of the ring. The ring has a radius a, and cross-sectional area (A). Moreover, the conductivity (κ) and specific heat (c_V) are known for gold. Find the rate that the temperature of the ring is increasing $\left(\dot{T} = \frac{d}{dt}T\right)$ over time, as a function of the magnetic field and its time derivatives. You may assume that $A \ll \pi a^2$.

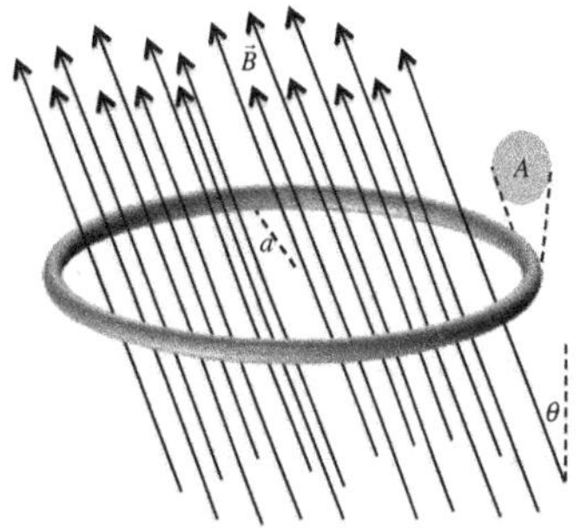

Thought Experiment 9.4: The Induced Electric Field in Free Space

Consider now an imaginary loop of wire surrounding a magnetic field. For this imaginary loop, the relationship between the electric field and the magnetic flux must also hold:

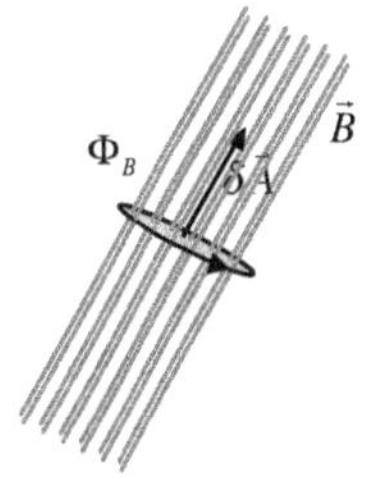

$$\sum_{\text{around loop}} \mathbb{V} = \oint_{\text{path}} \vec{E}\cdot d\vec{\ell} = -\frac{d\Phi_B}{dt}.$$

The rate of change of flux through the imaginary loop can be written as:

$$\frac{d\Phi_B}{dt} = \frac{d}{dt}\left(\oint_{\text{surface}} \vec{B}\cdot d\vec{A}\right) = \oint_{\text{surface}} \left(\frac{d\vec{B}}{dt}\right)\cdot d\vec{A}.$$

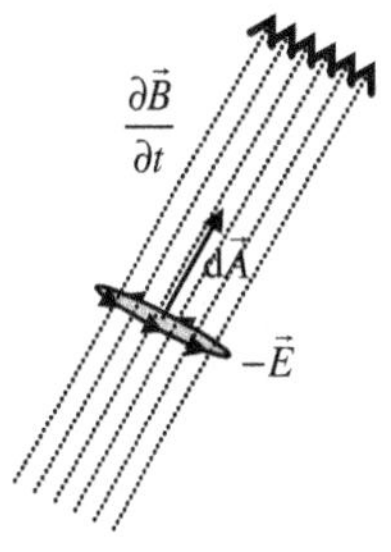

We now take the limit as the area of the loop becomes infinitesimally small:

$$\lim_{\delta A\to 0}\left(\frac{1}{\delta A}\frac{d\Phi_B}{dt}\right) = \lim_{\delta A\to 0}\frac{1}{\delta A}\oint_{\text{surface}}\frac{d\vec{B}}{dt}\cdot d\vec{A} = \frac{\partial\vec{B}}{\partial t}\cdot\delta\hat{A}.$$

This is the definition of the curl, perpendicular to the loop, so:

$$\left(\vec{\nabla}\times\vec{E}\right)\cdot\hat{n} \equiv \lim_{\delta A\to 0}\frac{\oint_{\text{path}}\vec{E}\cdot d\vec{\ell}}{\delta A_{\hat{n}}},$$

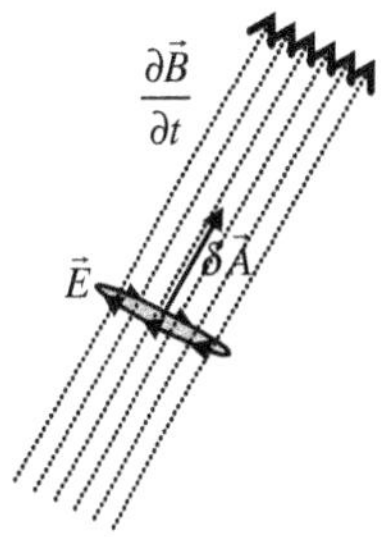

where $\hat{n}$ is again the unit vector in the direction of $\delta\vec{A}$, which is normal to the surface of integration using the right hand rule.

Including all three directions, we can write the curl of the electric field as:

$$\vec{\nabla}\times\vec{E} = \lim_{\delta A\to 0}\frac{\oint_{\text{path}}\vec{E}\cdot d\vec{\ell}}{\delta A}\delta\hat{A} = \lim_{A\to 0}\left(\frac{-\frac{d\Phi_B}{dt}}{\delta A}\right)\delta\hat{A} = -\frac{\partial B}{\partial t}\delta\hat{A} = -\frac{\partial\vec{B}}{\partial t}.$$

Note that the time derivative becomes a partial derivative because the position vector is constant over an infinitesimally small area.

We ought to mention an important caveat here. We assume in this thought experiment that the strength of the magnetic field varies while the field direction remains constant. This assumption is made only for purposes of clarity, however, and can be relaxed. Like the area vector, the electric and magnetic fields are vector quantities so this relationship must hold independently for the vector components in all three orthogonal directions.

Fundamental Law 7: Faraday's Law

$$\vec{\nabla}\times\vec{E} = -\frac{\partial\vec{B}}{\partial t}$$

Thought Experiment 9.5: A Loop Inside a Changing Magnetic Field

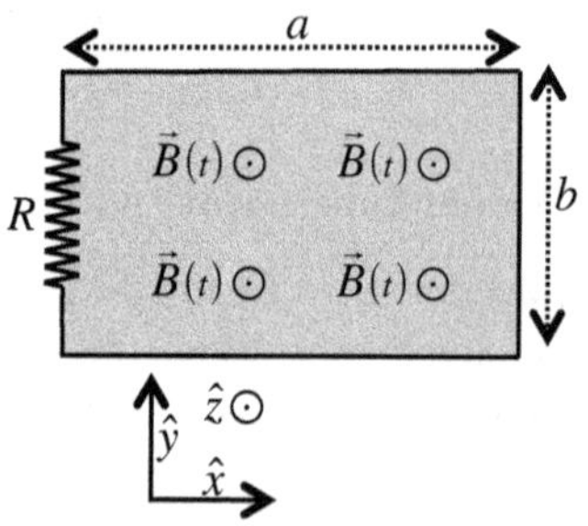

Consider a rectangular loop of wire with dimensions $a\times b$, which is attached to a resistor with resistance R. The loop is perpendicular to a time-dependent uniform magnetic field $\vec{B}(t) = B_z(t)\hat{z}$.

If Faraday's law is correct, we should be able to predict the current through the loop, and then measure this with a galvanometer.

First we consider the magnetic flux, Φ_B, through the wire loop:

$$\Phi_B = \int_{\text{surface}} \vec{B}\cdot d\vec{A} = \vec{B}\cdot(ab\hat{z}) = abB_z.$$

Then the sum of the voltages around the loop is:

$$\sum_{\text{around loop}}\mathbb{V} = \oint_{\text{path}}\vec{E}\cdot d\vec{\ell} = -\frac{d\Phi_B}{dt} = -\frac{d}{dt}(ab\,B_z) = -ab\frac{dB_z}{dt}.$$

This is the work per charge around the loop, or simply the voltage across the resistor, so the current in the loop can be found using Ohm's law:

$$I = -\frac{ab}{R}\frac{dB_z}{dt},$$

where positive current flows counter-clockwise around the loop. Faraday conducted many experiments like this one to verify his law.

Now consider the power dissipated through the resistor:

$$P = I\,\mathbb{V} = \left(\frac{ab}{R}\frac{dB_z}{dt}\right)\left(ab\frac{dB_z}{dt}\right) = \frac{a^2b^2}{R}\left(\frac{dB_z}{dt}\right)^2.$$

Checking our units, we have:

$$[P]=\left[\frac{a^2b^2}{R}\left(\frac{dB}{dt}\right)^2\right]=\left(\frac{m^4}{\Omega}\right)\left(\frac{T}{s}\right)^2=\left(\frac{m^4C^2}{Js}\right)\left(\frac{J}{Cm^2}\right)^2=\frac{J}{s}=W.$$

Problem 9.4: The Vector Potential in a Current Loop

Consider the wire loop in Thought Experiment 9.5 above (p. 276).

a) Find the average vector potential inside the wire.

b) Find the average electric field inside the wire.

c) Find the voltage across the resistor and the current through the circuit.

d) Show that this is mathematically identical to the method in Thought Experiment 9.5.

Problem 9.5: Induced Voltage in a Coil

Consider a coil or radius a with N loops of wire attached to a load resistor R. An external, cylindrically symmetrical, magnetic field interacts with the coil. The electric field is given by:

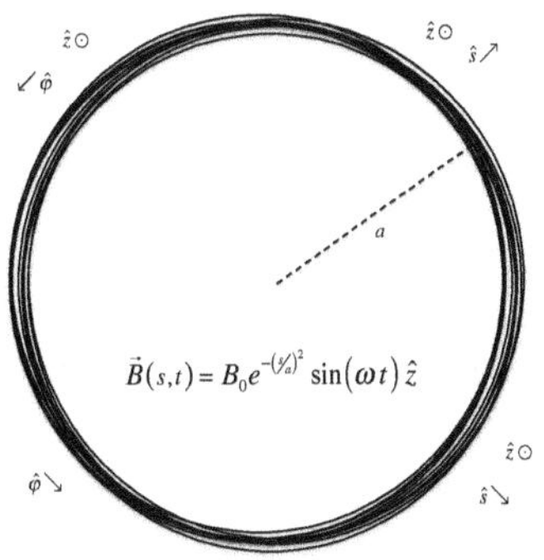

$$\vec{B}=B_0e^{-(s/a)^2}\sin(\omega t)\,\hat{z},$$

where the $\hat{z}$ axis is the axis of the coil.

a) What is the vector potential at the location of the coil?

b) What is the induced electric field inside wire?

c) What is the voltage across the load resistor?

d) What is the time-averaged power dissipated by the load resistor?

e) Where does this energy come from to power the resistor?

Problem 9.6: A Coil in a Capacitor

Consider a small toroidal coil of wire sandwiched between two large capacitor plates. The capacitor has an area A, gap spacing h and is initially at a voltage V_0. The coil has an overall radius b, and N loops of wire each will a small coil radius a.

Find the voltage across the coil as a function of time.

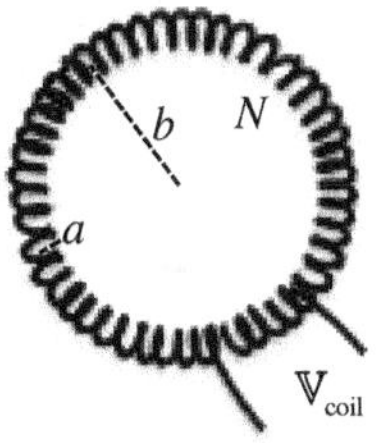

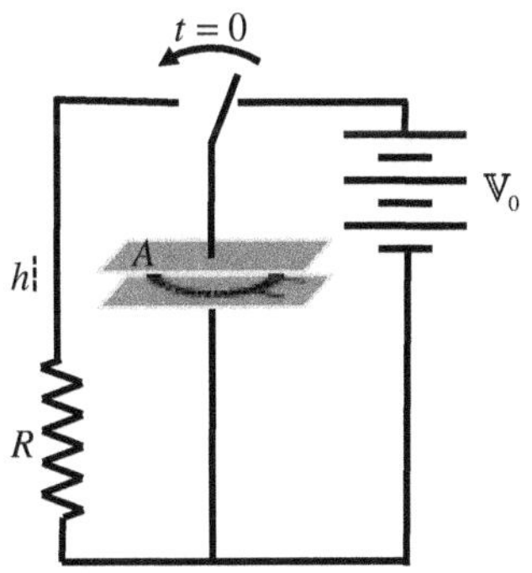

Problem 9.7: The Current Density in a Conducting Wire

Consider a cylindrical conducting wire with a steady current I and a uniform conductivity κ.

(a) Show that the curl of the current density vanishes, which is called *irrotational flow.*

(b) Show that, as long as the conductivity is uniform, the current density $\vec{J} = J(s)\hat{z}$ is uniform throughout the wire.

(c) If the conductivity is nonuniform, so $\kappa = \kappa(s)$, show that: $\frac{1}{J}\frac{\partial J}{\partial s} = \frac{1}{\kappa}\frac{\partial \kappa}{\partial s}$.

(d) In practice, the conductivity decreases with rising temperatures. If one part of the wire gets a little hotter than another, does this lead to increased current flowing through the hot part, and it in turn heating up more? Or, does less current flow through the hot part, cooling it off, thus mitigating the temperature variations? Explain.

Example 9.1: Magnetic Heating of a Coin

A coin is placed perpendicular to a harmonically varying magnetic field. Given the conductivity of the coin (κ), and that the magnetic field is given by the relation:

$$\vec{B} = B_0 \sin(\omega t)\hat{z},$$

how much energy per time per volume is being added to the coin? Which part is heating up the fastest, the center or the rim?

Our strategy for this problem involves: (1) deriving the expression for the work done by an electric field in a resistive medium, (2) finding the induced electric field, via Faraday's law, and (3) combining these to find the heating as a function of the changing magnetic field.

(1) First we recall that the magnetic force is always perpendicular to velocity of the charged particles, so this force will not directly do work on them.

Imagine that an electric field $\vec{E}$ is applied to a current of current density $\vec{J}$. How much work is done by the field on the charged fluid per time per volume?

We begin by considering the work done on a fluid element when it moves a distance δz in the $\hat{z}$ direction as:

$$\delta W = \int_z^{z+\delta z} \vec{F}\cdot d\vec{\ell} = \int_z^{z+\delta z} \left(\vec{E}\delta Q\right)\cdot d\vec{\ell} = \delta Q \int_z^{z+\delta z} \left(\vec{E}\cdot d\vec{\ell}\right) \approx E_z \delta z \delta Q.$$

Now, the current density is the rate of charge flow per area, and is given by:

$$\vec{J} = \left(\frac{dQ}{dt\, dA}\right)\hat{z} \approx \left(\frac{\delta Q}{\delta t\, \delta A}\right)\hat{z}.$$

Solving for the charge $(\delta Q = J_z \delta t \delta A)$ and substituting it in the work equation we find:

$$\delta W \approx E_z \delta z \left(J_z \delta t \delta A\right) = E_z J_z \left(\delta z \delta A\right)\delta t = \left(\vec{E}\cdot\vec{J}\right)\delta V \delta t.$$

Finally we can find the specific power done by the electric field, as:

$$\frac{dW}{dV\, dt} = \frac{\text{power}}{\text{volume}} = \lim_{\delta V \delta t \to 0} \frac{\delta W}{\delta V\, \delta t} = \vec{E}\cdot\vec{J}.$$

(2) Next, taking advantage of the azimuthal symmetry, we draw an imaginary circular loop, of radius s, and apply Faraday's law to this loop:

$$\vec{\nabla}\times\vec{E} = -\frac{\partial \vec{B}}{\partial t}$$

$$\oint_{\text{surface}} \left(\vec{\nabla}\times\vec{E}\right)\cdot d\vec{A} = \oint_{\text{surface}} \left(-\tfrac{\partial}{\partial t}\vec{B}\right)\cdot d\vec{A}$$

$$\oint_{\text{surface}} \left(\lim_{\delta A\to 0} \tfrac{1}{\delta A} \oint_{\text{path}} \vec{E}\cdot d\vec{\ell}\right)\cdot dA = \oint_{\text{surface}} \left(-\tfrac{\partial}{\partial t}\vec{B}\right)\cdot d\vec{A}$$

$$\oint_{\text{path}} \vec{E}\cdot d\vec{\ell} = \oint_{\text{surface}} \left(-\tfrac{\partial}{\partial t}\vec{B}\right)\cdot d\vec{A}$$

$$2\pi s E = -\pi s^2 B_0 \omega \cos(\omega t),$$

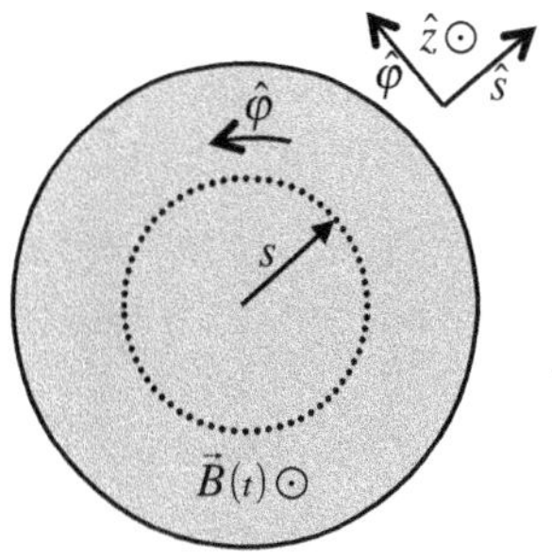

So, the electric field, with direction, is:

$$\vec{E} = -\tfrac{1}{2} s B_0 \omega \cos(\omega t)\hat{\phi}.$$

(3) Now we combine our results, to find the heating rate per volume:

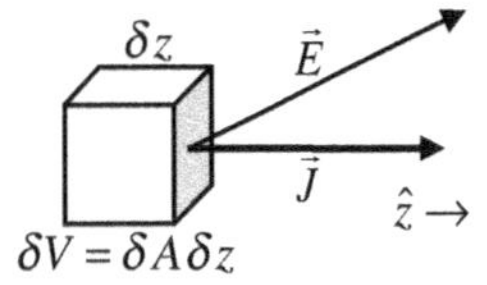

$$\frac{du}{dt} = \vec{E}\cdot\vec{J} = \kappa E^2 = \kappa\left(\tfrac{1}{2} s B_0 \omega \cos(\omega t)\right)^2.$$

Averaging over a full period, and recalling that $\langle \cos^2(\omega t)\rangle = \tfrac{1}{2}$, we find that the time-averaged heating rate, per volume, is given by:

$$\left\langle \frac{du}{dt} \right\rangle = \tfrac{1}{8}\kappa B_0^2\, \omega^2 s^2.$$

Notice how the electric field, and therefore the current density, is proportional to the frequency of the oscillatory magnetic field. Note also that the rim is heated more quickly than the rest of the coin. Good conductors of electricity tend to be also good conductors of heat, so heat flows from the rim toward the center of the coin.

Now we will check the consistency of our work by solving for the curl of the electric field:

$$\vec{\nabla}\times\vec{E} = \vec{\nabla}\times\left(-\tfrac{1}{2} s B_0 \omega \cos(\omega t)\hat{\phi}\right) = -\tfrac{1}{2} B_0 \omega \cos(\omega t)\vec{\nabla}\times\left(s\hat{\phi}\right)$$

$$= -\tfrac{1}{2} B_0 \omega \cos(\omega t)\left(-\frac{\partial s}{\partial z}\hat{s} + \frac{1}{s}\frac{\partial (s^2)}{ds}\hat{z}\right) = -B_0 \omega \cos(\omega t)\hat{z} = -\frac{\partial \vec{B}}{\partial t}.$$

Problem 9.8: A Spinning Hoop

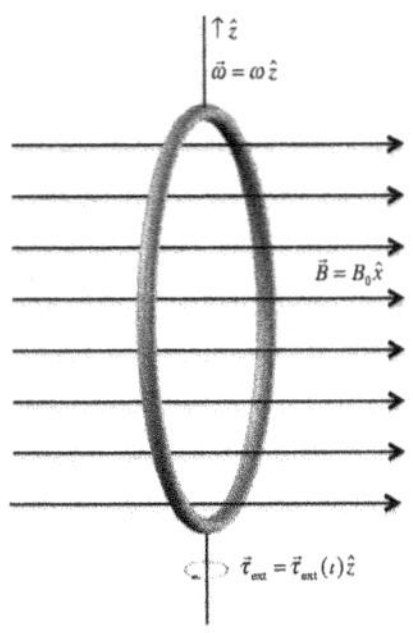

Consider a hoop of mass M, radius a, and total resistance R, which is spinning vertically with a constant angular velocity $\vec{\omega} = \omega\hat{z}$ in a constant and uniform magnetic field $\vec{B} = B_0\,\hat{x}$, as shown.

a) Find the induced voltage around the hoop as a function of time.

b) What is the induced current in the hoop as a function of time?

c) What is the power lost due to resistive heating as a function of time?

d) In order for the hoop not to slow down there must be an external driving torque $\vec{\tau}_{\text{ext}}(t)$. What external torque must be applied as a function of time in order for the angular velocity to be constant?

e) What must be the torque on the hoop, as a function of time, due to the interaction with the magnetic field?

f) Express this torque on the hoop in terms of the current around the hoop, the area vector of the hoop, and the magnetic field. Show that the dipole moment of the hoop is simply $\vec{m} = I\,\vec{A}$.

Example 9.2: The Brush Dynamo-Electric Machine

In 1879 C.F. Brush filed a patent for a dynamo-electric machine.[127] He introduced his invention as follows:

> In most dynamo-electric machines before the public, the bobbins of wire on the armature surround one or more cores of soft iron, to the changing magnetism of which the whole (in some) or part (in others) of the effect produced is due. This changing magnetism of the cores, is a source of great loss of driving power, which loss appears as heat in the cores. This is caused not only by the changing magnetism, but by the induction of currents in the iron itself due to its motion in the magnetic field. Thus not only is a large portion of the driving power wasted, but the field of force is largely diverted from its proper function.
>
> In other machines, wherein the armature carries no moving iron, the field of force is necessarily so large that much of it cannot be utilized, and the length of moving conductor on the armature is so great as to cause much resistance, and consequent waste of current, attended with heat. If now, a very great concentration of magnetic field can be attained, without diverting any of it from its proper function, then a rapid motion of a short armature conductor may develop a high electromotive force; and there being no changing magnetism or "local" induced currents, very nearly the whole of the driving power may be realized as available current. I have fully accomplished these important results in the apparatus I am about to describe.

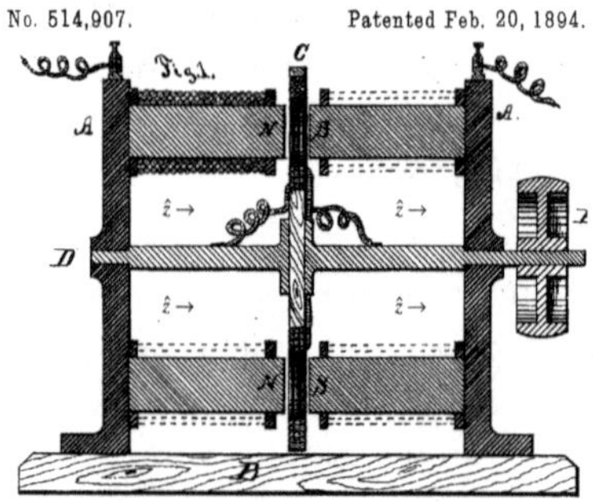

In the figures we overlay two of his figures as viewed from the left: one showing the positive direction of the coil windings in the rotating armature, and the other showing the fixed magnets. We rotated the windings figure by an angle of $\pi/4$ between the two diagrams, to show how the voltage is reversed. In his labeling, "S" means the magnetic field is from left to right in his Fig. 1, or out of the page in his Fig. 2 as it is a cut-away viewed from the left.

In this example, we are going to estimate the voltage produced by this AC generator, as a function of the number of windings per loop N, the area of each loop $\vec{A}$, the peak magnetic field B_0, the frequency of rotation ω_0, and the radius to the middle or each loop a.

To begin, we consider the system in a co-rotating reference frame, where the coils of wire are fixed and the magnetic field is considered to be rotating. From the vantage point of one of the wire coils, say above a south pole, the magnetic field will appear to oscillate with a an angular frequency four times the driving frequency, or:

[127] Brush, C.F., 1879, "Dynamo-Electric Machine," U.S. Patent No. 514907, awarded February 20, 1894.

$$\vec{B} \approx -B_0 \cos(\varphi)\,\hat{z} = -B_0 \cos(\omega t)\,\hat{z} \approx -B_0 \cos(4\,\omega_0\, t)\,\hat{z}.$$

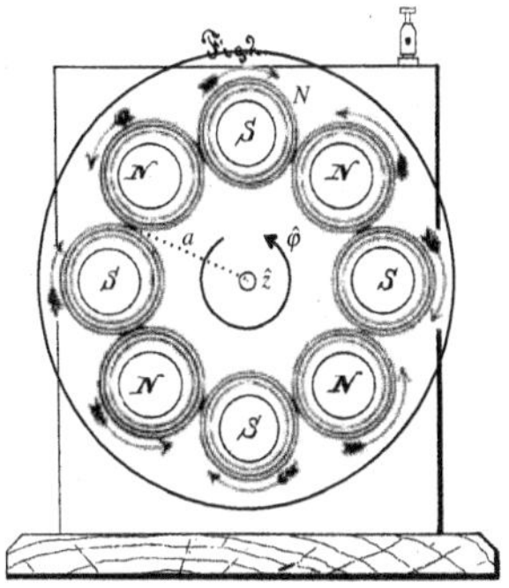

Now considering a single loop in the left-hand diagram about a south pole, it has an area of $\vec{A} = -A\,\hat{z}$, and so we can apply Faraday's law as:

$$\vec{\nabla} \times \vec{E} = -\frac{\partial \vec{B}}{\partial t} \quad \rightarrow \quad \int_{\text{surface}} \left(\vec{\nabla} \times \vec{E}\right) \cdot \mathrm{d}\vec{A} = -\int_{\text{surface}} \frac{\partial \vec{B}}{\partial t} \cdot \mathrm{d}\vec{A}.$$

So, the voltage around one loop is:

$$\mathbb{V}_{\text{loop}} = \oint_{\text{path}} \vec{E} \cdot \mathrm{d}\vec{\ell} = -\int_{\text{surface}} \frac{\partial}{\partial t}\left(-B_0 \cos(4\,\omega_0\, t)\,\hat{z}\right) \cdot \mathrm{d}\vec{A} = 4\,\omega_0\, B_0\, A \sin(4\,\omega_0\, t).$$

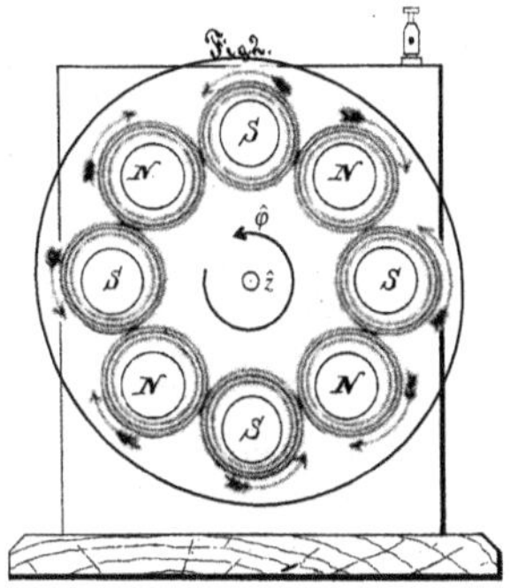

All eight loops are arranged in series, and configured such that their voltages add, so the voltage output will be:

$$\mathbb{V} = 8\mathbb{V}_{\text{loop}} = 32 N \omega_0\, B_0\, A \sin(4\,\omega_0\, t) = 8 N \omega\, B_0\, A \sin(\omega t).$$

So, for example, if one of these were built to produce $60\,\text{Hz}$ A.C. at $16 \times 120\text{V}{=}1.92\text{kV}$ peak-to-peak, then the designers would need to make sure that:

$$\frac{\mathbb{V}_{\text{pp}}}{4\omega_0} = 8 N\, B_0\, A = \frac{1920\text{V}}{(2\pi \cdot 60\,\text{Hz})} \approx 5.1\,\text{V} \cdot \text{s} \approx 5.1\,\text{Wb} \approx 5.1\,\text{T} \cdot \text{m}^2 \ .$$

Now let us imagine that this machine is expected to produce 500 kilowatts of power, then the wire will need to be thick enough to have an average current flowing through it of:

$$I_{\text{rms}} = \frac{\mathbb{P}}{\mathbb{V}_{\text{rms}}} = \sqrt{2}\,\frac{\mathbb{P}}{\mathbb{V}_{\text{pp}}} \approx \sqrt{2}\,\frac{500\,\cancel{\text{k}}\text{W}}{1.92\,\cancel{\text{k}}\text{V}} \approx 370\,\text{A} \ .$$

Alternatively, with slightly more effort, this same generator could be wired with each coil in parallel, rather than series. Would parallel wiring be a good design choice?

All other things being equal, parallel wiring would increase the length of wire by eight times to produce the same voltage, but reduce the current flowing also by eight times so the same total power were generated. Resistance is proportional to wire length, so we would expect the ratio of energy loss due to heat in the wires to be:

$$\frac{\mathbb{P}_{\text{lost parallel}}}{\mathbb{P}_{\text{lost series}}} = \frac{I_{\text{p}}^2 R_{\text{p}}}{I_{\text{s}}^2 R_{\text{s}}} = \left(\frac{I_{\text{p}}}{I_{\text{s}}}\right)^2 \left(\frac{R_{\text{p}}}{R_{\text{s}}}\right) = \left(\frac{I_{\text{p}}}{I_{\text{s}}}\right)^2 \left(\frac{l_{\text{p}}}{l_{\text{s}}}\right) = \left(\frac{1}{8}\right)^2 \left(\frac{8}{1}\right) = \frac{1}{8}$$

which argues in favor of wiring the coils in parallel, on the other hand using thicker wires in the series circuit would also reduce the energy losses as $R \approx l/\kappa A$. Keeping the total power losses the same, the ratio of wire areas would need to be:

$$\frac{\mathbb{P}_{\text{lost parallel}}}{\mathbb{P}_{\text{lost series}}} = \left(\frac{I_{\text{p}}}{I_{\text{s}}}\right)^2 \left(\frac{R_{\text{p}}}{R_{\text{s}}}\right) = \left(\frac{I_{\text{p}}}{I_{\text{s}}}\right)^2 \left(\frac{l_{\text{p}}}{l_{\text{s}}}\right) \left(\frac{A_{\text{p}}}{A_{\text{s}}}\right) \quad \text{and} \quad \left(\frac{A_{\text{s}}}{A_{\text{p}}}\right) = \left(\frac{I_{\text{p}}}{I_{\text{s}}}\right)^2 \left(\frac{l_{\text{p}}}{l_{\text{s}}}\right) = \left(\frac{1}{8}\right)^2 \left(\frac{8}{1}\right) = \frac{8}{1}.$$

So to keep the power output the same, both forms of wiring would use the same total mass of copper. As it turns out, thicker wires only help you up to a point when producing AC current, due to the *skin depth*, which will discuss in a later chapter, so it is usually better to use more thinner wires than fewer thicker ones.

Problem 9.9: The Wallace Machine

The Wallace Machine was a nineteenth century generator consisting of two horseshoe magnets. "Each magnet has a rotating armature of twenty-five bobbins, on which the wire is wound quadruply."[128]

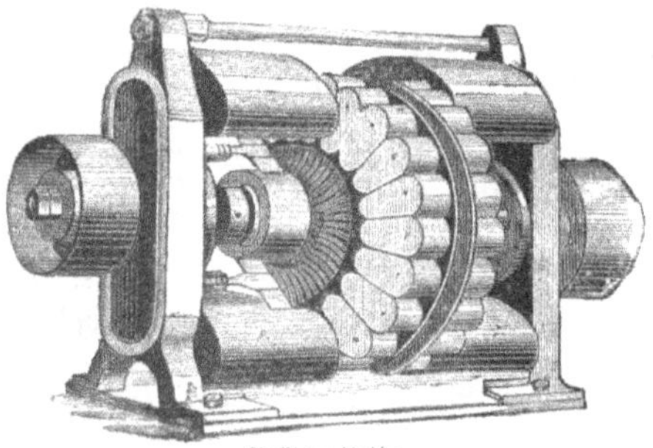

The Wallace Machine.

We will call the inner and outer radii of the side of each bobbin a and b respectively. The magnetic field between the horseshoe magnet poles is B and the angular frequency the crank is turned is ω.

(a) Find the electric field inside the wire as a function of the radial distance s from the axis.

(b) If a test charge q moves radially outward from $s = a$ to $s = b$ inside a single wire segment, how much work is done on it?

(c) Using the illustration and the quote, estimate how many such wire segments N are in the presence of the magnetic field at any given time.

(d) Using the picture, make a reasonable guess as to the total power generated by this machine. It has an angular crank speed of 60 Hz (i.e. $60 \times 2\pi \, ^{rad}/_{s}$) and ten amperes of current flowing.

(e) What torque must be applied in order to maintain the power you found in part (d)?

Problem 9.10: Faraday's Wheel

Michael Faraday set up a round copper plate in a strong magnetic field, as shown in his Fig. 7. The circuit was closed through a galvanometer, so he could measure the current that flows through the wires.

Faraday observed the following:

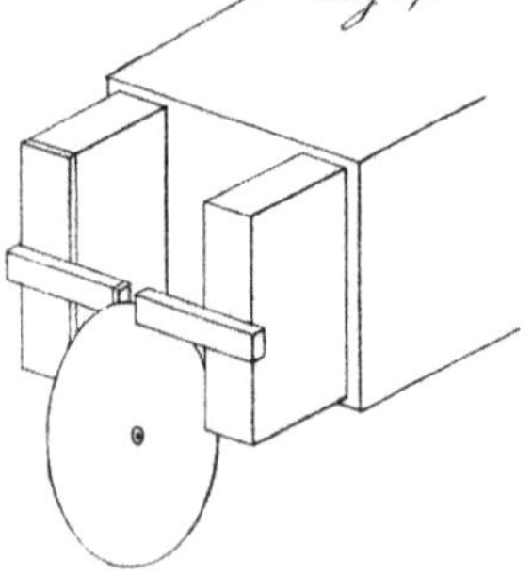

> All these arrangements being made, the copper disc was adjusted as in fig. 7, the small magnetic poles being about half an inch apart, and the edge of the plate inserted about half their width between them. One of the galvanometer wires was passed twice or thrice loosely round the brass axis of the plate, and the other to a conductor, which itself was retained by the hand in contact with the amalgamated edge of the disc at the part immediately between the magnetic poles. Under these circumstances, all was quiescent, and the galvanometer exhibited no effect. But the instant the plate moved, the galvanometer was influenced, and by revolving the plate quickly the needle could deflect

---

[128] Henry Frith, Marvels of Electricity and Magnetism, (London: Ward, Lock & Co). Figure from page 126. The publication date is unknown, but it was likely sometime around 1890 based on the content of the text.

90° or more.[129]

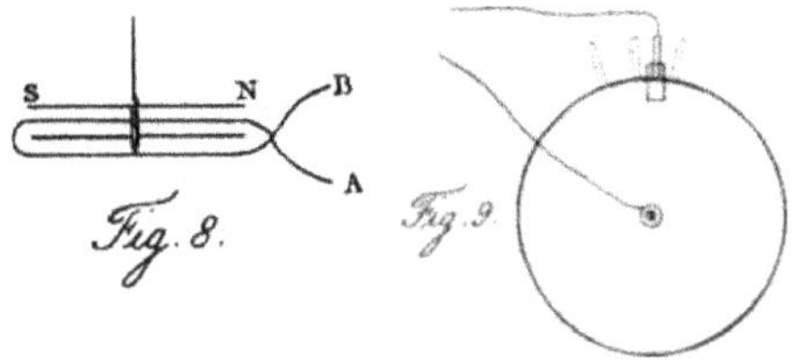

Faraday's Fig. 8 is a top view of the apparatus, while Fig. 9 shows the location of two wires he connected—one on the rotation axis of the disk and one attached to a sliding metal conductor on the disk's edge directly above the magnet.

To make this a quantitative problem we will assume that the disc has a mass M and radius a. We will also assume magnetic field $\vec{B}$ over a small cubic gap of dimensions $b \times b \times b$, and total resistance in the circuit R.

a) Of our assumed constant parameters of the system, b is the only one with a numeric measurement in the quote above. What is it in inches? Estimate a as well, assuming that Fig. 7 is drawn to scale.

b) Consider a small charge element δq in the disk and in the magnet. What is the electromagnetic force $\delta\vec{F}$ on the test charge as a function of the angular velocity ω?

c) What current is measured by the galvanometer, as a function of the angular velocity ω?

d) Find the torque on the disc due to current flow as a function of the angular velocity ω.

e) Find the power produced as a function of the angular velocity.

f) Let us now imagine that Faraday spins the wheel from rest with a constant torque τ_{ext}, which would be easy to do using a pulley and falling weight system. Express the angular velocity ω as a function of time. Plot this function.

Problem 9.11: Faraday's Generator and Capacitor

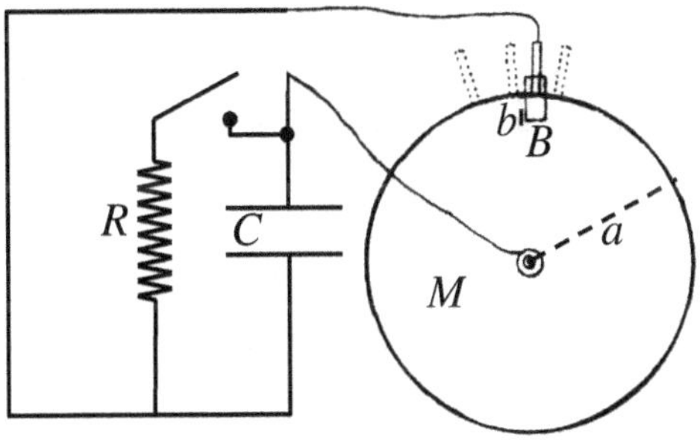

Reconsider Faraday's homopolar generator (Problem 9.10 above) except now we hook it up to a capacitor of known capacitance C, and assume that the generator is a frictionless flywheel with mass M and radius a. The magnetic field is B and covers a radial distance near the edge b.

a) At time zero, we give the wheel an initial fast angular velocity ω_0, calculate the charge on the capacitor Q, as a function of time.

b) After a long time, we close the switch to power a load resistor. Find another expression for the charge on the capacitor as a function of time. How does this differ from a simple RC circuit?

c) What is the frequency of the wheel as a function of time?

d) What is the rate of energy loss by the capacitor and the flywheel? Show that this equals the power output of the load resistor.

[129] Michael Faraday, Experimental Researches in Electricity, 3 vols. (London: Taylor, 1839, 1844; Taylor and Francis, 1855) **I**: 26, para. 88.

Problem 9.12: A Spinning Coil of Wire Surrounding a Magnet

Consider a small magnet with a wire coil of radius a with N loops of wire spinning at constant angular velocity perpendicular to the magnetic poles ($\vec{\omega} = \omega\hat{x}$). The total resistance in the circuit is R.

a) Show that the voltage across the coil is:

$$\mathbb{V}_{\text{coil}} = \frac{\mu_0 \omega m}{2a}\cos(\omega t).$$

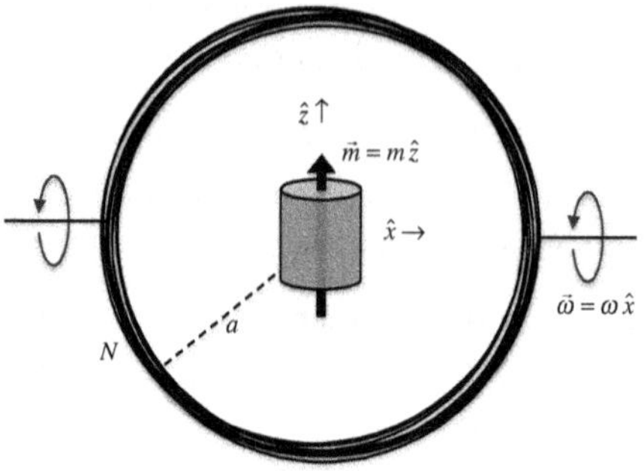

b) Clearly draw the system when the flux through the hoop is greatest and when it is zero.

c) Clearly draw the system when the voltage is greatest in the positive direction, when it is zero, and when it is most negative. Discuss how this relates to the flux.

d) What is the average power that this generator produces?

9.2 *Inductors*

Notice in the 1899 photograph showing Nicola Tesla in his Colorado lab on page 270, the large coil on the right that is sparking to a grounded metal sphere. This is an example of an inductor.

Inductors and electromagnets are essentially the same thing, except in the case of an inductor we care about how it affects the electric circuit, rather than the external magnetic field it produces. There are basically two kinds of inductors: self inductors, commonly called *inductors*, and mutual inductors, commonly called *transformers*. While a self inductor has only one coil of wire surrounding a steel core, a mutual inductor has two coils of wire usually with different numbers of turns.

In this section we first introduce the self inductor. Next, we use this as a toy model to find an expression for the magnetic energy density. Last we will introduce mutual inductors like Tesla's coil in the picture.

Thought Experiment 9.6: A Toroidal Inductor

Consider a torus of an iron alloy with a circumference l, and a cross-sectional area $\vec{A}$, so it has a total volume of $V = lA$. The steel torus is wrapped with n turns of wire per length as shown.

Next we attach it to a circuit with a known time-dependent current, $I(t)$, flowing through the circuit. Our goal is to predict the voltage across the inductor, $\mathbb{V}(t)$.

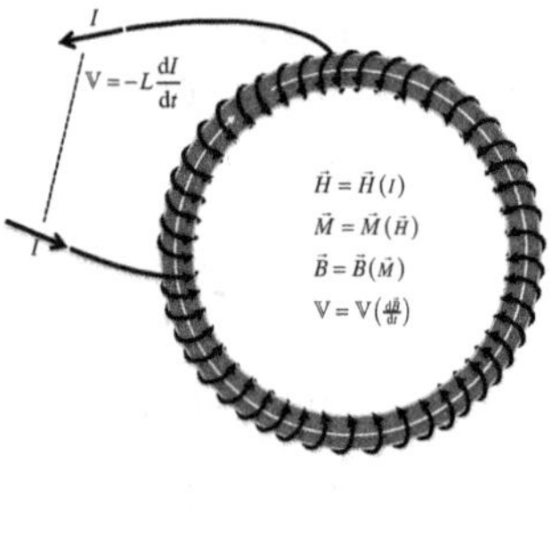

The magnetizing field, $\vec{H}$, inside the torus is given by:

$$\vec{H} = In\hat{A}.$$

In the diagram, and our sign convention (p. 24), $\vec{A}$ is pointing clockwise around the torus.

In Thought Experiment 8.14 (p. 263) we found that if the coercivity were small compared to the internal magnetization, we could write the magnetic moment per volume as:

$$M \approx M_I \tanh\left(\frac{\chi_M H}{M_I}\right).$$

and the magnetic field is approximately:

$$B = \mu_0 (H + M) = \mu_0 H + \mu_0 M_I \tanh\left(\frac{\chi_M H}{M_I}\right).$$

We can now write the magnetic flux through each loop as:

$$\Phi_B = \vec{B} \cdot \vec{A} = \mu_0 H\, A + \mu_0 M_I\, A \tanh\left(\frac{\chi_M H}{M_I}\right).$$

The magnetic flux through each loop is a function of time so it induces a voltage around each of the wire coils, according to Faraday's law, as:

$$\Delta \mathbb{V}_i(t) = -\frac{\mathrm{d}\Phi_B}{\mathrm{d}t} = -\mu_0\left(1 + \chi_M\left(1 - \tanh^2\left(\frac{\chi_M H}{M_I}\right)\right)\right)\left(\tfrac{\mathrm{d}}{\mathrm{d}t}\vec{H} \cdot \vec{A}\right).$$

Now, in the small field limit:

$$\Delta \mathbb{V}_i(t) \approx -\mu_0(1 + \chi_M)\left(\tfrac{\mathrm{d}}{\mathrm{d}t}\vec{H} \cdot \vec{A}\right) \approx -\mu\left(\tfrac{\mathrm{d}}{\mathrm{d}t}\vec{H} \cdot \vec{A}\right).$$

Notice that the permeability μ is a constant in the small field limit for an idealized magnetic material.

Recall that there were nl total loops, so the voltage is:

$$\mathbb{V}(t) = (nl)\Delta\mathbb{V}_i(t) = (nl)\left(-\mu \tfrac{\mathrm{d}}{\mathrm{d}t}\left(I n \hat{A} \cdot \vec{A}\right)\right) \approx -\left(\mu n^2 A l\right)\tfrac{\mathrm{d}}{\mathrm{d}t} I \approx -\left(\mu n^2 V\right)\tfrac{\mathrm{d}}{\mathrm{d}t} I.$$

Notice how everything inside the parentheses can be controlled by the manufacturer, therefore we define a quantity called *the inductance*, L, such that for and ideal inductor:

$$\mathbb{V}(t) = -L\frac{\mathrm{d}I}{\mathrm{d}t}, \quad \text{and:} \quad L = \mu n^2 V.$$

The circuit diagram symbol for an inductor is shown, as it is suppose to represent a coil of wire. The SI unit of inductance is called the henry, after the American physicist Joseph Henry (p. 271). The Henry is defined as:

$$1\,\mathrm{H} = 1\frac{\mathrm{Wb}}{\mathrm{A}} = 1\frac{\mathrm{T}\cdot\mathrm{m}^2}{\mathrm{A}} = 1\frac{\mathrm{V}\cdot\mathrm{s}}{\mathrm{A}} = 1\ \Omega\cdot\mathrm{s}.$$

It is extremely important to remember that the idea of inductance came from the assumed linearity of the ferromagnetic materials. In fact, they are not linear. While, in principle, inductors do not dissipate heat, in practice they do.

We now write the work done, per time, by the inductor on the rest of the circuit as:

$$\mathbb{P} = I\,\mathbb{V} = -LI\frac{\mathrm{d}I}{\mathrm{d}t}.$$

So, the total potential energy stored in an inductor, assuming that $I(t=0) = 0$, is given by:

$$\mathbb{U} = \int_0^t \mathbb{P}\,dt = -L\int_0^t I\frac{dI}{dt}dt = -L\int_0^I I\,dI = -\tfrac{1}{2}LI^2.$$

Thought Experiment 9.7: The Energy Density from Faraday's Law

Consider an inductor with n turns of wire per length and a steel core. What is the potential energy per volume in the core, as a function of the fields $\vec{B}$ and $\vec{H}$?

Inside the torus, the magnetizing field is $\vec{H} = In\hat{A}$.

Now, we can use Faraday's law to find the voltage drop as:

$$\vec{\nabla}\times\vec{E} = -\tfrac{\partial\vec{B}}{\partial t} \quad\rightarrow\quad \mathbb{V} = -\int_{\text{along wire}} \vec{E}\cdot d\vec{\ell} = -N\left(-\int \tfrac{\partial}{\partial t}\vec{B}\cdot d\vec{A}\right) = N\tfrac{\partial}{\partial t}\vec{B}\cdot\vec{A}.$$

Therefore the power that flows into the electromagnet is:

$$\mathbb{P} = I\mathbb{V} = (I)\left(N\tfrac{\partial}{\partial t}\vec{B}\cdot\vec{A}\right) = N\tfrac{\partial}{\partial t}\vec{B}\cdot I\vec{A} = (Al)\tfrac{\partial}{\partial t}\vec{B}\cdot\vec{H} = V\tfrac{\partial}{\partial t}\vec{B}\cdot\vec{H}.$$

So, the power and work per volume are:

$$\mathbb{p} = \frac{\mathbb{P}}{V} = \vec{H}\cdot\tfrac{\partial}{\partial t}\vec{B} \quad\rightarrow\quad \mathbb{w} = \int_0^t \mathbb{p}\,dt = \int_0^t \vec{H}\cdot\frac{d\vec{B}}{dt}dt = \int_0^B \vec{H}\cdot d\vec{B}.$$

Engineers plot ferromagnetic hysteresis curves in terms of the applied magnetic field, $\vec{H}$, and the induced magnetic field, $\vec{B}$, which allows them to interpret the curves thermodynamically in much the same way as pressure-volume diagrams are for gas dynamics.

To gain a more physical understanding, we use the constitutive relation to express this in terms of the two quantities that, in principle, could be measured in situ (recall Discussion 6.2, p. 195):

$$\mathbb{w} = \int_0^B \vec{H}\cdot d\vec{B} = \int_0^B \left(\tfrac{1}{\mu_0}\vec{B} - \vec{M}\right)\cdot d\vec{B} = \tfrac{1}{2\mu_0}B^2 - \int_0^B \vec{M}\cdot d\vec{B}.$$

The first term represents the energy density contained inside any magnetic field, including one in free space.

The second term represents the volume averaged magnetic energy benefit due to each domain's magnetic moment aligning with the local magnetic field it encounters. There is a catch! Aligning magnetic domains in ferromagnets is never friction free, so some of this work simply heats up the magnet. This is why stepping up, or down, the AC voltage always costs some energy.

In the linear regime, where $M \propto B$, the potential energy becomes:

$$\mathbb{u}_B = \tfrac{1}{2\mu_0}B^2 - \tfrac{1}{2}\vec{M}\cdot\vec{B}.$$

In case of a permanent magnet located in an increasing magnetic field:

$$\mathbb{u}_B = \tfrac{1}{2\mu_0}B^2 - \vec{M}\cdot\vec{B}.$$

Problem 9.13: The Energy Density of a Ferromagnet

Consider an electromagnet with a steel core with susceptibility χ_M, coercivity $H_C \to 0$, and intrinsic magnetization M_I.

a) Show that the energy density can be written:

$$\mathbb{u} = \tfrac{1}{2}\mu_0 H^2 + \vec{H}\cdot\vec{M} - \mu_0 \int_0^H \vec{M}\cdot \mathrm{d}\vec{H}.$$

b) Make the unitless variable substitution: $x = \dfrac{\chi_M H}{M_I}$, and express the energy density in terms of the same parameters and a unitless definite integral.

c) Rewrite the energy density in the form:

$$\mathbb{u}_B = \tfrac{1}{2\mu_0} B^2 - \xi \vec{M}\cdot\vec{B}.$$

And write ξ in terms of the unitless integral as best as you can.

d) Show that $\xi = \tfrac{1}{2}$ in the linear case and $\xi = 1$ for a permanent magnet.

e) What is ξ in the limit that $M \to M_I$.

Problem 9.14: Energy Cycle of a Hot Magnet

Consider a hot piece of magnetic material, with a temperature higher than its Curie temperature, that is completely unmagnetized. Call the potential energy per volume in this situation $\mathbb{u} = 0$.

a) Now you slowly turn up an external magnetic field $\vec{B}_{\text{external}} = \mu_0 H\,\hat{z}$, until its magnetization becomes $M = \chi_M H$.

What is the magnetic energy per volume, $\mathbb{u}_a$?

b) Keeping the external field in place, you cool the magnet below its Curie temperature, so that it becomes a permanent magnet with uniform magnetization $\vec{M} = M\,\hat{z}$. Now you twist the magnet 90°. What is the magnetic energy per volume, $\mathbb{u}_b$?

c) Now you turn off the external field. What is the magnetic energy per volume, $\mathbb{u}_c$?

d) Now you heat the magnet to above its Curie temperature again, and it loses its magnetization. Its potential energy density must be back to $\mathbb{u}_0$. Is this consistent with your analysis?

Explain this in terms of a classical toy model of magnetism, where matter is made up of a bunch of tiny permanent magnets that are free to rotate. Draw pictures for each situation, and discuss what makes them have different energy differences.

Problem 9.15: The Energy Density Inside of a Toroidal Inductor

In Thought Experiment 9.6 above we found the magnetic potential energy inside of a toroidal inductor. Use this to show that, in the linear limit, the magnetic potential energy density is:

$$\mathbb{u}_B = -\tfrac{1}{2}\vec{B}\cdot\vec{H}.$$

Problem 9.16: A Toroidal Inductor with an Iron Core

Consider the toroidal inductor shown in the diagram, which consists of N turns of wire wrapped around an iron torus of radius b, and cross-sectional radius a. The susceptibility of the iron is known and assumed to be large $\chi_M \gg 1$.

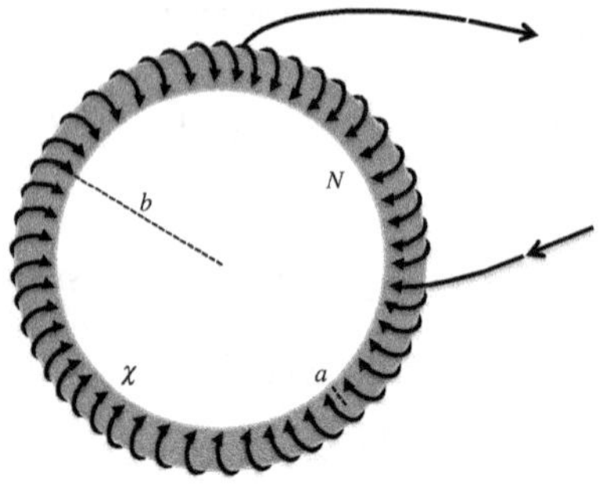

a) If a current I is flowing through the wire in the direction shown, what is the magnetic field inside the iron ring?

b) If the current increases at a rate $\dot{I} = \frac{dI}{dt}$, what is the voltage across the wires?

c) What is the inductance L?

d) If a current I flows and it changes at a rate $\dot{I}$, how much power flows into the inductor?

e) Integrate the power flowing into the inductor to find how much work it takes to go from no current flowing to having a current I flowing. Account for where the energy went.

Problem 9.17: Current Density Induced by a Toroidal Inductor

Consider a toroidal inductor with a radius b, cross-sectional radius $a \ll b$, N turns of wire, a core with magnetic susceptibility χ_M, and an AC currents $I = I_0 \sin(\omega t)$. A cylindrical metal rod, with electrical conductivity κ and radius $c \ll b$, is placed on the axis of the torus

a) Find the induced electric field, $\vec{E}$, in the rod?

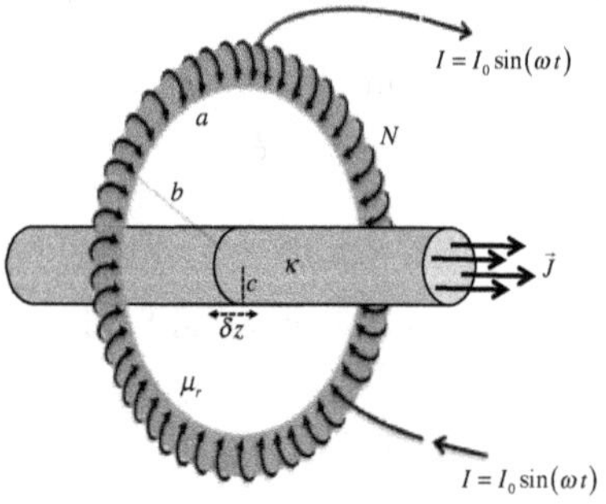

b) What is the current density inside the rod at the center of the torus? Assume no other electric fields in the wire.

c) How much energy is being dissipated per time per volume in the rod at the center of the loop?

d) What is the induced magnetic field, due to a small distance δz of the metal at the center of torus.

e) What is the induced voltage against the driving current due to the small section δz of the central wire? Is all the energy accounted for?

Problem 9.18: A Toroidal Inductor around an AC Current

Consider a long wire along the axis of a toroidal inductor. The wire carries a large AC current given by $I = I_0 \sin(\omega t)$. The inductor has N turns of wire surrounding an iron torus with magnetic susceptibility χ_M, overall radius b, cross-sectional radius a, and it is attached to a resistor R.

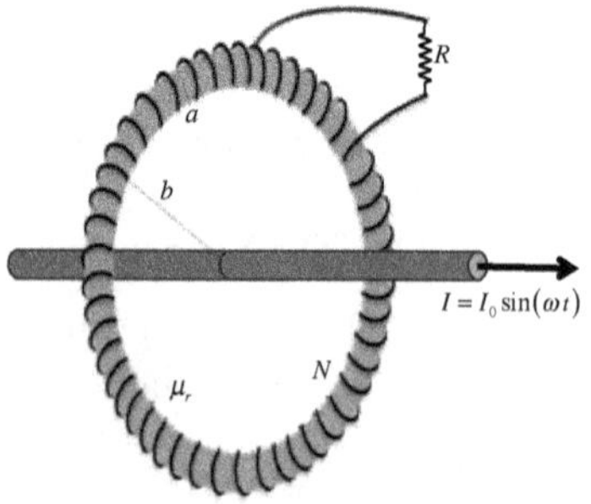

a) Find the voltage drop across the resistor as a function of time.

b) Find the current flowing through the resistor, also as a function of time.

c) Find the voltage drop in the power line due to the inductor's presence.

d) Report on how clamp-on AC current meters work and their use by electricians.

Problem 9.19: Electrical Steel

Electrical steel is the class of iron alloys used in electromagnetic devices such as AC transformers and electromagnets. Vendors provide plots of the induced magnetic field, B, as a function of the applied magnetic field $\mu_0 H$, for a sinusoidal applied magnetizing field $H = H_0 \sin(\omega t)$, with a particular amplitude and frequency. You can assume that the magnetization follows the form:

$$M = M_I \tanh\left(\frac{\chi_M (H \mp H_c)}{M_I}\right),$$

and that $\chi_M H_0 \ll M_I$ and $H_c \ll H_0$.

a) Find the energy lost per cycle per volume.

b) Find the magnetic energy density when fully magnetized.

c) What is the efficiency of the material as a function of driving frequency?

d) Look up the specifications for electrical steel, including its density, specific heat capacity, and thermal conductivity. Discuss the engineering considerations for it applications. Make sure you discuss how the material keeps from getting too hot.

e) Research the efficiency of the power grid in your country. Is it better, or worse, than in other countries? Why?

f) Research and estimate the total mass of electrical steel that is used in your country's power grid.

Thought Experiment 9.8: An LC Circuit

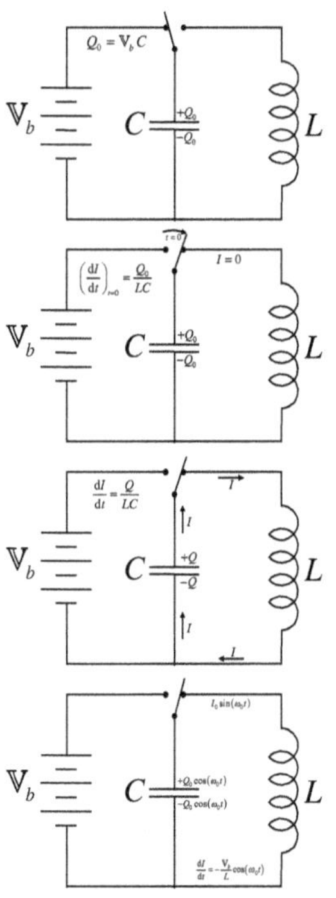

Consider a battery with a voltage V_b, a double-pole switch, a capacitor C and an inductor L. At time zero the switch is thrown, and the capacitor subsequently begins to discharge. What is the current through the circuit as a function of time?

Initially the capacitor is charged with an initial charge on the top plate of:

$$Q_0 = V_b C,$$

and of course a corresponding negative charge on the bottom plate so it remains net neutral. Charge conservation states that, at any time, the current flowing out of the capacitor is related to its charge by:

$$I = -\frac{dQ}{dt}.$$

After the switch is thrown, the voltage rise across the capacitor, plus the voltage drop across the inductor, must be zero because of Kirchhoff's loop rule:

$$\sum \Delta V_i = \frac{Q}{C} - \left(-L\frac{dI}{dt}\right) = 0$$

Or the voltage across the capacitor must be:

$$\mathbb{V} = \frac{Q}{C} = L\frac{\mathrm{d}}{\mathrm{d}t}\left(-\frac{\mathrm{d}Q}{\mathrm{d}t}\right) = -L\frac{\mathrm{d}^2 Q}{\mathrm{d}t^2}.$$

This is a 2nd order differential equation, so we write it in standard form as:

$$\frac{\mathrm{d}^2 Q}{\mathrm{d}t^2} + \frac{Q}{LC} = 0 .$$

First consider the units of LC:

$$[LC] = (\mathrm{H})\cdot(\mathrm{F}) = \left(\frac{\cancel{\mathrm{V}}\cdot \mathrm{s}}{\mathrm{A}}\right)\cdot\left(\frac{\mathrm{C}}{\cancel{\mathrm{V}}}\right) = \mathrm{s}^2 .$$

Recognizing that this will be an oscillatory system, we guess a solution with some characteristic angular frequency, which we will call ω_0, with units of inverse time. As the capacitor is fully charged at time $t = 0$, we will initially guess a solution in the form:

$$Q = Q_0 \cos(\omega_0 t).$$

Next we differentiate this:

$$\frac{\mathrm{d}Q}{\mathrm{d}t} = -\omega_0 Q_0 \sin(\omega_0 t) \quad \text{and} \quad \frac{\mathrm{d}^2 Q}{\mathrm{d}t^2} = -(\omega_0)^2 Q_0 \cos(\omega_0 t).$$

We now put this solution into the differential equation:

$$\frac{\mathrm{d}^2 Q}{\mathrm{d}t^2} + \frac{Q}{LC} = 0 \rightarrow \left(-(\omega_0)^2 \cancel{Q_0 \cos(\omega_0 t)}\right) + \frac{\left(\cancel{Q_0 \cos(\omega_0 t)}\right)}{LC} = 0$$

$$-(\omega_0)^2 + \frac{1}{LC} = 0 \rightarrow \omega_0 = \frac{1}{\sqrt{LC}}.$$

Notice since the time dependent terms cancel, our trial function is a solution to the differential equation if: $\omega_0 = \frac{1}{\sqrt{LC}}$.

Physically, this makes sense given that the units of ω_0 are radians per second, and the units of the product LC is seconds squared.

Finally, we find the current flowing around the loop as:

$$I = -\frac{\mathrm{d}Q}{\mathrm{d}t} = -(-\omega_0 Q_0 \sin(\omega_0 t)) = \omega_0 Q_0 \sin(\omega_0 t)$$

$$= \left(\frac{1}{\sqrt{LC}}\right)(\mathbb{V}_b C)\sin(\omega_0 t) = \mathbb{V}_b\sqrt{\frac{C}{L}}\sin(\omega_0 t).$$

As a check of our answer, let us now verify the initial conditions. At time zero, the current is zero, the charge is Q_0, and the rate of change of the current is:

$$\left(\frac{\mathrm{d}I}{\mathrm{d}t}\right)_{t=0} = \omega_0 \mathbb{V}_b\sqrt{\frac{C}{L}}\cos(0) = \frac{1}{\sqrt{L\cancel{C}}}\mathbb{V}_b\sqrt{\frac{\cancel{C}}{L}} = \frac{\mathbb{V}_b}{L},$$

all of these are consistent with applying the loop rule immediately as the switch is thrown.

As a second check, let us verify that the total energy in the system remains constant, by showing that the energy remains constant. First we write this total energy in terms of the capacitance, charge, inductance, and current:

$$\mathbb{U}_{\text{total}} = \tfrac{1}{2}\frac{Q^2}{C} + \tfrac{1}{2}LI^2 .$$

Now we include the solutions above and simplify:

$$\mathbb{U}_{\text{total}} = \tfrac{1}{2}\frac{\left(C\mathbb{V}_b\cos(\omega_0 t)\right)^2}{C} + \tfrac{1}{2}L\left(\sqrt{\frac{C}{L}}\,\mathbb{V}_b\sin(\omega_0 t)\right)^2 = \tfrac{1}{2}C\mathbb{V}_b^{\,2}\cos^2(\omega_0 t) + \tfrac{1}{2}\cancel{L}\frac{C}{\cancel{L}}\mathbb{V}_b^{\,2}\sin^2(\omega_0 t)$$
$$= \tfrac{1}{2}C\mathbb{V}_b^{\,2}\left(\cos^2(\omega_0 t) + \sin^2(\omega_0 t)\right) = \tfrac{1}{2}C\mathbb{V}_b^{\,2} .$$

Now, the total energy is no longer a function of time, but simply equals the potential energy stored in the capacitor at time zero.

On the face of it, this is analogous to the conversion between kinetic and potential energy in the simple harmonic motion of a mass on a spring. However, there is a key difference: both the electric (capacitor) and magnetic (inductor) energies are potential energy; the kinetic energy of the charge carriers is negligible.

Thought Experiment 9.9: Measuring Permeability

Recall that when we derived the inductance of the toroidal inductor (Thought Experiment 9.6), the magnetic and magnetizing fields were, approximately, directly proportional. We called the constant of proportionality, μ, the *permeability* of the ferromagnetic material:

$$\mu \equiv \frac{B}{H} .$$

With this assumption, we then derived the inductance in terms of controllable quantities:

$$L = \mu n^2 V ,$$

where n and V are the turns per length of the wire, and the volume of the core, respectively.

By making an LC circuit, and using an oscilloscope to measure the voltage as a function of time, one could measure the period T of these oscillations. The inductance could then be measured independently of the geometry of the actual inductor involved, from:

$$\omega_0 = \frac{2\pi}{T} = \frac{1}{\sqrt{LC}} , \quad \text{so:} \quad L = \frac{1}{C}\left(\frac{T}{2\pi}\right)^2 .$$

Therefore, the permeability of the material could be easily found from:

$$\mu = \frac{L}{n^2 V} .$$

Moreover, the inductance of most ferromagnetic materials is only approximately constant for relatively small currents through the coils. The inductance then decreases as the core becomes saturated, which is represented by the slope of the hysteresis curve. The specifications for manufactured inductors clearly indicate the range of currents where the inductance can safely be assumed to be approximately constant. The measured permeabilities of most iron alloys are on the order of magnitude of $\mu \sim 1\,\mathrm{mH/m}$.

Faraday and others found from their experimental work that non-magnetic materials have a small, but measurable, permeability of approximately a thousandth that of iron alloys. Moreover, all these materials had about the same permeability of $\mu \approx 1.257\,\mu\mathrm{H/m}$, differing only by values in the fifth, or higher, significant digit. These subtle differences are called paramagnetism and diamagnetism. For now, it is important to note that they have little practical effect.

Problem 9.20: An RLC Circuit

Consider a circuit with a capacitor C, inductor L, and resistor R all connected in series. At time $t = 0$, the voltage across the capacitor is V_0 and no current is flowing.

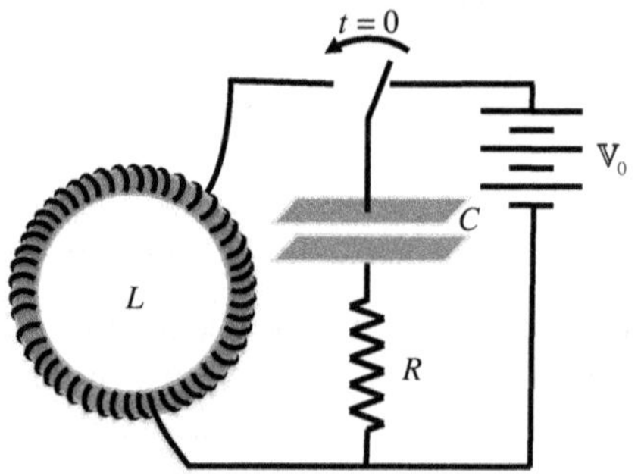

a) Derive an expression relating the charge on the capacitor to its time derivatives.

b) Solve for the charge on the capacitor, and the current flowing through the circuit, as a function of time.

c) What are the characteristic time constants governing the system in terms of the quantities R, L, and C? Discuss qualitatively what each one does.

d) Make a plot of the charge on the capacitor as a function of time, for the special cases where $R\sqrt{C/L} = \{\frac{1}{4}, \frac{1}{2}, 1, 2, 4\}$.

Problem 9.21: A Driven Series RLC Circuit

Consider a circuit with an alternating voltage source $\mathrm{V} = \mathrm{V}_0 \sin(\omega t)$, capacitor C, inductor L, and resistor R, all connected in series.

a) Express the voltage drop across each of the components in terms of the charge on the capacitor Q and its time derivatives

b) Use the loop rule to find a differential equation relating the charge to its time derivatives.

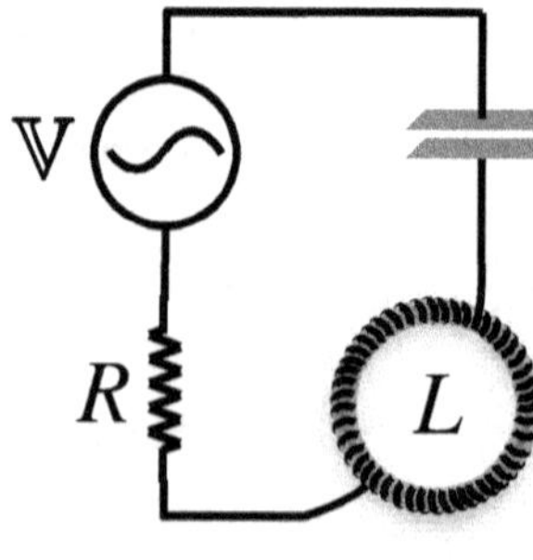

c) Assume that the current through the voltage source is in phase with the voltage and is of the form $I = I_0 \sin(\omega t)$. Now rewrite the differential equation in terms of the amplitude of the current I_0, and the quantities C, L, R, ω, and t.

d) Express the inductor energy in terms of I_0, L, ω, and t.

e) Express the capacitor energy in terms of I_0, C, ω, and t.

f) At any given time energy is conserved, so add up the total power added (+) and subtracted (-) from the system and set it equal to zero. Does the time cancel out as it must? Solve for the amplitude of the current I_0 in terms of the other quantities.

Thought Experiment 9.10: The Inductance of a Coaxial Cable

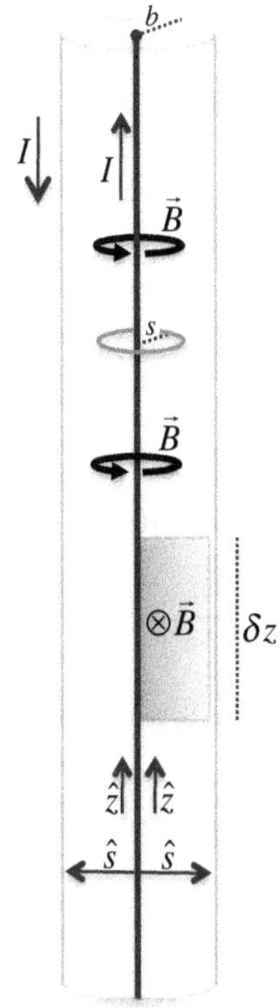

Consider our coaxial cable from Chapter 4. Our goal is to investigate the magnetic field inside the cable and calculate its inductance. We will assume we know the current I and its time derivatives. The radii of the central wire and the cable are a and b respectively. We assume that the material with has the same permeability and permittivity as free space.

We begin by applying the Maxwell-Ampère law:

$$\vec{\nabla}\times\vec{B}=\mu_0\vec{J}+\mu_0\varepsilon_0\tfrac{\partial}{\partial t}\vec{E}+\mu_0\tfrac{\partial}{\partial t}\cancel{\vec{P}}+\mu_0\vec{\nabla}\times\cancel{\vec{M}}.$$

And we pick an Ampèrian loop of radius s surrounding the central wire, but inside the outer sheath, so:

$$\oint_{\text{path}}\vec{B}\cdot\mathrm{d}\vec{\ell}=\mu_0\int_{\text{surface}}\vec{J}\cdot\mathrm{d}\vec{A}+\mu_0\varepsilon_0\int_{\text{surface}}\tfrac{\partial}{\partial t}\vec{E}\cdot\mathrm{d}\vec{A}\,.$$

Any charge difference between the central cable and outer sheath will produce an electric field in the radial direction, so no electric flux passes through our Ampèrian loop. Thus, the electric field inside the cable is the same as a long wire:

$$\vec{B}=\frac{\mu_0 I}{2\pi s}\hat{\varphi}\quad:\quad a\le s\le b\quad.$$

Now we apply Faraday's law: $\vec{\nabla}\times\vec{E}=-\tfrac{\partial}{\partial t}\vec{B}$, to a rectangular loop of constant φ, with dimensions $(b-a)\times\delta z$:

$$\oint_{\text{path}}\vec{E}\cdot\mathrm{d}\vec{\ell}=-\int_{\text{surface}}\frac{\partial\vec{B}}{\partial t}\cdot\mathrm{d}\vec{A}=-\frac{\mathrm{d}}{\mathrm{d}t}\int_{\text{surface}}\left(\frac{\mu_0 I}{2\pi s}\hat{\varphi}\right)\cdot(\mathrm{d}s\,\mathrm{d}z\hat{\varphi})\,.$$

Integrating over the surface yields:

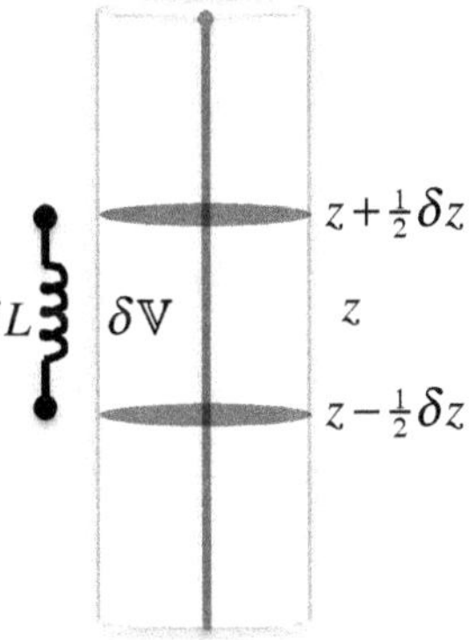

$$\oint_{\text{path}}\vec{E}\cdot\mathrm{d}\vec{\ell}=-\frac{\mathrm{d}}{\mathrm{d}t}\left(\frac{\mu_0 I}{2\pi}\int_{z-\delta z/2}^{z+\delta z/2}\int_a^b\frac{\mathrm{d}s}{s}\,\mathrm{d}z\right)=-\frac{\mathrm{d}}{\mathrm{d}t}\left(\frac{\mu_0 I}{2\pi}\ln\left(\frac{b}{a}\right)\int_{z-\delta z/2}^{z+\delta z/2}\mathrm{d}z\right),$$

which simplifies to:

$$\oint_{\text{path}}\vec{E}\cdot\mathrm{d}\vec{\ell}=-\frac{\mathrm{d}}{\mathrm{d}t}\left(\frac{\mu_0 I}{2\pi}\ln\left(\frac{b}{a}\right)\delta z\right)=-\left(\frac{\mu_0}{2\pi}\ln\left(\frac{b}{a}\right)\delta z\right)\frac{\mathrm{d}I}{\mathrm{d}t}\,.$$

Now we will evaluate the path integral, keeping in mind that $\vec{E}=E\,\hat{s}$:

$$\oint_{\text{path}}\vec{E}\cdot\mathrm{d}\vec{\ell}=\int_a^b\vec{E}\cdot\hat{s}\,\mathrm{d}s\Big|_{z-\delta z/2}+\cancel{\int_{z-\delta z/2}^{z+\delta z/2}\vec{E}\cdot\hat{z}\,\mathrm{d}z\Big|_b}+\int_b^a\vec{E}\cdot\hat{s}\,\mathrm{d}s\Big|_{z+\delta z/2}+\cancel{\int_{z+\delta z/2}^{z-\delta z/2}\vec{E}\cdot\hat{z}\,\mathrm{d}z\Big|_a}\,.$$

This path integral is simply the voltage difference in the $\hat{z}$ direction of the central wire:

$$\oint_{\text{path}} \vec{E}\cdot d\vec{\ell} = \left(\int_a^b \vec{E}\cdot\hat{s}\,ds \Big|_{z+\delta z/2} \right) - \left(-\int_a^b \vec{E}\cdot\hat{s}\,ds \Big|_{z-\delta z/2} \right) = \mathbb{V}\left(z+\delta z/2\right) - \mathbb{V}\left(z-\delta z/2\right) = \delta\mathbb{V}.$$

Now putting these together:

$$\delta\mathbb{V} = -\left(\frac{\mu_0}{2\pi}\ln\left(\frac{b}{a}\right)\delta z \right)\frac{dI}{dt}.$$

So, by the definition of the inductance δL, we have:

$$\delta L = -\frac{\delta\mathbb{V}}{dI/dt} = \frac{\mu_0}{2\pi}\ln\left(\frac{b}{a}\right)\delta z.$$

So the inductance per length is:

$$\mathtt{L} = \frac{\delta L}{\delta z} = \frac{\mu_0}{2\pi}\ln\left(\frac{b}{a}\right).$$

Revisiting the definition of the inductance, we use this result:

$$\delta\mathbb{V} = -\delta L\frac{dI}{dt} \quad\rightarrow\quad \frac{\delta\mathbb{V}}{\delta z} = -\mathtt{L}\frac{dI}{dt}.$$

Taking the limit as $\delta z \to 0$, this becomes a partial differential equation:

$$\frac{\partial\mathbb{V}}{\partial z} = -\mathtt{L}\frac{\partial I}{\partial t}.$$

This is one of Heaviside's telegrapher's equations, which we will discuss in more detail in Chapter 13 (p. 407).

Problem 9.22: The Inductance of Two Parallel Wires

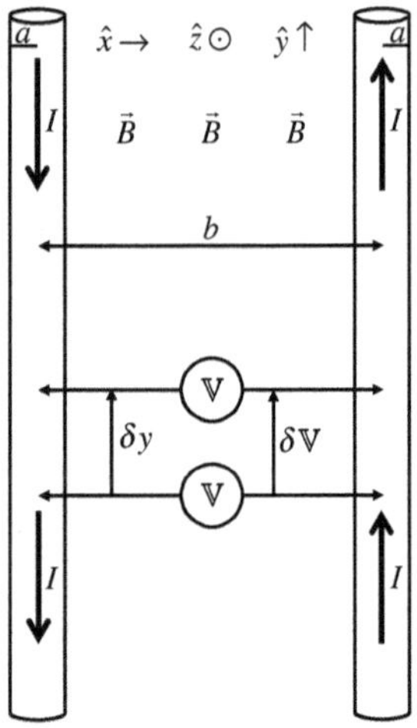

Consider two parallel wires of radius a carrying a current $+I(t)\,\hat{y}$ and $-I(t)\,\hat{y}$, which are separated by a distance b as in the diagram.

a) Find the magnetic field $\vec{B}_{z=0}(x)$ in the plane between the two wires.

b) The voltage across the wires becomes affected by the magnetic field. Find the amount that this voltage changes $\delta\mathbb{V}$ from position y to position $y+\delta y$ due to the magnetic field as a function of the current and its time derivatives?

c) What is the inductance per length $\mathtt{L}$ of the wires?

9.3 *Mutual Inductance*

In this section we introduce mutual inductance, where a time-varying current in one coil of wire creates a time-changing magnetic field, which induces an electric field in a different coil of wire. This not only is the basis for how transformers work, but also the physical origin of cross-talk among computers. As an example of mutual inductance, we consider the case in which two LC circuits are coupled. Next, we express the concept of inductance as a symmetrical matrix, where the diagonal terms are self-inductance and the off-diagonal are the mutual inductance terms.

While it would be natural to extend the concept of mutual inductance to express the relationship between any number of wire loops in any configuration, this will have to wait until Thought Experiment 11.20 (p. 360), after we have developed the law of Biot and Savart.

Thought Experiment 9.11: Mutual Inductance

Let us consider two long free space solenoids of different numbers of turns per length (n_1, n_2) and different cross-sections (A_1, A_2), but the same length h, with one solenoid located inside the other.

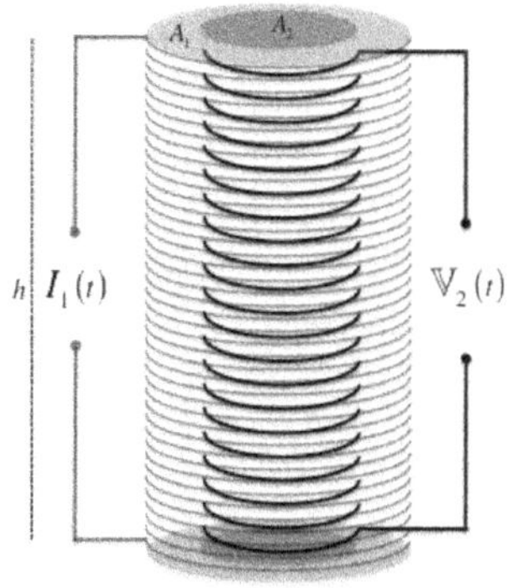

If we now we send a current $I_1(t)$ through one solenoid, what is the voltage $\mathbb{V}_2(t)$ across the other?

First, let's solve for the magnetic field inside the cylinder. From Thought Experiment 8.8 (p. 256), we see that the magnetic field is given by:

$$\vec{B}_1 = \mu_0 n_1 I_1(t)\hat{z}.$$

The magnetic flux through solenoid 2, because of magnetic field 1, is thus:

$$\Phi_{2B} = \int_{surface} \mu_0 n_1 I_1(t)\hat{z}\cdot d\vec{A} = \mu_0 n_1 I_1(t) A_2 .$$

Applying Faraday's law, we find that the induced voltage per loop is given by:

$$\frac{\mathbb{V}_2(t)}{n_2 h} = -\frac{\mathrm{d}\Phi_B}{\mathrm{d}t} = -\mu_0 n_1 A_2 \frac{\mathrm{d}I_1(t)}{\mathrm{d}t}.$$

So the output voltage should be:

$$\mathbb{V}_2(t) = -\mu_0 n_1 n_2 h A_2 \frac{\mathrm{d}I_1(t)}{\mathrm{d}t}.$$

Thus the mutual inductance between coil one and coil two is:

$$\mathbb{V}_2(t) = -L_{2,1}\frac{\mathrm{d}I_1(t)}{\mathrm{d}t} \quad \text{and} \quad L_{2,1} = \mu_0 n_1 n_2 h A_2 .$$

Now, if we reverse this, we can find the inverse relationship:

$$\vec{B}_2 = \mu_0 n_2 I_2(t)\hat{z} \quad \rightarrow \quad \Phi_{1B} = \int_{surface} \mu_0 n_2 I_2(t)\hat{z}\cdot d\vec{A} = \mu_0 n_2 I_2(t) A_2 \quad \rightarrow \quad \frac{\mathbb{V}_1(t)}{n_1 h} = -\mu_0 n_2 A_2 \frac{\mathrm{d}I_2(t)}{\mathrm{d}t}$$

$$\mathbb{V}_1(t) = -\mu_0 n_1 n_2 h A_2 \frac{\mathrm{d}I_2(t)}{\mathrm{d}t}.$$

Notice that the area of integration is still A_2, as it is on the inside, while every other index was reversed, so the mutual inductance is given by:

$$L_{12} = L_{21} = \mu_0 n_1 n_2 h A_2 .$$

And the self-inductance of each solenoid are:

$L_{11} = \mu_0 n_1^{\ 2} h A_1$ and $L_{22} = \mu_0 n_2^{\ 2} h A_2$.

Putting these together, we can write the induction as a matrix:

$$\mathbf{L} = \begin{pmatrix} L_{11} & L_{12} \\ L_{21} & L_{22} \end{pmatrix} = \begin{pmatrix} \mu_0 n_1^{\ 2} h A_1 & \mu_0 n_1 n_2 h A_2 \\ \mu_0 n_1 n_2 h A_2 & \mu_0 n_2^{\ 2} h A_2 \end{pmatrix}$$

So, for any currents $\begin{pmatrix} I_1(t) & I_2(t) \end{pmatrix}$ the corresponding induced voltages can now be written as:

$$\begin{pmatrix} \mathbb{V}_1(t) \\ \mathbb{V}_2(t) \end{pmatrix} = - \begin{pmatrix} L_{11} & L_{12} \\ L_{21} & L_{22} \end{pmatrix} \frac{\mathrm{d}}{\mathrm{d}t} \begin{pmatrix} I_1(t) \\ I_2(t) \end{pmatrix}$$

As a final remark, notice that conservation of energy implies that:

$$\mathbb{P} = I_1(t)\mathbb{V}_1(t) + I_2(t)\mathbb{V}_2(t) = \begin{pmatrix} I_1(t) & I_2(t) \end{pmatrix} \begin{pmatrix} \mathbb{V}_1(t) \\ \mathbb{V}_2(t) \end{pmatrix} = \frac{\mathrm{d}}{\mathrm{d}t} \left(\oint_{\text{volume}} \frac{B^2}{2\mu_0} \mathrm{d}V \right).$$

Example 9.3: A Coupled LC Circuit

Consider two LC circuits, as shown in the diagram, with known capacitances and inductances (both self and mutual), and a given initial voltage across one of the capacitors.

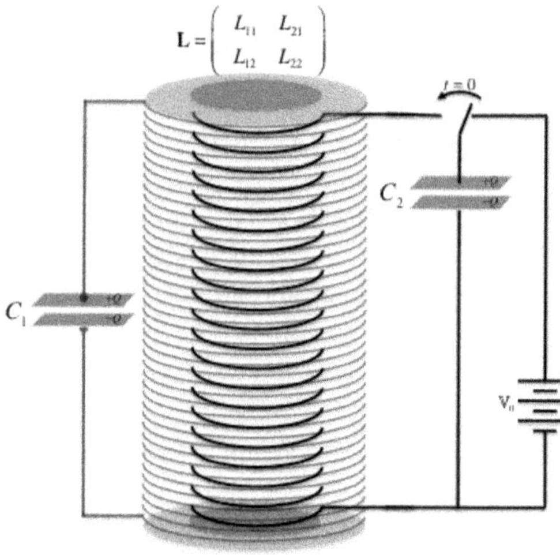

What is the charge on each capacitor, and the current in each circuit, as a function of time?

To answer this we begin with Kirchhoff's loop rule:

$$0 = \frac{Q_1(t)}{C_1} - L_{11}\frac{\mathrm{d}I_1(t)}{\mathrm{d}t} - L_{21}\frac{\mathrm{d}I_2(t)}{\mathrm{d}t}, \quad \text{and}$$

$$0 = \frac{Q_2(t)}{C_2} - L_{12}\frac{\mathrm{d}I_1(t)}{\mathrm{d}t} - L_{22}\frac{\mathrm{d}I_2(t)}{\mathrm{d}t}.$$

Now conservation of charge states:

$$\begin{pmatrix} I_1(t) \\ I_2(t) \end{pmatrix} = \frac{\mathrm{d}}{\mathrm{d}t} \begin{pmatrix} Q_1(t) \\ Q_2(t) \end{pmatrix}$$

So, our coupled differential equations can be written:

$$\begin{pmatrix} L_{11}C_1 & L_{21}C_1 \\ L_{12}C_2 & L_{22}C_2 \end{pmatrix} \frac{\mathrm{d}^2}{\mathrm{d}t^2} \begin{pmatrix} Q_1(t) \\ Q_2(t) \end{pmatrix} = \begin{pmatrix} Q_1(t) \\ Q_2(t) \end{pmatrix}.$$

Now, we invert this, so the two particular solutions can be found by solving for the eigenvalues (ω_A, ω_B) and eigenvectors, in the form of:

$$\begin{pmatrix} L_{11}C_1 & L_{21}C_1 \\ L_{12}C_2 & L_{22}C_2 \end{pmatrix}^{-1} \begin{pmatrix} Q_1 \\ Q_2 \end{pmatrix}_A = -\omega_A^{\,2} \begin{pmatrix} Q_1 \\ Q_2 \end{pmatrix}_A \text{ and } \begin{pmatrix} L_{11}C_1 & L_{21}C_1 \\ L_{12}C_2 & L_{22}C_2 \end{pmatrix}^{-1} \begin{pmatrix} Q_1 \\ Q_2 \end{pmatrix}_B = -\omega_B^{\,2} \begin{pmatrix} Q_1 \\ Q_2 \end{pmatrix}_B .$$

And taking into account the initial condition, the general solution becomes:

$$\begin{pmatrix} Q_1(t) \\ Q_2(t) \end{pmatrix} = a_1 \sin(\omega_1 t + \varphi_1) \begin{pmatrix} Q_1 \\ Q_2 \end{pmatrix}_1 + a_2 \sin(\omega_2 t + \varphi_2) \begin{pmatrix} Q_1 \\ Q_2 \end{pmatrix}_2 .$$

Problem 9.23: A Mutual Inductor

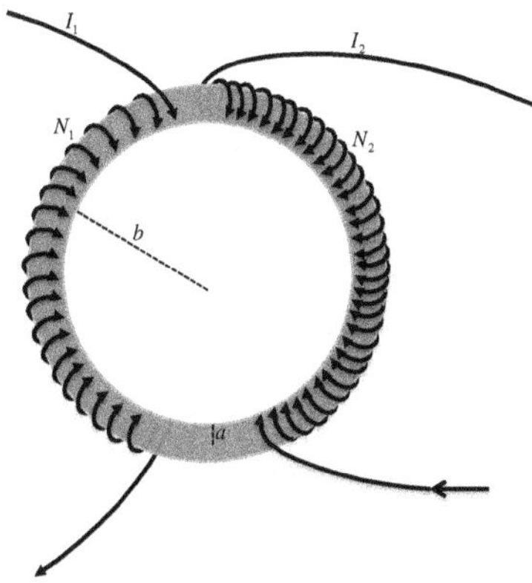

Consider a torus of soft iron with magnetic susceptibility χ_M, radius b, and cross-sectional radius a. The wires are wrapped N_1 and N_2 times and carry currents I_1 and I_2 , respectively, with positive directions as defined in the diagram. You can assume that the magnetic field is completely contained within the iron torus.

a) If: $I_1 = I_0 \sin(\omega t)$ and $I_2 = \frac{N_1}{N_2} I_0 \sin(\omega t + \phi)$,

what are the voltages on each side of the inductor?

b) What is the magnetic field inside the torus?

c) Account for all the energy that flows into, out of, and is contained in the inductor.

d) Now imagine that the magnetic material is strongly ferromagnetic, and has an intrinsic magnetization M_I, and a coercivity H_C. How will these affect the inductor?

e) People often say that to conserve energy you should unplug your AC to DC adaptors when they are not in use. Using physical arguments, make a case either for, or against, this. Is there a physical reason to unplug your transformers? If so, estimate the power lost for one of yours and the cost per month of keeping it pugged all the time.

Problem 9.24: Power Converters

A typical household wall outlet delivers sinusoidal power, so that $\mathbb{V} = \mathbb{V}_{pp} \sin(\omega t)$. The international standard for mains power is now $\mathbb{V}_{rms} \approx 230\,\mathrm{V}$ and $f = 50\,\mathrm{Hz}$, though in the Americas the standard is $\mathbb{V}_{rms} \approx 115\,V$ and $f = 60\,\mathrm{Hz}$. Depending on the country, however, the voltage standard may vary significantly.

a) Consider an appliance with a load resistance R. The average power it will use is $\mathbb{P}_{avg} = \frac{1}{R} \mathbb{V}_{rms}^2$. Find the relationship between $\mathbb{V}_{pp}$ and $\mathbb{V}_{rms}$.

b) The frequency of the power from a wall outlet is exactly 50 Hz or 60 Hz, but the voltage can vary significantly. Why is it so much more important for the electric company to provide power at exactly the standard frequency, when the voltage is much more important than the frequency for most household appliances? To answer this question, suppose that a company had two power stations, one that produced a voltage of $\mathbb{V}_0 \sin(\omega_0 t)$ and the other of $\mathbb{V}_0 \sin((\omega_0 + \delta\omega)t)$.

Write down the voltage as a function of time that they would receive, and plot it, assuming that $\delta\omega \approx 1\%$ of ω_0.

c) Design a one-component circuit to convert from the 230 V to the 115 V standard. Draw a circuit diagram, and write down the input and output voltages as a function of time. In your idealized case, there would be no power lost, however real power converters heat up. Why is this so? What design choices could you make to maximize the efficiency?

d) A common electrical device is a laptop computer power adaptor, to convert typical AC wall power to clean DC. First it must step down the voltage (part c), next it must rectify the current. Using 4 idealized diodes, design a circuit whose output voltage is the absolute value of the input voltage. Plot the input and output voltages as a function of time.

e) The last step in designing a good power converter is to clean up the output. This means creating as close to a constant output voltage as possible. Imagine that you can afford to include one inductor, two capacitors, and as many resistors as you need. Design a circuit to clean up the dirty DC, and return as close to a constant voltage as possible.

f) The Apple computer MagSafe® power converter can accept AC power with voltages ranging from 100 to 240 V, and returns very clean 14.85 V DC up to a maximum power of 45 W. They are also expensive to replace. Imagine that you are tasked by different company with the job of designing a cheaper replacement for the your domestic market only. Write a computer simulation program that accepts an input voltage and frequency at the your country's mains standard, applies each of your design components, and returns the output voltage as a function of time. Use this program to help you choose the right specifications for the transformer, inductor, capacitors and resistors. Report your design specifications, along with a plot of the output voltage as a function of time. Look up the price of each of your components, and estimate the cost of production. Given the budget constraints, will your power converter be good enough for an Apple computer owner who does not travel internationally?

9.4 *Maxwell's Equations in Potential Form*

> ... Maxwell's second relation in his equation of electric force in terms of two highly artificial quantities, a vector and scalar potential, say $\vec{\mathbb{A}}$ and $\mathbb{V}$, thus:
>
> $$\vec{E}_1 = -\frac{\partial\vec{\mathbb{A}}}{\partial t} - \vec{\nabla}\mathbb{V} \quad (2)$$
>
> ignoring impressed force for the present. From $\vec{\mathbb{A}}$ we get down to $\vec{H}_1$ again, thus,
>
> $$\text{curl}\,\vec{\mathbb{A}} = \vec{B} \qquad \vec{B} = \mu\vec{H}_1 \ ;$$
>
> $\vec{B}$ being the magnetic induction, and μ the inductivity.
>
> The equation (2) is arrived at through a rather complex investigation. ... Again—and this is an objection of some magnitude—the two potentials $\vec{\mathbb{A}}$ and $\mathbb{V}$, if given everywhere, are *not sufficient* to specify the state of the electromagnetic field. Try it; and fail. [130]

---

130 Oliver Heaviside, "On The Self-Induction of Wires," published in Philosophical Magazine August (1886), 118, republished in Electrical Papers, 2nd ed. (Bronx, NY: Chelsea Publishing, 1970) Vol. 2, 168. Heaviside calls the scalar potential P, for which we substituted $\mathbb{V}$, to keep our notation consistent.

The above 1886 quote from Oliver Heaviside shows the distain that many leading scientists and engineers had for the vector potential. Notice how Heaviside dismisses $\vec{\mathbb{A}}$, but writes the magnetic field as $\vec{H}$ rather than $\vec{B}$. With neither an aether nor moveable magnetic poles, many physicists today have a similarly low opinion of $\vec{H}$ as Heaviside had of the vector potential.

In this section, we reformulate Maxwell's field equations in terms of the vector and scalar potentials, which results in some beautiful mathematics. Keep in mind, however, that we use Maxwell's equations because they predict observable phenomena. While we no longer agree with the substance of Heaviside's argument, we laud his skeptical approach to problem solving.

Maxwell's equations, from simplest to most complex, are:

Peregrinus's Principle	$\vec{\nabla}\cdot\vec{B}=0$
Faraday's Law	$\vec{\nabla}\times\vec{E}=-\frac{\partial}{\partial t}\vec{B}$
Gauss's Law	$\vec{\nabla}\cdot\vec{E}=\frac{1}{\varepsilon_0}\rho-\frac{1}{\varepsilon_0}\vec{\nabla}\cdot\vec{P}$
Maxwell-Ampère Law	$\vec{\nabla}\times\vec{B}=\mu_0\vec{J}+\mu_0\varepsilon_0\frac{\partial}{\partial t}\vec{E}+\mu_0\varepsilon_0\frac{\partial}{\partial t}\vec{P}+\mu_0\vec{\nabla}\times\vec{M}$

Derivation 9.1: Peregrinus's Principle in Potential Form

In Chapter 7 we defined the vector potential such that, for any open surface, the magnetic flux through the surface is the line integral of the vector potential around the surface, or mathematically:

$$\Phi_B=\int_{\text{surface}}\vec{B}\cdot d\vec{A}=\int_{\text{path}}\vec{\mathbb{A}}\cdot d\vec{\ell}.$$

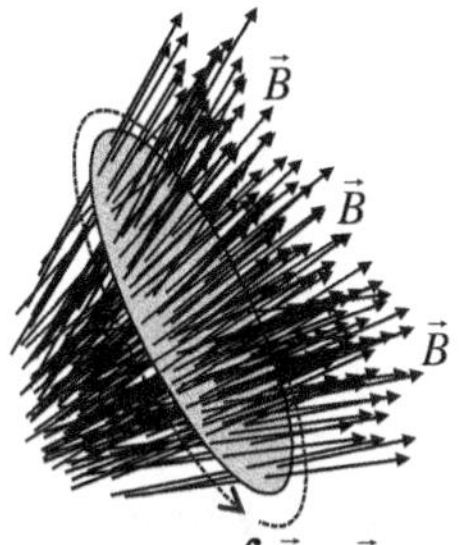

$$\Phi_B=\oint_{\text{path}}\vec{\mathbb{A}}\cdot d\vec{\ell}$$

And then we took the limit as the area of the surface goes to zero:

$$\vec{B}=\lim_{\delta A\to\infty}\frac{1}{A}\int_{\text{surface}}\vec{B}\cdot d\vec{A}=\lim_{\delta A\to\infty}\frac{1}{A}\int_{\text{path}}\vec{\mathbb{A}}\cdot d\vec{\ell}=\vec{\nabla}\times\vec{\mathbb{A}}.$$

Now, we take the divergence of the magnetic field:

$$\vec{\nabla}\cdot\vec{B}=\vec{\nabla}\cdot\vec{\nabla}\times\vec{\mathbb{A}}\equiv 0.$$

Thus, Peregrinus's principle in ensured by the very definition of the vector potential. Thus, if Peregrinus's principle were false, the vector potential would exist under false assumptions. Since no magnetic monopoles have ever been reproducibly found in nature, this is on sound footing.

Derivation 9.2: Faraday's Law in Potential Form

Back on page 103, we introduced the electrostatic potential, $\mathbb{V}$, as the potential energy per charge with respect to some zero point, which we called *ground*. Since for any conservative force, $\mathbb{U}=-\int_0^{\vec{r}}\vec{F}\cdot d\vec{\ell}$, we could define: $\mathbb{V}=\frac{\delta\mathbb{U}}{\delta q}=-\frac{1}{\delta q}\int_0^{\vec{r}}\delta\vec{F}\cdot d\vec{\ell}=-\int_0^{\vec{r}}\frac{1}{\delta q}\delta\vec{F}\cdot d\vec{\ell}=-\int_0^{\vec{r}}\vec{E}\cdot d\vec{\ell}$.

However, potential energy concepts only work for conservative forces, so this presupposes:

$$\oint_{\text{path}}\vec{E}\cdot d\vec{\ell}=0.$$

We now know that this is not always the case, because according to Faraday's law:

$$\oint_{\text{path}} \vec{E}\cdot \mathrm{d}\vec{\ell} = -\tfrac{\mathrm{d}}{\mathrm{d}t}\Phi_B .$$

So, what should we do? We find a quantity which is a conservative force field, and then integrate this. So, we write:

$$0 = \oint_{\text{path}} \vec{E}\cdot \mathrm{d}\vec{\ell} + \tfrac{\mathrm{d}}{\mathrm{d}t}\Phi_B = \oint_{\text{path}} \vec{E}\cdot \mathrm{d}\vec{\ell} + \tfrac{\mathrm{d}}{\mathrm{d}t}\oint_{\text{path}} \vec{\mathbb{A}}\cdot \mathrm{d}\vec{\ell} = \oint_{\text{path}} \left(\vec{E} + \tfrac{\partial}{\partial t}\vec{\mathbb{A}}\right)\cdot \mathrm{d}\vec{\ell}.$$

Therefore, the quantity $\vec{E} + \tfrac{\partial}{\partial t}\vec{\mathbb{A}}$ is a conservative force field, so we can use this to find the electric potential $\mathbb{V}$, which we now call the *scalar potential.* Following the same logic as before:

$$\mathbb{V} = \frac{\delta \mathbb{U}}{\delta q} = -\int_0^{\vec{r}} \left(\vec{E} + \tfrac{\partial}{\partial t}\vec{\mathbb{A}}\right)\cdot \mathrm{d}\vec{\ell}.$$

And, turning this around using the gradient function, we find:

$$\vec{E} + \tfrac{\partial}{\partial t}\vec{\mathbb{A}} = -\vec{\nabla}\mathbb{V} \quad \rightarrow \quad \vec{E} = -\vec{\nabla}\mathbb{V} - \tfrac{\partial}{\partial t}\vec{\mathbb{A}}.$$

To see what this has to do with Faraday's law, we now take the curl:

$$\vec{\nabla}\times\vec{E} = \vec{\nabla}\times\left(-\vec{\nabla}\mathbb{V} - \tfrac{\partial}{\partial t}\vec{\mathbb{A}}\right) = -\vec{\nabla}\times\vec{\nabla}\mathbb{V} - \tfrac{\partial}{\partial t}\vec{\nabla}\times\vec{\mathbb{A}} = 0 - \tfrac{\partial}{\partial t}\vec{B} = -\tfrac{\partial}{\partial t}\vec{B}.$$

As with Peregrinus's principle and the vector potential, Faraday's law is already incorporated into the mathematics of the scalar potential.

Discussion 9.1: Choice of Gauge

Remember Heaviside's complaint about the vector potential (p. 298). This was partially because he mistakenly considered $\vec{H}$ more fundamental than $\vec{B}$. But, the other part of his point is just as true today, which is that neither potential is uniquely defined mathematically.

Consider our definition of the vector potential, $\vec{B} = \vec{\nabla}\times\vec{\mathbb{A}}$. We could add any curl free vector field to $\vec{\mathbb{A}}$ and we would return the same magnetic field. This is called a freedom of *gauge*.

While the concept has been around since Maxwell, the term *gauge* arose later from a 1918 poor translation of the German term, *Eichinvarianz*, which means something closer to "measurement invariance," and has since become an important concept in modern physics. Therefore, when you have a choice of gauge, it is because each form of the vector potential will result in the same magnetic field, which is what you classically measure.

Thought Experiment 9.12: A Gauge Transformation

Consider any scalar function, ψ. Since $\vec{\nabla}\times\vec{\nabla}\psi = 0$, we can add its gradient to the vector potential without changing the magnetic field. To see this, consider a new vector potential $\vec{\mathbb{A}}'$, defined such that: $\vec{\mathbb{A}}' = \vec{\mathbb{A}} + \vec{\nabla}\psi$, so:

$$\vec{B}' = \vec{\nabla}\times\vec{\mathbb{A}}' = \vec{\nabla}\times\left(\vec{\mathbb{A}} + \vec{\nabla}\psi\right) = \vec{\nabla}\times\vec{\mathbb{A}} + \vec{\nabla}\times\vec{\nabla}\psi = \vec{\nabla}\times\vec{\mathbb{A}} = \vec{B}.$$

However, the scalar potential must change so the electric field remains the same. We can see this by calculating the new electric field $\vec{E}'$:

$$\vec{E}' = -\vec{\nabla}\mathbb{V}' - \tfrac{\partial}{\partial t}\vec{\mathbb{A}}' = -\vec{\nabla}\mathbb{V}' - \tfrac{\partial}{\partial t}\left(\vec{\mathbb{A}} + \vec{\nabla}\psi\right) = -\vec{\nabla}\mathbb{V}' - \vec{\nabla}\tfrac{\partial}{\partial t}\psi - \tfrac{\partial}{\partial t}\vec{\mathbb{A}} = -\vec{\nabla}\left(\mathbb{V}' + \tfrac{\partial}{\partial t}\psi\right) - \tfrac{\partial}{\partial t}\vec{\mathbb{A}}.$$

Notice that if $\mathbb{V} = \mathbb{V}' + \tfrac{\partial}{\partial t}\psi$, then $\vec{E}' = \vec{E}$, so for any valid set of potentials $\left(\vec{\mathbb{A}}, \mathbb{V}\right)$, we can pick a new set of valid potentials $\left(\vec{\mathbb{A}}', \mathbb{V}'\right)$ such that:

$$\vec{\mathbb{A}}' = \vec{\mathbb{A}} + \vec{\nabla}\psi \quad \text{and} \quad \mathbb{V}' = \mathbb{V} - \tfrac{\partial}{\partial t}\psi,$$

where $\psi = \psi(\vec{r}, t)$ is any scalar function.

Usually we choose a gauge to simplify our equations, for example in Chapter 7 we implicitly required that $\vec{\nabla} \cdot \vec{\mathbb{A}} = 0$, which is called the *Coulomb gauge*. However, for time dependent systems the convention is to pick another gauge, called the *Lorenz gauge*.

Definition 9.1: The Lorenz Gauge

The *Lorenz gauge*, first proposed by the Danish physicist Ludvig V. Lorenz,[131] requires that:

$$\vec{\nabla} \cdot \vec{\mathbb{A}} = -\mu_0 \varepsilon_0 \tfrac{\partial}{\partial t}\mathbb{V}.$$

In order for this to be a valid gauge constraint, we must show that an arbitrary set of potentials $\left(\vec{\mathbb{A}}', \mathbb{V}'\right)$ can be transformed to our potentials $\left(\vec{\mathbb{A}}, \mathbb{V}\right)$ with some gauge function ψ. Thus, we substitute in the two primed potentials above:

$$\begin{aligned}
\vec{\nabla} \cdot \left(\vec{\mathbb{A}}' - \vec{\nabla}\psi\right) &= -\mu_0 \varepsilon_0 \tfrac{\partial}{\partial t}\left(\mathbb{V}' + \tfrac{\partial}{\partial t}\psi\right) \\
\vec{\nabla} \cdot \vec{\mathbb{A}}' - \nabla^2\psi &= -\mu_0 \varepsilon_0 \tfrac{\partial}{\partial t}\mathbb{V}' - \mu_0 \varepsilon_0 \tfrac{\partial^2}{\partial t^2}\psi \\
\vec{\nabla} \cdot \vec{\mathbb{A}}' + \mu_0 \varepsilon_0 \tfrac{\partial}{\partial t}\mathbb{V}' &= \left(\nabla^2 - \mu_0 \varepsilon_0 \tfrac{\partial^2}{\partial t^2}\right)\psi.
\end{aligned}$$

Thus, for any valid set of potentials $\left(\vec{\mathbb{A}}', \mathbb{V}'\right)$ there exists some scalar function $\psi(\vec{r}, t)$, which can drive a gauge transformation to a new set of equally valid potentials $\left(\vec{\mathbb{A}}, \mathbb{V}\right)$ that conform to the Lorenz gauge constraint. Thus, we can apply the Lorenz gauge constraint *a priori*, and require it as a condition for all potentials.

Thought Experiment 9.13: Gauss's Law in Potential Form

In terms of the potentials the electric field is:

$$\vec{E} = -\vec{\nabla}\mathbb{V} - \tfrac{\partial}{\partial t}\vec{\mathbb{A}},$$

and its divergence is:

$$\vec{\nabla} \cdot \vec{E} = \vec{\nabla} \cdot \left(-\vec{\nabla}\mathbb{V} - \tfrac{\partial}{\partial t}\vec{\mathbb{A}}\right) = -\nabla^2\mathbb{V} - \tfrac{\partial}{\partial t}\left(\vec{\nabla} \cdot \vec{\mathbb{A}}\right) = -\nabla^2\mathbb{V} - \tfrac{\partial}{\partial t}\left(-\mu_0 \varepsilon_0 \tfrac{\partial}{\partial t}\mathbb{V}\right) = \left(\mu_0 \varepsilon_0 \tfrac{\partial^2}{\partial t^2} - \nabla^2\right)\mathbb{V}.$$

---

[131] Lorenz, L., "On the Identity of the Vibrations of Light with Electrical Currents," Philosophical Magazine, **34**, 287-301 (1847).

Thus, Gauss's law in potential form becomes:

$$\left(\mu_0\varepsilon_0 \tfrac{\partial^2}{\partial t^2} - \nabla^2\right)\mathbb{V} = \tfrac{1}{\varepsilon_0}\rho - \tfrac{1}{\varepsilon_0}\vec{\nabla}\cdot\vec{P}.$$

Thought Experiment 9.14: The Maxwell-Ampère Law in Potential Form

In terms of the potentials the magnetic and electric fields are given by:

$\vec{B} = \vec{\nabla}\times\vec{\mathbb{A}}$ and $\vec{E} = -\vec{\nabla}\mathbb{V} - \tfrac{\partial}{\partial t}\vec{\mathbb{A}}$.

By applying the Lorenz gauge, the curl of the magnetic field becomes:

$$\vec{\nabla}\times\vec{B} = \vec{\nabla}\times\vec{\nabla}\times\vec{\mathbb{A}} = \vec{\nabla}(\vec{\nabla}\cdot\vec{\mathbb{A}}) - \nabla^2\vec{\mathbb{A}} = \vec{\nabla}(-\mu_0\varepsilon_0\tfrac{\partial}{\partial t}\mathbb{V}) - \nabla^2\vec{\mathbb{A}} = \mu_0\varepsilon_0\tfrac{\partial}{\partial t}(-\vec{\nabla}\mathbb{V}) - \nabla^2\vec{\mathbb{A}}.$$

Maxwell-Ampère law is:

$$\vec{\nabla}\times\vec{B} = \mu_0\vec{J} + \mu_0\varepsilon_0\tfrac{\partial}{\partial t}\vec{E} + \mu_0\varepsilon_0\tfrac{\partial}{\partial t}\vec{P} + \mu_0\vec{\nabla}\times\vec{M} = \mu_0\vec{J} + \mu_0\varepsilon_0\tfrac{\partial}{\partial t}\left(-\vec{\nabla}\mathbb{V} - \tfrac{\partial}{\partial t}\vec{\mathbb{A}}\right) + \mu_0\varepsilon_0\tfrac{\partial}{\partial t}\vec{P} + \mu_0\vec{\nabla}\times\vec{M}.$$

Equating our terms, the Maxwell-Ampère law becomes:

$$\left(\mu_0\varepsilon_0\tfrac{\partial^2}{\partial t^2} - \nabla^2\right)\vec{\mathbb{A}} = \mu_0\vec{J} + \mu_0\varepsilon_0\tfrac{\partial}{\partial t}\vec{P} + \mu_0\vec{\nabla}\times\vec{M}.$$

Thought Experiment 9.15: The Vector Potential of a Long Wire

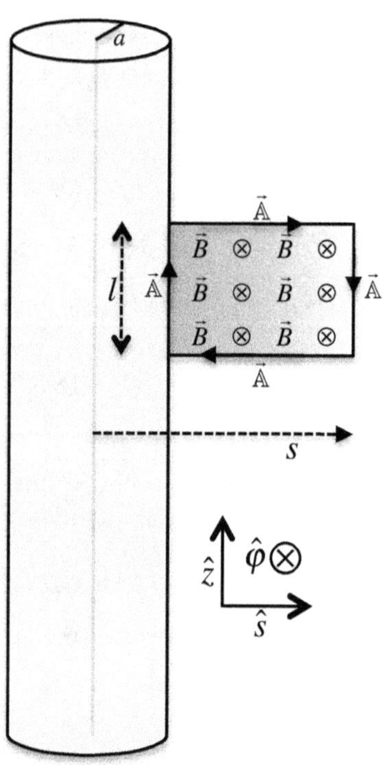

Consider a long wire of radius a that is carrying a steady current I in the $\hat{z}$ direction. From Thought Experiment 8.5 (p. 253), we know that the magnetic field surrounding it is:

$$\vec{B} = \frac{\mu_0 I}{2\pi s}\hat{\varphi}.$$

From the definition of the vector potential we have:

$$\vec{B} = \vec{\nabla}\times\vec{\mathbb{A}} \quad\rightarrow\quad \Phi_B = \oint_{\text{path}} \vec{\mathbb{A}}\cdot d\vec{\ell} = \oint_{\text{surface}} \vec{B}\cdot d\vec{A}.$$

Noticing both the azimuthal and axial symmetry, we integrate around an Ampèrian loop perpendicular to the wire, as in the picture:

$$\int_{-\frac{l}{2}}^{\frac{l}{2}}\int_0^s B\,ds\,dz = \int_{-\frac{l}{2}}^{\frac{l}{2}} \mathbb{A}_z(a)dz + \int_a^s \mathbb{A}_s(s)ds + \int_{\frac{l}{2}}^{-\frac{l}{2}} \mathbb{A}_z(s)dz + \int_s^a \mathbb{A}_s(s)ds$$

$$= \int_{-\frac{l}{2}}^{\frac{l}{2}} (\mathbb{A}_z(a) - \mathbb{A}_z(s))dz + \left(\int_a^s \mathbb{A}_s(s)ds - \int_a^s \mathbb{A}_s(s)ds\right).$$

Equating the integrands, we solve for the vector potential:

$$\mathbb{A}_z(s) = \mathbb{A}_z(a) - \int_a^s B\,ds = \mathbb{A}_z(a) - \int_a^s \left(\frac{\mu_0 I}{2\pi s}\right)ds = \mathbb{A}_z(a) - \frac{\mu_0 I}{2\pi}\ln\left(\frac{s}{a}\right).$$

Note that if we set $\mathbb{A}_z(a) = 0$, the vector potential becomes:

$$\vec{\mathbb{A}}(s)=\frac{\mu_0 I}{2\pi}\ln\left(\frac{a}{s}\right)\hat{z}\,.$$

Notice that the vector potential is parallel to the current.

Problem 9.25: The Long Wire

Consider the wire in Thought Experiment 9.15.

a) Show that $\vec{\nabla}\times\vec{\mathbb{A}}(s)=\frac{\mu_0 I}{2\pi s}\hat{\varphi}$.

b) Show that outside of the wire: $\left(\mu_0\varepsilon_0\frac{\partial^2}{\partial t^2}-\nabla^2\right)\vec{\mathbb{A}}=0$.

c) From the magnetic field, find the vector potential inside of the wire.

d) Show that inside of the wire: $\left(\mu_0\varepsilon_0\frac{\partial^2}{\partial t^2}-\nabla^2\right)\vec{\mathbb{A}}=\mu_0\vec{J}$.

Problem 9.26: Three Long Wires

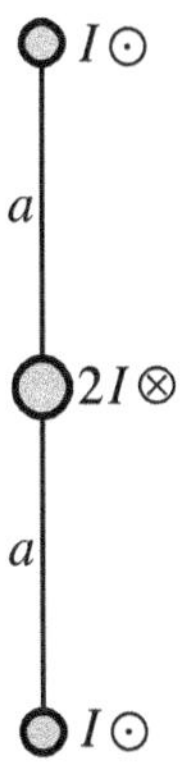

Three long wires are configured in a line as shown, with a current I flowing in the positive $\hat{z}$ directions for the wires located at positions $\pm a\hat{y}$, and a return current $2I$ flowing in the negative $\hat{z}$ direction down the $\hat{z}$ axis.

a) Find the surrounding vector potential $\vec{\mathbb{A}}(x,y)$.

b) Find the surrounding magnetic field $\vec{B}(x,y)$.

c) Make a contour plot of $\hat{z}\cdot\vec{\mathbb{A}}(x,y)=\mathbb{A}_z(x,y)$.

d) Make a vector or streamline plot of $\vec{B}(x,y)$.

Problem 9.27: Four Long Wires

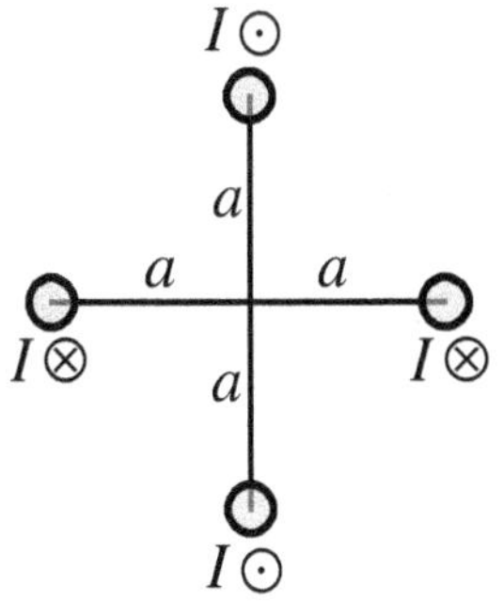

Four long wires are configured in a square as shown, with the current flowing in the positive $\hat{z}$ directions for the wires located at positions $\pm a\hat{y}$, and flowing in the negative $\hat{z}$ direction through the wires located at positions $\pm a\hat{x}$.

a) Find the surrounding vector potential $\vec{\mathbb{A}}(x,y)$.

b) Find the surrounding magnetic field $\vec{B}(x,y)$.

c) Make a contour plot of the $\hat{z}$ component of the vector potential as a function of x and y.

d) Make a vector plot of the magnetic field as a function of x and y.

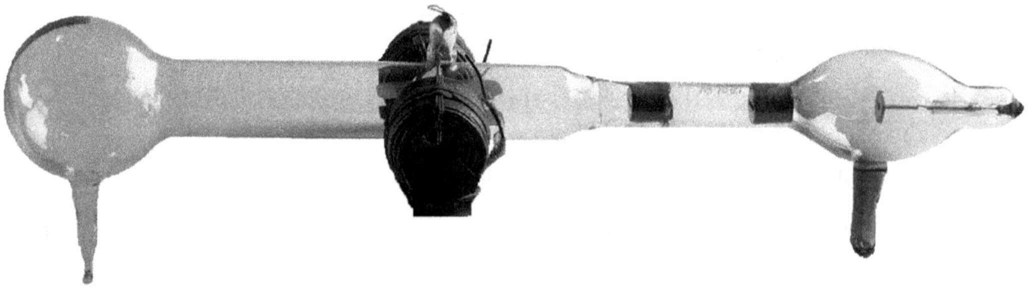

Plate 18: J.J. Thomson's Cathode Ray Tube[132]

[132]J.J. Thomson, "Cathode Rays," The London, Edinburgh, and Dublin Philosophical Magazine and Journal of Science, 5th series, **44** (1897), 293-315. Thomson's article is also published in The Electrician, **39** (1897), 104-109.

Chapter 10 The Electron

In a 1747 letter, Benjamin Franklin wrote

> … it is demonstrated and discovered, both here and in Europe, that the Electrical Fire is a real Element, or species of Matter, not created by the Friction, but collected only.[133]

So far we have, successfully, followed Franklin's lead and modeled this "Electrical Fire" as a conserved quantity that can flow from place to place. But what exactly comprises this electrical fluid? In fact, even after Maxwell's death, physicists still had not answered the question of whether Franklin was correct and there was one electrical fluid with different properties, or whether du Fay was correct and there were two different kinds of fluid that cancel one another out when they occur in equal amounts. The purpose of this chapter is tell the story of how this completely changed over the course of the thirty years from 1879 to 1909.

We begin this chapter with a novel experiment conducted by a young American graduate student, Edwin H. Hall, whose 1879 results showed a clear asymmetry in the flow of positive and negative charge. Most of his results suggested that negative charge flows through conductors, but some metals showed the opposite.

Meanwhile a number of scientists, especially the British chemist Sir William Crooks, began working with partially evacuated glass tubes containing electrodes. These would glow when connected to a high voltage circuit, and they also observed an asymmetry between the positive and negative terminals, as rays of light appeared to extend from only the negative terminal. These *cathode rays* could be deflected by a nearby magnet.

These experiments let to the hypothesis that there was some sort of negatively charged particle, called an electron. This led some physicists to start thinking of the electrical fluid as some sort of collection of negatively charged particles. This led Joseph John Thomson, Oliver Heaviside, and Hendrik A. Lorentz to finally consider what the electromagnetic force would be on a moving charged particle.

In 1897 J.J. Thomson successfully measured the radius of curvature of cathode rays when placed in a known magnetic field. At this time, the electron went from being just an idea, to being a particle with a measured charge to mass ratio.

This led to a flurry of different experiments, each with the goal of independently measuring the electron charge. By 1909 these varied techniques would agree, within only a few percent, on the value of charge of an electron. Physicists now knew the values of the most fundamental constants of nature to within a few percent.

So, who was right? Franklin or du Fay? Is there one fluid or two?

Protons are 1800 times more massive than electrons, yet carry the same magnitude of charge, so du Fay clearly guessed correctly. Electrons comprise his *resinous* fluid, and it exactly cancels out the heavy *vitreous* fluid of protons.

On the other hand, perhaps Franklin was right. Positive charge does not flow with respect to the matter, rather it *is* the matter. Negative charge is the excess of electrons, and positive charge is

---

[133] The Collected papers of Benjamin Franklin (New Haven: Yale University Press, 1961), 142. See also, J.L. Heilbron, Electricity in the 17th and 18th centuries: a study of Early Modern Physics, (Berkley: University of California Press, 1979), 330.

their absence. We do not talk about electrons and protons in semiconductors, but rather about electrons and holes. Under this picture, a century and a half after Franklin's arbitrary choice of positive charge, physicists found that he chose backwards.

10.1 *The Hall Effect*

> It must be carefully remembered, that the mechanical force which urges a conductor carrying a current across the lines of magnetic force, acts, not on the electric current, but on the conductor which carries it. If the conductor be a rotating disk or a fluid it will move in obedience to this force, and this motion may or may not be accompanied with a change of position of the electric current which it carries. But if the current itself be free to choose any path through a fixed solid conductor or a network of wires, then, when a constant magnetic force is made to act on the system, the path of the current through the conductors is not permanently altered, but after certain transient phenomena, called induction currents, have subsided, the distribution of the current will be found to be the same as if no magnetic force were in action. The only force which acts on electric currents is electromotive force, which must be distinguished from the mechanical force which is the subject of this chapter.[134]

The above quote from James Clerk Maxwell bothered Edwin Hall when he encountered it as a graduate student. How could the magnetic force act not on the electric current, but on the conductor which carries it? Would it not make more sense for the force to act, rather, on whatever charged fluid flowed? And, if it did, could not the fundamental question of early electricity be finally solved? Is there one electrical fluid or two?

Thought Experiment 10.1: The Magnetic Force on a Charged Fluid

Imagine an electrical fluid with a current density $(\vec{J})$ moving through a magnetic field $(\vec{B})$. What is the force per volume acting on the fluid?

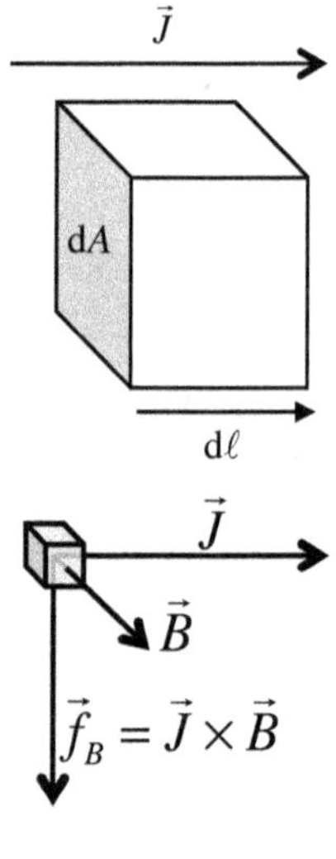

Assume a rectangular volume element with area δA and length $\delta\ell$, as shown. Now the force on a volume element of fluid is given by the law of Laplace as:

$$\delta\vec{F} = I\,\delta\vec{\ell}\times\vec{B} = \left(\vec{J}\cdot\delta\vec{A}\right)\left(\delta\vec{\ell}\times\vec{B}\right) = \left(\delta\vec{\ell}\cdot\delta\vec{A}\right)\left(\vec{J}\times\vec{B}\right).$$

The volume element is the surface area times the length:

$$\delta\vec{F} = \left(\vec{J}\times\vec{B}\right)\delta V\,.$$

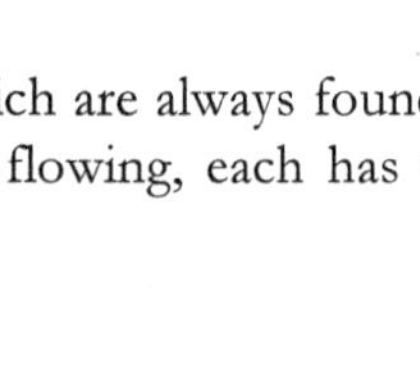

Therefore, the magnetic force density on a fluid is given by:

$$\vec{f}_B = \vec{J}\times\vec{B}\,.$$

Problem 10.1: The Magnetic Force and the Two Fluid Model

Imagine that there are two electrical fluids, on positive and one negative, which are always found in equal abundance in a conductive wire, or: $\rho_- = -\rho_+$. When current is flowing, each has a

---

134 From Maxwell's Electricity and Magnetism, Vol. II p. 144, as quoted by E.H. Hall in "On a New Action of the Magnet on Electric Current," American Journal of Mathematics, **2**, (1879) pp. 287-293, The Johns Hopkins University Press

different velocity: $\vec{v}_-$ and $\vec{v}_+$. Now consider a straight wire of cross-sectional area A with the current flowing in the $\hat{x}$ direction, located in an external magnetic $\vec{B} = B\hat{z}$.

a) Express the current, I, in terms of the variables introduced.

b) Find the force per length on the wire in terms of these same variables.

c) Imagine that equal amounts of positive and negative charge flows through the wire. What would the magnetic force, per volume, be on the positively charged fluid, and the negatively charged fluid, respectively? Are these forces in the same direction, or opposite directions?

d) Imagine that only positive charge flows, again, what are the respective body forces on the positively and negatively charged fluids?

e) Now if only the negative charge flows, what are the body forces?

f) Verify that in each case you obtain the same magnetic force, per length, on the wire, which is consistent with the law of Laplace.

Thought Experiment 10.2: The Electromagnetic Force on a Charged Fluid

Recall that the body force due to the electric field (p. 59) is given by:

$$\vec{f}_E = \rho\vec{E},$$

while the magnetic body force is given by:

$$\vec{f}_B = \vec{J} \times \vec{B}.$$

So, in the case of both electric and magnetic fields, the total body force should be the sum of these two forces.

The electromagnetic force per volume is given by:

$$\vec{f} = \rho\vec{E} + \vec{J} \times \vec{B}.$$

This law is simply a combination of Franklin's fluid model of electricity, Laplace's law, and Faraday's concept of the electric and magnetic fields. As with much of electrodynamics, James Clerk Maxwell was the first person to clearly string these concepts together in his Treatise on Electricity and Magnetism as quoted above.

Problem 10.2: The Electromagnetic Force on a Rod

Consider a rod of total charge q, that is moving longitudinally at a velocity $\vec{v}$, in a uniform electromagnetic field $\vec{B}$ and $\vec{E}$. Find the electromagnetic force on the rod? Show all of your work.

Thought Experiment 10.3: The Hall Effect

The Franklin fluid model of electricity had been around for almost a century by the 1870s, but it was still unknown whether positive, negative, or both charges together moved when current flowed. Graduate student Edwin Hall at Johns Hopkins University was the first to solve this mystery while being directed by professor Henry A. Rowland.[135]

Hall was interested in whether magnetic fields affect wires as a whole or only the substance inside of them. To investigate this, he ran current through a piece of gold leaf perpendicular to a magnetic field and measured the current resulting from a mutually perpendicular charge separation with a galvanometer.

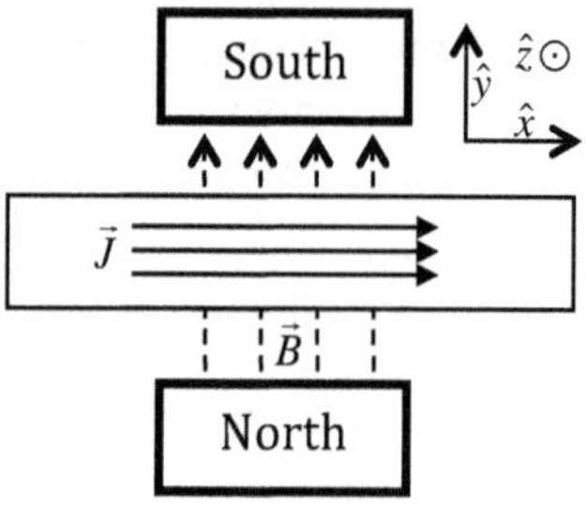

Consider the body force, $\vec{f}_B$, on a current carrying wire in a magnetic field, as shown in the diagram.

$$\vec{f}_B = \vec{J} \times \vec{B} = (J\hat{x}) \times (B\hat{y}) = J\,B\,\hat{z}\,.$$

There is a force in the $\hat{z}$ direction on any moving charged fluid, regardless of the charge. Therefore, Hall figured, if only one kind of charge can move, while the other were fixed, than perhaps he could measure an associated current.

Consider the three cases shown:

I. If only positive charge moves, then the positive charge would build up on the top plate.

II. If only negative charge moves, then a negative charge would be measured on the top plate.

III. If both negative and positive charge move, there would be no net charge buildup.

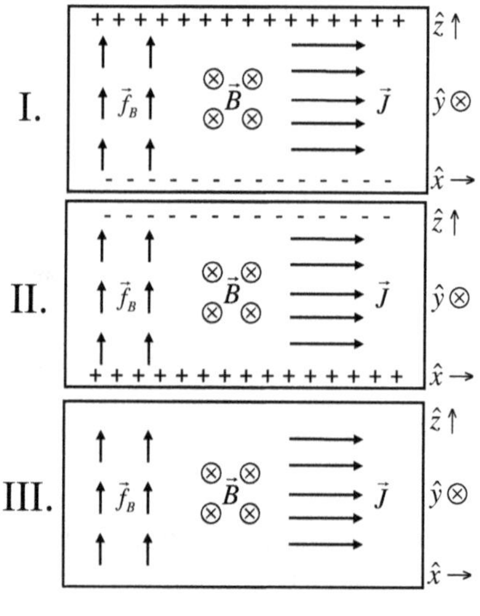

The plot shows Hall's measurement[136] of the small perpendicular current used to measure the charge separation, $\vec{f}_B$, as a function of the product of current and magnetic field $(\vec{J} \times \vec{B})$. Hall took these data by performing delicate measurements that were considered a remarkable feat of precision at the time, some eighteen years before the discovery of the electron.

Hall's results implied that it was mostly negative, rather than positive, charge that flowed in the gold foil. This behavior was subsequently confirmed for other monovalent metals such as silver, copper, lithium, sodium, and potassium. For divalent metals such as beryllium, zinc, and cadmium, however, it appears that the charge carriers are positive, and the simple *free charge carrier* model of metals that we have used here is not adequate. A quantum mechanical treatment is needed for these materials, as the *holes* appear to flow rather than the electrons.

---

[135] E. H. Hall, "On a New Action of the Magnet on Electric Currents," American Journal of Mathematics, **2** (1879), 287-292.

[136] Ibid., 290.

Problem 10.3: A Hall Effect Thruster

There is type of ion thruster for rockets that uses the Hall effect in order to make the light negatively charged fluid stay put, and expel the heavy positive fluid at high velocity.

Model a net neutral ionized plasma as two fluids with negative and positive charge densities, ρ_- and ρ_+, and negative and positive mass densities ρ_{M-} and ρ_{M+}. The plasma is placed in a cylindrically symmetrical chamber, with radius a, length h, and an axial electric field $\vec{E} = E\hat{z}$.

a) Show that to obtain a target ion velocity, $\vec{v}_+ = v_+\hat{z}$, you should design it with an axial electric field:

$$\vec{E} = \frac{\rho_{M+}v_+^2}{2\rho_+ h}\hat{z}.$$

b) Now we turn on a radial magnetic field $\vec{B} = B\hat{s}$. Show that in order for the electrons to orbit azimuthally inside the chamber, but sill allow the ions to leave, the magnetic field should be:

$$\vec{B} \approx \frac{v_+}{a}\sqrt{\frac{\rho_{M-}\rho_{M+}}{\rho_-\rho_+}}\hat{s}.$$

10.2 *The Lorentz Force*

Hall was not the only physicist honing in on the existence of a negatively charged electrical fluid. William Crooks developed a type of mostly evacuated gas discharge tube that glowed when he placed a high voltage across separated terminals. which he and others used to study the spectra of rarefied gasses. More importantly for our discussion, Crooks, as well as Eugen Goldstein, showed that rays emitted from the negative terminal, called *cathode rays*, bent under the influence of a magnet.[137]

In the spring of 1881, Hermann von Helmholtz delivered a lecture in England where he publicly argued that electricity is comprised of many tiny discrete particles.[138] He also invited Edwin Hall to visit his laboratory in Berlin, where Hall reproduced his experiment and made additional measurements of his current ratio, which he presented in England before returning to America. Now that particles may exist, it finally made sense to investigate the electromagnetic force on one, which is what J.J. Thomson attempted to do. Unfortunately he was off by a factor of two.[139]

Eight years after Thomson's attempt, Oliver Heaviside derived the correct expression for the magnetic force acting on a moving charge.[140] Meanwhile, Hendrik A. Lorentz independently discovered and published the same force law a few years later, having overlooked Heaviside's

---

[137] W. Crookes, "Contributions to Molecular Physics in High Vacua," Phil. Trans. (1879), 641-664; E. Goldstein, "On the Electrical Discharge in Rarefied Gases," parts 1 and 2, Philosophical Magazine, 5 **10** (1880), 173-190 and 234-247.

[138] H. von Helmholtz, "The Modern Development Of Faraday's Conception Of Electricity," The Faraday Lecture, delivered before the Fellows of the Chemical Society in London on April 5, 1881. Published in Popular Science Monthly, **19** (1881), 242-250.

[139] J.J. Thomson, "On the Electric and Magnetic Effects produced by the Motion of Electrified Bodies," Philosophical Magazine, 5 **11** (1881), 229-249.

[140] Oliver Heaviside, "On the Electromagnetic Effects due to the Motion of Electrification through a Dielectric," Philosophical Magazine, 5 **27** (1889), 324-339. Heaviside refers to Maxwell's result in this paper.

earlier work. The total electromagnetic force acting on such a charge is now generally referred to as the *Lorentz force*, despite the fact that Thomson and Heaviside found it first.

Hendrik Lorentz, a great and honorable man who was one of the leading physicists of the late nineteenth and early twentieth centuries, is given credit for his own work, but also that of others. Oliver Heaviside, on the other hand, is largely forgotten by physicists, despite his having developed much of what we now consider modern classical electrodynamics.[141] Why this happened is a historical question on which it is fun to speculate. Was it because of the contrast between Heaviside's rough personality, and Lorentz's likability? Was it because of the divergence of physics and electrical engineering at around the same time? In a somewhat comical lapse, physicists gave Lorentz credit for the Lorenz gauge (Definition 9.1, p. 301), because of the striking similarities of "Lorentz" and "Lorenz," and the similarities between them and the Lorentz transformations, which, by the way, were also found earlier by both George FitzGerald and Oliver Heaviside.

While setting the historical record straight is an important goal, there is a much more interesting question regarding the Lorentz force: why was it discovered and rediscovered so late in the development of electricity and magnetism? As we have just seen, applying a particle model of current to Maxwell's electromagnetic force leads naturally to a surprisingly simple force law. Our best guess is that the question did not seem relevant, as charge was conceived of as a continuous fluid during earlier periods. So, the moral for the young physicist: factors of two can be very important! Had J.J. Thomson not been off by a factor of two, the electromagnetic force on a charged particle would surely be called the "Thomson force."

Thought Experiment 10.4: The Lorentz Force

Once the idea of charged particles became commonplace, it was fairly straightforward to find the electromagnetic force on a charged particle moving through a magnetic field. Indeed, J.J. Thomson was motivated by Crookes's and Goldstein's experiments to seek a mathematical expression for such a magnetic force. (Thomson's result was nearly correct; unfortunately it included an erroneous coefficient of one-half. Several years later, Oliver Heaviside published the correct derivation of the magnetic force on a charged particle.)

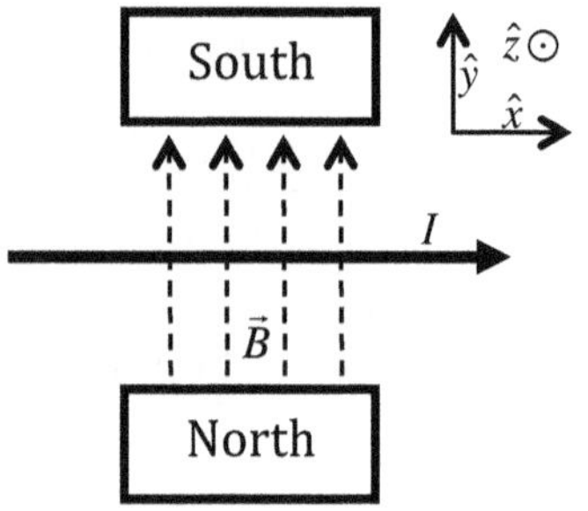

Imagine a long wire that carries a current I between the poles of a horseshoe magnet. The current flowing through the wire is moving perpendicular to the magnetic field. What force will the wire experience?

Again, let us model the interior of a wire as an electron sea with negative free electron charge density ρ_- which is moving at a velocity $\vec{v}_- = -v_-\hat{x}$. For each negatively charged electron there is a corresponding positively charged hole, so $\rho = \rho_+ + \rho_- = 0$ and the total current density is given by $\vec{J} = \rho_-\vec{v}_-$.

Now there will be a force per volume on the electron sea, which is given by:

$$\vec{f}_- = \rho_-\vec{E} + \vec{J}_- \times \vec{B} = \rho_-\vec{E} + \rho_-\vec{v}_- \times \vec{B} = \rho_-\left(\vec{E} + \vec{v}_- \times \vec{B}\right).$$

---

[141] For two interesting accounts of Heaviside's life and work, see Bruce J. Hunt, The Maxwellians (New York: Cornell University Press, 2005) and Paul Nahin, Oliver Heaviside: The Life, Work and Times of a Victorian Genius (Baltimore: Johns Hopkins University Press, 2002).

Now, because of the Thomson experiment, we know that the force per volume is the force per electron times the number of electrons per volume:

$$\vec{f}_e = \vec{F}_e n_e = \vec{F}_e \left(\frac{\rho_-}{q_e} \right),$$

where $q_e = -e$ is the charge of an electron.

By equating the expressions, we can now solve for the force on each electron as:

$$\vec{F}_e \left(\frac{\cancel{\rho_-}}{q_e} \right) = \cancel{\rho_-} \left(\vec{E} + \vec{v}_- \times \vec{B} \right)$$

$$\vec{F}_e = q_e \left(\vec{E} + \vec{v}_- \times \vec{B} \right).$$

Thus we generalize this result, to the force on any charged particle, as:

$$\vec{F}_B = q \left(\vec{E} + \vec{v} \times \vec{B} \right).$$

This is now commonly called the *Lorentz force.*

Fundamental Law 8: The Electromagnetic Force

In Thought Experiment 6.4 (p. 192) we found the force on a small magnet in a non-uniform magnetic field, and we discussed the magnetic moment of an electron in Thought Experiment 6.8 (p. 202). Putting this magnetic together with the Lorentz force, gives the force on a particle with a charge q and magnetic moment $\vec{m}$, which is under the influence of electric and magnetic fields $\vec{E}$ and $\vec{B}$. Therefore, we have another fundamental law:

$$\vec{F} = q \left(\vec{E} + \vec{v} \times \vec{B} \right) + \vec{\nabla} \left(\vec{m} \cdot \vec{B} \right).$$

Problem 10.4: Interplanetary Dust Grains

The table[142] shows calculated forces on typical interplanetary dust grains of different sizes. The force subscript L stands for Lorentz. The other forces are the gravitational force of the sun and the force of the sun's radiation.

Main forces on dust in interplanetary space at $5AU$, radius for spherical particles, absorbing particle, density $1000 kg/m^3$, $5V$, $v_{rel} \approx 400 km/s$, $B \approx 1nT$, $\alpha \approx 80°$.

mass (kg)	10^{-20}	10^{-17}	10^{-14}	10^{-11}	10^{-8}
radius (μm)	0.01	0.1	1	10	100
F_{grav} (N)	10^{-24}	10^{-21}	10^{-18}	10^{-15}	10^{-12}
F_{rad}/F_{grav}	0.5	2	0.5	0.05	$5 \cdot 10^{-3}$
charge-to-mass ratio (C/kg)	600	6	0.06	$6 \cdot 10^{-4}$	$6 \cdot 10^{-6}$
F_L/F_{grav}	2000	20	0.2	$2 \cdot 10^{-3}$	$2 \cdot 10^{-5}$

a) Given the data in the table, how long would it take each dust grain to orbit the assumed magnetic field of one nanotesla as stated in the table caption?

b) Estimate the radius of curvature of each of the 5 dust grains, assuming it is traveling at 400 km/s with respect to the magnetic field. Convert this to astronomical units.

c) In the model, the authors clearly assume some sort of constant charge density over all grain sizes. Did they assume a constant linear charge density, λ, surface charge density, σ, or volume charge density, ρ?

---

142 B. Grün and J. Svestka, *Physics of Interplanetary and Interstellar Dust*, Space Science Reviews, **78** (1996) 347-360, Kluwer Academic Publishers

d) What is the assumed mass density of the dust grains? Does this most compare to wood, water, rock, or metal?

e) Imagine that you broke a grain into two pieces, each with half the charge and half the mass. Would the electrostatic potential energy go up or down? Explain clearly.

f) Use Coulomb's law to estimate the repulsive force pulling apart the charges on a dust grain. How does this relate to the assumed size of each grain?

10.3 *The Discovery of the Electron*

Jean Baptiste Perrin, while still a graduate student at the École Normale Supérieure in Paris, discovered that the cathode rays emitted from heated negative terminals carried a negative charge. Perrin did this by shooting these cathode rays into a gold leaf electroscope and showing that the electroscope became negatively charged. For two years his interpretation was controversial, until J.J. Thomson published a series of detailed experiments confirming Perrin's conclusion. Thomson introduces his readers to Perrin's work as follows:

> [Perrin's] experiment proves that something charged with negative electricity is shot off from the cathode, traveling at right angles to it, and that this something is deflected by a magnet; it is open, however, to the objection that it does not prove that the cause of the electrification in the electroscope has anything to do with cathode rays. Now the supporters of the aethereal theory do not deny the electrified particles are shot off from the cathode; they deny, however, that these charged particles have any more to with the cathode rays than a rifle ball has to do with the flash when a rifle is fired. I have therefore repeated Perrin's experiment in a form which is not open to this objection.[143]

Next, Thomson built an experiment (picture on p. 304) where he showed that cathode rays could also be deflected by an electrostatic field. His diagram of the apparatus is shown, and he described it thus:

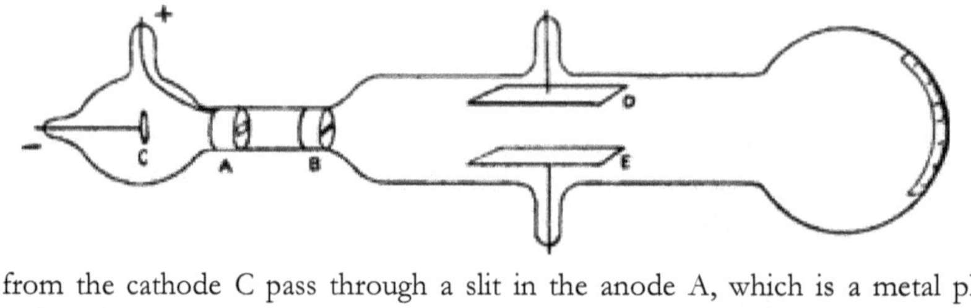

> The rays from the cathode C pass through a slit in the anode A, which is a metal plug fitting tightly into the tube and connected with the earth; after passing through a second slit in another earth-connected metal plug B, they travel between two parallel aluminum plates about 5 cm long by 2 broad and at a distance of 1.5 cm apart; they then fall on the end of the tube and produce a narrow well-defined phosphorescent patch.

With a sufficient vacuum inside the chamber, Thomson was able to measure the trajectory of the cathode rays, and to estimate the ratio of their mass to charge.

The only catch is that Thomson was unable to measure the velocity of the particles directly, but he could measure the accelerating electric field and the distance over which it acts. Therefore, the velocity of the cathode rays could be found as a function of charge and mass, by using the work-energy theorem:

---

[143] J.J. Thomson, "Cathode Rays," The London, Edinburgh, and Dublin Philosophical Magazine and Journal of Science, 5th series, **44** (1897), 293-315. Thomson's article is also published in The Electrician, **39** (1897), 104-109.

$$\mathbb{W} = \int q\vec{E}\cdot d\vec{l} = \tfrac{1}{2}mv^2 \quad \text{so} \quad v = \sqrt{\frac{q}{m}2\int \vec{E}\cdot d\vec{l}}\,,$$

where $\int \vec{E}\cdot d\vec{l}$ is easily measurable, as it is simply the voltage of the battery.

Thomson next investigated the motion of these cathode rays in the presence of a constant magnetic field $\vec{B}$. In this case, the rays were deflected perpendicularly to both the direction of motion and the magnetic field, and followed a circular path. The radius of curvature was found as follows:

> Suppose that the rays, when deflected by a magnet, strike against the glass at E, then, if R is the radius of the circular path of the rays,
>
> $$2R = \frac{\mathrm{CE}^2}{\mathrm{AC}} + \mathrm{AC}\,;$$
>
> thus, if we measure CE and AC we have the means of determining the radius of curvature of the path of the rays.

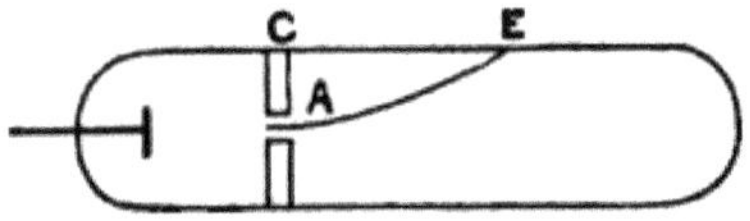

Let us now consider the direction and magnitude of the force on the particle in the constant magnetic field. We will use cylindrical coordinates so that the observed quantities are: $\vec{B} = B\hat{z}$, $\vec{v} = v\hat{\varphi}$, and $\vec{r} = R\hat{s}$, where the magnitude of each vector is a constant, but the direction of the velocity changes continuously.

The force on these particles is given by:

$$\vec{F} = q\vec{v}\times\vec{B} = m\vec{a}\,.$$

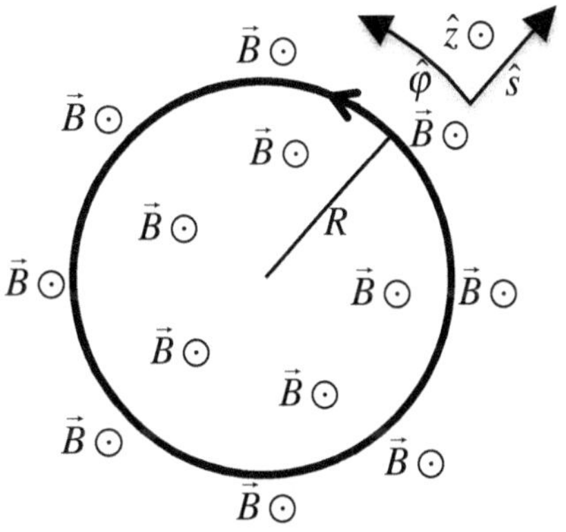

From this we can find the radius of curvature of the deflected particles, as follows:

$$qvB(\hat{\varphi}\times\hat{z}) = m\left(-\frac{v^2}{R}\hat{s}\right) \quad\rightarrow\quad RqB(\cancel{\hat{s}}) = -mv\,\cancel{\hat{s}}$$

$$R = -\frac{mv}{qB} = \frac{mv}{eB}\,.$$

For this to be the case q must be negative, so we write the charge as $q = -e$. Thomson did not measure the velocity directly, but the velocity is also a function of the mass to charge ratio. Solving for the mass to charge ratio as a function of measurable quantities, we get:

$$R = \frac{mv}{eB} = \frac{m}{eB}\sqrt{\frac{e}{m}\left(-2\int \vec{E}\cdot d\vec{l}\right)} \quad\rightarrow\quad \frac{m}{e} = \frac{R^2B^2}{-2\int \vec{E}\cdot d\vec{l}}\,.$$

Thomson investigated this further using cathodes of different metals, and unlike Hall's results, found comparable results for each one. He concluded:

> From these determinations we see that the value of m/e is independent of the nature of the gas, and that its value 10^{-7} is very small compared to the value of 10^{-4}, which is the smallest value of this quantity previously known, and which is the value for the hydrogen ion in electrolysis.

Our modern value, in SI units, for the mass to charge ratio is:

$m/e = 5.7 \times 10^{-12}$ kg/C.

J.J. Thomson's work identifying the cathode rays as beams of electrons earned him the 1906 Nobel Prize in Physics.

Problem 10.5: Thomson's Radius

We quote J.J. Thomson's formula for the radius of curvature of the electron path, given his two measured lengths AC and CE. Geometrically prove his relationship.

Problem 10.6: Thomson's m/e Ratio

According to the quote above, Thomson measured $m/e = 10^{-7}$, but he assumes his readers know what unit system he is working in. He does, however, say that for an hydrogen ion $m/e = 10^{-4}$.

a) Based on the quote from Thomson, what is ratio of the mass of a proton to an electron? How does this compare to the modern value?

b) After reading Appendix D (p. 719), convert the modern values of m/e for a proton, and an electron, to cgs units.

c) Do you think that Thomson was working in cgs units? Why, or why not?

Problem 10.7: The Ratio of Forces

The astute reader might ask why Thomson, or other experimentalists, could not separate out the charge from the mass by measuring the affect of gravity on the electrons—or if that was impossible, on ions instead. For example, why not inject the electrons into a vertical magnetic field and watch them spiral down due to gravity until balanced by a known electric field? Or, if that was too hard, why not measure the eccentricity of the path of vertically orbiting electrons, to find the perturbation of motion due to gravity. Should this not work because:

$\vec{F} = m\vec{g} + q(\vec{E} + \vec{v} \times \vec{B}) = m\vec{a}$?

As it turns out, this was impractical even for ions, which you will investigate here.

a) What electric field is required to balance the force of gravity of a proton? Express your answer in volts per meter.

b) The average random kinetic energy of a monatomic molecule is $\langle \mathbb{T} \rangle = \frac{3}{2} k_B T$, where k_B is the Boltzmann constant and T is the temperature. What would be the approximate ratio of magnetic force to weight of a proton with this kinetic energy here on Earth?

Problem 10.8: Jupiter's Magnetic Field

Around 1960 decametric radio emission was discovered with an intensity that had a periodicity of $9^h45^m30^s$, which is the rotation period of Jupiter. This Jovian emission also seemed to have louds bursts every so often. The figure (their Fig. 6) is from an early paper which (correctly) proposed to explain the radiation as cyclotron radiation from spiraling electrons, while (incorrectly) explaining the observed burst periodicities as electron's oscillating from pole to pole in Jupiter's ionosphere.[144]

---

[144] Eliis, G.R.A. "Cyclotron Radiation from Jupiter." *Australian Journal of Physics* 15 (1962): 344

a) Show that the angular velocity of an electron in a uniform magnetic field is given in vector form by: $\vec{\omega} = \frac{e}{m}\vec{B}$.

b) The decametric radio detections peaked at a radio frequency of 20 MHz. Assuming that the electron gyrofrequency is also 20 MHz, estimate the magnetic field surrounding Jupiter.

c) We now explain the bursts to be caused by one of Jupiter's moon's passing through Jupiter's magnetic field. Recalling the concept of *beat frequency* from introductory physics, and by clearly drawing pictures, determine which of Jupiter's 4 Galilean moons is most consistent with the two periods shown in the figure? The orbital periods, as seen from Earth, of Io, Europa, Ganymede, and Callisto are 1.769, 3.551, 7.155, and 16.69 days respectively.

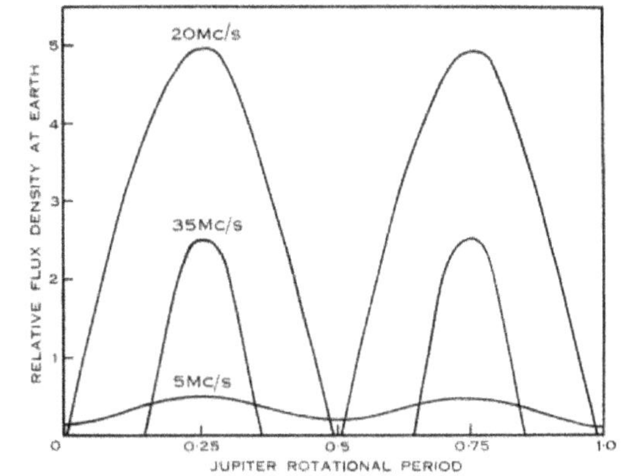

Fig. 6.—Variation of the integrated flux density of the radiation with rotation of Jupiter for different frequencies. Inclination of dipole axis 30°. Assumed mirror points of electrons 4 radii.

Example 10.1: Deflection of Electrons in an Oscilloscope

An oscilloscope consists of two pairs of parallel plates connected to an input signal. The electric field between the plates is proportional to the input signal.

An electron gun shoots electrons toward a screen, but they are deflected by the electric field in the region between the plates and so do not hit the middle of the screen.

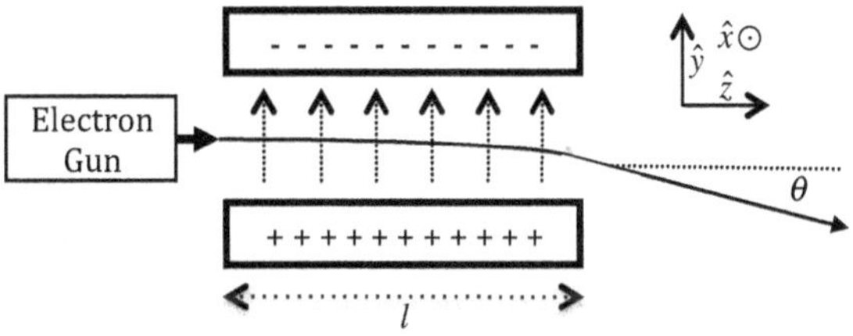

Given an initial velocity $\vec{v}_0 = v_0\hat{z}$, electric field $\left(\vec{E}\right)$ and rectangular plates of length l, by what angle (θ) is the electron deflected by the plates? Let:

$$\vec{E} = E\,\hat{y}.$$

The force on the electron is therefore given by

$$\vec{F} = q\vec{E} = (-e)(E\hat{y}).$$

Since $\vec{F} = m\vec{a}$, where $\vec{a}$ is the acceleration of the particle, we have that

$$\vec{a} = \left(\frac{-e}{m}\right)E\,\hat{y} = \frac{\mathrm{d}\vec{v}}{\mathrm{d}t}.$$

Integrating to obtain the velocity, we can write

$$\vec{v} = \vec{v}_0 + \int_0^t \vec{a}\,\mathrm{d}t = \vec{v}_0 + \vec{a}t,$$

because $\vec{a}$ is constant in this instance. The result is simply the equation for projectile motion.

Notice that the total time the electron spends between the plates is $t = l/v_0$. Finally, we can write the final velocity in terms of our known quantities:

$$\vec{v} = v_0\hat{z} + \left(\frac{(a)l}{v_0}\right)\hat{y} = v_0\hat{z} + \left(\frac{\left(-\frac{e}{m}E\right)l}{v_0}\right)\hat{y} \text{ so } \theta = \arctan\left(\frac{\sqrt{v_x^2+v_y^2}}{v_z}\right) = \arctan\left(\frac{eEl}{mv_0^2}\right).$$

Now we will check our units:

$$\left[\frac{eEl}{mv_0^2}\right] = \frac{(\mathrm{C})\left(\frac{\mathrm{N}}{\mathrm{C}}\right)(\mathrm{m})}{(\mathrm{kg})\left(\left(\mathrm{m}/\mathrm{s}\right)^2\right)} = \frac{\left(\frac{\mathrm{kg}\cdot\mathrm{m}}{\mathrm{s}^2}\right)(\mathrm{m})}{\mathrm{kg}\left(\mathrm{m}/\mathrm{s}\right)^2} = \text{unitless}.$$

Problem 10.9: An Oscilloscope

In the last example we discuss an oscilloscope with a single pair of plates. Real oscilloscopes have two inputs and two pairs of plates. The pair of plates perpendicular to the $\hat{y}$ direction is usually connected to the input voltage and another pair of plates perpendicular to the $\hat{x}$ direction is usually connected to a clock. You can assume an initial electron velocity $\vec{v}_0 = v_0\,\hat{z}$, a plate length l, and an electric field $\vec{E} = E_x\hat{x} + E_y\hat{y}$. Solve for the final velocity of the electrons.

Problem 10.10: A Mass Spectrometer

In a mass spectrometer, a sample of positive ions are accelerated by an electric field $\vec{E} = E\hat{z}$ over a distance $\vec{d} = d\hat{z}$ and injected into a magnetic field $\vec{B} = B\hat{x}$. After a 90° turn, the location of the particles is detected. We will assume that each atom is completely stripped of its electrons, so each has a charge $q = Ze$ and $m = A\mu$, where Z and A are the atomic number and mass respectively.

a) What is the velocity of the particles when they hit the detector? Express your answer in vector notation.

b) At what position are the particles detected?

c) What chemically interesting quantity can be measured using a mass spectrometer?

Problem 10.11: A Betatron Accelerator

In a betatron, electrons are first injected into an initial magnetic field $\vec{B} = B_0(s)\hat{z}$, with the appropriate speed so that they orbit with a radius R. By increasing the magnetic field $\vec{B} = B(s,t)\hat{z}$, the electrons speed up. However, the magnetic field must be chosen carefully so that the orbital radius will remain constant.

a) Given an initial magnetic field at the location of the electrons, $\vec{B} = \vec{B}(R,0)\,\hat{z}$, what must the injected velocity of the electrons be in order to be orbiting with a radius R?

b) Let $\overline{B}(t)$ represent the spatially averaged magnetic field through the face of the orbit, so by definition: $\overline{B}(t) \equiv \frac{1}{\pi R^2}\int_0^R B(s,t)\,2\pi s\,ds$. Find the acceleration of the electrons as a function of $\overline{B}(t)$ and its time derivatives.

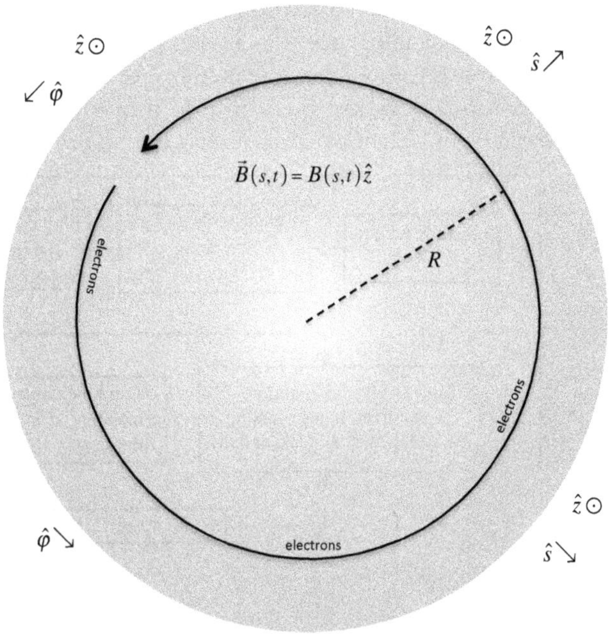

c) What is the velocity of the electrons as a function of time?

d) As the electrons accelerate, the magnetic field at the position of the electrons $\vec{B} = B(R,t)\,\hat{z}$ must also change in order to keep the electrons in an orbit with constant radius R. Find the magnetic field at the electron position, $\vec{B}(R,t)$, in terms of the average magnetic field $\overline{B}(t)$ through the orbital path.

e) Betatron accelerators began to be used in the late 1950s for treating cancer. Write a short report on their use today.

Discussion 10.1: Practically Solving Equations of Motion

It is often important to predict the motion of charged particles under the influence of only an electromagnetic field. Examples include electrons in an old television set, ions in the ionosphere, or dust grains in the interstellar medium. Whatever the situation, they are projectiles acted upon by known forces, so their motion can be predicted using Newton's laws of motion.

Consider first the familiar projectile motion problem: the flight of a ball. In your first physics course you ignored air resistance when solving this problem. As a first approximation this worked out well, but more importantly, it made the problem solvable with basic algebra. However, as sports fans well know, the forces between the air and the ball turn out to be crucial, especially when the ball is moving fast with a great deal of spin. So, how do we take these more difficult forces into account?

The drag force on projectiles is primarily due to momentum transfer from the ball to the air, while the sideways Magnus force is due to the difference in pressure from one side to the other of the ball much like lift on an airplane. Taking direction into account, the net force on the ball is:

$$\vec{F} \approx m\vec{g} - \alpha\,\rho_m\left(\pi R^2\right)|v|\,\vec{v} + \beta\,\rho_m\left(\pi R^3\right)\vec{\omega}\times\vec{v},$$

where m, $\vec{g}$, ρ_m, R, are the mass of the ball, the local gravitational field, the density of the air, and the radius of the ball respectively--while α (alpha) and β (beta) are unitless coefficients. The angular and linear velocities are denoted $\vec{\omega}$ and $\vec{v}$ respectively.

A reasonable way to attack a projectile motion problem such as this is to use the initial conditions and Newton's laws to estimate the acceleration of the ball. Assume constant acceleration and apply basic kinematics to find the future velocity and position. The drawback to this method is obvious: if your initial guess is wrong, or the acceleration is not constant, your estimate will be wrong. However, if only use the method for short time intervals, and nail down the first few time steps, it should work just fine—just do it over and over until the short time intervals add up to the whole flight. The flow chart shows this algorithm in detail.

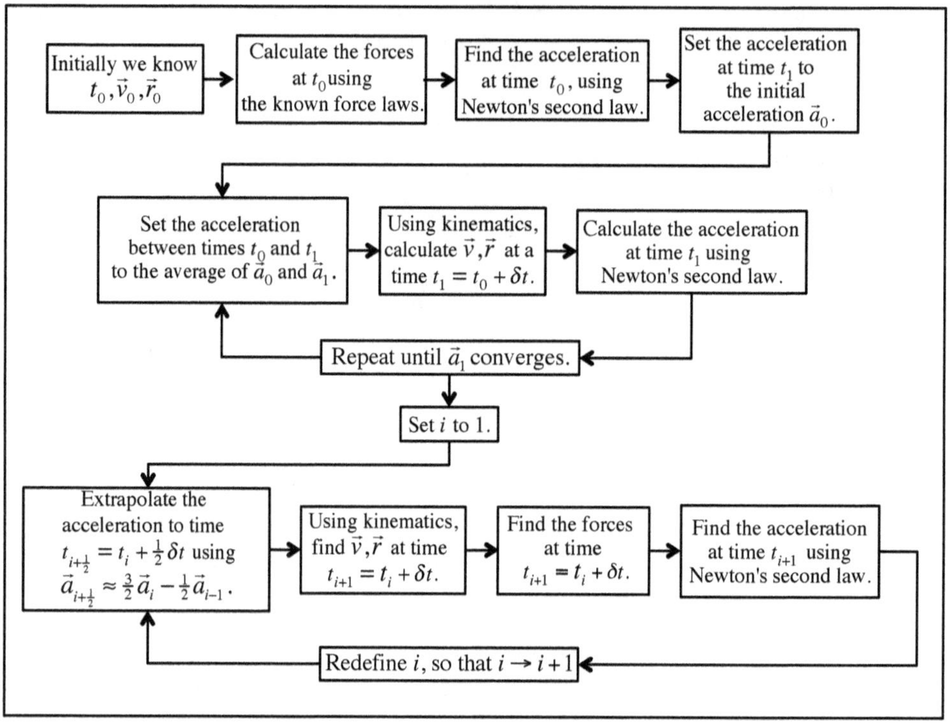

Thought Experiment 10.5: A Free Electron in an Electromagnetic Field

Consider the motion of a particle with charge q and mass m that is influenced by uniform and perpendicular magnetic and electric fields: $\vec{E} = E_x\hat{x} + E_y\hat{y}$ and $\vec{B} = B_z\hat{z}$. We wish to derive the equations of motion, so we can simulate the motion of a particle in the electric and magnetic fields.

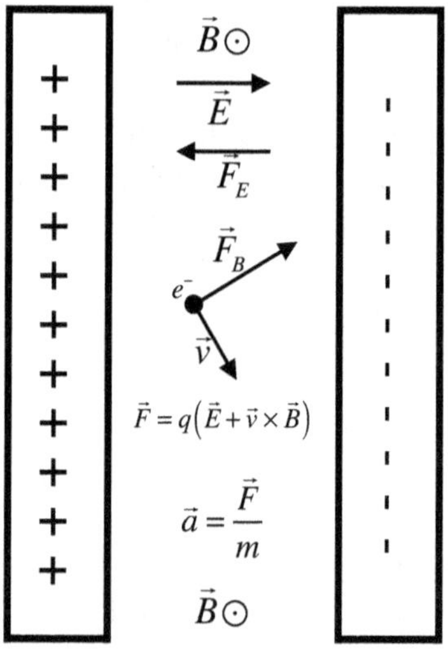

First, we apply the electromagnetic force and Newton's second law of motion:

$$\vec{a} = \frac{\vec{F}}{m} = \left(\frac{q}{m}\right)\left(\vec{E} + B_z(\vec{v} \times \hat{z})\right).$$

$$\vec{a} = \left(\frac{q}{m}\right)\begin{pmatrix} E_x\hat{x} + E_y\hat{y} \\ +B\left(v_x\hat{x}\times\hat{z} + v_y\hat{y}\times\hat{z} + v_z\hat{z}\times\hat{z}\right)\end{pmatrix}$$

$$= \left(\frac{q}{m}\right)\left((E_x + Bv_y)\hat{x} + (E_y - Bv_x)\hat{y}\right).$$

Now, we will express each component of the acceleration separately:

$$a_x = \frac{dv_x}{dt} = \tfrac{q}{m}\left(E_x + B_z\, v_y\right) \qquad \text{and} \qquad a_y = \frac{dv_y}{dt} = \tfrac{q}{m}\left(E_y - B\, v_x\right),$$

which are coupled first order differential equations.

While we can solve these particular equations analytically, our goal is to simulate the motion of the charged particle. The easiest way to accomplish this is to solve the equations numerically using the algorithm discussed above. The plot shows the motion of an electron inside an electric field of $\vec{E} = 10\,{}^{\text{kN}}\!/\!_{\text{C}}\,\hat{x}$ and a magnetic field picked so that it has a 10 ns gyroperiod, or:

$$\vec{B} = \left(\tfrac{m}{e}\right)\frac{2\pi}{10\,\text{ns}}\,\hat{z} \approx 3.6\,\text{mT}\,\hat{z}.$$

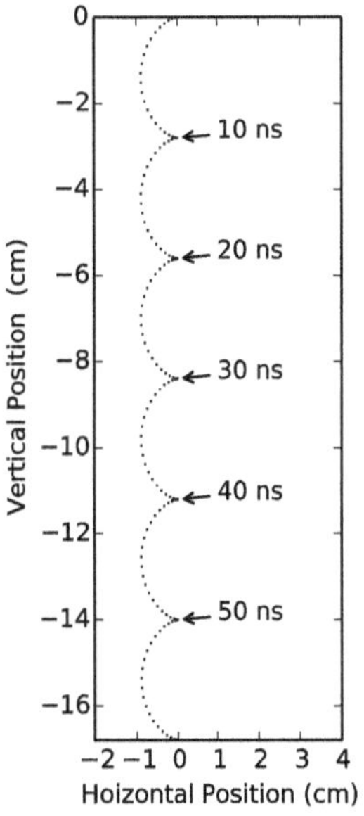

Notice that the electron appears to "roll" in the $-\hat{y}$ direction, with an average velocity of approximately:

$$\left\langle |v_y| \right\rangle \approx \frac{17\,\text{cm}}{60\,\text{ns}} \approx \frac{0.17\,\text{m}}{6\times 10^{-8}\,\text{s}} \approx 2.8\times 10^{6}\,{}^{\text{m}}\!/\!_{\text{s}}$$

$$\approx 2.8\times 10^{6}\,{}^{\text{N}}\!/\!_{\text{T}} \approx \frac{1\,{}^{\text{N}}\!/\!_{\text{C}}}{3.6\times 10^{-7}\,\text{T}} \approx \frac{10\,{}^{\text{N}}\!/\!_{\text{C}}}{3.6\,\text{mT}}.$$

This is suggestive that the average velocity would be given by the ratio of E to B, which is indeed the case, as we will analytically show later in this chapter. This velocity is often call the "E cross B" velocity, because charged particles will systematically move at an average velocity given by:

$$\left\langle \vec{v} \right\rangle \approx \frac{\vec{E}\times\vec{B}}{B^2}.$$

This result is very important in plasma physics, because positive and negative charges both move in the same direction.

Thought Experiment 10.6: A Free Particle in an Azimuthal Electric Field

Consider the motion of a particle with charge q and mass m, under the influence of an electric field given by the following relationship:

$$\vec{E} \propto \frac{1}{s}\hat{\varphi}.$$

What is the path of an electron, if it starts from rest at a position $\vec{r} = x_0\,\hat{x}$?

We apply the electromagnetic force to Newton's second law of Motion:

$$\vec{a} = \frac{\vec{F}}{m} = \frac{q\vec{E}}{m} = -\frac{e\vec{E}}{m} \propto -\frac{1}{s}\hat{\varphi} = -v_c^2\,\frac{\hat{\varphi}}{s},$$

where v_c^2 is a postive constant with units of velocity squared. We will think of v_c as a characteristic speed that is somehow defined by the problem. The acceleration in cartesian coordinates can now be written as:

$$\vec{a} = -\frac{v_c^2}{s}\hat{\varphi} = -\frac{v_c^2}{\sqrt{x^2+y^2}}\left(-\sin(\varphi)\hat{x}+\cos(\varphi)\hat{y}\right)$$

$$\vec{a} = \frac{v_c^2}{\sqrt{x^2+y^2}}\left(\left(\frac{y}{\sqrt{x^2+y^2}}\right)\hat{x} - \frac{x}{\sqrt{x^2+y^2}}\hat{y}\right) = \frac{v_c^2}{x^2+y^2}\left(y\hat{x}-x\hat{y}\right).$$

The acceleration components are simply a function of position, so we can simulate the motion by solving these two coupled second order differential equations:

$$\frac{\mathrm{d}^2x}{\mathrm{d}t^2} = \frac{v_c^2\,y}{x^2+y^2} \quad \text{and} \quad \frac{\mathrm{d}^2y}{\mathrm{d}t^2} = -\frac{v_c^2\,x}{x^2+y^2}.$$

The solution to this is shown in the figure for an electron that starts from rest at a position of one centimeter from the origin. The position of the particle, at unit time intervals, is shown as points on the plot, and the circles represent equal intervals of the electric field.

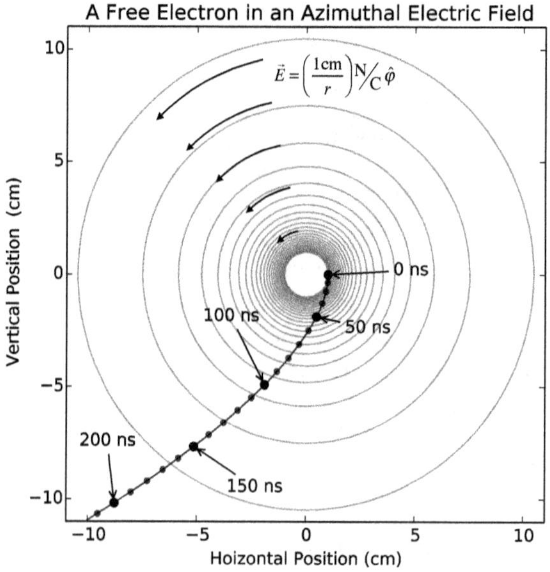

According to Ohm's law, which we discussed previously, the current in a conductor will always be parallel to the electric field, so the electron drift velocity is always anti-parallel to the electric field. If this were a normal conductor, then each electron would be moving mostly randomly, with a systemic velocity in the clockwise direction following the electric field lines.

A free particle behaves quite differently. In the case of even a very small electric field, the electron soon flies off. Electric fields are, therefore, unstable plasma containment fields, as free particles tend to fly away in even the most regular and modest electric fields.

Problem 10.12: A Magnetic Bottle

Consider a particle of mass m and positive charge q, interacting with a magnetic field given by:

$$\vec{B} = B_0\left(\left(1+\frac{z^2}{a^2}\right)\hat{z} - \frac{s\,z}{a^2}\hat{s}\right),$$

where B_0 and a are constants with units of magnetic field and length respectively.

a) Show that magnetic flux is conserved everywhere. If it were not, the given magnetic field would be unphysical.

b) Make a sketch of this magnetic field.

c) Solve for the force on the particle as a function of its position and velocity.

d) Find the equations of motion for this particle in cylindrical coordinates. You should solve for the acceleration $(\ddot{s},\ddot{\varphi},\ddot{z})$ as a function of the position (s,φ,z) and velocity $(\dot{s},\dot{\varphi},\dot{z})$. These will be three second order coupled differential equations.

e) Now, we will consider an equatorial orbit. What angular velocity ω_0 must the particle have if it is to remain orbiting at a fixed radial position $\vec{r}_0 = s_0\,\hat{s}$? In what direction is the orbit?

f) Assume now that this orbit is slightly perturbed, having a small initial axial velocity v_{z0}, so its initial velocity is given by $\vec{v}_0 = v_{z0}\,\hat{z} - s_0\,\omega_0\hat{\varphi}$. Assume that the radial time derivatives remain small ($\dot{s}_0 \approx 0, \ddot{s}_0 \approx 0$). What is the axial equation of motion? Solve for the position (z) as a function of time. Is this bounded? Explain.

g) Look up the Van Allen belts. Explain qualitatively what they are, how they work, and what they have to do with the aurora. Cite your references.

10.4 *The Elementary Charge*

J.J. Thomson measured the electron charge to mass ratio, but he did not separately determine the charge or the mass. Clearly electrons exist, and all of them are identical, so why should it be so difficult to measure the mass or the charge independently of the other?

Every experiment involving the electron, or ions for that matter, measured the acceleration and the electromagnetic field. Since the electromagnetic force is proportional to the charge, and the acceleration is inversely proportional to the mass, all measurements necessarily found the ratio of charge to mass: $\vec{a} = \frac{q}{m}(\vec{E} + \vec{v} \times \vec{B})$. One solution, therefore, is to balance the electromagnetic force with another known force. But how to do that on such a small scale?

Perhaps, if one electron of charge could be placed on a neutral object of known weight, then its motion could be balanced out by adjusting the electric field. This is the approach that the American experimentalist Robert A. Millikan took, but it was not the only result published in 1909. Millikan reviewed this in his 1911 follow-up paper.[145] For comparison, the modern value of one elementary charge is 4.803×10^{-10} electrostatic units[146] (1 nC = 2.998 esu).

> The value of *e* herewith obtained is in perfect agreement with the result reached by Regener[1] in his remarkably careful and consistent work on the counting of the number of scintillations produced by the particles emitted by a known amount of polonium and measuring the total charge carried by these same particles. His final value of this charge is 9.58×10^{-10}, and upon the assumption that this is twice the elementary charge—an assumption which seems to be justified by Rutherford's experiments[2]—he finds for *e* 4.79×10^{-10}, with a probable error of 3 per cent. Since the difference between this value and 4.89×10^{-10}, is but 2 per cent. the two results obviously agree within the limits of observational error.
>
> On the other hand, the present value of *e* is 4 per cent higher than the simple mean value which I previously obtained in work by a similar method upon drops of water and alcohol[3] and when the correction to Stokes's law is applied the difference becomes as high as 8 per cent. …

---

[145] R.A. Millikan, *The Isolation of an Ion, a Precision Measurement of it Charge, and the Correction of Stokes's Law.*, The Physical Review, **4** (1911) 349-397

[146] The electrostatic unit (esu) is defined in the centimeter, gram, second unit system by setting Coulomb's constant to unity. See Discussion 1.4 on p. 51.

[1] E. Regener, Sitz. Ber. d. k. Preuss. Acad. d. Wiss., XXXVII., p. 948, 1909.
[2] Rutherford, Phil. Mag., 17, p. 281, 1909.
[3] Millikan, Phil. Mag., 19, p. 209, 1909.

Thus, by 1909 multiple scientists had measured the elementary charge to within about 2% of the modern value. Millikan, and others, kept refining their methods, and critiquing each others methods, over the next few years.

The purpose of this section is to convey the physics necessary to understand the oil drop method of measuring the elementary charge, and that means learning some basic fluid dynamics. The bulk of Millikan's systematic error came from his calculation of the drag force on the oil drop, which required a knowledge of fluid dynamics to understand.

There is another reason to learn some fluid dynamics. Maxwell's electrodynamics is one large mathematical analogy to fluid dynamics. Much of this analogy is also physically based, such as current flowing through conductors, plasmas, and superconductors. Other parts, such as the analogy between magnetic fields and fluid velocity, is simply a mathematical analogy, or as Maxwell put it:

> In order to obtain physical ideas without adopting a physical theory we must make ourselves familiar with the existence of physical analogies. . . that partial similarity between the laws of one science and those of another which makes each of them illustrate the other.[147]

The final reason to include a taste of fluid dynamics is that it was excised from the undergraduate physics curriculum in the middle of the twentieth century to make room for modern physics, and is only now being added back in. This was a unique time in the history of physics, as it was after the fundamental physics of fluids was well established, but before there was the computational power to actually solve chaotic fluid problems. If fluids intrigue you, you may want to look into some fields of applied physics such as oceanography, meteorology, and aerospace engineering.

Discussion 10.2: Conserved Quantities and the Navier-Stokes Equation

All conserved quantities of charge must follow continuity equations. For example, the most fundamental equation in electricity is:

$$\vec{\nabla}\cdot\vec{J} = -\frac{\partial\rho}{\partial t}.$$

In fluid mechanics, the conservation of mass ensures that:

$$\vec{\nabla}\cdot\left(\rho_m\vec{v}\right) = -\frac{\partial\rho_m}{\partial t},$$

where $\vec{v}$ is the flow velocity. The conservation of energy is similar:

$$\vec{\nabla}\cdot\left(\vec{\mathbb{S}}+\mathbb{u}\vec{v}\right) = -\frac{\partial\mathbb{u}}{\partial t},$$

where, $\vec{v}$ is the flow velocity, $\mathbb{u}$ the energy density of the fluid, and $\vec{\mathbb{S}}$ is the radiated energy flux. Again, we can assume that in a steady state system whatever energy flows into a volume must also flow out, even if it changes form.

---

[147] J.C. Maxwell, "On Faraday's Lines of Force," Transactions of the Cambridge Philosophical Society, **10**, no. 1, (1856), 156.

And, finally, we must also conserve momentum. This is somewhat more complicated, as momentum is a vector. By analogy, we simply apply the same continuity equation to each component of the momentum:

$$\vec{\nabla}\cdot\left(\vec{P}_{x,y,z}+\mathbb{p}_{x,y,z}\,\vec{v}\right)=-\frac{\partial\,\mathbb{p}_{x,y,z}}{\partial t},$$

where, $\vec{P}_x$, $\vec{P}_y$, and $\vec{P}_z$ are the rate, per area, that each component of momentum flows out. This is deceptively simple, since each component of momentum can be transferred in any direction. This is the principle behind the *stress tensor*, which has units of pressure but may act either along a vector or sideways to it. Since momentum is mass times velocity, the above equation can be slightly simplified in tensor form as:

$$\vec{\nabla}\cdot\left(\vec{P}+\rho\,\vec{v}\circ\vec{v}\right)=-\frac{\partial(\rho\,\vec{v})}{\partial t},$$

where we have used the *outer product*, which produces a tensor from two vectors. Yes, this is very difficult, and so we will put off the idea of a generalized stress tensor until Chapter 17, where we generalize this to four dimensional space-time using Einstein's tensor notation.

When we simplify this even more, say for an incompressible liquid, we assume that the density is constant and, more importantly, that the stress tensor pushes the same in all directions. This isotropic stress tensor can now be characterized by only two parameters, pressure and viscosity. This is what is known as a *Newtonian fluid*, and the conservation of momentum becomes the fluid version of Newton's second law called the Navier–Stokes equations. As with most names in physics, this was developed by many scientists but named after only two of them: the eighteenth century French physicist Claude-Louis Navier and George Gabriel Stokes. Steady-state solutions to the Navier-Stokes equations primarily depend on the degree to which forces within the fluid can exchange momentum sideways, known as the viscosity η.

Thought Experiment 10.7: Newton's Second Law in a Fluid

Recall that Newton's second law states that the net force on an object causes a rate of change of momentum, or $\vec{F}_{\text{net}}=\frac{d}{dt}(m\vec{v})=m\vec{a}$. Now consider a rocket in space, so there are no external forces on it, but it can experience an apparent force (thrust) if it ejects mass. Integrating the ejecta's momentum flux around the surface of the rocket, the thrust, in terms of the ejecta's mass density and velocity, becomes:

$$\vec{F}_{\text{thrust}}=-\oint_{\text{surface}}\vec{v}_{\text{ejecta}}\left(\rho_{\text{ejecta}}\vec{v}_{\text{ejecta}}\cdot d\vec{A}\right)=-\oint_{\text{surface}}\rho_{\text{ejecta}}\vec{v}_{\text{ejecta}}\left(\vec{v}_{\text{ejecta}}\cdot d\vec{A}\right).$$

Now imagine that rather than a rocket, we have fluid with a given velocity field $\vec{v}(\vec{r})$ and density $\rho_m(\vec{r})$, the same function would be the apparent thrust that a given closed volume of fluid would experience simply due to inertia. So, we can write the thrust, per volume, at a given point as:

$$\vec{f}_{\text{thrust}}=-\lim_{\delta V\to 0}\frac{1}{\delta V}\oint_{\text{surface}}\rho_m\vec{v}\left(\vec{v}\cdot d\vec{A}\right)=-\left(\vec{v}\cdot\vec{\nabla}\right)\left(\rho_m\vec{v}\right).$$

This is a example of a directional derivative, because we are only concerned with a change in the vector field in a particular direction.

Now consider that a fluid has a constant density and is experiencing an external body force on it, $\vec{f}$. Newton's second law, including any thrust, would simply be $\vec{f}+\vec{f}_{\text{thrust}}=\rho_m\vec{a}$. In terms of our density and velocity fields this becomes:

$$\vec{f}=\rho_m\tfrac{\partial}{\partial t}\vec{v}+\rho_m\left(\vec{v}\cdot\vec{\nabla}\right)\vec{v}=\rho_m\left(\tfrac{\partial}{\partial t}\vec{v}+\left(\vec{v}\cdot\vec{\nabla}\right)\vec{v}\right)=\rho_m\left(\tfrac{\partial}{\partial t}+\left(\vec{v}\cdot\vec{\nabla}\right)\right)\vec{v}.$$

The left hand side of the total real external force per volume, while the right hand side represents rate of change of momentum per volume.

Definition 10.1: Viscosity

Viscosity is a property of a liquid that represents the shear drag force, per sideways area, from one layer to the next. Imagine a deep straight river moving in the $\hat{y}$ direction with a uniform depth h, and so the fluid velocity is simply a function of x. The force between any two parallel vertical slabs of water, each with length Δy, width δx, and height h will be proportional to the difference in their velocities or:

$$\vec{F}_{\text{of one slab on the next}}=-\eta\,\Delta y\,h\frac{\mathrm{d}v_y}{\mathrm{d}x}\hat{y}\quad\rightarrow\quad\vec{F}=-\eta\left(\frac{\mathrm{d}v_y}{\mathrm{d}x}\hat{z}\right)\times\left(\Delta y\,h\,\hat{x}\right),$$

where η is the *dynamic viscosity* and it is a property of matter that depends primarily on the strength of the electrical forces between molecules.

In vector calculus notation, we can write this as:

$$\delta\vec{F}=-\eta\left(\vec{\nabla}\times\vec{v}\right)\times\delta\vec{A}=-\eta\,\vec{\omega}\times\delta\vec{A}.$$

In the last expression, we introduced the *vorticity*, which is defined as the curl of the velocity field. In the deep river example, the vorticity is in the $\hat{z}$ direction.

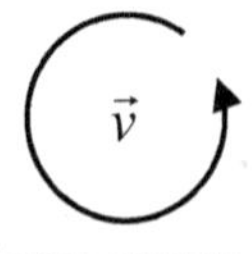

The laminar drag on an object is more complicated because it depends on the boundary layer and any turbulence that might be associated with this. However, if you assume the fluid is stationary at the surface, it is found by integrating the vorticity around the object, or with our standard outward sign convention:

$$\vec{F}_D=\eta\oint_{\text{surface}}\vec{\omega}\times\mathrm{d}\vec{A}.$$

In the deep river example, each vertical slab of water has two sides, one closer to the bank and other away from it. Taking into account the positive directions, we can write the net force on a slab of water as the difference from both sides:

$$\frac{F_y}{\Delta y\,\Delta z}=\left(-\eta\frac{\mathrm{d}v_y}{\mathrm{d}x}\right)_{\text{at }x-\frac{1}{2}\delta x}-\left(-\eta\frac{\mathrm{d}v_y}{\mathrm{d}x}\right)_{\text{at }x+\frac{1}{2}\delta x}.$$

Next we can find the force per volume as:

$$f_y=\lim_{\delta x\to 0}\frac{F_y}{\Delta y\,\Delta z\,\delta x}=\lim_{\delta x\to 0}\frac{1}{\delta x}\left(\left(\eta\frac{\mathrm{d}v_y}{\mathrm{d}x}\right)_{\text{at }x+\frac{1}{2}\delta x}-\left(\eta\frac{\mathrm{d}v_y}{\mathrm{d}x}\right)_{\text{at }x-\frac{1}{2}\delta x}\right)=\eta\frac{\mathrm{d}^2v_y}{\mathrm{d}x^2}.$$

Now, by extending this to three dimensions, this becomes:

$$\vec{f}_D = -\eta\left(\vec{\nabla}\times\vec{\omega}\right) = -\eta\left(\vec{\nabla}\times\left(\vec{\nabla}\times\vec{v}\right)\right).$$

The vector operator $\vec{\nabla}\times\vec{\nabla}\times$ is called the *double curl*, and notice that, as with any second derivative, it is evaluated from the right to left. The double curl comes up often in electrodynamics.

Problem 10.13: Water Through a Cylindrical Pipe

(a) Consider water steadily flowing through a long cylindrical pipe, so that its velocity is given by: $\vec{v} = v_z(s)\,\hat{z}$. Show that $\vec{\nabla}\times\vec{\nabla}\times\vec{v} = -\left(\frac{\partial^2}{\partial s^2}v_z + \frac{1}{s}\frac{\partial}{\partial s}v_z\right)\hat{z}$.

(b) Common sense suggests that the water moves fastest in the middle of the pipe, and slowest at the contact point with the pipe. Argue, using free-body diagrams, that this is expected for any positive viscosity.

(c) Imagine that the function velocity followed the following function of the speed of the water in the middle, v_0, and the pipe diameter, a: $\vec{v} = v_0\left(1-\left(\frac{s}{a}\right)^2\right)$. Show that the flow rate of the water would be given by: $I_m = \frac{1}{2}\left(\pi a^2\right)\rho_m v_0$.

(d) Given a viscosity, η, find the drag force per volume on the water.

(e) Imagine that this pipe carries water from a mountain reservoir to a city at a lower elevation. If the water pressure is constant throughout, find an expression for the downhill angle of incline.

(f) In 1837 New York City commenced work on damming the Croton River, located at an elevation of 200 feet above sea level and forty miles north of the city, with the hope of providing 20 million gallons per day of fresh water to New Yorkers. Calculate the minimum radius that a cylindrical pipe would need to be in order to transport the water. Express your answer in feet.

(g) They actually built a rectangular masonry aqueduct with dimensions of 7 feet × 8 feet × 40 miles. What is ratio the cross sectional area of this aqueduct to the minimum you calculated? Discuss your result.

(Note: at 70°F the dynamic viscosity of water is $2\times10^{-5}\,\frac{\text{s}\cdot\text{lb}}{\text{ft}^2}$ and its density is $62.3\frac{\text{lb}}{\text{ft}^3} = 8.0\,\frac{\text{lb}}{\text{gallon}}$.)

Thought Experiment 10.8: Irrotational Flow

Consider the drag body force on a fluid element:

$$\vec{f}_D = -\eta\left(\vec{\nabla}\times\left(\vec{\nabla}\times\vec{v}\right)\right).$$

As we discussed, this is a restorative force, meaning that it resists shear differences. In steady-state flow, the net force must be zero as there is no acceleration. Therefore the drag force must be equal and opposite to the external body force.

Now, if the viscosity is very large, like molasses or ketchup, we can write:

$$\left|\vec{\nabla}\times\left(\vec{\nabla}\times\vec{v}\right)\right| = \left|\vec{\nabla}\times\vec{\omega}\right| = \frac{f_D}{\eta} \to 0 \text{ for large viscosities.}$$

And, so we can write for any loop:

$$\oint_{\text{path}} \vec{\omega}\cdot d\vec{\ell} = \int_{\text{surface}} \vec{\nabla}\times\vec{\omega}\cdot d\vec{A},$$

Thus, highly viscous fluids will follow *irrotational flow*, or:

$$\vec{\omega} = \vec{\nabla}\times\vec{v} \approx 0.$$

Problem 10.14: Stirring a Glass of Water

(a) When you stir honey into a cup of tea, the tea swirls in circles for a while after you stop stirring. However, this is not the case for the honey. Use the definition of the viscosity, along with Newton's second law, to discuss why this is so.

(b) Take a tall glass of water and stir it at a constant rate with a chopstick around the outside. After a while the water in the glass is all moving in a steady manner. If your glass has a radius, a, and you keep stirring it at a speed $\vec{v}(s=a) = v_0\hat{\varphi}$, find the velocity, as a function of position, of the water. Is this what you observe?

(c) Find the centripetal acceleration of the water in the last problem, as a function of position, in ratio to the surface gravity, $g = 980\frac{\text{cm}}{\text{s}^2}$. Show that the height of the water should be given by the function: $z = h + \frac{2\pi^2 a^2}{gT^2} - \frac{2\pi^2 a^2}{gT^2}\left(\frac{a}{s}\right)^2$, where h is the height of the water's edge, a is the radius of the glass, and T is the period that you are stirring.

(d) Find an expression for the diameter of the circle you see at the bottom on the glass, and test it out experimentally.

(e) Imagine you fill the glass up to a height h_0, and then you stir it until it just reaches the brim at height h. Calculate the period, T, you need to stir it to keep this up. Test it out experimentally to check your work.

Problem 10.15: The Double Curl and the Vector Laplacian

(a) Using Cartesian coordinates, verify: $\vec{\nabla}\times\vec{\nabla}\times\vec{v} = \vec{\nabla}\left(\vec{\nabla}\cdot\vec{v}\right) - \nabla^2\vec{v}$.

(b) Find the vector Laplacian in cylindrical coordinates.

(c) Find the vector Laplacian in spherical coordinates.

(d) Argue that for a constant density fluid the viscosity body forces would be given by: $\vec{f}_{\text{D}} = \eta\nabla^2\vec{v}$.

Thought Experiment 10.9: Stoke's Law

Consider the example of a heavy metal ball, of density ρ_{ball} and radius a, that falls at a constant velocity $\vec{v} = -v_0\,\hat{z}$ through oil whose density is ρ_{oil}. We can determine the viscosity of the oil by finding the drag force opposing the downward force on the ball.

Next we will consider the same problem from the point of view of the ball, where the fluid is moving up. In spherical coordinates at large distances the velocity of fluid is:

$$\vec{v} = v_0\,\hat{z} = v_0\left(\cos(\theta)\hat{r} - \sin(\theta)\hat{\theta}\right).$$

However right on the surface the radial velocity must be zero.

Now we will make a guess, which may, or may not, be correct, that conforms to our boundary conditions, conserves, mass, and makes intuitive sense. Like the velocity of a liquid, the magnetic field is also divergence free. So, we will make our guess based on our experience with spherical magnets, and, in particular, follow the techniques developed in In Thought Experiment 7.1 (p. 213).

First we will guess, based on physical intuition, that the velocity field is simply the difference between the far velocity and a perfect dipole field:

$$\vec{v}_{\text{guess}} = v_0\left(\cos(\theta)\hat{r} - \sin(\theta)\hat{\theta}\right) - \left(\frac{C}{r^3}\right)\left(2\cos(\theta)\hat{r} + \sin(\theta)\hat{\theta}\right).$$

We normalize the dipole field by setting the radial surface velocity to zero:

$$\hat{r}\cdot\vec{v}_{\text{guess}}\Big|_{r=a} = 0 = v_0\cos(\theta) - C\left(\frac{2\cos(\theta)}{a^3}\right) \quad\rightarrow\quad C = \frac{v_0 a^3}{2}.$$

Thus our guess becomes:

$$\vec{v}_{\text{guess}} = v_0\left(\cos(\theta)\hat{r} - \sin(\theta)\hat{\theta}\right) - \left(\frac{v_0 a^3}{2r^3}\right)\left(2\cos(\theta)\hat{r} + \sin(\theta)\hat{\theta}\right) = v_0\cos(\theta)\left(1 - \frac{a^3}{r^3}\right)\hat{r} - v_0\sin(\theta)\left(1 + \frac{a^3}{2r^3}\right)\hat{\theta}.$$

Notice that our guess suggests that the velocity must increase near the surface of the ball, which must be the case in order to conserve mass, just like the speeding up of a river through a narrows.

However, right at the surface things often become complicated in a region called the *boundary layer*. In the limit of extremely high viscosities, very small spheres, or very slow velocities, we can simply assume that the boundary layer is thin. That said, mass still must be conserved in the boundary layer, so the divergence of the velocity on the sphere must still be zero.

At the top of the boundary layer, we will assume that:

$$\vec{v}_{\text{guess}} = v_0\cos(\theta)\left(1 - \frac{a^3}{a^3}\right)\hat{r} - v_0\sin(\theta)\left(1 + \frac{a^3}{2a^3}\right)\hat{\theta} = -\tfrac{3}{2}v_0\sin(\theta)\,\hat{\theta}.$$

Taking the divergence in the boundary layer, and setting it to zero, gives us:

$$0 = \frac{1}{a\sin(\theta)}\frac{\partial}{\partial\theta}\left(v_\theta\sin(\theta)\right) = \frac{1}{a\sin(\theta)}\left(\sin(\theta)\frac{\partial v_\theta}{\partial\theta} + v_\theta\cos(\theta)\right),$$

Now we use separation of variables to solve this:

$$\int\frac{dv_\theta}{v_\theta} = \int -\frac{d\theta}{\tan(\theta)} \quad\rightarrow\quad \ln v_\theta = \ln\left(\frac{1}{\sin(\theta)}\right) + const \quad\rightarrow\quad v_\theta \propto \frac{1}{\sin(\theta)}.$$

What this really means is that the cross-sectional surface area in the boundary layer increases toward the equator, so in order to conserve mass the boundary layer must slow down, not speed up as it appears to just above. Notice that this implies a perfectly small constriction at each pole.

In order to match everything up at the equator, we expect the boundary layer velocity to be:

$$\vec{v}_{\text{boundary}} = -\frac{3v_0}{2\sin(\theta)}\hat{\theta}.$$

Now we finally have a simpler problem, one of a sphere with a constant velocity fluid washing over it. In order to find the drag force we need to find the vorticity in the boundary layer, which is:

$$\vec{\nabla}\times\vec{v}_{\text{boundary}} = \tfrac{1}{r}\tfrac{\partial}{\partial r}\left(r v_\theta\right)\hat{\varphi} = -\frac{3v_0}{2a\sin(\theta)}\tfrac{\partial}{\partial r}\left(r\right)\hat{\varphi} = -\frac{3v_0}{2a\sin(\theta)}\hat{\varphi}.$$

Now recall the definition of the viscosity from page 324. Considering a patch of area on the ball's surface, $\delta\vec{A} = \delta A\hat{r}$, we can find the force on the object as:

$$\vec{F}_D = \eta \oint_{\text{surface}} \vec{\omega}\times d\vec{A}\,\vec{F}_D = \eta\int_0^{2\pi}\int_0^{\pi}\left(-\frac{3v_0}{2a\cancel{\sin(\theta)}}\hat{\varphi}\right)\times\left(a^2\cancel{\sin(\theta)}\,d\theta\, d\varphi\,\hat{r}\right) = -\tfrac{3}{2}av_0\eta\int_0^{2\pi}\int_0^{\pi}\left(\hat{\varphi}\times\hat{r}\right)d\theta\, d\varphi$$

$$= -\tfrac{3}{2}av_0\eta\int_0^{2\pi}\int_0^{\pi}\hat{\theta}\,d\theta\, d\varphi.$$

Since we are interested in the net force on the sphere, and we know that this is going to be in the direction of flow, we now write our unit vector, $\hat{\theta}$, in terms of the cylindrical unit vectors $\hat{s}$ and $\hat{z}$:

$$\vec{F}_D = \tfrac{3}{2}av_0\eta\int_0^{2\pi}\int_0^{\pi}\left(-\cos(\theta)\,\hat{s} + \sin(\theta)\,\hat{z}\right)d\theta\, d\varphi.$$

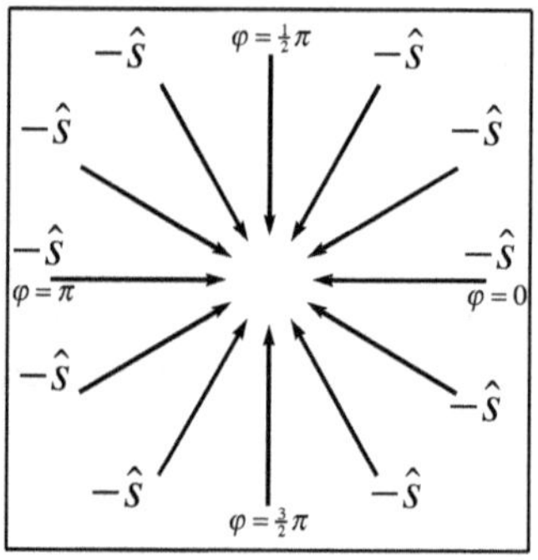

The direction of the unit vector $\hat{z}$ is fixed, so $\hat{z}$ is simply a constant of integration. However $\hat{s}$ is a function of the azimuthal angle, and it must be integrated with respect to the angle φ. Moreover, in a full circle, for every $-\hat{s}$ pointing in one fixed direction, there will be a corresponding unit vector $-\hat{s}$ that points in the opposite direction. This means that its sum over 2π will equal zero.

$$\vec{F}_D = \tfrac{3}{2}av_0\eta\left(\cancel{\int_0^{2\pi}\int_0^{\pi}\left(-\cos(\theta)\hat{s}\right)d\theta\, d\varphi} + \int_0^{2\pi}\int_0^{\pi}\left(\sin(\theta)\hat{z}\right)d\theta\, d\varphi\right) = \tfrac{3}{2}av_0\eta\hat{z}\left(2\pi\int_0^{\pi}\sin(\theta)\,d\theta\right) = 6\pi a v_0\eta\,\hat{z}.$$

This is called Stoke's law, after George Gabriel Stokes, and it is applicable to situations where there is a very high viscosity, or the spheres are very small. This does not apply to fast moving spheres through air, so do not apply it to ball sports. However, it does work well to model conductivity, as in the Heaviside model of Ohm's law.

Much of Millikan's 1911 and 1913 papers discussed the applicability of Stokes's law to find the drag on a falling drop. This was one reason he switched to using an oil drop between 1909 and 1911. He also empirically tested Stokes's law, and added his own correction to it, which was why it was important to measure the temperature and pressure inside of his chamber.

Problem 10.16: Stokes Law in the Millikan Experiment

Millikan wrote in 1913:[148]

> As is now well known, the oil-drop method rested originally upon the assumption of Stokes's law and gave the charge *e* on a given drop through the equation
>
> $$e_n = \frac{4}{3}\pi\left(\frac{9\eta}{2}\right)^{\frac{3}{2}}\left(\frac{1}{g(\sigma-\rho)}\right)^{\frac{1}{2}}\frac{(v_1+v_2)v_1^{\frac{1}{2}}}{F},$$
>
> in which η is the coefficient of viscosity of air, σ the density of the oil, ρ that of air, v_1 the speed of the descent of the drop under gravity and v_2 its speed of ascent under the influence of an electric field of strength F.

a) Draw a free body diagram when the drop is neutral and falling at a constant velocity v_1.

b) Find expressions for each of these forces, in terms of Millikan's measureable variables.

c) Find the size of the drop, a, in terms of Millikan's measureables.

d) Draw a free body diagram when the drop is charged and moving upward at velocity v_2.

e) Show that Millikan's expression is correct.

Problem 10.17: Conserving Mass in Stokes Law

In Thought Experiment 7.1 (p. 213), we calculated the total magnetic flux from a dipole field crossing the equator. In Thought Experiment 10.9 (p. 326), the premise is that fluid is flowing around a sphere. Well upstream of the sphere, and far downstream, the velocity is uniform.

(a) Calculate the total mass rate, I_m, without the sphere, as a function of v_0, ρ_m, and s.

(b) At the equator of the sphere, again calculate the total mass rate as a function of the same quantities as well as the sphere radius a.

(c) In our derivation of Stokes law (p. 326), we took advantage of the similarities between the magnetic field and the fluid velocity field to make an educated guess for the velocity field. Show that the total mass rates, I_m, that you found are equal in the limit of large distances s.

Problem 10.18: The Ratio of the Forces Between Electrons

(a) Find the ratio of the maximum attractive magnetic force to the electric force between two electrons, treated as magnetic dipoles, in terms of their known properties, fundamental constants, and their separation r.

(b) Show that the separation distance (in terms of the classical electron radius r_e) where these two forces will be equal is $\sqrt{\frac{24\pi\varepsilon_0}{\mu_0}\frac{\mu_e m_e}{e^3}r_e}$.

(c) Pretend that electrons are classical spheres of radius r_e. What would be the force ratio between them if they were stuck together?

---

[148] R.A. Millikan, "On the Elementary Electric charge and the Avogadro Constant," Physical Review, **2** (1913), 109-143.

Plate 19: Galileo's Cover Page

Chapter 11 Galilean Relativity in Electrodynamics

That a charge must move with a definite velocity to experience a magnetic force raises an interesting question. What happens if we transform to a reference frame in which the charge is at rest? Apparently, there is no magnetic force exerted on the charge in the rest frame of the charge. Yet the principle of relativity, first described by Galileo Galilei and then incorporated by Isaac Newton into his laws of motion, assures us that the laws of physics do not change whether you are traveling at a constant velocity or standing still.

Galileo used the principle of relativity in order to illustrate how the earth can move without our feeling the motion. As he (roughly) put it, imagine that you and a friend lock yourselves in a windowless cabin below the deck of a docked, motionless ship. Now look about the cabin. You observe water dripping from bottle that is suspended upside-down, fish swimming in all directions in an aquarium, and butterflies fluttering randomly. Throw something to your friend. It will require the same force no matter what direction you throw it in, so long as the distances of the throw are equal. Now imagine that the ship is cruising along at a constant velocity in calm waters. The water continues to drip as before. The fish and butterflies are equally content swimming or flying in all directions. There are no new challenges in throwing the ball. In fact, there is no way for you or your friend to tell whether or not you are stationary or moving uniformly with respect to the shore.[149] The earth is our cabin and though we are hurtling some 100,000 kilometers per hour relative to the sun, the butterflies do not seem to care.

Galileo's analysis of projectile motion can also be understood in light of the principle of relativity. Consider, once more, a ship sailing along a smooth sea at constant velocity. If one drops a heavy stone (so that air resistance is not a factor) from the top of the ship's mast, an observer on the ship would see it accelerate straight downward and hit the deck at the base of the mast. An observer standing on the shoreline sees the same stone move in a parabolic arc. In addition to the vertically accelerated motion of the stone, that person would also see the ship (and everything on it, including the falling stone) move horizontally together with the same speed.

Both observers, of course, are correct in describing the different scenes they saw, but the physics remains the same. For example, the time of flight for the stone—which we can derive by noting only the forces on the stone—remains the same, despite the different apparent flight paths. The laws of physics are consistent regardless of the frame of reference in which you choose to describe them.

As with falling bodies and butterflies, the physics for our test charge also will not change depending on the reference frame. Yet no magnetic force exists in the rest frame of a charge in the presence of a magnetic field, even though a magnetic force is exerted on that charge according to someone who moves uniformly with respect to it. How can this fact be reconciled with Galileo's principle of relativity? The ultimate answer to this conundrum comes from Einstein's special theory of relativity, which demonstrates that electric and magnetic fields are aspects of a more fundamental entity—the *electromagnetic field*. What appears to be only a magnetic field in one frame of reference will be observed as separate electric and magnetic fields in other frames of reference.

---

[149] Galileo Galilei, Dialogue Concerning the Two Chief World Systems: Ptolemy and Copernicus, (1632); (Translated by Stillman Drake, University of California Press, 1953), 186 – 187; Second Day).

Einstein's theory has its roots in electricity and magnetism; in fact, his first paper on the subject published in 1905 was titled *On the Electrodynamics of Moving Bodies.* It was his concern with questions related to the moving charge that led Einstein to develop this theory, and we will see just how successfully he clarified our understanding of electromagnetism later. We do not need the full machinery of special relativity to obtain at least a partial answer to the puzzle raised here, however. We will see that electric and magnetic fields transform in a simple way when measured by different moving observers and can be approximately related by the Galilean transformation from pre-Einstein physics. One consequence of this interrelationship of electric and magnetic fields observed in different frames of reference is that an electric field can be induced when a neutral metal bar moves through a given constant magnetic field.

We no longer think of electric and magnetic fields as separate entities, but instead as the unified electromagnetic field. Today, physicists adopt this concept of a *field* to mediate interactions among particles when describing all fundamental forces of nature. In light of this, we shall introduce the idea of electromagnetic fields in this chapter with an emphasis on the response of charges to already existing electric and magnetic fields.

11.1 *Galilean Relativity*

Galileo pioneered the concept that the laws of physics should be consistent whether they are applied in a stationary reference frame or in a frame moving in uniform translational motion relative to an observer. The *principle of relativity* is a fundamental axiom of all of physics.

Newton's laws of motion conform to the principle of relativity and are covariant under the simple Galilean transformation. To transform from one reference frame to another, the velocities simply add (or subtract) and all non-velocity-dependent quantities remain the same. It is common in classical mechanics to transform from one reference frame, analyze a problem, and transform back to the original frame. After covering a kinematic example of Galilean relativity, we discuss how to transform the current density.

Newton's second law of motion is preserved under Galilean transformations for forces that do not vary under Galilean transformation, such as Newton's gravitational force. But not all forces that arise in nature are Galilean invariant! The magnetic part of the Lorentz force does change under Galilean transformations, for example, so it is not Galilean invariant.

Problem 11.1: Newtonian Galilean Transforms

By convention, at time zero reference frames are assumed to be aligned. Use the Galilean transform relationship to show the following:

a) The position vector transforms as $\vec{r}' = \vec{r} - \vec{u}\,t$.

b) The acceleration is a Galilean invariant.

c) The momentum transforms as $\vec{p}' = \vec{p} - m\vec{u}$.

d) Newton's second law follows the principle of relativity.

e) The net force on an object in a Galilean invariant.

f) The drag force on a moving non-spinning sphere is given by $\vec{F}_{\text{drag}} = -k\,v^2\,\hat{v}$. Similarly, the wind force on the same object when it is stationary is $\vec{F}_{\text{wind}} = k\,v_{\text{wind}}^2\,\hat{v}_{\text{wind}}$. Show the sum of the two forces is related to the object's speed and the wind speed by:

$\vec{F}_{\text{wind}} + \vec{F}_{\text{drag}} = k|\vec{v}_{\text{wind}} - \vec{v}|(\vec{v}_{\text{wind}} - \vec{v})$. Is the wind force, or the drag force, a Galilean invariant? Is their sum?

Thought Experiment 11.1: The Current Density in a Neutral Medium

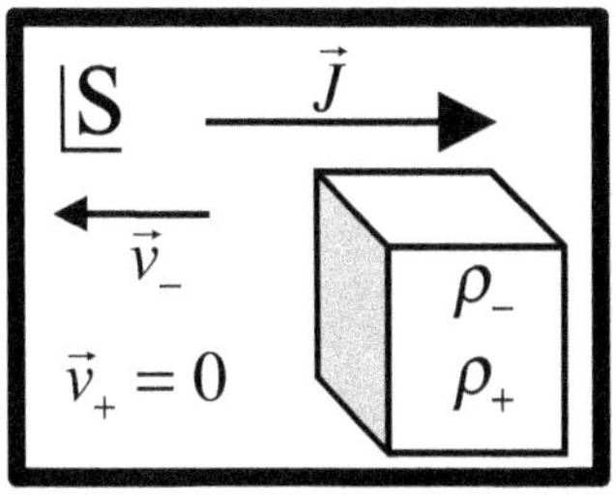

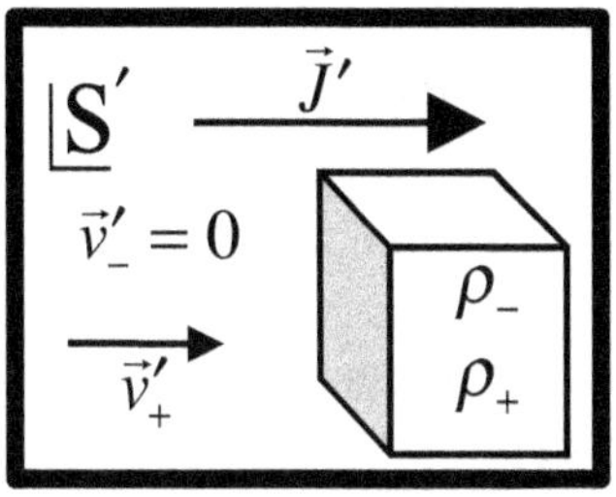

Does an observer drifting along with the electrons in a neutral conductor measure the same current density as we do in the rest frame of the conductor?

Consider both the positive and negative charges in the frame of the conductor, where:

$$\rho_- + \rho_+ = 0.$$

Since only the negative charge flows, we can write the corresponding current density as:

$$\vec{J} = \rho_+\vec{v}_+ + \rho_-\vec{v}_- = 0 + \rho_-\vec{v}_- = \rho_-\vec{v}_-.$$

Now consider the current density from the point of view of the observer in the moving reference frame:

$$\vec{J}' = \rho_+\vec{v}'_+ + \rho_-\vec{v}'_- = \rho_+\vec{v}'_+ + 0 = (-\rho_-)(-\vec{v}_-) = \vec{J}.$$

The answer to our question is a most definite "yes!" Note that from the point of view of the moving observer, $\vec{J}$ and $\vec{v}_+$ are parallel.

Thought Experiment 11.2: Galilean Transformation of the Current Density

In Chapter 1, we discovered that the current of a moving charged rod depends on the observer's reference frame, but the current in a neutral wire is frame independent. Here, we will show this in more general terms for a fluid of charge density ρ.

Using a two-fluid model of charge flow, we can write:

$$\rho = \rho_+ + \rho_-.$$

The net current density is related to the velocities of the positive and negative charge by:

$$\vec{J} = \rho_+\vec{v}_+ + \rho_-\vec{v}_-.$$

Now we transform to a reference frame $\lfloor\text{S}'$ moving at a speed $\vec{u}$ with respect to $\lfloor\text{S}$, so

$$\vec{v}'_+ = \vec{v}_+ - \vec{u} \text{ and } \vec{v}'_- = \vec{v}_- - \vec{u}.$$

So:

$$\vec{J}' = \rho_+\vec{v}'_+ + \rho_-\vec{v}'_- = \rho_+(\vec{v}_+ - \vec{u}) + \rho_-(\vec{v}_- - \vec{u}) = \rho_+\vec{v}_+ + \rho_-\vec{v}_- - (\rho_+ + \rho_-)\vec{u},$$

or simply:

$$\vec{J}' = \vec{J} - \rho\vec{u}.$$

Discussion 11.1: The Principle of Relativity

The last thought experiment touches on a significant theme in physics, which is that of transformations among reference frames. As we discussed in the introduction to this chapter, Galileo Galilei considered the thought experiment of someone in the hull of a ship moving uniformly through a calm sea to argue that all the laws of physics would be the same for someone moving at constant velocity as for someone not moving. There is only relative motion between objects, according to Galileo's argument, rather than absolute motion compared to an absolute reference frame.

As you have no doubt already experienced, this principle greatly simplifies problems in classical mechanics. Consider an elastic collision between a tennis ball and a tennis racquet. It is quite simple to transform to the rest frame of the racquet, reverse the direction of the ball, and then transform back to the original frame. Other common examples are those of boats in a current, and airplanes flying with respect to the prevailing winds. Each time you change reference frames, using common sense vector addition of velocities, you perform a Galilean transformation.

The one place in classical mechanics where there *appears* to be a preferred reference frame is wave motion through a medium, such as air or water. When we say that the speed of sound is approximately 343 m/s (for dry air at a temperature of 293 K), we mean that this is the speed at which sound waves travel for an observer at rest with respect to the air. Since we can deduce the speed of sound with respect to the medium by knowing such physical properties as its temperature, the speed of sound is characteristic of the medium itself.[150] Throughout the nineteenth century, electromagnetic waves were thought to behave like sound waves in this respect. Largely for this reason, scientists believed in the existence of a medium (i.e. the aether) which conveys the electromagnetic force and electromagnetic waves just as sound waves require the physical medium of air. As light is an electromagnetic wave, it seemed reasonable to presume that the characteristic propagation speed of the aether must be the speed of light.

Einstein argued in 1905 that the idea of a luminiferous aether is redundant at best. Such a concept is not necessary to explain light propagation and is incompatible with the principle of relativity. In fact, the main point of his paper, *On the Electrodynamics of Moving Bodies*, was that Galileo's *principle* of relativity must hold consistently throughout all of physics, regardless of whether we are talking about electrodynamical or mechanical phenomena. While Galileo's principle is correct, the simple mathematical rules that Galileo and Newton derived for transforming between inertial frames of reference turn out to be approximations, which are valid only for relative speeds that are small compared to the speed of light. The correct reference frame transformations for all systems are those that govern electrodynamics, not Newtonian mechanics.

As you continue your study of electrodynamics, keep in mind that the principle of relativity is not only a handy tool, but one of the most profound ideas in physics. Resolving the contradictions between this principle and physicists' conceptions of electromagnetism was one of the great early twentieth century breakthroughs in modern physics.

---

[150] For an observer moving with respect to the air, of course, the speed of sound is given by the usual velocity rule from the Galilean transformations. Chuck Yeager's breaking of the sound barrier depended on the relative velocity between his plane and the air, after all!

Fundamental Law 9: The Principle of Relativity

The laws governing the changes in the state of any physical system do not depend on which one of two coordinate systems, in uniform translational motion relative to each other, these changes of the state are referred to.[151]

Thus, the laws of nature are the same regardless of inertial reference frame.

Remember, however, that while the laws of physics apply equally in any constant velocity reference frame, you must still transform quantities when changing frames of reference.

Thought Experiment 11.3: Cycloid Motion of a Wheel

A car is driving at a constant speed v_0 and each wheel has a radius R. What is the velocity of a spot on the rim of the wheel as a function of time?

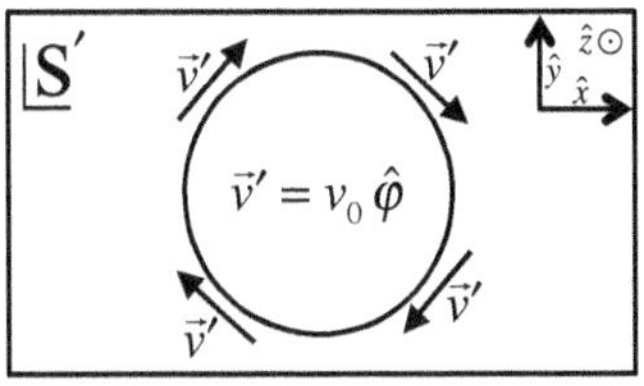

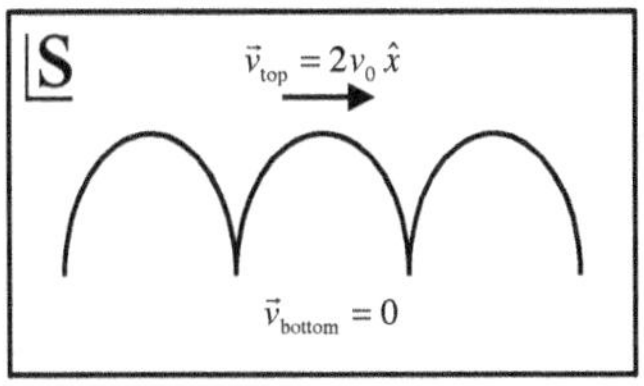

To answer this question we transform from the road reference frame $\lfloor s$ to a reference frame moving at constant velocity with the car $\lfloor s'$. In this frame the wheel is moving in a circle, so: $\vec{v}' = v_0\,\hat{\varphi}$. We denote the angular velocity by $\omega = \dfrac{v_0}{R}$, so that we have

$$\hat{\varphi}' = \hat{x}\sin(-\omega t) - \hat{y}\cos(-\omega t) = -\sin(\omega t)\hat{x} - \cos(\omega t)\hat{y},$$

$$\vec{v}' = v_0\left(-\sin(\omega t)\hat{x} - \cos(\omega t)\hat{y}\right) = -v_0\sin(\omega t)\hat{x} - v_0\cos(\omega t)\hat{y}.$$

Since the reference frame $\lfloor s'$ is moving with a velocity $\vec{u} = v_0\,\hat{x}$ with respect to $\lfloor s$, we can write:

$$\vec{v} = \vec{v}' + \vec{u} = \left(-v_0\sin(\omega t)\hat{x} - v_0\cos(\omega t)\hat{y}\right) + \left(v_0\,\hat{x}\right)$$
$$\vec{v} = v_0\left(1 - \sin(\omega t)\right)\hat{x} - v_0\cos(\omega t)\hat{y}.$$

Thought Experiment 11.4: The Electromagnetic Force in a Moving Frame

Consider a stationary charged particle in a magnetic field $\vec{B}$. There is no force on the particle.

Now let us transform to a reference frame $\lfloor s'$ moving at a constant velocity $(\vec{u})$. Since force is invariant under Galilean transformations, there must still be no force $\left(\vec{F}' = \vec{F} = 0\right)$.

Let us consider the Lorentz force in the moving reference frame. Since the particle moves with a velocity $\vec{v}' = -\vec{u}$, The force on the particle must be:

$$\vec{F}' = q\left(\vec{E}' + \vec{v}' \times \vec{B}'\right) = q\left(\vec{E}' - \vec{u} \times \vec{B}'\right).$$

Putting these two together, we find that:

$$\vec{F}' = 0 = q\left(\vec{E}' - \vec{u} \times \vec{B}'\right).$$

---

[151] A. Einstein, "On the Electrodynamics of Moving Bodies," Annalen der Physik **17** (1905), 891-921, translated by Anna Beck, 1989, Princeton Academic Press.

From this we conclude that the electric and magnetic fields transform when observed by a moving observer, and in this case $\vec{E}' = \vec{u} \times \vec{B}'$.

Thought Experiment 11.5: A Galilean Transformation of the Electric Field

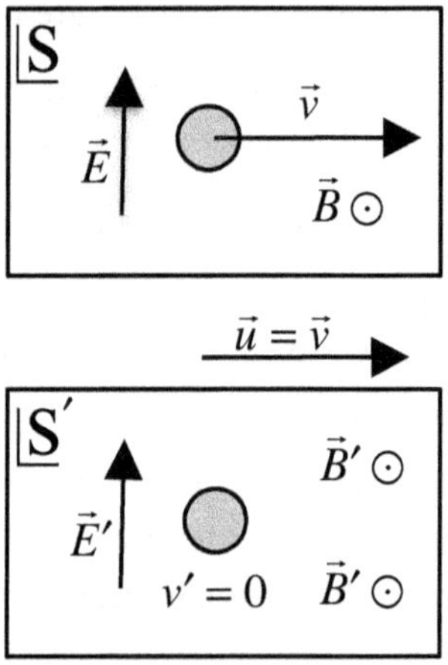

Consider a particle moving at a constant velocity $(\vec{v})$ in constant electric $(\vec{E})$ and magnetic $(\vec{B})$ fields, so:

$$\vec{F} = q\left(\vec{E} + \vec{v} \times \vec{B}\right).$$

Now consider the electromagnetic force in the rest frame of the particle $\lfloor S'$:

$$\vec{F}' = q\left(\vec{E}' + \vec{v}' \times \vec{B}'\right) = q\vec{E}'$$

Now in Newtonian physics $\vec{F}' = \vec{F}$, so:

$$\cancel{q}\vec{E}' = \cancel{q}\left(\vec{E} + \vec{v} \times \vec{B}\right)$$

$$\vec{E}' = \vec{E} + \vec{u} \times \vec{B}$$

Problem 11.2: Galilean Transformation of the Scalar Potential

By substituting in $\vec{E} = -\vec{\nabla}\mathbb{V} - \frac{\partial}{\partial t}\vec{\mathbb{A}}$, and making reasonable simplifying assumptions, show that the scalar potential transforms as:

$$\mathbb{V}' = \mathbb{V} - \vec{u} \cdot \vec{\mathbb{A}}.$$

Thought Experiment 11.6: The Magnetic Field is a Galilean Invariant

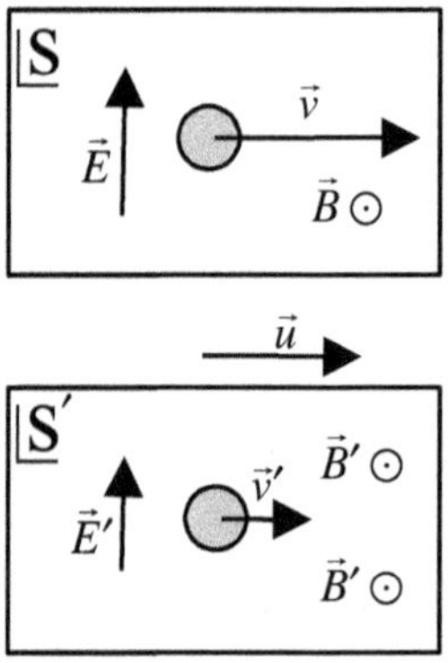

Now, let us look at the force on a particle moving at a velocity $\vec{v}$ in electric and magnetic fields $\vec{E}$ and $\vec{B}$ respectively. We will also consider the same problem in another inertial reference frame moving at a velocity $\vec{u}$. Thus:

$$\vec{F} = q\left(\vec{E} + \vec{v} \times \vec{B}\right) = \vec{F}' = q\left(\vec{E}' + \vec{v}' \times \vec{B}'\right) \quad \rightarrow \quad \vec{E}' + \vec{v}' \times \vec{B}' = \vec{E} + \vec{v} \times \vec{B}.$$

Substituting in $\vec{E}' = \vec{E} + \vec{u} \times \vec{B}$, we find that:

$$\left(\cancel{\vec{E}} + \vec{u} \times \vec{B}\right) + \vec{v}' \times \vec{B}' = \cancel{\vec{E}} + \vec{v} \times \vec{B} \quad \rightarrow \quad \vec{v}' \times \vec{B}' = \left(\vec{v} - \vec{u}\right) \times \vec{B}.$$

Thus the magnetic field must be a Galilean invariant—at least for now.

11.2 *The Breakdown of Galilean Relativity*

Until this chapter, we have not switched reference frames to solve problems, even when it might have made sense to do so. This is because electromagnetic quantities only approximately transform with Galilean transformations, so we avoided the issue by always keeping our analysis in the laboratory reference frame.

In this section, we will demonstrate that the Maxwell-Ampère law and the Lorentz force yield different Galilean transformations of the magnetic field. As long as the transformation velocities are small compared to the electromagnetic characteristic velocity, $c = \frac{1}{\sqrt{\mu_0 \varepsilon_0}}$, the differences do not matter. A number of physicists, including Oliver Heaviside, George Francis FitzGerald, and Hendrik Lorentz, derived a new set of transformations to be used for electrodynamics.

The paradox between Galileo's kinematic transformations, and the new electrodynamic transformations, bothered the young Albert Einstein. As we will discuss in Chapter 17, Einstein replaced the Galilean transformations with the Lorentz-FitzGerald transformations for basic kinematic calculations.

Thought Experiment 11.7: A Paradoxical Line Charge

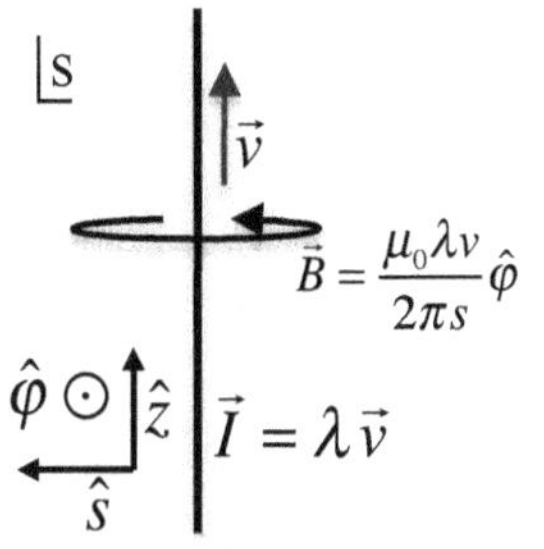

Consider a rod with a charge per length λ moving at a velocity $\vec{v}$ parallel to the rod in free space. From Gauss's and Ampère's laws, the electric and magnetic fields surrounding the wire are:

$$\vec{E} = \frac{\lambda}{2\pi\varepsilon_0 s}\hat{s}, \quad \vec{B} = \frac{\mu_0 \lambda v}{2\pi s}\hat{\varphi}.$$

Consider. now, the same rod from the point of view of an observer in a frame $\lfloor s'$ moving at velocity $\vec{u}$ with respect to the original frame $\lfloor s$. Let's use the Galilean transformations from Section 11.1 to express the electric and magnetic fields in $\lfloor s'$:

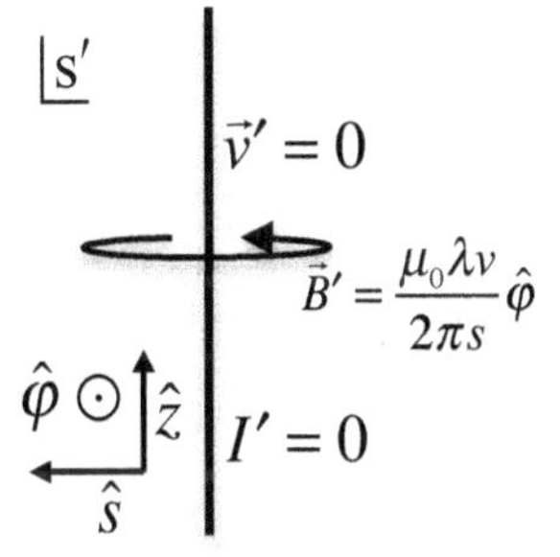

$$\vec{E}' = \frac{\lambda}{2\pi\varepsilon_0 s}\hat{s} + \vec{u} \times \frac{\mu_0 \lambda v}{2\pi s}\hat{\varphi} = \frac{\lambda}{2\pi s}\left(\frac{\hat{s}}{\varepsilon_0} + \mu_0 u v \hat{z} \times \hat{\varphi}\right) = \frac{\lambda}{2\pi\varepsilon_0 s}(1 - \mu_0 \varepsilon_0 u v)\hat{s},$$

$$\vec{B}' = \frac{\mu_0 \lambda v}{2\pi s}\hat{\varphi}.$$

So far, so good! Now let us consider the special case where $\lfloor S'$ is co-moving with the rod, so $\vec{u} = \vec{v} = v\hat{z}$, so we simply have a stationary rod with a charge per length λ:

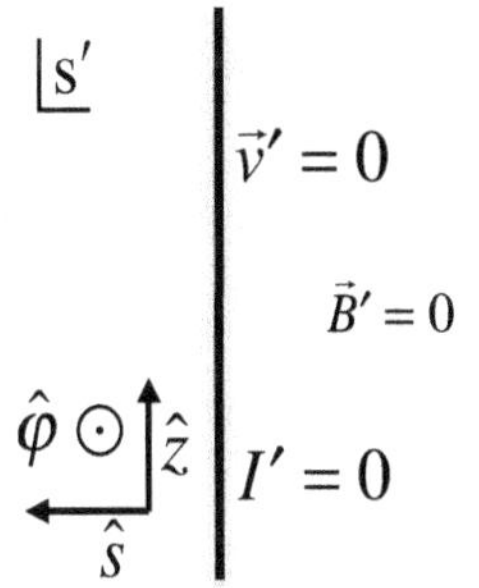

$$\vec{E}' = \frac{\lambda}{2\pi\varepsilon_0 s}\hat{s}, \quad \vec{B}' = 0.$$

We have arrived at a *contradiction*! The magnetic field should be the same regardless of the frame of reference. Yet, by applying the Galilean transformations we developed earlier, we have obtained a completely different answer for the magnetic field than before!

Most physicists of the day interpreted this as evidence for the existence of a preferred reference frame of the universe, which presumably contains an electrodynamic medium called *the aether*. We no longer believe in the existence of an aether, *per se*, but in 1965 radio astronomers Penzias and Wilson found did find a preferred reference frame of the universe, which is called *the cosmic microwave background radiation dipole anisotropy*.

Problem 11.3: A Moving Capacitor

Consider a large parallel plate capacitor, with a surface charge density on each plate of $\pm\sigma$ and plate separation h. The capacitor moves with a constant parallel velocity $\vec{v} = v\hat{x}$.

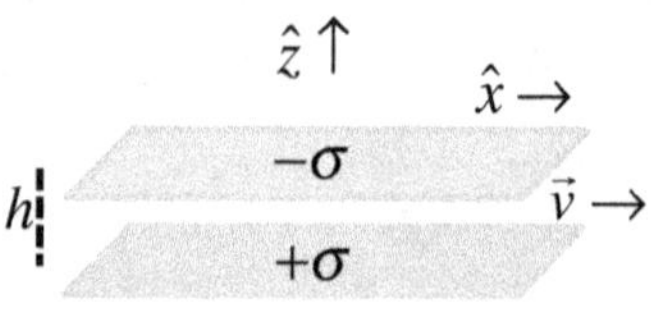

a) What is the associated current per perpendicular length associated with the top and bottom capacitor?

b) Use Ampère's law to find the magnetic field between the plates.

c) Write down the electric field in the rest frame of the capacitor, and transform this to the lab frame using the Galilean transformations

d) Find the magnetic field between the plates using the electric field in the laboratory frame.

e) Do your results from parts (b) and (d) agree? Discuss why or why not.

Thought Experiment 11.8: The Vector Potential Galilean Transformation

Consider now the potential formulation of Maxwell's equations, which for simplicity we will consider in free space. From Section 9.4 (p. 298), we can write them as:

The Lorenz Gauge	$\vec{\nabla}\cdot\vec{\mathrm{A}} = -\mu_0\varepsilon_0 \frac{\partial}{\partial t}\mathrm{V}$
Gauss's Law	$\left(\mu_0\varepsilon_0 \frac{\partial^2}{\partial t^2} - \nabla^2\right)\mathrm{V} = \frac{1}{\varepsilon_0}\rho - \frac{1}{\varepsilon_0}\vec{\nabla}\cdot\cancel{\vec{P}}$
The Maxwell-Ampère Law	$\left(\mu_0\varepsilon_0 \frac{\partial^2}{\partial t^2} - \nabla^2\right)\vec{\mathrm{A}} = \mu_0\vec{J} + \mu_0\varepsilon_0\frac{\partial}{\partial t}\cancel{\vec{P}} + \mu_0\vec{\nabla}\times\cancel{\vec{M}}$

Consider the Maxwell-Ampère law in the primed frame:

$$\left(\mu_0\varepsilon_0 \tfrac{\partial^2}{\partial t^2} - \nabla^2\right)\vec{\mathrm{A}}' = \mu_0\vec{J}' = \mu_0\left(\vec{J} - \rho\vec{u}\right) = \mu_0\vec{J} - \mu_0\rho\vec{u}.$$

Now we substitute for the current and charge densities:

$$\left(\mu_0\varepsilon_0 \tfrac{\partial^2}{\partial t^2} - \nabla^2\right)\vec{\mathrm{A}}' = \left(\mu_0\varepsilon_0 \tfrac{\partial^2}{\partial t^2} - \nabla^2\right)\vec{\mathrm{A}} - \mu_0\varepsilon_0\left(\mu_0\varepsilon_0 \tfrac{\partial^2}{\partial t^2} - \nabla^2\right)\mathrm{V}\vec{u} = \left(\mu_0\varepsilon_0 \tfrac{\partial^2}{\partial t^2} - \nabla^2\right)\left(\vec{\mathrm{A}} - \mu_0\varepsilon_0\mathrm{V}\vec{u}\right).$$

We now take the simplest solution as:

$$\vec{\mathrm{A}}' = \vec{\mathrm{A}} - \mu_0\varepsilon_0\mathrm{V}\vec{u}.$$

Problem 11.4: Transformation of the Charge Density

In this problem, you will test Galilean relativity against charge conservation.

a) From the potential formulation of Maxwell's equations, show that the charge density should transform as:

$$\rho' = \rho - \mu_0\varepsilon_0\vec{u}\cdot\vec{J}.$$

b) Show that, for time-dependent currents, this violates the conservation of charge.

Thought Experiment 11.9: The Galilean Transform of the Magnetic Field

We will take the curl of the vector potential to find the transform of the magnetic field:

$$\vec{B}' = \vec{\nabla} \times \vec{\mathbb{A}}' = \vec{\nabla} \times \left(\vec{\mathbb{A}} - \mu_0 \varepsilon_0 \mathbb{V} \vec{u} \right) = \vec{\nabla} \times \vec{\mathbb{A}} - \mu_0 \varepsilon_0 \vec{\nabla} \times \left(\mathbb{V} \vec{u} \right) = \vec{B} - \mu_0 \varepsilon_0 \vec{\nabla} \times \left(\mathbb{V} \vec{u} \right).$$

Now we will look at the last term:

$$\vec{\nabla} \times \left(\mathbb{V} \vec{u} \right) = \mathbb{V} \left(\vec{\nabla} \times \vec{u} \right) - \vec{u} \times \vec{\nabla} \mathbb{V} = -\vec{u} \times \left(-\vec{E} - \tfrac{\partial}{\partial t} \vec{\mathbb{A}} \right) = \vec{u} \times \left(\vec{E} + \tfrac{\partial}{\partial t} \vec{\mathbb{A}} \right).$$

Therefore:

$$\vec{B}' = \vec{B} - \mu_0 \varepsilon_0 \vec{u} \times \left(\vec{E} + \tfrac{\partial}{\partial t} \vec{\mathbb{A}} \right).$$

In steady state systems, this simplifies to:

$$\vec{B}' \approx \vec{B} - \mu_0 \varepsilon_0 \vec{u} \times \vec{E}.$$

Problem 11.5: The Magnetic Field Transformation

a) Reproduce the line charge analysis from Thought Experiment 11.7 (p. 337). Applying the new magnetic field transformation, discuss whether there is still a contradiction.

b) Show that $\vec{\nabla} \cdot \vec{B}' = 0$ still holds under the new Galilean transformation.

Problem 11.6: The Galilean Transformation of the Electromagnetic Force

a) Show that the time independent Galilean transformation of the electromagnetic force is:

$$\vec{F}' = \vec{F} - \mu_0 \varepsilon_0 (\vec{v} - \vec{u}) \times \left(\vec{u} \times q\vec{E} \right).$$

b) The second term contradicts Newtonian mechanics. At what velocity, $\vec{u}$, would the second term start becoming significant at the 1% level?

Discussion 11.2: Galilean Relativity and Modern Physics

No matter how we slice it, the Galilean transformations of the fields, when derived from their *effects*, contradict those same transformations derived from their *causes*. It was this contradiction of the Galilean transformations, when applied to electrodynamics, that eventually led to Albert Einstein's theory of special relativity.

The special theory of relativity does not aim at overthrowing classical physics so much as it attempts to preserve Galileo's principle of relativity and extend it to all of physics, especially electricity, magnetism, and optics. Classically, there is no limiting speed and no *a priori* reason why signals cannot be sent instantaneously from one location to another. In classical physics, moreover, the speed of any signal depends on the relative motion of the source and the receiver of that signal. But according to experiment, light signals just don't behave that way. Light traveling in a vacuum does so at a fixed finite speed that is independent of the motion of its source or observer. Einstein's attempt at reconciling this fact with the principle of relativity led him to posit a new set of transformation laws, known as the Lorentz-FitzGerald transformations, between inertial frames of reference. In the limit of low speeds (i.e. speeds which are much less than the speed of light), the Lorentz-FitzGerald transformations reduce to the Galilean transformations that we discussed earlier. Nevertheless, even for slow velocities, the results of applying the Galilean transformations are inconsistent with those derived from the laws governing the causes of magnetic fields. We will discuss this in more detail, though it requires Einstein's special theory of relativity to resolve fully these apparent inconsistencies.

Although the correct equations that connect one inertial reference frame to another are more complicated than the equations of Galilean relativity, the special theory of relativity preserves Galileo's original insight: the laws governing the changes in the state of any physical system do not depend on the choice of inertial reference frame. In fact, the special theory of relativity extends Galileo's principle from mechanics to all of physics.

Problem 11.7: The FitzGerald Transformations

The Irish physicist, George Francis FitzGerald, pointed out in 1889 that the Galilean relativity paradox could be avoided, if lengths in the aether foreshorten in the direction of motion, but not sideways, by a factor of $\sqrt{1-\mu_0\varepsilon_0 u^2}$.

a) At what speed would the object become flat?

b) Give a qualitative explanation of how lengths could shorten, from the point of view of nineteenth century aether theory. (How we explain it today is the topic of Chapter 17.)

c) Consider an object of charge q, and volume $V = \vec{A}\cdot\vec{h}$, transformed to another frame moving at a velocity $\vec{u}$ such that $\vec{u}\cdot\vec{h} = 0$. Show that FitzGerald's transformations of the charge and current densities must have been:

$$\rho' = \frac{\rho - \mu_0\varepsilon_0 \vec{u}\cdot\vec{J}}{\sqrt{1-\mu_0\varepsilon_0 u^2}}.$$

d) Show that, if the current density is perpendicular to the direction of transform, then:

$$\vec{J}'_\perp = \frac{\vec{J}_\perp}{\sqrt{1-\mu_0\varepsilon_0 u^2}}.$$

Table 11.1: The Galilean Transformations[152]		
Quantity	$\lfloor S$	$\lfloor S'$
Velocity of frame	0	$\vec{u}$
Velocity	$\vec{v}$	$\vec{v}-\vec{u}$
Electric field	$\vec{E}$	$\vec{E}+\vec{u}\times\vec{B}$
Magnetic field	$\vec{B}$	$\vec{B}-\mu_0\varepsilon_0\,\vec{u}\times\vec{E}$
Scalar Potential	V	$\mathrm{V}-\vec{u}\cdot\vec{\mathrm{A}}$
Vector Potential	$\vec{\mathrm{A}}$	$\vec{\mathrm{A}}-\mu_0\varepsilon_0\mathrm{V}\vec{u}$
Charge density	ρ	$\rho-\mu_0\varepsilon_0\,\vec{u}\cdot\vec{J}$
Current Density	$\vec{J}$	$\vec{J}-\rho\vec{u}$

152 These transformations are still not quite consistent with a Galilean transformation first to frame $\lfloor S'$, followed by a calculation of the electric and magnetic fields. However, they are close so long as: $\mu_0\varepsilon_0 u^2 << 1$.

Problem 11.8: Galileo's Cannon

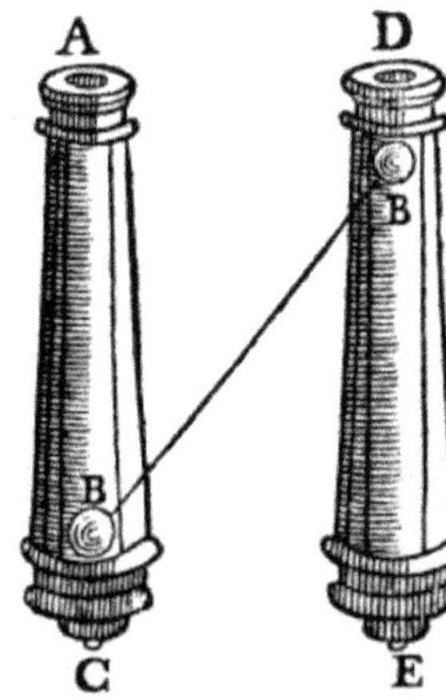

Galileo's illustration[153] shows a thought experiment involving a cannon that was used to argue for a moving earth. Consider a stationary cannon on the earth that is shot straight up in the air. The cannonball will come back down into the cannon. On the other hand, if the earth's surface is moving eastward at a velocity $\vec{u} = u\,\hat{x}$, then the cannon will be moving sideways too at the same speed. Thus, the cannonball will not be shot straight up, but rather at the angle drawn in the figure (B to B). Because everything else on the earth is also moving with the same velocity, Galileo argues, the cannonball would come back down into the cannon regardless of which reference frame is used to calculate its motion.

a) In the stationary reference frame, calculate the motion of both the ball (B) and the cannon as a function of time. Use Galileo's law of gravity, that the acceleration of the cannon ball is constant and downward, $\vec{a} = -g\,\hat{z}$. Show that the ball lands back into the cannon in this reference frame. You can assume that the earth is locally flat.

b) Consider now the Earth's magnetic field, which has horizontally northward (world-wide) and vertically downward (northern hemisphere) components, or: $\vec{B} = B_H\,\hat{y} - B_V\,\hat{z}$. Find the induced electric field in the stationary reference frame.

c) Find the force on a small charge q inside the cannonball as a function of time also in the stationary reference frame.

d) Show that the force on the same charge as calculated in the reference frame moving with the earth's surface is the same as your answer in part c.

Problem 11.9: A Moving Dielectric Rod

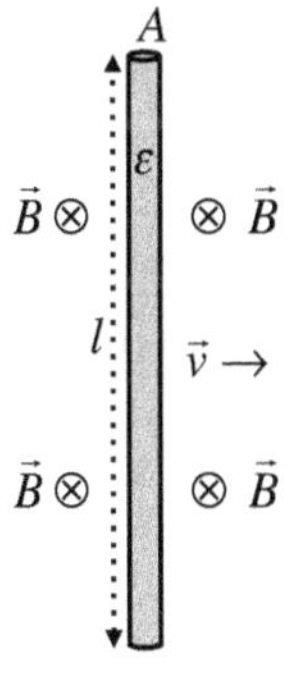

Consider a rod with permittivity ε, length l, and cross-sectional area A moving with a constant velocity $\vec{v}$ perpendicular to a uniform magnetic field $\vec{B}$.

a) What is the polarization of the rod?

b) What is the dipole moment of the rod?

c) If you model the rod as a single dumbbell dipole, what would be the charge $\pm q$ at each end?

Problem 11.10: A Velocity Selector

Consider a collimated beam of unknown charged particles. You wish to build a filter box to select only those particles with a particular velocity. To do this, you create perpendicular magnetic and electric fields, which are also perpendicular to the particle beam. If all goes well, only particles of a particular velocity will not be deflected by the electric and magnetic fields.

---

153 Galileo Galilei, Dialogue Concerning the Two Chief World Systems: Ptolemy and Copernicus, (1632); (Translated by Stillman Drake, University of California Press, 1953), 126; Second Day).

(a) What is the relationship between this velocity and the magnetic and electric fields? Draw a picture, and show your work.

(b) Now consider a new reference frame moving at the selected velocity v. What is the electric field in this reference frame?

(c) After passing through the fields, the beam passes through an opening at the end of box. Your task is to let only those particles within a velocity range $v \pm \delta v$ pass through the aperture. For a given aperture width a and path length b, solve for the electric and magnetic fields which are needed to select a velocity v within a tolerance δv.

Problem 11.11: A Rotating Dielectric Rod

Consider a dielectric rod with permittivity ε, length a, cross-sectional area A, is forced to rotate with constant angular velocity $\vec{\omega}$ perpendicular to a uniform magnetic field $\vec{B}$.

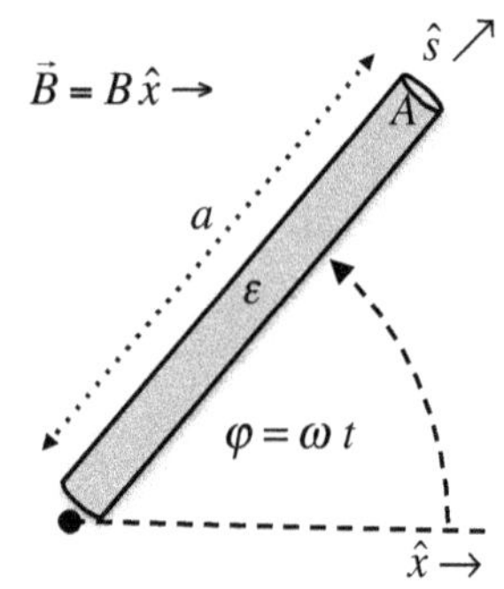

a) What is the electric field, in the reference frame of the rod, as a function of time and position s?

b) Find the polarization of the rod as a function of position and time.

c) What is the total induced electric dipole moment of the rod?

d) Find the total torque on the rod as a function of time.

Thought Experiment 11.10: The Free Particle Simulator

Recall Thought Experiment 10.5 (p. 318) where we considered a particle of mass m and positive charge q that is initially at rest with respect to an inertial reference frame $\lfloor \mathrm{S}$, and which is placed in an electric field $\vec{E} = E\hat{y}$ and a magnetic field $\vec{B} = B\hat{z}$. We algorithmically solved the problem in the laboratory reference frame, and we also found that the system appeared to roll along, as in Thought Experiment 11.3 (p. 335), with an average velocity:

$$\langle \vec{v} \rangle \approx \frac{\vec{E} \times \vec{B}}{B^2}.$$

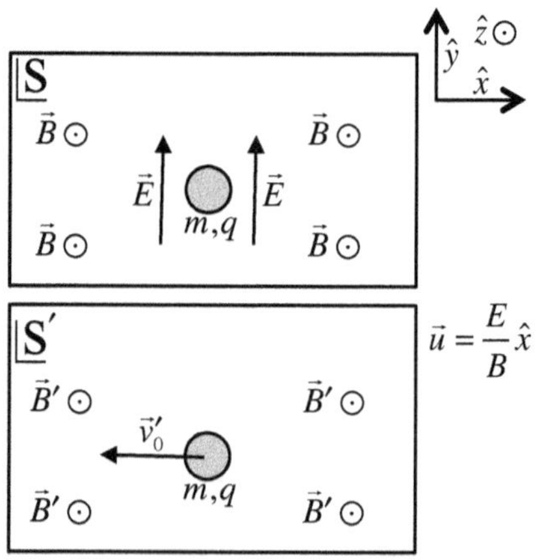

Now that we have some experience with this type of problem, we will find the motion of the particle analytically using the principles of Galilean relativity.

To solve this problem, first transform to a reference frame $\lfloor \mathrm{S}'$ that moves with uniform velocity $\vec{u}$ with respect to frame $\lfloor \mathrm{S}$, and in which there is no electric field:

$$\vec{E}' = 0 = \vec{E} + \vec{u} \times \vec{B} \quad \rightarrow \quad 0 = E\hat{y} + uB(\hat{x} \times \hat{z}) = E\hat{y} - uB\hat{y}$$

$$\vec{u} = \frac{E}{B}\hat{x}.$$

Next we will transform into the comoving reference frame $\lfloor \mathrm{S}'$, solve the problem, and transform back to the laboratory frame $\lfloor \mathrm{S}$.

In frame $\lfloor S'$ we can write the initial velocity of the particle $\vec{v}_0'$, the magnetic field $\vec{B}'$, and the electric field $\vec{E}'$:

$$\vec{v}_0' = \vec{v}_0 - \vec{u} = 0 - \frac{E}{B}\hat{x}, \quad \vec{B}' = \vec{B} = B\hat{z}, \quad \text{and} \quad \vec{E}' = \vec{E} + \vec{u}\times\vec{B} = 0.$$

Now the particle is in circular motion with constant speed $v' = E/B$ and gyration radius given by:

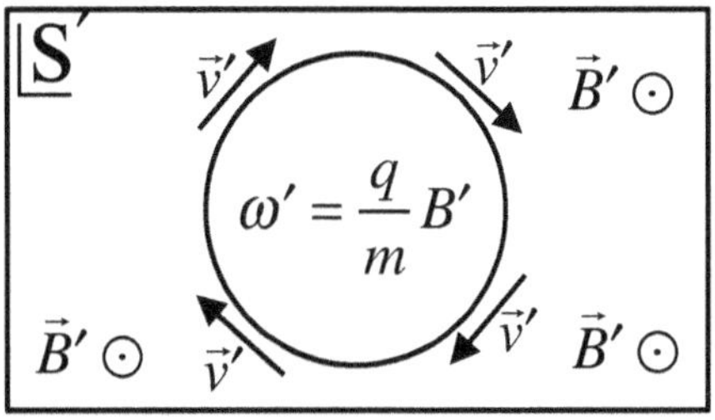

$$R' = \frac{mv'}{qB'} = \frac{mE}{qB^2}$$

and an angular velocity of:

$$\omega' = \frac{v'}{R'} = \frac{q}{m}B = \omega_g,$$

which is called the *gyrofrequency*. Thus the position, velocity, and acceleration are given by:

$$\vec{r}' = R'\left(-\sin(\omega' t)\hat{x} - \cos(\omega' t)\hat{y}\right),$$

$$\vec{v}' = R'\omega'\left(-\cos(\omega' t)\hat{x} + \sin(\omega' t)\hat{y}\right), \quad \text{and}$$

$$\vec{a}' = R'\omega'^2\left(\sin(\omega' t)\hat{x} + \cos(\omega' t)\hat{y}\right).$$

Transforming back to the lab reference frame:

$$\vec{r} = \vec{r}' + \vec{u}t = \left(ut - R'\sin(\omega' t)\right)\hat{x} - R'\cos(\omega' t)\hat{y} = \left(\frac{E}{B}t - \frac{mE}{qB^2}\sin(\omega_g t)\right)\hat{x} - \frac{mE}{qB^2}\cos(\omega_g t)\hat{y},$$

$$\vec{v} = \vec{v}' + \vec{u} = \left(u - R'\omega'\cos(\omega' t)\right)\hat{x} + R'\omega'\sin(\omega' t)\hat{y} = \frac{E}{B}\left(1 - \cos(\omega_g t)\right)\hat{x} + \frac{E}{B}\sin(\omega_g t)\hat{y}, \quad \text{and}$$

$$\vec{a} = \vec{a}' = \frac{qE}{m}\left(\sin(\omega' t)\hat{x} + \cos(\omega' t)\hat{y}\right) = \frac{qE}{m}\left(\sin(\omega_g t)\hat{x} + \cos(\omega_g t)\hat{y}\right).$$

This is the motion of a cycloid, as in Thought Experiment 11.3 on page 335.

Now, we will check our answer using just frame $\lfloor S$:

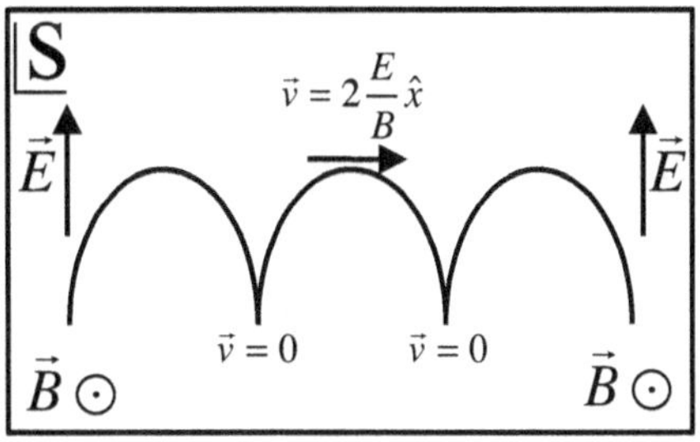

$$a_x \stackrel{?}{=} \tfrac{q}{m} B\, v_y$$

$$\frac{qE}{m}\sin(\omega_g t) = \tfrac{q}{m}\cancel{B}\left(\frac{E}{\cancel{B}}\sin(\omega_g t)\right)$$

$$a_y \stackrel{?}{=} \tfrac{q}{m}\left(E - B\, v_x\right)$$

$$\frac{qE}{m}\cos(\omega_g t) \stackrel{?}{=} \frac{q}{m}\left(E - \cancel{B}\left(\frac{E}{\cancel{B}}\left(1 - \cos(\omega_g t)\right)\right)\right)$$

$$\frac{qE}{m}\cos(\omega_g t) = \frac{qE}{m}\cos(\omega_g t)$$

Thus, our solution is consistent with solving the problem in the laboratory frame directly.

Problem 11.12: Shock Acceleration

The leading model for the origin of cosmic rays involves charged particles gaining energy by crossing a shock front. Consider the following scenario. A particle of mass m, charge q, and velocity $\vec{v}_0 = -v_0\,\hat{x}$ crosses into a fluid with a magnetic field $\vec{B} = B\,\hat{y}$ moving with a wind speed $\vec{v}_w = v_w\,\hat{x}$.

(a) Find the velocity of the particle after it has completed a 180° turn and left the fluid boundary.

(b) How much energy, ΔE, did the particle gain in the interaction?

(c) When the particle leaves the moving fluid it may also encounter another magnetic field and make another 180° turn. Thus, it could encounter the same moving fluid multiple times. Consider a particle initially with a velocity equal but opposite to the wind speed $(\vec{v}_0 = -\vec{v}_w)$. Plot its energy as a function of the number of times N that it encounters the moving magnetic fluid.

(d) Assuming both magnetic fields are equal, how much time will it take to increase the particle's energy by a factor of 100?

Problem 11.13: A Magnetic Bullet

Consider a magnetic bullet, with velocity $\vec{v} = v\,\hat{z}$. We will assume, for simplicity, that the dipole moment is parallel to the direction of motion, so the magnetic field in the reference frame of the bullet is given by the following relationship:

$$\vec{B} \propto \frac{3\vec{r}\,(\hat{z}\cdot\vec{r})}{r^5} - \frac{\hat{z}}{r^3}.$$

(a) Assume it is shot from a musket, so you can ignore spin. Find the induced electric field surround the bullet in the lab frame, and express your answer in spherical polar coordinates.

(b) Now assume that it was shot from a rifle, so it is spinning about an axis which coincides with both the magnetic moment and the direction of motion. Assume a known angular velocity $\vec{\omega} = \omega\,\hat{z}$. Again, find the induced electric field surround the bullet in the lab frame and express your answer in spherical polar coordinates.

11.3 *Motional EMF*

In Chapter 9 we discussed Michael Faraday's induction experiments, his corresponding law. Here we will discuss magnetic induction a different way, called *motional EMF*. By the end of this section, we will argue that the two are equivalent, and so you could choose which to use depending on the problem you are expected to solve.

Consider, for example, a neutral conducting wire that moves between the poles of a permanent magnet. In one frame of reference, that of the magnet, there is only a magnetic field. In the rest frame of the wire, however, there are both electric and magnetic fields. According to Ohm's law, this *induced electric field* will cause a current to flow through the conductor.

When one integrates this induced electric field around a loop, the result corresponds to the voltage gain around the loop due to this induced electric field. In practice, this voltage is offset by a voltage drop somewhere else, like a load resistor.

As we now have current flowing in a magnetic field, there will be a force on the wire, which will always oppose the motion of the wire. This fact, that the induced current always resists external motion, is called *Lenz's law*. We finish this section with the application of magnetic backpressure on a moving, conducting, and charged fluid.

Thought Experiment 11.11: The Induced Electric Field

Consider a large neutral metal bar with finite conductivity (κ) moving through a constant magnetic field $(\vec{B})$ at constant velocity $(\vec{v})$. What is the current density in the bar?

To see the effect on the bar, let us define our two reference frames as follows:

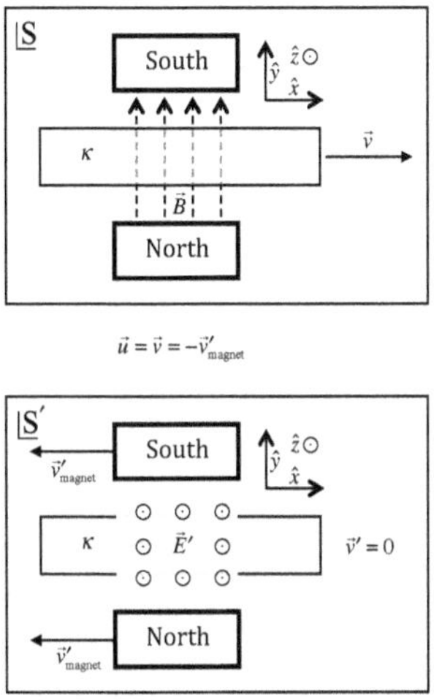

$\lfloor$S Frame of the Magnet

$\lfloor$S′ Frame of the Metal Bar

So we can see that $\vec{u}=\vec{v}$, because $v'=0$.

Now, we will transform the electric and magnetic fields to the bar.

$$\vec{B}'=\vec{B};\ \vec{v}'=\vec{v}-\vec{u}=0$$

$$\vec{E}'=\vec{E}+\vec{u}\times\vec{B}'=\vec{v}\times\vec{B}$$

Now we can apply Ohm's Law to the metal bar, so

$$\vec{J}'=\kappa\vec{E}'=\kappa\vec{v}\times\vec{B}.$$

Transforming back into the lab frame we find that

$$\vec{J}=\vec{J}'+\rho\vec{u}=\kappa\vec{v}\times\vec{B},$$

because the bar is net neutral.

Problem 11.14: An Azimuthal Magnetic Field

Consider an azimuthal magnetic field:

$$\vec{B}=\frac{\mu_0 I}{2\pi s}\hat{\varphi}.$$

(a) Verify that $\vec{\nabla}\cdot\vec{B}=0$ for this magnetic field.

(b) Consider an observer whose position as a function of time is given by: $\vec{r}=s_0\,\hat{s}+v_z t\,\hat{z}$. What is the electric field $\vec{E}'(t)$ in the reference frame $\lfloor$S′ of the observer?

(c) Now consider another observer moving in the radial direction, such that: $\vec{r}=v_s t\,\hat{s}$. What is the electric field $\vec{E}''(t)$ in the reference frame $\lfloor$S″ at the location of this observer?

Problem 11.15: An Airplane

Consider an airplane traveling due East at a constant speed $\vec{v}=v\hat{x}$. In the northern hemisphere, the magnetic field of the earth is northward and downward, so we will represent it by:

$\vec{B} = B_H \hat{y} - B_V \hat{x}$,

where B_H and B_V are the horizontal and vertical components of the magnetic field.

(a) Consider a charge q traveling with the aircraft, what is the force on this charge?

(b) Consider the reference frame of the plane, $\mathbb{S}'$. What is the induced electric field in this reference frame?

(c) Is the force acting on the charge the same, regardless of reference frame, as Galilean relativity would imply?

Problem 11.16: A Spinning Magnet

Consider a small magnet with magnetic moment $\vec{m} = m\hat{z}$ spinning about its magnetic axis with an angular velocity $\vec{\omega} = \omega\hat{z}$.

a) Express the vector potential of the magnet in its own reference frame.

b) Find the scalar potential in the laboratory frame, in polar coordinates.

c) Express the magnetic field of the magnet in its own reference frame.

d) Find the magnetic field in the laboratory frame, in polar coordinates.

e) The magnetic field at the earth's equator is about 40 microtesla. Estimate the magnetic moment of the earth, if it were a simple magnet.

f) The Earth has a natural electric field of about 150 volts per meter. Could this be predicted from a spinning magnet? Explain.

Problem 11.17: A Falling Bar

A cylindrical bar of mass m, conductivity κ, radius a, and length b is connected to two frictionless conducting vertical guides of height h so it is free to fall vertically. The whole apparatus is located in a perpendicular horizontal magnetic field. Each of the supports is connected to perfectly conducting wires, which are, in turn, connected the terminals of a switch.

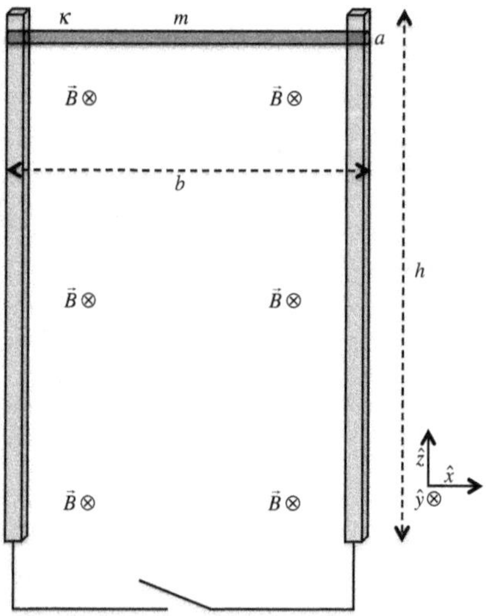

a) If the switch is open, find the position of the bar z as a function of time while it is falling from the height h under only the influence of the Earth's gravitational field $\vec{g} = -g\hat{z}$.

b) If the switch is closed and the field is strong, the bar will fall at a constant velocity. In this case, what is the position as a function of time?

c) If the switch is closed, and the field is not necessarily strong, the acceleration of the rod will depend on its velocity. In this case, find the equation of motion relating the downward speed v to the downward acceleration $a = \frac{dv}{dt}$. For convenience, write this in terms of the gravitational field g and a characteristic time τ, which you must define in terms of the given constant parameters.

d) Solve this equation for the position z as a function of time. In the limit that the magnetic field is small, does it agree with part (a)? In the limit that the field is large, does it agree with (b)?

Thought Experiment 11.12: Lenz's Law

In this thought experiment, we consider the forces on the same bar as in Thought Experiment 11.11 and find the acceleration of the bar, assuming there are no external forces.

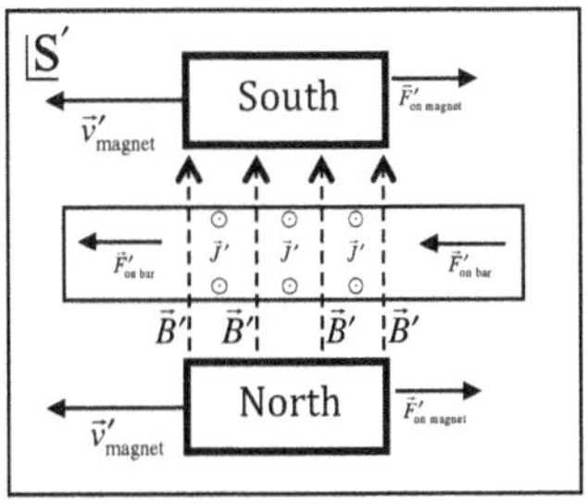

Now in the moving frame, we have a current-carrying bar inside a magnetic field, so there must be a body force on it.

$\vec{f}' = \vec{J}' \times \vec{B} = \vec{J} \times \vec{B}$,
because the body force is a Galilean invariant.

For a bar with mass density ρ_m, we can apply Newton's second law, and Ohm's law, to find the acceleration:

$$\vec{a} = \frac{\vec{F}}{m} = \frac{\frac{d\vec{F}}{dV}}{\frac{dm}{dV}} = \frac{\kappa\left(\vec{v} \times \vec{B}\right) \times \vec{B}}{\rho_m}.$$

Now we can simplify this using the "back cab" vector identity:

$$\vec{A} \times \left(\vec{B} \times \vec{C}\right) = \vec{B}\left(\vec{A} \cdot \vec{C}\right) - \vec{C}\left(\vec{A} \cdot \vec{B}\right).$$

So the acceleration becomes:

$$\vec{a} = -\frac{\kappa}{\rho_m}\vec{B} \times \left(\vec{v} \times \vec{B}\right) = -\frac{\kappa}{\rho_m}\left(\vec{v}\left(\vec{B} \cdot \vec{B}\right) - \vec{B}\left(\vec{B} \cdot \vec{v}\right)\right) = -\frac{\kappa}{\rho_m}\left(B^2 \vec{v}\right),$$

as $\vec{B} \perp \vec{v}$.

Thus the metal block slows down in the field. This makes sense, because frictional forces in the metal bar dissipate energy. The principle that the induced current is always in a direction that will resist the motion is commonly referred to as *Lenz's law*, after H. F. Emil Lenz who published it in 1834.[154]

Taking this idea one step further, the power dissipated in the bar per volume is given by

$$\mathbb{p} = \frac{d\mathbb{P}}{dV} = \frac{d\left(\vec{F} \cdot \vec{v}\right)}{dV} = \vec{f} \cdot \vec{v} = \left(-\kappa B^2 \vec{v}\right) \cdot \vec{v} = -\kappa B^2 v^2.$$

Notice that if the bar is connected to an open circuit, no current flows. Closing the circuit allows the current to flow freely, thus providing a braking force on the object. This is the basic principle behind magnetic braking in electric vehicles such as subway trains. Rather than the subway's kinetic energy being dissipated as heat with every stop, as would happen with a resistor, the subway's braking system returns the energy released in braking as electricity to the power grid, which makes it an extremely efficient form of urban transportation.

---

154 Lenz, Heinrich Emil, "Über die Bestimmung der Richtung der durch elektrodynamische Vertheilung erregten galvanischen Strömen," Annalen der Physik **31** (1834), 483-94.

Notice also that the more conductive the metal is, the more power is dissipated in the bar. This may seem odd, because one would think that there would be less friction. However, consider the limiting cases. If you move a non-conducting material through a magnetic field, you would expect that it would not be affected by the magnetic field at all. However, a superconducting body not only resists the motion, but the magnetic force remains even after the body halts because the currents that continue to flow within the superconductor, as we will discuss in Chapter 12.

Thought Experiment 11.13: Magnetic Backpressure

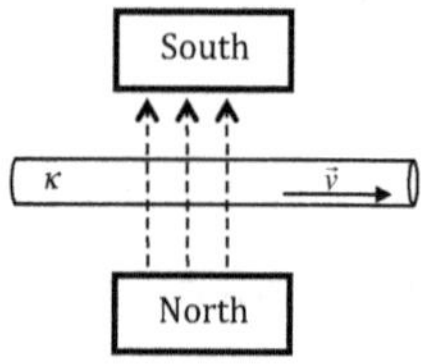

A fluid with conductivity κ and velocity $\vec{v}$ flows in a pipe of length l, perpendicular to a magnetic field $\vec{B}$. What is the backpressure on the fluid?

The force per volume on the fluid is:

$$\vec{f} = \kappa\left(\vec{v}\times\vec{B}\right)\times\vec{B} = -\kappa B^2\,\vec{v} + \kappa\left(\vec{v}\cdot\vec{B}\right)\vec{B}.$$

Again, the second term goes to zero if $\vec{v}$ is perpendicular to $\vec{B}$, as in the picture. Now let us look at a fluid element of length $\delta\vec{\ell}$ and of cross-sectional area $\delta\vec{A}$, so $\delta V = \delta\vec{A}\cdot\delta\vec{\ell}$. The force on the volume element is $\delta\vec{F} = -\kappa B^2\vec{v}\left(\delta\vec{A}\cdot\delta\vec{\ell}\right)$, so we write the magnitude of the force per area, in the backward direction, as:

$$\vec{P} = \frac{\delta\vec{F}}{\delta A} = -\int\kappa B^2\,\vec{v}\,d\ell.$$

In a pipe with constant length l traveling at a constant velocity v in a constant perpendicular magnetic field B, the backpressure is:

$$P_{\text{back}} = \kappa B^2 v l.$$

Problem 11.18: A Magic Pipe

In a magic trick, the magician allows an audience member to drop a slug through a hollow copper pipe, and it is seen to fall out the other end in a rather short time $t \approx \sqrt{\frac{2l}{g}}$. The magician then says a few magic words and makes a slight of hand replacement of the standard slug with a magnetic one. It then takes much longer for the magnetic slug to fall through the pipe this time—apparently defying gravity. The pipe has length l, conductivity κ, and inner and outer radii a and b respectively. Each slug has a mass M and length h, and it takes a time t for the magnetic slug to fall through the pipe.

a) What is rate of gravitational energy loss as the slug falls?

b) What is the power dissipated in the pipe?

c) Find the average electric field in the reference frame of the pipe.

d) Find the current density in the reference frame of the pipe.

e) Find the average magnetic field in the slug's reference frame.

f) Find the magnetic moment of the slug.

Problem 11.19: A Conducting Sphere in a Magnetic Field

Consider a solid sphere of mass M, radius a, and conductivity κ. This sphere is spinning about the $\hat{y}$ axis with an angular velocity $\vec{\omega} = \omega \hat{y}$ in a constant magnetic field $\vec{B} = -B\hat{x}$.

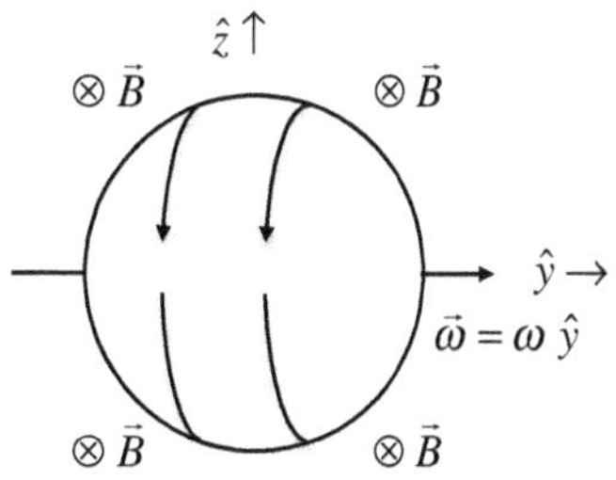

a) Find the induced electric field in the ball as a function of position, and express this in spherical coordinates.

b) Assuming that the current flows in circular loops, calculate the voltage around each loop. Express this as a function of r and θ.

c) Find the current density in the ball as a function of r and θ.

d) What is the total power generated by this ball?

e) Assuming that the ball started with an initial angular velocity ω_0, and there are no additional torques, what is the angular velocity as a function of time?

Problem 11.20: A Rotating Rod

A rod of length a, mass M, cross-sectional area A, and electrical conductivity κ rotates about one end with an angular frequency ω in a perpendicular constant and uniform magnetic field. The rod slides along a perfectly conducting ring as it rotates, which is connected in series to a load resistor R.

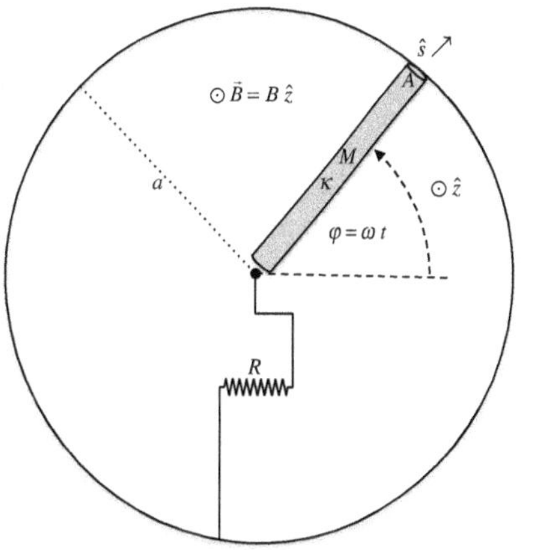

a) Find the induced electric field in the rod.

b) Find the induced voltage across the rod.

c) Find the total current flowing through the circuit.

d) Find the voltage across the resistor.

e) Find the net electric field inside the rod.

f) Find the total backward torque on the rod.

g) Assuming the rod started along the $\hat{x}$ axis with an initial angular velocity ω_0, find its angular position φ as a function of time.

Problem 11.21: A Falling Hoop of Metal

Consider a horizontal hoop of metal with radius a, conductivity κ, mass density ρ_m, and total mass M. It is dropped from a height h, with the axis of the hoop in the vertical ($\hat{z}$) direction, in a magnetic field given by:

$$\vec{B} = B_0\left(\frac{b}{s} + \frac{s}{2b}\right)\hat{s} - B_0\left(1 + \frac{z}{b}\right)\hat{z}.$$

a) What is the total current that flows through the hoop as a function of velocity?

b) What is the upward force as a function of hoop velocity?

c) What is the acceleration of the hoop as a function of hoop velocity?

d) What is the velocity of the hoop as a function of time?

e) What is the terminal velocity of the hoop?

Problem 11.22: Faraday's Waterloo Bridge Experiment

In 1832 Michael Faraday proposed that power could be generated by harnessing the flow of the Thames river directly,[155] which led to an unsuccessful experiment at the Waterloo bridge. The idea was to place two large vertical copper plates on either bank of the river and then connect them, and measure the current with a galvanometer. Assume that you know the conductivity of the river κ, the area of the plates A, and the width of the river w. London is in the Northern Hemisphere, so the Earth's magnetic field has a Northward horizontal component B_H, and a downward vertical component B_V (i.e. $\vec{B} = B_H\,\hat{y} - B_V\,\hat{z}$). The Thames flows approximately northeast under the Waterloo bridge, and we will assume that it flows twice as fast in the center than along the bank with a quadratic function, so: $\vec{v} = v_0\left(1 - \frac{2y^2}{w^2}\right)\hat{x}$, where x is the position from the middle of the river.

a) What is the voltage difference between the plates?

b) How much total current flows through the galvanometer if it has an internal resistance R?

c) Faraday tried unsuccessfully to measure the current flow through his galvanometer. Research and discuss why he may have been unsuccessful.

Thought Experiment 11.14: Motional EMF and Faraday's Law

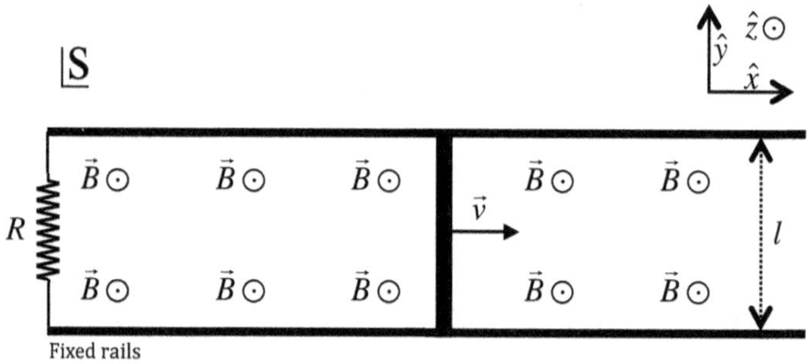

Consider a conducting bar that is being pulled along at a constant velocity $(\vec{v})$ inside a magnetic field $(\vec{B})$. Current flowing through the bar of length (l) can only complete a circuit by flowing through a load resistor (R).

a) How much power is dissipated from the resistor?

The force per charge in the moving wire is:

---

155 Michael Faraday, Experimental Researches in Electricity, 3 vols. (London: Taylor, 1839, 1844; Taylor and Francis, 1855) **I**: para., 188, 55.

$$\vec{F}/q = \cancel{\vec{E}} + \vec{v}\times\vec{B} = (v\hat{x})\times(B\hat{z}) = vB(\hat{x}\times\hat{z}) = -vB\hat{y}\ .$$

The work done, per charge, across the wire is therefore

$$\Delta \mathbb{V} = \mathbb{V}_{top} - \mathbb{V}_{bottom} = -\int_0^l \frac{\vec{F}}{q}\cdot d\vec{\ell} = -\int_0^l (-vB)\,dy = vBl\ .$$

Now applying the loop rule, the voltage drop across the resistor must also be $\Delta\mathbb{V}$ so the power dissipated at the resistor is:

$$P_{out} = I\Delta\mathbb{V} = \left(\frac{\Delta\mathbb{V}}{R}\right)(\Delta\mathbb{V}) = \frac{(\Delta\mathbb{V})^2}{R} = \frac{(vBl)^2}{R}\ .$$

b) What external force must be applied to the bar to move it at a constant velocity?

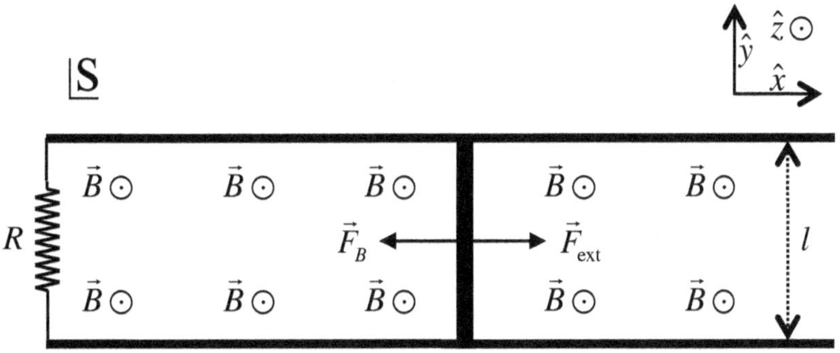

The premise of this problem is that the magnetic field is constant with time, so any energy which leaves the system via the load resistor must be offset by additional work on the system by the power entering the system. Thus $P_{in} = P_{out}$ and the net force on the bar must be zero, because it is moving at a constant velocity, so $\vec{F}_B = -\vec{F}_{ext}$.

Moreover we can write the power in and out as:

$$P_{in} = \vec{F}_{ext}\cdot\vec{v} = (-\vec{F}_B)\cdot\vec{v} \quad \text{and} \quad P_{out} = \frac{(vBl)^2}{R}.$$

Setting the powers equal to each other, we find:

$$(-\vec{F}_B)\cdot\vec{v} = \frac{(vBl)^2}{R} \quad\rightarrow\quad \vec{F}_B = \frac{-vB^2l^2}{R}\hat{x}.$$

So, conservation of energy implies that there must be a magnetic force on the bar which resists the motion.

c) Consider the same question, that of an external force applied to the bar using a force rather than an energy point of view.

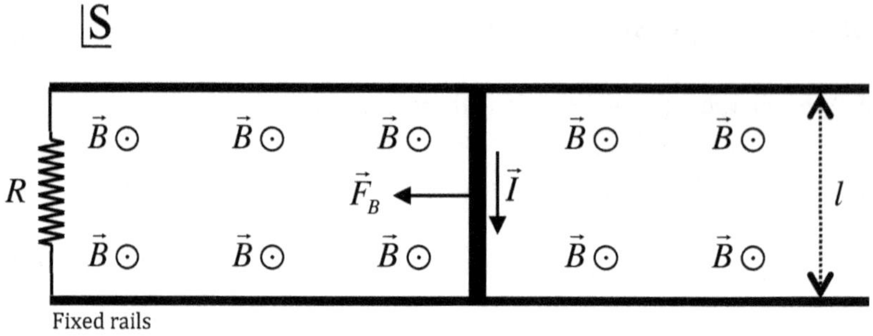

Applying the law of Laplace $\left(\frac{d\vec{F}_B}{d\ell} = \vec{I} \times \vec{B}\right)$ we find:

$$\vec{F}_B = \int_0^\ell \frac{d\vec{F}_B}{d\ell} d\ell = \int_0^\ell \vec{I} \times \vec{B} d\ell = \int_0^\ell (-I\hat{y}) \times (B\hat{z})\, d\ell = -IBl\hat{x}.$$

And applying Ohm's law we reproduce our result above:

$$\vec{F}_B = -\left(\frac{vBl}{R}\right) Bl\hat{x} = \frac{-vB^2l^2}{R}\hat{x}.$$

d) Now let us view the same problem in the reference frame of the bar.

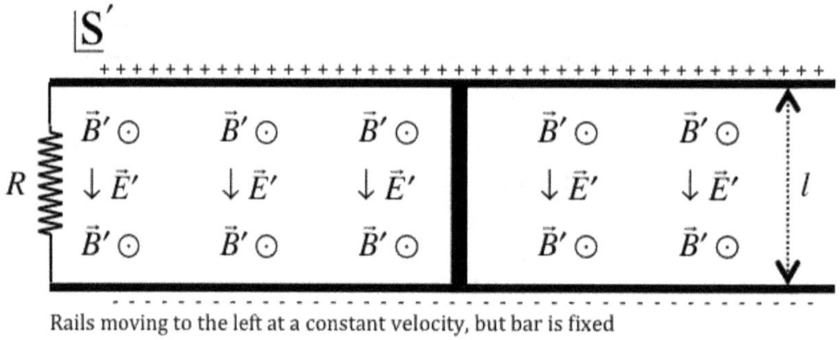

By applying the Galilean transformations to a primed reference frame that moves with the bar $(\vec{u} = \vec{v})$, we find an induced electric field in the bar. Thus in the reference frame of the bar:

$$\vec{v}' = 0, \quad \vec{B}' = \vec{B}, \quad \text{and} \quad \vec{E}' = 0 + \vec{u} \times \vec{B}.$$

So the voltage across the bar is:

$$\Delta V = -\int_0^l \vec{E}' \cdot d\vec{\ell} = -\int_0^l (\vec{u} \times \vec{B}) \cdot d\vec{\ell} = -vBl(\hat{x} \times \hat{z}) \cdot \hat{y} = vBl.$$

Thus, it does not matter whether we move the bar, or move the magnet in the opposite direction, the induced voltage across the bar is still the same.

e) Now we consider the magnetic flux, $\Phi_B = \int \vec{B} \cdot d\vec{A}$, through the loop of wire, and relate that to the voltage across the resistor:

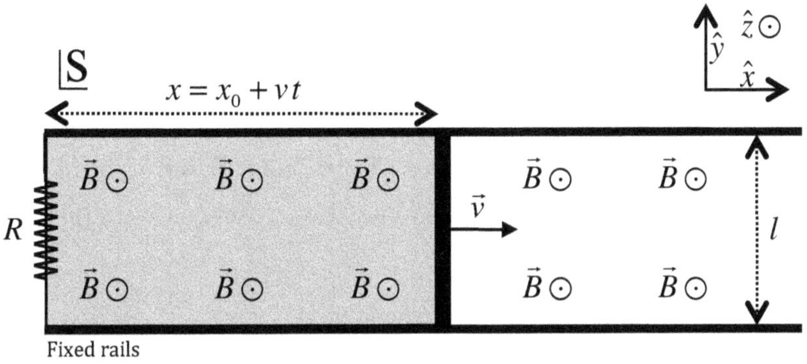

In this case, the magnetic flux through the loop is simply the product of the magnetic field and the area of the loop:

$$\Phi_B = (B\hat{z})\cdot(l(x_0+vt)\hat{z}) = Bl(x_0+vt) = (Blx_0)+(vBl)t\,.$$

And we notice something interesting: we already know that $\Delta\mathbb{V} = vBl$, so by taking the time derivative of both sides of this expression we obtain:

$$\frac{\mathrm{d}\Phi_B}{\mathrm{d}t} = vBl = \Delta\mathbb{V}.$$

f) In light of the previous discussion, we can reframe the description of induced voltage from an argument that involves the electromagnetic force to one that invokes the transfer of energy due to changing magnetic flux.

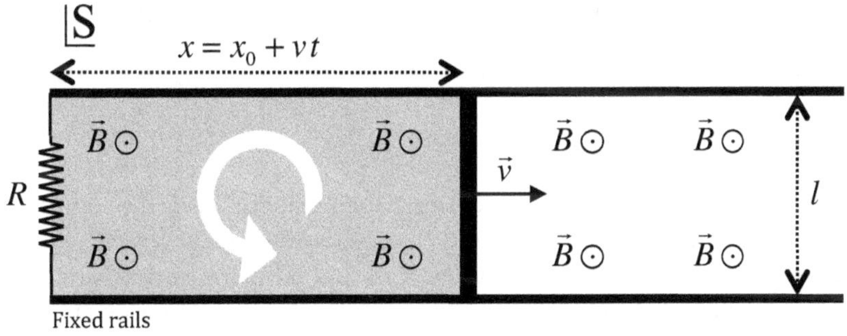

By integrating the electric field around the loop, we see that the work per charge around our loop is given by:

$$\sum_{\text{around loop}} \mathbb{V} = \oint_{\text{path}} \vec{E}\cdot \mathrm{d}\vec{\ell} = -\frac{\mathrm{d}\Phi_B}{\mathrm{d}t}\,.$$

This is Faraday's law, so we have shown that motional EMF and Faraday's law are equivalent physical concepts.

Problem 11.23: Motional EMF and the Vector Potential

Here you will repeat the analysis in the last two parts of Thought Experiment 11.14 above, except using the vector potential.

a) Express the vector potential in the laboratory frame in terms of the constant magnetic field. Use Cartesian coordinates.

b) Integrate the vector potential around the loop. Show that this equals the magnetic flux through the loop.

c) Find the scalar potential, also in Cartesian coordinates, in the frame of the moving wire.

d) Show that the potential difference from one side of the moving wire to the other agrees with our result from Thought Experiment 11.14.

Problem 11.24: A Moving Loop in a Changing Magnetic Field

Consider a square loop of wire, of dimensions $a \times b$, being forced to move at a constant velocity $\vec{v} = v\hat{x}$ in a magnetic field:

$$\vec{B} = B_0 \cos\left(\frac{\pi x}{a}\right)\sin(\omega t)\,\hat{z}\,.$$

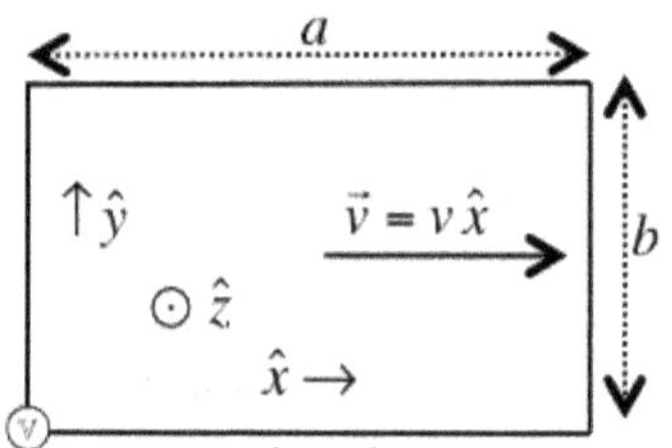

$$\vec{B} = B_0 \cos\left(\frac{\pi x}{a}\right)\sin(\omega t)\,\hat{z}$$

a) Find the vector potential in the laboratory reference frame.

b) Find the scalar potential in the moving reference frame.

c) Find the vector potential in the moving reference frame.

d) Find the electric field in the moving reference frame.

e) What does the voltmeter read as a function of time?

f) Consider the special case of driving frequencies, $\omega = n\dfrac{\pi v}{a}$, where n is an integer. Find the time-averaged voltmeter reading. Discuss.

11.4 *The Law of Biot and Savart*

One of the most useful relationships for calculating the magnetic field in practice is called the law of Biot and Savart after Jean-Baptiste Biot and his collaborator Felix Savart. As we discussed in Chapter 8, there was great competition between Biot and Ampère, and it is only fitting that one law is named after Ampère and the other Biot, with the more important one after Ampère.

That said, who specifically did what is much more complicated and jumbled in the actual history. A simplistic summary of the history reading would be that: Biot discovered the law of Laplace, Maxwell formulated Ampére's law, and Ampère and Savary developed the law of Biot-Savart. That said, Biot and Savart did make the first careful quantitative follow-up measurements to Ørsted's more qualitative observations.

Like motional EMF, the law of Biot-Savart is now best analyzed in term of reference frames and second order Galilean relativity. That is the approach we take in this section.

Thought Experiment 11.15: A Moving Point Charge

Consider a point charge Q moving at moderate velocity $\vec{v}$ through free space. Now transform to the rest frame $\lfloor S'$ of the charge, and use Coulomb's Law to write an expression for the electric field in the region surrounding the charge:

$$\vec{E}' = \frac{Q}{4\pi\varepsilon_0 r'^2}\hat{r}'.$$

Let's now transform back to the lab frame $\lfloor S$ and find the magnetic field in the vicinity of the moving charge.

$$\vec{B} = \vec{B}' + \mu_0\varepsilon_0\,\vec{u}\times\vec{E}'$$

$$= 0 + \mu_0 \not{\varepsilon_0}\,\vec{v}\times\left(\frac{Q}{4\pi\not{\varepsilon_0}r'^2}\hat{r}'\right)$$

$$= \frac{\mu_0}{4\pi r'^2}Q\vec{v}\times\hat{r}'.$$

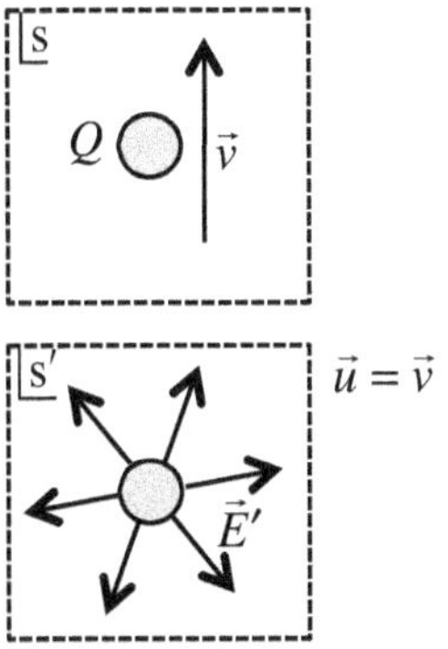

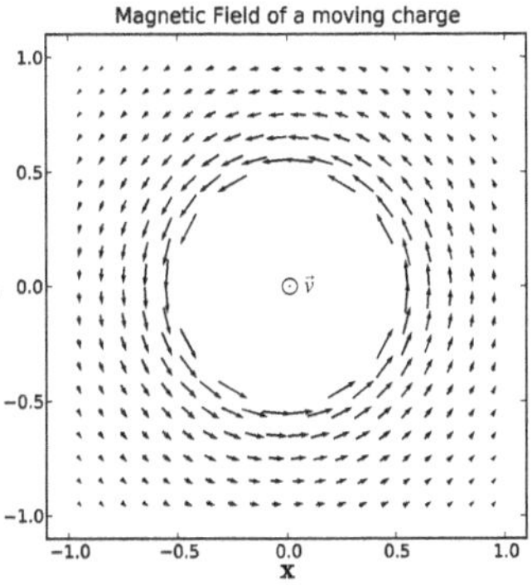

If $\vec{v} = v\hat{z}$, then:

$$\vec{B} = \frac{\mu_0}{4\pi r'^2}Qv\left(\hat{z}\times\hat{r}'\right) = \frac{\mu_0 Q v\sin\theta}{4\pi r'^2}\hat{\varphi}.$$

Since we used the Galilean transformation, this result is only an approximation. However, it holds quite well for speeds much slower than the speed of light.

Problem 11.25: The Force between Point Charges

Consider two point charges, one (q_1) located (temporarily) at the origin and another (q_2) located at a position $\vec{r}_2$. However, unlike in Chapter 2, we will assume that they are moving at velocities $\vec{v}_1$ and $\vec{v}_2$ respectively.

a) What is the electric field, due to charge #1, at the location of charge #2?

b) What is the magnetic field, due to charge #1, at the location of charge #2?

c) Show that for moderate speeds the force on charge #2 is given by:

$$\vec{F}_{1\text{on}2} = \tfrac{1}{4\pi\varepsilon_0}\frac{q_1 q_2}{r_2^2}\hat{r}_2 + \tfrac{\mu_0}{4\pi}\frac{q_1 q_2}{r_2^2}\left(\vec{v}_2\times\left(\vec{v}_1\times\hat{r}_2\right)\right).$$

d) Repeat the same arguments, but this time reverse which charge you consider #1 and #2. Show that:

$$\vec{F}_{2\text{on}1} = \tfrac{1}{4\pi\varepsilon_0}\frac{q_1 q_2}{r_1^2}\hat{r}_1 + \tfrac{\mu_0}{4\pi}\frac{q_1 q_2}{r_1^2}\left(\vec{v}_1\times\left(\vec{v}_2\times\hat{r}_1\right)\right).$$

c) Show that:

$$\vec{F}_{1\text{on}2} + \vec{F}_{2\text{on}1} = \tfrac{\mu_0}{4\pi}\frac{q_1 q_2}{r_2^2}\left(\hat{r}_2\times\left(\vec{v}_1\times\vec{v}_2\right)\right).$$

d) According to Newton's third law, $\vec{F}_{1\text{on}2} + \vec{F}_{2\text{on}1} = 0$. Under what conditions does you result agree with Newton's third aw? Under what conditions does it not?

e) What is the fractional difference from Newton's third aw, defined as: $\frac{\left|\vec{F}_{1\text{on}2}+\vec{F}_{2\text{on}1}\right|}{\left|\vec{F}_{1\text{on}2}-\vec{F}_{2\text{on}1}\right|}$? What is this numerically if $\vec{v}_1 = 3{,}000\,{}^{\text{km}}\!/_{\text{s}}\,\hat{x}$, $\vec{v}_2 = 3{,}000\,{}^{\text{km}}\!/_{\text{s}}\,\hat{y}$, and $\hat{r}_2 = \hat{x}$?

f) Assume now that Newton's law holds, but these equations are only an approximations for speeds such that $\mu_0\varepsilon_0 v_1 v_2 \ll 1$. Show that the average between the direct, and indirect, methods of calculating $\vec{F}_{1on2}$ gives:

$$\vec{F}_{1on2} = \tfrac{1}{4\pi\varepsilon_0}\frac{q_1q_2}{r_2^2}\hat{r}_2\left(1+\tfrac{1}{2}\mu_0\varepsilon_0\left(\vec{v}_1\left(\vec{v}_2\cdot\hat{r}_2\right)+\vec{v}_2\left(\vec{v}_1\cdot\hat{r}_1\right)\right)\right).$$

g) Show that there is no magnetic force between two particles moving at the same velocity. Explain why this also must be so given the principle of relativity.

h) What is the magnetic force between two particles moving directly away from each other? How about two particles moving directly toward each other?

i) Consider two particles separated by a distance ℓ along the $\hat{z}$ axis. Let $\hat{v}_1 = \cos\theta_1\,\hat{z}+\sin\theta_1\,\hat{x}$, and $\hat{v}_2 = \sin\theta_2\,\cos\varphi_2\,\hat{x}+\sin\theta_2\,\sin\varphi_2\,\hat{y}+\cos\theta_2\,\hat{z}$. Find the magnetic force between the particles in terms of their speeds and these three angles.

i) Calculate the angles, and draw the vectors, that produce a maximum magnetic force. Calculate the maximum ratio of the magnetic to electric forces given speeds v_1 and v_2.

Thought Experiment 11.16: The Law of Biot-Savart

We consider a wire carrying a current I, and focus on a charge element $\delta Q = Id\ell$ moving at drift velocity $\vec{v}$ in the wire. The magnetic field $\delta\vec{B}$ due to the current element is:

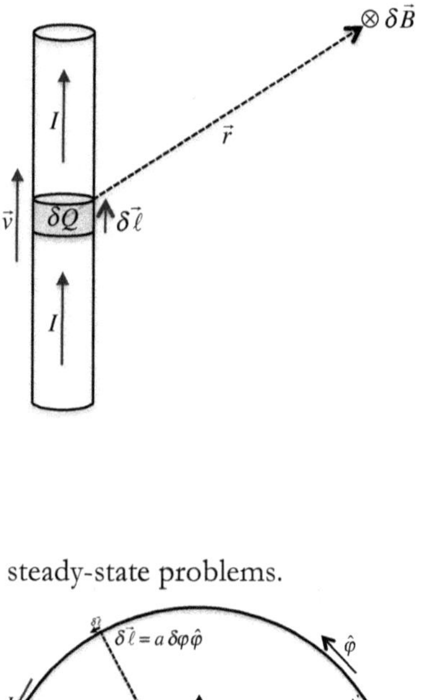

$$\delta\vec{B} = \frac{\mu_0}{4\pi}\frac{I\delta\vec{\ell}\times\hat{r}'}{r'^2}.$$

Where is the distance between the current element and the point at which $\delta\vec{B}$ is evaluated. In addition, a steady state of charge continuously flows into, and out of, a volume element at rest with respect to the wire. For steady currents, then, we can simply drop the prime in our velocity equation:

$$\delta\vec{B} = \frac{\mu_0 I\delta\vec{\ell}\times\hat{r}}{4\pi r^2}.$$

This is the *law of Biot-Savart*, and it is extremely useful for solving steady-state problems.

Thought Experiment 11.17: A Circular Loop of Wire

Now consider a single loop of wire, with a radius a and a current I flowing through it. Let's now find the magnetic field in the center of the loop using the law of Biot-Savart:

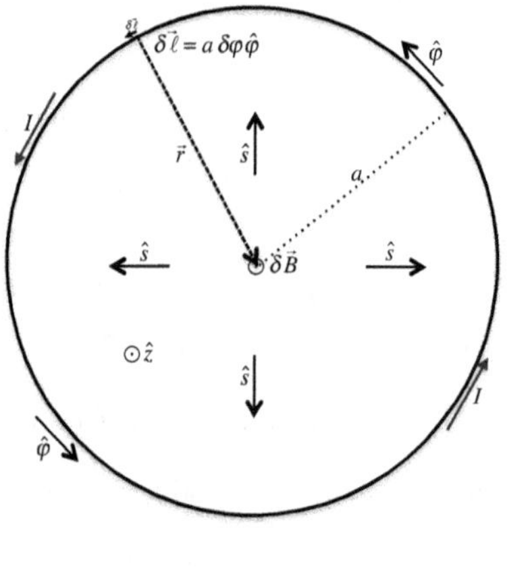

$$\delta\vec{B}_0 = \frac{\mu_0 I\left(a\delta\varphi\,\hat{\varphi}\right)\times\left(-\hat{s}\right)}{4\pi a^2} = \frac{\mu_0 I\hat{z}}{4\pi a}\delta\varphi$$

$$\vec{B}_0 = \int_0^{2\pi}\frac{\mu_0 I\hat{z}}{4\pi a}d\varphi = \frac{\mu_0 I}{2a}\hat{z}.$$

We discussed the magnetic moment of current loops already (p. 251), so we can write this in terms of the magnetic moment $\vec{m}$ as:

$$\vec{B}_0 = \frac{\mu_0 I}{2a}\hat{z} = \frac{\mu_0}{2\pi a^3}\left(I\pi a^2\hat{z}\right) = \frac{\mu_0 \vec{m}}{2\pi a^3}.$$

Problem 11.26: On Axis Field of a Wire Loop

Consider the circular loop of wire from Thought Experiment 11.17 above.

(a) Find the on-axis magnetic field as a function of the height z above the loop.

(b) In the small loop limit, show that this is consistent with the magnetic field of a small magnet with a magnetic moment $I\vec{A}$.

Problem 11.27: A Spiral Current

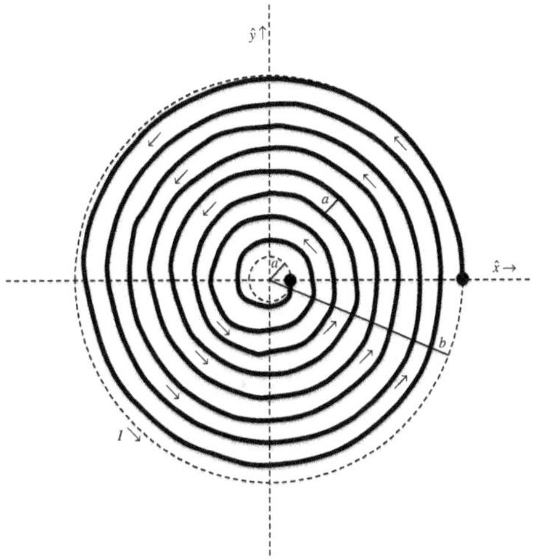

Consider a spiral coil of wire shown in the diagram with the current I flowing counterclockwise. We will assume that the circuit is completed with another wire some distance in the $\hat{z}$ direction, which can be ignored. The separation between each coil is a and the radius of the coil as whole is b.

Find the magnetic field $\vec{B}(0,0,z)$ on the $\hat{z}$ axis.

Problem 11.28: The Magnetic Field of a Short Solenoid

Find, using a stack of loops, the magnetic field in the center of a solenoid with a radius a, height h, current I, and n loops per length. Does it agree with our prior discussion in the limit that $h \gg a$?

Discussion 11.3: Ampère and the Law of Biot and Savart

After a series of experiments that Ampère conducted between 1820 and 1823, he presented a general expression for the force between two differential current elements, $I_1 d\vec{\ell}_1$ and $I_2 d\vec{\ell}_2$, based on the erroneous assumption that the force between the elements always acts along a line joining them. Later Maxwell showed that an infinite number of expressions for the force between differential current elements lead to the same result when integrated over the lengths of each wire:

$$\vec{F}_{12} = \frac{\mu_0}{4\pi}\int_{L_1}\int_{L_2}\frac{I_1 d\vec{\ell}_1 \times \left(I_2 d\vec{\ell}_2 \times \hat{r}_{12}\right)}{|r|^2}.$$

If we rewrite this to put it in the suggestive form:

$$\vec{F}_{12} = \frac{\mu_0}{4\pi}\int_{L_1} I_1 d\vec{\ell}_1 \times \int_{L_2}\frac{\left(I_2 d\vec{\ell}_2 \times \hat{r}_{12}\right)}{|r|^2}.$$

Comparison of this expression to the law of Laplace,

$$\vec{F} = \int_L I\,\mathrm{d}\vec{\ell} \times \vec{B},$$

suggests that we can write the magnetic field $\vec{B}_2$ due to a wire carrying current I_2 as:

$$\vec{B}_2(r) = \int_{L_2} \frac{\left(I_2 \mathrm{d}\vec{\ell}_2 \times \hat{r}'_{12}\right)}{|r'|^2},$$

where r is the distance between the current element and the point of observation at which is evaluated.

For a steady, uniform current, we can write

$$I_1 \vec{L}_1 = \int_{L_1} I_1 \mathrm{d}\vec{\ell}_1,$$

so that we have:

$$\vec{F}_{12} = \frac{\mu_0}{4\pi} I_1 \vec{L}_1 \times \int_{L_2} \frac{\left(I_2 \mathrm{d}\vec{\ell}_2 \times \hat{r}_{12}\right)}{|r|^2},$$

or,

$$\vec{F}_{12} = I_1 \vec{L}_1 \times \vec{B}_2 .$$

Thought Experiment 11.18: The Potentials of a Moving Point Charge

Again, consider a point charge Q moving at moderate velocity $\vec{v}$ through free space. Transforming to the rest frame $\lfloor S'$ of the charge, and use Coulomb's Law to write an expression for the electric potential:

$$V' = \frac{Q}{4\pi\varepsilon_0 r'}.$$

Let's now transform back to the lab frame $\lfloor s$ and find the potentials.

$$\vec{A} = \cancel{\vec{A}'} + \mu_0\varepsilon_0 V'\vec{u} = 0 + \mu_0 \cancel{\varepsilon_0}\, \vec{u}\left(\frac{Q}{4\pi \cancel{\varepsilon_0} r'}\right).$$

The scalar potential is:

$$V = V' + \vec{u}\cdot\cancel{\vec{A}'} = \frac{Q}{4\pi\varepsilon_0 r'}.$$

Since $\vec{u} = \vec{v}$:

$$\vec{A} = \frac{\mu_0 Q}{4\pi r'}\vec{v} \quad \text{and} \quad V = \frac{Q}{4\pi\varepsilon_0 r'}.$$

Notice that the distance is still in the particle frame, so it is the distance to the particle $(\vec{r}' = \vec{r} - \vec{v}t)$.

In fact, it is still not quite the distance to the where the particle is, but rather to where it was a very short time before, because information cannot travel instantaneously fast. As long as the particle is moving relatively slowly $\left(v^2\mu_0\varepsilon_0 \ll 1\right)$, and not accelerating, this will not matter.

Thought Experiment 11.19: Vector Potential of a Very Short Wire

In Thought Experiment 11.16 (p. 356) we found the magnetic field surrounding a current element, but often it is easier to find the vector potential of a current element, integrate this over the wire, the finally differentiate to obtain the magnetic field. We start with the Law of Biot and Savart:

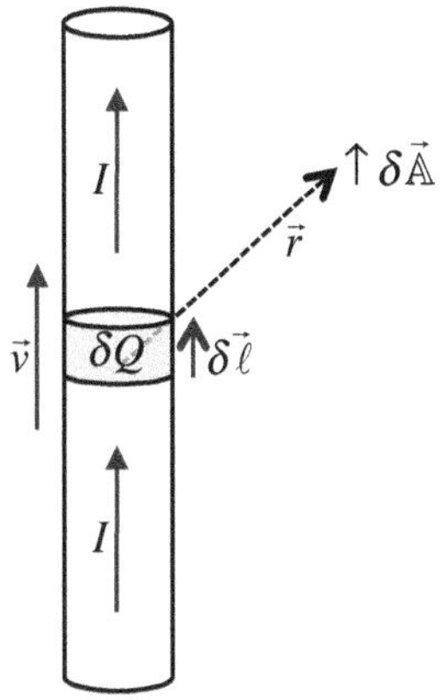

$$\delta \vec{B} = \vec{\nabla} \times \delta \vec{\mathbb{A}} = \frac{\mu_0 I \; \delta \vec{\ell} \times \hat{r}}{4 \pi r^2} \; .$$

Rendering the magnetic field in cylindrical coordinates, we can write:

$$\delta \vec{B} = \frac{\mu_0 I \; \delta \ell}{4 \pi} \frac{\hat{z} \times \vec{r}}{r^3} = \left(\frac{\mu_0 I \; \delta \ell}{4 \pi} \right) \frac{\hat{z} \times (s \hat{s} + z \hat{z})}{\left(\sqrt{s^2 + z^2} \right)^3} = \left(\frac{\mu_0 I \; \delta \ell}{4 \pi} \right) s \left(s^2 + z^2 \right)^{-\frac{3}{2}} \hat{\varphi} \; .$$

Relying upon our experience from Thought Experiment 9.15, we now assume that the vector potential is parallel to the current. The curl of $\delta \vec{\mathbb{A}}$ then becomes:

$$\vec{\nabla} \times \delta \vec{\mathbb{A}} = \vec{\nabla} \times (\delta \mathbb{A} \, \hat{z}) = \frac{1}{s} \frac{\partial \delta \mathbb{A}}{\partial \varphi} \hat{s} - \frac{\partial \delta \mathbb{A}}{\partial s} \hat{\varphi} \; .$$

Because of azimuthal symmetry, $\delta \vec{\mathbb{A}}$ cannot depend on φ, so this simplifies to:

$$\vec{\nabla} \times \delta \vec{\mathbb{A}} = - \frac{\partial \delta \mathbb{A}}{\partial s} \hat{\varphi} \; .$$

We can now equate the magnetic field $\delta \vec{B}$, and solve for the magnitude of the vector potential $\delta \mathbb{A}$:

$$\delta \vec{B} = \left(\frac{\mu_0 I \; \delta \ell}{4 \pi} \right) s \left(s^2 + z^2 \right)^{-\frac{3}{2}} \hat{\varphi} = \vec{\nabla} \times \delta \vec{\mathbb{A}} = - \frac{\partial \delta \mathbb{A}}{\partial s} \hat{\varphi}$$

$$\frac{\partial \delta \mathbb{A}}{\partial s} = - \left(\frac{\mu_0 I \; \delta \ell}{4 \pi} \right) s \left(s^2 + z^2 \right)^{-\frac{3}{2}}$$

$$\delta \mathbb{A} = \int \left(\frac{\mu_0 I \; \delta \ell}{4 \pi} \right) \left(- s \left(s^2 + z^2 \right)^{-\frac{3}{2}} \right) ds = \left(\frac{\mu_0 I \; \delta \ell}{4 \pi} \right) \int (2 s) \left(-\tfrac{1}{2} \right) \left(s^2 + z^2 \right)^{-\frac{3}{2}} ds$$

$$\delta \mathbb{A} = \left(\frac{\mu_0 I \; \delta \ell}{4 \pi} \right) \left(s^2 + z^2 \right)^{-\frac{1}{2}} = \frac{\mu_0 I \; \delta \ell}{4 \pi r}$$

And finally remember that we defined the $\hat{z}$ direction to the be in the same direction as $\delta \vec{\ell}$, so:

$$\delta \vec{\mathbb{A}} = \frac{\mu_0 I}{4 \pi r} \delta \vec{\ell} \; .$$

This is the vector potential version of the law of Biot-Savart. Again, unless solving for the magnetic field at one special point, it is usually easier to solve for the vector potential, and then take the curl, to obtain the magnetic field at all points of interest.

Problem 11.29: The Vector Potential of a Current Element

In Thought Experiment 11.18 (p. 358) we found the potentials of a moving charge, and in Thought Experiment 11.16 (p. 356) we found the law of Biot and Savart from the magnetic field of a moving charge.

Derive the law of Biot and Savart, in potential form, directly from the potentials of a moving charge.

Thought Experiment 11.20: The Inductance Matrix

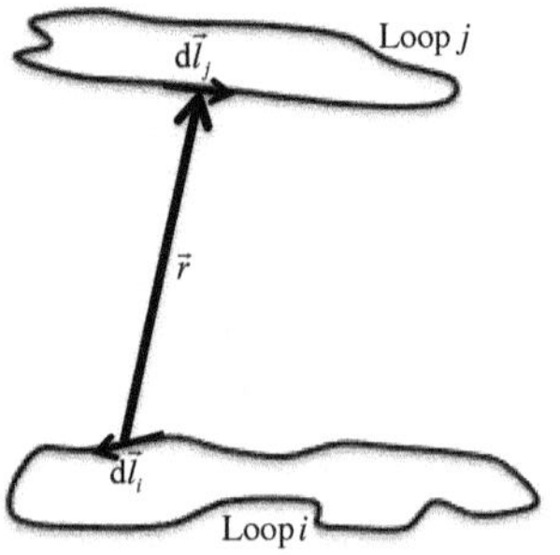

In Thought Experiment 9.11 (p. 295) we expressed the inductance of two coils as a matrix. The diagonal elements represented the self-inductance, and the off-diagonal elements representing the mutual inductance between the coils. In this thought experiment, we use the vector potential form of the law of Biot and Savart to generalize this to N loops of wires. Let us now consider any two loops of wire, i and j, and then calculate their mutual inductance, L_{ij}. to find a general expression for the mutual inductance.

First we find the vector potential due to loop i using the law of Biot-Savart (Thought Experiment 11.19), as follows:

$$\delta\vec{\mathbb{A}} = \frac{\mu_0 I}{4\pi r}\delta\vec{\ell} \quad \rightarrow \quad \vec{\mathbb{A}}_i = \frac{\mu_0 I_i}{4\pi}\oint_{\text{loop } i}\frac{d\vec{\ell}}{r}$$

Recall from Section 7.1 (p. 212) that flux through loop j due to loop i is simply:

$$\Phi_{ij} = \int_{\text{surface } j}\left(\vec{\nabla}\times\vec{\mathbb{A}}_i\right)\cdot d\vec{A} = \oint_{\text{loop } j}\vec{\mathbb{A}}_i\cdot d\vec{\ell}\,.$$

Now combining these results, the mutual magnetic flux is:

$$\Phi_{ij} = \oint_{\text{loop } j}\left(\frac{\mu_0 I_i}{4\pi}\oint_{\text{loop } i}\frac{d\vec{\ell}_i}{r_{ij}}\right)\cdot d\vec{\ell}_j = \frac{\mu_0 I_i}{4\pi}\oint_{\text{loop } j}\oint_{\text{loop } i}\frac{d\vec{\ell}_i\cdot d\vec{\ell}_j}{r_{ij}}\,.$$

Recall the definition of the mutual inductance between these two loops:

$$L_{ij} = -\frac{\mathbb{V}_{ij}}{dI_j/dt} = \frac{d\Phi_{ij}/dt}{dI_j/dt} = \frac{d\Phi_{ij}}{dI_j} = \frac{d}{dI_j}\left(\frac{\mu_0 I_i}{4\pi}\oint_{\text{loop } j}\oint_{\text{loop } i}\frac{d\vec{\ell}_i\cdot d\vec{\ell}_j}{r_{ij}}\right) = \frac{\mu_0}{4\pi}\oint_{\text{loop } j}\oint_{\text{loop } i}\frac{d\vec{\ell}_i\cdot d\vec{\ell}_j}{r_{ij}}\,.$$

This last expression is known as the Neumann formula, after the German physicist Franz Neumann who was also heavily involved in the development of the vector potential.

Problem 11.30: Inductance of two Loops

Consider two parallel circular coils of wire, each with N loops of radius a. The coils are separated by a distance $2a$.

a) Express the mutual inductance between the two coils in unitless integral form.

b) Numerically solve the integral to find an expression for the mutual inductance.

c) If one coil has a current $I = I_0\cos(\omega t)$ flowing through it, what would be the induced voltage in the other?

d) If the other had a load resistor R, what would be the current flowing through it?

e) How much power is transferred between the two coils?

Thought Experiment 11.21: Vector Potential of a Finite Length Wire

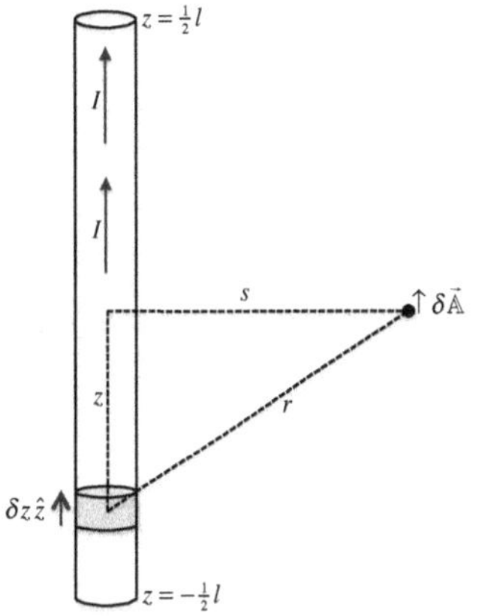

We now find the vector potential of a finite length wire using the potential version of the law of Biot-Savart obtained in Thought Experiment 11.19:

$$\delta\vec{\mathbb{A}} = \frac{\mu_0 I}{4\pi r}\delta\vec{\ell} = \frac{\mu_0 I}{4\pi}\frac{\delta z}{\sqrt{z^2+s^2}}\hat{z}\,.$$

Next we integrate over the wire, but because of even symmetry we can integrate from the middle out, and double our integral:

$$\vec{\mathbb{A}} = \frac{\mu_0 I}{4\pi}2\int_0^{l/2}\frac{\mathrm{d}z}{\sqrt{z^2+s^2}}\hat{z} = \frac{\mu_0 I}{2\pi}\left(\ln\left(z+\sqrt{z^2+s^2}\right)\Big|_0^{l/2}\right)\hat{z}$$

$$= \frac{\mu_0 I}{2\pi}\left(\ln\left(l/2+\sqrt{(l/2)^2+s^2}\right)-\ln\left(\sqrt{s^2}\right)\right)\hat{z} = \frac{\mu_0 I}{2\pi}\ln\left(\frac{l/2+\sqrt{(l/2)^2+s^2}}{s}\right)\hat{z}\,.$$

Notice that in this case the vector potential goes to zero as $s\to\infty$.

In the limiting case of a long wire, $l \gg s$, then:

$$\vec{\mathbb{A}} \approx \frac{\mu_0 I}{2\pi}\ln\left(l/s\right)\hat{z}\,,$$

which is simply off by a constant from Thought Experiment 9.15.

Thought Experiment 11.22: The Vector Potential of a Current Loop

Consider a single current loop of radius a. We will calculate the vector potential at a point $\vec{r}$ from the origin.

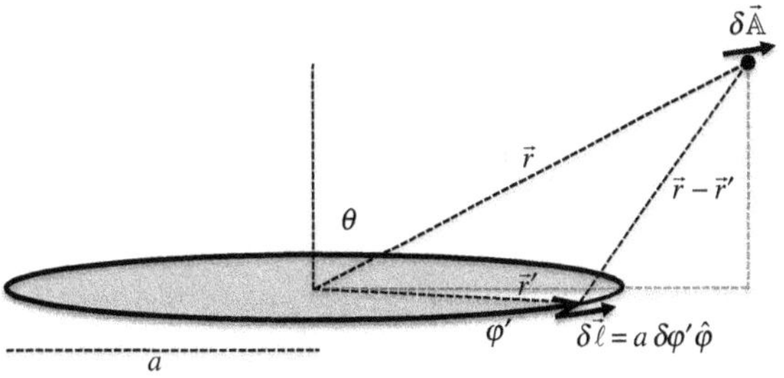

First we write the Biot-Savart law in potential form for a current element located at a position $\vec{r}'$:

$$\delta\vec{\mathbb{A}} = \frac{\mu_0 I\,\mathrm{d}\vec{\ell}}{4\pi|\vec{r}-\vec{r}'|}\,.$$

But first, we have a little mathematical prep-work to do. Recall the law of cosines:

$$|\vec{r}-\vec{r}'|^2 = r^2 + r'^2 - 2\vec{r}\cdot\vec{r}'\,,$$

and note that in cylindrical coordinates:

$$\vec{r}\cdot\vec{r}' = (s\hat{s}+z\hat{z})\cdot(a\cos\varphi'\,\hat{s}+a\sin\varphi'\,\hat{\varphi}).$$

So, in spherical coordinates we have that:

$$\vec{r}\cdot\vec{r}' = sa\cos\varphi' = ar\sin\theta\cos\varphi'$$

$$|\vec{r}-\vec{r}'|^2 = r^2+a^2-2ar\sin\theta\cos\varphi' = r^2\left(1+\left(\tfrac{a}{r}\right)^2-2\left(\tfrac{a}{r}\right)\sin\theta\cos\varphi'\right).$$

Now in spherical coordinates, we can write:

$$\delta\vec{\mathbb{A}} = \frac{\mu_0 I a\,\delta\varphi'}{4\pi r\sqrt{1+\left(\frac{a}{r}\right)^2-2\left(\frac{a}{r}\right)\sin\theta\cos\varphi'}}\hat{\varphi}'.$$

For the whole loop of wire, the vector potential becomes:

$$\vec{\mathbb{A}} = \frac{\mu_0 I a}{4\pi r}\int_0^{2\pi}\frac{\hat{\varphi}'\,d\varphi'}{\sqrt{1+\left(\frac{a}{r}\right)^2-2\left(\frac{a}{r}\right)\sin\theta\cos\varphi'}}.$$

This integral cannot be evaluated in terms of elementary functions, but it can be numerically integrated. We can, however, evaluate this analytically in the limit that $r \gg a$.

First we let $\xi = \left(\frac{a}{r}\right)^2-2\left(\frac{a}{r}\right)\sin\theta\cos\varphi'$, and expand around $\xi=0$ as:

$$\frac{1}{\sqrt{1+\xi}} \approx 1-\tfrac{1}{2}\xi+\tfrac{1}{8}\xi^2.$$

Next we substitute such that:

$$\frac{1}{\sqrt{1+\xi}} \approx 1-\tfrac{1}{2}\left[\left(\tfrac{a}{r}\right)^2-2\left(\tfrac{a}{r}\right)\sin\theta\cos\varphi'\right]+\tfrac{1}{8}\left[\left(\tfrac{a}{r}\right)^2-2\left(\tfrac{a}{r}\right)\sin\theta\cos\varphi'\right]^2$$

$$\approx 1+\left(\tfrac{a}{r}\right)\sin\theta\cos\varphi'-\tfrac{1}{2}\left(\tfrac{a}{r}\right)^2+\tfrac{1}{8}\left(2\left(\tfrac{a}{r}\right)\sin\theta\cos\varphi'\right)^2+\mathcal{O}\left(\left(\tfrac{a}{r}\right)^3\right).$$

Putting in a fixed coordinate system with the point of interest located on the $\hat{x}$ axis (i.e. $\varphi=0$), where $\hat{\varphi}'=-\sin\varphi'\,\hat{x}+\cos\varphi'\,\hat{y}$, yields:

$$\mathbb{A}_x \approx \frac{\mu_0 I a}{4\pi r}\int_0^{2\pi}\left(1-\tfrac{1}{2}\left(\tfrac{a}{r}\right)^2+\left(\tfrac{a}{r}\right)\sin\theta\cos\varphi'+\tfrac{1}{2}\left(\tfrac{a}{r}\right)^2\sin^2\theta\cos^2\varphi'\right)(-\sin\varphi')\,d\varphi'$$

$$\mathbb{A}_y \approx \frac{\mu_0 I a}{4\pi r}\int_0^{2\pi}\left(1-\tfrac{1}{2}\left(\tfrac{a}{r}\right)^2+\left(\tfrac{a}{r}\right)\sin\theta\cos\varphi'+\tfrac{1}{2}\left(\tfrac{a}{r}\right)^2\sin^2\theta\cos^2\varphi'\right)(\cos\varphi'\,\hat{y})\,d\varphi'.$$

Notice that when this expression is integrated around the circular loop, only the $\cos^2\varphi'$ term does not integrate to zero. Thus:

$$\vec{\mathbb{A}}(\vec{r}\gg a) \approx \frac{\mu_0 I a^2\sin\theta}{4\pi r^2}\hat{y}\int_0^{2\pi}\cos^2(\varphi')\,d\varphi' \approx \frac{\mu_0 I\pi a^2\sin\theta}{4\pi r^2}\hat{y}.$$

Now we will express this in terms of the magnetic moment of the current loop, which is given by $\vec{m}=I\vec{A}=\pi a^2 I\,\hat{z}$, so:

$$\vec{\mathbb{A}}(\vec{r} \gg a) \approx \frac{\mu_0 m \sin\theta}{4\pi r^2}\hat{y}.$$

Our point of interest is located at $\varphi = 0$, so we have the following unit vector relationship:

$$\hat{z}\times\hat{r} = \hat{z}\times(\sin\theta\,\hat{x} + \cos\theta\,\hat{z}) = \sin\theta\,\hat{z}\times\hat{x} = \sin\theta\,\hat{y}.$$

This allows us to take advantage of the vector nature of the magnetic moment, and to make a coordinate independent expression, as:

$$\vec{\mathbb{A}}(\vec{r} \gg a) \approx \frac{\mu_0 m \sin\theta}{4\pi r^2}\hat{y} \approx \frac{\mu_0 m \cancel{\sin\theta}}{4\pi r^2}\frac{(\hat{z}\times\hat{r})}{\cancel{\sin\theta}} \approx \frac{\mu_0}{4\pi}\frac{(m\hat{z})\times\hat{r}}{r^2} \approx \frac{\mu_0}{4\pi}\frac{\vec{m}\times\hat{r}}{r^2}.$$

This is a very important result, as this allows us to find the far away vector potential of anything with a magnetic moment.

Finally we will evaluate this analytically for the opposite limiting case, where $r \ll a$. In this case, we can write:

$$\vec{\mathbb{A}}(\vec{r} \ll a) \approx \frac{\mu_0 I a}{4\pi r}\int_0^{2\pi}\frac{\hat{\varphi}'\,\mathrm{d}\varphi'}{\sqrt{\cancel{1} + \left(\frac{a}{r}\right)^2 - 2\left(\frac{a}{r}\right)\sin\theta\cos\varphi'}} \approx \frac{\mu_0}{4\pi}I\int_0^{2\pi}\frac{\hat{\varphi}'\,\mathrm{d}\varphi'}{\sqrt{1 - 2\left(\frac{r}{a}\right)\sin\theta\cos\varphi'}}.$$

Now we let $\xi = -2\left(\frac{r}{a}\right)\sin\theta\cos\varphi'$, and again expand around $\xi = 0$, so:

$$\frac{1}{\sqrt{1+\xi}} \approx 1 - \tfrac{1}{2}\xi + \tfrac{1}{8}\xi^2 \approx 1 - \tfrac{1}{2}\left[-2\left(\tfrac{r}{a}\right)\sin\theta\cos\varphi'\right] + \tfrac{1}{8}\left[-2\left(\tfrac{r}{a}\right)\sin\theta\cos\varphi'\right]^2$$

$$\approx 1 + \left(\tfrac{r}{a}\right)\sin\theta\cos\varphi' + \tfrac{1}{2}\left(\tfrac{r}{a}\right)^2\sin^2\theta\cos^2\varphi'.$$

And again rendering this in fixed coordinates, and integrating, we have:

$$\mathbb{A}_x \approx \tfrac{\mu_0}{4\pi}I\int_0^{2\pi}\left(1 + \left(\tfrac{r}{a}\right)\sin\theta\cos\varphi' + \tfrac{1}{2}\left(\tfrac{r}{a}\right)^2\sin^2\theta\cos^2\varphi'\right)(-\sin\varphi')\,\mathrm{d}\varphi'$$

$$\mathbb{A}_y \approx \tfrac{\mu_0}{4\pi}I\int_0^{2\pi}\left(1 + \left(\tfrac{r}{a}\right)\sin\theta\cos\varphi' + \tfrac{1}{2}\left(\tfrac{r}{a}\right)^2\sin^2\theta\cos^2\varphi'\right)(\cos\varphi'\,\hat{y})\,\mathrm{d}\varphi'.$$

Again we integrate only the $\cos^2\varphi'$ term, so:

$$\vec{\mathbb{A}}(\vec{r} \ll a) \approx \frac{\mu_0 I r\sin\theta}{4\pi a}\hat{y}\int_0^{2\pi}\cos^2(\varphi')\,\mathrm{d}\varphi' \approx \frac{\mu_0 I \pi a^2 r\sin\theta}{4\pi a^3}\hat{y} \approx \frac{\mu_0 m r\sin\theta}{4\pi a^3}\hat{y} \approx \frac{\mu_0}{4\pi a^3}\vec{m}\times\vec{r}.$$

Problem 11.31: The Magnetic Field of a Current Loop

In Thought Experiment 11.22 (p. 361) we showed:

$$\vec{\mathbb{A}} = \frac{\mu_0 I a}{4\pi}\int_0^{2\pi}\frac{\mathrm{d}\varphi'}{r\sqrt{1 + \left(\frac{a}{r}\right)^2 - 2\left(\frac{a}{r}\right)\sin\theta\cos\varphi'}}\hat{\varphi}'.$$

We then solved this integral in the two limits: that $\frac{a}{r} \ll 1$ and $\frac{a}{r} \gg 1$.

a) Numerically integrate this in cylindrical coordinates over the range $0<\frac{s}{a}<2$ and $0<\frac{z}{a}<2$. Plot it as an image representing $\mathbb{A}_\varphi$.

b) Numerically take the curl of your result in part (a), to find both B_s and B_z. Plot the magnetic field as a vector plot on top of the image in part (a).

c) Verify that your results are consistent with the analytical results in the small and large limits.

Thought Experiment 11.23: The Magnetic Field of a Single Current Loop

Now that we have the vector potential or a current loop, we must take the curl of $\vec{\mathbb{A}}$ to find the magnetic field:

$$\vec{B}(\vec{r}\gg a)=\vec{\nabla}\times\vec{\mathbb{A}}=\frac{\mu_0}{4\pi}\vec{\nabla}\times\left(\frac{\vec{m}\times\hat{r}}{r^2}\right)=\frac{\mu_0}{4\pi}\vec{\nabla}\times\left(\frac{\vec{m}\times\vec{r}}{r^3}\right)$$

$$\vec{B}(\vec{r}\ll a)=\vec{\nabla}\times\vec{\mathbb{A}}=\frac{\mu_0}{4\pi a^3}\vec{\nabla}\times(\vec{m}\times\vec{r}).$$

Let's first evaluate the following:

$$\vec{\nabla}\times(\vec{m}\times\vec{r})=\vec{m}(\vec{\nabla}\cdot\vec{r})-(\vec{m}\cdot\vec{\nabla})\vec{r}=\vec{m}(3)-\left(m\frac{\partial}{\partial z}\right)\vec{r}=3\vec{m}-m\hat{z}=2\vec{m},$$

so near the center the magnetic field must be:

$$\vec{B}(\vec{r}\ll a)\approx\frac{\mu_0}{4\pi a^3}\vec{\nabla}\times(\vec{m}\times\vec{r})\approx\frac{\mu_0\vec{m}}{2\pi a^3}.$$

Now we will find the magnetic field for large distances:

$$\vec{B}(\vec{r}\gg a)\approx\frac{\mu_0}{4\pi}\left(\frac{1}{r^3}\vec{\nabla}\times(\vec{m}\times\vec{r})-(\vec{m}\times\vec{r})\times\left(\vec{\nabla}\frac{1}{r^3}\right)\right)=\frac{\mu_0}{4\pi r^3}\left(2\vec{m}+3\frac{(\vec{m}\times\vec{r})}{r}\times\hat{r}\right)$$

$$=\frac{\mu_0}{4\pi r^3}\left(2\vec{m}+3(\vec{m}\times\hat{r})\times\hat{r}\right)=\frac{\mu_0}{4\pi r^3}\left(2\vec{m}-3(\hat{r}\times\vec{m}\times\hat{r})\right)=\frac{\mu_0}{4\pi r^3}\left(2\vec{m}-3(\vec{m}-(\vec{m}\cdot\hat{r})\hat{r})\right).$$

Therefore,

$$\vec{B}(\vec{r}\gg a)\approx\frac{\mu_0}{4\pi r^3}\left(3(\vec{m}\cdot\hat{r})\hat{r}-\vec{m}\right).$$

It is especially convenient to write the magnetic field of a single current loop in cylindrical or spherical coordinates. In cylindrical coordinates, $\vec{m}=m\hat{z}$ and $\hat{r}=\frac{1}{r}(z\hat{z}+s\hat{s})$, so the magnetic field can be written as:

$$\vec{B}(s,z)=\begin{cases}\dfrac{\mu_0 m}{2\pi a^3}\hat{z} & r\ll a\\[2ex] \dfrac{\mu_0 m}{2\pi r^3}\left(\left(3\dfrac{sz}{r^2}\hat{s}\right)+\left(3\dfrac{z^2}{r^2}-1\right)\hat{z}\right) & r\gg a\end{cases}$$

In spherical coordinates, we have that: $\vec{B}(r,\theta)=\begin{cases}\dfrac{\mu_0 m}{2\pi a^3}\left(\cos\theta\,\hat{r}-\sin\theta\,\hat{\theta}\right) & r\ll a\\ \dfrac{\mu_0 m}{2\pi r^3}\left(\cos\theta\,\hat{r}+\tfrac{1}{2}\sin\theta\,\hat{\theta}\right) & r\gg a\end{cases}$

Thought Experiment 11.24: A Spinning Charged Spherical Shell

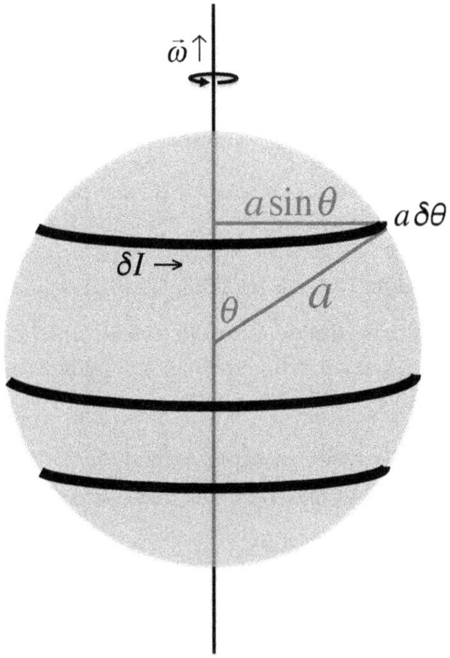

Consider a sphere with radius a, and a charge Q, that is uniformly distributed on the surface of the sphere, spinning at an angular velocity $\vec{\omega}$. What is the vector potential outside of the sphere?

First, note that the surface charge density is:

$$\sigma=\frac{Q}{4\pi a^2}.$$

Now, we will model the ball as a stack of current loops, each with charge element:

$$\mathrm{d}Q=\sigma\left(2\pi a\sin\theta\right)\left(a\,\mathrm{d}\theta\right)=2\pi a^2\sigma\sin\theta\,\mathrm{d}\theta,$$

so each current element is:

$$\mathrm{d}I=\frac{\mathrm{d}Q}{2\pi/\omega}=\omega a^2\sigma\sin\theta\,\mathrm{d}\theta.$$

The magnetic moment per current loop is:

$$\begin{aligned}\mathrm{d}\vec{m}&=A\,\mathrm{d}I\,\hat{z}=\left(\pi\left(a\sin\theta\right)^2\right)\left(\omega a^2\sigma\sin\theta\,\mathrm{d}\theta\right)\hat{z}\\&=\omega\sigma\pi a^4\sin^3\theta\,\mathrm{d}\theta\,\hat{z}=\tfrac{1}{4}Qa^2\omega\sin^3\theta\,\mathrm{d}\theta\,\hat{z}.\end{aligned}$$

The magnetic moment is linear and adds vectorally, so:

$$\vec{m}=\int_0^\pi \tfrac{1}{4}Qa^2\omega\sin^3\theta\,\mathrm{d}\theta\,\hat{z}=\tfrac{1}{4}Qa^2\omega\hat{z}\left(\int_0^\pi\sin^3\theta\,\mathrm{d}\theta\right).$$

The integral is equal to $\frac{4}{3}$, and $\vec{\omega}=\omega\hat{z}$, so the magnetic moment is simply:

$$\vec{m}=\tfrac{1}{3}Qa^2\vec{\omega}.$$

Notice the similarity with the angular momentum of a spherical shell of the same shape, which is given by:

$$\vec{L}=\tfrac{2}{3}Ma^2\vec{\omega},$$

so if the mass M and charge Q follow each other, then the ratio of the magnetic moment to the angular momentum is simply half the charge to mass ratio.

Finally we find the vector potential outside the sphere as:

$$\vec{\mathbb{A}} \approx \frac{\mu_0}{4\pi}\frac{\vec{m}\times\hat{r}}{r^2} = \frac{\mu_0 Q a^2}{12\pi}\frac{\vec{\omega}\times\hat{r}}{r^2}.$$

As it turns out, this is not really an approximation, which the reader can show by calculating the vector potential directly (see Problem 11.32).

Problem 11.32: The Vector Potential of a Charged Shell

Consider a spherical shell with uniform surface charge density σ, radius a and spinning with an angular velocity $\vec{\omega}$, as in Thought Experiment 11.24.

a) Show that inside the sphere, the vector potential is given by the relation:

$$\vec{\mathbb{A}}(\vec{r}) = \tfrac{1}{3}\mu_0 a\sigma\, \vec{\omega}\times\vec{r}\,, \quad \text{for } r \le a\,.$$

b) Show that, when $r \ge a$ this results is the vector potential of an *exact dipole*, not simply an approximate one.

Thought Experiment 11.25: The Classical Gyromagnetic Ratio

With the discovery of the electron, the Ampèrian model predicts that bound electrons should have a fixed ratio of their magnetic moment to their angular momentum, called the *gyromagnetic ratio* (γ_e).

Consider, for example, bound electrons moving in a circle of radius R with an angular velocity $\vec{\omega}$. Classically, these electrons have an angular momentum given by:

$$\vec{L} = \vec{r}\times\vec{p} = m_e\vec{\omega}R^2\,,$$

so the associated magnetic moment would be:

$$\vec{\mu} = I\vec{A} = \left(-\frac{e\omega}{2\pi}\right)\left(\pi R^2\right)\hat{\omega} = -\tfrac{1}{2}e\vec{\omega}R^2\,,$$

where we are using $\vec{\mu}$ instead of $\vec{m}$ for the magnetic moment to avoid confusion with mass. We therefore expect that for each bound electron, the ratio of its orbital magnetic moment to its orbital angular momentum would be given by:

$$\gamma_e = \frac{\mu_e}{L_e} = -\frac{e}{2m_e}.$$

This thought experiment was originally proposed by the British physicist Owen W. Richardson in 1908.[156] The actual gyromagnetic ratio of an electron is about twice this.

Problem 11.33: The Electron g Factor

Here you will model the electron classically as a uniform, in both mass and charge, spinning sphere with angular momentum $\vec{S} = \tfrac{1}{2}\hbar\,\hat{z}$, charge $Q = -e$, mass $M = m_e$, and magnetic moment $\vec{m} = -\mu_e\,\hat{z}$.

---

[156] O.W. Richardson, "A Mechanical Effect Accompanying Magnetization," Physical Review, (Series 1), **26**:3, (1908), 248-253.

a) Solve for the angular momentum of the sphere in terms of its mass, radius a, and angular velocity angular velocity $\vec{\omega}$ about an axis through its center.

b) Solve for the magnetic moment in terms of its charge, radius a, and angular velocity angular velocity $\vec{\omega}$.

c) Solve for the magnetic moment in terms of the measured parameters, assuming it were a classical solid spinning sphere.

d) Find the ratio of the measured, to classically predicted, magnetic moments.

e) This is customarily written with the letter g, so it is called the electron spin g-factor. Look this up to check your work.

Problem 11.34: Gyromagnetic Ratio of a Spinning Charged Sphere

Consider a uniform sphere whose mass is distributed uniformly throughout the sphere by volume, but the charge is distributed uniformly by area on the surface.

a-d) Repeat Problem 11.33, but with this sphere instead.

e) How well does this classical model predict the gyromagnetic ratio?

f) Do you think this is a better, or worse, toy model than the one Richardson proposed in 1908? Explain, not only based on the numbers, but also based on the physical concepts involved.

f) The *classical electron radius* is defined as: $r_e = \frac{\mu_0}{4\pi}\frac{e^2}{m_e} \approx 2.82\,\text{fm}$. Using this model, calculate its angular velocity ω and its linear velocity evaluated at the equator ωr_e. Comment on the physical plausibility of such a spinning sphere occurring in nature.

d) Research the nature of *electron spin*. How do modern scientists understand this concept, especially in light of your result from part c?

Problem 11.35: A Spinning Long Cylinder

Consider a long cylinder of radius a and uniform charge density ρ. The cylinder is spinning about its principal axis with an angular velocity ω.

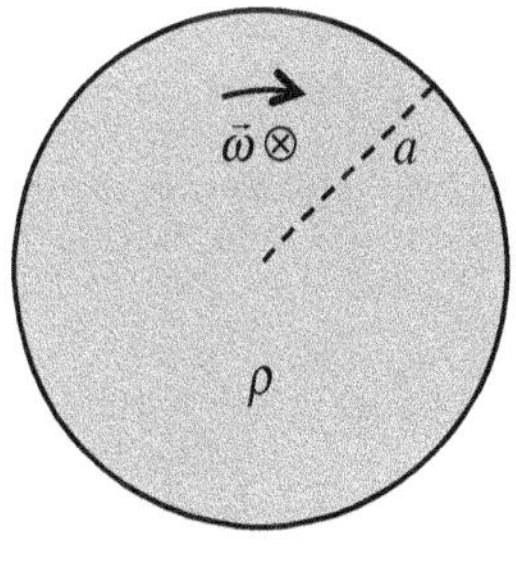

a) Find the current density as a function of azimuthal radius s.

b) Find the vector potential $\vec{\mathbf{A}}(s)$ inside and outside of the cylinder.

c) Find the magnetic field $\vec{B}(s)$ both inside and outside of the cylinder.

d) Is there a torque on the cylinder? Does it lose energy? If so, where does it go?

Problem 11.36: A Spinning Charged Disk

Consider a uniform disk of radius a and height $h \ll a$, with a total mass M and charge Q. The disk is spinning about its principal axis with an angular velocity ω.

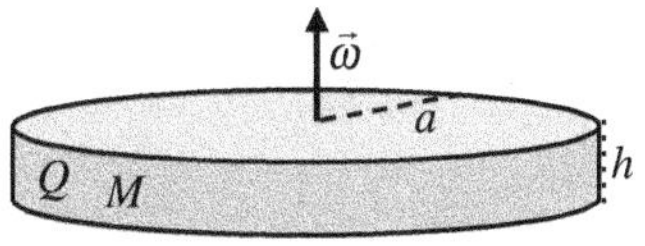

a) What is the total angular momentum of the disk?

b) Find the current density as a function of azimuthal radius s.

c) Find the on-axis magnetic field as a function of z, for all $|z| \geq \frac{1}{2}h$.

d) What is the magnetic moment of the disk?

e) The gyromagnetic ratio γ is defined as the ratio of the magnetic moment to the angular momentum of an object. What is the gyromagnetic ratio for this disk?

f) Find the vector potential surrounding the disk at large distances?

g) Find the magnetic field at large distances. Does this agree with the on-axis field?

h) Plot the magnitude of the vector potential as an image plot, and the magnetic field as a vector plot overlaid.

Problem 11.37: A Spinning Non-Uniform Charged Sphere

Consider a sphere of radius a, with an angular velocity ω, total mass M, and charge Q. The mass, however, is more dense toward the center of the sphere, while the charge is concentrated toward the edge. The mass and charge density profiles inside the sphere are given by:

$$\rho_M \propto a - r \quad \text{and} \quad \rho_Q \propto r \ .$$

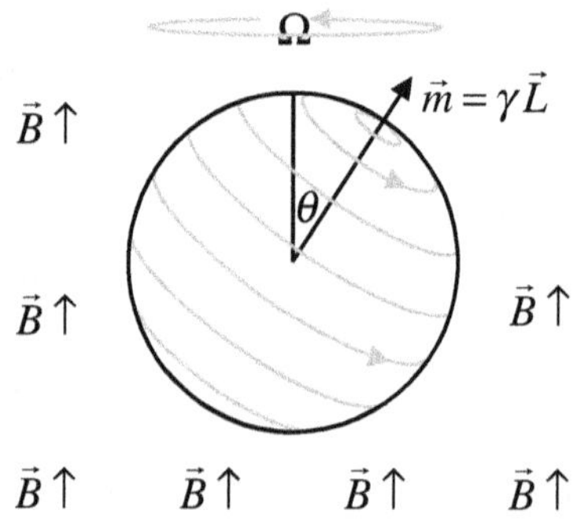

a) Normalize the density functions, and express them in terms of the total mass or charge, the radius a and the radial position r.

b) Find the current density as a function of position inside the sphere.

c) Find the total angular momentum of the sphere?

d) What is the magnetic moment of this sphere?

e) The gyromagnetic ratio γ is defined as the ratio of the magnetic moment to the angular momentum of an object. What is the gyromagnetic ratio for this sphere?

f) Find the vector potential surrounding the sphere.

g) Find the magnetic field surrounding the sphere.

Problem 11.38: A Precessing Top, MRI and NMR

Consider a spinning ball with a known gyromagnetic ratio γ and initial angular momentum $\vec{L}_0$, located in a constant magnetic field $\vec{B}$ as in the diagram.

a) Write down in vector notation the equations of motion in terms of the angular momentum, its time derivatives, the gyromagnetic ratio and the magnetic field.

b) Render these equations of motion in a fixed coordinate system, and solve these equations for the angular momentum as a function of time.

c) What is the precession frequency Ω in terms of B and γ?

d) A proton is a spin-½ particle, so for the hydrogen nucleus $\gamma = 2^{\mu_p}/_{\hbar}$, where $\mu_p = 1.411\times10^{-26}\,\frac{\text{J}}{\text{T}}$ is the proton magnetic moment. At what angular frequency will the proton precess in a one tesla magnetic field? What will be the frequency in MHz of the radio waves associated with the precession?

e) *Magnetic resonance imaging* utilizes this effect to map out body tissue. Report on modern MRI technology, the magnetic field strengths used, and the sensitivity and resolution of the machines. How do they know the position of the emitting hydrogen atoms?

f) Chemists also utilize this phenomenon of *nuclear magnetic resonance* to investigate unknown chemical samples and measure their chemical makeup. Report on how this works, and in particular why some nuclei are better candidates for NMR than others.

g) The electron magnetic moment is much larger than nuclear magnetic moments, so we would expect that this technique would be particularly successful at mapping electrons rather than nuclei. However the corresponding type of technique, *electron paramagnetic resonance*, is not nearly as widely used. Why is this? Report on this and what it is used for.

h) Geologists use this principle to measure the Earth's magnetic field; this is called a *proton magnetometer.* For typical Earth magnetic fields of about half of a gauss ($50\,\mu\text{T}$), what would be the frequency of proton precession? Report on these magnetometers, how they work, and whether they are commonly used.

i) Report on other forms of nuclear magnetic resonance technology, especially those that utilize the Earth's magnetic field. What other applications use this principle?

Thought Experiment 11.26: A Spinning Steel Rod

Consider a cylindrical rod of ferromagnetic material with a total mass m_{rod}, radius a, and height h, which is free to rotate about its primary axis. Initially, the bar is unmagnetized and not rotating. An external uniform magnetic field is applied, causing the bar to have a total magnetic moment $\vec{\mu}_{\text{rod}}$. What would we expect the bar's final angular velocity $\vec{L}_{\text{rod}}$ to be?

The total angular momentum would be zero initially, but the electrons will have a non-zero average angular momentum once the magnetic moment is induced. Since the electrons are coupled to the bar, the final total angular momentum of the bar becomes simply the net angular momentum of the electrons, or:

$$\vec{L}_{\text{rod}} = \frac{\vec{\mu}_{\text{rod}}}{\gamma_e}.$$

Thus, the classical Ampèrian loop model predicts that a steel rod will start rotating if placed in a strong magnetic field.

Discussion 11.4: The Einstein-de Haas and Barnett Effects

The nail in the coffin for the pole model were two experiments from 1915, each of which clearly demonstrated that magnetization is related to the angular momentum of the electron.

Albert Einstein and Wander J. de Haas devised an experiment, reminiscent of the Cavendish experiment, in which a cylindrical iron magnet is supported by a thin quartz fiber attached to a mirror. The magnet is suspended within a solenoid connected to a reversible DC power source. The current in the solenoid is sufficient to create a magnetic field strong enough to saturate the cylinder's magnetization in either direction. If Ampère's hypothesis is correct, this should create an angular displacement of the magnet that can be detected by deflection of a beam of light directed at the rotating mirror (see Thought Experiment 11.26).

Einstein and deHaas published their results in 1915, and these results confirmed the classical prediction we discussed above.

Meanwhile, the American physicist Samuel J. Barnett was investigating the converse effect, where spinning ferromagnetic materials become magnetized. Barnett also published his results in 1915.[157]

Barnett reasoned that electrons inside an object that rotates with an angular velocity Ω would experience a fictitious Coriolis force:

$$\vec{F}_{\text{cor}} = 2m_e\vec{v}_e \times \vec{\Omega}.$$

This means that whatever an electron does, so long as its motion is responsible for the magnetic effects of matter, would be just as if some fictitious magnetic field, $\vec{B}_{\text{cor}}$, were to exist such that the fictitious magnetic force would equal the Coriolis force. In a non-inertial reference frame, we can pretend there is a fictitious magnetic field in place of the Coriolis effect because both act in a similar manner to objects with a fixed charge to mass ratio. To see this, let's equate the Coriolis force on an electron moving with velocity $\vec{v}_e$ (with respect to the non-inertial reference frame) and the fictitious magnetic force on the same electron:

$$\vec{F}_{\text{cor}} = 2m_e\vec{v}_e \times \vec{\Omega} = q_e\vec{v}_e \times \vec{B}_{\text{cor}} \quad\rightarrow\quad q_e\vec{B}_{\text{cor}} = 2m_e\vec{\Omega} \quad\rightarrow\quad \vec{B}_{\text{cor}} = -2\frac{m_e}{e}\vec{\Omega}.$$

If the motion of electrons were responsible for magnetization, then spinning a ferromagnetic disk to high velocities would induce the same magnetization as if the disk were placed in a magnetic field. Moreover, because the two forces have exactly the same velocity dependence, the details of the electron motion should not matter. For example, imagine that the electrons were little balls of charge with classical spin; the results would be exactly the same, because the Coriolis effect would act on these in the same way as a magnetic field would, no matter how complicated the electron's motion otherwise.

Barnett conducted experiments where he spun a steel rod in a lathe at a fast rate, and measured the magnetization. The results clearly showed that the effect was real, and also that the effect is not strong enough to explain the magnetic moment of the earth. As you will find in Problem 11.39, the Barnett effect is applicable to spinning interstellar dust particles.

The measurements of the Einstein-deHaas and Barnett effects provided strong proof that magnetization was related to angular momentum, and thus drove the final nail into the coffin for any serious consideration of the pole model of matter.

The Einstein-deHaas and Barnett effects marked the end of the story for the pole model, but they uncovered a very odd puzzle about the electron. In 1918 Princeton students John Q. Stewart and Maurice Pate performed a follow-up Einstein-deHaas experiment showing that the electron gyromagnetic ratio is twice what is predicted classically,[158] or:

---

[157] S.J. Barnett, "Magnetization by Rotation," Phys. Rev., **6**:4, (1915), 239-270.

[158] J.Q. Stewart, "On the moment of momentum accompanying magnetic moment in iron and nickel," Phys. Rev., **11** (1918), 100-120.

$$\Gamma_e = \frac{\vec{\mu}_e}{\vec{L}_e} = -\frac{e}{m_e}.$$

We now measure this ratio as a unitless quantity called a g-factor, with:

$$g \equiv \left(\frac{2m}{q}\right)\frac{\vec{\mu}}{\vec{L}} = \left(\frac{2m}{q}\right)\Gamma,$$

so the electron g-factor is very close to 2.

As you will find in Problem 11.40, modern g-factors are not exactly 2, depending on the magnetic material involved.

The Barnett effect also depends on the same g-factor. The fictitious magnetic field due to rotation is now given by:

$$\vec{B}_{\text{cor}} = -\frac{2}{g}\frac{m_e}{e}\vec{\Omega} \approx -\frac{m_e}{e}\vec{\Omega}.$$

These experimental results baffled scientists in the early twentieth century.[159] Indeed, it was one of the first clues that hinted toward the quantum mechanical theory of *spin*, or more precisely *intrinsic angular momentum.* In 1928, the British physicist Paul Dirac formulated a relativistic quantum mechanical wave equation to describe the electron that provided a theoretical justification of spin as the consequence of uniting quantum mechanics and Einstein's special theory of relativity. The equation predicted, among other things, the existence of a new form of matter, anti-matter, and the anti-matter counterpart to the electron, the positron, was discovered in 1932. From his relativistic wave equation, which was later dubbed *the Dirac equation* in his honor, Dirac also predicted that the value of the g-factor is exactly $g_e = 2$. The modern value of the electron g-factor is actually $g_e = 2.0023193$, which differs from Dirac's relativistic quantum mechanical prediction of exactly two due to subtle effects explained by quantum electrodynamics.

Dirac then resurrected the pole model in 1931, when he predicted the existence of a quantum of magnetic monopole.[160] Dirac argued that if charge is quantized, so must be pole strength. Dirac also showed that the force holding two quanta of monopoles together must be about 4700 times stronger than that holding a proton and electron of the same distance, so it would be difficult to separate these monopoles if they did actually exist.

Despite much experimental effort, magnetic monopoles have yet to be reproducibly found in nature.[161] However, some very cold substances, called *spin ices* by analogy with crystalline water ice, appear to show a small gap between the poles of their intrinsic magnetic moments. The final upshot for Petrus Peregrinus is that his principle is still alive and well 750 years after the siege of Lucia.

---

[159] For a fascinating account of the Einstein-de Haas experiment and its aftermath, see P. Gallson, "Theoretical predispositions in experimental physics: Einstein and the gyromagnetic experiments, 1915-1925," Historical Studies in the Physical Sciences, **12**:2 (1982), 285-323.

[160] Dirac, P.A.M., "Quantised Singularities in the Electromagnetic Field," Proc. Roy. Soc. A., **133** (1931), 60-72.

[161] In 1982 one monopole was claimed to be found, but no others were reported from the same detector (B. Cabrera, "First Results from a Superconductive Detector for Moving Magnetic Monopoles," Phys. Rev. Lett., **48**, (1982) 1378–1381).

Problem 11.39: The Barnett Effect and Interstellar Dust Grains

Optical light from galaxies, such as the Whirlpool (M51), is slightly polarized azimuthally.[162]

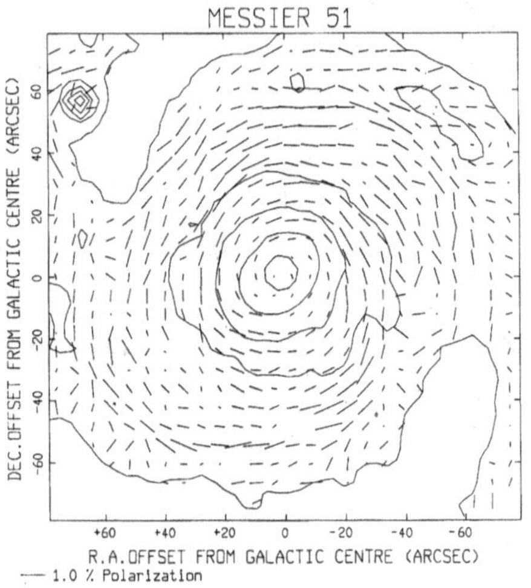

The *interstellar medium* is known to contain fine cigar-shaped particles that astronomers call *dust grains.* The most analogous substance on earth would be cigarette smoke, and like smoke dust grains absorb light. The grains, however, absorbing more perpendicularly polarized light, so the dust grains in the Whirlpool galaxy are aligned in the $\hat{s}$ direction; the polarization is in the $\hat{\varphi}$ direction; and we are looking at it from the $\hat{z}$ direction.

An early hypothesis was that dust grains were ferromagnetic and, like iron filings, would align with the galactic magnetic, however independent observations showed that in spiral galaxies the magnetic field is primarily azimuthal, so the early hypothesis must have been wrong. We now know that galaxy magnetic fields are about:

$$\vec{B}_G = B_G\hat{\varphi} \approx (1\,\mathrm{nT})\hat{\varphi}\,.$$

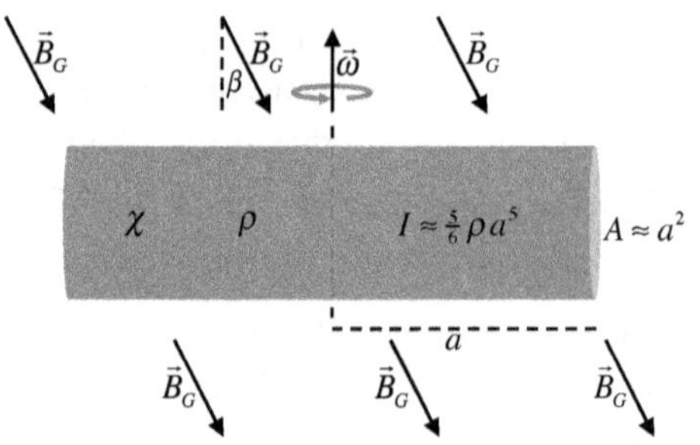

In this exercise you will model how dust grains align perpendicular to the galactic magnetic field.

Assume that a *typical* dust grain has the following parameters: $\rho \approx 3\,{}^{\mathrm{g}}\!/_{\mathrm{cm}^3}$, $A \approx a^2$, and $\chi \approx 10^{-4}$. Also assume that dust grains come in a whole range of sizes, a, but all have the same shape, density and magnetic susceptibility as the typical dust grain.

a) The average rotational kinetic energy should be related to the ambient temperature, and the spin rate, by: $\mathrm{T}_{\mathrm{spin}} = \frac{1}{2}k_B T = \frac{1}{2}I\omega^2$. If the typical dust grain have a rotation frequency of: $\frac{1}{2\pi}\omega \approx 40\,\mathrm{kHz}$, what is the corresponding local ambient temperature.

b) What is the fictitious magnetic field, B_{cor}, affecting the electrons in the typical dust grain? How does this this compare to the typical galactic field of a nanotesla?

c) Find the magnetization $\vec{M}$ as a function of the angular velocity $\vec{\omega}$.

d) For our typical dust grain, what will be its magnetic moment, $\vec{m}$, due to the Barnett effect? For a distribution of dust grains all with the same kinetic energy, $\mathrm{T}_{\mathrm{spin}}$, what will be the magnetic moment as a function of grain size a?

---

[162] Scarrott, S.M., Ward-Thomson, D. and Warren-Smith, R.F., "Evidence for a spiral magnetic field configuration in the galaxy M51," MNRAS, **224**, (1987), 299-305.

e) These dust grains are located in the galactic magnetic field, $\vec{B}_G$. Write down the torque on the dust grain in terms of the angle β and the magnetic moment $\vec{m}$.

f) This torque will cause the dust grain to precess about the magnetic field. Find the precession angular velocity Ω for the typical dust grain at an angle $\beta = 45°$. What is this as a function of grain size, again assuming constant kinetic energy?

g) To explain the observations, dust grains must align with the magnetic field: $\vec{\omega} \parallel \vec{B}_G$, which means that somehow the precessional energy must dissipate. What makes the field align is an area of active research, but one idea is that there are losses due to the changing magnetic field. In the reference frame of the dust grain, the magnetic field appears to be have a small time varying component given by: $\vec{B}_{\mathrm{AC}} = \vec{B}_G \sin\beta \sin(\Omega t)$.

Using Faraday's law, estimate the induced electric field in the dust grain.

h) Imagine that the grain has a conductivity like coal, $\kappa \approx 10\ \mathrm{Sm}^{-1}$, estimate the eddy current density inside the typical dust grain. From this estimate the magnetic moment due to these currents.

i) Compare the torque from the eddy currents to the precessional angular momentum, and estimate the typical time it would take for a dust grain to align with the field. The sun orbits our galaxy once every 230 million years; is the time you found long or short compared to this?

j) Astronomers observe extra diffuse light with a frequency of about 30 GHz, and some scientists claim that these microwaves come from tiny, very rapidly spinning, dust grains.[163] If you were to assume that these have the same rotational kinetic energy as the typical dust grain, how big would these dust grains be? How does this size compare to typical atomic spacing?

Problem 11.40: Measuring the g-factor of NiFe Film

In a 2006, National Institute of Standards and Technology (NIST) scientists published a measurement of the magneto mechanical g-factor of the alloy $Ni_{80}Fe_{20}$ using the Einstein-deHaas effect.[164]

A diagram of their set-up is shown in their FIG 1, which consists of a tiny flexible cantilever with dimensions $200\,\mu\mathrm{m} \times 20\,\mu\mathrm{m} \times 0.6\,\mu\mathrm{m}$. On the top of the beam, they spread a thin film, only 50 nanometers thick, of the $Ni_{80}Fe_{20}$ alloy. The effective modal mass of the tiny beam is $m_{\mathrm{mod}} = 1.86 \times 10^{-12}\,\mathrm{kg}$, and the beam has a quality factor of $Q = 24$. With this apparatus, very small deviations in the bending of the cantilever beam can be measured.

Notice that in their Fig. 1 the authors use the quantity $\vec{H}_{\mathrm{AC}}$ to denote the magnetizing field, which corresponds to $\vec{H}_{\mathrm{AC}} = n I_{\mathrm{AC}} \hat{x}$, where n is the number of turns per length of the solenoid and I_{AC} is the current through the coils.

---

163 B.T. Drain, & A. Lazarian, "Electric Dipole Radiation from Spinning Dust Grains," The Astrophysical Journal, **508** (1998), 157-179.

164 T.M. Wallis, J. Moreland and P. Kabos, "Einstein-de Haas effect in a NiFe film deposited on a microcantilever," Applied Physics Letters, **89**, (2006), 122502.

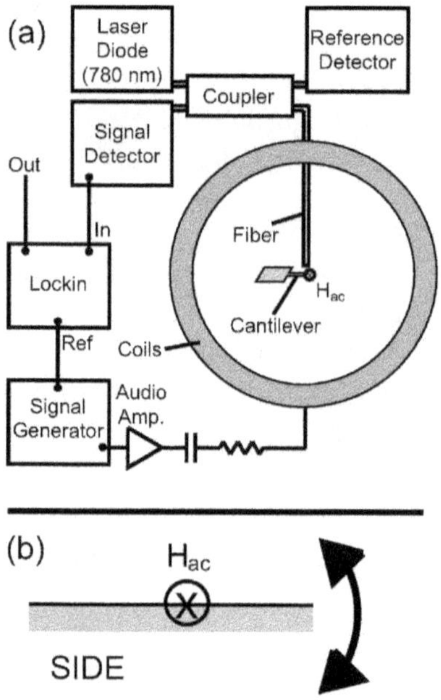

FIG. 1. (a) Apparatus for measurement of the Einstein–de Haas effect in a ferromagnetic film deposited on a microcantilever. (b) Side view of cantilever. The alternating magnetic field (H_{ac}) goes in and out of the page, as drawn. The resulting bending motion is indicated by black arrows.

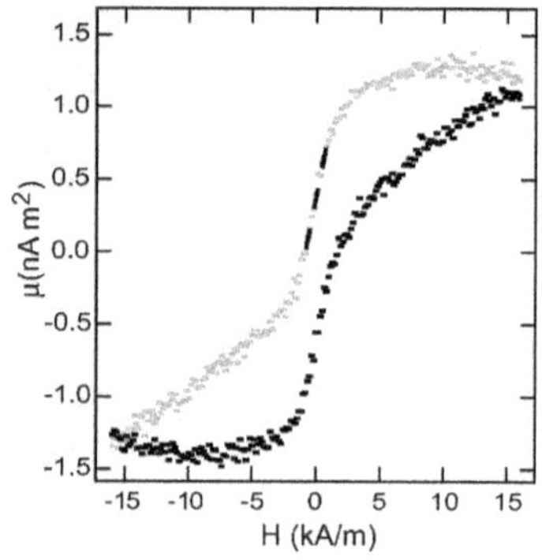

FIG. 2. Alternating gradient magnetometry of the film is shown. Black points correspond to the increasing field and gray points to the decreasing field. The dashed line is a linear fit to the region between −800 and 800 A/m and has a slope of $(4.57\pm0.21)\times10^{-13}$ m^3. Here, the linear fit is done on the curve corresponding to decreasing field; a linear fit to the increasing field (not shown) gives a consistent value of the slope.

a) Ferromagnetic substances do not follow a linear relationship between the applied field and the magnetization, so it was important to measure a hysteresis curve (their FIG. 2) to be able to manipulate the magnetic moment (μ) of the film. (Do not confuse the magnetic moment, their μ, with the permeability of the medium, our μ earlier in this chapter.)

From their plot, find the intrinsic magnetization and coercity of $Ni_{80}Fe_{20}$. Express these in units of amperes per meter and teslas respectively. Draw a diagram showing each value, and how you converted each set of units.

b) In a small regime with a low applied magnetic field, $|\mu_0 H| < 1\,\text{mT}$, the relationship between the magnetic moment and the magnetic field are approximately linear. This is shown in their FIG 2 with a dashed line, and a best fit slope of $(4.57\pm0.21)\times10^{-13}\,\text{m}^3$. What is the magnetic susceptibility, χ, of the material in this linear regime?

c) By driving an AC current through the coils, a sinusoidal field, $\vec{H}_{\text{AC}} = H_0 \sin(\omega t)\hat{x}$, is imparted at the location of the tiny beam. What is the magnetization of the film, $\vec{M}(t)$?

d) Find the total magnetic moment of the film as a function of the root means squared amplitude of the AC magnetizing field, H_0, the driving frequency ω, and time? One run of the experiment had a magnetizing field amplitude of $H_0 = 367\,{}^{\text{A}}\!/_{\text{m}}$. What is the change in magnetic moment, $\Delta\mu$, corresponding to twice the amplitude of the field?

e) The tiny beam has a deflection amplitude, z_0, which can be independently measured. In their FIG. 3, they plot this amplitude as a function of driving frequency $f = \frac{1}{2\pi}\omega$. The driving torque about the fulcrum has amplitude given by the beam equation when driven at resonance:

$$\tau_f = \frac{l_c m_{\text{mod}} \omega^2}{Q} z_0 .$$

From their Fig. 3, estimate the measured torque about the end of the beam.

f) Show that the torque about the center of the plate, which is causing the bending of the beam, is exactly half of the torque about the fulcrum. Draw a clear picture that illustrates this.

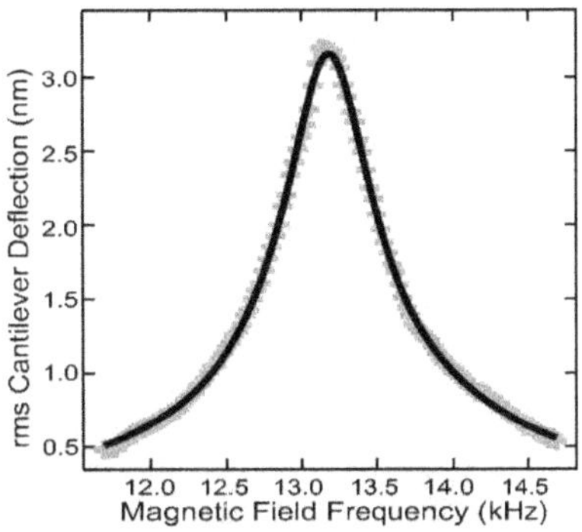

FIG. 3. Root mean square (rms) cantilever deflection is shown as a function of the frequency of the applied magnetic field [gray "plus" signs are measured data; solid black line is a fit of Eq. (4)]. The amplitude of the applied magnetic field (H) is 367 A/m, resulting in a change in magnetic moment of the film ($\Delta\mu$) of 0.335 ± 0.012 nA m^2.

g) Integrate the torque over half of a period to find the angular momentum difference from minimum to maximum, $\Delta\vec{J}$. Using your result in (d), find the gyromagnetic ratio Γ of the electrons in this NiFe film. Is this consistent with the paper's measured g-factor of $g' = 1.82$?

h) Good experimentalists make a number of measurements varying the independent variable. FIG. 4 is a plot of the magnetic moment difference of the film (independent variable), and the classically expected magnetic moment difference for each measured resonance deflection z (dependent variable). A least squares regression is fit to the points with a slope of 0.55 ± 0.03. What is the corresponding g-factor (g')?

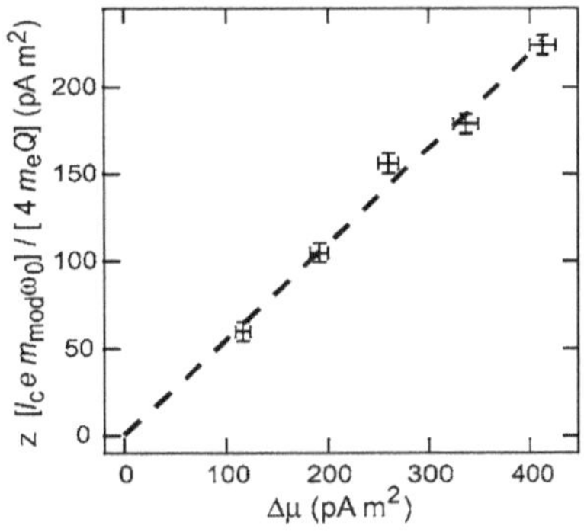

FIG. 4. Maximum rms deflection of the cantilever (i.e., deflection at mechanical resonance) is shown as a function of the change in the magnetic moment of the film. The vertical axis has been scaled so that the slope of a linear fit is $1/g'$. With $l_c = 200$ μm, $m_{\text{mod}} = 1.86\times10^{-12}$ kg, $f_0 = 13\,180$ Hz, and $Q = 24$, the fit gives $g' = 1.83$.

i) If you have not done so already, look up this paper,[165] and read it over. Look up and skim the abstracts of the more recent papers that cite it,[166] and discuss how the Einstein-deHaas effect is useful and relevant today.

[165] If you cannot download it from the journal, it should be posted at http://www.nist.gov/ as a publication on the topic of *electromagnetics*.

[166] You can use http://adsabs.harvard.edu/ to find citations to the article.

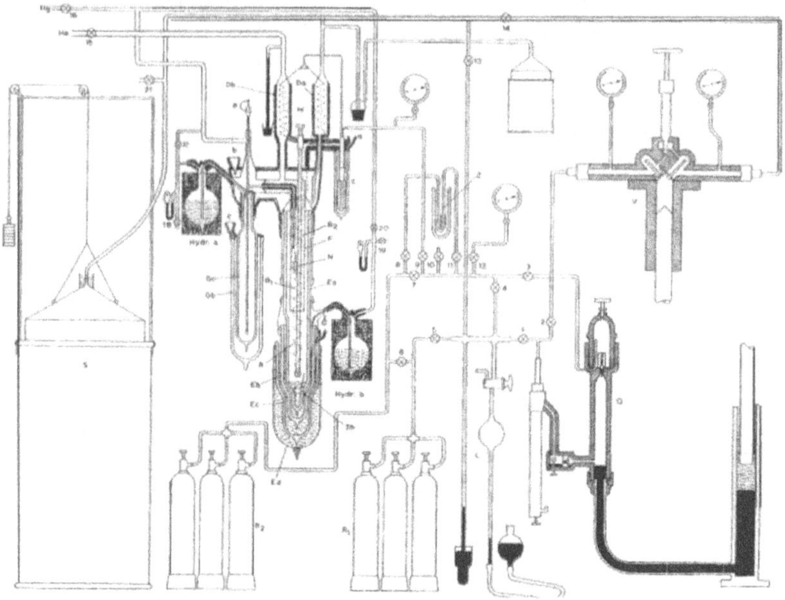

Plate 20: Kamerlingh Onnes's Cryogenics Lab[167]

[167] The top picture is of Paul Ehrenfest, Henrick Lorentz, Neils Bohr, and Kamerlingh Onnes. The bottom is a diagram from Onnes's 1913 Nobel lecture.

Chapter 12 Superconductors and Plasmas

Big whirls have little whirls
that feed on their velocity,
and little whirls have lesser whirls
and so on to viscosity.

--Lewis Fry Richardson[168] [169]

It takes completely different physics to model whitewater than flatwater, but why does this transition happen so suddenly, as when an ocean wave breaks into foam? In general, slow moving viscous fluids are extremely well understood, using laminar fluid dynamics which was developed over the course of centuries.

Turbulence has also been understood for centuries—but only at the level of the opening poem. Turbulent "whirls" are called *eddies*, or *eddy currents*, and when Arago and Foucault discovered similarly shaped electrical currents in conductors they assumed they had a similar cause. But, as Maxwell made clear, they had nothing to do with turbulent eddies. Rather it was quite the opposite. As Heaviside pointed out in 1883, modeling electricity as a laminar fluid nicely explains Ohm's law (p. 90). Also in 1883, Osborne Reynolds, a fellow Briton, developed a criterion, now called the *Reynolds number*, for determining whether a fluid would be fully laminar, fully turbulent, or somewhere in between. Putting their two arguments together (Thought Experiment 12.1), shows that normal conductors are not only laminar, but have Reynolds numbers a billion times below the turbulent threshold. Thus, except for the special phenomena covered in this chapter, the kinetic energy associated with the flow of charge can be completely ignored.

The most obvious area where electrodynamics and fluid mechanics merge, is the numerically intensive field of *magnetohydrodynamics (MHD)*, also called *plasma physics*. Usually MHD involves the flow of electrically conductive material, such as ionized gas or sea water. Charged particles move in circles in magnetic fields, inhibiting plasma flow perpendicular to the magnetic field, but allowing for free movement parallel to the field. Thus, either the material appears to become trapped along magnetic field lines, called *flux tubes*, or the magnetic field lines appear to bend with the matter, where the magnetic field is said to be *frozen in*. Ironically, the hotter a gas gets, the more likely it is to become ionized, and the more likely the magnetic field is to become frozen in.

To understand MHD, one has to understand hydrodynamics, which tends to become more difficult at higher Reynolds numbers. Recall that Millikan's dominant source of error in his oil drop experiment, for example, was whether Stokes law applied (p. 321). Stokes law, after all, relates the drag force on a sphere to its velocity, but only at low Reynolds numbers (p. 326). Thanks to this imperfect assumption, in hindsight other techniques, like Rutherford's for example, resulted in closer measurements of the elementary charge.

Predicting the weather is important, yet exceedingly difficult. Until the mid twentieth century, the best strategy was to simply fit periodic functions to the historical record, and record these in what is known as a *farmer's almanac*. For moderately high Reynolds number systems, especially one

---

168 Richardson, Lewis F. Weather Prediction by Numerical Process. 1st ed. (Cambridge University Press, 1922), retrieved from https://archive.org/.

169 This little ditty was based on a nursery rhyme, which, in-turn, came from a line in Jonathan Swift's 1733 poem On Poetry: a Rhapsody that reads: "So, naturalists observe, a flea / Has smaller fleas that on him prey / And these have smaller still to bite 'em, / And so proceed ad infinitum."

driven by the daily, monthly, and annual motion of the earth and the moon, these *ad hoc* methods used to outperform even those based on sound physical models like those of Lewis Richardson, who published the fun poem about whirls in his 1922 treatise on numerical weather prediction before defining them more rigorously using Gaussian surfaces. You may wonder why Richardson wrote about predicting the weather using numerical methods in 1922, a couple of decades before the computer was invented. Actually, applied mathematicians have used numerical algorithms to solve problems for millennia. Everyday arithmetic calculation, after all, are simply numerical algorithms. Computers did existed in 1922, but they were people with the job title "computer."

Today, the cost of computation dominates studies of turbulent systems, with the higher the Reynolds number the higher the cost (Problem 12.2). This is because the Reynolds number essentially measures the ratio of the largest, to the smallest, length scale of interest. Thus, the higher the Reynolds number, the higher the resolution required in a numerical simulation. That said, if the Reynolds number of an MHD system is very large, then one can assume *a priori* that whirling eddies, and a frozen in field, have already wound up the magnetic field lines as far as they can go, without dissipating energy. In these cases the energy is not only conserved, but also equally divided between kinetic energy and magnetic energy. This criterion of *equipartition* greatly simplifies the analysis, allowing for analytical studies of very low viscosity turbulent MHD systems, almost all of which occur at very high temperatures—with the stark exception of superconductivity.

In 1911 the Dutch experimentalist, Heike Kamerlingh Onnes, first observed that, at temperatures near absolute zero, the resistivity of liquid helium vanished. In 1933 German physicists Walther Meissner and Robert Ochsenfeld found that magnetic fields do not penetrate superconductors, in much the same way as the electric field goes to zero inside a normal conductor. This was confusing at the time, as is nicely outlined in the opening address to a 1935 conference on the topic by the Canadian physicist John Cunningham McLennan.

> The series of experiments on magnetic fields around supraconductors, commenced by Meissner and his co-workers, has recently opened a new method of attack on this problem. ... Meissner and his collaborators in Berlin began by studying the field distribution round cylinders, and inside a hollow cylinder, with a small test coil. ...
>
> In agreement with simple electromagnetic considerations, the supraconductor acts as if its magnetic permeability were zero, at any rate beyond a very thin surface layer. But when the transition to the supraconducting state takes place in the presence of a magnetic field, either by cooling in a steady external field or by reducing the field strength from an initial value greater than the critical field, the experiments agree only in that the result is never quite what was expected. However, certain experimental facts seem to stand out, at least for most metals which have been tested. First, the distribution of external field is spontaneously readjusted nearly to the distribution it would have if the induction within the body were everywhere zero. Second, the field inside a hollow is only slightly altered, if at all, in the transition, showing the curious phenomenon of an apparently isolated section of magnetic field in a stable state. Third, for a solid body varying amounts of the original magnetic flux are "locked in" in the course of the transition. The field thus locked in, like the field inside the hollow, cannot then be disturbed by any external influence so long as the body remains in the supraconducting state.[170]

---

[170] J.C. McLennan, et al., "A Discussion on Supraconductivity and Other Low Temperature Phenomena," Proceedings of the Royal Society of London A, **152** (1935), 1-46.

Notice that McLennan uses the word *magnetic field* to mean the magnetic field applied externally in the laboratory, $\vec{H} = \frac{1}{\mu_0}\vec{B}_0$, and the word *induction* to mean the magnetic field, $\vec{B}$, inside the "supraconductor," as in the Heaviside convention.

These experimental results led the brothers, Fritz and Heinz London (also in attendance), to come up with a classical model, which they referred to as *acceleration theory*, to replace Ohm's law.[171]

As the opening poem states, when stirred on large scales, "big whirls have little whirls that feed on their velocity, and little whirls have lesser whirls, and so on to viscosity." This *cascade* transfers kinetic energy from slowly moving large eddies to rapidly moving tiny eddies that dissipate the energy—the lower the viscosity the smaller the tiny eddies.

What happens when the electron sea has absolutely no viscosity? The faster the electrons move, $\vec{J}$, the greater the sideways magnetic force they encounter. On the other hand, the smaller the whirl, $\vec{\nabla}\times\vec{J}$, the stronger the magnetic field they induce. As is usually the case in a *stochastic* processes, these energies quickly even out. In any given volume, half of the energy resides in the magnetic field and the other half in the electrons. Thus, in superconductors, if current is whirling, there must be a magnetic field. Similarly, if there is no magnetic field, there must also be no current.

We commence the chapter with a review of viscosity and a discussion of laminar and turbulent fluid flow. Next we consider the magnetic fields in conductive fluids, such as the ionized plasma in the sun. Next we discuss early superconductivity experiments, followed by London acceleration theory. We finish the chapter with a short discussion of the growing field of high temperature superconductivity.[172]

12.1 ***Turbulence and the Reynolds Number***

Incompressible fluid flow tends to fall into one of three regimes (laminar, turbulent, or in between) depending on the dimensionless *Reynolds number* (R), named after Osborne Reynolds:[173]

$$R \approx \frac{\rho_m a v}{\eta},$$

where ρ_m is the fluid mass density, a the characteristic size of an object, v the velocity of the fluid, and η is the fluid's viscosity. A river, for example, may become turbulent after encountering a rock of radius a, with swirling *eddy currents* behind it. In general, the larger the Reynolds number, the more prone to turbulence a medium becomes. For example, molasses pouring from a jar would have $R \sim 1$, and a person in a pool would have $R \sim 10{,}000$. So, in order to model electricity as a fluid (electrons) flowing around stationary rocks (atoms), we must

---

[171] F. London, and H. London, "The Electromagnetic Equations of the Supraconductor," Proceedings of the Royal Society of London A, **149** (1935), 71-88.

[172] For further discussion of the behavior of a superconductor in a magnetic field, see, for example, K.B. Ma, et al., "Superconductor and magnet levitation devices," Review of Scientific Instruments **74**, no. 12, (2003), 4990-5017.

[173] O. Reynolds, "An Experimental Investigation of the Circumstances Which Determine Whether the Motion of Water Shall Be Direct or Sinuous, and of the Law of Resistance in Parallel Channels," Phil. Trans. R. Soc. Lond., **174** (1883), 935-982.

estimate the Reynolds number. If the Reynolds number is larger than a million or so, we will assume fully turbulent flow.

On the other hand, if the Reynolds number is less than unity, then we can assume that all motion stops as soon as external body forces cease to exist. Moreover, we can also assume that the kinetic energy associated with the flow is negligible, no matter what the current. In other words, all objects would move at a constant speed that is proportional to any external forces acting on them, just as Aristotle's physics predicted.

In Thought Experiment 12.1 below, we show that, due to the strength of the electrical forces between atoms and the electron sea and the small sizes of atoms, a typical current flowing through a copper wire has a Reynolds number of $R \approx 10^{-8}$, which is even smaller than the Reynolds number of one-celled animals or a person stuck in quicksand.[174]

In terms of a fluid model of electricity, we can therefore assume minimal shear from one streamline of current to the next.[175] This condition of minimal shear is called *irrotational flow*, and it combines with the continuity equation to uniquely solve the current density as a function of position, given only boundary conditions.

Plasma flow is also highly viscous, as the electromagnetic interaction length between the electrons is quite large. However, it is important not to be too sanguine when modeling plasmas in space, since their Reynolds numbers can be quite high due to the far distances between each particle.

Superconductors represent the opposite limit than normal conductors, since $R \to \infty$. With absolutely no viscosity, superconductors are actually quite understandable on classical scales. This is primarily because, without any dissipative forces, they must conserve kinetic energy.

Problem 12.1: The Reynolds Number in Everyday Life

(a) Estimate the Reynolds number of the New York aqueduct in Problem 10.13 (p. 325).

(b) The Reynolds number is often thought of as the ratio of two lengths. The largest scales of interest, represented by a, and the smallest scales that can affect the system. Find an expression for this viscosity scale length, λ.

(c) Research and discuss *the butterfly effect*. How are the Reynolds number and the butterfly effect related?

(c) Estimate the viscosity scale length for water in most everyday situations.

(d) Explain why white water looks white.

Thought Experiment 12.1: The Reynolds Number from Ohm's Law

In Thought Experiment 3.7 (p. 90), we discussed Heaviside's interpretation of Ohm's law as laminar drag in a viscous fluid, and we found the conductivity (κ) in terms of the viscosity (η):

---

174 E.M. Purcell, "Life at Low Reynolds Numbers," American Journal of Physics, **45** (1976), 3-11.

175 A *streamline* is a curve that is instantaneously tangent to the velocity vector of the flow. A streamline therefore indicates the direction in which a fluid element will travel at any point in time. In purely mathematical terms, a streamline is an *integral curve* of the *velocity field* of the fluid. The integral curves of electric and magnetic fields are called electric and magnetic *fieldlines*, for example.

$$\kappa = \frac{\rho_-^{\ 2}}{6\pi\eta a n_+} \quad \rightarrow \quad \eta = \frac{\rho_-^{\ 2}}{6\pi\kappa a n_+},$$

where a represents the size of the atoms and n_+ their number density.

We now use the expression for the Reynolds number:

$$R \approx \frac{\rho_m a v}{\eta} \approx \rho_m a v\left(\frac{6\pi\kappa a n_+}{\rho_-^{\ 2}}\right) \approx \left(\frac{6\pi\kappa a^2 \rho_m n_+}{\rho_-^{\ 2}}\right) v \approx \left(\frac{6\pi\kappa a^2 \rho_m n_+}{\rho_-^{\ 3}}\right) J.$$

Now we replace the mass density, and electron charge density, with the number density of electrons. Then the Reynolds number is given by:

$$R \approx \left(\frac{6\pi\kappa a^2 \rho_m n_+}{\rho_-^{\ 3}}\right) J \approx \left(\frac{6\pi\kappa a^2 m_e n_e n_+}{e^3 n_e^3}\right) J \approx \left(\frac{6\pi\kappa a^2 m_e n_+}{e^3 n_e^2}\right) J.$$

Now in the case of copper, there are two valence electrons per atom, so we can write R in terms of the total mass density ρ_{cu} of copper:

$$R \approx \left(\frac{6\pi\kappa a^2 m_e n_+}{e^3 (2n_+)}\right) J \approx \left(\frac{3\pi}{2}\right)\left(\frac{\kappa a^2 m_e}{e^3 n_+}\right) J \approx \left(\frac{3\pi}{2}\right)\left(\frac{\kappa a^2 A_{cu} m_p m_e}{e^3 \rho_{\text{cu}}}\right) J.$$

Putting in numbers for copper wire at maximum load, we have:

$$R \approx \frac{3\pi}{2}\,\frac{\left(6\times10^{7}\ \text{S}/\text{m}\right)\left(1.3\times10^{-10}\ \text{m}\right)^2 (63.5)\left(1.6\times10^{-27}\ \text{kg}\right)\left(9\times10^{-31}\ \text{kg}\right)\left(3\times10^{6}\ \text{A}/\text{m}^2\right)}{\left(1.6\times10^{-19}\ \text{C}\right)^3\left(9000\ \text{kg}/\text{m}^3\right)} \approx 3.5\times10^{-8}.$$

Now we check the units in our calculation:

$$[R] = \frac{\left(\text{S}/\text{m}\right)(\text{m})^2(\text{kg})(\text{kg})\left(\text{A}/\text{m}^2\right)}{(\text{C})^3\left(\text{kg}/\text{m}^3\right)} = \frac{\left(\frac{\text{C}^2\cdot\text{s}}{\text{kg}\cdot\text{m}^3}\right)(\text{m})^2(\text{kg})(\text{kg})\left(\frac{\text{C}}{\text{s}\cdot\text{m}^2}\right)}{(\text{C})^3\left(\text{kg}/\text{m}^3\right)} = \text{unitless}$$

So, even for copper wire at the maximum load, we can assume that the current follows the electric field, and that it stops when the electric field turns off.

Example 12.1: A Pump in a Pool and Numerical Viscosity

In Chapter 1 (p. 49) we discussed how to use the method of relaxation to ensure conservation of charge; in Chapter 5 we used relaxation methods to solve electrostatics problems (p. 174); and in Chapter 6 we used the same mathematical technique to model Faraday's magnetic field observations (p. 198). However, we argued that providing boundary conditions, and simply conserving charge, did not provide enough information to uniquely solve for the current density using the fluid model.

In this example, we consider an idealized pump in the middle of a square shallow pool. We set the mass flux to a uniform constant inside of pump, so $\vec{J}_m = J_0\hat{x}$. Our only other constraint is that we do not allow for mass to flow through the sides of the pump, or along the sides of the pool.

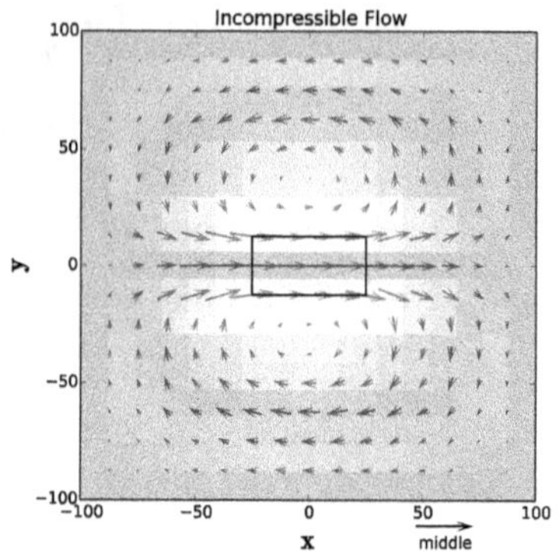

The figure shows our solution after convergence, and it appears reasonable. In fact, it looks exactly as we might expect from everyday experience.

As a physics model, on the other hand, it is somewhat lacking. Yes we did make sure that mass was conserved everywhere, but we did not include a viscosity. Why does it appear to work out so well?

The answer is quite simple, we actually did assume a viscosity due to the numerical resolution of our simulation, called the *numerical viscosity*.

Looking at the scale of the plot, you can see that the pool is about 100 cells from the center of the pump. Therefore, our simulation could not represent any physical system with Reynolds numbers bigger than about 100.

The time it takes a computer algorithm to run tends to be proportional to the number of calculations it must make. The number of calculations, in turn, monotonically increases with the total number of cells it must keep track of. Therefore, in order to simulate a hydrodynamic system with a given Reynolds number, R, in three dimensions, the number of cells to keep track of must scale as R^3. Moreover, each cell in three dimensions will depend on the 26 surrounding cells, which in turn depend on their surrounding cells, so the minimum number of iterations before convergence must be, at the very least, proportional to the maximum distance two cells can be apart, so at the very least the number of calculations would scale as R^4. Thus, the higher the Reynolds number, the much higher the computational cost. This is why it is so hard to predict the weather.

Problem 12.2: The Cost of a Hydrodynamic Simulation

a) How much memory is required to hold an $N \times N \times N$ three dimensional velocity field on a computer that stores each number in 32 bits? Express your answer in bytes, as a function of N.

b) What is the maximum Reynolds number of a simulation where each velocity field takes up 10 Mbytes of memory? What about 100 Mbytes? What about 1 Gbyte?

c) Imagine that you time your test code as it runs on your laptop and it takes about 1, 10, 80 and 640 milliseconds per iteration, with resolutions of 10, 20, 40, and 80 cells per side, respectively. How would the time, per iteration, of your code scale with Reynolds number if your ran it at even higher resolution on a computer with plenty of memory?

d) Imagine that you gave your program some criterion for convergence, and you found that the computer took 20, 80, 300, and 1300 iterations to converge for each of your test runs. How would the total time to run your code scale with Reynolds number if your ran it at even higher resolution on a computer with plenty of memory?

e) Now imagine that a computer lab at your school has 10 computers identical to your laptop, which you can use to run the code distributively with your laptop being the master computer and the other 10 each running at the same time. If your computer networks runs at 100 Mbs, estimate the total time it would take to run a $200 \times 200 \times 200$ cell simulation. Do you think it would run overnight? Clearly explain your assumptions about the code.

12.2 *Magnetic Fields in Plasmas*

Most of the universe is made of hydrogen, which is often fully ionized. The sun, and other stars, are mostly comprised of fully ionized hydrogen, because of the high temperature and intensity of the ambient light. The solar corona, just outside of the sun's photosphere, is particularly interesting. This is especially true when the solar magnetic field is particularly strong, call a *solar maximum*, because the plasma is both pushed by the sun's magnetic field and creates magnetic fields of its own. Moreover, when the radiation pressure exceeds the gravitational pull, this plasma blows away from the sun and can interact with the magnetic fields of planets, including Earth, causing the northern, and southern, lights when the particles hit the atmosphere near the magnetic poles.

When extremely massive stars implode, they subsequently explode in what is known as a *supernova*. Decades to centuries later, inhomogeneities in the surrounding *interstellar medium* can throw the plasma into turbulence, which, in turn, amplifies the surrounding magnetic field. Radio astronomers observe this magnetic field structure, because electrons radiate when accelerated and the greater the field the greater the acceleration (see Plate 27, p. 505). Curiously, the large scale magnetic field structure of many of these supernova remnants appear to be radial. These are not magnetic monopoles! The magnetic field, rather, becomes stretched and folded as plasma of high and low density exchange places due to the overall deceleration of the material.

Conductors have very small Reynolds numbers (Thought Experiment 12.1), but the Reynolds numbers in plasmas can vary greatly. So, like the flow of water, they can be either laminar or turbulent depending on the situation.

Since Coulomb forces among charged particles act over long distances, viscosities in plasmas are much higher than in even much denser gasses. On the other hand, stellar winds and shock waves travel fast over vast distances, and interact with large objects. These tendencies tend to cancel out, providing a whole range of Reynolds numbers, from very small to very large.

Magnetically, turbulent systems are most interesting because they magnify the magnetic field. This creates a transfer of kinetic to magnetic energy, until the magnetic field pushes back with enough force stop the turbulent cascade.

Thus, in plasmas, big whirls have lesser whirls and so on to magnetism. This, alas, does not rhyme, but it does make for interesting physics.

Thought Experiment 12.2: Magnetic Plasma Tubes

When a gas is ionized, creating a plasma, it becomes conductive as the electrons are now able to move freely. We model this as a net neutral fluid of conductivity κ and mass density ρ_m, which is moving at an initial velocity $\vec{v}_0$ through a constant magnetic field $\vec{B} = B\hat{z}$. What is the velocity, $\vec{v}$, as a function of time?

We begin by writing the body force as:

$$\vec{f} = \kappa\left(\vec{v}\times\vec{B}\right)\times\vec{B} = -\kappa B^2\vec{v} + \kappa\left(\vec{v}\cdot\vec{B}\right)\vec{B}.$$

Next, we render this into cylindrical coordinates:

$$\vec{f} = -\kappa B^2\left(v_s\hat{s} + v_z\hat{z}\right) + \kappa\left(v_z\right)B^2\hat{z} = -\kappa B^2 v_s\hat{s}.$$

Notice that the sideways force is opposite the velocity, but there is no net force parallel to the magnetic field, so the $\hat{z}$ component of the velocity remains constant. Applying Newton's second law, we can find an equation of motion:

$$\vec{a} = \frac{\vec{f}}{\rho_m} = -\frac{\kappa}{\rho_m} B^2 v_s \hat{s} .$$

We can solve this by separation of variables as:

$$\int_{v_{0x}}^{v_x} \frac{dv_s}{v_s} = -\frac{\kappa}{\rho_m} B^2 \int_0^t dt , \quad \text{so} \quad \ln\left(\frac{v_x}{v_{0x}}\right) = -\left(\frac{\kappa B^2}{\rho_m}\right) t .$$

Solving for the sideways velocity as a function of time, we obtain:

$$v_s = v_{0s} e^{-\left(\frac{\kappa B^2}{\rho_m}\right)t} = v_{0s} e^{-\left(\frac{t}{\tau}\right)} ,$$

where $\tau = \frac{\rho_m}{\kappa B^2}$ is the e-folding time.[176] Much after this time, the velocity will become parallel to the magnetic field.

In the solar corona, ejections of ionized gas often flow in the direction of the magnetic field, as shown in the picture.[177] You can see that each flux tube is significantly thinner than the radius of Earth (see the small Earth symbol), or about 1,000 km. Assuming a typical plasma speed of 100 km/s, this would imply that $\tau \ll 10\,\text{s}$.

Putting in rule of thumb quantities such as: $\rho_m \approx 10^{-11}\,\text{kg}/\text{m}^3$, $B \approx 10\,\text{mT}$, and $\kappa \approx 10^{-5}\,\text{S}/\text{m}$, implies:

$$\tau \approx \frac{\left(10^{-11}\,\text{kg}/\text{m}^3\right)}{\left(10^{-5}\,\text{S}/\text{m}\right)\left(10^{-2}\,\text{T}\right)^2} \approx 0.01 \frac{\left(\text{kg}/\text{m}^3\right)}{\left(\text{C}^2\cdot\text{s}/\text{kg}\cdot\text{m}^3\right)\left(\text{kg}/\text{C}\cdot\text{s}\right)^2} \approx 10\,\text{ms}.$$

This is consistent with the maximum time scale we estimated from the picture.

Problem 12.3: Solar Physics

Back in Problem 6.1 (p. 191) you found a relationship between the magnetic field and energy level splitting in atoms called the effect. Solar physicists use this to map out in detail the magnetic field in the sun, and use these measurements and others to model the current flow inside the sun. This is of particular interest, because the solar magnetic field changes on an 11 year cycle.

a) Research the solar cycle, and report on it.

b) Research how plasma physicists model solar magnetic field amplification, and report on it.

---

[176] The term *e-folding time* refers to the timescale for a quantity to decrease to $1/e$ of its previous value, or to increase by a factor of e. In certain contexts it may be identified with the "relaxation time" of a system or process.

[177] This is an ultraviolet negative picture of the solar corona taken by NASA's Solar Dynamics Observatory. The small cross represents the approximate size of Earth.

Problem 12.4: Deflection of a Salt Water Jet

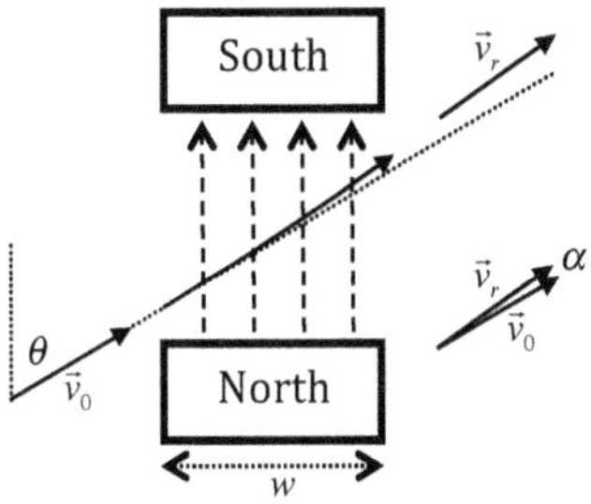

A jet of salt water with conductivity κ and mass density ρ_m is squirted through a horizontal magnetic field (strength B and width w) with an initial speed v_0 at an angle θ from the magnetic field (see diagram). By what angle α is the final velocity deflected from the initial velocity? You may assume that $\alpha << \theta$ and $\alpha << \frac{\pi}{2} - \theta$, and you may ignore gravity.

Problem 12.5: An Azimuthal Current Density

Consider stirring sugar into a tall cylindrical glass of iced tea clockwise with a straw. If you hold the straw vertically, and are very careful, you can make the flow azimuthal and independent of depth (i.e., $\vec{v} = -v(s)\hat{\varphi}$). Now imagine that rather than iced tea, the spinning fluid were negatively charged with a uniform charge density.

a) Draw pictures to clearly show that the current density can be written in cylindrical coordinates as $\vec{J} = J(s)\hat{\varphi}$.

b) If the glass has a height h and a radius a, show that its the total magnetic moment would be given in integral form by:

$$\vec{m} = \int_0^h \int_0^a \pi s^2 J_\varphi(s)\, \mathrm{d}s\, \mathrm{d}z .$$

c) Show that this can be written, in vector form, as a closed volume integral over the whole container as:

$$\vec{m} = \oint_{\text{volume}} \left(\tfrac{1}{2}\vec{r} \times \vec{J}\right) \mathrm{d}V .$$

d) Does this necessarily imply that all magnetization is caused by the equation $\vec{M} = \frac{1}{2}\vec{r} \times \vec{J}$, as Ampère argued. Clearly argue your point using words, equations, and diagrams.

Problem 12.6: The Magnetic Field in a Spinning Plasma

Consider a region of spinning ionized plasma that is flowing with a velocity profile in cylindrical coordinates of:

$$\vec{v} = v_0 \tanh\left(\tfrac{s}{R}\right)\hat{\varphi} .$$

Since protons are much more massive than electrons, you can imagine that there will be some centrifugal separation. For simplicity, model the charge density as a double step function that changes sign at two particular radii, so the plasma in the center of tube is negatively charged, while the outer part has a corresponding positive charge density, or:

$$\rho = \begin{cases} -\rho_0 : & s < a \ll R \\ +\rho_0 : & R \ll b - \frac{1}{2}\delta < s < b + \frac{1}{2}\delta \end{cases}, \quad \text{where } \rho_0 = \frac{2\varepsilon_0 m_p v_0^2}{e R^2} .$$

a) Show that the charge density has the correct dimensions.

b) Find δ in terms of a and b, such that the system is net neutral.

c) Find the current density as a function of s in terms of: v_0, R, a, and b.

d) Discuss qualitatively, with pictures, the direction of the magnetic field.

e) From Ampère's law find the surrounding magnetic field $\vec{B}(s)$.

f) Find the magnetic body force on the plasma: $\vec{f}_B(s)$.

g) Discuss all the body forces, and acceleration, on the spinning plasma, as viewed in (1) the laboratory frame, and (2) the comoving frame with the plasma. Discuss what it will do if no other forces act on it.

12.3 *The Meissner Effect*

As we discussed in the chapter introduction (p. 377), Walther Meissner and Robert Ochsenfeld, discovered, in 1933, that superconductors expel magnetic fields—in much the same way as normal conductors expel electric fields. In this section, we take this observation *a priori*, and use Peregrinus's principle to investigate the surrounding magnetic field structure. This type of analysis is similar to our discussion of Faraday's drawings of permanent magnets (p. 200).

Thought Experiment 12.3: The Meissner Effect

In this example, we consider Meissner's superconducting experiments within the context of the continuity of magnetic flux and simulate the magnetic field structure.

Let us consider a solid square superconductor with an initial internal magnetic field of $\vec{B}_0 = B_0\,\hat{x}$. Outside of the superconductor, $\vec{\nabla}\cdot\vec{B} = 0$ and $\vec{\nabla}\times\vec{B} \approx 0$, so by defining the *vector potential* $\mathbb{A}$ such that $\vec{\nabla}\times\vec{\mathbb{A}} \equiv \vec{B}$, we can solve these equations by finding $\vec{\mathbb{A}}$ initially, such that the magnetic field is constant. To simulate the transition to the superconducting state, we turn off the magnetic field inside the cylinder.

Initially, and near the boundaries, the magnetic field is constant, so:

$$\mathbb{A}_z = -\int B_0\,\mathrm{d}x = -B_0\,x + C.$$

As we plan on setting $\mathbb{A}_z = 0$ inside the cylinder, we will make sure that $\mathbb{A}_z = 0$ at the origin, so:

$$\mathbb{A}_z = -B_0\,y.$$

Once the conductor becomes a superconductor, we set $\mathbb{A}_z = 0$ inside the superconductor and apply the two dimensional method of relaxation. The figure shows our model of a superconductor inside of a constant magnetic field, with a normal conductor for comparison.

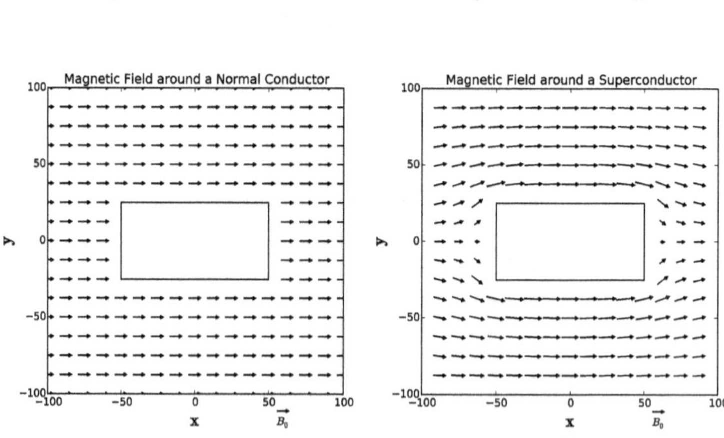

Problem 12.7: A Superconducting Sphere

A spherical superconductor, with radius a, inside of a large uniform magnetic field $\vec{B} = B_0\,\hat{z}$.

(a) Review the discussion of Stoke's law on page 326, and then make a plausible mathematical guess in spherical coordinates for the magnetic field outside the sphere after the superconductor has been cooled below its critical temperature.

(b) Show that your expression becomes simply $\vec{B} = B_0\,\hat{z}$ in the limit that $r \gg a$.

(c) Show that Peregrinus's principle holds everywhere outside the sphere.

(d) Draw a huge cylindrical Gaussian surface of radius $s \gg a$ and height $z_{\text{top}} - z_{\text{bottom}} \gg a$. Show that the entering magnetic flux equals the magnetic flux exiting the Gaussian surface.

(e) Now slice the surface into half, so you have a Gaussian surface of radius $s \gg a$ and height $\left(z_{\text{top}} - 0\right) \gg a$, with the bottom face located at the equatorial plane. Again, show that the entering magnetic flux equals the exiting magnetic flux.

(f) Draw a Gaussian surface at the surface of the sphere. Show that magnetic flux neither enters, nor exits, this surface.

(g) Show that the curl of the magnetic field is zero everywhere inside, and outside, of the sphere.

(h) Draw a wedged shaped Amperian loop, containing part of the surface of the sphere, with corners at the four points $\left(a \pm \tfrac{1}{2}\delta r, \theta \pm \tfrac{1}{2}\delta\theta, \varphi_0\right)$. Integrate the magnetic field around this loop. You may assume that $0 < \delta r \ll a$.

(i) Show that $\vec{J}\,\delta r = \dfrac{3\,B_0 \sin(\theta)}{2\mu_0}\hat{\varphi}$, where δr represents the thickness of the boundary layer?

Problem 12.8: Modeling the Meissner Effect

Consider a cylindrical superconductor of radius a and height h, located in a known uniform and constant external magnetic field $\vec{B}_{\text{ext}} = \mu_0 H\,\hat{z}$. According to the Meissner effect, the internal net magnetic field must be zero. For this problem, we will assume that there exists some thin boundary layer with thickness $\lambda \ll a$ that is still superconducting but also has a magnetic field.

a) Using Amperian loops, symmetry, and your experience from permanent magnets, argue that the magnetic field on the edge of the boundary layer is approximately double the external field.

b) Find the average magnetic field energy density, $\mathbb{u}_B$, within the boundary layer.

c) What is the current density, $\vec{J}$, that must flow around the superconductor in order to cancel out the external magnetic field? Express this in terms of a, h, λ, and H.

d) Find the electron velocity field in terms of a, h, λ, H, and the electron number density n_e.

d) Find the kinetic energy density, $\mathbb{u}_K$, of the electrons in the superconductor.

e) In many fluids undergoing random interactions, the energy per volume equilibrates. This is called equipartition, and it means that you can equate the two energy densities. Use this to solve for height of the boundary layer, λ.

f) Show that the thickness of the boundary layer is equal to the London penetration depth:

$$\lambda_L = \sqrt{\frac{m_e}{\mu_0 e^2 n}} \ ?$$

Thought Experiment 12.4: A Trapped Magnetic Field

In his talk (p. 377) John Cunningham McLennan discusses an experiment where his students drilled a cylindrical hole, of bore radius a and depth h, into a block of a superconducting material. They, then, inserted a magnetometer into the hole, and capped it, also with superconducting material. They, then, enclosed this inside a uniform magnetic field. At room temperature, the magnetometer detects the expected uniform field $\vec{B}_0$. After cooling the superconductor, and turning off the external magnetic field, the field remained inside the hole.

However, it follows from Peregrinus's principle, that the magnetic flux must be circulatory, and so it must somehow return. Later in his talk, McLennan discusses the supraconductivity of thin films and reports on findings by his students. (Note: the symbol μ denoted the length of one micron, until 1960 when it was changed to μm.)

> For thicknesses of the order of 10 to 1 μ, there is a slight lowering of the transition point, which may be due to strains set up by unequal contraction of the film and its support. But for a thickness of tin less than 1.0 μ, and of lead less than 0.8 μ, the transition point begins to decrease very rapidly until film thinner than 0.3 μ could not be made superconducting at 2° K, the lowest temperature at present attainable in Toronto.

It seems natural to conclude that the return magnetic flux is confined to a thin layer of thickness, δs, between 0.3 and 0.8 μm, making the return magnetic field extremely strong. To estimate the strength of this return field, we consider a hollow cylinder with hole radius of $a = 1\,\text{cm} = 10{,}000\,\mu\text{m}$ and thickness h. The total magnetic flux contained in the hole was found by using the maximum perpendicular area in the x-z plane:

$$\Phi_B = \int_{surface} \vec{B}\cdot d\vec{A} = B_0(2ah).$$

Now this same amount of flux must return within the thickness, δ, into the superconducting layer. Therefore the same flux must be given by:

$$\Phi_B = \int_{surface} \vec{B}\cdot d\vec{A} = B_r(h\delta).$$

Equating the flux values, we can now estimate the return magnetic field as:

$$\Phi_B = B_0(2ah) \approx B_r(h\delta) \quad \rightarrow \quad B_r \approx \frac{2a}{\delta} B_0 \approx 10{,}000\, B_0 .$$

As you can imagine, this trapping of the magnetic field inside of a hollow cylinder absolutely astonished scientists of the time.

Problem 12.9: A Spherical Hole

A superconductor is made with a spherical hole of radius a inside of it, and then cooled down inside of a uniform magnetic field $\vec{B}_0$.

a) Before cooling, what was the total magnetic flux through the equator of the hole?

b) Assuming the same magnetic field says inside the cavity after cooling, carefully draw what you presume to be the magnetic field lines, based on the experimental results of the McLennan team.

c) Show that the magnetic field, inside a boundary layer with a uniform thickness $h \ll a$, would be expressed as: $\vec{B} = \frac{a}{h} B_0 \sin^2(\theta)\,\hat{\theta}$.

12.4 *Acceleration Theory*

Fritz and Heinz London sought to replace Ohm's law with a classical model, which they referred to as *acceleration theory*.[178]

Like conductors, and unlike plasmas, only the electrons move. However, like plasmas, but unlike conductors, we must also take into account kinetic energy. This time, however, it is the electrons, rather than the protons, which retain their momentum. The Londons, therefore, modeled the inside of conductors as a gas of free electrons, but with the net charge density always being neutral.

Thought Experiment 12.5: The First London Equation

The charge carriers in a superconductor do not experience drag forces, so when a force acts on them they should accelerate as would free particles. Following Heaviside's derivation of Ohm's law (p. 294), we write for the body force:

$$\vec{f}_E = \rho_- \vec{E} \,.$$

The negatively charged fluid now experiences a net force, which accelerates it. In terms of the charge density ρ_- and the mass density ρ_m of the superconducting charge carriers, Newton's second law becomes:

$$\vec{f}_E = \rho_- \vec{E} = \rho_m \vec{a}_- = \rho_m \frac{\mathrm{d}\vec{v}_-}{\mathrm{d}t} = \left(\frac{\rho_m}{\rho_-}\right) \frac{\mathrm{d}\vec{J}}{\mathrm{d}t} \,.$$

Solving for the current density and grouping the constants, yields:

$$\frac{\mathrm{d}\vec{J}}{\mathrm{d}t} = \left(\frac{\rho_-^{\,2}}{\rho_m}\right)\vec{E} = \frac{1}{\Lambda}\vec{E} \,.$$

This is called *The First London Equation*, and in their words:

> This equation, which might replace Ohm's law for supraconductors, simply expresses the influence of the electric part of the Lorentz force on freely movable electrons of the mass m_e and charge e, the number per cm^3 being n_e. By definition the constant Λ must be positive. As a direct consequence of this equation stationary currents in supraconductors are possible when $\vec{E} = 0$.

Everyone knew the values for the mass and charge of an electron by 1913, so in 1935 the Londons wrote the superconductivity constant in terms of the number density of superconducting electrons as:

---

[178] F. London and H. London, "The Electromagnetic Equations of the Supraconductor," Proceedings of the Royal Society of London A, **149** (1935), 71-88.

$$\Lambda = \frac{\rho_m}{\rho_-^2} = \frac{\rho_m}{|\rho_-|^2} = \frac{n_e m_e}{n_e^2 e^2} = \frac{m_e}{n_e e^2}.$$

Now we check the units of Λ:

$$[\Lambda] = \frac{[\rho_m]}{[\rho_-^2]} = \frac{\left(\text{kg}/\text{m}^3\right)}{\left(\text{C}/\text{m}^3\right)^2} = \frac{[E]}{\left[\frac{dJ}{dt}\right]} = \frac{\text{N}/\text{C}}{\text{A}/\left(\text{s}\cdot\text{m}^2\right)} = \frac{[m_e]}{[n_e][e^2]} = \frac{\text{kg}}{\text{m}^{-3}\cdot\text{C}^2} = \frac{\text{kg}\cdot\text{m}^3}{\text{C}^2} = \left(\text{J}/\text{A}^2\right)\cdot\text{m} = \text{H}\cdot\text{m}.$$

Λ can be interpreted as the kinetic inductivity since it has the same relationship to inductance as resistivity has to resistance. Typical values for the kinetic inductivity of superconductors are on the order of: $\Lambda \sim 10^{-21}\,\text{H}\cdot\text{m}$.

Problem 12.10: The Kinetic Inductivity

Tin (Sn) and lead (Pb) have room temperature densities of 3.7 and 7.2 kg/liter and atomic masses of 119 and 201 g/mol respectively.

a) Find the number of atoms per Bohr volume $\left(\frac{4\pi}{3}a_0^3\right)$ for each element. You can assume that this is also the electron number density.

b) Show that the kinetic inductivity can be approximately written in terms of the Bohr volume as:

$$\Lambda \approx \mu_0\, a_0^2 \frac{\left(137^2 \div 3\right)}{n_e\left(\frac{4\pi}{3}a_0^3\right)} \approx 1500\frac{\mu_0}{a_0 n_e}.$$

c) Estimate each atom's kinetic inductivity expressed in terms of $\mu_0\, a_0^2$.

d) Estimate each atom's kinetic inductivity in SI units.

e) Estimate the London penetration depth, $\lambda_L = \sqrt{\frac{\Lambda}{\mu_0}}$, for each atom. Express this in terms of the Bohr radius and in SI units. How does it compare to the discussion by McLennan in Thought Experiment 12.4 (p. 388).

Problem 12.11: Electrons in a Superconductor

Consider a cylindrical superconductor of radius a and height h. Initially with no current flowing through it. A uniform magnetic field is then turned on with a characteristic rise time so it follows the following time dependence: $\vec{B} = \mu_0 H_0\left(1 - e^{-t/t_r}\right)\hat{z}$. For illustration, and simplicity, assume that each electrons first stays at a fixed distance, s, from the axis of symmetry and doe not interact with each other – at least at first.

a) What is the induced electric field inside the superconductor as a function of position and time?

b) What would be the electron velocity, $\vec{v}(s,t)$, be if each electron stayed in its particular orbit, and no energy were exchanged between the superconducting electrons?

c) What is the magnetic force, per electron, as a function of position and time? In what direction is this in?

c) What is the centripetal acceleration of each electron, again assuming that each one is independent and stays in its own orbit.

d) In a rotating reference frame, where each electron is stationary, the centripetal acceleration acts like a pseudo gravity, called the centrifugal force, where: $\vec{g}_{\text{effective}} = \frac{v^2}{s}\hat{s}$. In this reference frame, do the electrons experience a net force? If so, is it inward or outward?

Thought Experiment 12.6: The Second London Equation

We now take the London acceleration theory to the next level by investigating the magnetic part of the Lorentz force. Consider the same charged fluid, under the influence of a magnetic field $\vec{B} = B\hat{z}$. It will experience a perpendicular acceleration given by:

$$\vec{a}_\perp = \frac{\vec{f}}{\rho_m} = \frac{\rho_-}{\rho_m}\vec{v}\times\vec{B} \approx \left\langle \frac{\rho_- v B\sin\theta}{\rho_m} \right\rangle \approx \frac{\rho_-}{\rho_m} vB\langle \sin\theta \rangle \approx \frac{\rho_-}{2\rho_m} vB .$$

Next, we can equate this with the centripetal acceleration and solve for the radius, R, as a function of the magnetic field and current density, like so:

$$R \approx v^{\cancel{2}}\left(\frac{2\rho_m}{\rho_- \cancel{v} B}\right) \approx \frac{2\rho_m v}{\rho_- B} \approx \frac{2\rho_m v}{\rho_- B} \approx \frac{2\rho_m}{\rho_- B}\left(\frac{J}{\rho_-}\right) \approx 2\frac{\rho_m}{\rho_-^2}\left(\frac{J}{B}\right) \approx 2\Lambda\left(\frac{J}{B}\right).$$

Now we apply the definition of the curl of the current density around the $\hat{z}$ axis:

$$\left(\vec{\nabla}\times\vec{J}\right)_z \equiv \lim_{\delta A\to 0}\left(\frac{1}{\delta A}\oint_{\text{path}} \vec{J}\cdot d\vec{\ell}\right) = \lim_{R\to 0}\left(\frac{-J2\cancel{\pi}\cancel{R}}{\cancel{\pi}R^{\cancel{2}}}\right) \approx \lim_{J\to 0}\left(\frac{-\cancel{2}\cancel{J}}{\cancel{2}\Lambda\left(\cancel{J}/B\right)}\right) \approx -\frac{B}{\Lambda},$$

which in vector form can be written:

$$\vec{\nabla}\times\vec{J} = -\vec{B}/\Lambda .$$

This is known as *The Second London Equation.* In other words, the curl of the current density must be proportional to the internal magnetic field. Notice how the incompressibility of the current density is still maintained, as it is neutral on all scales of interest.

Discussion 12.1: The Development of Superconductivity

The London model of superconductivity is similar to the Heaviside model of conductivity in that it nicely explains the measured inductance of a superconductor but fails to explain the microphysics. By 1935 quantum physics was in full swing, so the Londons took their model one step further by deviating from classical physics in their second London equation. As you saw in Thought Experiment 12.6 (p. 391) this was an unnecessary leap of faith, because a direct application of the magnetic component of the Lorentz force would yield the same answer for a macroscopic toy model.

Our understanding of superconductivity has come a long way since 1935.[179] In 1952, John Bardeen, Leon N. Cooper, and J. Robert Schrieffer published a quantum model of

---

179 See Stephen Blundell, Superconductivity: A Very Short Introduction, (2009), Oxford University Press, ISBN 978-0-19-954090-7

superconductivity that is now called BCS theory.[180] Currently the field of superconductor physics is booming, as superconductors that work at ever higher temperatures are being found all the time.

As is the case with resistivity, viscosity can also disappear in cold materials, such as liquid helium. Normally the flow velocity in a pipe is proportional to the pressure difference, however in 1938 Pyotr Kapitza,[181] and independently Jack Allen and Don Misener,[182] showed that the fluid velocity is nearly independent of the pressure in cold helium (see their figure). In analogy to superconductivity, this is called superfluidity.

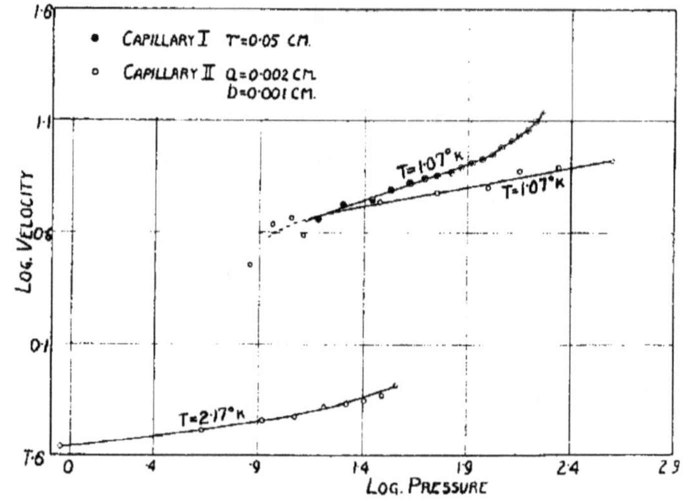

An effective toy model may apply the same pictures as quantum physics (e.g. a gas of free electrons) to explain some phenomenon, but its use of the limited concepts of classical physics means that it is heuristic at best.

12.5 *Magnetic Levitation*

According to Lenz's law all induced currents create reverse magnetic fields. This is exactly the opposite effect see between magnets and iron. In this section we apply London acceleration theory show how it can explain how magnets repel superconductors.

Thought Experiment 12.7: The Magnetic Force on a Superconductor

We will investigate the force between a given magnetic field and a superconductor in this thought experiment so that we can illustrate how to apply the electromagnetic force, and the London equations, to a superconductor.

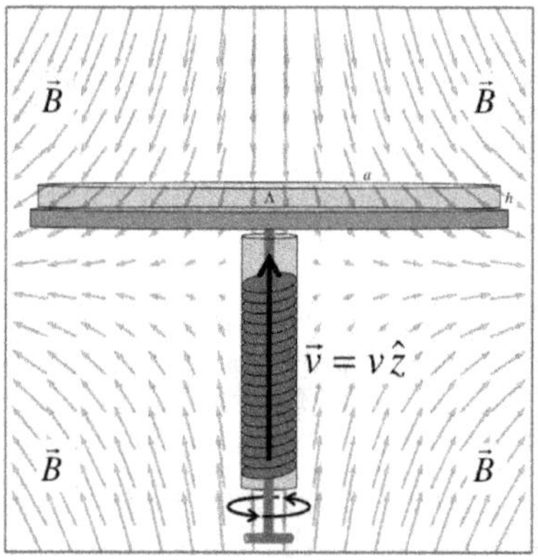

Consider a system of fixed magnets that are designed in such a way as to produce a magnetic field that can be represented by the simple function:

$$\vec{B} = K\left(s\,\hat{s} - 2z\,\hat{z}\right),$$

where K is a positive constant.

The divergence of any function must equal zero if it is to represent a magnetic field, so let's first check that our

---

[180] J. Bardeen, L.N. Cooper, and J.R. Schrieffer, "Microscopic Theory of Superconductivity," Physical Review, **106** (1952), 162-164.

[181] P. Kapiza, "Viscosity of Liquid Helium below the λ point," Nature, **141** (1938), 74.

[182] J.F. Allen and A.D. Meissner, "Flow of Liquid Helium II," Nature, **141** (1938), 75.

candidate function satisfies this requirement:

$$\vec{\nabla}\cdot\vec{B}=K\vec{\nabla}\cdot\left(s\,\hat{s}-2\,z\,\hat{z}\right)=K\left(\frac{1}{s}\frac{\partial\left(s^2\right)}{\partial s}-2\frac{\partial z}{\partial z}\right)=K\left(\frac{2\cancel{s}}{\cancel{s}}-2\right)=0.$$

Now consider a thin cylindrical superconductor with radius a and thickness $h \ll a$ placed on a lift that is elevated at a constant velocity $\vec{v}=v\hat{z}$.

What is the magnetic force on the superconductor?

To solve this problem we first calculate the body force on the negatively charged fluid, $\rho_- < 0$, due to the motion of the lift:

$$\vec{f}=\rho_-\vec{v}\times\vec{B}=\rho_- vB_s\left(\hat{z}\times\hat{s}\right)=\rho_- vB_s\hat{\varphi}=-\left|\rho_-\right|v\left(K\,s\right)\hat{\varphi}.$$

Notice something very interesting. The body force due the magnetic field, with the superconductor on a rising platform, has exactly the same form as if there were an azimuthal electric field. Therefore we can apply Newton's second law, and the Londons' acceleration theory, as:

$$\vec{f}=\rho_m\frac{\mathrm{d}}{\mathrm{d}t}\left(\frac{\vec{J}}{\rho_-}\right)=\left(\frac{\rho_m}{\rho_-}\right)\frac{\mathrm{d}\vec{J}}{\mathrm{d}t}=-\left(\frac{\rho_m}{\left|\rho_-\right|}\right)\frac{\mathrm{d}\vec{J}}{\mathrm{d}t}=-\left|\rho_-\right|v\left(K\,s\right)\hat{\varphi}.$$

We can therefore solve for the rate of change of the current density as:

$$\frac{\mathrm{d}\vec{J}}{\mathrm{d}t}=\frac{\left|\rho_-\right|^2}{\rho_m}\left(v\,K\,s\right)\hat{\varphi}=\frac{\left|\rho_-\right|^2}{\rho_m}\left(v\,K\,s\right)\hat{\varphi}=\frac{v\,K\,s}{\Lambda}\hat{\varphi},$$

where Λ is the Londons' kinetic inductivity.

Let us suppose scientists arrange this experiment such that when the material becomes superconducting $\left(t=0,\,\vec{J}=0\right)$, it is also at a vertical position $(z=0)$. Then the current density in the superconductor should be given by:

$$\vec{J}=\int_0^t\frac{v\,K\,s}{\Lambda}\hat{\varphi}\,\mathrm{d}t=\frac{K\,s}{\Lambda}\hat{\varphi}\int_0^t v\,\mathrm{d}t=\frac{K\,s}{\Lambda}\hat{\varphi}\int_0^t\frac{\mathrm{d}z}{\mathrm{d}t}\mathrm{d}t=\frac{K\,s}{\Lambda}\hat{\varphi}\int_0^{z_0}\mathrm{d}z=\frac{K\,s\,z_0}{\Lambda}\hat{\varphi}.$$

Thus the current density becomes azimuthal and is then simply a function of the distance that the superconductor traveled through the magnetic field, which we call $z_0 \equiv vt$.

Next we will consider the force on this current density, due to the same magnetic field as before:

$$\vec{f}=\vec{J}\times\vec{B}=\left(\frac{z_0\,K\,s}{\Lambda}\hat{\varphi}\right)\times\left(K\left(s\,\hat{s}-2\,z\,\hat{z}\right)\right),$$

and simplifying:

$$\vec{f}=\left(\frac{z_0\,K^2\,s^2}{\Lambda}\right)\left(\hat{\varphi}\times\hat{s}\right)+\left(\frac{2\,z_0\,K^2\,s\,z}{\Lambda}\right)\left(-\hat{\varphi}\times\hat{z}\right)=\left(\frac{z_0\,K^2\,s^2}{\Lambda}\right)\left(-\hat{z}\right)+\left(\frac{2\,z_0\,K^2\,s\,z}{\Lambda}\right)\left(-\hat{s}\right).$$

Now the charge cannot move in either of these directions while the body remains electrically neutral, so internal forces in the superconductor and its walls must counteract this force. We can now find the total force on the superconductor by integrating over the whole volume of the superconductor:

$$\vec{F} = \oint_{\text{volume}} \vec{f}\,\mathrm{d}V = \oint_{\text{volume}} \left(\frac{z_0 K^2 s^2}{\Lambda}\right)(-\hat{z}) + \left(\frac{2 z_0 K^2 s z}{\Lambda}\right)(-\hat{s})\,\mathrm{d}V$$

$$= \int_{z_0}^{z_0+h}\int_0^{2\pi}\int_0^{a} \left(\frac{z_0 K^2}{\Lambda}\right) s^2 (-\hat{z})\, s\,\mathrm{d}s\,\mathrm{d}\varphi\,\mathrm{d}z + \int_{z_0}^{z_0+h}\int_0^{2\pi}\int_0^{a} \left(\frac{2 z_0 K^2}{\Lambda}\right)(s z)(-\hat{s})\, s\,\mathrm{d}s\,\mathrm{d}\varphi\,\mathrm{d}z$$

$$= (-\hat{z})\left(\frac{z_0 K^2}{4\Lambda}\right)\int_{z_0+h}^{z_0+h}\mathrm{d}z\int_0^{2\pi}\mathrm{d}\varphi\int_0^{a} 4 s^3\,\mathrm{d}s + \left(\frac{2 z_0 K^2}{3\Lambda}\right)\int_{z_0+h}^{z_0+h} z\,\mathrm{d}z\ \cancel{\int_0^{2\pi}(-\hat{s})\,\mathrm{d}\varphi}\int_0^{a} 3 s^2\,\mathrm{d}s\,.$$

First notice that the vertical position of the center of the superconductor is vt, so the limits of z integration depend on time. Next, observe that since the superconductor is cylindrical, the limits of integration do not depend on each other, allowing us to separate the integrals.

For every $\hat{s}$ pointing in one fixed direction, there will be a corresponding unit vector $\hat{s}$ that points in the opposite direction, so its integral becomes zero. You can see the same thing mathematically by using fixed Cartesian coordinates:

$$\int_0^{2\pi}(-\hat{s})\,\mathrm{d}\varphi = -\hat{x}\cancel{\int_0^{2\pi}(\cos\varphi)\,\mathrm{d}\varphi} - \hat{y}\cancel{\int_0^{2\pi}(\sin\varphi)\,\mathrm{d}\varphi} = 0\,.$$

Therefore we can now write the resultant force on the superconductor as simply the first term:

$$\vec{F} = (-\hat{z})\left(\frac{z_0 K^2}{4\Lambda}\right)\int_{z_0}^{z_0+h}\mathrm{d}z\int_0^{2\pi}\mathrm{d}\varphi\int_0^{a} 4 s^3\,\mathrm{d}s = (-\hat{z})\left(\frac{z_0 K^2}{4\Lambda}\right)(h)(2\pi)(a^4) = \left(\frac{\pi z_0 K^2 h a^4}{2\Lambda}\right)(-\hat{z})\,.$$

Let us check the units:

$$[\vec{F}] = \left(\frac{[\cancel{\pi}][z_0][K]^2[h][a]^4}{[\cancel{2}][\Lambda]}\right)[\cancel{-\hat{z}}] = \left(\frac{\mathrm{m}\cdot(\mathrm{T}/\mathrm{m})^2\,\mathrm{m}\cdot\mathrm{m}^4}{\mathrm{H}\cdot\mathrm{m}}\right) = \frac{\mathrm{T}^2\cdot\mathrm{m}^3}{\mathrm{H}} = (\mathrm{T}^2\cdot\mathrm{m}^3)\left(\frac{\mathrm{A}}{\mathrm{T}\cdot\mathrm{m}^2}\right) = (\mathrm{T}\cdot\mathrm{m}\cdot\mathrm{A}) = \mathrm{N}.$$

Now, consider the implications of this thought experiment of a superconductor placed on a rising platform moving into a cylindrically symmetrical magnetic field.

First, notice that the force opposes the motion. This must be the case while the superconductor continues moving, because the kinetic energy of the superconducting fluid has increased and the force that keeps the lift moving at a constant velocity must have done the required work. Had the force been attractive, it would be unclear where the energy came from in order to circulate the electron sea. This is simply an application of Lenz's law, which states that forces due to induced currents must always oppose the motion that caused the induced current.

Let's note next that when the platform stops, the force remains and simply depends on the distance the platform traveled. While the superconductor needs to be moving in the first place, the azimuthal current remains once it is there.

Next we will consider the second London equation:

$$\vec{B} = -\Lambda \vec{\nabla} \times \vec{J} = -\Lambda \vec{\nabla} \times \left(\frac{K s z_0}{\Lambda} \hat{\varphi} \right) = -K z_0 \vec{\nabla} \times (s \hat{\varphi}) = -K z_0 \left(\frac{1}{s} \frac{\partial (s^2)}{\partial s} \hat{z} \right) = -K z_0 \left(\frac{\cancel{s}}{\cancel{s}} \right) \hat{z} = -2 K z_0 \hat{z}.$$

Notice that this is simply the $\hat{z}$ component of the magnetic field, so for a thin disk $h \ll a$ or $z \approx z_0$, the second London equation reproduced the magnetic field exactly as expected.

You might say that this is rather odd. During all of our analysis we used only the $\hat{s}$ component of the magnetic field; the $\hat{z}$ component was irrelevant, as we argued that the current would not be able to circulate in the $-\hat{s}$ direction. How, then, can the second London equation magically give us the $\hat{z}$ component of the magnetic field? The answer goes back to Petrus Peregrinus of Maricourt's thought experiment in 1269: the magnetic field is fundamentally divergence free, so the direction of each magnetic field line, of necessity, must depend on the directions of the other field lines. This is different from many—though not all—vectors in classical mechanics, whose different components are completely independent.

Another concern you may have with this result is Meissner's observation that the magnetic field appears to be repelled by superconductors. If this is so, then why is the magnetic field inside the superconductor not exactly equal to zero? Currents produce magnetic fields in their own right, which is something we have yet to take into account.

The final observation we want to leave you with is the idea of magnetic levitation. One of the most common superconductor demonstrations involves taking a small permanent magnet, setting it on top of a flat cylindrical superconductor and watching it float in mid air. This magnetic levitation demonstration shows that there is a repulsive force between the superconductor and the magnet, and, with a strong enough magnet, it can actually be greater than the magnet's weight! Imagine that instead of raising the superconductor into the magnetic field, rather we had lowered the magnet onto the superconductor. According to Newton's third law, the force on the magnet would be equal in magnitude, but opposite in direction, to the force on the superconductor. Thus, there would have been an upward force on the magnet.

Problem 12.12: Magnetic Levitation I

In a superconductivity demonstration a small magnet, with a magnetic moment $\vec{m} = m\hat{z}$, and mass M, dropped from a height h_0 directly over a superconductor with its magnetic moment up. It then sits in mid-air at a height h above a flat disk superconductor with a total volume V and radius a.

a) Express the induced electric field, $\vec{E}(s)$, at the location of the superconducting disk, as a function of the magnet's speed, v.

b) Estimate the current density inside of the superconductor, $\vec{J}(s)$, as a function of the position, Z, of the magnet above the disk and some kinetic inductivity Λ. Assume that the current only flows in the azimuthal direction.

c) Estimate the total magnetic force between the superconductor and the magnet, as a function of the magnet height, Z.

d) Find the kinetic inductivity, Λ, in terms of the observable parameters.

e) Find the magnetic field in the superconductor as a function of the height Z.

Problem 12.13: Magnetic Levitation II

Consider a small magnet whose surrounding magnetic field can be expressed by the following vector relationship:

$$\vec{B} \propto \frac{3\vec{r}\,(\hat{z}\cdot\vec{r})}{r^5} - \frac{\hat{z}}{r^3},$$

assuming that the direction of the magnetic moment is in the $\hat{z}$ direction.

In a superconductivity demonstration this magnet is brought toward a superconductor, from above, with magnetic moment up, and it sits in mid-air at a position z_0 above the superconductor.

The total mass of the magnet is M and the magnetic field strength, on-axis, at a position z_0, is B_0. The superconductor is a thin disk, with a height h and radius a.

a) Express the magnetic field, $\vec{B}$, in cylindrical coordinates, surrounding the magnet. Express your answer in terms of the coordinates s and z and the positive constants B_0 and z_0.

b) Find the superconductivity constant Λ of the superconductor in terms of M, B_0, z_0, h, a and the local gravitational field g.

12.6 *Type II Superconductors*

Studies of turbulence in incompressible fluids at high Reynolds number show that, over time, the kinetic energy cascades from large vortex radii, to smaller and smaller scales, until finally viscosity dissipates the energy. In a superconductor the viscosity is exactly zero, so you would expect the medium to become fully turbulent with a full cascade of vortices.

In higher temperature superconductors, called type II superconductors, there is direct evidence of these vortices, albeit seen in a regular pattern described by quantum physics. That said, on the intermediate scales they can be investigated with classical electrodynamics.

Unlike type I superconductors, type II superconductors do not expel the magnetic field, but rather they form vortices around magnetic flux tubes. This can give them additional stability, while at the same time allowing for superconductivity in the electron sea surrounding the vortices.

Discussion 12.2: Ampére's Current Loop Model of Magnetic Matter

André-Marie Ampère died long before the discovery of Type II superconductors, but one could argue that they are the natural consequence of his model of matter. Ampère imagined small current loops, perhaps on the atomic scale. In this model, a distinction was made between this hypothetical *bound current* and the large scale *free current.* Free current could be measured directly, while the bound current caused the magnetization.

While it makes sense conceptually, the Ampèrian model completely contradicts the simplest experiment involving ferromagnets. According to a classical Ampèrian loop model, there should be no ferromagnetic materials found in nature. Rather all materials should exhibit some sort of diamagnetism. In other words, a magnet should be repelled from your refrigerator door, defying some of your earliest experiences with magnetics!

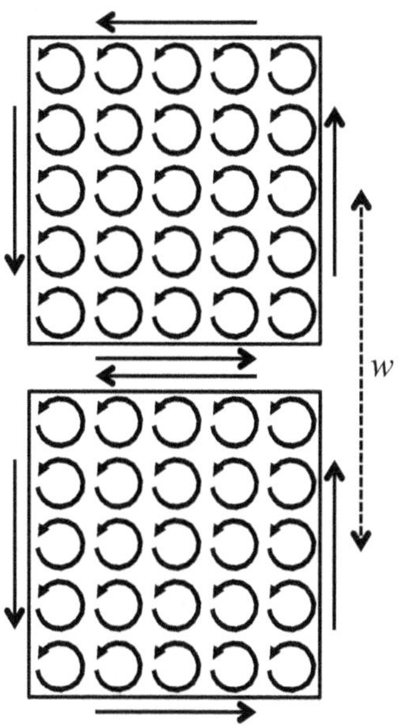

Consider a permanent magnet with a magnetic moment per volume $\vec{M} = M(x,y,z)\hat{z}$. Imagine a square Ampèrian loop of in the x-y plane of dimensions $l \times w$, with a finite height δz. Therefore the cause of the magnetic moment can be modeled as a loop of current δI with area $A = l \times w$, so:

$$\delta \vec{m} = \delta I\, A\hat{z} = \delta I(l \times w)\hat{z},$$

and the current per vertical length around the loop would be given by:

$$\frac{\delta I}{\delta z} = \frac{\delta m}{(l \times w)\delta z} = \frac{\delta m}{\delta V} = \vec{M} \cdot \hat{z} = M_z.$$

Now that we have established the current around a square Ampèrian loop, we can ask: what is the net current inside a magnetic material? To answer this, we consider adjacent loops, as in the figure. The current at the boundary must be the difference of the two associated currents, or:

$$\left.\frac{\delta I}{\delta z}\right|_{x,y,z} = \vec{M}_z(x, y + \tfrac{1}{2}w, z) - \vec{M}_z(x, y - \tfrac{1}{2}w, z) \approx \left(\frac{\partial M_z}{\partial y}\right) w.$$

Dividing by the width, and taking the limit of small cells, yields a current per area in the $\hat{x}$ direction of:

$$J_x = \lim_{w \to 0} \frac{\delta I}{w \delta z} = \frac{\partial M_z}{\partial y}.$$

If the magnetism vector is in the $\hat{y}$ direction, we can make the same argument, but with $J_x = -\frac{\partial}{\partial z} M_y$, so including both components yields:

$$J_x = \tfrac{\partial}{\partial y} M_z - \tfrac{\partial}{\partial z} M_y \quad \text{or} \quad \vec{J}_b = \vec{\nabla} \times \vec{M}.$$

This is referred to as the *bound current density*. In this model, the Maxwell-Ampère law can be written in terms of the sum of three currents:

$$\vec{\nabla} \times \vec{B} = \mu_0 \left(\vec{J}_f + \vec{J}_b + \vec{J}_D \right),$$

where Maxwell's displacement current, $\vec{J}_D = \frac{\partial}{\partial t}\left(\vec{P} + \varepsilon_0 \vec{E}\right)$, is also included.

Problem 12.14: Type II Superconductors and Vortex Lattices

Here you will investigate the vortex lattice, at least on the intermediate scale where classical electrodynamics could apply.

a) Show that the vortex diameter, λ, in a superconductor with azimuthal current density, $\vec{J} = J(s)\hat{\varphi}$, and a magnetic field, $\vec{B} = B\,\hat{z}$, is:

$$\lambda = \Lambda \frac{J}{B}.$$

b) Show that the kinetic energy density in a superconductor is given by:

$u_K = \frac{1}{2}\Lambda J^2$.

c) What would be the vortex diameter when the energy is equally distributed between kinetic and magnetic energy (equipartition)? How does this compare to $\lambda_L = \sqrt{\frac{\Lambda}{\mu_0}}$, the *London penetration depth.*

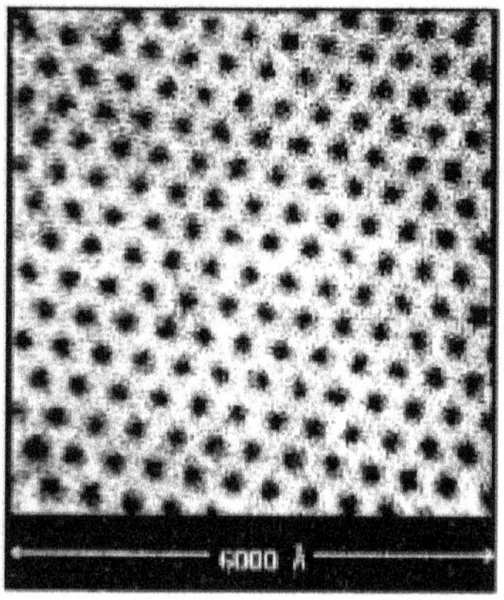

d) Niobium Selenide ($NbSe_2$) has a molecular weight of 251 g/mol and a mass density of 6.3 g/cm³. What is its London penetration depth?

e) The figure[183] shows Niobium Selenide vortices, observed using scanning tunneling microscopy with a field of view of 600 nm. How do the diameters of the vortices compare with λ_L?

f) Notice how the vortices in the superconductor form a regular lattice structure. When this happens, the amount of magnetic flux per lattice cell is given by the magnetic flux quantum, $\Phi_0 \equiv \frac{\pi\hbar}{e} \approx 2.07\,\text{fWb}$. Estimate the magnetic field in these vortices, and the vector potential around each vortex.

g) The study of high temperature superconductors is one of most dynamic areas of applied physics today. Research current developments in the field of high temperature superconductivity, the materials of interest, and where they are currently in use.

183 A. A. Abrikosov, "Type II Superconductors and the Vortex Lattice," Nobel Prize Lecture, December 8, 2003, Copyright The Nobel Foundation 2003

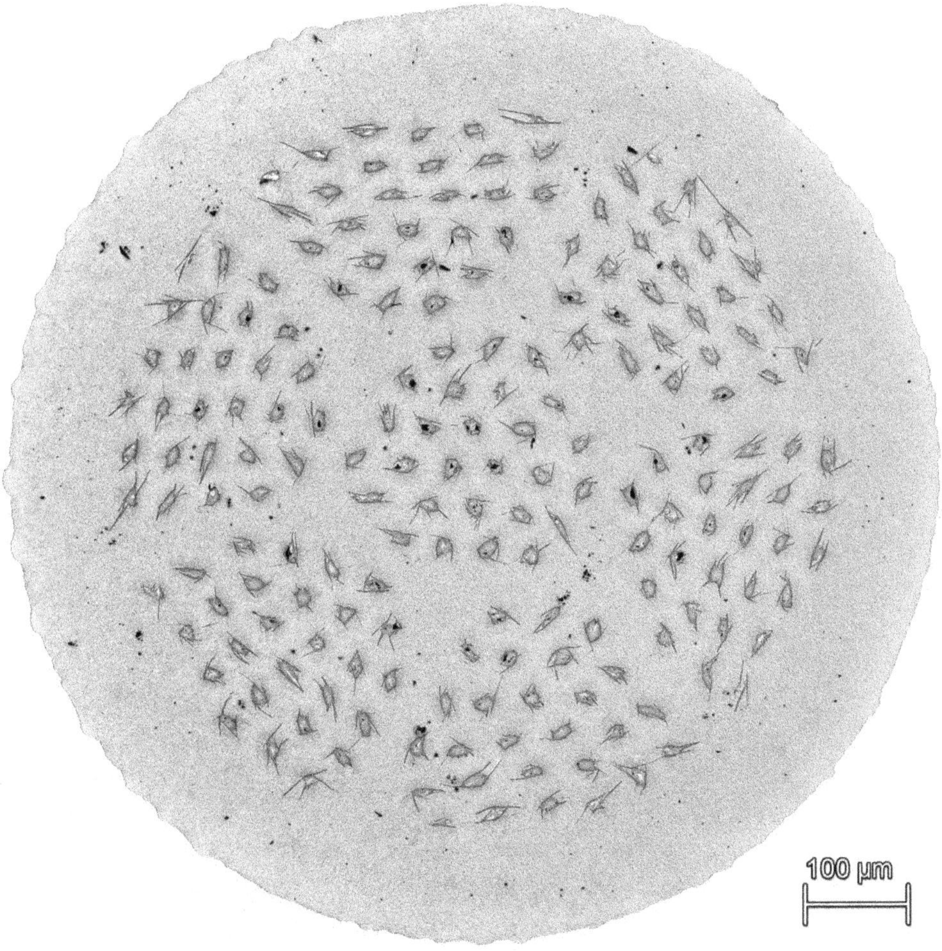

Plate 21: A High Temperature Superconductor[184]

[184] This is a scanning electron microscope image of a polished cross section of a Bi-2212 high-temperature superconductor 27×7 filament composite. Image by Jianyi Jiang of the Applied Superconducting Center at Florida State University, and downloaded from the National High Magnetic Field Laboratory (nationalmaglab.org).

Plate 22: Opticae Thesaurus[185]

185 Federico Risnero, Opticae thesaurus (Basile AE per Episcopios, 1572).

Part III: Light

The two previous Parts told two very different stories about the respective histories of electricity and magnetism. While knowledge of electricity seemed to progress monotonically over the course of time, leading scientists disagreed about the fundamental nature of magnetism for nearly a century. The history of optics took even more twists and turns than magnetism. The greatest minds of the last three centuries wrestled with its most fundamental question—what is light?

We have always been fascinated by optical phenomena, as the figure on the opposite page suggests. Armed with Euclidian geometry, the ancient Greeks investigated optics culminating in Ptolemy's optical treatise around the first century AD, where he argued that the eye emits a visual flux that probes external objects, called *emission theory*. Ptolemy also used the idea of refraction to explain whey the moon appears so large when it is near the horizon. Ptolemy's treatise was unfortunately lost, but not until after somebody translated part of it into Arabic.

Around the turn of the millennium, Ibn al-Haytham (also known as Alhacen and Alhazen) wrote the seven-volume treatise The Book of Optics, where he challenges both Ptolemy's emission theory of vision and his explanation for the moon illusion. In Alhacen's view, the eye only receives light and the moon is no larger when it is near the horizon than when it is high in the sky—he was right on both counts. Alhacen also conducted many optical experiments, and developed some of our known laws of optics, including the law of reflection.[186]

A few decades before Peregrinus performed his magnetism experiments, someone translated The Book of Optics from Arabic to Latin, and it became one of the first widely read scientific treatises of the middle ages. In the late sixteenth century, it was published in an anthology called Opticae thesaurus, which included the figure depicting famous optical phenomena of lore, including Noah's rainbow and Archimedes's death ray.[187] Now widely published, Alhacen's ideas permeated Europe, and found their way to the Netherlands, where someone invented a spyglass.

After learning about the Dutch spyglass Galileo designed and built his own telescope, which he pointing toward the sky. Among other things, he observed that Venus goes through the complete range of phases, with it appearing larger (and thus closer to Earth) when it is going through a crescent phase. More importantly, at other times it appears small (so it is far away) and almost full (so its day side is facing Earth). These observations were consistent with Venus orbiting the sun directly, but completely contradicted Ptolemy's geocentric model, which required Venus to always lie between the Sun and Earth.[188]

---

[186] See A. Mark Smith, "What is the History of Medieval Optics Really About?" Proceedings of the American Philosophical Society (2004) **148**, No. 2: 180-94. Also see Jim Al-Khalili, "In retrospect: Book of Optics" Nature (2015) **518**, 164-165.

[187]The Sicilian city-state of Syracuse famously fought off a vast Roman army for two years, primarily because of the ingenuity of the old Archimedes. While most of his inventions were mechanical in nature, Archimedes supposedly instructed the troops to use their shiny bronze shields to reflect sunlight on the attacking Roman ships—setting them ablaze. The television show *Mythbusters* (29-September-2004) found the "Archimedes Death Ray" to be implausible ("Busted"). However, the next year MIT students enrolled in Engineering 2.009 succeeded, so Mythbusters revisited the issue (25-January-2006), but again found it unlikely. They revisited the question four years later (8-December-2010), yet again with a null result.

[188] This insight did not originate with Galileo, or even with Copernicus. Centuries before Ptolemy, Aristarchus of Samos proposed a heliocentric universe. Unfortunately his treatise did not survive, but Archimedes of Syracuse discusses it in The Sand Reckoner, and Galileo mentioned it in his writings.

A spinning Earth made no physical sense to Galileo's contemporaries, who argued that a short jump would result in flying many feet to the west, so Galileo reinvented kinematics by proposing his principle of relativity. From that time on, all new physics had to work both terrestrially and universally, thus marrying the fields of physics and astronomy forever.

The development of the telescope, and the new science that it opened up, spurred a renewed interest in optics. This research included two contradictory theories of light: *the wave model* and *the particle model.*

In Traité de la Lumière,[189] Christiaan Huygens investigated light by analogy to sound, including accurately calculating the ratio of their speeds. This followed Robert Hooke's treatise on microscopy,[190] where he briefly discussed light as a wave that propagates through a medium he called *the aether.*

In Opticks or A Treatise of the Reflections, Refractions, Inflections and Colours of Light, Newton took a completely different tack to explain all the same phenomena—and more. According to Newton, light is not a wave, but rather composed of particles.

Opticks presents Newton's experiments with light in a manner that is clear, easy to understand, and written in English.[191] The bulk of Opticks contains carefully performed experiments that characterize properties of light. Newton's most startling experiments were those that investigated color.

Prior to Newton, it was widely believed that colored light contained some kind of impurity. Just as a dye-soaked piece of cloth imparts color to water flowing through it, so would a piece of stained glass lend color to light. Newton shocked the scientific community when he used two prisms to separate light into a rainbow, and then overlapped them again. He describes the experiment, referring to his figure shown, thus:

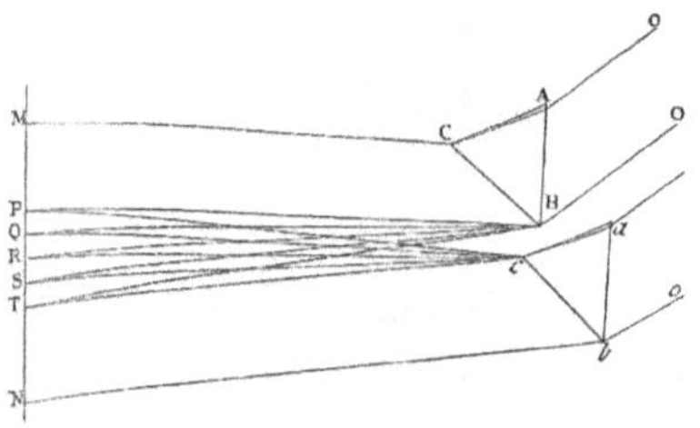

> Then let the Light trajected through them fall upon the Paper MN, distant about 8 or 12 Inches from the Prisms. And the Colours generated by the interior Limits B and C of the two Prisms, will be mingled at PT, and there compound white. For if either Prism be taken away, the Colours made by the other will appear in that Place PT, and when the Prism is restored to its Place again, so that its Colours may there fall upon the Colours of the other, the Mixture of them both will restore the Whiteness.[192]

Since red light refracts least, and blue (violet) light most, Newton theorized that the velocity of a light particle varies with its color. Newton looked to astronomy to verify this hypothesis. If his idea were correct, then blue would be the last color seen during an eclipse of Jupiter's moons, and

---

[189] C. Huygens, Treatise on Light, (1690), Translated by S. P. Thompson, Project Gutenberg E-book number 14725. The figure is from page 17 of the same source.

[190] R. Hooke, Micrographia: Some Physiological Descriptions of Minute Bodies Made by Magnifying Glasses with Observations and Inquiries Thereupon, (1665), London, Printed by Jo. Martyn, and Ja. Allestry, Printers to the Royal Society, Project Gutenberg E-book number 15491.

[191] Newton actually wrote three English editions (1704, 1717, and 1730), and a single Latin edition in 1706.

[192] Isaac Newton, Opticks or A Treatise on the Reflections, Refractions, Inflections and Colors of Light, 1730 Edition, Book One, Part II, Experiment 13.

red would be the first color observed when the satellite finally emerged from behind the planet. Unfortunately for Newton, no color changes in the eclipses of Jupiter's moons could be observed by him or any other astronomer.[193]

Astronomy soon did verify another prediction of Newton's particle model, and at the same time contradicted Huygens's wave model. In 1725, James Bradley and Samuel Molyneaux serendipitously observed a shift in the positions of stars due to the earth's motion. Just as rain appears to come toward your windshield, starlight appears to shift in a telescope throughout the year. This *aberration of starlight* is easily explained using Galilean relativity and the known velocity of light,[194] but only if light is a particle.

Newton's corpuscular theory of light began to unravel, however, in 1804 when the English polymath Thomas Young presented his famed double-slit experiment. This clearly demonstrated interference patterns of light, just as the Hooke-Huygens theory predicted.[195] When light from a distant source first passes through two open slits and then falls on the screen, there are regions on the screen that are dim but would be bright if only one slit were open. This fact is easily explained by assuming light travels as a wave, so that waves coming from each slit interfere either constructively or destructively.

Young's wave theory, however, did not immediately take off. Newton had already explained wavelike phenomena as ripples caused in the medium by moving light particles, similar to the ripples caused by rocks hitting water, or the wake produced by a passing boat. More importantly, Young's wave theory could explain neither Bradley's observations of stars nor experiments involving the polarization of light.

Augustin-Jean Fresnel successfully explained the results of optical polarization experiments by proposing that, rather than being a longitudinal wave like sound, light waves are transverse like the oscillations of a plucked string. By 1820, Fresnel's theory had passed every new experimental test, and thus became largely accepted. It still required, however, at least one lucky coincidence—Bradley's result only makes sense if the aether remains stationary with respect to the solar system.[196]

James Clerk Maxwell persuaded the rest of the physics community of the need for the aether when he derived a wave equation from his electromagnetic field equations. By the late nineteenth century, physicists the world over had accepted the aether's existence as established scientific fact. Maxwell's theory had passed every experimental test, especially the exacting experiments of Heinrich Hertz. William Thomson, who had just become Lord Kelvin a year earlier, concludes his preface to the 1893 English translation of Hertz's book on *Electric Waves* thusly:

> During the fifty-six years which has passed since Faraday first offended physical mathematicians with his curved lines of force, many workers and many thinkers have helped to build up the

---

[193] Alan Shapiro, "Newton's optics and atomism," in The Cambridge Companion to Newton, ed. I. Bernard Cohen and George Smith (Cambridge University Press, 2004), pp. 236-7.

[194] All measures of the speed of light at the time depended, in the same way, on the assumed length of one astronomical unit, so the uncertainty in the AU was not a problem.

[195] Thomas Young, "The Bakerian Lecture: Experiments and calculations relative to physical optics," Philosophical Transactions of the Royal Society of London (Royal Society of London) **94** (1804), 1–16.

[196] Young, Thomas, "The Bakerian Lecture: Experiments and calculations relative to physical optics," Philosophical Transactions of the Royal Society of London (Royal Society of London) **94** (1804), pp. 12-13. See also, E.T. Whittaker, A History of the Theories of Aether and Electricity (Dublin: Dublin University Press, 1910), p. 115.

> nineteenth-century school of *plenum*, one ether for light, heat, electricity, magnetism; and the German and English volumes containing Hertz's electrical papers, given to the world in the last decade of the century, will be a permanent monument of the splendid consummation now realized.[197]

By building a tabletop interferometer (his fig. 3, and p. 0), the American naval officer Albert Michelson tried, and failed, to measure the relative motion of the earth through the aether. Since his data (solid line) showed a clear discrepancy from theory (dotted line), Michelson made the bold conclusion we quoted in the overview of the book, that the "hypothesis of a stationary ether is thus shown to be incorrect." [198]

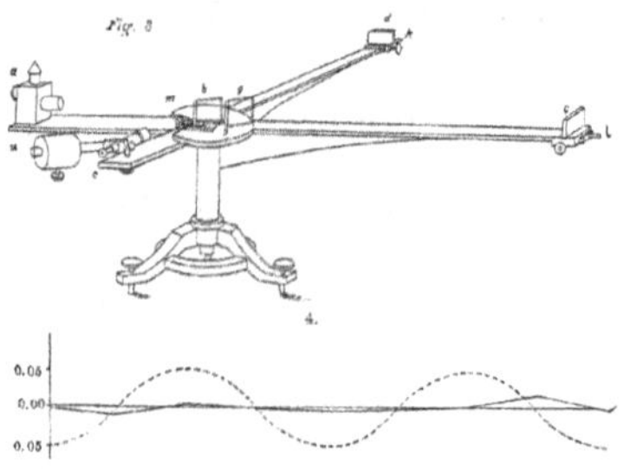

Michelson soon moved to a larger university, where he and his new colleague, Edward Morley, built a much more accurate optical interferometer, finally convincing the scientific community that the relative velocity between the earth and the aether is much less than the earth's orbital velocity.[199]

Recently Michelson interferometers, with fractional errors on the order of one part in nonillion (10^{30}), have finally succeeded in measuring tiny differences in light travel times. However they are not interpreted as changes in the speed of light, but rather as the detection of gravitational waves caused by coalescing black holes or neutron stars.[200]

By the turn of the century the problems with the theory of light came to the fore. In conjunction with the 1904 world's fair in St. Louis, the French physicist, J. Henri Poincaré, gave a lecture so brilliant we reproduce in full in Appendix C (p. 709). In his section on the principle of relativity, and its recent breakdown, he says this about Michelson's work:

> And then experiment, too, has taken upon itself to refute this interpretation of the principle of relativity; all the attempts to measure the velocity of the earth relative to the aether have led to negative results. Herein experimental physics has been more faithful to the principle than mathematical physics; the theorists would have dispensed with it readily in order to harmonize the other general points of view; but experimentation has insisted on confirming it. Methods were diversified; finally Michelson carried precision to its utmost limits; nothing came of it. It is precisely to overcome this stubbornness that today mathematicians are forced to employ all their ingenuity.
>
> Their task was not easy, and if Lorentz has succeeded, it is only by an accumulation of hypotheses. The most ingenious idea is that of local time.
>
> Let us imagine two observers, located at signal stations A and B, who wish to regulate their watches by means of optical signals. They exchange signals, but as they know that the transmission of light is not instantaneous, they are careful to cross them.
>
> . . .

---

197 H. Hertz, Electric Waves, trans. D. Jones (London: MacMillan and Co., 1893), p. xv.

198 Albert A. Michelson, "The Relative Motion of the Earth and the Luminiferous Ether", American Journal of Science, **22** (1881), 120-129.

199 Albert A. Michelson & Edward W. Morley, "On the Relative Motion of the Earth and the Luminiferous Ether", American Journal of Science **34** (1887), 333–345.

200 See the first detection article: B.P. Abbott, et al., "Observation of Gravitational Waves from a Binary Black Hole Merger", Physical Review Letters **116** (2016) 061102. Also see the textbook by Peter R. Saulson, Fundamentals of Interferometric Gravitational Wave Detectors (World Scientific, 1994, 2017)

> And, indeed, they mark the same hour at the same physical instant, but only if the two stations are stationary. Otherwise, the time of transmission will not be the same in the two directions, since the station A, for example, goes to meet the disturbance emanating from B, whereas station B flees before the disturbance emanating from A.
>
> Watches regulated in this way, therefore, will not mark the true time; they will mark what might be called the local time, so that one will gain on the other. It matters little, since we have no means of perceiving it. All the phenomena which take place at A, for example, will be behind time, but all just the same amount, and the observer will not notice it since his watch is also behind time; thus, in accordance with the principle of relativity he will have no means of ascertaining whether he is at rest or in absolute motion.
>
> Unfortunately this is not sufficient; additional hypotheses are necessary. We must admit that the moving bodies undergo a uniform contraction in the direction of the motion. One of the diameters of the earth, for example, is shortened by 1/200000000 as a result of our planet's motion, whereas the other diameter preserves its normal length. Thus we find the last minute differences accounted for.

It was around this same time that Albert Einstein led a book club where they read Poincaré. A year later, you can see Poincaré's influence in Einstein's 1905 paper *On the Electrodynamics of Moving Bodies*:

> Like every other electrodynamics, the theory to be developed is based on the kinematics of the rigid body, since assertions of each and any theory concern the relations between rigid bodies (coordinate systems), clocks, and electromagnetic processes.[201]

The connection that Einstein made was simple. Keep Galileo's principle of relativity and Maxwell's electrodynamics, but modify basic kinematics in order to make the union work. While the most startling results have to do with interpretations of time and length, the beauty of special relativity comes in its reinterpretation of electrodynamics.

With Einstein's theory of relativity, there is no longer a need for an aether. But with no medium, why does light act as an electromagnetic wave? What is it fundamentally? Does a field have any real meaning other than a way to put off harder physical questions, and advert action at a distance?

To address these questions, Einstein made the heuristic hypothesis that Newton was right after all, at least in principle, about light. This new theory of light, as a particle that follows ultra-relativistic kinematics, seemed to work. Einstein published his particle model of light just months before his special theory of relativity, and it was the only one of his papers that he conceded was truly "revolutionary."

What is the correct paradigm? Is light fundamentally a particle that often acts like a wave, or is it a wave that often acts as a particle? What is the solution to spooky action at a distance? Is the universe filled with some medium, be it called the aether, dark matter, dark energy, or something else? Some of these questions are still open. Keep them in mind as you study not only electrodynamics, but other scientific and philosophical fields as well.

---

[201] A. Einstein, "On the Electrodynamics of Moving Bodies," Annalen der Physik **17** (1905): 891-921, Translated by Anna Beck, in *The Collected Papers of Albert Einstein, Volume 2, The Swiss Years, 1900-1909.* John Stachel, ed. (Princeton: Princeton University Press). English Translation, ©1989 by the Hebrew University of Jerusalem.

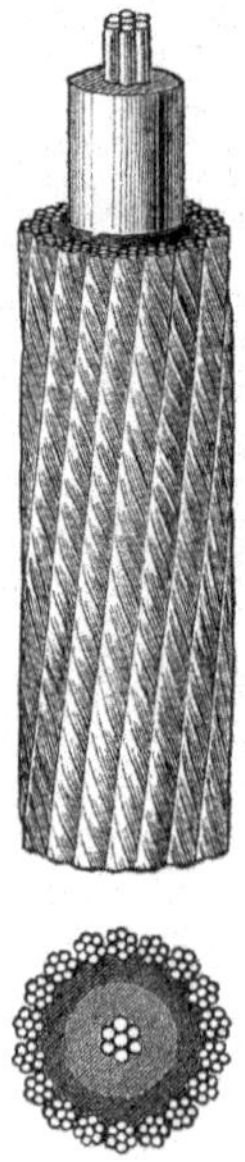

Plate 23: The First Atlantic Telegraph Cables[202]

[202] The top left figure is a diagram of the 1858 telegraph cable. Top right is an 1865 photo of an engineer laying cable. Bottom is a 1961 lithograph depicting the mid-atlantic rendezvous of the ships USS Niagra, HMS Valorous, HMS Gorgon, and HMS Agamemnon just before splicing together and laying the first transatalantic telegraph cable in 1858. The images are in the public domain and available from Wikimedia Commons.

Chapter 13 Transmission Lines

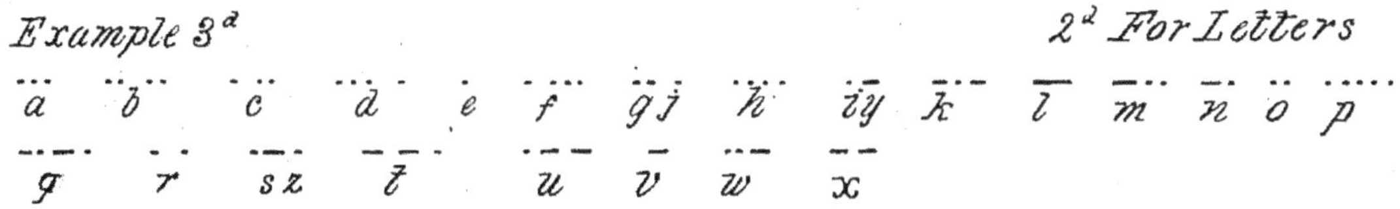

The first great digital information boom began when Samuel Morse patented the telegraph in 1840.[203] His primary concern was the encoding of binary information and a method for automating the communication, since the electrical components of the transmitter and receiver were relatively simple. The automation process of encoded *keys* (like those that fit in a lock) was soon dispensed with in lieu of manually tapping a button; also called a *key* (like on a piano), and telegraph engineers became quite adept at sending Morse code manually. Telegraph technology spread quickly, telegraph lines became longer, and expectations arose for increased rates of data telemetry.

Transmission lines, which carry high frequency signals over large distances, were especially important to the development of electrodynamics, as the telegraph was the emerging communication technology in the nineteenth century. In Germany, Franz Neumann and Wilhelm Weber derived a wave equation for a signal in a wire, and predicted that it would travel at the same speed as light travels in a vacuum. In Britain, Michael Faraday, William Thomson, and Oliver Heaviside made significant gains in implementing telegraph cable technology.

Of particular interest to scientists and engineers of that era was the question of how to string a telegraph cable under the ocean. Among the many issues that dogged early efforts in laying the trans-Atlantic cable, for example, was signal attenuation and distortion due to the conductivity of salt water. At the behest of the Electric Telegraph Company, Faraday, in 1854, witnessed an experimental demonstration of electrical signaling retardation in a long insulated cable submerged in the River Thames. He reasoned that the insulation acted as a dielectric placed between the copper wire and the slightly conducting water. Transmission of the electric current would therefore be limited because the cable acted like a capacitor that had to charge and discharge before the signal could be received at the other end.[204]

William Thomson soon developed a partial differential equation, identical in form to Fourier's equation describing the diffusion of heat, which modeled the behavior of electric current as it passed through a long cable of finite capacitance and resistance.[205] The solutions Thomson obtained for his equation allowed him to predict correctly that current pulses along the line would retard at a rate proportional to the square of the cable's length, thus limiting signal speed and causing poor performance. His expressed concerns about the first trans-Atlantic cable went unheeded, and the cable deteriorated only months after Queen Victoria exchanged the first telegraph messages with U.S. President James Buchanan in 1858. Thomson's early work was not in vain, however. He was subsequently appointed chief engineer for the company that

---

[203] S.F.B. Morse, "Improvement in the mode of communicating by signals by the application of electromagnetism," U.S. Patent No. 1647, awarded on June 20, 1840.

[204] M. Faraday, "On electric induction-Associated causes of current and static effects," Philosophical Magazine Series 4, **7**:44, (1854), 197-208.

[205] W. Thomson, "On the theory of the electric telegraph," Proceedings of the Royal Society **2**, (1855), 382.

successfully laid the transatlantic cable in 1867. His technical contributions—both theoretical and practical—were so crucial to the success of the project that Queen Victoria knighted him within the year.[206]

Gustav Kirchhoff, in his 1857 series of papers on the propagation of electric signals through telegraph lines, demonstrated "a very remarkable analogy between the propagation of electricity in a wire and the propagation of a wave in a [tense string]." He also found the propagation velocity to be "very nearly equal to the velocity of light in vacuo," although he failed to comment on the possible implications of this result.[207] Independently of Kirchhoff, Weber also performed a similar investigation, but his work wasn't published until shortly after Kirchhoff's. The remarkable implication of their analyses was that in a circuit of negligible resistivity, oscillating currents could be propagated along the wire with a constant velocity numerically equal to the velocity of light. Furthermore, this velocity was found to be independent of the nature of the conductors, of the cross section of the wire, and of the electric current density. This result of Kirchhoff and Weber is all the more remarkable as it came before Maxwell derived the wave equation from his introduction of the displacement current in Ampère's law in 1865.

Oliver Heaviside, a nephew of British inventor Charles Wheatstone (from whom the "Wheatstone bridge" gets its name), quit his first and only employment, as a telegraph operator, to move back to his parents' house and pursue research. Having chanced upon a copy of Maxwell's Treatise, Heaviside was inspired to devote the rest of his life to clarifying and applying Maxwell's theory. Heaviside was acquainted with the practical side of telegraphy through his job, and even published several early papers on elementary circuits and telegraph technology during that period. Despite, or perhaps because of, the fact that he was mostly self-taught, he made outstanding and profound contributions to electrodynamics, electrical engineering, and vector calculus in the 1870s and 1880s. Heaviside improved upon William Thomson's work by including inductance and modeling transmission cables as a long chain of coupled LC circuits.[208] And, although they had long been used in undersea telegraphy, Heaviside first explained, and then patented, the idea of a coaxial cable.[209]

We use the results of Heaviside's research to calculate the speed of transmission, the flux of energy, and the pressure exerted by electromagnetic fields along a coaxial cable. We will derive the *telegrapher's equations* for lossless transmission in Thought Experiment 13.4 (p. 415), for example, and show how these yield a wave equation for signal propagation along the cable. It is interesting to note that John Henry Poynting actually used the examples of a long straight wire and, later, of an underwater telegraph cable in discussing the energy flow by electromagnetic fields.[210]

---

[206] The Queen had also tried to knight Faraday, but he refused the honor.

[207] G. Kirchhoff, "On the motion of electricity in wires," Philosophical Magazine, **13** (1857), 393-412.

[208] O. Heaviside, "Electromagnetic induction and its propagation" in Electrical Papers by Oliver Heaviside in 2 vols. (Bronx, New York: Chelsea Publishing Company 1970), **I**: 492-56, **II**: 39-146.

[209] H. Griffiths, "Oliver Heaviside" in History of Wireless, ed. Tapan K. Sarkar et al. (Hoboken, New Jersey: J. Wiley & Sons, 2006), pp. 239-40; David Kraueter, British Radio and Television Pioneers: A Patent Bibliography (USA: Scarecrow Press), p. 66.

[210] J.H. Poynting, "Molecular electricity," Collected Scientific Papers (Cambridge: Cambridge University Press, 1920), pp. 269-298.

13.1 *RLC Circuits*

In this section we will further develop circuits involving resistors, inductors, and capacitors.

Thought Experiment 13.1: A Telegraph Signal Generator

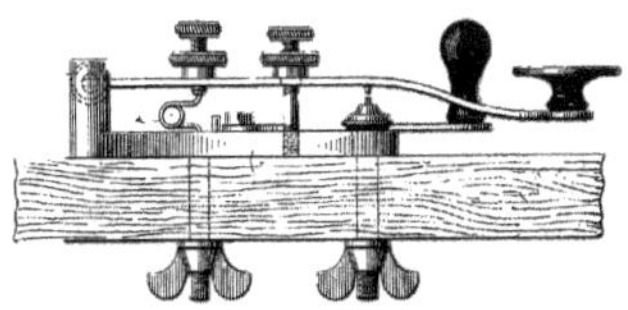

Old telegraph signal generators used batteries, Leyden jars, big inductor coils, and a spark gap. Throughout the nineteenth century the signal generators improved, until they were replaced with the Teletype machine in the first few decades of the twentieth century,[211] which was subsequently replaced with the networked digital computer we still use today. In this thought experiment, we will model the sending device as a circuit with a single capacitor, inductor, and resistor as in the diagram below. The capacitor is assumed to be fully charged to begin with, and the key is the open junction.

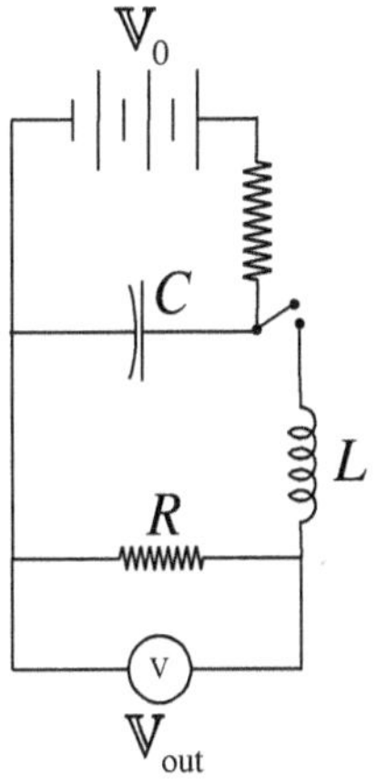

The telegraph operator pushes the key[212] at time zero. What is the voltage across the voltmeter as a function of time after each tap?

To solve this, we apply Kirchoff's loop rule as:

$$\frac{Q}{C} - L\frac{\mathrm{d}I}{\mathrm{d}t} - IR = 0 .$$

And we add conservation of charge, by noting that the current is related to the charge on the capacitor's anode via:

$$I = -\frac{dQ}{dt} .$$

So we can write expression for the sum of the voltages around the loop in terms of the charge on the capacitor's anode:

$$0 = Q + RC\frac{\mathrm{d}Q}{\mathrm{d}t} + LC\frac{\mathrm{d}^2Q}{\mathrm{d}t^2} .$$

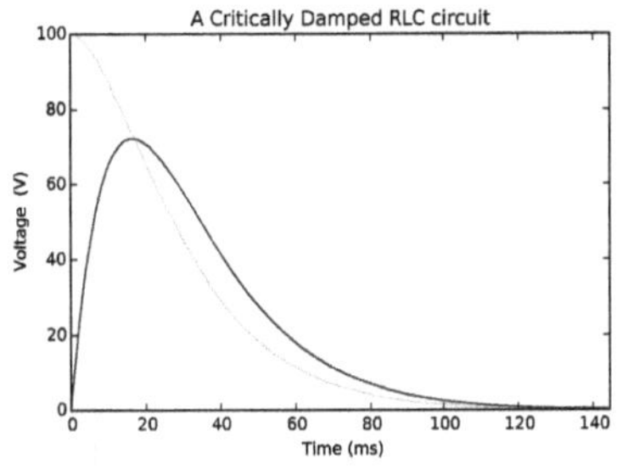

This is a second order differential equation, and it is similar to ones we've already solved. Notice that in the limit of no resistance the equation becomes that for an LC circuit, and if there is no inductance it simply describes an RC circuit. If the resistance to too high, the solution of this equation falls off exponentially, but if it is too low the system will oscillate with only a slight decrease over time. Between these two extremes of *overdamping* and *underdamping*, we have *critical damping*, which occurs when the resistance is given by:

$$R_c = 2\sqrt{L/C} \cdot$$

---

[211] Pearne, F.D., 1900, "System of Type-writing Telegraphs," U.S. Patent No. 674469, awarded May 21, 1901.

[212] Dickinson, O.A., 1895, "Telegraph Instrument," U.S. Patent No. 548969, awarded Oct 29, 1895.

While the damped oscillator has a standard solution, we nevertheless solved it numerically. The plot shows this solution for a critically damped system, with $\mathbb{V}_0 = 100\,\text{V}$, $L = 2.5\,\text{H}$, $C = 100\,\mu\text{F}$, and $R = 320\,\Omega$. The lighter curve shows the voltage across the capacitor, and the darker curve shows the voltage across the resistor, $\mathbb{V}_{\text{out}}$.

We also solved the equation first with increasing inductance, while holding the capacitance and resistance constant, and then by increasing capacitance while holding the inductance and resistance also constant. And, similarly, we obtained a solution while adjusting the resistance so the system is critically damped. Notice that the larger the capacitance, or the inductance, the further delayed the peak voltage across the resistor.

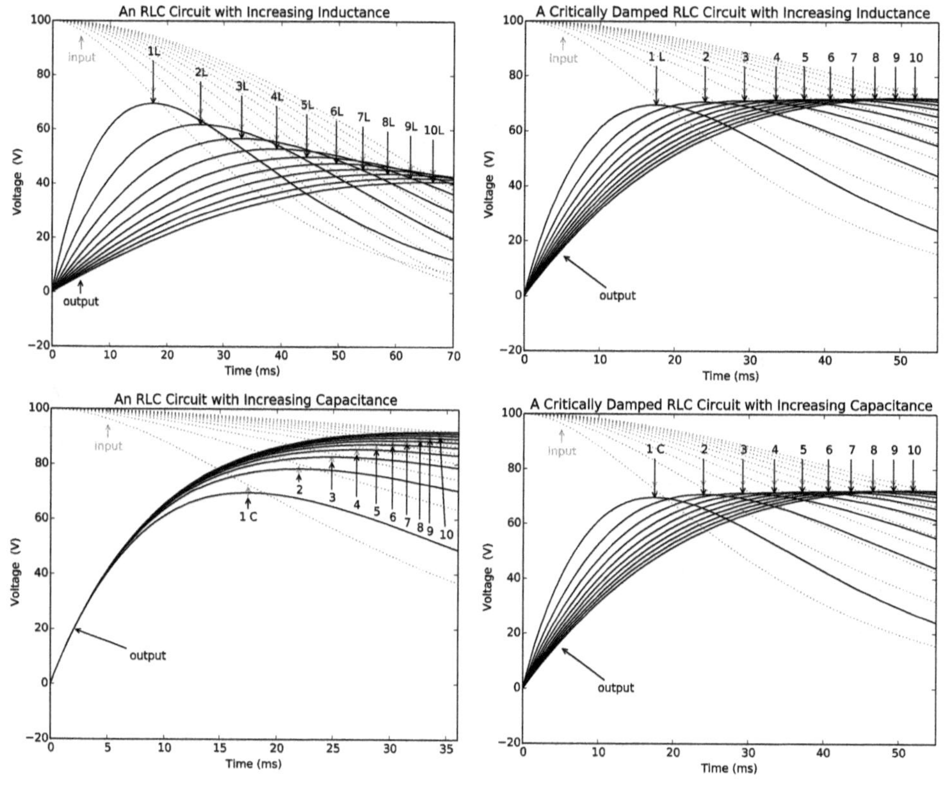

Problem 13.1: A Series RLC Circuit

Consider the RLC circuit in the diagram, where the switch is flipped at time zero.

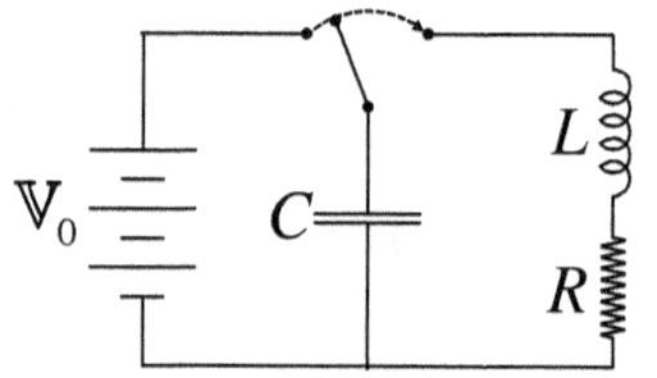

a) Write down the relationship among the charge on the capacitor and its time derivatives.

b) Imagine that $R = 0$, what would be the charge on the capacitor as a function of time?

c) Imagine that $L = 0$, what would be the charge on the capacitor as a function of time?

d) Solve for the charge on the capacitor as a function of time. Verify that this reverts to your prior results in the correct limits.

Problem 13.2: A Stack of Capacitors and Resistors

Consider a long line of capacitors and resistors wired in series, as in the diagram. Imagine that the input current rapidly changes.

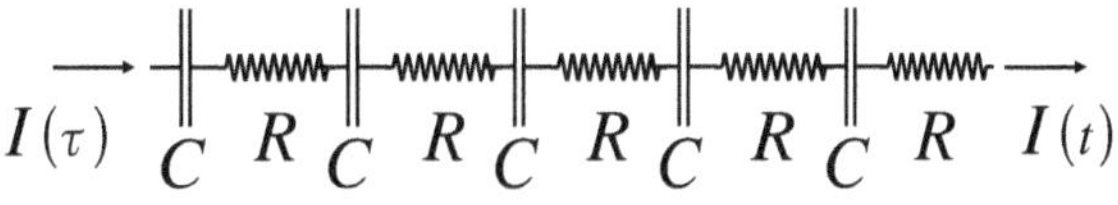

a) Consider a fast sinusoidal current at a particular node $I_n = I_0 \sin(\omega t)$. What would you expect the signal at the node just after to be? Clearly explain your reasoning.

b) If the current after n nodes is $I(t)$, because of an input signal $I(\tau)$, what is τ as a function of t? τ is called the *retarded time.*

c) Now imagine that each resistor to be cylindrical with a length δz, cross sectional area $\vec{A}$, and conductivity κ. Furthermore, assume that each capacitor also has a gap width δz, cross sectional area $\vec{A}$, and dielectrical permittivity ε. What would be the time delay Δt between nodes?

d) Imagine shocking a patient with a heart defibrillator. What would be the speed of the current wave assuming a permittivity of about 700 pF/m, a conductivity of about 3 S/m, and a node spacing of about 2 mm?

Thought Experiment 13.2: A Chain of LC Circuits

Consider Thought Experiment 13.1 above, and imagine that we took away the resistor and replaced it with the chain of identical RLC circuits. Thus, we will use our prior solution as the input signal $\mathbb{V}_0(t)$ and $I_0(t)$.

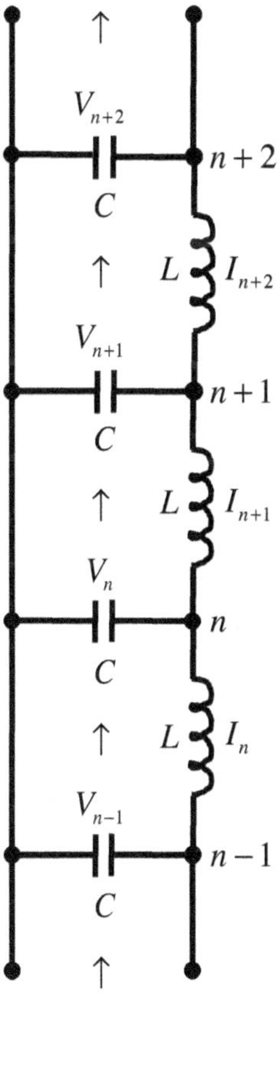

What is the voltage at each node down the line?

First we apply Kirchhoff's loop rule to the first loop, so the voltage across the first capacitor is:

$$\mathbb{V}_1(t) = \mathbb{V}_0(t) - L\frac{dI_0(t)}{dt},$$

and then applying the junction rule at node 1, we can find the current through the second inductor:

$$I_1(t) = I_0(t) - \frac{dQ_1(t)}{dt} = I_0(t) - C\frac{d\mathbb{V}_1(t)}{dt}.$$

And applying the loop rule around the second loop, and the junction rule at node 2, we have a similar relationship as before:

$$\mathbb{V}_2(t) = \mathbb{V}_1(t) - L\frac{dI_1(t)}{dt} \quad \text{and} \quad I_2(t) = I_1(t) - \frac{dQ_2(t)}{dt} = I_1(t) - C\frac{d\mathbb{V}_2(t)}{dt}.$$

We can then continue this many times so that the voltage at node n, and current leaving it, can be found from the loop and junction rules as:

$$\mathbb{V}_n = \mathbb{V}_{n-1} - L\frac{dI_{n-1}}{dt} \quad \text{and} \quad I_n(t) = I_{n-1}(t) - C\frac{d\mathbb{V}_n(t)}{dt}.$$

This is a *recursion relation*, which should make the problem relatively easy to solve on a computer since it is already a discrete relationship. We show the solution below for the first few nodes when a pulse is sent in one side but, as you can see, the solution quickly becomes numerically unstable. One reason is simply that taking multiple numerical derivatives can become unstable due to computer round-off error. The second reason for the instability is physical, and it has to do with the characteristic time $\Delta t = \sqrt{LC}$ of the circuit chain. So long as a periodic voltage is sent through one end with a characteristic period that is an integer multiple of the time Δt , each node will remain in phase. However, the circuit is inherently unstable to higher frequency input signals.

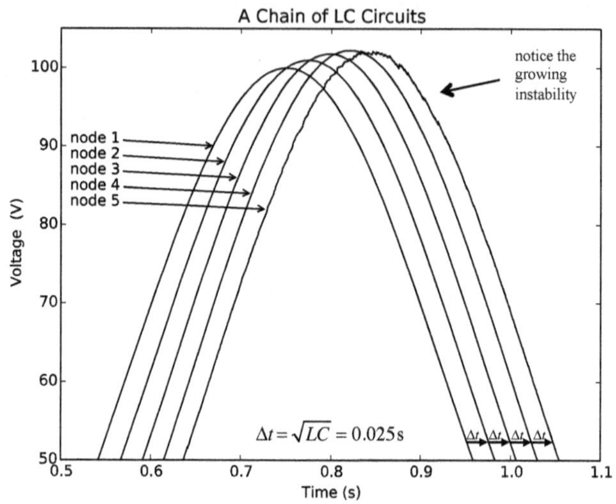

Since chains of LC circuits transmit signals at a particular frequency, they became useful for systems of automatic telegraphy. For example, Thomas Edison invented a system that was used during the 1870s, where long messages, such as news reports, were encoded onto perforated paper tape, and then sent at high telemetry rates of up to 1,000 bits per second, and received in a similar fashion on the other side.

Problem 13.3: An RLC Circuit Chain

Consider the chain of RLC circuits in the diagram.

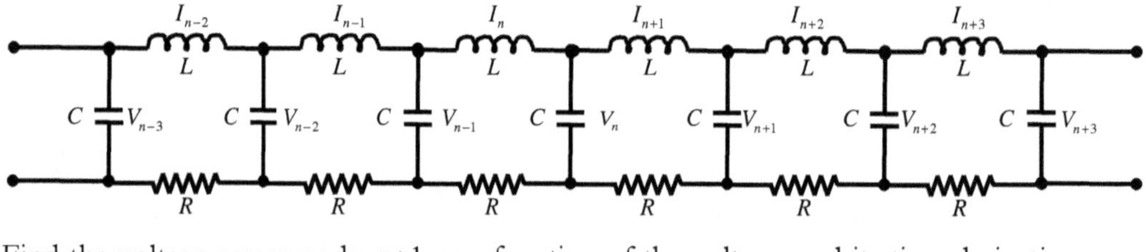

a) Find the voltage across node $n+1$ as a function of the voltage, and its time derivatives, across the prior node.

b) Find the output peak voltage in terms of the input peak voltage and the number of nodes?

c) After how many nodes will the voltage be halved?

d) Given a desired clocking period Δt and an unavoidable resistance R, what would be the best choice of inductance L and capacitance C?

e) If you are running this at a rate of 1000 nodes/second, what would be your choice for L and C if your resistance is ten ohms per node?

f) Plot the voltages of the first 100 nodes of this lossy transmission line for the five special cases where: $R\sqrt{C/L} = \{\tfrac{1}{4}, \tfrac{1}{2}, 1, 2, 4\}$.

13.2 *Continuous Transmission Lines*

In Thought Experiment 13.2 (p. 411) we showed how a chain of LC circuits would move a signal down the line at a rate of one register per $\sqrt{LC}$, so long as the clocking period is an even multiple of $\sqrt{LC}$. This is because the voltages and current obey the following *coupled* recursion relations:

$$\mathbb{V}_n - \mathbb{V}_{n-1} = -L\frac{dI_{n-1}}{dt} \quad \text{and} \quad I_n - I_{n-1} = -C\frac{d\mathbb{V}_n}{dt}.$$

But, what if we have a continuous cable? How then will a signal propagate down the line?

Oliver Heaviside, among others, described these signals with a pair of coupled first order differential equations, *the telegrapher's equations*. These equations, and their use, allowed the long distance telegraph to stretch all over the globe, including under water.

In this section we will describe these continuous cables with properties such as the resistance, capacitance and inductance per length, which, in this section, we will refer to as $\mathtt{R}$, $\mathtt{L}$, and $\mathtt{C}$.

Thought Experiment 13.3: An Ideal Coaxial Cable Transmission Line

In this thought experiment, we consider a continuous coaxial cable with a measured capacitance and inductance per length, $\mathtt{L}$ and $\mathtt{C}$, which we will model as a chain a small inductors and capacitors:

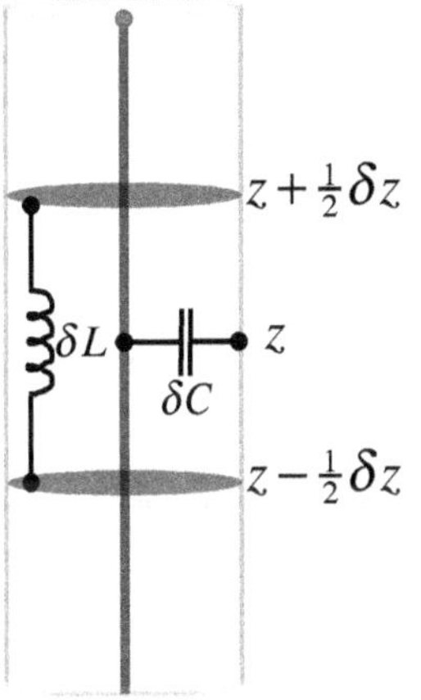

$$\delta\mathbb{V} \approx -\delta L\frac{dI}{dt} \approx -\mathtt{L}\delta z\frac{dI}{dt}$$
$$\delta I \approx -\delta C\frac{d\mathbb{V}}{dt} \approx -\mathtt{C}\delta z\frac{d\mathbb{V}}{dt}.$$

And taking the limit as $\delta z \to 0$, at a particular time, we have the following coupled differential equations:

$$\lim_{\delta z\to 0}\left.\frac{\delta\mathbb{V}}{\delta z}\right|_t = \frac{\partial\mathbb{V}}{\partial z} = -\mathtt{L}\frac{\partial I}{\partial t}$$
$$\lim_{\delta z\to 0}\left.\frac{\delta I}{\delta z}\right|_t = \frac{\partial I}{\partial z} = -\mathtt{C}\frac{\partial\mathbb{V}}{\partial t}.$$

These are the Heaviside telegrapher's equations, which we mentioned in the chapter introduction.

Recall that a discrete chain of circuits would replicate the voltage down the line, so long as the clocking period was an even multiple of the register period $\sqrt{LC}$. In the continuous limit, however, every period is a multiple of the register period, so the signal should be replicated down the line without this restriction. In this limit, the register period becomes a delay time per length:

$$\delta t = \sqrt{\delta L\delta C} = \sqrt{(\mathtt{L}\delta z)(\mathtt{C}\delta z)} = \delta z\sqrt{\mathtt{LC}}.$$

So the signal moves at a speed of:

$$v = \frac{\delta z}{\delta t} = \frac{1}{\sqrt{\mathtt{LC}}}.$$

If the signal is going to be replicated down the line, then the voltage and current, which are a function of both the position and time, must follow the same function of time as the input signal:

$$\mathbb{V}(t,z) = \mathbb{V}_{\text{in}}(t - z/v) = \mathbb{V}_{\text{in}}(\tau) \quad \text{and} \quad I(t,z) = I_{\text{in}}(t - z/v) = I_{\text{in}}(\tau),$$

where the time when the corresponding signal was at $z = 0$ is:

$$\tau = t - z/v,$$

which is called the *retarded time.*

While this seems reasonable, given our experience so far with transmission lines, how does this help us solve Heaviside's telegrapher's equations?

Since these equations contain partial derivatives, which we may want to solve using the chain rule, we begin by taking the partial derivatives of the retarded time:

$$\frac{\partial \tau}{\partial t} = \frac{\partial}{\partial t}\left(t - z/v\right) = 1 \quad \text{and} \quad \frac{\partial \tau}{\partial z} = \frac{\partial}{\partial z}\left(t - z/v\right) = -\frac{1}{v}\ .$$

Next we will apply our hypothesis, which is that the signal is characterized simply by functions of the retarded time.

First, let's use the chain rule to find each partial derivative in terms of the retarded time:

$$\frac{\partial \mathbb{V}}{\partial z} = \frac{\partial}{\partial z}\mathbb{V}_{\text{in}}(\tau) = \frac{d\mathbb{V}_{\text{in}}(\tau)}{d(\tau)}\frac{\partial(\tau)}{\partial z} = -\frac{1}{v}\frac{d\mathbb{V}_{\text{in}}}{dt}\bigg|_{t-z/v} = -\frac{1}{v}\frac{d}{d\tau}\mathbb{V}_{\text{in}}(\tau)$$

$$\frac{\partial \mathbb{V}}{\partial t} = \frac{\partial}{\partial t}\mathbb{V}_{\text{in}}(\tau) = \frac{\partial\mathbb{V}_{\text{in}}(\tau)}{\partial(\tau)}\frac{\partial(\tau)}{\partial t} = \frac{d\mathbb{V}_{\text{in}}}{dt}\bigg|_{t-z/v} = \frac{d}{d\tau}\mathbb{V}_{\text{in}}(\tau)$$

$$\frac{\partial I}{\partial t} = \frac{\partial}{\partial t}I_{\text{in}}(\tau) = \frac{\partial I_{\text{in}}(\tau)}{\partial(\tau)}\frac{\partial(\tau)}{\partial t} = \frac{dI_{\text{in}}}{dt}\bigg|_{t-z/v} = \frac{d}{d\tau}I_{\text{in}}(\tau)$$

$$\frac{\partial I}{\partial z} = \frac{\partial}{\partial z}I_{\text{in}}(\tau) = \frac{dI_{\text{in}}(\tau)}{d(\tau)}\frac{\partial(\tau)}{\partial z} = -\frac{1}{v}\frac{dI_{\text{in}}}{dt}\bigg|_{t-z/v} = -\frac{1}{v}\frac{d}{d\tau}I_{\text{in}}(\tau).$$

Notice the convenience that any function of two variables is now replaced by a function of a single variable.

Now, we can substitute these into the telegrapher's equations:

$$\frac{\partial \mathbb{V}}{\partial z} = -\mathbb{L}\frac{\partial I}{\partial t} \quad , \quad \frac{\partial I}{\partial z} = -\mathbb{C}\frac{\partial \mathbb{V}}{\partial t}$$

$$\left(-\frac{1}{v}\frac{d}{d\tau}\mathbb{V}_{\text{in}}(\tau)\right) = -\mathbb{L}\left(\frac{d}{d\tau}I_{\text{in}}(\tau)\right) \quad , \quad \left(-\frac{1}{v}\frac{d}{d\tau}I_{\text{in}}(\tau)\right) = -\mathbb{C}\left(\frac{d}{d\tau}\mathbb{V}_{\text{in}}(\tau)\right)$$

Multiplying the equations, we find:

$$\left(-\frac{1}{v}\cancel{\frac{d}{d\tau}\mathbb{V}_{\text{in}}(\tau)}\right)\left(-\frac{1}{v}\cancel{\frac{d}{d\tau}I_{\text{in}}(\tau)}\right) = \left(-\mathbb{L}\left(\cancel{\frac{d}{d\tau}I_{\text{in}}(\tau)}\right)\right)\left(-\mathbb{C}\left(\cancel{\frac{d}{d\tau}\mathbb{V}_{\text{in}}(\tau)}\right)\right) \quad \rightarrow \quad \frac{1}{v^2} = \mathbb{LC}.$$

This is the solution we surmised from the last thought experiment, namely that the signal speed is: $v = (\mathrm{LC})^{-\frac{1}{2}}$! Thus, our hypothesis is confirmed.

We now also divide the two equations, and find:

$$\frac{\left(\cancel{\frac{1}{\Delta}}\frac{d}{d\tau}\mathbb{V}_{in}(\tau)\right)}{\left(\cancel{\frac{1}{\Delta}}\frac{d}{d\tau}I_{in}(\tau)\right)} = \frac{-\mathrm{L}\left(\frac{d}{d\tau}I_{in}(\tau)\right)}{-\mathrm{C}\left(\frac{d}{d\tau}\mathbb{V}_{in}(\tau)\right)} \quad\rightarrow\quad \left(\frac{d}{d\tau}\mathbb{V}_{in}(\tau)\right)^2 = \left(\frac{\mathrm{L}}{\mathrm{C}}\right)\left(\frac{d}{d\tau}I_{in}(\tau)\right)^2.$$

If we take the square root on both sides, and then integrate, the result looks like Ohm's law:

$$|\mathbb{V}_{in}| = |I_{in}|\sqrt{\frac{\mathrm{L}}{\mathrm{C}}} \quad\rightarrow\quad |\mathbb{V}_{in}| = |I_{in}|Z.$$

The quantity, $Z = \sqrt{\frac{\mathrm{L}}{\mathrm{C}}}$, is called the *impedance of the line*, or simply the *impedance*. It acts as an effective resistance, so has units of ohms. Rather than the input power being dissipated as heat, however, it goes into the cable and eventually comes out on the other end. That is why it does not depend on the length of the cable, but rather on the material from which it is made.

If the other end of the cable is capped with a resistor, R, so that $R = \sqrt{\mathrm{L}/\mathrm{C}}$, then it will not reflect the signal. This is called *impedance matching*, and it is good practice when working with coaxial cable. Typically a telephone cable has a characteristic impedance of 100 ohms, while most television cable has an impedance of 75 ohms.

Thought Experiment 13.4: The Telegrapher's Wave Equation

We have solved Heaviside's telegrapher's equations (p. 409) for a signal passing down an ideal cable in a number of different ways. We solved them algorithmically for the finite case, then used this to guess the analytical solution, and then we verified that this was correct using the chian rule. It is often, however, easier to decouple these equations, so that we hae a partitial differential equation of either the voltage or current, but not both.

To decouple first order equations, we take the partial derivative with respect to position and time respectively:

$$\frac{\partial^2 \mathbb{V}}{\partial z^2} = -\mathrm{L}\frac{\partial^2 I}{\partial z \partial t} \quad \text{and} \quad \frac{\partial^2 I}{\partial t \partial z} = -\mathrm{C}\frac{\partial^2 \mathbb{V}}{\partial t^2}.$$

Since $\frac{\partial^2 I}{\partial z \partial t} = \frac{\partial^2 I}{\partial t \partial z}$,

we can substitue $-\mathrm{C}\frac{\partial^2 \mathbb{V}}{\partial t^2}$ for the mixed partial derivative $\frac{\partial^2 I}{\partial z \partial t}$, and write:

$$\frac{\partial^2 \mathbb{V}}{\partial z^2} = -\mathrm{L}\left(-\mathrm{C}\frac{\partial^2 \mathbb{V}}{\partial t^2}\right) = \mathrm{LC}\frac{\partial^2 \mathbb{V}}{\partial t^2}.$$

We could have done something similar to isolate the current instead:

$$\frac{\partial^2 \mathbb{V}}{\partial t\,\partial z} = -\mathrm{L}\frac{\partial^2 I}{\partial t^2} \quad \text{and} \quad \frac{\partial^2 I}{\partial z^2} = -\mathrm{C}\frac{\partial^2 \mathbb{V}}{\partial z\,\partial t} \quad \rightarrow \quad \frac{\partial^2 I}{\partial z^2} = \mathrm{LC}\frac{\partial^2 I}{\partial t^2}\,.$$

Partial differential equations in this form are called *wave equations*, as they govern such wave phenomena as tension waves, water waves, and sound waves. Since this is in one spatial dimension (distance along the cable), it is called *the one-dimensional wave equation.*

The solution to the one-dimensional wave equation must, naturally, be the same as the solution to the coupled first order equations we used to derive it. Therefore, as before, whatever signal is sent into the cable emerges later on the other end. Recall that we defined the retarded time to be the time when the corresponding signal was at zero position, so $\tau = t - z/v$.

We now show that we recover the same signal speed from this second order wave equation as we did before:

$$\frac{\partial^2 \mathbb{V}(\tau)}{\partial z^2} = \frac{\partial}{\partial z}\left(\frac{\partial}{\partial z}\mathbb{V}(\tau)\right) = \frac{\partial}{\partial z}\left(-\frac{1}{v}\mathbb{V}(\tau)\right) = \frac{1}{v^2}\mathbb{V}(\tau).$$

$$\frac{\partial^2 \mathbb{V}(\tau)}{\partial t^2} = \frac{\partial}{\partial t}\left(\frac{\partial}{\partial t}\mathbb{V}(\tau)\right) = \frac{\partial}{\partial t}\left(\mathbb{V}(\tau)\right) = \mathbb{V}(\tau).$$

and substituting into the wave equation:

$$\frac{\partial^2 \mathbb{V}}{\partial z^2} = \mathrm{LC}\frac{\partial^2 \mathbb{V}}{\partial t^2} \quad \rightarrow \quad \frac{1}{v^2}\mathbb{V}(\tau) = \mathrm{LC}\,\mathbb{V}(\tau) \quad \rightarrow \quad v = \frac{1}{\sqrt{\mathrm{LC}}}.$$

Thought Experiment 13.5: Heaviside's Lossy Transmission Line

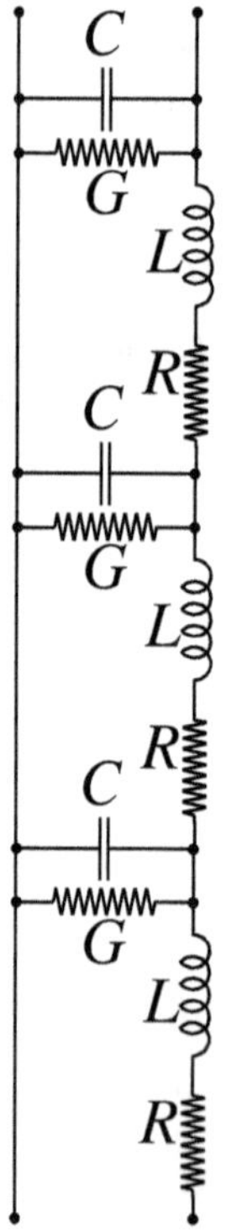

In Thought Experiment 13.2 we showed how a continuous chain of LC circuits would move a signal down the line at a signal speed given by:

$$v = (\mathrm{LC})^{-\frac{1}{2}},$$

and have an effective resistance, called the impedance:

$$Z = \sqrt{\frac{\mathrm{L}}{\mathrm{C}}}\,.$$

In this example, we extend this analysis to model real underwater cables of the late nineteenth century, that were used for telegraph and telephone transmission. For this purpose, we now introduce two new quantities, R and G, which are the resistance and leakage conductance per length, and whose SI units are be ohms per meter and siemens per meter, respectively.

As in the diagram, we model the cable as a continuous chain of circuits. In the ideal case, all parallel resistors would have infinite resistance, so there would be no leakage conductance $G = \mathrm{G}\delta z \approx 0$. Similarly, each series resistor would have infinite conductivity, so there would be no resistance $R = \mathrm{R}\delta z \approx 0$.

With these loss terms included, Heaviside's telegrapher's equations become:

$$\frac{\partial \mathbb{V}}{\partial z} = -\mathrm{R}I - \mathrm{L}\frac{\partial I}{\partial t} = -\left(\mathrm{R} + \mathrm{L}\frac{\partial}{\partial t}\right)I \quad \text{and} \quad \frac{\partial I}{\partial z} = -\mathrm{G} - \mathrm{C}\frac{\partial \mathbb{V}}{\partial t} = -\left(\mathrm{G} + \mathrm{C}\frac{\partial}{\partial t}\right)\mathbb{V}.$$

Notice that the Heaviside equations become much more complicated when losses are taken into account. The signals in a lossy cable still propagate at a speed $v = (\mathrm{LC})^{-\frac{1}{2}}$, but a signal now gets attenuated and it is frequency dependent as the time derivative may, or may not, come into play. One problem Heaviside tried to solve was how to limit signal distortion over long distances.

Consider only sinusoidal solutions at a particular angular frequency ω, which become attenuated due to the loss terms. Given the chain rule, the time derivatives of sinusoidal functions will pull out a factor of ω, and shift the phase forward by $\pi/2$. For mathematical convenience, Heaviside defined two unitless quantities that represent the relative importance of the resistance as compared to the inductance, and the leaking conductance relative to the capacitance:

$$f \equiv \frac{\mathrm{R}}{\omega \mathrm{L}} \quad \text{and} \quad g \equiv \frac{\mathrm{G}}{\omega \mathrm{C}} .$$

Heaviside argued that the leakage conductance is quite large for underwater cables, but for land lines the problem can be overcome by communicating at higher frequencies:

> Therefore on a landline of $1\,\mathrm{M\Omega}/\mathrm{km}$ megaohm per km insulation-resistance, and $0.1\,\mu\mathrm{F}/\mathrm{km}$ permittance, we have $g = 100/\omega$. Thus g is important at low frequencies and becomes a small fraction at high frequencies.

To see this we can write out the units as:

$$\mathrm{G} = \frac{1}{\mathrm{M\Omega}\cdot\mathrm{km}} = 10^{-6}\,\mathrm{S}/\mathrm{km} \quad \text{and} \quad g = \frac{1/\mathrm{M\Omega\,km}}{\omega 0.01\,\mu\mathrm{F}/\mathrm{km}} = \frac{1\,\mu\mathrm{S}}{\omega 0.01\,\mu\mathrm{F}} = \frac{100\,\mathrm{rad}/\mathrm{s}}{\omega}.$$

Heaviside did not suffer fools, as is clear from this aside:

> The Americans who went for low resistance had, I think, no idea of the important theoretical significance of the step they took, but did it because they wanted long-distance telephony, and because wires of high resistance would not go—a characteristically American way of doing things. Yet their action led the way to a rapid recognition of the sound practical merits of Maxwell's theory of dielectrics.[213]

Problem 13.4: 100 Ohm Twisted Pair Telephone Cable

Standard telephone line is rated at $Z_{\text{line}} = 100\,\Omega$. Typically two 22 gauge wires, each with a resistance per length of $\mathrm{R} = 53\,\frac{\mathrm{m\Omega}}{\mathrm{m}}$, are twisted together with a fixed separation that yields a capacitance per length of $50\,\frac{\mathrm{pF}}{\mathrm{m}}$.

a) What is the typical linear inductance of telephone line?

b) Telephones carry frequencies from about 30 to 3400 Hz, with most voice traffic between about 500 and 1000 Hz. At these frequencies find the average effective resistance $Z = \langle \frac{\mathbb{V}}{I} \rangle$?

---

[213] Oliver Heaviside, "On Telephone and Telegraph Circuits," (1887), published in Electrical Papers, 2nd ed. (Bronx, NY: Chelsea Publishing, 1970) Vol. 2, p. 341.

c) Technicians often terminate telephone line with either 600 or 900 ohm caps. Which should they pick for 22 gauge twisted pair wire?

d) What impedance should the technicians terminate the same line if the customer wanted 10 Mbaud internet service without a telephone?

Problem 13.5: The Characteristic Speed of a Signal in a Coaxial Cable

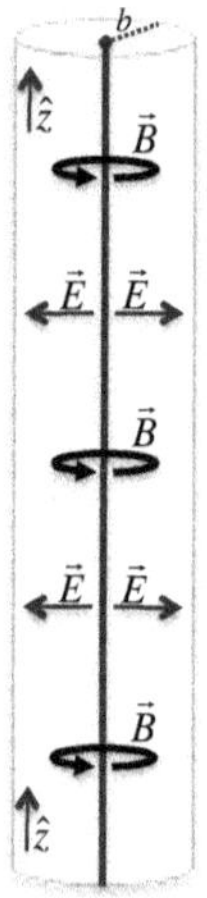

Consider a coaxial cable with a sleeve of iron infused insulating dielectric material between the inner and outer copper conductors. You can assume a coaxial cable of inner radius a and outer radius b, with an insulator of permittivity ε and permeability μ, and conductors with a large conductivity.

a) Show that the inductance per length L is: $L = \frac{1}{2\pi}\mu \ln(b/a)$.

b) Show that the capacitance per length C is: $C = \dfrac{2\pi\varepsilon}{\ln(b/a)}$.

c) Show that the propagation speed is independent of the geometry of the cable.

d) Find the propagation speed in the special case of a free space coaxial cable, and express this in units of kilometers per millisecond.

Problem 13.6: A Parallel Wire Transmission Line

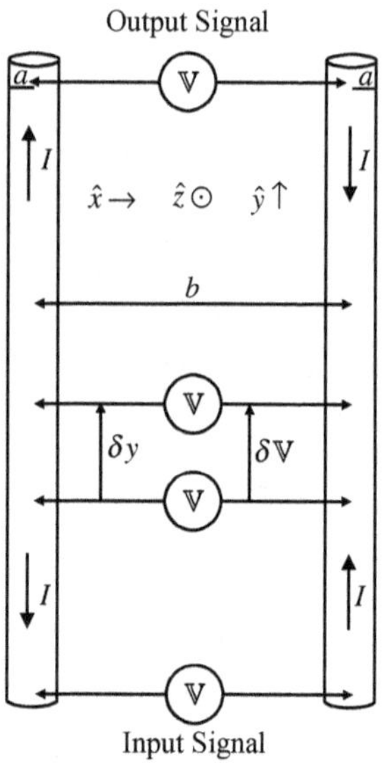

Consider the parallel transmission line shown with each wire having a radius a and a separation b. A high frequency voltage signal is applied to one end, and reproduced on the other end.

a) Rederive the capacitance per unit length that you found in Problem 13.5.

b) Rederive the inductance per unit length that you found in Problem 13.5.

c) Find the wave speed down the line. Express this in units of kilometers per millisecond.

d) Find the impedance of the line.

e) Show clearly that Heaviside's telegrapher's equations are applicable to these cables.

f) Consider that we can improve our model of a transmission line by including the finite conductivity κ of the wires. Rederive the telegrapher's equation for this slightly lossy transmission line.

g) How long could you string the cable before losing half the input voltage?

13.3 *Sinusoidally Driven Circuits*

The mathematics governing transmission lines and RLC circuits that are driven at a particular frequency ω can be greatly simplified using complex numbers. This is most obviously the case for circuits which are driven sinusoidally (although Fourier analysis allows us to analyze other driving inputs). The purpose of this section is to introduce you to the use of complex analysis, and also to persuade you to use it carefully. Since the goal of physics is to elucidate the fundamental laws of nature, in this text we mostly avoid such shortcuts.

On the other hand, sinusoidally driven circuits are so common, and the components are often contained inside of a *black box*, electrical engineers need a method to measure and model the systems relatively simply. Therefore, electrical engineers, including Oliver Heaviside, extended Ohm's law to AC power and telecommunication systems, by using the mathematical properties of both real and imaginary numbers.

Fundamentally, most applications of complex analysis in physics and engineering ultimately depend on the *Euler identity*:

$$e^{j\theta} = \cos\theta + j\sin\theta,$$

where we have used the electrical engineering notation $j = \sqrt{-1}$. You are no doubt used to expressing a complex number z as $z = x + jy$, where x and y are real numbers. Therefore, the complex number can be represented as a point (x, y) in *phase space*. By using the usual transformations, $x = r\cos\theta$ and $y = r\sin\theta$, and the Euler identity, we can also write $z = re^{j\theta}$, where $r = \sqrt{x^2 + y^2}$ and $\theta = \tan^{-1}(y/x)$. Thus, any complex number can be represented in either Cartesian or polar form, just like a two-dimensional vector.

Consider a circuit with a perfectly sinusoidal current flowing through it. As time zero is arbitrary, we could write this current as $I_0 \sin(\omega t)$, $I_0 \cos(\omega t)$, $I_0 \sin(\omega(t-t_0))$, or $I_0 \cos(\omega(t-t_0))$. Alternatively, we could write it as $I_{\cos}\cos(\omega t) + I_{\sin}\sin(\omega t)$, and use a basic property of trigonometric functions to show that $I_0^2 = I_{\cos}^2 + I_{\sin}^2$. Since the sine and cosine functions are mathematically orthogonal, we can treat them as components of two-dimensional vectors.

Using these techniques, the Kirchhoff junction rule can be easily extended to AC current, so long as we treat the cosine and sine functions as if they represented two unit vectors in *phase space*. Rather than $\hat{x}$ and $\hat{y}$, these pseudo-directions are actually the cosine and sine phase components, but are usually referred to as the *real* and the *imaginary* parts. Neither is more physical than the other. They are simply two orthogonal components in the phase space of a quantity called a *phasor*.

Definition 13.1: The Complex Impedance

The complex impedance, $\tilde{Z}$, of a circuit is measured by sending a current $I_0 \cos(\omega t)$ through the circuit, and measuring the resultant voltage $\mathbb{V} = \mathbb{V}_{\cos}\cos(\omega t) - \mathbb{V}_{\sin}\sin(\omega t)$. The complex impedance is defined as:

$$\tilde{Z} = \left(\frac{\mathbb{V}_{\cos}}{I_0}\right) + j\left(\frac{\mathbb{V}_{\sin}}{I_0}\right),$$

where $j = \sqrt{-1}$. The purpose of impedance is to act as an effective resistance in sinusoidal circuits. The *total impedance*, which is often just called *the impedance*, is given by adding the real and imaginary parts in quadrature:

$$Z = \sqrt{\left(\mathrm{Re}(\tilde{z})\right)^2 + \left(\mathrm{Im}(\tilde{z})\right)^2}.$$

The real part is called the *resistance* and the imaginary part the *reactance*.

Given his work in telegraphy, and his development of the telegrapher's equations, it perhaps comes as little surprise that Oliver Heaviside contributed greatly to the development of the complex impedance.

Example 13.1: The Impedance of Resistors

What is the complex impedance of a single resistor with resistance R driven at a frequency ω?

The voltage drop across the resistor is given by:

$$\mathbb{V} = IR = RI_0 \cos(\omega t).$$

Thus the complex impedance is:

$$\tilde{Z} = \left(\frac{\mathbb{V}_{\cos}}{I_0}\right) - j\left(\frac{\mathbb{V}_{\sin}}{I_0}\right) = \left(\frac{RI_0}{I_0}\right) - j\left(\frac{0}{I_0}\right) = R.$$

In the case of two resistors in series, the complex impedance, due to Kirchoff's loop rule, would be:

$$\mathbb{V} = IR_1 + IR_2 = I(R_1 + R_2) = (R_1 + R_2)(I_0 \cos(\omega t)).$$

Therefore the complex impedance is:

$$\tilde{Z} = \left(\frac{\mathbb{V}_{\cos}}{I_0}\right) - j\left(\frac{\mathbb{V}_{\sin}}{I_0}\right) = \left(\frac{(R_1 + R_2)I_0}{I_0}\right) - j\left(\frac{0}{I_0}\right) = R_1 + R_2.$$

Similarly, we can use Kirchoff's junction rule to find the impedance of two resistors in parallel:

$$I = I_0 \cos(\omega t) = I_1 \cos(\omega t) + I_2 \cos(\omega t) = \frac{\mathbb{V}}{R_1} + \frac{\mathbb{V}}{R_2} = \mathbb{V}\left(\frac{1}{R_1} + \frac{1}{R_2}\right).$$

Rearranging this we get:

$$\mathbb{V} = \frac{I_0}{\frac{1}{R_1} + \frac{1}{R_2}} \cos(\omega t) \quad \rightarrow \quad \frac{1}{\tilde{Z}} = \frac{1}{R_1} + \frac{1}{R_2} = \frac{1}{\tilde{Z}_1} + \frac{1}{\tilde{Z}_2}.$$

Example 13.2: The Impedance of an Inductor

What is the complex impedance of a single inductor with inductance L driven at a frequency ω?

The voltage drop across the inductor is given by:

$$\mathbb{V} = L\frac{dI}{dt} = L\frac{d}{dt}\left(I_0 \cos(\omega t)\right) = -L\omega I_0 \sin(\omega t).$$

So, applying the definition, the complex impedance is:

$$\tilde{Z} = \left(\frac{\mathbb{V}_{\cos}}{I_0}\right) - j\left(\frac{\mathbb{V}_{\sin}}{I_0}\right) = \left(\frac{0}{I_0}\right) - j\left(\frac{-L\omega I_0}{I_0}\right) = j\omega L.$$

Thus, the higher the frequency the higher the effective resistance.

Problem 13.7: Resistors and Inductors in Series and Parallel

a) Show that the impedance of two inductors in series add.

b) Show that the impedance of two inductors in parallel add in reciprocal.

c) Show that any combination of impedances of inductors and resistors add when hooked up in series.

d) Show that any combination of impedances of inductors and resistors add in reciprocal when hooked up in parallel.

Problem 13.8: The Impedance of a Capacitor

a) Show that the complex impedance of a capacitor is: $\tilde{Z} = \frac{1}{j\omega C}$.

b) Show that the impedance of two capacitors connected in series is simply the sum of the two impedances.

c) Show that the net impedance of two capacitors connected in parallel add in reciprocal.

Problem 13.9: The Impedance of RLC Circuits

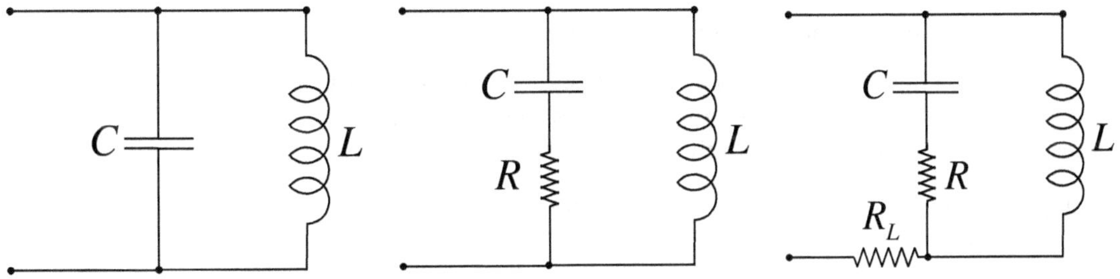

a) Find the complex impedance of each circuit shown, using the operational definition of impedance.

b) Find the complex impedance of each circuit shown, by the appropriate addition method. Show that they agree.

c) Find the total impedance of each circuit.

d) Discuss some uses for driven RLC circuits.

Problem 13.10: The Impedance of an RC Stack

In Problem 13.2 (p. 411) you analyzed an RC stack.

a) Find the complex impedance of this stack.

b) Find the total impedance of this stack.

c) Show that in the DC case this acts like an open circuit, and in the very high frequency case like just the resistors in series.

d) If you hooked this system up in series with a load resistor, R_L, and drove it with a constant AC voltage source at a frequency ω with peak-to-peak voltage V_0, what would be the power, as a function of time, on the load?

e) What would be the power dissipated by the resistors in the stack as a function of time?

f) How much energy is stored in the capacitors as a function of time?

g) Show that energy is conserved as a function of time.

Problem 13.11: The Impedance of an LC Chain

In Thought Experiment 13.2 (p. 411) we analyzed a chain of LC circuits.

a) Find a recursion relationship for the complex impedance of the line past node n as a function of the impedance past the node $n+1$.

b) If the chain is very long, adding one more node should not make much difference to one end, so they should be about equal. Find the complex impedance of the chain in the very long limit.

c) We stated before that the impedance of the line is $Z_{\text{line}} = \sqrt{\frac{L}{C}}$. Under what frequency ranges is this also the total impedance?

Problem 13.12: The Twisted Pair Telephone Cable

In Problem 13.4 (p. 417), you found the impedance of a long telephone cable as a function of driving frequency. Answer the same questions again, but now using complex impedance.

Discussion 13.1: Complex Amplitudes in Phase Space

A common mathematical technique when working with sinusoidal functions is to write the complex amplitudes of the current and voltage in much the same way as the impedance. In this method, we define the current with three parameters, rather than two, I_0, ω, and ϕ_0, such that:

$$I = I_0 \cos(\omega t - \phi_0) = (I_0 \cos(\phi_0))\cos(\omega t) + (I_0 \sin(\phi_0))\sin(\omega t) = I_{\cos}\cos(\omega t) + I_{\sin}\sin(\omega t).$$

Now we can define a complex current amplitude in the same way as we did the impedance:

$$\tilde{I}_\omega = I_{\cos} - jI_{\sin} = I_0(\cos\phi_0 - j\sin\phi_0) = I_0 e^{-\phi_0}.$$

To get the last term, we have used the Euler identity: $e^{-\phi_0} = \cos(\phi_0) + \sin(-\phi_0) = \cos\phi_0 - j\sin\phi_0$. Notice that this involves measuring *two* numbers at a particular frequency, which is easily done with an oscilloscope and a small shunt resistor. The trick is to keep in mind that complex numbers represent a vector in phase space at a particular driving frequency.

Once we have a complex current, we can return the peak-to-peak current by, essentially, taking the dot product with itself as we would with any vector. In polar coordinates, this would simply involve multiplying them together with the cosine of the mutual angle. But how do we do this simply?

First notice that there is not only one square root of negative one, but rather at least two, j and $-j$. So, when we multiply these vectors we must do it symmetrically. We do this using the complex conjugate, which we do by simply making the replacement: $j \to -j$. For example, the amplitude of the current is:

$$I_0 = \sqrt{\tilde{I}_\omega \cdot \tilde{I}_\omega} = \mathrm{Re}\sqrt{\tilde{I}_\omega^* \tilde{I}_\omega} = \mathrm{Re}\sqrt{\left(I_{\cos} - jI_{\sin}\right)\left(I_{\cos} + jI_{\sin}\right)} = \sqrt{I_{\cos}^2 + I_{\sin}^2} = I_0 .$$

We can, of course, do the same thing with the voltage, where:

$$\tilde{\mathbb{V}}_\omega = \mathbb{V}_{\cos} - j\mathbb{V}_{\sin} = \mathbb{V}_0\left(\cos\phi - j\sin\phi\right) = \mathbb{V}_0 e^{-j\phi}.$$

Again, notice that this involves measuring the amplitude and phase of the voltage at a particular frequency, which can be done with an oscilloscope.

Since the average power down the line is:

$$\langle \mathbb{P} \rangle = \langle I\,\mathbb{V} \rangle = \langle I_0 \cos(\omega t - \phi_0)\mathbb{V}_0 \cos(\omega t - \phi) \rangle = \tfrac{1}{2} I_0 \mathbb{V}_0 \langle \cos(\phi - \phi_0) \rangle,$$

it is also half of the dot product in phase space, so:

$$\langle \mathbb{P} \rangle = \tfrac{1}{2}\tilde{I}_\omega \cdot \tilde{\mathbb{V}}_\omega = \tfrac{1}{2}\mathrm{Re}\left(\tilde{\mathbb{V}}_\omega \tilde{I}_\omega^*\right) = \tfrac{1}{2} I_0 \mathbb{V}_0 \,\mathrm{Re}\left(e^{i(\phi-\phi_0)}\right) = \tfrac{1}{2} I_0 \mathbb{V}_0 \cos(\phi - \phi_0).$$

If you are going to use these techniques it is important to understand the underlying assumptions behind them. These techniques only work because we assume sinusoidal solutions, but they can be very handy especially when writing computer code, as modern computers support complex arithmetic.

So far, we've eschewed the use of complex numbers to avoid obscuring the physics, but there are many applications where these techniques greatly simplify the mathematics involved in classical electrodynamics.

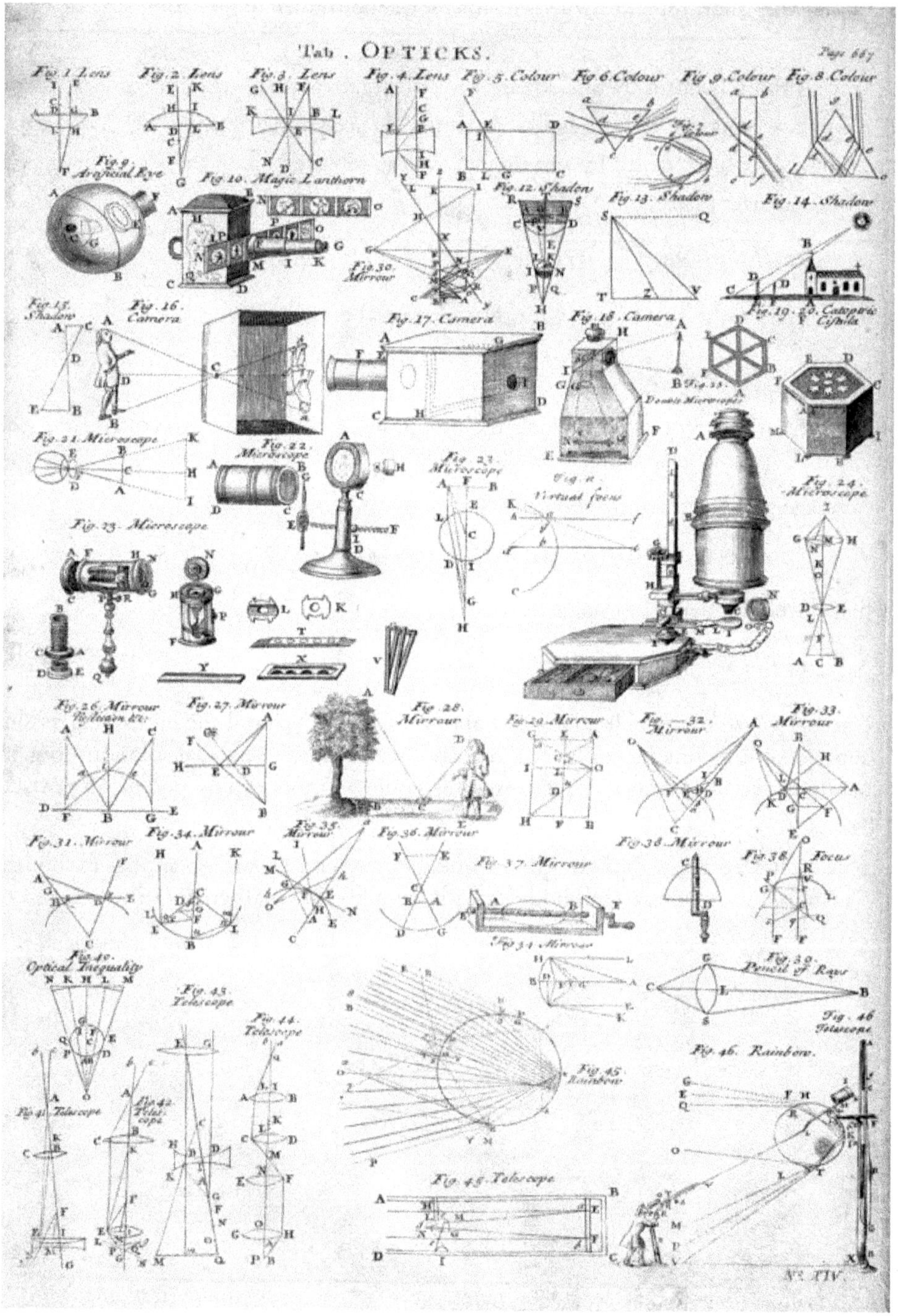

Plate 24: Newton's Opticks[214]

---

214 Ephraim Chambers, <u>Cyclopaedia: or, an Universal Dictionary of Arts and Sciences; …</u>, (London, 1728), Vol. II, Page 667

Chapter 14 Light in an Optical Medium

> As everyone knows, the aether played a great part in the physics of the nineteenth century; but in the first decade of the twentieth, chiefly as a result of the failure of attempts to observe the earth's motion relative to the aether, and the acceptance of the principle that such attempts must always fail, the word "aether" fell out of favor, and it became customary to refer to interplanetary spaces as 'vacuous"; the vacuum being conceived as mere emptiness, having no properties except that of propagating electromagnetic waves. But with the development of quantum electrodynamics, the vacuum has come to be regarded as the seat of the "zero-point" oscillations of the electromagnetic field, of the "zero-point" fluctuations of electric charge and current, and of a "polarisation" corresponding to a dielectric constant different from unity. It seems absurd to retain the name "vacuum" for an entity so rich in physical properties, and the historical word "aether" may fitly be retained.[215] — E. T. Whittaker

Before the nineteenth century, Isaac Newton's particle model and Christiaan Huygens's wave model could both explain most optical phenomena, but with opposite predictions for the speed of light in matter.

Both Newton and Huygens explained Snell's law of refraction as a sudden change in speed at the optical interface, but with one crucial difference: Newton's light particles should speed up, not slow down, when entering a more optically dense medium.[216] As Newton explained, attractive forces would accelerate a light particle upon entering a medium, and slow it again when exiting. As the force components parallel to the air-glass interface cancel, that component of the velocity would be unaffected, thus bending the trajectory. In this way, Newton derived Snell's law of refraction, with a clear and understandable mechanism. Huygens, on the other hand, noticed that a wave's speed also depends on the medium involved, but in the opposite sense—the waves must slow down to follow Snell's law.

In 1669, Rasmus Bartholin discovered double-refraction, where a crystal, such as calcite, splits a beam of light as it refracts. How a medium could have two different wave speeds was a puzzle. Furthermore, how could a longitudinal wave carry enough information to account for double refraction? A particle, however, could possess mass, velocity, and the ability to spin about an axis, favoring Newton's particle model of light.

Also in the 1660s, the Jesuit priest Francesco Maria Grimaldi observed the diffraction of light. This would seem to greatly favor Huygens, but Newton explained it as ripples in a local medium, caused by the impact of his light corpuscles. Thus, the two models remained deadlocked, until 1725 when James Bradley observed the *aberration of starlight* (see pp. 474 and 580), where the position of stars shifts due to the motion of the Earth. Bradley's observation could be easily explained with Newton's theory, providing definitive evidence favoring the particle model. However, that all changed with the turn of the nineteenth century.

Thomas Young's 1804 double slit experiment (p. 517) came down definitively on the side of light being a wave. In the years after he published the results, Young attempted to explain all of

---

[215] E.T. Whittaker, A History of the Theories of Aether and Electricity, Vols. 1&2 (1951; rpt. New York: Dover Publications, 2017). Despite Whittaker's argument, modern scientists have dropped the word *aether* from their lexicon, but they retain *vacuum* in the sense he describes.

[216] Isaac Newton, Principia, Mathematical Principles of Natural Philosophy: A New Translation, trans. I Bernard Cohen and Anne Whitman (Berkeley: University of California Press, 1999), Book I, Section 14, Proposition 94, p. 622.

optical phenomena in terms of the wave model. Although he had much success, he could still not explain the effect that some crystals have on light rays.

When light passes through two identical crystals, such as tourmaline, the intensity of the emerging light varies as the second crystal is rotated. The transmitted light is brightest when the axes of the two crystals are aligned, but is completely extinguished when the axes are oriented perpendicular to each other. Astonishingly, when a third crystal is placed between the two perpendicular ones, some light, again, passes through the last crystal.

Newton explained these results as due to an asymmetry in the shape of the particles. Just as a pick axe can only pass through prison bars aligned one way, only particles aligned parallel to the crystal lattice can pass. Light, the logic went, behaves like little dipoles, thus the origin of the confusing term *polarization* of light.

During Napoleon's reign, investigations by Étienne-Louis Malus continued to dog Young's wave theory of light.[217] Unpolarized light passing through a *birefringent* crystal, such as Iceland spar, separates into two rays, producing two images. While *double refraction* was known to Newton and Huygens, Malus observed something interesting when observing sunlight reflected from the windows of the Luxembourg Palace. The two images of the reflected setting sun dramatically differed, and, moreover, as he rotated the crystal the images alternately brightened and dimmed.

Augustin-Jean Fresnel saved the wave model by proposing that light is not longitudinal, but rather transverse like the waves on a string.[218] Fresnel, and his mentor François Jean Dominique Arago, showed that beams of perpendicularly polarized light do not interfere, which led, with a hint from Young, to Fresnel's transverse wave theory of light.[219]

Siméon Denis Poisson, who vigorously opposed the wave theory of light, pointed out what he saw as a damning feature of Fresnel's work. Poisson showed that Fresnel's theory predicted a bright spot in the center of the shadow of a circular screen, which, according to Poisson, was certainly an absurd result.[220] Arago performed the experiment and did indeed find the impossible bright spot, which is now known, in a nice bit of historical irony, as *the Poisson spot*.[221]

By the middle of the nineteenth century the transverse wave theory seemed the only possible way of accounting for such diverse optical phenomena as reflection, refraction, diffraction, and polarization. Light is a wave, nearly everyone agreed. Then, how and through what does it propagate?

Michael Faraday's experiments on capacitors with dielectrics had suggested a possible propagation mechanism. Since a slight displacement of charge can push the next atom down the line, perhaps this is how light propagates too? If space were filled with matter, then this process would convey the electric force in much the same manner as small elastic deformations convey

---

217 E-L. Malus, "Théorie de la Double Réfraction," Mémoires présentés à l'Institut des sciences par divers savants, **2** (1811), 303-508. See also, Buchwald, Jed Z., The Rise of the Wave Theory of Light, (Chicago: University of Chicago Press, 1989), 54-64.

218 Augustin Fresnel, "Memoirs on the Diffraction of Light," The Wave Theory of Light – Memoirs by Huygens, Young and Fresnel, (New York: American Book Company, 1900), 79–145.

219 Augustin Fresnel, "On the Action of Rays of Polarized Light upon Each Other," The Wave Theory of Light – Memoirs by Huygens, Young and Fresnel (American Book Company, 1900), 145–156.

220 A. Fresnel, Complete Works: Theory of Light, 3 vols., (Paris: Imprimiere Impériale, 1866, 1868, 1870), **1:** 366-372.

221 A. Fresnel, Complete Works: Theory of Light, 3 vols., (Paris: Imprimiere Impériale, 1866, 1868, 1870), **1:** 369.

the normal force through your chair. There was a problem since all these sound-like mechanisms suggested compression waves, not the transverse wave of Fresnel's theory.

On the other hand, both longitudinal, and transverse, waves can travel through an elastic solid. For example, earthquakes produce both longitudinal p-waves and transverse s-waves. Perhaps, thought George Gabriel Stokes, light simply consists of transverse waves propagating through the aether, analogous to s-waves traveling through the earth. As he put it:

> Undoubtedly, it does violence to the ideas that we should have been likely to form *a priori* of the nature of the aether, to assert that it must be regarded as an elastic solid in treating of the vibrations of light. When, however, we consider the wonderful simplicity of the explanations of the phenomena of polarization when we adopt the theory of transversal vibrations, and the difficulty, which to me at least appears quite insurmountable, of explaining these phenomena by any vibrations due to the condensation and rarefaction of an elastic fluid such as air, it seems reasonable to suspend our judgment, and be content to learn from phenomena the existence of forces which we should not beforehand have expected.[222,223]

One problem that bedeviled the elastic solid model of the aether was how to explain the existence of a medium that would be rigid enough to support transverse waves, yet fluid enough to allow planets and stars to glide through unhindered. This difficulty stimulated investigations of elastic solids and how they respond to various forces.

Stokes's model of light propagating through an elastic solid fascinated a precocious teen studying mathematics at the University of Edinburgh. At the age of 18, James Clerk Maxwell was finally deemed old enough to present a paper on his own to the local chapter of the Royal Society. His paper, titled "On the Equilibrium of Elastic Solids," tackled this question and was reported by the secretary with the following description:[224]

> This paper commenced by pointing out the insufficiency of all theories of elastic solids, in which the equations do not contain two independent constants deduced by experiments. One of these constants is common to liquids and solids, and is called the modulus of *cubical* elasticity [μ]. The other is peculiar to solids, and is here called the modulus of *linear* elasticity [m]. The equations of Navier, Poisson, and Lamé and Clapeyron, contain only one coefficient; and Professor G. G. Stokes of Cambridge, seems to have formed the first theory of elastic solids which recognized the independence of cubical and linear elasticity, although M. Cauchy seems to have suggested a modification of the old theories, which made the ratio of linear and cubical elasticity the same for all substances. Professor Stokes has deduced the theory of elastic solids from that of the motion of fluids, and his equations are identical with those of this paper, which are deduced from the following assumptions.
>
> In an element of an elastic solid, acted on by three pressures at right angles to on another, as long as the compressions do not pass the limits of perfect elasticity—
>
> 1st, The sum of the *pressures*, in the three rectangular axes, is proportional to the sum of the compressions in those axes.
>
> 2d, The difference of the pressures in two axes at right angles to one another, is proportional to the difference of the compressions in those axes.
>
> Or, in symbols:—

---

222 G.G. Stokes, "On the constitution of the luminiferous aether," Philosophical Magazine **29**, 3rd series, (1848), 6-10.

223 David B. Wilson, "George Gabriel Stokes on Stellar Aberration and the Luminiferous Ether," The British Journal for the History of Science, **6**, No. 1 (Jun., 1972), 57-72.

224 In the quote that follows, Maxwell uses the term "cubical elasticity" [μ] to denote what we now call ***bulk modulus***.

$$1.\ (P_1 + P_2 + P_3) = 3\mu\left(\frac{\delta x}{x} + \frac{\delta y}{y} + \frac{\delta z}{z}\right) \qquad 2. \begin{cases} (P_1 - P_2) = m\left(\frac{\delta x}{x} - \frac{\delta y}{y}\right) \\ (P_2 - P_3) = m\left(\frac{\delta y}{y} - \frac{\delta z}{z}\right) \\ (P_3 - P_1) = m\left(\frac{\delta z}{z} - \frac{\delta x}{x}\right) \end{cases}$$

> μ being the modulus of *cubical*, and m that of *linear* elasticity.
>
> These equations are found to be convenient for the solution of problems, some of which were given in the latter part of the paper.

The paper concluded with a conjecture, that as the quantity ω, (which expresses the relation of inequality of pressure in a solid to the doubly-refracting force produced) is probably a function of m; the determination of these quantities for different substances might lead to a more complete theory of double refraction, and extend our knowledge of the laws of optics.[225]

After that, Maxwell attended university at Cambridge. At the time, students sat for a grueling eight-day examination in order to graduate, where each question would be more difficult than the one before. A prize was given to the student who scored highest on the final 17 questions, which in 1854 were written by Professor George Gabriel Stokes. Much to Maxwell's delight, the last question on the exam read:

> Plane polarized light is transmitted, in a direction parallel to the axis of the crystal, across a thick plate of quartz cut perpendicular to the axis, and the emergent light, limited by a screen with a slit, is analyzed by a Nicol's prism combined with an ordinary prism; describe the appearance presented as the Nicol's prism is turned round, and from the phenomena deduce the nature of the action of quartz on polarized light propagated in the direction of the axis.[226]

Maxwell did not get the top score on the overall exam, but he did tie for the Smith prize for the best score on the final 17 questions.

Maxwell continued being fascinated by the relationship between elasticity and optics, and by 1856 had already landed an academic job. He moved to northern Scotland, where he taught, got married, and worked on problems in optics—and also showed that Saturn's rings were neither fluid or solid, but rather made of small objects. Then Maxwell caught another lucky break; he lost his job.

Maxwell landed on his feet at King's College London, where he took advantage of the intellectually stimulating environment and immersed himself in the works of Faraday, Ampère, Gauss, and William Thomson, in order to develop a mathematical theory of electricity and magnetism. In a sense, the fundamental laws that govern electricity and magnetism where understood by the time Maxwell took up his examination of them. Coulomb's law allowed one to calculate and predict the electrostatic interactions among charges, while Ampère's force law between currents and the formula credited to Biot and Savart could be used to calculate magnetic fields in the case of steady currents. Finally, Faraday's discovery of electromagnetic induction demonstrated how a changing magnetic field induces an electric field.

---

225 James Clerk Maxwell, Esq. (Communicated by the Secretary), "On the Equilibrium of Elastic Solids," in Proceedings of The Royal Society of Edinburgh, Vol. 2, December 1844 to April 1850, Neill and Co., Edinburgh, 1851.

226 George Gabriel Stokes, Esq. M.A., The Smith Prize Exam of 1854, from the website of the James Clerk Maxwell Foundation (http://www.clerkmaxwellfoundation.org).

The field concept was still in its infancy, but investigators such as Faraday and William Thomson used it with increasing success, especially in the laboratory. In formulating his own approach to the subject of electricity and magnetism, Maxwell emphasized his understanding of Faraday's idea of fields conveyed through the mathematics of fluids and solids:

> As I proceeded with the study of Faraday, I perceived that his method of conceiving phenomena was also a mathematical one, though not exhibited in the conventional form of mathematical symbols. I also found that these methods were capable of being expressed in the ordinary mathematical forms, and thus compared with those of the professed mathematicians.
>
> For instance, Faraday, in his mind's eye, saw lines of force traversing all of space, where the mathematicians saw only centers of force attracting at a distance...When I translated what I considered to be Faraday's ideas into a mathematical form...I found that several of the most fertile methods of research discovered by the mathematicians could be expressed much better in terms of ideas derived from Faraday than in their original form.[227]

His work ultimately vindicated Faraday's field conception of electromagnetism, and also confirmed Faraday's intuitive hunch that light was somehow connected with "vibrations of ... lines of force."[228] Maxwell achieved this by adding an insight of his own to the equations which govern electrical and magnetic phenomena.

Maxwell's insight was the converse of Faraday's discovery of 1831 that a changing magnetic field induces an electric field. Based on theoretical arguments arising from questions of symmetry and current continuity, Maxwell predicted that a changing electric field induces a magnetic field. With this insight, Maxwell was able to develop a fully electromagnetic theory of light.

Maxwell discovered that self-propagating electromagnetic waves would travel through space at a constant speed, which happened to equal the previously measured speed of light. Like others, Maxwell felt that the electromagnetic waves required a medium through which to travel. In the course of his work, he returned to his early research project and the work of his former professor—now good friend and colleague—George Stokes.

Maxwell applied Stokes's model of the aether as his basis for light. This allowed him to imagine the electric and magnetic fields in much the same way as shear stress and strain are related in perfectly elastic solids. In his model, the permittivity of a dielectric was analogous to the same inverse linear elasticity modulus m that Maxwell presented when he was eighteen:

$$\varepsilon \sim \frac{1}{m},$$

and the magnetic permeability provided the inertia, like the mass density in a mechanical wave:

$$\mu \sim \rho_m .$$

Since the transverse wave speed through an elastic medium was well known to be proportional to the square root of the ratio of the elasticity m to the mass density ρ_m, then by analogy:

$$v = \sqrt{\frac{m}{\rho_m}} \sim \sqrt{\frac{\frac{1}{\varepsilon}}{\mu}} \sim \frac{1}{\sqrt{\varepsilon\mu}} .$$

---

227 From the preface to J.C. Maxwell, Treatise on Electricity and Magnetism, vol. 1, 3rd ed. (1873; rpt. New York: Dover, 1954).

228 Faraday, "Thoughts on Ray Vibrations," 1844, Phil. Magazine; "On the physical character of the lines of magnetic force," 1852, Phil. Magazine. See page 102 of this chapter for more details.

Maxwell stated this result in his 1862 paper "On Physical Lines of Force":

> The velocity of transverse undulations in our hypothetical medium, calculated from the electromagnetic experiments of M.M. Kohlrausch and Weber, agrees so exactly with the velocity of light calculated from the optical experiments of M. Fizeau, that we can scarcely avoid the inference that *light consists in the transverse undulations of the same medium which is the cause of electric and magnetic phenomena.*[229]

Although the argument that Maxwell presented in his 1862 paper was suggestive, it was far from conclusive. In 1865, Maxwell derived an explicit wave equation for both the electric and magnetic fields, which predicted that those electric and magnetic fields are transverse waves that propagate at a speed in free space equal to the speed of light.

In 1873, he published A Treatise on Electricity and Magnetism containing a full mathematical description of the behavior of electric and magnetic fields in matter, the predictions of which have been fully confirmed by experiment.

The title of Maxwell's final chapter is "Theories of Action at a Distance." In it he criticizes the time-dependent theories of Gauss and Weber:

> There appears to be, in the minds of these eminent men, some prejudice, or *a priori* objection, against the hypothesis of a medium in which the phenomena of radiation of light and heat and the electric actions at a distance take place. It is true that at one time those who speculated as to the causes of physical phenomena were in the habit of accounting for each kind of action at a distance by means of a special aethereal fluid, whose function and property it was to produce these actions. They filled all space three or four times over with aethers of different kinds, the properties of which were invented merely to 'save appearances,' so that more rational enquirers were willing rather to accept not only Newton's definite law of attraction at a distance, but even the dogma of Cotes [in the preface to Newton's Principia, 2nd edition], that action at a distance is one of the primary properties of matter, and that no explanation can be more intelligible than this fact. Hence the undulatory theory of light has met with much opposition, directed not against its failure to explain the phenomena, but against its assumption of the existence of a medium in which light is propagated.[230]

His "constant aim in this treatise" was that "we ought to endeavor to construct a mental representation of all the details of" the aether.[231]

14.1 *Maxwell's Field Equations*

We have presented a number of fundamental laws of nature, all of which have been thoroughly tested experimentally. Some of these laws, such as the conservation of charge and Peregrinus's principle, still appear to always hold. Others, however, only apply in contexts where classical physics is valid, and are no longer considered fundamental in light of contemporary theories such as quantum electrodynamics.

As we discussed in the historical introduction, scientists such as Laplace, Biot, Poisson, Faraday, Thomson, and Maxwell were interested in the nature of matter, and the mechanism of how it mediates between cause and effect. The *Maxwellians*, those scientists such as Hertz, Heaviside,

---

[229] J.C. Maxwell, "On Lines of Physical Force," Philosophical Magazine, **21** & **23** (1862), Part III.

[230] J.C. Maxwell, A Treatise on Electricity and Magnetism, vol. 2, 3rd ed. (1873; rpt. New York: Dover, 1954), Art. 865, p. 492.

[231] Ibid., Art. 866, p. 493.

Gibbs, and Lorentz who later applied (and refined) Maxwell's work to explain electromagnetic phenomena, including optical phenomena, were also concerned about the nature of matter. But none of these scientists, from Laplace to the Maxwellians, could foresee the rise of the twentieth century quantum theories of matter and radiation that would bring about a revolutionary change in our understanding of the natural world.

Maxwell developed a set of field equations, now called *Maxwell's equations*, from which (assuming certain properties of matter) all principles of electrodynamics could subsequently be derived. Does this mean, therefore, that Maxwell's equations are more fundamental than other equations of electrodynamics? The answer is somewhat subtle (see Discussion 15.1, p. 479).

Maxwell's formulation of electromagnetism involved four force fields, two electric and two magnetic, which have had different interpretations at different times. In this chapter, we use this four-field approach to develop the propagation of electromagnetic waves through matter. We should also remember that Maxwell (and the Maxwellians) understood free space to contain a linear medium, the aether, which they believed was instrumental in relating cause to effect.

Thought Experiment 14.1: The Electric Displacement Vector

Consider the continuity equation, $\vec{\nabla}\cdot\vec{J}=-\frac{\partial}{\partial t}\rho$, and Gauss's law, $\vec{\nabla}\cdot\vec{E}=\frac{\rho}{\varepsilon_0}-\frac{1}{\varepsilon_0}\vec{\nabla}\cdot\vec{P}$. Both of these equations involve the charge density and the divergence of various vector fields. Thus, if we introduce a new vector field, $\vec{D}$, which we define such that its divergence is the free charge density $\left(\vec{\nabla}\cdot\vec{D}\equiv\rho\right)$, then Gauss's law becomes:

$$\vec{\nabla}\cdot\vec{E}=\tfrac{1}{\varepsilon_0}\rho-\tfrac{1}{\varepsilon_0}\vec{\nabla}\cdot\vec{P}=\tfrac{1}{\varepsilon_0}\left(\vec{\nabla}\cdot\vec{D}\right)-\tfrac{1}{\varepsilon_0}\vec{\nabla}\cdot\vec{P}=\vec{\nabla}\cdot\left(\tfrac{1}{\varepsilon_0}\left(\vec{D}-\vec{P}\right)\right).$$

Notice that our original definition is non-unique, as we could add a divergence free vector field, so we have some freedom of choice. Therefore, we can refine our definition to make $\vec{E}$ proportional to $\vec{D}$ in a linear medium, or simply:

$$\vec{D}\equiv\varepsilon_0\vec{E}+\vec{P}\approx\varepsilon\vec{E}.$$

Since the Maxwellians interpreted the permittivity as a property of matter, akin to elasticity, they called this the electric constitutive relation, as *constitutive* means restorative. They also interpreted free space as a dielectric medium, so they interpreted $\vec{D}$ in much the same way as we now interpret $\vec{P}$, which is why $\vec{D}$ is called the electric displacement. As Oliver Heaviside wrote: "$\vec{D}$, the displacement, is the charge (+ or -, according to the end)." [232]

Expressed in terms of the displacement vector, the continuity equation becomes:

$$\vec{\nabla}\cdot\vec{J}=-\tfrac{\partial}{\partial t}\rho=-\tfrac{\partial}{\partial t}\left(\vec{\nabla}\cdot\vec{D}\right)=\vec{\nabla}\cdot\left(-\tfrac{\partial}{\partial t}\vec{D}\right),$$

which simplifies to:

$$\vec{\nabla}\cdot\left(\vec{J}+\tfrac{\partial}{\partial t}\vec{D}\right)=0.$$

---

[232] Oliver Heaviside, "Notes on Nomenclature," published in The Electrician Sept. 4 (1885), 311, republished in Electrical Papers, vol. 2, 2nd ed. (Bronx, NY: Chelsea Publishing, 1970), 25.

Notice that this result does not simplify anything, as we have merely hidden the complexity in the definition of $\vec{D}$. Nevertheless, this form of the continuity equation did suggest to Maxwell and his colleagues that there was more to $\vec{D}$ than simply a mathematical definition.

Thought Experiment 14.2: Conservation of Charge and the Magnetizing Field

So far we have mostly used the magnetizing field $\vec{H}$ to represent the magnetic field external to a system. However, we could have defined it in terms of the continuity equation, in much the same way as we defined the vector potential in terms of Peregrinus's principle.

Since $\vec{\nabla}\cdot\left(\vec{J}+\frac{\partial}{\partial t}\vec{D}\right)=0$, there is a vector field, $\vec{H}$, with the property that:

$$\vec{\nabla}\times\vec{H}=\vec{J}+\tfrac{\partial}{\partial t}\vec{D}.$$

You can show that this is simply the Maxwell-Ampère law (see Problem 14.1 below).

Definition 14.1: The Auxiliary Fields

The auxiliary fields, $\vec{D}$ and $\vec{H}$, are now defined through the following relations:

$$\vec{D}\equiv\varepsilon_0\vec{E}+\vec{P} \quad\text{and}\quad \vec{H}\equiv\tfrac{1}{\mu_0}\vec{B}-\vec{M}.$$

Problem 14.1 : The Maxwell-Ampère Law in Auxiliary Form

Show that $\vec{\nabla}\times\vec{H}=\vec{J}+\frac{\partial}{\partial t}\vec{D}$ is mathematically equivalent to:

$$\vec{\nabla}\times\vec{B}=\mu_0\vec{J}+\mu_0\varepsilon_0\tfrac{\partial}{\partial t}\vec{E}+\mu_0\varepsilon_0\tfrac{\partial}{\partial t}\vec{P}+\mu_0\vec{\nabla}\times\vec{M}.$$

Discussion 14.1: Maxwell's Four-field Approach

The Maxwellians doggedly linked causes to effects, which enabled them to explain electromagnetic phenomena in terms of fields, with matter (including the aether) as the intermediary. With no *conceptual* distinction between the permittivity and permeability of free space, and that of any other material, the four-field approach made the most sense.

In terms of the four fields, $\vec{E}$, $\vec{D}$, $\vec{B}$, and $\vec{H}$, Maxwell's equations become:

$$\vec{\nabla}\cdot\vec{B}=0,\quad \vec{\nabla}\times\vec{E}=-\tfrac{\partial\vec{B}}{\partial t},\quad \vec{\nabla}\cdot\vec{D}=\rho,\quad\text{and}\quad \vec{\nabla}\times\vec{H}=\vec{J}+\tfrac{\partial}{\partial t}\vec{D}.$$

As shown in the solid arrows on the left side of the flow chart, the separation of charge (ρ) creates the electric displacement $(\vec{D})$, which affects the medium, which creates the electric field $(\vec{E})$, which, in turn, separates charge. The other arrows show common causal relationships.

Since the downfall of the aether, the four-field approach to electrodynamics has lost favor among physicists, especially theorists. Electrical engineers and applied physicists, on the other hand, retain the four field approach in much of their work. Therefore, an understanding of the four-field approach is key not only to understanding the history of science, but also to understand current practice is physics and electrical engineering.

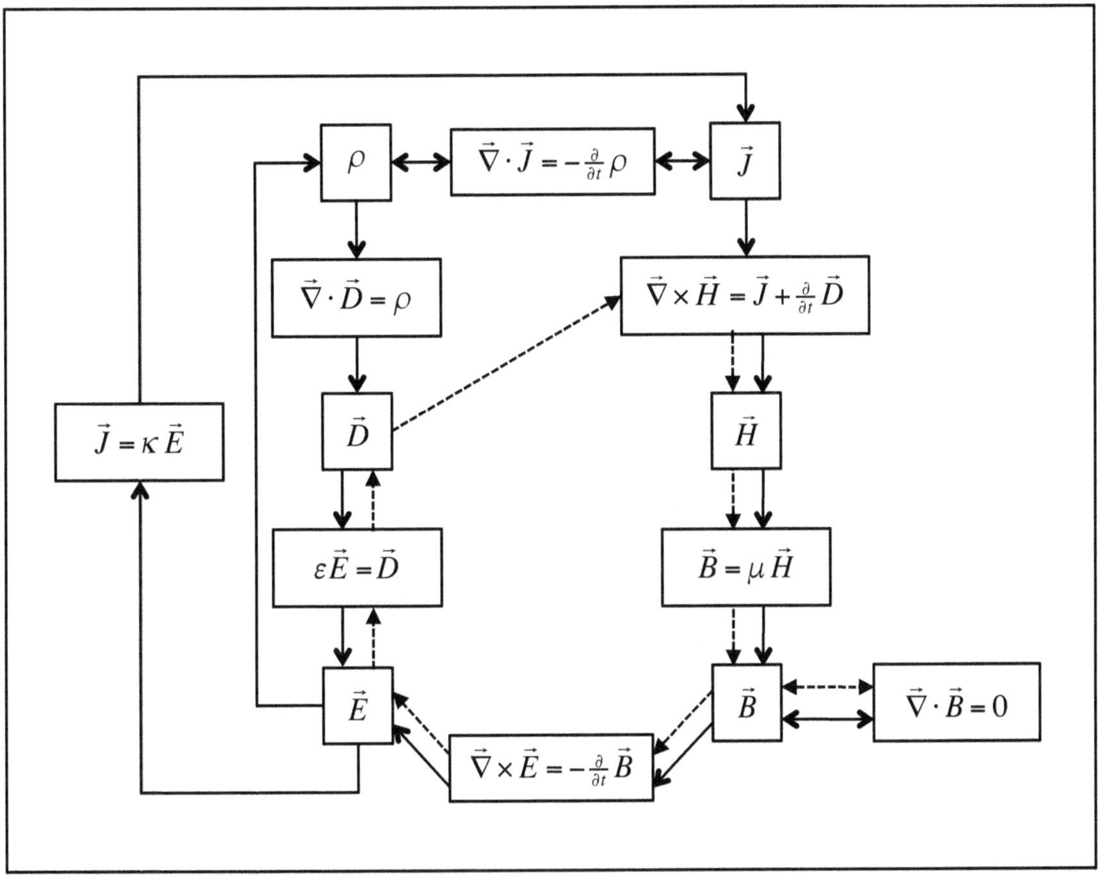

Problem 14.2: A Dielectric Capacitor

Imagine a simple parallel plate capacitor with a linear dielectric sandwiched in the middle.

a) Draw the electric displacement between the capacitor plates.

b) Draw the electric polarization vector inside the capacitor. Is this in the same, or the opposite, direction as the electric displacement? Clearly explain.

c) Given a dielectric permittivity ε, gap with h, plate area A, and plate charge $\pm Q$, find expressions for the electric displacement, and electric polarization, inside of the dielectric.

d) In the high permittivity limit, find the polarization of the dielectric in terms of the electric displacement.

Problem 14.3: A Steel Inductor

Consider an inductor made from a simple steel torus wrapped with a coil. Assume that you know the permeability and intrinsic magnetization of the steel torus, and assume no frictional losses. You also know any geometrical information, such as the circumference and cross-section of the steel torus, and the number of loops of wire in the inductor.

a) Draw the magnetizing field $\vec{H}$ inside the torus, and express it in terms of the current I flowing through the coils.

b) Draw the magnetization vector $\vec{M}$ inside the inductor. Is this in the same, or the opposite, direction as the magnetizing field? Clearly explain.

c) Express the magnetization in terms of the current.

d) Express the induced magnetic field, $\vec{B}$, also as a function of the current. Express this in general, and in the linear limit.

e) What physicists now call the magnetic field, $\vec{B}$, used to be called the *magnetic induction*, and still is by some electrical engineers. Explain why this makes sense in the context of inductors with large permeabilities.

f) As we have already discussed, physicists used to call $\vec{H}$ the magnetic field, but gradually switched once it became clear that $\vec{B}$, not $\vec{H}$, represented the field that could, in principle, be measured *in situ*. Now imagine that free space were made up of a perfectly linear paramagnetic medium with permeability μ_0, explain why both fields would seem to be equally fundamental. Also explain why $\vec{H}$ is easier to measure in inductor experiments.

Thought Experiment 14.3: A Cylindrical Electromagnetic Wave Algorithm

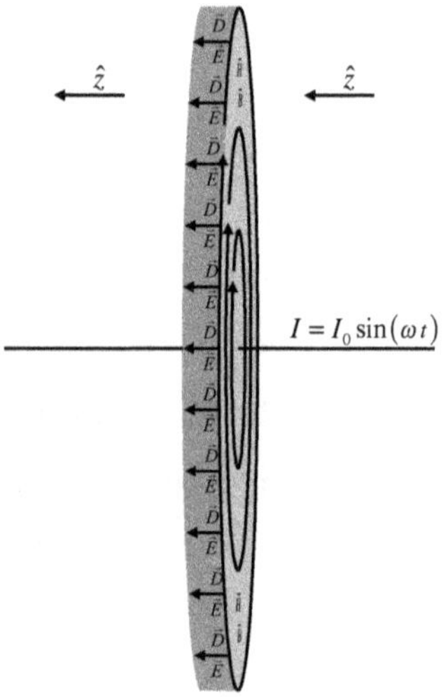

Consider a cylindrical disk with infinite radius but finite height h, made of iron infused glass, so we can assume that the permittivity and permeability are $\varepsilon \gg \varepsilon_0$ and $\mu \gg \mu_0$. In the middle of the ends are placed two circular conductive capacitor plates each of area πa^2, which are connected to a high frequency alternating current source $I(t) = I_0 \sin(\omega t)$.

The purpose of this thought experiment is to find all four-fields in turn, as we outlined in the flow chart on p. 432.

We will first consider the electric displacement vector from Gauss's law:

$$\vec{\nabla}\cdot\vec{D} \equiv \rho \quad \rightarrow \quad \oint_{\text{surface}} \vec{D}\cdot \mathrm{d}\vec{A} = Q \; .$$

Thus, $\vec{D}$ between the plates is:

$$\vec{D}(t,s) = \left(\frac{Q}{\pi a^2}\right)\hat{z} = \left(\frac{\hat{z}}{\pi a^2}\right)\int I\,\mathrm{d}t = -\left(\frac{I_0}{\omega \pi a^2}\right)\cos(\omega t)\,\hat{z} = \varepsilon\vec{E}(t,s).$$

Now we can find $\vec{H}$ from:

$$\vec{\nabla}\times\vec{H} = \cancel{\vec{J}} + \frac{\partial\vec{D}}{\partial t} = \left(\frac{I_0\hat{z}}{\cancel{\omega}\,\pi a^2}\right)\left(\cancel{\omega}\cos(\omega t)\right) = \left(\frac{I_0\sin(\omega t)\hat{z}}{\pi a^2}\right) = \frac{I}{\pi a^2}\hat{z}.$$

This is simplified by integrating over the capacitor surface:

$$\oint_{\text{path}} \vec{H}\cdot\mathrm{d}\vec{\ell} = \int_{\text{surface}} \left(\cancel{\vec{J}} + \frac{\partial\vec{D}}{\partial t}\right)\cdot\mathrm{d}\vec{A} = \int_{\text{surface}} \left(I/\pi a^2\right)\hat{z}\cdot\mathrm{d}\vec{A} = I\,.$$

This is a nice example of Maxwell's *displacement current*; the time derivative of the electric displacement vector, integrated over a perpendicular area, is simply the current driving the capacitor. Thus, in a linear medium, the electromagnetic field at the edge of the cylinder defined by the capacitor is:

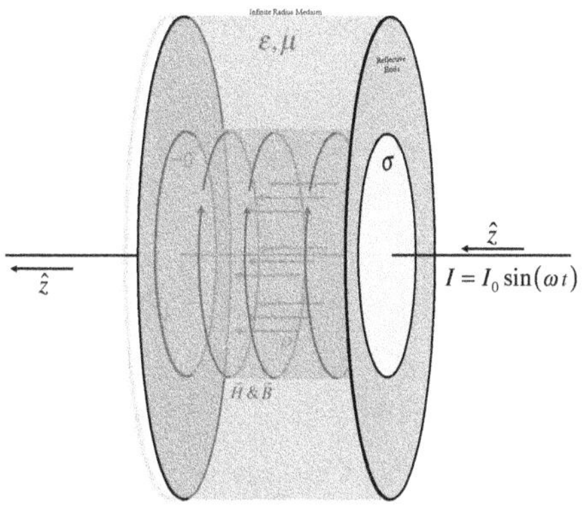

$$\vec{E}(t,a) = -\left(\frac{I_0}{\omega \pi \varepsilon a^2}\right)\cos(\omega t)\,\hat{z}$$

$$\vec{B}(t,a) = \frac{\mu I_0}{2\pi a}\sin(\omega t)\,\hat{\varphi}\,.$$

We now know the magnetic field at the edge of the capacitor, which we can refer to as *the surface of the source.*

Now comes the interesting part! We will now start with this electromagnetic field at the source, and then apply Maxwell's equations, in turn, to propagate the fields forward in time and outward in space.

The algorithm displayed in the diagram shows how the four-field model applies to this cylindrical signal. Notice how the Maxwellian four-field vision imagines that the fields propagate in space through the two equations involving the curl of the fields, and in time through the two constitutive relations. Recall that ε and μ were analogous to the elasticity and inertial modulus of the medium. As the wave propagates, because of the symmetry, the electric field and the electric displacement remain axial, while the magnetic field and magnetizing field remain azimuthal.

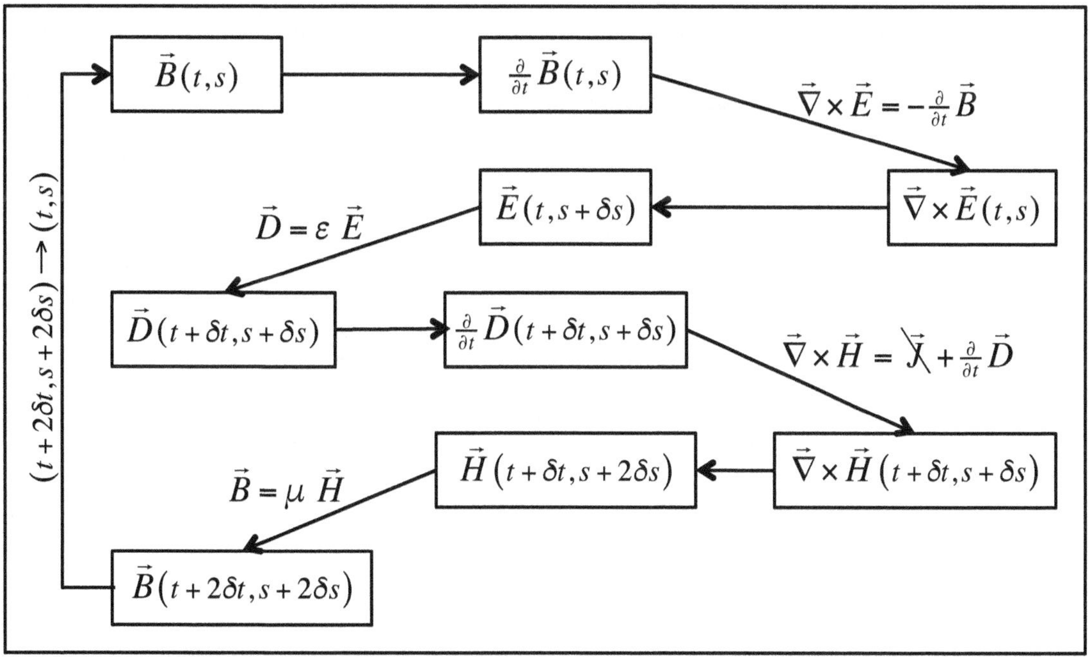

Let's begin our algorithm with the magnetic field, since we just calculated it. We will start with Faraday's law:

$$\tfrac{\partial}{\partial t}\vec{B}(a,t) = \frac{\mu I_0}{2\pi a}\tfrac{\partial}{\partial t}\sin(\omega t)\,\hat{\varphi} = \frac{\mu I_0 \omega}{2\pi a}\cos(\omega t)\,\hat{\varphi} = -\vec{\nabla}\times\vec{E}.$$

Now we will draw an Amperian loop from the edge of the capacitor out:

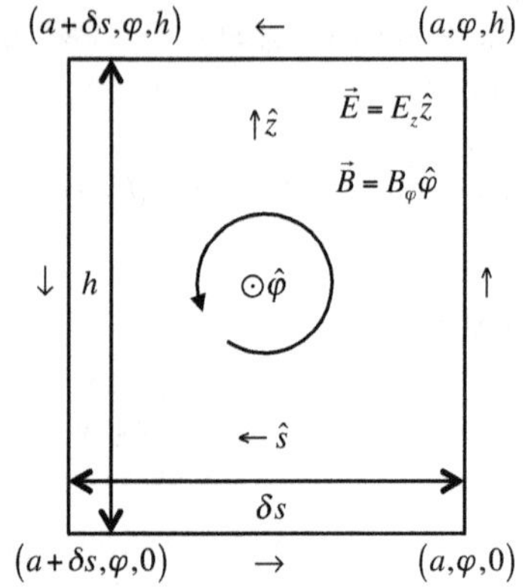

$$\oint_{\text{path}} \vec{E}\cdot d\vec{\ell} = \int_{a+\delta s}^{a} \cancel{\vec{E}\cdot\hat{s}}\Big|_{z=0} ds + \int_{0}^{h} \vec{E}\cdot\hat{z}\Big|_{s=a-\delta s} dz + \int_{a}^{a+\delta s} \cancel{\vec{E}\cdot\hat{s}}\Big|_{z=h} ds + \int_{h}^{0} \vec{E}\cdot\hat{z}\Big|_{s=a+\delta s} dz$$
$$\approx -h\left(E_z(a+\delta s) - E_z(a)\right).$$

And from Faraday's Law:

$$\oint_{\text{path}} \vec{E}\cdot d\vec{\ell} = \int_{\text{surface}} \left(\vec{\nabla}\times\vec{E}\right)\cdot d\vec{A} \approx \left(\vec{\nabla}\times\vec{E}\right)\cdot\left(\vec{h}\times\delta\vec{s}\right)$$
$$= -\delta s\frac{\mu I_0 \omega h}{2\pi a}\cos(\omega t).$$

Equating these, we can propagate the electric field spatially:

$$\vec{E}(t,a+\delta s) = \vec{E}(t,a) + \delta s\frac{\mu I_0 \omega}{2\pi a}\cos(\omega t)\hat{z} = I_0\cos(\omega t)\left(\delta s\frac{\mu\omega}{2\pi a} - \frac{1}{\omega\varepsilon\pi a^2}\right)\hat{z}.$$

We now imagine that it might take some time to polarize the material, so:

$$\vec{D}(t+\delta t, a+\delta s) = \varepsilon\vec{E}(t, a+\delta s).$$

Next we find the time derivative of the electric displacement and the curl of the magnetizing field:

$$\tfrac{\partial}{\partial t}\vec{D}(t+\delta s, a+\delta s) = I_0\varepsilon\sin(\omega t)\left(\delta s\frac{\mu\,\omega}{2\pi a} - \frac{1}{\varepsilon\,\omega\pi a^2}\right)\hat{z} = \vec{\nabla}\times\vec{H}(t+\delta s, a+\delta s).$$

We now find $\vec{H}$, from its curl, by drawing a wedge shaped loop around the $\hat{z}$ axis:

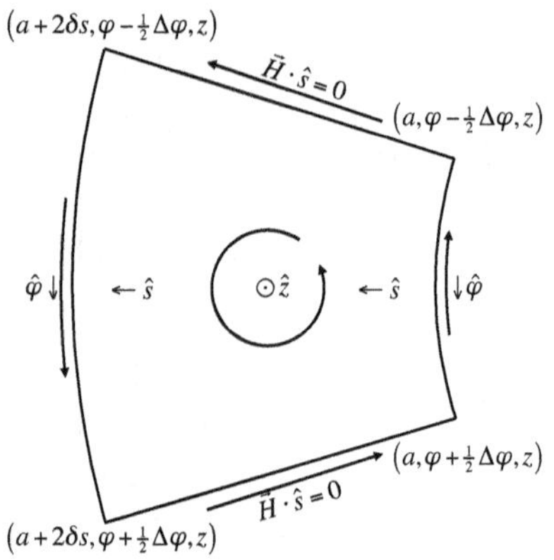

$$\oint_{\text{path}} \vec{H}\cdot d\vec{\ell} = (a+2\delta s)\left(H(a+2\delta s)\right)\Delta\varphi$$
$$-a\left(H(a)\right)\Delta\varphi.$$

The area of the wedge is the difference in the areas of two circles, scaled by the angle, so:

$$\int_{\text{surface}} \tfrac{\partial}{\partial t}\vec{D}\cdot d\vec{A} \approx \left(\tfrac{\partial}{\partial t}\vec{D}\right)\left(\frac{\Delta\varphi}{2\pi}\left(\pi(a+2\delta s)^2 - \pi a^2\right)\right).$$

Combining equations, and solving for $\vec{H}$, we find:

$$H(t+\delta t, a+2\delta s) = (2\delta s)\left(\frac{a+\delta s}{(a+2\delta s)}\right)\left(\tfrac{\partial}{\partial t}D(t+\delta s, a+\delta s)\right) + \left(\frac{a}{a+2\delta s}\right)H(t+\delta s, a).$$

In the small increment limit, we could discard the second order terms. However it does not cost too much computationally, so we would probably want to keep it to make the algorithm more robust.

And, again, we imagine that the inherent inductance (permeability) causes some delay in creating the induced magnetic field:

$$\vec{B}(t+2\delta t, a+2\delta s) = \mu H(t+\delta s, a+2\delta s).$$

By going back and substituting in for the electric field, and putting it in vector notation, we can now write:

$$\vec{B}(t+2\delta t, a+2\delta s) = (2\delta s\,\mu\varepsilon)\left(\frac{a+\delta s}{a+2\delta s}\right)\left(\tfrac{\partial}{\partial t}\vec{E}(t, a+\delta s)\times\hat{s}\right) + \left(\frac{a}{a+2\delta s}\right)\vec{B}(t,a).$$

Now that we are outside of the capacitor, we should be able to continue this process over and over.

Notice that we took two temporal derivatives analytically, and two spatial integrals numerically. Computational algorithms are particularly robust when adding and multiplying, however they can become unstable when dividing by small numbers. On the other hand, analytical differentiation is much easier than integration. Therefore, it is often good practice when solving differential equations to take derivatives analytically, but compute integrals numerically.

Consider now that this wave is propagating is the $\hat{s}$ direction at some speed v. Experience with other cylindrical waves, like ripples in water, suggests that the speed is primarily a function of the medium involved. In the case of transmission lines (Section 13.2), the wave speed was $v = (\mathtt{LC})^{-\frac{1}{2}}$, where $\mathtt{L}$ and $\mathtt{C}$ were the inductance and capacitance per length respectively. Therefore, we expect that the medium has its own wave speed, and that this speed is somehow related to the capacitance per length. Moreover, every capacitance we have calculated has been proportional to the permittivity, and every inductance has been proportional to the permeability. Putting these ideas together, and applying some dimensional analysis, suggests a medium speed of $v = (\mu\varepsilon)^{-\frac{1}{2}}$, so we must constrain the time step to be related to the grid size by: $\delta t = \delta s\sqrt{\mu\varepsilon}$.

Problem 14.4: The Cylindrical Wave Algorithm

In Thought Experiment 14.3 above we went through a single cycle of all four of Maxwell's equations to propagate the fields outward. In this problem you will investigate this yourself.

a) Continue the process shown in Thought Experiment 14.3 above until you have developed robust recursion relations for both the electric and magnetic fields. Keep terms up to second order in δs.

b) Imagine that the capacitor contained a strong dielectric embedded with iron, say with a permeability $\mu = 1000\mu_0$, and permittivity $\varepsilon = 10\varepsilon_0$. Calculate its characteristic speed. Express this as a percentage of the speed of light.

c) Write a computer program to solve this numerically.

Example 14.1: The Cylindrical Wave Solved Differentially

We, in Thought Experiment 14.3, and you, in Problem 14.4, investigated a cylindrical disk with infinite radius. In this example we solve the same problem, but this time analytically in the limit of far distances. We begin with the electric and magnetic fields at the surface of the source:

$$\vec{E}(t,a) = -\left(\frac{I_0}{\omega \pi \varepsilon a^2}\right)\cos(\omega t)\,\hat{z} \quad \text{and} \quad \vec{B}(t,a) = \frac{\mu I_0}{2\pi a}\sin(\omega t)\,\hat{\varphi}\,.$$

Recall in Thought Experiment 13.4 (p. 415) we decoupled the telegrapher's equations, so as to solve for only one variable. We now do the same thing by writing down the four relevant equations in cylindrical coordinates as:

$$\begin{array}{llll} \vec{\nabla}\times\vec{H} = \frac{\partial \vec{D}}{\partial t} & \vec{B} \approx \mu \vec{H} & \vec{\nabla}\times\vec{E} = -\frac{\partial \vec{B}}{\partial t} & \vec{D} \approx \varepsilon \vec{E} \\ \frac{1}{s}\frac{\partial}{\partial s}\left(s H_\varphi\right) = \frac{\partial}{\partial t} D_z & B_\varphi \approx \mu H_\varphi & -\frac{\partial}{\partial s} E_z = -\frac{\partial}{\partial t} B_\varphi & D_z \approx \varepsilon E_z\,. \end{array}$$

Taking the derivatives of each of Maxwell's equations gives us:

$$\begin{array}{ll} \frac{\partial}{\partial s}\frac{1}{s}\frac{\partial}{\partial s}\left(s H_\varphi\right) = \frac{\partial}{\partial s}\frac{\partial}{\partial t} D_z & \frac{\partial}{\partial s}\frac{\partial}{\partial s} E_z = \frac{\partial}{\partial s}\frac{\partial}{\partial t} B_\varphi \\ \frac{\partial}{\partial t}\frac{1}{s}\frac{\partial}{\partial s}\left(s H_\varphi\right) = \frac{\partial}{\partial t}\frac{\partial}{\partial t} D_z & \frac{\partial}{\partial t}\frac{\partial}{\partial s} E_z = \frac{\partial}{\partial t}\frac{\partial}{\partial t} B_\varphi\,. \end{array}$$

Next we rearrange to solve for the magnetic field:

$$\frac{\partial}{\partial s}\frac{1}{s}\frac{\partial}{\partial s}\left(s \frac{1}{\mu} B_\varphi\right) = \frac{\partial}{\partial s}\frac{\partial}{\partial t}\left(\varepsilon E_z\right) \quad , \quad \frac{\partial}{\partial t}\frac{\partial}{\partial s} E_z = \frac{\partial}{\partial t}\frac{\partial}{\partial t} B_\varphi$$

$$\frac{\partial}{\partial s}\frac{1}{s}\frac{\partial}{\partial s}\left(s B_\varphi\right) = \mu\varepsilon \frac{\partial}{\partial s}\frac{\partial}{\partial t} E_z = \mu\varepsilon \frac{\partial}{\partial t}\frac{\partial}{\partial s} E_z = \mu\varepsilon \frac{\partial}{\partial t}\frac{\partial}{\partial t} B_\varphi\,.$$

Therefore, we see that that:

$$\begin{aligned} \mu\varepsilon \frac{\partial^2}{\partial t^2} B_\varphi &= \frac{\partial}{\partial s}\frac{1}{s}\frac{\partial}{\partial s}\left(s B_\varphi\right) = \frac{\partial}{\partial s}\frac{1}{s}\left(B_\varphi + s\frac{\partial}{\partial s} B_\varphi\right) = \frac{\partial}{\partial s}\left(\frac{1}{s} B_\varphi\right) + \frac{\partial}{\partial s}\frac{\partial}{\partial s} B_\varphi \\ &= \left(-\frac{1}{s^2} B_\varphi + \frac{1}{s}\frac{\partial}{\partial s} B_\varphi + \frac{\partial^2}{\partial s^2} B_\varphi\right) = \left(\frac{\partial^2}{\partial s^2} B_\varphi + \frac{1}{s}\frac{\partial}{\partial s} B_\varphi - \frac{1}{s^2} B_\varphi\right). \end{aligned}$$

This is a second order partial differential equation, which we will solve by guessing a similar solution as for the telegrapher's equations. But here we consider the wave behavior analogous to ripples which decline in amplitude with distance from the source.

First, notice that the characteristic velocity of the medium is:

$$v = \frac{1}{\sqrt{\mu\varepsilon}}\,.$$

So we will now guess a self-similar solution, which only depends on the retarded time $\tau \equiv t - s/v_c = t - s\sqrt{\mu\varepsilon}$. Knowing we will use the chain rule, we find the partial derivatives of the retarded time as:

$$\frac{\partial \tau}{\partial t} = 1 \quad , \quad \frac{\partial \tau}{\partial s} = -\frac{1}{v} = -\sqrt{\mu\varepsilon}\,.$$

Next, we guess a particular (not general) solution in the form of:

$$B_\varphi(s,\tau) = S(s)\sin(\omega\tau)\ .$$

Taking the partial derivatives with respect to time and spatial coordinate, s, yields:

$$\tfrac{\partial^2}{\partial t^2}B_\varphi(s,\tau) = \tfrac{\partial^2}{\partial t^2}S(s)\sin(\omega\tau) = S(s)\tfrac{\partial^2}{\partial t^2}\sin(\omega\tau) \ = -\omega^2 S(s)\sin(\omega\tau)$$

$$\begin{aligned}
\tfrac{\partial}{\partial s}B_\varphi(s,\tau) &= \tfrac{\partial}{\partial s}S(s)\sin(\omega\tau) = \sin(\omega\tau)\tfrac{\partial}{\partial s}S(s) + S(s)\tfrac{\partial}{\partial s}\sin(\omega\tau) \\
&= \sin(\omega\tau)\tfrac{\partial}{\partial s}S(s) + \omega S(s)\cos(\omega\tau)\tfrac{\partial\tau}{\partial s} \\
&= \sin(\omega\tau)\tfrac{\partial}{\partial s}S(s) - \tfrac{1}{v_c}\omega S(s)\cos(\omega\tau)
\end{aligned}$$

$$\begin{aligned}
\tfrac{\partial^2}{\partial s^2}B_\varphi(s,\tau) &= \tfrac{\partial}{\partial s}\tfrac{\partial}{\partial s}S(s)\sin(\omega\tau) = \tfrac{\partial}{\partial s}\left(\sin(\omega\tau)\tfrac{\partial}{\partial s}S(s) + S(s)\tfrac{\partial}{\partial s}\sin(\omega\tau)\right) \\
&= \tfrac{\partial}{\partial s}\left(\sin(\omega\tau)\tfrac{\partial}{\partial s}S(s) + \omega S(s)\cos(\omega\tau)\tfrac{\partial\tau}{\partial s}\right) = \tfrac{\partial}{\partial s}\left(\sin(\omega\tau)\tfrac{\partial}{\partial s}S(s) - \tfrac{1}{v_c}\omega S(s)\cos(\omega\tau)\right) \\
&= \sin(\omega\tau)\tfrac{\partial^2}{\partial s^2}S(s) + \omega\cos(\omega\tau)\tfrac{\partial}{\partial s}S(s)\left(-\tfrac{1}{v_c}\right) - \tfrac{1}{v_c}\omega\cos(\omega\tau)\tfrac{\partial}{\partial s}S(s) - S(s)\left(\tfrac{\omega}{v_c}\right)^2\sin(\omega\tau) \\
&= \sin(\omega\tau)\tfrac{\partial^2}{\partial s^2}S(s) - S(s)\left(\tfrac{\omega}{v_c}\right)^2\sin(\omega\tau) - \tfrac{2}{v_c}\omega\cos(\omega\tau)\tfrac{\partial}{\partial s}S(s).
\end{aligned}$$

So, if the following criterion holds, we have a solution:

$$0 = \tfrac{\partial^2}{\partial s^2}B_\varphi + \tfrac{1}{s}\tfrac{\partial}{\partial s}B_\varphi - \tfrac{1}{s^2}B_\varphi - \mu\varepsilon\tfrac{\partial^2}{\partial t^2}B_\varphi$$

$$\begin{aligned}
0 = {} & \sin(\omega\tau)\tfrac{\partial^2}{\partial s^2}S(s) - S(s)\left(\tfrac{\omega}{v_c}\right)^2\sin(\omega\tau) - \tfrac{2}{v_c}\omega\cos(\omega\tau)\tfrac{\partial}{\partial s}S(s) \\
& + \tfrac{1}{s}\sin(\omega\tau)\tfrac{\partial}{\partial s}S(s) - \tfrac{1}{s}\tfrac{1}{v_c}\omega S(s)\cos(\omega\tau) \\
& - \tfrac{1}{s^2}S(s)\sin(\omega\tau) \\
& - \mu\varepsilon\left(-\omega^2 S(s)\sin(\omega\tau)\right)
\end{aligned}$$

$$\begin{aligned}
0 = {} & \sin(\omega\tau)\left(\tfrac{\partial^2}{\partial s^2}S(s) \cancel{- S(s)\left(\tfrac{\omega}{v_c}\right)^2} + \tfrac{1}{s}\tfrac{\partial}{\partial s}S(s) - \tfrac{1}{s^2}S(s) \cancel{+ \tfrac{1}{v_c^2}\omega^2 S(s)}\right) \\
& + \cos(\omega\tau)\left(-\tfrac{2}{v_c}\omega\tfrac{\partial}{\partial s}S(s) - \tfrac{1}{s}\tfrac{1}{v_c}\omega S(s)\right)
\end{aligned}$$

Notice that the sine and cosine functions are orthogonal to one another, so both conditions must be satisfied:

$$\begin{aligned}
0 &= \sin(\omega\tau)\left(\tfrac{\partial^2}{\partial s^2}S(s) + \tfrac{1}{s}\tfrac{\partial}{\partial s}S(s) - \tfrac{1}{s^2}S(s)\right) \\
0 &= \cos(\omega\tau)\left(-\tfrac{\omega}{v_c}\right)\left(2\tfrac{\partial}{\partial s}S(s) + \tfrac{1}{s}S(s)\right).
\end{aligned}$$

We solve the simpler one:

$$0 = \left(2\tfrac{\partial}{\partial s}S(s) + \tfrac{1}{s}S(s)\right) \quad\rightarrow\quad \frac{dS(s)}{S(s)} = -\tfrac{1}{2}\frac{ds}{s}$$

$$\ln S(s) = \ln s^{-\frac{1}{2}} + C \quad\rightarrow\quad \ln S(s) = -\tfrac{1}{2}\ln s + C \quad\rightarrow\quad S(s) \propto s^{-\frac{1}{2}}.$$

Finally, we have a solution for the magnetic field in the form of:

$$\vec{B}(s,\tau) = S(s)\sin(\omega\tau)\hat{\varphi} = \left(B_a\sqrt{a}\right)\frac{\sin(\omega\tau)}{\sqrt{s}}\hat{\varphi}.$$

As we know the solution at the surface of the source, we normalize the solution there, and we can now write:

$$\vec{B}(s,\tau) = B_a \sin(\omega\tau)\sqrt{\frac{a}{s}}\,\hat{\varphi} = I_0 \sin(\omega\tau)\frac{\mu}{2\pi a}\sqrt{\frac{a}{s}}\,\hat{\varphi} = I_0 \sin(\omega\tau)\frac{\mu}{2\pi}\sqrt{\frac{1}{a\,s}}\,\hat{\varphi}.$$

Now let's check our units:

$$[B] = [I_0]\,\cancel{\sin(\omega\tau)}\,\frac{[\mu]}{\cancel{2\pi}[a]}\sqrt{\frac{\cancel{a}}{\cancel{s}}}\,\cancel{\hat{\varphi}} = \frac{\mathrm{A}\cdot\mathrm{H}/\mathrm{m}}{\mathrm{m}} = \mathrm{T}$$

Next, we can use Faraday's law to find the electric field:

$$\vec{\nabla}\times\vec{E} = -\tfrac{\partial}{\partial s}E_z\,\hat{\varphi} = -\tfrac{\partial}{\partial t}\vec{B} = -\frac{\mu I_0}{2\pi\sqrt{a\,s}}\tfrac{\partial}{\partial t}\sin(\omega\tau)\,\hat{\varphi} = -\frac{\mu\omega I_0}{2\pi\sqrt{a\,s}}\cos(\omega\tau)\,\hat{\varphi}$$

$$E_z = \frac{\mu I_0}{2\pi\sqrt{a\,s}}\int \omega\cos(\omega\tau)\tfrac{\partial s}{\partial\tau}\,d\tau = -v\frac{\mu I_0}{2\pi\sqrt{a\,s}}\sin(\omega\tau) = -\frac{1}{\sqrt{\mu\varepsilon}}\frac{\mu I_0}{2\pi\sqrt{a\,s}}\sin(\omega\tau)$$

$$\vec{E} = -\sqrt{\frac{\mu}{\varepsilon}}\frac{I_0}{2\pi\sqrt{a\,s}}\sin(\omega\tau)\,\hat{z}.$$

Notice that we can factor out the wave velocity, so this simplifies to:

$$\vec{E} = v\frac{\mu I_0}{2\pi\sqrt{a\,s}}\sin(\omega\tau)\,\hat{z} = -(v\,\hat{s})\times\left(\frac{\mu I_0}{2\pi\sqrt{a\,s}}\sin(\omega\tau)\hat{\varphi}\right) = -\vec{v}\times\vec{B}.$$

where $\vec{v} = (\mu\varepsilon)^{-\frac{1}{2}}\,\hat{s}$ is the velocity of the wave.

In Part I we solved Laplace's equation with cylindrical symmetry, where the solution contained a special function called a *Bessel function*, after the astronomer who used them in his successful quest to measure the distance to a nearby star. Given the similarities between the wave equation and Laplace's equation, we might have expected this solution to contain a Bessel function. This would indeed have been the case had we not assumed solutions in the far field limit.

Thought Experiment 14.4: The Energy of a Cylindrical Wave

Consider the energy density of the same cylindrical wave:

$$u = \tfrac{\varepsilon}{2}E^2 + \tfrac{1}{2\mu}B^2 = \tfrac{\varepsilon}{2}\left(-\sqrt{\frac{\mu}{\varepsilon}}\frac{I_0}{2\pi\sqrt{a\,s}}\sin(\omega\tau)\right)^2 + \tfrac{1}{2\mu}\left(I_0\sin(\omega\tau)\frac{\mu}{2\pi}\sqrt{\frac{1}{a\,s}}\right)^2$$

$$= \left(\frac{\cancel{\varepsilon}}{2}\frac{\mu}{\cancel{\varepsilon}} + \frac{\mu^{\cancel{2}}}{2\cancel{\mu}}\right)\left(\frac{I_0}{2\pi\sqrt{a\,s}}\sin(\omega\tau)\right)^2 = \frac{1}{4\pi^2}\frac{\mu}{a\,s}\left(I_0\sin(\omega\tau)\right)^2.$$

Notice that the energy density is separated between the electric and magnetic fields in equal parts. When different forms of energy are distributed evenly this is called *equipartition*, as you may remember from our discussion of superconductivity from Part II.

Now, consider the flux of energy from the energy continuity equation:

$$\vec{\nabla}\cdot\vec{\mathbb{S}} = -\tfrac{\partial}{\partial t}\mathbb{u} = -\frac{I_0^2}{4\pi^2}\frac{\mu}{a\,s}\tfrac{\partial}{\partial t}\left(\sin^2(\omega\tau)\right)$$

$$\cancel{\tfrac{1}{s}}\tfrac{\partial}{\partial s}\left(s\mathbb{S}_s\right) = -\frac{I_0^2}{4\pi^2}\frac{\mu}{a\cancel{s}}\tfrac{\partial}{\partial \tau}\left(\sin^2(\omega\tau)\right)\cancel{\tfrac{\partial\tau}{\partial t}}$$

$$s\mathbb{S}_s = -\frac{I_0^2}{4\pi^2}\frac{\mu}{a}\int \tfrac{\partial}{\partial\tau}\left(\sin^2(\omega\tau)\right)\left(\tfrac{\partial s}{\partial\tau}\mathrm{d}\tau\right) = -\frac{I_0^2}{4\pi^2}\frac{\mu}{a}(-v_c)\left(\sin^2(\omega\tau)\right)$$

$$\vec{\mathbb{S}} = \frac{I_0^2}{4\pi^2}\frac{\mu}{a\,s}\frac{1}{\sqrt{\mu\varepsilon}}\sin^2(\omega\tau)\,\hat{s} = \frac{I_0^2}{4\pi^2}\frac{1}{a\,s}\sqrt{\frac{\mu}{\varepsilon}}\sin^2(\omega\tau)\,\hat{s}.$$

Next we factor this expression in terms of the electric and magnetic fields:

$$\vec{\mathbb{S}} = \frac{1}{\mu}\left(-\sqrt{\frac{\mu}{\varepsilon}}\frac{I_0}{2\pi\sqrt{a\,s}}\sin(\omega\tau)\,\hat{z}\right)\times\left(\frac{\mu I_0}{2\pi\sqrt{a\,s}}\sin(\omega\tau)\sqrt{\frac{1}{a\,s}}\,\hat{\varphi}\right) = \tfrac{1}{\mu}\vec{E}\times\vec{B}.$$

The electromagnetic energy flux is called the *Poynting vector*, after the English physicist John Henry Poynting. The name is also catchy, because the Poynting vector always points in the direction of the energy propagation.

We consider next the total power of the radiation by drawing a cylindrical Gaussian surface of radius s and height equal to the plate separation h:

$$\mathbb{P} = \oint_{\text{surface}} \vec{\mathbb{S}}\cdot\mathrm{d}\vec{A} = \oint_{\text{surface}}\left(\frac{I_0^2}{4\pi^2}\frac{1}{a\,s}\sqrt{\frac{\mu}{\varepsilon}}\sin^2(\omega\tau)\,\hat{s}\right)\cdot\mathrm{d}\vec{A} = \frac{I_0^2}{4\pi^2}\frac{1}{a\cancel{s}}\sqrt{\frac{\mu}{\varepsilon}}\sin^2(\omega\tau)\left(2\pi\cancel{s}\right)h$$

$$= \frac{h}{2\pi a}\sqrt{\frac{\mu}{\varepsilon}}\;I_0^2\sin^2(\omega\tau).$$

From the power, we can find voltage across the plates:

$$\mathbb{V} = \frac{\mathbb{P}}{I} = \frac{\frac{h}{2\pi a}\sqrt{\frac{\mu}{\varepsilon}}\;I_0^2\sin^2(\omega\tau)}{I_0\sin(\omega\tau)} = \frac{h}{2\pi a}\sqrt{\frac{\mu}{\varepsilon}}\;I_0\sin(\omega\tau).$$

And the impedance of the source is:

$$Z = \frac{\mathbb{V}}{I} = \frac{\frac{h}{2\pi a}\sqrt{\frac{\mu}{\varepsilon}}\;\cancel{I_0\sin(\omega\tau)}}{\cancel{I_0\sin(\omega\tau)}} = \frac{h}{2\pi a}\sqrt{\frac{\mu}{\varepsilon}} = \frac{h}{2\pi a}Z_c.$$

Notice that the impedance is on the order of $Z_c = \sqrt{\frac{\mu}{\varepsilon}}$, called the *characteristic impedance* of the medium. When the medium is free space, the characteristic impedance is naturally called the *impedance of free space*, and is now considered a fundamental constant defined as:

$$Z_0 \equiv \sqrt{\frac{\mu_0}{\varepsilon_0}} \approx 377\,\Omega.$$

Problem 14.5: A Cylindrical Wave in the Intermediate Limit

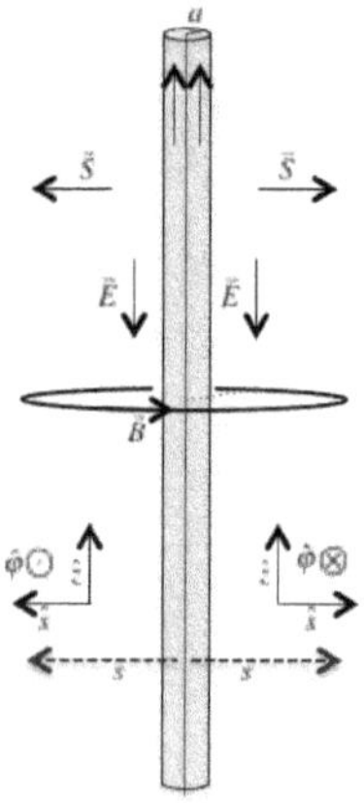

Consider a long wire emitting an electromagnetic wave of the form $\vec{B}_a = B_0 \sin(\omega t)\hat{\varphi}$ at the surface of the wire.

a) Use conservation of energy to find the Poynting vector as a function of position.

b) Find the electric and magnetic fields a distance s from the wire. Consider only distances, s, that are much longer than the wavelength of the light but much shorter than the length of the wire (i.e. $l \gg s \gg \frac{2\pi c}{\omega}$).

c) Report on practical examples of devices that could be modeled as a cylindrical electromagnetic wave.

14.2 *Light in a Linear Medium*

As early as the last millennium scholars such as Alhacen (p. 401) established, through experiment and careful observation, the rules by which light reflects and refracts as it passes from air to glass or water.

In this section we introduce some solutions to Maxwell's equations inside of a linear dielectric medium. Why do we limit ourselves to a linear medium? The answer is simply simplicity. Nonlinear media are terribly complicated, and the natural world is often reasonably well described by a linear approximation.

The change in speed, as light crosses from one dielectric medium to another, is key in our understanding of basic optical phenomena. From this speed ratio we derive Snell's law of refraction. Then we use the concept of retarded time to derive *Fermat's principle*, which states that the time of flight of light between two points is an extremum of the time with respect to any possible path it could take.

Derivation 14.1: The Wave Equation in a Linear Medium

In Thought Experiment 13.4 (p. 415) we introduced the one dimensional wave equation for a signal down a transmission line, and in Example 14.1 (p. 438) we introduced the two dimensional cylindrical wave equation. Here we will derive the general wave equation in a linear medium.

We begin with the four field representation of Maxwell's equations:

$$\vec{\nabla}\cdot\vec{B}=0,\quad \vec{\nabla}\times\vec{E}=-\tfrac{\partial\vec{B}}{\partial t},\quad \vec{\nabla}\cdot\vec{D}=\rho,\quad \text{and } \vec{\nabla}\times\vec{H}=\vec{J}+\tfrac{\partial}{\partial t}\vec{D}.$$

Our goal is to isolate one of these four fields. In our case here, we choose the electric field $\vec{E}$. We begin by taking the curl of Faraday's law:

$$\vec{\nabla}\times\vec{\nabla}\times\vec{E}=\vec{\nabla}\times\left(-\tfrac{\partial\vec{B}}{\partial t}\right)=-\tfrac{\partial}{\partial t}\left(\vec{\nabla}\times\vec{B}\right).$$

Next we consider the Maxwell-Ampere law, along with the constitutive relations $\vec{B}\approx\mu\vec{H}$ and $\vec{D}\approx\varepsilon\vec{E}$, which apply only in a uniform linear medium. Therefore:

$$\vec{\nabla}\times\vec{B}=\vec{\nabla}\times\left(\mu\vec{H}\right)=\mu\left(\vec{\nabla}\times\vec{H}\right)=\mu\left(\vec{J}+\tfrac{\partial}{\partial t}\vec{D}\right)=\mu\vec{J}+\tfrac{\partial}{\partial t}\mu\varepsilon\vec{E}.$$

And putting these together we have:

$$\vec{\nabla}\times\vec{\nabla}\times\vec{E} = -\tfrac{\partial}{\partial t}\left(\vec{\nabla}\times\vec{B}\right) = -\mu\left(\tfrac{\partial}{\partial t}\vec{J}\right) - \mu\varepsilon\tfrac{\partial}{\partial t}\left(\tfrac{\partial}{\partial t}\vec{E}\right) = -\mu\tfrac{\partial}{\partial t}\vec{J} - \mu\varepsilon\tfrac{\partial^2}{\partial t^2}\vec{E}.$$

Now we apply the double curl identity, and Gauss's law, so that:

$$\vec{\nabla}\times\vec{\nabla}\times\vec{E} = \vec{\nabla}\left(\vec{\nabla}\cdot\vec{E}\right) - \nabla^2\vec{E} = \vec{\nabla}\left(\tfrac{1}{\varepsilon}\rho\right) - \nabla^2\vec{E}.$$

And putting these last two together we find:

$$\nabla^2\vec{E} = \tfrac{1}{\varepsilon}\vec{\nabla}\rho + \mu\tfrac{\partial}{\partial t}\vec{J} + \mu\varepsilon\tfrac{\partial^2}{\partial t^2}\vec{E}.$$

Solutions to this equation can be quite complicated, but in particular cases they can be quite simple. For example, in static situations this just becomes Poisson's equation, which we solved in Part I.

In the special case of a neutral linear medium with no conductivity, such as glass (or free space), this simplifies to:

$$\nabla^2\vec{E} = \mu\varepsilon\tfrac{\partial^2}{\partial t^2}\vec{E} = \tfrac{1}{v^2}\tfrac{\partial^2}{\partial t^2}\vec{E},$$

where $v = \frac{1}{\sqrt{\mu\varepsilon}}$, is the propagation speed of the wave in the medium.

Problem 14.6: The Wave Speed in an Insulative Linear Medium

Consider an electromagnetic wave propagating through a linear non-conductive neutral medium:

a) Derive the wave equation in terms of the magnetic field $\vec{B}$.

b) Show that the magnetic wave speed is also: $v = (\mu\varepsilon)^{-\frac{1}{2}}$.

c) Common transparent substances, such as glass, water, and plastic, are dielectrics but non-magnetic. Show that the wave speed in these media is:

$$v = c\sqrt{\frac{\varepsilon_0}{\varepsilon}} = \frac{c}{\sqrt{K}},$$

where $K = \frac{\varepsilon}{\varepsilon_0}$ is called the *relative permittivity* or the *dielectric constant.*

Thought Experiment 14.5: The Plane Wave Solution in a Linear Medium

Light spreads out radially from a source, rarely traveling in only one direction. However, if the source is far away compared to the size of our detector, we can assume that it simply moves in one direction. This is called *the plane wave approximation*, and it greatly simplifies the wave equation. This means that we can again apply the principle of retarded time, but also assume a particular propagation direction $\hat{k} = \hat{z}$, so we go from having four independent spatial and temporal variables $(x, y, z, \text{ and } t)$ to having only the retarded time $(\tau = z - vt)$.

Since we expect to apply the chain rule, we take partial derivatives of the retarded time (τ):

$$\tfrac{\partial}{\partial t}\tau = \tfrac{\partial}{\partial t}\left(t - \tfrac{1}{v}z\right) = 1 \quad \text{and} \quad \tfrac{\partial}{\partial z}\tau = \tfrac{\partial}{\partial z}\left(t - \tfrac{1}{v}z\right) = -\tfrac{1}{v}.$$

Next we notice that for any function, $f(t,z) = f(\tau)$:

$\frac{\partial}{\partial t}f(\tau)=\left(\frac{\partial}{\partial\tau}f(\tau)\right)\left(\frac{\partial\tau}{\partial t}\right)=\frac{\partial}{\partial\tau}f(\tau)$ and $\frac{\partial}{\partial z}f(\tau)=\left(\frac{\partial}{\partial\tau}f(\tau)\right)\left(\frac{\partial\tau}{\partial z}\right)=-\frac{1}{v}\frac{\partial}{\partial\tau}f(\tau)$.

And, for good measure, let us find the second derivatives too:

$\frac{\partial^2}{\partial t^2}f(\tau)=\frac{\partial^2}{\partial\tau^2}f(\tau)$ and $\frac{\partial^2}{\partial z^2}f(\tau)=-\frac{1}{v}\frac{\partial}{\partial\tau}\left(-\frac{1}{v}\frac{\partial}{\partial\tau}f(\tau)\right)=\frac{1}{v^2}\frac{\partial^2}{\partial\tau^2}f(\tau)$.

Next we apply this solution to Maxwell's wave equation:

$$\nabla^2\vec{B}(\tau)=\mu\varepsilon\frac{\partial^2}{\partial t^2}\vec{B}(\tau) \quad , \quad \nabla^2\vec{B}(\tau)=\mu\varepsilon\frac{\partial^2}{\partial t^2}\vec{B}(\tau)$$
$$\frac{1}{v^2}\frac{\partial^2}{\partial\tau^2}\vec{B}=\mu\varepsilon\frac{\partial^2}{\partial\tau^2}\vec{B} \quad , \quad \frac{1}{v^2}\frac{\partial^2}{\partial\tau^2}\vec{B}=\mu\varepsilon\frac{\partial^2}{\partial\tau^2}\vec{B}$$
$$\frac{1}{v^2}=\mu\varepsilon \quad , \quad \frac{1}{v^2}=\mu\varepsilon.$$

Applying Faraday's law we can relate the two fields by:

$$\vec{\nabla}\times\vec{E}=-\frac{\partial}{\partial t}\vec{B} \rightarrow \frac{\partial\tau}{\partial z}\frac{\partial}{\partial\tau}\left(\hat{z}\times\vec{E}\right)=-\frac{\partial\tau}{\partial t}\frac{\partial}{\partial\tau}\vec{B} \rightarrow -\frac{1}{v}\frac{\partial}{\partial\tau}\left(\hat{z}\times\vec{E}\right)=-\frac{\partial}{\partial\tau}\vec{B},$$

and integrating with respect to τ :

$$\vec{B}=\frac{1}{v}\hat{z}\times\vec{E}.$$

To invert this, we cross both sides by $\vec{v}$, and apply the back cab rule:

$$\vec{v}\times\vec{B}=v\hat{z}\times\left(\frac{1}{v}\hat{z}\times\vec{E}\right)=\hat{z}\times\left(\hat{z}\times\vec{E}\right)=\left(\hat{z}\left(\hat{z}\cdot\vec{E}\right)-\vec{E}\left(\hat{z}\cdot\hat{z}\right)\right)=-\vec{E}$$
$$\vec{E}=-\vec{v}\times\vec{B}.$$

We will now find the Poynting vector, in terms of the magnetic field:

$$\vec{\mathbb{S}}=\frac{1}{\mu}\vec{E}\times\vec{B}=\frac{1}{\mu}\left(-\vec{v}\times\vec{B}\right)\times\vec{B}=\frac{1}{\mu}\vec{B}\times\left(\vec{v}\times\vec{B}\right)=\frac{1}{\mu}\left(\vec{v}\left(\vec{B}\cdot\vec{B}\right)-\vec{B}\left(\vec{B}\cdot\vec{v}\right)\right)=\frac{1}{\mu}B^2\vec{v}.$$

Similarly, we could also write this in terms of the electric field as:

$$\vec{\mathbb{S}}=\frac{1}{\mu}B^2\vec{v}=\frac{1}{\mu}\left(\frac{1}{v}E\right)^2\vec{v}=\frac{\mu\varepsilon}{\mu}E^2\vec{v}=\varepsilon E^2\vec{v}.$$

We now average the two versions of the solution, which yields:

$$\vec{\mathbb{S}}=\frac{1}{2}\left(\frac{1}{\mu}B^2+\varepsilon E^2\right)\vec{v}=\mathrm{u}\vec{v}.$$

Finally notice that the direction of propagation was arbitrarily assigned to the $\hat{z}$ direction. In general, we can replace $\hat{z}$ with a general propagation unit vector $\hat{k}$ $\left(\vec{v}=v\hat{k}\right)$.

Problem 14.7: The Wave Vector

Since waves follow sinusoidal solutions of a particular frequency and wavelength, it is often convenient to write the solutions in terms of the wave vector: $\vec{k}=\omega\frac{\vec{v}}{v^2}=\frac{\omega}{v}\hat{v}=\frac{\omega}{v}\hat{k}$. Show the following for plane waves:

a) $\omega\tau=\omega t-\vec{k}\cdot\vec{r}$, if the wave had past through the origin.

b) $\omega=\vec{v}\cdot\vec{k}$

c) $\vec{k}=-\vec{\nabla}(\omega\tau)=\vec{\nabla}(-\omega\tau)$

d) $\vec{\nabla} f(\tau) = -\dfrac{\vec{k}}{\omega}\dfrac{df}{d\tau}$ e) $\vec{\nabla} f(\tau) = -\frac{1}{\omega}\vec{k}\frac{d}{d\tau} f(\tau)$

f) $\vec{\nabla}\cdot\vec{F}(\tau) = -\frac{1}{\omega}\vec{k}\cdot\frac{d}{d\tau}\vec{F}(\tau)$ g) $\vec{\nabla}\times\vec{F}(\tau) = -\frac{1}{\omega}\vec{k}\times\frac{d}{d\tau}\vec{F}(\tau)$

h) $\nabla^2 f(\tau) = \frac{1}{\omega^2}k^2\frac{d^2}{d\tau^2} f(\tau)$ i) $k\lambda = 2\pi$

14.3 *The Refraction of Light*

As early as the last millennium scholars such as Alhacen (p. 401) established, through experiment and careful observation, the rules by which light reflects and refracts as it passes from air to glass or water.

The change in speed, as light crosses from one dielectric medium to another, is key in our understanding of basic optical phenomena. From this speed ratio we derive Snell's law of refraction. Then we use the concept of retarded time to derive *Fermat's principle*, which states that the time of flight of light between two points is an extremum of the time with respect to any possible path it could take.

Definition 14.2: The Index of Refraction

The *index of refraction* is now defined as the ratio of the wave speed in free space to that in the medium:

$$n \equiv \frac{c}{v}.$$

However, the concept of an index of refraction had already existed and was measurable using Snell's law of refraction, which allowed for a test of Maxwell's theory, or in his own words:

> There are not transparent media for which the magnetic capacity differs from that of air more than by a very small fraction. Hence the principle part of the difference between these media must depend on their dielectric capacity. According to our theory, therefore, the dielectric capacity of a transparent medium should be equal to the square of its index of refraction.

Thus,

$$n = \sqrt{\left(\frac{\varepsilon}{\varepsilon_0}\right)} = \sqrt{K}.$$

Maxwell then continues to compare the index of refraction of melted paraffin to the dielectric constant of the solid version:

> The only dielectric of which the capacity has been hitherto determined with sufficient accuracy in paraffin, for which in the solid form MM. Gibson and Barclay found K=1.975.
>
> Dr. Gladstone has found the following values of the index of refraction of melted paraffin, for the [spectral] lines A, D and H [see table], from which I find the index of refraction for waves of infinite length would be about 1.422.

Temperature	A	D	H
54°C	1.4306	1.4357	1.4499
57°C	1.4294	1.4343	1.4493

> The square root of K is 1.405.

However, there is a catch! The permittivities of real materials depend on the frequency of the oscillating electric field, so this was not the strongest of Maxwell's arguments.[233]

Problem 14.8: Dispersion of Light in Glass

The index of refraction of glass, in the optical range, is about:

$$n \approx 1.5 + \left(\frac{4\,\text{nm}}{\lambda}\right)^2 .$$

a) Write this relationship in terms of the angular frequency ω.

b) By extrapolating this to the DC limit, what would Maxwell's theory predict the relative permittivity, $K = \frac{1}{\varepsilon_0}\varepsilon$, to be?

c) The measured low frequency permittivity of glass is about $4.5\,\varepsilon_0$. How does this compare to what you found?

d) How might you have defended Maxwell's theory given these data?

Thought Experiment 14.6: Snell's Law

Consider light with incident energy flux $\vec{S}_I$ in free space that strikes a transparent dielectric, such as glass, at an angle θ_I from the normal.

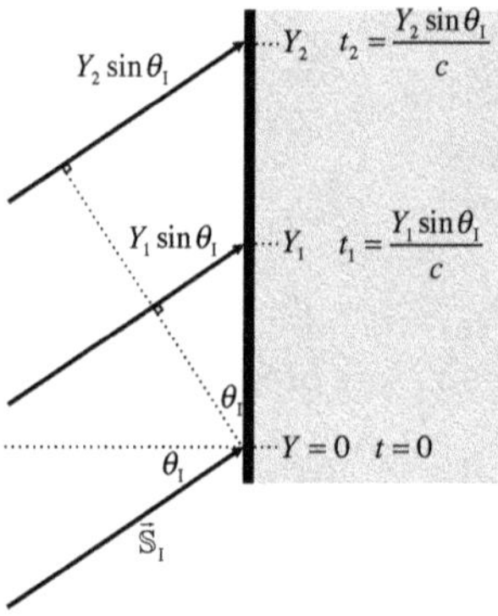

Let zero time be defined when a particular wavefront hits a position $Y = 0$ on the surfaced, as in the diagram. From the geometry, the same wavefront must travel further by a distance $Y \sin\theta$ before it strikes at position Y. Looking at the diagram, we can see that the same wavefront hits the surface at time:

$$t = \tfrac{1}{c} Y \sin(\theta_I).$$

After striking the glass, the light will travel a distance:

$$v\Delta t = vt - Y\sin(\theta_R).$$

where Y was the position where the light hit the glass, and Δt is the time it propagated inside the glass.

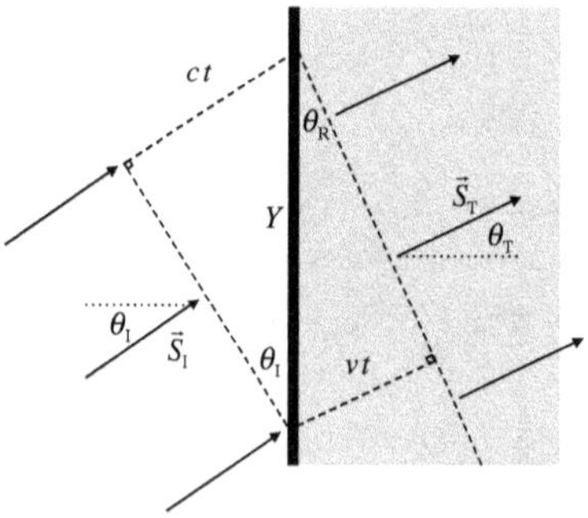

Consulting the diagram, notice that the total travel time of the light is the sum of the time before, and after, striking the glass. Thus, the retarded time τ is:

$$\begin{aligned}\tau &= t - (\text{time in air} + \text{time in glass}) \\ &= t - \left(\tfrac{1}{c} Y \sin(\theta_I) + \tfrac{1}{v}\left(vt - Y\sin(\theta_R)\right)\right) \\ &= Y\left(-\tfrac{1}{c}\sin(\theta_I) + \tfrac{1}{v}\sin(\theta_R)\right).\end{aligned}$$

A given wavefront will have constant retarded time, regardless of where it hits the glass, so:

---

[233] C.F. Bohren, "Maxwell's Good Fortune: The Visible Refractive Index and Static Relative Permittivity of Paraffin," American Journal of Physics, **85**, 165 (2017).

$$\frac{d\tau}{dY} = 0 = \left(-\tfrac{1}{c}\sin(\theta_I) + \tfrac{1}{v}\sin(\theta_R)\right).$$

Therefore:

$$\frac{\sin(\theta_I)}{c} = \frac{\sin(\theta_R)}{v}.$$

Generalizing this to the interface between any two optical media, with wave speeds, $v = c/n$, yields:

$$n_1 \sin\theta_1 = n_2 \sin\theta_2 .$$

This is called Snell's law, after the Dutch astronomer Willebrord Snellius .

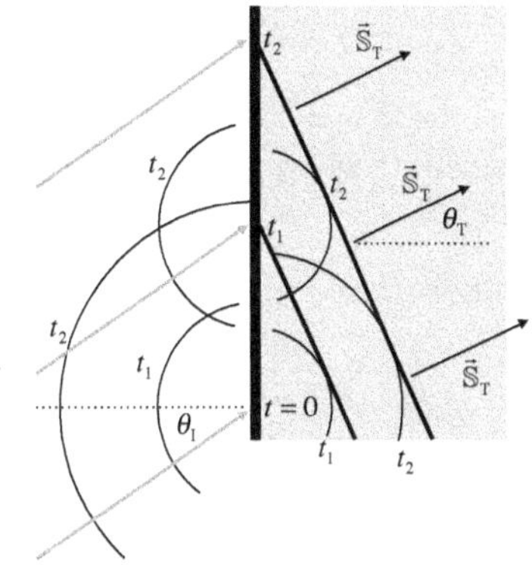

Problem 14.9: Making a Lens

A lens is made by pouring melted glass into a cylindrical container that is placed on a slowly rotating turntable while the glass cools. What is the focal length of the lens in terms of the index of refraction n and the turntable's angular velocity ω?

Problem 14.10: A Rectangular Prism

An isotropic light source is placed at point A and detected at point B, which is located at a position $a\hat{x} + b\hat{y}$ with respect to point A. A rectangular piece of unknown glass, whose thickness is half the horizontal separation, is placed between the source and the detector as shown in the diagram.

a) Show that a ray of light exiting the glass is parallel to the light entering it (i.e. $\theta_3 = \theta_1$).

b) You now slide a thin stick up the side of the prism until it casts a shadow over Point B, and measure its position h_0. Find the relative index of refraction, $n = \frac{n_2}{n_1}$, in terms of these three measured distances using Snell's law.

c) Find the total time that it takes light to travel from point A to point B, in terms of the same three measured lengths.

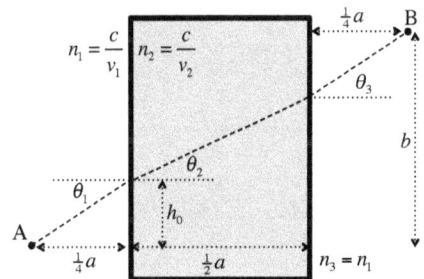

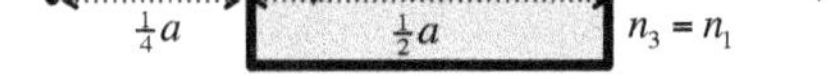

Thought Experiment 14.7: Fermat's Principle

Hero of Alexandria came up with a principle that light always takes the shortest distance between two points, so long as there is no change in propagation media. The purpose of this Thought Experiment is to introduce a similar principle later devised by Pierre de Fermat in the mid-seventeenth century, which states that light always takes the shortest time-path between two points.

A modern interpretation of Fermat's Principle is more general: light must take a path that is a local extremum in the travel time between two points.

We show this principle using the retarded time:

$\tau = t - (\text{travel time}) = t - \Delta t.$

For the wave crests to add, at any given time t, the retarded time τ must be the same. Therefore, as a function of the path taken, the difference $\Delta t = t - \tau$ must be local extremum, and therefore the light travel time must also be an extremum.

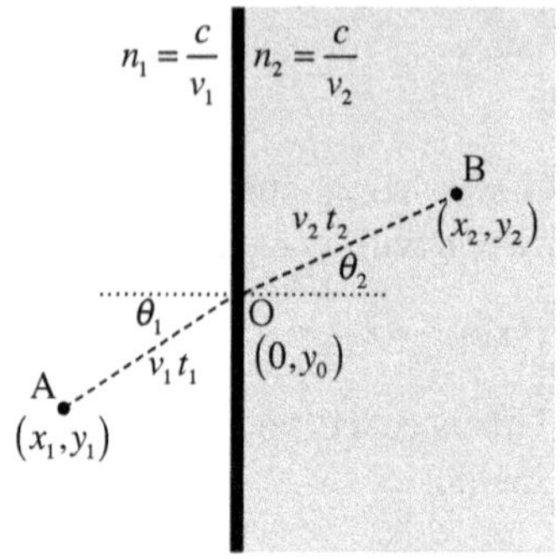

And, formally, the travel time is given in path integral form by:

$$\Delta t = \int_A^B \tfrac{1}{v}\left(\hat{k}\cdot d\vec{\ell}\right) = \int_A^B \tfrac{n}{c}\left(\hat{k}\cdot d\vec{\ell}\right) = \tfrac{1}{c}\int_A^B n\left(\hat{k}\cdot d\vec{\ell}\right).$$

The trick is to find the path where Δt is either a minimum or a maximum.

To illustrate this principle, we first consider a source of light located at point A inside one transparent medium, which is observed at a location B inside a second medium. What path should the light take to get from A to B in the least amount of time?

To quantify this question, we assign the positions of the points A, B and O as in the diagram. Our goal is find y_0, the location of point O. We first find the travel time as:

$$\Delta t = \frac{\sqrt{x_1^2 + (y_0 - y_1)^2}}{v_1} + \frac{\sqrt{x_2^2 + (y_2 - y_0)^2}}{v_2},$$

and then vary y_0 to minimize Δt, by taking the derivative:

$$\frac{d\Delta t}{dy_0} = -\frac{1}{2}\frac{1}{v_1\sqrt{x_1^2 + (y_0 - y_1)^2}}2(y_0 - y_1) + -\frac{1}{2}\frac{1}{v_2\sqrt{x_2^2 + (y_2 - y_0)^2}}2(y_2 - y_0)(-1),$$

and this becomes:

$$\begin{aligned}\frac{d\Delta t}{dy_0} &= -\frac{(y_0 - y_1)}{v_1\sqrt{x_1^2 + (y_0 - y_1)^2}} + \frac{(y_2 - y_0)}{v_2\sqrt{x_2^2 + (y_2 - y_0)^2}} \\ &= -\frac{\sin\theta_1}{v_1} + \frac{\sin\theta_2}{v_2} = -\frac{n_1\sin\theta_1}{c} + \frac{n_2\sin\theta_2}{c}.\end{aligned}$$

Notice that setting this zero simply reproduces Snell's Law:

$n_1 \sin\theta_1 = n_2 \sin\theta_2$.

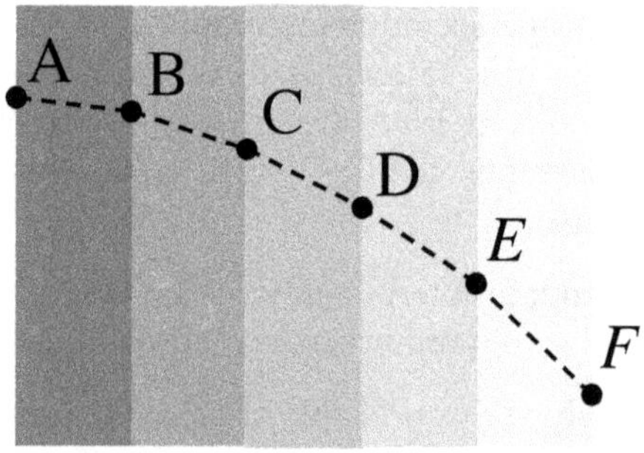

Now consider a series of evenly spaced flat pieces of optical material, and a ray of light traveling from a point on one edge to a point on the other, as in the diagram. Snell's law implies that point B is located along the shortest time path between point A and C, as we have shown. Moreover, point C is located at the shortest time path between points B and D, and likewise for any three points. Therefore, points B, C, D, and E all must lie along the shortest time path between points A and F. Finally, in the limit that the slices become infinitesimally thin, one can generalize Fermat's principle to any distribution of optical media.

Fermat's principle can confuse the thoughtful student for philosophical reasons. Fermat's principle makes it easy to consider light as a smart driver, taking the fast highways and avoiding the slow side streets. But how would a ray of light know beforehand which path to take? Such an interpretation would imbue the light ray with a sort of agency, and that makes no sense. Philosophically, *least action principles,* such as Fermat's principle appear to assign final causes to physical events. Thus statements that a particle "picks out a path of least action," or that the light ray "chooses the path that takes the least time to traverse," have an anthropomorphic air, which most scientists eschew. After all, few biologists believe that natural selection unfolds according to some pre-designed end!

While the way in which Fermat's principle is presented may appear teleological, we have also shown here that Fermat's principle follows directly from the concept of retarded time. No anthropomorphication is necessary to derive or understand Fermat's principle.

Example 14.2: A Spherical Lens

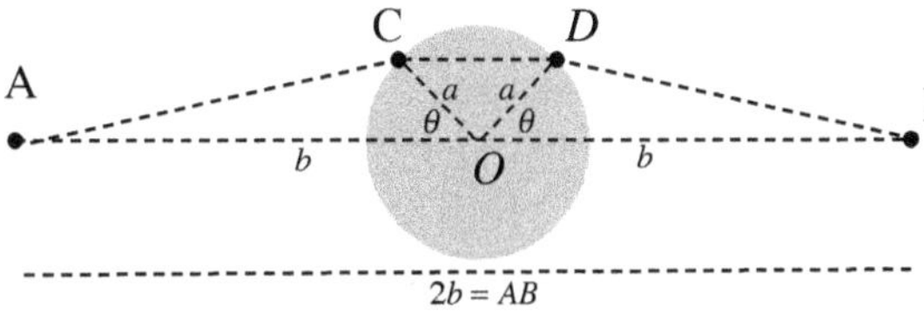

Consider a point source at a location A and a detector located at a position B, as in the diagram, with a spherical lens, of refractive index n, exactly in between. The radius of lens is a and the distance between the source and the middle of the lens is b. What is the angle θ of the point where the light that will end up at point B strikes the sphere?

We will use Fermat's principle to solve this problem, so we will find the total time of flight for the light from A to B, via points C and D.

Using the law of cosines, and basic trigonometry, we can find that:

$$AC = \sqrt{a^2 + b^2 - 2ab\cos\theta},$$

$$CD = 2a\cos\theta,$$

$$DB = \sqrt{a^2 + b^2 - 2ab\cos\theta}.$$

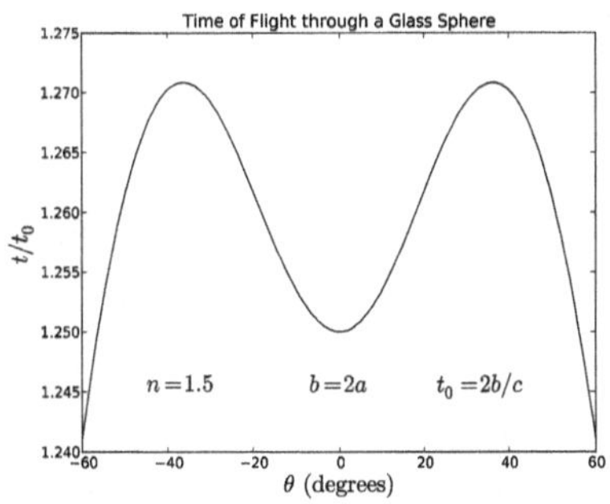

So the total time of flight is given by:

$$\Delta t = \frac{2}{c}\sqrt{a^2 + b^2 - 2ab\cos\theta} + \frac{n}{c}2a\cos\theta.$$

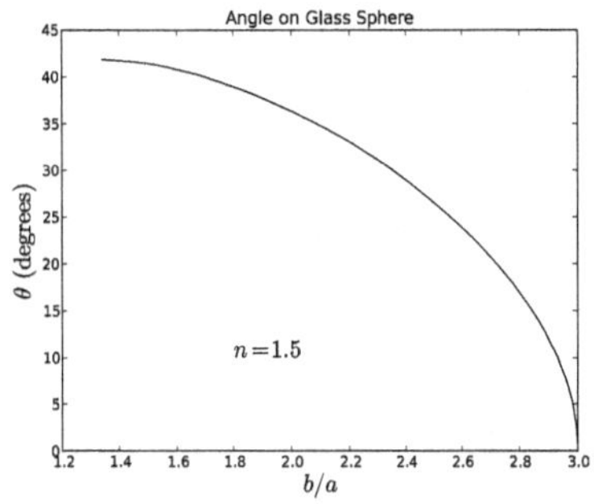

The upper plot shows the travel time as a function of the angle θ. Notice that it is a minimum when $\theta = 0$, as expected. However, with these particular input parameters it is also a maximum at an angle of slightly less than 40°.

Next we will find the extrema taking the derivative:

$$\frac{d\Delta t}{d\theta} = \frac{2}{c}\frac{ab\sin\theta}{\sqrt{a^2 + b^2 - 2ab\cos\theta}} - \frac{n}{c}2a\sin\theta.$$

Next we factor and set it to zero:

$$\left(\frac{2a}{c}\right)(\sin\theta)\left(\frac{b}{\sqrt{a^2+b^2-2ab\cos\theta}}-n\right)=0$$

One extremum must take place when $\theta=0$, and the other when:

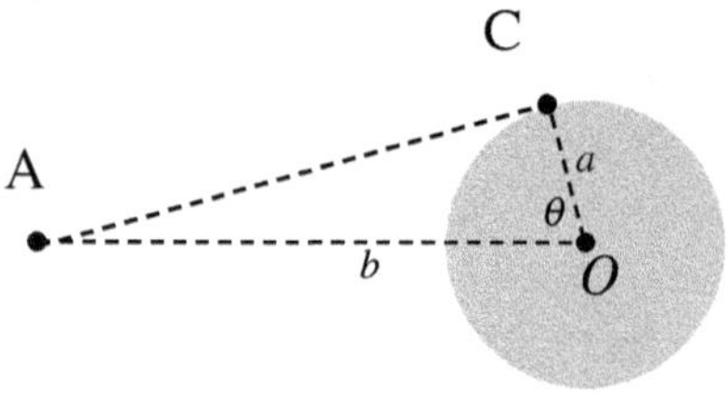

$$\theta=\arccos\left(\frac{a}{2b}+\frac{b}{2a}\left(1-\frac{1}{n^2}\right)\right).$$

This angle is plotted as a function of b/a for a unit glass sphere.

Reflecting on this solution, notice that this is simply a function of the ratio of the two lengths, which makes sense given that we are solving for an angle.

Problem 14.11: The Spherical Lens

a) Show that Example 14.2 above only gives a nontrivial solution if:

$$\frac{1}{\sqrt{1-\left(\frac{a}{b}\right)^2}}\le n\le\frac{1}{1-\frac{a}{b}},$$

b) What is this range if $b=2a$?

c) What is this range if $b=3a$?

d) What is the acceptable range of values for b/a, when $n=1.5$?

Problem 14.12: A Rectangular Prism Revisited

Consider Problem 14.10 (p. 447). Imagine, hypothetically, that every ray of light emitted from source A actually focused on source B. Show that the ray that takes the shortest time for go from point A to point B is the actual ray taken.

Problem 14.13: An Exponential Atmosphere

The index of refraction of Earth's atmosphere can be written as a function of elevation z above sea level by the Gladstone-Dale law as: $n=1+(n_0-1)e^{-z/h_0}$, where $h_0\approx 8\,\text{km}$ is the density scale height of the atmosphere and $n_0\approx 1.0002718$ is the red (633 nm) refractive index at standard temperature (20 C) and pressure (1 atm).

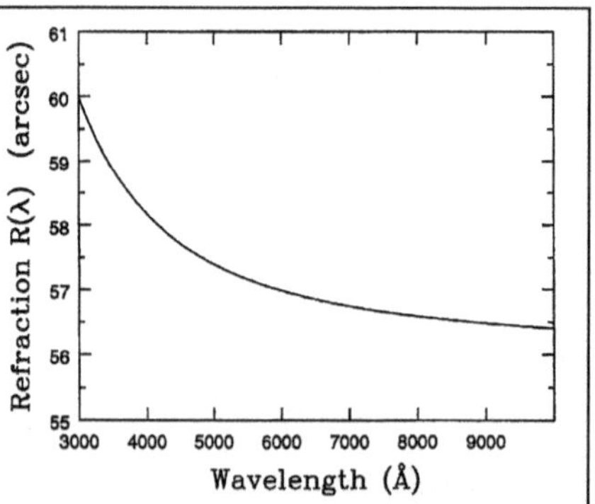

FIG. 2—Atmospheric refraction as a function of wavelength is shown over the wavelength interval λλ3000–10,000 Å, at a zenith distance of 45°, and ambient conditions characterized by temperature 15 °C, atmospheric pressure 760 mm, and no water-vapor pressure. Refraction is stronger at shorter wavelengths.

Imagine that a very powerful He-Ne laser (633 nm) is pointed at an angle θ_0 from directly overhead, called the *zenith angle*.

a) Write a computer program, or use a spreadsheet, to model the atmosphere as a series of horizontal slabs each with thickness δz. Calculate, and plot, the change in zenith angle as a function of height above sea level using Snell's law at each slab, so that:

$$\theta(z+\delta z) = \arcsin\left(\frac{n(z)}{n(z+\delta z)}\sin(\theta(z))\right).$$

b) The measured refraction by the total atmosphere, at a 45° zenith angle, is shown in the figure[234] as a function of wavelength. Does your model give similar results?

c) What is the total refraction of the atmosphere when observing something on the horizon, such as the sunset over the ocean. By how long could this shift the expected time of sunrise or sunset?

d) Based on your model, find the angular diameter of the moon when it is viewed near the horizon. (Its zenith angular diameter is 30 arcminutes.)

e) The moon looks much bigger when close to the horizon than it does after it rises. Ptolemy argued that this is due to refraction, but Alhacen argued that it is an optical illusion (see p. 401). Who was right? Is the moon illusion better explained using physics or psychology?

14.4 *Fresnel's Theory of Light*

Recall (pp. 403, 426) that Augustin-Jean Fresnel successfully explained the results of optical polarization experiments by proposing that light waves are transverse like the oscillations of a plucked string. His work, then, led to a set of equations governing the relative amount of reflection and transmission of light at an optical surface.

In this section, we briefly introduce linearly polarized light, followed by a discussion of the Fresnel equations for light at a dielectric surface. This allows us to find the reflection and transmission coefficients, and to find the angle of reflection that produces linearly polarized light.

Thought Experiment 14.8: Linearly Polarized Light

Consider a monochromatic plane wave travelling in the $\hat{z}$ direction. Using Cartesian coordinates, and choosing our point of interest to lie along the z axis, we can write the electric field as:

$$\vec{E}(z,t) = E_x\cos(\omega\tau)\hat{x} + E_y\cos(\omega\tau - \vartheta)\hat{y},$$

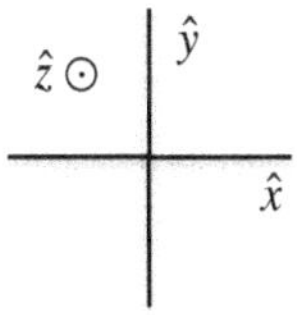

where τ represents the *retarded time*, $\tau = t - \frac{1}{c}z$. In the plane wave approximation the zero position does not need to be at the location of the source, but rather can be anywhere the wave has been or will be in the future.

If the two components of the electric field are in phase, so that $\vartheta = 0$, the electric field is:

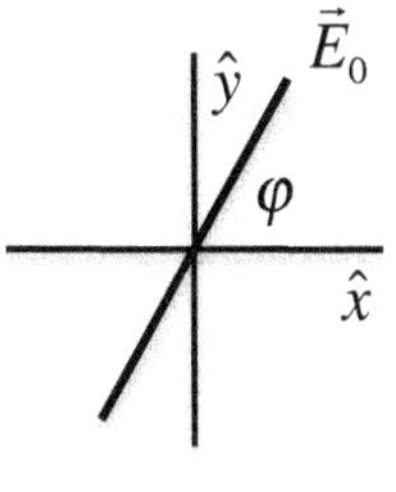

$$\vec{E}(z,t) = E_x\cos(\omega\tau)\hat{x} + E_y\cos(\omega\tau)\hat{y}.$$

This can be expressed more simply as $\vec{E}(z,t) = \vec{E}_0\cos(\omega\tau)$,

where $\vec{E}_0 = E_x\hat{x} + E_y\hat{y}$. Thus the electric field vector has amplitude

[234] Stone, R.C., "An Accurate Method for Computing Atmospheric Refraction," Publications of the Astronomical Society of the Pacific, **108** (1996), 1051-1058.

$\sqrt{E_x^2 + E_y^2}$ and the angle that $\vec{E}_0$ makes with respect to the $\hat{x}$ axis is given by:

$$\varphi = \arctan\left(\frac{E_y}{E_x}\right).$$

Notice that there is a redundancy in the *polarization angle* φ, because a difference of π in the angle can be absorbed into an overall phase shift also of π. That is, since $\sin(\theta + \pi) = -\sin\theta$, and $\cos(\theta + \pi) = -\cos\theta$, we have that $\tan(\theta + \pi) = \tan\theta$. We therefore usually restrict the range of the polarization angle to values such that $0 \le \varphi \le \pi$.

Thought Experiment 14.9: Boundary Conditions at an Optical Surface

The surface between two transparent media is important, as Snell's law clearly shows. Here we will investigate this more by applying each of Maxwell's equations, in turn, to the surface between two media of permittivities: $\varepsilon_1 = n_1^2 \varepsilon_0$ and $\varepsilon_2 = n_2^2 \varepsilon_0$.

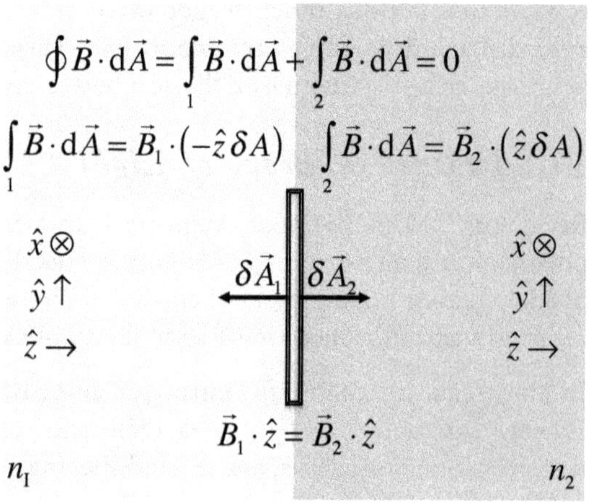

The four Maxwell's equations, in a linear non-conducting medium, are:

$$\vec{\nabla} \cdot \vec{B} = 0 \qquad \vec{\nabla} \cdot \left(\varepsilon \vec{E}\right) \approx 0$$

$$\vec{\nabla} \times \vec{E} = -\tfrac{\partial}{\partial t}\vec{B} \quad \vec{\nabla} \times \left(\tfrac{1}{\mu}\vec{B}\right) \approx \tfrac{\partial}{\partial t}\left(\varepsilon \vec{E}\right).$$

We will apply each equation, in turn, to a planar boundary, starting with:

$$\oint \vec{B} \cdot \mathrm{d}\vec{A} = 0.$$

$$\oint \vec{B} \cdot \mathrm{d}\vec{A} = \int_1 \vec{B} \cdot \mathrm{d}\vec{A} + \int_2 \vec{B} \cdot \mathrm{d}\vec{A} = 0$$

$$\int_1 \vec{B} \cdot \mathrm{d}\vec{A} = \vec{B}_1 \cdot (-\hat{z}\delta A) \qquad \int_2 \vec{B} \cdot \mathrm{d}\vec{A} = \vec{B}_2 \cdot (\hat{z}\delta A)$$

$\hat{x} \otimes$ $\hat{x} \otimes$

$\hat{y} \uparrow$ $\delta\vec{A}_1 \ \delta\vec{A}_2$ $\hat{y} \uparrow$

$\hat{z} \rightarrow$ $\hat{z} \rightarrow$

$$\vec{B}_1 \cdot \hat{z} = \vec{B}_2 \cdot \hat{z}$$

n_1 n_2

If we draw a thin Gaussian surface on the boundary, as in the diagram, then the component normal to the surface must be continuous, or:

$$\vec{B}_1^{\perp} = \vec{B}_2^{\perp}.$$

Now we will apply the curl of the magnetizing field to the surface by drawing a long/thin loop at the boundary. In the limit that the boundary is infinitesimally thin, then the area through the loop goes to zero:

$$\oint_{\text{path}} \tfrac{1}{\mu}\vec{B} \cdot \mathrm{d}\vec{\ell} = \int_{\text{surface}} \left(\varepsilon \tfrac{\partial}{\partial t}\vec{E}\right) \cdot \mathrm{d}\vec{A} = 0.$$

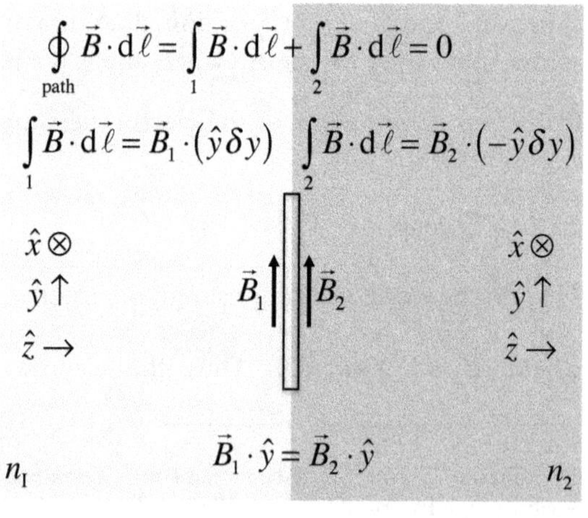

Thus the two parallel components of the magnetic field must be equal:

$$\vec{B}_1^{\|} \approx \vec{B}_2^{\|}.$$

Next we consider the boundary conditions for

$$\oint_{\text{path}} \vec{B} \cdot \mathrm{d}\vec{\ell} = \int_1 \vec{B} \cdot \mathrm{d}\vec{\ell} + \int_2 \vec{B} \cdot \mathrm{d}\vec{\ell} = 0$$

$$\int_1 \vec{B} \cdot \mathrm{d}\vec{\ell} = \vec{B}_1 \cdot (\hat{y}\delta y) \quad \int_2 \vec{B} \cdot \mathrm{d}\vec{\ell} = \vec{B}_2 \cdot (-\hat{y}\delta y)$$

$\hat{x} \otimes$ $\hat{x} \otimes$

$\hat{y} \uparrow$ $\vec{B}_1 \ \ \vec{B}_2$ $\hat{y} \uparrow$

$\hat{z} \rightarrow$ $\hat{z} \rightarrow$

$$\vec{B}_1 \cdot \hat{y} = \vec{B}_2 \cdot \hat{y}$$

n_1 n_2

the electric field:

$$\oint_{\text{surface}} \varepsilon \vec{E}\cdot \mathrm{d}\vec{A} = 0.$$

Setting $\varepsilon = n^2\varepsilon_0$, we can integrate around the surface:

$$\varepsilon_0 \int_1 n_1{}^2 \vec{E}\cdot \mathrm{d}\vec{A} + \varepsilon_0 \int_2 n_2{}^2 \vec{E}\cdot \mathrm{d}\vec{A} = 0.$$

Finally we can equate the perpendicular components as:

$$n_1^2 \vec{E}_1^{\perp} \approx n_2^2 \vec{E}_2^{\perp}.$$

Lastly we apply Faraday's law to find the parallel components of the electric field.

$$\oint_{\text{path}} \vec{E}\cdot \mathrm{d}\vec{\ell} = \int_{\text{surface}} \tfrac{\partial}{\partial t}\vec{B}\cdot \mathrm{d}\vec{A}.$$

Again, as the width of the loop approaches zero, the area approaches zero, so the parallel electric field condition becomes:

$$\vec{E}_1^{\parallel} \approx \vec{E}_2^{\parallel}.$$

These boundary conditions are quite powerful in the study of optics, as we will see in the next few examples.

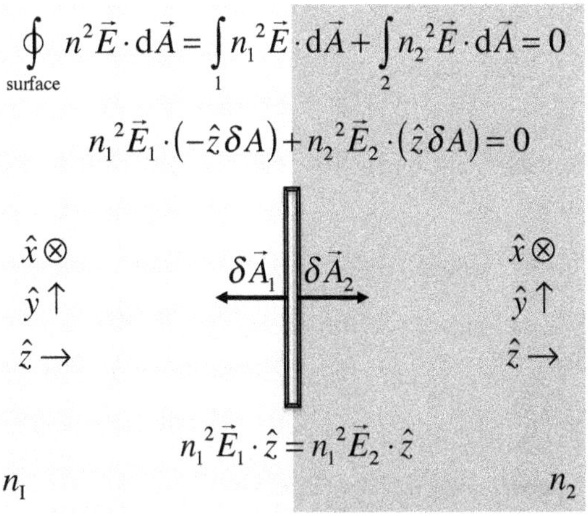

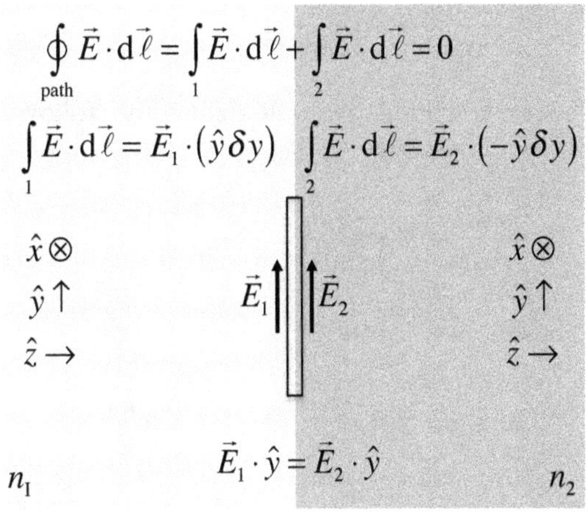

Thought Experiment 14.10: The Fresnel Equations

When light hits a boundary, it may be reflected, transmitted, or both. What fraction of the energy continues in each of these direction?

Consider now a large flat interface between two dielectric boundaries $\varepsilon_1, \varepsilon_2 = n_1^2\varepsilon_0, n_2^2\varepsilon_0$. The two rays of constructive interference that can be emitted at a flat interface are the reflected ray and the refracted, or transmitted, ray. The laws of reflection and refraction hold between the angles of incidence and the angle of reflection, as reviewed in the diagram.

First we apply the conservation of energy:

$$\oint_{\text{surface}} \vec{\mathbb{S}}\cdot \mathrm{d}\vec{A} = -\oint_{\text{volume}} \frac{\mathrm{d}\mathbb{u}}{\mathrm{d}t}\cdot \mathrm{d}V.$$

If we draw a very thin Gaussian surface, and we assume that the energy inside the surface is constant, we can write this as:

$$\vec{\mathbb{S}}_{\mathrm{I}}\cdot\delta\vec{A}_1 + \vec{\mathbb{S}}_{\mathrm{R}}\cdot\delta\vec{A}_1 + \vec{\mathbb{S}}_{\mathrm{T}}\cdot\delta\vec{A}_2 = 0 \;\rightarrow\; \vec{\mathbb{S}}_{\mathrm{I}}\cdot(-\hat{z}) + \vec{\mathbb{S}}_{\mathrm{R}}\cdot(-\hat{z}) + \vec{\mathbb{S}}_{\mathrm{T}}\cdot\hat{z} = 0$$

$$\vec{\mathbb{S}}_{\mathrm{I}}\cdot\hat{z} = \vec{\mathbb{S}}_{\mathrm{R}}\cdot(-\hat{z}) + \vec{\mathbb{S}}_{\mathrm{T}}\cdot\hat{z} \;\rightarrow\; \left|\vec{\mathbb{S}}_{\mathrm{I}}\cdot\hat{z}\right| = \left|\vec{\mathbb{S}}_{\mathrm{R}}\cdot\hat{z}\right| + \left|\vec{\mathbb{S}}_{\mathrm{T}}\cdot\hat{z}\right|.$$

Next we define the reflection and transmission coefficients as:

$$R \equiv \frac{\left|\vec{\mathbb{S}}_R \cdot \hat{z}\right|}{\left|\vec{\mathbb{S}}_I \cdot \hat{z}\right|} \quad T \equiv \frac{\left|\vec{\mathbb{S}}_T \cdot \hat{z}\right|}{\left|\vec{\mathbb{S}}_I \cdot \hat{z}\right|} .$$

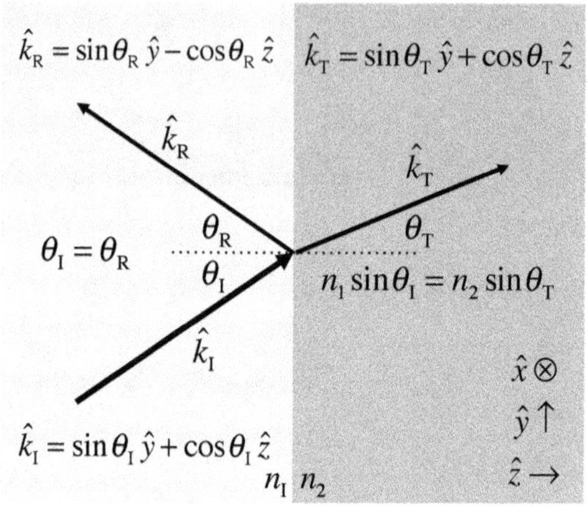

So we can write the conservation of energy simply as:

$$R + T = 1 .$$

We can also write the energy continuity equation as:

$$\mathbb{S}_I\left(\hat{k}_I \cdot \hat{z}\right) = \mathbb{S}_R \cdot \left(-\hat{k}_R \cdot \hat{z}\right) + \mathbb{S}_T \cdot \left(\hat{k}_T \cdot \hat{z}\right)$$

$$\mathbb{S}_I \cos\theta_I = \mathbb{S}_R \cos\theta_R + \mathbb{S}_T \cos\theta_T$$

$$\mathbb{S}_I = \mathbb{S}_R \frac{\cos\theta_R}{\cos\theta_I} + \mathbb{S}_T \frac{\cos\theta_T}{\cos\theta_I} .$$

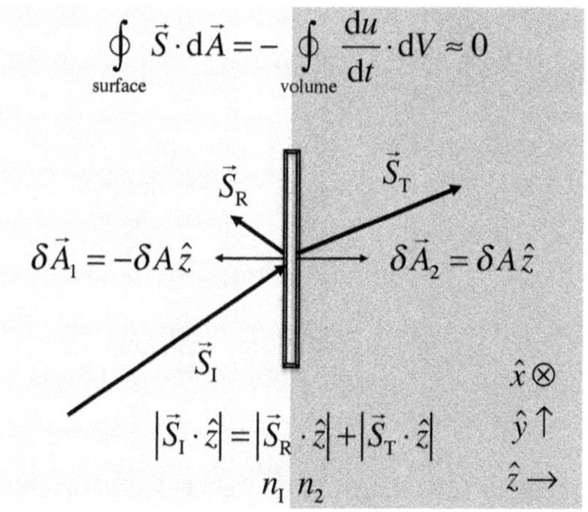

To save writing, we will define the following known constants:

$$\alpha \equiv \frac{\cos\theta_T}{\cos\theta_I} \quad \beta \equiv \frac{\sin\theta_I}{\sin\theta_T} = \frac{n_2}{n_1} .$$

Thus in terms of these:

$$\mathbb{S}_I = \mathbb{S}_R + \alpha\mathbb{S}_T , \quad R = \frac{\mathbb{S}_R}{\mathbb{S}_I} , \text{ and } \quad T = \alpha\frac{\mathbb{S}_T}{\mathbb{S}_I} .$$

Our goal in this example is to find R and T in terms of the angle of incidence and the indices of refraction. In order to do this, we will need to apply the boundary conditions from Thought Experiment 14.9. These are summarized as follows:

$$n_1^2 \vec{E}_1^{\perp} = n_2^2 \vec{E}_2^{\perp} \quad \vec{E}_1^{\parallel} = \vec{E}_2^{\parallel} \quad \vec{B}_1 = \vec{B}_2$$

So the electric field components are related by the following matrix equation:

$$\vec{E}_I + \vec{E}_R = \begin{pmatrix} 1 & 0 & 0 \\ 0 & 1 & 0 \\ 0 & 0 & \left(n_2/n_1\right)^2 \end{pmatrix} \vec{E}_T = \begin{pmatrix} 1 & 0 & 0 \\ 0 & 1 & 0 \\ 0 & 0 & \beta^2 \end{pmatrix} \vec{E}_T .$$

And the magnetic field components are:

$$\vec{B}_I + \vec{B}_R = \vec{B}_T .$$

For illustration, we will look at the simplest possible case—that of normal incidence, where $\alpha = 1$, $\hat{k} = \hat{z}$, and the $\hat{x}$ and $\hat{y}$ directions are in the directions of the electric and magnetic fields respectively.

First we write the magnetic field in terms of the electric field as:

$$\vec{B} = \frac{1}{v}\hat{k}\times\vec{E} = \frac{n}{c}\hat{k}\times\vec{E} = \frac{n}{c}E\left(\hat{k}\times\hat{x}\right),,$$

where $\hat{k} = \hat{z}$ for the incident and transmitted wave, but $\hat{k} = -\hat{z}$ for the reflected wave. Thus the boundary conditions imply:

$$E_{\mathrm{I}}\cancel{\hat{x}} + E_{\mathrm{R}}\cancel{\hat{x}} = E_{\mathrm{T}}\cancel{\hat{x}}$$

$$\frac{n_1}{\cancel{c}}E_{\mathrm{I}}\cancel{\hat{y}} - \frac{n_1}{\cancel{c}}E_{\mathrm{R}}\cancel{\hat{y}} = \frac{n_2}{\cancel{c}}E_{\mathrm{T}}\cancel{\hat{y}}.$$

Writing this in terms of $\beta = {}^{n_2}\!/_{n_1}$ this becomes:

$$E_{\mathrm{I}} + E_{\mathrm{R}} = E_{\mathrm{T}} \quad \text{and} \quad E_{\mathrm{I}} - E_{\mathrm{R}} = \beta E_{\mathrm{T}}.$$

We can now solve for each electric field as:

$$E_{\mathrm{T}} = \frac{2}{1+\beta}E_{\mathrm{I}} \quad \text{and} \quad E_{\mathrm{R}} = \frac{1-\beta}{1+\beta}E_{\mathrm{I}}.$$

Notice that if $n_2 > n_1$, the reflected wave will have a phase shift of 180°.

The Poynting vector is given by:

$$\vec{\mathbb{S}} = \frac{1}{\mu_0}\vec{E}\times\vec{B} = \frac{n}{\mu_0 c}E^2\hat{k},$$

so the energy continuity equation can be written as:

$$\mathbb{S}_{\mathrm{I}} = \mathbb{S}_{\mathrm{R}} + \mathbb{S}_{\mathrm{T}}$$

$$\frac{n_1}{\mu_0 c}E_{\mathrm{I}}^2 = \frac{n_1}{\mu_0 c}E_{\mathrm{R}}^2 + \frac{n_2}{\mu_0 c}E_{\mathrm{T}}^2$$

$$1 = \left(\frac{E_{\mathrm{R}}}{E_{\mathrm{I}}}\right)^2 + \beta\left(\frac{E_{\mathrm{T}}}{E_{\mathrm{I}}}\right)^2.$$

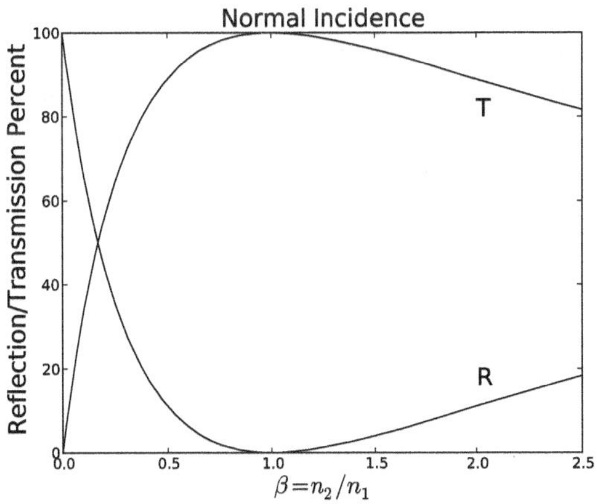

The reflection and transmission coefficients at normal incidence are therefore simply given by:

$$R = \left(\frac{1-\beta}{1+\beta}\right)^2$$

$$T = \beta\left(\frac{2}{1+\beta}\right)^2.$$

Now we will return to the more general case, where there are two polarizations to consider: that where the electric field is perpendicular to the plane of incidence ($\vec{E} = E\hat{x}$), and that where the magnetic field is perpendicular to the plane of incidence ($\vec{B} = B\hat{x}$). These are called *s* and *p* respectively, because of the German words for perpendicular (*senkrecht*) and parallel (*parallel*).

Consider s-polarization first, the case where $\vec{E} = E\hat{x}$ for each wave, so the magnetic field is given by:

$$\vec{B} = \frac{1}{v}\hat{k}\times\vec{E} = \frac{n}{c}\hat{k}\times\vec{E} = \frac{n}{c}\left(\cancel{\left(k_y E_z - k_z E_y\right)\hat{x}} + \hat{k}_z E_x \hat{y} - \hat{k}_y E_x \hat{z}\right).$$

Referring to the top figure, this implies that:

$\vec{B}_\mathrm{I} = \frac{n_1}{c}\left(\cos\theta_\mathrm{I}\,\hat{y} - \sin\theta_\mathrm{I}\,\hat{z}\right)E_\mathrm{I}$, $\vec{B}_\mathrm{R} = \frac{n_1}{c}\left(-\cos\theta_\mathrm{R}\,\hat{y} - \sin\theta_\mathrm{R}\,\hat{z}\right)E_\mathrm{R}$, and

$\vec{B}_\mathrm{T} = \frac{n_2}{c}\left(\cos\theta_\mathrm{T}\,\hat{y} - \sin\theta_\mathrm{T}\,\hat{z}\right)E_\mathrm{T}$.

Now we can apply the boundary conditions to the interface:

$E_\mathrm{I} + E_\mathrm{R} = E_\mathrm{T}$ and $\vec{B}_\mathrm{I} + \vec{B}_\mathrm{R} = \vec{B}_\mathrm{T}$.

The $\hat{y}$ component of the magnetic field, the boundary condition becomes:

$$\frac{n_1}{c}\left(\cos\theta_\mathrm{I}\,\hat{y}\right)E_\mathrm{I} + \frac{n_1}{c}\left(-\cos\theta_\mathrm{R}\,\hat{y}\right)E_\mathrm{R} = \frac{n_2}{c}\left(\cos\theta_\mathrm{T}\,\hat{y}\right)E_\mathrm{T}$$

$$E_\mathrm{I}\cos\theta_\mathrm{I} - E_\mathrm{R}\cos\theta_\mathrm{R} = \frac{n_2}{n_1}E_\mathrm{T}\cos\theta_\mathrm{T} \quad\rightarrow\quad E_\mathrm{I} - E_\mathrm{R} = \beta\alpha\,E_\mathrm{T}.$$

So, the electric fields, in terms of α and β (p. 454), become:

$$E_\mathrm{T} = \frac{2}{1+\beta\alpha}E_\mathrm{I} \quad\text{and}\quad E_\mathrm{R} = \frac{1-\beta\alpha}{1+\beta\alpha}E_\mathrm{I}.$$

And the transmission and reflection coefficients then are:

$$T_s = \alpha\left(\frac{2}{1+\beta\alpha}\right)^2 \quad\text{and}\quad R_s = \left(\frac{1-\beta\alpha}{1+\beta\alpha}\right)^2.$$

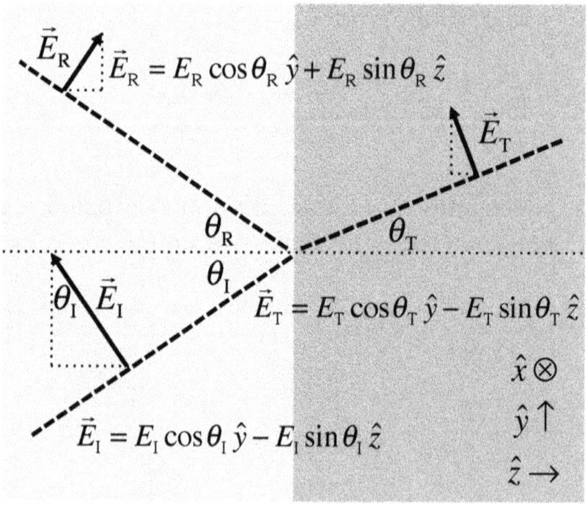

Now we look at the opposite special case, p-polarization, where the electric field has no $\hat{x}$ component, where the following relations hold:

$$\vec{E}_\mathrm{I} = E_\mathrm{I}\cos\theta_\mathrm{I}\,\hat{y} - E_\mathrm{I}\sin\theta_\mathrm{I}\,\hat{z}$$
$$\vec{E}_\mathrm{R} = E_\mathrm{R}\cos\theta_\mathrm{R}\,\hat{y} + E_\mathrm{R}\sin\theta_\mathrm{R}\,\hat{z}$$
$$\vec{E}_\mathrm{T} = E_\mathrm{T}\cos\theta_\mathrm{T}\,\hat{y} - E_\mathrm{T}\sin\theta_\mathrm{T}\,\hat{z}$$

Now the electric field boundary conditions are:

$$E_\mathrm{I}\cos\theta_\mathrm{I} + E_\mathrm{R}\cos\theta_\mathrm{R} = E_\mathrm{T}\cos\theta_\mathrm{T}$$
$$-E_\mathrm{I}\sin\theta_\mathrm{I} + E_\mathrm{R}\sin\theta_\mathrm{R} = -\beta^2 E_\mathrm{T}\sin\theta_\mathrm{T}.$$

And in terms of α and β (p. 454):

$E_\mathrm{I} + E_\mathrm{R} = \alpha E_\mathrm{T}$ and $-E_\mathrm{I} + E_\mathrm{R} = -\beta E_\mathrm{T}$.

Solving for the reflected and transmitted electric fields, we have:

$$E_\mathrm{T} = \left(\frac{2}{\alpha+\beta}\right)E_\mathrm{I} \quad\text{and}\quad E_\mathrm{R} = \left(\frac{\alpha-\beta}{\alpha+\beta}\right)E_\mathrm{I}.$$

Now we will calculate the magnitude of each Poynting vector:

$$\mathbb{S}_{\mathrm{I}} = \frac{1}{\mu_0}\vec{E}_{\mathrm{I}} \times \vec{B}_{\mathrm{I}} = \frac{1}{\mu_0}E_{\mathrm{I}}\,B_{\mathrm{I}} = \frac{E_{\mathrm{I}}^2}{\mu_0 v_1} = \frac{n_1}{Z_0}E_{\mathrm{I}}^2$$

$$\mathbb{S}_{\mathrm{R}} = \frac{1}{\mu_0}\vec{E}_{\mathrm{R}} \times \vec{B}_{\mathrm{R}} = \frac{1}{\mu_0}E_{\mathrm{R}}\,B_{\mathrm{R}} = \frac{E_{\mathrm{R}}^2}{\mu_0 v_1} = \frac{n_1}{Z_0}E_{\mathrm{R}}^2$$

$$\mathbb{S}_{\mathrm{T}} = \frac{1}{\mu_0}\vec{E}_{\mathrm{T}} \times \vec{B}_{\mathrm{T}} = \frac{1}{\mu_0}E_{\mathrm{T}}\,B_{\mathrm{T}} = \frac{E_{\mathrm{T}}^2}{\mu_0 v_1} = \frac{n_2}{Z_0}E_{\mathrm{T}}^2$$

And, in terms of α and β (p. 454), we have:

$$\mathbb{S}_{\mathrm{R}} = \left(\frac{\alpha-\beta}{\alpha+\beta}\right)^2 S_{\mathrm{I}} \quad \text{and} \quad \mathbb{S}_{\mathrm{T}} = \frac{n_2}{n_1}\left(\frac{2}{\alpha+\beta}\right)^2 \mathbb{S}_{\mathrm{I}} = \beta\left(\frac{2}{\alpha+\beta}\right)^2 \mathbb{S}_{\mathrm{I}} \ .$$

Again, we will check energy conservation with the continuity equation:

$$\mathbb{S}_{\mathrm{I}} = \mathbb{S}_{\mathrm{R}} + \alpha\,\mathbb{S}_{\mathrm{T}}$$

$$\mathbb{S}_{\mathrm{I}} = \left(\frac{\alpha-\beta}{\alpha+\beta}\right)^2 \mathbb{S}_{\mathrm{I}} + \alpha\beta\left(\frac{2}{\alpha+\beta}\right)^2 \mathbb{S}_{\mathrm{I}}$$

$$1 = \left(\frac{\alpha-\beta}{\alpha+\beta}\right)^2 + \alpha\beta\left(\frac{2}{\alpha+\beta}\right)^2$$

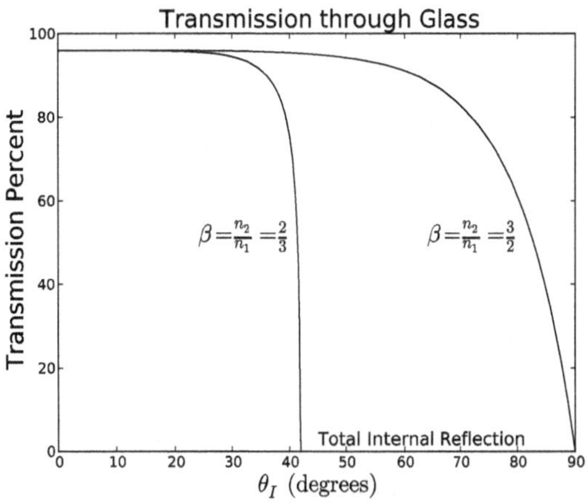

And we can write down the transmission and reflection coefficients as:

$$R_p = \left(\frac{\alpha-\beta}{\alpha+\beta}\right)^2$$

$$T_p = \alpha\beta\left(\frac{2}{\alpha+\beta}\right)^2$$

And finally, we can verify that $R_p + T_p = 1$ algebraically:

$$R_p + T_p = \left(\frac{\alpha-\beta}{\alpha+\beta}\right)^2 + \alpha\beta\left(\frac{2}{\alpha+\beta}\right)^2 = \frac{\alpha^2 - 2\alpha\beta + \beta^2 + 4\alpha\beta}{(\alpha+\beta)^2} = 1$$

Now let us consider unpolarized light. In this case, half of the incident light will be *s* polarized and the other half will be *p* polarized. Therefore, the reflection and transmission coefficients will be the average of the *s* and *p* values. The transmission of a glass-vacuum interface is plotted in the figure. Notice that when $\theta_{\mathrm{I}} > \arcsin\beta$ no light is transmitted from a more optically dense medium to a less optically dense medium. This is called *total internal reflection*, and it is what keeps light inside optical fibers.

Thought Experiment 14.11: Brewster's Angle

Consider unpolarized light incident on a surface at a normal angle θ_{I}, with an index of refraction ratio $\beta = {}^{n_2}\!/\!_{n_1}$. In Thought Experiment 14.10, we found that the reflection and transmission coefficients are not only functions of the angle of incidence and the indices of refraction, but are also a function of the polarization of the incident light. Thus, if unpolarized light lands on a dielectric interface, half of the light will be governed by one Fresnel equation and half by the other. Therefore, the reflected and transmitted light will become somewhat polarized.

The goal of this example is to calculate the polarization of the reflected light as a function of the angle of incidence and the index of refraction ratio.

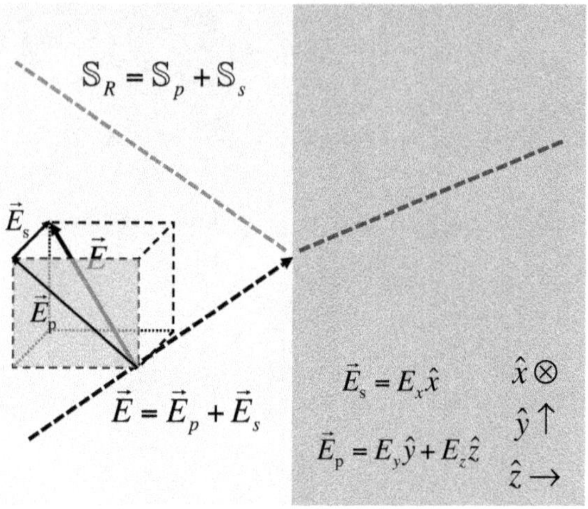

Recall the German names of the two types of polarization: *parallel* and *senkrecht*. Since our goal is to find the polarization of the reflected light, we define $\mathbb{S}_p$ and $\mathbb{S}_s$ to be the magnitudes of the reflected Poynting vectors where the electric field is parallel (*parallel*), or perpendicular (*senkrecht*), to the plane of the light rays, respectively.

In terms of α and β (p. 454), these components of the reflected wave are:

$$\mathbb{S}_s = \left(\frac{1-\beta\alpha}{1+\beta\alpha}\right)^2 \mathbb{S}_I \text{ and } \mathbb{S}_p = \left(\frac{\alpha-\beta}{\alpha+\beta}\right)^2 \mathbb{S}_I .$$

We will define the fractional polarization of the reflected light as:

$$Pol = \frac{|\mathbb{S}_s - \mathbb{S}_p|}{\mathbb{S}_R} = \frac{|\mathbb{S}_s - \mathbb{S}_p|}{\mathbb{S}_s + \mathbb{S}_p} = \frac{\left|\left(\frac{1-\beta\alpha}{1+\beta\alpha}\right)^2 - \left(\frac{\alpha-\beta}{\alpha+\beta}\right)^2\right|}{\left(\frac{1-\beta\alpha}{1+\beta\alpha}\right)^2 + \left(\frac{\alpha-\beta}{\alpha+\beta}\right)^2}$$

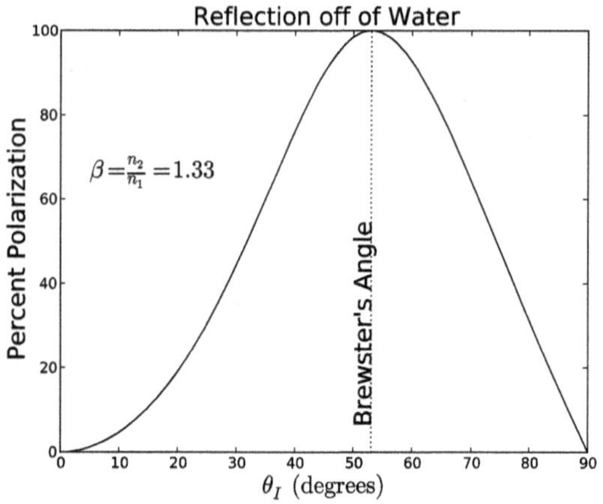

Now consider sunlight reflecting off water, and the advantage of polarized sunglasses especially when boating. The plot shows the percent polarization as a function of the incidence angle.

Notice that when $\tan\theta_I = {}^{n_2}\!/_{n_1}$, the reflected light is 100% polarized. This is called Brewster's angle, as shown in the figure for the air-water interface.

Problem 14.14: Brewster's Experiment

Shortly after Malus observed Luxembourg Palace at sunset, and published his findings, the Scottish physicist Sir David Brewster discovered that complete polarization of partially reflected light, at the interface of a transparent medium, occurs when the reflected light ray is perpendicular to the refracted ray. Show that the reflected and transmitted rays are perpendicular at Brewster's angle.

Problem 14.15: Dashboard Glare

When something is placed on the dashboard of a car, you clearly see its reflection through the windshield, which can be distracting to the driver. On the other hand, wearing polarized sunglasses can significantly reduce this problem.

a) Calculate Brewster's angle for the air-glass surface.

b) Place a light colored object, like a map, on the dashboard of a car. Measure the angle of incidence, and reflection, of this image. How close is it to Brewster's angle?

c) Using a polarizing sheet, or polarized sunglasses, measure the angle where the image of the map disappears. Rotate the polarizer 90°, and verify that the glare is just as bright as with no polarizing glasses.

d) Sometimes glare from a side window becomes distracting. Will polarized sunglasses help with this?

14.5 *Reflection off of Good Conductors*

Why are metals shiny and reflective? Light moving through an ionized gas becomes attenuated; why? In this section, we will consider several thought experiments that show how Maxwell's theory of light can answer these questions.

We begin by considering a toy model for an antenna that would produce a plane wave: a large thin conducting slab that is driven by an alternating current and therefore emits radiation. We then use this model to derive the characteristic impedance not of the conductor itself, but rather of the optical medium surrounding it, which we will assume is free space.

Next, we consider reflection off a very good, and very thin, conductor. We also present boundary conditions appropriate for the high conductivity limit, and apply each of Maxwell's equations in the standard way. Finally, we derive the law of angles of reflection and Huygens's principle for reflective surfaces.

Thought Experiment 14.12: Radiation from a Large Conducting Sheet

Imagine an infinitely large conducting sheet with a uniform AC current density passing through it. This is a toy model for an antenna that could produce a plane wave.

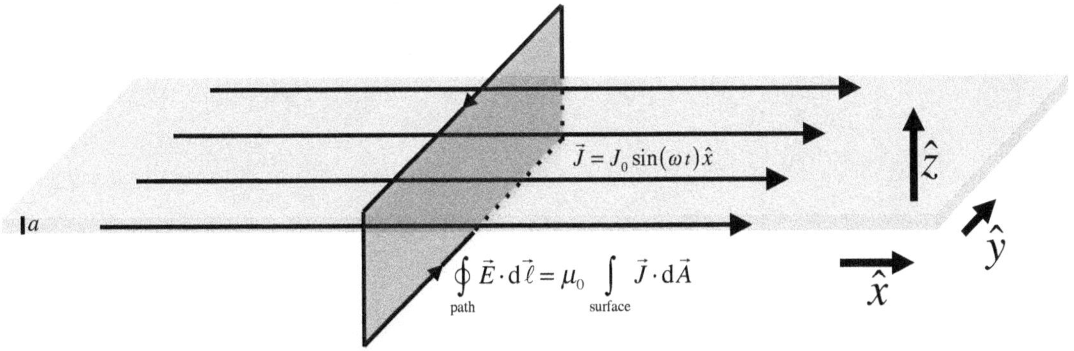

Consider a large conducting sheet in the x-y plane, with a small thickness a, and a current density given by:

$\vec{J} = J_0 \sin(\omega t)\hat{x}$.

We draw an Ampèrian loop to find the induced magnetic field, as:

$$\left(|B_{\text{top}}| + |B_{\text{bottom}}|\right)\cancel{\Delta x} = \mu_0 J a \cancel{\Delta x} \quad \rightarrow \quad |B_{\text{top}}| = |B_{\text{bottom}}| = \tfrac{1}{2}\mu_0 J a$$

$$\vec{B}_{\text{top}} = -\tfrac{1}{2}\mu_0 J a \hat{y} \quad \text{and} \quad \vec{B}_{\text{bottom}} = \tfrac{1}{2}\mu_0 J a \hat{y}.$$

Light travels at a finite speed, so the magnetic field is a function of the current as it was before, not as it its now. This *retarded time* is simply $t - \frac{1}{c}|z|$, so for each side of the thin plate the magnetic field is given by:

$$\vec{B}(z,t) = \begin{cases} \vec{B}_{top}(t - \frac{1}{c}|z|) = -\frac{1}{2}\mu_0 a J_0 \sin(\omega(t - \frac{1}{c}z))\,\hat{y} & : z > 0 \\ \vec{B}_{\text{bottom}}(t - \frac{1}{c}|z|) = \frac{1}{2}\mu_0 a J_0 \sin(\omega(t + \frac{1}{c}z))\,\hat{y} & : z < 0 \end{cases}.$$

Or we can write this more compactly as:

$$\vec{B} = -\tfrac{|z|}{z}\tfrac{1}{2}\mu_0 a J(t - \tfrac{1}{c}|z|)\hat{y} = -\tfrac{|z|}{z}\tfrac{1}{2}\mu_0 a J_0 \sin(\omega(t - \tfrac{1}{c}|z|))\hat{y}.$$

For there to be electromagnetic radiation, this magnetic field must induce an electric field outside of the conducting sheet. So, we apply Faraday's law outside of the conductor:

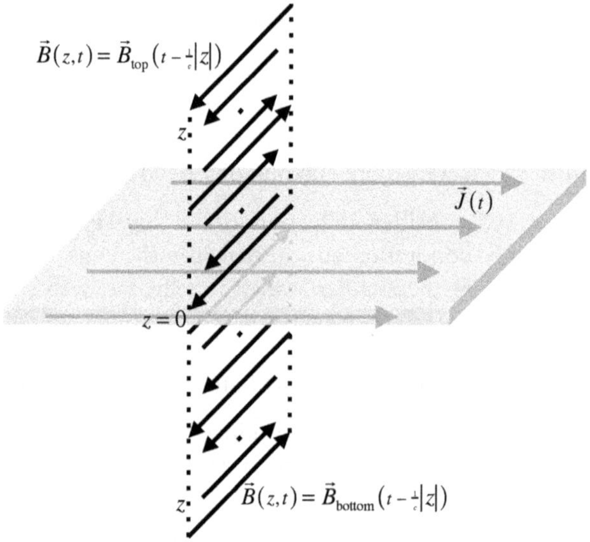

$$\vec{\nabla} \times \vec{E} = -\frac{\partial \vec{B}}{\partial t}$$

$$\vec{\nabla} \times \vec{E} = \frac{\partial}{\partial t}\left(-\tfrac{|z|}{z}\tfrac{1}{2}\mu_0 a J_0 \sin(\omega(t - \tfrac{1}{c}|z|))\hat{y}\right).$$

Differentiating with respect to time:

$$\vec{\nabla} \times \vec{E} = -\tfrac{|z|}{z}\tfrac{1}{2}\mu_0 a\omega J_0 \cos(\omega(t - \tfrac{1}{c}|z|))\hat{y}$$

Now applying the curl in Cartesian coordinates, and integrating, we have:

$$\left(\vec{\nabla} \times \vec{E}\right)_y = \frac{\partial E_x}{\partial z} - \cancel{\frac{\partial E_z}{\partial x}}$$

$$\frac{\partial E_x}{\partial z} = -\tfrac{1}{2}\tfrac{|z|}{z}\mu_0 a\omega J_0 \cos(\omega(t - \tfrac{1}{c}|z|))$$

$$E_x = -\tfrac{1}{2}\cancel{\tfrac{|z|}{z}}\mu_0 a \cancel{\omega} J_0\left(-\sin(\omega(t - \tfrac{|z|}{c}))\right)\left(\frac{c}{\cancel{\omega}}\right)\left(-\cancel{\tfrac{|z|}{z}}\right)$$

$$\vec{E} = -\tfrac{1}{2}c\mu_0 a J_0 \sin(\omega(t - \tfrac{1}{c}|z|))\hat{x}.$$

Now we consider the Poynting vector:

$$\vec{\mathbb{S}} = \tfrac{1}{\mu_0}\vec{E}\times\vec{B} = \tfrac{1}{\mu_0}\left(-\tfrac{1}{2}\mu_0 a c J_0 \sin\left(\omega\left(t-\tfrac{1}{c}|z|\right)\right)\hat{x}\right)\times\left(-\tfrac{|z|}{z}\tfrac{1}{2}\mu_0 a J_0 \sin\left(\omega\left(t-\tfrac{1}{c}|z|\right)\right)\hat{y}\right)$$
$$\vec{\mathbb{S}} = \tfrac{c}{\mu_0}\tfrac{|z|}{z}\left(\tfrac{1}{2}\mu_0 a\, J_0 \sin\left(\omega\left(t-\tfrac{1}{c}|z|\right)\right)\right)^2 \hat{z}.$$

Since energy is carried away from the radiating plane there must be an energy source driving the current in the conductor. To investigate this further, we apply the continuity equation inside the sheet:

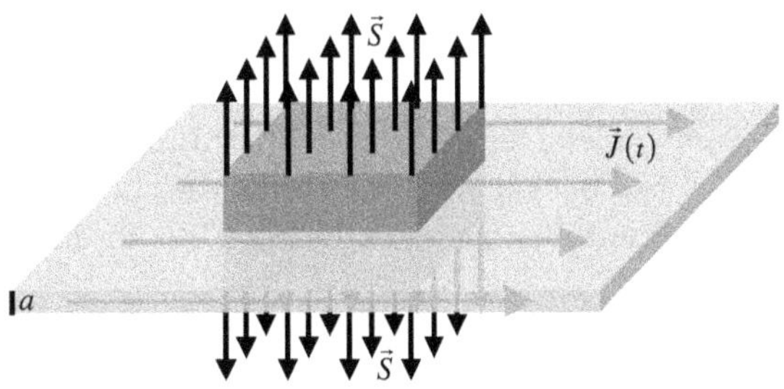

$$\vec{\nabla}\cdot\vec{\mathbb{S}} = -\vec{E}\cdot\vec{J}.$$

Integrating around a Gaussian surface containing an area of surface δA, and corresponding volume in the metal sheet of $\delta V = a\delta A$, we can solve for the electric field inside the conductor that is needed to produce the radiated energy:

$$2\tfrac{c}{\mu_0}\left(\tfrac{1}{2}\mu_0 a\, J_0 \sin(\omega t)\right)^2 \delta A = -\vec{E}\cdot\left(J_0\sin(\omega t)\hat{x}\right)\delta V$$
$$\vec{E} = -\tfrac{1}{2}\mu_0 c a J_0 \sin(\omega t)\hat{x} = -\left(\tfrac{1}{2}\mu_0 c a\right)\vec{J}.$$

Now let us consider the voltage required to drive the current in a sheet of dimensions $l\times l$:

$$\begin{aligned}\mathbb{V} &= -\int_0^l \vec{E}\cdot \mathrm{d}\vec{\ell} \\ &= -\int_0^l \left(-\left(\tfrac{1}{2}\mu_0 c a\right)\vec{J}\right)\cdot \mathrm{d}\vec{\ell} \\ &= \tfrac{1}{2}\mu_0 c \int_0^l J a\,\mathrm{d}x = \tfrac{1}{2}\mu_0 c J a l.\end{aligned}$$

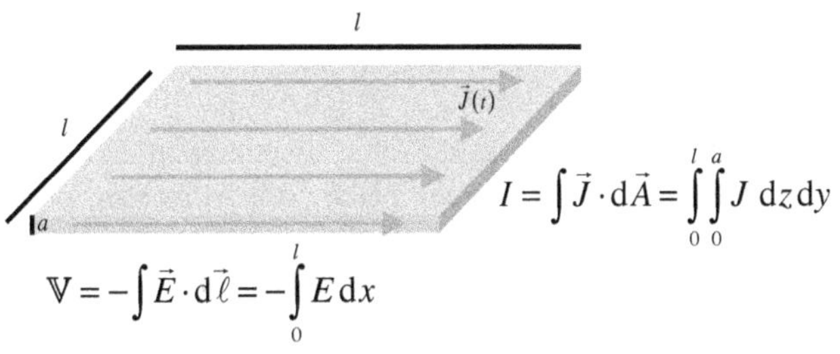

And if the large plate is square, then we can write the current as:

$$I = \int \vec{J}\cdot \mathrm{d}\vec{A} = \int_0^l\int_0^a J\,\mathrm{d}z\mathrm{d}y = J a l.$$

Thus the current and voltage are proportional, so we can find the impedance Z, or effective resistance, of the large square sheet as:

$$Z \equiv \frac{\mathbb{V}}{I} = \frac{\tfrac{1}{2}\mu_0 c \cancel{Jal}}{\cancel{Jal}} = \tfrac{1}{2}\mu_0 c = \tfrac{1}{2}\frac{\mu_0}{\sqrt{\mu_0\varepsilon_0}} = \tfrac{1}{2}\sqrt{\frac{\mu_0}{\varepsilon_0}} = \tfrac{1}{2}Z_0 \approx 188\,\Omega.$$

Recall that $Z_0 \approx 377\,\Omega$ is called *the impedance of free space.*

Thought Experiment 14.13: How Light Reflects off a Conducting Surface

Consider a plane wave incident on a conducting plane, with conductivity κ. Letting $z = 0$ at the surface boundary, the incident wave is:

$$\vec{E}_{z\le 0}(z,t) = E_0 \sin\left(\omega t - \tfrac{\omega}{c}z\right)\hat{x},$$

so at the surface:

$$\vec{E}(t) = E_0 \sin(\omega t)\hat{x}.$$

Inside the conductor, this electric field causes current to flow, so according to Ohm's law:

$$\vec{J}(t) = \kappa E_0 \sin(\omega t)\hat{x}.$$

We now ask what we mean by the surface. Is it some magical infinitesimal boundary, or does it have a real thickness? Let's model the surface as we did in Thought Experiment 14.12, as a large plane antenna with a small thickness a, so the solution for the reflected wave ($z<0$) becomes:

$$\vec{B}_{\text{reflection}} = \tfrac{1}{2}\mu_0 a\kappa E_0 \sin(\omega(t+z/c))\,\hat{y} \quad \text{and} \quad \vec{E}_{\text{reflection}} = -\tfrac{1}{2}c\mu_0 a\kappa E_0 \sin(\omega(t+z/c))\,\hat{x}.$$

Consider the implications of these equations. Notice first that the reflected magnetic field is in the same direction as the incident magnetic field, but the direction of the electric field vector is reversed. It is this reversal that leads to the 180° phase shift when light reflects off mirrors.

To conserve energy, the amplitude of the reflected wave must be less than, or equal to, the amplitude of the incoming wave. The interaction length a, therefore, must satisfy the inequality:

$$E_0 \geq \tfrac{1}{2}c\mu_0 a\kappa E_0, \text{ so that: } a \leq \frac{2}{Z_0\kappa},$$

where we have used $Z_0 = \mu_0 c$. For silver, this length is less than an angstrom, which is less than the size of the atom itself. We will discuss interaction lengths in more detail in the next section (p. 466).

Now that we understand the mechanism of reflection, and that it takes place in a very thin layer at the metal surface, we can apply Faraday's law to the boundary. Consulting the diagram, we see that:

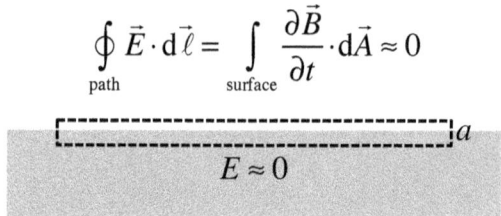

$$\oint_{\text{path}} \vec{E}\cdot d\vec{\ell} = \int_{\text{surface}} \frac{\partial \vec{B}}{\partial t}\cdot d\vec{A} \approx 0$$

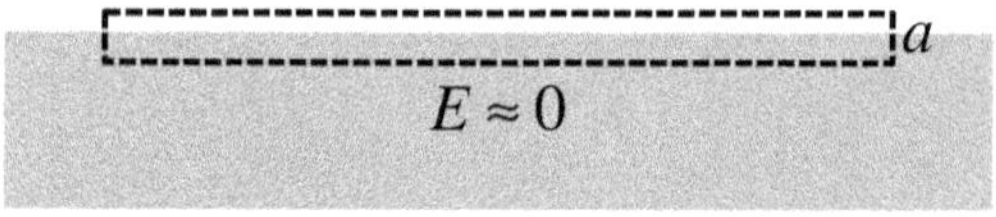

$$\oint_{\text{path}} \vec{E}\cdot d\vec{\ell} = \tfrac{\partial}{\partial t}\int_{\text{surface}} \vec{B}\cdot d\vec{A} \approx \tfrac{\partial}{\partial t}(a\delta x B_{z=0})$$

$$\left(\vec{E}_{\text{incident}} + \vec{E}_{\text{reflected}}\right)\delta x - \cancel{\vec{E}_{z>0}\,\delta x} \approx a\,\delta x \tfrac{\partial}{\partial t}(B_{z=0}).$$

Now, if the interaction distance, the frequency, or magnetic field, is sufficiently small, then the right hand term becomes approximately zero. This will be the case for good conductors, in the low frequency limit. When such a condition holds, we will have the following relation:

$$\vec{E}_{\text{incident}} + \vec{E}_{\text{reflected}} \approx a\tfrac{\partial}{\partial t}(B_{z=0}) \approx 0 \quad \rightarrow \quad \vec{E}_{\text{reflected}} \approx -\vec{E}_{\text{incident}}.$$

In other words, the surface of a good conductor has 100% reflectivity in the low frequency limit. This is, in fact, what is observed in nature, where metals such as silver and aluminum are very reflective and, when not tarnished, will reflect with close to 100% reflectivity at frequencies much lower than the inverse relaxation time or $\omega \ll Z_0 c\kappa$, which is typically in the near ultraviolet for good conductors.

Notice, also, how many metals, such as gold and copper, have a reddish shine to them. This is because the metals reflect red light efficiently as discussed in this thought experiment, but the reflectivity is significantly lower at blue frequencies. We therefore expect silver to be whiter in color than gold, due to its higher conductivity. Then, one might ask, what about aluminum or platinum? Aluminum and platinum are both highly conductive, but not as conductive as either copper or gold, yet are white in color while copper and gold are redder. One would expect, from our argument here, that copper would be whiter than aluminum. As it turns out, when considering light in the blue end of the optical spectrum, and distances on the scale of atoms themselves, classical physics breaks down and quantum mechanics takes over, so not only the particular assumptions of our model, but also classical physics, are too limited to correctly account for this.

Thought Experiment 14.14: Huygens's Principle

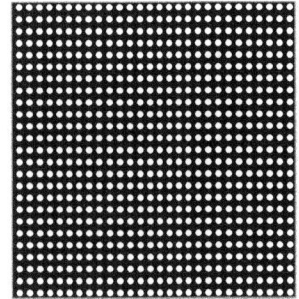

In the last two thought experiments we argued that reflection is due to the electric field inducing a current, and then that current radiating light. Therefore, every point on the surface of a mirror acts as its own light source.

As an illustration of this, consider normal incident light bouncing off a grid of shiny nails hammered into a back piece of wood. Thus, the light is reflected from small points of radius a separated by a two-dimensional grid of spacing b, as in the diagram. Each of these nail heads will be its own light source, while we will assume the light is absorbed in the regions between the nail points.

We also assume that the size of each head is large enough to reflect light, but otherwise small, so that each nail's head is a source of a spherical wave, which can be expressed as:

$$\vec{E}(\vec{r},t) = \vec{E}_a \frac{a}{|\vec{r}-\vec{R}|} \cos\left(\omega t - \tfrac{\omega}{c}|\vec{r}-\vec{R}|\right),$$

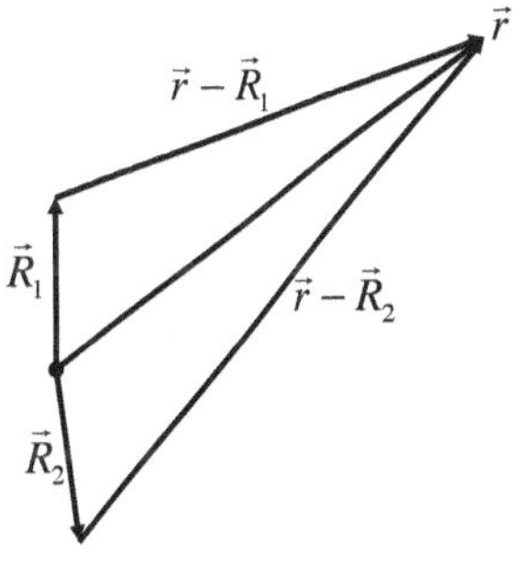

where $\vec{r}$ is measured with respect to the origin, such as the center of the grid of nails, and $\vec{R}$ is the position of the source with respect to the same origin. Thus, the electric field at a given location is simply the sum of the electric field from each nail, which is easy to calculate using a computer to model the finite grid.

Notice how there will be constructive interference between two identical sources at positions $\vec{R}_1$ and $\vec{R}_2$ when the following criterion holds:

$$\tfrac{\omega}{c}|\vec{r}-\vec{R}_1| - \tfrac{\omega}{c}|\vec{r}-\vec{R}_2| = N(2\pi) ,$$

where N is any integer. The condition for constructive interference is therefore:

$$|\vec{r}-\vec{R}_1| - |\vec{r}-\vec{R}_2| = N\lambda .$$

Now, we will investigate the limit as $b \to 0$. In this case, the grid becomes simply a continuous conductive plane, and we revert to the solution we found in Thought Experiment 14.13. Therefore, the reflected plane wave can be expressed as the infinite sum of spherical waves, each one originating at a different point on the plane.

This can be generalized to include the statement that any wavefront can be expressed as an infinite sum of spherical waves. This is called *Huygens's principle*, and it is extremely useful in the field of geometrical optics. Huygens's principle not only allows mirrors to be considered sources of light in and of themselves, but also allows for objects that block light to be expressed as the absence of a source.[235]

Example 14.3: The Law of Reflection

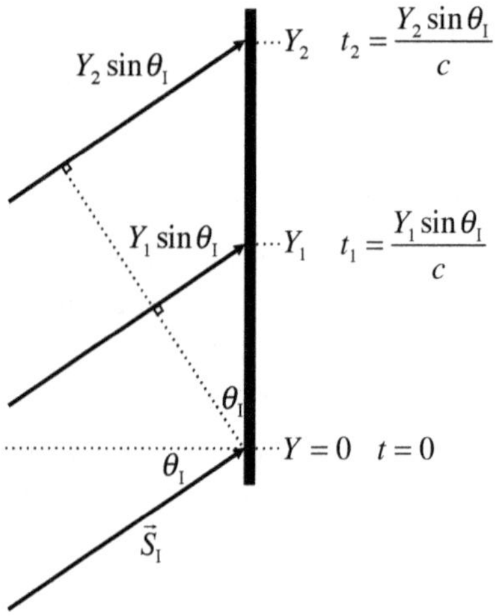

Now we consider a plane wave incident on a mirror at an angle θ_I from the normal. Each point on the mirror becomes its own spherical wave source, according to Huygens's principle. Due to the finite speed of light, however, the incoming wavefront is not incident on all the points of the mirror at the same time. There is a delay due to the extra length the wavefront must travel, as shown in the diagram.

In the diagram we placed the mirror in the x-y plane, while the incident light propagates in the direction:

$$\hat{k}_I = \hat{y}\sin\theta_I + \hat{z}\cos\theta_I .$$

Next we apply Huygens's principle, and note that the reflected wave from a small area $\delta X \delta Y$ must represent a spherical wave. However, each wave is not emitted at the same time, due to the delay we just mentioned.

Putting these ideas together, we can write the retarded time for a reflected wave that had bounced off of a point (X,Y) on the mirror, but given the translational symmetry we can ignore the $\hat{x}$ dimension, so:

$$\tau = t - \tfrac{1}{c}\left(Y \sin\theta_I + \sqrt{(y-Y)^2 + (z)^2}\right).$$

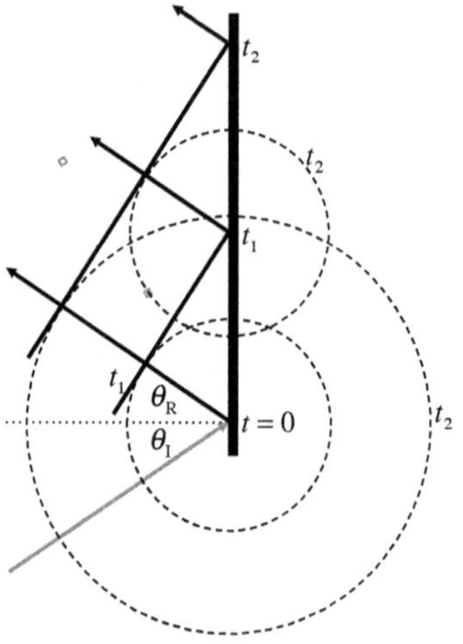

We will have constructive interference when the retarded time is the same regardless of where the light struck the mirror, so we solve for $\frac{d\tau}{dY}$ and set it to zero, yielding:

$$\frac{d\tau}{dY} = -\tfrac{1}{c}\left(\sin\theta_I - \frac{(y-Y)}{\sqrt{(y-Y)^2 + (z)^2}}\right) = 0.$$

Drawing these spherical wave crests, as in the diagram, we see that this is the case when:

$$\sin\theta_R = \sin\theta_I \quad \rightarrow \quad \theta_R = \theta_I ,$$

which corresponds to one of the most basic rules of ray optics.

---

[235] This last assertion is related to *Babinet's principle*, which says that the unperturbed plane wave field incident on an opaque body equals the sum of the wave field from an opaque body and that from a hole of the same size and shape. See, for example, S. Ganci, "Fraunhofer diffraction by a thin wire and Babinet's principle," Am. J. Phys., **73** (2004), 83-4.

Example 14.4: The Poisson Spot

As we discussed in the introduction (p. 426), in 1818, Poisson came up with a famous thought experiment designed to challenge the wave model of light. Here is a brief, modern version of Poisson's argument.

Consider a plane wave shining on an opaque circular disk of radius a. If we consider every point on the wavefront as a source of a spherical wave there will be constructive interference at the center, because the z axis is equidistant from the edge of circular aperture.

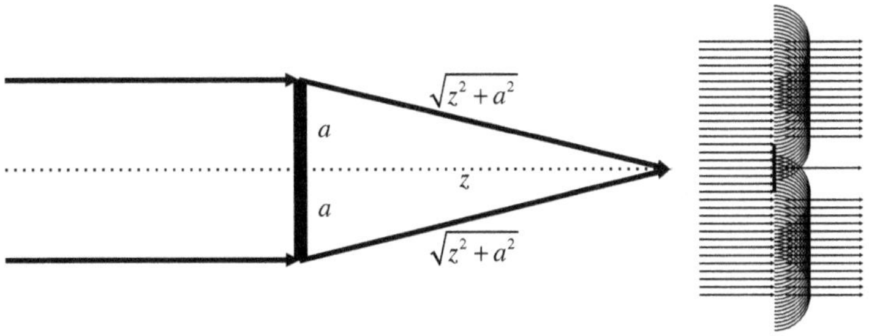

Poisson argued that this was absurd. However, Arago demonstrated in a series of experiments that the spot Poisson predicted actually does exist.

Discussion 14.2: Huygens's Principle and Refraction

In Thought Experiment 14.14 (p. 463) we discussed Huygens's principle in the context of reflection and diffraction. We showed that each encounter with a conducting medium created its own spherical wave, and the interference of these spherical waves defines the resulting wavefront. A plane wave could, also, then be characterized as a continuous distribution of spherical wave sources. If, as in Example 14.4 (p. 465), part of a wave were blocked, the resulting wave would then propagate again as a distribution of spherical waves.

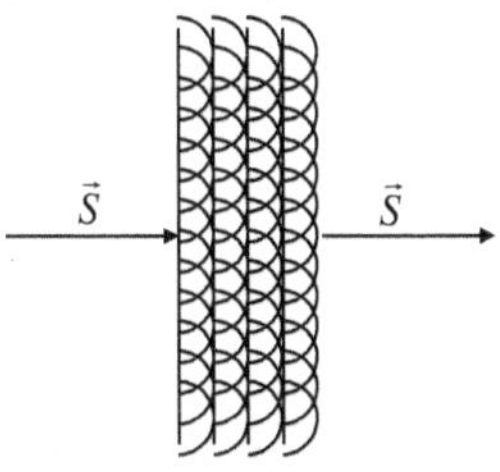

Consider a situation where the speed of light in the medium is a continuous function of position, as in Thought Experiment 14.7 (p. 447). Since each wavelet's radius is also a function of the position $\delta\ell = v(\vec{r})\delta t$, then the wavefront will bend so that the wave front remains tangent to all the wavelets, as in bottom diagram.

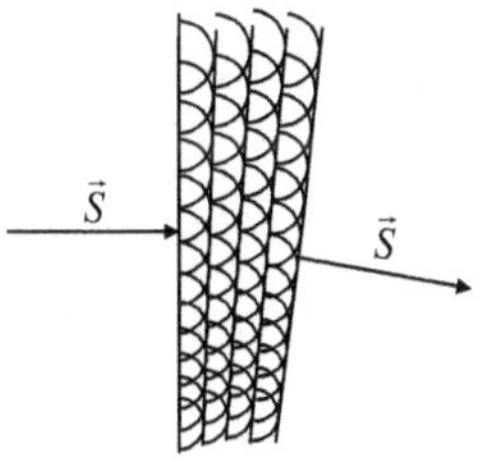

Notice that, regardless of how we think of it, the light bends because of a perpendicular gradient in the speed of light.

Problem 14.16: The Principles of Fermat and Huygens

Derive Fermat's principle (p. 447) from Huygens's principle.

14.6 *The Propagation of Light in a Conductor*

In Thought Experiment 14.13 we discussed the mechanism for reflection off of a good conductor, so we could explain how mirrors work. Here we will investigate how light propagates inside of a conductive medium. This will lead us to the idea of the penetration depth of light, called the *skin depth*, which was first found for a spherical conductor by the British physicist Horace Lamb[236] and later expanded upon by Oliver Heaviside.

According to Ohm's law, electric fields drive currents in conductors. At low frequencies, these currents either complete a circuit or die out within a very short time, called *the relaxation time.* At extremely high frequencies in a conductor, and at modest frequencies in a plasma, the acceleration, rather than electron velocity, is proportional to the driving electric field, causing the electrons to move in simple harmonic motion. This provides a phase shift between the electric and magnetic fields, which is not seen in a wave through free space.

Thought Experiment 14.15: A Plane Wave in a Conductive Medium

Consider an electromagnetic wave propagating through a neutral medium with a finite uniform conductivity κ. Again we write down the four Maxwell's equations, but we lift the restrictions that there are no currents, that the charge density is zero, and that we know the wave speed. Recalling the general wave equation from Derivation 14.1 (p. 442), we start with:

$$\nabla^2\vec{E} = \tfrac{1}{\varepsilon_0}\vec{\nabla}\rho + \mu_0 \tfrac{\partial}{\partial t}\vec{J} + \mu_0\varepsilon_0 \tfrac{\partial^2}{\partial t^2}\vec{E}.$$

Since the current is flowing perpendicular to the direction of propagation, the divergence of the current density is zero, so because of the continuity equation the charge density must be constant, and neutral, so: $\vec{\nabla}\rho = 0$.

We will now rewrite the wave equation in terms of the speed of light and the impedance of free space, $Z_0 = \mu_0 c$, so:

$$\nabla^2\vec{E} = \tfrac{1}{c} Z_0 \tfrac{\partial}{\partial t}\vec{J} + \tfrac{1}{c^2}\tfrac{\partial^2}{\partial t^2}\vec{E}.$$

Since the wave includes a first order term, we expect that it will become attenuated. Therefore we guess that the electric field follows the form:

$$\vec{E} = \vec{E}_0 e^{-z/\delta}\cos(\omega\tau).$$

And, according to Ohm's law:

$$\vec{J} = \kappa\vec{E}_0 e^{-z/\delta}\cos(\omega\tau).$$

Remember that $\frac{\partial\tau}{\partial t} = 1$ and $\frac{\partial\tau}{\partial z} = -\frac{\omega}{v} = -\frac{1}{\lambdabar}$, where $\lambdabar = \frac{c}{\omega} = \frac{1}{k}$ is the *reduced wavelength*, which is in length units per radian. Next we calculate, and simplify each term in the wave equation:

$$\nabla^2\vec{E} = \tfrac{\partial^2}{\partial z^2}\vec{E}_0 e^{-z/\delta}\cos(\omega\tau) = \vec{E}_0 e^{-z/\delta}\left[\left(\tfrac{1}{\delta^2} - \tfrac{1}{\lambdabar^2}\right)\cos(\omega\tau) - \left(\tfrac{2}{\lambdabar\delta}\right)\sin(\omega\tau)\right],$$

$$\mu_0\tfrac{\partial}{\partial t}\vec{J} = \tfrac{1}{c}Z_0\kappa\tfrac{\partial}{\partial t}\vec{E} = \tfrac{1}{c}Z_0\kappa\vec{E}_0 e^{-z/\delta}\tfrac{\partial}{\partial t}\cos(\omega\tau) = \left(-\tfrac{\omega}{c}Z_0\kappa\right)\left(\vec{E}_0 e^{-z/\delta}\right)\sin(\omega\tau),$$

and

$$\mu_0\varepsilon_0\tfrac{\partial^2}{\partial t^2}\vec{E} = \tfrac{1}{c^2}\tfrac{\partial^2}{\partial t^2}\vec{E}_0 e^{-z/\delta}\cos(\omega\tau) = \left(\vec{E}_0 e^{-z/\delta}\right)\left(-\tfrac{\omega^2}{c^2}\right)\cos(\omega\tau).$$

---

236 H. Lamb, On Electrical Motions in a Spherical Conductor," in Philosophical Transactions, **172**, 519 (1883), by the Royal Society

Canceling the common factor of $\vec{E}_0 e^{-z/\delta}$, and simplifying, this becomes:

$$\left(\tfrac{1}{\delta^2}-\tfrac{1}{\bar{\lambda}^2}+\left(\tfrac{\omega}{c}\right)^2\right)\cos(\omega\tau)+\left(\tfrac{\omega}{c}Z_0\kappa-\tfrac{2}{\bar{\lambda}\delta}\right)\sin(\omega\tau)=0.$$

Since the sine and cosine terms are orthogonal, we will have two valid equations.

$$\tfrac{1}{\delta^2}-\tfrac{1}{\bar{\lambda}^2}+\left(\tfrac{\omega}{c}\right)^2=0 \quad\text{and}\quad \tfrac{\omega}{c}Z_0\kappa-\tfrac{2}{\bar{\lambda}\delta}=0.$$

Now we have two equations, and two unknowns. Solving each for the inverse reduced wavelength, also called the wave number, gives us:

$$k=\tfrac{1}{\bar{\lambda}}=\sqrt{\tfrac{1}{\delta^2}+\tfrac{\omega^2}{c^2}}=\tfrac{1}{2c}Z_0\kappa\omega\delta.$$

Rearranging and factoring together terms with units of length, or inverse length, we obtain:

$$\tfrac{1}{4}\left(\tfrac{\omega}{c}\right)(Z_0\kappa)^2\delta^4-\left(\tfrac{\omega}{c}\right)\delta^2-\left(\tfrac{c}{\omega}\right)=0.$$

Now we apply the quadratic equation to find the skin depth squared:

$$\delta^2=\frac{\left(\frac{\omega}{c}\right)\pm\sqrt{\left(\frac{\omega}{c}\right)^2-4\frac{1}{4}\left(\frac{\omega}{c}\right)(Z_0\kappa)^2\left(-\frac{c}{\omega}\right)}}{2\frac{1}{4}\left(\frac{\omega}{c}\right)(Z_0\kappa)^2}=\frac{\left(\frac{\omega}{c}\right)\pm\sqrt{\left(\frac{\omega}{c}\right)^2+(Z_0\kappa)^2}}{\frac{1}{2}\left(\frac{\omega}{c}\right)(Z_0\kappa)^2}.$$

Only one solution generates positive values, so:

$$\delta=\sqrt{\frac{\left(\frac{\omega}{c}\right)+\sqrt{\left(\frac{\omega}{c}\right)^2+(Z_0\kappa)^2}}{\frac{1}{2}\left(\frac{\omega}{c}\right)(Z_0\kappa)^2}}.$$

In special case of high conductivity, but low frequency:

$$\delta\Big|_{Z_0\kappa\gg\frac{\omega}{c}}=\sqrt{\frac{\left(\frac{\omega}{c}\right)+\sqrt{\cancel{\left(\frac{\omega}{c}\right)^2}+(Z_0\kappa)^2}}{\frac{1}{2}\left(\frac{\omega}{c}\right)(Z_0\kappa)^2}}\approx\sqrt{\frac{\cancel{\left(\frac{\omega}{c}\right)}+(Z_0\kappa)}{\frac{1}{2}\left(\frac{\omega}{c}\right)(Z_0\kappa)^2}}\approx\sqrt{\frac{\cancel{(Z_0\kappa)}}{\frac{1}{2}\left(\frac{\omega}{c}\right)(Z_0\kappa)^{\cancel{2}}}}\approx\sqrt{\frac{2c}{\omega Z_0\kappa}}.$$

Notice that the lower the frequency, the bigger is the skin depth in good conductors. And, now, consider the poor conductor limit:

$$\delta\Big|_{Z_0\kappa\ll\frac{\omega}{c}}=\sqrt{\frac{\left(\frac{\omega}{c}\right)+\sqrt{\left(\frac{\omega}{c}\right)^2+\cancel{(Z_0\kappa)^2}}}{\frac{1}{2}\left(\frac{\omega}{c}\right)(Z_0\kappa)^2}}\approx\sqrt{\frac{\left(\frac{\omega}{c}\right)+\sqrt{\left(\frac{\omega}{c}\right)^2}}{\frac{1}{2}\left(\frac{\omega}{c}\right)(Z_0\kappa)^2}}\approx\sqrt{\frac{2\cancel{\left(\frac{\omega}{c}\right)}}{\frac{1}{2}\cancel{\left(\frac{\omega}{c}\right)}(Z_0\kappa)^2}}\approx\sqrt{\frac{4}{(Z_0\kappa)^2}}\approx\frac{2}{Z_0\kappa}.$$

So there is no frequency dependence in poor conductors. Notice that the skin depth in intermediate conductors is always between these two limits.

Problem 14.17: The Reduced Wavelength

a) Derive the general expression for the reduced wavelength $\bar{\lambda}$ of an electromagnetic wave of angular frequency ω in a conductor κ.

b) What is the reduced wavelength in the low conductivity limit?

c) What is the reduced wavelength in the high conductivity, limit?

d) Find the ratio of the reduced wavelength to skin depth in each case.

Problem 14.18: The Wave Speed in a Conductor

The wave speed can be expressed as $v = \omega ƛ$, and the index of refraction is defined such that $n = \frac{c}{v}$.

a) Derive the general expression for the index of refraction, n, in a conductor. Make sure to simplify your result.

b) Find the wave speed in the poor conductor limit. What does it depend on? Explain.

c) Find the wave speed in the high conductivity limit. What does it depend on? Explain.

d) Calculate the speed and wavelength of a 1.4 GHz radio wave inside of a copper wire.

e) One way that geologists measure ground conductivity is by measuring radio wave speed as a function of frequency, called the *dispersion*. Research these techniques and what they can measure. Report on your findings.

f) Radio astronomers measure the density of free electrons in the *interstellar medium* using dispersion. However, astronomers cannot place a transmitter and receiver on opposite sides of the Milky Way Galaxy. Rather they must rely on natural radio sources, such as *pulsars*, and measure their dispersion. Research and report on how they measure dispersion and how it relates to the quantity of ionized gas in our Galaxy.

Problem 14.19: Good Conductors Driven at Ultra High Frequencies

Imagine an extremely high frequency wave traveling through an excellent conductor. It may no longer be reasonable to assume no electron inertia.

We have worried about electron inertia before. In Chapter 12 we discussed the London brothers' classical acceleration theory of superconductors, where we quantified the inertial term as the *kinetic inductivity*, defined as in Thought Experiment 12.5 (p. 389):

$$\Lambda = \frac{m_e}{n_e e^2},$$

where n_e is the number density of the electron sea. We also found a related characteristic length, called the London penetration depth:

$$\lambda_L = \sqrt{\tfrac{1}{\mu_0}\Lambda} = \sqrt{\tfrac{c}{Z_0}\Lambda}.$$

However, this problem is not about superconductors, but about very good normal conductors driven at extremely high frequencies. So, you will apply the same wave equation as we did in Thought Experiment 14.5, but you will assume a phase lag between the electric field and the current density. Thus:

$$\vec{E} = \vec{E}_0 e^{-z/\delta}\cos(\omega\tau) \text{ and } \vec{J} = \vec{J}_0 e^{-z/\delta}\cos(\omega\tau - \phi).$$

a) By applying the wave equation, show that these criteria must hold:

$$\left(\frac{1}{\delta^2} - \frac{1}{ƛ^2} + \frac{\omega^2}{c^2}\right) = \mu_0\kappa\omega\sin(\phi) \text{ and } \frac{2}{ƛ\delta} = \mu_0\kappa\omega\cos(\phi).$$

b) Now assume that the characteristic time relating to inertial effects is the time it would take the wave to travel the distance of the London penetration depth. Therefore our expected phase lag would be:

$$\phi \approx \omega\left(\frac{\lambda_L}{v}\right) \approx \frac{\lambda_L}{\lambda\!\!\!^-}.$$

Assuming a small phase lag, so $\phi \approx \sin\phi$, solve for the phase lag as a function of the frequency, conductivity, and the London penetration depth, but not as a function of the skin depth or the reduced wavelength.

c) Find the driving frequency where $\phi \approx 1$ radian in copper. See Thought Experiment 12.1 (p. 380) for the relevant data.

Example 14.5: The Magnetic Field Inside a Conductor

In this example, we find the magnetic field of a wave in a conductor by applying the Maxwell-Ampère law:

$$\vec{\nabla}\times\vec{B} = \tfrac{1}{c}Z_0\vec{J} + \tfrac{1}{c^2}\tfrac{\partial}{\partial t}\vec{E}.$$

Let us now reconsider the plane wave solution inside a good conductor, with a solution for the electric field in the form:

$$\vec{E} = E_0 e^{-z/\delta}\cos\left(\omega t - z/\lambda\!\!\!^-\right)\hat{x} \quad \text{and} \quad \tfrac{\partial}{\partial t}\vec{E} = -\omega E_0 e^{-z/\delta}\sin\left(\omega t - z/\lambda\!\!\!^-\right)\hat{x}.$$

We will assume that the charges respond quickly, so:

$$\vec{J} = \kappa E_0 e^{-z/\delta}\cos\left(\omega t - z/\lambda\!\!\!^-\right)\hat{x}.$$

The wave is traveling in the $\hat{z}$ direction, and the electric field and current density are assumed to be in the $\hat{x}$ direction, so the magnetic field should be in the $\hat{y}$ direction. Therefore we will guess a magnetic field in the form:

$$\vec{B} = B_0 e^{-z/\delta}\cos\left(\omega t - z/\lambda\!\!\!^- - \phi\right)\hat{y}.$$

First we take the curl of our guess of the magnetic field:

$$\vec{\nabla}\times\vec{B} = -\tfrac{\partial}{\partial z}B_y = -\tfrac{\partial}{\partial z}\left(B_0 e^{-z/\delta}\cos\left(\omega t - z/\lambda\!\!\!^- - \phi\right)\right)\hat{x} = -B_0 e^{-z/\delta}\left(-\tfrac{1}{\delta}\cos\left(\omega t - z/\lambda\!\!\!^- - \phi\right) + \tfrac{1}{\lambda\!\!\!^-}\sin\left(\omega t - z/\lambda\!\!\!^- - \phi\right)\right)\hat{x}.$$

According to the Maxwell-Ampere law:

$$\vec{\nabla}\times\vec{B} = \tfrac{1}{c}Z_0\vec{J} + \tfrac{1}{c^2}\tfrac{\partial}{\partial t}\vec{E} = \tfrac{1}{c}Z_0\kappa E_0 e^{-z/\delta}\cos\left(\omega t - z/\lambda\!\!\!^-\right)\hat{x} - \tfrac{1}{c^2}E_0 e^{-z/\delta}\omega\sin\left(\omega t - z/\lambda\!\!\!^-\right)\hat{x}.$$

Using $\tau = \omega t - z/\lambda\!\!\!^-$ for the retarded time, we can write our guess of the magnetic field in the form:

$$\vec{\nabla}\times\vec{B} = B_0 e^{-z/\delta}\left(\left(\tfrac{1}{\delta}\sin(\phi) - \tfrac{1}{\lambda\!\!\!^-}\cos(\phi)\right)\sin(\omega\tau) + \left(\tfrac{1}{\lambda\!\!\!^-}\sin(\phi) - \tfrac{1}{\delta}\cos(\phi)\right)\cos(\omega\tau)\right)\hat{x}$$

Since the sine and cosine are orthogonal, we can equate each as:

$$B_0\left(\tfrac{1}{\delta}\sin(\phi) - \tfrac{1}{\lambda\!\!\!^-}\cos(\phi)\right) = -\tfrac{1}{c^2}E_0\,\omega$$
$$B_0\left(\tfrac{1}{\lambda\!\!\!^-}\sin(\phi) - \tfrac{1}{\delta}\cos(\phi)\right) = \tfrac{1}{c}Z_0\kappa E_0$$

Solving this system of equations for B_0 and ϕ will allow us to solve for the magnetic field. The most interesting one being ϕ, the phase shift.

In the low frequency, high conductivity, limit, this becomes:

$$\frac{\sin(\phi)}{\delta} - \frac{\cos(\phi)}{\lambda} = -\tfrac{1}{c^2}\omega\frac{E_0}{B_0} \approx 0 \quad \rightarrow \quad \frac{\sin(\phi)}{\delta} \approx \frac{\cos(\phi)}{\lambda}, \quad \text{so} \quad \tan(\phi) \approx \frac{\delta}{\lambda}.$$

In the low conductivity limit, this becomes:

$$\frac{\sin(\phi)}{\lambda} - \frac{\cos(\phi)}{\delta} = \tfrac{1}{c}Z_0\kappa\frac{E_0}{B_0} \approx 0 \quad \rightarrow \quad \frac{\sin(\phi)}{\lambda} \approx \frac{\cos(\phi)}{\delta}, \quad \text{so} \quad \tan(\phi) \approx \frac{\lambda}{\delta}.$$

Problem 14.20: The Magnetic Field Inside a Conductor

In Example 14.5 above, we discussed the magnetic field of a plane wave inside of a conductor. In this problem, you will follow-up on our analysis.

a) Show that, in the low conductivity limit, the magnetic field and the electric field are approximately in phase. Discuss why this is important.

b) Show that, in the high conductivity limit, the magnetic field lags the electric field by 45°.

c) Find the magnetic field phase angle, ϕ, in terms of the conductivity and angular frequency in general. Show that it is always between zero and 45°, and reverts to your prior answers in those limits.

d) Show that in the low conductivity, high frequency, limit: $B_0 = \tfrac{1}{c}E_0$.

e) Find the amplitude of the magnetic field, B_0, in terms of the electric field amplitude, conductivity, and angular frequency, in the high conductivity (low frequency) limit.

Problem 14.21: Skin Depth and Reflectivity

In Thought Experiment 14.13 (p. 461) we argued that when light reflects off of a conductor, the incident and reflected electric fields of are related by:

$$\left|\vec{E}_{\text{incident}} + \vec{E}_{\text{reflected}}\right| \approx a\tfrac{\partial}{\partial t}\left(B_{z=0}\right) \approx 0,$$

where a was the interaction length. This also showed that the reflected wave is about 180° out of phase.

a) In the high conductivity limit, estimate $a\tfrac{\partial}{\partial t}\left(B_{z=0}\right)$ using the skin depth for the interaction length.

b) Show that our estimated phase shift of π was off by: $\Delta\phi \approx \sqrt{\dfrac{2\omega}{cZ_0\kappa}}$.

Plate 25: Polishing a Gemini Mirror[237]

[237] A 1999 photograph of the Gemini North mirror after its first aluminum coating. The Gemini Observatory consists of twin 8.1-meter diameter telescopes located on mountains in Hawaii and Chile. This photo was downloaded from the Gemini web site at gemini.edu.

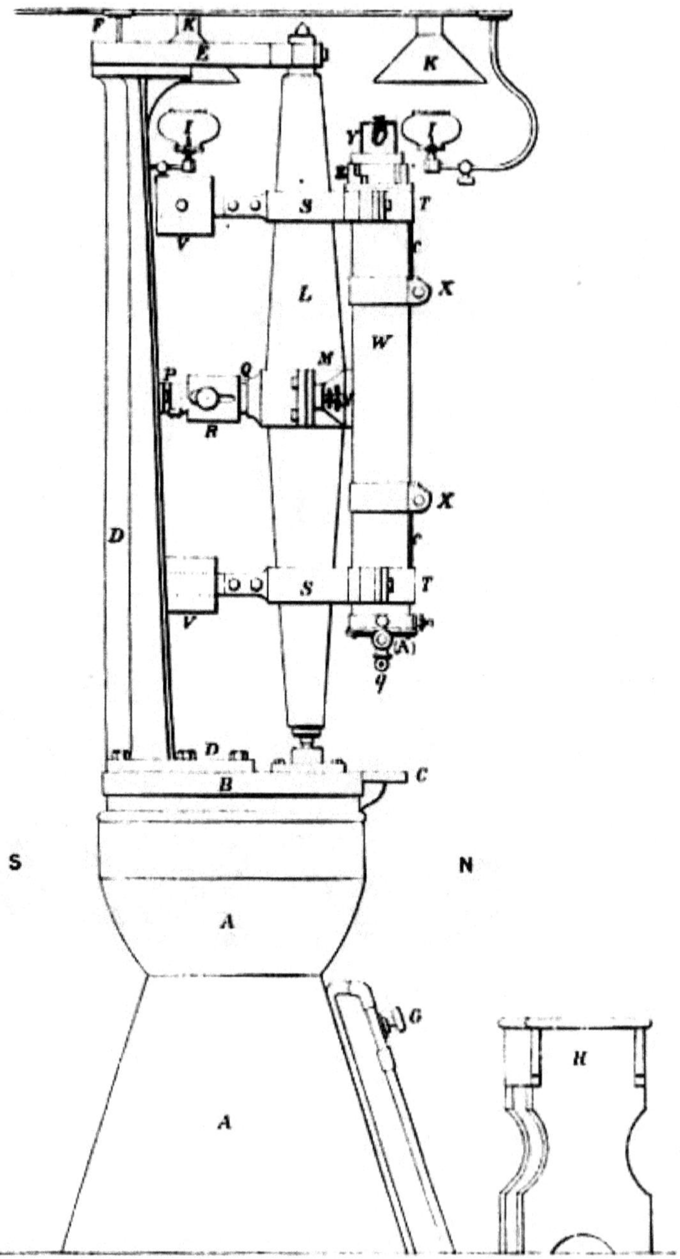

Plate 26: George Airy's Water Filled Telescope[238]

[238] G.B. Airy, "On the Supposed Alteration in the Amount of Astronomical Aberration of Light, Produced by the Passage of Light through a Considerable Thickness of Refracting Medium," Proceedings of the Royal Society of London **20** (1871), 35-39.

Chapter 15 Light in Free Space

How do you measure velocity without a clock? As long as you measure something traveling a certain distance, in a certain time, you always need some sort of clock. You can measure the speed of sound simply by watching and listening to the same event a known distance away, just as you can estimate the distance to a thunderstorm by the time difference between lightning and the thunder. But this needs a clock.

On the other hand, you could measure the temperature of the air, and from that estimate the ratio of the pressure to the density of the air. The sound speed is also given by the formula $v_s = \sqrt{P/\rho_m}$, where P is the pressure and ρ_m is the density of air. You do not need a clock to measure either the pressure or density of the air. Thus, you can measure the speed of sound either kinematically or by measuring the fluid properties of air.

When Maxwell showed a similar relationship between the permittivity ε and the permeability μ, and the speed of light:

$$v_{\text{light}} = 1/\sqrt{\mu\varepsilon},$$

the situation seemed resolved once and for all. Maxwell's result seemed to confirm the idea that there must be an all-pervasive medium through which light travels. Thus, many scientists reasoned, the speed of light in free space is really the speed that light travels with respect to the aether. But, what if light does not require a medium to propagate? How then could we either measure the speed of light using a clock, or not using a clock, without running afoul of Galileo's principle of relativity? On the one hand, if we put our laboratory into motion ε_0 and μ_0 should not change, but on the other hand velocities measured with clocks would.

A primary theme of ours, up to this point, has been the impressive success, and ultimate downfall, of the Galilean transformation laws that appear to be a necessary consequence of the principle of relativity. The arbitrary choice of inertial reference frame was employed successfully in both Newtonian mechanics and electrodynamics, until scientists ran into contradictory results. Recall the spectacular failure of the Galilean transformations (Section 11.2, p. 336), and how it forced us to modify the transformation laws between inertial reference frames.

In order to make these connections in the first place, however, the speed of light had to be measured. Light travels so fast that we are tempted to ask, how could anyone without a very accurate modern clock ever measure it? Yet it was first accurately measured in 1676, long before the invention of nanosecond clocks! The ideas, though not the actual measurement, originated with Galileo.

He argued in his 1628 Dialogues that light travels at a finite speed, and was the earliest scientist on record to suggest a measurement for the speed of light. He proposed that two people with lanterns, standing at a great distance apart, could measure the time required to signal one another by having one person uncover their lantern and the other uncovering their lantern in response. That way, they would not need a pair of synchronized clocks—one clock would do.

Despite practicing to minimize delay due to reaction time, this method did not show light to take a measurable time to travel.[239] Thus, light must travel faster than the distance of the lanterns divided by human perception time, or faster than about 100 miles per second. Galileo did, however, discover a clock in space—Jupiter's four large moons. While Galileo named them after his patron, they were renamed Io, Europa, Ganymede, and Callisto, after various love interests of the Greek god Zeus (a.k.a. Jupiter).

Working at the Paris observatory, the Danish astronomer Ole Rømer made detailed measurements of the orbits of the Galilean moons over a number of years, recording the time each moon disappears behind Jupiter, and reemerges on the other side. The time from one disappearance to the next takes Io about 42 hours, and double this for Europa, and double again for Ganymede. Callisto's period is not an even factor of two greater than Ganymede, but it is not too far off. This led to the first measurement of the speed of light in 1676.

Io, however, does not appear to orbit at exactly the same rate, but rather it appears to orbit faster when the Earth approaches Jupiter, and slower when receding. Rømer knew that the true orbital period of Io should have nothing to do with the relative positions of the Earth and Jupiter, and he therefore concluded that this must be due to the finite speed of light.

Rømer estimated that when the Earth is nearest to Jupiter, eclipses of Io should occur about twenty-two minutes earlier compared to when Earth was farthest away. So, Rømer estimated that light takes about 11 minutes to travel the radius of the earth's orbit, which was refined, over the years, to the 8.3 minutes per astronomical unit we are familiar with today. This speed was confirmed in 1727, when James Bradley calculated the speed of light using a different astronomical method, that of *stellar aberration*.[240]

Bradley was not trying to measure stellar aberration, but rather attempting to triangulate the distance from the earth to a star in the constellation Draco by measuring its annual apparent shift in the sky. Bradley did measure such an angular shift, but with a totally different functional dependence than he expected. The shift depended on the perpendicular *velocity*, rather than the perpendicular *position*, of the earth compared to the star. Bradley also found that all stars showed the same effect, so the shift must be a property of the light rather than the star.

On a calm rainy day, raindrops fall straight down. However, from the perspective of a moving car, the rain appears to fall at an angle. The faster you travel, the more slanted the rain appears. The angle of the rain only depends on the ratio of your horizontal velocity to the terminal velocity of the raindrops. If you were to look at the car's speedometer, and measure the angle of the rain from the side window, you could easily calculate the terminal velocity of a raindrop.

The same idea of falling rain applies to stellar aberration. Since Bradley knew the speed of the earth's orbit around the sun, and the angle of stellar aberration, he could calculate the speed light travels. This, for all practical purposes, established the modern value of 7.2 astronomical units per hour for the speed of light.

There was a catch. The distance from the earth to the sun was still very uncertain. While Bradley could accurately compare measurements of the speed of light, he could not express it in standard units, such as miles per hour, to anywhere near the same accuracy. In fact, Bradley himself had

---

[239] Foschi, R. and Leone, M, "Galileo, measurement of the velocity of light, and the reaction times", Perception 38, no. 8 (2009), 1251-9.

[240] James Bradley, "An account of a new discovered motion of the fixed stars", Phil Trans Roy Soc. **35** (1727), 637-61.

made the best measurement of the astronomical unit, just a few years earlier, with an uncertainty of about 30%. For the next 120 years, each improved measure of the astronomical unit resulted in a corresponding improvement in the value of the speed of light.

Bradley's discovery of stellar aberration not only corroborated Rømer's measurement of the speed of light, but also greatly supported Isaac Newton's particle model, over the wave model of Robert Hooke[241] and Christiaan Huygens.[242] Bradley could easily explain his result using Newton's light corpuscles, because simple vector addition of the velocities of earth and a light particle gave the correct angles.[243]

On the other hand, the Hooke-Huygens wave theory predicted that the angle would depend on not only the velocities of light and Earth, but also on the velocity of the medium through which light supposedly traveled. Just as our raindrop analogy requires it to be a calm day, the wave model could only explain stellar aberration if—by extraordinary coincidence—the medium were stationary with respect to the sun. Thus, Newton's corpuscular model dominated until Young's double slit experiment (p. 425), and Fresnel's transverse wave theory of light.

Bradley's observation was the hardest for Fresnel to explain. Perhaps, Fresnel thought, aether is partially dragged along in a transparent medium, like glass or water, with refractive index n. If the velocity of the aether then changed by a factor of $\left(1-\frac{1}{n^2}\right)$, Bradley's observation would also become consistent with a wave model of light.

Fresnel's hypothesis made a specific prediction: if measurements were made with a water-filled telescope the aberration would be unaffected by the presence of the water. There were no water filled telescopes at the time, but by mid-century the speed of light could finally be measured in the laboratory allowing for further testing of Fresnel's formula. The first laboratory measurement of the speed of light was done independently by two French physicists: A. Hippolyte L. Fizeau and Jean Bernard Léon Foucault, whose results Maxwell also cited (p. 479).

The apparatus designed by Fizeau consisted of a light source and a rotating toothed wheel. The rotating wheel had gaps between the teeth, or gears, through which light could pass. The light was sent out through one place in the teeth, reflected by a mirror placed 8 km away, and if the wheel was rotating fast enough, returned through a different gap. The speed of light could subsequently be calculated using the distance from mirror to wheel, the speed of wheel rotation, and the spacing between the teeth of the wheel. The results of Fizeau's experiment allowed him to calculate a value of 313,300 km/s for the speed of light.[244]

Foucault altered Fizeau's method by replacing the rotating toothed wheel with a rotating mirror. A light source was then shone onto the rotating mirror where it was reflected onto a distant concave fixed mirror. The light then reflected from the fixed mirror back to the rotating mirror, and returned to the light source. If the rotating mirror was spinning at high speeds, the returning

---

[241] R. Hooke, Micrographia: Some Physiological Descriptions of Minute Bodies Made by Magnifying Glasses with Observations and Inquiries Thereupon, (1665), London, Printed by Jo. Martyn, and Ja. Allestry, Printers to the Royal Society, Project Gutenberg E-book number 15491.

[242] C. Christian Huygens, Treatise on Light, translated by S. P. Thomson, (Chicago: University of Chicago Press, 1912).

[243] James Bradley, "An account of a new discovered motion of the fixed stars," Phil Trans Roy Soc. **35** (1727), 646-9.

[244] Fizeau, "Sur un expérience relative á la vitesse de propagation de la lumière", Comptes Rendus, (1849), **29** (1849), p. 90. An English translation is available in W.F. Magie, Source Book in Physics, (Cambridge: Harvard University Press, 1935), 341-42.

light hit it at a slightly different place causing the returning beam to be shifted from its original path. The speed of light could then be measured by taking into account the speed of mirror rotation, angle of the shift, and the distance between the rotating and fixed mirror.

Foucault used a small steam turbine to spin a mirror at the rate of 800 rotations per second, which he calibrated using a tuning fork and train whistle.[245] A light beam was reflected from it to another mirror 9 meters away. When it returned, 60 nanoseconds later, the mirror had rotated a small amount, causing the return beam to be deflected a little below the source. When the mirror is at any other angle, the light beam is reflected elsewhere in the room and lost. Foucault continually increased the accuracy of this method over the years, and his final measurement in 1862 determined that light travels at 299,796 km/s.[246]

To test the prediction of Newton's corpuscular theory that light should move faster in water than in air, Foucault introduced a tube of water, 3 meters in length, between the rotating and fixed mirrors. If Newton's prediction were correct, and the light speed in water is greater than in air, then the return light beam should arrive in less than 60 nanoseconds, and its path would be deflected closer to the source. Instead Foucault found that by introducing the water-filled tube, the light path deflected farther from the source. This showed that light travels more slowly in water than in air, in complete disagreement with the prediction of Newton's particle model. Foucault's experiment convinced the majority of scientists that Newton's theory had to be abandoned.

Their most astonishing experiment, however, was Fizeau's running water experiment of 1851. Fizeau set up two glass tubes, each about 5 mm in diameter and 1.5 m in length. He then circulated water through them at a speed of about 7 m/s in opposite directions.

Fizeau then built an interferometer by splitting a beam of light, and sending the rays down the two tubes, so one parallel ray traveled upstream and the other downstream. When the light rays were recombined, Fizeau could detect interference fringe shifts as a function of water velocity, in complete agreement with Fresnel's aether drag hypothesis. Fizeau concludes his paper thus:

> The success of the experiment seems to me to render the adoption of Fresnel's hypothesis necessary, or at least the law which he found for the expression of the alteration of the velocity of light by the effect of motion of a body; for although that law being found true may be a very strong proof in favour of the hypothesis of which it is only a consequence, perhaps the conception of Fresnel may appear so extraordinary, and in some respects so difficult, to admit, that other proofs and a profound examination on the part of geometricians will still be necessary before adopting it as an expression of the real facts of the case. — *Competes Rendus*, Sept. 29, 1851.[247]

So, the hypothesis of Fresnel became well-established, but it still seemed too neat a coincidence for Fresnel's factor, or *drag coefficient*, to miraculously cancel the effect of motion through the aether. Scientists wondered if they could measure the velocity of the earth relative to the aether. But how?

---

[245] Foucault's speed of light experiment in on display at the Musée des arts et métiers in Paris.

[246] Foucault, "Détermination expérimentale de la vitesse e la lumière; parallaxe du Soleil," Comptes Rendus, **55** (1862), 501-3, and 792-96. English translation is available in W.F. Magie, Source Book in Physics, (Cambridge: Harvard University Press, 1935) 343-44.

[247] Hippolyte Fizeau, "The Hypotheses Relating To The Luminous Æther, And An Experiment Which Appears To Demonstrate That The Motion Of Bodies Alters The Velocity With Which Light Propagates Itself In Their Interior," Philosophical Magazine, Series 4, vol. **2** (1851), 568-573.

James Clerk Maxwell turned to astronomy for a method to determine the solar system's possible motion through the aether by considering observations of the eclipses of Jupiter's moons, similar to those Rømer used to calculate the speed of light. By comparing the velocity of light when Jupiter is seen from the earth at nearly opposite points of the ecliptic, Maxwell believed that the velocity of the solar system relative to the aether could be found. He wrote that:

> The only practicable method of determining directly the relative velocity of the aether with respect to the solar system is to compare the values of the velocity of light deduced from the observation of the eclipses of Jupiter's satellites when Jupiter is seen from the earth at nearly opposite points of the ecliptic.[248]

Maxwell's method would measure an effect that is of first order in the velocity of the earth (or solar system) with respect to the aether. That is, the time delay he sought is approximately given by:

$$\Delta t = \frac{d_{Earth-Sun}}{(c-v)} - \frac{d_{Earth-Sun}}{(c+v)} = \frac{2(d_{Earth-Sun})v}{c^2 - v^2} \approx \frac{2(d_{Earth-Sun})v}{c^2} = \frac{2v}{c}t_0,$$

where $(d_{Earth-Sun})$ is the diameter of earth's orbit, v is the speed of the earth through the aether, and t_0 is the light-crossing time required for this distance (approximately 16 minutes). So, if $v = 30\ \text{km/s}$, for example, one would expect a time delay of about $\Delta t \approx 0.2\ \text{s}$. Unfortunately, it was impossible to detect such a short time delay over the course of half a year.[249]

Sir George Airy, in 1871, attempted to measure the velocity of this *aethereal wind*, by repeating Bradley's aberration measurement of the star γ-Draconis, but using a water-filled telescope (shown).[250] So, how would this detect the motion of the Earth?

Light moves slower in a medium by a factor of $1/n$, where n is the medium's index of refraction. Therefore, if light from a star moves through a telescope filled with water instead of air, it will take n times longer for the light to travel the length of the telescope.

When filled with water, the telescope should, therefore, be tilted more from the vertical to keep the star in view. According to a simple analysis using Snell's law, the difference between the aberration angles measured under these two different circumstances ought to equal $(n^2 - 1)v/c$, where v is the speed of the earth through the aether.

Airy failed to measure any velocity change, even after observing over the course of two years. The aberration angle was the same whether the telescope was filled with air or water. Instead, Fresnel's hypothesis of a partial drag of the light by the water itself appeared to explain perfectly Airy's null result.

---

248 J.C. Maxwell, "Ether,"Encyclopedia Britannica, Ninth Edition **8**: 568–572. By carefully reading Maxwell's quote above, we can derive the expression for Δt by considering the effect of measuring light from Jupiter 6 years apart (the period of Jupiter's orbit is 12 Earth years) and assuming the velocity to be directed along the line joining Earth and Jupiter at those antipodal points.

249 J.C. Maxwell, "On a Possible Mode of Detecting a Motion of the Solar System through the Luminiferous Ether," Nature **21** (1880), 314-315.

250 G.B. Airy, "On the Supposed Alteration in the Amount of Astronomical Aberration of Light, Produced by the Passage of Light through a Considerable Thickness of Refracting Medium," Proceedings of the Royal Society of London **20** (1871), 35-39.

Maxwell then made another point about the laboratory experiments that measured the speed of light. In each of these, such as those of Fizeau and Foucault, light beams retrace their paths. So any velocity v the earth may have relative to the aether would affect the time of this round trip by an amount of second order in v.

To see this, consider the problem a riverboat traveling upstream to a destination, and downstream on the return trip. The total round trip will still take longer than if on a calm lake, and if L is the one way distance, then the total round trip travel time for the light would be:

$$t = \frac{L}{v_{\text{ship}} + v_{\text{water}}} + \frac{L}{v_{\text{ship}} - v_{\text{water}}} = \frac{2Lv_{\text{ship}}}{v_{\text{ship}}^2 - v_{\text{water}}^2} \approx \frac{2L}{v_{\text{ship}}}\left(1 + \frac{v_{\text{water}}^2}{v_{\text{ship}}^2}\right).$$

In our analogy, the light is the ship and the aether is the river, so:

$$\Delta t = t - t_0 \approx \frac{2L}{c}\left(1 + \frac{v^2}{c^2}\right) - \frac{2L}{c} \approx \frac{2L}{c}\frac{v^2}{c^2} \approx \frac{v^2}{c^2}t_0,$$

where t_0 is the time required for light to make the round trip with no wind. If, for example, the speed of the Earth relative to the aether were on the order of its orbital speed about the Sun, then $v^2 / c^2 \approx 10^{-8}$. Maxwell noted that the small size of the time delay would be extremely difficult to detect, at least with the technology of his day.

This led Albert A. Michelson to build his perpendicular interferometer, which had produced interference fringes between perpendicular beams of light. As the apparatus rotated, leading him to the too bold conclusion that there was no aether (p. 404).

Michelson then joined forces with Edward Morley, and in 1887 they again attempted to measure the fringe shift to within one part in 200 (p. 481). Once again, the expected shift was not detected.[251] Even more than Airy's observation before it, this null result took nearly every scientist by surprise.

The Irish physicist George Francis FitzGerald, using a recent result of Oliver Heaviside's, suggested an explanation. Heaviside had shown that the electric field of a charged sphere moving uniformly at speed v is compressed by a factor of $1/\sqrt{1-v^2/c^2}$ in the direction of its motion.[252] Since it seemed likely that electromagnetic forces between molecules hold material bodies together, FitzGerald reasoned, would it be possible that all bodies suffer the same contraction when moving through the aether?[253]

FitzGerald's hypothesis drew scant attention until Lorentz discovered it independently in 1892.[254] Lorentz then made a straightforward calculation to show that the results of the Michelson-Morley

---

[251] A.A. Michelson, "The Relative Motion of the Earth and the Luminiferous Ether," Am. J. Sci., **122** (1881), 120-129; A.A. Michelson and E.W. Morley, "On the Relative Motion of the Earth and the Luminiferous Ether" Am. J. Sci., **134** (1887), 333-345.

[252] Oliver Heaviside, "On the Electromagnetic Effects due to the Motion of Electrification through a Dielectric," Philosophical Magazine, Series 5, vol. **27** (167) (1889), 324-339.

[253] G.F. Fitzgerald, "The Ether and Earth's Atmosphere," Science, **13** (1889), 390.

[254] H.A. Lorentz, 'The Relative Motion of the Earth and the Aether," Amsterdam, Zittingsverlag Akad. v. Wet., **1** (1892), 74.

aether-wind experiments would be explained if the length of the interferometer arm were contracted in its direction of motion by this same factor.

Yet, this explanation of Michelson and Morley's null result appeared to be even more *ad hoc* than using Fresnel's theory of partial drag to explain Airy's stellar aberration experiment. After all, Fizeau's laboratory measurements had seemingly confirmed Fresnel's hypothesis long before Airy filled his telescope with water. The nature of molecular forces, on the other hand, was not a settled question at the time. Moreover, since all bodies would be subject to this effect, it was difficult to see how the contraction could be measured. At the close of the nineteenth century, scientists' understanding of electrodynamics was therefore hampered by confusion regarding the optics of moving bodies.

15.1 *The Wave Equation in Free Space*

James Clerk Maxwell wrote down the equations of electrodynamics, including his modification of Ampère's law, and was able to derive a wave equation for the propagation of electromagnetic radiation through a medium. These coupled equations of Maxwell combine to form a three-dimensional wave equation that also describe how the fields propagate through a vacuum, so Maxwell concluded that free space must be filled with a substance, the aether, through which these waves could propagate. This also fits nicely with Maxwell's philosophy of cause and effect via contact, as there is no need for an action at a distance.

In this section we rederive Maxwell's wave equation from the free space field equations, and show that the speed of light is the same in any inertial reference frame. This constancy of the speed of light makes perfect sense in an aether, in much the same way that sound travels at a particular speed in air. This is one reason that Maxwell's aether model of light dominated the late nineteenth century. At the end of this section, we present experiments that attempted to measure the relative between Earth and the aether, and their conclusion, no matter how odd it may seem.

Discussion 15.1: Maxwell's Field Equations in Free Space

In Chapter 14 we developed Maxwell's equations in a medium, and showed how the speed of light relates to the permittivity and permeability of the medium. Since the permittivity and permeability of free space can be determined experimentally (without a clock), we can use these to calculate the speed of light in free space as:

$$c = \frac{1}{\sqrt{\mu_0 \varepsilon_0}} \approx \frac{1}{\sqrt{\left(1.26\,{}^{\mu\text{H}}\!/\!_{\text{m}}\right)\left(8.85\,{}^{\text{pF}}\!/\!_{\text{m}}\right)}} \approx \frac{1}{\sqrt{11\times10^{-18}\,{}^{\text{HF}}\!/\!_{\text{m}^2}}} \approx \frac{1}{3.3\,{}^{\text{ns}}\!/\!_{\text{m}}} \approx 0.30\,{}^{\text{m}}\!/\!_{\text{ns}}.$$

Maxwell presented a table of independently measured values of $c = \frac{1}{\sqrt{\mu_0 \varepsilon_0}}$ (the ratio of electric units) and of the speed of light as measured kinematically, and wrote the conclusion that follows:

Velocity of Light (mètres per second)		Ratio of Electric Units (mètres per second)	
Fizeau	314000000	Weber	310740000
Aberration, &c., and Sun's Parallax	308000000	Maxwell	288000000
Foucault	298360000	Thomson	282000000

> It is manifest that the velocity of light and the ratio of the units are quantities of the same order of magnitude. Neither of them can be said to be determined as yet with such as degree of accuracy as to enable us to assert that the one is greater or less than the other. It is to be hoped that, by further experiment, the relation between the magnitudes of the two quantities may be more accurately determined.
>
> In the meantime our theory, which asserts that these two quantities are equal, and assigns a

physical reason for equality, is certainly not contradicted by the comparison of these results such as they are.[255]

Now we consider the basic laws of electrodynamics in matter to be *Maxwell's field equations*, and they are expressed in the four-field version as:

$$\vec{\nabla}\cdot\vec{B}=0,\quad \vec{\nabla}\cdot\vec{D}=\rho,\quad \vec{\nabla}\times\vec{H}=\vec{J}+\tfrac{\partial}{\partial t}\vec{D},\quad \text{and}\quad \vec{\nabla}\times\vec{E}=-\tfrac{\partial}{\partial t}\vec{B}.$$

The Maxwellians pictured free space as a perfectly linear medium, which allowed the free space equations to be represented as:

$$\vec{\nabla}\cdot\vec{B}=0,\quad \vec{\nabla}\cdot\vec{E}=\rho/\varepsilon_0,\quad \vec{\nabla}\times\vec{B}=\mu_0\vec{J}+\mu_0\varepsilon_0\tfrac{\partial}{\partial t}\vec{E},\quad \text{and}\quad \vec{\nabla}\times\vec{E}=-\tfrac{\partial}{\partial t}\vec{B}.$$

Recall our discussion of the characteristic impedance of a medium (p. 441). Moreover, recall that by combining our fundamental constants, μ_0 and ε_0, we can define the *impedance of free space* as:

$$Z_0\equiv\sqrt{\frac{\mu_0}{\varepsilon_0}}\approx\sqrt{\frac{1.26\,^{\mu\text{H}}/_{\text{m}}}{8.85\,^{\text{pF}}/_{\text{m}}}}\approx\sqrt{0.14\tfrac{\mu\text{H}}{\text{pF}}}\approx\sqrt{0.14\tfrac{\mu\text{H}}{\text{pF}}}\approx 377\,\Omega.$$

Making the following substitutions:

$$Z_0=\sqrt{\frac{\mu_0}{\varepsilon_0}}\;,\qquad c=\frac{1}{\sqrt{\mu_0\,\varepsilon_0}}\;,\qquad \mu_0=\frac{Z_0}{c}\;,\qquad \varepsilon_0=\frac{1}{c\,Z_0}\;,$$

Maxwell's equations in free space become:

$$\vec{\nabla}\cdot\vec{B}=0,\quad \vec{\nabla}\cdot\vec{E}=c\,Z_0\,\rho,\quad \vec{\nabla}\times\vec{B}=\tfrac{Z_0}{c}\vec{J}+\tfrac{1}{c^2}\tfrac{\partial}{\partial t}\vec{E},\quad \text{and}\quad \vec{\nabla}\times\vec{E}=-\tfrac{\partial}{\partial t}\vec{B}.$$

Despite the overwhelming success of Maxwell's equations in predicting natural phenomena, Maxwell had a fundamental physical misconception about the "physical reason for equality" between the "ratio of electrical units" and the "velocity of light." Maxwell understood the constitutive relations between $\vec{D}\,\&\,\vec{E}$ and $\vec{H}\,\&\,\vec{B}$ to be properties of matter including the aether, which is why we still use the constants μ_0 and ε_0 today.

We now explain the propagation of electromagnetic waves throughout space without any need for Maxwell's aether, so theoretical physicists usually interpret the free space field equations as fundamental laws of nature in their own right, unless they are applying a more modern theory such as quantum electrodynamics.

Electrical engineering, on the other hand, has mostly retained Maxwell's four-field formulation, with its empirical constitutive relations. This makes a great deal of historical sense, given the parallel development of electrodynamics and engineering in the nineteenth century.

Since at least the time of Aristotle, the goal of physics has been to find the universal laws of nature. Maxwell was a true physicist in this philosophical sense. If Maxwell were alive today, he would probably say that he got it wrong, since his conclusion about the necessity of the aether has been debunked. On the other hand, others would surely disagree, touting the myriad of technological successes that have resulted from Maxwell's theory. Indeed, Heinrich Hertz, whose

---

[255] J.C. Maxwell, Treatise on Electricity and Magnetism, vol. 2, 3rd ed., (1873; rpt. New York: Dover, 1954), Art. 787, p. 436.

experiments verified the existence of Maxwell's electromagnetic waves, wrote, "Maxwell's theory is Maxwell's system of equations."[256]

Maxwell valued physical understanding, so he might be astonished that the equations he formulated have long since transcended the physical model of an aether. It is interesting, and perhaps a little ironic, that Maxwell's equations are now used as the starting point from which to mathematically derive all of electromagnetic theory. This irony is crystalized in the humorous tee shirts that say: "And God said:

$$\vec{\nabla}\cdot\vec{B}=0 \quad \vec{\nabla}\cdot\vec{D}=\rho \quad \vec{\nabla}\times\vec{E}=-\frac{\partial\vec{B}}{\partial t} \quad \vec{\nabla}\times\vec{H}=\vec{J}+\frac{\partial\vec{D}}{\partial t},$$

and there was light."

Of course, Maxwell was no mystic, nor was he addicted to a particular heuristic model of any natural phenomenon. He might point out, as a modern physicist certainly would, that his fundamental equations are not some arbitrary invention. All four equations express fundamental experimental facts, albeit in the language of field theory. For example, Gauss's law is equivalent to the inverse square law that Coulomb verified in his laboratory.

Thought Experiment 15.1: The Field Equations in a Vacuum

Here we will investigate Maxwell's equations in a complete vacuum, without any charges, currents, electric dipoles, or magnetic moments. So:

$$\vec{\nabla}\cdot\vec{B}=0,\quad \vec{\nabla}\cdot\vec{E}=0,\quad \vec{\nabla}\times\vec{B}=\tfrac{1}{c^2}\tfrac{\partial}{\partial t}\vec{E},\quad \text{and}\quad \vec{\nabla}\times\vec{E}=-\tfrac{\partial}{\partial t}\vec{B}.$$

We now combine Maxwell's equations to solve for the electric field:

$$\vec{\nabla}\times\vec{\nabla}\times\vec{E}=\vec{\nabla}\times\left(-\tfrac{\partial}{\partial t}\vec{B}\right) \;\rightarrow\; \vec{\nabla}\left(\vec{\nabla}\cdot\vec{E}\right)-\nabla^2\vec{E}=-\tfrac{\partial}{\partial t}\left(\vec{\nabla}\times\vec{B}\right)$$

$$-\nabla^2\vec{E}=-\tfrac{\partial}{\partial t}\left(\tfrac{1}{c^2}\tfrac{\partial}{\partial t}\vec{E}\right) \;\rightarrow\; \nabla^2\vec{E}=\tfrac{1}{c^2}\tfrac{\partial^2}{\partial t^2}\vec{E}.$$

In Problem 15.1, you will derive a similar equation for the magnetic field.

Notice that these are three dimensional wave equations, which propagate at a speed $v=c$. Moreover, notice an odd thing. To measure a speed requires two events that are separated by time and space—a ruler and a clock. However, the permeability and permittivity of free space can be measured in a stationary laboratory. If the laboratory were moving, therefore, either the permittivity and permeability must change, or the speed of light must be the same. Both conclusions are counterintuitive.

Problem 15.1: The Magnetic Field Wave Equation

Show from Maxwell's equations in a vacuum that the general wave equation for the magnetic field is: $\nabla^2\vec{B}=\tfrac{1}{c^2}\tfrac{\partial^2}{\partial t^2}\vec{B}$.

Discussion 15.2: The Michelson-Morley Experiment

---

[256] Heinrich Hertz, Electric Waves, trans. D.E. Jones (London: MacMillan and Co., 1893), 21.

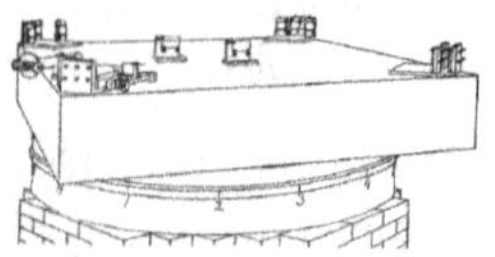

After Albert Michelson failed in his first attempt to measure differences in the speed of light because of the relative motion of the earth through the aether, he tried again with the help of his senior colleague, Edward Morley, to measure the Earth's velocity with respect to the aether using what is now known as a *Michelson interferometer.* In Problem 15.2 you will have a chance to work through the physical principles, but in short the following are their experimental results:

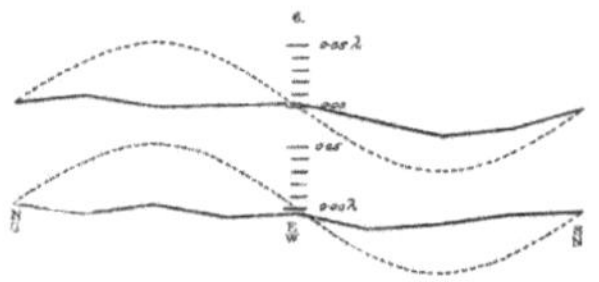

> The results of the observations are expressed graphically in fig. 6. The upper is the curve for the observations at noon, and the lower that for the evening observations. The dotted curves represent *one-eighth* of the theoretical displacements. It seems fair to conclude from the figure that if there is any displacement due to the relative motion of the earth and the luminiferous ether, this cannot be much greater than 0.01 of the distance between the fringes.
>
> Considering the motion of the earth in its orbit only, this displacement should be $2D(\frac{v}{c})^2 = 2D\times10^{-8}$. The distance D was about eleven meters, or 2×10^7 wavelengths of yellow light; hence the displacement to be expected was 0.4 fringe. The actual displacement was certainly less than the twentieth part of this, and probably less than a fortieth part. But since the displacement is proportional to the square of the velocity, the relative velocity of the earth and ether is probably less than one-sixth the earth's orbital velocity, and certainly less than one-fourth.
>
> In what proceeds, only the orbital motion of the earth is considered. If this is combined with the motion of the solar system, concerning which but little is known with certainty, the result would have to be modified; and it is just possible that the resultant velocity at the time of the observations was small though the chances are much against it. The experiment will therefore be repeated at intervals of three months, and thus all uncertainty will be avoided.
>
> It appears, from all that precedes, reasonably certain that if there be any relative motion between the earth and the luminiferous ether, it must be small; quite small enough entirely to refute Fresnel's explanation of aberration. Stokes has given a theory of aberration which assumes the ether at the earth's surface to be at rest with regard to the latter, and only requires in addition that the relative velocity have a potential; but Lorentz shows that these conditions are incompatible. Lorentz then proposes a modification which combines some ideas of Stokes and Fresnel, and assumes that existence of a potential, together with Fresnel's coefficient. If now it were legitimate to conclude that the ether were at rest with regard the earth's surface, according to Lorentz there could not be a velocity potential, and his own theory also fails.[257]

A clear experimental result that contradicts an existing theory, but does not confirm the predictions of any other known theory, presents a novel puzzle just waiting to be solved. If scientists already have a reasonable theory to explain something, why discard it *unless* there is an alternate theory that explains both everything the first theory does *and* the new observations? Experiments appear to contradict physical theories all the time, and usually either the experiment is irreproducible, or there is another relatively mundane explanation. For most scientists, it is usually better to just wait and see what comes up. After all, extraordinary conclusions require extraordinary evidence.

Michelson and Morley's null measurement of the aether wind was eventually explained by Albert Einstein. Einstein also explained all the other results that had been interpreted as supporting the aether theory. Moreover, his explanation made specific predictions about kinematics and classical

[257] A.A. Michelson and E.W. Morley, "On the Relative Motion of the Earth and the Luminiferous Ether" Am. J. Sci., **134** (1887), 333-345.

mechanics, which were subsequently tested. In every case, his new theory was consistent with experiment.

At this point, we therefore introduce a new fundamental law. This law was hinted at by the Michelson-Morley experiment, but proposed by Einstein in 1905. This law states that the speed of light in free space is a constant, regardless of how one measures it. Or, as Einstein put it:

Fundamental Law 10: The Constancy of the Speed of Light

> Every ray of light moves in the co-ordinate system "at rest" with the definite velocity v independent of whether this ray of light is emitted by a body at rest or in motion.[258]

Discussion 15.3: The Road to Relativity

Fundamental Law 10 means that no matter what reference frame you are in, the speed of light in free space will always have the same value. This completely contradicts what we would expect from Galilean relativity. Consider a train coming toward you at night with its headlight on. If the speed of light is $v = c$ with respect to the train, then should it not be $c + v_{\text{train}}$ with respect to you? After all, if a passenger is walking toward the engine at a speed v_{p}, she is approaching you at a speed $v_{\text{p}} + v_{\text{train}}$. Why should light behave differently?

Maxwell had answered that there must exist a medium through which light waves travel. The sound of a train whistle, after all, travels at the same speed, with respect to the air, regardless of how fast the train is moving.

However, the idea expressed by Fundamental Law 1 is different, as Michelson and Morley's null result makes very clear. If there is a medium through which light travels, it cannot be detected by any experiment, and the speed of light must be a universal constant.

So, what about Galileo's idea of relativity that we introduced earlier? If there is some fundamental speed, whose value is universal and frame-independent, then how can the laws of physics be the same regardless of reference frame?

We showed in Part II that the Galilean transformations were inconsistent with electrodynamics, but we did not conclude that the principle of relativity is no longer applicable to the electromagnetic field. The problem was partially solved in the late 1800s, when Oliver Heaviside and George FitzGerald found that, if electromagnetic lengths contract with velocity, the Michelson and Morley results could be easily explained. This led to a new set of field transformations, developed by FitzGerald, and then expanded upon by Lorentz. Chapter 17 covers Einstein's theory of relativity and these Lorentz-FitzGerald transformations in detail.

However, neither Heaviside, nor FitzGerald, nor Lorentz supposed that the aether did not actually exist. The question for them was: what property of the aether made lengths appear to contract? After all, hadn't the spooky action at a distance demon been slain once and for all? If there is no aether, something other than action at a distance must take its place.

Thought Experiment 15.2: The Speed of Light in a Moving Frame

Consider light propagating in a vacuum in reference frame $\lfloor \text{s}$. According to the wave equation that we derived from Maxwell's equations, the wave is propagating at a speed $v = c$.

---

[258] From "On the Electrodynamics of Moving Bodies," by Albert Einstein (1905), translated by Anna Beck, ©1989 by the Hebrew University of Jerusalem.

Now consider a reference frame $\lfloor s'$ moving at constant velocity $\vec{u}$ with respect to frame $\lfloor s$ in the same direction the light is traveling. The laws of physics must be equally valid in this reference frame as the first, so the electric and magnetic fields must also follow the vacuum wave equations:

$$\nabla^2 \vec{E}' = \tfrac{1}{c^2}\frac{\partial^2 \vec{E}'}{\partial t^2} \quad , \quad \nabla^2 \vec{B}' = \tfrac{1}{c^2}\frac{\partial^2 \vec{B}'}{\partial t^2} .$$

This wave is also propagating at the same speed, $v' = c$.

According to Galilean relativity, however, the speed of the wave, as seen in the new frame, must be: $v' = v - u$. Therefore we have encountered, yet again, a contradiction to Galilean relativity!

Problem 15.2: The Michelson-Morley Experiment

In the late nineteenth century the idea that light is an electromagnetic wave became well established, so it was assumed there must be some medium through which light waves propagate. This medium was called the aether, and the propagation speed of light must therefore be a property of the aether. The Earth moves on its axis, travels around the sun, and the sun travels around the galaxy, so the Earth must be moving with respect to the aether, creating an *aethereal wind*. Thus it should be possible to devise an experiment to measure this aethereal wind velocity $\vec{v}_w$. This is just what American physicist Albert Michelson set out to do in 1881, and again in 1887 with the help of Edward Morley.

Michelson and Morley set up an optical bench on a turntable. On it they arranged for a beam of monochromatic light, with angular frequency ω, to be split, travel equal distances d in perpendicular directions, recombine and finally be detected. From this they could very precisely measure the aethereal wind velocity $\vec{v}_w$.

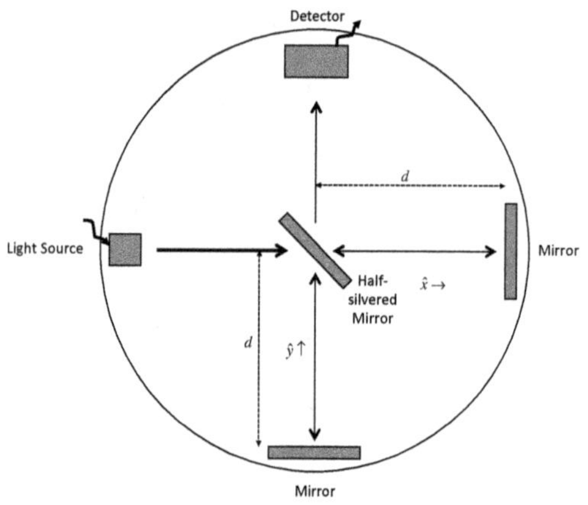

Assume for simplicity that the wind is blowing in the $\hat{x}$ direction, so, according to Galilean relativity, the speed of light in the $\hat{y}$ direction is simply c, while the speed of light in the $\hat{x}$ direction would be $c \pm v_w$, depending on whether it is traveling upwind or downwind.

a) Solve for the light travel time difference Δt between the two paths.

b) For a given measured phase difference ϕ, find the expected aethereal wind velocity.

c) Michelson and Morley never measured any phase differences, so the aethereal wind speed was clearly too slow to detect. Research and report on the most popular explanations at the time for this null result.

d) Thomas Kuhn was arguably the most influential historian and philosopher of science of the twentieth century. If you have not done so already, peruse Kuhn's book The Structure of

Scientific Revolutions,[259] and discuss how the idea of the aether fits into the concepts of *a paradigm*, *normal science*, and *incommensurability*.

15.2 *The Inverse Square Law*

Light originates from sources such as light bulbs and stars. Since energy must be conserved, the brightness must decrease with the distance squared. Plane waves, however, do not follow the inverse square law. In this sense they are unphysical, but since most detectors are small compared to the distance to sources, the plane wave approxmation usually holds.

For example, in the case of an astronomical source, the plane wave approximation holds at the location of Earth. On the other hand, the inverse square law is used all the time to relate the brightness, luminosity, and distance of astronomical objects.

Thought Experiment 15.3: The Inverse Square Law of Light

A small light source isotropically emits at a given power, called *luminosity*, $\mathbb{P}(t) = \frac{dU}{dt}$. What is the energy flux $\vec{\mathbb{S}}(r,t)$ we would measure at a distance r from the source?

Conservation of energy demands that:

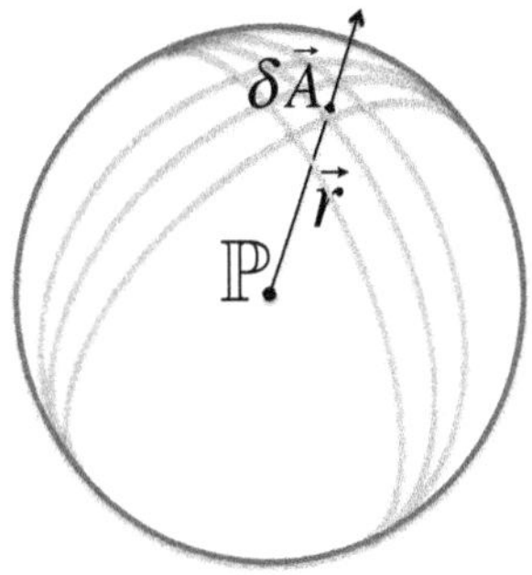

$$\vec{\nabla}\cdot\vec{\mathbb{S}} = -\tfrac{\partial}{\partial t}\mathfrak{u}_{\text{total}},$$

so:

$$\oint_{\text{Volume}} \vec{\nabla}\cdot\vec{\mathbb{S}}\,dV = \oint_{\text{Volume}} \left(-\tfrac{\partial}{\partial t}\mathfrak{u}_{\text{total}}\right)dV = \mathbb{P}$$

and drawing a spherical Gaussian surface of radius r centered on the source, we have:

$$\vec{\mathbb{S}} = \frac{\mathbb{P}}{4\pi r^2}\hat{r}.$$

While this relationship works well for a steady light source, when the source varies with time we must also take into account the time it takes light to travel from the source to the observer. The power the observer measures at time t is really the value of the object's luminosity when it was emitted at an earlier time $\tau = t - \frac{1}{c}r$, so to be more precise:

$$\oint_{\text{Surface}} \vec{\mathbb{S}}(t)\cdot d\vec{A} = \mathbb{P}\left(t - \frac{1}{c}r\right) \quad\to\quad \vec{\mathbb{S}}(t) = \frac{\mathbb{P}\left(t-\frac{r}{c}\right)}{4\pi r^2}\hat{r}.$$

This is the inverse square law of light. If we measure the energy flux $\vec{\mathbb{S}}(t)$, and know the distance between observer and source, we can determine the prior luminosity of the source.

Problem 15.3: The Solar Constant, Albedo, the Seasons, and Sea Ice

The solar radiation flux as measured just above Earth's atmosphere, $\vec{\mathbb{S}}(1\text{ AU}) = 1.36\frac{\text{kW}}{\text{m}^2}\hat{r}$, is called the solar constant. The fraction of this light that is reflected is called the *albedo* α. A white object has $\alpha = 1$, while a black object has $\alpha = 0$. Something is considered *gray* if its albedo is

[259] Kuhn, Thomas S., The Structure of Scientific Revolutions, (Chicago: The University of Chicago Press, 1962).

independent of wavelength. On a clear day, typical surface albedos are roughly 6%, 25%, and 60% for ocean, land, and ice, respectively.

a) Find the power per surface area imparted to a particular location on Earth, called the *heating rate*, as a function of the solar constant, the albedo, and the angle from the normal of sunlight, called the *solar zenith angle.*

b) What are the rates of heating on a clear day, in the following capital cities: Ottawa (45°N), Washington (39°N), Mexico City (19°N), Quito (0°), Lima (12°S), and Santiago (33°S)? Calculate this at noon on March 21st, June 21st, September 21st, and December 21st. Plot your results appropriately.

c) The Arctic Ocean is usually completely covered with ice each March when the sun reemerges. This ice melts for the next six months, until the sun sets in September. Find the heating rate of the Arctic Ocean on a sunny day as a function of the sea ice fraction, f, and the solar zenith angle.

d) Consider the whole Arctic Ocean, which as a surface area of 14 million square kilometers. Let $f_{\max}$ and $f_{\min}$ be the fraction of sea ice in March and September, which have historically been about 1.0 and 0.5 respectively. Estimate the total solar energy absorbed by the Arctic Ocean from April to September as a function of $f_{\max}$ and $f_{\min}$? Make a plot of the total energy as a function of $f_{\min}$, assuming that the whole ocean freezes every winter. Clearly state your simplifying assumptions.

Problem 15.4: Dwarf planets

Dwarf planets are continuously being discovered beyond the orbit of Neptune. To understand why they are often quite faint, you will investigate how bright they are as a function of their radius a, albedo α, distance $r_{\odot}$ from the sun, and distance $r_{\oplus}$ from the earth.

a) The solar radiation flux, as measured just above Earth's atmosphere, is called the solar constant $\left(\vec{\mathbb{S}}(1\text{ AU}) = 1.36\frac{\text{kW}}{\text{m}^2}\,\hat{r}\right)$. Find the equivalent solar energy per area as a function of distance from the sun as measured in astronomical units $\left(\frac{r}{\text{AU}}\right)$.

b) Find the power per surface area, imparted to a particular location on a planet, as a function of the distance from the sun, the solar constant, and the angle between the sun and the surface normal (the *solar zenith angle*).

c) Find the total power reflected back toward the sun. Assume that the light is reflected isotropically up, but not down, from each point on the dwarf planet's surface.

d) Of the solar system objects now called *dwarf planets*, Ceres was the first to be discovered (1801). It is orbiting the sun with a semi-major axis of 2.8 AU; its measured physical radius is about 470 km; and its average albedo is about 9%. Estimate the total power per area, as seen from Earth, when Ceres appears opposite the sun (*opposition*), when it appears near the sun (*conjunction*), and when it is at right angles to the sun (*quadrature*).

e) Discovered in 1930, Pluto orbits the sun with a semi-major axis of about 40 AU. Its physical radius is about 1200 km, and it appears about $\frac{1}{1000}$ as bright as Ceres at quadrature. Estimate the average albedo of Pluto.

Problem 15.5: The Cosmic Distance Ladder

The Small Magellanic cloud is a nearby dwarf galaxy that orbits the Milky Way. In 1912, Henrietta Leavitt of the Harvard College Observatory published a relationship between the period and the brightness of a particular kind of variable star called *Cepheid variables*.[260]

The prototypical Cepheid variable star is Delta Cephei, which is located about 1000 light years from the Earth, has a pulsation period of 5.4 days, and a visual magnitude range from 4.37 to 3.48.

The magnitude system has been used by astronomers for over two millennia. Since the mid 1800s, the magnitude difference between any two objects $(m_2 - m_1)$ has been defined in terms of their energy flux ratio $\left(S_2/S_1\right)$, measured using the same technique, such that:

$$m_2 - m_1 = -2.5\log_{10}\left(S_2/S_1\right),$$

with the bright star, Vega, defined to always have zero magnitude[261].

Leavitt's figures show the maximum, and minimum, magnitude as a function of period. Her Fig. 1 plots the period in days, while her Fig. 2 plots the base 10 logarithm of the same period.

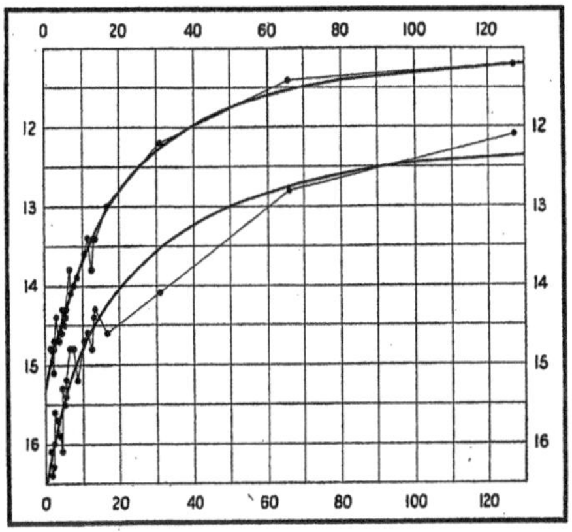

FIG. 1.

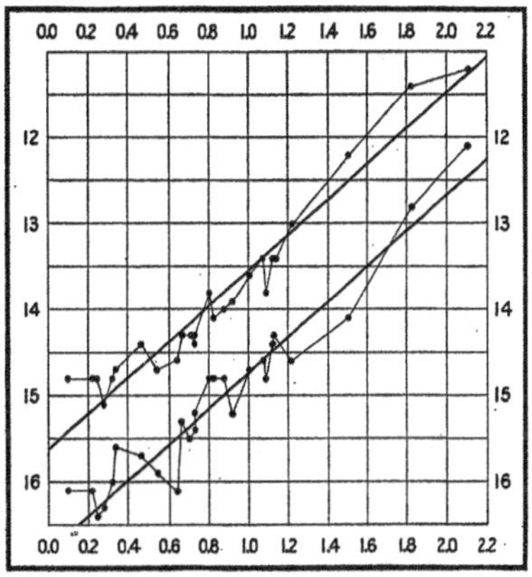

FIG. 2.

a) By reading the graph, find the magnitude range that δ-Cephei star would have if it were located in the Small Magellanic cloud.

b) What is the ratio of energy flux between δ-Cephei and a corresponding star in the Small Magellanic cloud?

c) According to these data, how far away is the small Magellanic cloud in light years? How does this compare to the modern distance of 200 thousand light years?

d) In 1925 Edwin Hubble published measurements of the magnitudes of stars in the brightest spiral nebulae in the constellation of Andromeda (M31).[262] At this time it was not known

---

[260] H.S. Leavitt, "Periods of 25 Variable Stars in the Small Magellanic Cloud," (Communicated by E.G. Pickering) Harvard College Observatory Circular, **173**, (1912).

[261] Astronomers no longer use one primary calibration star, but since the magnitude system is backward compatible, Vega's magnitude is still approximately zero.

[262] E. Hubble, "Cepheids in Spiral Nebulae," Popular Astronomy, **33** (1925), p. 252.

whether the spiral nebulae were inside our own galaxy, or external galaxies called *island universes.* Using Hubble's data, plot the same two quantities as in Leavitt's Fig 2, and fit a regression line. Find the ratio of the distance to M31 to the distance to the small Magellanic cloud.

CEPHEIDS IN M 31.

Var. No.	Period in Days	Log. P	Photographic Magnitude Max.
5	50.17	1.70	18.4
7	45.04	1.65	18.15
16	41.14	1.61	18.6
9	38	1.58	18.3
1	31.41	1.50	18.2
12	22.03	1.34	19.0
13	22	1.34	19.0
10	21.5	1.33	18.75
2	20.10	1.30	18.5
17	18.77	1.28	18.55
18	18.54	1.27	18.9
14	18	1.26	19.1

e) Putting your two results together, find the distance to M31.

f) Research and report on the primary reason for the discrepancy between Hubble's distance and the modern value of 2.5 million light years.

15.3 *Conserved Quantities in Plane Waves*

Sources of light tend to be finite in size. As these sources radiate, the light spreads out from a point resulting in the inverse square law (p. 485). However, in practice, sources of light are usually far away compared to the size of the detector. In such cases, it is most convenient to treat the wavefront as if it were part of an infinite plane, as we mostly did in the last chapter.

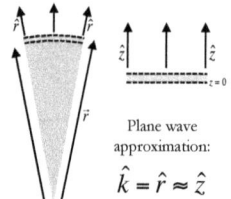

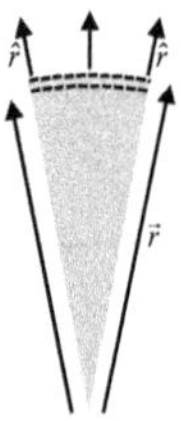

Plane wave approximation: $\hat{k} = \hat{r} \approx \hat{z}$

While widely applicable, the plane wave approximation can give unphysical results when overgeneralized to sources at finite distances. While a real wave expands out from its source, a plane wave stays the same size, so the inverse square law does not apply. Rather, in the plane wave approximation, all time-averaged conserved quantities remain uniform with position z.

In this section we will investigate conserved quantities associated with plane waves in free space, such as energy, momentum, and angular momentum.

Thought Experiment 15.4: The Energy Density

Consider light from a far away source traveling through a small volume element $dV = c\,dt\,dA$. The energy density is, therefore, given by:

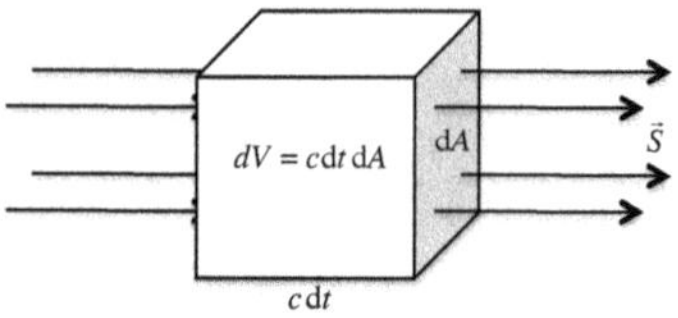

$$\mathbb{u} = \frac{d\mathbb{U}}{dV} = \frac{d\mathbb{U}}{c\,dt\,dA} = \frac{\mathbb{S}}{c}.$$

This works because of the definition of the Poynting vector (p. 440), as the energy per unit time per unit area.

Thought Experiment 15.5: Radiation Pressure

Consider a plane wave travelling through, but being absorbed by, a poorly conducting medium κ. Our goal is to find the force on the medium.

Recall Thought Experiment 14.15 (p. 466) where we found that the fields decay exponentially in the poor conductor limit, with attenuation length δ:

$$\delta = \frac{2}{\kappa}\sqrt{\frac{\varepsilon_0}{\mu_0}} = \frac{2}{Z_0\kappa}.$$

And, in Example 14.5 (p. 469) we found that in the poor conductor limit the magnetic field, so in terms of the distance and retarded time:

$$\vec{E} = \vec{E}_0 e^{-z/\delta}\cos(\omega\tau) \quad \text{and} \quad \vec{B} = \vec{B}_0 e^{-z/\delta}\cos(\omega\tau).$$

Now the time-averaged body force on the absorber is given by:

$$\vec{f} = \langle \vec{J}\times\vec{B}\rangle = \kappa\langle \vec{E}\times\vec{B}\rangle = \kappa\langle \vec{E}_0\cos(\omega\tau)e^{-z/\delta}\times\vec{B}_0\cos(\omega\tau)e^{-z/\delta}\rangle = \mu_0\kappa\langle \vec{\mathbb{S}}_{z=0}\rangle e^{-2z/\delta}.$$

So, in the low frequency and low conductivity limit (total absorption):

$$\frac{d\vec{F}}{dA} = \int_0^\infty \vec{f}\,dz = \kappa\frac{Z_0}{c}\langle \vec{\mathbb{S}}_{z=0}\rangle\int_0^\infty e^{-2z/\delta}\,dz = \kappa\frac{Z_0}{c}\langle \vec{\mathbb{S}}_{z=0}\rangle\left(\tfrac{1}{2}\delta\right) = \frac{1}{c}\langle \vec{\mathbb{S}}_{z=0}\rangle.$$

This is called *radiation pressure*, and is especially important in the study of very bright objects such as stars. Notice that the radiation pressure is equal to the energy density.

Problem 15.6: Solar Sails

A common futuristic navigation idea involves solar sails, which are highly reflective lightweight surfaces attached to spaceships.

a) The total radiative power output of the sun $(L_\odot)$ is called its *luminosity*. Use the energy continuity equation to write the solar energy flux $(\vec{\mathbb{S}}_\odot)$ as a function of heliocentric position $\vec{r}$.

b) Given a spaceship of mass (m), what must the area A of the sail be so that the force away from the sun balances the gravitational force toward the sun? Express your answer in terms of the mass of the ship and the mass and luminosity of the sun, which are $M_\odot = 2\times10^{30}\,\text{kg}$ and $L_\odot = 4\times10^{26}\,\text{W}$.

c) Gold has a density of about $20\,{}^{\text{g}}\!/_{\text{cm}^3}$. What is the maximum thickness of a gold foil sail that would be accelerated by the sun's radiation pressure? Are solar powered sails made of Gold foil a viable option for space travel?

d) Consider a spherical spaceship with radius a and mass density ρ. Find the acceleration of the ship as a function of the distance from the sun. How big of a ship would have no acceleration?

e) Models of star formation often involve a phase where the new star blows away the smallest particles, while leaving the bigger ones in place. What is this size cutoff for a sun-like star and for rock-like $(\rho \approx 2.6\,{}^{\text{g}}\!/_{\text{cm}^2})$ material. Would fine desert sand $(a \approx 200\,\mu\text{m})$ blow away?

Problem 15.7: Energy Density and Radiation Pressure

Consider pressure washing a house with a hose of water. The water hits the house with a velocity v and sprays sideways.

a) Using classical mechanics and Gaussian surfaces, show that the pressure on the house is given by $P = \rho v^2$, this is called *ram pressure.*

b) Find the kinetic energy density of the water, again in terms of the velocity and mass density.

c) Assume, for the sake of argument, that Newton was correct and light is made of a beam of particles governed by his laws. What would the relationship be between the radiation pressure and the energy density of light traveling in one direction? How does this compare to what we have found using Maxwell's equations?

Thought Experiment 15.6: The Momentum Density

Recall our derivation of radiation pressure (p. 488). Since force is simply the rate of momentum, we can use a similar method to find the momentum of an electromagnetic wave.

The radiation pressure is:

$$\frac{d\vec{F}}{dA} = \frac{\frac{d\vec{p}}{dt}}{dA} = \frac{d\vec{p}}{dAdt} = \frac{c\,d\vec{p}}{dAcdt} = c\frac{d\vec{p}}{dV},$$

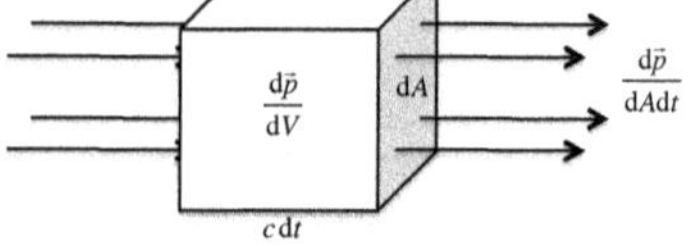

where $\vec{F}$ is the force exerted by the light on a small element dA of area, c is the speed of light, and $\vec{p}$ is the momentum in a small volume dV of the attenuator.

So, the momentum density of an electromagnetic field is given by:

$$\vec{\mathbb{p}} = \frac{d\vec{p}}{dV} = \frac{1}{c}\frac{d\vec{F}}{dA} = \frac{1}{c^2}\vec{\mathbb{S}}.$$

Thought Experiment 15.7: Circular Polarization

Now that we have discussed how electromagnetic waves carry momentum, a natural question arises: can they carry angular momentum? Before we dismiss the idea as far-fetched, let's consider *circular polarization.*

Consider the linearly polarized monochromatic light we discussed in Thought Experiment 14.8 (p. 452). However, assume that the two components are 90° out of phase, but with equal relative amplitudes, so:

$$\vec{E}(z,t) = E_0\left(\cos(\omega\tau)\hat{x} + \sin(\omega\tau)\hat{y}\right),$$

where, again, $\tau = t - \frac{1}{c}z$ represents the retarded time.

Notice that the angle of the electric field vector at a given position z is no longer independent of time, but it can be written:

$$\varphi(t) = \arctan\left(\frac{\sin(\omega\tau)}{\cos(\omega\tau)}\right) = \arctan\left(\tan(\omega\tau)\right) = \omega\tau.$$

Thus, at any given location on the z axis, the polarization angle appears to rotate to the *left* from the point of view of an observer looking at the source. We have expanded $\varphi(t)$ to show explicitly that the field vector appears to rotate uniformly with an angular velocity ω.

If the phase shift were, rather, in the other direction then:

$$\vec{E}(z,t) = E_0\left(\cos(\omega\tau)\hat{x} - \sin(\omega\tau)\hat{y}\right).$$

And the direction of the electric field vector is:

$$\varphi(z,t) = \arctan\left(\frac{-\sin(\omega\tau)}{\cos(\omega\tau)}\right) = -\omega\tau .$$

So, again, from the point of view of the observer the electric field vector now appears to rotate to the right, so it is called *right circular polarization.*

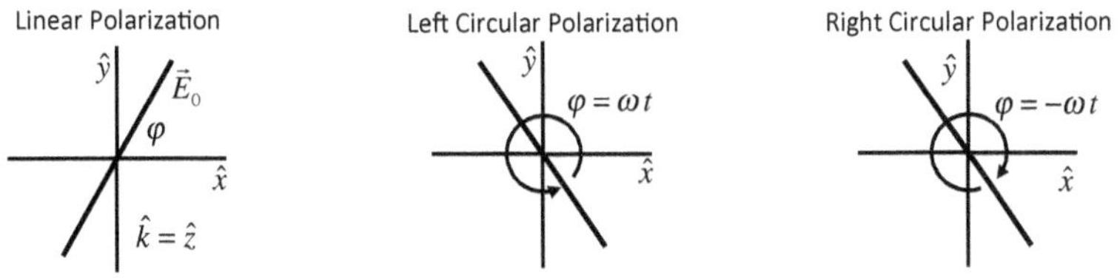

This sign convention can be confusing, because it is backwards from the right hand rule. To make matters worse, not all fields use the same convention. We suggest saying "left circular polarization" or "right circular polarization" when using our convention, and saying "right hand circular polarization" or "left hand circular polarization" when using the other.

Thought Experiment 15.8: The Angular Momentum of Light

Consider a free particle of mass m and charge q, initially at rest at the origin of our coordinate system. As a circularly polarized monochromatic wave (p. 490) passes by, the acceleration of the particle for (left/right) circular polarization is:

$$\vec{a} = \frac{\vec{F}}{m} = \frac{q}{m}\vec{E} = \frac{q}{m}E_0\left(\cos(\omega t)\hat{x} \pm \sin(\omega t)\hat{y}\right)$$

where "+" indicates left polarization, and "−" indicates right polarization.

Let us find the velocity and the position of the charge for a very short time t after the field is turned on. The magnetic field can be ignored because the particle starts at rest and the time under consideration is sufficiently brief. The velocity and position of the particle are therefore given by:

$$\vec{v} = \int_0^t \vec{a}\,dt = \frac{q}{m}E_0\int_0^t\left(\cos(\omega t)\hat{x} \pm \sin(\omega t)\hat{y}\right)dt = \frac{qE_0}{m\omega}\left(\sin\omega t\,\hat{x} \pm (1-\cos\omega t)\hat{y}\right)$$

$$\vec{r} = \int_0^t \vec{v}\,dt = \frac{qE_0}{m\omega}\int_0^t\left(\sin\omega t\,\hat{x} \pm (1-\cos\omega t)\hat{y}\right)dt$$

$$= \frac{qE_0}{m\omega^2}\left((1-\cos\omega t)\hat{x} \pm (\omega t - \sin\omega t)\hat{y}\right).$$

Evaluating these for small values of ωt yields:

$$\vec{a} = \frac{qE_0}{m}\left(\cos\omega t\,\hat{x} \pm \sin\omega t\,\hat{y}\right),\quad \vec{v} = \frac{qE_0}{m\omega}\left(\sin\omega t\,\hat{x} \pm \left(1-\cos\omega t\right)\hat{y}\right),$$

and $\vec{r} = \dfrac{qE_0}{m\omega^2}\left(1-\cos\omega t\right)\hat{x}.$

The rate of work done on the particle is:

$$P = q\vec{E}\cdot\vec{v} = \frac{q^2E_0^{\;2}}{m\omega}\left(\cos\omega t\,\hat{x} \pm \sin\omega t\,\hat{y}\right)\cdot\left(\sin\omega t\,\hat{x} \pm \left(1-\cos\omega t\right)\hat{y}\right)$$

$$= \frac{q^2E_0^{\;2}}{m\omega}\left(\cancel{\cos\omega t\sin\omega t} + \sin\omega t - \cancel{\cos\omega t\sin\omega t}\right) = \frac{q^2E_0^{\;2}}{m\omega}\sin\omega t \approx \frac{q^2E_0^{\;2}}{m}t.$$

We next find the torque on the particle for small times:

$$\vec{\mathrm{T}} = \vec{r}\times q\vec{E} = \frac{qE_0}{m\omega^2}\left(1-\cos\omega t\right)\hat{x}\times qE_0\left(\cos(\omega t)\hat{x} \pm \sin(\omega t)\hat{y}\right) \approx \pm\frac{q^2E^2{}_0}{m\omega}t\,\hat{z}.$$

Now notice that the ratio of angular momentum to energy lost is:

$$\frac{d\vec{\mathrm{L}}}{dU} = \frac{\dfrac{d\vec{\mathrm{L}}}{dt}}{\dfrac{dU}{dt}} = \frac{\vec{\mathrm{T}}}{P} = \pm\frac{\hat{z}}{\omega}.$$

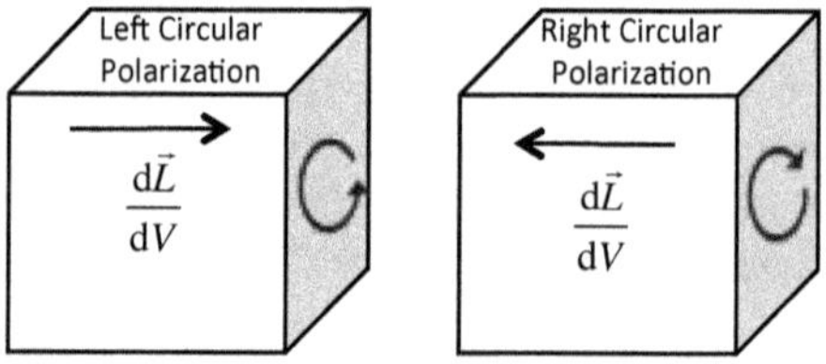

So, we can say that monochromatic light has an angular momentum density which is given by:

$$\vec{\mathbb{L}} = \frac{d\vec{\mathrm{L}}}{dV} = \frac{d\vec{\mathrm{L}}}{dA\,c\,dt} = \pm\frac{\vec{\mathbb{S}}}{\omega c}.$$

The angular momentum vector is in the direction of propagation for *Left Circular Polarization*, and in the direction of the source for *Right Circular Polarization*, as shown in the figure.

Problem 15.8: Monochromatic Plane Waves

Show that the following relations hold for a monochromatic plane wave in free space. Recall the wave vector, $\vec{k} = \frac{\omega}{v}\hat{k} = \frac{1}{\lambda}\hat{k}$:

a) $\vec{E}(\vec{r},t) = \vec{E}_0\cos\left(\vec{k}\cdot\vec{r} - \omega t\right)$ is a solution to the wave equation.

b) $\vec{B}(\vec{r},t) = \dfrac{1}{c}\hat{k}\times\vec{E}(\vec{r},t)$ is also a solution to the wave equation.

c) The instantaneous Poynting vector is: $\vec{\mathbb{S}} = \frac{1}{Z_0}E^2\,\hat{k}$.

d) The time average energy flux of the wave is: $\left\langle\vec{\mathbb{S}}\right\rangle = \frac{1}{2}\frac{1}{Z_0}E_0^{\;2}\,\hat{k}.$

e) The time average momentum density is: $\left\langle\dfrac{d\vec{p}}{dV}\right\rangle = \frac{1}{2}\varepsilon_0^{\;3/2}\mu_0^{\;1/2}E_0^{\;2}\,\hat{k}.$

f) The time average energy density is: $\langle \mathbb{u} \rangle = \frac{1}{2}\varepsilon_0 E_0^{\,2}\hat{k}$.

g) The monochromatic plane wave solution is often assumed in situations where it is clearly not valid, such as sources of light that are not monochromatic. Why is this so? What did the mathematical work of Joseph Fourier have to do with resolving this problem?

Problem 15.9: The Photon Model of Light

In 1905 Albert Einstein published a paper in which he suggested that light was actually made up of particles, called photons, and that the energy of a photon was proportional to the associated frequency where $\mathrm{E} = \hbar\omega$. In aggregate many photons would act exactly as if there were classical monochromatic radiation. In this problem you will assume that there are a great many photons per volume, n, and that they are all traveling at a velocity $\vec{v} = c\,\hat{k}$.

a) Express the energy density of light in terms of the number density of photons and their angular frequency ω.

b) Express the Poynting vector in terms of the number density of photons and their angular frequency ω.

c) Using your answer for part (b), find the momentum of each photon. Express this in terms of the monochromatic wave vector $\vec{k} = \omega/_c\,\hat{k}$.

d) Again using your answer for part (b) find the angular momentum of each photon in a circularly polarized wave.

e) Using your result for part (c), and Newton's second law, derive an expression for the radiation pressure in terms of the photon density n and the frequency ω.

15.4 *Measuring Light with a Photometer*

We often seek to characterize light measured from a distant source, especially in astronomy. If the frequency is small enough to allow us to use an antenna to make measurements, we can obtain the electric field as a function of time. Usually, however, we measure the energy imparted from the wave onto some type of photometer that is sensitive to a range of frequencies but measures only the total energy imparted in a given time.

For example, the cone cells in your eye measure the energy of light with wavelengths of between 400-500 nm, 450-630 nm and 500-700 nm for S (short), M (medium) and L (long) type cones respectively. This allows us to observe the energy flux as a function of frequency $\vec{\mathbb{S}}(\omega)$, which gives us color vision. Of course, our eyes also measure the energy of the light as a function of time (to within the limits of the temporal resolution of our eyes), and as a function of angle (to within the eye's angular resolution). Our eyes are not sensitive to the polarization state of light, but we can use polarized glasses to filter out one polarization state for such purposes as reducing glare or viewing a 3D movie.

In this section, we introduce a series of measureable quantities, which can then be used to fully characterize incident light, including its polarization state. Some thought experiments will involve antennas while others will involve photometers. The key difference is that the response time of an antenna is faster than the period of the wave, so the electric field can be amplified first, and the signal can be measured. In the case of a photometer, such as our eye, the energy is measured but not the electric field because the response time of the photometer is much slower than the period of the wave.

Thought Experiment 15.9: The Photometer

Consider light that is a collected via a lens with an area ΔA, then sent through a color filter of bandpass $\Delta\omega$, and then through a shutter that opens for a time Δt. This process describes how the energy (ΔU) of a light wave is measured with a photometer.

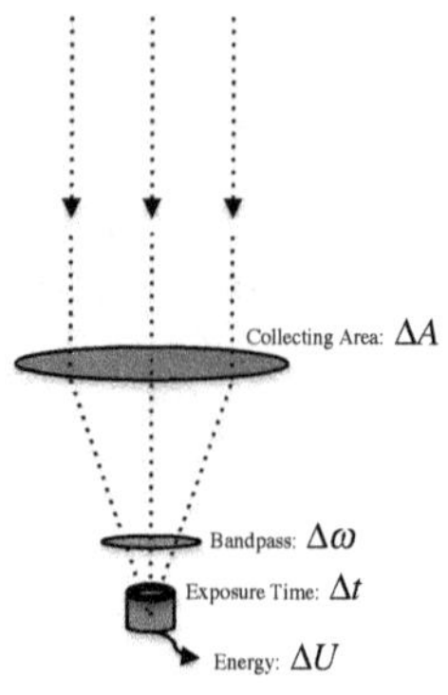

How much energy is actually detected is directly proportional to the brightness b of the source, the area of the collector, the bandpass of the filter and the exposure time. Thus:

$$\Delta U = b\,\Delta A\,\Delta\omega\,\Delta t .$$

The brightness of the incident light is determined from our photometric measurement of the energy ΔU:

$$b = \frac{\Delta U}{\Delta A\,\Delta\omega\,\Delta t}.$$

In the limit of short exposure times and narrow bandpasses, we can write:

$$b = \frac{\Delta U}{\Delta A\,\Delta\omega\,\Delta t} \approx \left\langle \frac{d\mathbb{S}}{d\omega} \right\rangle \equiv \mathbb{S}_{\omega} .$$

Thus, the time-averaged Poynting vector per angular frequency is directly measureable with a light filter and a photometer.

In the SI unit system, this has units of:

$$[\mathbb{S}_{\omega}] = \left[\frac{d\mathbb{S}}{d\omega}\right] = \frac{\text{J}}{(\text{m}^2)(\text{rad}/\text{s})(s)} = \frac{\text{W}}{(\text{m}^2)(\text{rad}/\text{s})}.$$

If we convert from angular to common frequency units, $\omega = 2\pi\nu$, so:

$$\mathbb{S}_{\nu} = \frac{d\mathbb{S}}{d\nu} = \frac{d\mathbb{S}}{d\omega}\frac{d\omega}{d\nu} = 2\pi\frac{d\mathbb{S}}{d\omega} = 2\pi\,\mathbb{S}_{\omega} \approx \frac{\Delta \text{U}}{\Delta A\,\Delta\nu\,\Delta t}.$$

Depending on the field, the power per area per time is called different things. For example, radio astronomers call $\vec{\mathbb{S}}_{\nu}$ *flux density* and often measure it in units named after Karl Jansky, the founder of radio astronomy, such that $1\,\text{Jy} \equiv 10^{-26}\,\frac{\text{W}}{\text{Hz}\,\text{m}^2}$..

Problem 15.10: S.I. Units of Perceived Brightness

Brightness has historically been measured using the naked eye in comparison to a candle. As with other units, scientists wanted to compare brightness in a more standardized manner, so they made a *standard candle.* The standard candle has now become an S.I. base unit, defined as:

> The candela is the luminous intensity, in a given direction, of a source that emits monochromatic radiation of frequency 540×10^{12} hertz and that has a radiant intensity in that direction of $1/683$ watt per steradian.

a) By drawing a spherical Gaussian surface of radius r, find the total power output of an isotropic 540 THz source with a *luminous intensity* of one candela. Recall that 4π steradians equal a full sphere.

b) The unit of the total visual light output of a source, called *luminous flux*, is the lumen. Again, by drawing a spherical Gaussian surface, derive the relationship between *luminous intensity* and *luminous flux*.

c) The unit of visual brightness, called *illuminance*, is the lux. The brightness of a one candela source, at a distance of one meter, is one lux. One lux is also one lumen per square meter. Using drawings and Gaussian surfaces, show that these two definitions are equivalent.

d) Recall that the term *flux* means different things in different branches of physics. Astronomers use the term to mean the amount of energy per area per time, however that is not the way it is used by lighting designers. Research the term, and compare the way it is used by engineers, as in *luminous flux*, to the way that astronomers use the term, as in *flux density*. Compare and contrast these with the way it is used in electricity and magnetism, as in the terms *electric flux* and *magnetic flux*.

Thought Experiment 15.10: Measuring the Angle of Linear Polarization

To determine the polarization of light using a photometer, we need a filter that can be rotated through an angle φ so that we can measure the relative brightness of the light as a function of angle. To start things simply, we will demonstrate how to measure the polarization angle of light with an incident beam assumed to be linearly polarized (p. 451).

Consider a monochromatic, linearly polarized, plane wave:

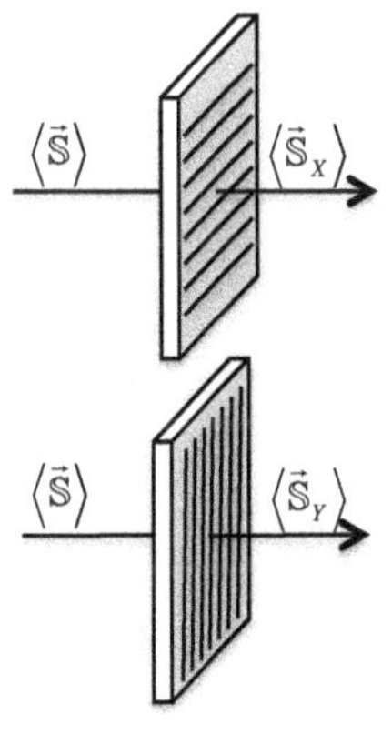

$$\vec{E}(z,t) = E_x \cos(\omega\tau)\hat{x} + E_y \cos(\omega\tau)\hat{y},$$

where $\tau = t - \frac{1}{c}z$ is the retarded time, and E_x and E_y are constants. As before, the polarization angle is given by:

$$\varphi = \arctan\left(\tfrac{E_y}{E_x}\right).$$

A photometer measures the energy per time, averaged over multiple cycles of the wave, so it can measure the time-averaged Poynting vector:

$$\langle \vec{\mathbb{S}} \rangle = \left\langle \tfrac{1}{\mu_0} \vec{E} \times \vec{B} \right\rangle = \left\langle \tfrac{c}{Z_0} \vec{E} \times \vec{B} \right\rangle = \tfrac{1}{Z_0}\left(E_x^2 + E_y^2\right)\hat{z}\left\langle \cos^2(\omega\tau) \right\rangle.$$

Since:

$$\left\langle \cos^2(\omega\tau) \right\rangle = \frac{1}{T}\int_0^T \cos^2(\omega\tau)\,dt = \frac{1}{2},$$

we obtain for the time-averaged energy flux of the wave:

$$\langle \vec{\mathbb{S}} \rangle = \tfrac{1}{2} Z_0 \left(E^2{}_x + E^2{}_y\right)\hat{z}.$$

After passing through a polarizer aligned with the $\hat{x}$ direction, the resulting electric field would be:

$$\vec{E}_X(z,t) = \hat{x} \cdot \vec{E}(z,t) = E_x \cos(\omega\tau)\hat{x}.$$

So, the average energy flux after passing through the polarizer is:

$$\left\langle \vec{\mathbb{S}}_X \right\rangle = \left\langle \frac{1}{Z_0} \left| \vec{E}_x(z,t) \right|^2 \right\rangle = \frac{1}{2}\frac{1}{Z_0} E_x^2 \, \hat{z}.$$

Thus, the fraction of the energy that passes through the polarizer is given by:

$$\frac{\left\langle \mathbb{S}_X \right\rangle}{\left\langle \mathbb{S} \right\rangle} = \frac{\frac{1}{Z_0}\frac{1}{2} E_x^2}{\frac{1}{Z_0}\frac{1}{2}\left(E_x^2 + E_y^2\right)} = \frac{E_x^2}{E_x^2 + E_y^2}.$$

If we rotate the polarizer by 90°, and make the same measurement, we find the fraction of light that is passed by a polarizer aligned in the $\hat{y}$ direction:

$$\frac{\left\langle \mathbb{S}_Y \right\rangle}{\left\langle \mathbb{S} \right\rangle} = \frac{E_y^2}{E_x^2 + E_y^2}.$$

Now, taking the ratio the two fluxes gives:

$$\frac{\left\langle \mathbb{S}_Y \right\rangle}{\left\langle \mathbb{S}_X \right\rangle} = \left(\frac{E_y}{E_x} \right)^2.$$

The polarization angle of the incident light can therefore be expressed in terms of measureable quantities as:

$$\varphi = \arctan\left(\sqrt{\frac{\left\langle \mathbb{S}_Y \right\rangle}{\left\langle \mathbb{S}_X \right\rangle}} \right).$$

And the incident flux is given simply by:

$$\left\langle \vec{\mathbb{S}} \right\rangle = \left\langle \vec{\mathbb{S}}_X \right\rangle + \left\langle \vec{\mathbb{S}}_Y \right\rangle.$$

The observer must already know that the incident beam is linearly polarized for this experiment to yield meaningful results. More measurements must be made to measure the polarization of *unknown* light using a photometer, as we will demonstrate below in Thought Experiment 15.12.

Thought Experiment 15.11: Measuring Circular Polarization

In Chapter 14 we discussed that dielectric materials slow down light. A material with an anisotropic dielectric constant will slow down light anisotropically, so that polarized light travels faster when the electric field is aligned with the fast axis, rather than the slow axis, of the material. Such *birefringent* materials (e.g. calcite, quartz, or Iceland spar) can be used, (at least in the case of monochromatic light) to make a quarter-wave plate, which delays the polarization on the slow axis by ¼ of a period compared to the fast axis. In this thought experiment, we will investigate how to measure circular polarization with a quarter-wave plate and a polarizer.

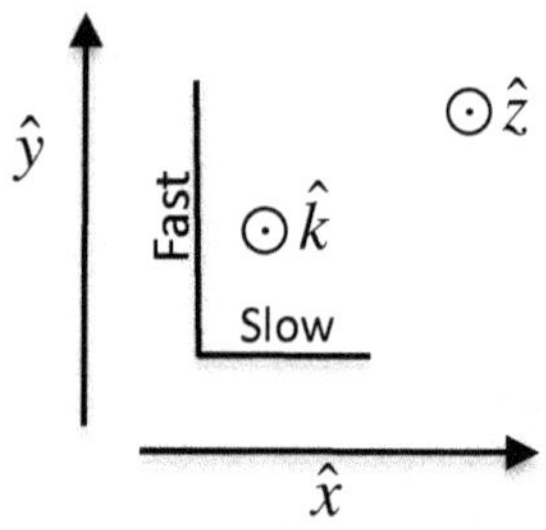

First consider incident circularly polarized light, with an electric field:

$$\vec{E}(z,t) = E_0\left(\cos(\omega\tau)\hat{x} + \sin(\omega\tau)\hat{y}\right) \qquad \text{(Left Circular Polarization)}$$

$$\vec{E}(z,t) = E_0\left(\cos(\omega\tau)\hat{x} - \sin(\omega\tau)\hat{y}\right) \qquad \text{(Right Circular Polarization).}$$

First we will can calculate the time-averaged Poynting vector for each, as:

$$\left\langle \vec{\mathbb{S}}_{CP} \right\rangle = \left\langle \frac{1}{\mu_0}\vec{E}\times\vec{B} \right\rangle = \left\langle \frac{E_0^2}{Z_0}\left(\cos^2(\omega\tau)+\sin^2(\omega\tau)\right) \right\rangle \hat{z} = \frac{E_0^2}{Z_0}\hat{z}.$$

After passing through a quarter-wave plate with the fast axis in the $\hat{y}$ direction, the $\hat{x}$ electric field is delayed by an additional ¼ period so that we have:

$$\vec{E}(z,t) = E_0\left(\cos\left(\omega\tau - \tfrac{\pi}{2}\right)\hat{x} \pm \sin(\omega\tau)\hat{y}\right) = E_0\left(\sin(\omega\tau)\hat{x} \pm \sin(\omega\tau)\hat{y}\right)$$
$$= E_0 \sin(\omega\tau)\left(\hat{x} \pm \hat{y}\right).$$

So, circular polarized light becomes linearly polarized with an angle:

$$\varphi_L = \arctan(1) = \tfrac{\pi}{4} = 45^\circ, \quad \text{or}$$
$$\varphi_R = \arctan(-1) = \tfrac{3\pi}{4} = 135^\circ,$$

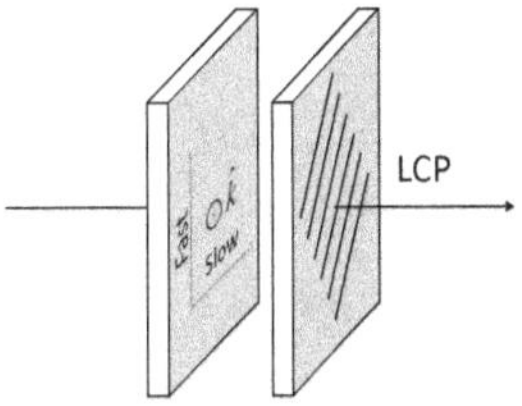

depending on whether the light was left or right circularly polarized.

A circular polarization filter can therefore be made by passing unknown light through a quarter wave plate, followed by a polarizer that is rotated by 45° for left circularly polarized light, or 135° for right circularly polarized light.

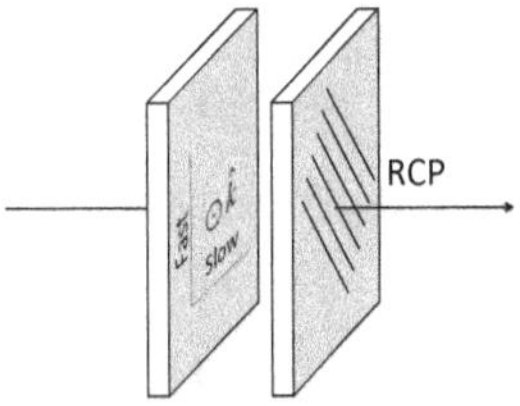

Noticing that the time-averaged energy flux, after going through the quarter wave plate, has not changed for either kind of circular polarization:

$$\left\langle \vec{\mathbb{S}}_{CP} \right\rangle = \left\langle \frac{E_0^2}{Z_0}\left(\cos^2\left(\omega\tau - \tfrac{\pi}{2}\right)+\sin^2(\omega\tau)\right) \right\rangle \hat{z} = \frac{E_0^2}{Z_0}\left\langle\left(2\sin^2(\omega\tau)\right)\right\rangle \hat{z} = \frac{E_0^2}{Z_0}\hat{z}.$$

Thought Experiment 15.12: Stokes Parameters

Consider a source of light, which may, or may not, be linearly polarized. It could, in fact, have multiple incoherent sources of light of different linear polarizations. Our goal is to measure the fractional polarization of the light, as well as the angle of polarization if it is at least partially polarized.

As in Thought Experiment 15.10, we assume that we have access to a photometer with a polarizing filter that can be rotated. But rather than making two measurements at right angles, we will make four measurements at angles 45° apart. Moreover, as in Thought Experiment 15.11, we will use an insertable quarter-wave plate so that left and right circularly polarized light can also be measured. Thus our photometer will be used to make six different measurements to fully characterize the unknown light.

For convenience, we define the flowing quantities:

$$\mathbf{I} = \langle \mathbb{S}_{0°} \rangle + \langle \mathbb{S}_{90°} \rangle = \langle \mathbb{S}_{45°} \rangle + \langle \mathbb{S}_{135°} \rangle = \langle \mathbb{S}_{LCP} \rangle + \langle \mathbb{S}_{RCP} \rangle$$
$$\mathbf{Q} = \langle \mathbb{S}_{0°} \rangle - \langle \mathbb{S}_{90°} \rangle$$
$$\mathbf{U} = \langle \mathbb{S}_{45°} \rangle - \langle \mathbb{S}_{135°} \rangle$$
$$\mathbf{V} = \langle \mathbb{S}_{L} \rangle - \langle \mathbb{S}_{R} \rangle$$

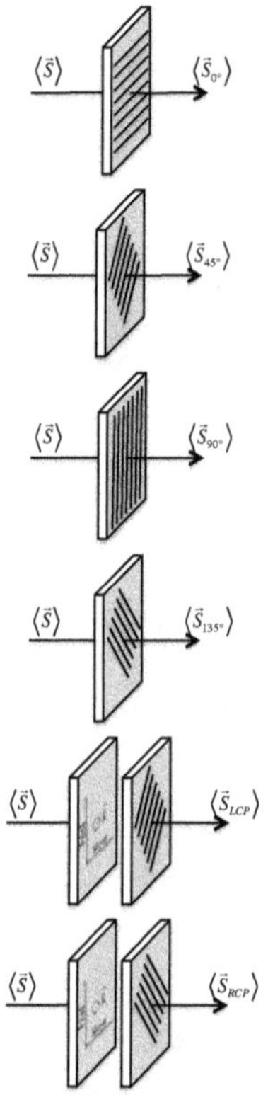

These are called *Stokes parameters* after the British physicist and mathematician George Gabriel Stokes who first introduced them in the nineteenth century.[263] By measuring all four Stokes parameters, the polarization of an unknown light source can be fully characterized.

A monochromatic coherent light wave will have the property that: $\mathbf{I}^2 = \mathbf{Q}^2 + \mathbf{U}^2 + \mathbf{V}^2$.

However, most natural sources of light have random phases and polarization angles. In this case:

$\langle \mathbb{S}_{0°} \rangle = \langle \mathbb{S}_{90°} \rangle = \langle \mathbb{S}_{45°} \rangle = \langle \mathbb{S}_{135°} \rangle = \langle \mathbb{S}_{LCP} \rangle = \langle \mathbb{S}_{RCP} \rangle$, so $\mathbf{Q} = \mathbf{U} = \mathbf{V} = 0$, and the light is said to be *unpolarized*. Thus, it is always the case that:

$$\mathbf{I}^2 \geq \mathbf{Q}^2 + \mathbf{U}^2 + \mathbf{V}^2.$$

The polarized flux is thus defined as:

$$\mathbf{P} = \sqrt{\mathbf{Q}^2 + \mathbf{U}^2 + \mathbf{V}^2}$$

and it often written as a percentage of the total flux:

$$\%\text{Polarization} = \frac{\mathbf{P}}{\mathbf{I}} \times 100\%.$$

Example 15.1: Stokes Parameters of Linear Polarized Light

Clearly, $\langle \mathbb{S}_{LCP} \rangle = \langle \mathbb{S}_{RCP} \rangle$, so $\mathbf{V} = 0$. We therefore concern ourselves with the Stokes parameters $\mathbf{I}, \mathbf{Q}$ and $\mathbf{U}$.

Let us now consider light that is linearly polarized along some angle φ with respect to the $\hat{x}$ axis, so that in Cartesian coordinates:

$$\vec{E}(z,t) = E_0 \cos(\omega\tau)\left(\cos(\varphi)\hat{x} + \sin(\varphi)\hat{y}\right)$$
$$\vec{B}(z,t) = \tfrac{1}{c} E_0 \cos(\omega\tau)\left(-\sin(\varphi)\hat{x} + \cos(\varphi)\hat{y}\right).$$

The time-averaged values of the energy fluxes measured at 0° and 90° are therefore:

$$\langle \mathbb{S}_{0°} \rangle = \left\langle \tfrac{1}{\mu_0} \hat{x} \cdot \left(\vec{E} \times \vec{B}\right) \right\rangle = \left\langle \tfrac{1}{Z_0} \left| E_0 \cos(\omega\tau)\cos(\varphi)\hat{x} \right|^2 \right\rangle = \tfrac{1}{2Z_0} E_0^2 \cos^2 \varphi$$
$$\langle \mathbb{S}_{90°} \rangle = \left\langle \tfrac{1}{\mu_0} \hat{y} \cdot \left(\vec{E} \times \vec{B}\right) \right\rangle = \left\langle \tfrac{1}{Z_0} \left| E_0 \cos(\omega\tau)\sin(\varphi)\hat{y} \right|^2 \right\rangle = \tfrac{1}{2Z_0} E_0^2 \sin^2 \varphi.$$

Similarly, we can write this in terms of the 45° rotated coordinates:

---

263 G.G. Stokes, "On the Composition of Streams of Polarized Light from different Sources," Transactions of the Cambridge Philosophical Society, **9** (1852), 399-416.

$$\begin{pmatrix} \hat{x}' \\ \hat{y}' \end{pmatrix} = \begin{pmatrix} \cos\left(\varphi - \frac{\pi}{4}\right) & \sin\left(\varphi - \frac{\pi}{4}\right) \\ -\sin\left(\varphi - \frac{\pi}{4}\right) & \cos\left(\varphi - \frac{\pi}{4}\right) \end{pmatrix} \begin{pmatrix} \hat{x} \\ \hat{y} \end{pmatrix}$$

so that the electric field is

$$\vec{E}(z,t) = E_0 \cos(\omega\tau)\left(\cos\left(\varphi - \tfrac{\pi}{4}\right)\hat{x} + \sin\left(\varphi - \tfrac{\pi}{4}\right)\hat{y}\right)$$

and the fluxes measured at 45° and 135° are given by

$$\langle \mathbb{S}_{45^\circ} \rangle = \left\langle \tfrac{1}{Z_0} \left| E_0 \cos(\omega\tau)\cos\left(\varphi - \tfrac{\pi}{4}\right)\hat{x} \right|^2 \right\rangle = \tfrac{1}{2Z_0} E_0^2 \cos^2\left(\varphi - \tfrac{\pi}{4}\right)$$

$$\langle \mathbb{S}_{135^\circ} \rangle = \left\langle \tfrac{1}{Z_0} \left| E_0 \cos(\omega\tau)\sin\left(\varphi - \tfrac{\pi}{4}\right)\hat{y} \right|^2 \right\rangle = \tfrac{1}{2Z_0} E_0^2 \sin^2\left(\varphi - \tfrac{\pi}{4}\right)$$

Now, the Stokes parameters can be written:

$$\begin{pmatrix} \mathbf{I} \\ \mathbf{Q} \\ \mathbf{U} \\ \mathbf{V} \end{pmatrix} = \frac{E_0^2}{2Z_0} \begin{pmatrix} 1 \\ \cos^2\varphi - \sin^2\varphi \\ \cos^2\left(\varphi - \frac{\pi}{4}\right) - \sin^2\left(\varphi - \frac{\pi}{4}\right) \\ 0 \end{pmatrix} = \frac{E_0^2}{2Z_0} \begin{pmatrix} 1 \\ \cos(2\varphi) \\ \sin(2\varphi) \\ 0 \end{pmatrix}.$$

The polarization angle is then determined by:

$$\varphi = \tfrac{1}{2}\arctan\left(\tfrac{\mathrm{U}}{\mathrm{Q}}\right).$$

Problem 15.11: Half Wave Plates

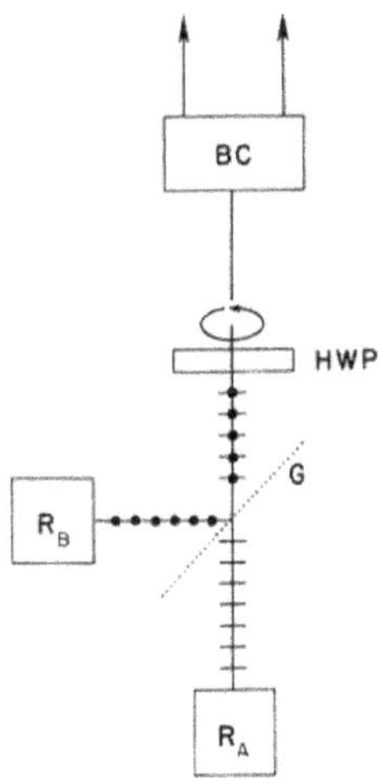

Often infrared polarimeters that are mounted on astronomical telescopes are made with a fixed diffraction grating and a rotatable half wave plate. In one case[264] light from the telescope first goes through a beam chopper (BC) that subtracts the sky from the source, followed by a rotatable half wave plate (HWP), and finally into a pair of receivers that measure two orthogonal polarizations at the same time. Without the half wave plate, receivers A and B are aligned with the $\hat{x}$ and $\hat{y}$ axes respectively.

Presumably, this polarimeter had been tested in the laboratory with a linearly polarized light source of intensity $\mathbf{I}$ and polarization aligned with the $\hat{x}$ axis.

a) Without the half wave plate receiver A reads $\mathbf{I}$ and receiver B reads zero. What would each receiver read with the half wave plate inserted and its fast axes aligned with the $\hat{x}$ axis?

b) What does each one read with the fast axis at an angle of 45°?

c) Make a plot of the signal in each receiver as a function of the angle of the fast axis.

---

264 G. Novak, C.R. Predmore, and P.F. Goldsmith, "Polarization of the $\lambda = 1.3$ millimeter Continuum Radiation from the Kelinmann-Low Nebula," The Astrophysical Journal, **355** (1990), 166-171.

d) Repeat the previous parts for a test beam that was linearly polarized along the $\hat{y}$ axis.

e) Write Stokes parameter $\mathbf{Q}$ in terms of the signal of each receiver when the fast axis of the half wave plate is at each of the following angles: 0°, 45°, 90°, and 135° with respect to $\hat{x}$ axis.

f) Write Stokes parameter $\mathbf{U}$ in terms of the signal of each receiver when the fast axis of the half wave plate is at each of the following angles: 22.5°, 67.5°, 112.5°, and 157.5° with respect to the $\hat{x}$ axis.

g) The half wave plate was designed to be rotated in 30° increments, rather than 22.5° increments. Why do you think they chose to do it this way?

Example 15.2: An Example of Elliptical Polarization

Consider monochromatic light with an electric field vector:

$$\vec{E}(z,t) = E_0\left(\cos\left(\omega\tau - \phi/2\right)\hat{x} + \cos\left(\omega\tau + \phi/2\right)\hat{y}\right).$$

What would be the resulting four measureable Stokes parameters for an elliptically polarized ray of light?

We will first consider the measurements of $\langle S_{0^\circ}\rangle$ and $\langle S_{90^\circ}\rangle$:

$$\langle S_{0^\circ}\rangle = \left\langle \tfrac{1}{Z_0}\left|E_0\cos\left(\omega\tau - \phi/2\right)\hat{x}\right|^2\right\rangle = \tfrac{1}{2Z_0}E_0^2$$

$$\langle S_{90^\circ}\rangle = \left\langle \tfrac{1}{Z_0}\left|E_0\cos\left(\omega\tau + \phi/2\right)\hat{y}\right|^2\right\rangle = \tfrac{1}{2Z_0}E_0^2.$$

so the first two Stokes parameters are given by:

$\mathbf{I} = \tfrac{1}{Z_0}E_0^2$ and $\mathbf{Q} = 0$.

Now, to measure $\mathbf{U}$, we must rotate our coordinate system by 45°, thus:

$$\begin{pmatrix}\hat{x}\\ \hat{y}\end{pmatrix} = \begin{pmatrix}\cos\pi/4 & -\sin\pi/4\\ \sin\pi/4 & \cos\pi/4\end{pmatrix}\begin{pmatrix}\hat{x}'\\ \hat{y}'\end{pmatrix} = \tfrac{\sqrt{2}}{2}\begin{pmatrix}1 & -1\\ 1 & 1\end{pmatrix}\begin{pmatrix}\hat{x}'\\ \hat{y}'\end{pmatrix} = \tfrac{\sqrt{2}}{2}\begin{pmatrix}\hat{x}' - \hat{y}'\\ \hat{x}' + \hat{y}'\end{pmatrix}.$$

The electric field expressed in the rotated system of coordinates is:

$$\begin{aligned}\vec{E}(z,t) &= E_0\tfrac{\sqrt{2}}{2}\left(\cos\left(\omega\tau - \phi/2\right)(\hat{x}' - \hat{y}') + \cos\left(\omega\tau + \phi/2\right)(\hat{x}' + \hat{y}')\right)\\ &= E_0\sqrt{2}\left(\tfrac{1}{2}\left(\cos\left(\omega\tau + \phi/2\right) + \cos\left(\omega\tau - \phi/2\right)\right)\hat{x}' + \tfrac{1}{2}\left(\cos\left(\omega\tau + \phi/2\right) - \cos\left(\omega\tau - \phi/2\right)\right)\hat{y}'\right)\\ &= E_0\sqrt{2}\left(\left(\cos(\omega\tau)\cos\left(\phi/2\right)\right)\hat{x}' - \left(\sin(\omega\tau)\sin\left(\phi/2\right)\right)\hat{y}'\right).\end{aligned}$$

We therefore have for the measureable fluxes:

$$\langle S_{45^\circ}\rangle = \left\langle \tfrac{1}{Z_0}\left|E_0\sqrt{2}\left(\sin\left(\omega\tau - \tfrac{\phi}{2}\right)\cos\left(\tfrac{\phi}{2}\right)\hat{x}'\right)\right|^2\right\rangle = \tfrac{1}{Z_0}E_0^2\cos^2\left(\tfrac{\phi}{2}\right)$$

$$\langle S_{135^\circ}\rangle = \left\langle \tfrac{1}{Z_0}\left|E_0\sqrt{2}\left(\sin\left(\tfrac{\phi}{2}\right)\cos\left(\omega\tau - \tfrac{\phi}{2}\right)\right)\right|^2\right\rangle = \tfrac{1}{Z_0}E_0^2\sin^2\left(\tfrac{\phi}{2}\right).$$

Thus we can now find $\mathbf{I}$ and $\mathbf{U}$:

$$\mathbf{I} = \tfrac{1}{Z_0} E_0^2 \cos^2\left(\tfrac{\phi}{2}\right) + \tfrac{1}{Z_0} E_0^2 \sin^2\left(\tfrac{\phi}{2}\right) = \tfrac{1}{Z_0} E_0^2$$
$$\mathbf{U} = \tfrac{1}{Z_0} E_0^2 \cos^2\left(\tfrac{\phi}{2}\right) - \tfrac{1}{Z_0} E_0^2 \sin^2\left(\tfrac{\phi}{2}\right) = \tfrac{1}{Z_0} E_0^2 \cos(\phi).$$

We now see that the expression for $\mathbf{I}$ is in agreement with our result in the $\hat{x},\hat{y}$ coordinate system.

Next we consider how we measure circularly polarized light. We first delay the $\hat{x}$ with the quarter wave plate, followed by filtering the linearly polarized light by 45°.

We being with our incident electric field of:

$$\vec{E}(z,t) = E_0\left(\cos\left(\omega\tau - \phi/2\right)\hat{x} + \cos\left(\omega\tau + \phi/2\right)\hat{y}\right).$$

After the quarter wave plate, the electric field is now:

$$\vec{E}(z,t) = E_0\left(\cos\left(\omega\tau - \phi/2 - \tfrac{\pi}{2}\right)\hat{x} + \cos\left(\omega\tau + \phi/2\right)\hat{y}\right) = E_0\left(\sin\left(\omega\tau - \phi/2\right)\hat{x} + \cos\left(\omega\tau + \phi/2\right)\hat{y}\right).$$

Next we rotate the axes by 45° as we did above:

$$\begin{aligned}\vec{E}(z,t) &= E_0 \tfrac{1}{\sqrt{2}}\left(\sin\left(\omega\tau - \phi/2\right)(\hat{x}' - \hat{y}') + \cos\left(\omega\tau + \phi/2\right)(\hat{x}' + \hat{y}')\right) \\ &= E_0 \tfrac{1}{\sqrt{2}}\left(\left(\cos\left(\omega\tau + \phi/2\right) + \sin\left(\omega\tau - \phi/2\right)\right)\hat{x} + \left(\cos\left(\omega\tau + \phi/2\right) - \sin\left(\omega\tau - \phi/2\right)\right)\hat{y}\right).\end{aligned}$$

Next we evaluate the expression:

$$\cos\left(\omega\tau + \phi/2\right) \pm \sin\left(\omega\tau - \phi/2\right) = \left(\cos(\omega\tau) \mp \sin(\omega\tau)\right)\left(\cos\left(\phi/2\right) \pm \sin\left(\phi/2\right)\right).$$

So the electric field becomes:

$$\vec{E}(z,t) = E_0 \tfrac{1}{\sqrt{2}}\begin{pmatrix}\left(\cos(\omega\tau) - \sin(\omega\tau)\right)\left(\cos\left(\phi/2\right) + \sin\left(\phi/2\right)\right)\hat{x} \\ +\left(\cos(\omega\tau) + \sin(\omega\tau)\right)\left(\cos\left(\phi/2\right) - \sin\left(\phi/2\right)\right)\hat{y}\end{pmatrix}$$

Next we find the Poynting vector for each component:

$$\langle \mathbb{S}_L \rangle = \left\langle \tfrac{1}{Z_0}\left|E_0 \tfrac{1}{\sqrt{2}}\left(\cos(\omega\tau) - \sin(\omega\tau)\right)\left(\cos\left(\phi/2\right) + \sin\left(\phi/2\right)\right)\right|^2 \right\rangle$$
$$\langle \mathbb{S}_R \rangle = \left\langle \tfrac{1}{Z_0}\left|E_0 \tfrac{1}{\sqrt{2}}\left(\cos(\omega\tau) + \sin(\omega\tau)\right)\left(\cos\left(\phi/2\right) - \sin\left(\phi/2\right)\right)\right|^2 \right\rangle$$

Factoring out variables without time dependence, we have:

$$\langle \mathbb{S}_L \rangle = \tfrac{E_0^2}{2Z_0}\left(\cos\left(\phi/2\right) + \sin\left(\phi/2\right)\right)^2 \left\langle \left(\cos(\omega\tau) - \sin(\omega\tau)\right)^2 \right\rangle = \tfrac{E_0^2}{2Z_0}\left(\cos\left(\phi/2\right) + \sin\left(\phi/2\right)\right)^2$$
$$\langle \mathbb{S}_R \rangle = \tfrac{E_0^2}{2Z_0}\left(\cos\left(\phi/2\right) - \sin\left(\phi/2\right)\right)^2 \left\langle \left(\cos(\omega\tau) + \sin(\omega\tau)\right)^2 \right\rangle = \tfrac{E_0^2}{2Z_0}\left(\cos\left(\phi/2\right) - \sin\left(\phi/2\right)\right)^2.$$

Finally, we can write $\mathbf{I}$ and $\mathbf{V}$:

$$\mathbf{I} = \langle \mathbb{S}_L \rangle + \langle \mathbb{S}_R \rangle = \tfrac{E_0^2}{Z_0}\left(\cos^2\left(\phi/2\right) + \sin^2\left(\phi/2\right)\right) = \tfrac{E_0^2}{Z_0}$$
$$\mathbf{V} = \langle \mathbb{S}_L \rangle - \langle \mathbb{S}_R \rangle = \tfrac{E_0^2}{Z_0}\left(2\cos\left(\phi/2\right)\sin\left(\phi/2\right)\right) = \tfrac{E_0^2}{Z_0}\left(\sin(\varphi)\right).$$

We can now write the four Stokes parameters as:

$$\begin{pmatrix} \mathbf{I} \\ \mathbf{Q} \\ \mathbf{U} \\ \mathbf{V} \end{pmatrix} = \tfrac{1}{Z_0} E_0^2 \begin{pmatrix} 1 \\ 0 \\ \cos(\phi) \\ \sin(\phi) \end{pmatrix}.$$

Problem 15.12: Elliptical Polarization and Stokes Parameters

Consider light that is coherent and linearly polarized with an electric field vector of $\vec{E} = E_0 \cos(\omega\tau)\hat{x}$.

First it passes through a linear polarizer at angle φ from the $\hat{x}$ axis. Next it passes through a birefringent plate of thickness Δz. The $\hat{x}$ component of the electric field is relatively delayed by a time:

$$\Delta t = \tfrac{\Delta z}{v_x} - \tfrac{\Delta z}{v_y} = \tfrac{\Delta z}{c}\left(n_x - n_y\right).$$

After this, the 4 Stokes parameters are measured in a laboratory.

a) Find all 4 Stokes parameters as a function of φ and Δt.

b) What are these in the special case where $\varphi = 45°$? How similar are they to Example 15.2? Explain.

c) What are these in the special case of $\Delta t = 0$? How similar are they to Example 15.1? Explain.

d) Under what conditions is the resulting light circularly polarized?

e) Under what conditions is the resulting light linearly polarized?

f) Show that the light is always 100% polarized.

Problem 15.13: Coherence

Most of the discussion in this chapter has involved *coherent* light. However, in the real world, independent light sources will produce light of random phases and with a range of frequencies. While at any instant the electromagnetic field vectors add, over the course of a real observation only the sum of the energy in the light can really be measured, so we have *incoherent* light. This leads to the question of how to measure the coherence of light with a photometer, where we add up the total energy over some integration time. This is not as easy as it seems, and the topic is addressed in this problem.

In Thought Experiment 15.12 we argued that, for all coherent light, the Stokes parameters have the property $\mathbf{I}^2 = \mathbf{Q}^2 + \mathbf{U}^2 + \mathbf{V}^2$.

a) Show that property $\mathbf{I}^2 = \mathbf{Q}^2 + \mathbf{U}^2 + \mathbf{V}^2$ holds for a monochromatic elliptically polarized wave with an electric field vector given by:

$$\vec{E}(z,t) = E_0\left(\cos\left(\omega\tau - \phi/2\right)\hat{x} + \cos\left(\omega\tau + \phi/2\right)\hat{y}\right)..$$

b) If you measure the four Stokes parameters from a distant source, and it has the property that $\mathbf{I}^2 > \mathbf{Q}^2 + \mathbf{U}^2 + \mathbf{V}^2$, is the source coherent? Discuss why, why not, or why you do not know.

Problem 15.14: The Poincaré Sphere

In Thought Experiment 15.12 we argued that for coherent light the Stokes parameters have the property $\mathbf{P}^2 = \mathbf{Q}^2 + \mathbf{U}^2 + \mathbf{V}^2$. This is reminiscent of our three dimensional vector $\vec{r}$ in Cartesian coordinates where $r^2 = x^2 + y^2 + z^2$, and in a similar manner we can write the polarized intensity in spherical coordinates with the following analogy with the position vector:

$$\mathbf{P} \leftrightarrow r \quad \mathbf{Q} \leftrightarrow x \quad \mathbf{U} \leftrightarrow y \quad \mathbf{V} \leftrightarrow z \quad 2\psi \leftrightarrow \varphi \quad 2\chi \leftrightarrow \theta$$

This is called the *Poincaré sphere*, after the French mathematical physicist, and philosopher of science, J. Henri Poincaré.

a) Find the parameters $\mathbf{Q},\mathbf{U},\mathbf{V}$ as a function of the polarized flux $\mathbf{P}$ and the two Poincaré sphere angles ψ,χ.

b) Find the angles two angles ψ,χ in terms of the Stokes parameters $\mathbf{P},\mathbf{Q},\mathbf{U},\mathbf{V}$.

c) Referring back to Example 15.1, we discussed the angle φ of linearly polarized light. Find the angle of linear polarization in terms of the Poincaré sphere angles.

d) In Example 15.2 we discussed elliptically polarized light angled at 45 degrees from the $\hat{x}$ axis. We found a phase angle ϕ that spanned the space between circular polarization $\left(|\phi| = \frac{\pi}{2}\right)$ and linear polarization $(\phi = 0)$. For this example find ϕ in terms of the Poincaré sphere angles.

e) For an unknown light source, you measure all four Stokes parameters $\mathbf{I},\mathbf{Q},\mathbf{U},\mathbf{V}$, and from these you calculate the Poincaré sphere angles. Model the polarized flux as it were elliptically polarized. Clearly draw the ellipse drawn out by the electric field vector, and label the position angle of the ellipse correctly.

f) The eccentricity of an ellipse is defined as $e \equiv \sqrt{1-\left(\frac{b}{a}\right)^2}$, where a and b are the semi-major and semi-minor axes respectively. Find the eccentricity of the electric field ellipse in terms of the Poincaré sphere angles.

Problem 15.15: Mueller Matrices

In Thought Experiment 15.12 we eventually wrote Stokes parameters as a 4×1 matrix $\mathbf{S}$, referred to as the Stokes vector. In an optics lab each device such as a polarizer or wave plate can have an associated 4×4 Mueller matrix $\mathbf{M}$. Using linear algebra, the resultant Stokes parameters can be calculated as $\mathbf{S} = \mathbf{M}\mathbf{S}_0$.

a) Consider unpolarized light that passes through a polarizer aligned with the $\hat{x}$ direction. The resultant light is linearly polarized also in the $\hat{x}$ direction. Show that the associated Mueller matrix for a horizontal polarizer is:

$$\mathbf{M}_{\hat{x}} = \frac{1}{2}\begin{pmatrix} 1 & 1 & 0 & 0 \\ 1 & 1 & 0 & 0 \\ 0 & 0 & 0 & 0 \\ 0 & 0 & 0 & 0 \end{pmatrix}.$$

b) Similarly show that the matrix of a vertical polarizer is:

$$\mathbf{M}_{\hat{y}} = \frac{1}{2}\begin{pmatrix} 1 & -1 & 0 & 0 \\ -1 & 1 & 0 & 0 \\ 0 & 0 & 0 & 0 \\ 0 & 0 & 0 & 0 \end{pmatrix}.$$

c) Similarly show that the matrix of a 45 degree polarizer is:

$$\mathbf{M}_{45^\circ} = \frac{1}{2}\begin{pmatrix} 1 & 0 & 1 & 0 \\ 0 & 0 & 0 & 0 \\ 1 & 0 & 1 & 0 \\ 0 & 0 & 0 & 0 \end{pmatrix}.$$

d) In a common introductory physics demonstration two polarizers are set up, one vertical and the other horizontal, and the students clearly see that no light passes through them. However when a 45° polarizer is sandwiched in-between, they can see through the stack of all 3 polarizers. Using the above Mueller matrices: (i) show that no light is transmitted through the two polarizers, and (ii) find the fraction of light that passes through the stack of three polarizers.

e) A Mueller matrix of a quarter wave plate is:

$$\begin{pmatrix} 1 & 0 & 0 & 0 \\ 0 & 1 & 0 & 0 \\ 0 & 0 & 0 & 1 \\ 0 & 0 & -1 & 0 \end{pmatrix}.$$

What are the directions of the slow and fast axes?

f) Find the Mueller matrix of a half wave plate.

g) Show using Mueller matrices that linearly polarized light at an angle of 45° becomes left circularly polarized after passing through a quarter wave plate.

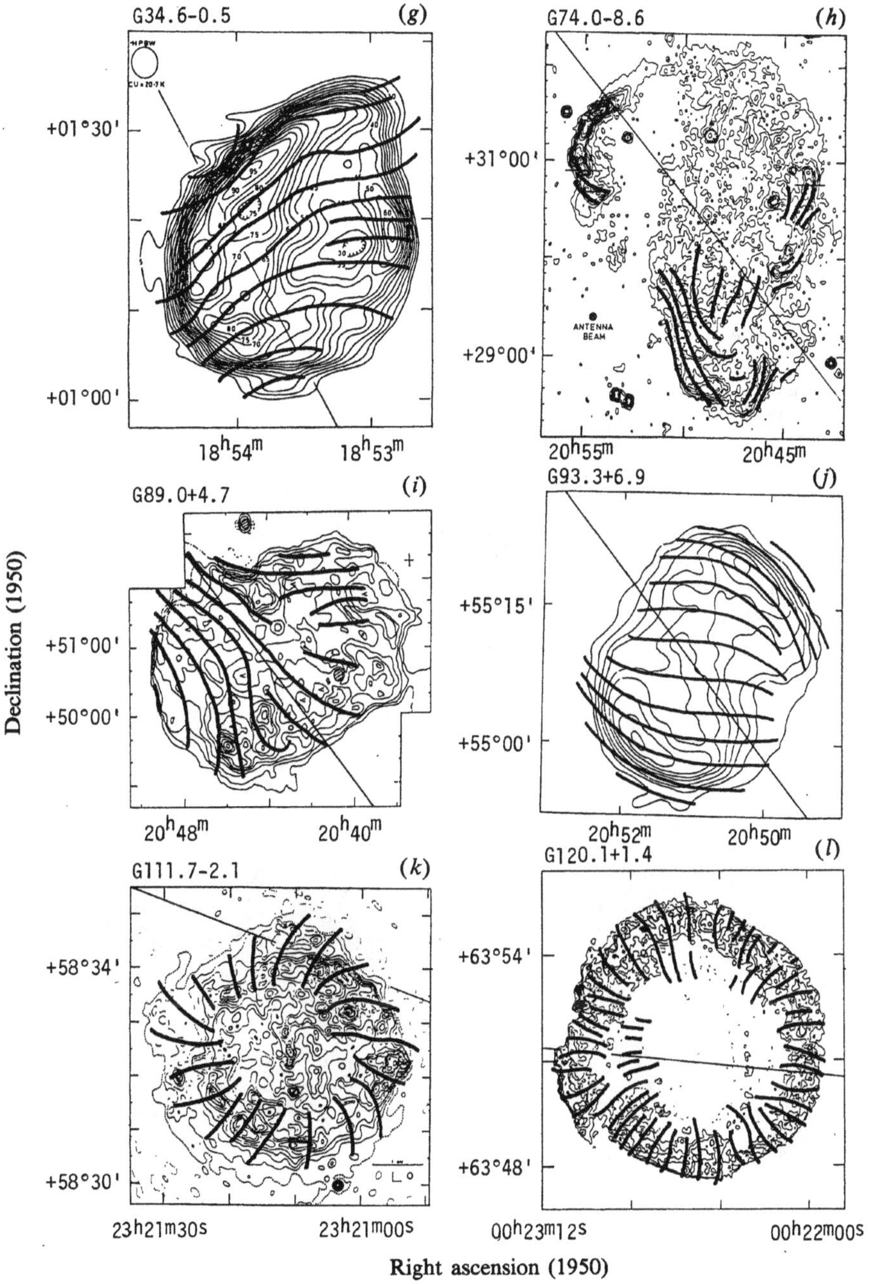

Plate 27: Magnetic Fields in Supernova Remnants[265]

---

[265] Radio intensity contour plots of selected supernova remnants with the direction perpendicular to the measured polarization angle φ overlaid. As this emission is caused by synchrotron radiation (p. 631) these lines show the direction of the projected magnetic field. Notice the radial magnetic field structure in the bottom two supernova remnants, however we know that $\vec{\nabla}\cdot\vec{B}=0$! Explain how this can be. (Hint, see p. 381.) The figure is from D.K. Milne, "An Atlas of Supernova Remnant Magnetic Fields," The Austrailian Journal of Physics, **40** (1987), pp. 771-787.

Plate 28: Karl Jansky's Antenna[266] and the Bell Labs Horn-Reflector Antenna[267]

---

[266] See Karl G. Jansky, "Electrical Phenomena that Apparently are of Interstellar Origin," Popular Astronomy, **41** (1933), pp. 548-555. The photo is from the website of the National Radio Astronomy Observatory.

[267] See A.A. Penzias and R.W. Wilson, "A Measurement of Excess Antenna Temperature at 4080 Mc/s.," The Astrophsical Journal, **142** (1965), pp. 419-421. The photo is figure 2 from: A.B. Crawford, D.C. Hogg, and L.E. Hunt, "Project Echo—Horn-Reflector for Space Communication," NASA Technical Note, **D-1131** (1961).

Chapter 16 Sources of Electromagnetic Radiation

Maxwell's electromagnetic theory of radiation predicts that a changing electric field induces a changing magnetic field, as described by Ampere's law, which in turn induces a changing electric field in accordance with Faraday's law. Heinrich Hertz experimentally confirmed Maxwell's theory in 1881 by generating and detecting radio waves in the laboratory, and demonstrating that these waves behaved like visible light, exhibiting properties such as reflection, refraction, diffraction, and interference.[268] Maxwell's theory and Hertz's experiments led directly to the development of modern radio, radar, television, electromagnetic imaging, and wireless communications. For example, the two antennas shown were developed by radio engineers at Bell Labs in 1933 and 1961 respectively. They were used to discover the very first radio waves from space, and the cosmic microwave background radiation, which are two of the greatest astronomical discoveries of all time.

We begin this chapter by discussing sources of light and the spherical wave equation. A straightforward analysis of a spherically symmetric light wave allows us to expand on the concept of *retarded time*. Later in the chapter we will apply the finite speed at which information travels to two different phenomena—radio antennas and moving charged particles. As it turns out, the key insight is to use the potential formulation of electrodynamics, while carefully accounting for the time of information travel between source and observer. This was accomplished, at the end of the nineteenth century, by the French engineer Alfred-Marie Liénard and the German physicist Emil Johann Wiechert, who independently derived the correct expressions for the potentials and the fields of a point charge in arbitrary motion. In fact, it is their expression, rather than those of Coulomb and Biot, that is the correct one for charges undergoing arbitrary motion.

We begin our discussion of antennas with the simplest antenna of them all, the Hertzian dipole radiator, or simply the *Hertzian dipole*, and derive the scalar and vector potentials, and electric and magnetic fields, of the oscillating dipole in the limit of large distances (i.e. the *radiation zone* or *far field region*) and long wavelengths (compared to the size of the dipole). From these we calculate the radiated power, Poynting vector, radiation resistance, and beam pattern of a Hertzian dipole antenna. We then show how the dipole can be used as both a receiver and transmitter, and discuss the weak and strong reciprocity theorems in the context of antennas. Hertz constructed his dipole antenna to test a key prediction of Maxwell's theory—the existence of electromagnetic waves—while John William Strutt (Lord Rayleigh) modeled the molecules of the Earth's atmosphere as tiny Hertzian dipoles that oscillate in response to an electromagnetic field. In this way, Rayleigh derived his famous law of scattering and explained the blue sky during the day and the red sun at dawn and dusk (p. 532).

In the case of antennas, we focus on the change in currents as the source of radiation in the far field, and we examine key results of antenna theory and single dish radio telescopes. From there, we continue with antenna theory in earnest. It is here that we come to understand why an engineer like Alfred-Marie Liénard would follow such a painstaking line of theoretical research. After calculating beam patterns of both whip and dish antennas, we will discuss how to use arrays of dish antennas to make detailed maps of faint radio sources, such as those extragalactic jets that exhibit superluminal motion.

---

[268] F.K. Vreeland and Henri Poincare, Maxwell's theory and wireless telegraphy, Part 1: Maxwell's theory and Hertzian oscillations, (New York: McGraw Publishing Co., 1904), 31-45. See also Heinrich Hertz, Electric Waves, trans. D.E. Jones (London: MacMillan and Co., 1893).

While it is frequently necessary to solve problems where the radiation is produced by time changing currents, it is sometimes the case that the radiation is being emitted by a single charge, or small number of charges, undergoing acceleration. In that case, we want to solve for the potentials and fields of such a charge (or charges). The result for arbitrarily moving point charges is called the Liénard-Wiechert potentials. Although Liénard and Wiechert derived these potentials prior to Einstein's discovery of special relativity, the Liénard-Wiechert potentials and the fields derived from them, are perfectly consistent with relativity.

Wiechert, unlike Liénard, was what we would now call a particle physicist. He, along with Hendrik Lorentz, Joseph Larmor, Henri Poincaré, Max Abraham, Paul Langevin, and others, worked on the relationship between the mass of the newly discovered electron and theories of the aether. Despite working under the false premise that the aether exists, these scientists developed many of the important relations we now know to be a consequence of special relativity. These included, among others, the relativistically correct potentials of a point charge, which we will discuss in Section 17.4 (p. 617) after developing relativitic electrodynamics.

16.1 *Spherical Waves*

That accelerated charges, and changing currents, emit radiation is key to our understanding of phenomena from radio antennas to the acceleration of cosmic rays, and is the main theme of this chapter. The salient fact that binds all of these phenomena together is the finite speed at which light travels.

Consider a far away source of light. The observer does not see the source as it is now, but rather as it had been. This is the same idea as *lookback time* in astronomy. Just as it takes approximately eight minutes for light to reach Earth from the Sun's surface, and the time delay between the closest other star (Proxima Centauri) and Earth is 4.2 years, we also see distant galaxies as they were billions of years ago. We use exactly this same concept for everyday purposes, such as calculating the beam pattern of a radio antenna or knowing which way to point your satellite television receiver.

Recall that ever since Maxwell demonstated that light is electromagnetic, it made perfect sense that all *electromagnetic* information would travel at the speed of light. Einstein's theory implies that *no* information can travel faster than light.

Thought Experiment 16.1: An Idealized Spherical Wave

Consider a small sphere of radius a emitting light with electric and magnetic fields $\vec{E}_a(t)$ and $\vec{B}_a(t)$ respectively. What are the electric and magnetic fields as a function of time and distance?

In this example, we will assume that the observer is very far from the source, that the source radiates equally in all directions, and that it is coherent light. We will later show that these assumptions are, in practice, mutually exclusive.

From conservation of energy, the Poynting vector $\vec{\mathbb{S}}$, on the surface of the source $(r=a)$, is simply related to the total power emitted $\mathbb{P}$ by:

$$\vec{\mathbb{S}}_a(t) = \frac{\mathbb{P}(t)}{4\pi a^2}\hat{r} = \frac{c}{Z_0}\vec{E}_a(t)\times\vec{B}_a(t).$$

It will take some time for this light to propagate, so the Poynting vector observed at a distance r will also take place at a time $\frac{|r-d|}{c} \approx \frac{r}{c}$ later, thus:

$$\vec{\mathbb{S}}(r,t) = \frac{\mathbb{P}\left(t-\frac{r}{c}\right)}{4\pi r^2}\hat{r} = \frac{1}{\mu_0}\vec{E}(r,t)\times\vec{B}(r,t).$$

From this, it seems reasonable that the propagating fields would be:

$$\vec{E}(r,t) = \frac{a}{r}\vec{E}_a\left(t-\frac{r}{c}\right) \quad \text{and} \quad \vec{B}(r,t) = \frac{a}{r}\vec{B}_a\left(t-\frac{r}{c}\right),$$

since they conserve energy:

$$\vec{\mathbb{S}}(r,t) = \frac{1}{\mu_0}\left(\frac{a}{r}\vec{E}_a\left(t-\frac{r}{c}\right)\right)\times\left(\frac{a}{r}\vec{B}_a\left(t-\frac{r}{c}\right)\right) = \frac{a^2}{r^2}\left(\vec{E}_a\left(t-\frac{r}{c}\right)\times\vec{B}_a\left(t-\frac{r}{c}\right)\right) = \frac{a^2}{r^2}\vec{\mathbb{S}}\left(a,t-\frac{r}{c}\right).$$

Now we will apply the free space wave equations:

$$\nabla^2\vec{E} = \frac{1}{c^2}\frac{\partial^2\vec{E}}{\partial t^2} \quad \text{and} \quad \nabla^2\vec{B} = \frac{1}{c^2}\frac{\partial^2\vec{B}}{\partial t^2}.$$

First recall that the vector Laplacian in spherical coordinates is not simply found by applying the scalar Laplacian to each coordinate. However, by making our assumption of radially propagating light, we can assume that neither the electric nor the magnetic fields have a radial component. By letting $\tau = t - \frac{1}{c}r$, and applying the chain rule for derivatives, the vector Laplacian of the electric field simplifies to:

$$\begin{aligned}
\nabla^2\vec{E} &= \frac{1}{r^2}\frac{\partial}{\partial r}\left(r^2\frac{\partial}{\partial r}\left(\frac{a}{r}\vec{E}_a(\tau)\right)\right) - \frac{a\vec{E}_a(\tau)}{r^3\sin^2(\theta)} \\
&= \frac{1}{r^2}\frac{\partial}{\partial r}\left(r^2\left(\frac{a}{r}\frac{\partial}{\partial r}\vec{E}_a(\tau) - \frac{a}{r^2}\vec{E}_a(\tau)\right)\right) - \frac{a\vec{E}_a(\tau)}{r^3\sin^2(\theta)} \\
&= \frac{1}{r^2}\frac{\partial}{\partial r}\left(-\frac{ar}{c}\frac{\partial}{\partial\tau}\vec{E}_a(\tau) - \vec{E}_a(\tau)\right) - \frac{a\vec{E}_a(\tau)}{r^3\sin^2(\theta)} \\
&= \frac{1}{r^2}\left(\frac{ar}{c^2}\frac{\partial^2}{\partial\tau^2}\vec{E}_a(\tau) - \frac{a}{c}\frac{\partial}{\partial\tau}\vec{E}_a(\tau) + \frac{a}{c}\frac{\partial}{\partial\tau}\vec{E}_a(\tau)\right) - \frac{a\vec{E}_a(\tau)}{r^3\sin^2(\theta)} \\
&= \frac{a}{rc^2}\frac{\partial^2}{\partial\tau^2}\vec{E}_a(\tau) - \frac{a\vec{E}_a(\tau)}{r^3\sin^2(\theta)}.
\end{aligned}$$

Now the time side of the wave equation simplifies to:

$$\frac{1}{c^2}\frac{\partial^2\vec{E}}{\partial t^2} = \frac{1}{c^2}\frac{\partial^2}{\partial t^2}\left(\frac{a}{r}\vec{E}_a\left(t-\frac{r}{c}\right)\right) = \frac{a}{rc^2}\frac{\partial^2}{\partial t^2}\left(\vec{E}_a\left(t-\frac{r}{c}\right)\right) = \frac{a}{rc^2}\frac{\partial^2}{\partial\tau^2}\vec{E}_a(\tau).$$

In the far limit, but only in the far limit, these are equal. Thus, our radial solution only applies when $r \gg a$.

Problem 16.1: A Spherical Magnetic Wave

Show that in the far limit $\vec{B}(r,t)=\frac{a}{r}\vec{B}_a\left(t-\frac{r}{c}\right)$ is a solution to Maxwell's equations in free space.

Discussion 16.1: Retarded Time

The important quantity in all this analysis is what is commonly called the *retarded time* (τ), which is simply the time that the information left the object being observed. This is not simply the time delay, but rather the time that the information left the source.

Imagine, by way of analogy, that on January 15 you receive a postcard from a friend vacationing in Paris, which is dated January 1. In this example, the current time is January 15, the retarded time is January 1, and the time delay would be two weeks. More importantly, your friend was in Paris when she sent you the postcard on January 1, but she may not be there on January 15 when you read her note. Then, on January 22, you receive a postcard from London dated January 15.

So, from her point of view, she was in Paris and two weeks later she was in London. However, from your point of view, it seemed that only one week passed between her visits to Paris and London. Unlike post cards, light always travels at the same speed. Moreover, nothing can travel faster than light, so we need not worry about receiving information in a different order that it was sent.

In this chapter, we will use a primed coodinate system to denote the location of the sender of the information, and an unprimed coordinate system for the observer. As your friend was the one sending the postcards, $\vec{r}'(\text{Jan 1})$ refers to Paris, and $\vec{r}'(\text{Jan 15})$ refers to London. However, the two correspondng times (t) from your point of view, as the observer, are rather January 15 for Paris and January 22 for London. But, on January 15 you were home, at location $\vec{r}(\text{Jan 15})$, reading about Paris.

Next we will investigate how the displacment vector, $\vec{R}$, that each postcard traveled, is related to the retarded time. In the case of the first postcard, it needed to go from Paris to your house, which we will label with the time your friend sent it, as follows:

$$\vec{R}(\text{Jan 1})=\vec{r}(\text{Jan 15})-\vec{r}'(\text{Jan 1}) \qquad \text{The postcard of Paris}$$
$$\vec{R}(\text{Jan 15})=\vec{r}(\text{Jan 22})-\vec{r}'(\text{Jan 15}) \qquad \text{The postcard of London.}$$

In the case of a moving particle, rather, the retarded time is that particular time, earlier than the time at which the measurement is being made, which allows for information traveling at the speed of light to pass from the source to the point of observation. The retarded time can therefore be found for a given observation time, and known particle trajectory, from:

$$\tau=t-\tfrac{1}{c}R(\tau),$$

where t is the current time at the point of observation, R is the distance between the source charge and observer, and c is the speed of signal propagation, which in this case is the speed of light in a vacuum.

At this point we run into our first small complication. As R is a function of the retarded time, and the retarded time is, in turn, a function of R, it is not obvious how to solve this relationship. The algebra will depend on the details of the specific particle motion $\vec{r}'(\tau)$. In this chapter, we will solve this relationship for a few simple systems, such as oscillating charges.

In practice, however, it is far easier to solve implicit functions, such as the one shown in the figure below, iteratively using a computer. This way, we can define the particle's position as a function of time. Then, for each detection by the observer, we simply guess the retarded time, calculate the time delay $\frac{1}{c}R(\tau)$, and calculate a new retarded time τ. Once we get the same value for τ out as we put in, we know we have finally guessed right. As computers can do this very quickly, it is relatively straightforward to actually solve for τ as a function of t.

Consider a particle moving toward an observer at a constant speed of $\frac{3}{4}c$, as shown in the figure. Now imagine that at regular intervals the particle emits a pulse of light. We will also assume that, as with the postcards, each pulse of light contained information regarding the particle's location. As the observer will use his own clock to measure the particle, he would measure the particle to be moving much faster than it really was, for the same reason that you thought that it took your friend only one week, rather than the two weeks she actually took, to travel from Paris to London. This would be the common mistake of confusing the particle's actual velocity of:

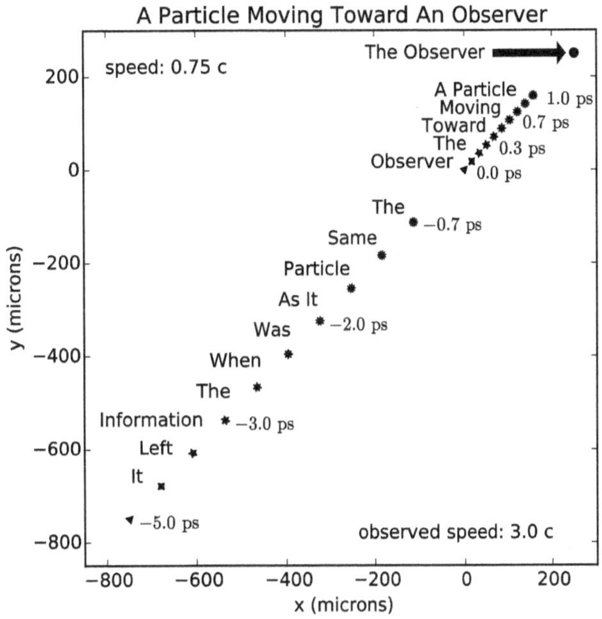

$$\vec{v}' = \frac{d\vec{r}'}{d\tau},$$

with the naïvely observed velocity $\frac{d}{dt}\vec{r}'$.

Sometimes something appears to travel sideways faster than the speed of light. This phenomenon, called *superluminal motion*, is observed in extragalactic radio galaxies when a blob of material moves toward, but not directly toward, Earth at a sufficiently high velocity. Of course, this is simply because we are using the observer's clock, along with the sender's ruler. When radio astronomers correctly take the time-delay into account, they can measure the actual velocity of the material $\frac{d}{d\tau}\vec{r}'$, which is always less than the speed of light.

Thought Experiment 16.2: Electric and Magnetic Field Interdependence

Consider a spherical wave that satisfies Maxwell's equations. What is the relationship between the electric and magnetic fields?

Recall from Thought Experiment 16.1 (p. 508) that:

$$\vec{E}(r,t) = \frac{a}{r}\vec{E}_a(\tau) \quad \text{and} \quad \vec{B}(r,t) = \frac{a}{r}\vec{B}_a(\tau),$$

where $\tau = t - \frac{r}{c}$ is the retarded time.

Applying Faraday's law we have:

$$\vec{\nabla}\times\vec{E} = -\frac{\partial\vec{B}}{\partial t} \quad\rightarrow\quad \vec{\nabla}\times\left(\frac{\cancel{a}}{r}\vec{E}_a(\tau)\right) = -\frac{\partial}{\partial t}\left(\frac{\cancel{a}}{r}\vec{B}_a(\tau)\right).$$

This is expressed in polar coordinates as:

$$\frac{1}{r}\left(-\frac{\partial}{\partial r}\left(\cancel{r}\frac{\hat{\varphi}\cdot\vec{E}_a(\tau)}{\cancel{r}}\right)\hat{\theta} + \frac{\partial}{\partial r}\left(\cancel{r}\frac{\hat{\theta}\cdot\vec{E}_a(\tau)}{\cancel{r}}\right)\hat{\varphi}\right) = -\frac{1}{r}\frac{\partial}{\partial t}\left(\vec{B}_a(\tau)\right) .$$

If we apply the chain rule:

$$\frac{\partial}{\partial r} = \frac{\partial\tau}{\partial r}\frac{\partial}{\partial\tau} \quad\text{and}\quad \frac{\partial}{\partial t} = \frac{\partial\tau}{\partial t}\frac{\partial}{\partial\tau},$$

we have that:

$$\frac{1}{r}\left(-\frac{\partial}{\partial r}\left(\cancel{r}\frac{\hat{\varphi}\cdot\vec{E}_a(\tau)}{\cancel{r}}\right)\hat{\theta} + \frac{\partial}{\partial r}\left(\cancel{r}\frac{\hat{\theta}\cdot\vec{E}_a(\tau)}{\cancel{r}}\right)\hat{\varphi}\right) = -\frac{1}{r}\frac{\partial}{\partial t}\left(\vec{B}_a(\tau)\right),$$

$$\left(-\hat{\varphi}\cdot\frac{\partial\vec{E}_a(\tau)}{\partial\tau}\frac{\partial\tau}{\partial r}\right)\hat{\theta} + \left(\hat{\theta}\cdot\frac{\partial\vec{E}_a(\tau)}{\partial\tau}\frac{\partial\tau}{\partial r}\right)\hat{\varphi} = -\frac{\partial\vec{B}_a(\tau)}{\partial\tau}\frac{\partial\tau}{\partial t},$$

so that

$$\left(-\hat{\varphi}\cdot\frac{\partial\vec{E}_a(\tau)}{\partial\tau}\left(-\frac{1}{c}\right)\right)\hat{\theta} + \left(\hat{\theta}\cdot\frac{\partial\vec{E}_a(\tau)}{\partial\tau}\left(-\frac{1}{c}\right)\right)\hat{\varphi} = -\frac{\partial\vec{B}_a(\tau)}{\partial\tau},$$

multiply both sides by the speed of light, to get:

$$\left(\hat{\varphi}\cdot\frac{\partial\vec{E}_a(\tau)}{\partial\tau}\right)\hat{\theta} - \left(\hat{\theta}\cdot\frac{\partial\vec{E}_a(\tau)}{\partial\tau}\right)\hat{\varphi} = -c\frac{\partial\vec{B}_a(\tau)}{\partial\tau}.$$

We therefore have that:

$$\frac{\partial}{\partial\tau}\left[\left(\hat{\varphi}\cdot\vec{E}_a(\tau)\right)\hat{\theta} - \left(\hat{\theta}\cdot\vec{E}_a(\tau)\right)\hat{\varphi}\right] = \frac{\partial}{\partial\tau}\left[-c\vec{B}_a(\tau)\right].$$

Integrating both sides with respect to τ yields:

$$\left(\hat{\varphi}\cdot\vec{E}_a(\tau)\right)\hat{\theta} - \left(\hat{\theta}\cdot\vec{E}_a(\tau)\right)\hat{\varphi} = -c\vec{B}_a(\tau),$$

This can be written as

$$\vec{E}_a(\tau)\times\hat{r} = -c\vec{B}_a(\tau),$$

so the magnetic field at $r = a$ is therefore:

$$\vec{B}_a(\tau) = \frac{\hat{r}}{c}\times\vec{E}_a(\tau) .$$

We can scale the fields by a factor of r^{-1} to get the general relationship:

$$\frac{a}{r}\vec{B}_a(\tau) = \frac{\hat{r}}{c}\times\frac{a}{r}\vec{E}_a(\tau) \quad\rightarrow\quad \vec{B}(r,t) = \tfrac{1}{c}\hat{r}\times\vec{E}(r,t).$$

In this case, the wave propagates in the $\hat{r}$ direction. As a check, we will compute the Poynting vector:

$$\mathbb{S} = \tfrac{1}{\mu_0}\vec{E}\times\vec{B} = \tfrac{1}{\mu_0}\vec{E}\times\left(\tfrac{1}{c}\hat{r}\times\vec{E}\right) = \tfrac{1}{\mu_0 c}\vec{E}\times\left(\hat{r}\times\vec{E}\right) = \tfrac{1}{Z_0}\left(\hat{r}\left(\vec{E}\cdot\vec{E}\right) - \vec{E}\cancel{\left(\vec{E}\cdot\hat{r}\right)}\right) = \tfrac{1}{Z_0}E^2\hat{r}.$$

And similarly in terms of the magnetic field:

$$\vec{\mathbb{S}} = \frac{c}{\mu_0}\left(\vec{B}\times\hat{r}\right)\times\vec{B} = \frac{c}{\mu_0}\left(\hat{r}\left(\vec{B}\cdot\vec{B}\right) - \vec{B}\cancel{\left(\vec{B}\cdot\hat{r}\right)}\right) = \frac{c^2}{Z_0}B^2\,\hat{r}.$$

Thought Experiment 16.3: A Beamed Spherical Wave

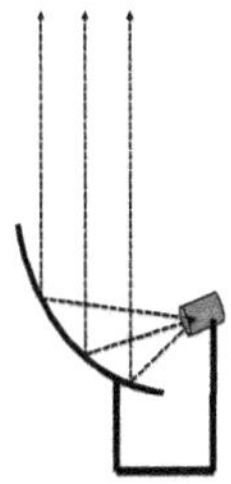

The radio waves in an ideal dish antenna are beamed in a cone with a Gaussian beam pattern. Find the Poynting vector and the electric field measured by the antenna in terms of the emitted power.

Setting $\hat{z}$ along the optical axis,[269] the Poynting vector is:

$$\vec{\mathbb{S}} \propto \frac{e^{-\left(\frac{\theta}{\alpha}\right)^2}}{r^2}\hat{r} = \frac{K(\tau)}{r^2}e^{-\left(\frac{\theta}{\alpha}\right)^2}\hat{r}.$$

Now we need to determine the time dependent coefficient $K(\tau)$. If we assume a steady-state solution and apply energy conservation, the emitted power is:

$$\mathbb{P} = \oint_{\text{volume}} \vec{\nabla}\cdot\vec{S}\,\mathrm{d}V = \oint_{\text{surface}} \vec{S}\cdot\mathrm{d}\vec{\mathrm{A}}$$

$$\mathbb{P}(\tau) = \oint_{\text{surface}}\left(\frac{K(\tau)}{\cancel{r^2}}e^{-\left(\frac{\theta}{\alpha}\right)^2}\hat{r}\right)\cdot\left(2\pi\cancel{r^2}\sin\theta\,\mathrm{d}\theta\,\hat{r}\right)$$

$$\mathbb{P}(\tau) = 2\pi\int_0^{\pi} K(\tau)e^{-\left(\frac{\theta}{\alpha}\right)^2}\sin\theta\,\mathrm{d}\theta$$

If the size of the beam is small enough that $\alpha \ll 1$ radian, then we need only integrate where $\sin\theta \approx \theta$:

$$\mathbb{P}(\tau) \approx 2\pi K(\tau)\int_0^{\infty}\theta e^{-\left(\frac{\theta}{\alpha}\right)^2}\mathrm{d}\theta = -\pi\alpha^2 K(\tau)\int_0^{\infty}-2\left(\frac{\theta}{\alpha}\right)e^{-\left(\frac{\theta}{\alpha}\right)^2}\mathrm{d}\left(\frac{\theta}{\alpha}\right) = \pi\alpha^2 K(\tau)$$

$$K(\tau) = \frac{\mathbb{P}(\tau)}{\pi\alpha^2}.$$

So, the Poynting vector is given by:

$$\vec{\mathbb{S}} = \frac{\mathbb{P}(\tau)}{\pi\alpha^2 r^2}e^{-\left(\frac{\theta}{\alpha}\right)^2}\hat{r},$$

and in terms of the electric field:

$$\vec{\mathbb{S}} = \tfrac{1}{\mu_0}\vec{E}\times\vec{B} = \sqrt{\tfrac{\varepsilon_0}{\mu_0}}\,E^2\,\hat{r} = \tfrac{1}{Z_0}E^2\,\hat{r}.$$

[269] The *optical axis* is the axis along which light propagates through an optical system such as a lens or a mirror. It usually coincides with the geometrical axis of rotational symmetry of the system. The optical and geometric axes of a parabolic dish, such as the one that comprises the radio telescope located at Green Bank, West Virginia (p. 177) do indeed coincide. Do not confuse the optical axis with the *optic* axis, by the way. The latter refers to a special property of crystals involving birefringence.

Solving for the electric field:

$$E = \sqrt{Z_0 \mathbb{S}} = \sqrt{Z_0}\sqrt{\left(\frac{\mathbb{P}(\tau)}{\pi\alpha^2 r^2} e^{-\left(\frac{\theta}{\alpha}\right)^2}\right)} = \sqrt{\frac{Z_0}{\pi}}\sqrt{\mathbb{P}(\tau)}\left(\frac{e^{-\frac{1}{2}\left(\frac{\theta^2}{\alpha^2}\right)}}{\alpha}\right)\frac{1}{r}.$$

Therefore, light emanating far from the source will follow the spherical wave solution, except that it will be attenuated away from the optical axis.

Example 16.1: A Monochromatic Light Source

Consider monochromatic light from a hypothetical point source, such as the radial antenna shown in the figure. The electric field is given by:

$$\vec{E}(r,t) = \frac{a}{r}\vec{E}_a\left(t - \frac{r}{c}\right) = \frac{a}{r}\vec{E}_a(\tau).$$

If we assume that the fields are sinusoidal in nature, we could write the monochromatic solution as:

$$\vec{E}(r,t) = \frac{a}{r}\left(E_\theta \sin(\omega\tau)\hat{\theta} + E_\varphi \sin(\omega\tau + \vartheta)\hat{\varphi}\right).$$

The magnetic field can be also obtained:

$$\vec{B}(r,t) = \frac{\hat{r}}{c}\times\vec{E}(r,t) = \frac{\hat{r}}{c}\times\frac{a}{r}\left(E_\theta \sin(\omega\tau)\hat{\theta} + E_\varphi \sin(\omega\tau+\vartheta)\hat{\varphi}\right)$$

$$\vec{B}(r,t) = \frac{a}{cr}\left(E_\theta \sin(\omega\tau)\hat{\varphi} - E_\varphi \sin(\omega\tau+\vartheta)\hat{\theta}\right).$$

Notice that four parameters characterize the wave: the amplitudes E_θ and E_φ, the relative phase ϑ between the two directions, and the angular frequency ω. The size of the source, a, is not independent of the amplitudes, but we have kept it so the units of the amplitude make sense.

Note that because of the linearity of the electric and magnetic fields, any waveform can be expressed as a sum of monochromatic light of different frequencies. Detectors are often sensitive to light in a given range, or *band*, of frequencies so it makes sense to sort light by frequency.

Finally, the Poynting vector of the light wave is given by:

$$\begin{aligned}\vec{\mathbb{S}} &= \tfrac{1}{\mu_0}\vec{E}\times\vec{B}\\ &= \tfrac{1}{\mu_0}\tfrac{a}{r}\left(E_\theta \sin(\omega\tau)\hat{\theta} + E_\varphi \sin(\omega\tau+\vartheta)\hat{\varphi}\right)\times\tfrac{1}{c}\tfrac{a}{r}\left(E_\theta \sin(\omega\tau)\hat{\varphi} - E_\varphi \sin(\omega\tau+\vartheta)\hat{\theta}\right)\\ &= \tfrac{1}{Z_0}\left(\frac{a}{r}\right)^2\left(E^2_\theta \sin^2(\omega\tau) + E^2_\varphi \sin^2(\omega\tau+\vartheta)\right)\hat{r}.\end{aligned}$$

This result agrees with the inverse square law for light, and should come as no surprise since we used the inverse square law to obtain the $\frac{1}{r}$ dependence of the electric and magnetic fields produced by spherically symmetric source in the first place.

Example 16.2: Two Source Interference

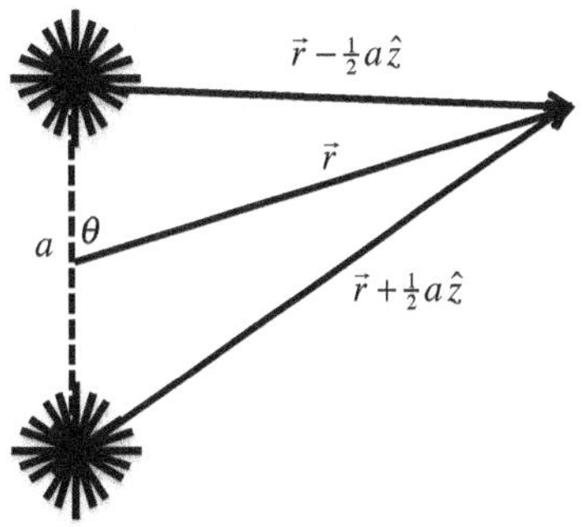

Consider now two antennas each emitting monochromatic light, such as a double tower radio transmitter. How is the power transmitted away from the towers?

First we write down the electric field vector due to each source, and simply add them together.

$$\vec{E} = \frac{r_o}{\left|\vec{r}-\frac{1}{2}a\hat{z}\right|}\vec{E}_{r_0}\left(t-\frac{\left|\vec{r}-\frac{1}{2}a\hat{z}\right|}{c}\right) + \frac{r_o}{\left|\vec{r}+\frac{1}{2}a\hat{z}\right|}\vec{E}_{r_0}\left(t-\frac{\left|\vec{r}+\frac{1}{2}a\hat{z}\right|}{c}\right),$$

where r_0 and $\vec{E}_{r_0}$ are evaluated at the surface of the antenna.

If the point of interest is equidistant from both transmitters, than the light will take the same time to arrive at the observer. This will be the case when the position is in the plane of symmetry, so that $\vec{r} = s\hat{s}$ in cylindrical coordinates, so:

$$\vec{E}(r,t) = \frac{2r_o}{\sqrt{s^2+\frac{1}{4}a^2}}\vec{E}_{r_0}\left(t-\frac{1}{c}\sqrt{s^2+\frac{1}{4}a^2}\right) = \frac{2r_o\vec{E}_{r_0}}{\sqrt{s^2+\frac{1}{4}a^2}}\sin\left(\omega t-\frac{\omega}{c}\sqrt{s^2+\frac{1}{4}a^2}\right).$$

Thus, in the plane between the two transmitters, the amplitudes of the electric fields simply add.

On the other hand, if the observer is closer to one source the signal from one source will be received before that of the other. Thus in cylindrical coordinates, the observed electric field is:

$$\vec{E}(r,t) = r_o\vec{E}_{r_0}\left(\frac{\sin\left(\omega t-\frac{1}{c}\omega\sqrt{s^2+z^2-az+\frac{1}{4}a^2}\right)}{\sqrt{s^2+z^2-az+\frac{1}{4}a^2}} + \frac{\sin\left(\omega t-\frac{1}{c}\omega\sqrt{s^2+z^2+az+\frac{1}{4}a^2}\right)}{\sqrt{s^2+z^2+az+\frac{1}{4}a^2}}\right).$$

And, the Poynting vector then becomes:

$$\mathbb{S} = \frac{1}{\mu_0}\vec{E}\times\vec{B} = \frac{1}{Z_0}E^2\hat{r} = \frac{1}{Z_0}\left(r_0E_{r_0}\right)^2\left(\frac{\sin\left(\omega t-\frac{\omega}{c}\sqrt{s^2+z^2-az+\frac{1}{4}a^2}\right)}{\sqrt{s^2+z^2-az+\frac{1}{4}a^2}} + \frac{\sin\left(\omega t-\frac{\omega}{c}\sqrt{s^2+z^2+az+\frac{1}{4}a^2}\right)}{\sqrt{s^2+z^2+az+\frac{1}{4}a^2}}\right)^2.$$

For large distances, and substituting $r=\sqrt{s^2+z^2}$:

$$\vec{E}(r,t) \approx \frac{2r_o\vec{E}_{r_0}}{r}\sin(\omega t)\cos\left(\frac{1}{2c}\omega\left(\sqrt{r^2+az}-\sqrt{r^2-az}\right)\right).$$

Substituting a normalization constant, $\mathbb{S}_0$, the Poynting vector at large distances becomes:

$$\mathbb{S} \approx \mathbb{S}_0\left(\frac{r_0}{r}\right)^2\sin^2(\omega t)\cos^2\left(\frac{1}{2c}\omega\left(\sqrt{r^2+az}-\sqrt{r^2-az}\right)\right).$$

We see that we have constructive interference if:

$$\frac{1}{2c}\omega\left(\sqrt{r^2+az}-\sqrt{r^2-az}\right) = \pi n,$$

where n is any integer.

Notice that $\lambda = \dfrac{2\pi c}{\omega}$, so if the difference in path length is an integral number of wavelengths we have constructive interference. For large distances, we can take a Taylor expansion to make the approximation:

$$\sqrt{r^2 \pm az} \approx r \pm \frac{az}{2r}.$$

Therefore:

$$n\lambda \approx \left(r + \frac{az}{2r}\right) - \left(r - \frac{az}{2r}\right) \approx \frac{az}{r} \quad \rightarrow \quad \frac{z}{r} \approx \frac{n\lambda}{a}.$$

Now in cylindrical and spherical coordinates, this becomes:

$$\frac{z}{\sqrt{z^2 + s^2}} \approx \frac{n\lambda}{a} \quad \rightarrow \quad \cos\theta \approx \frac{n\lambda}{a}.$$

Therefore, there will be constructive interference in the equatorial plane at any broadcast wavelength, but there will also be constructive interference emanating at particular angles from the sources. However, if the two sources are closer together than a wavelength, the only constructive interference will be on the equator.

Problem 16.2: A Diffraction Grating

A common piece of laboratory equipment is a diffraction grating. This is either a clear piece of glass (transmission grating) or a mirror (reflection grating) that is scratched with evenly spaced lines.

Consider a transmission grating with n scratches per length with monochromatic light of angular frequency ω incident on it from a distant source with an average energy per area of $\vec{\mathbb{S}}$ as in the diagram.

Model this as a row of cylindrical wave sources spaced evenly apart (recall Problem 14.5, p. 442).

a) Write the electric field as a summation of all of the electric field vectors. Assume a normalization constant E_a for the electric field at some distance from each hole a.

b) Consider two adjacent slits. At what angle $\theta = \arctan\left(\dfrac{y}{z}\right)$ is the first maximum?

Is this where there is constructive interference for the other slits as well?

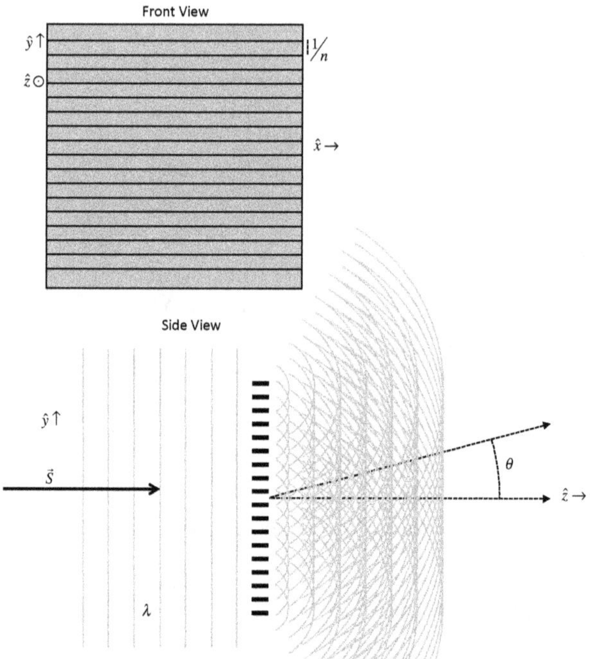

c) Numerically calculate, and plot, the intensity as a function of position y that would fall on a screen a distance z from the grating. Pick a distance z that is large compared to the slit spacing but small compared to the diffraction grating as a whole. Use at least 100 slits in your calculation, and assume a slit spacing equal to the wavelength of the light ($\lambda n = 1$).

d) Spectrometers often contain diffraction gratings to separate the light by frequency. When taking a spectrum of a faint source, you would expect that much of the light goes straight through the diffraction grating without being separated by frequency. How do astronomers design their spectrographs in order to keep light from being wasted? Research and discuss qualitatively the different designs that are currently in use at the largest observatories.

Problem 16.3: Young's Experiment

Consider the experiment by Thomas Young discussed in the introduction. In his experiment Young shined light through two holes, which effectively made two coherent point sources of light. You will now analyze this experiment and find the energy flux $\langle \vec{\mathbb{S}} \rangle$ as a function of position.

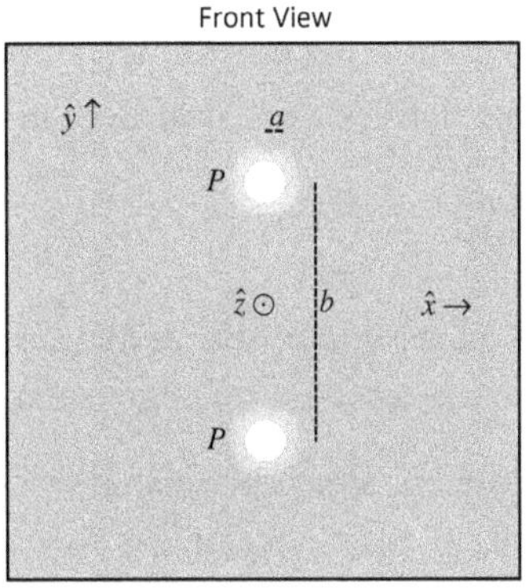

As in the picture, you will consider monochromatic light of frequency ω, with total power $\mathbb{P}$ emanating from each hole of radius a and separation b.

a) Write down the resultant electric field at a position $\vec{r}$ as measured from half way between the two holes.

b) What is the time-averaged Poynting vector $\langle \vec{\mathbb{S}} \rangle$ as a function of the position $\vec{r}$?

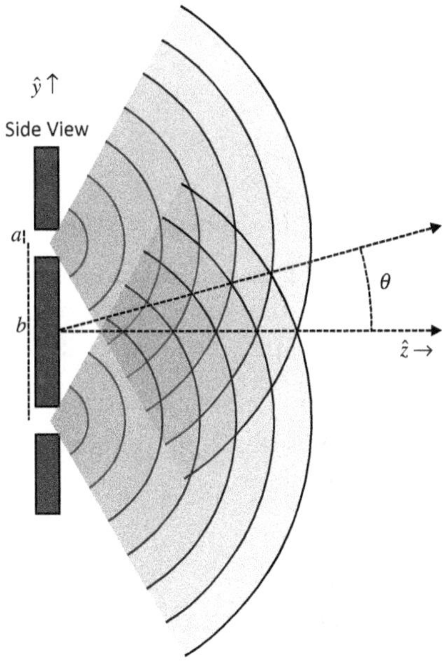

c) Render this intensity in the Cartesian coordinates shown, and make the assumption that $z \gg b$. What is $\langle \vec{\mathbb{S}} \rangle$ as a function of (x,y,z)?

d) Find the intensity $\langle \vec{\mathbb{S}}(s,\varphi,z) \rangle$ in cylindrical coordinates.

e) Plot the intensity as it would appear on a flat screen located at a far position $z \gg s$.

d) What is the angle θ of the first maximum, as in the diagram?

f) Report on this experiment, when it was performed, and the impact it had on our understanding of the nature of light.

Thought Experiment 16.4: A Moving Light Source

Consider a spherical wave emitted by a source a distance r from an observer, whose frame of reference $\lfloor\underline{s}$ is at rest with respect to the source. The electric field seen by the observer in $\lfloor\underline{s}$ is given by:

$$\vec{E}(r,t) = \frac{a}{r}\vec{E}_a(\tau),$$

where $\vec{E}_a$ is the electric field at the surface of the source $(r = a)$. The field $\vec{E}_a$ is evaluated at the delayed time $\tau = t - \frac{r}{c}$, which is the time the light was emitted by the source according to the observer who finally receives the light signal at time t in frame $\lfloor\underline{s}$.

Now consider a reference frame $\lfloor\underline{s'}$ that moves at a constant velocity $\vec{u}$ directly away from the source. The delayed time according to an observer in frame $\lfloor\underline{s'}$ is also given by:

$$\tau' = t - \frac{r + ut}{c},$$

where the plus sign in the numerator indicates that the light must travel farther to reach an observer in the receding reference frame $\lfloor\underline{s'}$ than for an observer in frame $\lfloor\underline{s}$.

Now at what rate does the time at $r = a$ appear to change according to an observer in $\lfloor\underline{s'}$? This is found by taking the partial derivative of τ' with respect to t:

$$\frac{\partial \tau'}{\partial t} = 1 - \frac{u}{c}$$

Any light signal viewed by an observer moving away from the source will appear to be in "slow motion" by a factor of $1 - \frac{1}{c}u$.

Example 16.3: Measuring the Velocity of a Monochromatic Light Source

Many natural light sources produce spectral line emission, which is light of a particular frequency. By measuring the frequency (or wavelength) of the observed light, the velocity of the source can be determined provided that we know the frequency at which the light is emitted.

Consider a monochromatic source that emits light at a known specific frequency ω_0, and whose emitted light is observed by a spectrometer to have a frequency of ω. What is the velocity of this source of light?

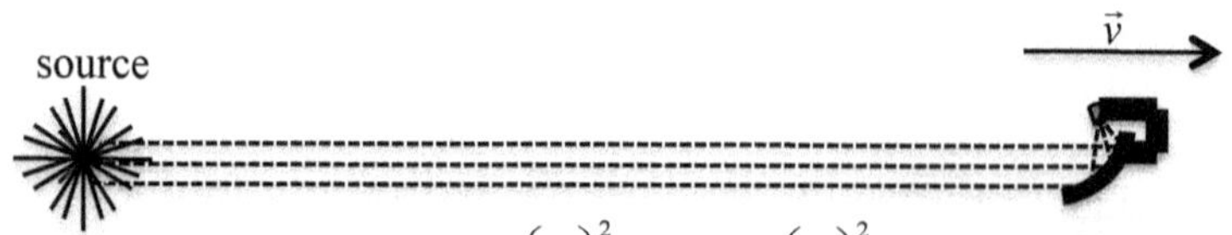

$\vec{S}_a(t) = S_0 \cos^2(\omega_0 t)\,\hat{r}$ $\quad$ $\vec{S}(r,t) = \left(\frac{a}{r}\right)^2 \vec{S}_a\!\left(t - \frac{r}{c}\right) = \left(\frac{a}{r}\right)^2 S_0 \cos^2\!\left(\omega_0\left(t - \frac{r}{c}\right)\right)\hat{r}$

Any sideways component of the velocity is not measurable, so we will solve for the velocity along the line of sight. In a reference frame with the source at the origin, we can write the energy flux at the surface of the source $(r = a)$ as:

$\vec{\mathbb{S}}_a(t) = \mathbb{S}_0 \cos^2(\omega_0 t)\,\hat{r}$.

The Poynting vector at a distance r from the source is given by:

$$\vec{\mathbb{S}}(r,t) = \left(\frac{a}{r}\right)^2 \vec{\mathbb{S}}_a\left(t-\frac{r}{c}\right) = \left(\frac{a}{r}\right)^2 \mathbb{S}_0 \cos^2(\omega_0(t-\tfrac{r}{c}))\,\hat{r} .$$

The detector is moving away at a velocity v in the frame of the source, so the distance is increasing linearly, i.e. $r = r_0 + vt$. So:

$$\vec{\mathbb{S}}(r,t) = \left(\frac{a}{r_0+vt}\right)^2 \mathbb{S}_0 \cos^2\left(\omega_0\left(t-\tfrac{r_0+vt}{c}\right)\right)\hat{r} .$$

Notice that the phase can now be written as:

$$\omega_0\left(t-\frac{r_0+vt}{c}\right) = \omega_0\left(t-\frac{vt}{c}-\frac{r_0}{c}\right) = \omega_0\left(1-\frac{v}{c}\right)t - \omega_0\frac{r_0}{c}$$

So, the observed frequency is given by:

$$\omega = \omega_0\left(1-\frac{v}{c}\right),$$

from which we can easily solve for the velocity v of the source along the line of sight:

$$v = c\left(1-\frac{\omega}{\omega_0}\right).$$

This result is valid only for speeds $v \ll c$.

This effect was first discovered by the Austrian astronomer Christian Doppler in 1842, who applied it erroneously to the colors of binary stars. Modern astronomers, however, do measure the velocities of binary stars, and many other objects, this way.

16.2 *Hertzian Dipole Radiation*

We will consider radiation mechanisms in the final sections of this chapter, and since our analysis relies heavily on the concept of the retarded time, you may wish to review Discussion 16.1 (p. 510). In the context of a charged particle emitting electromagnetic radiation, the retarded time is the time that the light was emitted, rather than when it is observed, and is given by the following formula:

$$\tau = t - \tfrac{1}{c}\left|\vec{r}-\vec{r}'\right| .$$

In this equation, t represents the time at the point of interest (i.e. the observer), $\vec{r}$ the position of the observer, and $\vec{r}'$ the position of the source—so $\left|\vec{r}-\vec{r}'\right|$ represents the distance the light traveled. In most applications thus far, we have assumed that the position of interest has plenty of time to find out about the current element, or, to put it mathematically, that $\tau \approx t$. We are now relaxing this assumption.

This approach seems straightforward, and we have already used it to illustrate the idea of a radiation source and particular solutions to the wave equation. In practice, however, it is easy to oversimplify an equation and miss a key phenomenon. In particular, it becomes very important to pay careful attention to which surface and volume elements correspond to the source and which correspond to the observer.

Since capital letters often represent different quantities (e.g. v and V represent speed and volume respectively), it is traditional to represent the source as a primed coordinate system and the observer as unprimed. This way all other derivative quantities can also be either primed or unprimed. Under this convention, the retarded time is given by:

$$\tau = t - \frac{\left|\vec{r}-\vec{r}'\right|}{c}.$$

Also, it is much easier to start with the potential representation rather than the fields. Starting with the fields, as we did earlier with the transmission line, requires a full series of iterations since the equations are coupled. Starting with the potentials, on the other hand, simplifies this problem at the start, and the correct fields can then be derived from the potentials evaluated at the retarded time.

As the physicists Alfred-Marie Liénard and Emil Wiechert introduced the retarded time independently, the corresponding point charge potentials are called the *Liénard -Wiechert potentials.*

Definition 16.1: The Retarded Potentials

Coulomb's law and the law of Biot-Savart, when written in potential form with the retarded time included, are together called the Liénard-Wiechert potentials. For a continuous distribution of charge:

$$\mathbb{V}(\vec{r},t) = \frac{Z_0 c}{4\pi} \int_{\text{volume}} \frac{\rho(\vec{r}',\tau)}{\left|\vec{r}-\vec{r}'\right|} dV' = \frac{Z_0 c}{4\pi} \int_{\text{volume}} \frac{\rho\left(\vec{r}',t-\frac{1}{c}\left|\vec{r}-\vec{r}'\right|\right)}{\left|\vec{r}-\vec{r}'\right|} dV'$$

$$\vec{\mathbb{A}}(\vec{r},t) = \frac{Z_0}{4\pi c} \int_{\text{volume}} \frac{\vec{J}(\vec{r}',\tau)}{\left|\vec{r}-\vec{r}'\right|} dV' = \frac{Z_0}{4\pi c} \int_{\text{volume}} \frac{\vec{J}\left(\vec{r}',t-\frac{1}{c}\left|\vec{r}-\vec{r}'\right|\right)}{\left|\vec{r}-\vec{r}'\right|} dV'$$

Problem 16.4: The Spherical Wave Solution in Potential Form

We exhibited the spherical wave solution in potential form, but we did not verify that it was in fact a solution to Maxwell's equations.

a) Show that the spherical wave solution for the potentials are, in fact, solutions to the wave equation.

b) Show, in spherical coordinates, that the potential form of the wave equation does in fact yield the spherical wave solution for the electromagnetic fields.

c) Solve for the Poynting vector in terms of the potentials for a spherical wave and simplify it.

Thought Experiment 16.5: The Hertzian Dipole

The dipole radiator, first considered by Hertz in connection with his famous experiments that confirmed key predictions of Maxwell's theory, is perhaps the simplest example of radiation due to time varying charge. Consider an electric dipole whose charge, and therefore dipole moment, is a sinusoidal function of time:

$$Q(t) = Q_0 \cos(\omega t),$$

so that

$$\vec{p}(t) = Q(t)\,\vec{d} = Q_0\cos(\omega t)\vec{d} = \vec{p}_0\cos(\omega t)\,,$$

where $p_0 = Q_0 d$ is the static dipole moment.

We could construct such an arrangement with two small conducting spheres connected by a long thin metal wire of negligible resistance and capacitance, as shown in the figure. The current flowing upward through the connecting wire is then:

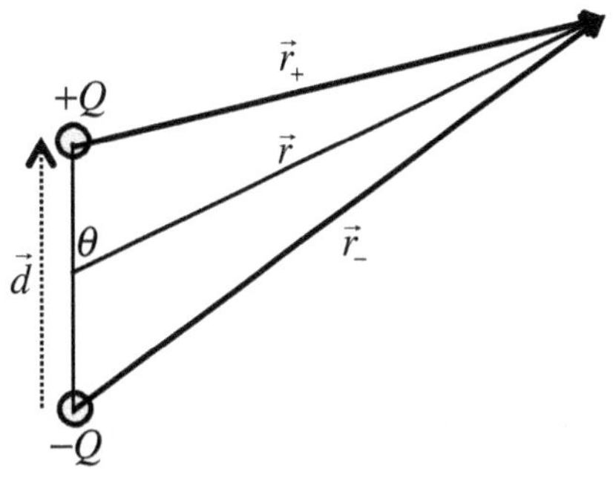

$$I = \frac{\mathrm{d}Q}{\mathrm{d}t} = -\omega Q_0\sin(\omega t) = -I_0\sin(\omega t)\,.$$

The retarded scalar potential is given by:

$$\mathbb{V}(\vec{r},t) = \frac{Q}{4\pi\varepsilon_0 r_+} + \frac{-Q}{4\pi\varepsilon_0 r_-}$$

$$\mathbb{V}(\vec{r},t) = \frac{Q_0\cos\left[\omega\left(t-\frac{r_+}{c}\right)\right]}{4\pi\varepsilon_0 r_+} - \frac{Q_0\cos\left[\omega\left(t-\frac{r_-}{c}\right)\right]}{4\pi\varepsilon_0 r_-}\,,$$

where

$$\vec{r}_+ = \vec{r} - \tfrac{1}{2}\vec{d} \text{ and } \vec{r}_- = \vec{r} + \tfrac{1}{2}\vec{d}\,.$$

Note that there is now a phase difference in addition to the amplitude difference we found in the static case. We write the ratio of r to $r_\pm$ as:

$$\frac{r}{r_\pm} = \left(1 + \left(\frac{d}{2r}\right)^2 \mp \frac{d\cos(\theta)}{r}\right)^{-\frac{1}{2}},$$

and expand the right hand side in a power series to obtain:

$$\frac{r}{r_\pm} = 1 \pm \frac{d\cos(\theta)}{2r} + \left(\frac{d}{2r}\right)^2\left(\frac{3\cos^2(\theta)-1}{2}\right) + \ldots.$$

We assume that the source is far enough away from the point of observation that $r \gg d$. We therefore retain only the two leading terms of the expansion:

$$r_\pm \approx r \mp \tfrac{1}{2}d\cos(\theta)\,.$$

The argument of the trigonometric functions in the above expression for $\mathbb{V}$ is called the phase $\phi_\pm$, and can be written as:

$$\phi_\pm = \omega\tau_\pm = \omega\left(t - \tfrac{1}{c}r_\pm\right) \approx \omega\left(t - \tfrac{1}{c}r \pm \tfrac{1}{c}\tfrac{1}{2}d\cos(\theta)\right).$$

The trigonometric functions in the numerator of $\mathbb{V}$ are therefore given by:

$$\cos\left(\omega\left(t - \tfrac{1}{c}r_\pm\right)\right) \approx \cos\left(\omega\left(t - \tfrac{1}{c}r\right) \pm \tfrac{d}{2c}\omega\cos(\theta)\right).$$

Since $\cos(\alpha \pm \beta) = \cos(\alpha)\cos(\beta) \mp \sin(\alpha)\sin(\beta)$, this is also:

$$\cos\left(\omega\left(t - \tfrac{1}{c} r_{\pm}\right)\right) \approx \cos\left(\omega\left(t - \tfrac{1}{c} r\right)\right)\cos\left(\tfrac{d}{2c}\omega\cos(\theta)\right) \mp \sin\left(\omega\left(t - \tfrac{1}{c} r\right)\right)\sin\left(\tfrac{d}{2c}\omega\cos(\theta)\right),$$

which can also be written as:

$$\cos[\phi_{\pm}] \approx \cos[\omega\tau]\cos[\omega\tfrac{d}{2c}\cos(\theta)] \mp \sin[\omega\tau]\sin[\omega\tfrac{d}{2c}\cos(\theta)].$$

Let us generalize our discussion for a moment to consider an arbitrary system of radiating charges with diameter d. Now imagine that in the center of the source is some sort of *cause* that is driving the oscillation, with the *effect* being the movement of charge. Thus, the minimum time between cause and effect must be given by:

$$\Delta t_{\text{source}} \geq \frac{d}{2c}.$$

The quantity, $t_c \equiv \frac{1}{c} d$, is called the *coherence time* of the source.

The coherence time indicates by how much an electromagnetic wave, traveling from one part of the radiating system, is delayed compared to the waves radiated by other parts of the system. In other words, the term $\frac{d}{c}$, or t_c, determines the time it takes for the electromagnetic wave to pass through the system of charges. In the case of the dipole, it is obviously just the time it takes for light to travel from one charge to the other.

If we assume, for simplicity's sake, that the velocity of the charges equals v, then the charges move through a distance $\frac{1}{c} vd$ in a time given by t_c. The delay inside the system is negligible when this distance is small compared to the size d of the system. If $\frac{1}{c} vd \ll d$, or $v \ll c$, the charges do not move very much as the wave passes through the system.

For the system to remain unchanged during that time, however, the charges' *velocities* must also remain approximately the same, because the vector potential $\vec{\mathbb{A}}$ depends on the currents (i.e. the velocities of the charges), and $\vec{\mathbb{A}}$ is needed to determine both the electric and magnetic fields at the observation point.

For example, consider charges oscillating with angular frequency ω; the wavelength of the resulting light is therefore given by $\lambda = 2\pi c / \omega$. The phase of the charge oscillations changes by ωt_c in the coherence time t_c, and this change must be small compared to 2π. If the delay inside the system is negligible, then the size of the system must be small compared with the wavelength of radiated light. Finally, the assumptions we make here hold, provided the following two inequalities are satisfied:

$v \ll c$ and $d \ll \lambda$.

We can avoid having to include details of how the charge distribution changes within the dipole by requiring that the propagation time for such changes be less than the time it takes light to travel across the dipole, or:

$T \gg t_c$,

where $T = 2\pi / \omega$. This is tantamount to requiring that that $\lambda \gg d$.

Since $\lambda = 2\pi c/\omega$, this also means that $\omega t_c \ll 1$, so the small angle approximation must hold in these two cases:

$$\cos\left[\tfrac{d}{2c}\omega\cos(\theta)\right] \approx 1\text{, and } \sin\left[\tfrac{d}{2c}\omega\cos(\theta)\right] \approx \tfrac{d}{2c}\omega\cos(\theta).$$

Thus,

$$\cos\left[\omega\left(t - \tfrac{1}{c}r_\pm\right)\right] \approx \cos\left[\omega\left(t - \tfrac{1}{c}r\right)\right] \mp \frac{\omega d}{2c}\cos(\theta)\sin\left[\omega\left(t - \tfrac{1}{c}r\right)\right].$$

As we did before, let us now denote the retarded time with $\tau = t - \frac{1}{c}r$. Let's now substitute this result into our expression for the retarded scalar potential:

$$\mathbb{V}(\vec{r},t) \approx \frac{Q_0}{4\pi\varepsilon_0}\left[\left(\frac{1}{r}+\frac{d}{2r^2}\right)\left(\cos(\omega\tau)-\frac{\omega d}{2c}\sin(\omega\tau)\cos\theta\right)-\left(\frac{1}{r}-\frac{d}{2r^2}\right)\left(\cos(\omega\tau)+\frac{\omega d}{2c}\sin(\omega\tau)\cos\theta\right)\right].$$

Carrying through the multiplication, we have that:

$$\mathbb{V}(\vec{r},t) \approx -\frac{Q_0}{4\pi\varepsilon_0 r}\left(\frac{\omega d}{c}\right)\sin(\omega\tau)\cos\theta + \frac{Q_0 d}{4\pi\varepsilon_0 r^2}\cos\theta\cos(\omega\tau).$$

We can write this in a more compact form by recalling that the static dipole moment is $p_0 = Q_0 d$, and retaining only the two highest order terms:

$$\mathbb{V}(\vec{r},t) \approx \frac{p_0\cos\theta}{4\pi\varepsilon_0 r}\left(-\left(\frac{\omega}{c}\right)\sin(\omega\tau)+\frac{1}{r}\cos(\omega\tau)\right).$$

This is consistent with the static potential, because as $\omega \to 0$:

$$\mathbb{V}(\vec{r},t) \to \frac{p_0\cos\theta}{4\pi\varepsilon_0 r^2}.$$

The cosine term in the parentheses above, then, is just the electrostatic potential multiplied by a time-dependent term that varies sinusoidally.

What about the sine term, which falls off as r^{-1}? This term gives the fields responsible for the radiation detected at r. To see this, we must also compute the vector potential:

$$\vec{\mathbb{A}}(\vec{r},t) = \frac{\mu_0}{4\pi}\int_{-d/2}^{d/2}\frac{I(\omega\tau)}{r}\hat{z}\,dz' = -\frac{\mu_0\,\omega p_0}{4\pi r}\sin(\omega\tau)\hat{z}.$$

Since $\hat{z} = \cos\theta\,\hat{r} - \sin\theta\,\hat{\theta}$, this can be written as:

$$\vec{\mathbb{A}}(\vec{r},t) = -\frac{\mu_0\,\omega p_0}{4\pi r}\sin(\omega\tau)\left(\cos\theta\,\hat{r} - \sin\theta\,\hat{\theta}\right).$$

Now, the electric field is given by:

$$\vec{E}(\vec{r},t) = -\vec{\nabla}\mathbb{V} - \tfrac{\partial}{\partial t}\vec{\mathbb{A}}.$$

We first compute the gradient of the scalar potential:

$$\vec{\nabla}\mathbb{V}(\vec{r},t) = \hat{r}\frac{\partial \mathbb{V}}{\partial r} + \hat{\theta}\frac{1}{r}\frac{\partial \mathbb{V}}{\partial \theta} + \hat{\phi}\frac{1}{r\sin\theta}\frac{\partial \mathbb{V}}{\partial \phi},$$

and:

$$\vec{\nabla}\mathbb{V}(\vec{r},t) = \frac{p_0}{4\pi\varepsilon_0}\left(\hat{r}\frac{\partial}{\partial r} + \hat{\theta}\frac{1}{r}\frac{\partial}{\partial \theta}\right)\left(-\frac{\omega\cos(\theta)}{cr}\sin(\omega\tau) + \frac{\cos(\theta)}{r^2}\cos(\omega\tau)\right).$$

Thus,

$$\vec{\nabla}\mathbb{V}(\vec{r},t) = \frac{p_0}{4\pi\varepsilon_0}\left\{2\frac{\omega}{c}\frac{\cos\theta}{r^2}\sin(\omega\tau) + \frac{\omega^2}{c^2}\frac{\cos\theta}{r}\cos(\omega\tau) - 2\frac{\cos\theta}{r^3}\cos(\omega\tau)\right\}\hat{r}$$
$$+ \frac{p_0}{4\pi\varepsilon_0}\left\{\frac{\omega}{c}\frac{\sin\theta}{r^3}\sin(\omega\tau) - \frac{\sin\theta}{r^3}\cos(\omega\tau)\right\}\hat{\theta}.$$

Clearly the terms which vary as r^{-2} and r^{-3} tend to zero faster than the r^{-1} term. If we consider distances for which $r \gg c/\omega$, which is known as the *radiation zone*, then we retain only the first order term:

$$\vec{\nabla}\mathbb{V}(\vec{r},t) \approx \frac{p_0}{4\pi\varepsilon_0}\frac{\omega^2}{c^2}\frac{\cos\theta}{r}\cos(\omega\tau)\hat{r}.$$

The partial derivative of the vector potential with respect to time is

$$\frac{\partial \vec{\mathbb{A}}(\vec{r},t)}{\partial t} = -\frac{\mu_0\omega^2 p_0}{4\pi}\frac{1}{r}\cos(\omega\tau)\left(\cos\theta\,\hat{r} - \sin\theta\,\hat{\theta}\right),$$

or, since $\mu_0 = \varepsilon_0^{-1}c^{-2}$,

$$\frac{\partial \vec{\mathbb{A}}(\vec{r},t)}{\partial t} = -\frac{p_0}{4\pi\varepsilon_0}\frac{\omega^2}{c^2}\frac{1}{r}\cos(\omega\tau)\left(\cos\theta\,\hat{r} - \sin\theta\,\hat{\theta}\right).$$

And the electric field, for $r \gg c/\omega$, is therefore given by

$$\vec{E}(\vec{r},t) = -\vec{\nabla}\mathbb{V} - \frac{\partial \vec{\mathbb{A}}}{\partial t} = -\frac{p_0}{4\pi\varepsilon_0}\frac{\omega^2}{c^2}\frac{\sin\theta}{r}\cos(\omega\tau)\,\hat{\theta}.$$

We leave it as an exercise for you to show that

$$\vec{B}(\vec{r},t) = \nabla \times \vec{\mathbb{A}} = -\frac{p_0}{4\pi\varepsilon_0}\frac{\omega^2}{c^3}\frac{\sin\theta}{r}\cos(\omega\tau)\,\hat{\phi}.$$

The instantaneous Poynting vector is:

$$\vec{\mathbb{S}}(\vec{r},t) = \frac{1}{\mu_0}\vec{E}(r,t) \times \vec{B}(r,t),$$

so for the oscillating dipole we have that:

$$\vec{\mathbb{S}}(\vec{r},t) = \frac{1}{\mu_0}\frac{p_0^2}{16\pi^2\varepsilon_0^2}\frac{\omega^4}{c^5}\left(\frac{\sin^2(\theta)}{r^2}\right)\cos^2(\omega\tau).$$

Since $\mu_0 \varepsilon_0^2 = \frac{1}{Z_0 c^3}$, this simplifies to:

$$\vec{\mathbb{S}}(\vec{r},t) = \frac{Z_0}{4\pi}\frac{p_0^2\,\omega^4}{c^2}\left(\frac{\sin^2(\theta)}{4\pi r^2}\right)\cos^2(\omega\tau) \ .$$

Notice that there is no emission along the dipole axis, where $\sin\theta = 0$.

Since the time average of the square of the cosine equals one-half:

$$\left\langle \cos^2(\omega\tau)\right\rangle = \frac{2\pi}{\omega}\int_0^{2\pi}\cos^2(\omega\tau)\,\mathrm{d}t = \frac{1}{2},$$

we can find the average energy flux as:

$$\left\langle \vec{\mathbb{S}}(\vec{r},t)\right\rangle = \frac{Z_0}{4\pi}\frac{p_0^2\,\omega^4}{c^2}\left(\frac{\sin^2(\theta)}{4\pi r^2}\right)\left\langle\cos^2(\omega\tau)\right\rangle\hat{r} = \frac{Z_0}{32\pi^2}\frac{p_0^2\,\omega^4}{c^2}\left(\frac{\sin^2(\theta)}{r^2}\right)\hat{r}.$$

The energy flux depends on $\sin^2(\theta)$, where θ is the angle between the observer's line of sight and the dipole axis.

The total radiated power is then just the surface integral over a sphere of radius centered at the origin of the dipole:

$$\left\langle \mathbb{P}\right\rangle = \oint\left\langle \vec{\mathbb{S}}(\vec{r},t)\right\rangle\cdot\mathrm{d}\vec{A} = \int_0^{2\pi}\int_0^{\pi}\left\langle \mathbb{S}\right\rangle r^2\sin\theta\,\mathrm{d}\theta\,\mathrm{d}\phi = \frac{Z_0}{32\pi^2}\frac{p_0^2\,\omega^4}{c^2}\int_0^{2\pi}\int_0^{\pi}\left(\frac{\sin^2(\theta)}{\cancel{r^2}}\right)\cancel{r^2}\sin\theta\,\mathrm{d}\theta\,\mathrm{d}\phi,$$

which, upon evaluating the integral, becomes:

$$\left\langle \mathbb{P}\right\rangle = \frac{Z_0}{12\pi}\frac{p_0^2\,\omega^4}{c^2} = \frac{Z_0}{12\pi}\left(\frac{p_0\,\omega^2}{c}\right)^2.$$

The units of $\mathbb{P}$, Z_0, and $\frac{1}{c}p_0\,\omega^2$ are simply watts, ohms and amperes respectively.

Example 16.4: The Radiation Resistance of a Dipole Antenna

Consider a linear antenna feed that is driven by an oscillating current, as in the diagram. We know the driving frequency ω, the total length of the antenna a, and the amplitude of the current I_0. We will assume that $\omega a \ll c$.

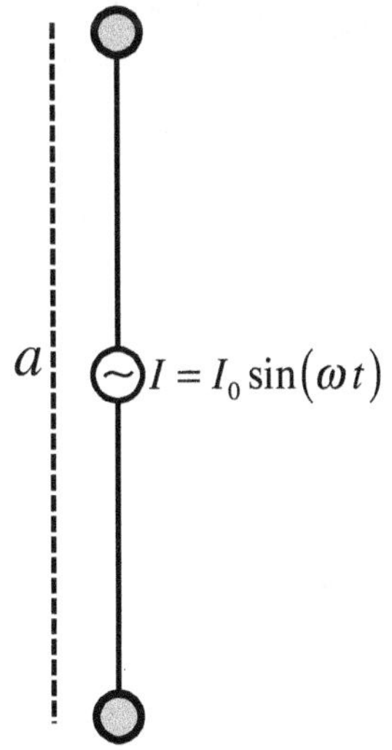

First let us find the total power emitted by the antenna. As this is simply a Hertzian dipole, we can apply our result from Thought Experiment 16.5 as long as we can find the dipole moment $\vec{p}$ as a function of time:

$$\vec{p}(t) = Q\vec{d} = a\hat{z}\int I_0\sin(\omega t)\,\mathrm{d}t = \left(-\frac{aI_0}{\omega}\right)\cos(\omega t)\,\hat{z},$$

so the total power is given by:

$$\langle \mathbb{P} \rangle = \frac{Z_0}{12\pi}\left(\frac{p_0\,\omega^2}{c}\right)^2 = \frac{Z_0}{12\pi}\left(-\frac{aI_0}{\omega}\frac{\omega^2}{c}\right)^2 = \frac{Z_0}{12\pi}a^2\,I_0^2\,\frac{\omega^2}{c^2}.$$

This power must have come from somewhere, so it must also be the average power from the signal generator. In terms of the root mean square current, $I_{\text{rms}} = \frac{\sqrt{2}}{2} I_0$, this can be written:

$$\langle \mathbb{P} \rangle = \frac{Z_0}{12\pi}a^2\left(\sqrt{2}I_{\text{rms}}\right)^2\frac{\omega^2}{c^2} = \frac{2\pi}{3}Z_0\left(a^2\left(\frac{\omega}{2\pi}\right)^2\right)\left(I_{\text{rms}}\right)^2 = \frac{2\pi}{3}\left(\frac{a}{\lambda}\right)^2 Z_0\left(I_{\text{rms}}\right)^2.$$

Now we can express this power loss as an equivalent resistance, called the *radiation resistance*, as it is proportional to the current squared, which is then:

$$R_{\text{rad}} \equiv \frac{\langle \mathbb{P} \rangle}{\langle I^2 \rangle} = \frac{2\pi}{3}\frac{1}{\varepsilon_0 c}\left(\frac{a}{\lambda}\right)^2 = \frac{2\pi}{3}\left(\frac{a}{\lambda}\right)^2 Z_0 \sim 2Z_0\left(\frac{a}{\lambda}\right)^2.$$

Example 16.5: The Beam Pattern of a Hertzian Dipole Antenna

First recall the hypothetical spherical coherent point source we discussed in Thought Experiment 16.1 (p. 508). In practice, single point source antennas do not exist. However, incoherent isotropic sources of light are extremely common, as there are about 100 billion stars in our galaxy and nearly 10 sextillion stars in the universe.

Antennas, unlike stars, do not emit their light isotropically. In fact, most antennas are purposely directional, especially for applications in fields such as satellite communications and radio astronomy. One simple measure of the directionality is called the *antenna gain*, and a more sophisticated measure is called the *beam pattern*. These are measures of the ratio of the intensity of the radiation in a particular direction to that of an equivalent hypothetical isotropic antenna with the same overall power.

Consider a Hertzian dipole that radiates at an angular frequency ω with a total radiated power $\mathbb{P}$. From Thought Experiment 16.5, the Poynting vector as a function of the polar angle and distance from the source, and the total power, are given by:

$$\langle \vec{\mathbb{S}}(\vec{r},t) \rangle = \frac{Z_0 p_0^2}{8\pi}\frac{\omega^4}{c^2}\left(\frac{\sin^2(\theta)}{4\pi r^2}\right)\hat{r} \quad \text{and} \quad \langle \mathbb{P} \rangle = \frac{Z_0}{12\pi}\left(\frac{p_0\,\omega^2}{c}\right)^2 = p_0^2\,\frac{Z_0\omega^4}{12\pi c^2}.$$

Inverting the second expression, and substituting it into the first for p_0^2, we can write this in terms of the total power, as:

$$\langle \vec{\mathbb{S}}(\vec{r},t) \rangle = \left(\langle \mathbb{P} \rangle \frac{12\not{\pi}\,\not{c^2}}{\not{Z_0}\,\not{\omega^4}}\right)\left(\frac{\not{Z_0}}{8\not{\pi}}\frac{\not{\omega^4}}{\not{c^2}}\left(\frac{\sin^2(\theta)}{4\pi r^2}\right)\hat{r}\right) = \frac{3}{2}\left(\frac{\langle \mathbb{P} \rangle}{4\pi r^2}\right)\sin^2(\theta)\hat{r}.$$

Notice that the term in parentheses is simply the equivalent isotropic energy flux, so the beam pattern is given in spherical polar coordinates by:

$$G(\theta,\varphi) = \tfrac{3}{2}\sin^2(\theta).$$

The antenna gain is simply 1.5, as this is the maximum value of G. Electrical engineers often express such ratios as the gain of an antenna in a logarithmic scale in units of decibels. For example, the gain of a Hertzian dipole can be expressed as:

$$G_0 = 1.5 = 10\log_{10}(1.5)\,\mathrm{dB} \approx 1.76\,\mathrm{dB}\,.$$

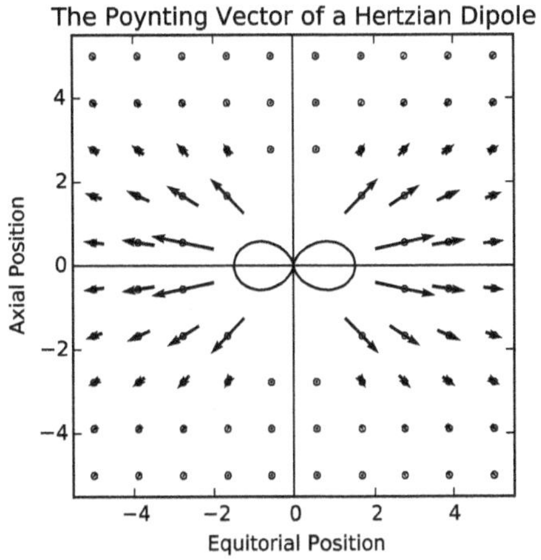

The beam pattern is independent of azimuthal angle in this example. An antenna with an axially symmetrical beam pattern (i.e. $G(\theta,\varphi) = G(\theta)$) is said to be *omnidirectional.*

The arrows in the plot show the energy flux, and the figure eight in the middle represents a *polar plot* of the beam pattern. In beam pattern plots, the radial coordinate represents $G(\theta,\varphi)$, so an isotropic emitter would have a beam pattern of the unit circle.

As you can see, this antenna would be optimized for radiation in the equatorial plane.

Example 16.6: The Power per Solid Angle

We have now found the Poynting vector for a monochromatic Hertzian dipole, and defined the beam pattern as the ratio of this to the equivalent Poynting vector from a hypothetical isotropic emitter. By combining the last two examples, we could characterize an antenna by the use of two quantities: (1) the total power output, and (2) the beam pattern. In this example, we will put these together and use the notion of a solid angle to find the power per solid angle.

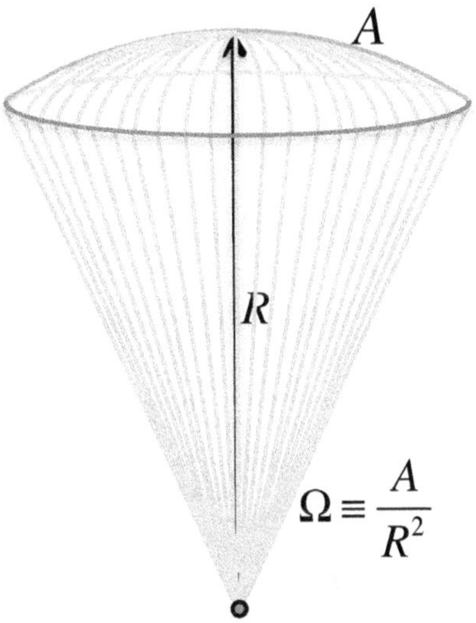

First, consider the time-averaged energy per area of a monochromatic Hertzian dipole at a distance r from the source:

$$\left\langle \vec{\mathbb{S}}(\vec{r},t) \right\rangle = \frac{p_0^2}{32\pi^2\varepsilon_0}\frac{\omega^4}{c^3}\left(\frac{\sin^2(\theta)}{r^2}\right)\hat{r} = \frac{3}{2}\left(\frac{\langle \mathbb{P}\rangle}{4\pi r^2}\right)\sin^2(\theta)\hat{r}\,.$$

Now consider the definition of solid angle, Ω. Just as an angle in a plane is defined as the ratio of the arc length to the radius of the corresponding circle, the solid angle is the ratio of the area of a wedge of a sphere to the corresponding radius squared. Therefore, the time-averaged power per solid angle for a radially emitting source is given by:

$$\mathbb{P}_\Omega \equiv \left\langle \frac{\delta \mathbb{P}}{\delta \Omega} \right\rangle = \frac{\langle \vec{\mathbb{S}}\rangle \cdot \delta \vec{A}}{\delta \Omega} = \frac{\left(\langle \mathbb{S}\rangle \hat{r}\right)\cdot\left(r^2 \delta\Omega\, \hat{r}\right)}{\delta\Omega} = r^2 \langle \mathbb{S}\rangle\,.$$

In the case of the Hertzian dipole, this is simply:

$$\mathbb{P}_\Omega = r^2\langle \mathbb{S}\rangle = \frac{p_0^2}{32\pi^2\varepsilon_0}\frac{\omega^4}{c^3}\sin^2(\theta) = \langle \mathbb{P}\rangle \frac{3}{8\pi}\sin^2(\theta)\,.$$

Let us check that this expression is consistent with the total power. Recall that we can render the differential solid angle in spherical coordinates as $d\Omega = \sin(\theta)d\varphi d\theta$, so we can write:

$$\langle \mathbb{P} \rangle = \oint \mathbb{P}_\Omega d\Omega = \int_0^\pi \int_0^{2\pi} \left(\langle \mathbb{P} \rangle \frac{3}{8\pi} \sin^2(\theta) \right) \left(\sin(\theta) d\varphi d\theta \right) = \langle \mathbb{P} \rangle \frac{3}{8\pi} 2\pi \int_0^\pi \sin^3(\theta) d\theta = \langle \mathbb{P} \rangle \tfrac{3}{4} \left(\tfrac{4}{3} \right) = \langle \mathbb{P} \rangle .$$

Finally, we can relate the power per solid angle $\mathbb{P}_\Omega$ to the total radiated power $\langle \mathbb{P} \rangle$ and the beam pattern $G(\theta,\varphi)$:

$$\mathbb{P}_\Omega = \tfrac{1}{4\pi} \langle \mathbb{P} \rangle \tfrac{3}{2} \sin^2(\theta) = \tfrac{1}{4\pi} \langle \mathbb{P} \rangle G(\theta,\varphi).$$

Thought Experiment 16.6: A Receiving Hertzian Dipole Antenna

Consider the conducting small dipole antenna in Example 16.4 (p. 525), but now acting as a receiver instead of a transmitter. To accomplish this, we replace the signal generator with an amplifier of equivalent resistance R.

Our goal, here, is to find the power dissipated at the resistor as a function of the Poynting vector of the incoming radiation $\vec{\mathbb{S}} = -\mathbb{S}\hat{r}$.

First, we assume that the incoming signal is a monochromatic signal, from a distance source, that is linearly polarized parallel to the antenna. Then we can write the incoming electric field at the location of the small antenna as: $\vec{E} = E_0 \sin(\omega t)\hat{\theta}$,

which means that the voltage across the antenna is:

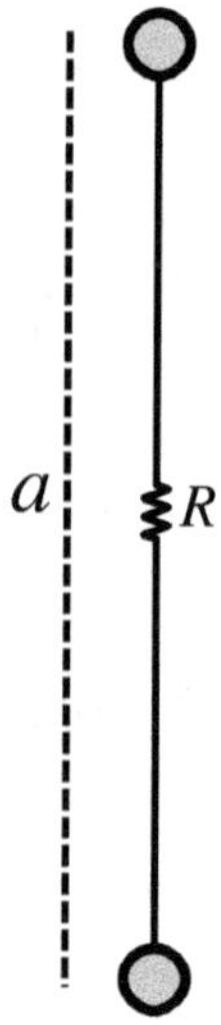

$$\mathbb{V} = \vec{E} \cdot (a\hat{z}) = E a \sin\theta = E_0 a \sin(\omega t) \sin\theta .$$

Therefore, the current in the wire is given by:

$$I = \frac{\mathbb{V}}{R} = \frac{\mathbb{V}_0}{R} \sin(\omega t) = \frac{E_0 a}{R} \sin(\theta) \sin(\omega t),$$

so the total *received* power as a function of time is:

$$\mathbb{P}_{\text{in}} = I\mathbb{V} = \frac{\mathbb{V}^2}{R} = \frac{\mathbb{V}_0^2}{R} \sin(\omega t) = \frac{E_0^2 a^2}{R} \sin^2(\theta) \sin^2(\omega t).$$

But this, in turn, acts exactly like the transmitting dipole antenna, so it must retransmit the signal, with a total emitted power given by:

$$\mathbb{P}_{\text{rad}} = \frac{2\pi}{3} Z_0 \left(\frac{a}{\lambda} \right)^2 I^2 = \frac{2\pi}{3} Z_0 \left(\frac{a}{\lambda} \right)^2 \left(\frac{E_0 a}{R} \sin(\theta) \sin(\omega t) \right)^2 .$$

Therefore, the net power received by the antenna is given by:

$$\begin{aligned} \mathbb{P}_{\text{net}} &= \mathbb{P}_{\text{in}} - \mathbb{P}_{\text{rad}} \\ &= \frac{E_0^2 a^2}{R} \sin^2(\theta) \sin^2(\omega t) - \frac{2\pi}{3} Z_0 \left(\frac{a}{\lambda} \right)^2 \frac{E_0^2 a^2}{R^2} \sin^2(\theta) \sin^2(\omega t). \end{aligned}$$

This simplifies to:

$$\mathbb{P}_{\text{net}} = E_0^2\, a^2 \sin^2(\theta)\sin^2(\omega t)\left(\frac{1}{R} - \frac{2\pi}{3}\left(\frac{a}{\lambda}\right)^2 \frac{Z_0}{R^2}\right) = E_0^2\, a^2 \sin^2(\theta)\sin^2(\omega t)\left(\frac{1}{R} - \frac{R_{\text{rad}}}{R^2}\right).$$

where R_{rad} represents the radiation resistance (p. 525) of the same antenna.

Notice that we can write this receiving power in terms of the magnitude of the incoming Poynting vector, $\mathbb{S} = \frac{1}{\mu_0}\left|\vec{E}\times\vec{B}\right| = \frac{1}{\mu_0 c}E^2 = \frac{1}{Z_0}E^2$, as:

$$\mathbb{P}_{\text{net}} = Z_0 \mathbb{S}\, a^2 \sin^2(\theta)\left(\frac{1}{R} - \frac{R_{\text{rad}}}{R^2}\right).$$

Next, we define the effective area A_{eff} of a receiving antenna to be the ratio of the power received to the incoming power per area. Thus:

$$A_{eff} \equiv \frac{\mathbb{P}_{\text{net}}}{\mathbb{S}} = a^2 \sin^2(\theta) Z_0\left(\frac{1}{R} - \frac{R_{\text{rad}}}{R^2}\right).$$

Notice that this has exactly the same angular dependence as the beam, so we can write the effective area in terms of the antenna gain, as:

$$A_{eff} = \tfrac{2}{3}a^2\left(\tfrac{3}{2}\sin^2(\theta)\right)Z_0\left(\frac{1}{R} - \frac{R_{\text{rad}}}{R^2}\right) = \tfrac{2}{3}a^2\, G(\theta,\varphi) Z_0\left(\frac{1}{R} - \frac{R_{\text{rad}}}{R^2}\right).$$

As it turns out, the effective area and the gain always have the same angular dependence regardless of the antenna design. This is called the *weak reciprocity theorem*, which we will soon discuss. But before doing so, let's first consider the fact that an electrical engineer will pick the resistance, R, to maximize the effective area. We can solve for this optimal resistance by setting the first derivative of A_{eff} equal to zero and solving for R:

$$\frac{\mathrm{d}A_{eff}}{\mathrm{d}R} = \left(\tfrac{2}{3}Z_0 a^2\, G(\theta,\varphi)\right)\left(-\frac{1}{R^2} + 2\frac{R_{\text{rad}}}{R^3}\right) = 0$$

$$R = 2R_{\text{rad}}.$$

As with coaxial cables in Chapter 13, this is called *impedance matching*. In this ideal case, the maximum effective area is given by:

$$A_{eff} = \tfrac{2}{3}Z_0 a^2\, G(\theta,\varphi)\left(\frac{1}{(2R_{\text{rad}})} - \frac{R_{\text{rad}}}{(2R_{\text{rad}})^2}\right) = \frac{Z_0 a^2}{6R_{\text{rad}}}G(\theta,\varphi).$$

And, finally, the effective area of the Hertzian dipole is given by:

$$A_{eff} = \frac{\cancel{Z_0}\,\cancel{a^2}\,G(\theta,\varphi)}{6\left(\frac{2\pi}{3}\cancel{Z_0}\left(\frac{\cancel{a}}{\lambda}\right)^2\right)} = \frac{\lambda^2}{4\pi}G(\theta,\varphi).$$

As it turns out, this is always true for impedance-matched reception antennas, regardless of their particular design, so long as the polarization of the received signal matches the equivalent polarization of the radiated signal. This *strong reciprocity theorem* is extremely useful because the radiation pattern is usually easier to calculate than the effective area.

Discussion 16.2: Strong Reciprocity Theorem and CPT Symmetry

Imagine that our universe suddenly changed in the following three ways: (1) every charge is reversed, (2) all directions are flipped, and (3) time moves backwards. Would the fundamental laws of physics still be valid? If the answer were "yes," then we would say that the laws of physics follow charge, parity, and time (CPT) symmetry.

Every known physical law, except for those involving statistical mechanics, conforms to the principle of CPT symmetry. Moreover, violations to CPT symmetry outside the normal margin of error have never been observed in nature despite a great deal of experimental work.[270]

Consider a transmitting dipole antenna driven by a current and changing dipole moment. How would a CPT transformation change these?

$$I=\frac{\mathrm{d}Q}{\mathrm{d}t}\rightarrow\frac{\mathrm{d}(-Q)}{\mathrm{d}(-t)}=\frac{\mathrm{d}Q}{\mathrm{d}t}=I \quad \text{and} \quad \vec{p}=Q\vec{d}\rightarrow(-Q)(-\vec{d})=\vec{p},$$

so the current and dipole moment would be unchanged. The resulting Poynting vector must then also be the same:

$$\vec{\mathbb{S}}(\vec{r},t)\rightarrow\vec{\mathbb{S}}(-\vec{r},-t)=\vec{\mathbb{S}}(\vec{r},t).$$

But what is $\vec{\mathbb{S}}(-\vec{r},-t)$? It would be a mirror image radio antenna that was receiving instead of transmitting. The other way around is true too. A radio telescope designed to be sensitive in a particular direction can be used to transmit a narrow beam.

There are two catches to all this: one is polarization, and the other is entropy. We will consider polarization first.

Imagine we have two identical antennas that emit linearly polarized waves, such as in a pair of toy walkie-talkies used by a pair of children. As long as each child points the antenna up, the transmitting walkie-talkie will transmit a vertically polarized wave, and the receiving antenna will be sensitive to the same wave.

Now consider two identical helical antennas pointing at each other. If each one emits left-polarized light, the antenna will be sensitive to right-polarized light. So, the two will not be able to communicate, as the parity of the light is also reversed. The solution, of course, is to use mirror image transmitters and receivers.

Entropy, on the other hand, is more problematic than polarization. All areas of physics that involve statistical mechanics measure the arrow of time, and radio antennas are no exception. As we discussed when we derived the spherical wave equation, theoretically one could have a large spherical shell as a source of coherent light, but they don't exist naturally. So, while transmitters emit spherical waves, receivers rarely catch them.

Problem 16.5: A Baseball Scout with a Radar Gun

Consider a baseball scout with a radar gun a distance z_0 behind home plate clocking the pitch speed. The radar gun has a single antenna with an on-axis gain of G and emits a total power $\mathbb{P}_0$ at a wavelength λ. The ball has a radius a and a conductivity κ.

---

[270] Kostelecky and Russell, *Data Tables for Lorentz and CPT Violation*, Reviews of Modern Physics **83**, 11 (2011) and subsequent updates.

a) What is the on-axis power per solid angle emitted by the radar gun?

b) What is the on-axis power per area a distance z in front of home plate?

c) How much power per solid angle, $\delta\mathbb{P}_{ball\_\Omega}$, is reflected in a small surface area δA on the surface of the baseball as a function of angle theta prime? You can assume that $a \ll z_0$.

d) What is the power per area per solid angle reflected off of the baseball's surface back toward the radar gun. Assuming that each point on the ball's surface is a point dipole antenna.

e) What is the reflected power per area at the location of the radar gun?

f) What is the ratio of the power emitted, to the power received, by the radar gun? Is this a function of the antenna gain?

g) Discuss how the reversibility of time affects, and does not affect, this problem.

Example 16.7: The Cross Section of a Dielectric Sphere

Consider monochromatic light of wavelength λ incident on a nonconducting sphere of radius a and dielectric constant $K = \frac{1}{\varepsilon_0}\varepsilon$. What is the cross section of interaction between the light and the sphere?

First consider the special case where the radius of the sphere is large compared to the wavelength. Such is the case for sunlight incident on the Earth: the total power received from the sun is due to the amount of light actually intercepted at the Earth's surface. Thus, we would say that the cross section of the Earth to sunlight is simply the projected area, or physical cross section, of Earth imagined as a sphere of radius a:

$$\sigma_{\oplus} = \pi a^2 = \pi\left(6371\,\text{km}\right)^2 \approx 1.3\times10^{14}\,\text{m}^2 .$$

Some of the incident light is reflected, and the rest is absorbed, but in thermal equilibrium the total power of light incident on the Earth must be balanced by the light radiated back into space.

While it would be silly to do so, we could imagine determining the size of the Earth by measuring the total power radiated into space by the Earth, and comparing it to the energy flux of light incident on earth. Then we could say that the cross section of Earth would simply be:

$$\sigma_{\oplus} = \frac{\mathbb{P}_{\text{Radiated by Earth}}}{\mathbb{S}_{\text{Incident Sunlight}}} \approx \frac{1.8\times10^{17}\,\text{W}}{1{,}400\,{}^{\text{W}}\!/_{\text{m}^2}} \approx 1.3\times10^{14}\,\text{m}^2 \approx \pi\left(6300\,\text{km}\right)^2 .$$

For light hitting the Earth, this cross section is the physical size of Earth. While this is trivial for a large opaque object, notice that the cross section of a large transparent object is always smaller than its cross-sectional area. For example, the cross section of the Earth for incident neutrinos from space is much smaller, because only a small fraction of the neutrinos actually interact with the Earth.

We now change tack to consider tiny objects which are much smaller than the wavelength of the incident light ($a \ll \lambda$). More specifically, the cross-section of a small dielectric sphere, which is irradiated by monochromatic light, may have little to do with the physical cross section πa^2.

First consider a monochromatic plane wave of light with an electric field, at the location of the sphere, given by:

$$\vec{E} = E_0 \sin(\omega t)\hat{z},$$

which will, in turn, induce a dipole moment per volume. As we discussed in Part I, for a linear medium this is:

$$\vec{P} = (\varepsilon - \varepsilon_0)\vec{E} ,$$

so the total dipole moment is given by:

$$\vec{p} = \tfrac{4}{3}\pi a^3 \vec{P} \approx \tfrac{4}{3}\pi a^3 (\varepsilon - \varepsilon_0)\vec{E} \approx \tfrac{4}{3}\pi a^3 (\varepsilon - \varepsilon_0) E_0 \sin(\omega t)\hat{z}.$$

The total radiated dipole radiation is therefore:

$$\langle \mathbb{P} \rangle \approx \frac{Z_0 p_0^2}{12\pi} \frac{\omega^4}{c^2} \approx \frac{\left(\tfrac{4}{3}\pi a^3 (\varepsilon - \varepsilon_0) E_0\right)^2}{12\pi \varepsilon_0} \frac{\omega^4}{c^3} \approx \frac{\tfrac{16}{9}\pi^2 a^6 (\varepsilon - \varepsilon_0)^2 E_0^2}{12\pi \varepsilon_0} \frac{\omega^4}{c^3}$$

$$\approx \frac{8\pi}{27} a^6 \left(\frac{\varepsilon - \varepsilon_0}{\varepsilon_0}\right)^2 \left(\frac{E_0^2}{2Z_0}\right) \left(\frac{\omega}{c}\right)^4 \approx \frac{8\pi}{27} a^6 \left(\frac{\varepsilon - \varepsilon_0}{\varepsilon_0}\right)^2 \left(\frac{\omega}{c}\right)^4 \langle \mathbb{S} \rangle .$$

The cross section to incident light will be:

$$\sigma \equiv \frac{\langle \mathbb{P} \rangle}{\langle \mathbb{S} \rangle} \approx \frac{8\pi^5}{27} a^6 \left(\frac{\varepsilon - \varepsilon_0}{\varepsilon_0}\right)^2 \left(\frac{2\pi}{\lambda}\right)^4 \approx \frac{3}{2} \left(\frac{4\pi a}{3\lambda}\right)^4 \left(\frac{\varepsilon - \varepsilon_0}{\varepsilon_0}\right)^2 \left(\pi a^2\right),$$

and the ratio of the effective cross section to the physical cross section is related to the diameter of the particles and the wavelength by:

$$\left(\frac{\sigma}{\pi a^2}\right) \propto \left(\frac{a}{\lambda}\right)^4 .$$

This only holds if $a \ll \lambda$, and the ratio is simply unity if $a \gg \lambda$.

Discussion 16.3: Why the Sky is Blue and Sunsets are Red

The Poynting vector and the total radiated power depend on the fourth power of the frequency, in other words the inverse of the fourth power of the radiated wavelength. John William Strutt, also Lord Rayleigh, first derived this to explain the blue color of the sky as due to the scattering of light by air molecules. Sunlight passing through the atmosphere causes the molecules to act as tiny electric dipole radiators that oscillate, and hence emit radiation at various frequencies (or wavelengths). Because this scattering is more effective at high frequencies (short wavelengths), blue light is preferentially scattered and the daytime sky appears blue. For example, the scattering at a wavelength of $400\,\mathrm{nm}$ is more than 9 times greater than that at a wavelength of $700\,\mathrm{nm}$ for equal incident intensity at the same angle.

This same Rayleigh scattering also explains the red sky at sundown, since the blue light is scattered and the red light is transmitted. At sunrise and sunset, sunlight travels a long path through the atmosphere to reach our eyes, by which point most of the blue light has been removed. The result is that the light that our eyes receive at sunrise and sunset is dominated by red light, which is scattered least.

On the other hand, clouds appear white or gray, no matter what time of day it is. This is as expected, since the typical droplet is larger than the typical wavelength of sunlight.

Ludvig Lorenz and Gustav Mie modeled the scattering by particles of approximately the same size as the wavelength, but rather by the sum of terms in an infinite series.[271] The radiation emitted by *Mie scattering* does not have a symmetrical pattern, unlike the case with Rayleigh scattering, but instead tends to be more intense in the forward direction (e.g. away from the Sun).

Problem 16.6: Mie Scattering

So far we have discussed scattering by large $(a \gg \lambda)$ and small $(a \ll \lambda)$ objects, but what if the scatterer is comparable to the wavelength?

a) Write a short computer function that uses the Rayleigh scattering formula to calculate the effective cross-section as a function of the wavelength and size of the scatterer.

b) Make a log-log plot of $(\sigma/\pi a^2)$ as a function of (λ/a) in the two limiting cases. Apply the Rayleigh formula when $a < \frac{1}{10}\lambda$ and the physical size when $a > 10\lambda$.

c) Consider now the intermediate solution. Think up and apply a simple ad-hoc interpolation between the two limits. Show this also on your plot.

d) Research the topic of Mie scattering. Discuss how scattering off of particles of intermediate size differs from your interpolation. What are the key physical principles?

e) Research current uses of Mie scattering theory. Discuss one application that you find most interesting.

16.3 *Antenna Theory*

In this section, we apply Liénard-Wiechert potential theory to antennas of finite size. We relax the requirement that the source is much smaller than the wavelength of the light it produces, but retain the assumption that the observer is very far away from the source.

Antennas are driven by currents in conductors, so we will assume that it is only the current, rather than the charge, that is important in producing the far radiation field. Thus, in this section, we need only concern ourselves with the vector, rather than the scalar, potential.

In the far limit, the resulting radiation must follow an outwardly expanding spherical wave solution. But, also in the far limit, we can assume that the outgoing wave is locally planar. When we consider a receiving antenna, we assume an incoming plane wave. Thus, other than the dependence on source distance, the processes of transmission and reception are identical and follow the same reciprocity relationship as the Hertzian dipole antenna.

Thought Experiment 16.7: Simple Plane Waves in Potential Form

In a vacuum, Maxwell's equations can be written as:

$$\left(\nabla^2 - \frac{1}{c^2}\frac{\partial^2}{\partial t^2}\right)\begin{pmatrix} \mathbb{V} \\ \vec{\mathbb{A}} \end{pmatrix} = \begin{pmatrix} 0 \\ 0 \end{pmatrix}.$$

We now recognize each of these as conforming to a wave equation. First consider the simplest linearly polarized plane wave solution propagating in the $\hat{z}$ direction:

[271] H.C. van de Hulst, Light scattering by small particles, (John Wiley & Sons, 1957; rpt. New York: Dover, 1981).

$$\vec{E}(\vec{r},t) = E\left(t-\tfrac{z}{c}\right)\hat{x} = E_0 \sin(\omega(t-\tfrac{1}{c}z))\hat{x} = E_0 \sin(\omega\tau)\hat{x}$$
$$\vec{B}(\vec{r},t) = B\left(t-\tfrac{z}{c}\right)\hat{y} = \tfrac{1}{c}E_0 \sin(\omega(t-\tfrac{1}{c}z))\hat{y} = \tfrac{1}{c}E_0 \sin(\omega\tau)\hat{y}.$$

Let's investigate the vector potential through the relationship:

$$\vec{B} = \vec{\nabla}\times\vec{\mathbb{A}} = \tfrac{1}{c}E_0 \sin(\omega\tau)\hat{y} = \left(\tfrac{\partial}{\partial z}\mathbb{A}_x - \tfrac{\partial}{\partial x}\mathbb{A}_z\right)\hat{y}.$$

Planar symmetry suggests that the vector potential should not be a function of the sideways position, so we suppose that in this simple case:

$$\vec{\mathbb{A}} = \hat{x}\int \tfrac{1}{c}E_0 \sin(\omega\tau)\mathrm{d}z = \tfrac{1}{\omega}E_0\hat{x}\int\left(\tfrac{\omega}{c}\right)\sin(\omega t - \tfrac{\omega}{c}z)\mathrm{d}z = \tfrac{1}{\omega}E_0 \cos(\omega\tau)\hat{x}.$$

We will now differentiate this expression to obtain the electric field:

$$\vec{E} = -\vec{\nabla}\mathbb{V} - \tfrac{\partial}{\partial t}\vec{\mathbb{A}} = -\vec{\nabla}\mathbb{V} - \tfrac{\partial}{\partial t}\left(\tfrac{1}{\omega}E_0 \cos(\omega\tau)\hat{x}\right) = -\vec{\nabla}\mathbb{V} + E_0 \sin(\omega\tau)\hat{x}.$$

Thus, the simplest plane wave solution in potential form is:

$$\mathbb{V} = 0 \quad \text{and} \quad \vec{\mathbb{A}} = \tfrac{1}{\omega}E_0 \cos(\omega\tau)\hat{x}.$$

Thus, the vector potential is in the same direction as the electric field, but 90° out of phase with it.

So, the Poynting vector of one polarization is:

$$\left\langle\vec{S}\right\rangle = \tfrac{1}{\mu_0}\left\langle\vec{E}\times\vec{B}\right\rangle = \tfrac{1}{\mu_0 c}E_0^2\left(\left\langle\sin(\omega\tau)\right\rangle\right)^2\hat{z} = \tfrac{\omega^2}{Z_0}\left(\left\langle\tfrac{1}{\omega}E_0\cos(\omega\tau)\right\rangle\right)^2\hat{z}$$
$$= \tfrac{\omega^2}{Z_0}\left\langle\mathbb{A}^2\right\rangle = \tfrac{\omega^2}{2Z_0}\mathbb{A}_0^2.$$

Thought Experiment 16.8: Spherical Waves in Potential Form

In Thought Experiment 16.1 (p. 508), we found a particular spherical wave solution as:

$$\vec{\mathbb{S}}(r,t) = \frac{\mathbb{P}(\tau)}{4\pi r^2}\hat{r} = \frac{c}{Z_0}\vec{E}(r,t)\times\vec{B}(r,t).$$

And, by applying the definition of the power per solid angle, we now write:

$$\mathbb{P}_\Omega(t-\tfrac{r}{c}) = \mathbb{P}_\Omega(\tau) = r^2\mathbb{S}(r,t) = \frac{r^2}{\mu_0}\vec{E}(r,t)\times\vec{B}(r,t) = \frac{cr^2}{Z_0}\vec{E}(r,t)\times\vec{B}(r,t).$$

Since the source is far away the wave is approximately planar, so we can express the relationship between the magnetic and electric fields as mutually perpendicular to the each other and the direction of propagation:

$$\vec{B}(r,t) = \tfrac{1}{c}\hat{r}\times\vec{E}(r,t) \quad , \quad \vec{E}(r,t) = -c\hat{r}\times\vec{B}(r,t) .$$

In terms of the magnetic field, we can write the Poynting vector as:

$$\vec{\mathbb{S}}(r,t) = \frac{1}{\mu_0}\left(-c\hat{r}\times\vec{B}(r,t)\right)\times\vec{B}(r,t) = \frac{c}{\tfrac{1}{c}Z_0}\vec{B}(r,t)\times\left(\hat{r}\times\vec{B}(r,t)\right) = \frac{c^2}{Z_0}B^2(r,t)\hat{r}.$$

In terms of the vector potential this becomes:

$$\vec{\nabla}\times\vec{\mathbb{A}}=\vec{B}(r,t)=\frac{1}{r}\left(\frac{1}{\sin\theta}\frac{\partial\mathbb{A}_r}{\partial\varphi}-\frac{\partial(r\mathbb{A}_\varphi)}{\partial r}\right)\hat{\theta}+\frac{1}{r}\left(\frac{\partial(r\mathbb{A}_\theta)}{\partial r}-\frac{\partial\mathbb{A}_r}{\partial\varphi}\right)\hat{\varphi}\ .$$

Thus, we can write the far Poynting vector in terms of the far vector potential, in spherical coordinates, as:

$$\vec{\mathbb{S}}(r,t)=\frac{c^2}{Z_0}B^2(r,t)\hat{r}=\frac{c^2}{Z_0}\left(\frac{1}{r}\left(\frac{1}{\sin\theta}\frac{\partial\mathbb{A}_r}{\partial\varphi}-\frac{\partial(r\mathbb{A}_\varphi)}{\partial r}\right)\hat{\theta}+\frac{1}{r}\left(\frac{\partial(r\mathbb{A}_\theta)}{\partial r}-\frac{\partial\mathbb{A}_r}{\partial\theta}\right)\hat{\varphi}\right)^2\hat{r}$$

$$=\frac{1}{r^2}\frac{c^2}{Z_0}\left(\left(\frac{1}{\sin\theta}\frac{\partial\mathbb{A}_r}{\partial\varphi}-\frac{\partial(r\mathbb{A}_\varphi)}{\partial r}\right)^2+\left(\frac{\partial(r\mathbb{A}_\theta)}{\partial r}-\frac{\partial\mathbb{A}_r}{\partial\theta}\right)^2\right)\hat{r}.$$

The power per solid angle is therefore given by:

$$\mathbb{P}_\Omega(r,t)=\frac{c^2}{Z_0}\left(\left(\frac{1}{\sin\theta}\frac{\partial\mathbb{A}_r}{\partial\varphi}-\frac{\partial(r\mathbb{A}_\varphi)}{\partial r}\right)^2+\left(\frac{\partial(r\mathbb{A}_\theta)}{\partial r}-\frac{\partial\mathbb{A}_r}{\partial\theta}\right)^2\right).$$

The antenna produces a spherical wave, but we are only concerned with the wave that propagates radially, so usually $\mathbb{A}_r=0$, and this simplifies to:

$$\mathbb{P}_\Omega(r,t)\approx\frac{c^2}{Z_0}\left(\left(-\frac{\partial(r\mathbb{A}_\varphi)}{\partial r}\right)^2+\left(\frac{\partial(r\mathbb{A}_\theta)}{\partial r}\right)^2\right).$$

Thought Experiment 16.9: The Poynting Vector in the Radiation Zone

First we will consider the plane-wave approximation, so the electric field in the direction of propagation is zero $\left(E_\parallel=\vec{E}\cdot\hat{k}=0\right)$. We now write the far vector potential as:

$$\vec{E}=-\vec{\nabla}\mathbb{V}-\tfrac{\partial}{\partial t}\vec{\mathbb{A}}=\cancel{\vec{E}_\parallel}+\vec{E}_\perp=\cancel{\left(-\vec{\nabla}_\parallel\mathbb{V}-\tfrac{\partial}{\partial t}\vec{\mathbb{A}}_\parallel\right)}-\left(\cancel{\vec{\nabla}_\perp\mathbb{V}}+\tfrac{\partial}{\partial t}\vec{\mathbb{A}}_\perp\right)\approx-\tfrac{\partial}{\partial t}\vec{\mathbb{A}}_\perp\,.$$

The relationship between the plane wave magnetic and electric fields is:

$$\vec{B}(r,t)=\tfrac{1}{c}\hat{k}\times\vec{E}(\vec{r},t)\quad,\quad\vec{E}(r,t)=-c\hat{k}\times\vec{B}(\vec{r},t)\quad,$$

so the Poynting vector can be expressed in terms of the vector potential as:

$$\vec{\mathbb{S}}=\tfrac{1}{\mu_0}\vec{E}\times\vec{B}=\tfrac{c}{Z_0}\left(\vec{E}\right)\times\left(\tfrac{1}{c}\hat{k}\times\vec{E}\right)=\tfrac{1}{Z_0}\hat{k}\left(\vec{E}\cdot\vec{E}\right)=\tfrac{1}{Z_0}\hat{k}\left|-\tfrac{\partial}{\partial t}\vec{\mathbb{A}}_\perp\right|^2=\tfrac{1}{Z_0}\left|\tfrac{\partial}{\partial t}\vec{\mathbb{A}}_\perp\right|^2\hat{k}.$$

A quick unit check shows that dimensionally this comes out correctly:

$$\left[\vec{\mathbb{S}}\right]=\frac{\mathrm{W}}{\mathrm{m}^2}=\frac{\mathrm{V}^2}{\Omega\mathrm{m}^2}=\frac{\left(\mathrm{V}/\mathrm{m}\right)^2}{\Omega}=\frac{\left(\mathrm{N}/\mathrm{C}\right)^2}{\Omega}=\frac{\left(\frac{\mathrm{T\,m}}{\mathrm{s}}\right)^2}{\Omega}=\frac{\left[\left|\tfrac{\partial}{\partial t}\vec{\mathbb{A}}_\perp\right|^2\right]}{\left[Z_0\right]}.$$

Therefore, in the radiation limit, if we can find the time derivative of the vector potential, we can also find the flux of energy that is being radiated.

In the radiation limit, the solution to Maxwell's vacuum equations:

$$\left(\nabla^2-\frac{1}{c^2}\frac{\partial^2}{\partial t^2}\right)\begin{pmatrix}\mathbb{V}\\ \vec{\mathbb{A}}\end{pmatrix}=\begin{pmatrix}0\\ 0\end{pmatrix},$$

is the spherical wave solution from Problem 16.4 (p. 520):

$$\begin{pmatrix}\mathbb{V}(r,t)\\ \mathbb{A}_r(r,t)\\ \vec{\mathbb{A}}_\perp(r,t)\end{pmatrix}=\begin{pmatrix}\left(\frac{r_0}{r}\right)\mathbb{V}(r_0,\tau)\\ \left(\frac{r_0}{r}\right)\mathbb{A}_r(r_0,\tau)\\ \left(\frac{r_0}{r}\right)\vec{\mathbb{A}}_\perp(r_0,\tau)\end{pmatrix},$$

where the retarded time was simply approximated by:

$$\tau=t-\tfrac{1}{c}\left|\vec{r}-\vec{r}'\right|\approx t-\tfrac{1}{c}r\,.$$

Now, we will calculate the Poynting vector, noting that $\hat{k}=\hat{r}$:

$$\vec{\mathbb{S}}(r,t)=\tfrac{1}{\mu_0}\vec{E}\times\vec{B}=\tfrac{1}{Z_0}\left|\tfrac{\partial}{\partial t}\vec{\mathbb{A}}_\perp\right|^2\hat{k}=\tfrac{1}{Z_0}\left|\tfrac{\partial}{\partial\tau}\vec{\mathbb{A}}_\perp\tfrac{\partial}{\partial t}\tau\right|^2\hat{r}=\tfrac{1}{Z_0}\left|\tfrac{\partial}{\partial\tau}\vec{\mathbb{A}}_\perp\right|^2\left(\tfrac{\partial\tau}{\partial t}\right)^2\hat{r}\,.$$

Since the vector potential is inversely proportional to the radial distance:

$$\vec{\mathbb{S}}(r,t)=\tfrac{1}{Z_0}\left|\tfrac{\partial}{\partial\tau}\left(\tfrac{r_0}{r}\right)\vec{\mathbb{A}}_\perp(r_0,\tau)\right|^2\left(\tfrac{\partial\tau}{\partial t}\right)^2\hat{r}=\tfrac{1}{Z_0}\left(\tfrac{r_0}{r}\right)^2\left|\tfrac{\partial}{\partial\tau}\vec{\mathbb{A}}_\perp(r_0,\tau)\right|^2\left(\tfrac{\partial\tau}{\partial t}\right)^2\hat{r}\,.$$

Finally, we find the emitted power per solid angle:

$$\mathbb{P}_\Omega(\tau)=r^2\mathbb{S}(r,t)=\tfrac{1}{Z_0}\left|r_0\tfrac{\partial}{\partial\tau}\vec{\mathbb{A}}_\perp(r_0,\tau)\right|^2\left(\tfrac{\partial\tau}{\partial t}\right)^2=\tfrac{1}{Z_0}\left|r_0\tfrac{\partial}{\partial\tau}\vec{\mathbb{A}}_\perp(r_0,\tau)\right|^2\left(\tfrac{1}{c^2}v_r^2\right).$$

Thought Experiment 16.10: An Omnidirectional Antenna

Consider a dipole antenna driven by an AC current source, as in Example 16.4, but unlike before we will not necessarily assume that the antenna length is much smaller than the wavelength. In fact, we will imagine that the two are comparable, or even that the length of the antenna is longer than the wavelength. However, we will still assume that the observer is very far away from the antenna.

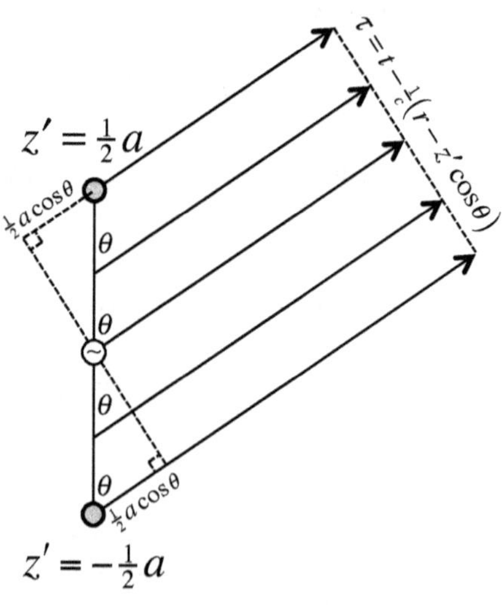

When a signal nears the end of a transmission line, it is usually reflected unless it is terminated properly. Therefore, we will assume that the current follows the form of a standing wave pattern, with an antinode at the signal generator, so:

$$I=I_0\cos\left(2\pi\tfrac{z'}{\lambda}\right)\sin(\omega t)\,.$$

Now an element of the vector potential is given by:

$$\delta\vec{\mathbb{A}}(\vec{r},t)=\frac{\mu_0 I(z',\tau)\delta z'}{4\pi r}\hat{z}=\frac{\mu_0 I_0\cos\left(2\pi\frac{z'}{\lambda}\right)\sin\left(\omega\left(t-\frac{r}{c}+\frac{1}{c}z'\cos\theta\right)\right)\delta z'}{4\pi r}\hat{z},$$

where: $\tau = t - \frac{r}{c} + \frac{z'\cos\theta}{c}$.

Therefore the vector potential differential is given by:

$$\delta\vec{\mathbb{A}}(\vec{r},t) = \frac{\mu_0 I_0 \cos\left(2\pi\frac{z'}{\lambda}\right)\sin\left(\omega t - \omega\frac{r}{c} + \omega\frac{1}{c}z'\cos\theta\right)\delta z'}{4\pi r}\hat{z}.$$

Next we will define the unitless quantities $\phi \equiv \omega\left(t - \frac{r}{c}\right)$, $w \equiv \frac{z'}{\lambda}$, and $N \equiv 2\frac{a}{\lambda}$ as the number of half wavelengths along with wire. So, in terms of these variables, we have:

$$\delta\vec{\mathbb{A}}(\vec{r},t) = \left(\frac{Z_0 I_0 \lambda}{4\pi c r}\hat{z}\right)\cos(2\pi w)\sin(\phi + 2\pi w\cos\theta)\,\delta w,$$

such that $-\frac{1}{2}a < z' < \frac{1}{2}a$, or that $-\frac{1}{4}N < w < \frac{1}{4}N$. Therefore:

$$\vec{\mathbb{A}} = \left(\frac{Z_0\lambda}{4\pi c r}\hat{z}\right)\int_{-\frac{1}{4}N}^{+\frac{1}{4}N}\left(I_0\cos(2\pi w)\right)\sin(2\pi w\cos\theta + \phi)\,\mathrm{d}w.$$

Notice that this integral does something very interesting. It transforms the current as a function of position on the wire to the vector potential as a function of the angle of the radiation. Integrals of this type are called *Fourier integrals*, and the two quantities, in this case the current and the vector potential, which are related to one another through these integrals, are called a *Fourier pair*. The variables w and $\cos\theta$ are said to be *conjugate* to one another. To evaluate this integral we will rewrite the integrand using the product formula:

$$\tfrac{1}{2}\sin\left(2\pi w(1+\cos\theta)+\phi\right) - \tfrac{1}{2}\sin\left(2\pi w(1-\cos\theta)-\phi\right).$$

Now we evaluate the integral:

$$\int_{-\frac{1}{4}N}^{+\frac{1}{4}N}\tfrac{1}{2}\sin\left(2\pi w(1\pm\cos\theta)\pm\phi\right)\mathrm{d}w = -\frac{\cos\left(2\pi w(1\pm\cos\theta)\pm\phi\right)}{4\pi(1\pm\cos\theta)}\Bigg|_{-\frac{1}{4}N}^{+\frac{1}{4}N} = \pm 2\frac{\sin(\phi)\sin\left(\frac{\pi}{2}N(1\pm\cos\theta)\right)}{4\pi(1\pm\cos\theta)}.$$

Thus, the vector potential becomes:

$$\vec{\mathbb{A}} = -\sin(\phi)\left(\frac{Z_0 I_0\lambda}{4\pi^2 c r}\hat{z}\right)\left(\frac{\sin\left(\frac{\pi}{2}N(1+\cos\theta)\right)}{2(1+\cos\theta)} + \frac{\sin\left(\frac{\pi}{2}N(1-\cos\theta)\right)}{2(1-\cos\theta)}\right) = -\sin(\phi)\left(\frac{Z_0 I_0\lambda}{4\pi^2 c r}\hat{z}\right)F(\theta).$$

We then find a common denominator, solve for the function $F(\theta)$, and simplify:

$$\begin{aligned}
F(\theta) &= \left(\frac{(1-\cos\theta)\sin\left(\frac{\pi}{2}N(1+\cos\theta)\right)}{2(1+\cos\theta)(1-\cos\theta)} + \frac{(1+\cos\theta)\sin\left(\frac{\pi}{2}N(1-\cos\theta)\right)}{2(1-\cos\theta)(1+\cos\theta)}\right)\\
&= \frac{1}{1-\cos^2\theta}\left(\tfrac{1}{2}(1-\cos\theta)\sin\left(\tfrac{\pi}{2}N(1+\cos\theta)\right) + \tfrac{1}{2}(1+\cos\theta)\sin\left(\tfrac{\pi}{2}N(1-\cos\theta)\right)\right)\\
&= \frac{1}{\sin^2\theta}\left(\sin\left(\tfrac{\pi}{2}N\right)\cos\left(\tfrac{\pi}{2}N\cos\theta\right) - \cos\theta\cos\left(\tfrac{\pi}{2}N\right)\sin\left(\tfrac{\pi}{2}N\cos\theta\right)\right).
\end{aligned}$$

Now, the Poynting vector in the far limit is:

$$\left\langle \vec{\mathbb{S}} \right\rangle = \tfrac{1}{\mu_0}\left\langle \vec{E} \times \vec{B} \right\rangle = \tfrac{1}{2Z_0}\omega^2 \mathbb{A}_0^2 \hat{r} = \tfrac{(2\pi)^2 c^2}{2Z_0\lambda^2}\left|\hat{r} \times \vec{\mathbb{A}}_0\right|^2 \hat{r} = \tfrac{(2\pi)^2 c^2}{2Z_0\lambda^2}\left\langle \left(-\sin(\phi)\left(\frac{\mu_0 I_0 \lambda}{4\pi^2 r}\right) F(\theta)\left|\hat{r} \times \hat{z}\right| \right)^2 \right\rangle \hat{r}$$

$$= \frac{1}{2Z_0}\left(-\frac{\cancel{2}\cancel{\pi} c}{\cancel{\lambda}} \frac{\mu_0 I_0 \cancel{\lambda}}{2^{\cancel{2}}\pi^{\cancel{2}} r} \right)^2 \left\langle \sin^2(\phi(t)) \right\rangle \hat{r} = \left(\frac{Z_0}{\pi}\right)\left(\frac{I_0^2}{2}\right)\frac{1}{4\pi r^2}\left(F(\theta)\sin(\theta)\right)^2 \hat{r} .$$

So, we have explicitly that:

$$\left\langle \vec{\mathbb{S}} \right\rangle = \left(\frac{I_0^2 Z_0}{2\pi}\right)\left(\frac{\sin\left(\frac{\pi}{2}N\right)\cos\left(\frac{\pi}{2}N\cos\theta\right) - \cos\theta\cos\left(\frac{\pi}{2}N\right)\sin\left(\frac{\pi}{2}N\cos\theta\right)}{\sin\theta}\right)^2 \left(\frac{\hat{r}}{4\pi r^2}\right).$$

The power per solid angle is therefore given by:

$$\mathbb{P}_\Omega = r^2\left\langle \vec{\mathbb{S}} \cdot \hat{r} \right\rangle = \left(\tfrac{1}{2}I_0^2\right)\left(\tfrac{Z_0}{4\pi^2}\right)\left(\frac{\sin\left(\frac{\pi}{2}N\right)\cos\left(\frac{\pi}{2}N\cos\theta\right) - \cos\theta\cos\left(\frac{\pi}{2}N\right)\sin\left(\frac{\pi}{2}N\cos\theta\right)}{\sin\theta}\right)^2$$

We integrate this expression to obtain the total emitted power:

$$\left\langle \mathbb{P} \right\rangle = \oint_\Omega \left\langle \mathbb{P}_\Omega \right\rangle d\Omega = \tfrac{1}{8\pi^2} Z_0 I_0^2 \int_0^\pi \left(\frac{\sin\left(\frac{\pi}{2}N\right)\cos\left(\frac{\pi}{2}N\cos\theta\right) - \cos\theta\cos\left(\frac{\pi}{2}N\right)\sin\left(\frac{\pi}{2}N\cos\theta\right)}{\sin\theta}\right)^2 (2\pi\sin\theta\, d\theta)$$

$$= \left(\tfrac{1}{2}I_0^2\right)\frac{Z_0}{2\pi}\int_0^\pi \frac{1}{\sin\theta}\left(\sin\left(\tfrac{\pi}{2}N\right)\cos\left(\tfrac{\pi}{2}N\cos\theta\right) - \cos\theta\cos\left(\tfrac{\pi}{2}N\right)\sin\left(\tfrac{\pi}{2}N\cos\theta\right)\right)^2 d\theta .$$

Note that the radiation resistance is given by:

$$R_{\text{rad}} = \frac{\left\langle \mathbb{P} \right\rangle}{\left(\frac{1}{2}I_0^2\right)} = \frac{Z_0}{2\pi}\int_0^\pi \frac{1}{\sin\theta}\left(\sin\left(\tfrac{\pi}{2}N\right)\cos\left(\tfrac{\pi}{2}N\cos\theta\right) - \cos\theta\cos\left(\tfrac{\pi}{2}N\right)\sin\left(\tfrac{\pi}{2}N\cos\theta\right)\right)^2 d\theta.$$

The beam pattern of the antenna is:

$$G = \frac{\left\langle \mathbb{P}_\Omega \right\rangle}{\frac{\left\langle \mathbb{P} \right\rangle}{4\pi}} = \frac{2\left(\frac{\sin\left(\frac{\pi}{2}N\right)\cos\left(\frac{\pi}{2}N\cos\theta\right) - \cos\theta\cos\left(\frac{\pi}{2}N\right)\sin\left(\frac{\pi}{2}N\cos\theta\right)}{\sin\theta}\right)^2}{\int_0^\pi \frac{1}{\sin\theta}\left(\sin\left(\frac{\pi}{2}N\right)\cos\left(\frac{\pi}{2}N\cos\theta\right) - \cos\theta\cos\left(\frac{\pi}{2}N\right)\sin\left(\frac{\pi}{2}N\cos\theta\right)\right)^2 d\theta}.$$

In the figure on the next page are a number of beam patterns found by numerical integration. The dotted curve is the equivalent Hertzian dipole beam, and notice how similar it is to the half-wave dipole. Notice that the opening angles of the primary beams decrease with the length of the antenna. The rule of thumb is that this opening angle, in radians, is approximately:

$$\text{opening angle} \approx \frac{\lambda}{a}.$$

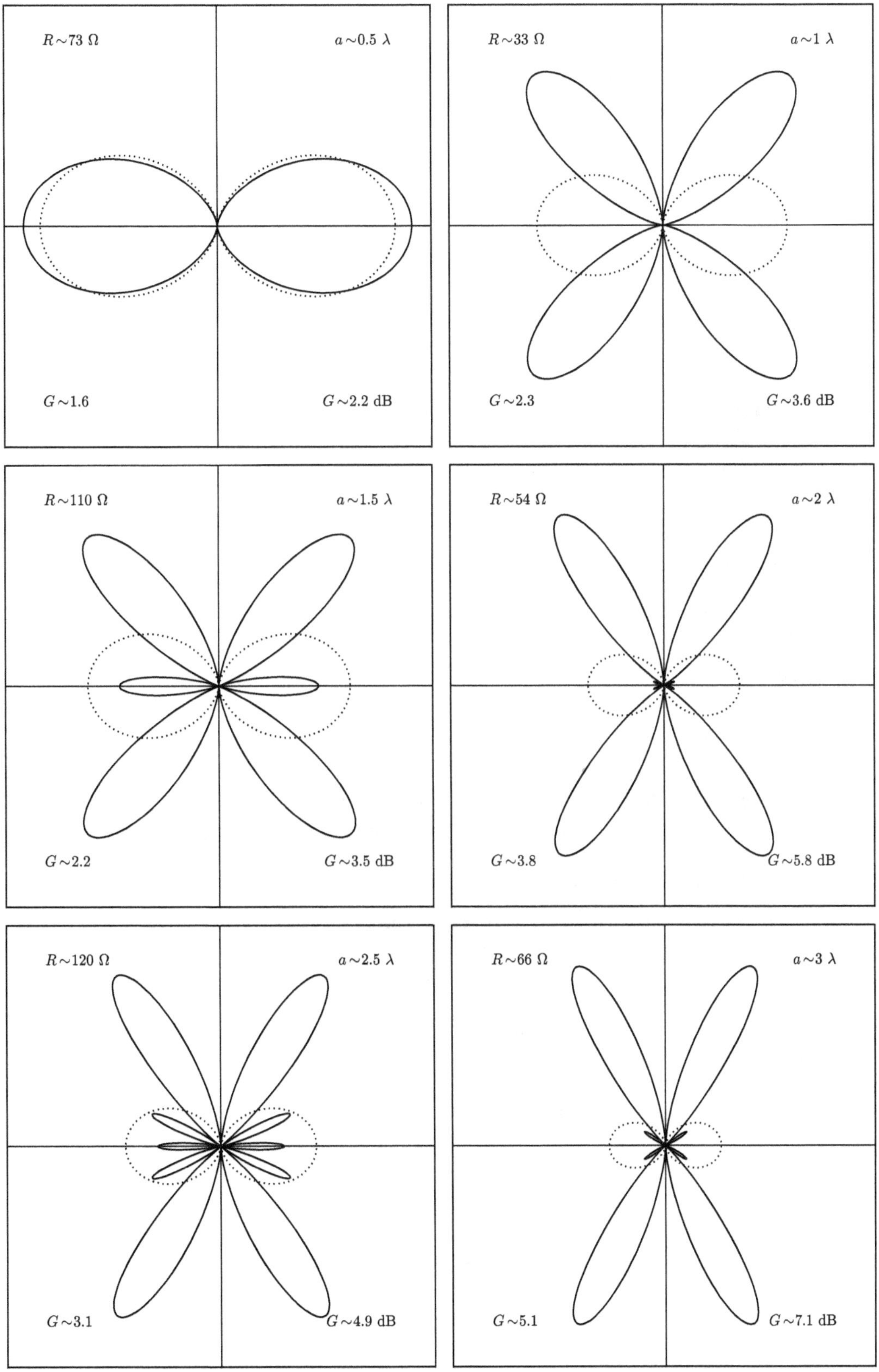
R∼73 Ω
a∼0.5 λ
G∼1.6
G∼2.2 dB
R∼33 Ω
a∼1 λ
G∼2.3
G∼3.6 dB
R∼110 Ω
a∼1.5 λ
G∼2.2
G∼3.5 dB
R∼54 Ω
a∼2 λ
G∼3.8
G∼5.8 dB
R∼120 Ω
a∼2.5 λ
G∼3.1
G∼4.9 dB
R∼66 Ω
a∼3 λ
G∼5.1
G∼7.1 dB

Problem 16.7: Car Radios

Consider a dipole antenna with a length of $a = \frac{1}{2}\lambda$, driven at an angular frequency $\omega = 2\pi c/\lambda$, as a special case of the antenna in Thought Experiment 16.10.

a) For a half wave antenna $(a = \frac{1}{2}\lambda)$, assume that the current in the wire is driven in the middle so the current follows the form:

$I = I_0 \cos(\pi \frac{z}{a}) \sin(\omega t)$.

Find the radiation resistance, and the gain, of a half wave dipole antenna. Show that these are consistent with those found numerically in Thought Experiment 16.10.

b) The most common coaxial cable transmission line to a radio antenna has a characteristic impedance of 75 ohms. Why is this so?

c) A classic car antenna is driven at one end, however, due to the reflectivity of the metal car, has the same beam pattern as an antenna driven in the middle with twice the actual length. The FM dial goes from about 88 MHz to about 108 MHz, with channels approximately every 0.1 MHz. What is the optical frequency for a 75 cm antenna to receive?

d) AM radio broadcasts from 540 to 1700 kHz on the radio dial. Compare the beam pattern of the same 75 cm car antenna when it is used to receive AM as compared to FM radio. Discuss which equation you would use for each and why.

e) Find the radiation resistance of the 75 cm car antenna when listening to AM as a function of its broadcast frequency.

f) AM radio channels are assigned in evenly spaced (10 kHz) frequency increments. Radio receivers are tunable LC circuits, where the radio dial controls one of the components. This could be either the inductor or the capacitor, but one is used much more than the other. Which one, the capacitor or the inductor, is fixed? Which one is tunable? Why is this so?

g) FM radio was an improvement over AM radio. Why was this so, but also why was it more difficult to build FM radios? Conduct some outside historical research, and cite your sources.

Problem 16.8: A Terminated Omnidirectional Antenna

Consider an antenna of length a, driven from one end at a frequency ω. Now, assume that the ends are terminated so that there is no reflected wave, so no standing waves occur. Thus the current is simply given by the formula: $I = I_0 \sin(\omega t - 2\pi \frac{z}{\lambda})$.

Reproduce Thought Experiment 16.10 (p. 536) for this antenna, including all of the calculations and the plots of the beam patterns.

Example 16.8: A Parabolic Dish Antenna

Recall that a parabola is a shape such that any point along the parabola is equidistant from a fixed point called *the focus*, and a fixed straight line called *the directrix*. (It is assumed that focus does not lie along the directrix). Thus, the path length from the focus to a far away source on the optical axis is simply the distance, z, from the directrix. By placing a small half-wave antenna, called *the feed*, at the focus, an AC surface current flows at the reflective surface.

In an ideal parabolic antenna, the dish is 100% reflective, and uniformly illuminated by the feed, so we can imagine that a circle of radius a, located at the directrix, has a uniform current per perpendicular length

$$\vec{K} = \int \vec{J}_\perp \, \mathrm{d}z' = K_0 \sin(\omega\tau).$$

Therefore an element of the vector potential, in polar coordinates with the $\hat{z}$ axis the optical axis, the vector potential becomes:

$$\tau = t - \tfrac{1}{c}r - \tfrac{1}{c}x'\sin\theta = t - \tfrac{1}{c}r - \tfrac{1}{c}s'\cos\varphi\sin\theta$$

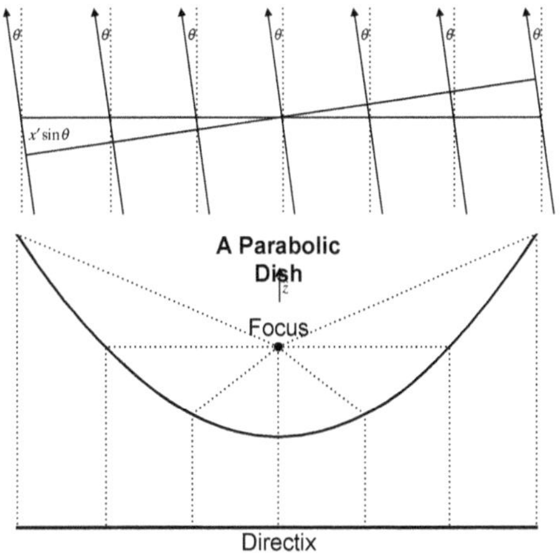

$$\begin{aligned}\delta\vec{\mathbb{A}}(\vec{r},t) &= \frac{\mu_0 \delta I(\tau)\delta y'}{4\pi r}\hat{y} \\ &= \frac{\mu_0 K_0 \sin\left(\omega\left(t-\frac{r}{c}-\frac{1}{c}x'\sin\theta\right)\right)\delta x'\delta y'}{4\pi r}\hat{y},\end{aligned}$$

which in polar coordinates we can write:

$$\delta\vec{\mathbb{A}}(\vec{r},t) = \frac{\mu_0 K_0 \sin\left(\omega\left(t-\frac{r}{c}\right)-\frac{\omega}{c}s'\sin\theta\right)2\pi s'\delta s'\cos\theta}{4\pi r}\hat{y}.$$

Next we will define the unitless quantities $\phi \equiv \omega\left(t-\frac{r}{c}\right)$, $p \equiv \frac{s'}{\lambda}$, and $R \equiv \frac{a}{\lambda}$:

$$\delta\delta\vec{\mathbb{A}}(\vec{r},t) = \left(\frac{Z_0 K_0 \lambda^2}{c4\pi r}\right)\cos\theta\sin(\phi - 2\pi p\sin\varphi'\sin\theta)\, p\delta\varphi'\delta p\,\hat{y}$$

$$\begin{aligned}\delta\vec{\mathbb{A}}(\vec{r},t) &= \left(\frac{Z_0 K_0 \lambda^2}{2cr}\right)\cos\theta\, p\delta p\,\hat{y}\int_0^{2\pi}\left(\sin(\phi)\cos(2\pi p\sin\varphi'\sin\theta) - \cancel{\cos(\phi)\sin(2\pi p\sin\varphi'\sin\theta)}\right)\mathrm{d}\varphi', \\ &= \left(\frac{Z_0 K_0 \lambda^2}{2cr}\right)\cos\theta\, p\delta p\,\hat{y}\sin(\phi)2\pi J_0\left(\left|2\pi p\sin\theta\right|\right).\end{aligned}$$

In this case, J_0 is the zeroth order Bessel function, so:

$$\vec{\mathbb{A}}(\vec{r},t) = \left(\frac{Z_0 K_0 \lambda^2}{2cr}\right)\cos\theta\,\hat{y}\sin(\phi)\int_0^R\left(2\pi p J_0\left(\left|2\pi p\sin\theta\right|\right)\mathrm{d}p\right),$$

and making the variable substitution:

$$\xi = 2\pi\sin\theta\, p \quad \text{so} \quad \mathrm{d}p = \frac{\mathrm{d}\xi}{2\pi\sin\theta},$$

$$\begin{aligned}\vec{\mathbb{A}}(\vec{r},t) &= \left(\frac{Z_0 K_0 \lambda^2}{2cr}\right)\frac{\cos\theta}{2\pi\sin^2\theta}\hat{y}\sin(\phi)\int_0^{2\pi\sin\theta R}\left(\xi J_0(|\xi|)\mathrm{d}\xi\right) \\ &= \left(\frac{Z_0 K_0 \lambda^2}{2cr}\right)\frac{\cos\theta}{\cancel{2\pi}\sin^{\cancel{2}}\theta}\hat{y}\sin(\phi)\cancel{2\pi}\,\cancel{\sin\theta}\,R\,J_1(2\pi\sin\theta R) \\ &= \left(\frac{Z_0 K_0 \lambda^2}{2cr}\right)\frac{\cos\theta}{\sin\theta}R\,J_1(2\pi\sin\theta R)\sin(\phi)\hat{y}.\end{aligned}$$

And substituting in for $\hat{y}$:

$$\vec{\mathbb{A}}_{\perp}(\vec{r},t) = \left(\frac{Z_0 K_0 \lambda^2}{2cr}\right)\frac{\cos\theta}{\sin\theta} R\, J_1(2\pi\sin\theta R)\sin(\phi)\left(\cos\theta\,\sin\varphi\,\hat{\theta} + \cos\varphi\,\hat{\varphi}\right)$$

$$= \left(\frac{Z_0 K_0 \lambda^2}{2cr} R\right)\left(\cos\theta\sqrt{\cos^2\theta\,\sin^2\varphi + \cos^2\varphi}\right)\frac{J_1(2\pi\sin\theta R)}{\sin\theta}\sin(\phi)$$

and the time-averaged Poynting vector is:

$$\left\langle\vec{\mathbb{S}}\right\rangle = \tfrac{\omega^2}{2Z_0}\mathbb{A}_0^2 = \tfrac{1}{2}\left(\tfrac{2\pi c}{\lambda}\right)^2\tfrac{1}{Z_0}\left(\tfrac{Z_0 K_0 \lambda^2}{2cr} R\left(\cos\theta\sqrt{\cos^2\theta\,\sin^2\varphi + \cos^2\varphi}\right)\frac{J_1(2\pi\sin\theta R)}{\sin\theta}\right)^2$$

$$= \tfrac{1}{2}Z_0 K_0^2\left(\frac{\pi^2 a^2}{r^2}\right)\left(\cos^2\theta\left(\cos^2\theta\,\sin^2\varphi + \cos^2\varphi\right)\right)\left(\frac{J_1(2\pi\sin\theta R)}{\sin\theta}\right)^2.$$

In general, this does not vary much with azimuthal angle, so in the plots we average over the angle φ and present the beam pattern (top). Notice how similar it is to a Gaussian beam (dotted), with a full width half maximum (FWHM) opening angle given by $\Delta\theta \approx \lambda/_{2a} \approx \lambda/_{D}$. The bottom plots show the logarithmic gain as a function of polar angle, and again the Gaussian beam is plotted in a dotted line. Notice how, on a logarithmic scale, the side lobes become pronounced, but they are hardly noticeable on a linear scale.

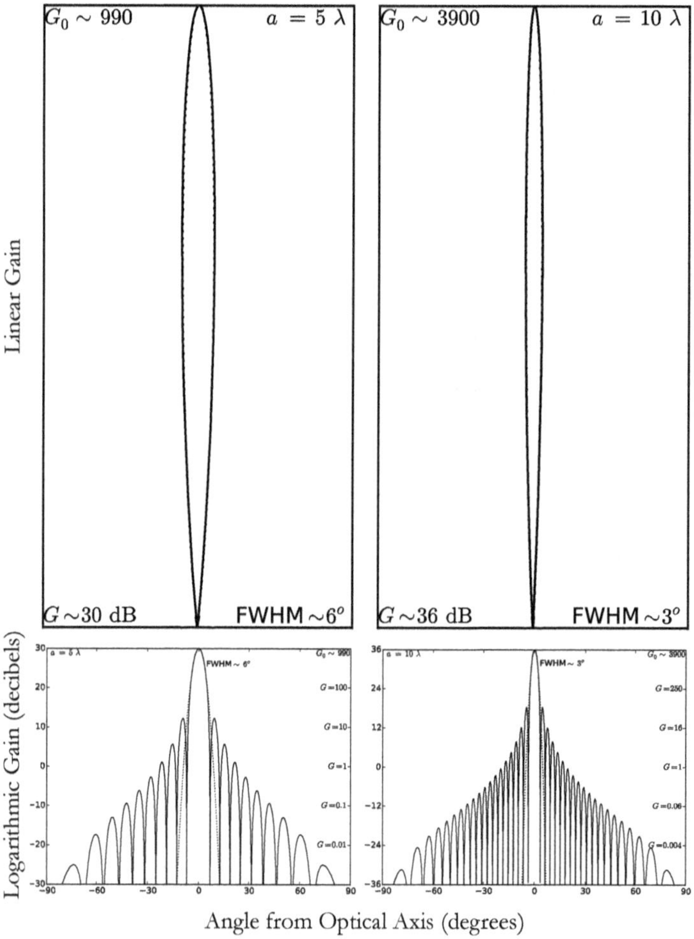

Recall that radio dishes can be used for either receiving or transmitting radio waves. In the case of a parabolic dish whose diameter is significantly larger than one wavelength, we expect that the collecting area will be simply the physical area when observing on-axis. This implies that for the peak gain, G_0, we have that:

$$A_{\text{eff}} = \frac{\lambda^2}{4\pi}G_0 \lesssim \left(\pi a^2\right), \quad \text{so} \quad G_0 \lesssim 4\pi^2\frac{a^2}{\lambda^2} \approx \left(\frac{2\pi a}{\lambda}\right)^2.$$

In practice, however, feeds need to be secured somehow. If the feed obstructs the beam it can greatly increase the side lobes. This is why in modern radio dishes, such as the small ones used for satellite television, only one side of the paraboloid is constructed. Nevertheless, these antennas are still parabolic, with the feed at the focus as shown in the picture of the Robert C. Byrd 100 meter radio telescope in Green Bank, West Virginia with a parabola overlaid.

Thought Experiment 16.11: Radio Interferometry

Often parabolic radio antennas are used for mapping faint objects, such as astronomical sources. While it may seem small, the beam size of a radio antenna is very large compared to the resolution of cameras attached to optical telescopes. However, radio astronomers routinely map out astronomical objects to resolutions of arcseconds at a wavelength of 20 cm (Plate 31, 639). How do they do that, when the telescope would need to have a diameter of about 30 km in order to have a beam that small. The solution is to correlate the signals from many telescopes, with the goal of recreating the energy as a function of angle: $\langle \vec{\mathbb{S}}(\theta,\varphi)\rangle = \langle \mathbb{S}(\theta,\varphi)\rangle \hat{k} = \langle \mathbb{S}(\theta,\varphi)\rangle(-\hat{r})$.

Consider two radio telescopes, call them antenna 1 and antenna 2, at positions of $\vec{r}_1'$ and $\vec{r}_2'$, and both pointing in a direction $\hat{z}$, so they would have a relative distance of $(\vec{r}_2' - \vec{r}_1')$. We will also assume that each produces a signal of $I_1(t)$ and $I_2(t)$ respectively.

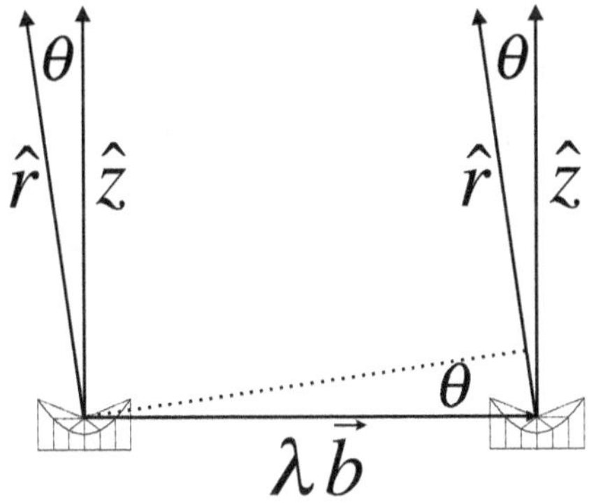

It is customary to work in units of wavelength, so we will define the unitless baseline as:

$$\vec{b}_{12} \equiv \frac{\hat{z}\times(\vec{r}_2' - \vec{r}_1')\times\hat{z}}{\lambda} = -\vec{b}_{21},$$

and to render the baselines into Cartesian coordinates, in terms of two unitless variables, defined such that $\vec{b} = u\,\hat{x} + v\,\hat{y}$. All modern interferometers can compensate for differences of position along the line of sight, with an overall delay time, $\frac{1}{c}(\vec{r}'\cdot\hat{z})$, so we can work in two dimensions without losing generality.

However, the ultimate goal is to map out the sky to great detail, so while signals coming from the $\hat{z}$ direction will be combined in phase, $\tau = t$, a signal coming from a direction $\hat{r}$ must travel a distance $\hat{k}\cdot(\vec{r}_2'-\vec{r}_1')=(-\hat{r})\cdot(\vec{r}_2'-\vec{r}_1')=-\lambda(\hat{r}\cdot\vec{b}_{12})$ further to arrive at telescope 2 than at telescope 1. This corresponds to a travel time difference of $t_{21}=-\frac{\lambda}{c}(\hat{r}\cdot\vec{b}_{12})=\frac{\lambda}{c}(\hat{r}\cdot\vec{b}_{21})$. The corresponding angular phase difference is $\phi_{21}=\omega t_{21}=\frac{\omega\lambda}{c}(\hat{r}\cdot\vec{b}_{21})=2\pi(\hat{r}\cdot\vec{b}_{21})$. Thus, by measuring the relative phases of the light entering each antenna, sources can be mapped out.

However, as astronomical sources are faint, it is imperative that the radio signal be averaged over time to increase the signal to noise ratio. The solution is to cross correlate the signals from each pair of antennas. This is done both in phase, and 90° out of phase, to gain the most information. Observing at a frequency ω, the two time-averaged cross-correlations will be defined as: $\langle I_1(t)I_2(t)\rangle$ and $\langle I_1(t)I_2(t-\frac{\pi}{2}\frac{1}{\omega})\rangle$. These two signals can be thought of as a vector in phase space, so we define a complex quantity, called the *visibility function*, where we assign the phase correlated power as the real part, and the out of phase correlated flux as the imaginary component. Thus:

$$\mathbb{S}_{12}\equiv\frac{\langle I_1(t)I_2(t)\rangle+i\langle I_1(t)I_2(t-\frac{\pi}{2\omega})\rangle}{\frac{1}{2}\left(\langle I_1(t)I_1(t)\rangle+\langle I_2(t)I_2(t)\rangle\right)},$$

where we use the astronomical notation of $i=\sqrt{-1}$, rather than the electrical engineering notation we used in Chapter 13.

We arbitrarily labeled each antenna. Had we, rather, reversed the labels "1" and "2," the real part of the visibility function would have stayed the same. However, when calculating the imaginary component, we would have delayed the other telescope by a quarter of a period, resulting in a shift of half a period from our prior result. Therefore, when reversed, the imaginary component would become negative. Thus we have:

$$\mathbb{S}_{21}=\mathbb{S}_{12}^*,$$

where $\mathbb{S}_{12}^*$ represents the complex conjugate of $\mathbb{S}_{12}$. If there are n antennas in the array, for any given exposure time, the visibility function is an $n\times n$ matrix with the property that the transpose is equal to the complex conjugate—called a *Hermitian matrix*.

As the earth turns, the position of each antenna, $\vec{r}_i'$, will also change in a predictable manner. So, by observing an object throughout the course of a day, a modern radio interferometer can measure the visibility function as a function of baseline. As $\vec{b}_{21}=-\vec{b}_{12}$, the visibility function has the symmetry property that:

$$\mathbb{S}(\vec{b})=\mathbb{S}^*(-\vec{b}),$$

so it is an *Hermitian function*, which will become very important soon. Thus, the technical purpose of an array of radio telescopes is to measure both the real and imaginary parts of the visibility, as a function of the projected baselines in units of wavelength, $\mathbb{S}(u,v)$.

An astronomer, on the other hand, is interested in creating maps of the sky—or more specifically finding the incoming power per area from a particular direction $\hat{r}$. So, with the visibility function, $\mathbb{S}(u,v)$, how do we map the sky brightness $\vec{\mathbb{S}}=-\mathbb{S}(\hat{r})\hat{r}=\mathbb{S}(\theta,\varphi)(-\hat{r})$? We take advantage of the reciprocity theorem.

Recall from the last two examples that we found the vector potential by integrating the current distribution, or:

$$\vec{\mathbb{A}}_{\text{outgoing}} = \frac{Z_0}{4\pi c} \int_{\text{volume}} \frac{\vec{J}(\vec{r}',\tau)}{|\vec{r}-\vec{r}'|} \mathrm{d}V' \quad \text{and} \quad \vec{\mathbb{A}}_{\text{incoming}} \propto \int_{\text{surface}} \vec{J}(\vec{r}',\tau)\,\mathrm{d}A' \propto \int_{\text{surface}} \vec{J}(\vec{r}',t-\tfrac{1}{c}\vec{r}'\cdot(-\hat{r}))\,\mathrm{d}A'.$$

As the wave is periodic, we can rewrite this as:

$$\mathbb{A}(\hat{r},t) \propto \int_{\text{surface}} \vec{J}_0(\vec{r}')\sin\left(\omega t+\tfrac{1}{c}\omega\vec{r}'\cdot\hat{r}\right)\mathrm{d}A',$$

which we called a Fourier integral, or commonly known as a two dimensional *Fourier transform* when written in complex notation as:

$$\mathbb{A}(\hat{r},t) \propto \int_{\text{surface}} J_0(\vec{r}')\,e^{-i\left(\omega t+\frac{1}{c}\omega\vec{r}'\cdot\hat{r}+\phi\right)}\mathrm{d}A' \propto e^{-i(\omega t+\phi)} \int_{\text{surface}} J_0(\vec{r}')\,e^{-2\pi i\left(\frac{1}{\lambda}\vec{r}'\right)\cdot\hat{r}}\mathrm{d}A', \quad \text{so} \quad \mathbb{A}(\hat{r}) \propto \mathrm{FT}\left(J_0(\vec{r}'/\lambda)\right).$$

Fourier transforms make a domain change on a function that exhibits periodicity. This inverse Fourier transform is almost the same except for changing the sign of the complex exponent:

$$J_0(\vec{r}'/\lambda) \propto \int_{\text{surface}} \mathbb{A}(\hat{r})\,e^{2\pi i\left(\frac{1}{\lambda}\vec{r}'\right)\cdot\hat{r}}\mathrm{d}\hat{A}.$$

The most common Fourier transform changes functions of time to functions of frequency, which is how radio spectrometers work. In our case we are converting our function of the position of the detector (in units of wavelength) to a function of the direction of the incoming light.

Next, recall, we found the outgoing Poynting vector by squaring the vector potential, so we will do the same thing here to find the incoming Poynting vector, so:

$$\langle\mathbb{S}(\hat{r})\rangle \propto \left\langle(\mathbb{A}(\hat{r}))^2\right\rangle \propto \left\langle\left(\mathrm{FT}\left(J_0(\vec{r}'/\lambda)\right)\right)^2\right\rangle.$$

Fourier transform theory is very well developed, and the Fourier transform has a number of desirable properties. The most practical of these is that computers can perform Fourier transforms on large data sets quickly using what is known as the *Fast Fourier Transform* (FFT) algorithm. The Fourier transform has another very desirable property, which is that the product of two Fourier transforms is the Fourier transform of their *convolution*.

Consider two functions, f and g, of two variables, u and v. Their convolution is by definition:

$$f \otimes g = \int f(x,y)\,g(x-x',y-y')\,\mathrm{d}x'\,\mathrm{d}y',$$

and under Fourier transform theory the following identity holds:

$$\mathrm{FT}(f)\,\mathrm{FT}(g) = \mathrm{FT}(f\otimes g).$$

Thus we can finally calculate the incoming Poynting vector as:

$$\langle\mathbb{S}(\hat{r})\rangle \propto \left\langle\left(\mathrm{FT}\left(J_0(\vec{r}'/\lambda)\right)\right)^2\right\rangle \propto \left\langle\mathrm{FT}\left(J_0(\vec{r}'/\lambda)\right)\cdot\mathrm{FT}\left(J_0(\vec{r}'/\lambda)\right)\right\rangle \propto \left\langle\mathrm{FT}\left(J_0\otimes J_0\right)\right\rangle.$$

And we can now write:

$$J_0 \otimes J_0 = \int\!\!\int J_0(x_1',y_1')\cdot J_0(x_1'-x_2',y_1'-y_2')\,\mathrm{d}x_2'\,\mathrm{d}y_2',$$

and making the variable substitution

$$u=\frac{x_1'-x_2'}{\lambda} \quad , \quad v=\frac{y_1'-y_2'}{\lambda} \quad , \quad x_1'=x_2'+\lambda u \quad , \quad y_1'=y_2'+\lambda v \quad .$$

$$(J_0 \otimes J_0)(u,v)=\int\int J_0(x_2'+\lambda u,y_2'+\lambda v)\cdot J_0(\lambda u,\lambda v)\mathrm{d}x_2'\,\mathrm{d}y_2'$$

Now recall that the result of a radio interferometric observation was the visibility function, which is a two dimensional array of the cross-correlation of the signal between each combination of telescopes. So for any pair of antennas, both the baseline, $\vec{b}$, and the visibility, $\mathbb{S}$, were measured. Notice, moreover, that the current convolution at a given measured baseline is simply proportional to the correlated current output of the two antennas:

$$(J_0 \otimes J_0)\propto\langle I_1 I_2\rangle .$$

But what about the phase information? Where would this come in?

To investigate this, we need to introduce another property of Fourier transforms. The inverse Fourier transform is simply the complex conjugate of the forward Fourier transform. Now consider a real sky brightness distribution, $\mathbb{S}(\hat{r})$, and apply the inverse Fourier transform. Unlike the sine transform, the Fourier transform of a real function need not also be real. However, it must be a Hermitian function—the real component is even, while the imaginary component is odd. Thus, in order to measure a real brightness distribution in Fourier space, we must guarantee that its Fourier transform is not necessarily real, but rather Hermitian. The cross-correlation function is, by definition, Hermitian, and it best measures the current density distribution. Therefore we can finally see that the sky brightness is likely proportional to the visibility function.

$$\langle\mathbb{S}(\hat{r})\rangle\propto\mathrm{FT}(S(\vec{b}))\propto\iint S(\vec{b})e^{-2\pi i(\hat{r}\cdot\vec{b})}\mathrm{d}^2\vec{b} .$$

And, in Cartesian coordinates, we can write:

$$\hat{r}\cdot\vec{b}=(\sin\theta\cos\varphi\hat{x}+\sin\theta\sin\varphi\hat{y}+\cos\theta\hat{z})\cdot(u\hat{x}+v\hat{y})=u\sin\theta\cos\varphi+v\sin\theta\sin\varphi=ul+vm ,$$

where the directional sine from the phase center east is l and north is m. Thus, the final image of the sky, $\langle\mathbb{S}(l,m)\rangle$, can be found by:

$$\langle\mathbb{S}(l,m)\rangle\propto\iint\mathbb{S}(u,v)e^{-2\pi i(ul+vm)}\,\mathrm{d}u\,\mathrm{d}v .$$

In practice, of course, the antennas can be further apart or closer together, and do not sample everywhere, so it is important to interpolate between absent points so that $\mathbb{S}(u,v)$ is a smooth function. The further apart the telescopes are the higher resolution the resultant image can become. On the other hand, information can be lost on the larger scales if the short spacings are absent. This is rarely a problem if the source does not change over time, as the astronomer can conduct a future observation with different spacings or with a single dish telescope.

Plate 29 (p. 561) show a photograph of the Very Large Array. Notice that each telescope sits on a double pair of railroad tracks, so that they can be spaced either far apart or close together. The map shown in Plate 31 (p. 639) was produced by observing at both $\lambda=6\mathrm{cm}$ and $\lambda=20\mathrm{cm}$, at gathering data at multiple array configurations.

Problem 16.9: Sea Interferometry

A single radio telescope observes a large solid angle of the sky, so it is often hard to separate out objects of interest. Post World War II Australian radio astronomers overcame this problem using a sea interferometer.

A sea interferometer consists of a radio antenna located on a cliff overlooking the ocean. The light from a source near the horizon reflects off the ocean and interferes with itself. As the source rises, the altitude angle α increases, changing the interference pattern over time. The confusing objects have different altitude angles, so they do not contribute to the source interference pattern.

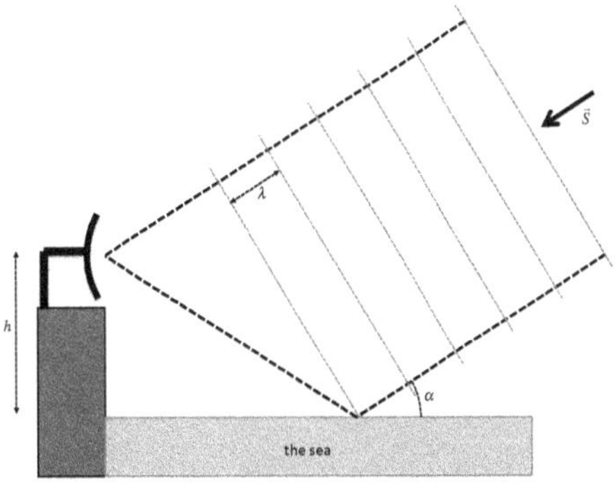

a) Given the angular frequency of the light ω, bandpass $\Delta\omega$, antenna collecting area A ,and the radio flux density b, find the power received by the telescope if it were pointing at the source directly. Call this power P_0.

b) For a given antenna height above the ocean h and altitude of the source α, power you would expect to receive P in terms of P_0. Note that at the reflection on the surface of the ocean, the phase of the light is shifted by half a period.

c) Graph the power received as a function of the altitude angle. What is the angular period $\Delta\alpha$ from constructive to constructive interference?

d) The graph[272] shows observations taken at a frequency of $\omega = 2\pi \times 100\,\text{MHz}$ and a height of 240 feet above the ocean. Estimate the rate at which Cygnus A was rising (i.e. $^{d\alpha}/_{dt}$) during the observation shown.

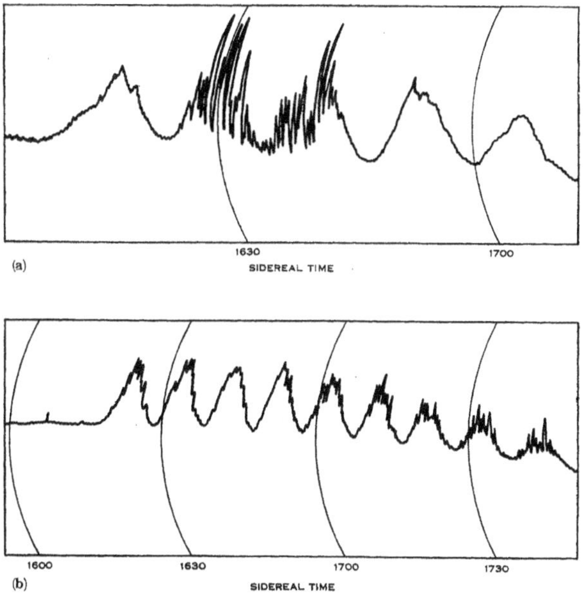

Fig. 1.—Typical records of the scintillations on interference patterns of the source in Cygnus at rising. (*a*) May 19, 1948; (*b*) June 11, 1949.

e) Report on the history of our understanding of the radio source Cygnus A. What do we believe it to be now and what were the crucial early observations that give us that understanding? Find a map of Cygnus A created by a modern radio interferometer.

f) Report on modern radio interferometry. What are the most sensitive radio interferometers in the world today? How many

[272] J. Bolton, O. Slee and G. Stanley, "Galactic Radiation at Radio Frequencies. VI. Low Altitude Scintillations of the Discrete Sources," Australian Journal of Physics, **6** (1953), p. 434.

antennas do they have? Which have the highest angular resolution? What is the highest frequency that it makes sense to map the sky using radio interferometry rather than another technique such a direct imaging.

Problem 16.10: Self-calibration

a) Consider a two antenna interferometer. Show that the phase difference between these two elements has the property: $\phi_{12} + \phi_{21} = 2n\pi$, where $n = 0$, if the phases are restricted to the range $-\pi < \phi < \pi$.

b) Consider a three antenna interferometer. Clearly argue that the phases around any closed loop should add such that: $\phi_{12} + \phi_{23} + \phi_{31} = 2n\pi$, where $n = -1$, 0, or 1.

c) Consider an interferometer with N antennas. Argue that the phases around any loop of antennas should add to $2\pi n$. What are the restrictions on the integer n?

d) Discuss how modern radio astronomers could use this property to calibrate their data *ex post facto*. They call this *self-calibration.*

e) Discuss why self-calibration is most effective when the phase errors are already small.

Discussion 16.4: Complex Numbers in Electrodynamics

In Chapter 13 we investigated AC circuits containing resistors, capacitors and inductors in series and parallel to introduce you to electromagnetic waves. These circuits are all driven sinusoidally and their solutions also come out to be oscillatory, so electrical engineers model them by analogy to DC circuits. Each component could be thought of as having a complex effective resistance called the *complex impedance*, which we represented with a capitol Z and measured in ohms. To find the overall impedance of a circuit, you simply add those components in series, and reciprocally add those in parallel, as one would do to find an effective resistance. Each factor of $j = i = \sqrt{-1}$ represents a 90° phase shift, greatly simplifying AC circuit theory.

There was a cost, however, of using the complex impedance. It greatly obscured the physics of oscillating circuits. We no longer had to consider the electric and magnetic fields. We just pretended that everything was an effective resistor, and we got the right answer. Since this is a textbook on the physics of electrodynamics, we have avoided using complex numbers them even when they simplify the algebra.

Many physical systems are described by differential equations. While solutions to first order differential equations must involve exponential growth, or decay, the solutions to second order equations can either be exponential or oscillatory. Sometimes there are competing factors that make it difficult to tell from the outset which will dominate. If you assume an oscillatory system, but you get an imaginary phase angle or frequency, it usually means that the system in unstable. In this text book, we have stressed physical understanding of the differential equations we encounter over formal methods of solution. In fact, often we have favored numerical solutions, even where analytical solutions exist, simply because we would not have expected you to already know which differential equations are solvable in closed form and which are not.

The prior example of a radio interferometer is probably, however, the first example you have encountered where measured physical quantities *inherently* represent complex numbers. This is because the *out of phase* visibility measurement is no less physically real than the *in phase* measurement, although it is represented by an imaginary number. The mathematics of what is known as *phase space*, where quantities are represented by sums of sines and cosines, can be considered much like two dimensional vector quantities where the sine and cosine terms represent two orthogonal components. This was why we needed two independent measurements in order to specify one visibility point.

The idea of an inherently complex phase space becomes very important in the study of quantum mechanics, as the momentum and position are often represented as imaginary and real representations of complex wave functions. So, complex numbers only become more and more important has you study physics at higher levels.

Problem 16.11: Mapping 3C 279 with the Very Long Baseline Array

The orbits of material that accretes onto a black hole can become unstable, and then be ejected along the two poles perpendicular to the accretion disk. Astronomers observe these jets coming out of active galactic nuclei. Moreover, sometimes the jets contain features, such as blobs of material, that can be followed over time as in the figure of the jet of quasar 3C 279.[273]

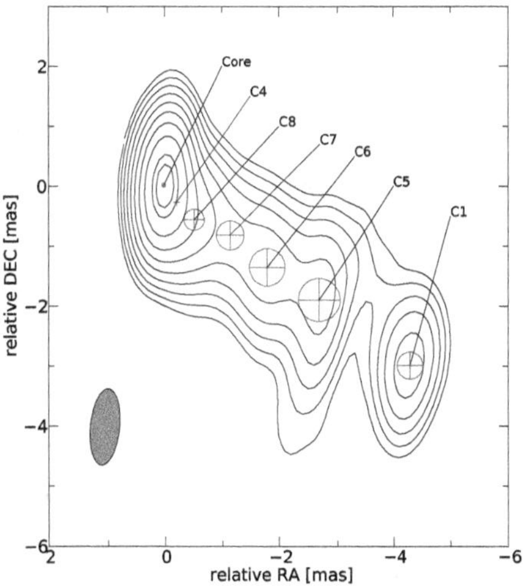

Their figure shows various features contained in the jet of 3C 279 as of June 15th 2003, as mapped using the Very Long Baseline Array (VLBA) at an observing frequency of 15 GHz. The VLBA is a radio interferometer with ten 25 meter antennas that range from the US Virgin Islands to Hawaii. The longest baselines range from about 8,000 km East/West and about 2,000 km North/South. The gray oval shows the size and shape of the synthesized beam pattern of the antenna array, which is about one milliarcsecond (mas) in angular diameter. By way of comparison, the resolution of the Hubble Space Telescope is only about 100 mas.

a) What is the wavelength of light being observed?

b) Based on the area of the gray oval, estimate: the solid angle of the beam in units of $(\text{mas})^2$, the same solid angle in steradians, and the gain of the antenna array.

c) The physical area of each antenna is simply $\frac{\pi}{4}D^2$, which is approximately the effective area of the system for sensitivity calculations. What is the beam size of each antenna? This is called the *primary beam*.

d) Imagine a giant parabolic dish 8,000 km in diameter, what would its beam size be? What about a 2,000 km parabolic dish? How do these hypothetical beam sizes compare to the synthesized beam of the array?

e) Explain how the coordinate system on the Earth compares to the fixed x,y,z coordinate system with $\hat{z}$ toward the location of the origin of the image. Find the corresponding values for the maximum east-west, and north-south, baseline to wavelength ratios: $u_{\max}$ and $v_{\max}$.

---

[273] S.D. Bloom, C.M. Fromm, and E. Ros, "The Accelerating Jet of 3C 279," The Astronomical Journal, **145**, article 12 (2013)

f) The map represents the function $\mathbb{S}(l,m)$, with north vertically up and east to the left. Use a ruler to measure the angular position, $(l,m)_{C1}$, from the core to feature C1. Find this in milliarcseconds and radians.

g) Calculate is the inverse Fourier transform of a point source at a single location, (l,m). Find both the real and imaginary components.

h) These data were collected by correlating each antenna with each other, and measuring the complex visibility function $\mathbb{S}(u,v)$. The team found $\mathbb{S}(l,m)$ by taking the Fourier transform of $\mathbb{S}(u,v)$. Draw two identical sets of axes, with the origin in the middle, and a labeled scale of u_{max} and v_{max}, in units of $M\lambda$. Based on the location of C1, sketch the real, and imaginary, components of the complex visibility function $\mathbb{S}(u,v)$ on these axes.

Problem 16.12: Superluminal Motion in the Jets of 3C 279

The distance, d, to 3C 279 (see Problem 16.11 above) is about 6 billion light years from Earth. Their paper also showed a graph of the angular sideways angular distance between the core and feature C1 over time.

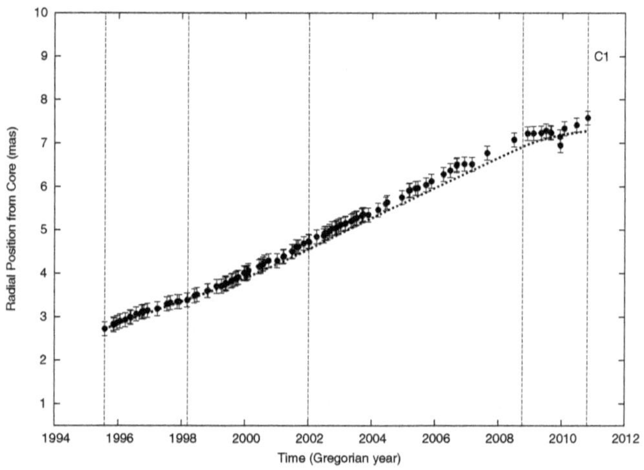

a) Using a ruler, draw a line through the data in the figure and find its slope and x-intercept. This slope, called proper motion, is the angular velocity the feature appears to move in units of milliarcseconds per year. The x-intercept is the date that we would have observed the blob leave the core, which you can call t_0.

b) Using the distance to the source, find the apparent tangential velocity in units of light-years per year, and call it $\frac{1}{c}v_\perp$. Discuss why this is called *superluminal motion*?

c) C1 is most likely moving radially away from the core at an angle θ with respect to the line of sight. Furthermore, it must be actually be moving at a speed close to light $(v' \lesssim c)$ since it appears to be moving sideways faster than light. Draw a diagram of this model and the position, $\vec{r}'$, of blob C1, and show that the retarded time is related to the actual time by: $\tau = t - \frac{1}{c}(d - r'\cos\theta)$.

d) Show that this physical speed is related to the observed tangential velocity by the relationship:

$$v' = \frac{v_\perp}{\sin\theta + \frac{1}{c}v_\perp \cos\theta}.$$

e) Given the observed tangential velocity, plot the physical speed as a function of the inclination angle θ. What is the smallest possible speed that feature C1 could actually be moving?

Problem 16.13: The Core of the Quasar 3C 279

Astronomers measure the brightness of objects, called quasars, that change over short time scales. During 2006 and 2007, a large collaboration of astronomers simultaneous measured the brightness of the core of the quasar 3C 279 as a function of time.[274]

a) The graph shows the correlation function between the brightness of radio observations at one time, versus those on a prior day. From this graph, estimate the maximum coherence time for the source.

b) Given this result, estimate the maximum radius of the emitting region. Is this closer to the size of a star, a solar system, or the distance between neighboring stars in a galaxy?

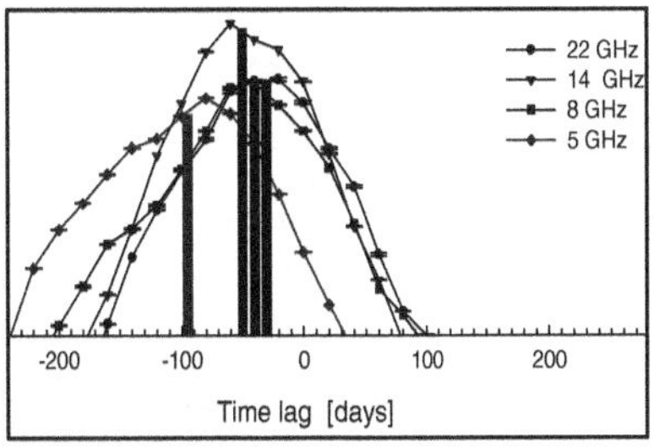

c) The next plot shows their log-log plot of the brightness of the core as a function of frequency from radio waves to x-rays. Estimate from the graph the total area under each curve. This total area represents the quantity:

$$\int \log(\nu F_\nu)\, d(\log \nu).$$

d) The units on a log-log plot are confusing, as differences in the log represent ratios in the actual quantity. In this plot, the total flux of the source in units of $\text{Jy} \cdot \text{Hz}$, would be given by the following equation:

$$\mathbb{S}_{\text{tot}} = \int \mathbb{S}_\nu \, d\nu .$$

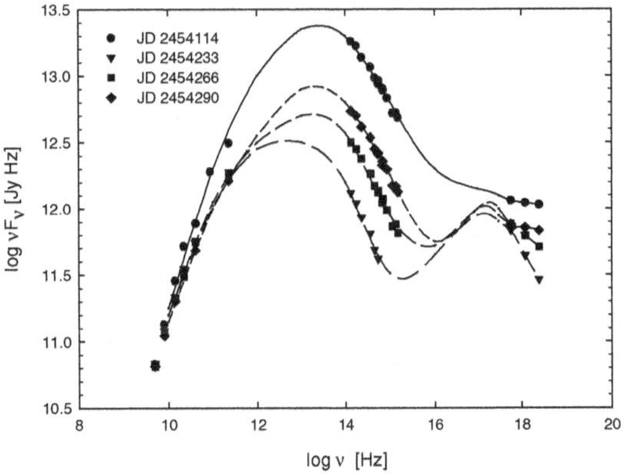

Estimate the total flux of the core of 3C 279, and convert it to SI units. Note: $1\,\text{Jy} \equiv 10^{-26} \frac{\text{W}}{\text{m}^2\text{Hz}}$, which is a unit used by radio astronomers. The unit is named Karl Guthe Jansky, who was an American radio engineer. In 1931, he discovered radio waves from the center of the Milky Way, and thus founded the field of radio astronomy.

d) The distance to 3C 279 is about 6 billion light years. If 3C 279 were magically moved to the same distance as the sun (8 light minutes), how many times brighter than the sun would it be?

e) What is the total power output of 3C 279? Express your answer in ratio to the sun's total luminosity ($L_\odot$).

f) Next estimate the ratio of the power emitted per unit volume of 3C 279, which is called the *emissivity* by astronomers. Express your answer in both watts per cubic meter and in ratio to the average emissivity of the sun $\left(\varepsilon_\odot = \dfrac{L_\odot}{\frac{4\pi}{3} R_\odot^3}\right)$.

---

[274]V.M. Larionov, et al., "Results of WEBT, VLBA and RXTE monitoring of 3C 279 during 2006–2007," Astronomy and Astrophysics, (2008), **492**, pp. 389-400.

g) Assume that the jets are powered by a disk of material orbiting the black hole, and heating up due to various frictional and tidal forces. At a given distance R from a black hole of mass M_{BH}, what is the total energy per mass ($\mathbb{V}_m \equiv {}^{\delta E}\!/\!_{\delta m}$) of circularly orbiting material?

h) The radius of the event horizon of a black hole, R_{BH}, is defined as the position where the escape velocity equals the speed of light. Assuming Newtonian gravity, what is the relationship between the radius of the event horizon and the mass, M_{BH}, of the black hole?

i) Like a dam on a dried up river, black holes without material falling into them do not produce power. On the other hand, when material falls into a black hole about half of the mass goes into the black hole, and about half of the mass leaves the system as some form of energy or matter based on Einstein's equation $\mathbb{E} = mc^2$. Assuming that most of this mass is actually converted to electromagnetic radiation with frequencies between 10^9 and 10^{18} Hz, estimate the total mass infall rate $\dot{m}$. Express your answer in both kilograms per second and solar masses per year.

16.4 *Radiation from Point Charges*

That accelerated point charges emit energy in the form of electromagnetic radiation wasn't appreciated when Maxwell first published his field equations, because charged point particles weren't thought to be real. This quickly changed in 1897 when J.J. Thomson discovered the electron. That same year, the Irish physicist Joseph Larmor derived an expression for the rate at which an electron should emit radiation.[275] Larmor's famous formula, which asserts that the rate of power loss is directly proportional to the square of the electron's charge and its acceleration, was originally derived to explain the Zeeman effect by assuming a model in which electrons move in circular or elliptical orbits about what Larmor termed "a fixed center." Nevertheless, Larmor's result holds good for any nonrelativistic classical charged particle. We will cover radiation from particles traveling close to the speed of light in Section 17.4 (p. 617), and we will discuss the quantum mechanical interpretation of radiation emission in Chapter 18 (p. 651). In this section, we derive Larmor's formula by using the retarded potentials and by way of analogy to the Hertzian dipole. This is similar in spirit to the way that Larmor obtained the result in his original paper.

Max Abraham and Hendrik Lorentz applied Larmor's work to show that the energy loss predicted by Larmor is equivalent to a damping term or *radiation reaction* force acting on the electron. The electromagnetic field emitted by the electron in the form of radiation acts back on its emitter by opposing the motion of the accelerating electron. This recoil force of the emitted field on the charged particle (that emitted the field) causes the charge to decelerate ever so slightly and therefore to emit radiation and lose energy, which is just what the damping term does in a classical harmonic oscillator. We will use conservation of energy and momentum to derive a simple expression for the radiation reaction force. We also study the interaction of light and matter by considering the scattering of light by a slowly moving electron and calculating the cross-section for such scattering.

---

[275] Joseph Larmor, "On the Theory of the Magnetic Influence on Spectra; and on the Radiation from moving Ions," Philosophical Magazine, Series 5, vol. **44**, 271 (1897), 503-512.

Thought Experiment 16.12: The Larmor Formula

Consider a non-relativistic particle of charge q, which is moving at a speed that is itself a function of time. What is the power per solid angle radiated by the particle?

First we will consider the vector potential as:

$$\vec{\mathbb{A}}(\vec{r},t)=\frac{Z_0}{4\pi c}\int_{\text{volume}}\frac{\vec{J}(\vec{r}',\tau)}{|\vec{r}-\vec{r}'|}\mathrm{d}V'=\frac{Z_0}{4\pi c}\int_{\text{volume}}\frac{\rho\vec{v}'(\vec{r}',\tau)}{|\vec{r}-\vec{r}'|}\mathrm{d}V'=\frac{Z_0 q}{4\pi c}\frac{\vec{v}'(\vec{r}',\tau)}{|\vec{r}-\vec{r}'|},$$

where again r' is the position of the particle from the origin, and $\vec{v}'=\frac{\mathrm{d}}{\mathrm{d}\tau}\vec{r}'$ is the velocity of the particle.

As with the antennas, we first consider a distant stationary observer at a position $\vec{r}$, where the observed radiation will be approximately a spherical wave whose wavefront can be locally approximated by a plane wave. Thus we can apply the result from Thought Experiment 16.9 (p. 535) as:

$$\vec{\mathbb{S}}(r,t)=\frac{1}{Z_0}\left|\frac{r_0}{r}\frac{\partial}{\partial t}\vec{\mathbb{A}}_\perp(r_0,t)\right|^2\hat{r}\Bigg|_{t=\tau}=\frac{1}{Z_0 r^2}\left|\left[r_0\frac{\partial}{\partial t}\vec{\mathbb{A}}_\perp(r_0,t)\right]_{\text{RET}}\right|^2\hat{r}$$

$$\mathbb{P}_\Omega(t)=r^2\mathbb{S}(r,t)=\frac{1}{Z_0}\left|r_0\frac{\partial}{\partial t}\vec{\mathbb{A}}_\perp(r_0,t)\right|^2\Bigg|_{t=\tau}=\frac{1}{Z_0}\left|\left[r_0\frac{\partial}{\partial t}\vec{\mathbb{A}}_\perp(r_0,t)\right]_{\text{RET}}\right|^2.$$

Similarly, the perpendicular component of the vector potential is:

$$\vec{\mathbb{A}}_\perp(\vec{r},t)=\left(\frac{Z_0 q}{4\pi c}\frac{\vec{v}'(\vec{r}',t)}{|\vec{r}-\vec{r}'|}\right)_\perp\Bigg|_{t=\tau}\approx\frac{Z_0 q}{4\pi r c}\left[\vec{v}'_\perp(\vec{r}',t)\right]_{\text{RET}}.$$

And taking its derivative as a function of time, we have:

$$\frac{\partial}{\partial t}\vec{\mathbb{A}}_\perp(\vec{r},t)=\frac{Z_0 q}{4\pi r c}\frac{\partial}{\partial t}\vec{v}'_\perp(\vec{r}',t)=\frac{Z_0 q}{4\pi r c}\left[\vec{a}'_\perp(\vec{r}',t)\right]_{\text{RET}}=\frac{Z_0 q}{4\pi r c}\vec{a}'_\perp(\vec{r}',\tau).$$

Where $a'_\perp(\vec{r}',\tau)$ is the acceleration of the particle at the retarded time, which we will simply denote as $\vec{a}_\perp$ from now on.

Putting the last equations together, we can now write:

$$\vec{\mathbb{S}}(r,t)=\frac{Z_0 q^2}{(4\pi)^2 r^2 c^2}|\vec{a}_\perp|^2\hat{r}=\frac{Z_0 q^2}{(4\pi)^2 r^2 c^2}|\hat{r}\times\vec{a}\times\hat{r}|^2\hat{r}.$$

Now consider the special case where the acceleration is along the $\hat{z}$ axis:

$$\vec{\mathbb{S}}(r,t)=\frac{Z_0 q^2}{(4\pi)^2 r^2 c^2}|\hat{r}\times a\hat{z}\times\hat{r}|^2\hat{r}=\frac{Z_0 q^2 a^2}{(4\pi)^2 r^2 c^2}\sin^2\theta\,\hat{r}.$$

Now we notice something very interesting. This beam pattern has the same angular dependence as the Hertzian dipole!

To find the total power, we integrate this over a sphere of constant radius:

$$\mathbb{P} = \oint \vec{\mathbb{S}} \cdot d\vec{A} = \int_0^{\pi} \int_0^{2\pi} \left(\frac{Z_0 q^2 a^2}{(4\pi)^2 \cancel{r^2} c^2} \sin^2\theta \right) \left(\cancel{r^2} \sin(\theta) d\varphi \, d\theta \right) = \left(\frac{Z_0 q^2 a^2}{8\pi c^2} \right) \int_0^{\pi} \sin^3(\theta) d\theta = \frac{Z_0 q^2 a^2}{6\pi c^2}.$$

The last expression is called the *Larmor Formula*,[276] and it famously shows that in its own frame of reference, the power radiated by a charged particle is proportional to its acceleration squared.

Problem 16.14: Radiation from an Oscillating Particle

For a particle in simple harmonic motion, the acceleration is related to the position by $a_z = -\omega^2 z$, so $qa = -q\omega^2 z = -\omega^2 p_z$, where p_z is the dipole moment.

a) Show that the amplitudes of the acceleration and dipole moment are related by $a_0 = \frac{\omega^2}{q} p_0$.

b) Show that the radiation pattern is consistent with that of a Hertzian dipole (p. 520): $\left\langle \vec{\mathbb{S}}(r,t) \right\rangle = \frac{Z_0 \omega^4 p_0^2}{2(4\pi)^2 c^2} \frac{\sin^2\theta}{r^2} \hat{r}$.

c) Integrate this expression over a full sphere, and compare it to the Larmor formula. Show that they are consistent with each other.

Thought Experiment 16.13: The Radiation Reaction Force

Consider now a slowly moving particle of charge q_1 and mass m_1. Something else, call it object 2, applies a force $\vec{F}_{21}$ on the particle. What is the net force on particle 1?

This seems like a silly question, since it is clearly just $\vec{F}_{21}$! But remember that conservation of momentum and energy are even more fundamental than Newton's idea of force. Since the particle carries charge, we expect it to radiate when accelerated, and we already know that light carries both energy and momentum. And, finally, Newton's second law says that $\sum \vec{F}_{ext} = \frac{d\vec{p}}{dt}$, so just like there is a force on a rocket when it expels its fuel, perhaps there is a force on the electron as it expels light. Thus, the answer to our silly question may not be so obvious after all!

Let us now call this force, due to the radiation, $\vec{F}_{R1}$. So the net force on particle 1 would be:

$$\vec{a}_1 = \frac{\vec{F}_{21} + \vec{F}_{R1}}{m_1}.$$

Now the total power radiated by particle 1 is given by:

$$\mathbb{P}_1 = \frac{Z_0 q_1^2 a_1^2}{6\pi c^2} = \frac{Z_0}{6\pi c^2} \left(\frac{q_1}{m_1} \right)^2 \left| \vec{F}_{21} + \vec{F}_{R1} \right|^2.$$

But this must also equal the work done on particle 1 minus its change in kinetic energy, so:

$$\mathbb{P}_1 = \vec{F}_{21} \cdot \vec{v}_1 - \tfrac{d}{dt}\mathbb{T} = \vec{F}_{21} \cdot \vec{v}_1 - \tfrac{d}{dt}\left(\tfrac{1}{2} m_1 \vec{v} \cdot \vec{v} \right) = \vec{F}_{21} \cdot \vec{v}_1 - m_1 \vec{v}_1 \cdot \vec{a}_1 = \vec{F}_{21} \cdot \vec{v}_1 - \cancel{m_1} \vec{v}_1 \cdot \left(\tfrac{1}{\cancel{m_1}} \left(\vec{F}_{21} + \vec{F}_{R1} \right) \right) = -\vec{v}_1 \cdot \vec{F}_{R1}.$$

---

[276] J. Larmor, "On a dynamical theory of the electric and luminiferous medium," Philosophical Transactions of the Royal Society **190**, (1897) 205–300.

Now we can equate this last result to the Larmor expression:

$$\mathbb{P}_1 = -\left(\vec{v}_1 \cdot \vec{F}_{\mathrm{R1}}\right) = \frac{Z_0\, q_1^{\,2} a_1^{\,2}}{6\pi\, c^2},$$

and solve for the *Radiation Reaction Force* of a slowly moving particle:

$$\vec{F}_{\mathrm{R1}} = -\frac{Z_0\, q_1^{\,2}}{6\pi\, c^2}\left(\frac{a_1}{v_1}\right)^2 \vec{v}_1 .$$

This force, as per the derivation, opposes the direction of motion. Notice something else. In the special case where the ratio of the magnitudes of the acceleration to the velocity is constant, such as circular motion, this can be approximated as a standard damping force.

As with many results in electrodynamics, this force runs afoul of Galilean relativity. While force is a Galilean invariant, velocity certainty is not, so how can a fundamental force depend on velocity? Contact forces, such as air drag, routinely depend on velocity, but not fundamental forces. Thus, if there is an aether our result makes perfect sense. In the post-aether era, however, physicists would naturally look to generalize the result differently even if doing so ran afoul of Newton's laws of motion.

Discussion 16.5: The Abraham-Lorentz Force

In this chapter we have discussed the development of *electron theory*. This was an early attempt at what we now would call particle physics, in which physicists try to unify the ideas of fields and particles. Now this is done using quantum fields, but then it was done classically using the electromagnetic field. Physicists, such as Hendrik Lorentz, attempted to better understand the electron by analyzing the nearby fields produced by electrons and their self-energy.

At the time, atomic spectral lines were understood as due to the resonance frequencies of electrons in atoms. The obvious place to investigate this electron theory would, therefore, be the simple harmonic oscillator.

In the case of an electron in one-dimensional simple harmonic motion, we can write the average power radiated from the Larmor formula as:

$$\langle \mathbb{P} \rangle = \frac{Z_0}{12\pi}\frac{p_0^2\,\omega^4}{c^2}.$$

Now, on average, the power must also be $\langle \mathbb{P} \rangle = \langle -\vec{F}\cdot\vec{v} \rangle$, thus the radiation reaction force for a charged particle in simple harmonic motion could be written:

$$F = -\frac{Z_0\, q^2}{6\pi\, c^2} x_0 \omega^3 \cos(\omega t),$$

if the position of the charged particle follows $x_0 \sin(\omega t)$.

As you can clearly see, the last expression is simply the third derivative of the position, thus the radiation reaction force is often written as:

$$\vec{F} = \frac{Z_0\, q^2}{6\pi\, c^2}\frac{\mathrm{d}^3\vec{r}}{\mathrm{d}t^3},$$

which is called the Abraham-Lorentz force.[277]

This expression, too, has its problems. Consider a charged object undergoing constant acceleration. According to the Larmor formula this object will lose energy, but according to the radiation reaction force formula no work would be done. From the viewpoint of Newtonian physics, this makes absolutely no sense! Clearly, this result could not be fundamental either.

The problem was shelved for the next thirty years, since physicists were preoccupied with developing relativity and quantum mechanics. In 1938 Paul Dirac picked it back up again—albeit now in a fully relativistic manner. Like the Abraham-Lorentz force, Dirac's relativistic version is proportional to the second derivative of the relativistic momentum with respect to the proper time. Thus, the Abraham-Lorentz force is often touted as the one force that is directly proportional to the jerk of the motion.

Problem 16.15: The Radiation Reaction Force and Momentum

Consider a particle of mass m and charge q at the origin at time zero.

a) If the particle is momentarily at rest at the origin, but experiencing an externally applied force $\vec{F}_{\text{ext}} = F_{\text{ext}}\,\hat{z}$, what is the emitted Poynting vector $\vec{\mathbb{S}}$ as a function of position and time?

b) Find the momentum density of the light, also as a function of position and time.

c) Find the total momentum of the light contained in a spherical shell with a radius $r = ct$ and small thickness δr. Show that this is zero.

d) Now we will change the problem, and assume that the particle has its own momentum $\vec{p}$, but is still experiencing an externally applied force. Now what is the Poynting vector $\vec{\mathbb{S}}$ of the radiated light as a function of position and time? What is its momentum density?

e) Again, find the total momentum of the light contained in a spherical shell with a radius $r = ct$ and small thickness δr. This will not be zero.

f) How much total momentum departed from the particle during the time interval $\delta t = \frac{1}{c}\delta r$?

g) Use this result to derive the radiation-reaction force on the particle. Show that this agrees with the result from Thought Experiment 16.13.

Problem 16.16: Cyclotron Radiation

Consider a non-relativistic electron with an initial velocity $\vec{v}\big|_{t=0} = v_0\,\hat{y}$ located within a constant, and uniform, magnetic field $\vec{B} = B\,\hat{z}$.

a) Find the acceleration, $\vec{a}_0$, of the electron at time zero only due to the magnetic field.

b) Define the origin such that the initial position is located at $\vec{r} = s_0\hat{x}$, where s_0 is the initial radius of curvature. What is s_0 in terms of the initial velocity and the magnetic field?

---

[277] The final chapter of J.D. Jackson's, Classical Electrodynamics, nicely discusses the Abraham-Lorentz model. See also, F. Rohrlich, "The dynamics of a charged sphere and the electron," Am. J. Phys., **65** (11) (1997), 1051-1056. Rohrlich provides historical context for the Abraham-Lorentz model, including a discussion of contributions by other physicists such as Larmor, Heaviside, Poincare, and Dirac to the classical model of the electron.

c) Show, to a reasonable approximation, that $(\tfrac{q}{v})^2 \approx (\tfrac{e}{m}B)^2$. Justify any assumptions you make.

d) Write down the acceleration vector (both components) as a function of velocity and position.

e) Find the speed of the electron as a function of time.

f) What is the radius of the orbit as a function of time? Does it increase or decrease with time?

g) Find the radial velocity as a function of time.

h) Find the kinetic energy of the electron as a function of time, and take its time derivative to find the rate that energy is lost from electron.

i) Find the total power radiated away as a function of time.

j) Show that energy is conserved in this system.

Example 16.9: The Thomson Cross Section

Consider the interaction of light with a free charged particle of known mass and charge. What is the effective cross-sectional area of the particle?

The fields of the light wave are related to each other and the propagation unit vector by:

$$\vec{B} = \tfrac{1}{c}\hat{k}\times\vec{E} \quad , \quad \vec{E} = -c\,\hat{k}\times\vec{B} \ ,$$

so in terms of the electric field we can write the force on the electron as:

$$\vec{F} = q\left(\vec{E}+\vec{v}\times\vec{B}\right) = q\left(\vec{E}+\tfrac{1}{c}\vec{v}\times\left(\hat{k}\times\vec{E}\right)\right) \approx q\vec{E} \ ,$$

so long as the velocity of the electron remains small. Similarly, we will also ignore the radiation reaction force, and find the acceleration of the electron simply as:

$$\vec{a} \approx \tfrac{q}{m}\vec{E} \ .$$

Applying the Larmor formula, we find the total power radiated by the electron as:

$$\mathbb{P} = \frac{Z_0 q^2 a^2}{6\pi c^2} = \frac{Z_0 q^2}{6\pi c^2}\left(\tfrac{q}{m}E\right)^2 = \frac{Z_0 q^4}{6\pi m^2 c^2}E^2 = \frac{Z_0 q^4}{6\pi m^2 c^2}\left(Z_0\mathbb{S}\right) = \frac{Z_0^2 q^4}{6\pi m^2 c^2}\mathbb{S} \ .$$

The effective area of a particle is called the cross-section, σ, which is:

$$\sigma = \frac{\mathbb{P}}{\mathbb{S}} = \frac{Z_0^2 q^4}{6\pi m^2 c^2} = \frac{1}{6\pi}\left(\frac{Z_0 q^2}{mc}\right)^2 = \frac{8\pi}{3}\left(\frac{Z_0 q^2}{4\pi c\, m}\right)^2 .$$

We now see that the larger the mass, the smaller the cross-section. In a primarily hydrogen plasma, such as the sun, the protons and electrons are in equal proportion. This means that the scattering of light off of electrons will dominate the scattering of light off of protons. This cross section of electrons is called the *Thomson cross-section*, σ_T, and given by:

$$\sigma_T = \frac{8\pi}{3}\left(\frac{Z_0 e^2}{4\pi c\, m_e}\right)^2 = \tfrac{8}{3}\pi r_e^2 = 6.652458734\times 10^{-29}\,\mathrm{m}^2 \approx 0.65\times 10^{-28}\,\mathrm{m}^2,$$

where the constant r_e is called the *classical radius of the electron*, which is the radius an electron would have if its electrostatic self-energy were equal to its rest mass energy.

Nuclear physicists use a unit of area called a *barn*, which is $10^{-28}\,\text{m}^2$. The name is a joke: shooting something that big would be like hitting the side of a barn.

Problem 16.17: The Eddington Luminosity

If a massive star is luminous enough, it can blow off its atmosphere because the outward force due to radiation pressure exceeds the gravitational force. (Recall the solar sail on p. 489.) You can assume that the number density of electrons equals the number density of protons, which you can call n.

a) Find the outward body force of the radiation on the free electrons, as a function of the distance from the center of the star, and the star's luminosity.

b) Find the inward gravitational body force on the free protons as a function of the distance from the center of the star, the and star's mass.

c) The *Eddington Luminosity* represents the largest luminosity that a star can have before it blows off its atmosphere. Express this as a function of the stellar mass. Notice that this is nether a function of number density nor radius.

d) Express your answer in solar units (see p. 489). What is the Eddington luminosity of a one solar mass star? What about a ten solar mass star?

e) Here you compared the outward force on a free electron with the inward force on a free proton. This seems like comparing apples to oranges. Explain why that was actually the right thing to do.

f) Research the Eddington luminosity. Who was Arthur Eddington, and what are some examples of objects that have a luminosity at, or above, their Eddington luminosity? What does this have to do with the mass-luminosity relationship for main sequence stars?

Example 16.10: Higher Order Thomson Scattering

Now we will redo the last example, but keep the terms we ignored before. Again, the force on the electron due to the incident radiation is:

$$\vec{F}_{\text{Fields}} = q\left(\vec{E} + \vec{v}\times\vec{B}\right) = q\left(\vec{E} + \tfrac{1}{c}\vec{v}\times\left(\hat{k}\times\vec{E}\right)\right) = q\left(\vec{E} + \tfrac{1}{c}\hat{k}\left(\vec{v}\cdot\vec{E}\right) - \vec{E}\left(\vec{v}\cdot\hat{k}\right)\right),$$

Similarly, we will write the radiation reaction force as:

$$\vec{F}_{\text{RR}} = -\frac{Z_0 q^2}{6\pi c^2}\left(\frac{a}{v}\right)^2 \vec{v} = -\frac{2}{3}\left(\frac{Z_0 q^2}{4\pi c m}\right)\left(\frac{m}{c}\right)\left(\frac{a}{v}\right)^2 \vec{v} = -\tfrac{2}{3} m_e r_e \left(\frac{a}{v}\right)^2 \frac{\vec{v}}{c},$$

Thus the net force on the electron is:

$$\vec{F} = q\left(\vec{E} + \tfrac{1}{c}\hat{k}\left(\vec{v}\cdot\vec{E}\right) - \vec{E}\left(\vec{v}\cdot\hat{k}\right)\right) - \tfrac{2}{3} m_e r_e \left(\tfrac{a}{v}\right)^2 \tfrac{1}{c}\vec{v}\,.$$

The acceleration of the electron is therefore:

$$\vec{a} = \tfrac{q}{m_e}\left(\vec{E} + \tfrac{1}{c}\hat{k}\left(\vec{v}\cdot\vec{E}\right) - \vec{E}\left(\vec{v}\cdot\hat{k}\right)\right) - \tfrac{2}{3} r_e \left(\tfrac{a}{v}\right)^2 \tfrac{1}{c}\vec{v}\,.$$

Once we have an equation of motion, it is straight-forward to simulate the electron's motion. That said, let us now make a few simplifying assumptions to get a handle on this. As the electron is engaged in simple harmonic motion in the direction perpendicular to $\hat{k}$, with a frequency ω, $a \approx \omega v$ because:

$$a_\perp \approx a_0 \sin(\omega t + \phi) \approx \tfrac{dv}{dt} \text{ and } v_\perp \approx -\tfrac{1}{\omega} a_0 \cos(\omega t + \phi) \text{ so: } \tfrac{a}{v} \approx \tfrac{|a_\perp|}{|v_\perp|} \approx \omega \left|\tan(\omega t + \phi)\right| \approx \omega .$$

The acceleration, then, becomes:

$$\vec{a} \approx \tfrac{q}{m_e}\left(\vec{E}\left(1 - \tfrac{1}{c}\vec{v}\cdot\hat{k}\right) + \tfrac{1}{c}\hat{k}\left(\vec{v}\cdot\vec{E}\right)\right) - \tfrac{2}{3}\tfrac{1}{c} r_e \omega^2 \vec{v} ,$$

so the radiation-reaction force acts much like a laminar drag factor with a drag constant proportional to the driving frequency squared.

Notice that the electron velocity is primarily perpendicular to the direction of the light propagation, so the acceleration in the direction of the light wave would be given by:

$$a_\parallel \approx \tfrac{q}{m_e c}\left(\vec{v}\cdot\vec{E}\right) - \tfrac{2}{3}\tfrac{1}{c} r_e \omega^2 v_\parallel ,$$

and for an electron $q = -e$.

Now, the acceleration perpendicular to the direction of propagation is:

$$\begin{aligned} a_\perp &\approx -\tfrac{q}{m_e} E_0 \sin(\omega t + \phi)\left(1 - \tfrac{1}{c} v_\parallel\right) - \tfrac{2}{3}\tfrac{1}{c} r_e \omega^2 \tan^2(\omega t + \phi)\left(\tfrac{1}{\omega} a_0 \cos(\omega t + \phi)\right) \\ &\approx -a_0\left(1 - \tfrac{1}{c} v_\parallel\right)\sin(\omega t + \phi) - a_0 \sin(\omega t + \phi)\tfrac{2}{3}\tfrac{1}{c} r_e \omega \tan(\omega t + \phi) \\ &\approx -a_0 \sin(\omega t + \phi)\left(1 - \tfrac{1}{c} v_\parallel + \tfrac{2}{3}\tfrac{\omega}{c} r_e \tan(\omega t + \phi)\right). \end{aligned}$$

Now let us look at the acceleration term parallel to the direction of motion again, which is dominated by a term of $\vec{v}\cdot\vec{E}$, which we can now approximate by:

$$\begin{aligned} a_\parallel &\approx -\tfrac{e}{m_e c}\vec{v}\cdot\vec{E} \approx -\tfrac{e}{m_e c} v_\perp E \approx -\tfrac{e}{m_e c} v_\perp\left(-\tfrac{m_e}{e} a_\perp\right) \approx \tfrac{1}{c} v_\perp a_\perp \\ &\approx \tfrac{1}{c}\left(-\tfrac{1}{\omega} a_0 \cos(\omega t + \phi)\right)\left(a_0 \sin(\omega t + \phi)\left(1 - \tfrac{1}{c} v_\parallel + \tfrac{2}{3}\tfrac{\omega}{c} r_e \tan(\omega t + \phi)\right)\right) \\ &\approx a_0^2 \tfrac{1}{\omega c}\left(\sin(\omega t + \phi)\cos(\omega t + \phi)\right)\left(1 - \tfrac{1}{c} v_\parallel + \tfrac{2}{3}\tfrac{\omega}{c} r_e \tan(\omega t + \phi)\right) \\ \left\langle a_\parallel \right\rangle &\approx a_0^2 \tfrac{1}{\omega c}\left\langle \tfrac{2}{3}\tfrac{\omega}{c} r_e \sin^2(\omega t + \phi)\right\rangle \approx a_0^2 \tfrac{2}{3}\tfrac{r_e}{c^2}\left\langle \sin^2(\omega t + \phi)\right\rangle \approx \tfrac{1}{3} a_0^2 \tfrac{r_e}{c^2} \end{aligned}$$

So, there is a net force in the opposite direction of the light propagation. If we make the approximation that $a_0 \approx \tfrac{e}{m_e} E_0$, then we can write:

$$a_0^2 \approx \left(\tfrac{e}{m_e}\right)^2 E_0^2 \approx \left(\tfrac{e}{m_e}\right)^2 E_0^2 \approx \left(\tfrac{e}{m_e}\right)^2\left(2Z_0\langle\mathbb{S}\rangle\right) \approx \left(\tfrac{2e^2 Z_0}{m_e^2}\right)\langle\mathbb{S}\rangle .$$

And therefore the average parallel acceleration would become:

$$\left\langle a_\parallel \right\rangle \approx \tfrac{1}{3}\tfrac{r_e}{c^2}\left(\tfrac{2e^2 Z_0}{m_e^2}\right)\langle\mathbb{S}\rangle \approx \tfrac{8\pi}{3}\tfrac{r_e}{m_e c}\left(\tfrac{e^2 Z_0}{4\pi m_e c}\right)\langle\mathbb{S}\rangle \approx \tfrac{1}{m_e c}\tfrac{8\pi r_e^2}{3}\langle\mathbb{S}\rangle \approx \tfrac{\sigma_T}{m_e c}\langle\mathbb{S}\rangle .$$

The parallel component of the acceleration is directly proportional to the energy flux. Another way to look at this is that the momentum carried by the light exerts a pressure. As the electron has an effective cross-section of σ_T, the force per electron would simply be given by the pressure multiplied by the area, which is the result we have obtained.

Problem 16.18: Free Electrons in the Sun

In Example 16.10, we derived the acceleration for an electron in a radiation field to be as follows:

$$\vec{a} = \frac{q_e}{m_e}\left(\vec{E} + \frac{1}{c}\hat{k}\left(\vec{v}\cdot\vec{E}\right) - \vec{E}\left(\vec{v}\cdot\hat{k}\right)\right) - \frac{2}{3}r_e\left(\frac{a}{v}\right)^2\frac{1}{c}\vec{v}.$$

Adding a constant gravitational field, $\vec{g}$, term, we now have:

$$\vec{a} = \frac{q_e}{m_e}\left(\vec{E} + \frac{1}{c}\hat{k}\left(\vec{v}\cdot\vec{E}\right) - \vec{E}\left(\vec{v}\cdot\hat{k}\right)\right) - \frac{2}{3}r_e\left(\frac{a}{v}\right)^2\frac{1}{c}\vec{v} + \vec{g}.$$

Your goal is to simulate the electron's motion as applied to the sun.

a) Render the equation above into Cartesian Coordinates, assuming that $\hat{k} = \hat{y}$, $\vec{E} = E_0 \sin(\omega t)\hat{x}$, and $\vec{g} = -g\hat{y}$. Simplify to find equations for a_x and a_y as a function of v_x, v_y, a/v, and t.

b) As round off errors can add up, it is often important to work in units near unity. The most elegant way to this is to use *natural units*. Let us therefore work in units where the following constants are set to the following values: $e = m_e = c = r_e = 1$. For example, we can now write our unit for charge as $[q_e] = e$, and we would therefore write for an electron $q_e = -1e$ but code into a computer simply q = -1. Combine these quantities to make expressions for the unit mass, time, length, velocity, acceleration, angular frequency, and electric field. What are each of these, including the unit charge, in the SI?

c) Factor out of your equations of motion (part a) the unit acceleration, and write each parameter that you can control $(E_0, \omega, \text{and } t)$ in ratio to its own unit quantity.

d) We have many examples in this text where we numerically solve equations of motion. Diagram the algorithm you will use to iteratively solve for the position and velocity of the electron as a function of time.

e) Code up your simulation, and run it for values consistent with the sun's surface. In other words, express your user input values for the energy flux and frequency in solar units, with the default values being:

$$\vec{\mathbb{S}} = 1\left(\frac{L_\odot}{4\pi R_\odot^2}\right)\hat{y}, \quad \omega = 1\left(\frac{2\pi c}{500\,\text{nm}}\right), \quad \text{and} \quad \vec{g} = 1\left(\frac{GM_\odot}{R_\odot^2}\right)(-\hat{y}).$$

f) Run your simulation, and plot the position and velocity of the electron at is position. You can do this with separate plots, or you can make a single quiver plot that shows both the position and velocity. Does the electron tend to go up or down over time? Discuss how this may relate to the solar wind. Are there other forces at work in the sun that we have ignored?

g) So far you have ignored the Coulomb interaction force between electrons and protons. Now, assume that this Coulomb force is large, so that packets of gas are net neutral. Given that the sun is about 75% hydrogen and about 25% helium, find the average mass per electron.

h) Rerun your simulation with this much larger effective mass per electron. Now does the surface plasma go up or down? Explain.

i) Recall the Eddington Luminosity from Problem 16.17 (p. 558). Now simulate the gas on the surface of an Eddington Luminosity star. Does the gas act as you would expect it to? Explain.

Plate 29: The Very Large Array[278]

[278] The Very Large Array radio interferometer (Thought Experiment 16.11, p. 543) is located in the United States near the village of Magdalena, New Mexico, and consists of 28 parabolic dish antennas (Example 16.8, p. 540). Each antenna is 25 meters in diameter, weighs 230 tons, and sits on a double pair of railroad tracks. At any given time, twenty seven antennas observe a single source, where the signals from each pair of antennas are combined, both in and out of phase, before measuring their correlated power. These complex visibility data (Discussion 16.4, p. 547) are then expressed as a function of the projected baseline, which is the distance between the antennas in ratio to the wavelength being observed. The spacing between adjacent antennas is sometimes as close as 100 meters, as in the picture, to as far as 4 kilometers, allowing for a broad range of angular resolution at centimeter wavelengths. When these data are calibrated (Problem 16.10, p. 547), their Fourier transform represents a map of the sky (Plate 31, p. 639). Each antenna observes both left and right polarization (Thought Experiment 15.11, p. 496), allowing for maps in all four Stokes parameters (Thought Experiment 15.12, p. 497) that, in the right circumstances, can map the magnetic field direction (Plate 27, p. 505). This photo was downloaded from the website of The National Radio Astronomy Observatory, and it was taken from the top of an antenna by B. Saxton.

Plate 30: Participants in the 1911 Solvay Conference

Standing from left to right: Robert Goldschmidt, Max Planck, Heinrich Rubens, Arnold Sommerfeld, Frederick Lindemann, Maurice de Broglie, Martin Knudsen, Friedrich Hasenöhrl, Georges Hostelet, Édouard Herzen, James Jeans, Ernest Rutherford, Kamerlingh Onnes, Albert Einstein and Paul Langevin.

Seated from left to right: Walther Nernst, Marcel Brillouin, Ernest Solvay,[279] Hendrik Lorentz, Emil Warburg, Jean Perrin, Wilhelm Wien, Marie Curie and Henri Poincaré.

[279]If you look carefully, you will see that Solvay's picture is pasted onto the group photo.

Chapter 17 Special Relativity

Should the local power grid be run using alternating or direct current? On the one hand, it is easier to store direct current in batteries. On the other hand, it is easier to step up, or down, the voltage of alternating current. While a large company may have the means to invest in both standards, a small company must choose one, or the other. The power engineering firm, Einstein & Cie, was one such company.

Founded in 1880 by the brothers Jakob and Hermann Einstein, Einstein & Cie built electric generators in Munich. Their business took off throughout the 1880s as DC generators remained in high demand. Unfortunately, these successful years also coincided with the invention of practical closed-coil transformers and AC induction motors. By the mid 1890s, the city of Munich had switched to AC power, and Einstein & Cie went bankrupt. Jacob and Hermann moved their families to two different Italian cities to found new firms, while Hermann's teenage son, Albert, stayed in Munich supposedly to finish his studies at the prestigious Luitpold Gymnasium. This did not last long, as Albert found the formal instruction uninteresting and soon joined his parents in Italy a few months later, without earning his diploma. This gave him much more freedom to follow his interests, and after about a year he moved to Switzerland to continue his studies—but this time in mathematical physics.

After earning a degree, but still working on his doctoral thesis, Albert could not find a teaching position. He did, however, use his technical background to secure a well-paying job working in a Swiss patent office. During this time, Einstein appeared to be at his most happy and productive. His patent office job was not too demanding, which allowed him to devote thought to questions of natural philosophy. At the same time, Einstein was a member of a small book group that read current works on the philosophy of science. Among these works was Henri Poincaré's book, Science and Hypothesis, which outlined many of the inconsistencies in theoretical physics at that time.

Einstein's miracle year came in 1905 when he published four papers that forever altered physics. Einstein's first paper made the radical proposal that light is not a wave after all, but rather behaves as if it were a particle. This made little sense to other physicists of the day, especially since it appeared to contradict everything nineteenth century scientists had learned about interference. Yet Einstein made testable predictions in this paper, and their later experimental confirmation by Robert A. Millikan would pull down the edifice of Maxwell's wave theory of light. Einstein's second 1905 paper gave a statistical proof that atoms and molecules really exist based on an analysis of Brownian motion, and it was this paper that would become his doctoral dissertation on the determination of Avogadro's number and the size of molecules. Einstein's third and fourth papers are the subject of this chapter.

Albert Einstein's paper *On The Electrodynamics of Moving Bodies* answered the question that dogged theoretical physics throughout the late nineteenth century: "What is the nature of the aether?" Einstein's solution was simple: there is no aether. It simply does not exist. Instead, he asserted, Galileo's principle of relativity (but not his transformation equations) is equally true for electrodynamics and mechanics.

The two fundamental postulates that Einstein presented as the foundation of his theory are:

> The laws governing the changes of the state of any physical system do not depend on which one of two coordinate systems in uniform translational motion relative to one another these changes are referred to.

> Every ray of light moves in the co-ordinate system "at rest" with the definite velocity V independent of whether this ray of light is emitted by a body at rest or in motion. [280]

The first postulate holds that electrodynamics, like mechanics, ought to conform to the principle of relativity. There is no privileged inertial frame, such as the rest frame of the aether, to which all laws of electrodynamics must be referred. All inertial frames, i.e. frames moving uniformly with respect to one another, are equally correct for describing electrodynamics as well as mechanics. Far from overthrowing Galileo or Newton, Einstein was extending the principle of relativity to all of physics. No violation of this principle has ever been found.

Einstein's second postulate is our Fundamental Law 10 (p. 483), which asserts that light propagates in a vacuum at the same speed $(V = c)$ regardless of the motion of the source or the observer. At the time, it was simply an inspired guess suggested by the form of Maxwell's wave equation (see Discussion 15.3, p. 483).

This assumption would, at first sight, appear to conflict with the principle of relativity. If a light source and an observer are in uniform relative motion with a constant velocity v, then the Galilean transformations demand that the speed of light measured by the observer is not simply equal to c, but instead equals $c \pm v$. Yet, this is exactly what Einstein's second postulate denies! If Einstein's assertions are true, then the second postulate must be reconciled with the principle of relativity and a new set of transformations must supplant the Galilean transformations. As we mentioned in Chapter 15 (pp. 478 and 483), such a set of transformations, called the *Lorentz-FitzGerald transformations*, does indeed exist and we will see how these both preserve Maxwell's theory and allow Einstein's two postulates to coexist.

Other than the Michelson-Morley experiment, there was scant direct experimental evidence for Einstein's postulate of the universality of c. However, his theory as a whole did explain such diverse phenomena as the Doppler effect, Fresnel's drag coefficient, and stellar aberration.

Einstein's second postulate has been confirmed many times since, which have put tiny upper limits on the value of a hypothetical k parameter $\left(k = \frac{|c'-c|}{v}\right)$. These experiments include: de Sitter's 1913 observations[281] of the light received from binary stars,[282] a 1964 pion decay measurement, $k < 1\times10^{-4}$,[283] and a 1977 observation of x-ray sources in binary star systems, $k < 2\times10^{-9}$.[284] We may, therefore, be confident that the speed of light is independent of the observer's reference frame.

---

[280] From "On the Electrodynamics of Moving Bodies," by Albert Einstein (1905), translated by Anna Beck, ©1989 by the Hebrew University of Jerusalem.

[281] W. de Sitter, "On the constancy of the velocity of Light," Proceedings of the Royal Netherlands Academy of Arts and Science, **16** (1913), pp. 395-396.

[282] J. G. Fox pointed out in the 1960s that light emitted by binary stars must pass through surrounding gases making de Sitter's conclusion premature because his measurements would be independent of the original motion of the binary stars (Am. J. Phys. **30** (1962), p. 297 and J. Opt. Soc. **57** (1967), p. 967).

[283] T. Alveger, et al. Phys. Letters, **12** (1964), 260.

[284] K. Brecher, "Is the speed of light independent of the velocity of the source?", Physical Review Letters, **39** (17) (1977), pp. 1051–1054.

Rather than focusing on the aether, which Einstein discarded as an unnecessary abstraction, his special theory of relativity focuses instead on what can, at least in principle, be observed. Consider this key passage from Einstein's introduction to his original paper:

> Like every other electrodynamics, the theory to be developed is based on the kinematics of the rigid body, since assertions of each and any theory concern the relations between rigid bodies (coordinate systems), clocks, and electromagnetic processes.[285]

This need to base theory on measurement led Einstein to ask what we mean by saying two events are simultaneous, and to demand an operational definition of simultaneity: one that stresses the use of clocks, rigid measuring rods, and light signals. In this manner, the old assumption of an absolute time that flows evenly and universally was revealed to be nothing more than metaphysical prejudice.

The aether, which had been retained from an overly mechanistic model of electromagnetic interactions, also proved unnecessary. Over and over, a predicted effect of such an aether had been contradicted by experiment, and every time this happened scientists resorted to more exotic refinements of the aether model. With no compelling evidence that it even existed, the aether—like the notion of absolute time—seemed to arise from an uncritical acceptance of common sense. The aether was a superfluous notion at best, Einstein argued, and the science of electricity and magnetism could dispense with it.

Even matter's very existence, at least as something distinct from energy, came into doubt with Einstein's relativity. Einstein's fourth 1905 paper addressed the question of relativistic dynamics and the equivalence of mass and potential energy.

We begin this chapter with Einstein's postulates, and use a series of thought experiments in order to illustrate the theory of special relativity and relativistic electrodynamics. In Section 17.1, we not only show that measurements of length and time depend on the observer's frame of reference, but that the order of events can also be relative. This interdependence of space and time naturally leads to denoting the time and location of an event with a single 4-vector, making the Lorentz-FitzGerald transformations mathematically similar to a rotation in spacetime.

In Section 17.2, we reproduce our most fundamental thought experiments from early in this text—but with a completely relativistic approach. We begin with a third axiom, that charge is both conserved and independent of reference frame. Just as we derived the electrodynamic Galilean transformations by considering moving rods, capacitors, and inductors, we now derive the corresponding electrodynamic Lorentz-FitzGerald transformations. Moreover, we also use simple moving capacitors and inductors to derive the relativistic energy and momentum relations, including the famous equation $\mathbb{E}_0 = mc^2$.

Finally, in Section 17.3, we introduce Einstein summation notation that is common in both special and general relativity. This leads to a discussion of polar and axial vectors, and covariant and contravariant tensors. It is in this form that Maxwell's set of equations most elegantly displays the interdependence between the electric and magnetic fields, and the unique nature of light as a wave without a medium.

---

[285] A. Einstein, "On the Electrodynamics of Moving Bodies," Annalen der Physik **17** (1905), 891-921, Translated by Anna Beck, in *The Collected Papers of* Albert Einstein, Volume 2, English Translation, ©1989 by the Hebrew University of Jerusalem

17.1 *Relativistic Kinematics*

Einstein's theory is rooted in electrodynamics, but the combination of his two postulates requires us to revise drastically our understanding of time and space. The principle of relativity, and the existence of a universal signaling speed c, force us to abandon naïve notions about simultaneity because we must use light signals to synchronize time measurements.

The concept is similar to that of retarded time from Chapter 16, but with an additional constraint. Recall the superluminal quasar jet from Problem 16.12 (p. 550), for example. Since it was travelling toward us, in our reference frame less time passed between each observation than in the reference frame of the source. Imagine that at the same time the blob of material left the core an identical blob was ejected in the opposite direction. At any given time here on Earth, t, the retarded time of each blobs, τ_1 and τ_2 and would be increasingly different. So, in the reference frame of the quasar's core, $\tau_1 = \tau_2$, but the times are not in our frame, and definitely not in the reference frame of each blob.

By considering clever thought experiments, Einstein showed that in certain circumstances events can occur in a different order to different observers. In one reference frame, for example, one event occurs before another, but it another the order is reversed. This relativity of the ordering of events raises a disturbing question about cause and effect: is it possible to transform to a frame where an effect can affect a cause? The resolution to this paradox turns out to be both simple and subtle. As long as the two events occur close together in time and far apart in space, such that light could not travel between them, then it does not matter which one happens first. Thus, Einstein considers what is metaphysically important (cause and effect), even if requires us to completely disregard any sort of common sense.

This does not mean that anything goes, as implied by postmodern popular culture. Mathematical logic must still hold, and, in the slow limit, Einstein's kinematics must revert to Galileo's. In one's own reference frame, and those moving slowly compared to light speed, common sense must hold. In relativity theory, common sense[286] quantities are called *proper.* Clocks and rulers, for example, measure *proper time* and *proper length* respectively.

We begin this section with a series of classic thought experiments, each of which would be trivially obvious under Galilean relativity. For example, Thought Experiment 17.1 proves the obvious, that everyone agrees on perpendicular lengths (p. 568). Each successive thought experiment, however, shows odder and odder results. Eventually, we derive the Lorentz-FitzGerald transformations for spacetime, and the relativistic equivalent of simple Galilean velocity addition.

The magnitude of a proper vector is simply its magnitude, which will not change under coordinate rotation. For example, the position vector, $\vec{r}$, has the same length whether rendered in Cartesian, cylindrical, or spherical coordinates, which is obvious under the rules of Euclidian geometry. We conclude this section by defining the position 4-vector, $(ct,\vec{r})$, which specifies both an object's time and location. In Thought Experiment 17.7 (p. 580), we show that, similarly, the 4-position's magnitude remains constant under any 4-dimensional rotation. There is a catch, however. For this 4-dimensional magnitude to have physical meaning, we must use a form of non-Euclidian geometry called *Minkowski space*, after Hermann Minkowski (1864-1909). When we do, a 4-rotation about the time axis simply represents a traditional rotation of spatial

[286] By "common sense" we mean "as measured at rest," so one's *proper velocity* is, by definition, zero.

coordinates, but a rotation about any spatial axis represents a Lorentz-FitzGerald transformation of reference frame.

Problem 17.1: Willem de Sitter's Argument

Soon after Einstein published his theory of relativity, Willem de Sitter wrote a paper arguing that observations of binary stars directly confirmed the constancy of the speed of light.[287]

Consider a star in a binary star system located a far distance ℓ from Earth. The star is in a uniform circular orbit with speed v and period T. Now consider two positions in its motion: when the star is moving directly away from an observer on Earth, and when the star is moving directly toward the observer.

a) Show that, if Galilean addition of velocities were correct, and the velocity of light were not constant, then a binary star would appear at both sides of its orbit at the same time when:

$$\ell = T\left(\frac{c^2 - v^2}{v}\right) \approx \frac{c^2 T}{v}.$$

Argue that this would be the farthest away that a binary star system could possibly be from Earth, and still be observed as such by astronomers.

b) Starlight contains spectral lines, making it relatively easy to measure velocity via the Doppler effect. The plot from a 1910 spectroscopic study[288] of the bright eclipsing binary star Algol shows the Doppler velocity (in km/s) as a function of time. Based on this plot, approximately what is the farthest away that Algol could possibly be if Galilean relativity were true? Express your answer in light years.

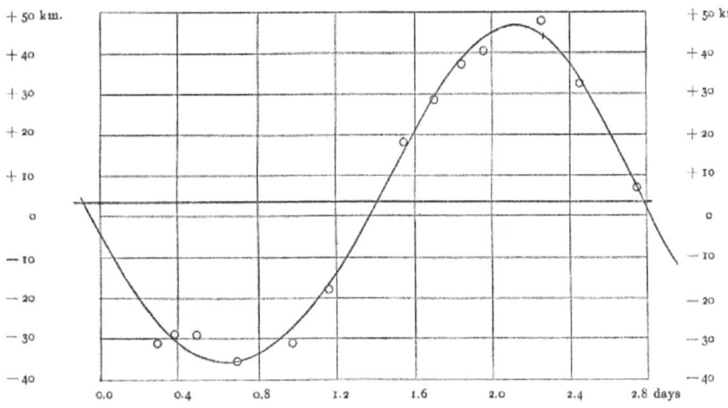

Velocity Curve from Curtiss' Observations.

c) Algol's distance was, at the time, estimated to be 23 light years[289], which is about a fifth of the modern value of 90 light years. Moreover, many dimmer Algol-like stars had been discovered by 1913, which were presumably much farther away. Discuss what this said about the constancy of the speed of light as of 1913.

[287] W. de Sitter, "On the constancy of the velocity of Light," Proceedings of the Royal Netherlands Academy of Arts and Science, **16** (1913), pp. 395-396.

[288] F. Schlesinger, F. and R.H. Curtiss, "The orbit of Algol from observations made in 1906 and 1907," Publications of the Allegheny Observatory, **1** (1910), pp. 25-32.

[289] J.E. Gore, "On the Probable Distance of Algol," Journal of the British Astronomical Association, **2** (1892), p. 443.

Thought Experiment 17.1: The Invariance of Perpendicular Length

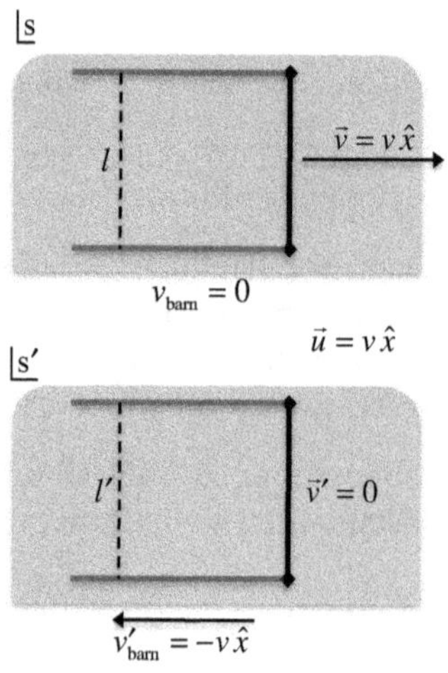

Consider a rod with paint brushes on either end, moving along at a velocity $\vec{v} = v\hat{x}$, painting two parallel lines on the side of a barn that is stationary in reference frame ⌊s. Now imagine another observer moving along with the rod at a speed $\vec{u} = v\hat{x}$, whose reference frame we will call ⌊s′.

Later the barn is inspected, with the rod nearby and at rest. Which is bigger, the spacing of the lines or the rod?

Let the *proper length* of the rod be l_0, and the observed lengths of the rod be l and l' in reference frames ⌊s and ⌊s′ respectively.

In frame ⌊s, the barn is stationary, so the spacing of the lines on the barn when inspected later must be l.

In frame ⌊s′, the rod is stationary, so the length of the rod is, by definition, its proper length. Thus, $l' = l_0$.

Now we consider three hypothetical cases:

(i) Moving perpendicular lengths appear shorter than stationary lengths.

(ii) Moving perpendicular lengths appear longer than stationary lengths.

(iii) Moving perpendicular lengths appear the same as stationary lengths.

And, upon comparison the following must follow for each case:

Case (i): According to frame ⌊s, the, now stationary, rod would be longer than the spacing of the stripes $(l_0 > l)$, because it made the stripes when it was moving. However, according to frame ⌊s′ the rod would be shorter than the stripe spacing $(l_0 < l)$, because the barn, not the rod, was moving when the stripes were painted $(l' < l)$. Thus, case (i) must be false, because it violates the principle of relativity.

Case (ii): According to frame ⌊s, the rod would be shorter than the spacing of the stripes $(l_0 < l)$, because it made the stripes when it was moving. However, according to frame ⌊s′ the rod would be longer than the stripe spacing $(l_0 > l)$, because the barn, not the rod, was moving when the stripes were painted. Thus, case (ii) must also be false, because it too violates the principle of relativity.

Therefore, if moving perpendicular lengths are neither longer nor shorter than stationary perpendicular lengths, then they must be equal. Thus, we conclude that case (iii) must hold, and lengths perpendicular to the direction of the velocity must not change under a Lorentz-FitzGerald transformation.

Thought Experiment 17.2: A Light Clock

Consider a clock made from two plane mirrors with a separation a, and a pulse of light that bounces back and forth between the mirrors. Every time the light bounces off of a mirror, it emits a flash of light in all directions. The time between flashes is simply the separation distance divided by the speed of light.

Now if we transform into a reference frame moving parallel to the plates, at a velocity $\vec{u} = u\hat{x}$, the light must travel farther between plates, but still at the same speed because of Fundamental Law 10 (p. 483), which is:

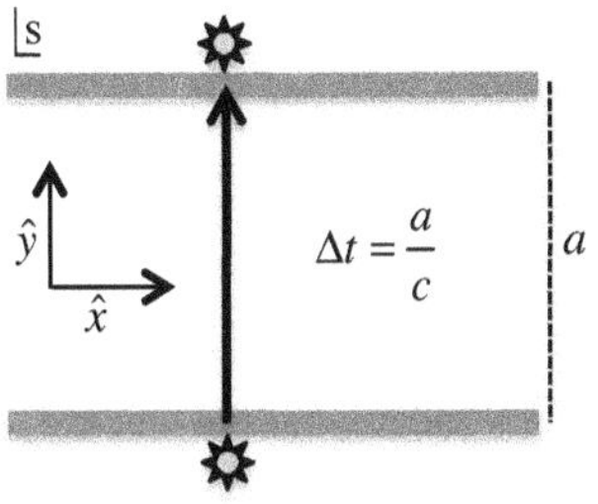

$$d' = \sqrt{(u\Delta t')^2 + (a')^2} = c\,\Delta t'$$

In Thought Experiment 17.1, we showed that perpendicular lengths are invariant to changes in reference frame, so $a = a'$ and:

$$d' = c\Delta t' = \sqrt{(a)^2 + (u\Delta t')^2} = \sqrt{(c\,\Delta t)^2 + (u\,\Delta t')^2}\,.$$

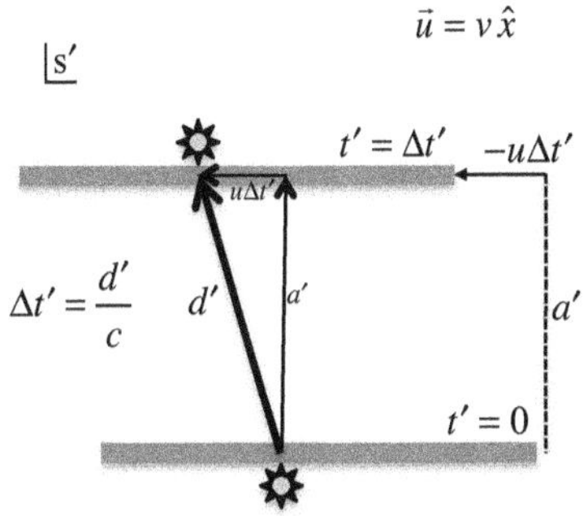

Now solving for $\Delta t'$ we have:

$$(\Delta t')^2 = (\Delta t)^2 + (\tfrac{1}{c}u\,\Delta t')^2$$

$$(\Delta t')^2\left(1 - \tfrac{1}{c^2}u^2\right) = (\Delta t)^2$$

$$\Delta t' = \frac{\Delta t}{\sqrt{1 - \frac{1}{c^2}u^2}}\,.$$

This is called *time dilation.*

Notice also that if we switched the time Δt with $\Delta t'$ we do not have the same relationship. How can that be, given the principle of relativity? Every clock, or ruler, is at rest in some reference frame. Our clock's *rest frame* is $\lfloor S$, so the time it measures must be the *proper time* t_0.

In general, a clock moving at speed v will always appear to tick slower compared to a stationary clock by a factor of $\dfrac{1}{\sqrt{1 - \frac{1}{c^2}v^2}}$.

Problem 17.2: Mork from Ork

In 1957 Mork visits Wisconsin from the planet Ork, leaving in early October bringing a clock with him. After traveling home at 90% the speed of light, he stays for twelve years on before his boss, Orson, assigns him to visit Earth on a long term anthropological mission, giving him a new ship that can travel at 99% the speed of light. On September 14 of 1978, he arrives in Colorado, where he meets and befriends a 21-year-old woman named Mindy.

a) How far away is Ork from Earth in light years?

b) When Mork meets Mindy, what does Mork's clock say?

Problem 17.3: Muon Decay

The half-life of μ-mesons (a.k.a. *muons*), when stopped, is measured to be 2.3μs. Such μ-mesons are produced in the Earth's atmosphere by cosmic ray reactions, so they must traverse a distance of at least 5 km before muon observatories on Earth detect them.

a) Find the time it would take for a muon to travel 5 km if it were traveling at almost the speed of light. Discuss how this is possible when their lifetime is only about 2.3 μs.

b) Now consider 100,000 muons traveling at a speed of 0.98c toward a muon observatory that is 5 km from the place the muons were formed. Calculate the muon half-life in the rest frame of an observer on Earth.

c) How many muons make it to the observatory before decaying?

d) If it were not for the special theory of relativity, what would your answer have been for part (b)?

e) Now consider this from the muon's point of view. Find the travel time, in the muon's rest frame, between when it was created and when it was detected in the observatory.

f) Using your results from part e, again how many muons make it to the observatory before decaying?

g) Research muon decay experiments, and discuss their importance in the development of 20th century physics. Also report on the reactions that produce the muons in the atmosphere and what they have discovered about the origin of cosmic rays.

Thought Experiment 17.3: Length Contraction

Consider a ruler of length l moving at velocity $\vec{v}$ with respect to a standard laboratory photogate attached to a clock. Clearly, the ruler will block the photogate for a length of time $t = \frac{l}{v}$.

In the rest frame of the ruler, which we call $\lfloor s'$, the clock moves at a velocity $\vec{v}'_{\text{clock}} = -\vec{v}$, so the time would be related to the proper time of the clock by:

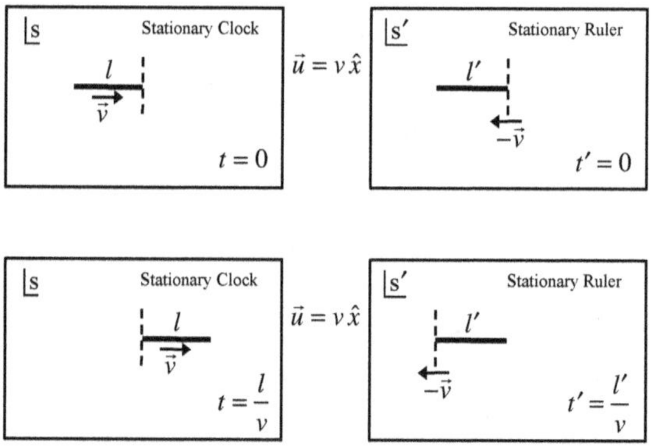

$$t' = \frac{t_0}{\sqrt{1-\left(-v/c\right)^2}} = \frac{t_0}{\sqrt{1-\left(v/c\right)^2}}.$$

In the primed reference frame the length of the ruler must be simply the distance the clock travels over a time interval t'. Therefore:

$$l' = v'_{\text{clock}}\, t' = v t' = \frac{vt}{\sqrt{1-\left(\frac{1}{c}v\right)^2}} = \frac{l}{\sqrt{1-\left(\frac{1}{c}v\right)^2}}.$$

Since the primed frame is the rod's rest frame, l' must be its proper length. We can, therefore, write the length in the unprimed frame as:

$$l = l_0\sqrt{1-\left(\frac{1}{c}v\right)^2}.$$

Thus moving objects are shorter than stationary objects in the direction of motion, which is called *length contraction*.

Thought Experiment 17.4: The Order of Events

Consider a ruler whose proper length is l_0, and is moving at a velocity $\vec{v}$, with respect to a clock attached to two photogates, which are also separated by a distance l_0.

In the reference frame of the clocks, the ruler has a length:

$$l = l_0\sqrt{1-\left(\frac{v}{c}\right)^2}\ .$$

In the reference frame of the ruler, the clock separation is similarly:

$$l' = l_0\sqrt{1-\left(\frac{v}{c}\right)^2}\ .$$

Next we consider four events, A-D, as pictured.

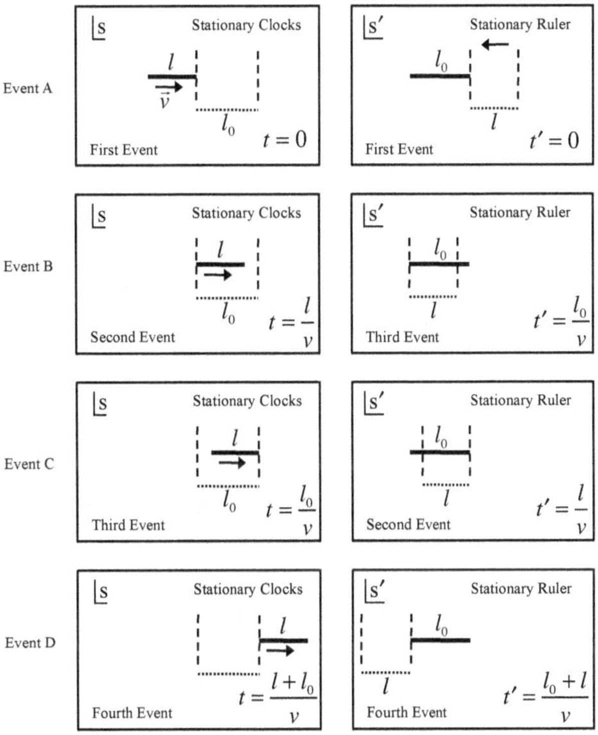

Event A: The front of the ruler crosses the first clock. Let this be the origin and time zero, so $t_A = t'_A = 0$.

Event B: The back of the ruler crosses the first clock. This is the second event in the reference frame of the clocks because the ruler is shorter than the clock spacing. However, this is the third event in the reference frame of the ruler because the ruler is longer than the clock spacing.

Event C: The front of the ruler crosses the second clock. This event takes place after Event B in the clock frame, but before Event B in the ruler frame.

Event D: The back of the ruler crosses the second clock.

Thus, the order of events need not be the same from one reference frame to the next!

This raises the question, what if Event B somehow caused Event C to happen? How could these be different in each one's own respective reference frame? Clearly they could not. How can we resolve this paradox?

For Event B to affect Event C, it would need to send it a message, and this takes time. In the clock's reference frame, the minimum communication distance, $c\Delta t$, in frame $\lfloor S$, is:

$$c\Delta t = c\left(t_C - t_B\right) = c\left(\frac{l_0}{v}-\frac{l}{v}\right) = \frac{c}{v}\left(l_0 - l_0\sqrt{1-\left(\frac{v}{c}\right)^2}\right) = l_0\frac{c}{v}\left(1-\sqrt{1-\left(\frac{v}{c}\right)^2}\right).$$

For causality to hold the communication distance $c\left(t_C - t_B\right)$ must be shorter than the physical separation $\Delta x = l_0$. Because of the square root, we will compare the squares of these two distances. Therefore, if $(\Delta x)^2-(c\Delta t)^2>0$, then there is no paradox. Now, $l=l_0\sqrt{1-\left(v/c\right)^2}$, so substituting for l:

$$(\Delta x)^2-(c\Delta t)^2=l_0^{\,2}\left(1-\left(\tfrac{c}{v}\right)^2\left(1-\sqrt{1-\left(\tfrac{v}{c}\right)^2}\right)^2\right).$$

Let us first look at the non-relativistic limit, where $v\ll c$:

$$(\Delta x)^2-(c\Delta t)^2\approx l_0^{\,2}\left(1-\left(\tfrac{c}{v}\right)^2\left(1-\sqrt{1-\cancel{\left(\tfrac{v}{c}\right)^2}}\right)^2\right)\approx l_0^{\,2}\left(1-\left(\tfrac{c}{v}\right)^2\left(1-1\right)^2\right)\approx l_0^{\,2}>0$$

And if $v\approx c$, then:

$$(\Delta x)^2-(c\Delta t)^2\approx l_0^{\,2}\left(1-(1)^2\left(1-\sqrt{1-(1)^2}\right)^2\right)\approx 0$$

For good measure, we will evaluate it at the intermediate velocity $\frac{v}{c}=\frac{3}{5}$:

$$(\Delta x)^2-(c\Delta t)^2=l_0^{\,2}\left(1-\left(\frac{5}{3}\right)^2\left(1-\sqrt{1-\left(\frac{3}{5}\right)^2}\right)^2\right)=l_0^{\,2}\left(1-\frac{25}{9}\left(1-\sqrt{\frac{16}{25}}\right)^2\right)=l_0^{\,2}\left(1-\frac{\cancel{25}}{9}\left(\frac{1}{\cancel{25}}\right)\right)=\frac{8}{9}l_0^{\,2}>0.$$

Therefore, while the order of events is different in the two frames, causality is preserved because a signal traveling at the speed of light could not move from one clock to the other in time to change the other event.

Problem 17.4: Car and the Garage

A car of rest-length 3m moves at a very high speed through a garage of rest-length 2m. The doors of the garage are open at both ends. According to an observer at rest with respect to the garage, there is one instant in which the car just barely fits in the garage. That is, at this one instant, the ends of the car simultaneously coincide with the garage door openings, according to the observer with respect to the garage.

a) Determine the speed of the car.

b) What is the length of the garage according to the driver of the car?

c) What is the length of the car, according to the driver?

d) Will there ever be a time when, according to the driver, the car is entirely in the garage? Discuss why or why not.

Problem 17.5: Superman and the Farmer

Suppose a farmer and Superman take a ladder, and measure it against a barn, and find that they are exactly the same length. Superman, then, picks up the ladder, and runs with it horizontally through the barn.

Is it possible for the Farmer to close the barn doors, simultaneously, and catch superman in the barn before he smashes through the exit door?

Analyze, and discuss, the situation from the point of view of the farmer first. Next, discuss it from the point of view of Superman.

What can the two agree on, and what do they not agree on. Is there any experiment that can be conducted to reconcile their differences?

Thought Experiment 17.5: The Lorentz-FitzGerald Transformations

We have shown previously that the Galilean transformations are invalid for objects moving at high speeds, and we have found some interesting results by applying Einstein's postulates to clocks, photogates and rulers. In this thought experiment, we derive the general Lorentz-FitzGerald transformation between two reference frames whose origins coincide at time zero and move at a relative velocity $\vec{u} = u\hat{x}$. These will relate the Cartesian spacetime coordinates t,x,y,z in frame $\lfloor\text{s}$ to those in frame $\lfloor\text{s}'$.

First recall from Thought Experiment 17.1 (p. 568) that perpendicular lengths are reference frame invariant. Therefore, we conclude that:

$$y' = y \quad \text{and} \quad z' = z.$$

Moreover, Galilean relativity must hold for slow velocities:

$$t' \underset{u \ll c}{\approx} t \quad \text{and} \quad x' \underset{u \ll c}{\approx} x - ut.$$

so the Lorentz-FitzGerald transformations must revert to these if $u \ll c$.

Notice that in Galilean relativity $\frac{\partial}{\partial t}x' = -u$. We will now guess a parallel structure for the time so that: $\frac{\partial}{\partial x}t' \propto -u$.

We now realize that unlike in Galilean relativity, the Lorentz-FitzGerald transformations depend on a fundamental constant c, which must be the same in all reference frames. Using dimensional analysis we include factors of the speed of light so the units work out. Thus:

$$\left[\frac{\mathrm{d}t'}{\mathrm{d}x}\right] = \frac{\mathrm{s}}{\mathrm{m}} = \left[\frac{-u}{c^2}\right].$$

Notice that when $u \ll c$, this ratio becomes very small. Therefore, we guess a linear transformation between the two frames of the form:

$$t' = A\left(t - \frac{u}{c^2}x\right), \quad x' = B(x - ut), \quad y' = y, \quad \text{and} \quad z' = z,$$

where A and B are unitless functions of the relative velocity u, that have the property $A(u=0) = 1$ and $B(u=0) = 1$. Our goal now is to find these functions A and B.

Let us now consider that the origin of frame $\lfloor\text{s}$ is moving backwards at a speed u, from the point of view of frame $\lfloor\text{s}'$, thus:

$$\left.\frac{\mathrm{d}x'}{\mathrm{d}t'}\right|_{x=0} = -u.$$

So, applying our assumed transformations above, we see that if $x = 0$: $t' = At$ and $x' = -But = -\frac{B}{A}ut'$.

Therefore:

$$\left.\frac{dx'}{dt'}\right|_{x=0} = -u = -\frac{B}{A}u \, .$$

Thus, $B = A$, so now the Lorentz-FitzGerald transformations become:

$$t' = A\left(t - \frac{u}{c^2}x\right), \quad x' = A(x - ut), \quad y' = y, \quad \text{and} \quad z' = z \, .$$

Now, we still need to find the function A.

Next, consider a spherical light pulse that is emitted at the origin of the two reference frames $\lfloor$s and $\lfloor$s′ at time zero in both frames. According to Einstein's second postulate, the speed of light is the same in both reference frames. Thus, in both reference frames the radius of the wave front is expanding at the speed of light, so:

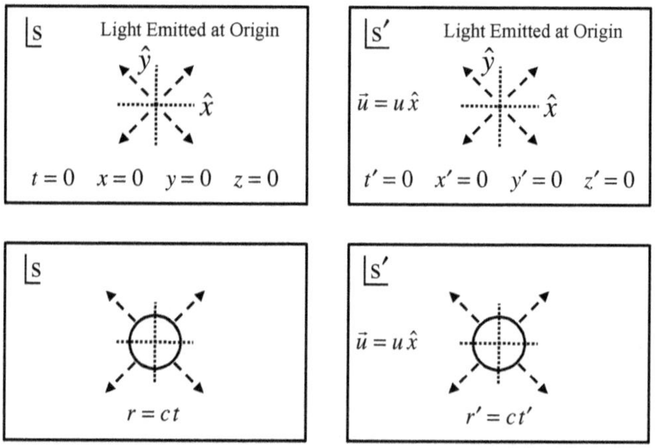

$$r = \sqrt{x^2 + y^2 + z^2} = ct$$
$$r' = \sqrt{x'^2 + y'^2 + z'^2} = ct' \, .$$

By squaring both sides, we see that:

$$x^2 + y^2 + z^2 - c^2t^2 = 0$$
$$x'^2 + y'^2 + z'^2 - c^2t'^2 = 0 \, .$$

Subtracting the equations:

$$\left(x^2 - x'^2\right) + \left(y^2 - y'^2\right) + \left(z^2 - z'^2\right) - c^2\left(t^2 - t'^2\right) = 0 \, .$$

Substituting for x' and t', we see that:

$$\left(x^2 - (Ax - Aut)^2\right) - c^2\left(t^2 - \left(At - A\tfrac{1}{c^2}ux\right)^2\right) = 0 \, .$$

We first expand out each term:

$$\left(x^2 - (Ax - Aut)^2\right) - c^2\left(t^2 - \left(At - A\tfrac{1}{c^2}ux\right)^2\right) = 0$$
$$x^2 - \left(A^2x^2 - 2A^2uxt + A^2u^2t^2\right) - c^2t^2 + c^2\left(A^2t^2 - 2A^2\tfrac{1}{c^2}uxt + A^2\tfrac{1}{c^4}u^2x^2\right) = 0$$
$$x^2 - A^2x^2 + 2A^2uxt - A^2u^2t^2 - c^2t^2 + c^2A^2t^2 - 2A^2uxt + A^2\tfrac{1}{c^2}u^2x^2 = 0 \, .$$

We now cancel opposite terms, and combine quadratic terms in t and x:

$$\left(x^2 - A^2x^2 + A^2\frac{u^2}{c^2}x^2\right) + \left(\cancel{2A^2uxt} - \cancel{2A^2uxt}\right) - \left(A^2u^2t^2 + c^2t^2 - c^2A^2t^2\right) = 0$$

Finally by factoring we find:

$$\left(1 - A^2 + A^2\tfrac{1}{c^2}u^2\right)x^2 - \left(1 - A^2 + A^2\tfrac{1}{c^2}u^2\right)c^2t^2 = 0.$$

Notice that this equation is satisfied, for all positions and times, if:

$$1 - A^2 + A^2\tfrac{1}{c^2}u^2 = 0.$$

Now, solving for A, we find:

$$1 - A^2\left(1 - \tfrac{1}{c^2}u^2\right) = 0 \quad \rightarrow \quad A^2\left(1 - \tfrac{1}{c^2}u^2\right) = 1 \quad \rightarrow \quad A = \frac{1}{\sqrt{1 - \tfrac{1}{c^2}u^2}}.$$

Substituting back into the original equation, we find the Lorentz-FitzGerald transformation formulas to be:

$$t' = \frac{t - \tfrac{1}{c^2}ux}{\sqrt{1 - \tfrac{1}{c^2}u^2}}, \quad x' = \frac{x - ut}{\sqrt{1 - \tfrac{1}{c^2}u^2}}, \quad y' = y, \quad \text{and} \quad z' = z.$$

Often it is saves writing to define the *relativistic gamma factor* (or the *Lorentz factor*) as a function of velocity as:

$$\gamma(u) \equiv \frac{1}{\sqrt{1 - \tfrac{1}{c^2}u^2}},$$

which is shown along with the low velocity approximation (p. 599):

$$\gamma(u) \approx 1 + \tfrac{1}{2c^2}u^2.$$

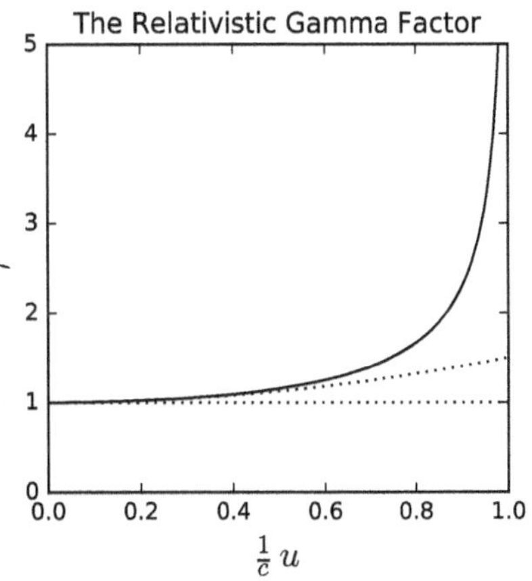

Note that there will frequently be more than one velocity to consider in a problem, so it is important to know to which velocity each gamma refers.

The Lorentz-FitzGerald transformations can also be written as a matrix:

$$\begin{pmatrix} ct' \\ x' \\ y' \\ z' \end{pmatrix} = \begin{pmatrix} \gamma(u) & -\frac{u}{c}\gamma(u) & 0 & 0 \\ -\frac{u}{c}\gamma(u) & \gamma(u) & 0 & 0 \\ 0 & 0 & 1 & 0 \\ 0 & 0 & 0 & 1 \end{pmatrix} \begin{pmatrix} ct \\ x \\ y \\ z \end{pmatrix}.$$

Notice that the transformation is a 4×4 unitless matrix. This is often referred to as a *boost*, to distinguish it from a *rotation* or a *translation*. The group of all boosts, translations, and rotations is referred to as the Poincaré group in honor of Henri Poincaré.

Also notice that when $u \ll c$ our boost is simply a Galilean transformation, which we can also write in matrix notation as:

$$\begin{pmatrix} ct' \\ x' \\ y' \\ z' \end{pmatrix} \approx \begin{pmatrix} 1 & 0 & 0 & 0 \\ -\frac{u}{c} & 1 & 0 & 0 \\ 0 & 0 & 1 & 0 \\ 0 & 0 & 0 & 1 \end{pmatrix} \begin{pmatrix} ct \\ x \\ y \\ z \end{pmatrix}.$$

Example 17.1: A Relativistically Moving Light Source

Now let us consider a ray of light having been emitted by a source at position $-x_0$ with respect to an observer at the origin in reference frame $\lfloor$s, and moving in the positive $\hat{x}$ direction. The position of the light is:

$$x = ct - x_0 .$$

Now we will consider a reference frame $\lfloor$s$'$, moving at a constant velocity $\vec{u} = u\hat{x}$ directly away from the source, and whose origin was aligned with that of $\lfloor$s at time zero. The new observer's time is given by:

$$t' = \left(t - \frac{u}{c^2}x\right)\gamma(u).$$

The rate at which time appears to change for the moving observer, as compared to the stationary observer, is:

$$\frac{dt'}{dt} = \left(\frac{dt}{dt} - \frac{u}{c^2}\frac{dx}{dt}\right)\gamma(u).$$

The new observer is observing the motion of a ray of light, which moves at the speed of light, so:

$$\frac{dt'}{dt} = \left(\frac{dt}{dt} - \frac{u}{c^2}\frac{dx}{dt}\right)\gamma(u) = \left(1 - \frac{u}{c^2}c\right)\gamma(u) = \left(1 - \frac{u}{c}\right)\gamma(u).$$

This reverts to the non-relativistic Doppler effect (p. 518) when $\gamma(u) \approx 1$. Notice that at no point in this example do we assume that light is an electromagnetic wave—simply that it is traveling at the speed of light. Therefore, this relationship must apply to anything moving at light speed!

Problem 17.6: The Relativistic Doppler Effect

The speed of light propagation is c in all inertial reference frames, but the frequency and wavelength can differ from one reference frame to the next. The purpose of this problem is to derive the relativistic Doppler effect, and investigate some of its implications.

Consider a monochromatic plane wave, with an electric field in reference frame $\lfloor$s:

$$\vec{E} = \vec{E}_0 \sin\left(\omega t - \tfrac{\omega}{c}x\right).$$

a) Show that the phase of an electromagnetic plane wave, $\left(\omega t - \tfrac{\omega}{c}x\right)$, is a Lorentz invariant. Transform the electric field into a reference frame $\lfloor$s$'$ moving at a velocity $\vec{u} = u\hat{x}$. Show that it can be written in a form:

$$\vec{E}' = \vec{E}_0' \sin\left(\omega' t' - \tfrac{\omega'}{c}x'\right) = \vec{E}_0' \sin\left(\omega t - \tfrac{\omega}{c}x\right).$$

b) Find the relationship between $\vec{E}_0'$ and $\vec{E}_0$. Discuss how this is related to the brightness of a moving source. Is a source moving toward, or away from, the observer brighter?

c) Apply Lorentz-FitzGerald transformations to the time and position, and your result in (a), to show that:

$$\omega' = \omega\sqrt{\frac{1-u/c}{1+u/c}}.$$

d) This is the relativistic Doppler effect, and it is often used to calculate the velocity of fast moving astronomical sources. Consider a source that emits light of a wavelength, λ_0, but we observe the light to have a wavelength λ. Find the relationship between the radial velocity v_r of the source and the two wavelengths involved. Make sure that your answer conforms to astronomical convention, which dictates that a source moving away from Earth has a positive velocity.

e) By 1917 the American astronomer Vesto Slipher had photographed the spectra of 15 spiral galaxies, which included a spectral line associated with ionized calcium that has a laboratory wavelength of 3940Å. Slipher measured the same calcium line in a number of distant galaxies located in the direction of the constellation Corona Borealis, which each had an observed wavelength of 4750Å. What is the velocity of the Corona Borealis galaxy cluster?

f) Show that your answer in part (d) reverts to the common non-relativistic Doppler effect when the speed is much less than the speed of light:

$$\frac{v_r}{c} \approx \frac{\lambda-\lambda_0}{\lambda_0}.$$

Would it have made any difference which formula Slipher used to find the velocity of the Corona Borealis galaxy?

g) Slipher found a most remarkable result, which was controversial in its day. He found that every galaxy was moving away from Earth, with the exception of those in the direction of the constellation Andromeda. Moreover, galaxies clustered together in the sky had similar velocities as each other. In the 1920s, Edwin Hubble measured the brightness of variable stars in these same galaxies to estimate their distances. He correlated the distances with Slipher's velocity results, to establish the famous linear relationship between the recession velocity and distance to galaxies known as Hubble's law.

h) Look up Hubble's 1929 papers using the search engine located at http://adsabs.harvard.edu. Also look up papers by V. Slipher between the years 1917 and 1929. Does Hubble correctly give credit to Slipher for the use of his published data in his 1929 papers? In your opinion, did Hubble cross the line into unethical behavior? Had he given Slipher due credit, do you think the expansion of the universe would be credited to Hubble, Slipher, or both jointly?

Thought Experiment 17.6: Addition of Velocities

Consider an object initially at the origin moving at a velocity $\vec{v} = v_x\hat{x} + v_y\hat{y} + v_z\hat{z}$ in a reference frame $\lfloor \underline{S}$. What is its velocity $\vec{v}'$ in a reference frame $\lfloor \underline{S'}$ moving at velocity $\vec{u} = u\hat{x}$ with respect to $\lfloor \underline{S}$?

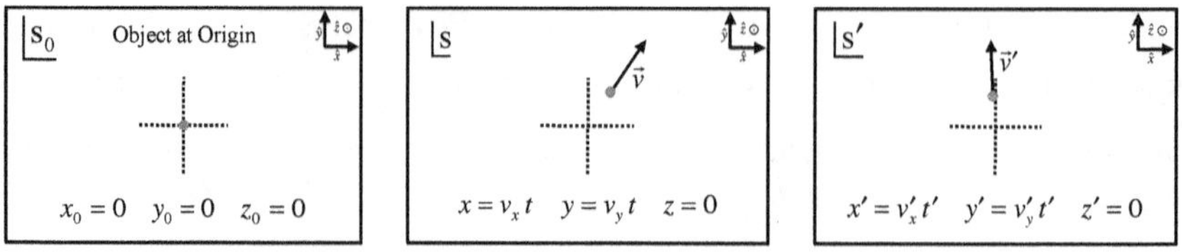

Our game plan is to solve for $\vec{r}'$ as a function of t', so we can differentiate to find $\vec{v}'$. To do this, we first boost frame $\lfloor s$ to frame $\lfloor s'$ by the relative velocity $u\hat{x}$:

$$\begin{pmatrix} ct' \\ x' \\ y' \\ z' \end{pmatrix} = \begin{pmatrix} \gamma(u) & -\frac{u}{c}\gamma(u) & 0 & 0 \\ -\frac{u}{c}\gamma(u) & \gamma(u) & 0 & 0 \\ 0 & 0 & 1 & 0 \\ 0 & 0 & 0 & 1 \end{pmatrix} \begin{pmatrix} ct \\ v_x t \\ v_y t \\ v_z t \end{pmatrix} = \begin{pmatrix} \gamma(u)ct - \frac{u}{c}\gamma(u)v_x t \\ -\frac{u}{c}\gamma(u)ct + \gamma(u)v_x t \\ v_y t \\ v_z t \end{pmatrix}.$$

Solving for t we can write:

$$ct' = \gamma(u)ct - \tfrac{u}{c}\gamma(u)v_x t = \gamma(u)\left(c - \tfrac{1}{c^2}uv_x\right)t \quad \rightarrow \quad t = \frac{ct'}{\gamma(u)\left(c - \frac{1}{c}uv_x\right)}.$$

Solving for x', substituting for t, and differentiating, we find:

$$x' = \gamma(u)(v_x - u)t = \cancel{\gamma(u)}(v_x - u)\frac{ct'}{\cancel{\gamma(u)}\left(c - \frac{u}{c}v_x\right)},$$

$$v_x' = \frac{d}{dt'}\left(\frac{(v_x - u)ct'}{\left(c - \frac{u}{c}v_x\right)}\right) = \frac{(v_x - u)c}{\left(c - \frac{u}{c}v_x\right)} = \frac{(v_x - u)}{\left(1 - \frac{1}{c^2}uv_x\right)}.$$

Similarly, solving for y' yields:

$$y' = v_y t = v_y \frac{ct'}{\gamma(u)\left(c - \frac{u}{c}v_x\right)} \quad \text{and} \quad v_y' = \frac{dy'}{dt'} = v_y \frac{\sqrt{1 - u^2/c^2}}{\left(1 - \frac{1}{c^2}uv_x\right)}.$$

And, solving for z' yields:

$$z' = v_z t = v_z \frac{ct'}{\gamma(u)\left(c - \frac{u}{c}v_x\right)} \quad \text{and} \quad v_z' = \frac{dz'}{dt'} = v_z \frac{\sqrt{1 - u^2/c^2}}{\left(1 - \frac{1}{c^2}uv_x\right)},$$

or in terms of unit vectors:

$$\vec{v}' = \frac{(v_x - u)}{\left(1 - \frac{1}{c^2}uv_x\right)}\hat{x} + v_y \frac{\sqrt{1 - \frac{1}{c^2}u^2}}{\left(1 - \frac{1}{c^2}uv_x\right)}\hat{y} + v_z \frac{\sqrt{1 - \frac{1}{c^2}u^2}}{\left(1 - \frac{1}{c^2}uv_x\right)}\hat{z} = \frac{(v_x - u)\hat{x} + \sqrt{1 - u^2/c^2}\left(v_y\hat{y} + v_z\hat{z}\right)}{1 - \frac{1}{c^2}uv_x}.$$

Now, combining these results, we can generalize the solution for parallel and perpendicular components as:

$$v_\parallel' = \frac{v_\parallel - u}{1 - \frac{1}{c^2}uv_\parallel} \quad \text{and} \quad v_\perp' = v_\perp \frac{\sqrt{1 - \frac{1}{c^2}u^2}}{1 - \frac{1}{c^2}uv_\parallel}.$$

Problem 17.7: Fizeau's Experiment

As discussed already in the text, in 1851 Hippolyte Fizeau measured the speed of light in flowing water. His experiment involved splitting a ray of light with a half-silvered mirror, and sending one beam upstream, and the other downstream, through a tube of water. Once the beams complete their respective round-trip paths, they are combined again with the same half-silvered mirror and sent to an eyepiece. Fizeau could observe the interference patterns as a function of the water speed through the glass pipes.

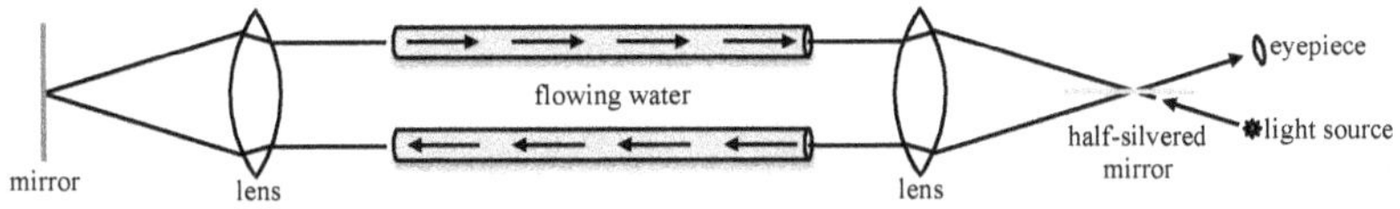

In this experiment, the index of refraction of water, n, the water speed v_w, and the total length of the tubes l are all known, and Fizeau could measure the time difference between the two optical paths Δt from the interference fringes.

a) Show that the difference in the apparent index of refraction for the upstream and downstream light could be found experimentally by:

$$n_u - n_d = \frac{c\,\Delta t}{l}.$$

b) Consider a reference frame $\lfloor S'$ that is stationary with respect to the water. In this reference frame the light must travel at a speed $v'_{\text{light}} = c/n$ inside each tube, regardless of the direction. Using Galilean relativity, as one would have done in 1851, show that the difference in apparent upstream and downstream indices of refraction, $n_u - n_d$, would be approximately given by:

$$n_u - n_d \approx \frac{2n^2}{c} v_w \ .$$

c) Fizeau did not find the expected relationship from simple Galilean relativity, but instead he found a different relationship, which follows:

$$n_u - n_d = 2\frac{v_w}{c} \ .$$

Show that his empirical relationship implies that the speed of light in a moving medium is given by:

$$v_{\text{light}} = \frac{c}{n} \pm \left(1 - \frac{1}{n^2}\right) v_w \ ,$$

for the downstream and upstream paths. This is Fizeau's formula for the speed of light in a moving medium. Fresnel, Fizeau and other nineteenth century physicists struggled to explain this in terms of the aether and light being dragged by the medium.

d) Fizeau's result was one of the clues that guided Einstein as he worked out the electrodynamics of moving bodies. To see how this helped, reanalyze part (b) above relativistically. Show that special relativity predicts Fizeau's experimental result.

Problem 17.8: Stellar Aberration

Recall Bradley's 1727 measurement of stellar aberration, where he pointed a telescope at a single star, and measured it moving slightly throughout the year. He found that this motion was a function of the Earth's perpendicular motion with respect to the starlight. Bradley explained the observations nicely using the Newtonian particle theory of light, and it is best understood by analogy to driving in the rain. Even if the rain is falling vertically, you observe it hitting your windshield at an angle.

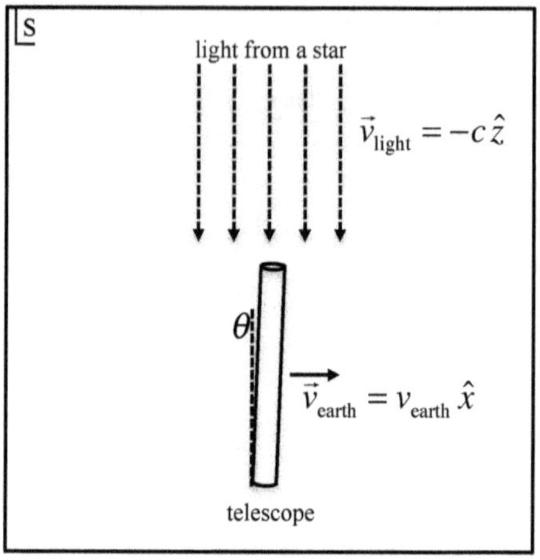

$$\vec{u} = -\vec{v}_{earth} = -v_{earth}\hat{x}$$

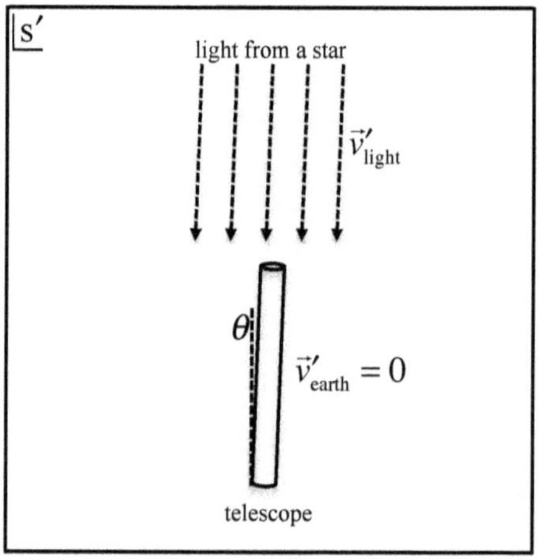

a) Show that under Galilean relativity and Newtonian optics, one would expect Bradley's result of:

$$\tan\theta = \frac{v_{earth}}{c}.$$

b) Bradley measured a maximum aberration angle of 20 seconds of arc. Knowing that the earth is a distance of 150 million km from the sun[290], and orbits once a year, estimate the speed of light. Is this a reasonable number?

c) Now consider light as a wave propagating in the supposed aethereal medium. Show that Bradley's result could only be explained if the sun were at rest compared to the aether.

d) Show, using velocity addition, that the correct expression for the aberration of a star located normal to the ecliptic plane is:

$$\tan\theta = \frac{v_{earth}}{c}\frac{1}{\sqrt{1-\left(v_{earth}/c\right)^2}}.$$

e) To what accuracy must the angle, θ, be measured in order to differentiate between Bradley's formula and the modern formula? Express your answer in arcseconds. Could this be measured in the 1800s?

f) That the sun is stationary compared with the aether, but the earth is not, seems an odd coincidence. In the history of astronomy, every time the observer must be in a special place for a theory to work, it turned out to be wrong (e.g. the geocentric universe). Did the light dragging model rely on this coincidence? Did Lorentz's aether theory? Did relativity theory?

Thought Experiment 17.7: The Magnitude of the Spacetime 4-Vector

Vectors have a magnitude, which is constant under rotation and translation, for example the magnitude of the position vector is:

---

290 At the time Bradley had yet to measure the distance to the sun, which he did later by carefully observing Mars as the Earth turned.

$$r = \sqrt{x^2 + y^2 + z^2}\,.$$

Note that under a rotation this remains constant. Under a translation, the origin must be shifted as well, and the scalar quantity becomes:

$$r = |\vec{r}' - \vec{r}_0'| = \sqrt{(x' - x_0')^2 + (y' - y_0')^2 + (z' - z_0')^2}\,.$$

Under a boost, however, the magnitude of the vector is clearly not preserved. This is even true for Galilean transformations, where a boost in the $\hat{x}$ direction would give:

$$r' = \sqrt{(x - ut)^2 + y^2 + z^2} \neq \sqrt{x^2 + y^2 + z^2}\,.$$

On the other hand, there is a quantity, much like a vector magnitude, which is preserved under Lorentz boosts. To get an understanding of this, let us first consider the light pulse from Thought Experiment 17.5 (p. 573). In the case of light emitted at the 4-origin, we found that:

$$0 = -(ct)^2 + x^2 + y^2 + z^2 = -(ct')^2 + x'^2 + y'^2 + z'^2\,.$$

In Thought Experiment 17.4 (p. 571), we found that to maintain a causal relationship between the 4-origin and another point in spacetime, the distance between these points must be smaller than the light travel time: $r \le ct$. By squaring this expression, we see that in order for one event to possibly affect another the following must be true:

$$-(ct)^2 + x^2 + y^2 + z^2 \le 0\,.$$

Because the quantity on the left hand side is less than or equal to zero, we call it a *timelike* spacetime interval. So long as the interval satisfies this inequality, we can always transform to a frame where two events occur at the same place (i.e. $x = 0, y = 0, z = 0$), but are separated in time. It would make no sense for two events to be timelike in one reference frame and *spacelike* $(r > ct)$ in another. So if $r \le ct$ in one inertial reference frame, then $r' \le ct'$ in any other inertial reference frame.

Thus, we will define the 4-magnitude of the position from the origin in spacetime as:

$$\|ct, \vec{r}\|^2 \equiv -(ct)^2 + r^2\,.$$

While we write this as a magnitude squared to remind us of our units, this quantity is not positive definite and cannot therefore arise from the ordinary algebra of real numbers nor is it compatible with Euclidean geometry. In fact, if there is a possible causal relationship between any two points then it must be the case that $\|ct, \vec{r}\|^2 \le 0$.

In matrix notation, we can write the 4-magnitude as the inner product of a 4-vector in the following manner:

$$\|ct, \vec{r}\|^2 \equiv (ct, \vec{r}) \bullet (ct, \vec{r}) \equiv \begin{pmatrix} ct & x & y & z \end{pmatrix} \begin{pmatrix} -1 & 0 & 0 & 0 \\ 0 & 1 & 0 & 0 \\ 0 & 0 & 1 & 0 \\ 0 & 0 & 0 & 1 \end{pmatrix} \begin{pmatrix} ct \\ x \\ y \\ z \end{pmatrix} = -(ct)^2 + x^2 + y^2 + z^2\,.$$

The middle tensor is called the *metric* $(\ddot{g})$ and it defines the type of geometry used. Geometry that obeys this type of 4-vector algebra is called *Minkowski space.*

Einstein's theory of gravity (*general relativity*) tells us that the metric can differ from this in the presence of gravitational fields. However, the deviation is very slight, except in extreme situations like near a black hole or when considering the universe as a whole.

17.2 *Relativistic Electrodynamics*

The special theory of relativity forces us to recast kinematics and accept radically counterintuitive notions about space and time. Yet, it would be slightly misleading to emphasize the revolutionary nature of Einstein's theory at the expense of recognizing the many ways that relativity preserves and extends fundamental principles from pre-relativistic physics. After all, Galileo formulated the principle of relativity, and Einstein simply extended this principle from mechanics to encompass all of physics.

It is even truer to say that relativity preserves, and extends, the fundamental ideas of classical electricity and magnetism. It is certainly the case that applying relativity to electrodynamics forces us to jettison such concepts as a privileged inertial frame, and with it notions of the luminiferous aether and light as a mechanical wave, just as we must also dispense with our deeply ingrained notions about simultaneity and absolute time in kinematics. But the physical insight that we gain by applying special relativity to electrodynamics clearly outweighs our loss of such metaphysical prejudices. The Galilean transformations for electric and magnetic fields already imply the interdependence of these fields. Yet, they also lead to inconsistencies, which led investigators such as FitzGerald, Heaviside, and Lorentz to repair them in an *ad hoc* manner. Relativity theory simply kept the new Lorentz-FitzGerald transformations for electrodynamics, thus preserving all of the fundamental laws of electrodynamics, including Maxwell's field equations and the conservation of charge.

We begin this section with our most fundamental law, that charge is conserved in all places at all times. Using length contraction and time dilation, we develop the Lorentz-FitzGerald transformations for the charge and current densities, which we can now write as a single covariant 4-current. This naturally leads us to express Maxwell's field equations as a single equation relating the 4-potential to the 4-current.

Next we turn our attention to a series of thought experiments involving idealized rods, capacitors, and inductors, which we use to gain a broader understanding of relativistic electrodynamic phenomena. Just as in non-relativistic electrodynamics, parallel plate capacitor thought experiments lead naturally from charge to voltage to energy density.

Next we express the electromagnetic energy of a moving capacitor in terms of its potential energy and velocity. By subtracting the potential energy, we then write the kinetic energy in terms of the proper energy and the velocity. In the non-relativistic limit, this simplifies to a quadratic function of the velocity, which would be equal to $\frac{1}{2}mv^2$ if the mass is related to the proper energy by $m = \frac{1}{c^2}\mathbb{E}_0$. We finish this section by generalizing the conservation laws of mass, energy and momentum, into a unified conservation of 4-momentum.

Thought Experiment 17.8: The Lorentz Invariance of Charge

Consider a particle of mass m and charge q under the influence of an electric field. The velocity of the charge is now changing with time. However, charge is conserved, so the charge must be constant. Therefore, charge must be invariant under a Lorentz-FitzGerald transformation.

Thought Experiment 17.9: The Charge Density

Consider a box with dimensions $l_0 \times w_0 \times h_0$ with a uniformly distributed charge Q_0, so the charge density is:

$$\rho_0 = \frac{Q_0}{l_0 w_0 h_0}.$$

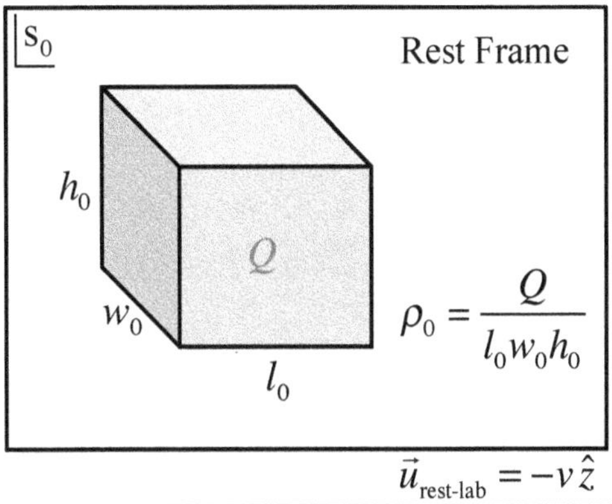

Now consider the same box of charge moving at a velocity $\vec{v} = v\hat{z}$ in a reference frame $\lfloor S$. In this reference frame only the length in the $\hat{z}$ direction is contracted. All perpendicular lengths and the charge are Lorentz invarients, thus:

$l = l_0, w = w_0, h = h_0\sqrt{1 - \frac{1}{c^2}v^2}$, so:

$$\rho = \frac{Q}{lwh} = \frac{Q_0}{l_0 w_0 \left(h_0\sqrt{1-\frac{1}{c^2}v^2}\right)} = \frac{Q_0}{l_0 w_0 h_0}\frac{1}{\sqrt{1-\frac{1}{c^2}v^2}}$$

$$= \frac{\rho_0}{\sqrt{1-\frac{1}{c^2}v^2}} = \rho_0\gamma(v).$$

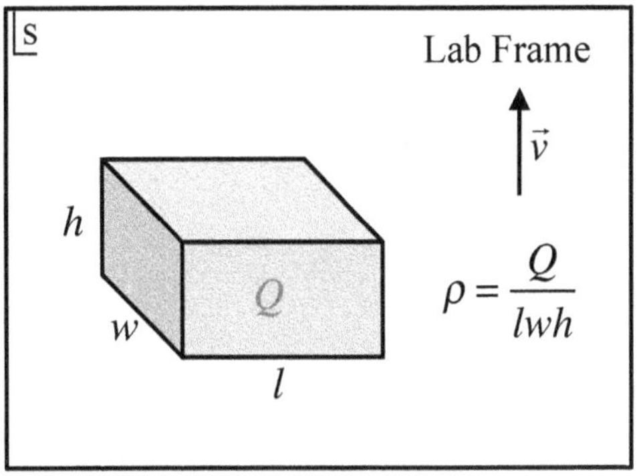

Now, in order to find the Lorentz-FitzGerald transformation, we now compare this to yet another reference frame, moving in the same direction at velocity $\vec{u}$, with respect to frame $\lfloor S_0$.

The charge density in the new reference frame is, therefore, given by:

$$\rho' = \frac{Q}{l'w'h'} = \rho_0\gamma(v').$$

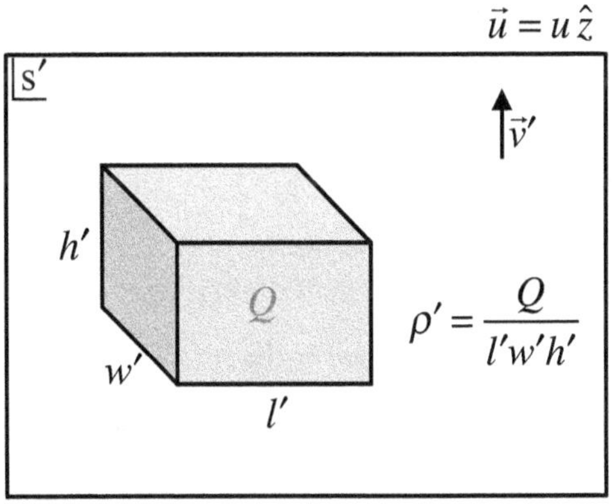

Now we will use the addition of parallel velocities (p. 577):

$$v'_{\parallel} = \frac{v_{\parallel} - u}{1 - \frac{1}{c^2}uv_{\parallel}}.$$

We can now write the charge density as:

$$\rho' = \rho_0\gamma(v') = \frac{\rho_0}{\sqrt{1 - \frac{1}{c^2}\left(\frac{v-u}{1-\frac{1}{c^2}uv}\right)^2}}.$$

We now expand and simplify:

$$\rho' = \frac{\rho_0}{\sqrt{1-\frac{1}{c^2}\frac{v^2-2uv+u^2}{\left(1-\frac{1}{c^2}uv\right)^2}}} = \frac{\rho_0}{\sqrt{\frac{\left(1-\frac{1}{c^2}uv\right)^2}{\left(1-\frac{1}{c^2}uv\right)^2}-\frac{1}{c^2}\frac{v^2-2uv+u^2}{\left(1-\frac{1}{c^2}uv\right)^2}}}$$

$$= \frac{\rho_0}{\sqrt{\frac{1-2\frac{1}{c^2}uv+\left(\frac{1}{c^2}uv\right)^2}{\left(1-\frac{1}{c^2}uv\right)^2}-\frac{\frac{1}{c^2}v^2-2\frac{1}{c^2}uv+\frac{1}{c^2}u^2}{\left(1-\frac{1}{c^2}uv\right)^2}}} = \frac{\rho_0\left(1-\frac{1}{c^2}uv\right)}{\sqrt{1+\left(\frac{1}{c^2}uv\right)^2-\frac{1}{c^2}v^2-\frac{1}{c^2}{}_0u^2}}$$

$$= \frac{\rho_0\left(1-\frac{1}{c^2}uv\right)}{\sqrt{\left(1-\frac{1}{c^2}v^2\right)\left(1-\frac{1}{c^2}u^2\right)}} = \rho_0\gamma(v)\left(1-\frac{1}{c^2}uv\right)\gamma(u).$$

Therefore, the density of charge moving at a velocity v transforms as:

$$\rho' = \rho\left(1-\frac{1}{c^2}uv\right)\gamma(u).$$

Problem 17.9: The Charge Density Lorentz-FitzGerald Transformation

Show the following relationship for the Lorentz-FitzGerald transformation for the charge density:

$$\rho' = \frac{\rho-\frac{1}{c^2}\vec{u}\cdot\vec{J}}{\sqrt{1-\frac{1}{c^2}u^2}} = \left(\rho-\frac{1}{c^2}\vec{u}\cdot\vec{J}\right)\gamma(u).$$

Thought Experiment 17.10: Transforming the Current Density

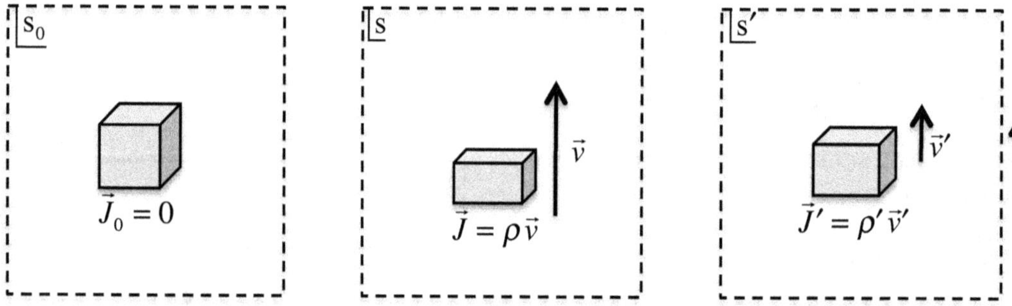

Here we will find the Lorentz-FitzGerald transformation of the current density,

First we will imagine a charge element that, in its own rest frame, has a charge $\delta Q_0 = \rho_0 \delta V_0$.

Now, imagine that in frame $\lfloor S$, the charge element is moving at a velocity $\vec{v} = v\hat{z}$. The current density is then given by:

$$\vec{J} = \rho\vec{v}.$$

In Problem 17.10 you will use a similar process as in Thought Experiment 17.9 to show that:

$$\vec{J}' = \frac{\vec{J}-\rho\vec{u}}{\sqrt{\left(1-\frac{1}{c^2}u^2\right)}} = \left(\vec{J}-\rho\vec{u}\right)\gamma(u).$$

Notice that the Lorentz-FitzGerald transformation of the current density is consistent with the Galilean transformation so long as $u^2 \ll c^2$.

Problem 17.10: The Lorentz-FitzGerald Current Density Transformation

Show that the Lorentz-FitzGerald transformation of the current density follows the form:

$$\vec{J}' = \frac{\vec{J} - \rho\vec{u}}{\sqrt{\left(1 - \frac{1}{c^2}u^2\right)}} = \left(\vec{J} - \rho\vec{u}\right)\gamma(u).$$

Thought Experiment 17.11: The 4-Current

In Problem 17.9 and Problem 17.10 you found that the charge and current transform, interdependently, as:

$$\rho' = \left(\rho - \tfrac{1}{c^2}\vec{u}\cdot\vec{J}\right)\gamma(u) \quad \text{and} \quad \vec{J}' = \left(\vec{J} - \rho\vec{u}\right)\gamma(u).$$

Putting both of these in current density units $\left(\mathrm{A}/\mathrm{m}^2\right)$, we can write:

$$\begin{pmatrix} c\rho' \\ J'_x \\ J'_y \\ J'_z \end{pmatrix} = \begin{pmatrix} \gamma(u) & -\frac{u}{c}\gamma(u) & 0 & 0 \\ -\frac{u}{c}\gamma(u) & \gamma(u) & 0 & 0 \\ 0 & 0 & 1 & 0 \\ 0 & 0 & 0 & 1 \end{pmatrix} \begin{pmatrix} c\rho \\ J_x \\ J_y \\ J_z \end{pmatrix}.$$

Notice that this transforms with the same Lorentz boost matrix that we introduced in Thought Experiment 17.5 (p. 573), giving it properties similar to the spacetime 4-vector. Thus, we can write its 4-magnitude as:

$$\left\|c\rho,\vec{J}\right\|^2 \equiv \left(c\rho,\vec{J}\right)\bullet\left(c\rho,\vec{J}\right) \equiv \begin{pmatrix} c\rho & J_x & J_y & J_z \end{pmatrix} \begin{pmatrix} -1 & 0 & 0 & 0 \\ 0 & 1 & 0 & 0 \\ 0 & 0 & 1 & 0 \\ 0 & 0 & 0 & 1 \end{pmatrix} \begin{pmatrix} c\rho \\ J_x \\ J_y \\ J_z \end{pmatrix} = -(c\rho)^2 + J_x^{\,2} + J_y^{\,2} + J_z^{\,2}.$$

Now, in terms of the proper charge density, this is:

$$\left\|c\rho,\vec{J}\right\|^2 = -\left(c\rho_0\gamma(v)\right)^2 + \left(\rho_0\gamma(v)\vec{v}\right)^2 = -\rho_0\gamma^2(v)\left(c^2 - v^2\right) = -\rho_0 c^2\left(\frac{1}{1-\frac{1}{c^2}v^2}\right)\left(1-\tfrac{1}{c^2}v^2\right) = -\rho_0 c^2.$$

Since this result is independent of the velocity, the 4-vector magnitude is a Lorentz invariant. We therefore have that:

$$-(c\rho_0)^2 = \left\|c\rho,\vec{J}\right\|^2 = \left\|c\rho',\vec{J}'\right\|^2 = -(c\rho)^2 + J_x^{\,2} + J_y^{\,2} + J_z^{\,2} = -(c\rho')^2 + J_x'^2 + J_y'^2 + J_z'^2.$$

Thought Experiment 17.12: The Continuity Equation in 4-Vectors

In this example, we will express the conservation of charge in terms of the 4-current discussed in Thought Experiment 17.11.

We begin by writing the continuity equation, and rearranging it:

$$\vec{\nabla}\cdot\vec{J}=-\frac{\partial\rho}{\partial t}\rightarrow\frac{\partial\rho}{\partial t}+\vec{\nabla}\cdot\vec{J}=0\rightarrow\frac{\partial(c\rho)}{\partial(ct)}+\vec{\nabla}\cdot\vec{J}=0.$$

And in 4-matrix notation this can be written:

$$\begin{pmatrix}-\frac{\partial}{\partial(ct)} & \frac{\partial}{\partial x} & \frac{\partial}{\partial y} & \frac{\partial}{\partial z}\end{pmatrix}\begin{pmatrix}-1 & 0 & 0 & 0\\ 0 & 1 & 0 & 0\\ 0 & 0 & 1 & 0\\ 0 & 0 & 0 & 1\end{pmatrix}\begin{pmatrix}c\rho\\ J_x\\ J_y\\ J_z\end{pmatrix}=0.$$

Now we define the 4-vector del operator (written as a square) where:

$$\vec{\Box}\equiv\left(-\frac{\partial}{\partial ct},\vec{\nabla}\right).$$

Therefore, the continuity equation simply becomes:

$$\vec{\Box}\bullet\left(c\rho,\vec{J}\right)=0.$$

Thought Experiment 17.13: The 4-Potential

Recall that Maxwell's equations in potential form is:

$$\left(\nabla^2-\mu_0\varepsilon_0\frac{\partial^2}{\partial t^2}\right)\begin{pmatrix}\mathbf{V}\\ \vec{\mathbf{A}}\end{pmatrix}=\begin{pmatrix}-\rho/\varepsilon_0\\ -\mu_0\vec{J}\end{pmatrix}.$$

We now write this so that each matrix has consistent units:

$$\left(\nabla^2-\mu_0\varepsilon_0\frac{\partial^2}{\partial t^2}\right)\begin{pmatrix}\frac{1}{c}\mathbf{V}\\ \vec{\mathbf{A}}\end{pmatrix}=\begin{pmatrix}-\rho/c\varepsilon_0\\ -\mu_0\vec{J}\end{pmatrix}=-\mu_0\begin{pmatrix}\rho/c\mu_0\varepsilon_0\\ \vec{J}\end{pmatrix}$$

$$\left(\nabla^2-\frac{\partial^2}{\partial(ct)^2}\right)\begin{pmatrix}\frac{1}{c}\mathbf{V}\\ \vec{\mathbf{A}}\end{pmatrix}=-\mu_0\begin{pmatrix}c\rho\\ \vec{J}\end{pmatrix}.$$

Defining the 4-Potential as $\left(\frac{1}{c}\mathbf{V},\vec{\mathbf{A}}\right)$, we can express Maxwell's equations in 4- matrix notation:

$$\begin{pmatrix}-\frac{\partial}{\partial ct} & \frac{\partial}{\partial x} & \frac{\partial}{\partial y} & \frac{\partial}{\partial z}\end{pmatrix}\begin{pmatrix}-1 & 0 & 0 & 0\\ 0 & 1 & 0 & 0\\ 0 & 0 & 1 & 0\\ 0 & 0 & 0 & 1\end{pmatrix}\begin{pmatrix}-\frac{\partial}{\partial ct}\\ \frac{\partial}{\partial x}\\ \frac{\partial}{\partial y}\\ \frac{\partial}{\partial z}\end{pmatrix}\begin{pmatrix}\frac{1}{c}\mathbf{V}\\ \mathbf{A}_x\\ \mathbf{A}_y\\ \mathbf{A}_z\end{pmatrix}=-\mu_0\begin{pmatrix}c\rho\\ J_x\\ J_y\\ J_z\end{pmatrix}.$$

We now see that this operator can simply be thought of as a 4-vector version of the Laplacian, which we usually write as the square of a vector operator, therefore:

$$\left\| -\frac{\partial}{\partial ct}, \vec{\nabla} \right\|^2 \left(\tfrac{1}{c}\mathbb{V}, \vec{\mathbb{A}} \right) = -\mu_0 \left(c\rho, \vec{J} \right).$$

This operator is called the d'Alembertian after the French mathematician Jean-Batiste le Rond d'Alembert. Now we can write the d'Alembertian as $\Box^2 = \vec{\Box} \bullet \vec{\Box}$, in much the same way as the Laplacian is $\nabla^2 = \vec{\nabla} \cdot \vec{\nabla}$. Finally, we can write Maxwell's equations in the following compact form:

$$\Box^2 \left(\tfrac{1}{c}\mathbb{V}, \vec{\mathbb{A}} \right) = -\mu_0 \left(c\rho, \vec{J} \right).$$

Thought Experiment 17.14: The Parallel Electric Field Transformation

Consider a parallel plate capacitor in its rest frame with a charge $\pm Q$ on plates of dimension $l \times w$, so the surface charge density $\pm\sigma_0$ is given by:

$$\sigma_0 = \frac{Q}{l_0 w_0}.$$

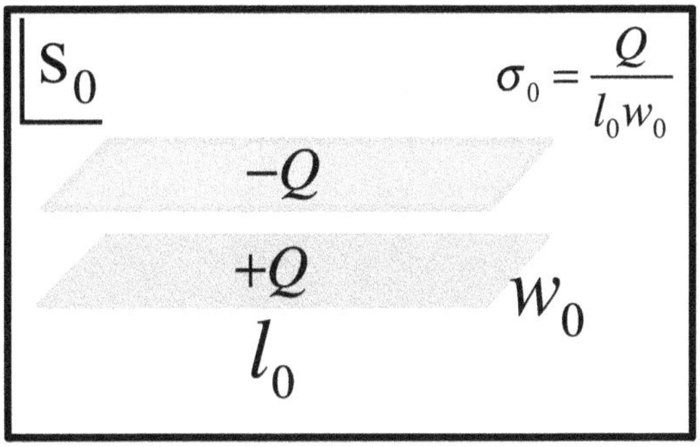

Taking the $\hat{z}$ direction to be normal to the plates, the electric field is given by:

$$\vec{E}_0 = \tfrac{1}{\varepsilon_0} \sigma_0 \hat{z}.$$

Now let us consider another reference frame $\lfloor S$, where the capacitor is moving upward with velocity $\vec{v} = v\hat{z}$. Notice that because l_0 and w_0 are both perpendicular to the direction of motion, they are unaffected by the length contraction, so $\sigma = \sigma_0$ and the electric field is given by:

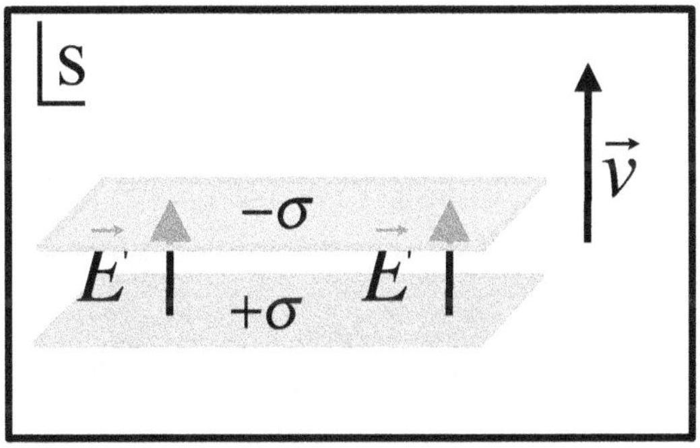

$$\vec{E} = \frac{\sigma}{\varepsilon_0}\hat{z} = \frac{\sigma_0}{\varepsilon_0}\hat{z} = \vec{E}_0.$$

Now consider yet another reference frame, $\lfloor S'$, with a relative velocity $\vec{u} = u\hat{z}$.

Since this is also moving in the $\hat{z}$ direction with respect to the rest frame, the electric field is also unchanged.

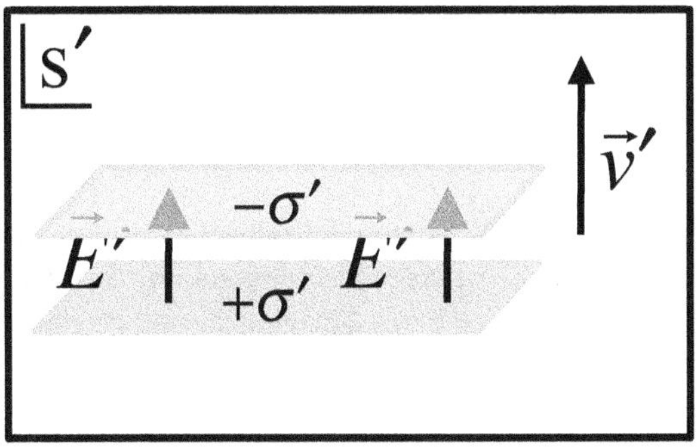

$$\vec{E}' = \frac{\sigma'}{\varepsilon_0}\hat{z} = \frac{\sigma_0}{\varepsilon_0}\hat{z} = \vec{E}_0 = \vec{E} \quad \rightarrow \quad E'_{\parallel} = E_{\parallel}.$$

Thus, the parallel electric field is a Lorentz-FitzGerald invariant.

Thought Experiment 17.15: The Parallel Magnetic Field Transformation

Consider a long solenoid with n_0 turns per length, with each loop carrying a current I_0, in its rest frame $\lfloor S_0$. We showed in Chapter 5 that the magnetic field inside the solenoid is given by:

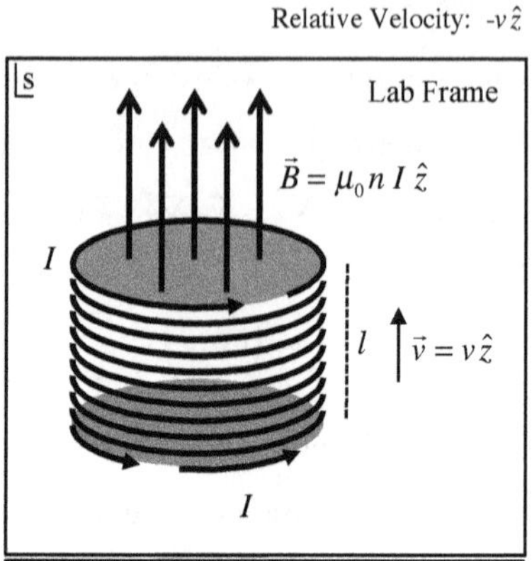

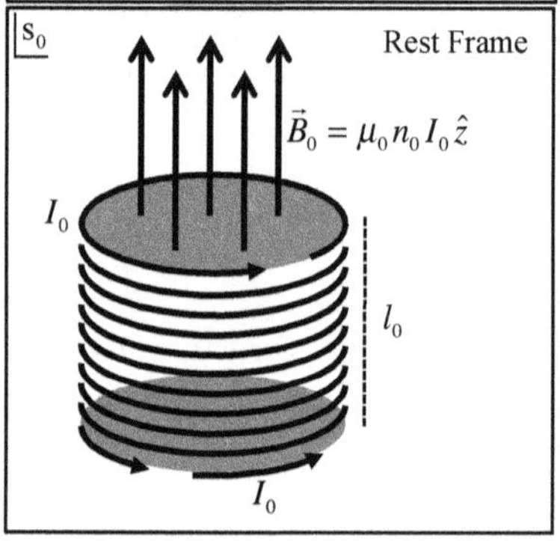

$$\vec{B}_0 = \mu_0 n_0 I_0 \hat{z}.$$

Now consider another reference frame $\lfloor S$, where the solenoid is moving with velocity $\vec{v} = v\hat{z}$. The magnetic field must be given by:

$$\vec{B} = \mu_0 n I \hat{z}.$$

We can count the total number of turns N, so it is clearly a Lorentz invariant. Therefore the number of turns per parallel length must be related to the number of turns per proper length by:

$$n = \frac{N}{l} = \frac{N}{l_0\sqrt{1-\frac{1}{c^2}v^2}} = n_0\gamma(v).$$

Now let us consider the current I flowing through the wire:

$$I = \frac{dQ}{dt} = \frac{dQ}{dt_0}\frac{dt_0}{dt} = \frac{dQ}{dt_0}\frac{\cancel{dt_0}}{\left(\cancel{dt_0}\Big/\sqrt{1-\frac{1}{c^2}v^2}\right)} = I_0\sqrt{1-\frac{1}{c^2}v^2} = \frac{I_0}{\gamma(v)}.$$

In this case, the proper reference frame is the same for both the ruler $\left(\frac{1}{n_0}\right)$ and clock $\left(\frac{1}{I_0}\right)$, so the magnetic field inside the solenoid is given by:

$$\vec{B} = \mu_0 n I \hat{z} = \mu_0\left(n_0\cancel{\gamma(v)}\right)\left(\frac{I_0}{\cancel{\gamma(v)}}\right)\hat{z} = \mu_0 n_0 I_0 \hat{z} = \vec{B}_0.$$

Now, consider the magnetic field in a new reference frame $\lfloor S'$ where the solenoid is moving at a velocity $\vec{v}' = v'\hat{z}$, so the magnetic field is given by:

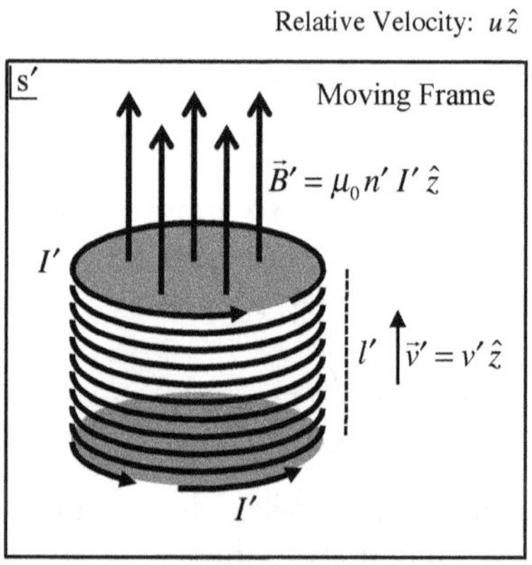

$$\vec{B}' = \mu_0 n' I' \hat{z} = \mu_0\left(n_0\cancel{\gamma(v')}\right)\left(\frac{I_0}{\cancel{\gamma(v')}}\right)\hat{z} = \mu_0 n_0 I_0 \hat{z} = \vec{B}_0 = \vec{B}.$$

Thus, the parallel magnetic field is invariant under a Lorentz-FitzGerald transformation, in much the same way as the electric field was from Thought Experiment 17.15, so we can write:

$$\vec{B}'_{\parallel} = \vec{B}_{\parallel}.$$

Thought Experiment 17.16: A Moving Line Charge

Consider a long line charge, moving at velocity $\vec{v} = v\hat{z}$, where $\hat{z}$ is the direction parallel to the line. In its rest frame, the line charge has a linear charge density of λ_0. What is its linear charge density in the lab frame $\lfloor S$? What are the surrounding electric and magnetic fields?

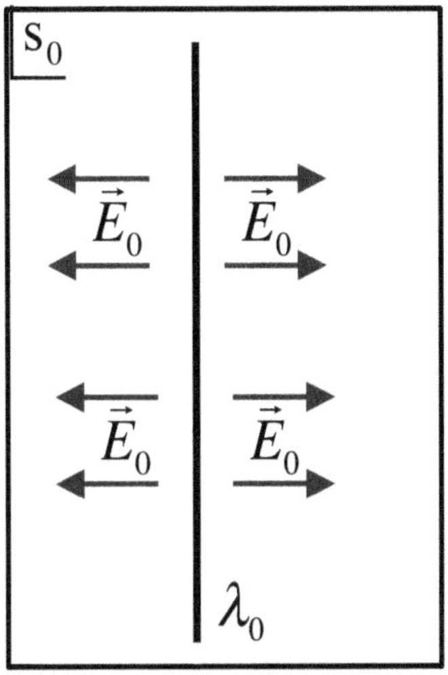

We begin by noticing that charge is conserved, so the linear charge density will increase due to length contraction, or

$$\delta z = \delta z_0 \sqrt{1-\left(v/c\right)^2},$$

$$\lambda = \frac{\mathrm{d}Q}{\mathrm{d}z} = \frac{\mathrm{d}Q}{\mathrm{d}z_0}\frac{\mathrm{d}z_0}{\mathrm{d}z} = \lambda_0 \frac{1}{\sqrt{1-\left(v/c\right)^2}} = \lambda_0 \gamma(v).$$

Now, applying Gauss's law, the electric field is:

$$\oint_{\text{surface}} \vec{E}\cdot \mathrm{d}\vec{A} = \oint_{\text{volume}} \frac{\rho}{\varepsilon_0}\mathrm{d}V \quad \rightarrow \quad E\,2\pi s\,\delta z = \frac{\lambda}{\varepsilon_0}\delta z$$

$$\vec{E} = \frac{\lambda}{2\pi\varepsilon_0 s}\hat{s} = \frac{\lambda_0\gamma(v)}{2\pi\varepsilon_0 s}\hat{s}.$$

From charge conservation, the current is given by $\vec{I} = \lambda\vec{v}$, so we now apply Ampère's Law to find the magnetic field:

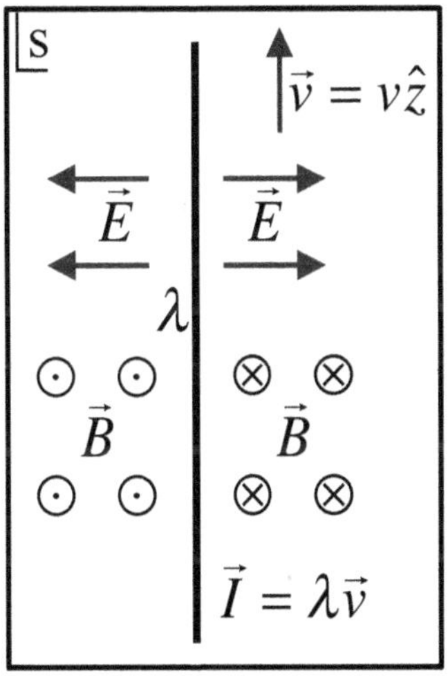

$$\oint_{\text{path}} \vec{B}\cdot \mathrm{d}\vec{\ell} = \int_{\text{surface}} \mu_0 \vec{J}\cdot \mathrm{d}\vec{A} = \mu_0 I$$

$$\vec{B} = \frac{\mu_0 I}{2\pi s}\hat{\varphi} = \frac{\mu_0 \lambda v}{2\pi s}\hat{\varphi} = \frac{\mu_0\lambda_0 v\gamma(v)}{2\pi s}\hat{\varphi}.$$

And, notice that the ratio of the magnetic to the electric field strengths is simply given by:

$$\frac{B}{E} = \frac{\dfrac{\mu_0 \lambda_0 v}{2\pi s\sqrt{1-\mu_0\varepsilon_0 v^2}}}{\dfrac{\lambda_0}{2\pi\varepsilon_0 s\sqrt{1-\mu_0\varepsilon_0 v^2}}} = \mu_0\varepsilon_0 v = \frac{v}{c^2}.$$

Or, in vector notation we can write:

$$\vec{B} = \mu_0\varepsilon_0\,\vec{v}\times\vec{E} = \tfrac{1}{c^2}\vec{v}\times\vec{E}.$$

Thought Experiment 17.17: The Perpendicular Field Transformations

Consider a line charge moving at velocity $\vec{v} = v\hat{z}$ in reference frame $\lfloor S$, so from Thought Experiment 17.16 the surrounding electromagnetic fields are:

$$\vec{E} = \frac{\lambda}{2\pi\varepsilon_0 s}\hat{s} = \frac{\lambda_0\gamma(v)}{2\pi\varepsilon_0 s}\hat{s} \quad \text{and} \quad \vec{B} = \frac{\mu_0\lambda v}{2\pi s}\hat{\varphi} = \frac{\mu_0\lambda_0 v\gamma(v)}{2\pi s}\hat{\varphi}.$$

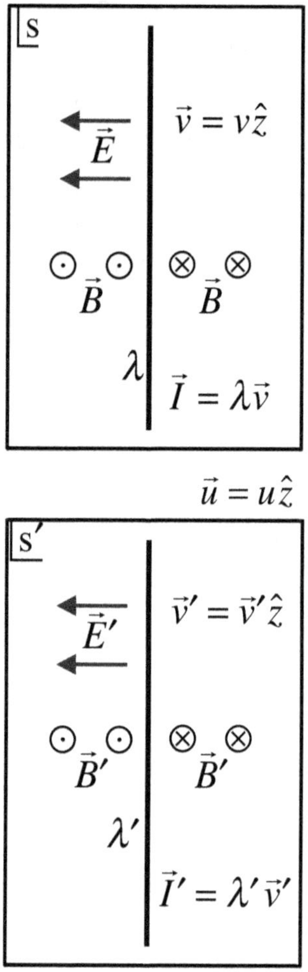

And, the same line charge in another inertial frame, $\lfloor s'$, has electric and magnetic fields:

$$\vec{E}' = \frac{\lambda'}{2\pi\varepsilon_0 s}\hat{s} = \frac{\lambda_0\gamma(v')}{2\pi\varepsilon_0 s}\hat{s} \quad \text{and} \quad \vec{B}' = \frac{\mu_0\lambda' v'}{2\pi s}\hat{\varphi} = \frac{\mu_0\lambda_0 v'\gamma(v')}{2\pi s}\hat{\varphi}.$$

Notice that the coordinate s is invariant, because it is perpendicular to the transformation velocity $\vec{u} = u\hat{z}$.

We now relate v and v' with the parallel velocity addition formula:

$$v'_{\parallel} = \frac{v_{\parallel} - u}{1 - \frac{1}{c^2}uv_{\parallel}}.$$

Now, our goal is to find the electric and magnetic fields in frame $\lfloor s'$ as a function of the corresponding fields in frame $\lfloor s$ and the frame velocity $\vec{u}$.

We begin by considering the quantity v'^2 from addition of velocities:

$$v'^2 = \left(\frac{v-u}{1-\frac{1}{c^2}uv}\right)^2 = \frac{v^2 - 2uv + u^2}{\left(1-\frac{1}{c^2}uv\right)^2} = \frac{\left(c^2 - 2uv\right) - \left(c^2 - v^2 - u^2\right)}{\left(1-\frac{1}{c^2}uv\right)^2}.$$

By grouping terms that look like Lorentz-FitzGerald factors, we have:

$$\frac{v'^2}{c^2} = 1 - \frac{\left(1-\frac{1}{c^2}v^2\right)\left(1-\frac{1}{c^2}u^2\right)}{\left(1-\frac{1}{c^2}uv\right)^2} = 1 - \frac{1}{\left(1-\frac{1}{c^2}uv\right)^2\gamma(v)\gamma(u)}.$$

We can now write the Lorentz factor for the primed frame as:

$$\gamma(v') = \frac{1}{\sqrt{1-\frac{1}{c^2}v'^2}} = \left(1-\frac{1}{c^2}uv\right)\gamma(u)\gamma(v).$$

Now we can write the electric field in frame $\lfloor s'$ as:

$$\vec{E}' = \frac{\lambda_0\gamma(v')}{2\pi\varepsilon_0 s}\hat{s} = \frac{\lambda_0\gamma(v)}{2\pi\varepsilon_0 s}\left(1-\frac{1}{c^2}uv\right)\gamma(u)\hat{s} = \left(\frac{\lambda}{2\pi\varepsilon_0 s}\right)\gamma(u)\left(1-\frac{1}{c^2}uv\right)\hat{s} = \gamma(u)\left(\frac{\lambda}{2\pi\varepsilon_0 s}\hat{s} - \mu_0\cancel{\varepsilon_0}uv\frac{\lambda}{2\pi\cancel{\varepsilon_0}s}\hat{s}\right)$$

$$= \gamma(u)\left(\frac{\lambda}{2\pi\varepsilon_0 s}\hat{s} + (u\hat{z})\times\left(\frac{\mu_0\lambda v}{2\pi s}\hat{\varphi}\right)\right) = \gamma(u)\vec{E} + \vec{u}\times\vec{B} = \gamma(u)\left(\vec{E} + \vec{u}\times\vec{B}\right).$$

And, following a similar argument for the magnetic field we find:

$$\vec{B}' = \frac{\mu_0\lambda_0 v'\gamma(v')}{2\pi s}\hat{\varphi} = \frac{\mu_0\lambda_0\gamma(v)}{2\pi s}\left(\frac{v-u}{\cancel{1-\frac{1}{c^2}uv}}\right)\cancel{\left(1-\frac{1}{c^2}uv\right)}\gamma(u)\hat{\varphi} = \gamma(u)\left(\frac{\mu_0\lambda v}{2\pi s}\hat{\varphi} - \mu_0\varepsilon_0(u\hat{z})\times\left(\frac{\cancel{\mu_0}\lambda}{2\pi\cancel{\mu_0}\varepsilon_0 s}\hat{s}\right)\right)$$

$$= \gamma(u)\left(\vec{B} - \frac{1}{c^2}\vec{u}\times\vec{E}\right).$$

Now we can generalize the perpendicular field transformations to:

$\vec{E}'_\perp = \left(\vec{E}_\perp + \vec{u} \times \vec{B}_\perp\right)\gamma(u)$ and $\vec{B}'_\perp = \left(\vec{B}_\perp - \frac{1}{c^2}\vec{u} \times \vec{E}_\perp\right)\gamma(u)$.

Problem 17.11: Motional EMF and Relativity

Consider a wire of length l moving at a constant velocity $\vec{v}$ perpendicular to a magnetic field $\vec{B}$. In Thought Experiment 11.14 (p. 350), we analyzed this system in the reference frame of the magnet $\lfloor S$, and the reference frame of the bar $\lfloor S'$, and we used Galilean relativity to find the voltage drop in the rest frame of the wire, which is usually called *motional EMF*. We then extended this argument to the magnetic flux, and derived Faraday's Law. In this problem you are to clearly reproduce this thought experiment, but rather than apply the Galilean transformations, apply the Lorentz-FitzGerald transformations. Discuss clearly which laws are the same, and which are different.

Problem 17.12: A Point Charge Moving Parallel to a Line Charge

Consider a long wire, at rest, with a linear charge density λ, and a small charge q moving parallel to the wire at a constant velocity $\vec{v}$, always remaining a distance h from the wire. The electric force of the small charge is balanced by the force of an unknown magnetic field $\vec{B}$.

a) Find $\vec{B}$ as a function of the quantities λ, q, h and $\vec{v}$.

b) Show that the electric and magnetic fields are related by: $c^2\vec{B} = \vec{v} \times \vec{E}$.

c) Now consider the same system in the rest frame of the charge, $\lfloor S'$. Find the corresponding quantities λ', q', and h' in terms of λ, q, h and $\vec{v}$.

d) Find the electric field on the charge, $\vec{E}'$, in terms of λ', q', and h'.

e) Find the magnetic force on the charge, $\vec{B}'$, in terms of λ', q', h' and $\vec{v}'_{\text{wire}}$, where $\vec{v}'_{\text{wire}}$ is the velocity of the wire in the rest frame of the small charge.

f) Apply the Lorentz field transformations directly to $\vec{E}$ and $\vec{B}$, and express $\vec{E}'$ and $\vec{B}'$ in terms of the electric field $\vec{E}$ and the velocity $\vec{v}$.

g) Express the electric and magnetic fields in reference frame $\lfloor S'$ in terms of the physical quantities from reference frame $\lfloor S$: λ, q, h and $\vec{v}$? Do this two different ways: (1) by first transforming λ, q, h, and then calculating the electric and magnetic fields; and (2) by first calculating the electric and magnetic fields and then transforming them to the new reference frame. Show that you obtain the same results.

Problem 17.13: A Moving Capacitor

Consider a parallel plate capacitor is at rest in a reference frame $\lfloor S$. It has dimensions $L \times L \times d$ and the electric field between the plates is given by: $\vec{E} = E\hat{z}$. There is no magnetic field in the rest frame of the capacitor.

a) Find the electric and magnetic fields in a laboratory reference frame, $\lfloor s'$, where the capacitor moves with velocity $\vec{v}' = v'\hat{x}$.

b) What is the surface charge density, σ, in the rest frame of the capacitor?

c) Transform the surface charge density to the laboratory frame.

d) What is the potential difference in both the rest frame, $\mathbb{V}$, and the lab frame, $\mathbb{V}'$? What is the relationship between them?

e) Calculate the quantities, $E^2 - cB^2$ and $\vec{E}\cdot\vec{B}$, in each reference frame.

f) Consider now the potential differences, $\mathbb{V}$ and $\mathbb{V}'$. Calculate from these, the electric field in each reference frame.

g) Here you have found some quantities different ways. Discuss the various ways that you have confirmed the principle of relativity while doing this problem.

Problem 17.14: Two Lorentz Invariants

a) Show that, given an electric field $\vec{E}$ and magnetic field $\vec{B}$, the quantity $E^2 - c^2B^2$ is invariant under any arbitrary Lorentz-FitzGerald transformation.

b) Now argue that if $E = cB$ in any inertial reference frame, $E' = cB'$ in any other inertial frame.

c) Consider now the dot product of the electric and magnetic fields $\vec{E}\cdot\vec{B}$. Show that this is also a Lorentz invariant.

d) Now argue that if the electric and magnetic fields are perpendicular in any inertial reference frame, they will be perpendicular in all inertial reference frames.

e) Show that if, and only if, $E < cB$ and $\vec{E}\times\vec{B} = 0$, there exists some inertial reference frame with no electric field (i.e. $\vec{E}' = 0$). Discuss how this relates to cycloidal motion?

f) Show that for a given electromagnetic field there only exists an inertial frame with no magnetic field (i.e. $\vec{B}' = 0$), if $cB < E$ and $\vec{E}\times\vec{B} = 0$. How does this relate to parabolic motion?

g) Calculate these new Lorentz invariant quantities for light traveling in a vacuum. Discuss.

Problem 17.15: Making Fields Parallel and Perpendicular

In some inertial reference frame, $\lfloor s$, there are electric and magnetic fields $\vec{E}$ and $\vec{B}$, which are neither parallel nor perpendicular.

a) Find another inertial reference frame, $\lfloor s'$, moving at a velocity $\vec{u}$ with respect to $\lfloor s$, such that the magnetic fields are parallel.

b) Is there one in which they are perpendicular? Discuss.

Problem 17.16: Magnetic Moment Transformations

A small magnetic moment $\vec{m}$, at rest, produces a vector potential:

$$\vec{\mathbb{A}} = \frac{\mu_0}{4\pi}\frac{\vec{m}\times\vec{r}}{r^3}.$$

If $\vec{m}$ moves with a constant velocity $\vec{v}$, such that $v \ll c$, show that there is also a scalar potential that appears to come from a fictitious electric dipole moment:

$\vec{p}' = \frac{\vec{v} \times m'}{c^2}$. (Hint: Find $\mathbb{V}'$, and then consider the limit that $v \ll c$.)

Problem 17.17: Light From an Arbitrarily Moving Source

A source of light, which emits light isotropically at frequency ω' in its own rest frame $\lfloor \underline{s}'$, travels with constant velocity $\vec{v} = \vec{v}_{\text{source}}$ with respect to an inertial frame at rest with respect to Earth, $\lfloor \underline{s}$.

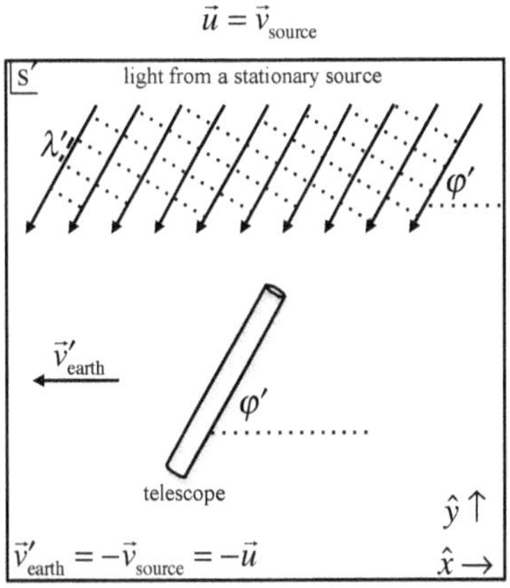

We will now call $\vec{r}'$ the position of a particular wave crest in the rest frame of the source, then the electric field of the wave can be written as:

$$\vec{E}'(\vec{r}', t') = \frac{a'}{|\vec{r}' - \vec{r}_0'|} \vec{E}_a' \sin(\omega' t' - \tfrac{\omega'}{c}|\vec{r}' - \vec{r}_0'|),$$

where $\vec{r}_0'$ is the position of the source from the origin and t' is the time in the same reference frame.

a) At large distances show that the spherical wavefronts can be written as a plane wave, in the form of:

$$\vec{E}'(\vec{r}', t') = \vec{E}_0' \sin(\omega' t' + \tfrac{\omega'}{c}\cos\varphi' x' + \tfrac{\omega'}{c}\sin\varphi' y'),$$

where the angle φ' is defined as toward the source as in the diagram.

Consider the same situation from the reference frame of Earth $\lfloor \underline{s}$, as in the top two figures. The received light will appear as a plane wave traveling in a single direction $\hat{k}(\varphi)$, where φ is the source velocity angle as shown in the diagram, so $\hat{k} = -\cos\varphi\,\hat{x} - \sin\varphi\,\hat{y}$.

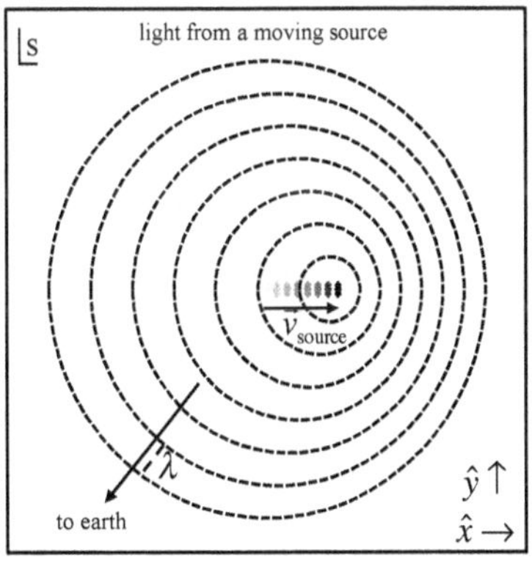

So we can express the observed electric field as a function of the observed angle φ, amplitude of $\vec{E}_0$, angular frequency $\omega = 2\pi c/\lambda$, and the coordinates x and y, as:

$$\vec{E} = \vec{E}_0 \sin(\omega t + \tfrac{\omega}{c}\cos\varphi\, x + \tfrac{\omega}{c}\sin\varphi\, y).$$

Notice that we assume it is a plane wave, and we can set the origin to a convenient local point.

b) Transform the electric field into the reference frame $\lfloor \underline{s}'$ moving at a velocity $\vec{u} = v\hat{x}$. Show that the phase of an electromagnetic plane wave, $\omega t + \frac{\omega}{c}\cos\varphi\,\hat{x} + \frac{\omega}{c}\sin\varphi\,\hat{y}$, is a Lorentz invariant.

c) Using the Lorentz-FitzGerald transformations for the position and time, show that the electric field in the rest-frame of the source, $\lfloor \underline{s}'$, can be written as:

$$\vec{E}'(\vec{r}',t') = \vec{E}'_0 \sin\left(\frac{\omega'(1-\frac{v}{c}\cos\varphi')t}{\sqrt{1-(\frac{v}{c})^2}} + \frac{\omega'(\cos\varphi'-\frac{v}{c})x}{c\sqrt{1-(\frac{v}{c})^2}} + \frac{\omega'\sin\varphi' y}{c} \right) .$$

d) From this, show that the following relations must hold true:

$$\omega\cos\varphi = \omega'\frac{\cos\varphi'-\frac{v}{c}}{\sqrt{1-(\frac{v}{c})^2}}, \quad \omega\sin\varphi = \omega'\sin\varphi', \quad \text{and} \quad \omega = \omega'\frac{1-\frac{v}{c}\cos\varphi'}{\sqrt{1-(\frac{v}{c})^2}} .$$

e) In Problem 17.8 (p. 580) you derived the formula for stellar aberration, in the special case that the source was located perpendicular to the motion of Earth. Now show that the more general formula is:

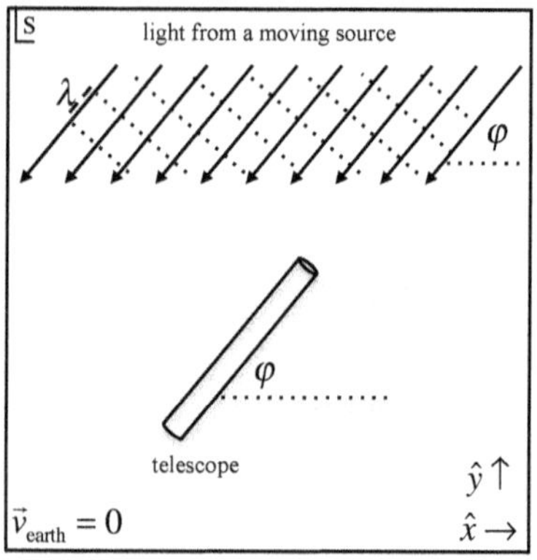

$$\tan\varphi = \frac{\sin\varphi'\sqrt{1-(\frac{v}{c})^2}}{\cos\varphi'+\frac{v}{c}} .$$

Also derive the inverse expression for $\tan\varphi'$?

f) Clearly show that in the limit that $v \ll c$, these expressions conform to Bradley's measurement for the aberration of starlight.

g) Derive the relativistic Doppler effect relation:

$$\omega = \omega'\frac{1-\frac{v}{c}\cos\varphi'}{\sqrt{1-(\frac{v}{c})^2}} .$$

Solve for the velocity as a function of the emitted, and observed, frequencies—and the angle of motion in the observer's reference frame φ. What component of the velocity can be measured with the Doppler effect?

h) You have derived three equations in part (d), with three unknowns (φ,φ',v). Thus, if one could measure the frequency of a spectral line, presumably the space velocity could be determined. Albert Einstein, in 1907, suggested that the apparent wavelength of light, emitted by a fast moving beam of atoms, might be measured at right angles to the source velocity. However, there is a practical difficulty testing the *transverse Doppler effect.* Any deviation from $\varphi = 90°$ would cause the first order factor (i.e. $1+\frac{v}{c}\cos\varphi$) to overwhelm the second order effect, making it a useless tool, so far, for gathering astronomical space velocities.

In their attempt to test Einstein's theory, H.E. Ives and G.R. Stilwell circumvented this difficulty in 1938,[291] and published a follow-up study in 1941[292], that fully confirmed the existence of the second order effects predicted by Einstein. Use the search engine at *http://adsabs.harvard.edu* to look up their two papers to see how they did this.

Prepare a report describing their work and how they confirmed the second order Doppler effects. Can current astronomers use this effect to measure the transverse velocities?

---

291 H.E. Ives and G.R. Stilwell, 1938, J. Opt. Soc. Am, **28**, 215.

292 H.E. Ives and G.R. Stilwell, 1941, J. Opt. Soc. Am, **31**, 369.

Thought Experiment 17.18: The Potentials of a Moving Rod

Imagine a long wire of rest linear charge density λ_0 and radius a. The electric field from Gauss's law is:

$$\vec{E}_0 = \frac{\lambda_0 \hat{s}}{2\pi\varepsilon_0 s} \quad ; s > a.$$

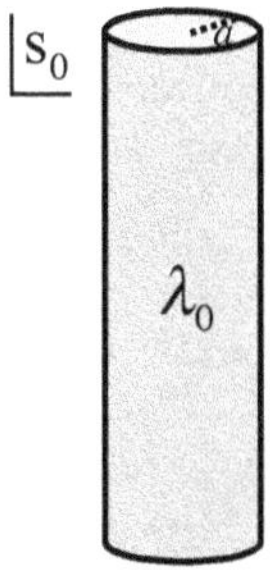

The scalar potential is therefore:

$$\mathbb{V}_0 = -\int \vec{E}\cdot d\vec{l} = -\int_a^s \frac{\lambda_0}{2\pi\varepsilon_0 s} ds = -\frac{\lambda_0}{2\pi\varepsilon_0} \ln\left(\frac{s}{a}\right),$$

and with stationary charges, $\vec{\mathbb{A}}_0 = 0$

Now we will consider a frame $\lfloor \mathrm{S}$, moving at a velocity $\vec{u} = -v\hat{z}$, with respect to $\lfloor \mathrm{S}_0$. Now the new linear charge density is given by:

$$\lambda = \frac{\lambda_0}{\sqrt{1-\frac{1}{c^2}v^2}} = \lambda_0 \gamma(v).$$

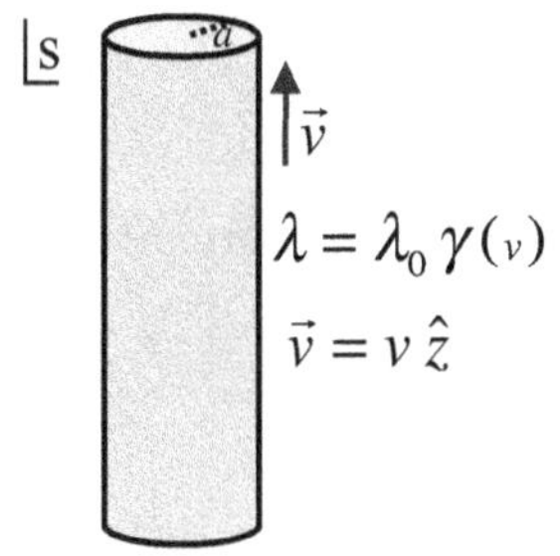

Recalling that perpendicular lengths are invariant, we can simply write the scalar potential in frame $\lfloor \mathrm{S}$ as:

$$\mathbb{V} = -\frac{\lambda}{2\pi\varepsilon_0}\ln\left(\frac{s}{a}\right) = -\frac{\lambda_0\gamma(v)}{2\pi\varepsilon_0}\ln\left(\frac{s}{a}\right) = \mathbb{V}_0\gamma(v) = \frac{\mathbb{V}_0}{\sqrt{1-\frac{1}{c^2}v^2}}.$$

Now, the rod has a velocity in frame $\lfloor \mathrm{S}$, so there is a current associated with the rod's motion:

$$\vec{I} = \lambda\vec{v},$$

and a corresponding vector potential:

$$\vec{\mathbb{A}}(s) = -\frac{\mu_0 I}{2\pi}\ln\left(\frac{s}{a}\right)\hat{z} = -\frac{\mu_0}{2\pi}\ln\left(\frac{s}{a}\right)\lambda\vec{v} = -\frac{\mu_0}{2\pi}\ln\left(\frac{s}{a}\right)\frac{\vec{v}}{\sqrt{1-\frac{1}{c^2}v^2}}.$$

Thus we see that for a moving long charged rod, the vector potential can be found from the scalar potential in the rest frame, as:

$$\vec{\mathbb{A}}(s) = \frac{\mu_0\varepsilon_0}{2\pi\varepsilon_0}\ln\left(\frac{s}{a}\right)\frac{\vec{v}}{\sqrt{1-\frac{1}{c^2}v^2}} = \mu_0\varepsilon_0\mathbb{V}_0(s)\frac{\vec{v}}{\sqrt{1-\frac{1}{c^2}v^2}} = \frac{\vec{v}\gamma(v)}{c^2}\mathbb{V}_0(s).$$

Thought Experiment 17.19: The Transformation of the Potentials

We showed earlier that Maxwell's equations are inconsistent with Galilean relativity, and in Thought Experiment 17.5 (p. 573) that spacetime can be transformed using the Lorentz boost matrix. Moreover, we showed in Thought Experiment 17.11 (p. 585) that the 4-current transforms the same way as the spacetime 4-vector, and we showed in Thought Experiment 17.13 (p. 586) that it makes sense to represent the potentials as a 4-vector when expressing Maxwell's equations. Here we find the Lorentz-FitzGerald transformation of the potentials and write this in 4-vector notation.

Consider the rod from Thought Experiment 17.18 (p. 595) in reference frame $\lfloor s$, the potentials are given by:

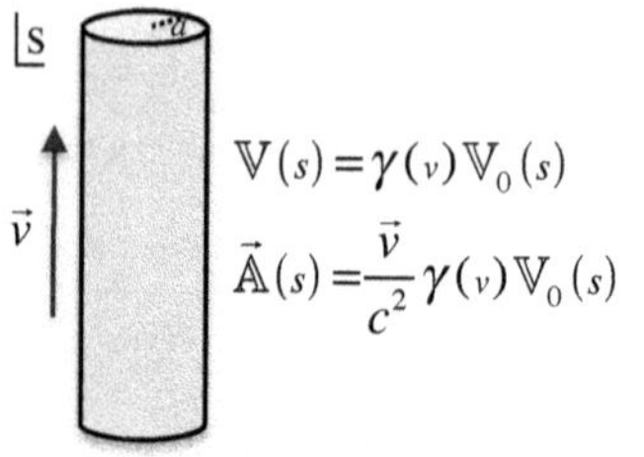

$$\mathbb{V} = \mathbb{V}_0\gamma(v) \quad \text{and} \quad \vec{\mathbb{A}} = \frac{\vec{v}}{c^2}\mathbb{V}_0(s)\gamma(v).$$

Now, in another reference frame moving at a velocity $\vec{u} = u\hat{z}$ with respect to frame $\lfloor s$, and where the velocity of the rod is $\vec{v}' = v'\hat{z}$, the potentials must be given by:

$$\mathbb{V}' = \mathbb{V}_0\gamma(v') \quad \text{and} \quad \vec{\mathbb{A}}' = \tfrac{1}{c^2}\vec{v}'\mathbb{V}_0(s)\gamma(v').$$

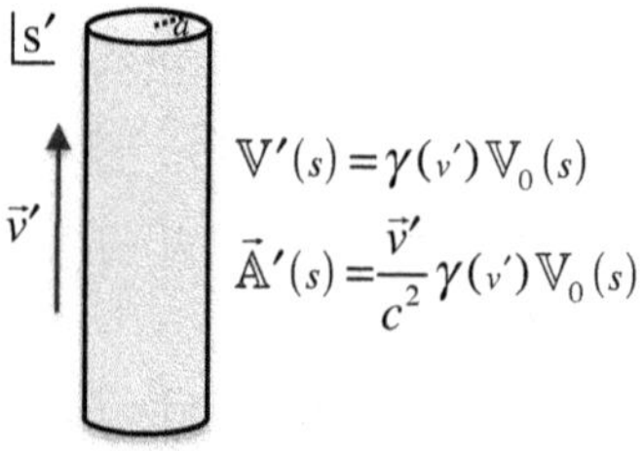

Following the argument from Thought Experiment 17.17 (p. 589), we can write the velocity and relativistic gamma factor in the new frame as:

$$v' = \frac{v-u}{1-\frac{1}{c^2}uv} \quad \text{and} \quad \gamma(v') = \frac{1}{\sqrt{1-\frac{1}{c^2}v'^2}} = \left(1-\tfrac{1}{c^2}uv\right)\gamma(v)\gamma(u).$$

So, the transformed potentials can be written as:

$$\mathbb{V}' = \mathbb{V}_0\gamma(v') = \mathbb{V}_0\gamma(v)\gamma(u)\left(1-\tfrac{1}{c^2}uv\right) = \left(\mathbb{V}_0(s)\gamma(v) - \tfrac{1}{c^2}uv\mathbb{V}_0(s)\gamma(v)\right)\gamma(u)$$

$$\vec{\mathbb{A}}' = \tfrac{1}{c^2}\left(\frac{v-u}{\cancel{1-\frac{1}{c^2}uv}}\hat{z}\right)\mathbb{V}_0\gamma(v)\gamma(u)\left(\cancel{1-\tfrac{1}{c^2}uv}\right) = \tfrac{1}{c^2}\mathbb{V}_0(s)\gamma(v)\gamma(u)(v-u)\hat{z}$$

$$= \left(\tfrac{1}{c^2}v\mathbb{V}_0(s)\gamma(v)\hat{z} - \tfrac{1}{c^2}u\mathbb{V}_0(s)\gamma(v)\hat{z}\right)\gamma(u).$$

Now, substituting to get rid of v, yields:

$$\mathbb{V}' = \left(\mathbb{V} - u\mathbb{A}_{\parallel}\right)\gamma(u) \quad \text{and} \quad \mathbb{A}'_{\parallel} = \left(\mathbb{A}_{\parallel} - \tfrac{1}{c^2}u\mathbb{V}\right)\gamma(u).$$

Recall the parallel structure to the spacetime transformations from Thought Experiment 17.5 (p. 573), where for a relative velocity in the $\hat{z}$ direction:

$$t' = \left(t - \frac{u}{c^2}x\right)\gamma(u), \quad x' = x, \quad y' = y, \text{ and } \quad z' = (z - ut)\gamma(u).$$

Now using the same boost matrix, we can write the potential transform as:

$$\begin{pmatrix} \frac{1}{c}\mathbb{V}' \\ \mathbb{A}'_x \\ \mathbb{A}'_y \\ \mathbb{A}'_z \end{pmatrix} = \begin{pmatrix} \gamma(u) & 0 & 0 & -\frac{u}{c}\gamma(u) \\ 0 & 1 & 0 & 0 \\ 0 & 0 & 1 & 0 \\ -\frac{u}{c}\gamma(u) & 0 & 0 & \gamma(u) \end{pmatrix} \begin{pmatrix} \frac{1}{c}\mathbb{V} \\ \mathbb{A}_x \\ \mathbb{A}_y \\ \mathbb{A}_z \end{pmatrix} = \begin{pmatrix} \gamma(u)\frac{1}{c}\mathbb{V} - \frac{u}{c}\gamma(u)\mathbb{A}_z \\ \mathbb{A}_x \\ \mathbb{A}_y \\ -\frac{u}{c}\gamma(u)\frac{1}{c}\mathbb{V} + \gamma(u)\mathbb{A}_z \end{pmatrix}.$$

The perpendicular components of the vector potential are Lorentz invariants because the parallel magnetic field is an invariant. This can be understood from the definition of the vector potential as $\vec{B} = \vec{\nabla} \times \vec{\mathbb{A}}$:

$$\left(\vec{\nabla}\times\vec{\mathbb{A}}\right)_z = B_z = B'_z = \left(\vec{\nabla}\times\vec{\mathbb{A}}'\right)_z \quad\rightarrow\quad \frac{\partial \mathbb{A}_y}{\partial x} - \frac{\partial \mathbb{A}_x}{\partial y} = \frac{\partial \mathbb{A}'_y}{\partial x} - \frac{\partial \mathbb{A}'_x}{\partial y}.$$

Clearly any multiplicative factors to the Lorentz-FitzGerald transformation would be retained. Thus, in order for the equation above to always be true:

$$\mathbb{A}'_\perp = \mathbb{A}_\perp .$$

Now we take the analogy with the spacetime 4-vector one more step, and realize that the 4-magnitude must also be a Lorentz invariant, because both 4-vectors transform the same way, so

$$\left\|\tfrac{1}{c}\mathbb{V},\vec{\mathbb{A}}\right\|^2 = \left(\tfrac{1}{c}\mathbb{V},\vec{\mathbb{A}}\right)\bullet\left(\tfrac{1}{c}\mathbb{V},\vec{\mathbb{A}}\right) = -\tfrac{1}{c^2}\mathbb{V}^2 + \mathbb{A}^2_x + \mathbb{A}^2_y + \mathbb{A}^2_z = \mathbb{A}^2 - \tfrac{1}{c^2}\mathbb{V}^2,$$

which you will show in Problem 17.18 below.

Problem 17.18: The 4-magnitude of the 4-Potential Vector

In Thought Experiment 17.19, the potentials were given as:

$$\mathbb{V}' = \left(\mathbb{V} - u\mathbb{A}_\parallel\right)\gamma(u), \quad \mathbb{A}'_\parallel = \left(\mathbb{A}_\parallel - \tfrac{1}{c^2}u\mathbb{V}\right)\gamma(u), \text{ and } \mathbb{A}'_\perp = \mathbb{A}'_\perp = 0.$$

By comparing two arbitrary reference frames, show explicitly that the 4-magnitude of the 4-potential vector is a Lorentz invariant.

Problem 17.19: Maxwell's Equations

Show that Maxwell's equations, in free space, are invariant under the Lorentz-FitzGerald transformations. You may want to express Maxwell's equations in Cartesian coordinates and use the chain rule of differentiation.

Problem 17.20: The Fields of a Point Charge

a) Derive the electric and magnetic fields for a uniformly moving point charge moving at velocity $\vec{v} = v\hat{z}$, by first writing the Cartesian components of its electric and magnetic fields in its own rest frame $\underline{|\mathrm{S}'}$, and then transforming them back to the lab frame using the spacetime coordinate transformations.

b) Show that Gauss's Law is satisfied by a moving point charge.

c) Verify your answer in part (a) by, once again, writing the Cartesian components of the electric and magnetic fields, and yet again transforming them. However, rather than using the spacetime Lorentz-FitzGerald transformations as you did in (a), explicitly use the field transformations instead.

d) Convert the electric field to polar coordinates. Does it satisfy Coulomb's law in the non-relativistic limit?

e) Convert the magnetic field to polar coordinates. Show that, in the non-relativistic limit, it satisfies the law of Biot-Savart.

f) Make vector plots of the electric and magnetic fields of a point charge moving at a speed of $v = 0.5c$.

Problem 17.21: Parallel Wires

Consider two long stationary parallel rods separated by a distance of $2a$, with equal and opposite linear charge densities of $+\lambda$ and $-\lambda$.

a) What are the electric and magnetic fields at a point exactly half way between the two wires when they are at rest?

b) Now consider a reference frame $\lfloor s'$ moving at a relative velocity of $\vec{u}$, parallel to the wires. Determine the linear charge density, λ', of the wires in $\lfloor s'$.

c) Find the current I' associated with each wire, in frame $\lfloor s'$.

d) Use the results of (c) to find the magnetic field halfway in-between the wires.

e) Transform the fields you calculated in part (a) into frame $\lfloor s'$. Does your answer agree with your result in part (d)?

Thought Experiment 17.20: Relativistic Energy

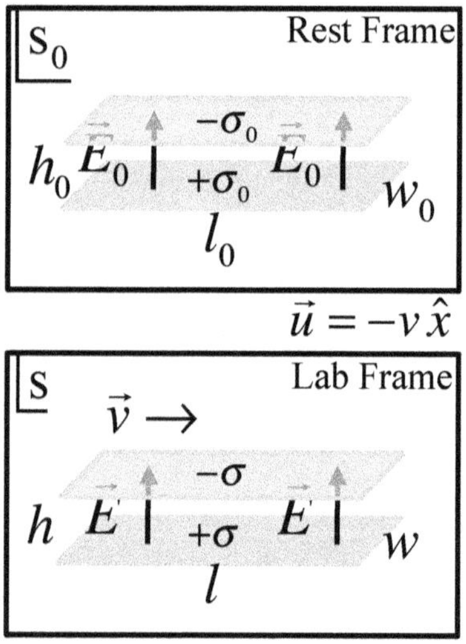

Consider a parallel plate capacitor at rest with charge $\pm Q_0$ and a voltage across the plates $\mathbb{V}_0$. The energy contained in the capacitor is:

$$\mathbb{E}_0 = \tfrac{1}{2}Q\mathbb{V}_0 .$$

The voltage is related to the charge and dimensions of the capacitor by:

$$\mathbb{V}_0 = \frac{Q_0 h_0}{\varepsilon_0 l_0 w_0} .$$

Now, we transform to a frame in which the capacitor is moving sideways at a velocity $\vec{v} = v\hat{x}$, so the voltage is now:

$$\mathbb{V} = \frac{Qh}{\varepsilon_0 lw} = \frac{Q_0 h_0}{\varepsilon_0 \left({}^{l_0}\!/_{\gamma(v)} \right) w_0} = \frac{Q_0 h_0}{\varepsilon_0 l_0 w_0}\gamma(v) = \mathbb{V}_0 \gamma(v).$$

Thus the total energy $(\mathbb{E})$ contained in the capacitor is related to the proper energy $(\mathbb{E}_0)$ by the relationship:

$$\mathbb{E} = \tfrac{1}{2}Q\mathbb{V} = \tfrac{1}{2}Q\mathbb{V}_0\gamma(v) = \mathbb{E}_0\gamma(v).$$

The kinetic energy $(\mathbb{T})$ contained in the capacitor must, therefore, be the difference between the energy of the moving capacitor and the energy it would have if it were at rest:

$$\mathbb{T} = \mathbb{E} - \mathbb{E}_0 = \mathbb{E}_0\gamma(v) - \mathbb{E}_0 = \mathbb{E}_0\left(\gamma(v) - 1\right).$$

How does this result relate to our non-relativistic understanding of kinetic energy expressed as $\mathbb{T} = \tfrac{1}{2}mv^2$? To answer this question we Taylor expand the relativistic kinetic energy about the velocity $v = 0$, beginning with the Lorentz factor itself (see graph on p. 575):

$$\gamma(v) \approx \gamma(0) + \left.\frac{\mathrm{d}\gamma(v)}{\mathrm{d}v}\right|_{v=0} u + \frac{1}{2}\left.\frac{\mathrm{d}^2\gamma(v)}{\mathrm{d}v^2}\right|_{v=0} v^2 + \ldots.$$

Taking the first two derivatives of the Lorentz factor gives:

$$\frac{\mathrm{d}}{\mathrm{d}v}\gamma(v) = \frac{\mathrm{d}}{\mathrm{d}v}\left(1-\tfrac{1}{c^2}v^2\right)^{-\frac{1}{2}} = -\tfrac{1}{2}\left(1-\tfrac{1}{c^2}v^2\right)^{\frac{1}{2}}\left(-\tfrac{2}{c^2}v\right) = \tfrac{1}{c^2}v\sqrt{1-\tfrac{1}{c^2}v^2} = \tfrac{1}{c^2}v\gamma^{-1}(v)$$

and

$$\frac{\mathrm{d}^2}{\mathrm{d}v^2}\gamma(v) = \frac{\mathrm{d}}{\mathrm{d}v}\tfrac{1}{c^2}v\gamma^{-1}(v) = \tfrac{1}{c^2}\gamma^{-1}(v) - \tfrac{1}{c^2}v\gamma^{-2}(v)\frac{\mathrm{d}\gamma(v)}{\mathrm{d}v} = \tfrac{1}{c^2}\gamma^{-1}(v) - \tfrac{1}{c^4}v^2\gamma^{-3}(v).$$

For low velocities, the Lorentz-FitzGerald factor can be approximated by:

$$\gamma(v) \approx \gamma(0) + \cancel{\left(\tfrac{1}{c^2}0\gamma^{-1}(0)\right)v} + \tfrac{1}{2}\left(\tfrac{1}{c^2}\gamma^{-1}(0) - \cancel{\tfrac{1}{c^4}0^2\gamma^{-3}(0)}\right)v^2 + \ldots \approx 1 + \tfrac{1}{2c^2}v^2 + \ldots$$

Returning to the kinetic energy of the moving capacitor:

$$\mathbb{T} = \mathbb{E} - \mathbb{E}_0 = \mathbb{E}_0\left(\gamma(v) - 1\right) \approx \mathbb{E}_0\left(\cancel{1} + \tfrac{1}{2c^2}v^2 - \cancel{1}\right) \approx \tfrac{1}{2}\left(\tfrac{1}{c^2}\mathbb{E}_0\right)v^2 \approx \tfrac{1}{2}mv^2.$$

Thus, the mass associated with the electric fields must be given by:

$$m = \tfrac{1}{c^2}\mathbb{E}_0 \quad \rightarrow \quad \mathbb{E}_0 = mc^2!$$

Einstein's relationship $\mathbb{E}_0 = mc^2$ is universal. It does not just apply to the electric field, but any change in the rest energy of an object implies a change in its mass. Maxwell's equations of electrodynamics, when combined with the correct transformations between inertial reference frames, predict that a charged capacitor will have more mass than an uncharged capacitor, by what is, in practice, too small a fraction of any normal capacitor's mass to measure. For example, a $6.0\,\mu\mathrm{F}$ capacitor that is charged to 10 volts carries a charge of $0.6\times10^{-4}\,\mathrm{C}$ and therefore stores energy in the amount of 3×10^{-4} joules. This charged capacitor has a mass only 3.3×10^{-21} kg greater than when uncharged. Similarly, you have about a trillionth again more mass in Denver than in Death Valley. We do not notice this because the mass change is negligibly small.

Now the total energy of an object can be written as:

$$\mathbb{E} = mc^2\gamma(v) = mc^2 + \mathbb{T}.$$

In the non-relativistic limit $(v \ll c)$, this becomes:

$$\mathbb{E} = mc^2\gamma(v) \approx mc^2 + \tfrac{1}{2}mv^2.$$

And in the ultra-relativistic limit $(v \approx c)$, it becomes:

$$\mathbb{E} = mc^2\gamma(v) \approx \mathbb{T}.$$

Discussion 17.1: Mass and Potential Energy

The relationship $\mathbb{E}_0 = mc^2$ solves a nagging metaphysical question about potential energy, its arbitrary zero point. In introductory mechanics, we learned that energy is stored when work is done against a conservative force. But where is it stored?

Worse still, we also learned that energy is some sort of fundamental quantity, but also that potential energy has an arbitrary zero point. How can this be? The astute student comes out quite confused. After all, when we first learn about temperature the zero point was arbitrary on the Celsius or Fahrenheit scales. However, later we learned about absolute zero, and the Kelvin temperature scale. Why would potential energy be any different? Special relativity solves this problem, as changes in potential energy correspond to changes in its mass.

We can also attribute a proper energy to a collection of particles either by transforming to a frame in which the particles are at rest or, if they are in relative motion with one another, to the center of mass frame where the total momentum is zero. In the case of a collection of particles in relative motion, Einstein's relationship actually makes no distinction between potential and kinetic energy. For example, the mass of an atom of a particular element is related to the total energy of that atom, as measured in the center of mass rest frame. Therefore, the rest mass energy of a composite body may be thought of as that body's *total internal energy*.

Philosophically this puts a limit on the amount of energy we could possibly extract from something, as its total internal energy is simply $\mathbb{E}_0 = mc^2$. In practice, however, this amount of energy was so enormous that it actually led scientists to imagine extracting even more energy from natural resources than had ever been considered before.

Also around the turn of the twentieth century, other physicists were investigating radioactivity and nuclear energy. They discovered that differences in nuclear binding energy were on the order of 1% of mc^2, which, though small, was large enough to be measured.

Problem 17.22: The Magnetic Field Energy

Consider a stationary solenoid of total length l_0 with current I_0 flowing through trough $n_0 l_0$ coils of wire, so n_0 represents the number of turns per length. Now consider the same solenoid, but it is now moving at a velocity $\vec{v} = v\hat{z}$, where $\hat{z}$ is, as usual, along the axis of symmetry.

a) How many turns per length, n, does the moving solenoid have?

b) What is the current I flowing through the solenoid?

c) What is the total magnetic energy, $\mathbb{E}_0$, in the solenoid when it is at rest?

d) Transform this magnetic field to find both the magnetic, and electric, fields of the solenoid in motion.

e) Find the total electromagnetic energy, $\mathbb{E}$, when it is in motion.

f) Show that $\mathbb{E} = \mathbb{E}_0\gamma(v)$.

g) Reconsider the solenoid at rest. If its mass with no current flowing is m_0, what would be its mass with current I_0 flowing through it? What is the fractional mass difference of a tightly packed coil of 1 mm diameter copper wire with 100 A of current? (Copper's density is 9 g/cm³.)

Thought Experiment 17.21: Relativistic Momentum

Consider a parallel plate capacitor at rest with charge $\pm Q_0$, voltage across the plates $\mathbb{V}_0$, and with dimensions $l_0 \times w_0 \times h_0$. The electric field between the capacitor plates must be:

$$\vec{E}_0 = \frac{\mathbb{V}_0}{h_0}\hat{z} = \frac{Q_0}{\varepsilon_0 l_0 w_0}\hat{z}.$$

Now, we transform to a frame in which the capacitor is moving sideways at a velocity $\vec{v} = v\hat{x}$, so their relative velocity is $\vec{u} = -v\hat{x}$. The electric and magnetic fields must now be given by:

$$\vec{E} = \frac{\vec{E}_0 + \vec{u}\times\vec{B}_0}{\sqrt{1-\frac{1}{c^2}u^2}} = \vec{E}_0\gamma(u) = \vec{E}_0\gamma(v) \quad \text{and} \quad \vec{B} = \frac{\vec{B}_0 - \frac{1}{c^2}\vec{u}\times\vec{E}_0}{\sqrt{1-\frac{1}{c^2}u^2}} = -\tfrac{1}{c^2}\vec{u}\times\vec{E}_0\gamma(u) = \tfrac{1}{c^2}\left(\vec{v}\times\vec{E}_0\right)\gamma(v).$$

We now write down the momentum density in the field in terms of the Poynting vector:

$$\frac{\mathrm{d}\vec{p}}{\mathrm{d}V} = \frac{1}{c^2}\vec{S} = \mu_0\varepsilon_0\,\vec{S} = \cancel{\mu_0}\,\varepsilon_0\frac{1}{\cancel{\mu_0}}\vec{E}\times\vec{B}.$$

We now multiply by the volume to find the total momentum of the fields:

$$\vec{p} = lwh\frac{\mathrm{d}\vec{p}}{\mathrm{d}V} = lwh\,\varepsilon_0\,\vec{E}\times\vec{B} = \left(\frac{l_0}{\cancel{\gamma(v)}}\right)w_0h_0\,\varepsilon_0\left(\vec{E}_0\,\cancel{\gamma(v)}\right)\times\left(\tfrac{1}{c^2}\vec{v}\times\vec{E}_0\gamma(v)\right) = l_0w_0h_0\,\varepsilon_0\tfrac{1}{c^2}\vec{v}\left(\vec{E}_0\cdot\vec{E}_0\right)\gamma(v)$$

$$= \left(l_0w_0h_0\right)\left(\varepsilon_0 E^2{}_0\right)\left(\tfrac{1}{c^2}\vec{v}\gamma(v)\right) = \tfrac{1}{c^2}\left(l_0w_0h_0\right)u_E\vec{v}\gamma(v) = \tfrac{1}{c^2}\mathbb{E}_0\vec{v}\gamma(v) = m\vec{v}\gamma(v).$$

This relationship not only holds for electrodynamics, but for mechanics as well, which underscores the universal nature of Einstein's theory.

Example 17.2: A Charged Particle in a Constant Electric Field

Consider a particle of mass m and charge q initially at rest in a constant electric field $\vec{E} = E\hat{x}$. Find the velocity as a function of time.

We solve this problem using the electromagnetic force and Newton's second law, as:

$$\vec{F} = q\vec{E} = \frac{\mathrm{d}\vec{p}}{\mathrm{d}t},$$

and we solve for $\vec{p}$ as:

$$\vec{p} = q\,\vec{E}\,t = q\,E\,t\,\hat{x}.$$

So, clearly the momentum of the particle is proportional to time. Now to find the velocity, we write the momentum in terms of the velocity as:

$$\vec{p} = m\,\vec{v}\,\gamma(v) = m\,v\,\gamma(v)\,\hat{x}.$$

Now, inverting the equation and solving for the speed we have:

$$p = mv\gamma(v) \quad \rightarrow \quad p^2 = m^2v^2\gamma^2(v) = \frac{m^2v^2}{1-\frac{1}{c^2}v^2}$$

$$p^2 - \tfrac{1}{c^2}p^2v^2 = m^2v^2 \quad \rightarrow \quad m^2v^2 + \tfrac{1}{c^2}p^2v^2 = p^2$$

$$m^2v^2\left(1+\frac{p^2}{m^2c^2}\right) = p^2 \quad \rightarrow \quad v = \frac{p}{m\sqrt{1+\frac{p^2}{m^2c^2}}}.$$

Clearly, if the momentum is small, then the velocity is simply $\vec{p}/m$, however if $p \gg mc$ then:

$$\underset{p \gg mc}{v} \approx \frac{p}{m\sqrt{\cancel{1}+\frac{p^2}{m^2c^2}}} \approx \frac{\cancel{p}}{\cancel{m}\left(\frac{\cancel{p}}{\cancel{m}c}\right)} \approx c.$$

Thus no matter how much momentum is imparted by the electric field, the particle will never obtain a speed at or above the speed of light.

Finally, the velocity as a function of time is given by:

$$v = \frac{p}{m\sqrt{1+\frac{p^2}{m^2c^2}}} = \frac{qEt}{m\sqrt{1+\frac{(qEt)^2}{m^2c^2}}} = \frac{qEt}{m}\frac{1}{\sqrt{1+\left(\frac{qEt}{mc}\right)^2}}.$$

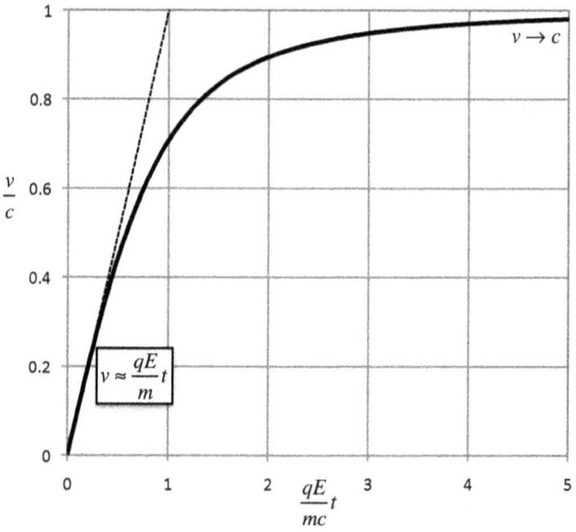

No matter how long the particle is accelerated by the electric field, its speed will never quite equal the speed of light.

Notice also how important the mass is in the velocity of the particle. For a given momentum, the smaller the mass the faster the particle travels. In fact, taking the limit of a massless particle yields the following result:

$$\lim_{m\to 0} v = \lim_{m\to 0}\frac{p}{m\sqrt{\cancel{1}+\frac{p^2}{m^2c^2}}} = \lim_{m\to 0}\frac{\cancel{p}}{\cancel{m}\left(\frac{\cancel{p}}{\cancel{m}c}\right)} = c.$$

Therefore, if a particle has no mass it is always traveling at the speed of light. The corollary is also true. If something is moving at exactly the speed of light, and it carries momentum, it must have no mass. Even in classical physics, light conforms to this definition: it carries momentum and energy, and travels at the speed of light. Therefore, treating light like a massless particle does not violate special relativity.

Thought Experiment 17.22: Transformation of Energy and Momentum

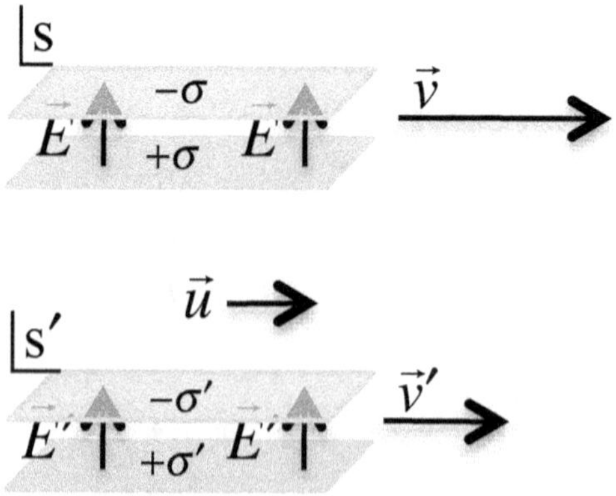

In Thought Experiment 17.20 (p. 598) and Thought Experiment 17.21 (p. 600), we showed that energy and momentum of a body moving at a velocity $\vec{v} = v\hat{x}$ is given by:

$\mathbb{E} = mc^2\gamma(v)$ and $\vec{p} = m\vec{v}\gamma(v)$.

And, of course, in a transformed reference frame, the energy of and momentum are:

$\mathbb{E}' = mc^2\gamma(v')$ and $\vec{p}' = m\vec{v}'\gamma(v')$.

Our goal is to use the addition of parallel

velocities (p. 577) in order to solve for the transformed energy and momentum in terms of the velocity in the laboratory frame.

The addition of parallel velocities, and the corresponding addition of Lorentz factors (p. 589), give us:

$$v' = \frac{v-u}{1-\frac{1}{c^2}uv} \quad \text{and} \quad \gamma(v') = \frac{1}{\sqrt{1-\frac{1}{c^2}v'^2}} = \left(1-\tfrac{1}{c^2}uv\right)\gamma(u)\gamma(v).$$

Therefore the transformed energy and momentum are:

$$\mathbb{E}' = mc^2\gamma(v') = mc^2\left(1-\tfrac{1}{c^2}uv\right)\gamma(u)\gamma(v)$$

$$\vec{p}' = m\vec{v}'\gamma(v') = m\left(\frac{v-u}{\cancel{1-\frac{1}{c^2}uv}}\right)\left(\cancel{1-\tfrac{1}{c^2}uv}\right)\gamma(u)\gamma(v)\,\hat{x}.$$

Now, in terms of $\mathbb{E}$ and $\vec{p}$, for a parallel boost:

$$\mathbb{E}' = \left(\left(mc^2\gamma(v)\right) - u\left(mv\gamma(v)\right)\right)\gamma(u) = \left(\mathbb{E} - \vec{u}\cdot\vec{p}\right)\gamma(u)$$

$$\vec{p}' = \left(\left(mv\gamma(v)\hat{x}\right) - \tfrac{1}{c^2}u\hat{x}\left(mc^2\gamma(v)\right)\right)\gamma(u) = \left(\vec{p} - \tfrac{1}{c^2}\vec{u}\,\mathbb{E}\right)\gamma(u).$$

Notice that this follows a similar form as the 4-position, 4-current density, and the 4-potential. Therefore, we will put the energy into momentum units, and write a 4-momentum $\left(\frac{1}{c}\mathbb{E},\vec{p}\right)$, and its transform as:

$$\begin{pmatrix} \frac{1}{c}\mathbb{E}' \\ p'_x \\ p'_y \\ p'_z \end{pmatrix} = \begin{pmatrix} \gamma(u) & -\frac{u}{c}\gamma(u) & 0 & 0 \\ -\frac{u}{c}\gamma(u) & \gamma(u) & 0 & 0 \\ 0 & 0 & 1 & 0 \\ 0 & 0 & 0 & 1 \end{pmatrix} \begin{pmatrix} \frac{1}{c}\mathbb{E} \\ p_x \\ p_y \\ p_z \end{pmatrix} = \begin{pmatrix} \gamma(u)\frac{1}{c}\mathbb{E} - \frac{u}{c}\gamma(u)p_x \\ -\frac{u}{c}\gamma(u)\frac{1}{c}\mathbb{E} + \gamma(u)p_x \\ p_y \\ p_z \end{pmatrix}.$$

Since this transform is the same as our other 4-vectors, its 4-magnitude must also be a Lorentz invariant:

$$\left\|\tfrac{1}{c}\mathbb{E},\vec{p}\right\|^2 = \left(\tfrac{1}{c}\mathbb{E},\vec{p}\right)\bullet\left(\tfrac{1}{c}\mathbb{E},\vec{p}\right) = \begin{pmatrix} \frac{1}{c}\mathbb{E} & p_x & p_y & p_z \end{pmatrix} \begin{pmatrix} -1 & 0 & 0 & 0 \\ 0 & 1 & 0 & 0 \\ 0 & 0 & 1 & 0 \\ 0 & 0 & 0 & 1 \end{pmatrix} \begin{pmatrix} \frac{1}{c}\mathbb{E} \\ p_x \\ p_y \\ p_z \end{pmatrix}$$

$$= -\tfrac{1}{c^2}\mathbb{E}^2 + p_x^2 + p_y^2 + p_z^2 = p^2 - \tfrac{1}{c^2}\mathbb{E}^2 = \vec{p}\cdot\vec{p} - \tfrac{1}{c^2}\mathbb{E}^2 = p^2 - \tfrac{1}{c^2}\mathbb{E}^2.$$

Now we will investigate this quantity further, by expanding it in terms of the mass and velocity:

$$\left\|\tfrac{1}{c}\mathbb{E},\vec{p}\right\|^2 = -\frac{\mathbb{E}^2}{c^2} + p^2 = -\frac{\left(mc^2\gamma(v)\right)^2}{c^2} + \left(mv^2\gamma(v)\right)^2$$

$$= m^2\left(-c^2+v^2\right)\gamma^2(v) = -m^2c^2\left(\cancel{1-\tfrac{1}{c^2}v^2}\right)\left(\frac{1}{\cancel{1-\frac{1}{c^2}v^2}}\right) = -m^2c^2.$$

Notice, again, that this is simply $\left\|\tfrac{1}{c}\mathbb{E},\vec{p}\right\|^2$ in the rest frame of the object. Now equating the magnitudes, we have the simple relationship:

$$p^2 - \frac{\mathbb{E}^2}{c^2} = -m^2c^2 \quad \rightarrow \quad \mathbb{E}^2 = m^2c^4 + p^2c^2 .$$

Problem 17.23: The Kinetic Energy and Momentum

a) Show that, in general, the kinetic energy is given by:

$$\mathbb{T} = mc^2\left(\sqrt{1+\left(\tfrac{p}{mc}\right)^2} - 1\right).$$

b) Show that, in the non-relativistic limit, $\mathbb{T} \approx \frac{p^2}{2m}$.

c) Show that in the super-relativistic limit $\mathbb{T} \approx pc$.

d) Make a plot of $\frac{1}{mc^2}\mathbb{T}$ as a function of $\frac{1}{mc}p$.

Problem 17.24: A Dogfight in Space

A spaceship with mass m_1 is approaching the stern of an enemy ship m_2. A combat journalist, in an inertial reference frame located far away and to the side, films the interaction. When preparing her news story, she measures the speeds of the two ships to be v_1 and v_2.

a) Find the momentum of each ship, according to the journalist. What is the total momentum of the system?

b) Ship 1 launches a torpedo from its bow toward the stern of ship 2. The torpedo has a mass m_3, and leaves the launch bay with a speed v_3'. At what speed, v_3, does the journalist see the torpedo move?

c) At what speed, v_3'', does the crew of ship 2 see the torpedo approach?

d) Ship 2 catches the torpedo in a tractor beam, and safely pulls it into a cargo bay in its stern. From the journalist's point of view, what is the final velocity of each ship after the interaction?

Problem 17.25: The Velocity 4-Vector

The 4-velocity vector can be defined by the relation:

$${}^4\vec{v} \equiv \frac{{}^4\vec{p}}{m} = \frac{\left(\tfrac{1}{c}\mathbb{E},\vec{p}\right)}{m} = \left(\tfrac{1}{mc}\mathbb{E},\tfrac{1}{m}\vec{p}\right).$$

a) Show that the velocity 4-vector is then simply given by:

$${}^4\vec{v} = \frac{(c,\vec{v})}{\sqrt{1-\left(\tfrac{v}{c}\right)^2}}.$$

b) Find the 4-magnitude of the 4-velocity, and show that it is independent of reference frame.

c) Boost the velocity 4-vector to derive the correct rules for velocity addition that we found in Thought Experiment 17.6 (p. 577).

Problem 17.26: The Relativistic Child-Langmuir Law

In the early days of radio technology vacuum tubes were used in much the same ways as diodes and transistors are now. This problem is designed to walk you through the relativistic Child-Langmuir law, which is a relationship between current and voltage for vacuum tubes in much the same way as Ohm's law is for resistors.

Consider a pair of parallel plates in a vacuum tube, with electrons flowing from the negative plate (cathode) to the positive plate (anode). The anode is held at a voltage $\mathbb{V}_0$ with respect to the cathode. The gap between the plates is h, and the surface area of each plate is A.

a) Assume that the electrons start from rest at the cathode and accelerate toward the anode. Show that the velocity of the electrons, after passing through a potential difference $\mathbb{V}$, will be given by:

$$v = c\frac{\sqrt{\mathbb{V}/\mathbb{V}_e\left(2+\mathbb{V}/\mathbb{V}_e\right)}}{\left(1+\mathbb{V}/\mathbb{V}_e\right)},$$

where $\mathbb{V}_e = mc^2/e = 511\,\text{kV}$ is the rest energy potential of an electron. Also show that, in the limit that $\mathbb{V} \ll \mathbb{V}_e$, we reproduce the non-relativistic limit of $v = \sqrt{\frac{2e\mathbb{V}}{m_e}}$.

b) Show that the charge density in the gap, ρ, is related to the local electric potential $\mathbb{V}$, and the total current I, by:

$$\rho = \frac{I}{Ac}\sqrt{\frac{\mathbb{V}_e}{2\mathbb{V}}}\frac{\left(1+\mathbb{V}/\mathbb{V}_e\right)}{\sqrt{\left(1+\mathbb{V}/2\mathbb{V}_e\right)}}$$

c) Show that the electric potential in the gap is governed by the following differential equation:

$$\frac{d^2\mathbb{V}}{dz^2} = \frac{Z_0 I}{A}\sqrt{\frac{\mathbb{V}_e}{2\mathbb{V}}}\frac{\left(1+\mathbb{V}/\mathbb{V}_e\right)}{\sqrt{\left(1+\mathbb{V}/2\mathbb{V}_e\right)}}.$$

d) Making the unitless substitution $\eta \equiv \mathbb{V}/\mathbb{V}_e$, show that this is equivalent to:

$$\frac{d^2\eta}{dz^2} = \frac{Z_0 I}{A\mathbb{V}_e}\sqrt{\frac{1}{2\eta}}\frac{(1+\eta)}{\sqrt{\left(1+\frac{1}{2}\eta\right)}},$$

And show that:

$$\left(\frac{d\eta}{dz}\right)^2 = \frac{Z_0 I}{A\mathbb{V}_e}2\sqrt{\eta(\eta+2)}.$$

e) Show that the current can be given by the following expression:

$$I = \frac{V_e}{2Z_0}\frac{A}{h^2}\left(\int_0^{\mathbb{V}_0/\mathbb{V}_e}\frac{d\eta}{\sqrt[4]{\eta(\eta+2)}}\right)^2.$$

f) Find a relationship between the current and the voltage in the non-relativistic limit, where $\mathbb{V}_0 \ll \mathbb{V}_e$.

g) Show that at extremely high voltages, $\mathbb{V}_0 \gg \mathbb{V}_e$, this simply follows Ohm's law. What is the resistance in the super-relativistic case?

Problem 17.27: Proton Collisions and the Higgs Boson

Consider a proton of mass m_p moving with a kinetic energy $\mathbb{T}$, that is approaching another proton at rest.

a) Nonrelativistically show that the total kinetic energy of the system, in the center-of-mass rest frame, is exactly half that of the initial proton in the lab frame ($\mathbb{T}_{cm} = \frac{1}{2}\mathbb{T}$). Recall that the center-of-mass rest frame has, by definition, zero total momentum.

b) Use the Lorentz invariance of the Energy-Momentum 4-vector to find the total kinetic energy in the center of mass frame $\mathbb{T}_{cm}$ as a function of the total energy in the lab frame $\mathbb{T}$.

c) Show that your result in (b) reverts to your result in (a) when: $\mathbb{T} \ll m_p c^2$.

d) Plot the available energy $\mathbb{T}_{cm}$ as a function of the initial kinetic energy $\mathbb{T}$ and the proton mass $m_p c^2 = 0.938\,\text{GeV}$. How much energy would it take to produce 1 TeV of available center of mass energy?

e) Plot the efficiency, $\mathbb{T}_{cm}/\mathbb{T}$, as a function of the initial kinetic energy in units of GeV. At low energies it should be 50%, because of part (a). Does it become more or less efficient at high energies?

f) The Large Hadron Collider, at CERN, actually accelerates two beams of protons in opposite directions. If $\mathbb{T}$ is the kinetic energy of each proton, what is the center of mass energy budget when they collide? Show that, for the non-relativistic case, it is 4 times greater than if there were only one beam hitting a stationary target.

g) In 2013, scientists at CERN reported discovering a new particle with a mass of about 125 GeV. What is the minimum kinetic energy per proton to have produced a particle of this mass?

h) Report on the discovery of the Higgs Boson, how it fits in with the standard model, and the experiments that have measured its mass.

Problem 17.28: Pion Decay

A π^+-meson, or pion, is created in a high-energy collision of a cosmic ray with the atmosphere 200 km above sea level. The pion travels vertically downward towards the surface of the earth at a speed of 0.99c. It disintegrates 26 ns after its creation, as measured in its own rest frame. The pion rest mass is $m_{\pi^+} = 140\,\frac{\text{MeV}}{c^2}$.

a) In the pion's rest frame, how far did it travel in 26 ns?

b) At what altitude above sea level did the pion disintegrate?

c) Calculate the radius of curvature in the Earth's 25 μT magnetic field. How does this compare to the radius of the Earth?

d) How much kinetic energy does the pion have with respect to Earth?

e) Pions decay into muons and neutrinos. In this process they must conserve charge, and another quantity called lepton number, so our positive pion must decay into an antimuon (μ^+) and a neutrino (ν^0). The rest mass of the antimuon is $m_{\mu^+} = 106\,\frac{\text{MeV}}{c^2}$, and the neutrino mass is less than an electron volt per c^2. In the rest frame of the pion, calculate the energies of the antimuon and the neutrino.

f) Assuming a sideways decay in its rest frame, calculate the kinetic energy of the antimuon and the neutrino with respect to Earth. In the Earth's reference frame, was energy conserved?

Example 17.3: Relativistic Doppler Beaming

Consider a ray of light moving toward the origin with a Poynting vector $\vec{\mathbb{S}} = \mathbb{S}\hat{x}$. Next, consider a reference frame $\lfloor\underline{s'}$ moving away from the source at a velocity $\vec{u} = u\hat{x}$. What is the Poynting flux $\vec{\mathbb{S}}$ in the new reference frame?

First we imagine a box, with cross-section $A = w \times h$ and length l. The energy and momentum of a volume element of light are:

$$\mathbb{E} = \tfrac{1}{c}\mathbb{S}Al \quad \text{and} \quad \vec{p} = \tfrac{1}{c^2}\mathbb{S}Al = \tfrac{1}{c}\mathbb{E}\hat{x}.$$

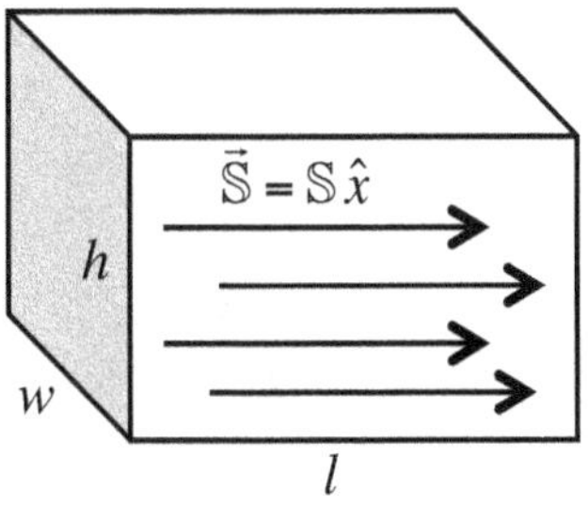

We begin by writing the energy-momentum 4-vector in terms of the energy, and transforming it to frame $\lfloor\underline{s'}$:

$$\begin{pmatrix} \frac{1}{c}\mathbb{E}' \\ p_x' \\ p_y' \\ p_z' \end{pmatrix} = \begin{pmatrix} \gamma(u) & -\frac{u}{c}\gamma(u) & 0 & 0 \\ -\frac{u}{c}\gamma(u) & \gamma(u) & 0 & 0 \\ 0 & 0 & 1 & 0 \\ 0 & 0 & 0 & 1 \end{pmatrix} \begin{pmatrix} \frac{1}{c}\mathbb{E} \\ \frac{1}{c}\mathbb{E} \\ 0 \\ 0 \end{pmatrix} = \begin{pmatrix} \frac{1}{c}\mathbb{E}\left(1-\frac{u}{c}\right)\gamma(u) \\ \frac{1}{c}\mathbb{E}\left(1-\frac{u}{c}\right)\gamma(u) \\ 0 \\ 0 \end{pmatrix}.$$

Next we define two events, separated by a time Δt and distance l, such that event 1 represents the observation of the light at the leading face of the box, and event 2 the trailing face. Clearly the leading corner is further along than the trailing corner, so its position is greater. Moreover, the leading edge is observed first, so we find the time and position differences in both reference frames as:

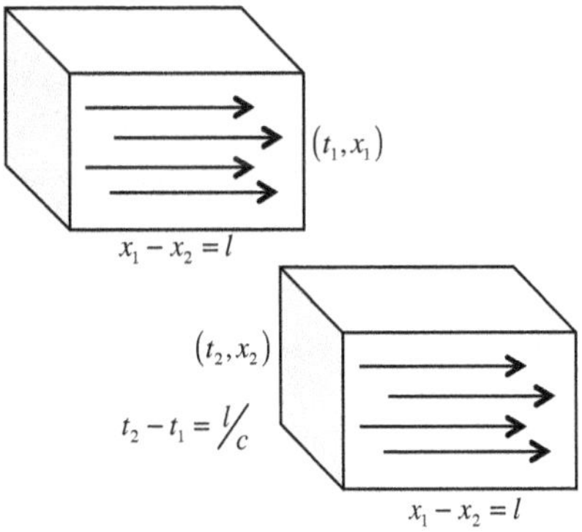

$$t_2 - t_1 = l/c, \quad x_2 - x_1 = -l, \quad t_2' - t_1' = l'/c, \quad \text{and} \quad x_2' - x_1' = -l'.$$

The 4-displacement vectors are therefore $(l, -l, 0, 0)$ and $(l', -l', 0, 0)$, which are related as follows:

$$\begin{pmatrix} l' \\ -l' \\ 0 \\ 0 \end{pmatrix} = \begin{pmatrix} \gamma(u) & -\frac{u}{c}\gamma(u) & 0 & 0 \\ -\frac{u}{c}\gamma(u) & \gamma(u) & 0 & 0 \\ 0 & 0 & 1 & 0 \\ 0 & 0 & 0 & 1 \end{pmatrix} \begin{pmatrix} l \\ -l \\ 0 \\ 0 \end{pmatrix} = \begin{pmatrix} l\gamma(u) + l\frac{u}{c}\gamma(u) \\ -l\frac{u}{c}\gamma(u) - l\gamma(u) \\ 0 \\ 0 \end{pmatrix}.$$

Therefore, the length of the box in frame $\lfloor S'$ is: $l' = l\left(1+\tfrac{1}{c}u\right)\gamma(u)$.

Now, we take the ratio of the energy contained in the box between the two frames, and solve for the ratio of the Poynting fluxes:

$$\frac{\mathbb{E}'}{\mathbb{E}} = \frac{\cancel{\tfrac{1}{c}}\mathbb{S}'\cancel{A'}l'}{\cancel{\tfrac{1}{c}}\mathbb{S}\cancel{A}l} = \frac{\mathbb{S}'l'}{\mathbb{S}l} \quad \text{and} \quad \frac{\mathbb{S}'}{\mathbb{S}} = \frac{\mathbb{E}'}{\mathbb{E}}\frac{l}{l'}.$$

Now, applying the results from our transformations above:

$$\frac{\mathbb{S}'}{\mathbb{S}} = \frac{\mathbb{E}'}{\mathbb{E}}\frac{l}{l'} = \left(\frac{\tfrac{1}{c}\cancel{\mathbb{E}}\left(1-\tfrac{1}{c}u\right)\cancel{\gamma(u)}}{\cancel{\mathbb{E}}}\right)\left(\frac{\cancel{l}}{\cancel{l}\left(1+\tfrac{1}{c}u\right)\cancel{\gamma(u)}}\right) = \frac{1-\tfrac{1}{c}u}{1+\tfrac{1}{c}u}.$$

This is often referred to as *Doppler beaming* or the *headlight effect*, in that the brightness of a relativistic light source is extremely dependent on the velocity toward the observer. This can be seen in quasars with bidirectional jets, as the radio interferometer map[293] shows. The approaching jet is quite clear, but the receding jet cannot be seen at all.

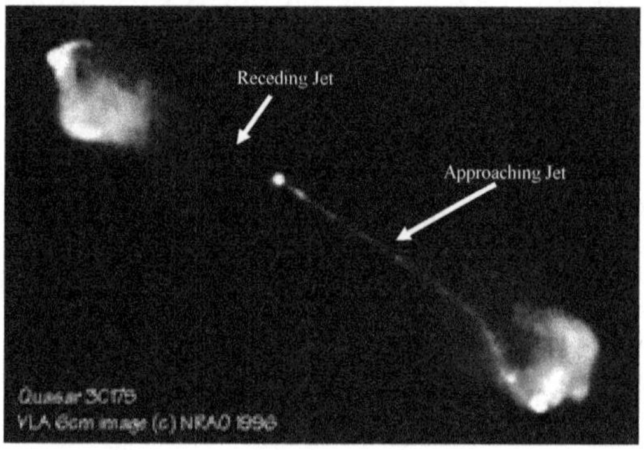

In the super-relativistic limit, the flux densities of the receding, and approaching, sources become:

$$\left.\frac{\mathbb{S}'}{\mathbb{S}}\right|_{u\approx c} = \frac{1-\tfrac{1}{c}u}{1+\tfrac{1}{c}u} \approx \frac{1-\tfrac{1}{c}u}{2} \approx 0.$$

$$\left.\frac{\mathbb{S}'}{\mathbb{S}}\right|_{u\approx -c} = \frac{1-\tfrac{1}{c}u}{1+\tfrac{1}{c}u} = \left(1-\tfrac{1}{c}u\right)^2\gamma^2(u) \approx 4\gamma^2(u).$$

17.3 *The Electromagnetic Field Tensor*

Albert Einstein altered our understanding of notions such as space, time, and simultaneity with the special theory of relativity. He also extended Maxwell's unification of electricity, magnetism, and optics by replacing the separate electric and magnetic fields of pre-relativity physics with a unified quantity, the *electromagnetic field*, to describe Maxwellian electrodynamics.

In this section, we will introduce the electromagnetic field tensor, which allows us to write Maxwell's equations in a way that makes their Lorentz covariance obvious by inspection. Of course, Maxwell's equations are already Lorentz covariant, so the reader may wonder what is gained from recasting them in an admittedly slick mathematical form. The covariant formulation of classical electromagnetism allows us to write laws of classical electromagnetism (in particular, Maxwell's equations and the Lorentz force) in a form that is manifestly invariant under Lorentz-FitzGerald transformations. These Lorentz invariant expressions make it easy to prove that the laws of classical electromagnetism take the same form in any inertial coordinate system, and also provide an especially convenient way to translate the fields and forces from one frame to another.

---

[293]Picture from: Deep VLA Imaging of Twelve Extended 3CR Quasars, by Alan H. Bridle, David H. Hough, Colin J. Lonsdale, Jack O. Burns and Robert A. Laing, The Astronomical Journal, **108**, (1994) 766-820.

Discussion 17.2: Einstein Notation

Until this section, we have used a 4-vector notation that we believe to be straightforward for the reader. However, this is not the notation that was used by Einstein, nor is it the notation used professionally by scholars of general relativity theory. Here we introduce the reader to *Einstein Notation*, which is algebraically flexible and clear, as each vector or tensor component is written explicitly as a summation of other components.

Recall the 4-magnitude of the spacetime 4-vector we discussed in Thought Experiment 17.7:

$$\|ct,\vec{r}\|^2 = -(ct)^2 + x^2 + y^2 + z^2 ,$$

which is Lorentz invariant. It is convenient to write the components of the spacetime 4-vector using the following notation:

$$x^\mu = \left(x^0, x^1, x^2, x^3\right) \equiv \left(ct, x, y, z\right).$$

The superscript μ ranges over the four values 0, 1, 2, and 3. If we write the components of the Lorentz-FitzGerald transformation matrix (see Thought Experiment 17.5 and Thought Experiment 17.6) as Λ^μ_ν, where μ is the row number and ν the column number of the array element, then the components, x'^μ, of the transformed spacetime 4-vector are:

$$x'^\mu = \sum_{\nu=0}^{3} \Lambda^\mu_\nu x^\nu \quad \leftrightarrow \quad \begin{pmatrix} ct' \\ x' \\ y' \\ z' \end{pmatrix} = \begin{pmatrix} \gamma(u) & -\frac{u}{c}\gamma(u) & 0 & 0 \\ -\frac{u}{c}\gamma(u) & \gamma(u) & 0 & 0 \\ 0 & 0 & 1 & 0 \\ 0 & 0 & 0 & 1 \end{pmatrix} \begin{pmatrix} ct \\ x \\ y \\ z \end{pmatrix}.$$

For example:

$$x'^0 = \Lambda^0_0 x^0 + \Lambda^0_1 x^1 + \Lambda^0_2 x^2 + \Lambda^0_3 x^3 = \gamma x^0 - \gamma \tfrac{1}{c} u x^1 + 0 \cdot x^2 + 0 \cdot x^3 ,$$

and

$$ct' = \gamma\left(ct - \tfrac{1}{c} u x\right).$$

It is standard, and quite convenient, to introduce the *Einstein summation convention*, whereby repeated indices imply sums over the appropriate range. The index over which the sum is taken is repeated in the product of Λ^μ_ν and x^ν, so the Einstein convention allows us to write:

$$\Lambda^\mu_\nu x^\nu = \sum_{\nu=0}^{3} \Lambda^\mu_\nu x^\nu , \text{ so that we have: } x'^\mu = \Lambda^\mu_\nu x^\nu .$$

Now it is easy to see that:

$$\frac{\partial x'^\mu}{\partial x^\nu} = \Lambda^\mu_\nu ,$$

so the transformation may be rewritten as:

$$x'^\mu = \frac{\partial x'^\mu}{\partial x^\nu} x^\nu .$$

We then say that x^{μ} transforms as a *contravariant* vector. The displacement vector $\Delta \underline{x} = (c\Delta t, \Delta x, \Delta y, \Delta z)$ also transforms as a contravariant vector, as do its time derivatives such as velocity and acceleration.

The metric tensor for flat spacetime, or Minkowski space, is so special that we denote it by $\ddot{\eta}$ instead of $\ddot{g}$, which one usually writes for any general metric tensor, and write its components as:

$$\eta^{\mu\nu} = \eta_{\mu\nu} = \begin{pmatrix} -1 & 0 & 0 & 0 \\ 0 & 1 & 0 & 0 \\ 0 & 0 & 1 & 0 \\ 0 & 0 & 0 & 1 \end{pmatrix}.$$

As we said before, the product $\underline{x} \cdot \underline{x}$ is affected the metric tensor:

$$\underline{x} \cdot \underline{x} = \eta_{\mu\nu}\, x^{\mu} x^{\nu} = \sum_{\mu=0}^{3} \sum_{\nu=0}^{3} \eta_{\mu\nu}\, x^{\mu} x^{\nu} = \eta_{00} x^{0} x^{0} + \eta_{11} x^{1} x^{1} + \eta_{22} x^{2} x^{2} + \eta_{33} x^{3} x^{3} = -c^2 t^2 + x^2 + y^2 + z^2 .$$

The off-diagonal components of the metric, such as η_{12}, for example, were not expressly written since $\eta_{\mu\nu} = 0$ when $\mu \neq \nu$. Again, we used the Einstein summation convention to write $\eta_{\mu\nu} x^{\mu} x^{\nu}$ for the double sum $\sum_{\mu=0}^{3} \sum_{\nu=0}^{3} \eta_{\mu\nu}\, x^{\mu} x^{\nu}$. Since μ and ν are repeated twice in the expression, we sum over both indices. It is always the case that the repeated indices to be summed are one-up, and one-down, as in the expression above. An index is never repeated more than twice; for example, the expression $\eta_{\nu\nu}\, x^{\nu}$ is not defined.

We have used the Minkowski metric to form the inner product between two 4-vectors, but we can also use it to form another set of components of the 4-vector $\underline{x}$, which we denote by x_{μ}:

$$x_{\mu} = (x_0, x_1, x_2, x_3) = (-ct, x, y, z),$$

where the subscripted components differ from the superscripted ones only in the sign of the time component. The subscripted variables form the components of a *covariant* vector, x_{μ}. The relationship between the two types, covariant and contravariant, can be easily displayed with the use of the metric tensor:

covariant: $x_{\mu} = \eta_{\mu\nu}\, x^{\nu} = \sum_{\nu=0}^{3} \eta_{\mu\nu}\, x^{\nu}$

contravariant: $x^{\mu} = \eta^{\mu\nu}\, x_{\nu} = \sum_{\mu=0}^{3} \eta^{\mu\nu}\, x_{\nu}$.

Definition 17.1: The Kronecker and Levi-Cevita Functions

In summation notation, there are two common symbols that arise with regularity for writing three dimensional dot and cross products, which have the symbols $\delta_{i,j}$ and $\varepsilon_{i,j,k}$ respectively. Each is defined such that:

$$\delta_{i,j} \equiv \begin{cases} 0 \text{ if } i \neq j \\ 1 \text{ if } i = j \end{cases} \quad \text{and} \quad \varepsilon_{i,j,k} \equiv \begin{cases} -1 & \text{if } (i,j,k) = (1,3,2),(2,1,3),(3,2,1). \\ 1 & \text{if } (i,j,k) = (1,2,3),(2,3,1),(3,1,2). \\ 0 & \text{if any two indices are equal.} \end{cases}$$

These are referred to as the *Kronecker delta function* and the *Levi-Cevita cyclical permutation symbol* respectively. They are named after the German mathematician Leopold Kronecker and the Italian Mathematician Tullio Levi-Civita.

Thus the dot and cross products of two 3-vectors can we written as:

$$\vec{A}\cdot\vec{B} = \sum_{i=1}^{3}\sum_{j=1}^{3} A_i B_j \delta_{i,j} = A_i B_j \delta_{i,j} \quad \text{and} \quad \left(\vec{A}\times\vec{B}\right)_k = \sum_{i=1}^{3}\sum_{j=1}^{3} A_i B_j \varepsilon_{i,j,k} = A_i B_j \varepsilon_{i,j,k}.$$

Notice our use of the Einstein summation convention, which assumes summation over the appropriate range for repeated indices. For brevity, we will omit the commas in the indices of the Kronecker delta and the Levi-Cevita symbol and simply write $\delta_{i,j} = \delta_{ij}$ and $\varepsilon_{i,j,k} = \varepsilon_{ijk}$.

For example, the work from the origin due to a force and the i-th component of the angular momentum are given by:

$$W = \delta_{ij}\, x_j\, F_i \quad \text{and} \quad L_i = \varepsilon_{ijk}\, x_j\, p_k.$$

Problem 17.29: The Triple Product of Three Vectors

a) Consider that a vector $\vec{A}$ can be expressed in terms of a set of orthogonal unit vectors $\hat{e}_i$ and suitable components A_i:

$$\vec{A} = \sum_{i=1}^{3} A_i \hat{e}_i,$$

which in Einstein notation is simply written $\vec{A} = A_i \hat{e}_i$. Because the unit vectors are specified to be orthogonal, their scalar product can be written as $\hat{e}_i \cdot \hat{e}_j = \delta_{ij}$, where δ_{ij} is the Kronecker delta defined above.

Use this fact, and the Einstein summation convention, to find explicitly the scalar product of two 3-vectors, $\vec{A}$ and $\vec{B}$, in Cartesian coordinates.

b) If we calculate the cross product between the 3-vectors $\vec{A}$ and $\vec{B}$ following the method of part (a), we are lead to the following expression:

$A \times B = A_i B_j \hat{e}_i \times \hat{e}_j$, where $1 \le i \le 3$ and $1 \le j \le 3$.

Show by *explicit* calculation that if we want to obtain the usual expression for the cross product, then the following identity must hold:

$$\hat{e}_i \times \hat{e}_j = \varepsilon_{ijk}\, \hat{e}_k,$$

where ε_{ijk} is the Levi-Cevita symbol defined above.

c) Write the triple product $\vec{C}\cdot\left(\vec{A}\times\vec{B}\right)$ in terms of the components of $\vec{A}$, $\vec{B}$ and $\vec{C}$, and the Kronecker delta and Levi-Cevita symbols. Can you simplify it further so that the Kronecker delta is no longer in the final expression?

d) Prove the identity: $\vec{A}\cdot\left(\vec{B}\times\vec{C}\right)=\vec{B}\cdot\left(\vec{C}\times\vec{A}\right)=\vec{C}\cdot\left(\vec{A}\times\vec{B}\right)$.

Problem 17.30 A Useful Identity

a) Whenever two vector products occur, two Levi-Cevita symbols will appear, and will frequently have an index in common over which we are supposed to sum. For example, suppose we must sum the product $\varepsilon_{ink}\varepsilon_{jmk}$ over the repeated index k where $1\le k\le 3$. It turns out that such a sum can be expressed in terms of Kronecker deltas and is given by the following handy identity:

$\varepsilon_{ink}\varepsilon_{jmk}=\delta_{ij}\delta_{nm}-\delta_{im}\delta_{nj}$. Verify this identity by substituting specific numbers for the indices and evaluating the appropriate Levi-Cevita symbols and Kronecker deltas.

b) Use the identity from part (a) to prove that:

$$\vec{A}\times\left(\vec{B}\times\vec{C}\right)=\left(\vec{A}\cdot\vec{C}\right)\vec{B}-\left(\vec{A}\cdot\vec{B}\right)\vec{C}=\vec{B}\left(\vec{A}\cdot\vec{C}\right)-\vec{C}\left(\vec{A}\cdot\vec{B}\right).$$

This is sometimes referred to as the "Bac(k) Cab" rule.

Discussion 17.3: Polar & Axial Vectors and The Faraday Tensor

We have so far neglected a subtle difference between electric and magnetic fields. They are different kinds of vectors. Vectors whose scalar products remain unchanged by all orthogonal transformations of coordinates, such as translations, rotations, and reflections, are *polar* vectors. Obvious examples of a polar vector are displacement, velocity, force, and the electric field. Of course, if any, or all, of the coordinate axes changes sign, the corresponding vector components change sign. For example, we have the following reflection transformation rule for position and momentum:

$x_i \to x_i' = -x_i$, and $p_i \to p_i' = -p_i$.

But this is not the case for the cross product of two polar vectors. The angular momentum vector, $\vec{L}=\vec{r}\times\vec{p}$, whose components are $L_i=\varepsilon_{ijk}x_j p_k$, transforms according to a different reflection rule:

$L_i \to L_i' = \varepsilon_{ijk}x_j' p_k' = \varepsilon_{ijk}x_j p_k = L_i$.

Vectors that transform according to this rule, i.e. that transform like a vector formed by the cross product of two polar vectors, are called *axial* vectors. Examples of axial vectors include angular momentum, torque, and the magnetic field. Axial vectors transform just like polar vectors under translations and rotations of the coordinate system, but under any changes involving reflections of the coordinates the relative sign between axial and polar vectors is reversed. The scalar product of an axial vector and a polar vector is not a true scalar, but a psuedoscalar: it changes sign under an inversion of the axes. Physical vector relations can therefore equate only vectors of the same kind.

Axial vectors cannot form the spatial components of a four-vector, but the components of an axial vector formed from the cross product of two polar vectors can be expressed by two indices, one from each one of the polar vectors. If, for example, we have that $\vec{L} = \vec{r} \times \vec{p}$, then the components of $\vec{L}$ can be labeled in a right-handed coordinate system:

$L_{jk} = x_j\, p_k - x_k\, p_j = -L_{kj}$, where $L_{12} = L_z$, $L_{23} = L_x$, and $L_{13} = L_y$.

For example, Faraday's law:

$$\vec{\nabla} \times \vec{E} = -\tfrac{\partial}{\partial t}\vec{B},$$

is written in this notation as:

$$\frac{\partial E_j}{\partial x^i} - \frac{\partial E_i}{\partial x^j} = -\frac{\partial}{\partial t} B_{ij},$$

where $1 \le i \le 3$ and $1 \le j \le 3$. The magnetic field is derived from the standard vector potential via $\vec{B} = \vec{\nabla} \times \vec{\mathbb{A}}$, or in this notation as:

$$B_{ij} = \frac{\partial \mathbb{A}_j}{\partial x^i} - \frac{\partial \mathbb{A}_i}{\partial x^j}.$$

On the other hand, the electric field is polar since it is derived from the scalar and vector potential via:

$$E_i = -\frac{\partial \mathbb{V}}{\partial x^i} - \frac{\partial \mathbb{A}_i}{\partial t}.$$

If we look at the expressions for the magnetic and electric fields, we can introduce a four dimensional tensor $F_{\mu\nu}$ that is derived from the 4-potential we introduced in Thought Experiment 17.13 (p. 586):

$$\mathcal{A}_\mu = \left(\tfrac{1}{c}\mathbb{V}, \vec{\mathbb{A}}\right),$$

and implicitly summing from 0 to 3:

$$F_{\mu\nu} = \frac{\partial \mathcal{A}_\nu}{\partial x^\mu} - \frac{\partial \mathcal{A}_\mu}{\partial x^\nu}.$$

The contravariant tensor is, therefore, given by

$$F^{\mu\nu} = \frac{\partial \mathcal{A}^\nu}{\partial x_\mu} - \frac{\partial \mathcal{A}^\mu}{\partial x_\nu}, \text{ where: } \mathcal{A}^\mu = \left(-\tfrac{1}{c}\mathbb{V}, \vec{\mathbb{A}}\right).$$

Explicitly, the components of the covariant field tensor are:

$$F_{\mu\nu} = \begin{pmatrix} 0 & E_x/c & E_y/c & E_z/c \\ -E_x/c & 0 & -B_z & B_y \\ -E_y/c & B_z & 0 & -B_x \\ -E_z/c & -B_y & B_x & 0 \end{pmatrix}.$$

The components of the corresponding contravariant tensor are:

$$F^{\mu\nu} = \begin{pmatrix} 0 & -E_x/c & -E_y/c & -E_z/c \\ E_x/c & 0 & -B_z & B_y \\ E_y/c & B_z & 0 & -B_x \\ E_z/c & -B_y & B_x & 0 \end{pmatrix}.$$

The electromagnetic field tensor is also called the *Faraday tensor.*

Next let us consider the two inhomogeneous equations in free space:

$\vec{\nabla}\cdot\vec{E} = \frac{1}{\varepsilon_0}\rho = \mu_0 c^2 \rho$ and $\vec{\nabla}\times\vec{B} - \frac{1}{c^2}\frac{\partial}{\partial t}\vec{E} = \mu_0 \vec{J}$.

which can be written in the form:

$$\frac{\partial F^{\mu\nu}}{\partial x^\nu} = \mu_0 j^\mu,$$

where j^μ is the 4-current that we defined on page 585:

$$j^\mu = (c\rho, \vec{J}).$$

Note that the left side of the expression above is just the covariant divergence of the field tensor, so it is like a four dimensional Gauss's law.

Next the covariant form of the two homogeneous equations: $\vec{\nabla}\cdot\vec{B} = 0$ and $\vec{\nabla}\times\vec{E} = -\frac{\partial}{\partial t}\vec{B}$ is:

$$\frac{\partial F_{\mu\nu}}{\partial x^\eta} + \frac{\partial F_{\nu\eta}}{\partial x^\mu} + \frac{\partial F_{\eta\mu}}{\partial x^\nu} = 0.$$

The left side of this expression vanishes if any index equals another. For example, if $\mu = 1$, $\nu = 1$, and $\eta = 3$, we get:

$$\frac{\partial F_{11}}{\partial x^3} + \frac{\partial F_{13}}{\partial x^1} + \frac{\partial F_{31}}{\partial x^2} = 0 ,$$

which is:

$$\frac{\partial(0)}{\partial z} + \frac{\partial(B_y)}{\partial x} + \frac{\partial(-B_y)}{\partial x} = 0.$$

If, on the other had, we consider the case where $\mu = 1$, $\nu = 2$, and $\eta = 3$, then we have:

$$\frac{\partial F_{12}}{\partial x^3} + \frac{\partial F_{23}}{\partial x^1} + \frac{\partial F_{31}}{\partial x^2} = 0, \quad \text{or,} \quad -\frac{\partial B_z}{\partial z} - \frac{\partial B_x}{\partial x} - \frac{\partial B_y}{\partial y} = 0,$$

which is just $-\vec{\nabla}\cdot\vec{B} = 0$.

When $\mu = 0$, $\nu = 2$, and $\eta = 3$, we obtain:

$$\frac{\partial F_{02}}{\partial x^3} + \frac{\partial F_{23}}{\partial x^0} + \frac{\partial F_{30}}{\partial x^2} = 0, \quad \text{or,} \quad \frac{\partial(E_y/c)}{\partial z} - \frac{\partial B_x}{\partial(ct)} - \frac{\partial(E_z/c)}{\partial y} = 0.$$

Rearranging terms and multiplying both sides by c yields:

$$\frac{\partial E_y}{\partial z}-\frac{\partial E_z}{\partial y}=\frac{\partial B_x}{\partial t}\text{, or, }\frac{\partial E_z}{\partial y}-\frac{\partial E_y}{\partial z}=-\frac{\partial B_x}{\partial t},$$

which is simply the x component of $\nabla\times\vec{E}=-\frac{\partial}{\partial t}\vec{B}$.

The expression $\frac{\partial F_{\mu\nu}}{\partial x^\eta}+\frac{\partial F_{\nu\eta}}{\partial x^\mu}+\frac{\partial F_{\eta\mu}}{\partial x^\nu}=0$ actually represent 64 equations, all but four of which are redundant under a permutation of the indices. For example, if we consider $\mu=3$, $\nu=0$, and $\eta=2$, we obtain:

$$\frac{\partial F_{30}}{\partial x^2}+\frac{\partial F_{02}}{\partial x^3}+\frac{\partial F_{23}}{\partial x^0}=0,$$

which translates to:

$$\frac{\partial(-E_z/c)}{\partial y}+\frac{\partial(E_y/c)}{\partial z}+\frac{\partial(-B_x)}{\partial(ct)}=0.$$

This is exactly the result we obtained for the case of $\mu=0$, $\nu=2$, and $\eta=3$. We can achieve some economy by introducing the so-called *dual field tensor* $G^{\mu\nu}$, (also with magnetic field units) which is obtained from $F^{\mu\nu}$ by replacing the entries $\vec{E}/c$ with $\vec{B}$ and $\vec{B}$ with $-\vec{E}/c$:

$$G^{\mu\nu}=\begin{pmatrix} 0 & B_x & B_y & B_z \\ -B_x & 0 & -E_z/c & E_y/c \\ -B_y & E_z/c & 0 & -E_x/c \\ -B_z & -E_y/c & E_x/c & 0 \end{pmatrix}.$$

The sourceless Maxwell equations,

$\vec{\nabla}\cdot\vec{B}=0$, and $\vec{\nabla}\times\vec{E}+\frac{\partial}{\partial t}\vec{B}=0$,

can now be written as the four scalar equations:

$$\frac{\partial G^{\mu\nu}}{\partial x^\nu}=0.$$

(Don't forget that we sum over the repeated index ν !)

Thought Experiment 17.23: Transforming the Field Tensor

It is natural to ask how the electromagnetic field tensor transforms from one frame to another. First notice that for any contravariant vector A^ν:

$$A'^\mu=\Lambda^\mu_\nu A^\nu,$$

and, therefore,

$$F'_{\mu\nu} = \frac{\partial A'_\nu}{\partial x'^\mu} - \frac{\partial A'_\mu}{\partial x'^\nu} = \frac{\partial\left(\Lambda^\rho_\nu A_\rho\right)}{\partial x'^\mu} - \frac{\partial\left(\Lambda^\sigma_\mu A_\sigma\right)}{\partial x'^\nu} = \Lambda^\rho_\nu \frac{\partial A_\rho}{\partial x'^\mu} - \Lambda^\sigma_\mu \frac{\partial A_\sigma}{\partial x'^\nu},$$

where Λ^μ_ν is the entry in row μ and column ν of the Lorentz-FitzGerald transformation matrix:

$$\Lambda(u) = \begin{pmatrix} \gamma(u) & -\frac{1}{c}u\gamma(u) & 0 & 0 \\ -\frac{1}{c}u\gamma(u) & \gamma(u) & 0 & 0 \\ 0 & 0 & 1 & 0 \\ 0 & 0 & 0 & 1 \end{pmatrix}.$$

Use of the chain rule yields

$$F'_{\mu\nu} = \Lambda^\rho_\nu \frac{\partial A_\rho}{\partial x'^\mu} - \Lambda^\sigma_\mu \frac{\partial A_\sigma}{\partial x'^\nu} = \Lambda^\rho_\nu \left(\frac{\partial A_\rho}{\partial x^\sigma}\right)\left(\frac{\partial x^\sigma}{\partial x'^\mu}\right) - \Lambda^\sigma_\mu \left(\frac{\partial A_\sigma}{\partial x^\rho}\right)\left(\frac{\partial x^\rho}{\partial x'^\nu}\right).$$

Recall that $\frac{\partial x'^\mu}{\partial x^\nu} = \Lambda^\mu_\nu$, which means that:

$$\frac{\partial x^\nu}{\partial x'^\mu} = \left(\Lambda^\mu_\nu\right)^{-1} = \Lambda^\nu_\mu(-\beta) = -\Lambda^\nu_\mu = -\Lambda^\mu_\nu,$$

where $\left(\Lambda^\mu_\nu\right)^{-1}$ is the inverse of the Lorentz-FitzGerald transformation matrix Λ^μ_ν. Therefore:

$$F'_{\mu\nu} = \Lambda^\rho_\nu \Lambda^\sigma_\mu \left(\frac{\partial A_\sigma}{\partial x^\rho} - \frac{\partial A_\rho}{\partial x^\sigma}\right) = \Lambda^\rho_\nu \Lambda^\sigma_\mu F_{\rho\sigma}.$$

From here we can derive the electromagnetic field transformations that we found in the thought experiments on pages 587, 588, and 589.

Take, for example, $\mu = 0$ and $\nu = 0$:

$$F'_{01} = \Lambda^\rho_1 \Lambda^\sigma_0 F_{\rho\sigma} = \Lambda^0_1 \Lambda^\sigma_0 F_{0\sigma} + \Lambda^1_1 \Lambda^\sigma_0 F_{1\sigma} + \Lambda^2_1 \Lambda^\sigma_0 F_{2\sigma} + \Lambda^3_1 \Lambda^\sigma_0 F_{3\sigma},$$

after summing over σ, the only non-vanishing terms left are:

$$F'_{01} = \Lambda^0_1 \Lambda^1_0 F_{01} + \Lambda^1_1 \Lambda^0_0 F_{10}, \quad \text{or:} \quad -E'_x = \left(\tfrac{1}{c^2}u^2 - 1\right)\gamma^2 E_x = -\frac{\left(1 - \frac{1}{c^2}u^2\right)}{1 - \frac{1}{c^2}u^2} E_x,$$

so that $E'_x = E_x$.

The remaining cases are left as an exercise for you.

Problem 17.31: Properties of the Electromagnetic Tensor

a) Show that the components of the electric and magnetic field vectors can be written as:

$$E_i = cF^{0i} \text{ and } B_i = -\frac{1}{2}\varepsilon_{ijk}F^{jk},$$

where $1 \le i \le 3$ and ε_{ijk} is the Levi-Cevita symbol.

b) Calculate the product $F_{\mu\nu}F^{\mu\nu}$ and comment on the physical significance of your result. (Don't forget the Einstein summation convention.)

c) Calculate the product of the Faraday tensor and its dual: $G_{\mu\nu}F^{\mu\nu}$. Comment on the physical significance of your result.

d) Use the Faraday tensor to explicitly calculate the fields under a Lorentz- FitzGerald transformation between inertial frames $\lfloor\underline{s}$ and $\lfloor\underline{s'}$:

$$F'_{\mu\nu} = \Lambda^{\rho}_{\nu}\Lambda^{\sigma}_{\mu}\, F_{\rho\sigma}\,.$$

e) Demonstrate, by explicit evaluation, that Maxwell's four equations can be written as the two equations:

$$\frac{\partial F^{\mu\nu}}{\partial x^{\nu}} = \mu_0 j^{\mu}\,, \text{ and } \frac{\partial G^{\mu\nu}}{\partial x^{\nu}} = 0\,,$$

where j^{μ} is the current 4-vector defined in Thought Experiment 17.13.

17.4 *Radiation from Relativistic Point Charges*

Recall from the introduction to Chapter 16 (p. 507) that, at the end of the nineteenth century, the French engineer Alfred-Marie Liénard and the German physicist Emil Johann Wiechert independently derived the correct expressions for the potentials and the fields of a point charge in arbitrary motion. In this section, we use relativistic electrodynamics to derive the Liénard-Wiechert potentials, and then employ them to calculate the electric and magnetic fields, and the angular distribution of the power radiated by accelerated point charges.

We first compute the potentials for a moving charge by applying the appropriate Lorentz transformation to the potential four vector in the charge's rest fame. (If the charge is accelerated, then we apply the transformations to the momentarily co-moving rest frame of the charge.) Of course, we must take care to correct for the finite speed at light travels from the charge (i.e. the light source) to an observer, so we calculate the position and velocity of the charge at the retarded time, τ. If $\vec{r}$ is the observer's position at time t, and $\vec{r}'(\tau)$ is the position of the charge at the retarded time τ, then the following condition must be satisfied:

$$|\vec{r} - \vec{r}'(\tau)| = c(t - \tau)\,.$$

The Liénard-Wiechert potentials must be evaluated at the retarded time, but we can easily solve this equation for τ and express the potentials and fields in terms of the observer's position and present time for the special case of uniformly moving point charge. As we will see later in the section, things are trickier for the case of accelerated motion, for which one will need to know the trajectory of the source charge.

We demonstrate that a point charge in uniform motion emits no radiation by calculating the Poynting vector and integrating it over all of space. Of course, this result is consistent with relativity: a charge at rest does not radiate energy, so it should not radiate when observed in a different inertial frame.

We carefully calculate the electric and magnetic fields of an accelerated charge, and then investigate how which parts of these fields contribute to the radiation, and which do not. From there we derive the non-relativistic and fully relativistic Larmor formulas for radiative power loss, discuss the angular distribution of the power radiated by accelerated charges both parallel and perpendicular to their instantaneous velocity, the latter case is referred to as synchrotron radiation. We provide a brief, and mostly qualitative, discussion of the power spectrum produced by synchrotron radiation.

Thought Experiment 17.24: The Potentials of a Moving Particle

Consider a reference frame with a stationary particle with a surrounding electrostatic potential $\mathbb{V}_0(t_0,\vec{r}_0)$. This need not be a point charge, as it could rather be a dipole, quadrupole, or any other charge distribution. However, the charges must not be moving with respect to each other, and there may not be any intrinsic magnetic moments.

Now let us consider this from the point of view of a reference frame $\lfloor\underline{\mathrm{S}}$ that is moving at a constant velocity $\vec{u}=-\vec{v}$, as in this frame the particle will be moving at velocity $\vec{v}$, which we will assume is in the z direction. As we discussed in Chapter 8, we can now apply a boost to this particle collection as which gives us the following 4-potential in frame $\lfloor\underline{\mathrm{S}}$:

$$\begin{pmatrix} \frac{1}{c}\mathbb{V} \\ \mathbb{A}_x \\ \mathbb{A}_y \\ \mathbb{A}_z \end{pmatrix} = \begin{pmatrix} \gamma(-v) & 0 & 0 & -\frac{(-v)}{c}\gamma(-v) \\ 0 & 1 & 0 & 0 \\ 0 & 0 & 1 & 0 \\ -\frac{u}{c}\gamma(-v) & 0 & 0 & \gamma(-v) \end{pmatrix} \begin{pmatrix} \frac{1}{c}\mathbb{V}_0 \\ 0 \\ 0 \\ 0 \end{pmatrix} = \begin{pmatrix} \gamma(v)\frac{1}{c}\mathbb{V} \\ 0 \\ 0 \\ \frac{v}{c}\gamma(v)\frac{1}{c}\mathbb{V} \end{pmatrix}$$

Or, if we write it more generally, we have the potentials in the laboratory reference frame:

$$\mathbb{V}=\mathbb{V}_0(t_0,\vec{r}_0)\gamma(v) \quad \text{and} \quad \vec{\mathbb{A}}=\frac{\vec{v}}{c^2}\mathbb{V}_0(t_0,\vec{r}_0)\gamma(v)..$$

There, of course, is a catch. The position of the particle needs also to be measured in the laboratory frame. Thus, we must do a reverse boost, $\vec{u}=\vec{v}$, as to find the position and time in the proper frame:

$$\begin{pmatrix} ct_0 \\ x_0 \\ y_0 \\ z_0 \end{pmatrix} = \begin{pmatrix} \gamma(v) & 0 & 0 & -\frac{v}{c}\gamma(v) \\ 0 & 1 & 0 & 0 \\ 0 & 0 & 1 & 0 \\ -\frac{v}{c}\gamma(v) & 0 & 0 & \gamma(v) \end{pmatrix} \begin{pmatrix} ct \\ x \\ y \\ z \end{pmatrix} = \gamma(v) \begin{pmatrix} ct-\frac{v}{c}z \\ x \\ y \\ -\frac{v}{c}ct+z \end{pmatrix}.$$

Or, again more generally:

$$t_0=\gamma(v)t-\gamma(v)\frac{1}{c^2}\vec{v}\cdot\vec{r} \quad , \quad \vec{r}_0=\gamma(v)\vec{r}-\gamma(v)\frac{1}{c}\vec{v}t \quad ,$$

Now we will take this one step further, and notice that the information from the particle propagates at the speed of light. Thus, always, $r=ct$. Another way to look at the same thing is to take the 4-magnitude of the 4-position vector, and set it equal to zero, or: $0=\vec{r}\cdot\vec{r}-(ct)^2$, thus $\sqrt{\vec{r}\cdot\vec{r}}=ct=r$.

We can now write the following:

$$t_0 = \gamma(v)\left(r - \tfrac{1}{c^2}\vec{v}\cdot\vec{r}\right) \quad , \quad \vec{r}_0 = \gamma(v)\left(\vec{r} - \tfrac{1}{c}\vec{v}\,r\right) \quad ,$$

Now we will use the triple product rule to write this as:

$$\vec{r}_0 = \gamma(v)\left(\vec{r} - \tfrac{1}{c}\vec{v}\left(\frac{\vec{r}\cdot\vec{r}}{r}\right)\right) = \gamma(v)\left(\vec{r} - \tfrac{1}{c}\vec{r}\,\frac{(\vec{v}\cdot\vec{r})}{r}\right) = \gamma(v)\vec{r}\left(1 - \tfrac{1}{c}(\vec{v}\cdot\hat{r})\right).$$

And, from the 4-magnitude argument, we can write:

$$r^2 - (ct)^2 = r_0^2 - (ct_0)^2 = 0\,.$$

Therefore, we can now write the potentials of a collection of charge moving at a velocity $\vec{v}$ as:

$$\mathbb{V} = \gamma(v)\,\mathbb{V}_0\left(\gamma(v)\left(r - \tfrac{1}{c^2}\vec{v}\cdot\vec{r}\right), \gamma(v)\vec{r}\left(1 - \tfrac{1}{c}(\vec{v}\cdot\hat{r})\right)\right)$$
$$\vec{\mathbb{A}} = \gamma(v)\frac{\vec{v}}{c^2}\mathbb{V}_0\left(\gamma(v)\left(r - \tfrac{1}{c^2}\vec{v}\cdot\vec{r}\right), \gamma(v)\vec{r}\left(1 - \tfrac{1}{c}(\vec{v}\cdot\hat{r})\right)\right).$$

And, we see that the vector and scalar potentials are now related by:

$$\vec{\mathbb{A}} = \frac{\vec{v}}{c^2}\mathbb{V}\,.$$

We also see that the 4-magnitude of the potential vector is given by:

$$\left\|\tfrac{1}{c}\mathbb{V},\vec{\mathbb{A}}\right\|^2 = \mathbb{A}^2 - \tfrac{1}{c^2}\mathbb{V}^2 = \left(\tfrac{1}{c^2}v\mathbb{V}\right)^2 - \tfrac{1}{c^2}\mathbb{V}^2 = \left(\left(\tfrac{1}{c}v\right)^2 - 1\right)\tfrac{1}{c^2}\mathbb{V}^2 = -\left(\frac{1}{\gamma(v)}\right)^2\tfrac{1}{c^2}\mathbb{V}^2\,.$$

But in terms of the stationary potential, this becomes:

$$\left\|\tfrac{1}{c}\mathbb{V},\vec{\mathbb{A}}\right\|^2 = -\left(\frac{1}{\gamma(v)}\right)^2\tfrac{1}{c^2}\left(\gamma(v)\mathbb{V}_0\right)^2 = -\tfrac{1}{c^2}\left(\mathbb{V}_0\right)^2$$

Which makes sense, because this is consistent with the 4-magnitude in the rest frame of the particle, as we would expect.

Thought Experiment 17.25: The Constant Velocity Liénard-Wiechert Potentials

According to Coulomb's Law, the scalar potential of a point charge, located at the origin, is given by:

$$\mathbb{V}_0(t_0,\vec{r}_0) = \frac{Z_0 c}{4\pi}\frac{q}{r_0},$$

So the scalar potential of a charge moving at velocity $\vec{v}$ is:

$$\mathbb{V} = \gamma(v)\,\mathbb{V}_0\left(\gamma(v)\left(r - \tfrac{1}{c^2}\vec{v}\cdot\vec{r}\right), \gamma(v)\vec{r}\left(1 - \tfrac{1}{c}(\vec{v}\cdot\hat{r})\right)\right)$$
$$= \gamma(v)\frac{Z_0 c}{4\pi}\left(\frac{q}{\left|\gamma(v)\vec{r}\left(1 - \tfrac{1}{c}(\vec{v}\cdot\hat{r})\right)\right|}\right) = \cancel{\gamma(v)}\frac{Z_0 c}{4\pi}\left(\frac{q}{\cancel{\gamma(v)}\left(1 - \tfrac{1}{c}(\vec{v}\cdot\hat{r})\right)|\vec{r}|}\right) = \frac{Z_0 c q}{4\pi r}\left(\frac{1}{1 - \tfrac{1}{c}(\vec{v}\cdot\hat{r})}\right).$$

And therefore the vector potential is:

$$\vec{\mathbb{A}} = \frac{\vec{v}}{c^2}\mathbb{V} = \frac{\vec{v}}{c^{\cancel{2}}}\frac{Z_0 \cancel{c} q}{4\pi}\left(\frac{1}{r\left(1-\frac{1}{c}(\vec{v}\cdot\hat{r})\right)}\right) = \frac{Z_0 q}{4\pi c}\left(\frac{\vec{v}}{r\left(1-\frac{1}{c}(\vec{v}\cdot\hat{r})\right)}\right) .$$

These are the relativistically correct potentials of a point charge located at the origin. Notice that this is almost the same as Coulomb's law, but with one factor that becomes small if $v \ll c$.

However, recalling the analogy with the postcards in Chapter 16 (p. 510), we must be very careful about correcting for the travel time of light, and so we must calculate the position and velocity of the particle at the retarded time, τ, rather than at the current time.

Taken together, these expressions for the scalar and vector potentials of a point charge are called the *Liénard-Wiechert potentials*. We can write them in four-vector notation compactly as:

$$\left(\tfrac{1}{c}\mathbb{V},\vec{\mathbb{A}}\right) = \frac{Z_0 q}{4\pi}\left[\frac{\left(1,\frac{1}{c}\vec{v}\right)}{R-\frac{1}{c}\vec{v}\cdot\vec{R}}\right]_{\mathrm{RET}} .$$

Notice that we have just used the following standard notation: a bracket $[\ ]_{\mathrm{RET}}$ with the subscript "RET" indicates that the quantities inside the bracket are functions of time that must be evaluated at the retarded, time τ.

Example 17.4: The Potentials of a Uniformly Moving Point Charge

The trajectory of the point charge moving at uniform velocity $\vec{v}$ is given by

$\vec{r}'(t) = \vec{v}t$. (Remember that we are not using $\vec{r}'$ to denote the time derivative of $\vec{r}$, which is standard notation in some applications. In other words, $\vec{r}' \neq \dfrac{d\vec{r}}{dt}$!)

First, we must compute the retarded time τ using our standard formula:

$\left|\vec{r} - \vec{r}'(\tau)\right| = c(t-\tau)$.

Squaring both sides of the standard formula yields

$r^2 - 2\vec{r}\cdot\vec{v}\tau + v^2\tau = c^2t^2 - 2c^2t\tau + c^2\tau^2$.

We now collect like terms and solve for the delayed time τ :

$$\tau = \frac{\left(c^2t - \vec{r}\cdot\vec{v}\right) \pm \sqrt{\left(c^2t - \vec{r}\cdot\vec{v}\right) - \left(c^2 - v^2\right)\left(c^2t^2 - r^2\right)}}{c^2 - v^2} .$$

Since we are interested in the delayed rather than the advanced time, we choose the "-" sign in the numerator of the solution for τ.

Now we compute the value of the denominator, $R - \vec{\beta}\cdot\vec{R}$, in the Liénard-Wiechert potentials at the appropriate delayed time τ. To do so, note that $R = c\left(t-\tau\right)$, and $\vec{R} = \vec{r} - \vec{v}\tau = \vec{r} - c\vec{\beta}\tau$. Thus,

$$R - \vec{\beta}\cdot\vec{R} = c(t-\tau) - \vec{\beta}\cdot\vec{r} + c\beta^2\tau = c(t-\tau) - \frac{\vec{v}\cdot\vec{r}}{c} + \frac{v^2}{c}\tau ,$$

which we can rearrange and write as

$$R-\vec{\beta}\cdot\vec{R}=\frac{1}{c}\left\{(c^2t-\vec{r}\cdot\vec{v})-(c^2-v^2)\tau\right\}.$$

Now substitute the physically appropriate solution for the retarded time,

$\tau=\dfrac{\left(c^2t-\vec{r}\cdot\vec{v}\right)-\sqrt{\left(c^2t-\vec{r}\cdot\vec{v}\right)-\left(c^2-v^2\right)\left(c^2t^2-r^2\right)}}{c^2-v^2}$, into the expression for $R-\vec{\beta}\cdot\vec{R}$ to obtain the denominator in terms of the present time t and the field point $\vec{r}$:

$$R-\vec{\beta}\cdot\vec{R}=\frac{1}{c}\sqrt{(c^2t-\vec{r}\cdot\vec{v})^2+(c^2-v^2)(r^2-c^2t^2)}.$$

The Liénard-Wiechert potentials are therefore given by:

$$\mathbb{V}(\vec{r},t)=\frac{1}{4\pi\varepsilon_0}\frac{qc}{\sqrt{(c^2t-\vec{r}\cdot\vec{v})^2-(c^2-v^2)(r^2-c^2t^2)}},$$

and

$$\vec{\mathbb{A}}(\vec{r},t)=\frac{\mu_0}{4\pi}\frac{qc\vec{v}}{\sqrt{(c^2t-\vec{r}\cdot\vec{v})^2-(c^2-v^2)(r^2-c^2t^2)}}.$$

or in compact form:

$$\left(\tfrac{1}{c}\mathbb{V},\vec{\mathbb{A}}\right)=\frac{Z_0}{4\pi}\frac{q}{c}\frac{\left(c,\vec{v}\right)}{\sqrt{(c^2t-\vec{r}\cdot\vec{v})^2-(c^2-v^2)(r^2-c^2t^2)}}.$$

Example 17.5: The Electric Field of a Uniformly Moving Charge

We will now calculate the fields due to a charge moving with uniform velocity $\vec{v}$ from the Liénard-Wiechert potentials via:

$$\vec{E}(\vec{r},t)=-\vec{\nabla}\mathbb{V}-\frac{\partial\vec{\mathbb{A}}}{\partial t},\quad\text{and}\quad\vec{B}(\vec{r},t)=\vec{\nabla}\times\vec{\mathbb{A}}.$$

While laborious, the differentiation and requisite algebra are straightforward in the case of a uniformly moving charge because we have been able to write the potentials exclusively in terms of the field point $\vec{r}$ and the present time t. The gradient of $\mathbb{V}$ is easily determined if we generalize from one component:

$$\frac{\partial\mathbb{V}}{\partial x}=\left(\frac{Z_0\,q\,c}{4\pi}\right)\frac{\partial}{\partial x}\left\{\frac{c}{\sqrt{\left(c^2t-\vec{r}\cdot\vec{v}\right)^2+\left(c^2-v^2\right)\left(r^2-c^2t^2\right)}}\right\},$$

$$\frac{\partial\mathbb{V}}{\partial x}=\left(\frac{Z_0\,q\,c^2}{4\pi}\right)\frac{\left(\left(c^2t-\vec{r}\cdot\vec{v}\right)v_x+\left(c^2-v^2\right)x\right)}{\left[\left(c^2t-\vec{r}\cdot\vec{v}\right)^2+\left(c^2-v^2\right)\left(r^2-c^2t^2\right)\right]^{3/2}},$$

where v_x is the x-component of the velocity $\vec{v}$.

This result is easily generalized to:

$$\vec{\nabla}\mathbb{V} = \left(\frac{Z_0\, q\, c^2}{4\pi}\right)\frac{\left(\left(c^2t - \vec{r}\cdot\vec{v}\right)\vec{v} - \left(c^2 - v^2\right)\vec{r}\right)}{\left[\left(c^2t - \vec{r}\cdot\vec{v}\right)^2 + \left(c^2 - v^2\right)\left(r^2 - c^2t^2\right)\right]^{3/2}}.$$

It is also straightforward to show that:

$$\frac{\partial\vec{\mathbb{A}}}{\partial t} = -\left(\frac{Z_0\, q\, c^2}{4\pi}\right)\frac{\left(\left(c^2t - \vec{r}\cdot\vec{v}\right) + \left(c^2 - v^2\right)t\right)\vec{v}}{\left[\left(c^2t - \vec{r}\cdot\vec{v}\right)^2 + \left(c^2 - v^2\right)\left(r^2 - c^2t^2\right)\right]^{3/2}}.$$

Therefore, the electric field is given by

$$\vec{E} = \left(\frac{Z_0\, q\, c}{4\pi}\right)\frac{c\left(c^2 - v^2\right)\left(\vec{r} - \vec{v}t\right)}{\left[\left(c^2t - \vec{r}\cdot\vec{v}\right)^2 + \left(c^2 - v^2\right)\left(r^2 - c^2t^2\right)\right]^{3/2}}.$$

Discussion 17.4: More on the Electric Field of a Uniformly Moving Charge

The expression for the electric field due to a uniformly moving charge can be simplified if we define

$$\vec{\xi} \equiv \vec{r} - \vec{v}t.$$

Then we can show that

$$\left(c^2t - \vec{r}\cdot\vec{v}\right)^2 + \left(c^2 - v^2\right)\left(r^2 - c^2t^2\right) = \xi^2c^2\left(1 - \beta^2\sin^2\theta\right),$$

where θ is the angle between $\vec{\xi}$ and $\vec{v}$. (We leave the details of the algebra to you!) Thus,

$$\vec{E} = \frac{q}{4\pi\varepsilon_0}\frac{\left(1 - \beta^2\right)}{\left[1 - \beta^2\sin^2\theta\right]^{3/2}}\frac{\vec{\xi}}{\xi^3}.$$

What is interesting about this result is that $\vec{v}t$ is the present position of the charge, so the electric field points in the direction of charge's current location despite being caused by the charge when the latter is at the delayed location!

The magnetic field can be computed directly from the curl of the vector potential:

$$\vec{B} = \vec{\nabla}\times\vec{\mathbb{A}} = \vec{\nabla}\times\left(\frac{1}{c^2}\mathbb{V}\vec{v}\right) = \frac{1}{c^2}\left(\vec{\nabla}\mathbb{V}\right)\times\vec{v} + \frac{1}{c^2}\mathbb{V}\left(\cancel{\vec{\nabla}\times\vec{v}}\right) = -\frac{1}{c^2}\vec{v}\times\left(\vec{\nabla}\mathbb{V}\right) = \frac{1}{c^2}\vec{v}\times\vec{E}.$$

For velocities that are small compared to the speed of light, we recover the Coulomb and (naïve, at least for a point charge) Biot-Savart expressions:

$$\vec{E} = \frac{q}{4\pi\varepsilon_0}\frac{\vec{r}}{r^3}, \text{ and } \vec{B} = \frac{q\vec{v}\times\vec{r}}{4\pi\varepsilon_0 r^3},$$

as we expected. What about high velocities, for which $\beta \to 1$? First note both $\vec{E}$ and $\vec{B}$ become increasingly dependent on the angle between $\vec{\xi}$ and the direction of motion. For positions in the direction of motion, or in the exact opposite direction, $\vec{E}$ is decreased by a factor of $\left(1-\beta^2\right)$. On the other hand, for directions perpendicular to the direction of motion, the field strength is actually enhanced by a factor of $\left(1-\beta^2\right)^{-1/2}$. The net result is that the electric field lines are compressed perpendicular to the velocity. The magnetic field lines appear as closed circles surrounding the particle's trajectory. Now, as $v \to c$, the fields are nearly purely transverse and resemble the brief pulse of a plane wave, a resemblance that is sometimes exploited in practical problems via the so-called method of virtual radiation. You should not be misled, however, into believing that electric and magnetic fields of a rapidly moving, but unaccelerated, charge are what we mean by light or radiation; they are not.

We can see this by computing the Poynting vector $\vec{\mathbb{S}}$ for the uniformly moving charge, and then integrating over all of space. The vector product of $\vec{E}$ and $\vec{B}$ is

$$\vec{E}\times\vec{B} = \left(\frac{q}{4\pi\varepsilon_0}\right)^2 \frac{(1-\beta^2)}{(1-\beta^2\sin^2\theta)}\left(\frac{1}{\xi^6}\right)\vec{\xi}\times\left(\vec{v}\times\vec{\xi}\right) \propto \frac{\hat{\xi}\times\left(\hat{v}\times\hat{\xi}\right)}{\xi^4}.$$

Which means that if we calculate the total energy by integrating $\vec{\mathbb{S}}\cdot d\vec{\xi}$ over a sphere of radius ξ, we obtain the radiated power

$$\oint \vec{\mathbb{S}}\cdot d\vec{\xi} \propto \int \frac{d\xi}{\xi^4} \propto \xi^{-2},$$

which approaches zero in the limit as ξ becomes large. A uniformly moving charge does not radiate energy. A simpler way to see this is to consider things from the point of view of special relativity. Since all inertial frames are equivalent for describing electrodynamics, and since the charge moves at constant velocity, we can transform to a frame in which the charge is at rest. In such a frame, the charge clearly does not radiate energy, so no radiation occurs whether the charge is observed to be stationary or moving at a constant velocity.

Example 17.6 The Fields of a Uniformly Moving Point Charge Revisited

Yet another way of deriving the potentials and fields of a uniformly moving charge is to solve the inhomogeneous wave equation for the vector potential. Assume, for simplicity, that a charge moves with constant speed along the positive x-axis. Recall that we can use Faraday's law to eliminate $\vec{E}$ in the Maxwell-Ampère law, and write it in the form of a wave equation. This led us to Maxwell's equations in potential form, one of which is a wave equation, with source, for the vector potential $\vec{\mathbb{A}}$:

$$\frac{1}{c^2}\frac{\partial^2\vec{\mathbb{A}}}{\partial t^2} - \vec{\nabla}^2\vec{\mathbb{A}} = \frac{1}{\varepsilon_0 c^2}\rho\vec{v}.$$

Because the charge moves along the x-axis, we can assume that $\mathbb{A}_y = 0$ and $\mathbb{A}_z = 0$, so:

$$\frac{1}{c^2}\frac{\partial^2\mathbb{A}_x}{\partial t^2} - \vec{\nabla}^2\mathbb{A}_x = \frac{1}{\varepsilon_0 c^2}\rho v.$$

Let (x_1, y, z) denote coordinates in the rest frame of the charge, and are related to (x, y, z) by:

$$x_1 = x - vt.$$

Then we have that

$$\frac{\partial}{\partial x}=\frac{\partial}{\partial x_1} \text{ and } \frac{\partial}{\partial t}=-v\frac{\partial}{\partial x_1}.$$

The wave equation for $\mathbb{A}_x$ can therefore be written as

$$\left(1-\frac{v^2}{c^2}\right)\frac{\partial^2 \mathbb{A}_x}{\partial x_1{}^2}+\frac{\partial^2 \mathbb{A}_y}{\partial y^2}+\frac{\partial^2 \mathbb{A}_z}{\partial z^2}=-\frac{1}{\varepsilon_0 c^2}\rho v.$$

If we now make the following substitution

$$\chi=\left(1-\frac{v^2}{c^2}\right)^{-\frac{1}{2}} x_1,$$

then the chain rule for derivatives yields

$$\frac{\partial}{\partial x_1}=\left(1-\frac{v^2}{c^2}\right)^{-\frac{1}{2}}\frac{\partial}{\partial \chi}, \text{ and, thus, } \frac{\partial^2 \mathbb{A}_x}{\partial x_1{}^2}=\left(1-\frac{v^2}{c^2}\right)^{-1}\frac{\partial^2 \mathbb{A}_x}{\partial \chi^2}.$$

The wave equation can therefore be written as

$$\frac{\partial^2 \mathbb{A}_x}{\partial \chi^2}+\frac{\partial^2 \mathbb{A}_y}{\partial y^2}+\frac{\partial^2 \mathbb{A}_z}{\partial z^2}=-\frac{1}{\varepsilon_0 c^2}\rho v.$$

This is just Poisson's equation, whose standard solution is given by

$$\mathbb{A}_x=\frac{1}{4\pi\varepsilon_0 c}\iiint_{volume}\frac{\rho(\chi',y',z')v\,d\chi'dy'dz'}{\sqrt{(\chi'-\chi)^2+(y'-y)^2+(z'-z)^2}},$$

which can also be written as

$$\mathbb{A}_x=\frac{1}{4\pi\varepsilon_0 c}\iiint_{volume}\frac{\rho(x',y',z')v\,dx'dy'dz'}{\sqrt{(x'-x)^2+(1-v^2/c^2)\left[(y'-y)^2+(z'-z)^2\right]}}.$$

For a point particle of charge q, the integral gives

$$\mathbb{A}_x=\frac{1}{4\pi\varepsilon_0 c}\frac{qv}{\sqrt{x^2+(1-v^2/c^2)\left[y^2+z^2\right]}}.$$

If we note that

$$\sin^2\theta=\frac{y^2+z^2}{x_1{}^2+y^2+z^2}=\frac{y^2+z^2}{\xi^2},$$

then we may write $\mathbb{A}_x$ as:

$$\mathbb{A}_x=\frac{1}{4\pi\varepsilon_0 c}\frac{qv}{\xi^2\sqrt{1-\left(\frac{1}{c}v\right)^2\sin^2\theta}}=\frac{1}{4\pi\varepsilon_0 c}\frac{qv}{\xi^2\sqrt{1-\beta^2\sin^2\theta}}.$$

The magnitude of the magnetic field is therefore given by:

$$B = \frac{1}{4\pi\varepsilon_0 c}\frac{qv\sin\theta}{\xi^2\left[1-\beta^2\sin^2\theta\right]^{3/2}},$$

and the electric field, which is purely radial, is

$$\vec{E} = \frac{1}{4\pi\varepsilon_0}\frac{q\left(1-\beta^2\right)}{\xi^3\left[1-\beta^2\sin^2\theta\right]^{3/2}}\vec{\xi},$$

which are the results obtained in Discussion 17.4.

With hindsight we can see that the coordinate substitutions we made for reasons of mathematical convenience really amount to a Lorentz transformation in disguise:

$$\chi = \frac{x - vt}{\sqrt{1-\dfrac{v^2}{c^2}}}.$$

Example 17.7 The electric and magnetic fields of an accelerated particle

Calculating the fields involves taking derivatives of the denominator in the expressions for the Liénard-Wiechert potentials,

$$\mathbb{V}(\vec{r},t) = \frac{q}{4\pi\varepsilon_0\left[R-\vec{\beta}\cdot\vec{R}\right]_{\text{RET}}} \quad \text{and} \quad \vec{\mathbb{A}}(t,\vec{r}) = \frac{q}{4\pi\varepsilon_0 c}\left[\frac{\vec{\beta}}{R-\vec{\beta}\cdot\vec{R}}\right]_{\text{RET}} = \frac{1}{c}\left[\vec{\beta}\,\mathbb{V}(t,\vec{r})\right]_{\text{RET}},$$

which is tricky because the position $\vec{R} = \vec{r} - \vec{r}'(\tau)$ is evaluated at the delayed time τ, which in turn depends on the position: $\tau = t - R/c = t - |\vec{r} - \vec{r}'|/c$. We must therefore evaluate the time and space derivatives of $\mathbb{V}$ and $\vec{\mathbb{A}}$ carefully, especially because in this example we no longer assume that $\vec{v}$ is constant.

To calculate the fields, we differentiate the potentials with respect to the coordinates (x,y,z) at a constant time t rather than a constant delayed time τ, and we differentiate the vector potential with respect to t rather than τ:

$$E(\vec{r},t) = -\vec{\nabla}\mathbb{V} - \frac{\partial\vec{\mathbb{A}}}{\partial t}, \text{ and } B(\vec{r},t) = \vec{\nabla}\times\vec{\mathbb{A}}.$$

When we evaluate $\vec{\nabla}\mathbb{V}$, for example, we are simultaneously comparing values of the potential at closely separated points, yet these values actually originated at different times from the source charge as it moved along its trajectory. When we take the partial derivative of the vector potential with respect to time t, $\frac{\partial\vec{\mathbb{A}}}{\partial t}$, it is now the values of the field point coordinates (x,y,z) that are held constant. In this case, we compare successive values of the vector potential at a fixed field point over a time interval during which the source charge coordinates have changed. Now, the position of the charge is an explicit function of the delayed time τ : $\vec{r}' = \vec{r}'(\tau)$. This presents no problem for the special instance of uniform motion, where we can express the potentials, and hence the fields, entirely in terms of the present position and time. But it is not generally possible to express the potentials in terms of the present position for the case of accelerated charges, and this means that in order to solve for the fields at the present time and position, we must express $\vec{\nabla}\big|_t$ and $\partial/\partial t\big|_{(x,y,z)}$ in terms of $\partial/\partial\tau\big|_{(x,y,z)}$.

Now, the potential actually depends on the delayed time τ, which is a function of the present time and position:

$$\mathbb{V}(\vec{r},t) = \mathbb{V}(\vec{r},\tau[r,t]).$$

The electric field can therefore be written as:

$$\vec{E}(\vec{r},t) = -\vec{\nabla}\mathbb{V} - \frac{\partial\vec{\mathbb{A}}}{\partial t} = -\vec{\nabla}\mathbb{V}\Big|_\tau - \left(\left(\frac{\partial\mathbb{V}}{\partial\tau}\right)\Bigg|_{(x,y,z)}\right)\vec{\nabla}\tau - \frac{\partial\vec{\mathbb{A}}}{\partial\tau}\frac{\partial\tau}{\partial t},$$

where we have used the chain rule to write the derivatives on the right hand side of the equation. The subscript τ indicates differentiation with respect to field coordinates $\vec{r} = (x,y,z)$ while holding the delay time constant. Similarly, the subscript (x,y,z) implies that we are to differentiate with respect to the delay time τ while holding (x,y,z) constant.

If we evaluate the vector potential of the moving point charge in terms of the scalar potential, we have that

$$\vec{E}(\vec{r},t) = -\vec{\nabla}\mathbb{V}\Big|_\tau - \left(\left(\frac{\partial\mathbb{V}}{\partial\tau}\right)\Bigg|_{(x,y,z)}\right)\vec{\nabla}\tau - \frac{1}{c}\left(\mathbb{V}\frac{\partial\vec{\beta}}{\partial\tau}\Bigg|_{(x,y,z)} + \vec{\beta}\frac{\partial\mathbb{V}}{\partial\tau}\Bigg|_{(x,y,z)}\right)\frac{\partial\tau}{\partial t}.$$

To clarify, and to stress the fact that we are taking the derivative only with respect to the field coordinates, we will write $\vec{\nabla}\mathbb{V}\big|_\tau = \vec{\nabla}_r\mathbb{V}$. We also drop the subscript (x,y,z) from $\partial\mathbb{V}/\partial\tau$ for convenience, with the understanding that the field coordinates (x,y,z) are held constant when taking the partial derivative with respect to τ. The electric field is therefore,

$$\vec{E}(\vec{r},t) = -\vec{\nabla}_r\mathbb{V} - \left(\frac{\partial\mathbb{V}}{\partial\tau}\right)\vec{\nabla}\tau - \frac{1}{c}\left(\mathbb{V}\frac{\partial\vec{\beta}}{\partial\tau} + \vec{\beta}\frac{\partial\mathbb{V}}{\partial\tau}\right)\frac{\partial\tau}{\partial t}.$$

It is convenient to begin by evaluating the partial derivative of R with respect to τ while holding the field coordinates fixed:

$$\tfrac{\partial}{\partial\tau}R = \tfrac{\partial}{\partial t}R\Big|_{(x,y,z)}.$$

This will allow us to also determine the important quantities:

$$\tfrac{\partial}{\partial t}\tau,\ \vec{\nabla}R\text{, and }\vec{\nabla}\tau.$$

Since $\vec{R} = \vec{r} - \vec{r}'(\tau)$, we have that

$$R = \sqrt{(\vec{r}-\vec{r}')\cdot(\vec{r}-\vec{r}')} = \sqrt{r^2 - 2rr' + r'^2},$$

and so

$$\frac{\partial R}{\partial\tau} = \frac{\partial}{\partial\tau}\left\{\sqrt{r^2 - 2rr' + r'^2}\right\} = \frac{1}{\cancel{2}\sqrt{r^2 - 2rr' + r'^2}}\left\{-\cancel{2}r\frac{\partial r'}{\partial\tau} + \cancel{2}r'\frac{\partial r'}{\partial\tau}\right\}.$$

Since $\vec{v} = \dfrac{\partial \vec{r}'}{\partial \tau}$, this reduces to $\dfrac{\partial R}{\partial \tau} = \dfrac{-(r-r')v}{\sqrt{r^2 - 2rr' + r'^2}} = -\dfrac{\vec{R}\cdot\vec{v}}{R}$.

So we have the first of our necessary derivative identities:

$$\frac{\partial R}{\partial \tau} = -\frac{\vec{R}\cdot\vec{v}}{R}.$$

Since $R = c(t-\tau)$, the partial derivative of R with respect to the present time is given by:

$$\frac{\partial R}{\partial t} = \frac{\partial}{\partial t}\big(c(t-\tau)\big) = c\left(1 - \frac{\partial \tau}{\partial t}\right),$$

but the chain rule for derivatives also implies that:

$$\frac{\partial R}{\partial t} = \frac{\partial R}{\partial \tau}\frac{\partial \tau}{\partial t} = \left(-\frac{\vec{R}\cdot\vec{v}}{R}\right)\frac{\partial \tau}{\partial t}.$$

We can equate the two expressions for $\partial R/\partial t$:

$$c - c\left(\frac{\partial \tau}{\partial t}\right) = -\frac{\vec{R}\cdot\vec{v}}{R}\left(\frac{\partial \tau}{\partial t}\right),$$

and collect terms to solve for $\partial \tau / \partial t$:

$$\frac{\partial \tau}{\partial t} = \frac{1}{1 - \dfrac{\vec{R}\cdot\vec{\beta}}{R}} = \frac{R}{R - \vec{R}\cdot\vec{\beta}}.$$

We therefore have that:

$$\frac{\partial}{\partial t} = \frac{R}{\left(R - \vec{R}\cdot\vec{\beta}\right)}\frac{\partial}{\partial \tau}.$$

We next wish to determine $\vec{\nabla}R$. Again, since $R = c(t-\tau)$, we have that:

$$\vec{\nabla}R = -c\vec{\nabla}\tau = \left(\vec{\nabla}R\right)\Big|_{\tau} + \frac{\partial R}{\partial \tau}\vec{\nabla}\tau = \vec{\nabla}_r R + \frac{\partial R}{\partial \tau}\vec{\nabla}\tau = \frac{\vec{R}}{R} - \frac{\vec{R}\cdot\vec{v}}{R}\vec{\nabla}\tau,$$

where the subscripts indicate that the retarded time τ is held constant during differentiation with respect to the field coordinates. We have dropped the subscript (x,y,z) from $\partial R/\partial \tau$ for convenience, but it should be clear that the field coordinates are held constant when one takes the partial derivative with respect to τ.

We can solve expression we developed above

$$-c\vec{\nabla}\tau = \frac{\vec{R}}{R} - \frac{\vec{R}\cdot\vec{v}}{R}\vec{\nabla}\tau$$

for $\vec{\nabla}\tau$ to obtain two necessary identities:

$$\vec{\nabla}\tau = \frac{-\vec{R}}{Rc - \vec{R}\cdot\vec{v}} = \frac{-\vec{R}/c}{R - \vec{R}\cdot\vec{\beta}},$$

and:

$$\vec{\nabla}R = \frac{c\vec{R}}{Rc - \vec{R}\cdot\vec{v}} = \frac{\vec{R}}{R - \vec{R}\cdot\vec{\beta}} .$$

We now evaluate $\vec{\nabla}\mathbb{V}\big|_\tau$ and $\dfrac{\partial \mathbb{V}}{\partial \tau}$. To begin:

$$\vec{\nabla}_r\mathbb{V} = \vec{\nabla}_r\left\{\frac{q}{4\pi\varepsilon_0\left[R - \vec{R}\cdot\vec{\beta}\right]_{\text{RET}}}\right\} = \frac{-q}{4\pi\varepsilon_0}\left[\frac{\vec{\nabla}_r\left\{R - \vec{R}\cdot\vec{\beta}\right\}}{\left(R - \vec{R}\cdot\vec{\beta}\right)^2}\right]_{\text{RET}} = \frac{-q}{4\pi\varepsilon_0}\left[\frac{\left\{\vec{\nabla}_r R - \vec{\nabla}_r\left(\vec{R}\cdot\vec{\beta}\right)\right\}}{\left(R - \vec{R}\cdot\vec{\beta}\right)^2}\right]_{\text{RET}} .$$

The numerator on the right hand side of the expression for $\vec{\nabla}_r\mathbb{V}$ is easily determined as long as we keep in mind that the delay time is held constant when differentiating with respect to the field point coordinates:

$$\vec{\nabla}_r R - \vec{\nabla}_r\left(\vec{R}\cdot\vec{\beta}\right) = \frac{\vec{R}}{R} - \vec{\beta} .$$

Hence:

$$\vec{\nabla}_r\mathbb{V} = \frac{-q}{4\pi\varepsilon_0}\left[\frac{\left\{\left(\vec{R}/R\right) - \vec{\beta}\right\}}{\left(R - \vec{R}\cdot\vec{\beta}\right)^2}\right]_{\text{RET}} .$$

Now,

$$\frac{\partial \mathbb{V}}{\partial \tau} = \frac{-q}{4\pi\varepsilon_0}\left[\frac{\left(\partial R/\partial\tau\right) - \partial\left(\vec{\beta}\cdot\vec{R}\right)/\partial\tau}{\left(R - \vec{R}\cdot\vec{\beta}\right)^2}\right]_{\text{RET}} = \frac{-q}{4\pi\varepsilon_0}\left[\frac{-\left(\vec{R}\cdot\vec{v}/R\right) - \partial\left(\vec{\beta}\cdot\vec{R}\right)/\partial\tau}{\left(R - \vec{R}\cdot\vec{\beta}\right)^2}\right]_{\text{RET}} ,$$

and evaluating the second term in the numerator yields:

$$\frac{\partial\left(\vec{\beta}\cdot\vec{R}\right)}{\partial\tau} = \frac{\partial}{\partial\tau}\left(\vec{\beta}\cdot\vec{r} - \vec{\beta}\cdot\vec{r}'\right) = \left(\frac{\partial\vec{\beta}}{\partial\tau}\right)\cdot\vec{r} - \left(\frac{\partial\vec{\beta}}{\partial\tau}\right)\cdot\vec{r}' - \vec{\beta}\cdot\left(\frac{\partial\vec{r}'}{\partial\tau}\right) = \dot{\vec{\beta}}\cdot\vec{R} - c\beta^2 .$$

Thus,

$$\frac{\partial \mathbb{V}}{\partial \tau} = \frac{-q}{4\pi\varepsilon_0}\left[\frac{-\left(\vec{R}\cdot\vec{v}/R\right) - \vec{R}\cdot\dot{\vec{\beta}} + c\beta^2}{\left(R - \vec{R}\cdot\vec{\beta}\right)^2}\right]_{\text{RET}} = \frac{q}{4\pi\varepsilon_0}\left[\frac{c\hat{R}\cdot\vec{\beta} + \vec{R}\cdot\dot{\vec{\beta}} + c\beta^2}{\left(R - \vec{R}\cdot\vec{\beta}\right)^2}\right]_{\text{RET}} .$$

We can now find the electric field, since:

$$\vec{E}(\vec{r},t) = -\vec{\nabla}_r\mathbb{V} - \left(\tfrac{\partial}{\partial\tau}\mathbb{V}\right)\vec{\nabla}\tau - \tfrac{1}{c}\left(\mathbb{V}\tfrac{\partial}{\partial\tau}\vec{\beta} + \vec{\beta}\tfrac{\mathrm{d}}{\mathrm{d}\tau}\mathbb{V}\right)\tfrac{\partial}{\partial t}\tau .$$

Upon substituting the result we derived above for $\vec{\nabla}_r\mathbb{V}$, we obtain:

$$\vec{E}(\vec{r},t) = \frac{q}{4\pi\varepsilon_0}\left[\frac{\hat{R} - \vec{\beta} - \left\{c\hat{R}\cdot\vec{\beta} + \vec{R}\cdot\dot{\vec{\beta}} - c\beta^2\right\}\vec{\nabla}\tau}{\left(R - \vec{R}\cdot\vec{\beta}\right)^2} - \tfrac{1}{c}\left(\mathbb{V}\dot{\vec{\beta}} + \vec{\beta}\tfrac{\partial}{\partial\tau}\mathbb{V}\right)\tfrac{\partial}{\partial t}\tau\right]_{\text{RET}} .$$

Evaluating the derivatives $\vec{\nabla}\tau$ and $\frac{\partial}{\partial t}\tau$ yields:

$$\vec{E} = \frac{q}{4\pi\varepsilon_0}\left[\frac{\vec{R} - R\vec{\beta} + (\vec{R}\cdot\vec{\beta})\vec{\beta} + \frac{1}{c}(\vec{R}\cdot\dot{\vec{\beta}})\vec{R} - \beta^2\vec{R}}{(R - \vec{R}\cdot\vec{\beta})^3} - \frac{R\dot{\vec{\beta}}/c}{(R - \vec{R}\cdot\vec{\beta})^2} + \frac{-R(\hat{R}\cdot\vec{\beta})\vec{\beta} - \frac{1}{c}R(\vec{R}\cdot\dot{\vec{\beta}})\vec{\beta} + R\beta^2\vec{\beta}}{(R - \vec{R}\cdot\vec{\beta})^3}\right]_{\text{RET}}$$

With a little algebra, we can write this as:

$$\vec{E}(\vec{r},t) = \frac{q}{4\pi\varepsilon_0}\left[\frac{(\vec{R} - R\vec{\beta})(1-\beta^2)}{(R - \vec{R}\cdot\vec{\beta})^3} + \frac{(\vec{R}\cdot\dot{\vec{\beta}})(\vec{R} - R\vec{\beta})}{c(R - \vec{R}\cdot\vec{\beta})^3} - \frac{R\dot{\vec{\beta}}}{c(R - \vec{R}\cdot\vec{\beta})^2}\right]_{\text{RET}}.$$

Using the triple product identity, we can combine the last two terms:

$$\vec{R}\times\left\{(\vec{R} - R\vec{\beta})\times\dot{\vec{\beta}}\right\} = (\vec{R} - R\vec{\beta})(\vec{R}\cdot\dot{\vec{\beta}}) + R(R - \vec{R}\cdot\vec{\beta})\dot{\vec{\beta}},$$

and therefore write the electric field as the sum of two terms:

$$\vec{E}(\vec{r},t) = \frac{q}{4\pi\varepsilon_0}\left[\frac{(\vec{R} - R\vec{\beta})(1-\beta^2) + \frac{1}{c}\vec{R}\times\left\{(\vec{R} - R\vec{\beta})\times\dot{\vec{\beta}}\right\}}{(R - \vec{R}\cdot\vec{\beta})^3}\right]_{\text{RET}}.$$

The magnetic field is given by:

$$\vec{B}(\vec{r},t) = \vec{\nabla}\times\vec{\mathbb{A}}(\vec{r},t) = \frac{q}{4\pi\varepsilon_0}\vec{\nabla}\times\left(\left[\frac{\vec{\beta}}{R - \vec{\beta}\cdot\vec{R}}\right]_{\text{RET}}\right)$$

$$= \frac{q}{4\pi\varepsilon_0 c^2}\left[-\frac{\vec{R}\times\dot{\vec{\beta}}}{(R - \vec{R}\cdot\vec{\beta})^2} + \frac{c\vec{\beta}\times\vec{R}}{(R - \vec{R}\cdot\vec{\beta})^2}\left(\frac{1}{R} + \frac{(\hat{R}\cdot\vec{\beta}) + \frac{1}{c}(\vec{R}\cdot\dot{\vec{\beta}}) - \beta^2}{(R - \vec{R}\cdot\vec{\beta})}\right)\right]_{\text{RET}}$$

$$= \frac{q}{4\pi\varepsilon_0 c^2}\left[-\frac{\vec{R}\times\dot{\vec{\beta}}}{(R - \vec{R}\cdot\vec{\beta})^2} + \frac{c\vec{\beta}\times\vec{R}}{(R - \vec{R}\cdot\vec{\beta})^2}\left(1 - \beta^2 + \frac{1}{c}\vec{R}\cdot\dot{\vec{\beta}}\right)\right]_{\text{RET}}.$$

Using the same triple product formula as above, we obtain:

$$\vec{B}(\vec{r},t) = \frac{q}{4\pi\varepsilon_0 c^2}\left[\frac{c\vec{\beta}\times\vec{R}}{(R - \vec{R}\cdot\vec{\beta})^3}(1-\beta^2) + \frac{\vec{R}\times\left\{\vec{R}\times\left(\left\{\vec{R} - R\vec{\beta}\right\}\times\dot{\vec{\beta}}\right)\right\}}{R(R - \vec{R}\cdot\vec{\beta})^3}\right]_{\text{RET}}.$$

If we compare this result to the electric field, we obtain, as expected:

$$\vec{B}(\vec{r},t) = \frac{1}{c}\hat{R}\times\vec{E}(\vec{r},t).$$

The magnetic field is therefore perpendicular to both $\vec{R}$ and $\vec{E}$.

Discussion 17.5: Radiation and Fields of an Accelerated Charge

The expression for the electric field of an accelerated particle reduces to

$$\vec{E}(\vec{r},t)=\frac{q}{4\pi\varepsilon_0}\left[\frac{\left(\vec{R}-R\vec{\beta}\right)\left(1-\beta^2\right)}{\left(R-\vec{R}\cdot\vec{\beta}\right)^3}\right]_{\text{RET}},$$

when $\dot{\vec{\beta}}=0$. Thus the first term of $\vec{E}(\vec{r},t)$ that we obtained in the previous example represents the field due to a uniformly moving charge. That this term equals to what we have already derived for the case of uniform motion can be seen immediately once we recall our earlier result from Example 17.5:

$$R-\vec{\beta}\cdot\vec{R}=\frac{1}{c}\sqrt{(c^2t-\vec{r}\cdot\vec{v})^2+(c^2-v^2)(r^2-c^2t^2)},$$

and also note that:

$$\vec{R}-R\vec{\beta}=\vec{r}-\vec{r}'-c(t-\tau)\vec{\beta}=\vec{r}-c\tau\vec{\beta}-c(t-\tau)\vec{\beta}=\vec{r}-\vec{v}t.$$

Substitution of these two results into the first term of yields the result from Example 17.5:

$$\vec{E}=\frac{qc^3}{4\pi\varepsilon_0}\frac{\left(1-\beta^2\right)\left(\vec{r}-\vec{v}t\right)}{\left[\left(c^2t-\vec{r}\cdot\vec{v}\right)^2+\left(c^2-v^2\right)\left(r^2-c^2t^2\right)\right]^{3/2}}.$$

Notice that as R increases, the first term of $\vec{E}$ falls off as R^{-2}. Thus, as we pointed out before, this term does not contribute to the energy radiated away by the charge. Such radiation comes from the second term, which depends on the charge's acceleration. We will label this term as $\vec{E}_{\text{RAD}}$:

$$\vec{E}_{\text{RAD}}(\vec{r},t)=\frac{q}{4\pi\varepsilon_0}\left[\frac{\vec{R}\times\left\{\left(\vec{R}-R\vec{\beta}\right)\times\dot{\vec{\beta}}\right\}}{c\left(R-\vec{R}\cdot\vec{\beta}\right)^3}\right]_{\text{RET}}.$$

As R increases, $\vec{E}_{\text{RAD}}$ falls off as R^{-1}, so this term gives the only finite contribution from $\vec{E}$ to the energy flux as $R\to\infty$. The same holds true for the acceleration term of the magnetic field $\vec{B}$. We will now look at special cases of accelerated motion and radiation.

Example 17.8 Radiation from a slowly moving accelerated point charge

For the case in which the velocity of the charge is much less than the speed of light, the radiation terms of the electric and magnetic fields are approximately given by

$$\vec{E}_{\text{RAD}}(\vec{r},t)\approx\frac{q}{4\pi\varepsilon_0 c}\frac{\vec{R}\times\left(\vec{R}\times\dot{\vec{\beta}}\right)}{R^3}=\frac{q}{4\pi\varepsilon_0 c}\frac{\vec{R}\left(\vec{R}\cdot\dot{\vec{\beta}}\right)-R^2\dot{\vec{\beta}}}{R^3},$$

and:

$$\vec{B}_{\text{RAD}}(\vec{r},t)\approx\frac{q}{4\pi\varepsilon_0 c^2}\frac{\dot{\vec{\beta}}\times\vec{R}}{R^2}.$$

These terms resemble the fields for a static dipole of dipole moment $\vec{p} = qR\dot{\vec{\beta}}$. If, on the other hand, we compare these terms to the fields of the oscillating electric dipole, then we have that $\dot{\vec{p}} = qc\dot{\vec{\beta}} / \omega^2$.

The part of the Poynting vector that contributes to the radiation is given by

$$\vec{\mathbb{S}}_{\mathrm{RAD}} = \frac{1}{\mu_0}\vec{E}_{\mathrm{RAD}} \times \vec{B}_{\mathrm{RAD}} = \frac{1}{\mu_0}\vec{E}_{\mathrm{RAD}} \times \left(\frac{\vec{R} \times \vec{E}_{\mathrm{RAD}}}{cR}\right) = \frac{1}{\mu_0}\frac{\vec{R}E^2_{\mathrm{RAD}} - \vec{E}_{\mathrm{RAD}}\left(\vec{E}_{\mathrm{RAD}} \cdot \vec{R}\right)}{cR}.$$

Since $\vec{E}_{\mathrm{RAD}} \cdot \vec{R} = 0$, we have that:

$$\vec{\mathbb{S}}_{\mathrm{RAD}} = \frac{1}{\mu_0 c}\frac{\vec{R}E^2_{\mathrm{RAD}}}{R} = \frac{E^2_{\mathrm{RAD}}}{\mu_0 c}\hat{R},$$

where $\hat{R} = \dfrac{\vec{R}}{R}$. Now,

$$E^2_{\mathrm{RAD}} = \left(\frac{q}{4\pi\varepsilon_0 c}\right)^2 \left[\frac{R^2\dot{\beta}^2 - \left(\vec{R} \cdot \dot{\vec{\beta}}\right)^2}{R^4}\right].$$

If is θ the angle between $\vec{R}$ and $\dot{\vec{\beta}}$ (it amounts to the same thing if we define θ as the angle between $\vec{R}$ and $\vec{p}$), then we can write:

$$E^2_{\mathrm{RAD}} = \left(\frac{q}{4\pi\varepsilon_0 c}\right)^2 \frac{\dot{\beta}^2}{R^2}\left[1 - \cos^2\theta\right] = \left(\frac{q}{4\pi\varepsilon_0 c}\right)^2 \frac{\dot{\beta}^2}{R^2}\sin^2\theta,$$

and, thus, we have:

$$\vec{\mathbb{S}}_{\mathrm{RAD}} = \frac{1}{\mu_0 c}\left(\frac{q}{4\pi\varepsilon_0 c}\right)^2 \frac{\dot{\beta}^2}{R^2}\sin^2\theta\,\hat{R} = \frac{q^2}{16\pi^2\varepsilon_0 c}\frac{\dot{\beta}^2}{R^2}\sin^2\theta\,\hat{R}.$$

The Poynting vector is related to the differential power radiated per unit time per unit solid angle through the relationship

$$\frac{\mathrm{d}P}{\mathrm{d}\Omega} = \left(\vec{\mathbb{S}}_{\mathrm{RAD}} \cdot \hat{R}\right)R^2 = \frac{q^2}{16\pi^2\varepsilon_0 c}\dot{\beta}^2\sin^2\theta,$$

which gives the classic pattern that is also produced by dipole radiation.

We integrate to get the total radiated power

$$P = \oint \frac{\mathrm{d}P}{\mathrm{d}\Omega}\mathrm{d}\Omega = \frac{q^2}{16\pi^2\varepsilon_0 c}\dot{\beta}^2 \int_0^{2\pi}\mathrm{d}\phi \int_0^{\pi}\left(\sin^2\theta\right)\sin\theta\,\mathrm{d}\theta = \frac{q^2}{6\pi\varepsilon_0 c}\dot{\beta}^2.$$

This is, again, the famous *Larmor formula* for the power radiated by a nonrelativistic accelerated point charge.

Example 17.9: Radiation with Parallel Velocity and Acceleration

For the purpose of this discussion we relax the condition that the velocity be much less than the speed of light; the results we derive here hold for any velocity less than c. We are interested only in radiation so we consider just the radiation terms of the fields:

$$\vec{E}_{\text{RAD}}(\vec{r},t) = \frac{q}{4\pi\varepsilon_0}\left[\frac{\vec{R}\times\left\{\left(\vec{R}-R\vec{\beta}\right)\times\dot{\vec{\beta}}\right\}}{c\left(R-\vec{R}\cdot\vec{\beta}\right)^3}\right]_{\text{RET}} = \frac{q}{4\pi\varepsilon_0 c}\left[\frac{\vec{R}\times\left(\vec{R}\times\dot{\vec{\beta}}\right)}{\left(R-\vec{R}\cdot\vec{\beta}\right)^3}\right]_{\text{RET}},$$

since $\vec{\beta}\times\dot{\vec{\beta}}=0$ by assumption. The magnetic field is thus

$$\vec{B}_{\text{RAD}}(\vec{r},t) = \frac{q}{4\pi\varepsilon_0 c^2}\left[\frac{R\dot{\vec{\beta}}\times\vec{R}}{\left(R-\vec{R}\cdot\vec{\beta}\right)^3}\right]_{\text{RET}}.$$

Note that these expressions are nearly the same as the ones we obtained in the case of the slowly moving accelerated particle, save for the exact form of the denominator. The denominator in the case of the slowly moving accelerated charge differs from this one by a factor of:

$$\frac{R^3}{\left(R-\vec{R}\cdot\vec{\beta}\right)^3} = \frac{1}{\left(1-\beta\cos\theta\right)^3}.$$

The main effect of this factor is that the angular distribution of the radiation is increased in the forward direction (i.e. the direction of motion).

Using the well-known identity for triple vector products, $\vec{A}\times(\vec{B}\times\vec{C}) = \vec{B}(\vec{A}\cdot\vec{C}) - \vec{C}(\vec{A}\cdot\vec{B})$:

$$\vec{E}_{\text{RAD}}(\vec{r},t) = \frac{q}{4\pi\varepsilon_0}\left[\frac{\vec{R}\left(\vec{R}\cdot\dot{\vec{\beta}}\right)-\dot{\vec{\beta}}R^2}{c\left(R-\vec{R}\cdot\vec{\beta}\right)^3}\right]_{\text{RET}}, \quad \text{so:} \quad E^2_{\text{RAD}}(\vec{r},t) = \left(\frac{q}{4\pi\varepsilon_0}\right)^2\left[\frac{\dot{\beta}^2R^4-\left(\vec{R}\cdot\dot{\vec{\beta}}\right)^2R^2}{c^2\left(R-\vec{R}\cdot\vec{\beta}\right)^6}\right]_{\text{RET}}.$$

The numerator is identical with the expression we found in the previous example $R^2\dot{\beta}^2-\left(\vec{R}\cdot\dot{\vec{\beta}}\right)^2 = R^2\dot{\beta}^2\left(1-\cos^2\theta\right)$, while the denominator is given by $R-\vec{R}\cdot\vec{\beta} = R(1-\beta\cos\theta)$.

We therefore have that:

$$E^2_{\text{RAD}}(\vec{r},t) = \left(\frac{q}{4\pi\varepsilon_0 c}\right)^2\left[\frac{\dot{\beta}^2\left(1-\cos^2\theta\right)}{R^2(1-\beta\cos\theta)^6}\right]_{\text{RET}} = \left(\frac{q}{4\pi\varepsilon_0 c}\right)^2\left[\frac{\dot{\beta}^2\sin^2\theta}{R^2(1-\beta\cos\theta)^6}\right]_{\text{RET}},$$

and the Poynting vector is given by

$$\vec{\mathbb{S}}_{\text{RAD}} = \frac{E^2_{\text{RAD}}}{\mu_0 c}\hat{R} = \frac{1}{\mu_0 c}\left(\frac{q}{4\pi\varepsilon_0 c}\right)^2\left[\frac{\dot{\beta}^2\sin^2\theta}{R^2(1-\beta\cos\theta)^6}\hat{R}\right]_{\text{RET}}.$$

The amount of power radiated into a unit solid angle, and that crosses a surface at a distance R from the emitting charge at time t, is equal to the energy per unit time lost by the particle at time τ. The differential amount of energy radiated into a solid angle $d\Omega$ is

$$dU = -\left[\vec{\mathbb{S}}_{\text{RAD}} \cdot \hat{R}\right]_{\text{RET}} dt\, R^2 d\Omega = -\left(\vec{\mathbb{S}}_{\text{RAD}} \cdot \hat{R}\right)\frac{dt}{d\tau} d\tau R^2 d\Omega = -\left(\frac{E_{\text{RAD}}^2}{\mu_0 c}\right)\frac{dt}{d\tau} d\tau R^2 d\Omega \,,$$

where the minus sign indicates that the energy is lost by the charge to radiation.

Since:

$$\frac{dt}{d\tau} = 1 - \frac{\vec{\beta}\cdot\vec{R}}{R} = 1 - \beta\cos\theta \,,$$

we can write the rate at which energy is radiated into a solid angle as:

$$\frac{dU}{d\tau} d\Omega = -\frac{1}{\mu_0 c}\left(\frac{q}{4\pi\varepsilon_0 c}\right)^2 \frac{\dot{\beta}^2 \sin^2\theta}{\left(1-\beta\cos\theta\right)^5} d\Omega \,,$$

or:

$$\frac{dP}{d\Omega} = -\frac{1}{\mu_0 c}\left(\frac{q}{4\pi\varepsilon_0 c}\right)^2 \frac{\dot{\beta}^2 \sin^2\theta}{\left(1-\beta\cos\theta\right)^5} \,.$$

This result is generally useful for getting an estimate of the energy lost when a charged particle is decelerated, a process that physicists refer to as *bremsstrahlung* (from the German word for "braking radiation").

The angle θ_{max} for which the intensity is a maximum can be found by taking the derivative of the above expression and setting it equal to zero and solving for the angle. The result is:

$$\cos\left(\theta_{\text{max}}\right) = \frac{\sqrt{1+15\beta^2}-1}{3\beta} = \frac{4\left(1-15/(16\gamma^2)\right)^{1/2}-1}{3\left(1-1/\gamma^2\right)^{1/2}} \,,$$

or:

$$\theta_{\text{max}} = \cos^{-1}\left\{\frac{\sqrt{1+15\beta^2}-1}{3\beta}\right\} = \cos^{-1}\left\{\frac{4\left(1-15/(16\gamma^2)\right)^{1/2}-1}{3\left(1-1/\gamma^2\right)^{1/2}}\right\},$$

where, as usual, $\gamma^2 = \dfrac{1}{1-\beta^2}$.

For extremely high speeds, such that $\beta \to 1$, $\cos\left(\theta_{\text{max}}\right) \approx 1 - \frac{1}{2}\theta_{\text{max}}^2$, and the right hand side of the above expression can be written as an expansion in powers of $1/\gamma^2$:

$$1 - \frac{1}{2}\theta_{\text{max}}^2 \approx 1 - \frac{1}{8\gamma^2} \,.$$

Thus,

$$\theta_{\text{max}} \approx \frac{1}{2\gamma} \ll 1 \,.$$

A polar diagram of the radiation pattern would therefore show that the radiation vanishes along the axes defined by $\vec{\beta}$ and $\dot{\vec{\beta}}$, but has two lobes whose maxima lie at an approximate angle of $\theta_{\text{max}} \approx 1/(2\gamma)$ on either side of this axis. A three-dimensional view would show the radiation forming a mostly hollow cone centered on the $\vec{\beta}$ axis.

Compared to radiation from a charge at rest, the emission of a charge moving at high velocity is shifted in the direction of motion. This is often referred to as *beaming*.

Example 17.10 Synchrotron Radiation

Charged particles are accelerated perpendicular to their velocity when deflected by a prescribed magnetic field such as we have in the case of synchrotron devices for accelerating charges to high energies. In such a synchrotron accelerator, the guiding magnetic field, which bends the particles into a closed path, is time-dependent, being *synchronized* to a particle beam of increasing kinetic energy. In many astrophysical situations, charged particles gyrate about magnetic field lines and emit radiation similar to that produced in synchrotrons. For this reason, radiation emitted by charges in circular motion is referred to as *synchrotron radiation.* Electrons and positrons in the Crab Nebula, which is a young supernova remnant, orbit magnetic fields coupled, or "frozen," to the nebular plasma and emit synchrotron radiation at x-ray, optical and radio wavelengths.

Imagine a charged particle moving in a circular orbit of fixed radius ρ. Its acceleration $\dot{\vec{\beta}}$ is directed toward the center of the orbit, and its velocity vector $\vec{\beta}$ is tangentially directed along the orbit's circumference and is thus perpendicular to the acceleration. The angular frequency of the charge's motion is:

$\omega = \dfrac{c\beta}{\rho}$, and the acceleration is related to the frequency via $\dot{\beta} = \dfrac{\rho\omega^2}{c}$.

For definiteness, imagine that the orbit lies entirely in the $y-z$ plane, with directed $\vec{R}$ from the charge to the fixed point of observation. Then:

$\vec{\beta} \cdot \vec{R} = \beta R \cos\theta$, and $\dot{\vec{\beta}} \cdot \vec{R} = \dot{\beta} R \sin\theta \cos\varphi$.

The acceleration field is given by:

$$\vec{E}_{\text{RAD}}(\vec{r},t) = \frac{q}{4\pi\varepsilon_0}\left[\frac{\vec{R}\times\left\{\left(\vec{R}-R\vec{\beta}\right)\times\dot{\vec{\beta}}\right\}}{c\left(R-\vec{R}\cdot\vec{\beta}\right)^3}\right]_{\text{RET}},$$

and the Poynting vector is given by the square of the field:

$$\vec{\mathbb{S}}_{\text{RAD}} = \frac{E_{\text{RAD}}^2}{\mu_0 c}\hat{R}.$$

Now we want to express this in terms of the coordinate angles θ and φ, especially for the purposes of calculating the angular distribution of the Poynting vector and radiated power. We begin by defining the vector:

$$\vec{R}_\beta \equiv \vec{R} - R\vec{\beta},$$

so that:

$$\left\{\vec{R}\times\left(\vec{R}_\beta\times\dot{\vec{\beta}}\right)\right\}^2=\left\{\vec{R}_\beta\left(\vec{R}\cdot\dot{\vec{\beta}}\right)-\dot{\vec{\beta}}\left(\vec{R}\cdot\vec{R}_\beta\right)\right\}^2.$$

Upon expanding the right hand side, we obtain:

$$\left\{\vec{R}\times\left(\vec{R}_\beta\times\dot{\vec{\beta}}\right)\right\}^2=R_\beta^2\left(\vec{R}\cdot\dot{\vec{\beta}}\right)^2-2\left(\vec{R}\cdot\dot{\vec{\beta}}\right)\left(\dot{\vec{\beta}}\cdot\vec{R}_\beta\right)\left(\vec{R}\cdot\vec{R}_\beta\right)+\dot{\beta}^2\left(\vec{R}\cdot\vec{R}_\beta\right)^2.$$

From the definition of the vectors and the angles, we also have that:

$$R_\beta^2=\left(\vec{R}-R\vec{\beta}\right)^2=R^2-2R\vec{R}\cdot\vec{\beta}+R^2\beta^2=R^2\left(1-2R\beta\cos\theta+\beta^2\right),$$

$$\vec{R}\cdot\vec{R}_\beta=R\left(1-\beta\cos\theta\right),$$

and:

$$\dot{\vec{\beta}}\cdot\vec{R}_\beta=\dot{\vec{\beta}}\cdot\vec{R}-R\dot{\vec{\beta}}\cdot\vec{\beta}=\dot{\beta}R\sin\theta\cos\varphi,$$

since $\dot{\vec{\beta}}\cdot\vec{\beta}=0$.

Thus,

$$\left\{\vec{R}\times\left(\vec{R}_\beta\times\dot{\vec{\beta}}\right)\right\}^2=R^4\dot{\beta}^2\left\{\left(1-\beta\cos\theta\right)^2-\left(1-\beta^2\right)\sin^2\theta\cos^2\varphi\right\}.$$

Therefore, the Poynting vector is:

$$\vec{\mathbb{S}}_{\text{RAD}}=\frac{E_{\text{RAD}}^2}{\mu_0 c}\hat{R}=\frac{1}{\mu_0 c}\left(\frac{q}{4\pi\varepsilon_0 c}\right)^2\frac{\dot{\beta}^2\left\{\left(1-\beta\cos\theta\right)^2-\left(1-\beta^2\right)\sin^2\theta\cos^2\varphi\right\}}{R^2\left(1-\beta\cos\theta\right)^6}\hat{R}.$$

Repeating the analysis of the previous example, we obtain:

$$\frac{d\mathbb{P}}{d\Omega}=-\frac{1}{\mu_0 c}\left(\frac{q}{4\pi\varepsilon_0 c}\right)^2\frac{\dot{\beta}^2\left\{\left(1-\beta\cos\theta\right)^2-\left(1-\beta^2\right)\sin^2\theta\cos^2\varphi\right\}}{\left(1-\beta\cos\theta\right)^5}.$$

For $\beta\approx 1$, the radiation has a very sharply peaked intensity pattern in the direction of motion, which is often referred to as *synchrotron beaming*, and a backward lobe of lower intensity.

The total power radiated by the charged particle is given by:

$$\mathbb{P}=\oint\frac{dP}{d\Omega}d\Omega=\frac{q^2\dot{\beta}^2}{6\pi\varepsilon_0 c}\frac{1}{\left(1-\beta^2\right)^2},$$

which reduces to the Larmor formula when $\beta\ll 1$. It is clear that charged particles moving in a circular orbit at high speed will quickly radiate away their energy. We may write this result in terms of the angular frequency at which the particle gyrates:

$$\mathbb{P}=\frac{2}{3}\frac{r_e}{c}\frac{\rho^2\omega^4}{c^2}\gamma^4\left(m_e c^2\right)=\frac{2\omega^2 r_e}{3c}\beta^2\gamma^4 m_e c^2,$$

where we have now assumed the particle to be an electron or positron (i.e. $q = \mp e$), and r_e is the *classical electron radius* given by:

$$r_e = \frac{e^2}{4\pi\varepsilon_0 m_e c^2}.$$

We will not discuss the frequency spectrum radiated by accelerated charges, which is a subject that is covered in more advanced texts. In general, one computes the Fourier transform of the fields and then calculates the Poynting flux to obtain the radiated power as a function of frequency. This is important since detectors measure the frequency spectrum, and an understanding of the power spectra arising from different physical processes aids in an analysis of measured signals.

For example, a charged particle that is quickly decelerated emits radiation whose spectrum is independent of frequency at least up to a maximum, or cut-off, frequency, above which the spectrum exponentially falls to zero. This is the classic bremsstrahlung spectrum that comes up often in astrophysics and atomic physics. A charge radiates as it traverses a circular path, and then we expect the spectrum to be composed of discrete harmonics of the fundamental frequency of its motion.

We can get a brief idea of the physics involved in understanding the spectrum from a charged particle in circular motion by considering the real world example of a charged particle gyrating about a magnetic field, such as happens in the synchrotron emitting regions of the Crab Nebula. If we denote the prescribed magnetic field as $\vec{B}_0$, then the fundamental frequency of motion equals the cyclotron frequency, ω_B, which is given by:

$$\omega_0 = \omega_B = \frac{eB_0}{m_e\gamma}.$$

The radiated power, in ratio to the electron mass, is:

$$\frac{\mathbb{P}}{m_e c^2} = \frac{2}{3}\left(\frac{eB_0}{m_e}\right)^2\left(\frac{r_e}{c}\right)\beta^2\gamma^2.$$

Now, consider successive points along the charge's circular trajectory separated by the angular amount $\Delta\theta$:

$$\Delta\theta \approx \frac{1}{\gamma} \ll 1,$$

and a pulse that is emitted during a time $\Delta\tau = \Delta\theta/\omega_0$ is received by the observer in a time interval:

$$\Delta t = \frac{\Delta\theta}{\omega_0}(1-\beta) \approx \frac{\Delta\theta}{2\omega_0\gamma^2}.$$

For the relativistic case, where $\gamma \gg 1$, this time interval is approximately given by:

$$\Delta t \approx \frac{\Delta\theta}{2\omega_0\gamma^2}.$$

This means that the emitted frequency spectrum will contain harmonics that are multiples of the fundamental frequency up to a maximum of:

$$\Delta\omega \approx \gamma^3\omega_0 = \gamma^3\omega_B = \left(\frac{eB_0}{m_e}\right)\gamma^2 .$$

The electric field of a slowly gyrating charge varies sinusoidally with the same frequency, ω_B, with which the charge revolves about its force center. (Because of the nature of the magnetic part of the Lorentz force, the force center coincides with the location of the magnetic field line.) For nonrelativistic motion, then, the power spectrum consists of a single line centered on ω_B. As β increases, higher harmonics of the fundamental frequency contribute to the spectrum. That is, the spectrum consists of integer multiples of ω_B, since there is a periodicity in time intervals $T = 2\pi/\omega_B$. For mildly relativistic speeds, for example, there is radiation in the first harmonic of ω_B, i.e. at $2\omega_B$, as well as in the fundamental at ω_B. For extremely relativistic motion, $\beta \approx 1$, the electric field will be a series of repeated sharp pulses of approximate width $1/(\gamma^3\omega_B)$ separated by a time interval $\Delta t = 2\pi/\omega_B$. The resulting power spectra comprise such a large number of harmonics that the envelope of these harmonics approaches a smooth continuous function, which is characteristic of synchrotron radiation.[294]

Example 17.11 Radiation from an arbitrarily accelerated charge

Classically, we can use our earlier result for the field of an arbitrarily accelerated charge:

$$\vec{E}_{\text{RAD}}(\vec{r},t) = \frac{q}{4\pi\varepsilon_0}\left[\frac{\vec{R}\times\left\{\left(\vec{R}-R\vec{\beta}\right)\times\dot{\vec{\beta}}\right\}}{c\left(R-\vec{R}\cdot\vec{\beta}\right)^3}\right]_{\text{RET}}, \text{ and: } \frac{\mathrm{d}\mathbb{U}}{\mathrm{d}\tau}\delta\Omega = -\frac{q^2\vec{R}}{16\pi^2\varepsilon_0 c^2}\left[\frac{\vec{R}\times\left\{\left(\vec{R}-R\vec{\beta}\right)\times\dot{\vec{\beta}}\right\}}{\left(R-\vec{R}\cdot\vec{\beta}\right)^5}\right]\delta\Omega.,$$

We can now integrate over solid angle to obtain the radiated power:

$$\mathbb{P} = -\oint\frac{\mathrm{d}\mathbb{U}}{\mathrm{d}\tau}\mathrm{d}\Omega = \frac{q^2}{6\pi\varepsilon_0 c}\gamma^6\left[\dot{\beta}^2 - \left(\vec{\beta}\times\dot{\vec{\beta}}\right)^2\right].$$

Liénard followed just such a procedure when he obtained this result in 1898. The integration is straightforward, but tedious.

We can also derive Liénard's expression for the radiated power by noting the relativistic generalization of the Larmor formula we derived earlier:

$$\mathbb{P} = \frac{q^2}{6\pi\varepsilon_0 m^2 c^3}\left(\frac{\mathrm{d}p_\mu}{\mathrm{d}\tau}\frac{\mathrm{d}p^\mu}{\mathrm{d}\tau}\right), \text{ where: } p^\mu = \left(\frac{\mathbb{E}}{c},\vec{p}\right) = mc\gamma\left(1,\vec{\beta}\right) \text{ and } p_\mu = mc\gamma\left(-1,\vec{\beta}\right),$$

and the repeated use of upper and lower indices implies summation over the index μ. The proper time interval is related to the observer's time via $\mathrm{d}\tau = \mathrm{d}t/\gamma$.[295]

---

[294] See G. Rybicki and A. Lightman, Radiative Processes in Astrophysics (New York: John Wiley & Sons, 1979), pages 181-186, for more details on the synchrotron power spectrum.

[295] Heaviside derived the relativistic generalization of Larmor's formula prior to the discovery of relativity! See, O. Heaviside, "The waste of energy from a moving electron," Nature, **67**, (1902), 6-7.

So:

$$\left(\frac{\mathrm{d}p_\mu}{\mathrm{d}\tau}\frac{\mathrm{d}p^\mu}{\mathrm{d}\tau}\right)=\left(\frac{\mathrm{d}\vec{p}}{\mathrm{d}\tau}\right)^2-\left(\frac{\mathrm{d}(mc\gamma)}{\mathrm{d}\tau}\right)^2,$$

where:

$$\frac{\mathrm{d}}{\mathrm{d}\tau}(mc\gamma)=\frac{mc\vec{\beta}\cdot\dot{\vec{\beta}}}{\left(1-\beta^2\right)^{3/2}}=mc\gamma^3\vec{\beta}\cdot\dot{\vec{\beta}},$$

and:

$$\frac{\mathrm{d}\vec{p}}{\mathrm{d}\tau}=\frac{\mathrm{d}}{\mathrm{d}\tau}\left(mc\gamma\,\vec{\beta}\right)=\frac{mc\dot{\vec{\beta}}}{\left(1-\beta^2\right)^{1/2}}+\frac{mc\vec{\beta}\left(\vec{\beta}\cdot\dot{\vec{\beta}}\right)}{\left(1-\beta^2\right)^{3/2}}=mc\gamma\,\dot{\vec{\beta}}+mc\gamma^3\vec{\beta}\left(\vec{\beta}\cdot\dot{\vec{\beta}}\right).$$

Collecting terms, and recalling the vector identity:

$$\left(\vec{A}\times\vec{B}\right)\left(\vec{C}\times\vec{D}\right)=\left(\vec{A}\cdot\vec{C}\right)\left(\vec{B}\cdot\vec{D}\right)-\left(\vec{A}\cdot\vec{D}\right)\left(\vec{B}\cdot\vec{C}\right),$$

yields Alfred-Marie Liénard's result.

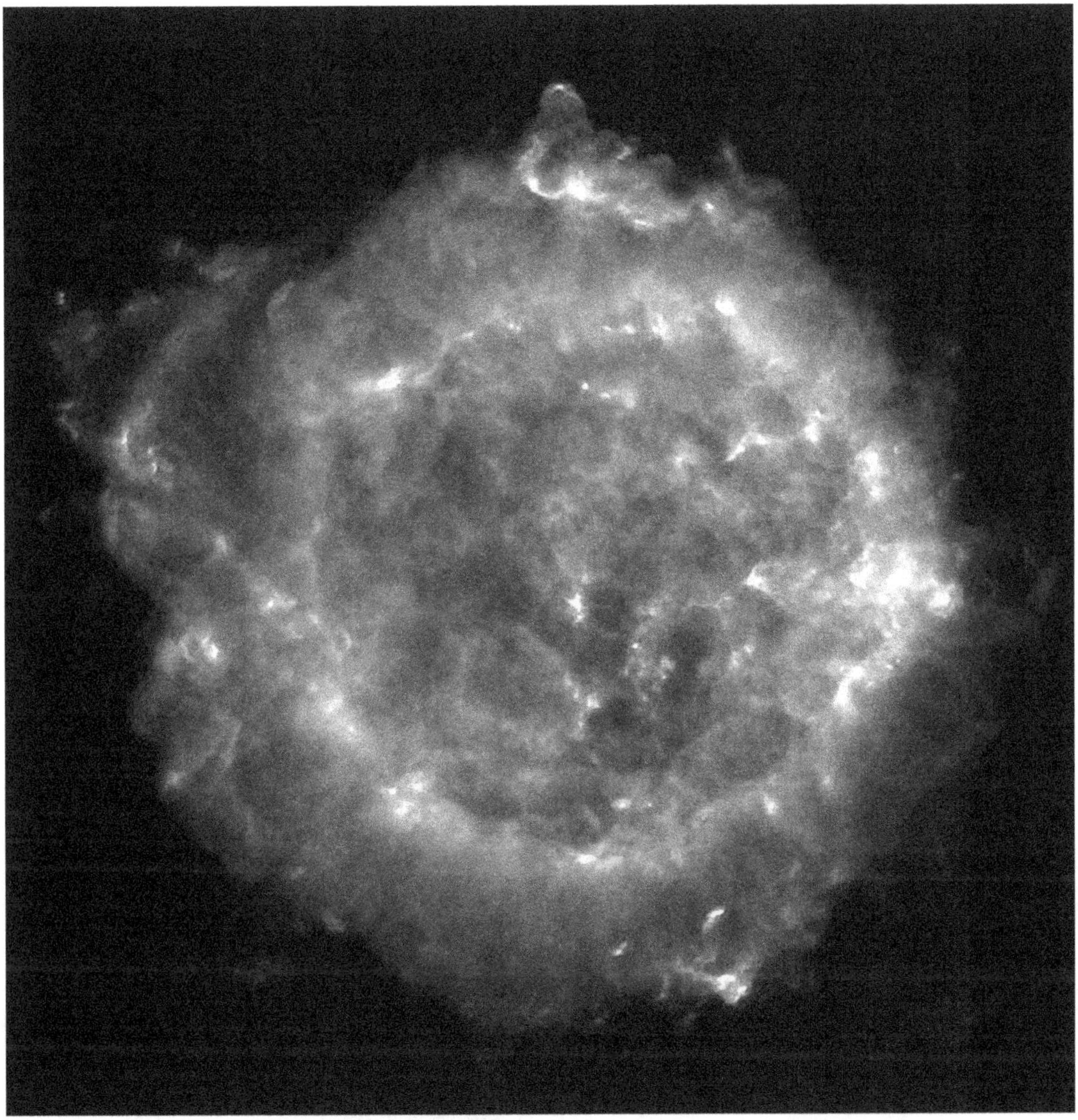

Plate 31: The Supernova Remnant Cassiopeia A[296]

[296] This Very Large Array (p. 561) radio map has a field of view of six arcminutes, which at that distance is about six light-years, and represents synchrotron emission by relativistic electrons interacting with the local magnetic field, which we know because the radio emission is somewhat linearly polarized (see Plate 27 panel k, p. 503). As we discussed in Example 17.10 (p. 630), the brightness is proportional to the square of the magnetic field, so you can interpret the brighter regions as representing areas of high magnetic field strength, which was presumably amplified in more turbulent regions as the magnetic field is frozen into the plasma (see Section 12.2, p. 381). This image was downloaded from the website of The National Radio Astronomy Observatory, and it was produced by L. Rudnick, T. Delaney, J. Keohane, & B. Koralesky, and composed by T. Rector.

Fig. 1.

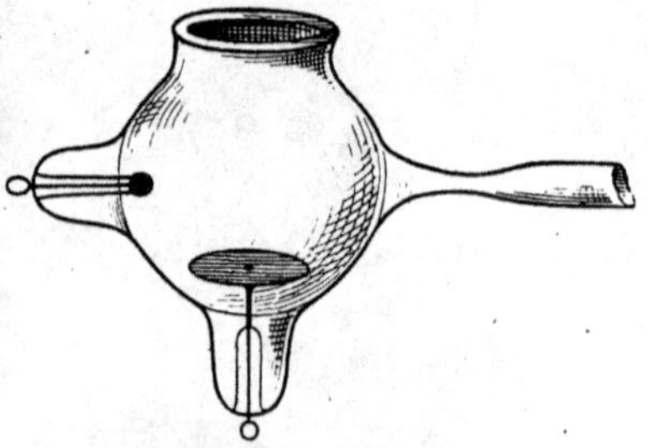

Fig. 2 (ca. natürl. Gr.).

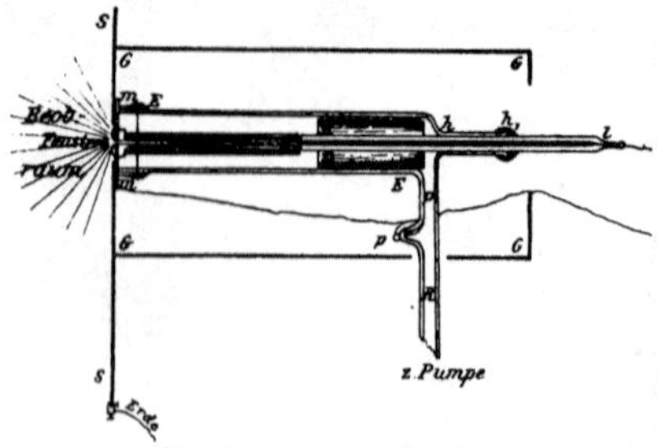

Fig. 3 (ca. ¹/₅ natürl. Gr.).

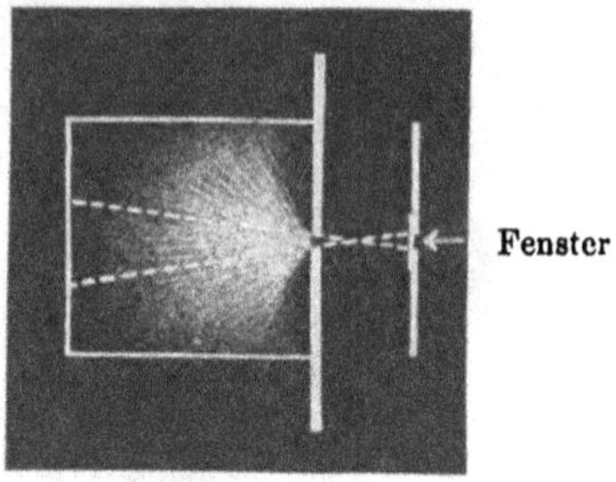

Fig. 4.

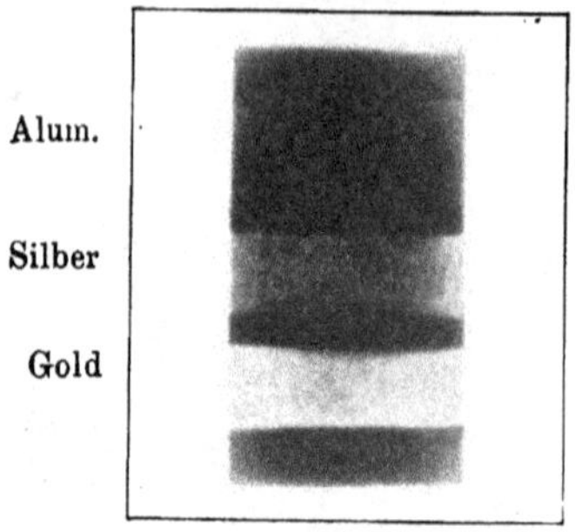

Fig. 7.
gleich dick (0,00071 mm)

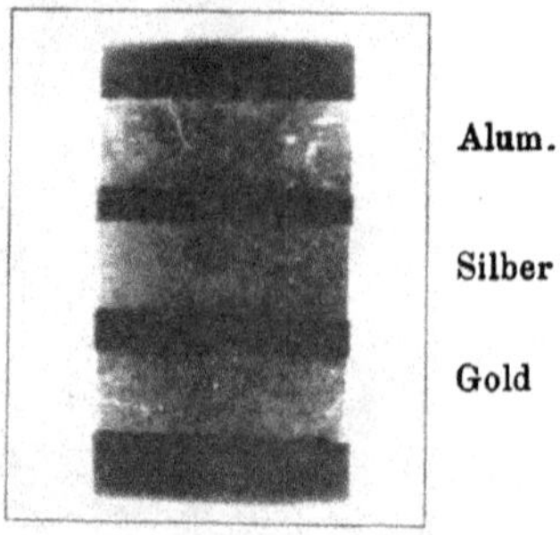

Fig. 8.
gleich schwer (0,75 mg/mm²)

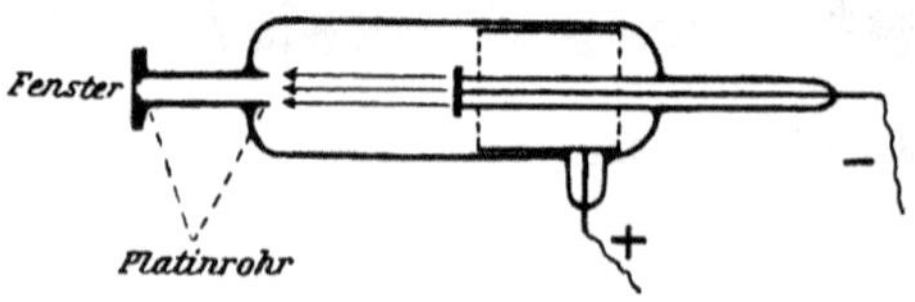

Fig. 9 (ca. ¹/₄ natürl. Gr.).

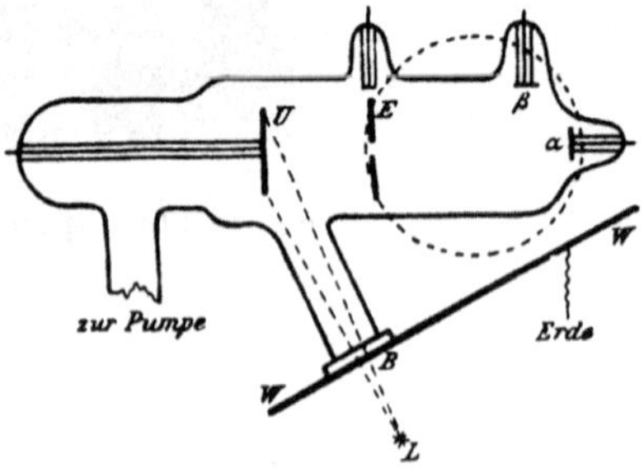

Fig. 10 (ca. ¹/₃ natürl. Gr.).

Plate 32: Phillip Lenard's Photoelectric Experiment[297]

[297]P. Lenard, Über Kathodenstrahlen Nobel-Vorlesung (Verlag von Johann Ambrosius Barth, 19). He won the 1905 Nobel prize for his work that took place around 1890.

Chapter 18 The Photon

> Whereas the main Business of natural Philosophy, is to argue from Phaenomena without feigning Hypotheses, and to deduce cause from effect until we come to the very first Cause, which certainly is not mechanical; and not only to unfold the Mechanism for the World, but chiefly to resolve these and such like Questions.[298]

Isaac Newton demonstrated with his famous seventeenth century prism experiment that white light is composed of the colors of the rainbow. Based on this, and a whole host of other experiments, he put forth a particle theory of light in his works Principia (1687) and Opticks (1704) that he used to explain rectilinear propagation, reflection, refraction, and optical dispersion. In his *corpuscular theory*, Newton rejected the wave model of Huygens, largely because the medium required for such a wave was incompatible with the motion of the planets. Instead he proposed that light consists of very small particles endowed with a tiny mass, and that these corpuscles can therefore experience force, impulse, and convey momentum. And, most importantly, Newton's corpuscle model clearly explained cause and effect—even across the vast distances of space.

James Bradley's 1727 discovery of stellar aberration supported Newton's corpuscular model, as it could easily be explained from vector addition of particle velocities, so Newton's theory of light dominated through the rest of the century—especially in the Anglosphere.

As the nineteenth century dawned, Newton's corpuscle theory set. This commenced in 1804 when Thomas Young presented his famous experiment in which light falls on an opaque screen after pasting through a system of two slits.[299] Young also suggested, in place of the idea that different colors were associated with different velocities or corpuscular masses, that each color has a specific wavelength.

After that, things just got worse and worse for the corpuscular theory of light. In 1862, Foucault measured the speed of light accurately in both air and water.[300] The result was decisive! Light moves slower in a denser medium, exactly as predicted by wave theory. Finally, Maxwell's 1864 paper, "Dynamical Theory of the Electromagnetic Field," provided a clear wave mechanism, and the question of the nature of light was totally solved.

In the last decade of the century, there were plenty of other problems to solve. Most of these had to do with the detailed nature of matter, including the medium that was assumed to pervade free space—the aether. Not only did the aether decide the nature of light, but also it solved the action at a distance problem once and for all, a point Lord Kelvin made clear in his preface to the 1893 edition of the English translation of Hertz's treatise that we have reproduced in Appendix B (p. 705). Things would soon change, and quickly too. In 1900, the elderly Kelvin changed his tune, and now spoke of two ominous dark clouds looming over physics. The first cloud shed doubt on the existence of the aether, and the second cloud shed doubt on "the Maxwell-Boltzmann doctrine regarding the partition of energy."

---

[298] Isaac Newton, Opticks or A Treatise on the Reflections, Refractions, Inflections and Colors of Light, Book III, Part I, Query 28

[299] Young, Thomas, "The Bakerian Lecture: Experiments and calculations relative to physical optics", Philosophical Transactions of the Royal Society of London (Royal Society of London) **94** (1804), 1–16.

[300] Foucault, Comptes Rendus, "Détermination expérimentale de la vitesse e la lumière; parallaxe du Soleil"(1862), **55**, 501-3, 792-96. English translation is available in W.F. Magie, Source Book in Physics, (Cambridge: Harvard University Press, 1935) 343-44.

In hindsight, Max Planck became the first heretic when he used energy quanta to derive the correct formula for the spectrum of blackbody radiation. Perhaps because Planck did not dwell on this radical assumption in his original papers of 1900 and 1901, few scientists recognized the complete break from what we now call *classical physics*. In fact, during the five or six years immediately following the publication of Planck's papers, only a handful of scientists publicly questioned Planck's result or the route by which he had obtained it. Among these scientists were Sir James Jeans, Lord Rayleigh, and Albert Einstein, all of whom pointed out that any consistent derivation that recognized energy equipartition should, instead, lead to the following result for the spectral energy density of radiation inside an oven with walls kept at a fixed temperature T:

$$\mathrm{u}(\omega,T)=\frac{\omega^2}{\pi^2c^3}k_BT.$$

Rayleigh, who had been critical of the equipartition theorem, showed that this formula failed at short wavelengths and suggested it might be due to a failure of the equipartition theorem at high frequencies. To remedy what Paul Ehrenfest would later refer to as "the ultraviolet catastrophe," Rayleigh proposed an exponential cutoff so it would agree with most experimental data. Despite including this plausible, but *ad hoc*, exponential factor, Rayleigh's modified formula failed to reproduce Wien's displacement law that $\lambda_{\max}T$ = constant, which was known to be in excellent accord with experiment. For this reason, Rayleigh's more general formula is largely forgotten. The name *Rayleigh-Jeans law* is attached to the above classical expression after Rayleigh, and Jeans who corrected a factor in Rayleigh's original equation.

The so-called Rayleigh-Jeans law might be called the *Rayleigh-Jeans-Einstein* law instead, since Einstein obtained the above expression for the thermal radiation energy density $\mathrm{u}(\omega,T)$ independently of Rayleigh and Jeans, and (unlike Rayleigh and Jeans) demonstrated explicitly that it diverges for high frequencies. He also argued that Planck's formula, though in excellent agreement with data at all wavelengths, could only be derived by assuming the discontinuous transfer of energy between the oven wall resonators and the radiation within the oven. In fact, Einstein demonstrated, had Planck maintained a consistently classical approach, he would have found the Rayleigh-Jeans expression for $\mathrm{u}(\omega,T)$, rather than the experimentally correct formula.

Einstein publicly revealed Planck's tentative heresy by showing that Planck's distribution function implied light should be conceived of as a collection of particles, or quanta—despite the previous century's worth of overwhelming evidence to the contrary. As Einstein puts it in his famous 1905 paper:

> It seems to me that the observations associated with blackbody radiation, fluorescence, the production of cathode rays by ultraviolet light, and other related phenomena connected with the emission or transformation of light are more readily understood if one assumes that the energy of light is discontinuously distributed in space. In accordance with the assumption to be considered here, the energy of a light ray spreading out from a point source is not continuously distributed over an increasing space but consists of a finite number of energy quanta which are localized at points in space, which move without dividing, and which can only be produced and absorbed as complete units.
>
> In the following I wish to present the line of thought and the facts which have led me to this point of view, hoping that this approach may be useful to some investigators in their research.[301]

---

[301] Albert Einstein, "Concerning a Heuristic Point of View Toward the Emission and Transformation of Light," Annalen der Physik **17** (1905), 132-148.

Einstein's proposal of light quanta faced strong opposition, even from those scientists who enthusiastically embraced relativity, and it is not difficult to see why this was the case. Although the particle hypothesis explains the photoelectric effect and blackbody radiation, it is inconsistent with the overwhelming number of experiments that appear to confirm the wave nature of light.

Einstein, of course, was well aware of the success of the wave theory, about which he wrote:

> The wave theory of light, which operates with continuous spatial functions, has worked well in the representation of purely optical phenomena and will probably never be replaced by another theory. It should be kept in mind, however, that the optical observations refer to time averages rather than instantaneous values. In spite of the complete experimental confirmation of the theory as applied to diffraction, reflection, refraction, dispersion, etc., it is still conceivable that the theory of light which operates with continuous spatial functions may lead to contradictions with experience when it is applied to the phenomena of emission and transformation of light.[302]

Einstein's hypothesis was actually well reasoned, and in retrospect his argument is quite convincing. After confirming that the Rayleigh-Jeans law agrees with Planck's distribution at low frequencies, Einstein considered the high frequency limit of Planck's law. In the region of high frequencies, for which $\hbar\omega \gg k_B T$, Planck's radiation law takes the following form:

$$\mathbb{u}(\omega,T) = \frac{\hbar\omega^3}{\pi^3 c^3} e^{-\hbar\omega/k_B T} .$$

Based on experimental data taken at high frequency, Wilhelm Wien had actually proposed this function (without explicit reference to the constant $\hbar$) as the correct expression for the spectral energy density of blackbody radiation at all frequencies. In the limit where this expression applies, Einstein demonstrated that the entropy of radiation does not behave like that of waves, but rather like particles of energy $\mathbb{E} = \hbar\omega$!

Where Planck had assumed only that the energy absorbed, or emitted, by the material oscillators of the cavity walls was restricted to integer multiples of $\hbar\omega$, Einstein now proposed that it was the radiation in the cavity whose energy was actually quantized. Not only that, but he broke entirely with the received physics of his day by suggesting a new *heuristic* viewpoint that would treat light as if it comprised particles whose energy is proportional to the frequency of that light.

Einstein adduced such phenomena as fluorescence and the photoionization of gases to argue for the plausibility of his heuristic light-quanta, which were later dubbed *photons.* Most significantly, he explained the ejection of electrons from the surface of an irradiated metal, which is called the *photoelectric effect*, which had been studied by Phillip Lenard whose figures from his 1905 Nobel Prize lecture are shown at the beginning of this chapter (p. 640).

In the end, it was the skeptical Robert Millikan who convinced the physics community to take the photon concept seriously by thoroughly testing the photoelectric effect. He eloquently describes his understanding of the relevant physics in his 1916 paper:

> Quantum theory was not originally developed for the sake of interpreting photoelectric phenomena. It was solely a theory as to the mechanism of absorption and emission of electromagnetic waves by resonators of atomic or subatomic dimensions. It had nothing whatever to say about the energy of an escaping electron or about the conditions under which such an electron could make its escape, and up to this day the form of the theory developed by its

---

302 Albert Einstein, "Concerning a Heuristic Point of View Toward the Emission and Transformation of Light," Annalen der Physik **17** (1905), 132-148.

author has not been able to account satisfactorily for the photoelectric facts presented here-with. We are confronted, however, by the astonishing situation that these facts were correctly and exactly predicted nine years ago by a form of quantum theory which has now been pretty generally abandoned.

It was in 1905 that Einstein made the first coupling of photo effects and with any form of quantum theory by bringing forward the bold, not to say the reckless, hypothesis of an electro-magnetic light corpuscle of energy $h\nu$, which energy was transferred upon absorption to an electron. This hypothesis may well be called reckless first because an electromagnetic disturbance which remains localized in space seems a violation of the very conception of an electromagnetic disturbance, and second because it flies in the face of the thoroughly established facts of interference. The hypothesis was apparently made solely because it furnished a ready explanation of one of the most remarkable facts brought to light by recent investigations, viz., that the energy with which an electron is thrown out of a metal by ultra-violet light or X-rays is independent of the intensity of the light while it depends on its frequency. This fact alone seems to demand some modification of classical theory or, at any rate, it has not yet been interpreted satisfactorily in terms of classical theory.[303]

Despite recklessly flying in the face of the thoroughly established facts of interference, Millikan had to conclude that Einstein's particle model correctly predicted the key facts of photoelectric emission. Or, as Millikan put it in his conclusions:

1. Einstein's photoelectric equation has been subjected to very searching tests and it appears in every case to predict exactly the observed results.

2. Planck's h has been photoelectrically determined with a precision of about .5 per cent. and is found to have the value $h = 6.57 \times 10^{-27}$ erg · s.

Now that Einstein's model had to be taken seriously, the interference problem became more acute.[304] In 1924, the French graduate student, Louis deBroglie, proposed his own bold and reckless hypothesis. If waves act like particles, perhaps particles act like waves?

Meanwhile two Americans, Clinton Davisson and Lester Germer, were working on the problem of using electrons as a probe to understand the atomic structure of nickel. Against expectations, they observed electron diffraction patterns, which confirmed deBroglie's hypothesis.

At nearly the same moment that physicists were coming to grips with the idea of *light quanta*, the analysis of Louis de Broglie and the electron diffraction experiments of Davisson and Germer forced them to consider the wave properties of electrons. The new quantum physics that eventually emerged from this crucible compels us to accept a fusion of the concepts of *wave* and *particle* when explaining light, atoms, and subatomic particles. Electrons are clearly detected as whole particles: even in the double slit experiment, a fraction of an electron is never observed. Yet, even when only one electron at a time is allowed to pass through the diffracting system, an interference pattern typical of classical waves appears after a sufficiently large number of electrons

---

303R.A. Millikan, "A Direct Photoelectric Determination of Planck's 'h'," Physical Review **7** (1916), 355-388.

304The results of the double slit experiment turn out to be the same whether several photons or only one photon passes through the double slit apparatus. The British physicist G.I. Taylor who conducted painstaking observations using an extremely dim light source first obtained evidence for this startling result. The resulting diffraction patterns that Taylor obtained were one reason Dirac famously wrote, "Each photon then interferes only with itself." See G.I. Taylor, "Interference fringes with feeble light," Proceedings of the Cambridge Philosophical Society **15** (1909), 114-115. For a thorough discussion of the quantum mechanical double slit experiment, see Richard Feynman, Robert Leighton and Matthew Sands, The Feynman Lectures on Physics (Addison-Wesley), Vol. III, pp. I-1-11. Feynman discusses the double slit experiment using matter (electrons) instead of light, but the pertinent physics is the same.

have passed through. Real particles in nature, such as electrons, do not behave like classical particles! In hindsight, it is clear that classically intuitive notions of particles and waves are abstracted from our everyday experiences with macroscopic systems such as rocks and ponds, and we shouldn't be surprised when these concepts do not literally describe the subatomic world of which we have no direct experience.

Oddly enough, Newton's theory of light was, at least in principle, right after all: light is composed of particles. Newton's own laws of motion, however, do not govern these photons. Rather, both matter and light must be governed by a whole new theory of *quantum mechanics* and *quantum electrodynamics.*

18.1 *Einstein and the Photoelectric Effect*

Though Rayleigh and Jeans questioned the route by which Planck obtained his formula for the blackbody spectrum, it was Albert Einstein and fellow physicist Paul Ehrenfest who first grasped the truly revolutionary nature of Planck's work. Einstein took Planck's results even further by suggesting that light is really composed of particles instead of waves. Here we give the outline of Einstein's earliest argument for the existence of light quanta, and his use of the photoelectric effect as the prime example of a physically observed phenomenon where the wave theory is inadequate to explain the basic facts.

Einstein first presented his heuristic idea that we ought to conceive of light as consisting of particles or "quanta" in the same paper in which he analyzed the photoelectric effect. As we mentioned in the introduction, Einstein inferred from the thermal properties of high frequency blackbody radiation that such radiation behaves as if it consisted of spatially localized, independent energy quanta. Most of the paper in which he introduced light quanta is devoted to a theoretical analysis that includes calculating the entropy of high frequency blackbody radiation, which obeys Wien's law (p. 668), as a function of volume. Einstein then compares his result to the volume dependence of the entropy for an ideal gas, and concludes:

> Monochromatic light of low density (i.e. within the range of validity of Wien's radiation formula) behaves thermodynamically as though it consisted of a number of independent energy quanta of magnitude $\hbar\omega$.[305]

Next, Einstein uses his concept of light quanta to explain the photoemission of electrons from metal surfaces, a phenomenon that had first been noted by Heinrich Hertz. Of all the amazing work Einstein put forth in 1905, such as his explanation of Brownian motion and his famous formula $\mathbb{E} = mc^2$, the light quanta was perhaps the most audacious.

In the thought experiments that follow, we develop the ideas behind Einstein's simple, yet radical, explanation of the photoelectric effect.

Thought Experiment 18.1: The Ejection of Electrons from a Metal Surface

In a metal the valence electrons are free to move about the interior, while remaining bound inside the metal as a whole. With an addition of energy, some of these electrons can be ejected from the conductor. Indeed, electrons can be removed from the surface of a metal by several different

---

[305]Albert Einstein, "Concerning a Heuristic Point of View Toward the Emission and Transformation of Light," Annalen der Physik **17** (1905), 132-148.

processes. No matter how the electrons are liberated, however, an amount of work, $\mathbb{W}$, must be done on each electron to overcome some surface potential energy called the *work function.*

In practice, the metal surface is the negative side (cathode) of a capacitor with some potential difference, so the freed electrons flow and the resulting current can be measured. The trick for most measurements of the work function is to increase the voltage enough for all the ejected electrons to be measured in the current, but not so high that the electric field itself overcomes the work function. Moreover, these measurements must be conducted inside an evacuated tube to eliminate collisions between photoelectrons and gas molecules, and also prevent oxidation of the metal surface itself. (Recall the vacuum tube in Problem 17.26 on p. 605.)

When the electric field itself causes the current to flow, this is called *field emission*, and it is used for electron microscopy because of the ability to focus an electron beam. When measuring the work function of a metal using an external electric field, care must be taken that the applied voltages not exceed the work function.

Cathode ray tubes, on the other hand, create a steady flow of free electrons by heating the cathode to a temperature on the order of:

$$T \approx \frac{\mathbb{W}}{k_B},$$

so the electrons effectively boil off of the surface. This is how J.J. Thomson produced cathode rays in his experiment of 1897 that revealed the existence of electrons. And, of course, it is the way old cathode ray tube TV sets and oscilloscopes worked. It was also the way old vacuum tube AC to DC converters worked. Thermionic emission is also used to measure the work functions for various metals.

Multiplier tubes have been used to measure just a few free electrons, for example in a Geiger counter. These devices take advantage of so-called *secondary emission*, where a particle transfers its kinetic energy to the metal surface. As long as the incoming particle's kinetic energy exceeds the work function, $\mathbb{T} > \mathbb{W}$, electrons can be ejected. If the kinetic energy is much bigger than the work function, many electrons can be ejected. So, with a series of voltage drops, a free electron is accelerated toward a metal plate at a higher voltage, thus ejecting many other electrons. Now, these become accelerated to the next higher voltage plate, ejecting more electrons. By the last plate (the anode), the number of secondary electrons add up to make a measureable current.

Finally, light can also impart enough energy to free an electron. This is called the *photoelectric effect.* For example, very sensitive short wavelength light detectors combine the photoelectric effect with a multiplier tube. In these *photomultiplier tubes*, the readout current is proportional to the intensity of the light. Classical physics predicts this proportionality nicely, but it fails to predict why these tubes are only sensitive to short wavelength light.

Thought Experiment 18.2: The Photoelectric Effect

The photoelectric effect occurs when electromagnetic radiation shines on a clean metal surface and, as a result, electrons are liberated from the surface. The light beam supplies an electron with an amount of energy that equals or exceeds the energy, called the *work function*, that binds the electron to the surface. An electron that receives energy equal to, or greater than, the work function is able to escape from the metal to which it is bound.

A detailed description of the photoelectric effect requires knowledge, based on experiment, of how the several variables involved in photoelectric emission are related. These variables are the frequency of light ω, power of the incident light beam on the metal surface $\mathbb{P}$, the photoelectric current I, the kinetic energy of an individual ejected electron $\mathbb{T}$, and the work function of the metal involved $\mathbb{W}$.

First we will consider the conservation of charge. Each electron has a negative charge, $-e$, so the number of electrons per time ejected is simply given by $\frac{1}{e}I$. Thus, conservation of energy gives us:

$$\mathbb{P} = \tfrac{1}{e}I(\mathbb{T} + \mathbb{W}).$$

Making reasonable assumptions about the kinetic energy of the free electrons, this implies that the current is proportional to the power of incident light. This is indeed the case, so long as the light is of a sufficiently high frequency. However, if the incident light is of long wavelength a curious thing is observed: there is no current flowing at all. Once the frequency exceeds some threshold frequency, ω_0, current flows and is proportional to the light intensity. In fact, even for extremely low intensities of short wavelength light, the current turns on within nanoseconds, whereas one would otherwise expect an electron current only after several hundred hours for such low intensities.

The existence of a *threshold frequency* for a given metal, a frequency below which no photoemission occurs however great the light intensity, is completely inexplicable from the point of view of classical wave theory. Rather, the wave theory predicts that the energy reaching the surface per unit time, and not the frequency, ought to determine when photoemission occurs.

To investigate this further, the German physicist Phillip Lenard, who had been an assistant to Hertz, measured the kinetic energy of the individual electrons by relating it to the maximum stopping potential in a phototube.[306] Lenard used an evacuated glass tube, with two electrodes, shown in his figures at the start of this chapter (p. 640). Lenard experimented with making windows (*fenster* in German) strong enough to hold a vacuum, but transparent to short-wavelength ultraviolet light.

Using a variable voltage source, he reversed the polarity of the electrodes so that the monochromatic light now shined on a positively charged metal anode. Despite the anode being positively charged, electrons were still ejected, but now they were repelled by the negatively charged cathode. Nevertheless, a few electrons could still reach the cathode to constitute the current flowing in the circuit. By increasing this backward voltage, $\mathbb{V}$, Lenard could make the current go to zero. The value of the voltage for which the current equals zero is called the *stopping potential*, which he found to be a function of frequency, but independent of light intensity. Lenard was also able to investigate the photoelectron velocity, and found a maximum kinetic energy, independent of the light intensity. This result completely contradicted Maxwell's wave theory of light!

"According to classical physics," one author has written, "light spreads out from a source in the form of waves of ever increasing radius and the energy of the waves is spread out more and more thinly with increasing distance from the source. It was, therefore, as incomprehensible that distant and nearby electrons should acquire the same energy from waves of light as that a

---

[306] P. Lenard, Über Kathodenstrahlen Nobel-Vorlesung (Verlag von Johann Ambrosius Barth, 19) He won the 1905 Nobel prize for his work that took place around 1890.

submarine, miles away from a depth charge, should sustain the same damage as one very near the site of the explosion."[307]

Discussion 18.1: Einstein's Heuristic Photoelectric Effect

Let us continue with the submarine analogy. Unlike a depth charge, torpedoes can damage two submarines at different distances equally—as long as they each hit the target, of course. Albert Einstein suggested that light behaves more like a collection of torpedoes than a blast wave from a depth charge. An electron will be emitted when a quantum of light, or photon, of sufficient energy strikes a metal surface.

Einstein begins his analysis of the photoelectric effect by assuming that the beam of light incident on the metal consists of photons, and that each photon has an energy that depends only on its frequency, or wavelength

$$\mathbb{E} = \hbar\omega = 2\pi\hbar\frac{c}{\lambda} = \frac{hc}{\lambda}.$$

According to Einstein, a single photon can only interact with a single electron at the metal surface; it cannot share its energy with several electrons. Since the photon travels at the speed of light, it must, according to the theory of relativity, have zero rest mass and an energy that is entirely kinetic. When a zero-mass particle ceases to move at speed c, it ceases to exist. So when a photon strikes an electron bound in a metal, it must relinquish all of its energy $\hbar\omega$ to the single electron it strikes. If the energy the electron gains from the photon exceeds the energy binding it to the metal surface, the excess energy becomes the kinetic energy of the photoelectron.

On the basis of quantum theory, Einstein gave a simple meaning to the energies of the photon and the photoelectron via the equation:

$$\hbar\omega = \mathbb{T} + \mathbb{W},$$

where $\hbar\omega$ is the photon energy, $\mathbb{T}$ the kinetic energy of the photoelectron, and $\mathbb{W}$ is the work function. This is referred to as *Einstein's equation*, which Robert Millikan verified ten years after Einstein first proposed it. When Einstein published his first paper on light quanta in 1905, no available experimental evidence indicated that the kinetic energy of the photoelectrons is directly proportional to the frequency of the incident light, let alone that the constant of proportionality is $\hbar$! For example, Lenard's experiments of 1902 could demonstrate only that the energy of the ejected electrons increases with light frequency, but it could produce the functional dependence of energy on frequency.

An electron bound with energy $\mathbb{W}$ can be released only if $\hbar\omega > \mathbb{W}$, or, equivalently, if the frequency is higher than the threshold: $\omega > \omega_0$. Thus, Einstein argued, the frequency of the electromagnetic radiation determines precisely the photon energy $(\mathbb{E} = \hbar\omega)$. Thus Einstein's quantum theory of light could easily explain Lenard's result that photoemission only occurs when light of sufficient frequency is incident on the metal surface. How does this quantum analysis square with Lenard's discovery that photoemission is independent of the incident light's intensity?

---

[307] A.E.E. McKenzie, The Major Achievements of Science, Volume 2 (Cambridge University Press, Cambridge, 1960, rpt: Touchstone, 1973), p. 318.

According to Einstein's quantum interpretation of the photoelectric effect, light intensity is simply the energy of each photon multiplied by the number of photons crossing a unit area per unit time. An increase in the intensity of the light means, therefore, a proportionate increase in the number of photons striking the metal surface. Therefore, if light consists of photons the current would be proportional to the intensity. Photoemission occurs with little delay because, even at the lowest intensity, an electron is emitted only if it is hit by a photon, which on stopping relinquishes all of its energy to the electron. An electron's release from the surface, therefore, does not depend on its gradually accumulating energy from incident waves.

Discussion 18.2: Millikan's Photoelectric Experiment

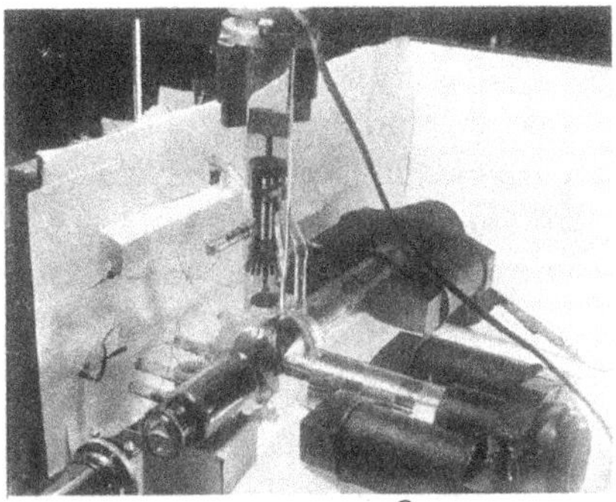

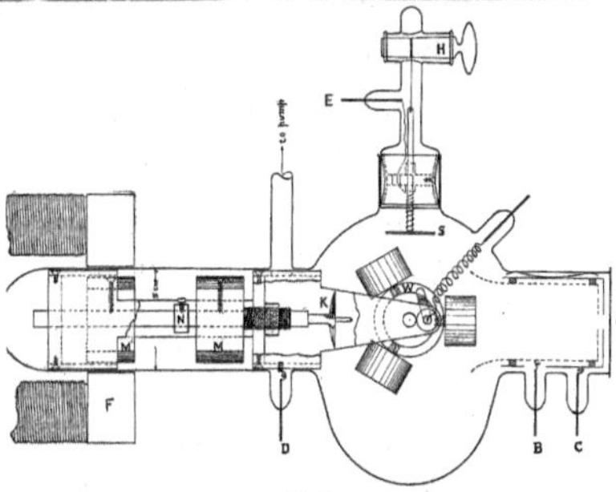

Fig. 2.

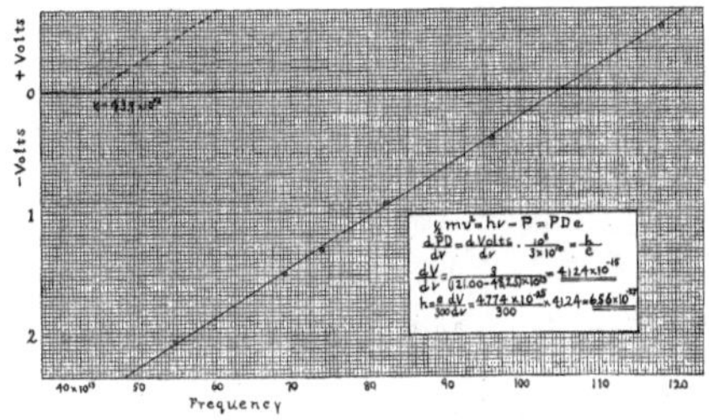

Physicists quickly accepted Einstein's theory of special relativity, but not his explanation of the photoelectric effect. In fact, he faced a good deal of opposition because, as we discussed in this chapter's introduction, his idea of light quanta completely flew in the face of all of classical electrodynamics. Wasn't this settled when Young's double slit experiment clearly showed the wave nature of light? Was Einstein really saying that Newton's corpuscular theory of light was right after all? How could Einstein explain interference phenomena with light quanta? What can it mean to talk about a "particle" whose energy depends on its "frequency"?

Perhaps we should not, with the benefit of twenty-first century hindsight, so quickly dismiss the skepticism of Einstein's co-workers. Einstein's proposal was indeed paradoxical, and there were solid grounds for doubting the existence of photons. Einstein's equation, $\hbar\omega = \mathbb{T} + \mathbb{W}$, which asserts that the kinetic energy $\mathbb{T}$ of the ejected electrons is proportional to the frequency ω of the incident radiation, and, furthermore, that the constant of proportionality is none other than Planck's constant $\hbar$, appeared to be pure theoretical conjecture. Prior to 1916, no experimental results had been announced to quantitatively verify, or refute, Einstein's proposed relation between frequency and photoelectron energy. That had to wait for the great American experimentalist, Robert Millikan.

Millikan, who had already achieved success by determining the amount of charge carried by an electron with his famous oil drop experiment, was so skeptical of Einstein's light quanta proposal that he expected to disprove Einstein's equation with experiment. Over the course of ten years, Millikan's team at the University of Chicago experimentally tested Einstein's theory by illuminating the clean surfaces of various metals with light at different frequencies and measuring the kinetic energy of the electrons emitted in each case.[308]

---

[308] R. A. Millikan, "A Direct Photoelectric Determination of Planck's 'h'," Physical Review **7** (1916), 355-388.

As alkali metals (e.g. lithium, sodium, or potassium) are quite reactive, the metal had to be freshly cut to avoid the effects of oxidation, so that only a pure metal surface would actually be presented to the incident light. Millikan devised an evacuated glass tube, inside of which the metal plate was mounted on a wheel. The wheel would move past a scraper knife and then into the path of monochromatic light whose frequency could be varied (see Millikan's Figure 2). Millikan jokingly referred to his apparatus as a barbershop, because he had to shave each metal in a vacuum to obtain a completely clean surface before irradiating it with light.

In Millikan's photoelectric apparatus, the metals to be studied were placed inside a vacuum and mounted on an electromagnetically controlled wheel, w. The three metals on the periphery of the wheel were cast cylinders of sodium, potassium, and lithium, and could be shaved as they moved past and adjustable rotary knife, K. The wheel was rotated freely until the freshly cut metal surface was perpendicular to the light beam entering at the point marked O in his diagram, so that the monochromatic light shined on each of the cleanly shaved metal plates in turn. Electrode C collected the photoelectrons, which in turn flowed, via the variable resistor and micro ammeter, back to the plate assembly. By adjusting the potentiometer, Millikan found the potential difference that was just large enough to stop the electrons moving round the circuit.

For each metal, Millikan measured the stopping voltage for several values of incident light frequency. His graph of the known incident light frequency versus stopping voltage, is shown in the figure. Millikan determined the slope to be equal to Planck's constant divided by the electron charge, and whose intercept was the work function per elementary charge, so the stopping voltage followed the equation:

$$\mathbb{V} = \left(\tfrac{h}{e}\right)\omega - \tfrac{1}{e}\mathbb{W}.$$

Millikan's value for Planck's constant agreed extremely well with the value Planck had derived from blackbody radiation, thus verifying Einstein's equation! Nevertheless, Millikan remarked near the end of his paper that he still regarded Einstein's hypothesis of light quanta as untenable. In fact, even as late as 1923, Millikan continued to express doubts in his acceptance speech for the Nobel Prize that was awarded, in part, for his work on the photoelectric effect:

> The general validity of Einstein's equation is, I think, now universally conceded, and to that extent the reality of Einstein's light quanta may be considered as experimentally established. But the conception of localized light quanta out of which Einstein got his equation must still be regarded as far from being established...until it can account for the facts of interference and the other effects which have seemed thus far to be irreconcilable with it, we must withhold our full assent.[309]

Problem 18.1: Millikan's Value of Planck's Constant

Consider Millikan's plot of the stoppage voltage vs. frequency above, which shows a remarkably linear relationship with a slope, according to Millikan, is given by $\frac{d\mathrm{V}}{d\nu} = 4.124\times10^{-15}\,\frac{\mathrm{V}}{\mathrm{Hz}}$. From this, he concluded the following:

$$h = \frac{e}{300}\frac{d\mathrm{V}}{d\nu} = \frac{4.774\times10^{-25}}{300}\times 4.124 = 6.56\times10^{-27}.$$

a) Show that his slope, $\frac{d\mathrm{V}}{d\nu}$, is a direct measurement of Planck's constant in units of electron volts per hertz. Find the percent difference from the current accepted value of 4.136×10^{-15}eV-s?

309 R. A. Millikan, "The electron and the light-quantum from experimental point of view" (1924) Nobel Prize lecture.

b) The erg is the unit of energy in the cgs system, where $1 \text{ erg} = 10^{-7}$ J. Convert Millikan's value from Planck's constant from cgs to SI units. Find the percent difference from the current value of $h = 6.626 \times 10^{-34}$ J-s.

c) Recall that the cgs unit system is based on electrostatic units, where Coulomb's constant is defined to be $K = 1\frac{\text{erg-cm}}{\text{esu}^2}$, where the *esu* is the cgs unit of charge. Use this to show that $1\,\text{esu} = 0.33356$ nC, given today's value of ε_0.

d) What value did Millikan assume for the elementary charge in esu? What is the modern value in the same units? What is the percent difference? Which contributed more to the error in Planck's constant, Millikan's measurement of the photoelectric effect or his value for the elementary charge.[310]

e) The number 300 in Millikan's equation represents a conversion factor between an SI unit and its electrostatic (cgs) equivalent. What is this unit? Which represents the bigger quantity, the SI unit or its cgs counterpart?

18.2 *Thermal Radiation and the Hydrogen Atom*

The study of the light that hot gases emit was well advanced by the middle of the nineteenth century. In 1802, the English scientist William Wollaston reported the existence of dark lines in the solar spectrum, which he observed by passing sunlight through a narrow slit and then through a prism. Projecting the light over a distance of ten to twelve feet, Wollaston saw the red, yellowish-green, green, blue and violet colors, as expected for the continuous spectrum that Newton produced in his experiments, but Wollaston also reported seven dark lines. Five of these dark lines he reported as being on the boundaries between two colors, but two lines were within the color boundaries (specifically yellowish-green and blue).[311]

A dozen years later the Bavarian physicist, Joseph von Fraunhofer, rediscovered these dark lines in the sun's spectrum while measuring the dispersive powers of various kinds of glass for light of different colors. Fraunhofer eventually mapped out the 574 thin black lines that he observed in the sun's spectrum, with eight of the most prominent lines being labeled A to K. Today, these lines are known as the *Fraunhofer lines*.[312] Fraunhofer also noticed that a bright "orange" line in the spectrum of the flame he was using was in the same position as one of the dark lines in the solar spectrum, which he had previously named the "D-line." This same line had been observed in flames from alcohol and sulfur as well as from candles. The D-line was found to consist of

---

310 Millikan's assumed elementary charge appears to be based on a paper he wrote in 1913 (Physical Review, Vol. II, No. 2, 109-137), where he addresses systematic errors in his measurements of the elementary charge due to uncertainties in the drag force on an oil drop. Around 1911, many researchers were measuring the elementary charge, and Millikan was neither the first nor, in hindsight, the most accurate. In his 1911 paper (Phys Rev., 32, pp. 349-397), he reviews the existing measurements, some of which are closer to the modern value than his own—particularly one by Ernest Rutherford.

311 William Hyde Wollaston, "A method of examining refractive and dispersive powers, by prismatic refraction," Philosophical Transactions of the Royal Society, **92** (1802): 365-380. Details of Wollaston's observation of the solar spectrum are given on pages 378-380 of his paper. Wollaston used a narrow slit no wider than 0.05 inches. Previously, investigators used circular openings or relatively wide slits, which gave spectra that were "blurred out" and therefore masked the presence of such lines.

312 A translation of Fraunhofer's original papers can be found in J.S. Ames, ed., Prismatic and diffraction spectra: Memoirs by Joseph von Fraunhofer, (New York and London: Harper and Brothers, 1899).

two closely spaced lines, subsequently called D_1 and D_2, and its position turned out to be in the yellow portion of the spectrum, not orange as Fraunhofer originally thought. Fraunhofer noted that in the spectrum of a candle flame two bright lines occur which coincide with the two dark lines D of the solar spectrum. So, it turned out, not all luminous bodies emit spectra like the rainbow pattern Newton observed with his prism.

Indeed, Léon Foucault discovered, in 1851, that he could make the Fraunhofer D lines in the solar spectrum appear darker by passing sunlight through sodium vapor.[313] These dark lines also showed up as bright lines in the light from a sodium lamp in Fraunhofer's original work. Heated sodium vapor, Foucault reasoned, emits light that forms these bright spectral lines. In contrast, cool sodium vapor absorbs light at these same wavelengths and generates dark spectral lines when the light originates from elsewhere and passes through the cool sodium vapor.

Investigators observed line spectra from the sun and certain elements in the laboratory, but no one undertook a systematic study of these characteristic colors emitted by a variety of samples until Gustav Kirchhoff and Robert Bunsen identified thousands of spectral lines and discovered the elements cesium and rubidium using their newly developed spectroscope and Bunsen's famous invention, the Bunsen burner. Sample material burning in the flame of a Bunsen burner can be clearly distinguished because the burner's flame emits so little light. Kirchhoff found that each element, when heated to incandescence, emits light of certain distinct wavelengths or frequencies. Kirchhoff divided the spectra he observed into three main types, and used this classification as the basis of his three laws of spectral analysis[314].

Kirchhoff's First Law, a luminous solid, or liquid (or a hot, ionized, opaque gas) emits light of all wavelengths or frequencies, which therefore results in a continuous spectrum. The visible part of this spectrum appears like the familiar rainbow with colors that change continuously from red to violet, just like the spectrum Newton obtained by passing sunlight through prisms. The continuous spectrum is nearly the same for all substances and varies in total brightness with the temperature of the solid or liquid. The way the intensity depends upon wavelength or frequency (e.g. color in the visible part of the spectrum) also varies slightly as the temperature is changed, but otherwise continuous spectra reveal little about the chemical composition of the emitting substances.

Kirchhoff's Second Law, a rarefied or low-pressure gas emits light whose spectrum contains bright lines, which are sometimes superimposed on a faint continuous spectrum. This is a bright line or emission spectrum. It appears as a set of distinct colors, separated by darker spaces. The wavelengths of specific lines are characteristic of the atoms of the gas. Because each type of atom or element in the periodic table has its own unique set of characteristic lines, the emission spectrum allows us to determine the chemical composition of the gas.

Kirchoff's Third Law, if a continuous spectrum from a luminous source passes through a cool, low-density gas, regions at certain wavelengths will be removed from the continuum causing dark lines (absorption lines) to appear. This absorption spectrum looks like a continuum with sharp lines blacked out at exactly the same position that would be occupied if the gas were emitting instead of absorbing. These absorption lines are therefore just as useful as emission lines in identifying the composition of the gas.

---

[313] J.B.L. Foucault, whose report in the French bulletin L'Insitut, Feb. 7, 1849, was translated and published by G. G. Stokes in Philosophical Magazine, **19** (1860): 194.

[314] Keep in mind that the term *Kirchhoff's laws* can refer to these laws of spectral analysis, or it can refer to his loop and junction rules of circuit theory, depending on the context.

The second and third laws of Kirchhoff and Bunsen's work were especially important for our understanding of matter, but it would be more than 60 years until the work of Niels Bohr began to shed light on the origin of these discrete spectral lines and their implications for the structure of matter. Nevertheless, the pair grasped the essential point of their results when they wrote that: "Spectrum analysis should become important for the discovery of hitherto unknown elements."[315] In fact, astronomers Norman Lockyer and Pierre Janssen independently identified an unknown element by using a spectroscope to observe the chromosphere of the Sun shortly after a solar eclipse in 1868. Both Lockyer and Janssen noticed a yellow line in their observed spectra that did not match any element known at the time. Lockyer ultimately determined its wavelength to be 587.62 nm, which is nearly equal to the wavelengths of Fraunhofer's "D" lines of sodium (589.592 nm for D_1 and 588.995 nm for D_2), and named the element *helium* for the location where it was found, the Sun (*helios* in Greek).[316]

Kirchhoff not only reported the coincidence of the wavelengths of spectrally resolved lines of absorption and of emission of visible light, he also observed that bright lines or dark lines were apparent depending on the temperature difference between emitter and absorber. This last result turned out to be of vital importance for thermal physics generally, but especially for the thermodynamics of radiation. For closely related to this work was the problem of so called cavity or *blackbody* radiation, which Kirchhoff introduced to the scientific community in 1860.[317]

If an enclosed cavity with perfectly absorbing (i.e. black) walls is held at a constant temperature, its interior will be filled with radiant energy of all wavelengths. If that radiation is in equilibrium, both within the cavity and its walls, then the rate at which energy is radiated across any surface or unit area does not depend upon the position or orientation of that surface. Makers of pottery and fine porcelain, such as eighteenth-century scientist and manufacturer Josiah Wedgwood, had noted that all objects in their kilns began to glow red after the oven reached a certain temperature (now known as the Draper temperature), regardless of their shape, size, or composition. Wedgwood invented a special type of thermometer, called the pyrometer, which allowed him to measure high temperatures in his furnaces without having to place a device in direct contact with the hot objects in his kiln.[318] In his first attempt, which anticipates modern pyrometry, Wedgwood tried to compare the colors inside the kiln with the tint of clay objects fired at known temperatures. Technical difficulties marred this early attempt at a modern pyrometer based on thermal radiation, so Wedgwood instead measured the shrinkage of pieces of clay as the standard for determining oven temperature.[319] Thomas Wedgwood, son of Josiah and a photography pioneer, made use of his "father's thermometer" to investigate how different bodies radiate with substances ground to a powder and heated to high temperatures in a crucible. He then irradiated small silver tubes, one of which was blackened so as to absorb incident light, that were placed in

---

315 G. Kirchhoff and R. Bunsen, "Chemical Analysis by Observation of Spectra," Ann. Phys. **110** (1860): 161-89.

316 N. Lockyer, "The Growth of Our Knowledge of Helium," Science Progress, vol.5, (1896), 249-285.

317 It is interesting to note that Kirchoff invented the terms "black body" and "blackbody radiation," which he introduced in the paper "On the relation between the radiating and absorbing powers of different bodies for light and heat," Philosophical Magazine, **20: 4**, (1860), 1–21. (The original German version may be found in Annalen der Physik und Chemie **109** (1860), 275–301).

318 J.A. Chaldecott, "Josiah Wedgwood (1730-95)—Scientist," Presidential Address, The British Journal For the History of Science, **8,** No. 1, (1975), 1-16; J. Wedgwood, Philosophical Transactions, **lxxii** (1782), 305-326, and Philosophical Transactions, **lxxiv** (1784), 558-584.

319 This method improved on Wedgwood's first attempt, but its usefulness also had its limits due to the properties of thermal expansion that were not well understood at the time. For more on this subject, see Chaldecott's paper.

an earthenware tube.[320] In the course of his experiments, the younger Wedgwood touched on questions that wouldn't be answered for at least another half century, such as, "Is not the light *emitted* [from a heated, blackened silver tube] equal to the light *received*?"

Kirchhoff argued that, for any body in thermal equilibrium with radiation, the emitted energy is proportional to the amount of energy absorbed.

An example of such a case would be the heated walls of a kiln or a furnace with its door closed and held at a constant temperature. The radiation within the walls of the furnace would be in thermal equilibrium when the radiation energy within the furnace is absorbed, exchanged, and reemitted many times over until the entire walls of the cavity (or furnace or kiln) are in thermal equilibrium. The radiation in thermal equilibrium with the walls of the furnace is similar to the radiation emitted by a black body, which is an ideal object that absorbs radiation of all frequencies or wavelengths and therefore would appear black. Actually, a black body at room temperature appears black because it radiates most of its energy at infrared wavelengths and so cannot be perceived by the human eye. At higher temperatures, black bodies glow with increasing intensity. Furthermore, colors shift from dull red to blindingly brilliant blue-white as the temperature increases. A black body emits the energy it absorbs in accord with Kirchhoff's observations; and the energy emitted is a function of the temperature of the black body and frequency of the emitted light, and independent of the size, shape, and chemical nature of the black body.

By the end of the nineteenth century, Maxwell's electromagnetic theory of light had been applied to explain long observed, and diverse, phenomena such as refraction, interference and diffraction, optical polarization and optical activity. In addition, Hertz had brilliantly confirmed the most basic predictions of this theory in the laboratory, and these experiments paved the way for the invention of modern radio. It seemed eminently reasonable for physicists such as Max Planck to suppose that Maxwell's theory would successfully explain the properties of cavity radiation. This is precisely what Planck sought to achieve when he began his program of theoretical research on blackbody radiation during the middle of the 1890s. Planck was already an expert in thermodynamics, especially on the role that entropy plays in physical processes, so he hoped he could explain the properties of cavity radiation by employing the established laws of thermodynamics and Maxwell's electromagnetic wave theory of light. Planck achieved some remarkable success with his program, and he derived an expression for the energy spectrum of blackbody radiation that appeared to account for the known facts. He soon revised his work, however, after experiments revealed that his earlier formula did not agree with measurements taken at longer wavelengths. In any case, his earlier derivation turned out to be specious, and he had to begin over from scratch. He first guessed the correct mathematical formula for the blackbody spectrum from the data, and then he set out to derive the expression from first principles. In the course of deriving the correct expression, Planck was forced to make a radical break with the physics of his time and assume that material systems, such as Hertzian dipoles or atoms, could not absorb and emit energy continuously, but only in discrete amounts or *quanta*.

Now our story turns to the remarkable farmer's son from the south island of New Zealand, Ernest Rutherford. After being educated in Christchurch, he moved to Cambridge to work with J.J. Thomson at the Cavendish Lab, where he investigated electromagnetic waves, the mobility of

---

[320] T. Wedgwood, "Experiments and observations on the production of light from different bodies, by heat and by attrition," Phil. Trans. Roy. Soc. London **82** (1792): 28-47; "Continuation of a paper on the production of light and heat from different bodies," Phil. Trans. Roy. Soc. London **82** (1792): 270-282.

ions in electric fields, and radioactivity. This work led to Thomson recommending him for a professorship at McGill University in Montreal. While in Canada, Rutherford continued to study radioactivity, and used it to measure the age of the earth. In 1907, Rutherford moved to Manchester England, where he analyzed the results of numerous scattering experiments, in which alpha particles (ionized helium nuclei) were hurled with great speed at thin gold foils. As a result, he concluded that the model of the atom that best fit facts was one for which negatively charged electrons orbit about a small positively charged nucleus, like the planets orbiting the sun (p. 655 below).

While Rutherford's planetary model neatly explained his gold foil experiment, it violently conflicted with Maxwell's electromagnetic theory. As we discussed in Section 16.4, an accelerating charge radiates energy in the form of electromagnetic waves. Since an electron traveling in a circular orbit is surely undergoing acceleration, it ought to lose energy as it traverses its path about the nucleus and eventually spiral into the nucleus. In fact, a straightforward calculation yields the result that the electron in a hydrogen atom should collide into the nucleus in about a nanosecond! If this were true, all atoms should have collapsed long ago, absent some mechanism that somehow saves or regenerates them.

Thought Experiment 18.3: Rutherford's Planetary Model of an Atom

In 1909 Ernest Rutherford supervised Hans Geiger and Ernest Marsden who carried out the famous gold foil experiment at the University of Manchester, which clearly showed that atoms are made up of mostly empty space with an extremely dense positively charged nucleus. This led Rutherford to propose that the atom is a little solar system, with the electrons orbiting the dense positive nucleus. In this thought experiment, we take this to its natural conclusion.

The electron of mass m_e revolves in a circle of radius r about the positive nucleus with a charge given by Ze where Z is the atomic number of the atom. From Newton's second law, the acceleration must be:

$$a = \frac{F}{m_e} = \frac{1}{m_e}\frac{Ze^2}{4\pi\varepsilon_0 r^2} = \frac{e^2}{4\pi\varepsilon_0 m_e c^2}\frac{Zc^2}{r^2} = r_e c^2 \frac{Z}{r^2}.$$

Where the *classical radius of the electron* is defined as:

$$r_e \equiv \frac{e^2}{4\pi\varepsilon_0 m_e c^2} \approx 2.818 \times 10^{-15}\,\text{m}.$$

And the electron's kinetic energy would then be given by:

$$\mathbb{T} = \tfrac{1}{2} m_e v^2 = \tfrac{1}{2} m_e \left(\frac{v^2}{r}\right) r = \tfrac{1}{2} m_e a r = \tfrac{1}{2} m_e r_e c^2 \frac{Z}{r^{\cancel{2}}} \cancel{r} = m_e c^2 \frac{Z r_e}{2r}.$$

The total energy of an electron is then:

$$\mathbb{E} = \mathbb{T} + \mathbb{U} = \frac{1}{2}\frac{Ze^2}{4\pi\varepsilon_0 r} - \frac{Ze^2}{4\pi\varepsilon_0 r} = -\frac{1}{2}\frac{Ze^2}{4\pi\varepsilon_0 r} = -\mathbb{T} = -m_e c^2 \frac{Z r_e}{2r}.$$

According to Larmor's formula, which we discussed in Chapter 16, an accelerated charged particle emits radiation of intensity proportional to the square of the particle's acceleration. We can use this fact to see if an orbiting electron emits sufficient radiation to be detectable and, more importantly, loose its energy quickly enough to become unstable over a reasonable amount of time.

Applying the Larmor formula, the total power radiated by an orbiting electron would be:

$$\mathbb{P} = \frac{e^2 a^2}{6\pi\varepsilon_0 c^3} = \frac{4}{6}\frac{m_e}{c}\left(\frac{e^2}{4\pi\varepsilon_0 m_e c^2}\right)\left(r_e c^2 \frac{Z}{r^2}\right)^2 = \frac{2m_e c^3 r_e^3}{3}\frac{Z^2}{r^4}.$$

Consider now the question of electron stability. If the electron is radiating energy, then it will eventually loose its kinetic energy, which would be a problem for the classical orbital model if it happens relatively quickly. The characteristic time that this will take place would be given by the ratio of the energy to the power radiated, or:

$$\tau = \left|\frac{\mathbb{E}}{\mathbb{P}}\right| = \left|\frac{-m_e c^2 \dfrac{Z r_e}{2r}}{\dfrac{2m_e c^3 r_e^3}{3}\dfrac{Z^2}{r^4}}\right| = \frac{3r^3}{4Zcr_e^2}$$

$$\tau_{\text{gold}} \sim \frac{\left(1.5\times10^{-10}\,\text{m}\right)^3}{4(79)\left(3\times10^{8}\,\text{m/s}\right)\left(3\times10^{-15}\,\text{m}\right)^2} \sim 4\,\text{ps}.$$

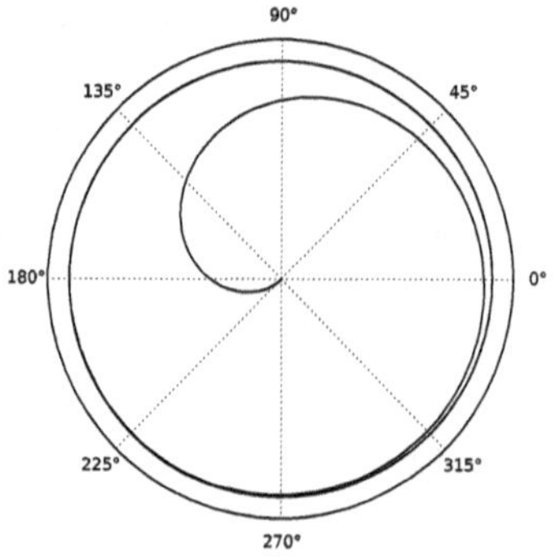

In other words, Rutherford devised a model that explained the experimental results, but made no sense from a classical physics point of view. The figure to the right is a simulation of the orbit of an electron in the atom: notice how it spirals into the nucleus. Clearly, classical physics combined with the planetary model cannot account for two of the most basic facts about atoms: (1) they are remarkably stable, and (2) they do not emit continuous radiation.

Problem 18.2: The Planetary Atom

Based on his gold foil experiment Ernest Rutherford came up with a toy model for the atom, which he knew was theoretically flawed. The point of this problem is for you to investigate this question yourself, and discover the obvious flaw that led Rutherford to hire a young theorist to sort it out. In this problem, assume that the hydrogen atom works much like a planet orbiting the sun in approximately circular orbits.

As you answer each part, express all your constants in terms of the classical electron radius, the mass of an electron, and the speed of light.

a) Find the acceleration of the electron as a function of its distance from the nucleus.

b) Find the power radiated by this electron, as a function of its distance from the nucleus.

c) What is the orbital velocity as a function of distance from the nucleus?

d) What is the radiation reaction force on the electron, as a function of the distance from the nucleus?

e) What is the ratio of the electrostatic force to the radiation reaction force? Is this big or small? If it is big, it validates your prior assumption that the acceleration and speed are dominated by the electrostatic force, and the orbit is approximately circular.

f) What is the azimuthal component of the acceleration, as a function of the distance to the nucleus?

g) You have been assuming, correctly, that $v_\varphi \gg v_r$, but that does not mean that v_r is necessarily zero. Knowing that:

$$v_r = \frac{dr}{dt} = \frac{dr}{dv_\varphi}\frac{dv_\varphi}{dt} = \frac{dr}{dv_\varphi}a_\varphi ,$$

find the electron's radial velocity as a function of radius.

h) Assuming that the electron starts in a circular orbit of radius r_0, find its radial position as a function of time.

i) The typical radius of a real hydrogen atom is called the *Bohr radius*, which is often written in terms of the classical electron radius as: $a_0 = \frac{r_e}{\alpha^2} \approx (137)^2 r_e$, where α is called the fine structure constant. Assuming that the electron starts in a circular orbit of $r_0 = a_0$, find the time it will take to reach the nucleus of the atom. You can assume that the nuclear radius is the *classical proton radius*, or $r_p = \left(\frac{m_e}{m_p}\right) r_e \approx \left(\frac{1}{1836}\right) r_e$.

Thought Experiment 18.4: The Bohr Hydrogen Atom

Niels Bohr, while working for Rutherford in 1913, combined Rutherford's atom with the quantum concepts of Planck and Einstein to fashion a workable model of the hydrogen atom.[321]

Following Rutherford, he assumed a planetary atom, but with a significant additional constraint. The electrons can only orbit stably in certain discrete allowed orbits, which Bohr called *stationary orbits* or *stationary states.* No other orbits are physically possible. Bohr's stationary orbits, now called *shells* or *energy levels,* are associated with definite energies.

Bohr then flatly contradicted Maxwell's electrodynamics, by postulating that, when in a stationary orbit, an electron will not continuously radiate energy. Rather, electrons can only gain and lose energy by making a transition from one allowed orbit to another, absorbing or emitting electromagnetic radiation with the right amount of energy to account for the difference $\Delta\mathbb{E}$. Thus, from Einstein's formula, the observed corresponding angular frequency of light, ω, would be related to the energy difference according to the relationship: $\Delta\mathbb{E} = \mathbb{E}_2 - \mathbb{E}_1 = \hbar\omega$, where $\hbar$ is Planck's constant divided by 2π.

Finally, in the limit of a distant electron, the new model must be in complete agreement with Rutherford's model. This is known as the *Bohr correspondence principle.*

Starting with Bohr's first assumption, we apply Rutherford's planetary model to find the total energy of a hydrogen atom:

$$\mathbb{E} = \mathbb{T} + \mathbb{U} = -\mathbb{T} = -m_e c^2 \frac{r_e}{2r} .$$

---

[321] N. Bohr, "On the Constitution of Atoms and Molecules Part I", Phil. Mag. (1913) **26** (151) 1-24; N. Bohr, "On the Constitution of Atoms and Molecules Part II", Phil. Mag. (1913) **26** (153) 476-502.

Bohr now quantizes the energy in terms on an integral quantum number, and relates the frequency of the emitted light to the energy difference. Thus, for two integer states, m and n:

$$\omega_{mn} = \frac{\mathbb{E}_m - \mathbb{E}_n}{\hbar}.$$

The Bohr correspondence principle, however, demands that the results of quantum physics agree with classical physics for large quantum numbers. For large values of n, in other words, the orbits become so large that the atom can be regarded as a macroscopic system for which classical physics yields the correct description. Therefore, the classical electron angular velocity, ω_e, must equal the angular frequency of the light emitted:

$$\omega_e = \lim_{n\gg 1} \omega_{n+1,n} = \tfrac{1}{\hbar} \lim_{n\gg 1} \left(\mathbb{E}_{n+1} - \mathbb{E}_n\right) = -\tfrac{1}{\hbar} \lim_{n\gg 1} \left(\mathbb{T}_{n+1} - \mathbb{T}_n\right).$$

Next we apply the definition of the derivative to find:

$$\omega_e \approx -\frac{1}{\hbar}\left(\frac{\mathbb{T}_{n+1} - \mathbb{T}_n}{(n+1)-n}\right) \approx -\frac{1}{\hbar}\left(\frac{d\mathbb{T}_n}{dn}\right).$$

Recall that the kinetic energy is $\mathbb{T} = \tfrac{1}{2} m_e v_e^2 = \tfrac{1}{2} m_e \omega_e^2 r^2$, so classically:

$$\omega_e = \frac{v}{r} = \frac{\sqrt{\dfrac{2\mathbb{T}}{m_e}}}{\left(\dfrac{m_e c^2 r_e}{2\mathbb{T}}\right)} = c\sqrt{\frac{2\mathbb{T}}{m_e c^2}}\left(\frac{2\mathbb{T}}{m_e c^2 r_e}\right) = \frac{c}{r_e}\left(\frac{2\mathbb{T}}{m_e c^2}\right)^{3/2}.$$

We can equate these in the continuous limit:

$$-\frac{1}{\hbar}\left(\frac{d\mathbb{T}_n}{dn}\right) \approx \frac{c}{r_e}\left(\frac{2\mathbb{T}}{m_e c^2}\right)^{3/2} \rightarrow \frac{d\mathbb{T}}{dn} \approx -\frac{\hbar c}{r_e}\left(\frac{2\mathbb{T}}{m_e c^2}\right)^{3/2}.$$

Now, by integration we find for large quantum numbers:

$$\int_{\mathbb{T}_m}^{\mathbb{T}_n} \mathbb{T}^{-3/2}\, d\mathbb{T} \approx -\frac{\hbar c}{r_e}\left(\frac{2}{m_e c^2}\right)^{3/2} \int_m^n dn$$

$$-2\left(\frac{1}{\sqrt{\mathbb{T}_n}} - \frac{1}{\sqrt{\mathbb{T}_m}}\right) \approx -\frac{\hbar c}{r_e}\left(\frac{2}{m_e c^2}\right)^{3/2} (n-m)$$

$$\frac{1}{\sqrt{\mathbb{T}_n}} - \frac{1}{\sqrt{\mathbb{T}_m}} \approx \frac{\sqrt{2}\,\hbar c}{r_e}\left(\frac{1}{m_e c^2}\right)^{3/2} (n-m).$$

Now, by distributing and equating each term we can write:

$$\frac{1}{\sqrt{\mathbb{T}_n}} \approx \frac{\sqrt{2}\,\hbar c}{r_e}\left(\frac{1}{m_e c^2}\right)^{3/2} n.$$

Solving for the kinetic energy we have:

$$\mathbb{T}_n = \left(\frac{m_e^3\, c^4\, r_e^2}{2\hbar^2}\right)\frac{1}{n^2} \approx \frac{13.6\,\text{eV}}{n^2}.$$

This expression for the kinetic energy of an electron corresponds to the classical approach in the far limit. That said, what matters most is its ability to be tested experimentally. If it works in the classical limit, but not at the ground state, then it would be of no use. To test it we must write it in terms of the total energy $\mathbb{E} = -\mathbb{T}$:

$$\mathbb{E}_n = -\left(\frac{m_e^3\, c^4\, r_e^2}{2\hbar^2}\right)\frac{1}{n^2} \approx \frac{-13.6\,\text{eV}}{n^2}.$$

Problem 18.3: The Balmer Formula

The Swiss physicist Johann Balmer Jr. came up with an empirical relationship to specify the optical emission lines of hydrogen, which states:

$$\lambda_{mn} \propto \frac{m^2}{m^2 - n^2}.$$

For the optical lines, $n = 2$ and m is any integer greater than 2.

a) The 3-2 transition is a red line called the *hydrogen alpha line* with a wavelength of $\lambda_{\text{H}_\alpha} \approx 656\,\text{nm}$. Show that Balmer's formula can be written as:

$$\lambda_{mn} = \tfrac{5}{9}\lambda_{\text{H}_\alpha}\frac{m^2}{m^2 - n^2}.$$

b) Derive Balmer's formula from Bohr's energy relationship.

c) Express the wavelength of the hydrogen alpha line in terms of fundamental constants, and calculate its value to six significant figures.

Problem 18.4: The Bohr Radius

a) From the Bohr model, show that an electrons orbital radius is:

$$r = \frac{1}{r_e}\left(\frac{\hbar}{m_e c}\right)^2 n^2.$$

b) The constant of proportionality, a_0, is called *the Bohr radius.* Show that:

$$a_0 = \frac{1}{r_e}\left(\frac{\hbar}{m_e c}\right)^2 = \frac{4\pi\varepsilon_0\hbar^2}{m_e e^2} \approx 52.9\,\text{pm} \approx 0.529\,\text{Å}.$$

c) Show that the energy can be written as:

$$\mathbb{E}_n = -\tfrac{1}{2}m_e c^2\frac{r_e}{a_0}\frac{1}{n^2}.$$

and calculate numerically the unitless ratio a_0/r_e.

Problem 18.5: The Orbital Angular Momentum

Clearly show that the orbital angular momentum of an electron, according to the Bohr model, is given simply by:

$$L_n = \hbar n.$$

Bohr derived this expression in his first paper on the atom, and used it as the quantization condition that restricts the allowed values of the orbital radii (i.e. stationary orbits) in all subsequent work.

Problem 18.6: The Bohr Atom and Fine Structure

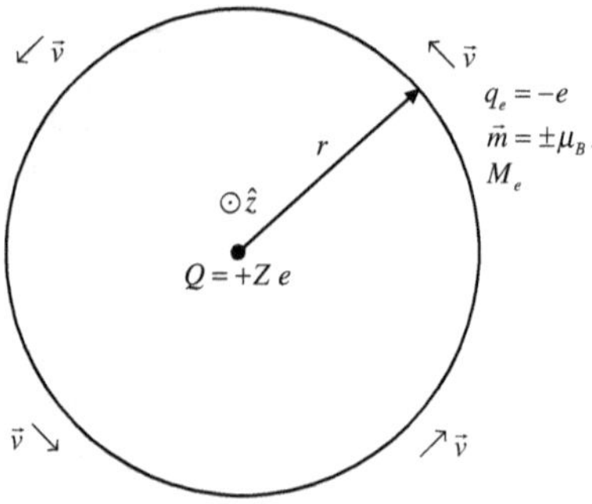

In the Bohr model the electrons orbit in circular planetary-style orbits, but each allowed orbit is quantized in angular momentum: $\vec{L} = n\hbar\,\hat{z}$. In this problem, you can assume a hydrogen-like atom with a nuclear charge $Q = +Ze$, an electron charge $q_e = -e$, and an electron mass M_e that is much less than the nuclear mass.

a) Write the velocity of the orbit in terms of known constants and the principle quantum number n.

b) Write the total energy, kinetic and potential, also in terms of known constants and the principle quantum number.

c) In the reference frame of the electron, the nuclear Coulomb electric field will appear as both an electric and magnetic field. Find the magnetic field in the electron rest frame, in terms of known constants and the principle quantum number.

d) The electron has its own magnetic moment, which is either aligned (up) or anti-aligned (down) with the magnetic field seen in the electron's rest frame. Thus the magnetic moment is given by $\vec{m} = \pm\mu_e\,\hat{z}$. Find the energy difference between a spin up and spin down electron, and express it in terms of the electron magnetic moment $\mu_e = -9.285\times 10^{-24}\,\frac{\text{N·m}}{\text{T}}$, known constants, and the principle quantum number.

e) In energy units of electron volts, calculate the difference in energy of the two spin states in the $n = 2$ doublet. This difference in energy is measured to be 4.28×10^{-5} electron volts. How close is this semi-classical model? Report a percent difference.

Thought Experiment 18.5: Spontaneous Emission in the Classical Limit

In this thought experiment we consider how the Bohr correspondence principle relates to the spontaneous emission of radiation.

Consider a hydrogen atom with a high quantum number, n, so that it is approximately classical in nature. As discussed in Thought Experiment 18.3 (p. 655), the power radiated is given by:

$$\mathbb{P} = \frac{2m_e c^3 r_e^3}{3r^4}.$$

Consider the system from a quantum mechanical point of view. Rather than a continuous rate of power radiated, we imagine an atom with a given probability rate of dropping from a higher energy level to a lower energy level.

The probability per unit time for an atom to transit from state n to state m by giving off radiation will be denoted by $A_{n,m}$. Thus we can write:

$$A_{nm} = \frac{\mathbb{P}}{\hbar\omega} \approx \frac{\mathbb{P}}{\hbar\omega_e}, \text{ for } n \gg 1, m < n.$$

Now the kinetic energy $\mathbb{T}$ and the orbital radius are related by:

$$\mathbb{T} = \tfrac{1}{2} m_e v_e^2 = \tfrac{1}{2} m_e \omega_e^2 r^2 = m_e c^2 \frac{r_e}{2r},$$

so the classical angular velocity of an orbiting electron is:

$$\omega_e = c\sqrt{\frac{r_e}{r^3}}.$$

So, the semi-classical probability rate for an energy state change is:

$$A_{nm} \approx \frac{\mathbb{P}}{\hbar\omega} \approx \frac{2m_e c^3 r_e^{\,3}}{3r^4}\frac{1}{\hbar\omega_e} \approx \frac{2m_e c^3 r_e^{\,3}}{3r^4}\frac{1}{\hbar c}\sqrt{\frac{r^3}{r_e}} \approx \frac{2m_e c^2}{3\hbar}\left(\frac{r_e}{r}\right)^{5/2}.$$

The mass of an electron is related to the Bohr radius and the classical radius of an electron by:

$$m_e c^2 = \frac{\hbar c}{\sqrt{r_e a_0}},$$

so:

$$A_{nm} \approx \frac{2m_e c^2}{3\hbar}\left(\frac{r_e}{r}\right)^{5/2} = \frac{2}{3\cancel{\hbar}}\frac{\cancel{\hbar} c}{\sqrt{r_e a_0}}\left(\frac{r_e}{r}\right)^{5/2} = \frac{2c}{3\sqrt{r_e a_0}}\left(\frac{r_e}{r}\right)^{5/2}.$$

We can also write this in terms of the initial quantum number n, keeping in mind that in terms of the Bohr radius:

$$r = a_0 n^2,$$

so the semi-classical probability rate can be expressed as:

$$A_{n,m} \approx \frac{2c}{3\sqrt{r_e a_0}}\left(\frac{r_e}{r}\right)^{5/2} = \frac{2c}{3\sqrt{r_e a_0}}\left(\frac{r_e}{a_0 n^2}\right)^{5/2} = \frac{2}{3}\frac{r_e^{\,2}}{a_0^3}\frac{c}{n^5} \sim \frac{\left(1\times 10^{10}\ ^1\!/_{\text{s}}\right)}{n^5} \sim \left(\frac{100}{n}\right)^5 \frac{1}{\text{s}}.$$

The astute reader has probably already asked the question, "Which quantum state m are we going to?"

On one hand, the states in the classical limit become continuous, so one would expect that the corresponding quantum jump would be from n to $n-1$. However, another argument can be made for it to correspond to the total power radiated by an electron in state n, in which case we would want to compare the classical limit to the sum of all transition rates to the lower quantum states. The plot shows the exact quantum mechanical solution[322] for both assumptions. Notice that our classical model overestimates the transition rate to the next lower energy level, but underestimates the total emission rate to all lower energy levels.

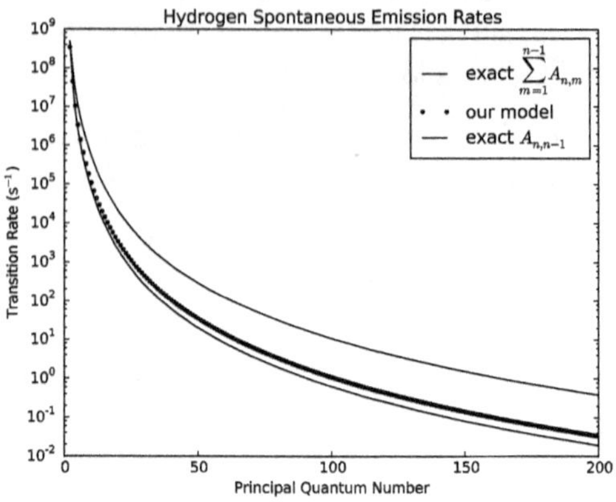

Problem 18.7: The Planetary Atom

Assume, in the sake of this problem, that the classical planetary model is correct and holds for hydrogen.

a) Show that the relationship between the power and the energy of an electron would be given by:

$$\mathbb{P} = \frac{128\pi\varepsilon_0}{3c^3 m_e^2 e^2}\mathbb{E}^4 = \frac{32}{3}\frac{c}{r_e\left(m_e c^2\right)^3}\mathbb{E}^4 .$$

b) From this equation, find the time a hydrogen atom with an initial energy $\mathbb{E}$ would take to reach the ground state of $\mathbb{E}_1 = -13.6\text{eV}$. What is this time for the following initial energies: -3.4 eV, -1.5 eV, -0.54 eV, and $-13.6/n^2$?

c) Find the orbital radius of an electron starting with a circular orbit of any initial energy, until it reaches the ground state, as a function of time. You can assume that it is always in a circular orbit.

Thought Experiment 18.6: The Driven Classical Atom

Consider a classical hydrogen atom, with its electron in a circular orbit with an initial angular velocity ω_0. The atom is subjected to a monochromatic electromagnetic plane wave of the same angular frequency ω_0 as the initial orbital frequency. We want to characterize the motion of the electron as a function of time.

First, we will make a free body diagram and write down the forces on the electron assuming that the electric field is driving the electron in phase $(\varphi_0 = 0)$, so that energy will be being added from the electric field to the electron.

There are three forces acting on the electron: the Coulomb force $\vec{F}_C$, the force from the driving electric field $\vec{F}_E$, and the radiation reaction force $\vec{F}_R$. We write each in turn as:

---

[322] The data used were from Kholupenko, Ivanchik, and Varshalovich found at http://www.ioffe.ru/astro/QC/CMBR/sp_tr.html, and published in Gravitation & Cosmology, v. 11(2005), No. 1-2, pp. 161-165.

$$\vec{F}_C = \frac{e^2}{4\pi\varepsilon_0 r^2}(-\hat{s}) = -m_e c^2 \frac{r_e}{r^2}\hat{s}$$
$$\vec{F}_E = -e\vec{E} = -eE_0 \sin(\omega_0 t)\hat{x}$$
$$\vec{F}_R = \frac{\mathbb{P}}{v}(-\hat{v}) = -\frac{2}{3}\frac{m_e r_e}{c}\left(\frac{a}{v}\right)^2 \vec{v}.$$

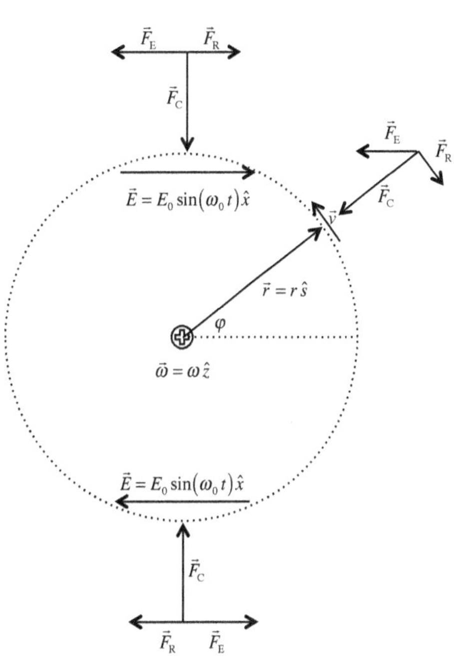

Notice that the radiation reaction force accounts for the radiation energy losses, so it is simply given by the Larmor formula divided by the speed. The total force in cylindrical coordinates is therefore:

$$\vec{F} = -m_e c^2 \frac{r_e}{r^2}\hat{s} - eE_0 \sin(\omega_0 t)\hat{x} - \frac{2}{3}\frac{m_e r_e}{c}\left(\frac{a}{v}\right)^2 \vec{v}.$$

Next we apply Newton's second law to find the acceleration:

$$\vec{a} = -c^2 \frac{r_e}{r^2}\hat{s} - \frac{e}{m_e}E_0 \sin(\omega_0 t)\hat{x} - \frac{2}{3}\frac{r_e}{c}\left(\frac{a}{v}\right)^2 \vec{v}.$$

And, finally, the acceleration in cylindrical coordinates becomes:

$$\vec{a} = -c^2 \frac{r_e}{r^2}\hat{s} - \frac{e}{m_e}E_0 \sin(\omega_0 t)\left(\cos(\varphi)\hat{s} - \sin(\varphi)\hat{\varphi}\right) - \frac{2}{3}\frac{r_e}{c}\left(\frac{a}{v}\right)^2 \left(v_s\hat{s} + v_\varphi\hat{\varphi}\right).$$

Notice that the net force is a function of the acceleration and speed, so that we will have two non-linear coupled second order differential equations. For this reason, we will solve this iteratively, using the following algorithm.

First we set our circular motion initial conditions, which are:

$$s = \sqrt[3]{\frac{c^2 r_e}{\omega_0^2}} \quad \varphi = \varphi_0 \quad v_s = 0 \quad v_\varphi = \omega_0 r \quad .$$

Then the acceleration can be found from Newton's second law:

$$\begin{pmatrix} a_s \\ a_\varphi \end{pmatrix} = \begin{pmatrix} -\frac{e}{m_e}E_0 \cos(\varphi)\sin(\omega_0 t) - c^2 \frac{r_e}{r^2} - \frac{2}{3}\frac{r_e}{c}\left(\frac{a}{v}\right)^2 v_s \\ \frac{e}{m_e}E_0 \sin(\varphi)\sin(\omega_0 t) - \frac{2}{3}\frac{r_e}{c}\left(\frac{a}{v}\right)^2 v_\varphi \end{pmatrix}.$$

Next we will find the time derivatives of the velocity components as:

$$\frac{\mathrm{d}}{\mathrm{d}t}\begin{pmatrix} v_s \\ v_\varphi \end{pmatrix} = \begin{pmatrix} a_s + \frac{v_\varphi^2}{r} \\ a_\varphi \end{pmatrix}.$$

Then after a short time δt, the new positions and velocities become:

$s(t+\delta t) \approx s + v_s \delta t + \frac{1}{2} \dot{v}_s \delta t^2 \quad v_s(t+\delta t) \approx v_s + \dot{v}_s \delta t$

$$\varphi(t+\delta t) \approx \varphi + \frac{v_\varphi}{s}\delta t + \frac{\dot{v}_\varphi \delta t^2}{2s} \quad v_\varphi(t+\delta t) \approx v_\varphi + \dot{v}_\varphi \delta t \ .$$

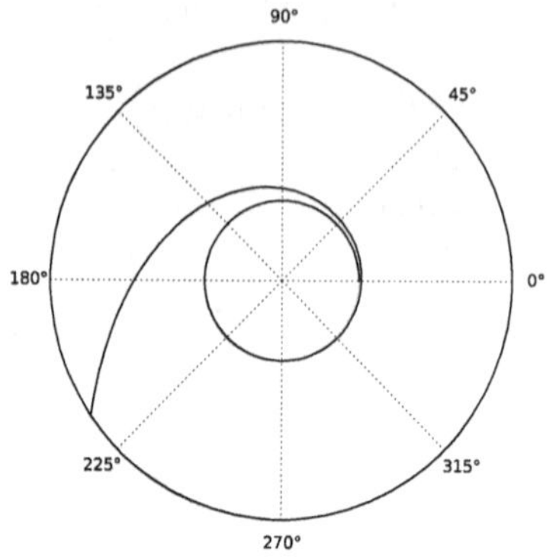

And, now, new accelerations can be calculated again using Newton's second law, as we repeat the last three steps.

The plot shows the position of the electron originally in a quantum state with $n = 10$, driven with an electric field of 10,000 N/C in phase with the electron's orbit. Notice that an electron driven in phase that gains energy increases its distance from the nucleus exponentially.

Thought Experiment 18.7: Classical Stimulated Emission and Absorption

Now we will imagine two cases, an atom driven in phase, and 180° out of phase. The driving frequency is assumed to be the same as the orbital frequency. We will then find the power absorbed, or radiated, as:

$$\frac{d\mathbb{E}}{dt} = \pm\mathbb{P}_{\text{driven}} - \mathbb{P}_{\text{spontaineous}} \ .$$

where $\mathbb{P}_{\text{spontaneous}}$ represents the power radiated due to spontaneous emission, and $\mathbb{P}_{\text{driven}}$ is the power either added (in phase) or subtracted (out of phase) by the driving radiation field. From this we can calculate $\mathbb{P}_{\text{driven}}$ as a function of the driving electric field E_0.

Looking at the driving power we have:

$$\mathbb{P}_{\text{driven}} = \left|\vec{F}_E \cdot \vec{v}\right| = F_\varphi v = eE_0 v \sin(\omega_0 t)\sin(\varphi) = eE_0 v \sin(\omega_0 t)\sin(\omega_0 t - \varphi_0).$$

The radial component of the force does not contribute to our expression for the power since we are assuming that the radius changes only slightly (i.e. it remains nearly constant and $v \approx v_\varphi$) in the limit of large quantum numbers. Now, we will assume that absorption happens in a very short time, δt, centered around $\varphi = \pm \pi/2$. To find the time that it takes to increase the energy by a single unit, $\Delta\mathbb{E} = \hbar\omega$, we choose our initial conditions so that $\varphi_0 = 0$ (i.e. in phase) and evaluate the following integral:

$$\Delta\mathbb{E} = \int_{-\frac{1}{2}\delta t}^{\frac{1}{2}\delta t} \mathbb{P}_{\text{driven}} dt = \frac{1}{\omega}\int_{\frac{\pi}{2}-\frac{1}{2}\omega\delta t}^{\frac{\pi}{2}+\frac{1}{2}\omega\delta t} eE_0 v \sin^2(\varphi) d\varphi = \frac{1}{2}\frac{eE_0 v}{\omega}(\omega\delta t + \sin(\omega\delta t))$$
$$\approx eE_0 v \delta t .$$

This means that, in the classical limit, the amount of time it takes to change energy levels is approximately:

$$\delta t \approx \frac{\Delta\mathbb{E}}{eE_0 v} = \frac{\hbar\omega}{eE_0 v} = \frac{\hbar}{eE_0 r}.$$

Recall that this very short time, δt, is centered at two phases $\varphi = \pm\frac{\pi}{2}$, which occur twice per orbit. Therefore, the probability that a jump will take place, during a given orbit, is going to be given by approximately twice the ratio of the period to δt.

So, on average, the power absorbed from the field is given by:

$$\langle \mathbb{P} \rangle \approx \left(2\frac{2\pi/\omega}{\delta t} \right) eE_0 v \approx \left(\frac{2\pi r}{2\pi/\omega} \right) eE_0 \left(\frac{4\pi}{\omega\delta t} \right) \approx \left(\frac{1}{\delta t} \right) 4\pi r eE_0 ,$$

which we can then write in terms of the energy density u_ω as:

$$\langle \mathbb{P} \rangle \approx \left(\frac{eE_0 r}{\hbar} \right) 4\pi r eE_0 \approx \frac{4\pi e^2 r^2}{\hbar} E_0^2 \approx \frac{4\pi e^2 r^2}{\hbar} \left(\frac{2u_\omega \delta\omega}{\varepsilon_0} \right).$$

Now we will define a rate for either absorption, or stimulated emission, and denote it as $B_{n,n\pm1}$:

$$B_{n,n\pm1} = \frac{\langle \mathbb{P} \rangle}{\hbar\omega u_\omega} \approx \frac{8\pi e^2 r^2}{\hbar^2 \varepsilon_0} \left(\frac{\delta\omega}{\omega} \right).$$

There are a few things to notice about this. First, the rate increases as the square of the radius of the orbit. This makes sense because the further out the electron is from the nucleus, the more susceptible it would be to perturbations from an external electric field. Notice, also, that it depends on the ratio of the bandwidth $\delta\omega$ to the frequency. For a driven damped oscillator this is related to the quality factor, Q_f, where:

$$\frac{\delta\omega}{\omega} = \frac{1}{Q_f} \approx \frac{\text{damping rate}}{\omega},$$

which can usually be assumed to be a constant on the order 10 to 100.

Problem 18.8: The Ratio of Einstein's A and B Coefficients

In the past thought experiments, we found the semi-classical rates for spontaneous emission and absorption/stimulated emission. Here you will find the ratio of the two, at least in the classical limit.

a) In the classical limit, show that the ratio of rates scales with radius and quantum number as:

$$\frac{A_{n,n-1}}{B_{n,n\pm1}} \propto \frac{1}{r^{9/2}} \propto \frac{1}{n^9}.$$

b) Like planets, the orbiting electron must follow Kepler's third law, $P^2 \propto r^3$. Use this to show that:

$$\frac{A_{n,n-1}}{B_{n,n\pm1}} \propto \frac{1}{P^3} \propto \omega^3.$$

Where P is the orbital period and ω is the orbital angular velocity.

c) Show that classically:

$$\frac{A_{n,n-1}}{B_{n,n\pm1}} = \left(\frac{Q_f}{48}\right)\left(\frac{\hbar\omega^3}{\pi^2c^3}\right) \approx \left(\frac{\hbar\omega^3}{\pi^2c^3}\right).$$

Thought Experiment 18.8: Einstein's Theory of Radiation

Einstein, in 1917, made use of Planck's quantum hypothesis to study the probabilities of radiative transitions for a system in equilibrium with electromagnetic radiation of energy density per frequency $\mathbb{u}_\omega(\omega)$. Consider an atomic system that consists of N atoms, with N_n atoms occupying energy level $\mathbb{E}_n$ and N_m atoms excited to energy $\mathbb{E}_m$, where we assume that $m > n$ and $\mathbb{E}_m > \mathbb{E}_n$. Radiation, whether a collection photons or waves, of energy $\hbar\omega = \mathbb{E}_m - \mathbb{E}_n$ can be absorbed by atoms in the state with energy $\mathbb{E}_n$, thus raising them to the $\mathbb{E}_m$ state; or, they can induce atoms in the $\mathbb{E}_n$ state to a quantum of energy as light $\hbar\omega$ and transit to the lower $\mathbb{E}_n$ state. The light is radiated by induced or *stimulated* emission, and is in phase with the radiation field. Einstein also assumed that spontaneous transitions can occur from $\mathbb{E}_m$ to $\mathbb{E}_n$. The probabilities for these three types of transitions are:

$A_{m,n} \equiv$ the probability per second for spontaneous emission,

$B_{n,m}\mathbb{u}_\omega(\omega)\delta\omega \equiv$ the probability per second for absorption,

$B_{m,n}\mathbb{u}_\omega(\omega)\delta\omega \equiv$ the probability per second for stimulated emission,

where A and B are constants and $\mathbb{u}_\omega(\omega)\delta\omega$ is the energy density of the radiation field over some resonance frequency range $\delta\omega$. Note that the constants $B_{n,m}$ and $B_{m,n}$ are multiplied by $\mathbb{u}_\omega(\omega)\delta\omega$, because both absorption and stimulated emission are induced by the radiation field. The transition rate for total emission is therefore given by:

$$\left[A_{m,n} + B_{m,n}\mathbb{u}_\omega(\omega)\delta\omega\right]N_m,$$

while that for absorption is

$$B_{n,m}\mathbb{u}_\omega(\omega)\delta\omega N_n.$$

At equilibrium these two rates must be equal (principle of detailed balance), so we have that

$$\frac{N_n}{N_m} = \frac{A_{m,n} + B_{m,n}\mathbb{u}_\omega(\omega)\delta\omega}{B_{n,m}\mathbb{u}_\omega(\omega)\delta\omega}.$$

From statistical mechanics, we can apply the *Boltzmann distribution*, which gives the probability of a state for a given energy, and hence the average number of particles in that particular state, as proportional to the Boltzmann factor, $\mathrm{e}^{-\mathbb{E}/k_BT}$, where T is the equilibrium temperature, and $\mathbb{E}$ is the energy of that particular state. The ratio of two equilibrium energy state populations depends on the energy differences between the states:

$$\frac{N_n}{N_m} = \frac{\mathrm{e}^{-\mathbb{E}_n/k_BT}}{\mathrm{e}^{-\mathbb{E}_m/k_BT}} = e^{(\mathbb{E}_m-\mathbb{E}_n)/k_BT} = e^{\hbar\omega/k_BT}.$$

Then,

$$e^{\hbar\omega/k_BT} B_{n,m} \mathrm{u}_\omega(\omega)\delta\omega = A_{m,n} + B_{m,n} \mathrm{u}_\omega(\omega)\delta\omega,$$

and the radiation energy density distribution is:

$$\mathrm{u}_\omega(\omega)\delta\omega = \frac{A_{m,n}}{B_{n,m} e^{\hbar\omega/k_BT} - B_{m,n}}\delta\omega.$$

At high temperatures it is reasonable to assume that $N_n \approx N_m$ and that $B_{m,n} u(\omega) \gg A_{m,n}$. Then, if this is this case, we get $B_{n,m} = B_{m,n}$, and

$$\mathrm{u}_\omega(\omega)\delta\omega = \frac{A_{m,n}}{B_{m,n}} \frac{\delta\omega}{e^{\left(\hbar\omega/k_BT\right)} - 1}.$$

Finally, if the ratio of Einstein's two coefficients represents a constant, then their ratio must be the same as in the classical limit, which you found in Problem 18.8 to be:

$$\frac{A_{m,n}}{B_{m,n}} = \frac{\hbar\omega^3}{\pi^2 c^3},$$

which would imply that in equilibrium the radiation field must have an energy density given by the following distribution:

$$\mathrm{u}_\omega(\omega)\delta\omega = \frac{\hbar\omega^3}{\pi^2 c^3} \frac{\delta\omega}{e^{\hbar\omega/k_BT} - 1}.$$

This is the famous Planck blackbody radiation law.

Problem 18.9: Thermal Hydrogen Emission

Atomic hydrogen is very common in space, so it often affects the radiation field surrounding it.

a) Taylor expand the Blackbody radiation law to find u_ω, in the low frequency limit where $\hbar\omega \ll k_B T$. This is called the *Raleigh-Jeans tail.*

b) When radio telescopes observe thermal emission from space, the brightness increases as the square of the frequency. However, this is not the case when optical telescopes observe atomic hydrogen. Explain.

c) When a bright hot star is observed behind a cooler cloud of atomic hydrogen, the star's spectrum shows dark *absorption lines.* On the other hand, when electricity runs through a hydrogen gas in the lab, its spectrum shows corresponding bright *emission lines.* Explain why this is.

Problem 18.10: Brightness Temperature

Brightness temperature is a concept that radio astronomers use to better understand extended astronomical sources. By pointing a radio dish at an object larger in angular size than the telescope's beam, a radio astronomer measures the *surface brightness,* which is simply the power per area per frequency per solid angle. If the radio waves were produced via a thermal process, then the temperature can be directly calculated from the surface brightness using the Rayleigh-Jeans law.

a) Consider a radio dish antenna of diameter D. For an unobstructed aperture, the diffraction limited opening angle θ where the telescope is most sensitive is given by:

$\theta \approx 1.22 \lambda/_D$.

What is the corresponding solid angle, Ω , that the radio dish observes?

b) An extended object is assumed to "fill the beam," which means that a telescope with a larger field of view (i.e. smaller diameter) will observe more of it. Consider an extended source of surface brightness Σ. What is the power per unit area per unit frequency that is observed by the telescope? This is called the *flux density*, and radio astronomers measure this in a unit called a jansky, where $1\,\mathrm{Jy} \equiv 10^{-26}\,\mathrm{Wm^{-2}Hz^{-1}}$. Is the flux density observed by a small diameter telescope larger or smaller than one by a smaller telescope? Explain.

c) Express the total power per frequency collected by the radio telescope as a function of the surface brightness of the source at that frequency, and the observation frequency $\nu = \frac{1}{2\pi}\omega$. Show that this is independent of the antenna diameter, but does depend on the frequency.

d) Consider a source of blackbody radiation at a temperature T . What is the power collected by a radio antenna from such a source? What does this depend on?

e) Most astronomical sources of thermal radiation are hot compared to the observing frequency $(k_B T \gg \hbar\omega)$, so the Rayleigh-Jeans temperature applies. In this limit, show that the power collected is simply proportional to the source temperature.

f) Radio astronomers often use a source of known temperature for calibration and noise measurements, so they use temperature units to measure the surface brightness of sources. This pseudo-temperature is called *brightness temperature.* Given a source with a measured brightness temperature, T_B, in units of kelvin, find the surface brightness in units of jansky per square arc minute.

g) In most radio antennas, the noise is dominated by the temperature of the antenna front end—before the signal is amplified. Thus the total random noise is also measured in temperature units and is called the antenna temperature. Some of the most significant survey work of extended sources has been done by small radio telescopes, while it is important to use a large telescope observe a radio point source. Research single-dish radio telescopes, and discuss the importance of each one.

Problem 18.11: Wien's Displacement Law

The color of a hot object is a function of its temperature alone. Using purely thermodynamic arguments, Wilhelm Wien first proposed an expression for the peak wavelength of light from a blackbody of a given temperature in 1893 before Planck commenced his work on blackbody radiation. Here you will find his expression from the Planck function.

a) Given the Planck distribution, find the value for the unitless factor

$$\frac{\hbar\omega_{\max}}{k_B T},$$

where $\omega_{\max}$ is the angular frequency at the peak of the distribution.

b) Find the equivalent relationship in term of the frequency υ in units of cycles per second (Hz), rather than the angular frequency.

c) Express this relationship in terms of degrees kelvin and the wavelength in nanometers, rather than the frequency.

d) The sun produces light with a peak wavelength in the middle of the visible spectrum, at about 500 nm. What is its surface temperature?

e) The Earth has an average temperature of about 15°C. At what peak wavelength does the Earth radiate?

Problem 18.12: Astronomical Colors

When hot objects, such as stars, are observed from afar their surface temperatures are usually measured based on the blackbody spectrum. There are a number of ways to do this, in practice. One common way, especially for star surveys, is to observe the stars with light filters that let in particular colors. For example, the Sloan Digital Sky Survey uses filters called u', g', r', i', and z', which are centered around the following wavelengths ranging from the ultraviolet to the infrared:

$\lambda_u = 3524\mathring{A}$, $\lambda_g = 4714\mathring{A}$, $\lambda_r = 6182\mathring{A}$, $\lambda_i = 7592\mathring{A}$, and $\lambda_z = 9003\mathring{A}$.

Each of these filers have corresponding bandwidths of:

$\Delta\lambda_u = 599\mathring{A}$, $\Delta\lambda_g = 1379\mathring{A}$, $\Delta\lambda_r = 1382\mathring{A}$, $\Delta\lambda_i = 1535\mathring{A}$ and $\Delta\lambda_z = 1370\mathring{A}$.

a) The star Vega has a surface temperature of 9,600 kelvin. Find the energy density of the light at Vega's surface per angular frequency centered at each of these filters.

b) If we define the total power from Vega that passes through the green filter as 1, estimate the power collected using each of the other filters. You can ignore sensitivity changes in the detector.

c) Astronomers use a system of measuring the brightness of stars which has been in existence for over 2000 years. In the 1800s, the magnitude system was standardized in such a way that Vega would, by definition, have a magnitude of zero using any filter or telescope. So that it would be backward compatible to historical star catalogues, a star's magnitude was defined by the formula:

$$m = -2.5\log_{10}\left(\frac{\mathbb{S}}{\mathbb{S}_{vega}}\right)$$

where $\mathbb{S}$ is the energy flux of the star and $\mathbb{S}_{vega}$ is the energy flux of Vega using the same technique.

Colors of stars are often measured using the color index, which is simply the difference between the magnitudes of two color filters. The color index of Vega is zero, by definition.

Calculate the u'-g', g'-r', r'-i' and the i'-z' color indices for a star with a surface temperature of 4000K.

d) Write a computer program to input each of these magnitudes, and fit a blackbody curve to the data. The program should return two quantities: (1) the surface temperature of the star, and (2) the ratio of the total integrated flux from the star to that of Vega. Download data from the Sloan Digital Sky Survey database to test your results.

Problem 18.13: Spectral Energy Distributions

Often natural sources of radiation follow powerlaw spectral energy distributions. Thus an antenna, telescope, x-ray, or gamma ray detector would measure a particular energy flux density, which could be expressed as: $\vec{\mathbb{S}}_\omega(\omega) \propto \omega^\alpha$, where $\vec{\mathbb{S}}_\omega$ is in the direction toward the telescope and it has dimensions of energy flux per frequency:

$$[\mathbb{S}_\omega] = \frac{\text{energy}}{\text{time} \times \text{area} \times \text{angular frequency}} \underset{(\text{in SI})}{=} \frac{\text{W}}{\text{m}^2 \cdot \text{rad}/\text{s}} = \frac{\text{J}}{\text{m}^2}.$$

a) A common unit in radio astronomy of the equivalent quantity, $\vec{\mathbb{S}}_\nu$, is called the jansky, after Karl Jansky, where: $1\,\text{Jy} \equiv 10^{-26}\,\frac{\text{W}}{\text{m}^2 \cdot \text{Hz}}$. Express the relationship between the $\vec{\mathbb{S}}_\omega$ and $\vec{\mathbb{S}}_\nu$. What is $\mathbb{S}_\omega$, in SI, for a one jansky source?

b) Show that if $\mathbb{S}_\omega(\omega) \propto \omega^\alpha$ then $\mathbb{S}_\nu(\nu) \propto \nu^\alpha$, where as above ν is the frequency rather than the angular frequency.

c) Now show that the corresponding quantity, but with the frequency measured in units of photon energy $\mathbb{E}$, also has the same powerlaw index α. What is $\mathbb{S}_\mathbb{E}$, in SI, for a one jansky source?

d) X-ray astronomers, on the other hand, usually count the number of photons that arrive in their detectors. This is useful as their errors are often dominated by counting statistics, rather than the antenna temperature as in radio astronomy. They, rather, measure the quantity $\vec{\mathbb{F}}_\mathbb{E}(\mathbb{E}) \propto \mathbb{E}^{-\Gamma}$, which has dimensions of photon flux per energy, or:

$$[\mathbb{F}_\mathbb{E}] = \frac{\text{photons}}{\text{time} \times \text{area} \times \text{photon energy}} \underset{(\text{in SI})}{=} \frac{\text{counts}}{\text{s} \cdot \text{m}^2 \cdot \text{J}}.$$

Derive a formula relating $\mathbb{F}_\mathbb{E}$ to $\mathbb{S}_\mathbb{E}$, and vice versa, in terms of the other quantity and $\mathbb{E}$.

e) What is the relationship between an x-ray astronomer's Γ and a radio astronomer's α?

f) A typical reference frequency in radio astronomy is one gigahertz, while a typical photon energy in x-ray astronomy is 1 kilo-electron-volt. What are these equivalent reference frequencies in units of: Hz, GHz, eV, and keV? What are the equivalent wavelengths of the corresponding light?

g) Recall from Thought Experiment 18.8 (p. 666), the Planck distribution function for the energy density of light $\mathbb{u}_\omega(\omega) = \frac{\hbar\omega^3}{\pi^2 c^3} \frac{1}{e^{\hbar\omega/k_BT} - 1}$.

If you were to observe an unresolved spherical opaque thermal source (like a star) of radius R, uniform surface temperature T, and located a distance r away from us, show that the expected spectral energy distribution would be given by:

$$\mathbb{S}_\omega(\omega) = \frac{\hbar\omega^3}{\pi^2 c^2} \left(\frac{R}{r}\right)^2 \frac{1}{e^{\hbar\omega/k_BT} - 1}.$$

h) Rewrite this Planck distribution, making the two substitutions: $\mathbb{E}_\gamma = \hbar\omega$ and $\mathbb{E}_T = k_B T$, express the quantity $\mathbb{S}_{\mathbb{E}}$ in terms of these two energies. What is the temperature equivalent (i.e. $\mathbb{E}_\gamma = \mathbb{E}_T$) of a frequency of 1 GHz. What about an energy of 1 keV?

i) All astronomical sources are hot enough that at radio wavelengths $\mathbb{E}_\gamma \ll \mathbb{E}_T$. If a source is thermal in nature, what would its measured powerlaw index be? This is how radio astronomers know if a source they are observing is thermal or not.

j) Even if there were a collection of sources at different temperatures, in the radio the single power-law index you found would still be observed. Explain why this is unique to radio astronomy, and why the power-law index for other astronomers would depend on distribution of temperatures.

18.3 *Does the Photon Exist?*

Of all Einstein's work from 1905, his conception of light quanta, later dubbed *photons*, was surely the most controversial. Planck felt the need to apologize for it when he recommended Einstein to his post at the Berlin Academy, and Bohr resisted accepting the notion until the bitter end when he *lost* the first Einstein-Bohr debate in 1924.

To aid in our understanding of what Einstein's particle model of light could explain, and what it could not explain, we will spend this section reanalyzing common physical scenarios from a light particle point of view. These examples will hopefully illuminate why Einstein's model was so profound, yet so controversial.

For the purposes of this section, define the *wave theory* of light to be the classical theory that we have covered so far in this book, so we can contrast it with the *particle theory* which we define as making the following assumptions:

- Light is made of real particles of zero mass, called photons.
- The kinetic energy per photon, of a beam of identical particles, is measured classically as the frequency of light. The relationship is simply given by $\mathbb{E} = \mathbb{T} = \hbar\omega$.
- Photons have momentum and angular momentum, but possess no independently measureable properties other than these two conserved quantities.[323]
- Otherwise, photons follow the laws of classical mechanics, including those of the special theory of relativity, as they were already understood.

When the year 1924 commenced, the particle theory appeared to be uniquely successful at explaining the photoelectric effect. It also explained blackbody radiation, and other phenomena involving the absorption and emission of light by matter, better than the wave theory. Still, leading physicists, such as Bohr, Millikan, and Planck, criticized the particle theory of light because it could scarcely explain such phenomena as diffraction and interference. By the end of the year, two decisive developments, one theoretical and the other experimental, had changed the minds of the majority of physicists. In the realm of theory, the Indian physicist S.N. Bose used

---

[323] Since the energy of a photon is given by $\mathbb{E} = pc = \hbar\omega$, a measurement of photon energy, momentum, or frequency are all equivalent.

Einstein's light quantum idea to derive Planck's blackbody spectrum, without the *ad hoc* classical assumptions that had marred previous derivations.

The experimental breakthrough came from the American physicist Arthur H. Compton, who scattered nearly monochromatic x-rays off the valence electrons[324] of carbon and found that the wavelength of the scattered light was quite different than that of the incident light, even at the lowest intensities. While this was inexplicable from the standpoint of Thomson scattering (p. 557), Compton scattering could easily be explained by assuming that light behaves as if it consists of particles that carry momentum and energy.

Compton's experiments turned the tide in favor of Einstein's light particle hypothesis. Before we get to that part of the story, however, we will review some of the basic properties of the photon, especially spin and momentum, and review the state of scientists' understanding just prior to the discovery of Compton scattering.

Thought Experiment 18.9: The Momentum and Speed of a Photon

In Einstein's particle model of light, the photon is real, but has no rest mass. A photon can carry other conserved quantities such as momentum.

Applying the relativistic energy, mass and momentum relationship:

$$\mathbb{E}^2 = m^2c^4 + p^2c^2 ,$$

Now, remember that the mass is zero, so the momentum is given by:

$$\vec{p} = \lim_{m\to 0}\tfrac{1}{c}\sqrt{\mathbb{E}^2 - m^2c^4}\;\hat{k} = \tfrac{1}{c}\mathbb{E}\,\hat{k} = \tfrac{1}{c}\hbar\omega\hat{k} = \hbar\vec{k}.$$

Thus, the vacuum wave vector measures the momentum per photon.

Next, we will investigate the speed that the photon must be traveling. To do this, we will write the relativistic velocity in terms of the momentum as:

$$\vec{p} = m\vec{v}\gamma(v).$$

Rearranging, and dropping the vectors, we have:

$$\frac{v}{\sqrt{1-\dfrac{v^2}{c^2}}} = \frac{p}{m} \to v^2 = \left(\frac{p}{m}\right)^2\left(1-\frac{v^2}{c^2}\right) = \left(\frac{p}{m}\right)^2 - \left(\frac{p}{mc}\right)^2 v^2 \to v^2\left(1+\left(\frac{p}{mc}\right)^2\right) = \left(\frac{p}{m}\right)^2 .$$

Again, there is no mass, so the velocity is:

$$v = \lim_{m\to 0}\frac{p/m}{\sqrt{1+\left(p/mc\right)^2}} = p\lim_{m\to 0}\frac{1}{\sqrt{m^2+\left(p/c\right)^2}} = p\frac{1}{\sqrt{\cancel{m^2}+\left(p/c\right)^2}} = \cancel{p}\frac{c}{\cancel{p}} = c ,$$

which is also what we found in Example 17.2 (p. 601). In fact, all massless particles travel at exactly the speed of light. No slower, and no faster.

---

324 Since the energy of one of Compton's photons was $\approx 20\,\mathrm{keV}$, which is many times greater than the energy binding a valence electron in carbon to its parent atom, the electrons may be considered as essentially free.

This also means that a single photon requires no time to be absorbed into an atom, because once it slows slightly it ceases to exist. This is also why a particular photon's measureable properties must be conserved quantities, because they can only be detected when the photon either becomes absorbed, or is emitted, from something whose difference in energy, momentum or angular momentum can be measured. Notice that this is quite different from Newton's corpuscle model, where light particles have mass and may merely bounce off surfaces.

Problem 18.14: The Energy and Speed of a Photon

In Thought Experiment 18.9 above, we showed that a massless particle must travel at exactly the speed of light. Make the same argument, except using energy, rather than momentum.

Problem 18.15: The Non-Relativistic Doppler Effect

Use the particle model of light to find the non-relativistic Doppler effect, which is used to find the velocity of a source, as:

$$\frac{v}{c}=\frac{\omega_0}{\omega}-1 \quad \text{(see p. 518 above)},$$

where ω_0 and ω represent the angular frequency emitted by the source, and received by the detector, respectively.

Problem 18.16: Ram and Radiation Pressure

The force of the wind, per area, is an example of *ram pressure*. In this problem, you will model radiation pressure in much the same way as you would ram pressure.

a) Consider a sailboat with a sail of area, A, that is running downwind at a velocity v and in a wind of speed v_w. What is the force of the air on the sailboat, assuming an air density ρ ?

b) Now consider a spaceship with a solar sail of area, A, moving directly away from the sun at a velocity $v \ll c$ in a radiation field $\vec{\mathbb{S}}$. What is the force of the light on the spaceship?

c) Use this analogy, and the particle model, to derive the radiation pressure formula we found in Thought Experiment 15.5 (p. 488).

Problem 18.17: The Recoil of an Atom

Consider an atom of mass m at rest. A photon is spontaneously emitted in the $\hat{x}$ direction, and is observed to have an energy $\mathbb{E}$.

a) What is the velocity of the atom after the photon is emitted?

b) What is the kinetic energy of the atom after the photon was emitted?

c) How much energy was lost by the atom in the decay process?

d) This analysis is valid so long as the mass of the atom did not change very much. Find the fractional change in mass of the atom.

e) Now consider the same situation, except we will assume that, rather than being free to move, the emitting atom is fixed into a lattice structure. What would be the energy of the photon emitted?

f) Consider an Iron-57 nuclear decay, with an energy difference of 14.4 keV. If the iron atom were free to move, how fast would the atom be moving after the emission? What would be the observed energy of the photon emitted?

g) In the field of Mössbauer spectroscopy, an Iron-57 source is physically moved with a velocity measured in $\frac{\text{mm}}{\text{s}}$. How does the recoil velocity compare to these velocities? Discuss whether Mössbauer spectroscopists need to take into account recoil in their analysis, and don't forget to cite any external references you use.

Thought Experiment 18.10: The Spin of a Photon

Consider a beam of coherent monochromatic light with a Poynting vector $\vec{\mathbb{S}}$, so that the energy density of the beam is given by $\mathbb{u} = \frac{1}{c}\mathbb{S}$. In the particle model, this is a beam of identical photons, each with energy $\mathbb{E}$, moving in the same direction. The number density n of photons is:

$$n = \frac{\mathbb{u}}{\mathbb{E}} = \frac{\mathbb{S}}{c\mathbb{E}} = \frac{\mathbb{S}}{\hbar c\omega} .$$

Now that we have a number density of photons, we can write down the momentum density $(\vec{\mathbb{p}})$ of the beam as:

$$\vec{\mathbb{p}} = n\vec{p} = \left(\frac{\mathbb{S}}{c\mathbb{E}}\right)\vec{p} = \left(\frac{\vec{\mathbb{S}}}{c\cancel{\mathbb{E}}}\right)\left(\frac{\cancel{\mathbb{E}}}{c}\right) = \frac{1}{c^2}\vec{\mathbb{S}} ,$$

which agrees with our result from Thought Experiment 15.6 (p. 490).

We will now turn this argument around to find the angular momentum of a particular photon. In Thought Experiment 15.8 (p. 491) we found the angular momentum density of a circularly polarized light wave to be:

$$\vec{\mathbb{L}} = \frac{d\vec{L}}{dV} = \frac{d\vec{L}}{dA(cdt)} = \pm\frac{\vec{\mathbb{S}}}{\omega c} .$$

However, if we now interpret this as a beam of identical photons each with an individual angular momentum $\vec{L}$, then:

$$\vec{L} = \frac{\vec{\mathbb{L}}}{n} = \frac{\pm\left(\dfrac{\vec{\mathbb{S}}}{\omega c}\right)}{\left(\dfrac{\mathbb{S}}{\hbar c\omega}\right)} = \pm\hbar\hat{k}$$

where, as usual, $\hat{k}$ is the unit vector in the direction of propagation.

Thus, the photon is a *spin 1* particle.

Problem 18.18: Electromagnetic Waves and Photons

We have derived a number of relationships between the Poynting vector of a coherent electromagnetic wave and various conserved quantities conveyed by the wave. In this problem you are to derive the same relationships using the particle model of light.

a) Considering light as a collection of particles, each travelling at the same speed c, show that: $\mathbb{u} = \frac{1}{c}\mathbb{S}$.

b) In Thought Experiment 15.6 (p. 490), we found the momentum density contained in a beam of light is given by the formula: $\vec{\mathbb{p}} = \frac{1}{c^2}\vec{\mathbb{S}}$. Show that it would hold under Einstein's particle model.

c) In Thought Experiment 15.8 (p. 491), we found that the angular momentum density contained in a circularly polarized monochromatic beam of light is given by the formula $\vec{\mathbb{L}} = \pm\frac{1}{\omega c}\vec{\mathbb{S}}$. Considering light as a collection of spin-1 particles, derive this same relationship.

Discussion 18.3: Successes of the Particle Model of Light

In general, the particle model of light can explain many phenomena that can be calculated using the wave theory. In some instances, the photon concept explains the facts more simply and convincingly than does the wave theory. Other phenomena often require a detailed knowledge of the matter with which the light interacts for the particle picture to be plausible. Here we will discuss a number of phenomena that are more convincingly explained by Einstein's photon concept. When Millikan published the results of his photoelectric experiment in 1916, only the photoelectric effect appeared to *completely* contradict the predictions of classical electrodynamics. In his original paper on the subject of light quanta, Einstein also adduced the Stokes rule for photoluminescence and the photoionization of gases as evidence for light quanta.

Stokes gave the name *fluorescence* to the phenomena of light emission by substances that have absorbed light, or other electromagnetic radiation.[325] Einstein referred to the same phenomenon as *photoluminescence*, presumably to stress the role of the incident light in causing emission, unlike *incandescence* in which light is emitted due to heating. From his extensive experiments with a variety of substances, including fluor-spar (fluorite) from which he derived the word *fluorescence*, Stokes formulated the following rule: *the frequency of the light emitted by a photoluminescent substance cannot exceed the frequency of the incident light.* We can easily explain this, along with the fact that photoluminescence is so prompt that it appears to occur instantaneously, by appealing to Einstein's particle model of light.

Let monochromatic light be transformed by photoluminescence into light of another frequency, and assume that both incident and generated light consist of photons whose energy is given by the relation $\mathbb{E} = \hbar\omega$. During the process, according to Einstein's analysis, a photon of frequency ω_1 is absorbed and a photon of frequency ω_2 is generated, possibly along with other light quanta of different frequencies, as well as other forms of energy such as heat, all of which can be generated simultaneously. Note that it does not matter through which intermediate processes the final result comes about. If the fluorescing substance isn't a continuous source of energy, then energy conservation demands that the energy of the generated photon cannot be larger than the generating photon's energy, i.e. $\hbar\omega_2 \leq \hbar\omega_1$, or $\omega_2 \leq \omega_1$. Einstein's photon concept, when combined with energy conservation, is all we need to derive Stokes' rule of fluorescence. (Although this rule generally holds true, an exception can occur if some the atoms are initially in excited energy states.)

---

[325] Stokes, G. G. (1852), "On the Change of the Refrangibility of Light," Philosophical Transactions of the Royal Society of London, **142** (1852), 463–562.

Einstein also explained the ionization of gas by ultraviolet light using his idea of light quanta. Assume that one absorbed photon ionizes just one gas molecule, Einstein argued in his original paper, then it follows that the ionization energy of a molecule cannot be larger than the energy of an absorbed photon. To support his interpretation, Einstein produced a simple calculation from the best data then available. In essence, photoionization is simply the photoelectric effect per atom.

In subsequent papers on the subject, Einstein argued that the light quantum hypothesis was required even to derive the correct blackbody spectrum from first principles. Einstein pointed out that Planck's derivation is fundamentally flawed because it used a classical result to work out the rate at which the oscillator radiates energy, on the one hand, while implicitly assuming that the material oscillators of the wall can only absorb and emit energy in discrete amounts. Einstein wrote:

> We must view the following proposition as the basis underlying Planck's theory of radiation: the energy of an elementary resonator can only assume values that are integral multiples of $\hbar\omega$; by emission and absorption, the energy of a resonator changes by jumps of multiples of $\hbar\omega$.[326]

Einstein was pragmatic in the face of this contradiction, and proposed the following way of dealing with it until a correct theory could be found:

> Although Maxwell's theory is not applicable to elementary resonators, the average energy of such a resonator in a radiation field is the same as one would compute from Maxwell's theory.

Einstein's suggestion was as bold and provisional as Bohr's model of the hydrogen atom, in which classical and quantum concepts were mixed together. Until the advent of the complete quantum theory that Heisenberg, Born, Dirac, and Schrödinger introduced in the last half of the 1920s, physicists were therefore stuck with two contradictory theories for explaining atoms and radiation. Planck, Einstein, Bohr, and others used both Maxwell's equations and classical mechanics in an entirely *ad hoc* way to solve problems within the theoretical framework of what later became known as the *old quantum theory*. Theoretically, this *old quantum theory* was unsustainable: neither Newton's mechanics nor Maxwell's wave theory could correctly explain all the observed phenomena of atoms or light.

Again and again, experience from the real world contradicted predictions of both theories. Before anyone else, Einstein probably understood the necessity of a new theory that must supersede Newtonian mechanics and Maxwellian electrodynamics. The earlier theories of Newton and Maxwell would now appear as limiting cases of this new theory, just as Newtonian mechanics is a reasonable approximation of special relativity for speeds much less than the speed of light. But this new theory would involve a break with past concepts even more radical than that required by relativity.

In 1906, Einstein presented the derivation of Planck's distribution that we discussed in Thought Experiment 18.8 (p. 666). He obtained the correct blackbody distribution, despite his use of classical Boltzmann statistics, because he assumed *equal probability* for *all* energy states, even those with different energies (i.e. $0, \hbar\omega, 2\hbar\omega, 3\hbar\omega, \ldots$). *Classically*, however, one would assume equal probabilities *only* for states having equal energy. The equipartition theorem, for example, is a consequence of assuming that all states with the same energy are equally likely to be populated.

---

[326] A. Einstein, "On the theory of light production and absorption," Annalen der Physik, **20** (1906), 199-206. (Translated by Anna Beck in volume 2, Doc. 34, 192-199 of The Collected Papers of Albert Einstein.)

But with the equipartition theorem and classical Boltzmann statistics, as Einstein demonstrated, one can obtain only the Rayleigh-Jeans law.

Now, rather than making the somewhat forced assumption that the frequency of each oscillator in the wall happens to be quantized the right way, one can consider a box of photons, similar to gas particles in a box, and derive a suitable distribution function. This was the model that Indian physicist Satyendra Nath Bose employed when, in 1924, he gave the first derivation of Planck's spectrum that did not rely on *ad hoc* classical assumptions.[327] Now Bose's argument is a standard part of any course on thermodynamics and statistical physics.

Bose's derivation of the blackbody spectrum was impressive because it relied only on quantum assumptions and Einstein's photon model of light. Impressive as they were, however, Millikan's confirmation of Einstein's equation for the photoelectric effect and Bose's derivation remained strong, but not conclusive, pieces of evidence in favor of regarding the photon as physically real. Only after Compton's experiments on the scattering of x-rays by free electrons were most scientists convinced that the particle model of light had to be correct.

Before discussing the Compton experiment, however, we must remind ourselves of why physicists had abandoned Newton's particle model over a century earlier.

Discussion 18.4: The Failure of the Particle Model of Light

A particle model of light has a very difficult time explaining physical optics. If photons do not interact with each other, how can they form interference patterns? It is relatively straightforward, after all, to explain interference as due to the constructive and destructive overlapping of transverse waves propagating toward a screen. It isn't at all obvious, however, how a particle model of light could explain interference patterns, especially if these light particles do not interfere with one another.

If light strikes a glass surface at an angle, for example, why would it not be absorbed and reradiated at a random angle, making the mirror appear white rather than specularly reflective? Similarly, if photons enter an optical medium, should they not be absorbed by atoms, and perhaps subsequently reradiated? And finally, why would particles sometime act in a way that magically reproduce the successful predictions of Maxwell's wave theory, but conveniently act like corpuscles when needed to? Clearly, Einstein's photons cannot be simply some kind of miniature, massless versions of baseballs and pebbles, even with relativistic kinematics.

The inadequacy of a particle model of light can perhaps best be illuminated using Young's classic double slit experiment, which you worked out in Problem 16.3 (p. 517).[328] Recall that this experiment, from a full century before Einstein's 1905 papers, involved shining light through two rectangular slits, which showed exactly the interference pattern predicted by the superposition of waves.

The experiment is illustrated in the figure showing the particle (left) and wave (right) predictions of Newton and Huygens. In order to exhaustively investigate this further, we label the slits (A

---

[327] S.N. Bose, "Planck's law and the light quantum hypothesis," Z. Phys., **26** (1924), 178-181. The English translation of Bose's paper appears in full here: O. Theimer and Budh Ram, "The beginning of quantum statistics," Am. J. Phys., **44** (1976), 1056-1057.

[328] The discussion follows that presented by E.H. Wichmann, Quantum Physics (Volume IV of the Berkeley Physics Course) (New York: McGraw-Hill, 1971), 166-168. Wichmann's text is one of the best sources of physical insights into elementary quantum physics.

and B), assume a frequency of the incident light ω, separation distance b, and long and narrow rectangular slits.

In the twentieth century version, Young's screen is replaced with a very sensitive photomultiplier tube. The intensity is proportional to the counting rate observed with the tube, and measured at various angles φ from the centerline.

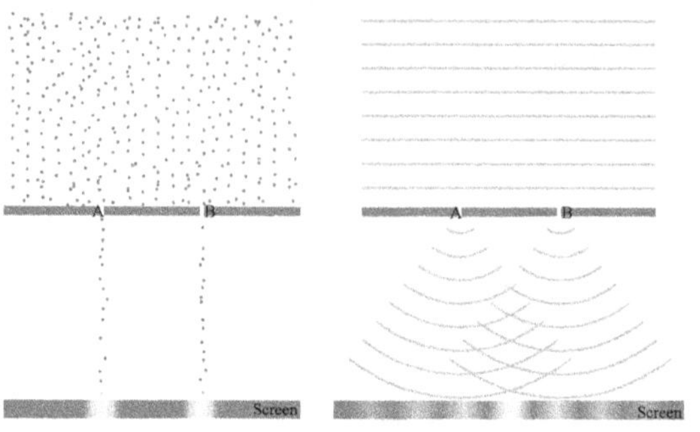

According to classical electromagnetic theory, the intensity distribution should display an interference pattern with alternating bright and dim angles. As you know from Problem 16.3, this is because, while the two cylindrical waves (p. 434) have equal amplitudes, but each wave is retarded in phase by an amount $\Delta\phi = \frac{1}{c}\omega\frac{1}{2}b\sin(\pm\varphi)$ relative to the center. When added together the combined wave becomes:

$$E_2 = \sqrt{\frac{a}{s}}E_0\cos\left(\omega t - \tfrac{1}{c}\omega s - \tfrac{1}{2c}\omega b\sin(-\varphi)\right) + \sqrt{\frac{a}{s}}E_0\cos\left(\omega t - \tfrac{1}{c}\omega s - \tfrac{1}{2c}\omega b\sin(\varphi)\right)$$

$$= 2\cos\left(\tfrac{1}{c}\omega b\sin(\varphi)\right)\left(\sqrt{\frac{a}{s}}E_0\cos\left(\omega t - \tfrac{1}{c}\omega s\right)\right) = 2\cos\left(\tfrac{1}{c}\omega b\sin(\varphi)\right)E_1(s,\varphi),$$

where a is the slit width, s is the cylindrically radial coordinate, and $E_1(s,\varphi)$ is the electric field with only one slit open. Therefore, the intensity of diffracted radiation is given by:

$$\vec{\mathbb{S}}_2(s,\varphi) = \tfrac{1}{\mu_0}\vec{E}\times\vec{B} = \tfrac{1}{Z_0}E^2\hat{s} = 4\cos^2\left(\tfrac{1}{c}\omega b\sin\theta\right)\vec{\mathbb{S}}_1(s,\varphi),$$

where $\vec{\mathbb{S}}_1$ is the intensity of the light with only one slit open.

The factor $4\cos^2\left(\tfrac{1}{c}\omega b\sin\varphi\right)$ is called the *interference term,* because it describes the interference that occurs between waves originating at slits A and B. Thus wave theory predicts high intensity in directions for which φ is an integer multiple of π, and zero intensity at $\varphi = \frac{\pi}{2} + n\pi$. This has been overwhelmingly verified by every experiment since 1800.

What would we expect if light is truly composed of indivisible light quanta that propagate "without dividing" (as Einstein puts it) through empty space? If photons behave like the particles of classical mechanics, then we would expect that if a photon goes through, say, slit A, it will not be influenced by the presence of slit B, and vice versa. Then the intensity with both slits open would be

$$\vec{\mathbb{S}}_2(s,\varphi) = \vec{\mathbb{S}}_A(s,\varphi) + \vec{\mathbb{S}}_B(s,\varphi) = 2\,\vec{\mathbb{S}}_1(s,\varphi),$$

which corresponds to the left hand figure above. This expression completely contradicts experience. So, we understand why leading scientists such as Millikan and Bohr were so skeptical of Einstein's proposed particle model of light.

We next discuss Compton scattering, which convinced most scientists that photons are indeed real. We will return to the double-slit experiment and see how one can make sense of the experimental results using Einstein's light quanta in the section after that.

Problem 18.19: The Double Slit Power

Show that, according to the wave model, the total power let through both slits is the sum of the power let though each slit: $\mathbb{P}_2 = \mathbb{P}_A + \mathbb{P}_B = 2\mathbb{P}_1$.

Discussion 18.5: Compton's Experiment

The scattering of light off other particles is a natural way to test whether the photon can explain observed phenomena. The case of light incident on a metal plate in Millikan's photoelectric effect experiment clearly demonstrates how the photon concept can explain observed facts far better than wave theory. On the other hand, as Einstein himself pointed out, Maxwell's classical wave theory of light is quite successful in explaining an amazingly diverse set of phenomena from interference to the scattering of light by matter. In Chapter 16, we discussed Thomson scattering of a free electron by an electromagnetic wave. We derived both the cross section for Thomson scattering (p. 557), and the Doppler shift (p. 518) that wave theory predicts will arise in the case of a high intensity incident beam of radiation.

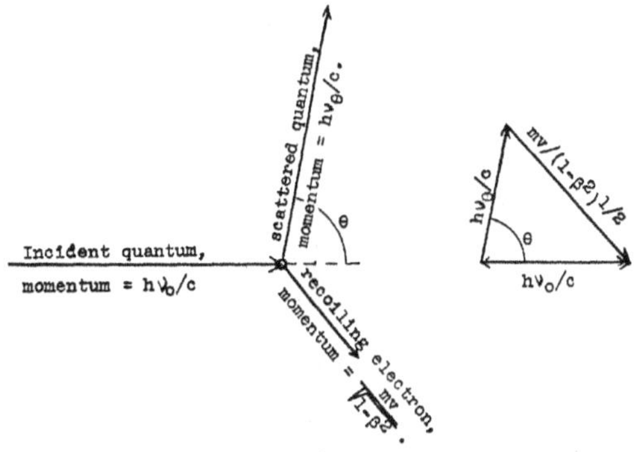

We now discuss the scattering of high-energy photons, such as x-rays, by relativistic electrons. The photon model makes a clear prediction in this case, and Arthur Compton set out to test it. This is called Compton scattering, and it clearly helped convince a majority of physicists that the photon is a real particle – albeit one with no rest mass.

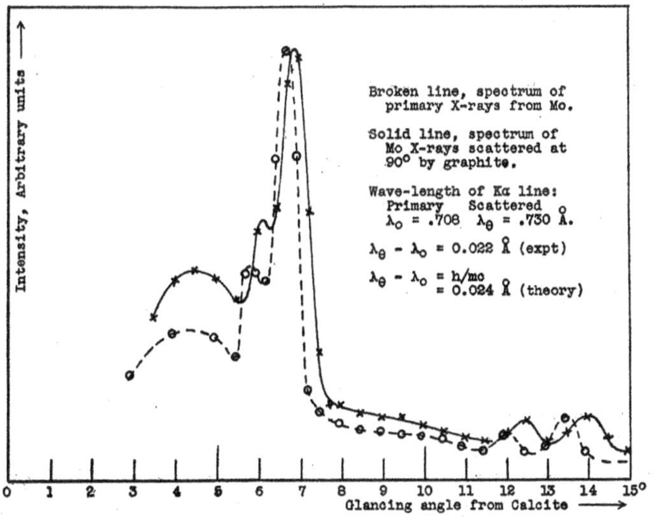

Compton's research on what we now call *Compton scattering* settled the question of the existence of photons, and from then on the question was how to develop a fully quantum theory of light that would reduce to Maxwell's theory under the right circumstances.

Of all the experiments that supported the particle nature of light, the Compton effect was the one that ultimately convinced the majority of scientists that Einstein's photon concept was correct. Early on, researchers using monochromatic x-rays noted that a scattered beam always contained a longer wavelength component in addition to the incident wavelength. In 1923, Compton scattered nearly monochromatic x-rays $(\lambda \approx 1\text{Å})$ off a target of carbon, a material known to contain many loosely bound electrons.[329] Compton measured the intensity of scattered light as a function of wavelength and scattering angle, θ. His results clearly demonstrated that the observed wavelength shift is independent of the source's wavelength, but strongly dependent on the scattering angle, θ. Compton explained his experimental results by modeling the interaction as the elastic collision of two particles. In particular, Compton applied relativistic energy and momentum conservation to the collision of an x-ray photon with a free electron to deduce the correct expression for the change in wavelength as a function of the angle, θ, through which the photon is scattered.

Compton's work provided strong evidence that light can undergo particle-like collisions with both atoms and unbound electrons. Later studies of recoil electrons, and their energies, added further confirmation to the predictions of the theory.[330] Compton's confirmation of Einstein's light quanta made it more, rather than less, urgent for physicists to reconcile photons with the successes of Maxwell's theory. Granted that photons are real, scientists asked, how can a particle model of light ever explain such obviously wave-like phenomena as diffraction and interference?

Thought Experiment 18.11: Compton Scattering

Consider an x-ray photon, whose initial energy and momentum are given by $\mathbb{E}_0 = \hbar\omega$ and $\vec{p}_0 = \frac{1}{c}\hbar\omega\hat{z}$, which is incident on a free electron at rest. Let $\vec{p}$ and $\vec{p}_e$ denote the momentum of the photon and free electron, respectively, after the collision. Since the initial momentum of the electron is zero, conservation of momentum in the scattering plane yields:

$$\vec{p}_0 = p_0\hat{x} = \vec{p} + \vec{p}_e = (p\sin\theta\,\hat{x} + p\cos\theta\,\hat{z}) + (p_e\sin\theta_e\,\hat{x} + p_e\cos\theta_e\,\hat{z}),$$

Similarly, energy conservation implies that:

$$\mathbb{E}_0 + m_e c^2 = \mathbb{E} + \mathbb{E}_e \quad \rightarrow \quad \mathbb{E}_e = m_e c^2 + \mathbb{E}_0 - \mathbb{E}.$$

The scattered electron's energy-momentum 4 vector is therefore:

$$^4\vec{p}_e = \left(\tfrac{1}{c}\mathbb{E}_e, \vec{p}_e\right) = \begin{pmatrix} \frac{1}{c}\mathbb{E}_e \\ p_e\sin\theta_e \\ 0 \\ p_e\cos\theta_e \end{pmatrix} = \begin{pmatrix} m_e c + \frac{1}{c}\mathbb{E}_0 - \frac{1}{c}\mathbb{E} \\ p\cos\theta \\ 0 \\ p\cos\theta - p_0 \end{pmatrix}.$$

Thus, its 4-magnitude must be:

$$\begin{aligned} \left\|{}^4\vec{p}_e\right\| &= -\left(m_e c + \tfrac{1}{c}\mathbb{E}_0 - \tfrac{1}{c}\mathbb{E}\right)^2 + (p\sin\theta)^2 + (p\cos\theta - p_0)^2 \\ &= -(m_e c)^2 - \tfrac{1}{c^2}(\mathbb{E}_0 - \mathbb{E})^2 - 2m_e(\mathbb{E}_0 - \mathbb{E}) + p^2 - 2p_0 p\cos\theta + p_0^2 \\ &= -(m_e c)^2 - \cancel{(p_0 - p)^2} - 2m_e c(p_0 - p) + \cancel{(p_0 - p)^2} + 2p_0 p(1 - \cos\theta). \end{aligned}$$

---

[329] Arthur H. Compton, "A Quantum Theory of the Scattering of X-rays by Light Elements," Physical Review, **21** (1923), 483-502.

[330] A.H. Compton and A.W. Simon, "Directed Quanta of Scattered X-rays," Physical Review, **26** (1925), 289-299.

In its rest frame, the 4-magnitude is simply $-(m_e c)^2$. Simplifying and solving for the momentum of the scattered photon gives:

$$-(m_e c)^2 = -(m_e c)^2 - 2m_e c(p_0 - p) + 2p_0 p(1-\cos\theta)$$
$$0 = -m_e c\, p_0 + m_e c\, p + p_0 p(1-\cos\theta)$$
$$p\left(m_e c + p_0(1-\cos\theta)\right) = m_e c\, p_0$$
$$p = \frac{m_e c\, p_0}{m_e c + p_0(1-\cos\theta)}.$$

We can also write this in terms of the photon energy as:

$$\frac{1}{\mathbb{E}} - \frac{1}{\mathbb{E}_0} = \frac{(1-\cos\theta)}{m_e c^2}.$$

Problem 18.20: The Compton Wavelength

a) Show that in Compton scattering:

$$\lambda = \lambda_0 - \tfrac{2\pi\hbar}{m_e c}(1-\cos\theta) = \lambda_0 - \tfrac{h}{m_e c}(1-\cos\theta) = \lambda_0 - \lambda_c(1-\cos\theta),$$

where the quantity $\lambda_c = 2\pi\hbar/m_e c = 2.43$ pm is called *the Compton wavelength*.

b) Unlike a bound electron, a free electron can neither emit nor absorb a photon: it can only *scatter* photons. Clearly explain why this must be so.

Problem 18.21: The Inverse Compton Effect

One model to explain astronomical gamma-ray production is called the *inverse Compton effect*, where an electron scatters off of a photon, giving the photon much more energy. Assuming this emission mechanism, a gamma-ray astronomer can estimate the energy of relativistic electrons in space.

a) Cosmic microwave background radiation (CMB) follows a blackbody curve, with a corresponding temperature of 2.7 K. What is the most common wavelength, λ_{CMB}, of the light from the CMB?

b) What is the ratio of energy between a 1 GeV gamma ray and a typical CMB photon?

c) A gamma ray is observed with energy $\mathbb{E}_\gamma$, how much total energy $\mathbb{E}_e$ did the initial electron have? Clearly explain your reasoning.

d) Research the field of gamma ray astronomy. Report on some astronomical objects that are believed to produce gamma rays via the inverse Compton effect?

Problem 18.22: The Relativistic Doppler Effect

Previously in this book we have shown that a moving light source will be observed at a different frequency due to the Doppler effect (see pp. 518, 576, & 673). Consider a source of monochromatic light, which emits at a wavelength λ_0, but is moving away from an observer at a velocity v, will be observed at a longer wavelength. In this problem you will show this same result, but under the photon interpretation.

a) For a photon in reference frame $\lfloor S_0$ with a characteristic wavelength of λ_0, write down the energy-momentum 4-vector. Factor out any quantities with units.

b) Perform the appropriate Lorentz boost, so that this same 4-vector is from the point of view of the observer.

c) Define the *redshift*, which astronomers call z, such that $z \equiv \frac{\lambda-\lambda_0}{\lambda_0}$. Show that the redshift can we written as: $z = (\gamma(v) - 1) + \frac{v}{c}\gamma(v) \approx \frac{v}{c}\Big|_{v \ll c}$.

d) How fast is a $z=1$ galaxy receding from us, neglecting any gravitational effects.

e) The distance to galaxies are observed to follow the relationship $d \approx v t_H$, where $t_H \approx 14.4\,\text{Billion Years}$ is called the *Hubble Time*. How many light years away would a $z=1$ galaxy be?

18.4 *Matter Waves*

As we have already discussed, electromagnetic radiation behaves in some circumstances like particles. Photons carry energy in discrete amounts, and are emitted and absorbed as though they were particles, yet they also exhibit wave-like behavior such as is observed when light travels through a Young double-slit apparatus. After Compton's x-ray scattering experiments, few scientists doubted the particle nature of light, but explaining how light could exhibit both particle and wave characteristics remained a puzzle.

Another remarkable fact, which emerged around the time that Compton published the results of his experiments, is that particles such as electrons and neutrons also display wave properties under certain experimental conditions. Experimentally it is found that atoms emit light at only discrete frequencies, just as a stretched string only emits sound waves at a discrete set of frequencies. Bohr's model of the atom, whatever its theoretical shortcomings, successfully accounted for the discrete line spectrum of the hydrogen atom. It might be supposed that a wavelike character of particles would provide the key to understanding atomic spectra. Just as discrete frequencies of a taut string are due to standing waves on the string, so the discrete frequencies of atomic spectra could be due to standing waves within the atoms. So reasoned Louis de Broglie in the early 1920's, whose ideas on the duality between light waves and matter were to have far-reaching consequences.

In this section we introduce de Broglie waves, and the idea that particles behave in a wave-like manner. We will discuss how the idea that particles act as waves came about, and that there must be some sort of probability function. This idea allowed for the easing of the obvious failures of the particle model of light. For, if electrons, which are truly particles, behave as waves then perhaps light is actually a particle and, just perhaps, all the wavelike properties of light really have to do with light being a particle, but a particle that is governed by a *probability density wave.*

Thought Experiment 18.12: de Broglie Waves

Louis de Broglie followed an interesting line of reasoning in 1924 when he postulated that all particles, such as electrons and photons, obey similar relations between conserved quantities like energy and momentum and a characteristic frequency or wavelength. Reasoning by analogy with classical wave packets, de Broglie assumed that each particle of speed v had an associated wave possessing some phase speed less than the speed of light c. This speed v_{phase} is related to v by a simple equation, $\vec{v}\cdot\vec{v}_{\text{phase}}=c^2$., which can be easily derived from the Lorentz invariance of a wave's phase. That a phase speed can be associated with a particle in a consistent way depends on some special features of the relativistic Lorentz transformations. We begin by recalling some results of the transformation properties of plane waves of any kind under Lorentz boosts between inertial frames.

Imagine a wave traveling in an arbitrary direction that is observed in two different inertial frames $\lfloor\underline{S}$ and $\lfloor\underline{S}'$. Imagine that $\lfloor\underline{S}'$ travels with respect to $\lfloor\underline{S}$ at velocity $\vec{u}=u\hat{x}$. Then the three components of the monochromatic wave vector $\vec{k}$ and the frequency ω transform between frames according to the following equations:

$$k'_x=\frac{1}{\sqrt{1-\frac{u^2}{c^2}}}\left(k_x-u\frac{\omega}{c^2}\right),\quad k'_y=k_y,\ k'_z=k_z,\text{ and } \omega'=\frac{1}{\sqrt{1-\frac{u^2}{c^2}}}\left(\omega-u\,k_x\right).$$

These may be expressed more compactly in vector form:

$$\vec{k}'=\gamma(u)\left(\vec{k}-\tfrac{1}{c^2}\vec{u}\,\omega\right)\quad\text{and}\quad \omega'=\gamma(u)\left(\omega-\vec{u}\cdot\vec{k}\right).$$

Note how these equations closely resemble the relativistic transformation laws for momentum and energy of a particle:

$$\vec{p}'=\gamma(u)\left(\vec{p}-\tfrac{1}{c^2}\vec{u}\,\mathbb{E}\right)\quad\text{and}\quad \mathbb{E}'=\gamma(u)\left(\mathbb{E}-\vec{u}\cdot\vec{p}\right).$$

Now for photons, we have that $\mathbb{E}=\hbar\omega$ and, since photons have zero rest mass, we also have $\vec{p}=\frac{1}{c}\mathbb{E}\,\hat{k}=\frac{1}{c}\hbar\omega\,\hat{k}=\hbar\,k$. If particles do indeed have waves associated with them, then a cursory look at the transformations for energy and momentum strongly suggest that the energy and momentum of a freely moving particle (i.e. a particle subject to no net force) must be of the form $\vec{p}\propto\vec{k}$ and $\mathbb{E}\propto\omega$. Let's try to determine the relationship between the phase velocity of the wave and the particle's velocity.

At first, it might be tempting to simply equate the phase velocity of the wave with the velocity of the particle by writing:

$$v_{phase}=\left|\vec{v}\right|=\frac{\left|\vec{p}\right|c^2}{\mathbb{E}}.$$

It is the *group velocity*, however, that we expect to be equal to the velocity of the particle since the group velocity represents the speed with which a wave transports energy. As we noted above, the speed of the particle and the phase speed are therefore related by $\vec{v}\cdot\vec{v}_{\text{phase}}=c^2$..

Now a wave packet generally obeys a dispersion relation of the form $\omega=\omega(\vec{k})$. The group velocity of such a wave is therefore given by:

$$v=\frac{\mathrm{d}\omega}{\mathrm{d}k}=\frac{\mathrm{d}\omega}{\mathrm{d}v}\frac{\mathrm{d}v}{\mathrm{d}k},$$

where we have invoked the chain rule in the last step. The energy of any particle of mass m that moves with speed v is given by

$$\mathbb{E} = mc^2\gamma(v).$$

We now assert, as a hypothesis, that the energy relation for photons is also true for material particles. Therefore:

$$\mathbb{E} = \hbar\omega = mc^2\gamma(v).$$

It therefore follows that:

$$\frac{dk}{dv} = \frac{1}{v}\frac{d\omega}{dv} = \left(\frac{m}{\hbar}\right)\left(1 - \frac{v^2}{c^2}\right),$$

which may be integrated to yield

$$\hbar k = m\int_0^v \left(1 - \frac{v'^2}{c^2}\right)^{-3/2} dv' = \frac{mv}{\sqrt{1 - (v/c)^2}} = mv\gamma(v) = p.$$

If de Broglie's hypothesis is correct, then we have the relations $\mathbb{E} = \hbar\omega$, and $\vec{p} = \hbar\vec{k}$ for matter waves associated with a particle. There should thus be a characteristic wavelength, known as the de Broglie wavelength, associated with every particle of mass m and momentum $p = |\vec{p}|$:

$$\lambda = \frac{2\pi\hbar}{p}.$$

For a non-relativistic particle, $\vec{p} \approx m\vec{v}$, the de Broglie wavelength is:

$$\lambda \approx \frac{2\pi\hbar}{mv} = \frac{2\pi\hbar}{\sqrt{2m\mathbb{T}}},$$

where $\mathbb{T}$ is the kinetic energy of the particle. This was a definite and testable prediction of de Broglie's bold hypothesis—one that would soon receive experimental confirmation.

Thought Experiment 18.13: The Davisson–Germer Experiment

De Broglie predicted that electrons would behave like waves with a wavelength $\lambda = 2\pi\hbar/p$. When these waves enter a crystal with regular lattice spacing, they should scatter and show interference, much as light does when incident on a diffraction grating. In the limiting case when the velocity of the particle is small compared to the speed of light, the de Broglie wavelength of a particle of mass m with kinetic energy $\mathbb{T}$ is:

$$\lambda \approx \frac{2\pi\hbar}{\sqrt{2m\mathbb{T}}} \approx \frac{2\pi\hbar}{mv}.$$

It is clear from this formula that to get the longest possible wavelength, so that it is relatively easy to measure, requires using a particle with the smallest practical mass. The most obvious candidate is the electron, which is the least massive of any elementary particle that carries electric charge and has a nonzero rest mass (neutrinos, which are electrically neutral, have much smaller rest masses than do electrons). If we substitute numbers into the above formula, we obtain the following useful expression:

$$\lambda \approx \frac{2\pi\hbar}{\sqrt{2m\mathbb{T}}} = \left(4.909...\times 10^{-19}\ \mathrm{m^2\,kg\,s^{-1}}\right)\sqrt{\frac{1}{\mathbb{T}}} = \sqrt{\frac{150.4\ \mathrm{eV}}{\mathbb{T}(\mathrm{eV})}}\,\mathrm{\AA},$$

where one Angstrom is denoted by $1\,\mathrm{\AA} = 1\times 10^{-10}\,\mathrm{m}$ and $\mathbb{T}(\mathrm{eV})$ is the particle's kinetic energy expressed in electron volts. If the kinetic energy of the electron is $150.4\ \mathrm{eV}$, for example, then the wavelength will be about $1\times 10^{-10}\,\mathrm{m} = 10\,\mathrm{nm}$, which is about the same order of magnitude as the lattice spacing in a typical crystal. Thus, one might follow the example of x-rays and use a crystal lattice as a diffraction grating.

The American physicists, Clinton J. Davisson and Lester H. Germer, demonstrated the existence of these matter waves in 1927 by observing diffraction peaks in a beam of electrons scattered from a sliced nickel crystal.[331] George P. Thomson, son of J.J. Thomson, independently replicated Davisson's results at about the same time by scattering electrons off gold foil.[332] For their groundbreaking work, Davisson and Thomson shared the Nobel Prize in 1937.

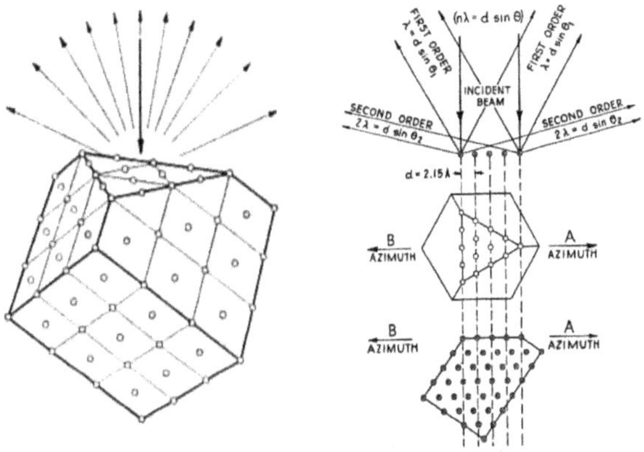

The basic set up for the Davisson-Germer experiment is shown in their diagram. A hot cathode emits electrons, which are accelerated through an electron gun, and then strike a crystal of nickel (Ni) and are diffracted back at an angle ϕ into a Faraday cup detector.[333] We may compute the electrons' momentum if we know the value of the accelerating potential, and the wavelength can be determined from the scattering angle ϕ and the lattice spacing of nickel. Hence, we may determine the value of $\hbar$ for electrons.

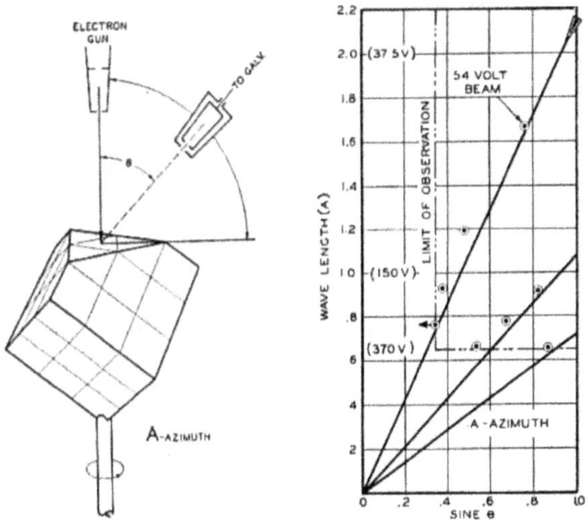

Planes of atoms in the crystal (known as Bragg planes) are regularly spaced and can produce a constructive interference pattern, if the so-called Bragg condition, $n\lambda = 2d\sin\theta = D\sin\phi$, where n is a positive integer, λ is the

---

[331] C.J. Davisson and L.H. Germer, "Diffraction of electrons by a crystal of nickel," Physical Review, **30** (1927), 705-740, and C.J. Davisson, "The Diffraction of Electrons by a Crystal of Nickel, " Bell System Technical Journal, **7** (1) (1927), 90-105.

[332] G.P. Thomson, "Experiments on the diffraction of cathode rays," Proc. of the Royal Soc., **117A** (1928), 600-609.

[333] A Faraday cup is a type of charged particle detector in which electrons simply strike a metal cup or plate, and a current is produced that is directly proportional to the number of electrons incident to the cup. Electron multiplier tubes can also be used as collectors in this experiment.

wavelength of the incident electrons, d is the spacing of atomic planes, θ is the angle between the incident ray and the atomic planes, and D is the distance between atoms in the crystal, is satisfied. This condition basically states that the reflected beams from the planes of atoms in the crystal will constructively interfere if the distance that the wave travels between two successive planes $(2d\sin\theta)$ amounts to an integral number of wavelengths.

The wavelength for the first order $(n=1)$ can therefore be determined from $\lambda_1 = D\sin\theta_1$. The experimental results are presented below in a polar plot of intensity for various accelerating voltages. It is seen that a diffraction maximum occurs for $\mathbb{V}=54V$ and $\theta_1 = 50^\circ$. The lattice spacing for nickel is known from x-ray diffraction experiments to be $D = 2.15\text{Å}$. Therefore,

$$\lambda_1 = 1.65\times10^{-10}\,\text{m},$$

which is in good agreement with the value predicted by de Broglie's relation of $\lambda_1 = 1.67\times10^{-10}\,\text{m}$.

Fundamental Law 11: Wave-Particle Duality

Fundamental particles follow probabilistic laws of physics, with an associated wave vector, $\vec{k}$, that is proportional to the particle's momentum $\vec{p}$, or:

$$\vec{p} = \hbar\vec{k}.$$

In the classical limit, the probability density function governing photons is proportional to the electric, or magnetic, field, of the associated electromagnetic wave.

Discussion 18.6: The Wave Nature of Electrons: What is Waving?

Given that electrons exhibit wavelike properties, the obvious question is what is it, exactly, that is waving? For example, in a mechanical wave it is the medium that actually moves. What is a matter wave? What does it mean when we speak of a wave being associated with a particle? How can we understand a picture of electrons and other particles that show evidence of interference and diffraction? Finally, if light is indeed a particle, how can we reconcile this with Young's double-slit experiment, which would appear to rule out a particle interpretation?

The answer begins with a simple fact: a fraction of an electron is never detected in any experiment. The detector in the Davisson-Germer experiment, or any similar experiment, always records a whole electron, by which we mean an entity with the electron's mass and charge. The same is true for photons: a photon of a specific frequency ω is never split in the sense that a photomultiplier detects only a fraction of the energy $\hbar\omega$. To see the relevance of this fact, let us return to the Young double-slit experiment, but instead consider its application to electrons.

If we substitute a monoenergetic beam of electrons for monochromatic light, and charged particle detectors for the photomultiplier tubes, we obtain results similar to what is observed in the case of light. Electrons passing through the double-slit apparatus display an interference pattern![334] With only one slit is open, we observe pretty much what we would expect for a beam of minute charged baseballs: a maximum intensity in the direction nearest that of the open slit,

---

[334] In fact, the most precise version of this experiment has been recently performed (2013) and the results are consistent with what we outline here. See R. Bach *et al*, "Controlled double-slit electron diffraction," New Journal of Physics, (2013) 15: 033018 for further details.

with a gentle roll off on either side of this maximum. With both slits open, however, we see a similar interference pattern as we did for light, or mathematically:

$$N_2(r,\theta) = \left(N_1(r,\theta)\right)4\cos^2\left(\tfrac{1}{2c}\omega b\sin\theta\right),$$

where $N_1 = N_1(r,\theta)$ is the number of electrons that are observed with only one slit open, b is the separation distance between the two slits, λ is the de Broglie wavelength for the electrons, and θ is the angle of observation. What accounts for this striking result?

Perhaps there is some mutual interaction that occurs between the electrons as they pass through the two slits? To test this hypothesis, we can lower the incident electron beam current until only one electron at a time passes through, and then record the hits and misses at various angles. There are two experimental possibilities: (1) the interference pattern goes away, and our count distribution follows:

$$N_2(r,\theta) = 2N_1(r,\theta);$$

or (2) the interference pattern remains as is:

$$N_2(r,\theta) = \left(2N_1(r,\theta)\right)\left(2\cos^2\left(\tfrac{1}{2c}\omega b\sin\theta\right)\right).$$

All experiments have been definitive; the interference pattern remains! Electrons clearly are not waves, at least in the classical sense, but they do seem to obey laws of motion which are somehow wave-like. Despite the fact that each electron yields a single count in a detector, both slits must somehow participate in the passage of each electron through the apparatus, since the diffraction pattern would otherwise not exist.

Next we could try to determine through which slit the single electron travels by placing a minimally invasive detector at each slit. When this is done, the interference pattern goes away, and possibility (1) is now observed. The very act of experimentally measuring which slit the electron goes through destroys the interference pattern, and the results change from possibility (1) to possibility (2). From the point of view of common sense (i.e. classical physics), this make no sense. Though electrons are detected as whole particles, they obviously do not behave classically. Likewise, photons do not behave classically either, even taking relativity into account.

Thus, while it may not be possible to predict the trajectory of a particular electron, we should be able to find the probability that an electron will be detected at each position. Recall that with light, we measure the relative intensity versus observation angle, so perhaps we can make a useful analogy here. The light intensity $\mathbb{S}$ at each point is proportional to the square of a wave amplitude vector E, which can be represented by the real, or imaginary, part of a complex phasor function:

$$\vec{E} = \vec{E}_0 e^{i(kx-\omega t)}.$$

By analogy, we define a wave amplitude function for an electron $\Psi(\vec{r},t)$ and give it the following interpretation: the modulus square of the wave amplitude $|\Psi|^2$ is proportional to the probability that the electron is located at position $\vec{r}$ at time t. For this reason, the wave function Ψ is often called the *probability amplitude*, which we can write as

$$\Psi(r,t) = \Psi_0\, e^{i(\vec{k}\cdot\vec{r}-\omega t)} = \Psi_0\, e^{i(\vec{r}\cdot\vec{p}-Et)/h},$$

where we have employed the relationships $p = \hbar k$ and $\mathbb{E} = \hbar\omega$. The probability amplitude function is inherently complex, unlike the real-valued electric and magnetic fields of classical electrodynamics.

To account for interference effects, we extend the analogy between light and electrons, by assuming that the principle of superposition is also valid for matter waves. Thus, when light is coherent, we add the amplitudes first, and then square the resultant amplitude to obtain the intensity:

$$\mathbb{S}_2 \propto \left(\vec{E}_{\mathrm{A}} + \vec{E}_{\mathrm{B}}\right)^2 \propto E_{\mathrm{A}}^2 + E_{\mathrm{B}}^2 + 2\vec{E}_{\mathrm{A}} \cdot \vec{E}_{\mathrm{B}}.$$

On the other hand, when light is incoherent, such as in the case of light from an incandescent light bulb, the resultant intensity is simply:

$$\mathbb{S}_2 \propto E_{\mathrm{A}}^2 + E_{\mathrm{B}}^2.$$

Recall Problem 18.19 (p. 679), where you showed that the same total power was transmitted in both cases.

Let's now apply the principle of superposition to the double slit experiment with electrons. Let Ψ_{A} be the probability amplitude for finding a particle at a particular location when only slit A is open, and Ψ_{B} the probability amplitude when only slit B is open. If we apply the principle of superposition to the electron double slit experiment with both slits open, then the probability amplitude with both slits open will be:

$$\Psi_2 = \Psi_{\mathrm{A}} + \Psi_{\mathrm{B}},$$

and the corresponding probability function of position is:

$$P_2 \propto |\Psi_2|^2 = |\Psi_{\mathrm{A}} + \Psi_{\mathrm{B}}|^2 = |\Psi_{\mathrm{A}}|^2 + |\Psi_{\mathrm{B}}|^2 + \Psi_{\mathrm{A}}^*\Psi_{\mathrm{B}} + \Psi_{\mathrm{B}}^*\Psi_{\mathrm{A}}.$$

The last two terms are interference terms, which depend upon the relative phases of the two waves.

We have discussed complex numbers in regard to impedance in Section 13.3 and radio interferometry in Thought Experiment 16.11. In both of these cases, we defined both the read and imaginary part of a two dimensional vector quantity, called a *phasor*, based on how it is measured. In neither of these cases were complex numbers inherent to understanding the physics, but rather they were a product of the sinusoidal nature of the topic being discussed. Quantum wave mechanics is mathematically similar, but physically different. Rather than measuring the real and imaginary, components of a complex quantity, we infer the complex wave function from measurements of mathematically real quantities.

We can shed further light on the idea of a probability wave if we return to electrodynamics and associate a probability amplitude with the photon.

Problem 18.23: Complex Identities

A complex number can be expressed either in Cartesian or polar form, as:

$$\Psi = \mathrm{Re}(\Psi) + i\,\mathrm{Im}(\Psi) = |\Psi| e^{i\phi}.$$

The complex conjugate of a complex number is defined such that the imaginary quantity becomes negative. Notice that that if $i^2=-1$, then also $(-i)^2=-1$. It is written with an asterisk and defined such that:

$$\Psi^*=\mathrm{Re}(\Psi)-i\,\mathrm{Im}(\Psi).$$

Using these two definitions, and Euler's equation $\left(e^{i\phi}=\cos\phi+i\sin\phi\right)$:

a) Show that if $\Psi=|\Psi|e^{i\phi}$ then $\Psi^*=|\Psi|e^{-i\phi}$.

b) Show that $|\Psi|^2=|\Psi^*|^2=\Psi\Psi^*$.

c) Show that $\tan\phi=\dfrac{\mathrm{Im}(\Psi)}{\mathrm{Re}(\Psi)}$.

d) Show that, for two complex numbers, $\Psi_A\Psi_B^*=\left(\Psi_A^*\Psi_B\right)^*$.

e) Show that $\Psi_A\Psi_B^*+\Psi_A^*\Psi_B$ is a real number and equal to $2\,\mathrm{Re}\left(\Psi_A^*\Psi_B\right)$, and also equal to $2\,\mathrm{Re}\left(\Psi_A\Psi_B^*\right)$.

f) Show that that $\Psi_A\Psi_B^*-\Psi_A^*\Psi_B$ is an imaginary number and equal to $2i\,\mathrm{Im}\left(\Psi_A^*\Psi_B\right)$, and also equal to $-2i\,\mathrm{Im}\left(\Psi_A\Psi_B^*\right)$.

g) Show that $\Psi_A\Psi_B^*+\Psi_A^*\Psi_B=2|\Psi_A||\Psi_B|\cos(\phi_A-\phi_B)$.

h) Prove the identity: $\cos(\alpha-\beta)=\cos(\alpha)\cos(\beta)+\sin(\alpha)\sin(\beta)$.

i) Prove the identity: $\sin(\alpha-\beta)=\sin(\alpha)\cos(\beta)-\cos(\alpha)\sin(\beta)$.

Thought Experiment 18.10: The Probability Density Function of Light

Let us now consider the possibility that light is a particle, just like the electron. How can we interpret the associated electric and magnetic fields? To investigate this, and related questions that arise, consider a classical monochromatic plane wave with an electric field:

$$\vec{E}=\vec{E}_0\sin\left(\omega t-\vec{k}\cdot\vec{r}\right),$$

where $\vec{k}=\frac{1}{c}\omega\hat{k}$ is the wave vector.

The instantaneous Poynting vector is therefore given by:

$$\vec{\mathbb{S}}(z,t)=\tfrac{1}{\mu_0}\vec{E}(z,t)\times\vec{B}(z,t)=\tfrac{1}{\mu_0}\vec{E}\times\left(\tfrac{1}{c}\hat{k}\times\vec{E}\right)=\tfrac{1}{Z_0}E^2\,\hat{k}=\tfrac{1}{Z_0}E_0^{\,2}\sin^2\left(\omega t-\vec{k}\cdot\vec{r}\right)\hat{k}.$$

We discussed above that if there are a large number of photons, at any instant in time, the photon density would be given by:

$$n=\frac{\mathbb{u}}{\mathbb{E}}=\frac{\mathbb{S}}{c\mathbb{E}}=\frac{E^2}{Z_0\hbar c\omega}=\frac{E_0^{\,2}}{Z_0\hbar c\omega}\sin^2\left(\omega t-\vec{k}\cdot\vec{r}\right).$$

What if there are just a few photons? Perhaps only one, but moving in the $\hat{k}$ direction at the speed of light? If such a photon were a classical particle, the probability of its coming into contact with something else in a given place $\vec{r}$ at a given time t would be unity if $\vec{r} = ct\hat{k}$, and zero if that were not the case, assuming you knew that it passed through the origin exactly at time zero.

To be more specific, consider a classical particle of a particular size. For the sake of argument, we will assume it to be a cube of volume $a \times a \times a$. So as long as the position is within a distance a of $ct\hat{k}$, in any direction, the classical particle would come in contact with the location $\vec{r}$.

Now we will put this concept to use and find the probability density function of the photon. In the case of a light source shining in the $\vec{k}$ direction, the probability that a photon would be found within a volume element $a \times a \times a$ at a given time t is:

$$P = n\delta V = na^3 = \frac{E_0^{\ 2}}{Z_0 \hbar c \omega} \sin^2\left(\omega t - \vec{k}\cdot\vec{r}\right) a^3 .$$

Of course, there is one catch with this. Sources of single photons have a given probability in a given time of emitting the photon (e.g. the Einstein A coefficient), and we do not know when that was until after it becomes detected at a given location at a given time.

Since the square of the electric field is proportional to the intensity of light, we can now put forward a way to interpret Young's double slit experiment using the particle model of light. The *probability* that a photon will strike the detector at a specific location is proportional to the classically calculated diffraction intensity.

We may, therefore, think of $\vec{E}(\vec{r},t)$ and $\vec{B}(\vec{r},t)$ as probability amplitudes, and the sum of their squares will no longer be interpreted as representing an energy density that pervades all of space. Instead, the square of a wave amplitude gives the probability that a photon of energy $\hbar\omega$ is located at some region of space. If a photon is detected, the energy delivered to the detector is equal to $\hbar\omega$. Therefore, the classical energy density integrated over some region, must be the product of the photon's energy with the probability of finding the photon in that region.

Suppose, for example, that we switch on a steady light source and keep it on for a long time. Because the source emits a large number of photons, the *average energy* that can be detected in some region is indeed equal to the classically computed energy in the region. This vindicates Einstein's suggestion that we use results of Maxwell's theory as valid statements about *average quantities*, and is consistent with the Bohr correspondence principle.

Now suppose the light source is nearly monochromatic and that it emits, on average, N photons per second of frequency ω. If we place a detector some distance away from the light source, we will detect some fraction of the total energy flux emitted by the source and, from our discussion above, the probability of a photon entering the detector is simply the following:

$$P = \frac{\mathbb{E}_{\text{Classical}}}{\hbar\omega} = \frac{1}{Z_0 \hbar c \omega} \int \left\langle \left|\vec{E}(\vec{r},t)\right|^2 \right\rangle \mathrm{d}V ,$$

where $\mathbb{E}_{\text{Classical}}$ is the classically computed energy, $\vec{E}$ is the electric field, and the angular brackets indicate a time average:

$$\langle E^2 \rangle = \frac{1}{T}\int_0^T E^2 \, dt \, .$$

For each photon emitted by the source we cannot predict whether or not a photon with actually be counted, but we can assert that the probability for this to happen is given by P. If the counter clicks, then the amount of energy transmitted from the source to the photodetector equals $\hbar\omega$. It follows that the average power transmitted from the light source to the detector, when the source is kept steady, is given by $\mathbb{P} = PN\hbar\omega = N\mathbb{E}_{\text{Classical}}$.

The classically calculated energy is proportional to the inverse square of the distance between source and detector. In classical field theory, we interpret this $1/r^2$ dependence as a decrease in the electromagnetic energy density due to the energy being spread out over an ever greater volume. With the particle model of light, we use the relation above to point out that the probability is also proportional to $1/r^2$. Since the counting rate of the detector equals the product of the probability P and number N of photons emitted per second, the counting rate itself falls off as $1/r^2$.

Problem 18.24: Photons and Probability Waves

a) Consider a monochromatic beam of light of frequency ω and energy flux $\vec{\mathbb{S}}$. What is the photon number density of this beam?

b) For the same monochromatic beam, what is the probability, $P_{A\cdot t}$, that a single photon will pass through an area δA within a time interval δt?

c) Classically, for the same monochromatic coherent light, the Poynting vector is given by $\vec{\mathbb{S}} = \frac{c}{Z_0}\vec{E}\times\vec{B} = \frac{1}{Z_0}E^2\hat{k}$. Express the probability that a single photon will pass through an area δA within a time interval δt, $P_{A\cdot t}$, in terms of the classical electric field E.

d) If we assume some sort of quantum field, Ψ, such that the probability per are per time, $P_{A\cdot t}$, is given by $P_{A\cdot t} = \Psi\Psi^*$, express the function Ψ in terms of the electric field. What would be its units?

Discussion 18.7: Photons, Interference, and Young's Experiment

Now let's return to Young's double slit apparatus, and ask, again, if we can somehow keep both slits open, and yet determine through which slit a photon travels.

Let's suppose we place detectors very close to the apertures, at a distance much less than the separation between the slits, so that we can determine which aperture the photon goes through. What do we find? The photon does indeed go through one slit or the other: one of the detectors records a hit; the other does not. The detector that records a hit receives the full energy of the photon, so the photon has not been split in any way as it traveled through the apparatus. Moreover, approximately equal numbers of photons go through each slit. There is catch, however. Since, the photon is a massless particle, it must be absorbed in order to be detected. Therefore, it is impossible to both measure it traveling through the slit and the diffraction pattern.

Massive particles, however, do not have this problem. We can measure add a detector that only absorbs a small fraction of the kinetic energy of a massive particle. What happens then? As we mentioned briefly with regard to electrons, this experiment has been performed using a variety of particles, including electrons, neutrons and atoms. The conclusion is always the same. Every attempt at determining which slit the particle goes through, while simultaneously trying to preserve the interference pattern, fails! If we measure which slit the particle goes through, we destroy the interference pattern in the process, even when only a tiny fraction of each a particle's energy is changed by the detector.

Here is the essence of wave-particle duality, and the necessity of adopting a probability interpretation of physical phenomena. If we arrange to detect through which aperture a particle, be it electron, photon, or neutron, passes, then the probability is simply 50% that the particle will travel through one slit and not the other. If, on the other had, we do not measure through which slit a particle comes, we get the interference pattern characteristic of double slit diffraction.

From a field point of view this makes some sense, since one could imagine the electron field traveling through both slits simultaneously. This, of course, raises the question of why electrons act like particles at all. A more complete answer began to emerge in the late 1920s and throughout the 1930s, and comes under the heading of *quantum field theory*. From this point of view, rather than particles exhibiting wave light properties, *quantum fields* exhibit particle-like behavior. In the next section, we briefly describe modern theories regarding the nature of electrons and photons.

18.5 *The New Electrodynamics*

With the advent of Heisenberg's matrix mechanics, Schrödinger's famous equation for the time evolution of particle wave functions, and Born's interpretation of these functions as probability amplitudes, the basic, non-relativistic, quantum theory of matter was completed by 1927. Electron spin was also introduced as an *ad hoc* explanation of the observed facts, and Heisenberg completed the foundations of the subject when he announced the *uncertainty principle*, also called *the principle of indeterminacy*. This latter idea completely altered physicists' conception of nature, and forced them to abandon the dream that one could, at least in principle, simultaneously measure such quantities as a system's position and momentum with infinite accuracy. In a certain sense, the uncertainty principle prescribes the limit for using such common sense notions as trajectories or waves in describing how the fundamental things of this world behave. Trajectories and waves do not really exist; they are convenient abstractions, fictions, or approximations, in describing reality. Heisenberg's principle of indeterminacy tells us to what extent we can make use of these notions before we run afoul on the shoals of subatomic reality.

As radical, and successful, as the new quantum mechanics was, however, it addressed only the behavior of *matter*, and even then material particles were handled non-relativistically. Schrödinger's equation for the time development of the probability amplitude Ψ,

$$i\hbar\frac{\partial\Psi}{\partial t}=\frac{-\hbar^2}{2m}\vec{\nabla}^2\Psi+\mathbb{U}(r,t)\Psi \quad ,$$

where $\mathbb{U}$ is the potential energy, accurately describes the behavior of electrons for situations that do not require the full machinery of relativity, but it is a partial differential equation that is second-order in the spatial coordinates and only first-order in time. Thus space and time do not enter on equal footing as required by relativity. In addition, the question of how to account for the particle-like properties of the electromagnetic field is left unanswered in the original quantum mechanics of Heisenberg and Schrödinger. In the standard quantum mechanics taught in undergraduate courses, radiation is treated semi-classically. This means that the photon is introduced as an auxiliary concept as needed (in energy transfers, for example), when one wishes to invoke $\mathbb{E} = \hbar\omega$ or $\vec{p} = \hbar\vec{k}$, but that otherwise Maxwell's wave theory is still assumed to be valid. While this is an acceptable approximation for a wide array of practical problems, it is hardly satisfactory from a theoretical perspective. Schrödinger's equation does not tell us how the atom transitions from one state to another, nor does it allow us to calculate the transition probabilities, such as the Einstein coefficients, that we need to determine the intensities of spectral lines.[335] We need a truly quantum theory of light to calculate these quantities correctly. Einstein's photon model was one of the main things that inaugurated the search for a complete and independent quantum theory, so it is certainly unsatisfactory to leave matters alone without addressing the quantization of light.

Dirac, Heisenberg, Pauli, and Fock, among others, began to address this question shortly after the standard non-relativistic quantum theory was completed. To obtain a coherent quantum theory of light requires a mathematically systemic way of *quantizing* the electromagnetic field.

Let's first outline the elements of a crude theory that allows us to account for some of the phenomena that we discussed in the previous sections of this chapter. We will then discuss how Dirac developed this theory further to fashion our modern theory of *quantum electrodynamics*. The main elements of a basic theory are: [336]

- Monochromatic light consists of discrete entities, or quanta, which we call photons, each of which carries energy $\hbar\omega$.

- Maxwell's equations correctly describe the propagation of a photon through space. This means that a photon can be defined by the electric and magnetic fields $\vec{E}$ and $\vec{B}$, which satisfy Maxwell's equations. Furthermore, each photon can be thought of as a kind of wave packet or disturbance that undergoes diffraction, when passing through an interference device such as a double slit apparatus.

- The sum $E^2 + B^2$ does not represent the energy density of a photon or electromagnetic field, but is instead proportional to the probability density that a photon will be detected if we try to observe it with a device such as a photomultiplier tube.

- Whenever a photon is detected, it delivers its entire energy $\hbar\omega$ to the detector. As we mentioned previously, if we keep a steady source of light turned on for a sufficiently long time, so that many, many photons are present, then the volume integral of $E^2 + B^2$ is proportional to the average energy of the light in that volume.

Two things that really distinguish this theory from classical physics are the apparent wave-particle duality displayed by the photon and the way that probability is applied to explain optical phenomena.

The first person to describe a successful procedure for quantizing the electromagnetic field was Dirac, who invented (or discovered?) quantum electrodynamics in 1927 by replacing the standard classical model of the electromagnetic field as an infinite set of classical, charged harmonic

---

[335] Standard non-relativistic quantum mechanics does provide a method for calculating transition probabilities, or transition rates, from an energy state to a *continuum* of energy states due to a perturbation. The general form of the formula applies to atomic transitions, nuclear decay, and scattering, and is known as Fermi's golden rule. But the method is not applicable to processes such as spontaneous emission.

[336] E.H. Wichmann, Quantum Physics (Volume IV of the Berkeley Physics Course) (New York: McGraw-Hill, 1971), 169-170.

oscillators, or *normal modes*, with quantum oscillators. According to Dirac's picture, photons are simply *excitations* of the quantum oscillators, which is analogous to the peak amplitude response of a driven harmonic oscillator near resonant frequency which acts as an excitation of the oscillator. His procedure was pretty straightforward, and the quantum theory of fields isn't necessarily difficult to understand conceptually, but the mathematical details can be very difficult.

In the 1940s, physicists such as Richard Feynman, Julian Schwinger, Sin-Itiro Tomonaga, and Freeman Dyson fixed many technical mathematical problems that afflicted quantum electrodynamics from its inception, but we will not discuss the details here. Despite technical difficulties, the quantum theory of electrodynamics scored notable successes. Dirac was able to calculate Einstein's coefficients using his new theory, for example, which was one of the earliest triumphs for quantum electrodynamics. Using quantum field theory, Wolfgang Pauli in 1940 proved the so-called *spin-statistics theorem*, which asserts that spin one-half particles obey Fermi-Dirac statistics, and therefore obey the Pauli exclusion principle, while particles with integer spin, such as photons, conform to the statistics of Bose and Einstein. Quantum electrodynamics is one the most rigorously tested, and therefore successful, theories in all of the physical sciences.

Thought Experiment 18.14: A Square Box of Light

Imagine that electromagnetic radiation fills a cubical box with sides of length L and volume $V = L^3$.

In the absence of source currents or charges, the vector potential $\vec{\mathbb{A}}$ satisfies the wave equation:

$$\vec{\nabla}^2\vec{\mathbb{A}} - \frac{1}{c^2}\frac{\partial^2}{\partial t^2}\vec{\mathbb{A}} = 0 .$$

For simplicity, we impose periodic boundary conditions at each face:

$\vec{\mathbb{A}}(x=0,y,z,t) = \vec{\mathbb{A}}(x=L,y,z,t)$, or $\vec{\mathbb{A}}(x,y=0,z,t) = \vec{\mathbb{A}}(x,y=L,z,t)$, etc.,

which sets the wave vector, $\vec{k} = k_x\hat{x} + k_y\hat{y} + k_z\hat{z}$, to:

$$k_x = \frac{2\pi}{L}n_x\ ,\ k_y = \frac{2\pi}{L}n_y\ ,\ k_z = \frac{2\pi}{L}n_z\ , \text{ where } n_x, n_y, n_z = 0, \pm 1, \pm 2, \ldots .$$

A particular solution to the wave equation takes the form:

$$\vec{\mathbb{A}}(\vec{r},t) \propto \left(\vec{\varepsilon}_m(\vec{k})\right)\cos(\omega t - \vec{k}\cdot\vec{r}) ,$$

where we note that $\vec{k}\cdot\vec{r} = k_x x + k_y y + k_z z$. The unit vectors $\vec{\varepsilon}_m(\vec{k})$ are linearly independent and mutually perpendicular to $\vec{k}$: $\vec{k}\cdot\vec{\varepsilon}_m(\vec{k}) = 0$, where $m = 1, 2$.. These are called the *circular polarization vectors*, and it is often convenient to express $\vec{\mathbb{A}}$ using $\vec{\varepsilon}_m(\vec{k})$ in place of the Cartesian unit vectors.

Problem 18.25: Blackbody Radiation and Wave-Particle Duality

From the Planck distribution function, $\mathbb{u}_\omega(\omega)$, on page 667:

a) Make a variable substitution to find the energy density per photon energy interval, $\mathbb{u}_{\mathbb{E}}(\mathbb{E})$. What are the dimensions of this quantity?

b) Find the number density of photons per photon energy interval $n_{\mathbb{E}}$. What are the dimensions of this quantity?

c) Consider a three-dimensional cube whose sides equals a single wavelength λ. How many photons, per photon energy interval, are contained in this box as a function of photon energy? What are the dimensions of this quantity?

d) Using your last result, find the energy bandpass, $\Delta\mathbb{E}$, which contains exactly one photon.

e) How much time does it take for a photon to traverse this box? Use the energy-time version of the Heisenberg Uncertainty Principle, $\delta\mathbb{E}\,\delta t \geq \frac{1}{2}\hbar$, to find the uncertainty in the energy as a function of the photon energy. Notice that this will be an inequality.

f) Finally, substitute $\Delta\mathbb{E}$, from part (d), in for $\delta\mathbb{E}$, from part (e), not the other way around. Find an inequality relationship relating the temperature to the photon energy, which corresponds to the Heisenberg Uncertainty Principle. Solve for the photon energy as a function of temperature.

g) Discuss what the inequality you found in part (f) means physically. How does this relate the so-called wave-particle duality?

Problem 18.26: Polarization Vectors

a) Show that, when $\hat{k} = \hat{z}$, the two linearly independent polarization unit vectors are related to the other two Cartesian unit vectors via:

$$\vec{\varepsilon}_1(\vec{k}) = \frac{-1}{\sqrt{2}}\left(\hat{x} + i\hat{y}\right) \text{ and } \vec{\varepsilon}_2(\vec{k}) = \frac{1}{\sqrt{2}}\left(\hat{x} - i\hat{y}\right).$$

b) Consider the classical plane wave, written in complex form, as:

$$\vec{\mathbb{A}}(\vec{r},t) = \mathbb{A}_0 \operatorname{Re}\left(\sum_{m=1}^{2}\left(\vec{\varepsilon}_m(\vec{k})\right)\cos(\omega t - \vec{k}\cdot\vec{r})\right).$$

Express this in Cartesian coordinates with no complex numbers.

c) Draw the electric field, looking at the source, in the two special cases where $\vec{\varepsilon}_1(\vec{k}) = 0$ and $\vec{\varepsilon}_2(\vec{k}) = 0$. Explain how one represents left circular polarization and the other represents right circular polarization (p. 490).

d) Most astronomical sources of radio waves are unpolarized, some are linearly polarized, and none (that we know of) are circularly polarized. Modern radio telescopes, however, use pairs of helical feeds, rather than linear feeds. Discuss how this is advantageous for calibration when observing a, somewhat, linearly polarized source.

Thought Experiment 18.15: Quantizing the Electromagnetic Field

The standing wave in the square box could also be thought of as photons in the box. In the classical limit these must match. By equating the time averaged classical electromagnetic energy density to the energy per volume of the photons, we could show that the proper normalization constant for the general solution in is given by:

$$\mathbb{A}_0 = \sqrt{2Z_0 \frac{N}{V}\frac{\hbar c}{\omega}}.$$

Now, for one quanta, and for each circular polarization, we therefore have:

$$\vec{\mathbb{A}}(\vec{r},t)=\sqrt{\frac{2Z_0\hbar c}{\omega V}}\,\vec{\varepsilon}_m(\vec{k})\cos(\omega t-\vec{k}\cdot\vec{r}).$$

Furthermore, since Euler's formula assures us that $\exp(ix)=\cos(x)+i\sin(x)$, we can also write the classical vector potential $\vec{\mathbb{A}}(\vec{r},t)$ in terms of $\exp(\pm i(\omega t-\vec{k}\cdot\vec{r}))$ for each wave vector $\vec{k}$:

$$\vec{\mathbb{A}}(\vec{r},t)=\sqrt{\frac{Z_0\hbar c}{2\omega V}}\,\varepsilon_m(\vec{k})\left(e^{i(\omega t-\vec{k}\cdot\vec{r})}+e^{-i(\omega t-\vec{k}\cdot\vec{r})}\right).$$

From this we could derive the electric and magnetic fields via:

$$\vec{E}(\vec{r},t)=-\frac{\partial}{\partial t}\vec{\mathbb{A}}(\vec{r},t) \text{ and } \vec{B}(\vec{r},t)=\vec{\nabla}\times\vec{\mathbb{A}}(\vec{r},t).$$

Except for the normalization constant, which we found by invoking the relationship $\mathbb{E}=\hbar\omega$ for a single photon, these results are perfectly consistent with classical electrodynamics.

Derivation 18.1: The Quantum Mechanical Equations of Motion

Here we derive a similar relationship, in general, for completeness, and for the benefit of the reader who is already familiar with formal quantum mechanics. In particular, we will show that the vector potential is simply the quantum mechanical wave function, Ψ, and that photons are simply eigenstates of that function.

So that we can generalize to more arbitrary wave functions, we expand the vector potential as an infinite Fourier series over the indices k_x, k_y, and k_z and treat the time-dependent part of $\vec{\mathbb{A}}$ as an unknown for which we must solve. We write:

$$\vec{\mathbb{A}}(\vec{r},t)=\sum_{\vec{k}}\sum_{m=1}^{2}\left[\vec{\varepsilon}_m(\vec{k})\sqrt{\frac{\hbar}{2\varepsilon_0\omega V}}\left(a(\vec{k},t)e^{i\vec{k}\cdot\vec{r}}+a^*(\vec{k},t)e^{-i\vec{k}\cdot\vec{r}}\right)\right].$$

The functions $a(\vec{k},t)$ or $a^*(\vec{k},t)$ are the vector potential amplitudes, or Fourier coefficients of $\vec{\mathbb{A}}$, that are as yet undetermined (the coefficient $a^*(\vec{k},t)$ is the complex conjugate of $a(\vec{k},t)$.) Then the wave equation for $\vec{\mathbb{A}}(\vec{r},t)$,

$$\vec{\nabla}^2\vec{\mathbb{A}}-\frac{1}{c^2}\frac{\partial^2}{\partial t^2}\vec{\mathbb{A}}=0,$$

gives equations that each coefficient, $a(\vec{k},t)$ or $a^*(\vec{k},t)$, must satisfy. To see that this is so, note that since:

$$\vec{\nabla}\left(e^{+i\vec{k}\cdot\vec{r}}\right)=i\vec{k}e^{+i\vec{k}\cdot\vec{r}},$$

and

$$\vec{\nabla}^2\left(e^{+i\vec{k}\cdot\vec{r}}\right)=(ik)(ik)\left(e^{+i\vec{k}\cdot\vec{r}}\right)=-k^2\left(e^{+i\vec{k}\cdot\vec{r}}\right),$$

we can use the wave equation for $\vec{\mathbb{A}}$ to write the following equations for the Fourier coefficients:

$$-k^2\, a(\vec{k},t) - \frac{1}{c^2}\frac{\partial^2 a(\vec{k},t)}{\partial t^2} = 0\,.$$

These equations are identical to those for a classical harmonic oscillator, whose solutions take the form $a(\vec{k},t) \propto e^{-i\omega(\vec{k})t}$, with the dispersion relationship $\omega = c|\vec{k}|$. Thus, these equations and their solutions are the equations of motion and solutions for a mechanical system having an infinite number of degrees of freedom. These can be taken as a classical analogy to the electromagnetic field in free space.

We can quantize this system by assuming that the amplitudes a are quantum mechanical operators, which satisfy appropriate commutation relations, and the amplitudes a^* are their adjoints. Thus, in quantum field theory, the vector potential now becomes a vector field operator:

$$\hat{\mathbb{A}}(\vec{r},t) = \sum_{\vec{k}}\sum_{m}\left[\vec{\varepsilon}_m(\vec{k})\sqrt{\frac{\hbar}{2\varepsilon_0\omega V}}\left(\hat{a}(\vec{k},t)e^{i\vec{k}\cdot\vec{r}} + \hat{a}^*(\vec{k},t)e^{-i\vec{k}\cdot\vec{r}}\right)\right],$$

in which the amplitudes, $\hat{a}(\vec{k},t)$ and $\hat{a}^*(\vec{k},t)$, of the $\vec{k}-th$ field mode satisfy the equations for a set of quantum, rather than classical, oscillators.

For photons, we can show that $\hat{a}$ and $\hat{a}^*$ are the lowering and raising operators that are used in treating the harmonic oscillator in standard, non-relativistic quantum mechanics. In this context, the raising operator $\hat{a}$ adds one photon while the lowering operator $\hat{a}^*$ removes a photon. Recall from a standard quantum course that the commutation relation for these operators is given by:

$$\left[\hat{a},\hat{a}^*\right] = \hat{a}\hat{a}^* - \hat{a}^*\hat{a} = 1\,.$$

Furthermore, if we denote a state, or *normalized eigenstate*, by the *ket vector* $|n\rangle$, where n is the number of photons present, we have the following eigenvalue relations:

$$\hat{a}^*|n\rangle = (n+1)^{1/2}|n+1\rangle,\ \text{and}\ \hat{a}^*|n\rangle = n^{1/2}|n-1\rangle\,.$$

The normalization factors $\sqrt{n+1}$ and $\sqrt{n}$ are derived for a quantum oscillator in any standard text on quantum theory.

We define the number operator $\hat{N} = \hat{a}^*\hat{a}$, which has the property that:

$$\hat{N}|n\rangle = \hat{a}^*(\hat{a}|n\rangle) = \hat{a}^*(n+1)^{1/2}|n+1\rangle = (n+1)^{1/2}(\hat{a}^*|n+1\rangle) = n|n\rangle\,.$$

Hence, $\hat{\mathbb{A}}(\vec{r},t)$ is now a vector field operator that obeys quantum physics. The fields are easily derived from $\hat{\mathbb{A}}(\vec{r},t)$:

$$\vec{E}(\vec{r},t) = i\sum_{\vec{k}}\sum_{m}\vec{\varepsilon}_m(\vec{k})\sqrt{\frac{\hbar}{2\varepsilon_0\omega V}}\left[\hat{a}*(k,t)\exp(-i\vec{k}\cdot\vec{r}) - \hat{a}(k,t)\exp(i\vec{k}\cdot\vec{r})\right]$$

$$\vec{B}(\vec{r},t) = \vec{\nabla}\times\hat{\mathbb{A}}(\vec{r},t) = i\left(\vec{k}\times\vec{\varepsilon}_m(\vec{k})\right)\sqrt{\frac{\hbar}{2\varepsilon_0\omega V}}\left[\hat{a}*(k,t)e^{-i\vec{k}\cdot\vec{r}} - \hat{a}(k,t)e^{i\vec{k}\cdot\vec{r}}\right].$$

The fields are also operators, which have a discrete energy spectrum:

$\hbar\omega(\vec{k})\left(n(\vec{k})+\frac{1}{2}\right)$, with $n(\vec{k})=0,1,2,\ldots,$

just like the quantum harmonic oscillator problem, where the frequency associated with the $\vec{k}-th$ mode is: $\omega(\vec{k})=c\left|\vec{k}\right|=c\sqrt{k_x^2+k_y^2+k_z^2}\geq 0$.

The integer $n(\vec{k})$ is the number of quanta, or photons, in the $k-th$ mode. Notice that even in the vacuum state, where $n(\vec{k})=0$, each mode has a ground state energy $\frac{1}{2}\hbar\omega(\vec{k})$. This is the *zero-point energy* that is characteristic of the quantum mechanical oscillator, and must exist if Heisenberg's uncertainty relation holds true.

Problem 18.27 : Mechanism of Light Propagation

Consider an electromagnetic wave with an energy flux vector $\vec{\mathbb{S}}$. Explain qualitatively the method in which the energy is transferred from one place to another. For each model, discuss not only the physical mechanism of the energy transfer, but also any implications regarding the ideas of action at a distance:

(a) Newton's corpuscular model of light,

(b) Huygens's wave model of light,

(c) Maxwell's transverse wave theory,

(d) Einstein's relativistic but classical model of light,

(e) Einstein's photon model of light, and

(f) modern theories of quantum electrodynamics.

Discussion 18.8: Is a Photon a Particle After All?

Is a photon a particle that mimics a propagating electromagnetic field when many photons propagate collectively? Alternatively, is the electromagnetic field the fundamental quantity with the photon simply representing discrete energy states?

From Thought Experiment 18.15 (p. 694) and Derivation 18.1 (p. 696) would say that the latter is the better interpretation. Each photon is an *excitation* of the field because the photon's energy elevates the energy of the field above the ground state. The simplest excited state is the *one-photon state*, in which just one mode is excited to its first excited state, so that the energy of the field (above the ground state energy) is just $\hbar\omega(\vec{k})$. Two photons appear when two modes are excited, and the energy is the sum $\hbar\omega(\vec{k})+\hbar\omega(\vec{k}')$, where the wave vector denotes the mode. If one mode is excited to its second excited state then the energy is $2\hbar\omega$, indicating the presence of two identical photons (each of energy $\hbar\omega$). Photons, then, are not really particles in any classical sense of the word; they are excitations of the quantized electromagnetic field.

If photons aren't really particles, it seems obvious to ask, then how do we explain the fact that a photomultiplier tube, photocell, or photographic screen, such as we might use in a double-slit experiment, appears to detect light quanta as point-like particles? Haven't we previously asserted that detectors observe only entire electrons and photons? The answer is found in the process of detection itself. The flashes seen on a photographic plate or the readout current from a photo detector are not the simple, single particle events they appear to be, but are instead the result of a sequence of events initiated by interactions of a single quantum with many atoms of the screen or the detector. Each photon interacts with a portion of a fluorescent screen, for example, and though this appears focused into a point image, it actually involves the response of a multitude of atoms and molecules. It is true that when a quantum field, such as the quantized electric field, interacts with matter, it delivers all of its energy as a single quantum, but it doesn't necessarily mean that the quantum is a point particle. In other words, the localization that is observed is characteristic of the detector, which is made of localized atoms, rather than the detected quanta.[337]

On the other hand, if we start with the premise that an electron is a particle, then what makes a photon any different? Is this argument not also true for an electron, or any other fundamental particle for that matter? Are they not all simply eigenstates of their governing field? Therefore, one could argue that light really is a particle after all.

This debate is largely moot because the question is based on a false dichotomy. In fact, a quantum mechanical system behaves neither like a classical particle nor a classical wave. It is not classical at all. Quantum mechanics does not make the distinction between particles and waves, as they are simply analogies for the benefit of macroscopic people.

Where you come down on the metaphysical debate about particles or fields probably depends on your style of doing physics. Are you, like many Continental physicists in the eighteenth century, persuaded by mathematical beauty? Are you more like Newton himself, and bothered by a lack of a mechanism relating cause to effect? Are you like Hertz, and care about observable predictions, but not the philosophical upshots? Are you like Einstein, and care deeply about philosophical inconsistencies?

Regardless of whether it should be called a particle, or a state of the electromagnetic field, a photon has the following particle-like properties:

1. It has a definite frequency ω and a definite wave vector $\vec{k}$.

2. Its energy is $\mathbb{E} = \hbar\omega$, and its momentum is $\vec{p} = \hbar\vec{k}$. The energy and momentum of a photon satisfy the dispersion relation $\mathbb{E} = |\vec{p}|c = pc$, which is characteristic of particles with zero rest mass.

3. The photon carries an angular momentum of $\vec{L} = \pm\hbar\hat{k}$, which correspond to the two states of circular polarization.

---

337 As Art Hobson puts it in his article "There are no particles, there are only fields," Am. J. Phys. **81**, (2013) 211–223, just because a big balloon interacts locally with a tiny needle doesn't mean that the balloon is as localized as the needle.

Plate 33: The Chandra X-ray Observatory[338]

[338] Photos of the Chandra X-ray Observatory. The top photo shows the whole mirror assembly, and the bottom photo shows the coaxial nested mirrors. The mirrors can only reflect X-rays at grazing angles, so the x-rays enter one end of the tube, skip off the conical mirrors, and exit the other end. For all but the very brightest objects, no more than one x-ray ever lands on the same camera pixel per clock cycle. Unlike in an optical camera, therefore, the energy per pixel represents the energy per photon. Chandra also has an optional diffraction gration for gathering high resolution spectra of bright poit sources. Photos are from www.nasa.gov, and information about Chandra, and access to data, can be found at chandra.harvard.edu and cxc.harvard.edu.

Part IV: Appendices

During the nineteenth century the separate fields of electricity, magnetism, and optics unified, ultimately resulting in Einstein's theory of relativity and the beginnings of quantum electrodynamics. Along the way there were many twists and turns, but also many historical gems that highlight the beauty of the process of unlocking nature's secrets. Many of these treasures worked there way into our storyline, but others simply did not make the final cut, including three of our favorite primary source documents—so we include them as the first three appendices of the seven shown in the table.

On the one hand, Michael Faraday discovered that magnetic fields can rotate the polarization angle of linearly polarized light through a medium (Faraday rotation), so we may assume that he made a strong connection between optics and magnetism. On the other hand, Faraday was a staunch empiricist, so he was reluctant to speculate on matters that he did not measure. Therefore, we were delighted to find a discussion on ray vibrations that is quite nuanced by Faraday, which we present in Appendix A.

Soon after becoming Lord Kelvin, William Thomson wrote a delightful preface to the English translation of Heinrich Hertz's treatise, which most clearly illustrates the Maxwellian view of how light propagates over long distances through space. Keep in mind as you read Appendix B, however, that a few years later Kelvin would give a series of lectures in America questioning the aether's very existence. This ability to be open-minded, even about a belief he held dearly, is clearly one of the things that made Kelvin such an excellent scientist.

Perhaps the most interesting documents we found was an English translation of a lecture Henri Poincaré gave in 1904 at the St. Louis World's fair. This is the best illustration we have found of the state of theoretical physics immediately before Albert Einstein's meraculous year, so we include it as Appendix C. This is not a coincidence, however, since the Einstein participated in a book group that closely studied Poincaré's work.

Appendix D explains the electrostatic, or Gaussian, system of units that are still common in many branches of physics. The next two appendices contain useful mathematical definitions and identities in the same style as the physical relations in the inside covers of the text.

We often found it difficult to keep straight the chronology of discoveries, and so we have added a timeline as Appendix G.

Finally, Appendix H is this book's index. However unlike most text books, our index only refers to historical figures. You should, however, refer to the long table of contents in the front of the book to quickly find more technical material.

Appendix A Faraday on Ray Vibrations

James Clerk Maxwell saw his work as a direct extension of that of Michael Faraday, and as such the later reader of Maxwell's comprehensive treatise on electricity and magnetism might assume Faraday strongly believed in an aether or some sort of medium that would convey all electromagnetic phenomena including light. In his writings, however, Faraday remained cautious about making conclusions that he could not support experimentally, though he sometimes speculated on more aethereal matters. Shortly after his discovery of Faraday rotation, which gave scientists their first clear evidence that light is somehow related to electricity and magnetism, Faraday wrote a letter titled "Thoughts on Ray-Vibrations" to his friend Richard Phillips, which shows that Faraday had a nuanced view of the aether and the nature of light as an electromagnetic phenomenon. For example, he considered an alternative to the aether theory as early as the 1840s. Regarding a lecture he had just given, Faraday wrote the following to his friend Phillips:

> The point intended to be set forth for consideration of the hearers was, whether it was not possible that vibrations which in a certain theory are assumed to account for radiation and radiant phenomena may not occur in the lines of force which connect particles, and consequently masses of matter together; a notion which as far as is admitted, will dispense with the aether, which in another view, is supposed to be the medium in which these vibrations take place.

Faraday's notions were based on a forerunner to the atomic hypothesis put forward by the physicist Roger Boscovich (1711-1787), who speculated that atoms were force centers devoid of matter. Faraday alludes to this in his letter to Phillips:

> You are aware of the speculation which I some time since uttered respecting that view of the nature of matter which considers its ultimate atoms as centres of force, and not as so many little bodies surrounded by forces, the bodies being considered in the abstract as independent of the forces and capable of existing without them...The consideration of matter under this view gradually led me to look at the lines of force as being perhaps the seat of vibrations of radiant phenomena.

Faraday compares the speed of light in a vacuum to the speed with which electrical effects travel in a wire, and suggests a connection between matter and radiation:

> Another consideration bearing conjointly on the hypothetical view both of matter and radiation, arises from the comparison of the velocities with which the radiant action and certain powers of matter are transmitted. The velocity of light through space is about 190,000 miles in a second; the velocity of electricity is, by the experiments of Wheatstone, shown to be as great as this, if not greater: the light is supposed to be transmitted by vibrations through an aether which is, so to speak, destitute of gravitation, but infinite in elasticity; the electricity is transmitted through a small metallic wire, and is often viewed as transmitted by vibrations also.
>
> That the electric transference depends on the forces or powers of the matter of the wire can hardly be doubted, when we consider the different conductibility of the various metallic and other bodies...The power of electric conduction (being a transmission of force equal in velocity to that of light) appears to be tied up in and dependent upon the properties of the matter, and is, as it were, existent in them.

He then compares the standard theories that build ordinary matter and aether out of material atoms, or "nuclei," with his and Boscovich's conception of such atoms as mere "centers of force" that are "devoid of matter":

> I suppose we may compare together the matter of the aether and ordinary matter (as, for instance, the copper of the wire through which the electricity is conducted), and consider them as alike in their essential constitution; i.e. either as both composed of little nuclei, considered in the abstract

as matter, and of force or power associated with these nuclei, or else both consisting of mere centres of force, according to Boscovich's theory and the view put forth in my speculation; for there is no reason to assume that the nuclei are more requisite in the one case than in the other. It is true that the copper gravitates and the aether does not, and that therefore the copper is ponderable and the aether is not; but that cannot indicate the presence of nuclei in the copper more than in the aether, for of all the powers of matter gravitation is the one in which the force extends to the greatest possible distance from the supposed nucleus, being infinite in relation to the size of the latter, and reducing the nucleus to a mere centre of force. The smallest atom of matter on the earth acts directly on the smallest atom of matter in the sun, though they are 95,000,000 miles apart; further, atoms which, to our knowledge, are at least nineteen times that distance, and indeed in cometary masses, far more, are in a similar way tied together by the lines of force extending from and belonging to each. What is there in the condition of the particles of the supposed aether, if there be even only one such particle between us and the sun, that can in subtlety and extent compare to this?

In experimental philosophy we can, by the phenomena presented, recognize various kinds of lines of force; thus there are the lines of gravitating force, those of electro-static induction, those of magnetic action, and others partaking of a dynamic character might be perhaps included. The lines of electric and magnetic action are by many considered as exerted through space like the lines of gravitating force. For my own part, I incline to believe that when there are intervening particles of matter (being themselves only centres of force), they take part in carrying on the force through the line, but that when there are none, the line proceeds through space. Whatever the view adopted respecting them may be, we can, at all events, affect these lines of force in a manner which may be conceived as partaking of the nature of a shake or lateral vibration. For suppose two bodies, A B, distant from each other and under mutual action, and therefore connected by lines of force, and let us fix our attention upon one resultant of force, having an invariable direction as regards space; if one of the bodies move in the least degree right or left, or if its power be shifted for a moment within the mass (neither of these cases being difficult to realise if A and B be either electric or magnetic bodies), then an effect equivalent to a lateral disturbance will take place in the resultant upon which we are fixing our attention; for, either it will increase in force whilst the neighboring results are diminishing, or it will fall in force as they are increasing.

It may be asked, what lines of force are there in nature which are fitted to convey such an action and supply for the vibrating theory the place of the aether? I do not pretend to answer this question with any confidence; all I can say is, that I do not perceive in any part of space, whether (to use the common phrase) vacant or filled with matter, anything but forces and the lines in which they are exerted. The lines of weight or gravitating force are, certainly, extensive enough to answer in this respect any demand made upon them by radiant phenomena; and so, probably, are the lines of magnetic force: and then who can forget that Mossotti has shown that gravitation, aggregation, electric force, and electro-chemical action may all have one common connection or origin; and so, in their actions at a distance, may have in common that infinite scope which some of these actions are known to possess? [339]

The view which I am so bold to put forth considers, therefore, radiation as a kind of species of vibration in the lines of force which are known to connect particles and also masses of matter together. It endeavors to dismiss the aether, but not the vibration. The kind of vibration which, I believe, can alone account for the wonderful, varied, and beautiful phenomena of polarization, is not the same as that which occurs on the surface of disturbed water, or the waves of sound in

---

[339] Faraday is referring to a theory advanced by the Italian physicist Ottaviano-Fabrizio Mossotti (1791–1863), which purportedly derived the gravitational and electrical force laws from a more general law. The theory ultimately proved unsuccessful, but Faraday considered it for a while. For a further discussion of Mossotti's theory, see James F. Woodward, "Early attempts at a Unitary Understanding of Nature," published in Old and New Questions in Physics, Cosmology, Philosophy, and Theoretical Biology, ed. Awyn van der Merwe, (New York: Plenum Press, 1983), 886-894 [885-908].

> gases or liquids, for the vibrations in these cases are direct, or to and from the centre of action, whereas the former are lateral. It seems to me, that the resultant of two or more lines of force is in an apt condition for that action which may be considered as equivalent to a lateral vibration; whereas a uniform medium, like the aether, does not appear apt, or more apt than air or water.
>
> The occurrence of a change at one end of a line of force easily suggests a consequent change at the other. The propagation of light, and therefore probably of all radiant action, occupies *time*; and, that a vibration of the line of force should account for the phenomena of radiation, it is necessary that such vibration should occupy time also. I am not aware whether there are any data by which it has been, or could be ascertained whether such a power as gravitation acts without occupying time, or whether lines of force being already in existence, such a lateral disturbance at one end as I have suggested above, would require time, or must of necessity be felt instantly at the other end.

Faraday explains how his conception of lines of force would obviate the need for a the hypothesized aether:

> The aether is assumed as pervading all bodies as well as space: in the view now set forth, it is the forces of the atomic centres which pervade (and make) all bodies, and also penetrate all space. As regards space, the difference is, that the aether presents successive parts of centres of action, and the present supposition only lines of action; as regards matter, the difference is, that the aether lies between the particles and so carries on the vibrations, whilst as respects the supposition, it is by the lines of force between the centres of the particles that the vibration is continued. As to the difference in intensity of action within matter under the two views, I suppose it will be very difficult to draw any conclusion, for when we take the simplest state of common matter and that which most nearly causes it to approximate to the condition of the aether, namely the state of the rare gas, how soon do we find in its elasticity and the mutual repulsion of its particles, a departure from the law, that the action is inversely as the square of the distance! [340]

---

[340] M. Faraday, "Experimental Researches in Electricity," Vol. III , 447-452, and M. Faraday, Philosophical Magazine, S.3, Vol XXVIII, N188, May 1846.

Appendix B Kelvin's Preface to Hertz's Treatise

Heinrich Hertz, in 1881, confirmed Maxwell's theory experimentally by generating and detecting radio waves in the laboratory and demonstrating that these waves behaved exactly like visible light, exhibiting properties such as reflection, refraction, diffraction, and interference.[341]

Not only did Maxwell's theory explain experiment, but it also made philosophical sense, as it answered the fundamental question of action at a distance:

> If something is transmitted from one particle to another at distance, what is its condition after it has left the one particle and before it has reached the other?[342]

The movement of real charge caused the medium to deform. This in turn caused the fields to react, which again caused the medium to deform, so on and so forth. Just like sound in air or waves on water. A charge in one location could now affect one in another location, using a mechanism that made causal sense. Spooky action at a distance would haunt physics no more, thanks to the heroic work of Faraday, Maxwell, both Thomsons, Heaviside, and other scrappy subjects of the British empire!

In 1893, Heinrich Hertz's detailed confirmation of Maxwell's theory came out in English, with a beautifully written, yet pompous, forward by the eminent Lord Kelvin,[343] which we reproduce in full here:

> To fully appreciate the work now offered to the English reading public, we must carry our minds back two hundred years to the time when Newton made known to the world the law of universal gravitation. The idea that the sun pulls Jupiter, and Jupiter pulls back against the sun with equal force, and that the sun, earth, moon, and planets all act on one another with mutual attractions seemed to violate the supposed philosophic principle that matter cannot act where it is not the explanation of the motions of the planets by a mechanism of crystal cycles and epicycles seemed natural and intelligible, and the improvement on this mechanism invented by Descartes in his vortices was no doubt quite satisfactory to some of the greatest of Newton's scientific contemporaries. Descartes's doctrine died hard among the mathematicians and philosophers of continental Europe; and for the first quarter of last century belief in universal gravitation was an insularity of our countrymen.
>
> Voltaire, referring to a visit which he made to England in 1727, wrote: "A Frenchman who arrives in London finds a great alteration in philosophy, as in other things. He left the world full; he finds it empty. At Paris you see the universe composed of vortices of subtle matter; at London we see nothing of the kind. With you it is the pressure of the moon which causes the tides of the sea; in England it is the sea which gravitates towards the moon.... You will observe also that the sun, which in France has nothing to do with the business, here comes in for a quarter of it. Among you Cartesians all is done by impulsion; with the Newtonians it is done by an attraction of which we know the cause no better."1 Indeed, the Newtonian opinions had scarcely any disciples in France till Voltaire asserted their claims on his return from England in 1728. Till then, as he himself says, there were not twenty Newtonians out of England.

---

341 F.K. Vreeland and Henri Poincaré, Maxwell's theory and wireless telegraphy, Part 1: Maxwell's theory and Hertzian oscillations, (New York: McGraw Publishing Co., 1904), 31-45.

342 Maxwell's 3rd ed. -- Article 866.

343 Heinrich Hertz, Researches on the Propagation of Electric Action with Finite Velocity Through Space, translated by D.E. Jones, forward written by W. Thomson (London: MacMillan and Co., 1893).

In the second quarter of the century sentiment and opinion in France, Germany, Switzerland, an Italy experienced a great change. 'The mathematical prize questions proposed by the French Academy naturally brought the two sets of opinions into conflict.' A Cartesian memoir of John Bernoulli was the one which gained the prize in 1730. It not infrequently happened that the Academy, as if desirous to show its impartiality, divided the prize between Cartesians and Newtonians. Thus, in 1734, the question being the cause of the inclination of the orbits of the planets, the prize was shared between John Bernoulli, whose memoir was founded on the system of vortices, and his son Daniel, who was a Newtonian. The last act of homage of this kind to the Cartesian system was performed in 1740, when the prize on the question of the tides was distributed between Daniel Bernoulli, Euler, Maclaurin, and Cavallieri; the last of whom had tied to amend and patch up the Cartesian hypothesis on this subject.

On the 4th February 1744 Daniel Bernoulli wrote as follows to Euler: "Uebrigens glaube ich, dass der Aether sowohl gravis versus solem, als die Luft versus terram sey, und kann Ihnen nicht Bergen, dass ich über diese Puncte ein völliger Newtonianer bin, und verwundere ich mich, dass sie den Principiis Cartesianis so lang adhäriren; es möchte wohl einige Passion vielleicht mit unterlaufen. Hat Gott Können eine animam, deren Natur uns unbegreiflich ist, erschaffen, so hat er auch können eine attractionem universalem materiae imprimiren, wenn gleich solche attraction supra captum ist, da hingegen die Principia Cartesiana allzeit contra captum etwas involviren."

Here the writer, expressing wonder that Euler had so long adhered to the Cartesian principles, declares himself a thorough-going Newtonian, not merely in respect to gravitation versus vortices, but in believing that matter may have been created simply with the law of universal attraction without the aid of any gravific medium or mechanism. But in this he was more Newtonian than Newton himself.

Indeed Newton was not a Newtonian, according to Daniel Bernoulli's idea of Newtonianism, for in his letter to Bentley of date 25th February 1692, he wrote: "That gravity should be innate, inherent, and essential to matter, so that one body may act upon another at a distance through a vacuum without the mediation of anything else, by and through which their action and force may be conveyed from one to another, is to me so great an absurdity that I believe no man who has in philosophical matters a competent faculty of thinking, can ever fall into it." Thus Newton, in giving out his great law, did not abandon the idea that matter cannot act where it is not. In respect, however, merely of philosophic thought, we must feel that Daniel Bernoulli was right; we can conceive the sun attracting Jupiter, and Jupiter attracting the sun, without any intermediate medium, if they are ordered to do so. But the question remains—Are they so ordered? Nevertheless, I believe all, or nearly all, his scientific contemporaries agreed with Daniel Bernoulli in answering this question affirmatively. Very soon after the middle of the eighteenth century Father Boscovich gave his brilliant doctrine (if infinitely improbable theory) that elastic rigidity of solids, the elasticity of compressible liquids and gases, the attractions of chemical affinity and cohesion, the forces of electricity and magnetism; in short, all the properties of matter except heat, which he attributed to a sulphureous essence, are to be explained by mutual attractions and repulsions, varying solely with distances, between mathematical points endowed also, each of them, with inertia. Before the end of the eighteenth century the idea of action-at-a-distance through absolute vacuum had become so firmly established, and Boscovich's theory so unqualifiedly accepted as a reality, that the idea of gravitational force or electric force or magnetic force being propagated through and by a medium, seemed as wild to the naturalists and mathematicians of one hundred years ago as action-at-a-distance had seemed to Newton and his contemporaries one hundred years earlier. But a retrogression from the eighteenth century school of science set in early in the nineteenth century.

Faraday, with his curved lines of electric force, and his dielectric efficiency of air and of liquid and solid insulators, resuscitated the idea of a medium through which, and not only through which but by which, forces of attraction or repulsion, seemingly acting at a distance, are transmitted. The long struggle of the first half of the eighteenth century was not merely on the question of a medium to serve for gravific mechanism, but on the correctness of the Newtonian law of gravitation as a matter of fact however explained. The corresponding controversy in the nineteenth century was very short, an fait soon became obvious that Faraday's idea of the transmission of electric force by a medium not only did not violate Coulomb's law of relation between force an distance, but that, if real, it must give a thorough explanation of that law. Nevertheless, after Faraday's discovery of the different specific inductive capacities of different insulators, twenty years passed before it was generally accepted in continental Europe. But before his death, in 1867, he had succeeded in inspiring the rising generation of the scientific world with something approaching to faith that electric force is transmitted by a medium call ether, of which, as had been believed by the whole scientific world for forty years, light and radiant heat are transverse vibrations. Faraday himself did not rest with this theory for electricity alone. The very last time I saw him at work in the royal Institution was in an underground cellar, which he had chosen for freedom from disturbance; and he was arranging experiments to test the time of propagation of magnetic force from an electromagnet through a distance of many yards of air to a fine steel needle polished to reflect light; but no result come from those experiments. About the same time or soon after, certainly not long before the end of his working time, he was engaged (I believe at the shot tower near Waterloo Bridge on the Surrey side) in efforts to discover relations between gravity and magnetism, which also led to no result.

Absolutely nothing has hitherto been done for gravity either by experiment or observation towards deciding between Newton and Bernoulli, as to the question of its propagation through a medium, and up to the present time we have no light, even so much as to point a way for investigation in that direction. But for electricity and magnetism Faraday's anticipations and Clerk-Maxwell's splendidly developed theory have been established on the sure basis of experiment by Hertz's work, of which his own most interesting account is now presented to the English reader by his translator, Professor D. E. Jones. It is interesting to know, as Hertz explains in his introduction, and it is very important in respect to the experimental demonstration of magnetic waves to which he was led, that he began his electric researches in a problem happily put before him thirteen years ago by Professor von Helmholtz, of which the object was to find by experiment some relation between electromagnetic forces an dielectric polarization of insulators, without, in the first place, any idea of discovering a progressive propagation of those forces through space.

It was by sheer perseverance in philosophical experimenting that Hertz was led to discover (VII., p. 107 below) a finite velocity of propagation of electromagnetic action, and then to pass on the electromagnetic waves in air and their reflection (VIII.), and to be able to say, as he says in a short reviewing sentence at the end of VIII.: "Certainly it is a fascinating idea that the processes in air which we have been investigating, represent to us on a million-fold larger scale the same processes which go on in the neighbourhood of a Fresnel mirror or between the glass plates used for exhibiting Newton's rings."

Professor Oliver Lodge has done well, in connection with Hertz's work, to call attention to old experiments, and ideas taken from them, by Joseph Henry, which came more nearly to an experimental demonstration of electromagnetic waves than anything that had been done previously. Indeed Henry, after describing experiments showing powerful enough induction due to a single spark from the prime conductor of an electric machine to magnetize steel needles at a distance of 30 feet in a cellar beneath with two floors and ceilings intervening, says that he is "disposed to adopt the hypothesis of an electrical plenum," and concludes with a short reviewing sentence, "It may be further inferred that the diffusion of motion in this case is almost comparable with that of a spark from a flint and steel in the case of light."

Professor Oliver Lodge himself did admirable work in his investigations regarding lightning rods, coming very near to experimental demonstration of electromagnetic waves; and he drew important lesson regarding "electrical surgings" in an insulated bar of metal "induced by Maxwell's and Heaviside's electromagnetic waves, " and many other corresponding phenomena manifested both in ingenious and excellent experiments devised by himself and in natural effects of lightening.

Of electrical surgings or waves in a short insulated wire, and of interference between ordinary and reflected waves, and positive electricity appearing where negative might have been expected, we hear first it seems in Herr von Bezold's "Researches on the Electric Discharge" (1870), which Hertz gives as the Third Paper in the present series, with interesting and ample recognition of its importance in relation to his own great work.

Readers of the present volume will, I am sure, be pleased if I call their attention to two papers by Prof. G. F. Fitzgerald which I heard myself at the meeting of th British Association at Southport in 1883. One of them is entitled, "On a Method of producing Electromagnetic Disturbances of comparatively Short Wave-length." The paper itself is not long, and I quote it here in full, as it appeared in the Report of the British Association, 1883; "This is by utilizing the alternating currents produced when an accumulator is discharged through a small resistance. It is possible to produce waves of as little as two metres wave-length, or even less." This was a brilliant and useful suggestion. Hertz, not knowing of it, used the method; and, making as little as possible of the "accumulator," got waves of as little as twenty-four centimetres wave-length in many of his fundamental experiments. The title alone of the other paper, "On the Energy lost by Radiation from Alternating Currents," is in itself a valuable lesson in the electromagnetic theory of light, or the undulatory theory of magnetic disturbance. The reader of the present volume will be interested in comparing it with the title of Hertz's Eleventh Paper; but I cannot refer to this paper without expressing the admiration and delight with which I see the words "rectilinear propagation," "polarization," "reflection," "refraction," appearing in it as sub-titles.

During the fifty-six years which have passed since Faraday first offended physical mathematicians with his curved lines of force, many workers and many thinkers have helped to build up the nineteenth-century school of plenum, one ether for light, heat, electricity, magnetism; and the German and English volumes containing Hertz's electrical papers, given to the world in the last decade of the century, will be a permanent monument of the splendid consummation now realized.

KELVIN.

Appendix C Henri Poincaré's 1904 Lecture

In 1904, the city of St. Louis hosted the world's fair with many grand events including the third modern Olympics and the International Congress of Arts and Sciences. Under the Normative Science Division, the section on Applied Mathematics contained two speakers who were invited to discuss the connection between natural philosophy and applied mathematics: Ludwig Boltzmann and Henri Poincaré.

Poincaré's talk is particularly insightful, as he gives a brief history of nineteenth century physics, followed by a discussion of current problems in theoretical physics and then by a discussion of the future of physics, which we only reproduce in part.

Poincaré beautifully explains a condensed history of early nineteenth century theoretical physics, beginning with a discussion of the Laplacian view of mathematical physics. He talks about the astronomical view of matter that all microscopic forces must follow gravitational-style force laws. While this method showed early promise, Poincaré makes it clear that its reliance on central force laws severely limited the scope and creativity of the Laplacians.

Next, Poincaré reviews the fundamental principles of physics, including the principle of relativity, the conservation of energy and the conservation of mass. He discusses how much more powerful this reasoning was, as compared to the central force methods of the Laplacians.

But there are now, as of 1904, contradictions in physics. All the principles cannot be true, and so there is a crisis in mathematical physics. Poincaré articulates this crisis beautifully, and he cites some recent ideas to resolve this crisis. All of the current investigations to which he refers foreshadow the work of Einstein. The section on the conservation of mass (*Lavoisier's Principle*), for example, discusses different ideas of mass, and different types of mass, all of which we now know was barking up the wrong tree. Einstein, after all, is just about to show that $E = mc^2$.

As you read Poincaré's talk, appreciate how well he understood the theoretical problems that were dogging classical physics at the turn of the century. Also keep in mind that of Einstein's famous four 1905 papers, three were accepted relatively quickly. These papers directly address Poincaré's present crisis. They solve the problem of *Brownian motion*, restore the principle of relativity as fundamental, and combine the conservation of mass with that of energy.

Moreover, notice the complete absence of any discussion of the problem of blackbody radiation or the photoelectric effect. These problems were the topic of Einstein's first 1905 paper, the one on the *photoelectric effect*. It is understandable that Einstein's contemporaries regarded this paper a reckless attempt to solve a non-problem with a sweeping and unsupported hypothesis. Later, of course, he would be awarded the Nobel Prize for his work on the photoelectric effect, precisely because he not only found a solution to a known problem, but also discovered the problem in the first place.

The Present and the Future of Mathematical Physics

Professor Henri Poincaré

(Talk in association with the 1904 world fair.)

What is the present state of mathematical physics? What are its problems? What is its future? Will the object and methods of this science appear in ten years as they appear to us now? Or are we to witness a far-reaching transformation? These are the questions we are forced to face today at the outset of our inquiry.

This is easy to ask, but difficult to answer. If we feel tempted to hazard a prediction, we should stop to think of the nonsense the most eminent scholars of a hundred years ago would have spoken in answer to the question of what this science would be in the nineteenth century. They would have thought themselves bold in their predictions, and after the event how timid we should have found them! Do not expect of me any kind of prophesy.

But if, like all prudent physicians, I refuse to give a prognosis, still I cannot deny myself a little diagnosis. Well, then, yes, there are symptoms of a serious crisis, which would seem to indicate that we may expect, presently, a transformation. However, there is no cause for great anxiety. We are assured that the patient will not die, and indeed we may hope that this crisis will be salutary, since the history of the past would seem to insure that. In fact, this crisis is not the first, and in order to understand it, we must recall those that have gone before. Allow me a brief historical sketch.

A Brief History of Mathematical Physics

Mathematical physics, as we are well aware, is an offspring of celestial mechanics, which gave it birth at the end of the eighteenth century---at the moment when it had, itself, attained its complete development. The child, especially during its first years, showed a striking resemblance to its mother. The astronomical universe consists of masses, undoubtedly of great magnitude, but separated by such immense distances that they appear to us as material points; these points attract each other in the inverse ratio of the squares of their distances, and this attraction is the only force which affects their motion.

The Physics of Central Forces: If we could measure all the details of the infinitesimal stars that are the atoms that make up all of matter, the spectacle thus disclosed would hardly differ from the one that the astronomer contemplates. There too we should see material points separated by intervals that are enormous in comparison with their dimensions, and describing orbits according to regular laws. Like the stars they attract each other---or they repel following similar laws. This attraction or repulsion, which is along the line joining them, depends only on the distance. The law according to which this force varies with the distance is perhaps not the law of Newton, but it is analogous thereto. Instead of the exponent 2, we probably have another exponent. From this diversity in the exponents proceeds all the diversity of the physical phenomena, the variety in qualities and sensations, all the world of color and sound which surrounds us; in a word, all of nature.

Such is the primitive, first half of the nineteenth century, conception in its utmost purity. Nothing remains but to inquire in the different cases, what value must be given to this exponent in order to account for all the facts. On this model, for example, Pierre Laplace constructed his beautiful theory of capillarity; he simply regards the latter as a special case of attraction, or, as he says, of universal gravitation, and no one is surprised to find it in the middle of one of the five volumes of his celestial mechanics.

More recently, Charles Briot believes he has laid bare the last secret of optics, when he has proved that the atoms of the aether attract each other in the inverse sixth power of the distance; and does not Maxwell, Maxwell himself, say somewhere that the atoms of a gas repel each other in the inverse ratio of the fifth power of the distance? We have the exponent 6 or 5, instead of the exponent 2; but it is always an exponent.

Among the theories of this period there is a single one that forms an exception, namely that of Joseph Fourier; here there are indeed atoms acting at a distance; they send each other heat, but they do not attract each other, they do not stir. From this point of view, Fourier's theory must have appeared imperfect and provisional to the eyes of his contemporaries, and even to himself.

This conception was not without greatness. It was alluring, and many of us have not given it up. These Laplacian holdouts know that the ultimate elements of things will not be attained except by disentangling with patience the complex skein furnished us by our senses; that progress should be made step by step without neglecting any intermediate portions; that our fathers were unwise in not wishing to stop at all the stations; but they still believe that when we once arrive at these ultimate elements, we shall meet again the majestic simplicity of celestial mechanics.

Nor has this conception been useless; it has rendered us a priceless service inasmuch as it has contributed to making more precise the fundamental concept of the physical law. Let me explain.

What did the ancients understand by a law? It was to them an internal harmony, static as it were, and unchangeable; or else a model which nature tried to imitate. To us a law is no longer that at all; it is a constant relation between a phenomenon of today and that of tomorrow; in a word, it is a differential equation.

Here we have the ideal form of the physical law; and, indeed, it is Newton's law that first gave it this form. If, later on, this form has become inured in physics, it has become so precisely by copying as far as possible this law of Newton. The similarity between the physics of atoms, and of astronomical bodies, is a result of our having used celestial mechanics as a model. Nevertheless, there came a day when the conception of central forces appeared no longer to suffice, and this is the first of the crises to which I referred a moment ago.

What was done? Abandoned was the thought of exploring the details of the universe, of isolating the parts of this vast mechanism, of analyzing one by one the forces that set them going. Rather we took certain general principles that have precisely the object of relieving us of this minute study. How is this possible?

The Physics of the Principles: Suppose we have before us any kind of machine; the part of the mechanism where the power is applied and the ultimate resultant motion alone are visible, while the transmissions, the intermediate gearing whereby the motion is communicated from one part to another, are hidden in the interior and escape our notice. We know not whether the transmissions are made by cog-wheels or by belts, by connecting-rods or other contrivances. Shall we say that it is impossible for us to learn anything about this machine unless we are allowed to take it apart?

You well know that such is not the case, and that the principle of the conservation of energy suffices to furnish us the most interesting feature. We can easily show that the last wheel turns ten times more slowly than the first, since these two wheels are visible; and we can conclude therefrom that a couple applied to the first will be in equilibrium with a couple ten times as great applied to the second. To obtain this result, it is unnecessary to look into the mechanism of this equilibrium, or to know how the forces balance in the interior of the machine.

In the case of the universe, the principle of the conservation of energy can render us the same service. This universe also is a machine, much more complicated than any in use in the industries, of which nearly all the parts are deeply hidden; but by observing the motion of those which we can see, we can by the aid of this principle draw conclusions which will remain valid no matter what the details of the invisible mechanism which actuates them.

The principle of the conservation of energy, or *Mayer's principle*, is certainly the most important, but it is not the only one; there are others from which we can derive the same advantage. These are:

Carnot's principle, or the principle of the dissipation of energy, or the second law of thermodynamics.

Newton's principle, or the principle of the equality of action and reaction.

The principle of relativity, according to which the laws of physical phenomena must be the same for a stationary observer as for one carried along in a uniform motion of translation, so that we have no means, and can have none, of determining whether or not we are being carried along in such a motion.

Lavoisier's principle, or the principle of the conservation of mass.

The principle of least action.

The application of these five or six general principles to the various physical phenomena suffices to teach us what we may reasonably hope to know about them. The most remarkable example of this new mathematical physics is without doubt Maxwell's electromagnetic theory of light. What is the aether? How are its molecules distributed? Do they attract or repel each other? Of these things we know nothing. But we do know that this medium transmits both optical and electrical disturbances; we know that this transmission must take place in conformity with the general principles of mechanics and that suffices to establish the equations of the electromagnetic field.

These principles are the boldly generalized results of experiment; but they appear to derive from their very generality a high degree of certainty. In fact, the greater the generality, the more frequent are the opportunities for verifying them. Such verifications, as they multiply, take the most varied and most unexpected forms, and leave in the end no room for doubt.

Such is the second phase of the history of mathematical physics, and we have not yet left it. Shall we say that the first has been useless, that for fifty years [1800-1850] science was on a wrong path and that there is nothing to do but to forget all that accumulation of effort which a vicious conception from the very beginning doomed to failure? By no means! Do you think the second period could have existed without the first?

The hypothesis of central forces contained all the principles. It involved them as necessary consequences. It involved the principle of the conservation of energy, as well as that of mass, and the equality of action and reaction, and the law of least action, which appeared to be sure, not as experimental facts, but as theorems.

It is the mathematical physics of our fathers that has gradually made us familiar with these various principles, and which has taught us to recognize them in the different garbs in which they are disguised. They have been compared with the results of experiment, where it has been found necessary to change their expression in order to make them conform to the facts; thus they have been extended and strengthened. In this way they came to be regarded as experimental truths. The conception of central forces then became a useless support, or rather an encumbrance, inasmuch as it imposed upon the principles its own hypothetical character.

The bonds then are not broken, because they were elastic; but they have been extended. Our fathers who established them have not labored in vain; and in the science of today we recognize the general features of the outline they traced.

The Present Crisis of Mathematical Physics

The New Crisis: Are we now about to enter upon a third period? Are we on the eve of a second crisis? Are these principles on which we have reared everything about to fall in their turn? This has recently become a vital question.

Hearing me speak thus, you are thinking without doubt of radium, that great revolutionary of the present day; and indeed I shall return to it presently. But there is something else. It is not merely the conservation of energy that is concerned; all the other principles are in equal danger, as we shall see by successively passing them in review.

Carnot's Principle: Let us begin with Carnot's principle. It is the only one that does not present itself as an immediate consequence of the hypothesis of central forces. Quite to the contrary, indeed, it appears, if not actually to contradict this hypothesis, at least not to be reconcilable with it without some effort. If physical phenomena were due exclusively to the motion of atoms the mutual attractions of which depend only on the distance, it would seem that all these phenomena should be reversible; if all the initial velocities were reversed, these atoms, if still subject to the same forces, should traverse their trajectories in the opposite direction, just as the earth would describe backward this same elliptical orbit that it now describes forward, if the initial conditions of its motion had been reversed. Thus, if a physical phenomenon is possible, the inverse phenomenon should be equally possible, and one should be able to retrace the course of time. Now, it is not so in nature, and this it is precisely that the principle of Carnot teaches us; heat may pass from a hot body to a cold; it is impossible to compel it to take the opposite route and to reestablish differences of temperature which have disappeared. Motion can be entirely destroyed and transformed into heat by friction; the converse transformation can only occur partially.

Efforts have been made to reconcile this apparent contradiction. If the world tends toward uniformity, it is not because its ultimate parts, though diversified at the start, tend to become less and less different; it is because moving at random they become mixed. To an eye which could distinguish all the elements, the variety would remain always as great; every grain of this powder retains its originality and does not fashion itself after its neighbors; but as the mixture becomes more and more perfect, our rough senses perceive only uniformity. That is why, for example, temperatures tend to equalize themselves, without its being possible to go back.

A drop of wine, let us say, falls into a glass of water; whatever the internal motion of the liquid, we shall soon see it assume a uniformly roseate hue, and from then on no possible shaking of the vessel would seem to be capable of again separating the wine and the water. Here, then, we have what may be the type of the irreversible phenomenon of physics: to hide a grain of barley in a great mass of wheat would be easy; to find it again and to remove it is practically impossible. All this has been explained by Maxwell and Boltzmann, but the man who has put it most clearly was Gibbs in a book that is too little read because it is a little difficult to read, in his Elements of Statistical Mechanics.[344]

To those who take this point of view, Carnot's principle is an imperfect principle, a sort of concession to the frailty of our senses. It is because our eyes are too coarse that we do not distinguish the elements of the mixture. It is because our hands are too coarse that we cannot compel them to separate. The imaginary demon of Maxwell, who can pick out the molecules one by one, would be quite able to constrain the world to move backwards. That it should return of its own accord is not impossible; it is only infinitely improbable; the chances are that we should wait a long time for that combination of circumstances that would permit a retrogression; but, sooner or later, they will occur, after years, the number of which would require millions of figures. These reservations, however, all remained theoretical; they caused little uneasiness and Carnot's principle preserved all of its practical value.

But now here is where the scene changes. The biologist, armed with his microscope, has for a long time noticed in his preparations certain irregular motions of small particles in suspension; this is known as *Brownian motion.* Brown believed at first that it was a phenomenon of life, but he soon saw that inanimate bodies hopped about with no less ardor than others; he then turned the matter over to the physicists. Unfortunately, the physicists did not become interested in the question for a long time. Light is concentrated, so they argued, in order to illuminate the microscopic preparation; light involves heat, and this causes differences in temperature and these produce internal currents in the liquid, which bring about the motions referred to.

Louis Georges Gouy had the idea of looking a little more closely, and thought he saw that this explanation was untenable; that the motion becomes more active as the particles become smaller, but that they are uninfluenced by the manner of lighting. If, then, these motions do not cease, or, rather, if they come into existence incessantly, without borrowing from any external source of energy, what must we think? We must surely not abandon on this account the conservation of energy; but we see before our eyes motion transformed into heat by friction and conversely heat changing into motion, and all without any sort of loss, since the motion continues forever. It is the contradiction of Carnot's principle. If such is the case, we no longer need the infinitely keen eye of Maxwell's demon in order to see the world move backward; our microscope suffices...

The Principle of Relativity: Let us consider the principle of relativity; this principle is not only confirmed by our daily experience, not only is it the necessary consequence of the hypothesis of central forces, but it appeals to our common sense with irresistible force. And yet it also is being fiercely attacked. Let us think of two electrified bodies; although they seem to be at rest, they are,

---

[344] Gibbs, Josiah Willard, Elementary Principles in Statistical Mechanics, ©1902 Yale University, published in 1914, Yale University Press, New Haven

both of them, carried along with the motion of the earth; Rowland has shown us that an electric charge in motion is equivalent to a current; these two charged bodies, then, are equivalent to two parallel currents in the same direction; these two currents should attract each other. By measuring this attraction we should be measuring the velocity of the earth; not its velocity relative to the sun and stars, but its absolute velocity.

I know what will be said; it is not its absolute velocity; it is its velocity relative to the aether. But, how unsatisfactory that is! Is it not clear that with this interpretation, nothing could be inferred from the principle? It could no longer teach us anything, simply because it would no longer fear any contradiction. Whenever we have succeeded in measuring anything, we would always be free to say that it is not the absolute velocity, and if it is not the velocity relative to the aether, it might always be the velocity relative to some new unknown fluid with which we might fill all space.

And then experiment, too, has taken upon itself to refute this interpretation of the principle of relativity; all the attempts to measure the velocity of the earth relative to the aether have led to negative results. Herein experimental physics has been more faithful to the principle than mathematical physics; the theorists would have dispensed with it readily in order to harmonize the other general points of view; but experimentation has insisted on confirming it. Methods were diversified; finally Michelson carried precision to its utmost limits; nothing came of it. It is precisely to overcome this stubbornness that today mathematicians are forced to employ all their ingenuity.

Their task was not easy, and if Lorentz has succeeded, it is only by an accumulation of hypotheses. The most ingenious idea is that of local time.

Let us imagine two observers, located at signal stations A and B, who wish to regulate their watches by means of optical signals. They exchange signals, but as they know that the transmission of light is not instantaneous, they are careful to cross them. When station B sees the signal from station A, its timepiece should not mark the same hour as that of station A at the moment the signal was sent, but this hour increased by a constant representing the time of transmission. Let us suppose, for example, that station A sends a signal at the moment when its timepiece marks the hour zero, and that station B receives it when its timepiece marks the hour t. The watches will be set, if the time t is the time of transmission, and in order to verify it, station B in turn sends a signal at the instant when its timepiece is at zero; station A must then see it when its timepiece is at t. Then the watches are regulated.

And, indeed, they mark the same hour at the same physical instant, but only if the two stations are stationary. Otherwise, the time of transmission will not be the same in the two directions, since the station A, for example, goes to meet the disturbance emanating from B, whereas station B flees before the disturbance emanating from A.

Watches regulated in this way, therefore, will not mark the true time; they will mark what might be called the local time, so that one will gain on the other. It matters little, since we have no means of perceiving it. All the phenomena which take place at A, for example, will be behind time, but all just the same amount, and the observer will not notice it since his watch is also behind time; thus, in accordance with the principle of relativity he will have no means of ascertaining whether he is at rest or in absolute motion.

Unfortunately this is not sufficient; additional hypotheses are necessary. We must admit that the moving bodies undergo a uniform contraction in the direction of the motion. One of the diameters of the earth, for example, is shortened by 1/200000000 as a result of our planet's motion, whereas the other diameter preserves its normal length. Thus we find the last minute differences accounted for.

Then there is still the hypothesis concerning forces. Forces, whatever their origin, weight as well as elasticity, will be reduced in a certain ratio in a world endowed with a uniform translatory motion; or rather that would happen for the components at right angles to the direction of translation; the parallel components will not change.

Let us then return to our example of the two electrified bodies; they repel each other; but at the same time, if everything is carried along in a uniform transition, they are equivalent to two parallel currents in the same direction, which attract each other. This electrodynamic attraction is, then, subtracted from the electrostatic repulsion, and the resultant repulsion is weaker than if the two bodies had been at rest. But since we must, in order to measure this repulsion, balance it by another force, and since all these other forces are reduced in the same ratio, we observe nothing. Everything, then, appears to be in order.

But have all doubts been dissipated? What would happen if we could communicate by signals other than those of light? If, after having regulated our watches by the optimal method, we wished to verify the result by means of these new signals, we should observe discrepancies due to the common translatory motion of the two stations.

And are such signals inconceivable, if we take the view of Laplace, that universal gravitation is transmitted with a velocity a million times as great as that of light? Thus the principle of relativity has in recent times been valiantly defended; but the very vigor of the defense shows how serious was the attack.

Newton's Principle: And now let us speak of the principle of Newton, concerning the equality of action and reaction. This principle is intimately connected with the preceding and it would seem that the fall of one would involve the fall of the other. Nor must we be surprised to find here again the same difficulties.

The electrical phenomena, it is thought, are due to displacements of small charged particles called electrons that are immersed in the medium we call the aether. The motions of these electrons produce disturbances in the surrounding aether; these disturbances are propagated in all directions with the aether velocity of light, and other electrons initially at rest are displaced when the disturbance reaches the portions of the aether in which they lie.

The electrons, then, act one upon the other, but this action is not direct; it takes place by mediation of the aether. Under these conditions, is it possible to have equality between action and reaction, at least for an observer who takes account only of the motion of matter, that is of the electrons, and who ignores that of the aether that he is unable to see? Evidently not: even if the compensation were exact, it could not be instantaneous. The disturbance is propagated with a finite velocity; it reaches the second electron, therefore, only after the first has long been reduced to rest. This second electron will, then, after an interval, be subjected to the action of the first, but will certainly not at that moment react upon it, since there is no longer anything in the neighborhood of this first electron that stirs.

The analysis of the facts will allow us to become more definite. Let us imagine, for example, a Hertzian oscillator such as those used in wireless telegraphy. It sends energy in all directions; but we may attach to it a parabolic mirror, as was done by Hertz with his smallest oscillators, so as to send all the energy produced in a single direction. What then will happen according to the theory? Why, the apparatus will recoil as though it were a cannon and the projected energy a ball, and that contradicts the principle of Newton, since our present projectile has no mass; it is not matter, it is energy. It is the same, moreover, in the case of a lighthouse having a reflector, since light is merely a disturbance in the electromagnetic field. This lighthouse would recoil; as though the light it sends forth were a projectile. What is the force that must produce this recoil? It is what is known as the Maxwell-Bartholdi pressure; it is very small, and to put it in evidence caused much trouble, even with the most sensitive radiometers; but it is sufficient for our purpose that it exists.

If all the energy issuing from our oscillator strikes a receiver, the latter will act as though it had received a physical shock, which in a sense will represent the compensation of the oscillator's recoil. The reaction will be equal to the action, but they will not be simultaneous; the receiver will advance, but not at the instant when the oscillator recoils. If the energy is propagated indefinitely without meeting a receiver, the compensation will never take place.

Shall we say that the space which separates the oscillator from the receiver, and which the disturbance must traverse in passing from one to the other, is not empty? Rather it is filled not only with aether, but with air, or even in inter-planetary space with some subtle, yet ponderable fluid. Under this hypothesis, this matter receives the shock, as does the receiver, at the moment the energy reaches it, and recoils, when the disturbance leaves it.

That would save Newton's principle, but it is not true. If the energy during its propagation remained always attached to some material substratum, this matter would carry the light along with it and Fizeau has shown, at least for the air, that there is nothing of the kind. Michelson and Morley have since confirmed this. We might also suppose that the motions of matter proper were exactly compensated by those of the aether, but that would lead us to the same considerations as those made a moment ago. The principle, if thus interpreted, could explain anything, since whatever the visible motions we could imagine hypothetical motions to compensate them. But if it can explain anything, it will allow us to foretell nothing. It therefore becomes useless.

And then the suppositions that must be made concerning the motions of the aether are not very satisfactory. If the electric charges were doubled, it would be natural to suppose that the velocities of the atoms of the aether also became twice as great, and for the compensation it would be necessary that the mean velocity of the aether become four times as great.

This is why I have for a long time thought that these consequences of the theory, which contradict Newton's principle, would some day be abandoned; and yet the recent experiments on the motion of the electrons emitted by radium seem rather to confirm them.

Lavoisier's Principle: I now come to Lavoisier's principle concerning the conservation of mass. This is certainly a principle that cannot be tampered with without shaking the science of mechanics. And still there are persons who think that it seems true to us only because in mechanics we consider only moderate velocities, and that it would cease to be so for bodies having velocities comparable with that of light. Now, such velocities are at present believed to have been realized; the cathode rays and those of radium would seem to be formed of very minute particles or electrons that move with velocities that are no doubt less than that of light, but which appear to be about one tenth or one third of it.

These rays can be deflected either by an electric or by a magnetic field, and by comparing these deflections it is possible to measure both the velocity of the electrons and their mass (or rather the ratio of their mass to their charge). But it was found that as soon as these velocities approached that of light a correction was necessary.

Since these particles are electrified, they cannot be displaced without disturbing the aether; to put them in motion, it is necessary to overcome a double inertia, that of the particle itself and that of the aether. The total or apparent mass that is measured is then composed of two parts: the real or mechanical mass of the particle and the electrodynamic mass representing the inertia of the aether.

Now, the calculations of Max Abraham and the experiments of Walter Kaufmann have shown that this mechanical mass property is nothing, and that the mass of the electrons, at least of the negative electrons, is purely of electrodynamic origin. This is what compels us to change our definition of mass; we can no longer distinguish between the mechanical mass and the electrodynamic mass, because then the first would have to vanish. There is no other mass than the electrodynamic inertia; but in this case, the mass can no longer be constant; it increases with the velocity; and indeed it depends on the direction, and a body having a considerable velocity will not oppose the same inertia to forces tending to turn it off its path that it opposes to those tending to accelerate or retard its motion.

There is indeed another resource: the ultimate elements of bodies are electrons, some with a negative charge, others with a positive charge. It is understood that the negative electrons have no mass; but the positive electrons, from what little is known of them, would seem to be much larger. They perhaps have besides their electrodynamic mass a true mechanical mass. The real mass of a body would then be the sum of the mechanical masses of its positive electrons, the negative electrons would not count; the mass defined in this way might still be constant.

Alas, this resource is also denied. Let us recall what we said concerning the principle of relativity and the efforts made to save it. And it is not simply a principle that is to be saved; the indubitable results of Michelson's experiments are involved. Lorentz, to account for these results, was obliged to suppose that all forces, whatever their origin, are reduced in the same ratio in a medium having a uniform translatory motion. But that is not sufficient; it is not enough that this should take place for the real forces, it must also be the same in the case of the forces of inertia. It is necessary, therefore—so he says—that the masses of all particles be influenced by a translation in the same degree as the electromagnetic masses of the electrons.

Hence, the mechanical masses must vary according to the same laws as the electrodynamic; they can then not be constant.

Do I need to remark that the fall of Lavoisier's principle carries with it that of Newton's? The latter implies that the center of gravity of an isolated system moves in a straight line; but if there no longer exists a constant mass, there no longer exists a center of gravity; indeed the phrase would be meaningless. This is why I said above that the experiments on cathode rays seemed to justify the doubts of Lorentz concerning Newton's principle.

From all these results, if they were to be confirmed, would issue a wholly new mechanics which would be characterized above all by this fact, that there could be no velocity greater than that of light,[345] any more than a temperature below that of absolute zero. For an observer, participating himself in a motion of translation of which he has no suspicion, no apparent velocity could surpass that of light, and this would be a contradiction, unless one recalls the fact that this observer does not use the same sort of timepiece as that used by a stationary observer, but rather a watch giving the "local time."

Here we are then face to face with a question, of which I shall confine myself to the mere statement. If there is no longer any mass what becomes of Newton's law? Mass has two aspects: it is at the same time a coefficient of inertia and an attracting mass entering as a factor into Newton's law of attraction. If the coefficient of inertia is not constant, can the attracting mass be constant? This is the question.

Mayer's Principle: The principle of the conservation of energy at least still remained and appeared more finely established. Shall I recall to your minds how it too was thrown into discredit? That event made more noise than the preceding; the journals are full of it. Ever since the first work of Becquerel, and above all after the Curies had discovered radium, it was seen that every radioactive substance was an inexhaustible source of radiation. Its activity seemed to continue without change through months and years. That is already a strain on the principles; these radiations in fact were energy, and from the same piece of radium came forth this energy and it came forth indefinitely. But these quantities of energy were too minute to be measured; at least that was the belief, and the matter caused little uneasiness.

The scene changed when Curie thought of placing the radium in a calorimeter. It was then seen that the quantity of heat continuously generated was very considerable.

The explanations advanced were numerous; but in a case of this kind it is not possible to say that an abundance of good does no harm: as long as one explanation has not displaced the others we cannot be sure that any one of them is good. For some time, however, one of these explanations seems to be gaining the upper hand and we may reasonably hope that we hold the key to the mystery.

Sir W. Ramsey has attempted to show that radium is transformed, that it contains an enormous amount of energy, but not an inexhaustible amount. The transformation of radium must then produce a million times as much heat as any known transformation; the radium would be exhausted in 1250 years; that is not long, but you see that we are at least sure of being bound to

[345] Because bodies would oppose an increasing inertia to the causes that would tend to accelerate their motion; and when approaching the velocity of light, this inertia would become infinite.

the present state of affairs for some hundreds of years. While we wait our doubts subsist.

The Future of Mathematical Physics

In the midst of such ruin, what remains standing? The principle of least action up to now is intact, and Larmor appears to think that it will long survive the others. It is in fact more vague and even more general.

In the presence of this general collapse of principles, what attitude should mathematical physics take? First of all, before becoming too excited, it is well to ask whether all this is really true. All this disparagement of principles is encountered only in the case of the infinitely small; the microscope is needed to see Brownian motion, the electrons are rather tiny, radium is very rare and never more than a few milligrams are together; and then we can ask whether by the side of the minute thing that was observed, there was not another minute thing which was not noticed and which counterbalanced the first.

The question is surely debatable, and apparently only experiment can solve it. We should merely have to turn the matter over to the experimenters and, while waiting for them definitely to settle the controversy, not to trouble ourselves with these disquieting problems, and to keep quietly at our work, as though the principles were still unchallenged. We certainly have enough to do without leaving the domain where they can be applied with all certainty; we have enough to keep us busy during this period of doubt.

And yet is it really true that we can do nothing to relieve science of these doubts? It must indeed be said that it has not been experimental physics alone that has brought them into existence; mathematical physics has contributed its share. It was the experimenters who saw radium emit energy; but the theorists were the ones who brought to light all the difficulties inherent in the propagation of light through a moving medium. Had it not been for the mathematical physicists, the problems with light propagation probably would not have been noticed.

The physical principles, then, have done their very best to embarrass us. It is only fitting that they should help us to extricate ourselves. They must be subject to a searching criticism of all the new conceptions that I have outlined today, or they must be abandoned after a loyal effort to save them.

Appendix D Electrodynamics in Gaussian Units

Physicists and engineers have historically used one of two consistent metric unit systems: units based on the meter, kilogram, and second, called *MKS*, or units based on the centimeter, gram, and second called *cgs*. In addition to these two standards, and one based on the foot, pound, and second, there are many other standards that are field specific. Regardless of unit system, however, all physical equations should, in principle, still work—except for those in electrodynamics.

Mechanical Units
$1\,\text{m} = 100\,\text{cm}$
$1\,\text{kg} = 1000\,\text{g}$
$1\,\text{s} = 1\,\text{s}$
$1\,\text{N} = 100{,}000\,\text{dynes}$
$1\,\text{J} = 10{,}000{,}000\,\text{ergs}$

During the nineteenth century, however, physicists defined the unit of charge differently. Continental physicists defined their charge unit electrostatically, by simply setting Coulomb's constant to unity. These are called *electrostatic units*. The electrostatic unit of charge is called the *Franklin*, the *esu*, or the *statcoulomb*. The unit of magnetic field is called the Gauss, but called the Oersted when referring to $\vec{H}$.

Electrodynamic Units

Quantity	Relation	In cgs base units
Q or q	$1\,\text{C} \approx 2.998\times10^{9}\,\text{Fr}$	$\text{Fr} = \frac{\sqrt{\text{cm}^3\text{g}}}{\text{s}}$
I	$1\,\text{A} \approx 2.998\times10^{9}\,\text{Fr}/\text{s}$	$\text{Fr}/\text{s} = \frac{\sqrt{\text{cm}^3\text{g}}}{\text{s}^2}$
$\mathbb{V}$	$1\,\text{V} \approx 33.36\,\text{erg}/\text{Fr}$	$\text{erg}/\text{Fr} = \frac{\sqrt{\text{cm g}}}{\text{s}}$
R	$1\,\Omega \approx 1.113\times10^{-12}\,\frac{\text{erg s}}{\text{Fr}^2}$	$\frac{\text{erg s}}{\text{Fr}^2} = \frac{\text{s}}{\text{cm}}$
$\vec{E}$	$1\,\text{V}/\text{m} \approx 3.336\times10^{-5}\,\text{G}$	$\text{G} = \frac{\sqrt{\text{g}}}{\text{s}\sqrt{\text{cm}}}$
$\vec{B}$	$1\,\text{T} = 10^{4}\,\text{G}$	$\text{G} = \frac{\sqrt{\text{g}}}{\text{s}\sqrt{\text{cm}}}$

By the twentieth century, virtually all physicists had standardized on a version of electrostatic cgs units called *Gaussian units*. This system was particularly elegant, as it required only one constant of nature, c. The hallmarks of Gaussian units are:

- Coulomb's constant is unity.
- All fields have the same unit.
- Velocities are often expressed in ratio to the speed of light.

When working with electric circuits, however, it made more sense to define a unit of current in terms of the magnetic force between two parallel wires. These are called *electromagnetic units*. Since magnetic forces between wires are so small, these units produce a huge unit of current. For example, the cgs unit of electromagnetic current is 10A, and the MKS unit is over 2000A. Once physicists reconciled electrostatic and electromagnetic units, they found the ratio of the charge unit to be a constant called c with a value of about 30 billion cm/s (see Maxwell's quote on p. 479).

Free Space Equations

	Gaussian Units (cgs)	**Rationalized Units (SI)**
Coulomb's Law	$\vec{F} = \frac{q_1 q_2}{r^2}\hat{r}$	$\vec{F} = \left(\frac{Z_0 c}{4\pi}\right)\frac{q_1 q_2}{r^2}\hat{r}$
Electromagnetic Force	$\vec{F} = q\left(\vec{E} + \frac{1}{c}\vec{v}\times\vec{B}\right)$	$\vec{F} = q\left(\vec{E} + \vec{v}\times\vec{B}\right)$
Peregrinus's Principle	$\vec{\nabla}\cdot\vec{B} = 0$	$\vec{\nabla}\cdot\vec{B} = 0$
Gauss's Law	$\vec{\nabla}\cdot\vec{E} = 4\pi\rho$	$\vec{\nabla}\cdot\vec{E} = Z_0 c\rho$
Faraday's Law	$\vec{\nabla}\times\vec{E} = -\frac{1}{c}\frac{\partial}{\partial t}\vec{B}$	$\vec{\nabla}\times\vec{E} = -\frac{\partial}{\partial t}\vec{B}$
The Maxwell-Ampere Law	$\vec{\nabla}\times\vec{B} = \frac{4\pi}{c}\vec{J} + \frac{1}{c}\frac{\partial}{\partial t}\vec{E}$	$\vec{\nabla}\times\vec{B} = \frac{Z_0}{c}\vec{J} + \frac{1}{c^2}\frac{\partial}{\partial t}\vec{E}$
The Larmor Equation	$\mathbb{P} = \frac{2}{3}\frac{1}{c^3}q^2a^2$	$\mathbb{P} = \frac{2}{3}\frac{Z_0}{4\pi c^2}q^2a^2$
The Vector Potential	$\vec{B} = \vec{\nabla}\times\vec{\mathbb{A}}$	$\vec{B} = \vec{\nabla}\times\vec{\mathbb{A}}$
The Electric Field and Potentials	$\vec{E} = -\vec{\nabla}\mathbb{V} - \frac{1}{c}\frac{\partial}{\partial t}\vec{\mathbb{A}}$	$\vec{E} = -\vec{\nabla}\mathbb{V} - \frac{\partial}{\partial t}\vec{\mathbb{A}}$
The Poynting Vector	$\vec{\mathbb{S}} = \frac{c}{4\pi}\vec{E}\times\vec{B}$	$\vec{\mathbb{S}} = \frac{c}{Z_0}\vec{E}\times\vec{B}$

To experimentalists, instrument calibration, and the size of the unit, are more important than elegant equations, so they defined a new unit of current, the *ampere*, to be $\frac{1}{10}$ of the cgs electromagnetic unit. Thus, operationally, the ampere is: the current needed to produce two dynes of force between meter-long parallel wires one centimeter apart (p. 247). This gave a standard current unit of reasonable size. When combined with the MKS unit of power (the watt), the ampere produced a convenient unit of electric potential (the volt), and electrical resistance (the ohm). This became the *MKSA* system of units, used by electrical engineers.

In the 1880s, Oliver Heaviside "rationalized" Maxwell's field equations, which included as the fundamental constants the two aethereal properties, ε_0 and μ_0, and doing away with extraneous factors of 4π. Electrical engineers followed Heaviside, but physicists did not.

Since 1889, global standards of measurement have been decided at the General Conferences on Weights and Measures (CGPM). After much debate between physicists and electrical engineers, the 1948 and 1960 meetings set Heaviside's rationalized MKSA units as the international standard (SI). Despite this vote, physicists did not quickly change practice. In fact, up until this century most graduate students were still taught electrodynamics using Gaussian units.

Often you will encounter electromagnetic equations that look sort of familiar, but do not contain the familiar constants, except for c. To translate a free space equation from Gaussian to SI, follow the following steps:

- Replace each unfamiliar factor of 4π with either the constant cZ_0, or $\frac{1}{c}Z_0$, depending on whether the context is electric, or magnetic, respectively.
- Replace each $\vec{B}$, or $\vec{\mathbb{A}}$, with $c\vec{B}$, or $c\vec{\mathbb{A}}$, respectively.
- Simplify and check the consistency of the units. Use dimensional analysis to include, or remove, necessary factors of the speed of light.

When working with matter, the translations become much more confusing. In the Gaussian system, all four fields have the same units, and you will find more unfamiliar factors of 4π. This is because the definition of the susceptibilities (χ_E, χ_M) is different—even though they are unitless quantities.

As a practical matter, if you find data that are given in non-SI units, convert them to SI units before doing your analysis. If you are required to present your results in non-SI units, you can always convert back when you are done. Keep in mind, however, the importance of standardized units, as they make results more understandable to scientists and engineers from different fields of study.

Constitutive Relations

Gaussian Units (cgs)	**Rationalized Units (SI)**
$\vec{D} = \vec{E} + 4\pi\vec{P} \approx \varepsilon\vec{E}$	$\vec{D} = \varepsilon_0\vec{E} + \vec{P} \approx \varepsilon\vec{E}$
$\vec{P} \approx \chi_E\vec{E}$	$\vec{P} \approx \chi_E\varepsilon_0\vec{E}$
$\varepsilon = 1 + 4\pi\,\chi_E$	$\varepsilon = \varepsilon_0(1+\chi_E)$
$\vec{H} = \vec{B} - 4\pi\vec{M} \approx \frac{1}{\mu}\vec{B}$	$\vec{H} = \frac{1}{\mu_0}\vec{B} - \vec{M} \approx \frac{1}{\mu}\vec{B}$
$\vec{M} \approx \chi_M\vec{H}$	$\vec{M} \approx \chi_M\vec{H}$
$\mu = 1 + 4\pi\,\chi_M$	$\mu = \mu_0(1+\chi_M)$
$\vec{\nabla}\cdot\vec{D} = 4\pi\rho$	$\vec{\nabla}\cdot\vec{D} = \rho$
$\vec{\nabla}\times\vec{H} = \frac{4\pi}{c}\vec{J} + \frac{1}{c}\frac{\partial}{\partial t}\vec{D}$	$\vec{\nabla}\times\vec{H} = \vec{J} + \frac{\partial}{\partial t}\vec{D}$

Appendix E Vector Calculus

Geometrical Definitions		Fundamental Theorems
$\vec{\nabla}\cdot\vec{F} \equiv \lim_{V\to 0}\frac{1}{V}\oint_{\text{surface}}\vec{F}\cdot\mathrm{d}\vec{A}$	The Divergence	$\oint_{\text{surface}}\vec{F}\cdot\mathrm{d}\vec{A} = \oint_{\text{volume}}\left(\vec{\nabla}\cdot\vec{F}\right)\mathrm{d}V$
$\vec{\nabla}f \equiv \lim_{\delta\vec{r}\to 0}\frac{\delta f}{\delta\vec{r}}$	The Gradient	$f(\vec{r})-f(\vec{r}_0) = \int_{\vec{r}_0}^{\vec{r}}\vec{\nabla}f\cdot\mathrm{d}\vec{\ell}$
$\left(\vec{\nabla}\times\vec{F}\right)_{\hat{n}} \equiv \lim_{A\to 0}\frac{1}{A}\oint_{\text{path}}\vec{F}\cdot\mathrm{d}\vec{\ell}\;\hat{n}$	The Curl	$\oint_{\text{path}}\vec{F}\cdot\mathrm{d}\vec{\ell} = \int_{\text{surface}}\left(\vec{\nabla}\times\vec{F}\right)\cdot\mathrm{d}\vec{A}$

$\vec{A}\cdot\vec{B} = A\,B_{\parallel}$

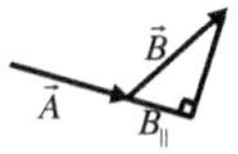

$\vec{A}\times\vec{B} = A\,B_{\perp}\,\hat{n}$

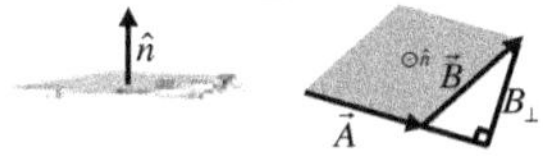

Vector Algebra Identities

$$A = \left|\vec{A}\right| = \sqrt{\vec{A}\cdot\vec{A}}$$

$$\vec{A}\cdot\vec{A} = A^2$$

$$\vec{A}\cdot\vec{B} = \vec{B}\cdot\vec{A}$$

$$\vec{A}\times\vec{A} = 0$$

$$\vec{A}\times\vec{B} = -\vec{B}\times\vec{A}$$

$$\vec{A}\times\vec{B}\times\vec{C} = \vec{A}\times\left(\vec{B}\times\vec{C}\right)$$

$$\vec{A}\times\left(\vec{B}\times\vec{C}\right) = \vec{B}\left(\vec{A}\cdot\vec{C}\right)-\vec{C}\left(\vec{A}\cdot\vec{B}\right)$$

$$\vec{A}\times\left(\vec{B}\times\vec{C}\right) = \vec{B}\times\left(\vec{A}\times\vec{C}\right)-\vec{C}\times\left(\vec{A}\times\vec{B}\right)$$

$$\vec{A}\cdot\left(\vec{B}\times\vec{C}\right) = \vec{B}\cdot\left(\vec{C}\times\vec{A}\right) = \vec{C}\cdot\left(\vec{A}\times\vec{B}\right)$$

$$\vec{A}\times\left(\vec{B}\times\vec{C}\right)+\vec{B}\times\left(\vec{C}\times\vec{A}\right)+\vec{C}\times\left(\vec{A}\times\vec{B}\right) = 0$$

$$\left(\vec{A}\times\vec{B}\right)\cdot\left(\vec{C}\times\vec{D}\right) = \left(\vec{A}\cdot\vec{C}\right)\left(\vec{B}\cdot\vec{D}\right) - \left(\vec{B}\cdot\vec{C}\right)\left(\vec{A}\cdot\vec{D}\right)$$

$$\left(\vec{A}\times\vec{B}\right)\times\left(\vec{C}\times\vec{D}\right) = \left(\vec{A}\cdot\left(\vec{B}\times\vec{D}\right)\right)\vec{C} - \left(\vec{A}\cdot\left(\vec{B}\times\vec{C}\right)\right)\vec{D}$$

Vector Calculus Identities

$$\vec{\nabla}\times\left(\vec{\nabla}f\right) = 0 \qquad \vec{\nabla}\cdot\left(\vec{\nabla}\times\vec{A}\right) = 0$$

$$\nabla^2 f = \vec{\nabla}\cdot\left(\vec{\nabla}f\right) \qquad \nabla^2\vec{A} = \vec{\nabla}\left(\vec{\nabla}\cdot\vec{A}\right)-\vec{\nabla}\times\left(\vec{\nabla}\times\vec{A}\right)$$

$$\vec{\nabla}f(g) = \tfrac{\mathrm{d}f}{\mathrm{d}g}\vec{\nabla}g \qquad \vec{\nabla}\cdot\vec{A}(f) = \vec{\nabla}f\cdot\left(\tfrac{\mathrm{d}}{\mathrm{d}f}\vec{A}\right)$$

$$\nabla^2 f(g) = \tfrac{\mathrm{d}f}{\mathrm{d}g}\nabla^2 g + \tfrac{\mathrm{d}^2 f}{\mathrm{d}g^2}\left|\vec{\nabla}g\right|^2 \qquad \vec{\nabla}\times\vec{A}(f) = \vec{\nabla}f\times\left(\tfrac{\mathrm{d}}{\mathrm{d}f}\vec{A}\right)$$

$$\vec{\nabla}\cdot\left(f\vec{A}\right) = f\left(\vec{\nabla}\cdot\vec{A}\right)+\vec{A}\cdot\left(\vec{\nabla}f\right)$$

$$\vec{\nabla}\cdot\left(\vec{A}\times\vec{B}\right) = \vec{B}\cdot\left(\vec{\nabla}\times\vec{A}\right)-\vec{A}\cdot\left(\vec{\nabla}\times\vec{B}\right)$$

$$\vec{\nabla}\times\left(f\vec{A}\right) = f\left(\vec{\nabla}\times\vec{A}\right)-\vec{A}\times\left(\vec{\nabla}f\right)$$

$$\vec{\nabla}\times\left(\vec{A}\times\vec{B}\right) = \left(\vec{B}\cdot\vec{\nabla}\right)\vec{A}-\left(\vec{A}\cdot\vec{\nabla}\right)\vec{B} + \vec{A}\left(\vec{\nabla}\cdot\vec{B}\right)-\vec{B}\left(\vec{\nabla}\cdot\vec{A}\right)$$

$$\vec{\nabla}\left(f\,g\right) = f\left(\vec{\nabla}g\right)+g\left(\vec{\nabla}f\right)$$

$$\vec{\nabla}\left(\vec{A}\cdot\vec{B}\right) = \vec{A}\times\left(\vec{\nabla}\times\vec{B}\right)+\vec{B}\times\left(\vec{\nabla}\times\vec{A}\right) + \left(\vec{A}\cdot\vec{\nabla}\right)\vec{B}+\left(\vec{B}\cdot\vec{\nabla}\right)\vec{A}$$

$$\nabla^2\left(f\,g\right) = f\left(\nabla^2 g\right)+2\left(\vec{\nabla}f\right)\cdot\left(\vec{\nabla}g\right)+g\left(\nabla^2 f\right)$$

$$\nabla^2\left(f\vec{A}\right) = f\left(\nabla^2\vec{A}\right)+2\left(\vec{\nabla}f\cdot\vec{\nabla}\right)\cdot\vec{A}+\vec{A}\left(\nabla^2 f\right)$$

$$\nabla^2\left(\vec{A}\cdot\vec{B}\right) = \vec{A}\cdot\left(\nabla^2\vec{B}\right)-\vec{B}\cdot\left(\nabla^2\vec{A}\right) + 2\vec{\nabla}\cdot\left(\left(\vec{B}\cdot\vec{\nabla}\right)\vec{A}+\vec{B}\times\left(\vec{\nabla}\times\vec{A}\right)\right)$$

Common Vector Derivatives

$$\vec{\nabla}\cdot\hat{r} = \tfrac{2}{r} \qquad \vec{\nabla}\cdot\vec{r} = 3 \qquad \vec{\nabla}\times\hat{\theta} = \tfrac{\hat{\varphi}}{r}$$

$$\vec{\nabla}\cdot\hat{s} = \tfrac{1}{s} \qquad \vec{\nabla}\times\hat{\varphi} = \tfrac{\hat{z}}{s} = \tfrac{\hat{r}}{r\tan\theta}-\tfrac{\hat{\theta}}{r}$$

$$\vec{\nabla}\cdot\hat{\theta} = \tfrac{1}{r\tan\theta} \qquad \vec{\nabla}\cdot\left(r^n\,\hat{r}\right) = \left(n+2\right)r^{n-1}$$

Common Curl and Divergence Free Vector Fields

$$\vec{F}\propto\frac{\hat{s}}{s} \qquad \vec{F}\propto\frac{\hat{\varphi}}{s} \qquad \vec{F}\propto\frac{\hat{r}}{r^2} \qquad \vec{F}\propto\frac{3\left(\hat{z}\cdot\hat{r}\right)-\hat{z}}{r^3}$$

$$\vec{\nabla}\times\vec{r} = 0 \qquad \vec{\nabla}\times f(r)\hat{r} = 0$$

$$\vec{\nabla}r = \hat{r} \qquad \vec{\nabla}z = \hat{z} \qquad \vec{\nabla}s = \hat{s}$$

Appendix F Curvilinear Coordinate Systems

Cartesian Coordinates

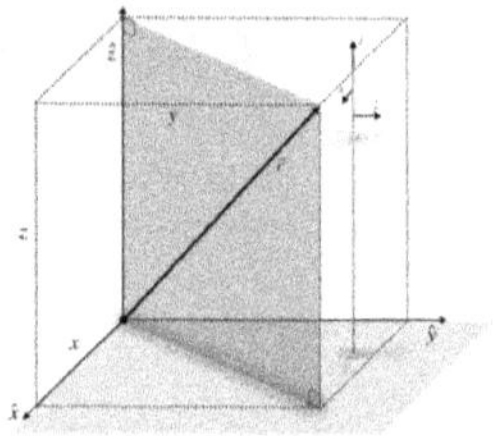

$$\hat{x}\times\hat{y}=\hat{z}$$
$$\vec{r}=x\hat{x}+y\hat{y}+z\hat{z}$$
$$r=\sqrt{x^2+y^2+z^2}$$
$$\delta\vec{\ell}=\delta x\,\hat{x}+\delta y\,\hat{y}+\delta z\,\hat{z}$$
$$\delta\vec{A}=\delta z\delta y\,\hat{x}+\delta y\delta z\,\hat{y}+\delta x\delta y\,\hat{z}$$
$$\delta V=\delta x\delta z\delta y$$
$$\vec{\nabla}\cdot\vec{F}=\tfrac{\partial}{\partial x}F_x+\tfrac{\partial}{\partial y}F_y+\tfrac{\partial}{\partial y}F_z$$
$$\vec{\nabla}f=\tfrac{\partial f}{\partial x}\hat{x}+\tfrac{\partial f}{\partial y}\hat{y}+\tfrac{\partial f}{\partial z}\hat{z}$$
$$\vec{\nabla}\times\vec{F}=\left(\tfrac{\partial}{\partial y}F_z-\tfrac{\partial}{\partial z}F_y\right)\hat{x}+\left(\tfrac{\partial}{\partial z}F_x-\tfrac{\partial}{\partial x}F_z\right)\hat{y}+\left(\tfrac{\partial}{\partial x}F_y-\tfrac{\partial}{\partial y}F_x\right)\hat{z}$$
$$x=s\cos\varphi=r\sin\theta\cos\varphi$$
$$y=s\sin\varphi=r\sin\theta\sin\varphi$$
$$z=z=r\cos\theta$$

Cylindrical Coordinates

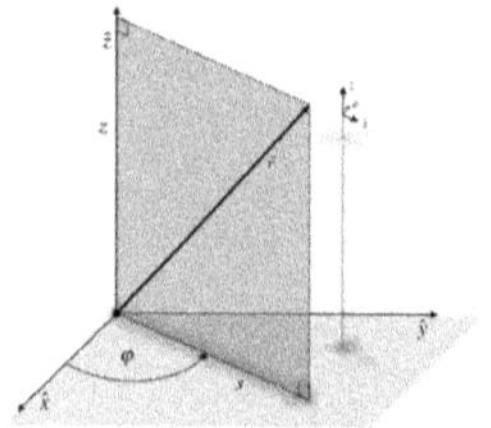

$$\hat{s}\times\hat{\varphi}=\hat{z}$$
$$\vec{r}=s\,\hat{s}+z\,\hat{z}$$
$$r=\sqrt{s^2+z^2}$$
$$\delta\vec{\ell}=\delta s\,\hat{s}+s\,\delta\varphi\,\hat{\varphi}+\delta z\,\hat{z}$$
$$\delta\vec{A}=s\,\delta\varphi\,\delta z\,\hat{s}+\delta z\delta s\,\hat{\varphi}+s\,\delta\varphi\,\delta s\,\hat{z}$$
$$\delta V=s\,\delta s\,\delta\varphi\,\delta z$$
$$\vec{\nabla}\cdot\vec{F}=\tfrac{1}{s}\tfrac{\partial}{\partial s}(sF_s)+\tfrac{1}{s}\tfrac{\partial}{\partial\varphi}F_\varphi+\tfrac{\partial}{\partial z}F_z$$
$$\vec{\nabla}f=\tfrac{\partial f}{\partial s}\hat{s}+\tfrac{1}{s}\tfrac{\partial f}{\partial\varphi}\hat{\varphi}+\tfrac{\partial f}{\partial z}\hat{z}$$
$$\vec{\nabla}\times\vec{F}=\left(\tfrac{1}{s}\tfrac{\partial}{\partial\varphi}F_z-\tfrac{\partial}{\partial z}F_\varphi\right)\hat{s}+\left(\tfrac{\partial}{\partial z}F_s-\tfrac{\partial}{\partial s}F_z\right)\hat{\varphi}+\left(\tfrac{1}{s}\tfrac{\partial}{\partial s}(sF_\varphi)-\tfrac{1}{s}\tfrac{\partial}{\partial\varphi}F_s\right)\hat{z}$$
$$s=r\sin\theta=\sqrt{x^2+y^2}$$
$$\varphi=\arctan\left(\tfrac{y}{x}\right)$$
$$z=r\cos\theta$$

Spherical Coordinates

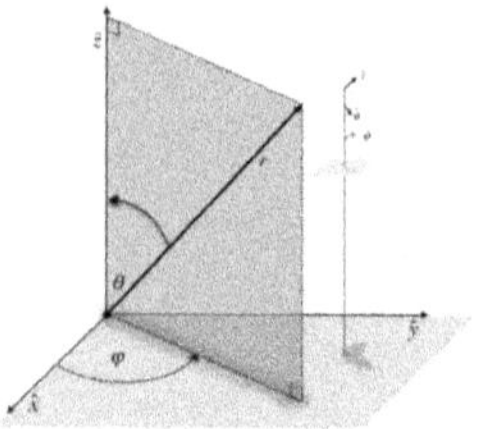

$$\hat{r}\times\hat{\theta}=\hat{\varphi}$$
$$\vec{r}=r\,\hat{r}$$
$$\delta\vec{\ell}=\delta r\,\hat{r}+r\,\delta\theta\,\hat{\theta}+r\sin\theta\,\delta\theta\,\hat{\varphi}$$
$$\delta\vec{A}=r^2\sin\theta\,\delta\theta\,\delta\varphi\,\hat{r}+r\sin\theta\,\delta r\,\delta\varphi\,\hat{\theta}+r\,\delta\theta\,\delta r\,\hat{\varphi}$$
$$\delta V=r^2\sin\theta\;\delta r\,\delta\theta\,\delta\varphi$$
$$\vec{\nabla}\cdot\vec{F}=\tfrac{1}{r^2}\tfrac{\partial}{\partial r}\left(r^2F_r\right)+\tfrac{1}{r\sin\theta}\tfrac{\partial}{\partial\theta}\left(\sin(\theta)F_\theta\right)+\tfrac{1}{r\sin\theta}\tfrac{\partial}{\partial\varphi}F_\varphi$$
$$\vec{\nabla}f=\tfrac{\partial f}{\partial r}\hat{r}+\tfrac{1}{r}\tfrac{\partial f}{\partial\theta}\hat{\theta}+\tfrac{1}{r\sin\theta}\tfrac{\partial f}{\partial\varphi}\hat{\varphi}$$
$$\vec{\nabla}\times\vec{F}=\tfrac{1}{r\sin\theta}\left(\tfrac{\partial}{\partial\theta}\left(\sin\theta\,F_\varphi\right)-\tfrac{\partial}{\partial\varphi}F_\theta\right)\hat{r}+\tfrac{1}{r}\left(\tfrac{1}{\sin\theta}\tfrac{\partial}{\partial\varphi}F_r-\tfrac{\partial}{\partial r}\left(rF_\varphi\right)\right)\hat{\theta}+\tfrac{1}{r}\left(\tfrac{\partial}{\partial r}\left(rF_\theta\right)-\tfrac{\partial}{\partial\theta}F_r\right)\hat{\varphi}$$
$$r=\sqrt{x^2+y^2+z^2}=\sqrt{s^2+z^2}$$
$$\theta=\arctan\left(\sqrt{\tfrac{x^2+y^2}{z^2}}\right)=\arctan\left(\tfrac{s}{z}\right)$$
$$\varphi=\arctan\left(\tfrac{y}{x}\right)=\varphi$$

$$\hat{x}=\cos\varphi\,\hat{s}-\sin\varphi\,\hat{\varphi}$$
$$\hat{y}=\sin\varphi\,\hat{s}+\cos\varphi\,\hat{\varphi}$$
$$\hat{z}=\hat{z}$$

$$\hat{s}=\cos\varphi\,\hat{x}+\sin\varphi\,\hat{y}$$
$$\hat{\varphi}=-\sin\varphi\,\hat{x}+\cos\varphi\,\hat{y}$$
$$\hat{z}=\hat{z}$$

$$\hat{s}=\sin\theta\,\hat{r}+\cos\theta\,\hat{\theta}$$
$$\hat{\varphi}=\hat{\varphi}$$
$$\hat{z}=\cos\theta\,\hat{r}-\sin\theta\,\hat{\theta}$$

$$\hat{r}=\sin\theta\,\hat{s}+\cos\theta\,\hat{z}$$
$$\hat{\theta}=\cos\theta\,\hat{s}-\sin\theta\,\hat{z}$$
$$\hat{\varphi}=\hat{\varphi}$$

$$\hat{x}=\sin\theta\cos\varphi\,\hat{r}+\cos\theta\,\cos\varphi\,\hat{\theta}-\sin\varphi\,\hat{\varphi}$$
$$\hat{y}=\sin\theta\sin\varphi\,\hat{r}+\cos\theta\,\sin\varphi\,\hat{\theta}+\cos\varphi\,\hat{\varphi}$$
$$\hat{z}=\cos\theta\,\hat{r}-\sin\theta\,\hat{\theta}$$

$$\hat{r}=\sin\theta\,\cos\varphi\,\hat{x}+\sin\theta\,\sin\varphi\,\hat{y}+\cos\theta\,\hat{z}$$
$$\hat{\theta}=\cos\theta\,\cos\varphi\,\hat{x}+\cos\theta\,\sin\varphi\,\hat{y}-\sin\theta\,\hat{z}$$
$$\hat{\varphi}=-\sin\varphi\,\hat{x}+\cos\varphi\,\hat{y}$$

Appendix G Historical Timeline

600 B.C. The Ancient Greek philosopher and mathematician Thales discovers that by rubbing amber ("ἤλεκτρον" or "elektron") with a piece of fur, he could attract feathers to the amber. The phenomenon was probably known well before Thales.

200 B.C. The Chinese invented the compass around the beginning of the Han Dyansty.

60 B.C. The Roman philosopher Lucretius discusses magnetism in his famous work On the Nature of Things (Latin: De Rerum Natura). Among other things, he attributes the attraction of iron to a loadstone as due to a stream of atoms given off by the magnet.

1000 Ibn al-Haytham writes the The Book of Optics.

1269 Petrus Peregrinus writes Epistola de magnete.

1581 Robert Norman presents the dip needle in The Newe Attractive.

1600 William Gilbert publishes On the Magnet, which introduced Peregrinus's work to modern Europe.

1632 Galileo publishes Dialogue Concerning the Two Chief World Systems: Ptolemy and Copernicus, with the principle of relativity.

1665 Robert Hooke publishes Micrographia, which reports observations of microscopic objects and suggests that light is a wave.

1676 Ole Rømer measures the speed of light, in astronomical units, by observing the Galilean moons of Jupiter.

1687 Isaac Newton publishes Philosophiæ Naturalis Principia Mathematica.

1690 Christiaan Huygens publishes Traité de la Lumière.

1704 Isaac Newton publishes the first edition of Opticks.

1725 James Bradley and Samuel Molyneaux discover the *aberration of starlight*, a shift in the positions of stars due to the earth's motion, supporting Newton's particle model of light.

1745 Pieter van Musschenbroek invents the Leyden jar

1754 Benjamin Franklin publishes Experiments and Observations on Electricity.

1757 Leonhard Euler publishes the continuity equation.

1776 The American revolution.

1785 Charles Coulomb publishes Mémoires des Académie des Sciences, which includes what is now known as Coulomb's law.

1789 The French Revolution begins.

1791 Luigi Galvani makes a dead frog's legs twitch with electricity.

1792 France declares war on Austria.

1799 Napoleon Bonaparte takes power in France.

1799 Simon-Pierre Laplace publishes his equation in the context of celestial mechanics.

1800 Alessandro Volta invents the voltaic pile.

1800 William Nicholson, Anthony Carlisle, and William Cruickshank separate hydrogen and oxygen from water using electric current.

1804 Thomas Young publishes his double-slit experiment.

1811 Étienne-Louis Malus discovers polarization of reflected sunlight, and publishes "Théorie de la Double Réfraction."

1812 Siméon Denis Poisson translates Lagrange's gravitational scalar potential to the electrostatic scalar potential.

1813 Carl Friedrich Gauss re-expresses Newton's law of gravity in terms of the mass enclosed by an arbitrary closed surface.

1814 David Brewster explains polarization off of a reflective surface.

1817 Augustin-Jean Fresnel proposes that light is a transverse wave.

1819 André-Marie Ampère demonstrates parallel current-carrying wires attracting, or repelling, when current flows in the same, or opposite, direction.

1819 François Arago observes the Poisson spot, supporting Fresnel's theory of light.

1819 Hans Christian Ørsted discovers that an electric current twists a compass needle, and maps the direction of the magnetic field.

1820 Jean-Baptiste Biot and Felix Savart make the first careful quantitative measurements of forces between magnets and current carrying wires, discovering the law of Laplace.

1821 André-Marie Ampère and Félix Savary discover the law of Biot and Savart.

1822 Joseph Fourier publishes The Analytical Theory of Heat.

1828 George Green introduces the term *potential* into gravity, electricity, and magnetism.

1830 Joseph Henry discovers electromagnetic induction.

1831 David Brewster publishes A Treatise on Optics.

1831 Michael Faraday discovers electromagnetic induction (independently of Henry).

1834 Emil Lenz discovers that the induced current in a moving conductor is always in a direction that results in a force opposing the velocity.

1838 Queen Victoria is coroneted.

1839 Joseph Henry publishes Contributions to Electricity and Magnetism.

1839 Michael Faraday publishes Experimental Researches in Electricity.

1840 Carl Friedrich Gauss and Wilhelm Weber publish an atlas of geomagnetism.

1840 Samuel Morse patents the telegraph and his code.

1842 Christian Doppler discovers the relationship among the emitted and received wavelengths, the wave speed in the medium, and the relative velocity of the source and detector.

1845 Franz Ernst Neumann introduces the vector potential.

1845 Michael Faraday discovers that, under the influence of a magnetic field parallel to the propagation direction, the polarization angle of light rotates when it passes through certain media. This is the first direct evidence that light is an electromagnetic phenomenon.

1846 Michael Faraday writes an intriguing, and nuanced, letter speculating on the nature of light as an electromagnetic wave.

1847 Georg Ohm publishes what became known as Ohm's Law.

1847 William Thomson finds the vector potential of a small magnet.

1848 Wilhelm Weber publishes his action at a distance theory.

1849 Hippolyte Fizeau measures the speed of light using a rotating toothed wheel and a mirror placed 8 km away.

1850 Gustav Kirchhoff expresses Ohm's Law in the modern form.

1851 Hippolyte Fizeau conducts his running water experiment, apparently confirming Fresnel's aether drag hypothesis.

1854 William Thomson derives Stokes's theorem of vector calculus.

1856 Rudolf Kohlrausch and Wilhelm Weber show that the ratio of electrostatic to electromagnetic units, c, numerically equals the speed of light.

1857 Gustav Kirchhoff and Wilhelm Weber model electricity in a telegraph wire as a wave with a speed close to the speed of light.

1858 The first transatlantic telegraph functions for three weeks.

1861 James Clerk Maxwell applies Gauss's law to electricity.

1861 James Clerk Maxwell derives the Maxwell-Ampère law.

1862 Léon Foucault measures the speed of light using a mirror rotating at 48,000 rpm, a tuning fork, and a train whistle.

1866 William Thomson oversees the operation of the first stable transatlantic telegraph cable.

1867 Ludvig Lorenz introduces a mathematical addendum to Weber's action at a distance theory, which we now call the Lorenz gauge.

1871 George Airy attempts, but fails, to measure the relative velocity of the earth and aether using a water-filled telescope.

1873 James Clerk Maxwell publishes A Treatise on Electricity and Magnetism.

1879 Edwin Hall discovers that usually negative, rather than positive, charge, predominantly flows through wires.

1880 Oliver Heaviside patents the coaxial cable.

1881 Albert Michelson presents his first paper where he fails to measure the aether wind.

1881 Heinrich Hertz confirms Maxwell's theory experimentally by generating and detecting radio waves.

1881 Hermann von Helmholtz argues that electricity is comprised of many tiny discrete particles.

1881 J.J. Thomson attempts to derive the magnetic force on a moving charged particle, but is off by a factor of two.

1883 Oliver Heaviside explains Ohm's Law with a fluid model.

1883 Horace Lamb derives the skin depth for a spherical conductor.

1883 Osborne Reynolds investigates the onset of turbulence.

1884 John Henry Poynting investigates the energy flow by electromagnetic fields in wires and telegraph cables.

1887 Albert Michelson and Edward Morley fail to detect the aethereal wind.

1889 George FitzGerald points out that lengths foreshorten in the direction of motion, but not sideways, anticipating both so-called Lorentz-FitzGerald contraction and the Lorentz-FitzGerald transformations.

1889 Oliver Heaviside derives the correct expression for the magnetic force acting on a moving charge.

1895 Guglielmo Marconi sends the first radio signal over 1.5 km.

1895 Pierre Curie publishes his thesis on magnetization.

1897 J.J. Thomson discovers the electron.

1897 Pieter Zeeman observes the splitting of a spectral line when the material was placed in a strong magnetic field.

1898 Alfred-Marie Liénard introduces the *retarded potentials*, which are the potentials due to an electric charge undergoing arbitrary motion.

1900 Emil Wiechert also derives the Liénard-Wiechert potentials.

1901 Queen Victoria dies.

1904 Henri Poincaré delivers "The Present and the Future of Mathematical Physics" at the St. Louis world's fair.

1905 Albert Einstein publishes "Concerning a Heuristic Point of View Toward the Emission and Transformation of Light."

1905 Albert Einstein publishes "On the Electrodynamics of Moving Bodies."

1906 Albert Einstein correctly derives the Planck distribution, despite his use of classical Boltzmann statistics.

1909 Hans Geiger and Ernest Marsden carry out the gold foil experiment, which demonstrates that atoms are made up of mostly empty space with a dense positively charged nucleus at the center.

1911 Ernest Rutherford proposes his planetary model of the atom.

1911 Kamerlingh Onnes discovers superconductivity.

1911 Robert Millikan, Ernest Rutherford, and others independently measure the elementary charge.

1913 Niels Bohr combines Rutherford's planetary model with Planck's energy quanta to fashion a workable model of the hydrogen atom, where electrons only orbit in particular stationary orbits.

1915 Albert Einstein and Wander J. de Haas measure the angular momentum associated with magnetization.

1916 Robert Millikan directly determines Planck's constant using the photoelectric effect, verifying Einstein's concept of light quanta.

1915 Samuel Barnett shows that spinning ferromagnetic materials become magnetized.

1918 John Stewart and Maurice Pate measure the electron gyromagnetic ratio to be twice what is predicted classically.

1922 Otto Stern and Walter Gerlach measure the quantized magnetic moment of an electron.

1923 Arthur Compton scatters x-rays off of the loosely bound electrons in a carbon target, showing that the photon exists.

1924 Louis de Broglie postulates that all particles have an associated wavelength related to their momentum.

1924 Satyendra Nath Bose derives Planck's spectrum without *ad hoc* classical assumptions, introducing Bose-Einstein statistics.

1927 Clint Davisson and Lester Germer observe diffraction peaks in a beam of electrons scattered from a sliced nickel crystal.

1927 George Thomson observes electron diffraction by scattering electrons off of gold foil.

1927 Paul Dirac quantizes the electromagnetic field.

1928 Paul Dirac unites quantum mechanics and special relativity, which results in a single particle theory of the electron and the famous equation that bears Dirac's name.

1928 Werner Heisenberg explains the energetics of the alignment of electron spins in ferromagnetic materials.

1931 Robert van de Graaff invents his electrostatic generator.

1933 Walther Meissner and Robert Ochsenfeld find that magnetic fields do not penetrate superconductors.

1935 Fritz and Heinz London develop acceleration theory to account for superconductivity.

1935 John Cunningham McLennan opens an international conference on superconductivity in London.

1935 Lev Landau and Evgeny Liftshitz explain magnetic domain structure in ferromagnetic materials.

1938 Jack Allen and Don Misener, and independently Pyotr Kapitza, discover superfluidity in cold helium.

1940 Wolfgang Pauli proves the *spin-statistics theorem*, which asserts that spin one-half particles obey Fermi-Dirac statistics, and therefore obey the Pauli exclusion principle.

1952 John Bardeen, Leon N. Cooper, and J. Robert Schrieffer develop a quantum model of superconductivity.

1959 Yakir Aharonov and David Bohm show that in quantum mechanics the potentials are more fundamental than the fields.

1962 Thomas Kuhn writes The Structure of Scientific Revolutions.

1965 Gordon Moore, publishes "Cramming More Components onto Integrated Circuits."

2016 Gravitational Waves are first detected using Michelson interferometers.

Appendix H Biographical Index

A

B

C

D

E

F

G

H

I

J

K

L

M

W

Y

Z

www.ingramcontent.com/pod-product-compliance
Lightning Source LLC
Chambersburg PA
CBHW051802200326
41597CB00025B/4659
9781949942002